D1465896

Element	Symbol	Atomic Number	Atomic Mass*
Actinium	Ac	89	[227]§
Aluminum	Al	13	26.98
Americium	Am	95	[243]
Antimony	Sb	51	121.8
Argon	Ar	18	39.95
Arsenic	As	33	74.92
Astatine	At	85	[210]
Barium	Ba	56	137.3
Berkelium	Bk	97	[247]
Beryllium	Be	4	9.012
Bismuth	Bi	83	209.0
Bohrium	Bh	107	[264]
Boron	B	5	10.81
Bromine	Br	35	79.90
Cadmium	Cd	48	112.4
Calcium	Ca	20	40.08
Californium	Cf	98	[251]
Carbon	C	6	12.01
Cerium	Ce	58	140.1
Cesium	Cs	55	132.90
Chlorine	Cl	17	35.45
Chromium	Cr	24	52.00
Cobalt	Co	27	58.93
Copernicium	Cn	112	[285]
Copper	Cu	29	63.55
Curium	Cm	96	[247]
Darmstadtium	Ds	110	[271]
Dubnium	Db	105	[262]
Dysprosium	Dy	66	162.5
Einsteinium	Es	99	[252]
Erbium	Er	68	167.3
Europium	Eu	63	152.0
Fermium	Fm	100	[257]
Flerovium	Fl	114	[289]
Fluorine	F	9	19.00
Francium	Fr	87	[223]
Gadolinium	Gd	64	157.3
Gallium	Ga	31	69.72
Germanium	Ge	32	72.59
Gold	Au	79	197.0
Hafnium	Hf	72	178.5
Hassium	Hs	108	[265]
Helium	He	2	4.003
Holmium	Ho	67	164.9
Hydrogen	H	1	1.008
Indium	In	49	114.8
Iodine	I	53	126.9
Iridium	Ir	77	192.2
Iron	Fe	26	55.85
Krypton	Kr	36	83.80
Lanthanum	La	57	138.9
Lawrencium	Lr	103	[260]
Lead	Pb	82	207.2
Livermorium	Lv	116	[293]
Lithium	Li	3	6.9419
Lutetium	Lu	71	175.0
Magnesium	Mg	12	24.31
Manganese	Mn	25	54.94
Meitnerium	Mt	109	[268]
Mendelevium	Md	101	[258]
Mercury	Hg	80	200.6
Molybdenum	Mo	42	95.94
Neodymium	Nd	60	144.2
Neon	Ne	10	20.18
Neptunium	Np	93	[237]
Nickel	Ni	28	58.69
Niobium	Nb	41	92.91
Nitrogen	N	7	14.01
Nobelium	No	102	[259]
Osmium	Os	76	190.2
Oxygen	O	8	16.00
Palladium	Pd	46	106.4
Phosphorus	P	15	30.97
Platinum	Pt	78	195.1
Plutonium	Pu	94	[244]
Polonium	Po	84	[209]
Potassium	K	19	39.10
Praseodymium	Pr	59	140.9
Promethium	Pm	61	[145]
Protactinium	Pa	91	[231]
Radium	Ra	88	226
Radon	Rn	86	[222]
Rhenium	Re	75	186.2
Rhodium	Rh	45	102.9
Roentgenium	Rg	111	[272]
Rubidium	Rb	37	85.47
Ruthenium	Ru	44	101.1
Rutherfordium	Rf	104	[261]
Samarium	Sm	62	150.4
Scandium	Sc	21	44.96
Seaborgium	Sg	106	[263]
Selenium	Se	34	78.96
Silicon	Si	14	28.09
Silver	Ag	47	107.9
Sodium	Na	11	22.99
Strontium	Sr	38	87.62
Sulfur	S	16	32.07
Tantalum	Ta	73	180.9
Technetium	Tc	43	[98]
Tellurium	Te	52	127.6
Terbium	Tb	65	158.9
Thallium	Tl	81	204.4
Thorium	Th	90	232.0
Thulium	Tm	69	168.9
Tin	Sn	50	
Titanium	Ti	22	
Tungsten	W	74	
Uranium	U	92	
Vanadium	V	23	
Xenon	Xe	54	
Ytterbium	Yb	70	
Yttrium	Y	39	
Zinc	Zn	30	
Zirconium	Zr	40	

*The values given here are to four significant figures where possible. §A value given in parentheses denotes the mass of the longest-lived isotope.

Chemistry

Volume II

Ninth Edition

Steven S. Zumdahl | Susan A. Zumdahl

CENGAGE
Learning·

Australia • Brazil • Japan • Korea • Mexico • Singapore • Spain • United Kingdom • United States

Chemistry: Volume II, Ninth Edition

Chemistry, 9th Edition
Steven S. Zumdahl | Susan A. Zumdahl

© 2014, 2010 Cengage Learning. All Rights Reserved.

Senior Manager, Student Engagement:

Linda deStefano

Janey Moeller

Manager, Student Engagement:

Julie Dierig

Marketing Manager:

Rachael Kloos

Manager, Production Editorial:

Kim Fry

Manager, Intellectual Property Project Manager:

Brian Methe

Senior Manager, Production and Manufacturing:

Donna M. Brown

Manager, Production:

Terri Daley

For product information and technology assistance, contact us at
Cengage Learning Customer & Sales Support, 1-800-354-9706
For permission to use material from this text or product,
submit all requests online at **cengage.com/permissions**
Further permissions questions can be emailed to
permissionrequest@cengage.com

This book contains select works from existing Cengage Learning resources and was produced by Cengage Learning Custom Solutions for collegiate use. As such, those adopting and/or contributing to this work are responsible for editorial content accuracy, continuity and completeness.

Compilation © 2014 Cengage Learning

ISBN-13: 978-1-305-01559-3
ISBN-10: 1-305-01559-2

WCN: 01-100-101

Cengage Learning
5191 Natorp Boulevard
Mason, Ohio 45040
USA

Cengage Learning is a leading provider of customized learning solutions with office locations around the globe, including Singapore, the United Kingdom, Australia, Mexico, Brazil, and Japan. Locate your local office at:
international.cengage.com/region.

Cengage Learning products are represented in Canada by Nelson Education, Ltd.
For your lifelong learning solutions, visit **www.cengage.com/custom.**
Visit our corporate website at **www.cengage.com.**

Printed in the United States of America

Brief Contents

Chapter 11

Properties of Solutions

Opals are formed from colloidal suspensions of silica when the liquid evaporates. (Horizon International/ Alamy)

Most of the substances we encounter in daily life are mixtures: Wood, milk, gasoline, champagne, seawater, shampoo, steel, and air are common examples. When the components of a mixture are uniformly intermingled—that is, when a mixture is homogeneous—it is called a *solution*. Solutions can be gases, liquids, or solids, as shown in Table 11.1. However, we will be concerned in this chapter with the properties of liquid solutions, particularly those containing water. As we saw in Chapter 4, many essential chemical reactions occur in aqueous solutions because water is capable of dissolving so many substances.

11.1 | Solution Composition

Because a mixture, unlike a chemical compound, has a variable composition, the relative amounts of substances in a solution must be specified. The qualitative terms *dilute* (relatively little solute present) and *concentrated* (relatively large amount of solute) are often used to describe solution content, but we need to define solution composition more precisely to perform calculations. For example, in dealing with the stoichiometry of solution reactions in Chapter 4, we found it useful to describe solution composition in terms of **molarity**, or the number of moles of solute per liter of solution (symbolized by M).

Other ways of describing solution composition are also useful. **Mass percent** (sometimes called *weight percent*) is the percent by mass of the solute in the solution:

$$\text{Mass percent} = \left(\frac{\text{mass of solute}}{\text{mass of solution}} \right) \times 100\%$$

Another way of describing solution composition is the **mole fraction** (symbolized by the lowercase Greek letter chi, X), the ratio of the number of moles of a given component to the total number of moles of solution. For a two-component solution, where n_A and n_B represent the number of moles of the two components,

$$\text{Mole fraction of component A} = X_A = \frac{n_A}{n_A + n_B}$$

Still another way of describing solution composition is **molality** (symbolized by m), the number of moles of solute per *kilogram of solvent*:

$$\text{Molality} = \frac{\text{moles of solute}}{\text{kilogram of solvent}}$$

A solute is the substance being dissolved. The solvent is the dissolving medium.

$$\text{Molarity} = \frac{\text{moles of solute}}{\text{liters of solution}}$$

When liquids are mixed, the liquid present in the largest amount is called the *solvent*.

In very dilute aqueous solutions, the magnitude of the molality and the molarity are almost the same.

Table 11.1 | Various Types of Solutions

Example	State of Solution	State of Solute	State of Solvent
Air, natural gas	Gas	Gas	Gas
Vodka, antifreeze	Liquid	Liquid	Liquid
Brass	Solid	Solid	Solid
Carbonated water	Liquid	Gas	Liquid
Seawater, sugar solution	Liquid	Solid	Liquid
Hydrogen in platinum	Solid	Gas	Solid

Interactive Example 11.1

Sign in at http://login.cengagebrain .com to try this Interactive Example in **OWL**.

Since molarity depends on the volume of the solution, it changes slightly with temperature. Molality is independent of temperature because it depends only on mass.

Various Methods for Describing Solution Composition

A solution is prepared by mixing 1.00 g ethanol (C_2H_5OH) with 100.0 g water to give a final volume of 101 mL. Calculate the molarity, mass percent, mole fraction, and molality of ethanol in this solution.

Solution

Molarity

The moles of ethanol can be obtained from its molar mass (46.07 g/mol):

$$1.00 \text{ g } C_2H_5OH \times \frac{1 \text{ mol } C_2H_5OH}{46.07 \text{ g } C_2H_5OH} = 2.17 \times 10^{-2} \text{ mol } C_2H_5OH$$

$$\text{Volume} = 101 \text{ mL} \times \frac{1 \text{ L}}{1000 \text{ mL}} = 0.101 \text{ L}$$

$$\text{Molarity of } C_2H_5OH = \frac{\text{moles of } C_2H_5OH}{\text{liters of solution}} = \frac{2.17 \times 10^{-2} \text{ mol}}{0.101 \text{ L}}$$

$$= 0.215 \ M$$

Mass Percent

$$\text{Mass percent } C_2H_5OH = \left(\frac{\text{mass of } C_2H_5OH}{\text{mass of solution}}\right) \times 100\%$$

$$= \left(\frac{1.00 \text{ g } C_2H_5OH}{100.0 \text{ g } H_2O + 1.00 \text{ g } C_2H_5OH}\right) \times 100\%$$

$$= 0.990\% \ C_2H_5OH$$

Mole Fraction

$$\text{Mole fraction of } C_2H_5OH = \frac{n_{C_2H_5OH}}{n_{C_2H_5OH} + n_{H_2O}}$$

$$n_{H_2O} = 100.0 \text{ g } H_2O \times \frac{1 \text{ mol } H_2O}{18.0 \text{ g } H_2O} = 5.56 \text{ mol}$$

$$\chi_{C_2H_5OH} = \frac{2.17 \times 10^{-2} \text{ mol}}{2.17 \times 10^{-2} \text{ mol} + 5.56 \text{ mol}}$$

$$= \frac{2.17 \times 10^{-2}}{5.58} = 0.00389$$

Molality

$$\text{Molality of } C_2H_5OH = \frac{\text{moles of } C_2H_5OH}{\text{kilogram of } H_2O} = \frac{2.17 \times 10^{-2} \text{ mol}}{100.0 \text{ g} \times \dfrac{1 \text{ kg}}{1000 \text{ g}}}$$

$$= \frac{2.17 \times 10^{-2} \text{ mol}}{0.1000 \text{ kg}}$$

$$= 0.217 \ m$$

See Exercises 11.29 through 11.31

Critical Thinking

You are given two aqueous solutions with different ionic solutes (Solution A and Solution B). What if you are told that Solution A has a greater concentration than Solution B by mass percent, but Solution B has a greater concentration than Solution A in terms of molality? Is this possible? If not, explain why not. If it is possible, provide example solutes for A and B and justify your answer with calculations.

Table 11.2 | The Molar Mass, Equivalent Mass, and Relationship of Molarity and Normality for Several Acids and Bases

Acid or Base	Molar Mass	Equivalent Mass	Relationship of Molarity and Normality
HCl	36.5	36.5	$1\,M = 1\,N$
H_2SO_4	98	$\dfrac{98}{2} = 49$	$1\,M = 2\,N$
NaOH	40	40	$1\,M = 1\,N$
$Ca(OH)_2$	74	$\dfrac{74}{2} = 37$	$1\,M = 2\,N$

The definition of an equivalent depends on the reaction taking place in the solution.

The quantity we call *equivalent mass* here traditionally has been called *equivalent weight*.

Oxidation–reduction half-reactions were discussed in Section 4.10.

Another concentration measure sometimes encountered is **normality** (symbolized by N). Normality is defined as the number of *equivalents* per liter of solution, where the definition of an equivalent depends on the reaction taking place in the solution. *For an acid–base reaction,* the equivalent is the mass of acid or base that can furnish or accept exactly 1 mole of protons (H^+ ions). In Table 11.2, note, for example, that the equivalent mass of sulfuric acid is the molar mass divided by 2, since each mole of H_2SO_4 can furnish 2 moles of protons. The equivalent mass of calcium hydroxide is also half the molar mass, since each mole of $Ca(OH)_2$ contains 2 moles of OH^- ions that can react with 2 moles of protons. The equivalent is defined so that 1 equivalent of acid will react with exactly 1 equivalent of base.

For oxidation–reduction reactions, the equivalent is defined as the quantity of oxidizing or reducing agent that can accept or furnish 1 mole of electrons. Thus 1 equivalent of reducing agent will react with exactly 1 equivalent of oxidizing agent. The equivalent mass of an oxidizing or reducing agent can be calculated from the number of electrons in its half-reaction. For example, MnO_4^- reacting in acidic solution absorbs five electrons to produce Mn^{2+}:

$$MnO_4^- + 5e^- + 8H^+ \longrightarrow Mn^{2+} + 4H_2O$$

Since the MnO_4^- ion present in 1 mole of $KMnO_4$ consumes 5 moles of electrons, the equivalent mass is the molar mass divided by 5:

Sign in at http://login.cengagebrain.com to try this Interactive Example in **OWL**.

$$\text{Equivalent mass of } KMnO_4 = \frac{\text{molar mass}}{5} = \frac{158\ g}{5} = 31.6\ g$$

Interactive Example 11.2

Calculating Various Methods of Solution Composition from the Molarity

The electrolyte in automobile lead storage batteries is a 3.75 M sulfuric acid solution that has a density of 1.230 g/mL. Calculate the mass percent, molality, and normality of the sulfuric acid.

Solution

What is the density of the solution in grams per liter?

$$1.230\,\frac{g}{mL} \times \frac{1000\ mL}{1\ L} = 1.230 \times 10^3\ g/L$$

What mass of H_2SO_4 is present in 1.00 L of solution?
We know 1 liter of this solution contains 1230. g of the mixture of sulfuric acid and water. Since the solution is 3.75 M, we know that 3.75 moles of H_2SO_4 is present per liter of solution. The number of grams of H_2SO_4 present is

$$3.75\ mol \times \frac{98.0\ g\ H_2SO_4}{1\ mol} = 368\ g\ H_2SO_4$$

A modern 12-volt lead storage battery of the type used in automobiles.

How much water is present in 1.00 L of solution?

The amount of water present in 1 liter of solution is obtained from the difference

$$1230. \text{ g solution} - 368 \text{ g H}_2\text{SO}_4 = 862 \text{ g H}_2\text{O}$$

What is the mass percent?

Since we now know the masses of the solute and solvent, we can calculate the mass percent.

> Mass percent $H_2SO_4 = \dfrac{\text{mass of H}_2\text{SO}_4}{\text{mass of solution}} \times 100\% = \dfrac{368 \text{ g}}{1230. \text{ g}} \times 100\%$

$$= 29.9\% \text{ H}_2\text{SO}_4$$

What is the molality?

From the moles of solute and the mass of solvent, we can calculate the molality.

> Molality of $H_2SO_4 = \dfrac{\text{moles H}_2\text{SO}_4}{\text{kilogram of H}_2\text{O}}$

$$= \dfrac{3.75 \text{ mol H}_2\text{SO}_4}{862 \text{ g H}_2\text{O} \times \dfrac{1 \text{ kg H}_2\text{O}}{1000 \text{ g H}_2\text{O}}} = 4.35 \ m$$

What is the normality?

Since each sulfuric acid molecule can furnish two protons, 1 mole of H_2SO_4 represents 2 equivalents. Thus a solution with 3.75 moles of H_2SO_4 per liter contains $2 \times 3.75 = 7.50$ equivalents per liter.

> The normality is 7.50 *N*.

See Exercise 11.37

11.2 | The Energies of Solution Formation

DDT

Dissolving solutes in liquids is very common. We dissolve salt in the water used to cook vegetables, sugar in iced tea, stains in cleaning fluid, gaseous carbon dioxide in water to make carbonated water, ethanol in gasoline to make gasohol, and so on.

Solubility is important in other ways. For example, because the pesticide DDT is fat-soluble, it is retained and concentrated in animal tissues, where it causes detrimental effects. This is why DDT, even though it is effective for killing mosquitos, has been banned in the United States. Also, the solubility of various vitamins is important in determining correct dosages. The insolubility of barium sulfate means it can be used safely to improve X rays of the gastrointestinal tract, even though Ba^{2+} ions are quite toxic.

What factors affect solubility? The cardinal rule of solubility is *like dissolves like*. We find that we must use a polar solvent to dissolve a polar or ionic solute and a nonpolar solvent to dissolve a nonpolar solute. Now we will try to understand why this behavior occurs. To simplify the discussion, we will assume that the formation of a liquid solution takes place in three distinct steps:

Polar solvents dissolve polar solutes; nonpolar solvents dissolve nonpolar solutes.

1. Separating the solute into its individual components (expanding the solute)
2. Overcoming intermolecular forces in the solvent to make room for the solute (expanding the solvent)
3. Allowing the solute and solvent to interact to form the solution

These steps are illustrated in Fig. 11.1. Steps 1 and 2 require energy, since forces must be overcome to expand the solute and solvent. Step 3 usually releases energy. In other words, Steps 1 and 2 are endothermic, and Step 3 is often exothermic. The en-

Figure 11.1 | The formation of a liquid solution can be divided into three steps: (1) expanding the solute, (2) expanding the solvent, and (3) combining the expanded solute and solvent to form the solution.

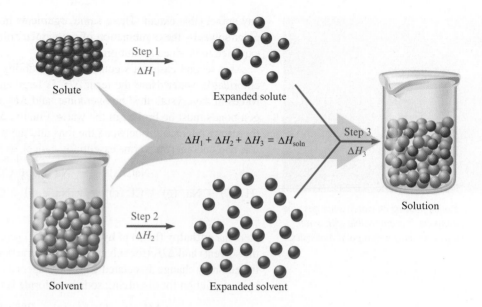

The enthalpy of solution is the sum of the energies used in expanding both solvent and solute and the energy of solvent–solute interaction.

thalpy change associated with the formation of the solution, called the **enthalpy (heat) of solution** (ΔH_{soln}), is the sum of the ΔH values for the steps:

$$\Delta H_{soln} = \Delta H_1 + \Delta H_2 + \Delta H_3$$

where ΔH_{soln} may have a positive sign (energy absorbed) or a negative sign (energy released) (Fig. 11.2).

To illustrate the importance of the various energy terms in the equation for ΔH_{soln}, we will consider two specific cases. First, we know that oil is not soluble in water. When oil tankers leak, the petroleum forms an oil slick that floats on the water and is eventually carried onto the beaches. We can explain the immiscibility of oil and water by considering the energy terms involved. Oil is a mixture of nonpolar molecules that interact through London dispersion forces, which depend on molecule size. We expect ΔH_1 to be small for a typical nonpolar solute, but it will be relatively large for the large oil molecules. The term ΔH_3 will be small, since interactions between the nonpolar solute molecules and the polar water molecules will be negligible. However, ΔH_2 will be large and positive because it takes considerable energy to overcome the hydrogen-bonding forces among the water molecules to expand the solvent. Thus ΔH_{soln} will be large and positive because of the ΔH_1 and ΔH_2 terms. Since a large amount of energy would have to be expended to form an oil–water solution, this process does not occur to

ΔH_1 is expected to be small for nonpolar solutes but can be large for large molecules.

Figure 11.2 | The heat of solution (a) ΔH_{soln} has a negative sign (the process is exothermic) if Step 3 releases more energy than that required by Steps 1 and 2. (b) ΔH_{soln} has a positive sign (the process is endothermic) if Steps 1 and 2 require more energy than is released in Step 3. (If the energy changes for Steps 1 and 2 equal that for Step 3, then ΔH_{soln} is zero.)

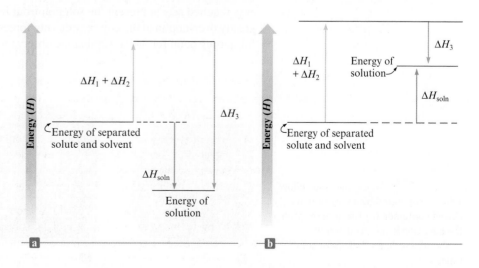

Gasoline floating on water. Since gasoline is nonpolar, it is immiscible with water, because water contains polar molecules.

any appreciable extent. These same arguments hold true for any nonpolar solute and polar solvent—the combination of a nonpolar solute and a highly polar solvent is not expected to produce a solution.

As a second case, let's consider the solubility of an ionic solute, such as sodium chloride, in water. Here the term ΔH_1 is large and positive because the strong ionic forces in the crystal must be overcome, and ΔH_2 is large and positive because hydrogen bonds must be broken in the water. Finally, ΔH_3 is large and negative because of the strong interactions between the ions and the water molecules. In fact, the exothermic and endothermic terms essentially cancel, as shown from the known values:

$$NaCl(s) \longrightarrow Na^+(g) + Cl^-(g) \qquad \Delta H_1 = 786 \text{ kJ/mol}$$
$$H_2O(l) + Na^+(g) + Cl^-(g) \longrightarrow Na^+(aq) + Cl^-(aq) \qquad \Delta H_{\text{hyd}} = \Delta H_2 + \Delta H_3$$
$$= -783 \text{ kJ/mol}$$

Here the **enthalpy (heat) of hydration** (ΔH_{hyd}) combines the terms ΔH_2 (for expanding the solvent) and ΔH_3 (for solvent–solute interactions). The heat of hydration represents the enthalpy change associated with the dispersal of a gaseous solute in water. Thus the heat of solution for dissolving sodium chloride is the sum of ΔH_1 and ΔH_{hyd}:

$$\Delta H_{\text{soln}} = 786 \text{ kJ/mol} - 783 \text{ kJ/mol} = 3 \text{ kJ/mol}$$

Note that ΔH_{soln} is small but positive; the dissolving process requires a small amount of energy. Then why is NaCl so soluble in water? The answer lies in nature's tendency toward higher probability of the mixed state. That is, processes naturally run in the direction that leads to the most probable state. For example, imagine equal numbers of orange and yellow spheres separated by a partition [Fig. 11.3(a)]. If we remove the partition and shake the container, the spheres will mix [Fig. 11.3(b)], and no amount of shaking will cause them to return to the state of separated orange and yellow. Why? The mixed state is simply much more likely to occur (more probable) than the original separate state because there are many more ways of placing the spheres to give a mixed state than a separated state. This is a general principle. *One factor that favors a process is an increase in probability.*

The factors that act as driving forces for a process are discussed more fully in Chapter 16.

But energy considerations are also important. *Processes that require large amounts of energy tend not to occur.* Since dissolving 1 mole of solid NaCl requires only a small amount of energy, the solution forms, presumably because of the large increase in the probability of the state when the solute and solvent are mixed.

The various possible cases for solution formation are summarized in Table 11.3. Note that in two cases, polar–polar and nonpolar–nonpolar, the heat of solution is expected to be small. In these cases, the solution forms because of the increase in the probability of the mixed state. In the other cases (polar–nonpolar and nonpolar–polar), the heat of solution is expected to be large and positive, and the large quantity of energy required acts to prevent the solution from forming. Although this discussion has greatly oversimplified the complex driving forces for solubility, these ideas are a useful starting point for understanding the observation that *like dissolves like*.

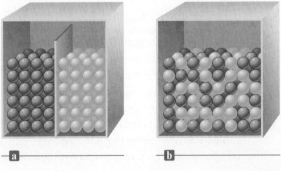

Figure 11.3 (a) Orange and yellow spheres separated by a partition in a closed container. (b) The spheres after the partition is removed and the container has been shaken for some time.

Table 11.3 | The Energy Terms for Various Types of Solutes and Solvents

	ΔH_1	ΔH_2	ΔH_3	ΔH_{soln}	Outcome
Polar solute, polar solvent	Large	Large	Large, negative	Small	Solution forms
Nonpolar solute, polar solvent	Small	Large	Small	Large, positive	No solution forms
Nonpolar solute, nonpolar solvent	Small	Small	Small	Small	Solution forms
Polar solute, nonpolar solvent	Large	Small	Small	Large, positive	No solution forms

Critical Thinking

You and a friend are studying for a chemistry exam. What if your friend tells you, "Since exothermic processes are favored and the sign of the enthalpy change tells us whether or not a process is endothermic or exothermic, the sign of ΔH_{soln} tells us whether or not a solution will form"? How would you explain to your friend that this conclusion is not correct? What part, if any, of what your friend says is correct?

Sign in at http://login.cengagebrain.com to try this Interactive Example in **OWL**.

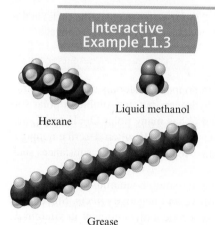

Hexane

Liquid methanol

Grease

Interactive Example 11.3

Differentiating Solvent Properties

Decide whether liquid hexane (C_6H_{14}) or liquid methanol (CH_3OH) is the more appropriate solvent for the substances grease ($C_{20}H_{42}$) and potassium iodide (KI).

Solution

Hexane is a nonpolar solvent because it contains C—H bonds. Thus hexane will work best for the nonpolar solute grease. Methanol has an O—H group that makes it significantly polar. Thus it will serve as the better solvent for the ionic solid KI.

See Exercises 11.43 through 11.45

11.3 | Factors Affecting Solubility

Structure Effects

In the last section we saw that solubility is favored if the solute and solvent have similar polarities. Since it is the molecular structure that determines polarity, there should be a definite connection between structure and solubility. Vitamins provide an excellent example of the relationship among molecular structure, polarity, and solubility.

Recently, there has been considerable publicity about the pros and cons of consuming large quantities of vitamins. For example, large doses of vitamin C have been advocated to combat various illnesses, including the common cold. Vitamin E has been extolled as a youth-preserving elixir and a protector against the carcinogenic (cancer-causing) effects of certain chemicals. However, there are possible detrimental effects from taking large amounts of some vitamins, depending on their solubilities.

Vitamins can be divided into two classes: *fat-soluble* (vitamins A, D, E, and K) and *water-soluble* (vitamins B and C). The reason for the differing solubility characteristics can be seen by comparing the structures of vitamins A and C (Fig. 11.4). Vitamin A, composed mostly of carbon and hydrogen atoms that have similar electronegativities, is

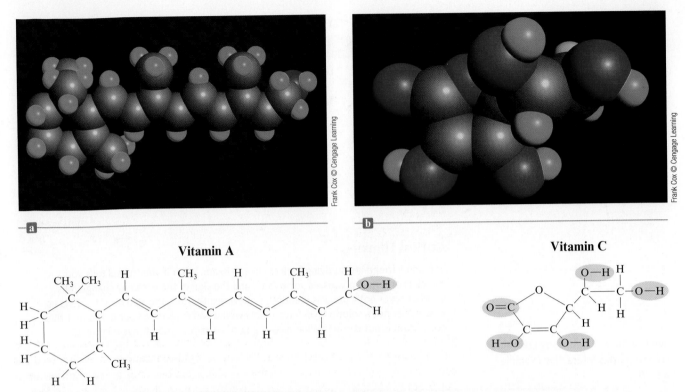

Figure 11.4 | The molecular structures of (a) vitamin A (nonpolar, fat-soluble) and (b) vitamin C (polar, water-soluble). The circles in the structural formulas indicate polar bonds. Note that vitamin C contains far more polar bonds than vitamin A Frank Cox (c) Cengage Learning.

virtually nonpolar. This causes it to be soluble in nonpolar materials such as body fat, which is also largely composed of carbon and hydrogen, but not soluble in polar solvents such as water. On the other hand, vitamin C has many polar O—H and C—O bonds, making the molecule polar and thus water-soluble. We often describe nonpolar materials such as vitamin A as *hydrophobic* (water-fearing) and polar substances such as vitamin C as *hydrophilic* (water-loving).

Because of their solubility characteristics, the fat-soluble vitamins can build up in the fatty tissues of the body. This has both positive and negative effects. Since these vitamins can be stored, the body can tolerate for a time a diet deficient in vitamin A, D, E, or K. Conversely, if excessive amounts of these vitamins are consumed, their buildup can lead to the illness *hypervitaminosis.*

In contrast, the water-soluble vitamins are excreted by the body and must be consumed regularly. This fact was first recognized when the British navy discovered that scurvy, a disease often suffered by sailors, could be prevented if the sailors regularly ate fresh limes (which are a good source of vitamin C) when aboard ship (hence the name "limey" for the British sailor).

Pressure Effects

While pressure has little effect on the solubilities of solids or liquids, it does significantly increase the solubility of a gas. Carbonated beverages, for example, are always bottled at high pressures of carbon dioxide to ensure a high concentration of carbon dioxide in the liquid. The fizzing that occurs when you open a can of soda results from the escape of gaseous carbon dioxide because under these conditions the pressure of CO_2 above the solution is now much lower than that used in the bottling process.

The increase in gas solubility with pressure can be understood from Fig. 11.5. Figure 11.5(a) shows a gas in equilibrium with a solution; that is, the gas molecules are entering and leaving the solution at the same rate. If the pressure is suddenly increased

Figure 11.5 | (a) A gaseous solute in equilibrium with a solution. (b) The piston is pushed in, which increases the pressure of the gas and the number of gas molecules per unit volume. This causes an increase in the rate at which the gas enters the solution, so the concentration of dissolved gas increases. (c) The greater gas concentration in the solution causes an increase in the rate of escape. A new equilibrium is reached.

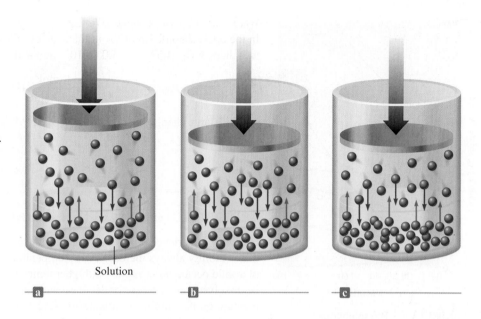

Solution

a   b   c

[Fig. 11.5(b)], the number of gas molecules per unit volume increases, and the gas enters the solution at a higher rate than it leaves. As the concentration of dissolved gas increases, the rate of the escape of the gas also increases until a new equilibrium is reached [Fig. 11.5(c)], where the solution contains more dissolved gas than before.

The relationship between gas pressure and the concentration of dissolved gas is given by **Henry's law**:

$$C = kP$$

where C represents the concentration of the dissolved gas, k is a constant characteristic of a particular solution, and P represents the partial pressure of the gaseous solute above the solution. In words, Henry's law states that *the amount of a gas dissolved in a solution is directly proportional to the pressure of the gas above the solution.*

Henry's law is obeyed most accurately for dilute solutions of gases that do not dissociate in or react with the solvent. For example, Henry's law is obeyed by oxygen gas in water, but it does *not* correctly represent the behavior of gaseous hydrogen chloride in water because of the dissociation reaction

$$HCl(g) \xrightarrow{H_2O} H^+(aq) + Cl^-(aq)$$

William Henry (1774–1836), a close friend of John Dalton, formulated his law in 1801.

Henry's law holds only when there is no chemical reaction between the solute and solvent.

Interactive Example 11.4

Sign in at http://login.cengagebrain.com to try this Interactive Example in **OWL**.

Carbonation in a bottle of soda.

Calculations Using Henry's Law

A certain soft drink is bottled so that a bottle at 25°C contains CO_2 gas at a pressure of 5.0 atm over the liquid. Assuming that the partial pressure of CO_2 in the atmosphere is 4.0×10^{-4} atm, calculate the equilibrium concentrations of CO_2 in the soda both before and after the bottle is opened. The Henry's law constant for CO_2 in aqueous solution is 3.1×10^{-2} mol/L · atm at 25°C.

Solution

What is Henry's law for CO_2?

$$C_{CO_2} = k_{CO_2}P_{CO_2}$$

where $k_{CO_2} = 3.1 \times 10^{-2}$ mol/L · atm.

What is the C_{CO_2} in the unopened bottle?
In the *unopened* bottle, $P_{CO_2} = 5.0$ atm and

❯ $C_{CO_2} = k_{CO_2}P_{CO_2} = (3.1 \times 10^{-2}$ mol/L · atm$)(5.0$ atm$) = 0.16$ mol/L

doug3437/iStockphoto.com

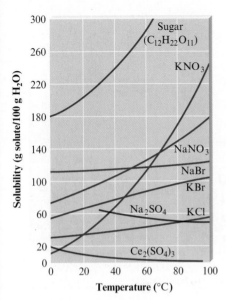

Figure 11.6 | The solubilities of several solids as a function of temperature. Note that while most substances become more soluble in water with increasing temperature, sodium sulfate and cerium sulfate become less soluble.

ΔH_{soln}° refers to the formation of a 1.0-M ideal solution and is not necessarily relevant to the process of dissolving a solid in a saturated solution. Thus ΔH_{soln}° is of limited use in predicting the variation of solubility with temperature.

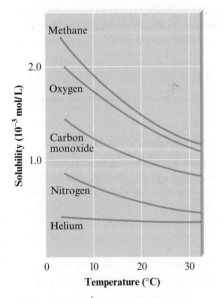

Figure 11.7 | The solubilities of several gases in water as a function of temperature at a constant pressure of 1 atm of gas above the solution.

What is the C_{CO_2} in the opened bottle?
In the *opened* bottle, the CO_2 in the soda eventually reaches equilibrium with the atmospheric CO_2, so $P_{CO_2} = 4.0 \times 10^{-4}$ atm and

$$\quad C_{CO_2} = k_{CO_2}P_{CO_2} = \left(3.1 \times 10^{-2} \frac{mol}{L \cdot atm}\right)(4.0 \times 10^{-4}\ atm) = 1.2 \times 10^{-5}\ mol/L$$

Note the large change in concentration of CO_2. This is why soda goes "flat" after being open for a while.

See Exercises 11.49 and 11.50

Temperature Effects (for Aqueous Solutions)

Everyday experiences of dissolving substances such as sugar may lead you to think that solubility always increases with temperature. This is not the case. The dissolving of a solid occurs *more rapidly* at higher temperatures, but the amount of solid that can be dissolved may increase or decrease with increasing temperature. The effect of temperature on the solubility in water of several solids is shown in Fig. 11.6. Note that although the solubility of most solids in water increases with temperature, the solubilities of some substances (such as sodium sulfate and cerium sulfate) decrease with increasing temperature.

Predicting the temperature dependence of solubility is very difficult. For example, although there is some correlation between the sign of ΔH_{soln}° and the variation of solubility with temperature, important exceptions exist.* The only sure way to determine the temperature dependence of a solid's solubility is by experiment.

The behavior of gases dissolving in water appears less complex. The solubility of a gas in water typically decreases with increasing temperature,† as is shown for several cases in Fig. 11.7. This temperature effect has important environmental implications because of the widespread use of water from lakes and rivers for industrial cooling. After being used, the water is returned to its natural source at a higher than ambient temperature (**thermal pollution** has occurred). Because it is warmer, this water contains less than the normal concentration of oxygen and is also less dense; it tends to "float" on the colder water below, thus blocking normal oxygen absorption. This effect can be especially important in deep lakes. The warm upper layer can seriously decrease the amount of oxygen available to aquatic life in the deeper layers of the lake.

The decreasing solubility of gases with increasing temperature is also responsible for the formation of *boiler scale*. As we will see in more detail in Chapter 14, the bicarbonate ion is formed when carbon dioxide is dissolved in water containing the carbonate ion:

$$CO_3^{2-}(aq) + CO_2(aq) + H_2O(l) \longrightarrow 2HCO_3^-(aq)$$

When the water also contains Ca^{2+} ions, this reaction is especially important—calcium bicarbonate is soluble in water, but calcium carbonate is insoluble. When the water is heated, the carbon dioxide is driven off. For the system to replace the lost carbon dioxide, the reverse reaction must occur:

$$2HCO_3^-(aq) \longrightarrow H_2O(l) + CO_2(aq) + CO_3^{2-}(aq)$$

This reaction, however, also increases the concentration of carbonate ions, causing solid calcium carbonate to form. This solid is the boiler scale that coats the walls of containers such as industrial boilers and tea kettles. Boiler scale reduces the efficiency of heat transfer and can lead to blockage of pipes (Fig. 11.8).

---

*For more information see R. S. Treptow, "Le Châtelier's Principle Applied to the Temperature Dependence of Solubility," *J. Chem. Ed.* **61** (1984): 499.
†The opposite behavior is observed for most nonaqueous solvents.

Figure 11.8 | A pipe with accumulated mineral deposits. The cross section clearly indicates the reduction in pipe capacity.

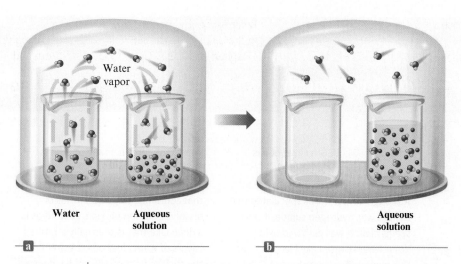

Figure 11.9 | An aqueous solution and pure water in a closed environment. (a) Initial stage. (b) After a period of time, the water is transferred to the solution.

11.4 | The Vapor Pressures of Solutions

Liquid solutions have physical properties significantly different from those of the pure solvent, a fact that has great practical importance. For example, we add antifreeze to the water in a car's cooling system to prevent freezing in winter and boiling in summer. We also melt ice on sidewalks and streets by spreading salt. These preventive measures work because of the solute's effect on the solvent's properties.

To explore how a nonvolatile solute affects a solvent, we will consider the experiment represented in Fig. 11.9, in which a sealed container encloses a beaker containing an aqueous sulfuric acid solution and a beaker containing pure water. Gradually, the volume of the sulfuric acid solution increases and the volume of the pure water decreases. Why? We can explain this observation if the vapor pressure of the pure solvent is greater than that of the solution. Under these conditions, the pressure of vapor necessary to achieve equilibrium with the pure solvent is greater than that required to reach equilibrium with the aqueous acid solution. Thus as the pure solvent emits vapor to attempt to reach equilibrium, the aqueous sulfuric acid solution absorbs vapor to try to lower the vapor pressure toward its equilibrium value. This process results in a net transfer of water from the pure water through the vapor phase to the sulfuric acid solution. The system can reach an equilibrium vapor pressure only when all the water is transferred to the solution. This experiment is just one of many observations indicating that the presence of a *nonvolatile solute lowers the vapor pressure of a solvent.*

> A nonvolatile solute has no tendency to escape from solution into the vapor phase.

We can account for this behavior in terms of the simple model shown in Fig. 11.10. The dissolved nonvolatile solute decreases the number of solvent molecules per unit volume and it should proportionately lower the escaping tendency of the solvent molecules. For example, in a solution consisting of half nonvolatile solute molecules and half solvent molecules, we might expect the observed vapor pressure to be half that of the pure solvent, since only half as many molecules can escape. In fact, this is what is observed.

Detailed studies of the vapor pressures of solutions containing nonvolatile solutes were carried out by François M. Raoult (1830–1901). His results are described by the equation known as **Raoult's law**:

$$P_{\text{soln}} = \chi_{\text{solvent}} P^0_{\text{solvent}}$$

where P_{soln} is the observed vapor pressure of the solution, χ_{solvent} is the mole fraction of solvent, and P^0_{solvent} is the vapor pressure of the pure solvent. Note that for a solution of

Chemical connections
The Lake Nyos Tragedy

On August 21, 1986, a cloud of gas suddenly boiled from Lake Nyos in Cameroon, killing nearly 2000 people. Although at first it was speculated that the gas was hydrogen sulfide, it now seems clear it was carbon dioxide. What would cause Lake Nyos to emit this huge, suffocating cloud of CO_2? Although the answer may never be known for certain, many scientists believe that the lake suddenly "turned over," bringing to the surface water that contained huge quantities of dissolved carbon dioxide. Lake Nyos is a deep lake that is thermally stratified: Layers of warm, less dense water near the surface float on the colder, denser water layers near the lake's bottom.

Under normal conditions the lake stays this way; there is little mixing among the different layers. Scientists believe that over hundreds or thousands of years, carbon dioxide gas had seeped into the cold water at the lake's bottom and dissolved in great amounts because of the large pressure of CO_2 present (in accordance with Henry's law). For some reason on August 21, 1986, the lake apparently suffered an overturn, possibly due to wind or to unusual cooling of the lake's surface by monsoon clouds. This caused water that was greatly supersaturated with CO_2 to reach the surface and release tremendous quantities of gaseous CO_2 that suffocated thousands of humans and animals before they knew what hit them—a tragic, monumental illustration of Henry's law.

Since 1986 the scientists studying Lake Nyos and nearby Lake Monoun have observed a rapid recharging of the CO_2 levels in the deep waters of these lakes, causing concern that another deadly gas release could occur at any time. Apparently the only way to prevent such a disaster is to pump away the CO_2-charged deep water in the two lakes. Scientists at a conference to study this problem in 1994 recommended such a solution, but it has not yet been funded by Cameroon.

© Thierry Orban/Corbis
Lake Nyos in Cameroon.

Pure solvent

**Solution with a
nonvolatile solute**

Figure 11.10 | The presence of a nonvolatile solute inhibits the escape of solvent molecules from the liquid and so lowers the vapor pressure of the solvent.

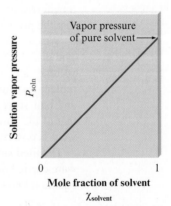

Vapor pressure of pure solvent →

Solution vapor pressure P_{soln}

Mole fraction of solvent $\chi_{solvent}$

Figure 11.11 | For a solution that obeys Raoult's law, a plot of P_{soln} versus $\chi_{solvent}$ gives a straight line.

half solute and half solvent molecules, $X_{solvent}$ is 0.5, so the vapor pressure of the solution is half that of the pure solvent. On the other hand, for a solution in which three-fourths of the solution molecules are solvent, $X_{solvent} = \frac{3}{4} = 0.75$, and $P_{soln} = 0.75P^0_{solvent}$. The idea is that the nonvolatile solute simply dilutes the solvent.

Raoult's law is a linear equation of the form $y = mx + b$, where $y = P_{soln}$, $x = X_{solvent}$, $m = P^0_{solvent}$, and $b = 0$. Thus a plot of P_{soln} versus $X_{solvent}$ gives a straight line with a slope equal to $P^0_{solvent}$ (Fig. 11.11).

Interactive Example 11.5

Sign in at http://login.cengagebrain.com to try this Interactive Example in OWL.

Calculating the Vapor Pressure of a Solution

Calculate the expected vapor pressure at 25°C for a solution prepared by dissolving 158.0 g common table sugar (sucrose, molar mass = 342.3 g/mol) in 643.5 cm³ of water. At 25°C, the density of water is 0.9971 g/cm³ and the vapor pressure is 23.76 torr.

Solution

What is Raoult's law for this case?

$$P_{soln} = X_{H_2O}P^0_{H_2O}$$

To calculate the mole fraction of water in the solution, we must first determine the number of moles of sucrose and the moles of water present.

What are the moles of sucrose?

$$\text{Moles of sucrose} = 158.0 \text{ g sucrose} \times \frac{1 \text{ mol sucrose}}{342.3 \text{ g sucrose}}$$

$$= 0.4616 \text{ mol sucrose}$$

What are the moles of water?

To determine the moles of water present, we first convert volume to mass using the density:

$$643.5 \text{ cm}^3 \text{ H}_2\text{O} \times \frac{0.9971 \text{ g H}_2\text{O}}{\text{cm}^3 \text{ H}_2\text{O}} = 641.6 \text{ g H}_2\text{O}$$

The number of moles of water is therefore

$$641.6 \text{ g H}_2\text{O} \times \frac{1 \text{ mol H}_2\text{O}}{18.02 \text{ g H}_2\text{O}} = 35.60 \text{ mol H}_2\text{O}$$

What is the mole fraction of water in the solution?

$$X_{H_2O} = \frac{\text{mol H}_2\text{O}}{\text{mol H}_2\text{O} + \text{mol sucrose}} = \frac{35.60 \text{ mol}}{35.60 \text{ mol} + 0.4616 \text{ mol}}$$

$$= \frac{35.60 \text{ mol}}{36.06 \text{ mol}} = 0.9873$$

> The vapor pressure of the solution is: $P_{soln} = X_{H_2O}P^0_{H_2O} = (0.9872)(23.76 \text{ torr})$
> $= 23.46 \text{ torr}$

Thus the vapor pressure of water has been lowered from 23.76 torr in the pure state to 23.46 torr in the solution. The vapor pressure has been lowered by 0.30 torr.

See Exercises 11.51 and 11.52

The phenomenon of the lowering of the vapor pressure gives us a convenient way to "count" molecules and thus provides a means for experimentally determining molar masses. Suppose a certain mass of a compound is dissolved in a solvent and the vapor pressure of the resulting solution is measured. Using Raoult's law, we can determine the number of moles of solute present. Since the mass of this number of moles is known, we can calculate the molar mass.

We also can use vapor pressure measurements to characterize solutions. For example, 1 mole of sodium chloride dissolved in water lowers the vapor pressure approximately twice as much as expected because the solid has two ions per formula unit, which separate when it dissolves. Thus vapor pressure measurements can give valuable information about the nature of the solute after it dissolves.

The lowering of vapor pressure depends on the number of solute particles present in the solution.

Interactive Example 11.6

Sign in at http://login.cengagebrain .com to try this Interactive Example in OWL.

Calculating the Vapor Pressure of a Solution Containing Ionic Solute

Predict the vapor pressure of a solution prepared by mixing 35.0 g solid Na_2SO_4 (molar mass = 142.05 g/mol) with 175 g water at 25°C. The vapor pressure of pure water at 25°C is 23.76 torr.

Solution

First, we need to know the mole fraction of H_2O.

$$n_{H_2O} = 175 \text{ g } H_2O \times \frac{1 \text{ mol } H_2O}{18.02 \text{ g } H_2O} = 9.71 \text{ mol } H_2O$$

$$n_{Na_2SO_4} = 35.0 \text{ g } Na_2SO_4 \times \frac{1 \text{ mol } Na_2SO_4}{142.05 \text{ g } Na_2SO_4} = 0.246 \text{ mol } Na_2SO_4$$

It is essential to recognize that when 1 mole of solid Na_2SO_4 dissolves, it produces 2 moles of Na^+ ions and 1 mole of SO_4^{2-} ions. Thus the number of solute particles present in this solution is three times the number of moles of solute dissolved:

$$n_{solute} = 3(0.246) = 0.738 \text{ mol}$$

$$\chi_{H_2O} = \frac{n_{H_2O}}{n_{solute} + n_{H_2O}} = \frac{9.71 \text{ mol}}{0.738 \text{ mol} + 9.72 \text{ mol}} = \frac{9.71}{10.458} = 0.929$$

Now we can use Raoult's law to predict the vapor pressure:

$$P_{soln} = \chi_{H_2O} P^0_{H_2O} = (0.929)(23.76 \text{ torr}) = 22.1 \text{ torr}$$

See Exercise 11.56

Nonideal Solutions

So far we have assumed that the solute is nonvolatile and so does not contribute to the vapor pressure over the solution. However, for liquid–liquid solutions where both components are volatile, a modified form of Raoult's law applies:

$$P_{TOTAL} = P_A + P_B = \chi_A P^0_A + \chi_B P^0_B$$

where P_{TOTAL} represents the total vapor pressure of a solution containing A and B, χ_A and χ_B are the mole fractions of A and B, P^0_A and P^0_B are the vapor pressures of pure A and pure B, and P_A and P_B are the partial pressures resulting from molecules of A and of B in the vapor above the solution (Fig. 11.12).

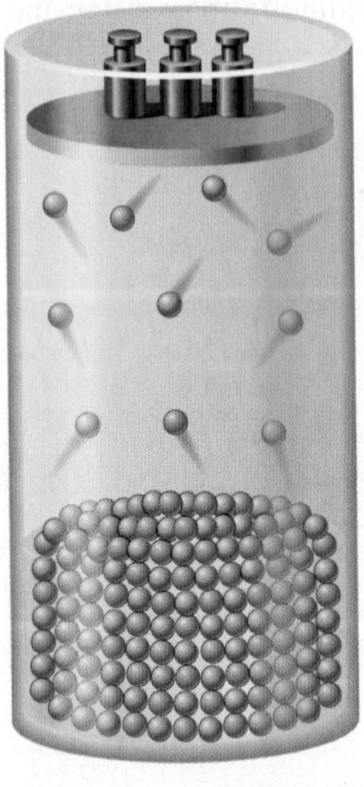

A liquid–liquid solution that obeys Raoult's law is called an **ideal solution**. Raoult's law is to solutions what the ideal gas law is to gases. As with gases, ideal behavior for solutions is never perfectly achieved but is sometimes closely approached. Nearly ideal behavior is often observed when the solute–solute, solvent–solvent, and solute–solvent interactions are very similar. That is, in solutions where the solute and solvent are very much alike, the solute simply acts to dilute the solvent. However, if the solvent has a special affinity for the solute, such as if hydrogen bonding occurs, the tendency of the solvent molecules to escape will be lowered more than expected. The observed

Figure 11.12 | When a solution contains two volatile components, both contribute to the total vapor pressure. Note that in this case the solution contains equal numbers of the components ⬤ and ⬤, but the vapor contains more ⬤ than ⬤. This means that component ⬤ is more volatile (has a higher vapor pressure as a pure liquid) than component ⬤.

vapor pressure will be *lower* than the value predicted by Raoult's law; there will be a *negative deviation from Raoult's law.*

When a solute and solvent release large quantities of energy in the formation of a solution, that is, when ΔH_{soln} is large and negative, we can assume that strong interactions exist between the solute and solvent. In this case we expect a negative deviation from Raoult's law, because both components will have a lower escaping tendency in the solution than in the pure liquids. This behavior is illustrated by an acetone–water solution, where the molecules can hydrogen-bond effectively:

$$CH_3 \atop CH_3 \diagdown C=O\,\cdots\,H-O \diagup ^H$$
$$\delta- \qquad \delta+$$

In contrast, if two liquids mix endothermically, it indicates that the solute–solvent interactions are weaker than the interactions among the molecules in the pure liquids. More energy is required to expand the liquids than is released when the liquids are mixed. In this case the molecules in the solution have a higher tendency to escape than expected, and *positive* deviations from Raoult's law are observed (Fig. 11.13). An example of this case is provided by a solution of ethanol and hexane, whose Lewis structures are as follows:

Ethanol   **Hexane**

The polar ethanol and the nonpolar hexane molecules are not able to interact effectively. Thus the enthalpy of solution is positive, as is the deviation from Raoult's law.

Finally, for a solution of very similar liquids, such as benzene and toluene (shown in margin), the enthalpy of solution is very close to zero, and thus the solution closely obeys Raoult's law (ideal behavior).

A summary of the behavior of various types of solutions is given in Table 11.4.

Strong solute–solvent interaction gives a vapor pressure lower than that predicted by Raoult's law.

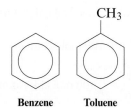

Benzene   **Toluene**

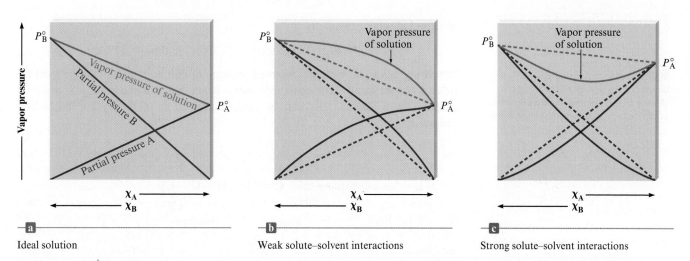

Ideal solution   Weak solute–solvent interactions   Strong solute–solvent interactions

Figure 11.13 | Vapor pressure for a solution of two volatile liquids. (a) The behavior predicted for an ideal liquid–liquid solution by Raoult's law. (b) A solution for which P_{TOTAL} is larger than the value calculated from Raoult's law. This solution shows a positive deviation from Raoult's law. (c) A solution for which P_{TOTAL} is smaller than the value calculated from Raoult's law. This solution shows a negative deviation from Raoult's law.

Table 11.4 | Summary of the Behavior of Various Types of Solutions

Interactive Forces Between Solute (A) and Solvent (B) Particles	ΔH_{soln}	ΔT for Solution Formation	Deviation from Raoult's Law	Example
A ↔ A, B ↔ B ≡ A ↔ B	Zero	Zero	None (ideal solution)	Benzene–toluene
A ↔ A, B ↔ B < A ↔ B	Negative (exothermic)	Positive	Negative	Acetone–water
A ↔ A, B ↔ B > A ↔ B	Positive (endothermic)	Negative	Positive	Ethanol–hexane

Interactive Example 11.7

Sign in at http://login.cengagebrain.com to try this Interactive Example in **OWL**.

Acetone

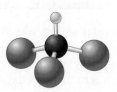

Chloroform

In this case the usually nonpolar C—H bond is strongly polarized by the three attached, highly electronegative chlorine atoms, thus producing hydrogen bonding.

Calculating the Vapor Pressure of a Solution Containing Two Liquids

A solution is prepared by mixing 5.81 g acetone (C_3H_6O, molar mass = 58.1 g/mol) and 11.9 g chloroform ($HCCl_3$, molar mass = 119.4 g/mol). At 35°C, this solution has a total vapor pressure of 260. torr. Is this an ideal solution? The vapor pressures of pure acetone and pure chloroform at 35°C are 345 and 293 torr, respectively.

Solution

To decide whether this solution behaves ideally, we first calculate the expected vapor pressure using Raoult's law:

$$P_{TOTAL} = \chi_A P_A^0 + \chi_C P_C^0$$

where A stands for acetone and C stands for chloroform. The calculated value can then be compared with the observed vapor pressure.

First, we must calculate the number of moles of acetone and chloroform:

$$5.81 \text{ g acetone} \times \frac{1 \text{ mol acetone}}{58.1 \text{ g acetone}} = 0.100 \text{ mol acetone}$$

$$11.9 \text{ g chloroform} \times \frac{1 \text{ mol chloroform}}{119 \text{ g chloroform}} = 0.100 \text{ mol chloroform}$$

Since the solution contains equal numbers of moles of acetone and chloroform, that is,

$$\chi_A = 0.500 \quad \text{and} \quad \chi_C = 0.500$$

the expected vapor pressure is

$$P_{TOTAL} = (0.500)(345 \text{ torr}) + (0.500)(293 \text{ torr}) = 319 \text{ torr}$$

Comparing this value with the observed pressure of 260. torr shows that the solution does not behave ideally. The observed value is lower than that expected. This negative deviation from Raoult's law can be explained in terms of the hydrogen-bonding interaction

$$\underset{\text{Acetone}}{\underset{CH_3}{\overset{CH_3}{\diagdown}}C=O}\cdots\underset{\text{Chloroform}}{H-\underset{Cl}{\overset{Cl}{\underset{|}{\overset{|}{C}}}}-Cl}$$

$\delta- \quad \delta+$

which lowers the tendency of these molecules to escape from the solution.

See Exercises 11.57, 11.58, 11.64, and 11.99

11.5 | Boiling-Point Elevation and Freezing-Point Depression

In the preceding section we saw how a solute affects the vapor pressure of a liquid solvent. Because changes of state depend on vapor pressure, the presence of a solute also affects the freezing point and boiling point of a solvent. Freezing-point depression, boiling-point elevation, and osmotic pressure (discussed in Section 11.6) are called **colligative properties**. As we will see, they are grouped together because they depend only on the number, and not on the identity, of the solute particles in an ideal solution. Because of their direct relationship to the number of solute particles, the colligative properties are very useful for characterizing the nature of a solute after it is dissolved in a solvent and for determining molar masses of substances.

Boiling-Point Elevation

Normal boiling point was defined in Section 10.8.

The normal boiling point of a liquid occurs at the temperature where the vapor pressure is equal to 1 atmosphere. We have seen that a nonvolatile solute lowers the vapor pressure of the solvent. Therefore, such a solution must be heated to a higher temperature than the boiling point of the pure solvent to reach a vapor pressure of 1 atmosphere. This means that *a nonvolatile solute elevates the boiling point of the solvent.* Figure 11.14 shows the phase diagram for an aqueous solution containing a nonvolatile solute. Note that the liquid/vapor line is shifted to higher temperatures than those for pure water.

As you might expect, the magnitude of the boiling-point elevation depends on the concentration of the solute. The change in boiling point can be represented by the equation

$$\Delta T = K_b m_{solute}$$

where ΔT is the boiling-point elevation, or the difference between the boiling point of the solution and that of the pure solvent, K_b is a constant that is characteristic of the solvent and is called the **molal boiling-point elevation constant**, and m_{solute} is the *molality* of the solute in the solution.

Values of K_b for some common solvents are given in Table 11.5. The molar mass of a solute can be determined from the observed boiling-point elevation, as shown in Example 11.8.

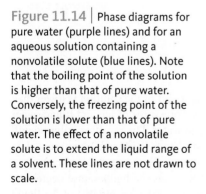

Figure 11.14 | Phase diagrams for pure water (purple lines) and for an aqueous solution containing a nonvolatile solute (blue lines). Note that the boiling point of the solution is higher than that of pure water. Conversely, the freezing point of the solution is lower than that of pure water. The effect of a nonvolatile solute is to extend the liquid range of a solvent. These lines are not drawn to scale.

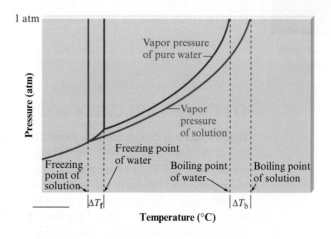

Table 11.5 | Molal Boiling-Point Elevation Constants (K_b) and Freezing-Point Depression Constants (K_f) for Several Solvents

Solvent	Boiling Point (°C)	K_b (°C · kg/mol)	Freezing Point (°C)	K_f (°C · kg/mol)
Water (H_2O)	100.0	0.51	0	1.86
Carbon tetrachloride (CCl_4)	76.5	5.03	−22.99	30.
Chloroform ($CHCl_3$)	61.2	3.63	−63.5	4.70
Benzene (C_6H_6)	80.1	2.53	5.5	5.12
Carbon disulfide (CS_2)	46.2	2.34	−111.5	3.83
Ethyl ether ($C_4H_{10}O$)	34.5	2.02	−116.2	1.79
Camphor ($C_{10}H_{16}O$)	208.0	5.95	179.8	40.

Sugar dissolved in water to make candy causes the boiling point to be elevated above 100°C.

Interactive Example 11.8

Sign in at http://login.cengagebrain.com to try this Interactive Example in **OWL**.

Calculating the Molar Mass by Boiling-Point Elevation

A solution was prepared by dissolving 18.00 g glucose in 150.0 g water. The resulting solution was found to have a boiling point of 100.34°C. Calculate the molar mass of glucose. Glucose is a molecular solid that is present as individual molecules in solution.

Solution

We make use of the equation

$$\Delta T = K_b m_{solute}$$

where

$$\Delta T = 100.34°C − 100.00°C = 0.34°C$$

From Table 11.5, for water $K_b = 0.51$. The molality of this solution then can be calculated by rearranging the boiling-point elevation equation to give

$$m_{solute} = \frac{\Delta T}{K_b} = \frac{0.34°C}{0.51°C \cdot kg/mol} = 0.67 \ mol/kg$$

The solution was prepared using 0.1500 kg water. Using the definition of molality, we can find the number of moles of glucose in the solution.

$$m_{solute} = 0.67 \ mol/kg = \frac{mol \ solute}{kg \ solvent} = \frac{n_{glucose}}{0.1500 \ kg}$$

$$n_{glucose} = (0.67 \ mol/kg)(0.1500 \ kg) = 0.10 \ mol$$

Thus 0.10 mole of glucose has a mass of 18.00 g, and 1.0 mole of glucose has a mass of 180 g (10×18.00 g). The molar mass of glucose is 180 g/mol.

See Exercise 11.66

Freezing-Point Depression

When a solute is dissolved in a solvent, the freezing point of the solution is lower than that of the pure solvent. Why? Recall that the vapor pressures of ice and liquid water are the same at 0°C. Suppose a solute is dissolved in water. The resulting solution will not freeze at 0°C because *the water in the solution has a lower vapor pressure than that of pure ice.* No ice will form under these conditions. However, the vapor pressure of ice decreases more rapidly than that of liquid water as the temperature decreases. Therefore, as the solution is cooled, the vapor pressure of the ice and that of the liquid

Melting point and *freezing point* both refer to the temperature where the solid and liquid coexist.

Figure 11.15 | (a) Ice in equilibrium with liquid water. (b) Ice in equilibrium with liquid water containing a dissolved solute (shown in red).

water in the solution will eventually become equal. The temperature at which this occurs is the new freezing point of the solution and is below 0°C. The freezing point has been *depressed*.

We can account for this behavior in terms of the simple model shown in Fig. 11.15. The presence of the solute lowers the rate at which molecules in the liquid return to the solid state. Thus for an aqueous solution, only the liquid state is found at 0°C. As the solution is cooled, the rate at which water molecules leave the solid ice decreases until this rate and the rate of formation of ice become equal and equilibrium is reached. This is the freezing point of the water in the solution.

Because a solute lowers the freezing point of water, compounds such as sodium chloride and calcium chloride are often spread on streets and sidewalks to prevent ice from forming in freezing weather. Of course, if the outside temperature is lower than the freezing point of the resulting salt solution, ice forms anyway. So this procedure is not effective at extremely cold temperatures.

The solid/liquid line for an aqueous solution is shown on the phase diagram for water in Fig. 11.14. Since the presence of a solute elevates the boiling point and depresses the freezing point of the solvent, adding a solute has the effect of extending the liquid range.

The equation for freezing-point depression is analogous to that for boiling-point elevation:

$$\Delta T = K_f m_{solute}$$

where ΔT is the freezing-point depression, or the difference between the freezing point of the pure solvent and that of the solution, and K_f is a constant that is characteristic of a particular solvent and is called the **molal freezing-point depression constant**. Values of K_f for common solvents are listed in Table 11.5.

Like the boiling-point elevation, the observed freezing-point depression can be used to determine molar masses and to characterize solutions.

Spreading salt on a highway.

Interactive Example 11.9

Sign in at http://login.cengagebrain.com to try this Interactive Example in OWL.

Ethylene glycol

Freezing-Point Depression

What mass of ethylene glycol ($C_2H_6O_2$, molar mass = 62.1 g/mol), the main component of antifreeze, must be added to 10.0 L water to produce a solution for use in a car's radiator that freezes at $-10.0°F$ ($-23.3°C$)? Assume the density of water is exactly 1 g/mL.

Solution

The freezing point must be lowered from 0°C to $-23.3°C$. To determine the molality of ethylene glycol needed to accomplish this, we can use the equation

$$\Delta T = K_f m_{solute}$$

The addition of antifreeze lowers the freezing point of water in a car's radiator.

where $\Delta T = 23.3°$ and $K_f = 1.86$ (from Table 11.5). Solving for the molality gives

$$m_{solute} = \frac{\Delta T}{K_f} = \frac{23.3°C}{1.86°C \cdot kg/mol} = 12.5 \text{ mol/kg}$$

This means that 12.5 moles of ethylene glycol must be added per kilogram of water. We have 10.0 L, or 10.0 kg, of water. Therefore, the total number of moles of ethylene glycol needed is

$$\frac{12.5 \text{ mol}}{kg} \times 10.0 \text{ kg} = 1.25 \times 10^2 \text{ mol}$$

The mass of ethylene glycol needed is

$$1.25 \times 10^2 \text{ mol} \times \frac{62.1 \text{ g}}{mol} = 7.76 \times 10^3 \text{ g (or 7.76 kg)}$$

See Exercises 11.69 and 11.70

Interactive Example 11.10

Sign in at http://login.cengagebrain.com to try this Interactive Example in **OWL**.

Determining Molar Mass by Freezing-Point Depression

A chemist is trying to identify a human hormone that controls metabolism by determining its molar mass. A sample weighing 0.546 g was dissolved in 15.0 g benzene, and the freezing-point depression was determined to be 0.240°C. Calculate the molar mass of the hormone.

Solution

From Table 11.5, K_f for benzene is 5.12°C · kg/mol, so the molality of the hormone is

$$m_{hormone} = \frac{\Delta T}{K_f} = \frac{0.240°C}{5.12°C \cdot kg/mol} = 4.69 \times 10^{-2} \text{ mol/kg}$$

The moles of hormone can be obtained from the definition of molality:

$$4.69 \times 10^{-2} \text{ mol/kg} = m_{solute} = \frac{\text{mol hormone}}{0.0150 \text{ kg benzene}}$$

or

$$\text{mol hormone} = \left(4.69 \times 10^{-2} \frac{mol}{kg}\right)(0.0150 \text{ kg}) = 7.04 \times 10^{-4} \text{ mol}$$

Since 0.546 g hormone was dissolved, 7.04×10^{-4} mole of hormone has a mass of 0.546 g, and

$$\frac{0.546 \text{ g}}{7.04 \times 10^{-4} \text{ mol}} = \frac{x}{1.00 \text{ mol}}$$

$$x = 776$$

Thus the molar mass of the hormone is 776 g/mol.

See Exercise 11.71

11.6 | Osmotic Pressure

Osmotic pressure, another of the colligative properties, can be understood from Fig. 11.16. A solution and pure solvent are separated by a **semipermeable membrane**, which allows *solvent but not solute* molecules to pass through. As time passes, the volume of the solution increases and that of the solvent decreases. This flow of solvent into the solution through the semipermeable membrane is called **osmosis**. Eventually the liquid levels stop changing, indicating that the system has reached equilibrium. Because the liquid levels are different at this point, there is a greater hydrostatic pressure on the solution than on the pure solvent. This excess pressure is called the **osmotic pressure**.

We can take another view of this phenomenon, as illustrated in Fig. 11.17. Osmosis can be prevented by applying a pressure to the solution. *The minimum pressure that stops the osmosis is equal to the osmotic pressure of the solution.* A simple model to explain osmotic pressure can be constructed as shown in Fig. 11.18. The membrane allows only solvent molecules to pass through. However, the initial rates of solvent transfer to and from the solution are not the same. The solute particles interfere with the passage of solvent, so the rate of transfer is slower from the solution to the solvent than in the reverse direction. Thus there is a net transfer of solvent molecules into the solution, which causes the solution volume to increase. As the solution level rises in the tube, the resulting pressure exerts an extra "push" on the solvent molecules in the solution, forcing them back through the membrane. Eventually, enough pressure develops so that the solvent transfer becomes equal in both directions. At this point, equilibrium is achieved and the levels stop changing.

Osmotic pressure can be used to characterize solutions and determine molar masses, as can the other colligative properties, but osmotic pressure is particularly useful because a small concentration of solute produces a relatively large osmotic pressure.

Experiments show that the dependence of the osmotic pressure on solution concentration is represented by the equation

$$\Pi = MRT$$

where Π is the osmotic pressure in atmospheres, M is the molarity of the solution, R is the gas law constant, and T is the Kelvin temperature.

A molar mass determination using osmotic pressure is illustrated in Example 11.11.

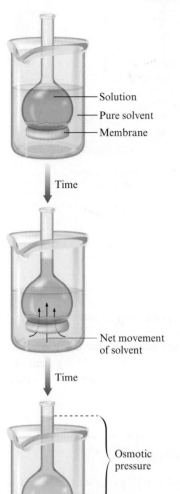

Solution
Pure solvent
Membrane

Time

Net movement of solvent

Time

Osmotic pressure

(at equilibrium)

Figure 11.16 | A tube with a bulb on the end that is covered by a semipermeable membrane. The solution is inside the tube and is bathed in the pure solvent. There is a net transfer of solvent molecules into the solution until the hydrostatic pressure equalizes the solvent flow in both directions.

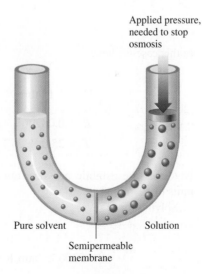

Applied pressure, needed to stop osmosis

Pure solvent   Solution

Semipermeable membrane

Figure 11.17 | The normal flow of solvent into the solution (osmosis) can be prevented by applying an external pressure to the solution. The minimum pressure required to stop the osmosis is equal to the osmotic pressure of the solution.

Figure 11.18 | (a) A pure solvent and its solution (containing a nonvolatile solute) are separated by a semipermeable membrane through which solvent molecules (blue) can pass but solute molecules (green) cannot. The rate of solvent transfer is greater from solvent to solution than from solution to solvent. (b) The system at equilibrium, where the rate of solvent transfer is the same in both directions.

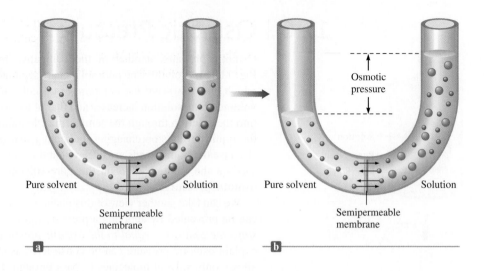

Critical Thinking

Consider the model of osmotic pressure as shown in Fig. 11.18. What if both sides contained a different pure solvent, each with a different vapor pressure? What would the system look like at equilibrium? Assume the different solvent molecules are able to pass through the membrane.

Interactive Example 11.11

Sign in at http://login.cengagebrain .com to try this Interactive Example in **OWL**.

Determining Molar Mass from Osmotic Pressure

To determine the molar mass of a certain protein, 1.00×10^{-3} g of it was dissolved in enough water to make 1.00 mL of solution. The osmotic pressure of this solution was found to be 1.12 torr at 25.0°C. Calculate the molar mass of the protein.

Solution

We use the equation

$$\Pi = MRT$$

In this case we have

$$\Pi = 1.12 \text{ torr} \times \frac{1 \text{ atm}}{760 \text{ torr}} = 1.47 \times 10^{-3} \text{ atm}$$

$$R = 0.08206 \text{ L} \cdot \text{atm/K} \cdot \text{mol}$$

$$T = 25.0 + 273 = 298 \text{ K}$$

Note that the osmotic pressure must be converted to atmospheres because of the units of R.

Solving for M gives

$$M = \frac{1.47 \times 10^{-3} \text{ atm}}{(0.08206 \text{ L} \cdot \text{atm/K} \cdot \text{mol})(298 \text{ K})} = 6.01 \times 10^{-5} \text{ mol/L}$$

Since 1.00×10^{-3} g protein was dissolved in 1 mL solution, the mass of protein per liter of solution is 1.00 g. The solution's concentration is 6.01×10^{-5} mol/L. This concentration is produced from 1.00×10^{-3} g protein per milliliter, or 1.00 g/L. Thus 6.01×10^{-5} mol protein has a mass of 1.00 g and

$$\frac{1.00 \text{ g}}{6.01 \times 10^{-5} \text{ mol}} = \frac{x}{1.00 \text{ mol}}$$

$$x = 1.66 \times 10^4 \text{ g}$$

Measurements of osmotic pressure generally give much more accurate molar mass values than those from freezing-point or boiling-point changes.

The molar mass of the protein is 1.66×10^4 g/mol. This molar mass may seem very large, but it is relatively small for a protein.

See Exercises 11.75 and 11.76

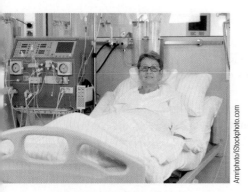

Patient undergoing dialysis.

The brine used in pickling causes the cucumbers to shrivel.

In osmosis, a semipermeable membrane prevents transfer of *all* solute particles. A similar phenomenon, called **dialysis**, occurs at the walls of most plant and animal cells. However, in this case the membrane allows transfer of both solvent molecules and *small* solute molecules and ions. One of the most important applications of dialysis is the use of artificial kidney machines to purify the blood. The blood is passed through a cellophane tube, which acts as the semipermeable membrane. The tube is immersed in a dialyzing solution (Fig. 11.19). This "washing" solution contains the same concentrations of ions and small molecules as blood but has none of the waste products normally removed by the kidneys. The resulting dialysis (movement of waste molecules into the washing solution) cleanses the blood.

Solutions that have identical osmotic pressures are said to be **isotonic solutions**. Fluids administered intravenously must be isotonic with body fluids. For example, if red blood cells are bathed in a hypertonic solution, which is a solution having an osmotic pressure higher than that of the cell fluids, the cells will shrivel because of a net transfer of water out of the cells. This phenomenon is called *crenation*. The opposite phenomenon, called *hemolysis,* occurs when cells are bathed in a hypotonic solution, a solution with an osmotic pressure lower than that of the cell fluids. In this case, the cells rupture because of the flow of water into the cells.

We can use the phenomenon of crenation to our advantage. Food can be preserved by treating its surface with a solute that gives a solution that is hypertonic to bacteria cells. Bacteria on the food then tend to shrivel and die. This is why salt can be used to protect meat and sugar can be used to protect fruit.

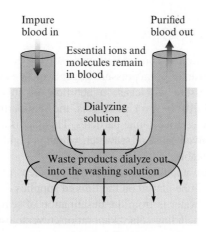

Impure blood in

Purified blood out

Essential ions and molecules remain in blood

Dialyzing solution

Waste products dialyze out into the washing solution

Figure 11.19 | Representation of the functioning of an artificial kidney.

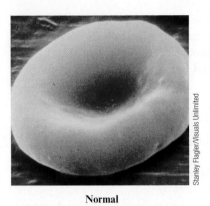

Normal

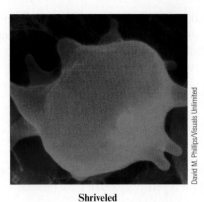

Shriveled

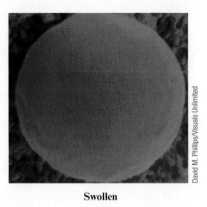

Swollen

Red blood cells in three stages of osmosis. (left) The normal shape of a red blood cell. (center) This cell has shrunk because water moved out of it by osmosis. (right) This cell is swollen with water that has moved into it by osmosis.

Interactive Example 11.12

Sign in at http://login.cengagebrain.com to try this Interactive Example in **OWL**.

Isotonic Solutions

What concentration of sodium chloride in water is needed to produce an aqueous solution isotonic with blood ($\Pi = 7.70$ atm at 25°C)?

Solution

We can calculate the molarity of the solute from the equation

$$\Pi = MRT \quad \text{or} \quad M = \frac{\Pi}{RT}$$

$$M = \frac{7.70 \text{ atm}}{(0.08206 \text{ L} \cdot \text{atm/K} \cdot \text{mol})(298 \text{ K})} = 0.315 \text{ mol/L}$$

This represents the total molarity of solute particles. But NaCl gives two ions per formula unit. Therefore, the concentration of NaCl needed is $\dfrac{0.315 \, M}{2} = 0.1575 \, M = 0.158 \, M$. That is,

$$NaCl \longrightarrow Na^+ + Cl^-$$

$$\underbrace{0.1575 \, M \quad 0.1575 \, M \quad 0.1575 \, M}_{0.315 \, M}$$

See Exercise 11.78

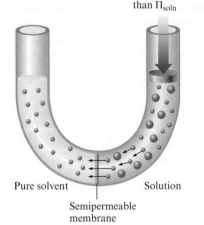

Pressure greater than Π_{soln}

Pure solvent   Solution

Semipermeable membrane

Figure 11.20 | Reverse osmosis. A pressure greater than the osmotic pressure of the solution is applied, which causes a net flow of solvent molecules (blue) from the solution to the pure solvent. The solute molecules (green) remain behind.

Reverse Osmosis

If a solution in contact with pure solvent across a semipermeable membrane is subjected to an external pressure larger than its osmotic pressure, **reverse osmosis** occurs. The pressure will cause a net flow of solvent from the solution to the solvent (Fig. 11.20). In reverse osmosis, the semipermeable membrane acts as a "molecular filter" to remove solute particles. This fact is applicable to the **desalination** (removal of dissolved salts) of seawater, which is highly hypertonic to body fluids and thus is not drinkable.

As the population of the Sun Belt areas of the United States increases, more demand will be placed on the limited supplies of fresh water there. One obvious source of fresh water is from the desalination of seawater. Various schemes have been suggested, including solar evaporation, reverse osmosis, and even a plan for towing icebergs from Antarctica. The problem, of course, is that all the available processes are

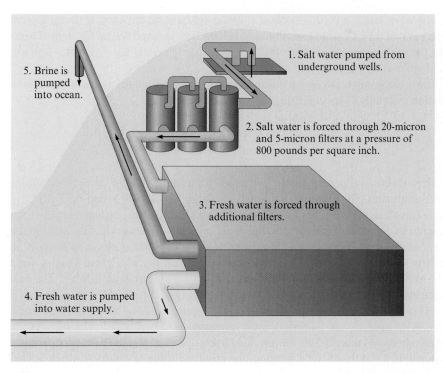

Figure 11.21 | (a) Residents of Catalina Island off the coast of southern California are benefiting from a desalination plant that can supply 132,000 gallons a day, or one-third of the island's daily needs. (b) Machinery in the desalination plant for Catalina Island.

expensive. However, as water shortages increase, desalination is becoming necessary. For example, the first full-time public desalination plant in the United States started operations on Catalina Island, just off the coast of California (Fig. 11.21). This plant, which can produce 132,000 gallons of drinkable water from the Pacific Ocean every day, operates by reverse osmosis. Powerful pumps, developing over 800 lb/in^2 of pressure, are used to force seawater through synthetic semipermeable membranes.

Catalina Island's plant may be just the beginning. The city of Santa Barbara opened a $40 million desalination plant in 1992 that can produce 8 million gallons of drinking water per day. The southern California city of Carlsbad opened a reverse osmosis desalination plant in 2012 that can produce 50 million gallons of drinking water daily from seawater. Desalination plants are also in the works for Huntington Beach, California, and Camp Pendleton, a military base just north of Carlsbad.

A small-scale, manually operated reverse osmosis desalinator has been developed by the U.S. Navy to provide fresh water on life rafts. Potable water can be supplied by this desalinator at the rate of 1.25 gallons of water per hour—enough to keep 25 people alive. This compact desalinator, which weighs only 10 pounds, can now replace the bulky cases of fresh water formerly stored in Navy life rafts.

11.7 | Colligative Properties of Electrolyte Solutions

As we have seen previously, the colligative properties of solutions depend on the total concentration of solute particles. For example, a 0.10-m glucose solution shows a freezing-point depression of 0.186°C:

$$\Delta T = K_f m = (1.86°C \cdot kg/mol)(0.100 \text{ mol/kg}) = 0.186°C$$

On the other hand, a 0.10-m sodium chloride solution should show a freezing-point depression of 0.37°C, since the solution is 0.10 m Na^+ ions and 0.10 m Cl^- ions. Therefore, the solution contains a total of 0.20 m solute particles, and $\Delta T = (1.86°C \cdot kg/mol)$ (0.20 mol/kg) = 0.37°C.

The relationship between the moles of solute dissolved and the moles of particles in solution is usually expressed using the **van't Hoff factor**, i:

$$i = \frac{\text{moles of particles in solution}}{\text{moles of solute dissolved}}$$

The *expected* value for i can be calculated for a salt by noting the number of ions per formula unit. For example, for NaCl, i is 2; for K_2SO_4, i is 3; and for $Fe_3(PO_4)_2$, i is 5. These calculated values assume that when a salt dissolves, it completely dissociates into its component ions, which then move around independently. This assumption is not always true. For example, the freezing-point depression observed for 0.10 m NaCl is 1.87 times that for 0.10 m glucose rather than twice as great. That is, for a 0.10-m NaCl solution, the observed value for i is 1.87 rather than 2. Why? The best explanation is that **ion pairing** occurs in solution (Fig. 11.22). At a given instant a small percentage of the sodium and chloride ions are paired and thus count as a single particle. In general, ion pairing is most important in concentrated solutions. As the solution becomes more dilute, the ions are farther apart and less ion pairing occurs. For example, in a 0.0010-m NaCl solution, the observed value of i is 1.97, which is very close to the expected value.

Ion pairing occurs to some extent in all electrolyte solutions. Table 11.6 shows expected and observed values of i for a given concentration of various electrolytes. Note that the deviation of i from the expected value tends to be greatest where the ions have multiple charges. This is expected because ion pairing ought to be most important for highly charged ions.

The colligative properties of electrolyte solutions are described by including the van't Hoff factor in the appropriate equation. For example, for changes in freezing and boiling points, the modified equation is

$$\Delta T = imK$$

where K represents the freezing-point depression or boiling-point elevation constant for the solvent.

For the osmotic pressure of electrolyte solutions, the equation is

$$\Pi = iMRT$$

Dutch chemist J. H. van't Hoff (1852–1911) received the first Nobel Prize in chemistry in 1901.

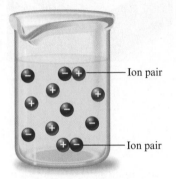

Figure 11.22 | In an aqueous solution a few ions aggregate, forming ion pairs that behave as a unit.

Table 11.6 | Expected and Observed Values of the van't Hoff Factor for 0.05 m Solutions of Several Electrolytes

Electrolyte	i (expected)	i (observed)
NaCl	2.0	1.9
$MgCl_2$	3.0	2.7
$MgSO_4$	2.0	1.3
$FeCl_3$	4.0	3.4
HCl	2.0	1.9
Glucose*	1.0	1.0

*A nonelectrolyte shown for comparison.

Chemical connections
The Drink of Champions—Water

In 1965, the University of Florida football team, the Gators, participated in a research program to test a sports drink formula containing a mixture of carbohydrates and electrolytes. The drink was used to help prevent dehydration caused by extreme workouts in the hot Florida climate. The Gators' success that season was in part attributed to their use of the sports drink formula. In 1967, a modified form of this formula was marketed with the name Gatorade. Today, Gatorade leads sales in sports drinks, but many other brands have entered a market where annual sales exceed $700 million!

During moderate- to high-intensity exercise, glycogen (a fuel reserve that helps maintain normal body processes) can be depleted within 60 to 90 minutes. Blood-sugar levels drop as the glycogen reserves are used up, and lactic acid (a by-product of glucose metabolism) builds up in muscle tissue causing fatigue and muscle cramps. Muscles also generate a large amount of heat that must be dissipated. Water, which has a large specific heat capacity, is used to take heat away from these muscles. Sweating and evaporative cooling help the body maintain a constant temperature, but at a huge cost. During a high-intensity workout in hot weather, anywhere from 1 to 3 quarts of water can be lost from sweating per hour. Sweating away more than 2% of your body weight—a quart for every 100 pounds—can put a large stress on the heart, increasing body temperature and decreasing performance. Excessive sweating also results in the loss of sodium and potassium ions—two very important electrolytes that are present in the fluids inside and outside cells.

All the major sports drinks contain three main ingredients—carbohydrates in the form of simple sugars such as sucrose, glucose, and fructose; electrolytes, including sodium and potassium ions; and water. Because these are the three major substances lost through sweating, good scientific reasoning suggests that drinking sports drinks should improve performance. But just how effectively do sports drinks deliver on their promises?

Recent studies have confirmed that athletes who eat a balanced diet and drink plenty of water are just as well off as those who consume sports drinks. A sports drink may have only one advantage over drinking water—it

For healthy athletes, drinking water during exercise may be as effective as drinking sports drinks.

tastes better than water to most athletes. And if a drink tastes better, it will encourage more consumption, thus keeping cells hydrated.

Since most of the leading sports drinks contain the same ingredients in similar concentrations, taste may be the single most important factor in choosing your drink. If you are not interested in any particular sports drink, drink plenty of water. The key to quality performance is to keep your cells hydrated.

Adapted with permission from "Sports Drinks: Don't Sweat the Small Stuff," by Tim Graham, *ChemMatters*, February 1999, p. 11.

Interactive Example 11.13

Sign in at http://login.cengagebrain.com to try this Interactive Example in **OWL**.

Osmotic Pressure

The observed osmotic pressure for a 0.10-*M* solution of $Fe(NH_4)_2(SO_4)_2$ at 25°C is 10.8 atm. Compare the expected and experimental values for *i*.

Solution

The ionic solid $Fe(NH_4)_2(SO_4)_2$ dissociates in water to produce 5 ions:

$$Fe(NH_4)_2(SO_4)_2 \xrightarrow{H_2O} Fe^{2+} + 2NH_4^+ + 2SO_4^{2-}$$

Thus the expected value for *i* is 5. We can obtain the experimental value for *i* by using the equation for osmotic pressure:

$$\Pi = iMRT \qquad \text{or} \qquad i = \frac{\Pi}{MRT}$$

where $\Pi = 10.8$ atm, $M = 0.10$ mol/L, $R = 0.08206$ L · atm/K · mol, and $T = 25 + 273 = 298$ K. Substituting these values into the equation gives

$$i = \frac{\Pi}{MRT} = \frac{10.8 \text{ atm}}{(0.10 \text{ mol/L})(0.08206 \text{ L} \cdot \text{atm/K} \cdot \text{mol})(298 \text{ K})} = 4.4$$

The experimental value for *i* is less than the expected value, presumably because of ion pairing.

See Exercises 11.87 and 11.88

11.8 | Colloids

Figure 11.23 | The Tyndall effect.

Mud can be suspended in water by vigorous stirring. When the stirring stops, most of the particles rapidly settle out, but even after several days some of the smallest particles remain suspended. Although undetected in normal lighting, their presence can be demonstrated by shining a beam of intense light through the suspension. The beam is visible from the side because the light is scattered by the suspended particles (Fig. 11.23). In a true solution, on the other hand, the beam is invisible from the side because the individual ions and molecules dispersed in the solution are too small to scatter visible light.

The scattering of light by particles is called the **Tyndall effect** and is often used to distinguish between a suspension and a true solution.

A suspension of tiny particles in some medium is called a **colloidal dispersion**, or a **colloid**. The suspended particles are single large molecules or aggregates of molecules or ions ranging in size from 1 to 1000 nm. Colloids are classified according to the states of the dispersed phase and the dispersing medium. Table 11.7 summarizes various types of colloids.

What stabilizes a colloid? Why do the particles remain suspended rather than forming larger aggregates and precipitating out? The answer is complicated, but the main factor seems to be *electrostatic repulsion*. A colloid, like all other macroscopic substances, is electrically neutral. However, when a colloid is placed in an electric field, the dispersed particles all migrate to the same electrode and thus must all have the same charge. How is this possible? The center of a colloidal particle (a tiny ionic crystal, a group of molecules, or a single large molecule) attracts from the medium a layer of ions, all of the same charge. This group of ions, in turn, attracts another layer of oppositely charged ions (Fig. 11.24). Because the colloidal particles all have an outer

Figure 11.24 | A representation of two colloidal particles. In each the center particle is surrounded by a layer of positive ions, with negative ions in the outer layer. Thus, although the particles are electrically neutral, they still repel each other because of their outer negative layer of ions.

Table 11.7 | Types of Colloids

Examples	Dispersing Medium	Dispersed Substance	Colloid Type
Fog, aerosol sprays	Gas	Liquid	Aerosol
Smoke, airborne bacteria	Gas	Solid	Aerosol
Whipped cream, soap suds	Liquid	Gas	Foam
Milk, mayonnaise	Liquid	Liquid	Emulsion
Paint, clays, gelatin	Liquid	Solid	Sol
Marshmallow, polystyrene foam	Solid	Gas	Solid foam
Butter, cheese	Solid	Liquid	Solid emulsion
Ruby glass	Solid	Solid	Solid sol

Chemical connections
Organisms and Ice Formation

The ice-cold waters of the polar oceans are teeming with fish that seem immune to freezing. One might think that these fish have some kind of antifreeze in their blood. However, studies show that they are protected from freezing in a very different way from the way antifreeze protects our cars. As we have seen in this chapter, solutes such as sugar, salt, and ethylene glycol lower the temperature at which the solid and liquid phases of water can coexist. However, the fish could not tolerate high concentrations of solutes in their blood because of the osmotic pressure effects. Instead, they are protected by proteins in their blood. These proteins allow the water in the bloodstream to be supercooled—exist below 0°C—without forming ice. They apparently coat the surface of each tiny ice crystal as soon as it begins to form, preventing it from growing to a size that would cause biologic damage.

Although it might at first seem surprising, this research on polar fish has attracted the attention of ice cream manufacturers. Premium quality ice cream is smooth; it does not have large ice crystals in it. The makers of ice cream would like to incorporate these polar fish proteins, or molecules that behave similarly, into ice cream to prevent the growth of ice crystals during storage.

Fruit and vegetable growers have a similar interest: They also want to prevent ice formation that damages their crops during an unusual cold wave. However, this is a very different kind of problem than keeping polar fish from freezing. Many types of fruits and vegetables are colonized by bacteria that manufacture a protein that *encourages* freezing by acting as a nucleating agent to start an ice crystal. Chemists have identified the offending protein in the bacteria and the gene that is responsible for making it. They

Blackfin icefish, Chaenocephalus aceratus.

have learned to modify the genetic material of these bacteria in a way that removes their ability to make the protein that encourages ice crystal formation. If testing shows that these modified bacteria have no harmful effects on the crop or the environment, the original bacteria strain will be replaced with the new form so that ice crystals will not form so readily when a cold snap occurs.

layer of ions with the same charge, they repel each other and do not easily aggregate to form particles that are large enough to precipitate.

The destruction of a colloid, called **coagulation**, usually can be accomplished either by heating or by adding an electrolyte. Heating increases the velocities of the colloidal particles, causing them to collide with enough energy that the ion barriers are penetrated and the particles can aggregate. Because this process is repeated many times, the particle grows to a point where it settles out. Adding an electrolyte neutralizes the adsorbed ion layers. This is why clay suspended in rivers is deposited where the river reaches the ocean, forming the deltas characteristic of large rivers like the Mississippi. The high salt content of the seawater causes the colloidal clay particles to coagulate.

The removal of soot from smoke is another example of the coagulation of a colloid. When smoke is passed through an electrostatic precipitator (Fig. 11.25), the suspended solids are removed. The use of precipitators has produced an immense improvement in the air quality of heavily industrialized cities.

Figure 11.25 | The Cottrell precipitator installed in a smokestack. The charged plates attract the colloidal particles because of their ion layers and thus remove them from the smoke.

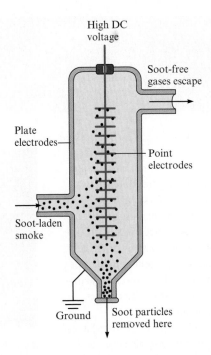

High DC voltage

Soot-free gases escape

Plate electrodes

Point electrodes

Soot-laden smoke

Ground

Soot particles removed here

For review

Key terms

Section 11.1
molarity
mass percent
mole fraction
molality
normality

Section 11.2
enthalpy (heat) of solution
enthalpy (heat) of hydration

Section 11.3
Henry's law
thermal pollution

Section 11.4
Raoult's law
ideal solution

Section 11.5
colligative properties
molal boiling-point elevation constant
molal freezing-point depression constant

Section 11.6
semipermeable membrane
osmosis
osmotic pressure

Solution composition

> Molarity (M): moles solute per liter of solution

> Mass percent: ratio of mass of solute to mass of solution times 100%

> Mole fraction (X): ratio of moles of a given component to total moles of all components

> Molality (m): moles solute per mass of solvent (in kg)

> Normality (N): number of equivalents per liter of solution

Enthalpy of solution (ΔH_{soln})

> The enthalpy change accompanying solution formation

> Can be partitioned into
>> The energy required to overcome the solute–solute interactions
>> The energy required to "make holes" in the solvent
>> The energy associated with solute–solvent interactions

Factors that affect solubility

> Polarity of solute and solvent
>> "Like dissolves like" is a useful generalization
> Pressure increases the solubility of gases in a solvent
>> Henry's law: $C = kP$
> Temperature effects
>> Increased temperature decreases the solubility of a gas in water
>> Most solids are more soluble at higher temperatures, but important exceptions exist

Key terms

dialysis
isotonic solution
reverse osmosis
desalination

Section 11.7
van't Hoff factor
ion pairing

Section 11.8
Tyndall effect
colloid (colloidal dispersion)
coagulation

Vapor pressure of solutions

> A solution containing a nonvolatile solute has a lower vapor pressure than a solution of the pure solvent

> Raoult's law defines an ideal solution:

$$P_{vapor}^{soln} = X_{solvent} P_{vapor}^{solvent}$$

> Solutions in which the solute–solvent attractions differ from the solute–solute and solvent–solvent attractions violate Raoult's law

Colligative properties

> Depend on the number of solute particles present
> Boiling-point elevation: $\Delta T = K_b m_{solute}$
> Freezing-point lowering: $\Delta T = K_f m_{solute}$
> Osmotic pressure: $\Pi = MRT$
> > Osmosis occurs when a solution and pure solvent are separated by a semipermeable membrane that allows solvent molecules to pass but not solute particles
> > Reverse osmosis occurs when the applied pressure is greater than the osmotic pressure of the solution
> Because colligative properties depend on the number of particles, solutes that break into several ions when they dissolve have an effect proportional to the number of ions produced
> > The van't Hoff factor i represents the number of ions produced by each formula unit of solute

Colloids

> A suspension of tiny particles stabilized by electrostatic repulsion among the ion layers surrounding the individual particles
> Can be coagulated (destroyed) by heating or adding an electrolyte

Review questions   Answers to the Review Questions can be found on the Student website (accessible from www.cengagebrain.com).

1. The four most common ways to describe solution composition are mass percent, mole fraction, molarity, and molality. Define each of these solution composition terms. Why is molarity temperature-dependent, whereas the other three solution composition terms are temperature-independent?

2. Using KF as an example, write equations that refer to ΔH_{soln} and ΔH_{hyd}. Lattice energy was defined in Chapter 8 as ΔH for the reaction $K^+(g) + F^-(g) \longrightarrow KF(s)$. Show how you would utilize Hess's law to calculate ΔH_{soln} from ΔH_{hyd} and ΔH_{LE} for KF, where $\Delta H_{LE} =$ lattice energy. ΔH_{soln} for KF, as for other soluble ionic compounds, is a relatively small number. How can this be since ΔH_{hyd} and ΔH_{LE} are relatively large negative numbers?

3. What does the axiom "like dissolves like" mean? There are four types of solute/solvent combinations: polar solutes in polar solvents, nonpolar solutes in polar solvents, and so on. For each type of solution, discuss the magnitude of ΔH_{soln}.

4. Structure, pressure, and temperature all have an effect on solubility. Discuss each of their effects. What is Henry's law? Why does Henry's law not work for HCl(g)? What do the terms *hydrophobic* and *hydrophilic* mean?

5. Define the terms in Raoult's law. Figure 11.9 illustrates the net transfer of water molecules from pure water to an aqueous solution of a nonvolatile solute. Explain why eventually all of the water from the beaker of pure water will transfer to the aqueous solution. If the experiment illustrated in Fig. 11.9 was performed using a volatile solute, what would happen? How do you calculate the total vapor pressure when both the solute and solvent are volatile?

6. In terms of Raoult's law, distinguish between an ideal liquid–liquid solution and a nonideal liquid–liquid solution. If a solution is ideal, what is true about ΔH_{soln}, ΔT for the solution formation, and the interactive forces within the pure solute and pure solvent as compared to the interactive forces within the solution? Give an example of an ideal solution. Answer the previous two questions for solutions that exhibit either negative or positive deviations from Raoult's law.

7. Vapor-pressure lowering is a colligative property, as are freezing-point depression and boiling-point elevation. What is a colligative property? Why is the freezing point depressed for a solution as compared to the pure solvent? Why is the boiling point elevated for a solution as compared to the pure solvent? Explain how to calculate ΔT for a freezing-point depression problem or a boiling-point elevation problem. Of the solvents listed in Table 11.5, which would have the largest freezing-point depression for a 0.50 molal solution? Which would have the smallest boiling-point elevation for a 0.50 molal solution?

   A common application of freezing-point depression and boiling-point elevation experiments is to provide a means to calculate the molar mass of a nonvolatile solute. What data are needed to calculate the molar mass of a nonvolatile solute? Explain how you would manipulate these data to calculate the molar mass of the nonvolatile solute.

8. What is osmotic pressure? How is osmotic pressure calculated? Molarity units are used in the osmotic pressure equation. When does the molarity of a solution approximately equal the molality of the solution? Before refrigeration was common, many foods were preserved by salting them heavily, and many fruits were preserved by mixing them with a large amount of sugar (fruit preserves). How do salt and sugar act as preservatives? Two applications of osmotic pressure are dialysis and desalination. Explain these two processes.

9. Distinguish between a strong electrolyte, a weak electrolyte, and a nonelectrolyte. How can colligative properties be used to distinguish between them? What is the van't Hoff factor? Why is the observed freezing-point depression for electrolyte solutions sometimes less than the calculated value? Is the discrepancy greater for concentrated or dilute solutions?

10. What is a colloidal dispersion? Give some examples of colloids. The Tyndall effect is often used to distinguish between a colloidal suspension and a true solution. Explain. The destruction of a colloid is done through a process called coagulation. What is coagulation?

Active Learning Questions

These questions are designed to be used by groups of students in class.

1. Consider Fig. 11.9. According to the caption and picture, water seems to go from one beaker to another.
   a. Explain why this occurs.
   b. The explanation in the text uses terms such as *vapor pressure* and *equilibrium.* Explain what these have to do with the phenomenon. For example, what is coming to equilibrium?
   c. Does all the water end up in the second beaker?
   d. Is water evaporating from the beaker containing the solution? If so, is the rate of evaporation increasing, decreasing, or staying constant?

   Draw pictures to illustrate your explanations.

2. Once again, consider Fig. 11.9. Suppose instead of having a nonvolatile solute in the solvent in one beaker, the two beakers contain different volatile liquids. That is, suppose one beaker contains liquid A ($P_{vap} = 50$ torr) and the other beaker contains liquid B ($P_{vap} = 100$ torr). Explain what happens as time passes. How is this similar to the first case (shown in the figure)? How is it different?

3. Assume that you place a freshwater plant into a saltwater solution and examine it under a microscope. What happens to the plant cells? What if you placed a saltwater plant in pure water? Explain. Draw pictures to illustrate your explanations.

4. How does ΔH_{soln} relate to deviations from Raoult's law? Explain.

5. You have read that adding a solute to a solvent can both increase the boiling point and decrease the freezing point. A friend of yours explains it to you like this: "The solute and solvent can be like salt in water. The salt gets in the way of freezing in that it blocks the water molecules from joining together. The salt acts like a strong bond holding the water molecules together so that it is harder to boil." What do you say to your friend?

6. You drop an ice cube (made from pure water) into a saltwater solution at 0°C. Explain what happens and why.

7. Using the phase diagram for water and Raoult's law, explain why salt is spread on the roads in winter (even when it is below freezing).

8. You and your friend are each drinking cola from separate 2-L bottles. Both colas are equally carbonated. You are able to drink 1 L of cola, but your friend can drink only about half a liter. You each close the bottles and place them in the refrigerator. The next day when you each go to get the colas, whose will be more carbonated and why?

9. Is molality or molarity dependent on temperature? Explain your answer. Why is molality, and not molarity, used in the equations describing freezing-point depression and boiling-point elevation?

10. Consider a beaker of salt water sitting open in a room. Over time, does the vapor pressure increase, decrease, or stay the same? Explain.

A blue question or exercise number indicates that the answer to that question or exercise appears at the back of this book and a solution appears in the *Solutions Guide,* as found on PowerLecture.

Solution Review

If you have trouble with these exercises, review Sections 4.1 to 4.3 in Chapter 4.

11. Rubbing alcohol contains 585 g isopropanol (C_3H_7OH) per liter (aqueous solution). Calculate the molarity.

12. What mass of sodium oxalate ($Na_2C_2O_4$) is needed to prepare 0.250 L of a 0.100-M solution?

13. What volume of 0.25 M HCl solution must be diluted to prepare 1.00 L of 0.040 M HCl?

14. What volume of a 0.580-M solution of $CaCl_2$ contains 1.28 g solute?

15. Calculate the sodium ion concentration when 70.0 mL of 3.0 M sodium carbonate is added to 30.0 mL of 1.0 M sodium bicarbonate.

16. Write equations showing the ions present after the following strong electrolytes are dissolved in water.

   a. HNO_3 f. NH_4Br
   b. Na_2SO_4 g. NH_4NO_3
   c. $Al(NO_3)_3$ h. $CuSO_4$
   d. $SrBr_2$ i. $NaOH$
   e. $KClO_4$

Questions

17. Rationalize the temperature dependence of the solubility of a gas in water in terms of the kinetic molecular theory.

18. The weak electrolyte $NH_3(g)$ does not obey Henry's law. Why? $O_2(g)$ obeys Henry's law in water but not in blood (an aqueous solution). Why?

19. The two beakers in the sealed container illustrated below contain pure water and an aqueous solution of a volatile solute.

Water   Aqueous
               solution

If the solute is less volatile than water, explain what will happen to the volumes in the two containers as time passes.

20. The following plot shows the vapor pressure of various solutions of components A and B at some temperature.

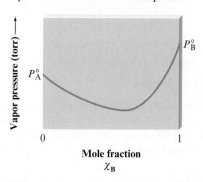

Mole fraction
χ_B

Which of the following statements is false concerning solutions of A and B?

   a. The solutions exhibit negative deviations from Raoult's law.

   b. ΔH_{soln} for the solutions should be exothermic.

   c. The intermolecular forces are stronger in solution than in either pure A or pure B.

   d. Pure liquid B is more volatile than pure liquid A.

   e. The solution with $\chi_B = 0.6$ will have a lower boiling point than either pure A or pure B.

21. When pure methanol is mixed with water, the resulting solution feels warm. Would you expect this solution to be ideal? Explain.

22. Detergent molecules can stabilize the emulsion of oil in water as well as remove dirt from soiled clothes. A typical detergent is sodium dodecylsulfate, or SDS, and it has a formula of $CH_3(CH_2)_{10}CH_2SO_4^-Na^+$. In aqueous solution, SDS suspends oil or dirt by forming small aggregates of detergent anions called *micelles*. Propose a structure for micelles.

23. For an acid or a base, when is the normality of a solution equal to the molarity of the solution and when are the two concentration units different?

24. In order for sodium chloride to dissolve in water, a small amount of energy must be added during solution formation. This is not energetically favorable. Why is NaCl so soluble in water?

25. Which of the following statements is(are) *true*? Correct the false statements.

   a. The vapor pressure of a solution is directly related to the mole fraction of solute.

   b. When a solute is added to water, the water in solution has a lower vapor pressure than that of pure ice at 0°C.

   c. Colligative properties depend only on the identity of the solute and not on the number of solute particles present.

   d. When sugar is added to water, the boiling point of the solution increases above 100°C because sugar has a higher boiling point than water.

26. Is the following statement true or false? Explain your answer. When determining the molar mass of a solute using boiling-point or freezing-point data, camphor would be the best solvent choice of all of the solvents listed in Table 11.5.

27. Explain the terms *isotonic solution, crenation,* and *hemolysis.*

28. What is ion pairing?

Exercises

In this section similar exercises are paired.

Solution Composition

29. A solution of phosphoric acid was made by dissolving 10.0 g H_3PO_4 in 100.0 mL water. The resulting volume was 104 mL. Calculate the density, mole fraction, molarity, and molality of the solution. Assume water has a density of 1.00 g/cm³.

30. An aqueous antifreeze solution is 40.0% ethylene glycol ($C_2H_6O_2$) by mass. The density of the solution is 1.05 g/cm³. Calculate the molality, molarity, and mole fraction of the ethylene glycol.

31. Common commercial acids and bases are aqueous solutions with the following properties:

	Density (g/cm³)	Mass Percent of Solute
Hydrochloric acid	1.19	38
Nitric acid	1.42	70.
Sulfuric acid	1.84	95
Acetic acid	1.05	99
Ammonia	0.90	28

Calculate the molarity, molality, and mole fraction of each of the preceding reagents.

32. In lab you need to prepare at least 100 mL of each of the following solutions. Explain how you would proceed using the given information.
 a. 2.0 *m* KCl in water (density of $H_2O = 1.00$ g/cm³)
 b. 15% NaOH by mass in water ($d = 1.00$ g/cm³)
 c. 25% NaOH by mass in CH_3OH ($d = 0.79$ g/cm³)
 d. 0.10 mole fraction of $C_6H_{12}O_6$ in water ($d = 1.00$ g/cm³)

33. A solution is prepared by mixing 25 mL pentane (C_5H_{12}, $d = 0.63$ g/cm³) with 45 mL hexane (C_6H_{14}, $d = 0.66$ g/cm³). Assuming that the volumes add on mixing, calculate the mass percent, mole fraction, molality, and molarity of the pentane.

34. A solution is prepared by mixing 50.0 mL toluene ($C_6H_5CH_3$, $d = 0.867$ g/cm³) with 125 mL benzene (C_6H_6, $d = 0.874$ g/cm³). Assuming that the volumes add on mixing, calculate the mass percent, mole fraction, molality, and molarity of the toluene.

35. A bottle of wine contains 12.5% ethanol by volume. The density of ethanol (C_2H_5OH) is 0.789 g/cm³. Calculate the concentration of ethanol in wine in terms of mass percent and molality.

36. Calculate the molarity and mole fraction of acetone in a 1.00-*m* solution of acetone (CH_3COCH_3) in ethanol (C_2H_5OH). (Density of acetone = 0.788 g/cm³; density of ethanol = 0.789 g/cm³.) Assume that the volumes of acetone and ethanol add.

37. A 1.37-*M* solution of citric acid ($H_3C_6H_5O_7$) in water has a density of 1.10 g/cm³. Calculate the mass percent, molality, mole fraction, and normality of the citric acid. Citric acid has three acidic protons.

38. Calculate the normality of each of the following solutions.
 a. 0.250 *M* HCl
 b. 0.105 *M* H_2SO_4
 c. 5.3 × 10⁻² *M* H_3PO_4
 d. 0.134 *M* NaOH
 e. 0.00521 *M* Ca(OH)₂

What is the equivalent mass for each of the acids or bases listed above?

Energetics of Solutions and Solubility

39. The lattice energy* of NaI is −686 kJ/mol, and the enthalpy of hydration is −694 kJ/mol. Calculate the enthalpy of solution per mole of solid NaI. Describe the process to which this enthalpy change applies.

40. **a.** Use the following data to calculate the enthalpy of hydration for calcium chloride and calcium iodide.

	Lattice Energy	ΔH_{soln}
$CaCl_2(s)$	−2247 kJ/mol	−46 kJ/mol
$CaI_2(s)$	−2059 kJ/mol	−104 kJ/mol

 b. Based on your answers to part a, which ion, Cl^- or I^-, is more strongly attracted to water?

41. Although $Al(OH)_3$ is insoluble in water, NaOH is very soluble. Explain in terms of lattice energies.

42. The high melting points of ionic solids indicate that a lot of energy must be supplied to separate the ions from one another. How is it possible that the ions can separate from one another when soluble ionic compounds are dissolved in water, often with essentially no temperature change?

43. Which solvent, water or carbon tetrachloride, would you choose to dissolve each of the following?
 a. KrF_2 **e.** MgF_2
 b. SF_2 **f.** CH_2O
 c. SO_2 **g.** $CH_2=CH_2$
 d. CO_2

44. Which solvent, water or hexane (C_6H_{14}), would you choose to dissolve each of the following?
 a. $Cu(NO_3)_2$ **d.** $CH_3(CH_2)_{16}CH_2OH$
 b. CS_2 **e.** HCl
 c. CH_3OH **f.** C_6H_6

45. For each of the following pairs, predict which substance would be more soluble in water.

a. NH_3 or PH_3

---

*Lattice energy was defined in Chapter 8 as the energy change for the process $M^+(g) + X^-(g) \rightarrow MX(s)$.

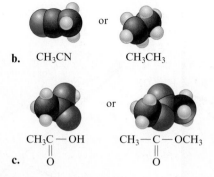

b. CH_3CN CH_3CH_3

c. CH_3C-OH $CH_3-C-OCH_3$
              $\|$ $\|$
              O O

46. Which ion in each of the following pairs would you expect to be more strongly hydrated? Why?

a. Na^+ or Mg^{2+} **d.** F^- or Br^-

b. Mg^{2+} or Be^{2+} **e.** Cl^- or ClO_4^-

c. Fe^{2+} or Fe^{3+} **f.** ClO_4^- or SO_4^{2-}

47. Rationalize the trend in water solubility for the following simple alcohols:

Alcohol	Solubility (g/100 g H_2O at 20°C)
Methanol, CH_3OH	Soluble in all proportions
Ethanol, CH_3CH_2OH	Soluble in all proportions
Propanol, $CH_3CH_2CH_2OH$	Soluble in all proportions
Butanol, $CH_3(CH_2)_2CH_2OH$	8.14
Pentanol, $CH_3(CH_2)_3CH_2OH$	2.64
Hexanol, $CH_3(CH_2)_4CH_2OH$	0.59
Heptanol, $CH_3(CH_2)_5CH_2OH$	0.09

48. In flushing and cleaning columns used in liquid chromatography to remove adsorbed contaminants, a series of solvents is used. Hexane (C_6H_{14}), chloroform ($CHCl_3$), methanol (CH_3OH), and water are passed through the column in that order. Rationalize the order in terms of intermolecular forces and the mutual solubility (miscibility) of the solvents.

49. The solubility of nitrogen in water is 8.21×10^{-4} mol/L at 0°C when the N_2 pressure above water is 0.790 atm. Calculate the Henry's law constant for N_2 in units of mol/L · atm for Henry's law in the form $C = kP$, where C is the gas concentration in mol/L. Calculate the solubility of N_2 in water when the partial pressure of nitrogen above water is 1.10 atm at 0°C.

50. Calculate the solubility of O_2 in water at a partial pressure of O_2 of 120 torr at 25°C. The Henry's law constant for O_2 is 1.3×10^{-3} mol/L · atm for Henry's law in the form $C = kP$, where C is the gas concentration (mol/L).

Vapor Pressures of Solutions

51. Glycerin, $C_3H_8O_3$, is a nonvolatile liquid. What is the vapor pressure of a solution made by adding 164 g glycerin to 338 mL H_2O at 39.8°C? The vapor pressure of pure water at 39.8°C is 54.74 torr and its density is 0.992 g/cm^3.

52. The vapor pressure of a solution containing 53.6 g glycerin ($C_3H_8O_3$) in 133.7 g ethanol (C_2H_5OH) is 113 torr at 40°C. Calculate the vapor pressure of pure ethanol at 40°C assuming that glycerin is a nonvolatile, nonelectrolyte solute in ethanol.

53. The normal boiling point of diethyl ether is 34.5°C. A solution containing a nonvolatile solute dissolved in diethyl ether has a vapor pressure of 698 torr at 34.5°C. What is the mole fraction of diethyl ether in this solution?

54. At a certain temperature, the vapor pressure of pure benzene (C_6H_6) is 0.930 atm. A solution was prepared by dissolving 10.0 g of a nondissociating, nonvolatile solute in 78.11 g of benzene at that temperature. The vapor pressure of the solution was found to be 0.900 atm. Assuming the solution behaves ideally, determine the molar mass of the solute.

55. A solution is made by dissolving 25.8 g urea (CH_4N_2O), a nonelectrolyte, in 275 g water. Calculate the vapor pressures of this solution at 25°C and 45°C. (The vapor pressure of pure water is 23.8 torr at 25°C and 71.9 torr at 45°C.)

56. A solution of sodium chloride in water has a vapor pressure of 19.6 torr at 25°C. What is the mole fraction of solute particles in this solution? What would be the vapor pressure of this solution at 45°C? The vapor pressure of pure water is 23.8 torr at 25°C and 71.9 torr at 45°C, and assume sodium chloride exists as Na^+ and Cl^- ions in solution.

57. Pentane (C_5H_{12}) and hexane (C_6H_{14}) form an ideal solution. At 25°C the vapor pressures of pentane and hexane are 511 and 150. torr, respectively. A solution is prepared by mixing 25 mL pentane (density, 0.63 g/mL) with 45 mL hexane (density, 0.66 g/mL).

a. What is the vapor pressure of the resulting solution?

b. What is the composition by mole fraction of pentane in the vapor that is in equilibrium with this solution?

58. A solution is prepared by mixing 0.0300 mole of CH_2Cl_2 and 0.0500 mole of CH_2Br_2 at 25°C. Assuming the solution is ideal, calculate the composition of the vapor (in terms of mole fractions) at 25°C. At 25°C, the vapor pressures of pure CH_2Cl_2 and pure CH_2Br_2 are 133 and 11.4 torr, respectively.

59. What is the composition of a methanol (CH_3OH)–propanol ($CH_3CH_2CH_2OH$) solution that has a vapor pressure of 174 torr at 40°C? At 40°C, the vapor pressures of pure methanol and pure propanol are 303 and 44.6 torr, respectively. Assume the solution is ideal.

60. Benzene and toluene form an ideal solution. Consider a solution of benzene and toluene prepared at 25°C. Assuming the mole fractions of benzene and toluene in the vapor phase are equal, calculate the composition of the solution. At 25°C the vapor pressures of benzene and toluene are 95 and 28 torr, respectively.

61. Which of the following will have the lowest total vapor pressure at 25°C?

a. pure water (vapor pressure = 23.8 torr at 25°C)

b. a solution of glucose in water with $X_{C_6H_{12}O_6} = 0.01$

c. a solution of sodium chloride in water with $X_{NaCl} = 0.01$

d. a solution of methanol in water with $X_{CH_3OH} = 0.2$ (Consider the vapor pressure of both methanol [143 torr at 25°C] and water.)

62. Which of the choices in Exercise 61 has the highest vapor pressure?

63. Match the vapor pressure diagrams with the solute–solvent combinations and explain your answers.

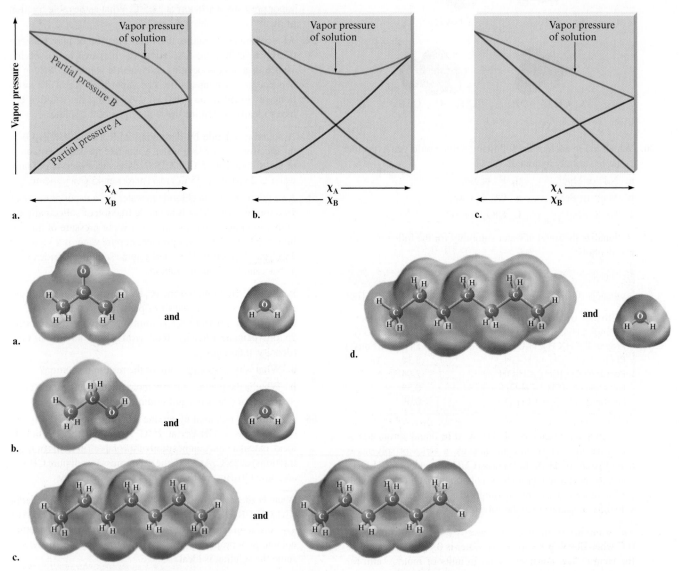

a.

b.

c.

64. The vapor pressures of several solutions of water–propanol $(CH_3CH_2CH_2OH)$ were determined at various compositions, with the following data collected at 45°C:

X_{H_2O}	Vapor Pressure (torr)
0	74.0
0.15	77.3
0.37	80.2
0.54	81.6
0.69	80.6
0.83	78.2
1.00	71.9

a. Are solutions of water and propanol ideal? Explain.

b. Predict the sign of ΔH_{soln} for water–propanol solutions.

c. Are the interactive forces between propanol and water molecules weaker than, stronger than, or equal to the interactive forces between the pure substances? Explain.

d. Which of the solutions in the data would have the lowest normal boiling point?

Colligative Properties

65. A solution is prepared by dissolving 27.0 g urea, $(NH_2)_2CO$, in 150.0 g water. Calculate the boiling point of the solution. Urea is a nonelectrolyte.

66. A 2.00-g sample of a large biomolecule was dissolved in 15.0 g carbon tetrachloride. The boiling point of this solution was determined to be 77.85°C. Calculate the molar mass of the biomolecule. For carbon tetrachloride, the boiling-point constant is 5.03°C · kg/mol, and the boiling point of pure carbon tetrachloride is 76.50°C.

67. What mass of glycerin ($C_3H_8O_3$), a nonelectrolyte, must be dissolved in 200.0 g water to give a solution with a freezing point of $-1.50°C$?

68. The freezing point of *t*-butanol is 25.50°C and K_f is 9.1°C · kg/mol. Usually *t*-butanol absorbs water on exposure to air. If the freezing point of a 10.0-g sample of *t*-butanol is 24.59°C, how many grams of water are present in the sample?

69. Calculate the freezing point and boiling point of an antifreeze solution that is 50.0% by mass of ethylene glycol ($HOCH_2CH_2OH$) in water. Ethylene glycol is a nonelectrolyte.

70. What volume of ethylene glycol ($C_2H_6O_2$), a nonelectrolyte, must be added to 15.0 L water to produce an antifreeze solution with a freezing point of $-25.0°C$? What is the boiling point of this solution? (The density of ethylene glycol is 1.11 g/cm³, and the density of water is 1.00 g/cm³.)

71. Reserpine is a natural product isolated from the roots of the shrub *Rauwolfia serpentina*. It was first synthesized in 1956 by Nobel Prize winner R. B. Woodward. It is used as a tranquilizer and sedative. When 1.00 g reserpine is dissolved in 25.0 g camphor, the freezing-point depression is 2.63°C (K_f for camphor is 40.°C · kg/mol). Calculate the molality of the solution and the molar mass of reserpine.

72. A solution contains 3.75 g of a nonvolatile pure hydrocarbon in 95 g acetone. The boiling points of pure acetone and the solution are 55.95°C and 56.50°C, respectively. The molal boiling-point constant of acetone is 1.71°C · kg/mol. What is the molar mass of the hydrocarbon?

73. **a.** Calculate the freezing-point depression and osmotic pressure at 25°C of an aqueous solution containing 1.0 g/L of a protein (molar mass = 9.0×10^4 g/mol) if the density of the solution is 1.0 g/cm³.

   b. Considering your answer to part a, which colligative property, freezing-point depression or osmotic pressure, would be better used to determine the molar masses of large molecules? Explain.

74. Erythrocytes are red blood cells containing hemoglobin. In a saline solution they shrivel when the salt concentration is high and swell when the salt concentration is low. In a 25°C aqueous solution of NaCl, whose freezing point is $-0.406°C$, erythrocytes neither swell nor shrink. If we want to calculate the osmotic pressure of the solution inside the erythrocytes under these conditions, what do we need to assume? Why? Estimate how good (or poor) of an assumption this is. Make this assumption and calculate the osmotic pressure of the solution inside the erythrocytes.

75. An aqueous solution of 10.00 g of catalase, an enzyme found in the liver, has a volume of 1.00 L at 27°C. The solution's osmotic pressure at 27°C is found to be 0.745 torr. Calculate the molar mass of catalase.

76. A 0.15-g sample of a purified protein is dissolved in water to give 2.0 mL of solution. The osmotic pressure is found to be 18.6 torr at 25°C. Calculate the protein's molar mass.

77. How would you prepare 1.0 L of an aqueous solution of sucrose ($C_{12}H_{22}O_{11}$) having an osmotic pressure of 15 atm at a temperature of 22°C? Sucrose is a nonelectrolyte.

78. How would you prepare 1.0 L of an aqueous solution of sodium chloride having an osmotic pressure of 15 atm at 22°C? Assume sodium chloride exists as Na^+ and Cl^- ions in solution.

Properties of Electrolyte Solutions

79. Consider the following solutions:

   0.010 *m* Na_3PO_4 in water
   0.020 *m* $CaBr_2$ in water
   0.020 *m* KCl in water
   0.020 *m* HF in water (HF is a weak acid.)

   a. Assuming complete dissociation of the soluble salts, which solution(s) would have the same boiling point as 0.040 *m* $C_6H_{12}O_6$ in water? $C_6H_{12}O_6$ is a nonelectrolyte.

   b. Which solution would have the highest vapor pressure at 28°C?

   c. Which solution would have the largest freezing-point depression?

80. From the following:

   pure water
   solution of $C_{12}H_{22}O_{11}$ (*m* = 0.01) in water
   solution of NaCl (*m* = 0.01) in water
   solution of $CaCl_2$ (*m* = 0.01) in water

   Choose the one with the

   a. highest freezing point.
   b. lowest freezing point.
   c. highest boiling point.
   d. lowest boiling point.
   e. highest osmotic pressure.

81. Calculate the freezing point and the boiling point of each of the following solutions. (Assume complete dissociation.)

   a. 5.0 g NaCl in 25 g H_2O
   b. 2.0 g $Al(NO_3)_3$ in 15 g H_2O

82. A water desalination plant is set up near a salt marsh containing water that is 0.10 *M* NaCl. Calculate the minimum pressure that must be applied at 20.°C to purify the water by reverse osmosis. Assume NaCl is completely dissociated.

83. Determine the van't Hoff factor for the following ionic solute dissolved in water.

84. Consider the following representations of an ionic solute in water. Which flask contains $MgSO_4$, and which flask contains NaCl? How can you tell?

85. Calculate the freezing point and the boiling point of each of the following aqueous solutions. (Assume complete dissociation.)

   a. 0.050 *m* $MgCl_2$

   b. 0.050 *m* $FeCl_3$

86. Calculate the freezing point and the boiling point of each of the following solutions using the observed van't Hoff factors in Table 11.6.

   a. 0.050 *m* $MgCl_2$

   b. 0.050 *m* $FeCl_3$

87. Use the following data for three aqueous solutions of $CaCl_2$ to calculate the apparent value of the van't Hoff factor.

Molality	Freezing-Point Depression (°C)
0.0225	0.110
0.0910	0.440
0.278	1.330

88. The freezing-point depression of a 0.091-*m* solution of CsCl is 0.320°C. The freezing-point depression of a 0.091-*m* solution of $CaCl_2$ is 0.440°C. In which solution does ion association appear to be greater? Explain.

89. In the winter of 1994, record low temperatures were registered throughout the United States. For example, in Champaign, Illinois, a record low of −29°F was registered. At this temperature can salting icy roads with $CaCl_2$ be effective in melting the ice?

   a. Assume *i* = 3.00 for $CaCl_2$.

   b. Assume the average value of *i* from Exercise 87.

   (The solubility of $CaCl_2$ in cold water is 74.5 g per 100.0 g water.)

90. A 0.500-g sample of a compound is dissolved in enough water to form 100.0 mL of solution. This solution has an osmotic pressure of 2.50 atm at 25°C. If each molecule of the solute dissociates into two particles (in this solvent), what is the molar mass of this solute?

Additional Exercises

91. The solubility of benzoic acid ($HC_7H_5O_2$),

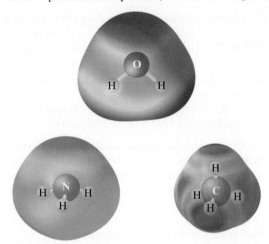

is 0.34 g/100 mL in water at 25°C and is 10.0 g/100 mL in benzene (C_6H_6) at 25°C. Rationalize this solubility behavior. (*Hint:* Benzoic acid forms a dimer in benzene.) Would benzoic acid be more or less soluble in a 0.1-*M* NaOH solution than it is in water? Explain.

92. Given the following electrostatic potential diagrams, comment on the expected solubility of CH_4 in water and NH_3 in water.

93. In a coffee-cup calorimeter, 1.60 g NH_4NO_3 was mixed with 75.0 g water at an initial temperature 25.00°C. After dissolution of the salt, the final temperature of the calorimeter contents was 23.34°C.

   a. Assuming the solution has a heat capacity of 4.18 J/g · °C, and assuming no heat loss to the calorimeter, calculate the enthalpy of solution (ΔH_{soln}) for the dissolution of NH_4NO_3 in units of kJ/mol.

   b. If the enthalpy of hydration for NH_4NO_3 is −630. kJ/mol, calculate the lattice energy of NH_4NO_3.

94. In Exercise 96 in Chapter 5, the pressure of CO_2 in a bottle of sparkling wine was calculated assuming that the CO_2 was insoluble in water. This was a bad assumption. Redo this problem by assuming that CO_2 obeys Henry's law. Use the data given in that problem to calculate the partial pressure of CO_2 in the gas phase and the solubility of CO_2 in the wine at 25°C. The Henry's law constant for CO_2 is 3.1×10^{-2} mol/L · atm at 25°C with Henry's law in the form $C = kP$, where C is the concentration of the gas in mol/L.

95. Explain the following on the basis of the behavior of atoms and/or ions.

   a. Cooking with water is faster in a pressure cooker than in an open pan.

   b. Salt is used on icy roads.

   c. Melted sea ice from the Arctic Ocean produces fresh water.

   d. $CO_2(s)$ (dry ice) does not have a normal boiling point under normal atmospheric conditions, even though CO_2 is a liquid in fire extinguishers.

   e. Adding a solute to a solvent extends the liquid phase over a larger temperature range.

96. The term *proof* is defined as twice the percent by volume of pure ethanol in solution. Thus, a solution that is 95% (by volume) ethanol is 190 proof. What is the molarity of ethanol in a 92 proof ethanol–water solution? Assume the density of ethanol, C_2H_5OH, is 0.79 g/cm³ and the density of water is 1.0 g/cm³.

97. At 25°C, the vapor in equilibrium with a solution containing carbon disulfide and acetonitrile has a total pressure of 263 torr and is 85.5 mole percent carbon disulfide. What is the mole fraction of carbon disulfide in the solution? At 25°C, the vapor pressure of carbon disulfide is 375 torr. Assume the solution and vapor exhibit ideal behavior.

98. For each of the following solute–solvent combinations, state the sign and relative magnitudes for ΔH_1, ΔH_2, ΔH_3, and ΔH_{soln} (as defined in Fig. 11.1 of the text). Explain your answers.

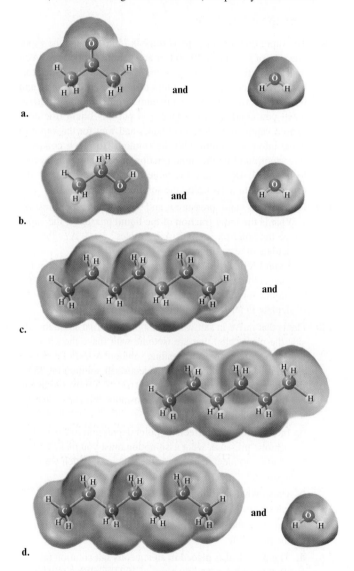

a.

b.

c.

d.

99. A solution is made by mixing 50.0 g acetone (CH_3COCH_3) and 50.0 g methanol (CH_3OH). What is the vapor pressure of this solution at 25°C? What is the composition of the vapor expressed as a mole fraction? Assume ideal solution and gas behavior. (At 25°C the vapor pressures of pure acetone and pure methanol are 271 and 143 torr, respectively.) The actual vapor pressure of this solution is 161 torr. Explain any discrepancies.

100. If the fluid inside a tree is about 0.1 M more concentrated in solute than the groundwater that bathes the roots, how high

will a column of fluid rise in the tree at 25°C? Assume that the density of the fluid is 1.0 g/cm³. (The density of mercury is 13.6 g/cm³.)

101. Thyroxine, an important hormone that controls the rate of metabolism in the body, can be isolated from the thyroid gland. When 0.455 g thyroxine is dissolved in 10.0 g benzene, the freezing point of the solution is depressed by 0.300°C. What is the molar mass of thyroxine? See Table 11.5.

102. If the human eye has an osmotic pressure of 8.00 atm at 25°C, what concentration of solute particles in water will provide an isotonic eyedrop solution (a solution with equal osmotic pressure)?

103. An unknown compound contains only carbon, hydrogen, and oxygen. Combustion analysis of the compound gives mass percents of 31.57% C and 5.30% H. The molar mass is determined by measuring the freezing-point depression of an aqueous solution. A freezing point of −5.20°C is recorded for a solution made by dissolving 10.56 g of the compound in 25.0 g water. Determine the empirical formula, molar mass, and molecular formula of the compound. Assume that the compound is a nonelectrolyte.

104. Consider the following:

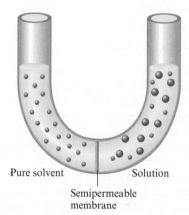

Pure solvent   Solution

Semipermeable membrane

What would happen to the level of liquid in the two arms if the semipermeable membrane separating the two liquids were permeable to

a. H_2O (the solvent) only?

b. H_2O and solute?

105. Consider an aqueous solution containing sodium chloride that has a density of 1.01 g/mL. Assume the solution behaves ideally. The freezing point of this solution at 1.0 atm is −1.28°C. Calculate the percent composition of this solution (by mass).

106. What stabilizes a colloidal suspension? Explain why adding heat or adding an electrolyte can cause the suspended particles to settle out.

107. The freezing point of an aqueous solution is −2.79°C.

a. Determine the boiling point of this solution.

b. Determine the vapor pressure (in mm Hg) of this solution at 25°C (the vapor pressure of pure water at 25°C is 23.76 mm Hg).

c. Explain any assumptions you make in solving parts a and b.

108. Specifications for lactated Ringer's solution, which is used for intravenous (IV) injections, are as follows to reach 100. mL of solution:

285–315 mg Na$^+$

14.1–17.3 mg K$^+$

4.9–6.0 mg Ca^{2+}

368–408 mg Cl$^-$

231–261 mg lactate, C$_3$H$_5$O$_3{}^-$

  a. Specify the amount of NaCl, KCl, CaCl$_2 \cdot$ 2H$_2$O, and NaC$_3$H$_5$O$_3$ needed to prepare 100. mL lactated Ringer's solution.

  b. What is the range of the osmotic pressure of the solution at 37°C, given the preceding specifications?

109. Patients undergoing an upper gastrointestinal tract laboratory test are typically given an X-ray contrast agent that aids with the radiologic imaging of the anatomy. One such contrast agent is sodium diatrizoate, a nonvolatile water-soluble compound. A 0.378-m solution is prepared by dissolving 38.4 g sodium diatrizoate (NaDTZ) in 1.60×10^2 mL water at 31.2°C (the density of water at 31.2°C is 0.995 g/cm^3). What is the molar mass of sodium diatrizoate? What is the vapor pressure of this solution if the vapor pressure of pure water at 31.2°C is 34.1 torr?

ChemWork Problems

These multiconcept problems (and additional ones) are found interactively online with the same type of assistance a student would get from an instructor.

110. A solution is prepared by dissolving 52.3 g cesium chloride in 60.0 g water. The volume of the solution is 63.3 mL. Calculate the mass percent, molarity, molality, and mole fraction of the CsCl solution.

111. The lattice energy of NaCl is −786 kJ/mol, and the enthalpy of hydration of 1 mole of gaseous Na$^+$ and 1 mole of gaseous Cl$^-$ ions is −783 kJ/mol. Calculate the enthalpy of solution per mole of solid NaCl.

112. For each of the following pairs, predict which substance is more soluble in water.

  a. CH$_3$NH$_2$ or NH$_3$

  b. CH$_3$CN or CH$_3$OCH$_3$

  c. CH$_3$CH$_2$OH or CH$_3$CH$_2$CH$_3$

  d. CH$_3$OH or CH$_3$CH$_2$OH

  e. (CH$_3$)$_3$CCH$_2$OH or CH$_3$(CH$_2$)$_6$OH

  f. CH$_3$OCH$_3$ or CH$_3$CO$_2$H

113. The normal boiling point of methanol is 64.7°C. A solution containing a nonvolatile solute dissolved in methanol has a vapor pressure of 556.0 torr at 64.7°C. What is the mole fraction of methanol in this solution?

114. A solution is prepared by mixing 1.000 mole of methanol (CH$_3$OH) and 3.18 moles of propanol (CH$_3$CH$_2$CH$_2$OH). What is the composition of the vapor (in mole fractions) at 40°C? At 40°C, the vapor pressure of pure methanol is 303 torr, and the vapor pressure of pure propanol is 44.6 torr.

115. The molar mass of a nonelectrolyte is 58.0 g/mol. Determine the boiling point of a solution containing 35.0 g of this compound and 600.0 g of water. The barometric pressure during the experiment was such that the boiling point of pure water was 99.725°C.

116. A 4.7×10^{-2} mg sample of a protein is dissolved in water to make 0.25 mL of solution. The osmotic pressure of the solution is 0.56 torr at 25°C. What is the molar mass of the protein?

117. A solid consists of a mixture of NaNO$_3$ and Mg(NO$_3$)$_2$. When 6.50 g of the solid is dissolved in 50.0 g of water, the freezing point of the solution is lowered by 5.23°C. What is the composition by mass of the solid mixture?

Challenge Problems

118. The vapor pressure of pure benzene is 750.0 torr and the vapor pressure of toluene is 300.0 torr at a certain temperature. You make a solution by pouring "some" benzene with "some" toluene. You then place this solution in a closed container and wait for the vapor to come into equilibrium with the solution. Next, you condense the vapor. You put this liquid (the condensed vapor) in a closed container and wait for the vapor to come into equilibrium with the solution. You then condense this vapor and find the mole fraction of benzene in this vapor to be 0.714. Determine the mole fraction of benzene in the original solution assuming the solution behaves ideally.

119. Liquid A has vapor pressure x, and liquid B has vapor pressure y. What is the mole fraction of the liquid mixture if the vapor above the solution is 30.% A by moles? 50.% A? 80.% A? (Calculate in terms of x and y.)

  Liquid A has vapor pressure x, liquid B has vapor pressure y. What is the mole fraction of the vapor above the solution if the liquid mixture is 30.% A by moles? 50.% A? 80.% A? (Calculate in terms of x and y.)

120. Plants that thrive in salt water must have internal solutions (inside the plant cells) that are isotonic with (have the same osmotic pressure as) the surrounding solution. A leaf of a saltwater plant is able to thrive in an aqueous salt solution (at 25°C) that has a freezing point equal to −0.621°C. You would like to use this information to calculate the osmotic pressure of the solution in the cell.

  a. In order to use the freezing-point depression to calculate osmotic pressure, what assumption must you make (in addition to ideal behavior of the solutions, which we will assume)?

  b. Under what conditions is the assumption (in part a) reasonable?

  c. Solve for the osmotic pressure (at 25°C) of the solution in the plant cell.

  d. The plant leaf is placed in an aqueous salt solution (at 25°C) that has a boiling point of 102.0°C. What will happen to the plant cells in the leaf?

121. You make 20.0 g of a sucrose (C$_{12}$H$_{22}$O$_{11}$) and NaCl mixture and dissolve it in 1.00 kg water. The freezing point of this solution is found to be −0.426°C. Assuming ideal behavior, calculate the mass percent composition of the original mixture, and the mole fraction of sucrose in the original mixture.

122. An aqueous solution is 1.00% NaCl by mass and has a density of 1.071 g/cm^3 at 25°C. The observed osmotic pressure of this solution is 7.83 atm at 25°C.

a. What fraction of the moles of NaCl in this solution exist as ion pairs?

b. Calculate the freezing point that would be observed for this solution.

123. The vapor in equilibrium with a pentane–hexane solution at 25°C has a mole fraction of pentane equal to 0.15 at 25°C. What is the mole fraction of pentane in the solution? (See Exercise 57 for the vapor pressures of the pure liquids.)

124. A forensic chemist is given a white solid that is suspected of being pure cocaine ($C_{17}H_{21}NO_4$, molar mass = 303.35 g/mol). She dissolves 1.22 ± 0.01 g of the solid in 15.60 ± 0.01 g benzene. The freezing point is lowered by $1.32 \pm 0.04°C$.

a. What is the molar mass of the substance? Assuming that the percent uncertainty in the calculated molar mass is the same as the percent uncertainty in the temperature change, calculate the uncertainty in the molar mass.

b. Could the chemist unequivocally state that the substance is cocaine? For example, is the uncertainty small enough to distinguish cocaine from codeine ($C_{18}H_{21}NO_3$, molar mass = 299.36 g/mol)?

c. Assuming that the absolute uncertainties in the measurements of temperature and mass remain unchanged, how could the chemist improve the precision of her results?

125. A 1.60-g sample of a mixture of naphthalene ($C_{10}H_8$) and anthracene ($C_{14}H_{10}$) is dissolved in 20.0 g benzene (C_6H_6). The freezing point of the solution is 2.81°C. What is the composition as mass percent of the sample mixture? The freezing point of benzene is 5.51°C and K_f is 5.12°C · kg/mol.

126. A solid mixture contains $MgCl_2$ and NaCl. When 0.5000 g of this solid is dissolved in enough water to form 1.000 L of solution, the osmotic pressure at 25.0°C is observed to be 0.3950 atm. What is the mass percent of $MgCl_2$ in the solid? (Assume ideal behavior for the solution.)

127. Formic acid (HCO_2H) is a monoprotic acid that ionizes only partially in aqueous solutions. A 0.10-M formic acid solution is 4.2% ionized. Assuming that the molarity and molality of the solution are the same, calculate the freezing point and the boiling point of 0.10 M formic acid.

128. You have a solution of two volatile liquids, A and B (assume ideal behavior). Pure liquid A has a vapor pressure of 350.0 torr and pure liquid B has a vapor pressure of 100.0 torr at the temperature of the solution. The vapor at equilibrium above the solution has double the mole fraction of substance A that the solution does. What is the mole fraction of liquid A in the solution?

129. In some regions of the southwest United States, the water is very hard. For example, in Las Cruces, New Mexico, the tap water contains about 560 μg of dissolved solids per milliliter. Reverse osmosis units are marketed in this area to soften water. A typical unit exerts a pressure of 8.0 atm and can produce 45 L water per day.

a. Assuming all of the dissolved solids are $MgCO_3$ and assuming a temperature of 27°C, what total volume of water must be processed to produce 45 L pure water?

b. Would the same system work for purifying seawater? (Assume seawater is 0.60 M NaCl.)

Integrative Problems

These problems require the integration of multiple concepts to find the solutions.

130. Creatinine, $C_4H_7N_3O$, is a by-product of muscle metabolism, and creatinine levels in the body are known to be a fairly reliable indicator of kidney function. The normal level of creatinine in the blood for adults is approximately 1.0 mg per deciliter (dL) of blood. If the density of blood is 1.025 g/mL, calculate the molality of a normal creatinine level in a 10.0-mL blood sample. What is the osmotic pressure of this solution at 25.0°C?

131. An aqueous solution containing 0.250 mole of Q, a strong electrolyte, in 5.00×10^2 g water freezes at −2.79°C. What is the van't Hoff factor for Q? The molal freezing-point depression constant for water is 1.86°C · kg/mol. What is the formula of Q if it is 38.68% chlorine by mass and there are twice as many anions as cations in one formula unit of Q?

132. Anthraquinone contains only carbon, hydrogen, and oxygen. When 4.80 mg anthraquinone is burned, 14.2 mg CO_2 and 1.65 mg H_2O are produced. The freezing point of camphor is lowered by 22.3°C when 1.32 g anthraquinone is dissolved in 11.4 g camphor. Determine the empirical and molecular formulas of anthraquinone.

Chapter 12

Chemical Kinetics

These wheelchair athletes generate a great amount of kinetic energy as they race through the mountains. (Simon Balson/Alamy)

The applications of chemistry focus largely on chemical reactions, and the commercial use of a reaction requires knowledge of several of its characteristics, including its stoichiometry, energetics, and rate. A reaction is defined by its reactants and products, whose identity must be learned by experiment. Once the reactants and products are known, the equation for the reaction can be written and balanced, and stoichiometric calculations can be carried out. Another very important characteristic of a reaction is its spontaneity. Spontaneity refers to the *inherent tendency* for the process to occur; however, it implies nothing about speed. *Spontaneous does not mean fast.* There are many spontaneous reactions that are so slow that no apparent reaction occurs over a period of weeks or years at normal temperatures. For example, there is a strong inherent tendency for gaseous hydrogen and oxygen to combine, that is,

$$2H_2(g) + O_2(g) \longrightarrow 2H_2O(l)$$

but in fact the two gases can coexist indefinitely at 25°C. Similarly, the gaseous reactions

$$H_2(g) + Cl_2(g) \longrightarrow 2HCl(g)$$
$$N_2(g) + 3H_2(g) \longrightarrow 2NH_3(g)$$

are both highly likely to occur from a thermodynamic standpoint, but we observe no reactions under normal conditions. In addition, the process of changing diamond to graphite is spontaneous but is so slow that it is not detectable.

To be useful, reactions must occur at a reasonable rate. To produce the 20 million tons of ammonia needed each year for fertilizer, we cannot simply mix nitrogen and hydrogen gases at 25°C and wait for them to react. It is not enough to understand the stoichiometry and thermodynamics of a reaction; we also must understand the factors that govern the rate of the reaction. The area of chemistry that concerns reaction rates is called **chemical kinetics**.

One of the main goals of chemical kinetics is to understand the steps by which a reaction takes place. This series of steps is called the *reaction mechanism.* Understanding the mechanism allows us to find ways to facilitate the reaction. For example, the Haber process for the production of ammonia requires high temperatures to achieve commercially feasible reaction rates. However, even higher temperatures (and more cost) would be required without the use of iron oxide, which speeds up the reaction.

In this chapter we will consider the main ideas of chemical kinetics. We will explore rate laws, reaction mechanisms, and simple models for chemical reactions.

12.1 | Reaction Rates

The kinetics of air pollution is discussed in Section 12.7.

To introduce the concept of the rate of a reaction, we will consider the decomposition of nitrogen dioxide, a gas that causes air pollution. Nitrogen dioxide decomposes to nitric oxide and oxygen as follows:

$$2NO_2(g) \longrightarrow 2NO(g) + O_2(g)$$

Suppose in a particular experiment we start with a flask of nitrogen dioxide at 300°C and measure the concentrations of nitrogen dioxide, nitric oxide, and oxygen as the nitrogen dioxide decomposes. The results of this experiment are summarized in Table 12.1, and the data are plotted in Fig. 12.1.

Note from these results that the concentration of the reactant (NO_2) decreases with time and the concentrations of the products (NO and O_2) increase with time (Fig. 12.2).

Table 12.1 | Concentrations of Reactant and Products as a Function of Time for the Reaction $2NO_2(g) \rightarrow 2NO(g) + O_2(g)$ (at 300°C)

Time (±1 s)	Concentration (mol/L)		
	NO₂	NO	O₂
0	0.0100	0	0
50	0.0079	0.0021	0.0011
100	0.0065	0.0035	0.0018
150	0.0055	0.0045	0.0023
200	0.0048	0.0052	0.0026
250	0.0043	0.0057	0.0029
300	0.0038	0.0062	0.0031
350	0.0034	0.0066	0.0033
400	0.0031	0.0069	0.0035

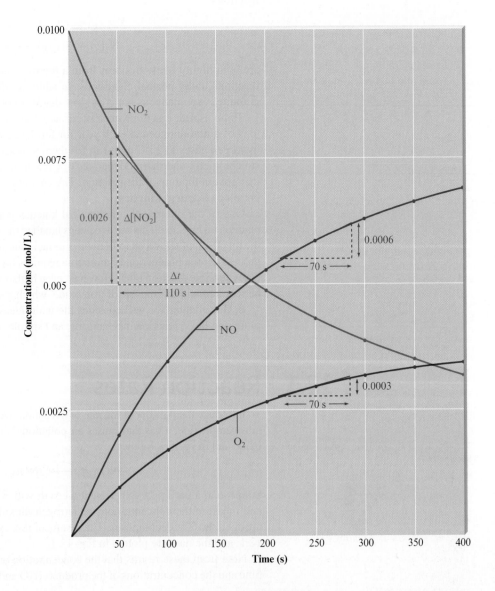

Figure 12.1 | Starting with a flask of nitrogen dioxide at 300°C, the concentrations of nitrogen dioxide, nitric oxide, and oxygen are plotted versus time.

Figure 12.2 | Representation of the reaction $2NO_2(g) \rightarrow 2NO(g) + O_2(g)$. (a) The reaction at the very beginning ($t = 0$). (b) and (c) As time passes, NO_2 is converted to NO and O_2.

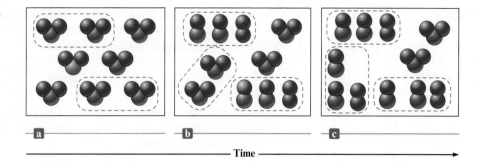

Chemical kinetics deals with the speed at which these changes occur. The speed, or *rate*, of a process is defined as the change in a given quantity over a specific period of time. For chemical reactions, the quantity that changes is the amount or concentration of a reactant or product. So the **reaction rate** of a chemical reaction is defined as the *change in concentration of a reactant or product per unit time*:

$$\text{Rate} = \frac{\text{concentration of A at time } t_2 - \text{concentration of A at time } t_1}{t_2 - t_1}$$

$$= \frac{\Delta[A]}{\Delta t}$$

[A] means concentration of A in mol/L.

where A is the reactant or product being considered, and the square brackets indicate concentration in mol/L. As usual, the symbol Δ indicates a *change* in a given quantity. Note that a change can be positive (increase) or negative (decrease), thus leading to a positive or negative reaction rate by this definition. However, for convenience, we will always define the rate as a positive quantity, as we will see.

Now let us calculate the average rate at which the concentration of NO_2 changes over the first 50 seconds of the reaction using the data given in Table 12.1.

$$\frac{\text{Change in }[NO_2]}{\text{Time elapsed}} = \frac{\Delta[NO_2]}{\Delta t}$$

$$= \frac{[NO_2]_{t=50} - [NO_2]_{t=0}}{50.\text{ s} - 0\text{ s}}$$

$$= \frac{0.0079\text{ mol/L} - 0.0100\text{ mol/L}}{50.\text{ s}}$$

$$= -4.2 \times 10^{-5}\text{ mol/L} \cdot \text{s}$$

The energy required for athletic exertion and the combustion of fuel in a race car both result from chemical reactions.

Note that since the concentration of NO_2 decreases with time, $\Delta[NO_2]$ is a negative quantity. Because it is customary to work with *positive* reaction rates, we define the rate of this particular reaction as

$$\text{Rate} = -\frac{\Delta[NO_2]}{\Delta t}$$

Appendix 1.3 reviews slopes of straight lines.

Since the concentrations of reactants always decrease with time, any rate expression involving a reactant will include a negative sign. The average rate of this reaction from 0 to 50 seconds is then

$$\text{Rate} = -\frac{\Delta[NO_2]}{\Delta t}$$
$$= -(-4.2 \times 10^{-5}\ \text{mol/L} \cdot \text{s})$$
$$= 4.2 \times 10^{-5}\ \text{mol/L} \cdot \text{s}$$

The average rates for this reaction during several other time intervals are given in Table 12.2. *Note that the rate is not constant but decreases with time.* The rates given in Table 12.2 are *average rates* over 50-second time intervals. The value of the rate at a particular time (the **instantaneous rate**) can be obtained by computing the slope of a line tangent to the curve at that point. Figure 12.1 shows a tangent drawn at $t = 100$ seconds. The *slope* of this line gives the rate at $t = 100$ seconds as follows:

$$\text{Slope of the tangent line} = \frac{\text{change in } y}{\text{change in } x}$$
$$= \frac{\Delta[NO_2]}{\Delta t}$$

But

$$\text{Rate} = -\frac{\Delta[NO_2]}{\Delta t}$$

Therefore,

$$\text{Rate} = -(\text{slope of the tangent line})$$
$$= -\left(\frac{-0.0026\ \text{mol/L}}{110\ \text{s}}\right)$$
$$= 2.4 \times 10^{-5}\ \text{mol/L} \cdot \text{s}$$

So far we have discussed the rate of this reaction only in terms of the reactant. The rate also can be defined in terms of the products. However, in doing so we must take

Table 12.2 | Average Rate (in mol/L · s) of Decomposition of Nitrogen Dioxide as a Function of Time*

$\dfrac{\Delta[NO_2]}{\Delta t}$	Time Period (s)
4.2×10^{-5}	$0 \rightarrow 50$
2.8×10^{-5}	$50 \rightarrow 100$
2.0×10^{-5}	$100 \rightarrow 150$
1.4×10^{-5}	$150 \rightarrow 200$
1.0×10^{-5}	$200 \rightarrow 250$

*Note that the *rate* decreases with time.

Los Angeles on a clear day, and on a day when air pollution is significant.

Jeff Hunter/The Image Bank/Getty Images

David McNew/Getty Images

into account the coefficients in the balanced equation for the reaction, because the stoichiometry determines the relative rates of consumption of reactants and generation of products. For example, in the reaction we are considering,

$$2NO_2(g) \longrightarrow 2NO(g) + O_2(g)$$

both the reactant NO_2 and the product NO have a coefficient of 2, so NO is produced at the same rate as NO_2 is consumed. We can verify this from Fig. 12.1. Note that the curve for NO is the same shape as the curve for NO_2, except that it is inverted, or flipped over. This means that, at any point in time, the slope of the tangent to the curve for NO will be the negative of the slope to the curve for NO_2. (Verify this at the point $t = 100$ seconds on both curves.) In the balanced equation, the product O_2 has a coefficient of 1, which means it is produced half as fast as NO, since NO has a coefficient of 2. That is, the rate of NO production is twice the rate of O_2 production.

We also can verify this fact from Fig. 12.1. For example, at $t = 250$ seconds,

$$\text{Slope of the tangent to the NO curve} = \frac{6.0 \times 10^{-4} \text{ mol/L}}{70. \text{ s}}$$

$$= 8.6 \times 10^{-6} \text{ mol/L} \cdot \text{s}$$

$$\text{Slope of the tangent to the } O_2 \text{ curve} = \frac{3.0 \times 10^{-4} \text{ mol/L}}{70. \text{ s}}$$

$$= 4.3 \times 10^{-6} \text{ mol/L} \cdot \text{s}$$

The slope at $t = 250$ seconds on the NO curve is twice the slope of that point on the O_2 curve, showing that the rate of production of NO is twice that of O_2.

The rate information can be summarized as follows:

Rate of consumption of NO₂		rate of production of NO		2(rate of production of O₂)
$-\dfrac{\Delta[NO_2]}{\Delta t}$	$=$	$\dfrac{\Delta[NO]}{\Delta t}$	$=$	$2\left(\dfrac{\Delta[O_2]}{\Delta t}\right)$

As we will see later, there is a certain type of rate that remains constant over time.

We have seen that the rate of a reaction is typically not constant. Most reaction rates change with time. This is so because the concentrations change with time (see Fig. 12.1).

Because the reaction rate changes with time, and because the rate is different (by factors that depend on the coefficients in the balanced equation) depending on which reactant or product is being studied, we must be very specific when we describe a rate for a chemical reaction.

12.2 | Rate Laws: An Introduction

Chemical reactions are *reversible*. In our discussion of the decomposition of nitrogen dioxide, we have so far considered only the *forward reaction,* as shown here:

$$2NO_2(g) \longrightarrow 2NO(g) + O_2(g)$$

However, the *reverse reaction* also can occur. As NO and O_2 accumulate, they can react to re-form NO_2:

$$O_2(g) + 2NO(g) \longrightarrow 2NO_2(g)$$

When gaseous NO_2 is placed in an otherwise empty container, initially the dominant reaction is

$$2NO_2(g) \longrightarrow 2NO(g) + O_2(g)$$

When forward and reverse reaction rates are equal, there will be no changes in the concentrations of reactants or products. This is called *chemical equilibrium* and is discussed fully in Chapter 13.

and the change in the concentration of NO_2 ($\Delta[NO_2]$) depends only on the forward reaction. However, after a period of time, enough products accumulate so that the reverse reaction becomes important. Now $\Delta[NO_2]$ depends on the *difference in the rates of the forward and reverse reactions.* This complication can be avoided if we study the rate of a reaction under conditions where the reverse reaction makes only a negligible contribution. Typically, this means that we must study a reaction at a point soon after the reactants are mixed, before the products have had time to build up to significant levels.

If we choose conditions where the reverse reaction can be neglected, the *reaction rate will depend only on the concentrations of the reactants.* For the decomposition of nitrogen dioxide, we can write

$$\text{Rate} = k[NO_2]^n \tag{12.1}$$

Such an expression, which shows how the rate depends on the concentrations of reactants, is called a **rate law**. The proportionality constant k, called the **rate constant**, and n, called the **order** of the reactant, must both be determined by experiment. The order of a reactant can be an integer (including zero) or a fraction. For the relatively simple reactions we will consider in this book, the orders will often be positive integers.

Note two important points about Equation (12.1):

1. The concentrations of the products do not appear in the rate law because the reaction rate is being studied under conditions where the reverse reaction does not contribute to the overall rate.
2. The value of the exponent n must be determined by experiment; it cannot be written from the balanced equation.

Before we go further we must define exactly what we mean by the term *rate* in Equation (12.1). In Section 12.1 we saw that reaction rate means a change in concentration per unit time. However, which reactant or product concentration do we choose in defining the rate? For example, for the decomposition of NO_2 to produce O_2 and NO considered in Section 12.1, we could define the rate in terms of any of these three species. However, since O_2 is produced only half as fast as NO, we must be careful to specify which species we are talking about in a given case. For instance, we might choose to define the reaction rate in terms of the consumption of NO_2:

$$\text{Rate} = -\frac{\Delta[NO_2]}{\Delta t} = k[NO_2]^n$$

On the other hand, we could define the rate in terms of the production of O_2:

$$\text{Rate}' = \frac{\Delta[O_2]}{\Delta t} = k'[NO_2]^n$$

Note that because $2NO_2$ molecules are consumed for every O_2 molecule produced,

$$\text{Rate} = 2 \times \text{rate}'$$

or

$$k[NO_2]^n = 2k'[NO_2]^n$$

and

$$k = 2 \times k'$$

Thus the value of the rate constant depends on how the rate is defined.

In this text we will always be careful to define exactly what is meant by the rate for a given reaction so that there will be no confusion about which specific rate constant is being used.

Types of Rate Laws

Notice that the rate law we have used to this point expresses rate as a function of concentration. For example, for the decomposition of NO_2 we have defined

$$\text{Rate} = -\frac{\Delta[NO_2]}{\Delta t} = k[NO_2]^n$$

The name *differential rate law* comes from a mathematical term. We will regard it simply as a label. The terms *differential rate law* and *rate law* will be used interchangeably in this text.

which tells us (once we have determined the value of *n*) exactly how the rate depends on the concentration of the reactant, NO_2. A rate law that expresses how the *rate depends on concentration* is technically called the **differential rate law**, but it is often simply called the **rate law**. Thus when we use the term *rate law* in this text, we mean the expression that gives the rate as a function of concentration.

A second kind of rate law, the **integrated rate law**, also will be important in our study of kinetics. The integrated rate law expresses how the *concentrations depend on time*. Although we will not consider the details here, a given differential rate law is always related to a certain type of integrated rate law, and vice versa. That is, if we determine the differential rate law for a given reaction, we automatically know the form of the integrated rate law for the reaction. This means that once we determine experimentally either type of rate law for a reaction, we also know the other one.

Which rate law we choose to determine by experiment often depends on what types of data are easiest to collect. If we can conveniently measure how the rate changes as the concentrations are changed, we can readily determine the differential (rate/concentration) rate law. On the other hand, if it is more convenient to measure the concentration as a function of time, we can determine the form of the integrated (concentration/time) rate law. We will discuss how rate laws are actually determined in the next several sections.

Why are we interested in determining the rate law for a reaction? How does it help us? It helps us because we can work backward from the rate law to infer the steps by which the reaction occurs. Most chemical reactions do not take place in a single step but result from a series of sequential steps. To understand a chemical reaction, we must learn what these steps are. For example, a chemist who is designing an insecticide may study the reactions involved in the process of insect growth to see what type of molecule might interrupt this series of reactions. Or an industrial chemist may be trying to make a given reaction occur faster. To accomplish this, he or she must know which step is slowest, because it is that step that must be speeded up. Thus a chemist is usually not interested in a rate law for its own sake but because of what it reveals about the steps by which a reaction occurs. We will develop a process for finding the reaction steps in this chapter.

> ### Let's Review | *Rate Laws: A Summary*
>
> ❯ There are two types of rate laws.
>
> **1.** The *differential rate law* (often called simply the *rate law*) shows how the rate of a reaction depends on concentrations.
>
> **2.** The *integrated rate law* shows how the concentrations of species in the reaction depend on time.
>
> ❯ Because we typically consider reactions only under conditions where the reverse reaction is unimportant, our rate laws will involve only concentrations of reactants.
>
> ❯ Because the differential and integrated rate laws for a given reaction are related in a well-defined way, the experimental determination of either of the rate laws is sufficient.
>
> ❯ Experimental convenience usually dictates which type of rate law is determined experimentally.
>
> ❯ Knowing the rate law for a reaction is important mainly because we can usually infer the individual steps involved in the reaction from the specific form of the rate law.

12.3 | Determining the Form of the Rate Law

The first step in understanding how a given chemical reaction occurs is to determine the *form* of the rate law. That is, we need to determine experimentally the power to which each reactant concentration must be raised in the rate law. In this section we will

Figure 12.3 | A plot of the concentration of N_2O_5 as a function of time for the reaction $2N_2O_5(soln) \rightarrow 4NO_2(soln) + O_2(g)$ (at 45°C). Note that the reaction rate at $[N_2O_5] = 0.90\ M$ is twice that at $[N_2O_5] = 0.45\ M$.

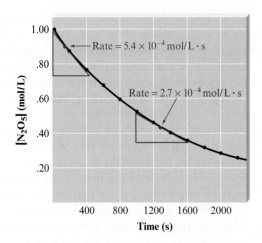

Table 12.3 | Concentration/Time Data for the Reaction $2N_2O_5(soln) \rightarrow 4NO_2(soln) + O_2(g)$ (at 45°C)

$[N_2O_5]$ (mol/L)	Time (s)
1.00	0
0.88	200
0.78	400
0.69	600
0.61	800
0.54	1000
0.48	1200
0.43	1400
0.38	1600
0.34	1800
0.30	2000

First order: rate = k[A]. Doubling the concentration of A doubles the reaction rate.

explore ways to obtain the differential rate law for a reaction. First, we will consider the decomposition of dinitrogen pentoxide in carbon tetrachloride solution:

$$2N_2O_5(soln) \longrightarrow 4NO_2(soln) + O_2(g)$$

Data for this reaction at 45°C are listed in Table 12.3 and plotted in Fig. 12.3. In this reaction the oxygen gas escapes from the solution and thus does not react with the nitrogen dioxide, so we do not have to be concerned about the effects of the reverse reaction at any time over the life of the reaction. That is, the reverse reaction is negligible at all times over the course of this reaction.

Evaluation of the reaction rates at concentrations of N_2O_5 of 0.90 M and 0.45 M, by taking the slopes of the tangents to the curve at these points (see Fig. 12.3), yields the following data:

$[N_2O_5]$	Rate (mol/L · s)
0.90 M	5.4×10^{-4}
0.45 M	2.7×10^{-4}

Note that when $[N_2O_5]$ is halved, the rate is also halved. This means that the rate of this reaction depends on the concentration of N_2O_5 to the *first power*. In other words, the (differential) rate law for this reaction is

$$\text{Rate} = -\frac{\Delta[N_2O_5]}{\Delta t} = k[N_2O_5]^1 = k[N_2O_5]$$

Thus the reaction is *first order* in N_2O_5. Note that for this reaction the order is *not* the same as the coefficient of N_2O_5 in the balanced equation for the reaction. This reemphasizes the fact that the order of a particular reactant must be obtained by *observing* how the reaction rate depends on the concentration of that reactant.

We have seen that by determining the instantaneous rate at two different reactant concentrations, the rate law for the decomposition of N_2O_5 is shown to have the form

$$\text{Rate} = -\frac{\Delta[A]}{\Delta t} = k[A]$$

where A represents N_2O_5.

Method of Initial Rates

The value of the initial rate is determined for each experiment at the same value of t as close to $t = 0$ as possible.

One common method for experimentally determining the form of the rate law for a reaction is the **method of initial rates**. The **initial rate** of a reaction is the instantaneous rate determined just after the reaction begins (just after $t = 0$). The idea is to

Table 12.4 | Initial Rates from Three Experiments for the Reaction $NH_4^+(aq) + NO_2^-(aq) \rightarrow N_2(g) + 2H_2O(l)$

Experiment	Initial Concentration of NH_4^+	Initial Concentration of NO_2^-	Initial Rate (mol/L · s)
1	0.100 M	0.0050 M	1.35×10^{-7}
2	0.100 M	0.010 M	2.70×10^{-7}
3	0.200 M	0.010 M	5.40×10^{-7}

determine the instantaneous rate before the initial concentrations of reactants have changed significantly. Several experiments are carried out using different initial concentrations, and the initial rate is determined for each run. The results are then compared to see how the initial rate depends on the initial concentrations. This allows the form of the rate law to be determined. We will illustrate the method of initial rates using the following equation:

$$NH_4^+(aq) + NO_2^-(aq) \longrightarrow N_2(g) + 2H_2O(l)$$

Table 12.4 gives initial rates obtained from three experiments involving different initial concentrations of reactants. The general form of the rate law for this reaction is

$$\text{Rate} = -\frac{\Delta[NH_4^+]}{\Delta t} = k[NH_4^+]^n[NO_2^-]^m$$

We can determine the values of n and m by observing how the initial rate depends on the initial concentrations of NH_4^+ and NO_2^-. In Experiments 1 and 2, where the initial concentration of NH_4^+ remains the same but the initial concentration of NO_2^- doubles, the observed initial rate also doubles. Since

$$\text{Rate} = k[NH_4^+]^n[NO_2^-]^m$$

we have for Experiment 1

$$\text{Rate} = 1.35 \times 10^{-7} \text{ mol/L} \cdot \text{s} = k(0.100 \text{ mol/L})^n(0.0050 \text{ mol/L})^m$$

and for Experiment 2

$$\text{Rate} = 2.70 \times 10^{-7} \text{ mol/L} \cdot \text{s} = k(0.100 \text{ mol/L})^n(0.010 \text{ mol/L})^m$$

The ratio of these rates is

$$\frac{\text{Rate 2}}{\text{Rate 1}} = \underbrace{\frac{2.70 \times 10^{-7} \text{ mol/L} \cdot \text{s}}{1.35 \times 10^{-7} \text{ mol/L} \cdot \text{s}}}_{2.00} = \frac{k(\cancel{0.100 \text{ mol/L}})^n(0.010 \text{ mol/L})^m}{k(\cancel{0.100 \text{ mol/L}})^n(0.0050 \text{ mol/L})^m}$$

$$= \frac{(0.010 \text{ mol/L})^m}{(0.0050 \text{ mol/L})^m} = (2.0)^m$$

Rates 1, 2, and 3 were determined at the same value of t (very close to $t = 0$).

Thus

$$\frac{\text{Rate 2}}{\text{Rate 1}} = 2.00 = (2.0)^m$$

which means the value of m is 1. The rate law for this reaction is first order in the reactant NO_2^-.

A similar analysis of the results for Experiments 2 and 3 yields the ratio

$$\frac{\text{Rate 3}}{\text{Rate 2}} = \frac{5.40 \times 10^{-7} \text{ mol/L} \cdot \text{s}}{2.70 \times 10^{-7} \text{ mol/L} \cdot \text{s}} = \frac{(0.200 \text{ mol/L})^n}{(0.100 \text{ mol/L})^n}$$

$$= 2.00 = \left(\frac{0.200}{0.100}\right)^n = (2.00)^n$$

The value of n is also 1.

We have shown that the values of n and m are both 1 and the rate law is

$$\text{Rate} = k[NH_4^+][NO_2^-]$$

This rate law is first order in both NO_2^- and NH_4^+. Note that it is merely a coincidence that n and m have the same values as the coefficients of NH_4^+ and NO_2^- in the balanced equation for the reaction.

The **overall reaction order** is the sum of n and m. For this reaction, $n + m = 2$. The reaction is second order overall.

The value of the rate constant k can now be calculated using the results of *any* of the three experiments shown in Table 12.4. From the data for Experiment 1, we know that

> Overall reaction order is the sum of the orders for the various reactants.

$$\text{Rate} = k[NH_4^+][NO_2^-]$$
$$1.35 \times 10^{-7}\ \text{mol/L} \cdot \text{s} = k(0.100\ \text{mol/L})(0.0050\ \text{mol/L})$$

Then

$$k = \frac{1.35 \times 10^{-7}\ \text{mol/L} \cdot \text{s}}{(0.100\ \text{mol/L})(0.0050\ \text{mol/L})} = 2.7 \times 10^{-4}\ \text{L/mol} \cdot \text{s}$$

Example 12.1

Determining a Rate Law

The reaction between bromate ions and bromide ions in acidic aqueous solution is given by the equation

$$BrO_3^-(aq) + 5Br^-(aq) + 6H^+(aq) \longrightarrow 3Br_2(l) + 3H_2O(l)$$

Table 12.5 gives the results from four experiments. Using these data, determine the orders for all three reactants, the overall reaction order, and the value of the rate constant.

Solution

The general form of the rate law for this reaction is

$$\text{Rate} = k[BrO_3^-]^n[Br^-]^m[H^+]^p$$

We can determine the values of n, m, and p by comparing the rates from the various experiments. To determine the value of n, we use the results from Experiments 1 and 2, in which only $[BrO_3^-]$ changes:

$$\frac{\text{Rate 2}}{\text{Rate 1}} = \frac{1.6 \times 10^{-3}\ \text{mol/L} \cdot \text{s}}{8.0 \times 10^{-4}\ \text{mol/L} \cdot \text{s}} = \frac{k(0.20\ \text{mol/L})^n(0.10\ \text{mol/L})^m(0.10\ \text{mol/L})^p}{k(0.10\ \text{mol/L})^n(0.10\ \text{mol/L})^m(0.10\ \text{mol/L})^p}$$

$$2.0 = \left(\frac{0.20\ \text{mol/L}}{0.10\ \text{mol/L}}\right)^n = (2.0)^n$$

Thus n is equal to 1.

Table 12.5 | The Results from Four Experiments to Study the Reaction $BrO_3^-(aq) + 5Br^-(aq) + 6H^+(aq) \rightarrow 3Br_2(l) + 3H_2O(l)$

Experiment	Initial Concentration of BrO_3^- (mol/L)	Initial Concentration of Br^- (mol/L)	Initial Concentration of H^+ (mol/L)	Measured Initial Rate (mol/L · s)
1	0.10	0.10	0.10	8.0×10^{-4}
2	0.20	0.10	0.10	1.6×10^{-3}
3	0.20	0.20	0.10	3.2×10^{-3}
4	0.10	0.10	0.20	3.2×10^{-3}

To determine the value of m, we use the results from Experiments 2 and 3, in which only $[Br^-]$ changes:

$$\frac{\text{Rate 3}}{\text{Rate 2}} = \frac{3.2 \times 10^{-3} \text{ mol/L} \cdot \text{s}}{1.6 \times 10^{-3} \text{ mol/L} \cdot \text{s}} = \frac{k(0.20 \text{ mol/L})^n(0.20 \text{ mol/L})^m(0.10 \text{ mol/L})^p}{k(0.20 \text{ mol/L})^n(0.10 \text{ mol/L})^m(0.10 \text{ mol/L})^p}$$

$$2.0 = \left(\frac{0.20 \text{ mol/L}}{0.10 \text{ mol/L}}\right)^m = (2.0)^m$$

Thus m is equal to 1.

To determine the value of p, we use the results from Experiments 1 and 4, in which $[BrO_3^-]$ and $[Br^-]$ are constant but $[H^+]$ differs:

$$\frac{\text{Rate 4}}{\text{Rate 1}} = \frac{3.2 \times 10^{-3} \text{ mol/L} \cdot \text{s}}{8.0 \times 10^{-4} \text{ mol/L} \cdot \text{s}} = \frac{k(0.10 \text{ mol/L})^n(0.10 \text{ mol/L})^m(0.20 \text{ mol/L})^p}{k(0.10 \text{ mol/L})^n(0.10 \text{ mol/L})^m(0.10 \text{ mol/L})^p}$$

$$4.0 = \left(\frac{0.20 \text{ mol/L}}{0.10 \text{ mol/L}}\right)^p$$

$$4.0 = (2.0)^p = (2.0)^2$$

Thus p is equal to 2.

The rate of this reaction is first order in BrO_3^- and Br^- and second order in H^+. The overall reaction order is $n + m + p = 4$.

The rate law can now be written

$$\text{Rate} = k[BrO_3^-][Br^-][H^+]^2$$

The value of the rate constant k can be calculated from the results of any of the four experiments. For Experiment 1, the initial rate is 8.0×10^{-4} mol/L $\cdot$ s and $[BrO_3^-] = 0.100$ M, $[Br^-] = 0.10$ M, and $[H^+] = 0.10$ M. Using these values in the rate law gives

$$8.0 \times 10^{-4} \text{ mol/L} \cdot \text{s} = k(0.10 \text{ mol/L})(0.10 \text{ mol/L})(0.10 \text{ mol/L})^2$$

$$8.0 \times 10^{-4} \text{ mol/L} \cdot \text{s} = k(1.0 \times 10^{-4} \text{ mol}^4/\text{L}^4)$$

$$k = \frac{8.0 \times 10^{-4} \text{ mol/L} \cdot \text{s}}{1.0 \times 10^{-4} \text{ mol}^4/\text{L}^4} = 8.0 \text{ L}^3/\text{mol}^3 \cdot \text{s}$$

Reality Check | Verify that the same value of k can be obtained from the results of the other experiments.

See Exercises 12.29 through 12.32

12.4 | The Integrated Rate Law

The rate laws we have considered so far express the rate as a function of the reactant concentrations. It is also useful to be able to express the reactant concentrations as a function of time, given the (differential) rate law for the reaction. In this section we show how this is done.

We will proceed by first looking at reactions involving a single reactant:

$$aA \longrightarrow \text{products}$$

all of which have a rate law of the form

$$\text{Rate} = -\frac{\Delta[A]}{\Delta t} = k[A]^n$$

We will develop the integrated rate laws individually for the cases $n = 1$ (first order), $n = 2$ (second order), and $n = 0$ (zero order).

First-Order Rate Laws

For the reaction

$$2N_2O_5(soln) \longrightarrow 4NO_2(soln) + O_2(g)$$

we have found that the rate law is

$$\text{Rate} = -\frac{\Delta[N_2O_5]}{\Delta t} = k[N_2O_5]$$

Since the rate of this reaction depends on the concentration of N_2O_5 to the first power, it is a **first-order reaction**. This means that if the concentration of N_2O_5 in a flask were suddenly doubled, the rate of production of NO_2 and O_2 also would double. This rate law can be put into a different form using a calculus operation known as integration, which yields the expression

$$\ln[N_2O_5] = -kt + \ln[N_2O_5]_0$$

Appendix 1.2 contains a review of logarithms.

where ln indicates the natural logarithm, t is the time, $[N_2O_5]$ is the concentration of N_2O_5 at time t, and $[N_2O_5]_0$ is the initial concentration of N_2O_5 (at $t = 0$, the start of the experiment). Note that such an equation, called the *integrated rate law,* expresses the *concentration of the reactant as a function of time.*

For a chemical reaction of the form

$$aA \longrightarrow \text{products}$$

where the kinetics are first order in [A], the rate law is

$$\text{Rate} = -\frac{\Delta[A]}{\Delta t} = k[A]$$

and the **integrated first-order rate law** is

$$\ln[A] = -kt + \ln[A]_0 \tag{12.2}$$

There are several important things to note about Equation (12.2):

An integrated rate law relates concentration to reaction time.

1. The equation shows how the concentration of A depends on time. If the initial concentration of A and the rate constant k are known, the concentration of A at any time can be calculated.
2. Equation (12.2) is of the form $y = mx + b$, where a plot of y versus x is a straight line with slope m and intercept b. In Equation (12.2),

$$y = \ln[A] \quad x = t \quad m = -k \quad b = \ln[A]_0$$

For a first-order reaction, a plot of ln[A] versus t is always a straight line.

Thus for a first-order reaction, plotting the natural logarithm of concentration versus time always gives a straight line. This fact is often used to test whether a reaction is first order. For the reaction

$$aA \longrightarrow \text{products}$$

the *reaction is first order in A if a plot of ln[A] versus t is a straight line.* Conversely, if this plot is not a straight line, the reaction is not first order in A.

3. This integrated rate law for a first-order reaction also can be expressed in terms of a *ratio* of [A] and $[A]_0$ as follows:

$$\ln\left(\frac{[A]_0}{[A]}\right) = kt$$

| Example 12.2 | First-Order Rate Laws I |

The decomposition of N_2O_5 in the gas phase was studied at constant temperature.

$$2N_2O_5(g) \longrightarrow 4NO_2(g) + O_2(g)$$

The following results were collected:

[N₂O₅] (mol/L)	Time (s)
0.1000	0
0.0707	50
0.0500	100
0.0250	200
0.0125	300
0.00625	400

Using these data, verify that the rate law is first order in $[N_2O_5]$, and calculate the value of the rate constant, where the rate $= -\Delta[N_2O_5]/\Delta t$.

Solution

We can verify that the rate law is first order in $[N_2O_5]$ by constructing a plot of $\ln[N_2O_5]$ versus time.

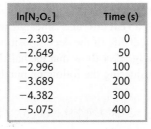

ln[N₂O₅]	Time (s)
−2.303	0
−2.649	50
−2.996	100
−3.689	200
−4.382	300
−5.075	400

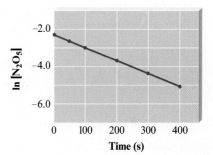

The values of $\ln[N_2O_5]$ at various times are given in the table above and shown in the plot of $\ln[N_2O_5]$. The fact that the plot is a straight line confirms that the reaction is first order in N_2O_5, since it follows the equation $\ln[N_2O_5] = -kt + \ln[N_2O_5]_0$.

Since the reaction is first order, the slope of the line equals $-k$, where

$$\text{Slope} = \frac{\text{change in } y}{\text{change in } x} = \frac{\Delta y}{\Delta x} = \frac{\Delta(\ln[N_2O_5])}{\Delta t}$$

Since the first and last points are exactly on the line, we will use these points to calculate the slope:

$$\text{Slope} = \frac{-5.075 - (-2.303)}{400.\text{ s} - 0\text{ s}} = \frac{-2.772}{400.\text{ s}} = -6.93 \times 10^{-3}\text{ s}^{-1}$$

$$k = -(\text{slope}) = 6.93 \times 10^{-3}\text{ s}^{-1}$$

See Exercise 12.37

| Example 12.3 | **First-Order Rate Laws II** |

Using the data given in Example 12.2, calculate $[N_2O_5]$ at 150 s after the start of the reaction.

Solution

We know from Example 12.2 that $[N_2O_5] = 0.0500$ mol/L at 100 s and $[N_2O_5] = 0.0250$ mol/L at 200 s. Since 150 s is halfway between 100 and 200 s, it is tempting to assume that we can simply use an arithmetic average to obtain $[N_2O_5]$ at that time. This is incorrect because it is $\ln[N_2O_5]$, not $[N_2O_5]$, that is directly proportional to t. To calculate $[N_2O_5]$ after 150 s, we use Equation (12.2):

$$\ln[N_2O_5] = -kt + \ln[N_2O_5]_0$$

The antilog operation means to exponentiate (see Appendix 1.2).

where $t = 150.$ s, $k = 6.93 \times 10^{-3}$ s^{-1} (as determined in Example 12.2), and $[N_2O_5]_0 = 0.1000$ mol/L.

$$\ln([N_2O_5]_{t=150}) = -(6.93 \times 10^{-3} \text{ s}^{-1})(150. \text{ s}) + \ln(0.100)$$
$$= -1.040 - 2.303 = -3.343$$
$$[N_2O_5]_{t=150} = \text{antilog}(-3.343) = 0.0353 \text{ mol/L}$$

Note that this value of $[N_2O_5]$ is *not* halfway between 0.0500 and 0.0250 mol/L.

See Exercise 12.37

Half-Life of a First-Order Reaction

The time required for a reactant to reach half its original concentration is called the **half-life of a reactant** and is designated by the symbol $t_{1/2}$. For example, we can calculate the half-life of the decomposition reaction discussed in Example 12.2. The data plotted in Fig. 12.4 show that the half-life for this reaction is 100 seconds. We can see this by considering the following numbers:

$[N_2O_5]$ (mol/L)	Δt (s)		Ratio of Concentrations
0.100	0		
		$\Delta t = 100$ s;	$\dfrac{[N_2O_5]_{t=100}}{[N_2O_5]_{t=0}} = \dfrac{0.050}{0.100} = \dfrac{1}{2}$
0.0500	100		
		$\Delta t = 100$ s;	$\dfrac{[N_2O_5]_{t=200}}{[N_2O_5]_{t=100}} = \dfrac{0.025}{0.050} = \dfrac{1}{2}$
0.0250	200		
		$\Delta t = 100$ s;	$\dfrac{[N_2O_5]_{t=300}}{[N_2O_5]_{t=200}} = \dfrac{0.0125}{0.0250} = \dfrac{1}{2}$
0.0125	300		

Note that it *always* takes 100 seconds for $[N_2O_5]$ to be halved in this reaction.

A general formula for the half-life of a first-order reaction can be derived from the integrated rate law for the general reaction

$$aA \longrightarrow \text{products}$$

If the reaction is first order in [A],

$$\ln\left(\frac{[A]_0}{[A]}\right) = kt$$

Figure 12.4 | A plot of $[N_2O_5]$ versus time for the decomposition reaction of N_2O_5.

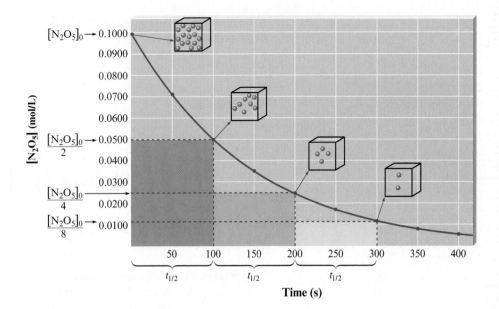

By definition, when $t = t_{1/2}$,

$$[A] = \frac{[A]_0}{2}$$

Then, for $t = t_{1/2}$, the integrated rate law becomes

$$\ln\left(\frac{[A]_0}{[A]_0/2}\right) = kt_{1/2}$$

or

$$\ln(2) = kt_{1/2}$$

Substituting the value of $\ln(2)$ and solving for $t_{1/2}$ gives

$$t_{1/2} = \frac{0.693}{k} \qquad (12.3)$$

For a first-order reaction, $t_{1/2}$ is independent of the initial concentration.

This is the *general equation for the half-life of a first-order reaction*. Equation (12.3) can be used to calculate $t_{1/2}$ if k is known or k if $t_{1/2}$ is known. Note that for a first-order reaction, *the half-life does not depend on concentration*.

Half-Life for a First-Order Reaction

A certain first-order reaction has a half-life of 20.0 minutes.

a. Calculate the rate constant for this reaction.

b. How much time is required for this reaction to be 75% complete?

Solution

a. Solving Equation (12.3) for k gives

$$k = \frac{0.693}{t_{1/2}} = \frac{0.693}{20.0 \text{ min}} = 3.47 \times 10^{-2} \text{ min}^{-1}$$

b. We use the integrated rate law in the form

$$\ln\left(\frac{[A]_0}{[A]}\right) = kt$$

If the reaction is 75% complete, 75% of the reactant has been consumed, leaving 25% in the original form:

$$\frac{[A]}{[A]_0} \times 100\% = 25\%$$

This means that

$$\frac{[A]}{[A]_0} = 0.25 \quad \text{or} \quad \frac{[A]_0}{[A]} = \frac{1}{0.25} = 4.0$$

Then

$$\ln\left(\frac{[A]_0}{[A]}\right) = \ln(4.0) = kt = \left(\frac{3.47 \times 10^{-2}}{\text{min}}\right)t$$

and

$$t = \frac{\ln(4.0)}{\dfrac{3.47 \times 10^{-2}}{\text{min}}} = 40. \text{ min}$$

Thus it takes 40. minutes for this particular reaction to reach 75% completion.

Let's consider another way of solving this problem using the definition of half-life. After one half-life the reaction has gone 50% to completion. If the initial concentration were 1.0 mol/L, after one half-life the concentration would be 0.50 mol/L. One more half-life would produce a concentration of 0.25 mol/L. Comparing 0.25 mol/L with the original 1.0 mol/L shows that 25% of the reactant is left after two half-lives. This is a general result. (What percentage of reactant remains after three half-lives?) Two half-lives for this reaction is 2(20.0 min), or 40.0 min, which agrees with the preceding answer.

See Exercises 12.38 and 12.49 through 12.52

Second-Order Rate Laws

For a general reaction involving a single reactant, that is,

$$aA \longrightarrow \text{products}$$

that is second order in A, the rate law is

Second order: rate = $k[A]^2$. Doubling the concentration of A quadruples the reaction rate; tripling the concentration of A increases the rate by nine times.

$$\text{Rate} = -\frac{\Delta[A]}{\Delta t} = k[A]^2 \tag{12.4}$$

The **integrated second-order rate law** has the form

$$\frac{1}{[A]} = kt + \frac{1}{[A]_0} \tag{12.5}$$

Note the following characteristics of Equation (12.5):

For second-order reactions, a plot of 1/[A] versus *t* will be linear.

1. A plot of $1/[A]$ versus t will produce a straight line with a slope equal to k.
2. Equation (12.5) shows how [A] depends on time and can be used to calculate [A] at any time t, provided k and $[A]_0$ are known.

When one half-life of the second-order reaction has elapsed ($t = t_{1/2}$), by definition,

$$[A] = \frac{[A]_0}{2}$$

Equation (12.5) then becomes

$$\frac{1}{\dfrac{[A]_0}{2}} = kt_{1/2} + \frac{1}{[A]_0}$$

$$\frac{2}{[A]_0} - \frac{1}{[A]_0} = kt_{1/2}$$

$$\frac{1}{[A]_0} = kt_{1/2}$$

Solving for $t_{1/2}$ gives *the expression for the half-life of a second-order reaction:*

$$t_{1/2} = \frac{1}{k[A]_0} \tag{12.6}$$

Example 12.5

Determining Rate Laws

Butadiene reacts to form its dimer according to the equation

$$2C_4H_6(g) \longrightarrow C_8H_{12}(g)$$

The following data were collected for this reaction at a given temperature:

When two identical molecules combine, the resulting molecule is called a dimer.

Butadiene (C_4H_6)

$[C_4H_6]$ (mol/L)	Time (± 1 s)
0.01000	0
0.00625	1000
0.00476	1800
0.00370	2800
0.00313	3600
0.00270	4400
0.00241	5200
0.00208	6200

a. Is this reaction first order or second order?

b. What is the value of the rate constant for the reaction?

c. What is the half-life for the reaction under the initial conditions of this experiment?

Solution

a. To decide whether the rate law for this reaction is first order or second order, we must see whether the plot of $\ln[C_4H_6]$ versus time is a straight line (first order) or the plot of $1/[C_4H_6]$ versus time is a straight line (second order). The data necessary to make these plots are as follows:

t (s)	$\dfrac{1}{[C_4H_6]}$	$\ln[C_4H_6]$
0	100	-4.605
1000	160	-5.075
1800	210	-5.348
2800	270	-5.599
3600	320	-5.767
4400	370	-5.915
5200	415	-6.028
6200	481	-6.175

Figure 12.5 (a) A plot of $\ln[C_4H_6]$ versus t. (b) A plot of $1/[C_4H_6]$ versus t.

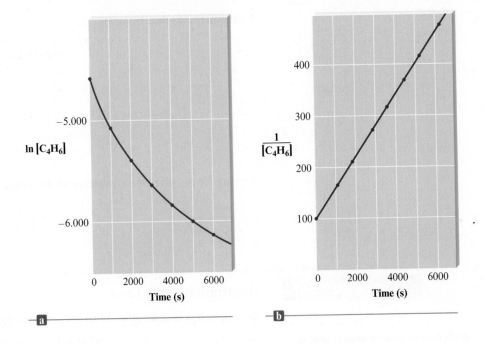

The resulting plots are shown in Fig. 12.5. Since the $\ln[C_4H_6]$ versus t plot [Fig. 12.5(a)] is not a straight line, the reaction is *not* first order. The reaction is, however, second order, as shown by the linearity of the $1/[C_4H_6]$ versus t plot [Fig. 12.5(b)]. Thus we can now write the rate law for this second-order reaction:

$$\text{Rate} = -\frac{\Delta[C_4H_6]}{\Delta t} = k[C_4H_6]^2$$

b. For a second-order reaction, a plot of $1/[C_4H_6]$ versus t produces a straight line of slope k. In terms of the standard equation for a straight line, $y = mx + b$, we have $y = 1/[C_4H_6]$ and $x = t$. Thus the slope of the line can be expressed as follows:

$$\text{Slope} = \frac{\Delta y}{\Delta x} = \frac{\Delta\left(\frac{1}{[C_4H_6]}\right)}{\Delta t}$$

Using the points at $t = 0$ and $t = 6200$, we can find the rate constant for the reaction:

$$k = \text{slope} = \frac{(481 - 100)\ \text{L/mol}}{(6200. - 0)\ \text{s}} = \frac{381}{6200.}\text{L/mol} \cdot \text{s} = 6.14 \times 10^{-2}\ \text{L/mol} \cdot \text{s}$$

c. The expression for the half-life of a second-order reaction is

$$t_{1/2} = \frac{1}{k[A]_0}$$

In this case $k = 6.14 \times 10^{-2}\ \text{L/mol} \cdot \text{s}$ (from part b) and $[A]_0 = [C_4H_6]_0 = 0.01000\ M$ (the concentration at $t = 0$). Thus

$$t_{1/2} = \frac{1}{(6.14 \times 10^{-2}\ \text{L/mol} \cdot \text{s})(1.000 \times 10^{-2}\ \text{mol/L})} = 1.63 \times 10^3\ \text{s}$$

The initial concentration of C_4H_6 is halved in 1630 s.

See Exercises 12.39, 12.40, 12.53, and 12.54

It is important to recognize the difference between the half-life for a first-order reaction and the half-life for a second-order reaction. For a second-order reaction, $t_{1/2}$ depends on both k and $[A]_0$; for a first-order reaction, $t_{1/2}$ depends only on k. For a first-order reaction, a constant time is required to reduce the concentration of the reactant by half, and then by half again, and so on, as the reaction proceeds. From Example 12.5 we can see that this is *not* true for a second-order reaction. For that second-order reaction, we found that the first half-life (the time required to go from $[C_4H_6] = 0.010\ M$ to $[C_4H_6] = 0.0050\ M$) is 1630 seconds. We can estimate the second half-life from the concentration data as a function of time. Note that to reach $0.0024\ M$ C_4H_6 (approximately $0.0050/2$) requires 5200 seconds of reaction time. Thus to get from $0.0050\ M$ C_4H_6 to $0.0024\ M$ C_4H_6 takes 3570 seconds ($5200 - 1630$). The second half-life is much longer than the first. This pattern is characteristic of second-order reactions. In fact, *for a second-order reaction, each successive half-life is double the preceding one* (provided the effects of the reverse reaction can be ignored, as we are assuming here). Prove this to yourself by examining the equation $t_{1/2} = 1/(k[A]_0)$.

Zero-Order Rate Laws

Most reactions involving a single reactant show either first-order or second-order kinetics. However, sometimes such a reaction can be a **zero-order reaction**. The rate law for a zero-order reaction is

$$\text{Rate} = k[A]^0 = k(1) = k$$

For a zero-order reaction, the rate is constant. It does not change with concentration as it does for first-order or second-order reactions.

The **integrated rate law for a zero-order reaction** is

$$[A] = -kt + [A]_0 \tag{12.7}$$

In this case a plot of $[A]$ versus t gives a straight line of slope $-k$ (Fig. 12.6).

The expression for the half-life of a zero-order reaction can be obtained from the integrated rate law. By definition, $[A] = [A]_0/2$ when $t = t_{1/2}$, so

$$\frac{[A]_0}{2} = -kt_{1/2} + [A]_0$$

or

$$kt_{1/2} = \frac{[A]_0}{2k}$$

Solving for $t_{1/2}$ gives

$$t_{1/2} = \frac{[A]_0}{2k} \tag{12.8}$$

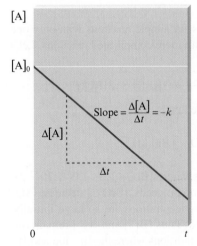

Figure 12.6 | A plot of $[A]$ versus t for a zero-order reaction.

Zero-order reactions are most often encountered when a substance such as a metal surface or an enzyme is required for the reaction to occur. For example, the decomposition reaction

$$2N_2O(g) \longrightarrow 2N_2(g) + O_2(g)$$

occurs on a hot platinum surface. When the platinum surface is completely covered with N_2O molecules, an increase in the concentration of N_2O has no effect on the rate, since only those N_2O molecules on the surface can react. Under these conditions, *the rate is a constant* because it is controlled by what happens on the platinum surface rather than by the total concentration of N_2O (Fig. 12.7). This reaction also can occur at high temperatures with no platinum surface present, but under these conditions, it is not zero order.

Figure 12.7 | The decomposition reaction $2N_2O(g) \rightarrow 2N_2(g) + O_2(g)$ takes place on a platinum surface. Although $[N_2O]$ is three times as great in (b) as in (a), the rate of decomposition of N_2O is the same in both cases because the platinum surface can accommodate only a certain number of molecules. As a result, this reaction is zero order.

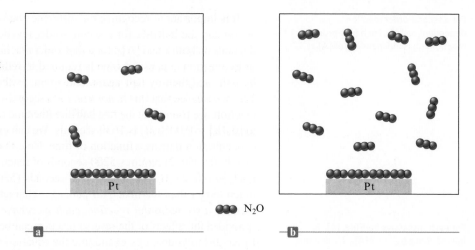

N_2O

a

b

Critical Thinking

Consider the simple reaction aA → products. You run this reaction and wish to determine its order. What if you made a graph of reaction rate versus time? Could you use this to determine the order? Sketch three plots of rate versus time for the reaction if it is zero, first, or second order. Sketch these plots on the same graph and compare them. Defend your answer.

Integrated Rate Laws for Reactions with More Than One Reactant

So far we have considered the integrated rate laws for simple reactions with only one reactant. Special techniques are required to deal with more complicated reactions. Let's consider the reaction

$$BrO_3^-(aq) + 5Br^-(aq) + 6H^+(aq) \longrightarrow 3Br_2(l) + 3H_2O(l)$$

From experimental evidence we know that the rate law is

$$\text{Rate} = -\frac{\Delta[BrO_3^-]}{\Delta t} = k[BrO_3^-][Br^-][H^+]^2$$

Suppose we run this reaction under conditions where $[BrO_3^-]_0 = 1.0 \times 10^{-3}\ M$, $[Br^-]_0 = 1.0\ M$, and $[H^+]_0 = 1.0\ M$. As the reaction proceeds, $[BrO_3^-]$ decreases significantly, but because the Br^- ion and H^+ ion concentrations are so large initially, relatively little of these two reactants is consumed. Thus $[Br^-]$ and $[H^+]$ remain *approximately constant*. In other words, under the conditions where the Br^- ion and H^+ ion concentrations are much larger than the BrO_3^- ion concentration, we can assume that throughout the reaction

$$[Br^-] = [Br^-]_0 \quad \text{and} \quad [H^+] = [H^+]_0$$

This means that the rate law can be written

$$\text{Rate} = k[Br^-]_0[H^+]_0^2[BrO_3^-] = k'[BrO_3^-]$$

where, since $[Br^-]_0$ and $[H^+]_0$ are constant,

$$k' = k[Br^-]_0[H^+]_0^2$$

The rate law

$$\text{Rate} = k'[\text{BrO}_3^-]$$

is first order. However, since this law was obtained by simplifying a more complicated one, it is called a **pseudo-first-order rate law**. Under the conditions of this experiment, a plot of $\ln[\text{BrO}_3^-]$ versus t will give a straight line where the slope is equal to $-k'$. Since $[\text{Br}^-]_0$ and $[\text{H}^+]_0$ are known, the value of k can be calculated from the equation

$$k' = k[\text{Br}^-]_0[\text{H}^+]_0^2$$

which can be rearranged to give

$$k = \frac{k'}{[\text{Br}^-]_0[\text{H}^+]_0^2}$$

Note that the kinetics of complicated reactions can be studied by observing the behavior of one reactant at a time. If the concentration of one reactant is much smaller than the concentrations of the others, then the amounts of those reactants present in large concentrations will not change significantly and can be regarded as constant. The change in concentration with time of the reactant present in a relatively small amount can then be used to determine the order of the reaction in that component. This technique allows us to determine rate laws for complex reactions.

Let's Review | *Rate Laws: A Summary*

1. To simplify the rate laws for reactions, we have always assumed that the rate is being studied under conditions where only the forward reaction is important. This produces rate laws that contain only reactant concentrations.

2. There are two types of rate laws.
   a. The *differential rate law* (often called the *rate law*) shows how the rate depends on the concentrations. The forms of the rate laws for zero-order, first-order, and second-order kinetics of reactions with single reactants are shown in Table 12.6.
   b. The *integrated rate law* shows how concentration depends on time. The integrated rate laws corresponding to zero-order, first-order, and second-order kinetics of one-reactant reactions are given in Table 12.6.

3. Whether we determine the differential rate law or the integrated rate law depends on the type of data that can be collected conveniently and accurately. Once we have experimentally determined either type of rate law, we can write the other for a given reaction.

4. The most common method for experimentally determining the differential rate law is the method of initial rates. In this method several experiments are run at different initial concentrations and the instantaneous rates are determined for each at the same value of t (as close to $t = 0$ as possible). The point is to evaluate the rate before the concentrations change significantly from the initial values. From a comparison of the initial rates and the initial concentrations, the dependence of the rate on the concentrations of various reactants can be obtained—that is, the order in each reactant can be determined.

5. To experimentally determine the integrated rate law for a reaction, concentrations are measured at various values of t as the reaction proceeds. Then the job is to see which integrated rate law correctly fits the data. Typically this is done visually by ascertaining which type of plot gives a straight line. A summary for one-reactant reactions is given in Table 12.6. Once the correct straight-line plot is found, the correct integrated rate law can be chosen and the value of k obtained from the slope. Also, the (differential) rate law for the reaction can then be written.

(continued)

Let's Review | *Rate Laws: A Summary (continued)*

6. The integrated rate law for a reaction that involves several reactants can be treated by choosing conditions such that the concentration of only one reactant varies in a given experiment. This is done by having the concentration of one reactant remain small compared with the concentrations of all the others, causing a rate law such as

$$\text{Rate} = k[A]^n[B]^m[C]^p$$

to reduce to

$$\text{Rate} = k'[A]^n$$

where $k' = k[B]_0^m[C]_0^p$ and $[B]_0 \gg [A]_0$ and $[C]_0 \gg [A]_0$. The value of n is obtained by determining whether a plot of [A] versus t is linear ($n = 0$), a plot of ln[A] versus t is linear ($n = 1$), or a plot of 1/[A] versus t is linear ($n = 2$). The value of k' is determined from the slope of the appropriate plot. The values of m, p, and k can be found by determining the value of k' at several different concentrations of B and C.

Table 12.6 | Summary of the Kinetics for Reactions of the Type aA → Products That Are Zero, First, or Second Order in [A]

	Order		
	Zero	**First**	**Second**
Rate law	$\text{Rate} = k$	$\text{Rate} = k[A]$	$\text{Rate} = k[A]^2$
Integrated rate law	$[A] = -kt + [A]_0$	$\ln[A] = -kt + \ln[A]_0$	$\dfrac{1}{[A]} = kt + \dfrac{1}{[A]_0}$
Plot needed to give a straight line	[A] versus t	ln[A] versus t	$\dfrac{1}{[A]}$ versus t
Relationship of rate constant to the slope of straight line	Slope $= -k$	Slope $= -k$	Slope $= k$
Half-Life	$t_{1/2} = \dfrac{[A]_0}{2k}$	$t_{1/2} = \dfrac{0.693}{k}$	$t_{1/2} = \dfrac{1}{k[A]_0}$

12.5 | Reaction Mechanisms

Most chemical reactions occur by a *series of steps* called the **reaction mechanism**. To understand a reaction, we must know its mechanism, and one of the main purposes for studying kinetics is to learn as much as possible about the steps involved in a reaction. In this section we explore some of the fundamental characteristics of reaction mechanisms.

Consider the reaction between nitrogen dioxide and carbon monoxide:

$$NO_2(g) + CO(g) \longrightarrow NO(g) + CO_2(g)$$

The rate law for this reaction is known from experiment to be

$$\text{Rate} = k[NO_2]^2$$

A balanced equation does not tell us *how* the reactants become products.

As we will see, this reaction is more complicated than it appears from the balanced equation. This is quite typical; the balanced equation for a reaction tells us the reactants, the products, and the stoichiometry but gives no direct information about the reaction mechanism.

Figure 12.8 | A molecular representation of the elementary steps in the reaction of NO_2 and CO.

Step 1

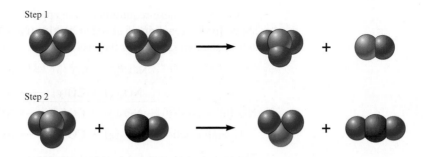

Step 2

For the reaction between nitrogen dioxide and carbon monoxide, the mechanism is thought to involve the following steps:

$$NO_2(g) + NO_2(g) \xrightarrow{k_1} NO_3(g) + NO(g)$$
$$NO_3(g) + CO(g) \xrightarrow{k_2} NO_2(g) + CO_2(g)$$

An intermediate is formed in one step and used up in a subsequent step and so is never seen as a product.

where k_1 and k_2 are the rate constants of the individual reactions. In this mechanism, gaseous NO_3 is an **intermediate**, a species that is neither a reactant nor a product but that is formed and consumed during the reaction sequence. This reaction is illustrated in Fig. 12.8.

Each of these two reactions is called an **elementary step**, *a reaction whose rate law can be written from its molecularity.* **Molecularity** is defined as the number of species that must collide to produce the reaction indicated by that step. A reaction involving one molecule is called a **unimolecular step**. Reactions involving the collision of two and three species are termed **bimolecular** and **termolecular**, respectively. Termolecular steps are quite rare, because the probability of three molecules colliding simultaneously is very small. Examples of these three types of elementary steps and the corresponding rate laws are shown in Table 12.7. Note from Table 12.7 that the rate law for an elementary step follows *directly* from the molecularity of that step. For example, for a bimolecular step the rate law is always second order, either of the form $k[A]^2$ for a step with a single reactant or of the form $k[A][B]$ for a step involving two reactants.

The prefix uni- means one, bi- means two, and ter- means three.

A unimolecular elementary step is always first order, a bimolecular step is always second order, and so on.

We can now define a reaction mechanism more precisely. It is a *series of elementary steps that must satisfy two requirements:*

1. The sum of the elementary steps must give the overall balanced equation for the reaction.
2. The mechanism must agree with the experimentally determined rate law.

The rate at which this colored solution enters the flask is determined by the size of the funnel stem, not how fast the solution is poured.

© Cengage Learning

Table 12.7 | Examples of Elementary Steps

Elementary Step	Molecularity	Rate Law
A → products	*Uni*molecular	Rate = $k[A]$
A + A → products	*Bi*molecular	Rate = $k[A]^2$
(2A → products)		
A + B → products	*Bi*molecular	Rate = $k[A][B]$
A + A + B → products	*Ter*molecular	Rate = $k[A]^2[B]$
(2A + B → products)		
A + B + C → products	*Ter*molecular	Rate = $k[A][B][C]$

To see how these requirements are applied, we will consider the mechanism given above for the reaction of nitrogen dioxide and carbon monoxide. First, note that the sum of the two steps gives the overall balanced equation:

$$NO_2(g) + NO_2(g) \longrightarrow NO_3(g) + NO(g)$$
$$NO_3(g) + CO(g) \longrightarrow NO_2(g) + CO_2(g)$$

$$\overline{NO_2(g) + NO_2(g) + NO_3(g) + CO(g) \longrightarrow NO_3(g) + NO(g) + NO_2(g) + CO_2(g)}$$

Overall reaction: $NO_2(g) + CO(g) \longrightarrow NO(g) + CO_2(g)$

The first requirement for a correct mechanism is met. To see whether the mechanism meets the second requirement, we need to introduce a new idea: the **rate-determining step**. Multistep reactions often have one step that is much slower than all the others. Reactants can become products only as fast as they can get through this slowest step. That is, the overall reaction can be no faster than the slowest, or rate-determining, step in the sequence. An analogy for this situation is the pouring of water rapidly into a container through a funnel. The water collects in the container at a rate that is essentially determined by the size of the funnel opening and not by the rate of pouring.

Which is the rate-determining step in the reaction of nitrogen dioxide and carbon monoxide? Let's *assume* that the first step is rate-determining and the second step is relatively fast:

$$NO_2(g) + NO_2(g) \longrightarrow NO_3(g) + NO(g) \quad \text{Slow (rate-determining)}$$
$$NO_3(g) + CO(g) \longrightarrow NO_2(g) + CO_2(g) \quad \text{Fast}$$

What we have really assumed here is that the formation of NO_3 occurs much more slowly than its reaction with CO. The rate of CO_2 production is then controlled by the rate of formation of NO_3 in the first step. Since this is an elementary step, we can write the rate law from the molecularity. The bimolecular first step has the rate law

$$\text{Rate of formation of } NO_3 = \frac{\Delta[NO_3]}{\Delta t} = k_1[NO_2]^2$$

Since the overall reaction rate can be no faster than the slowest step,

$$\text{Overall rate} = k_1[NO_2]^2$$

Note that this rate law agrees with the experimentally determined rate law given earlier. The mechanism we assumed above satisfies the two requirements stated earlier and *may* be the correct mechanism for the reaction.

How does a chemist deduce the mechanism for a given reaction? The rate law is always determined first. Then, using chemical intuition and following the two rules given on the previous page, the chemist constructs possible mechanisms and tries, with further experiments, to eliminate those that are least likely. *A mechanism can never be proved absolutely.* We can say only that a mechanism that satisfies the two requirements is *possibly* correct. Deducing mechanisms for chemical reactions can be difficult and requires skill and experience. We will only touch on this process in this text.

A reaction is only as fast as its slowest step.

Example 12.6

Reaction Mechanisms

The balanced equation for the reaction of the gases nitrogen dioxide and fluorine is

$$2NO_2(g) + F_2(g) \longrightarrow 2NO_2F(g)$$

The experimentally determined rate law is

$$\text{Rate} = k[NO_2][F_2]$$

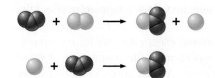

A suggested mechanism for this reaction is

$$NO_2 + F_2 \xrightarrow{k_1} NO_2F + F \qquad \text{Slow}$$

$$F + NO_2 \xrightarrow{k_2} NO_2F \qquad \text{Fast}$$

Is this an acceptable mechanism? That is, does it satisfy the two requirements?

Solution

The first requirement for an acceptable mechanism is that the sum of the steps should give the balanced equation:

$$NO_2 + F_2 \longrightarrow NO_2F + F$$
$$\underline{F + NO_2 \longrightarrow NO_2F}$$
$$2NO_2 + F_2 + \cancel{F} \longrightarrow 2NO_2F + \cancel{F}$$

Overall reaction: $\qquad 2NO_2 + F_2 \longrightarrow 2NO_2F$

The first requirement is met.

The second requirement is that the mechanism must agree with the experimentally determined rate law. Since the proposed mechanism states that the first step is rate-determining, the overall reaction rate must be that of the first step. The first step is bimolecular, so the rate law is

$$\text{Rate} = k_1[NO_2][F_2]$$

This has the same form as the experimentally determined rate law. The proposed mechanism is acceptable because it satisfies both requirements. (Note that we have not proved that it is *the correct* mechanism.)

See Exercises 12.61 and 12.62

Although the mechanism given in Example 12.6 has the correct stoichiometry and fits the observed rate law, other mechanisms may also satisfy these requirements. For example, the mechanism might be

$$NO_2 + F_2 \longrightarrow NOF_2 + O \qquad \text{Slow}$$
$$NO_2 + O \longrightarrow NO_3 \qquad \text{Fast}$$
$$NOF_2 + NO_2 \longrightarrow NO_2F + NOF \qquad \text{Fast}$$
$$NO_3 + NOF \longrightarrow NO_2F + NO_2 \qquad \text{Fast}$$

To decide on the most probable mechanism for the reaction, the chemist doing the study would have to perform additional experiments.

12.6 | A Model for Chemical Kinetics

How do chemical reactions occur? We already have given some indications. For example, we have seen that the rates of chemical reactions depend on the concentrations of the reacting species. The initial rate for the reaction

$$aA + bB \longrightarrow \text{products}$$

can be described by the rate law

$$\text{Rate} = k[A]^n[B]^m$$

where the order of each reactant depends on the detailed reaction mechanism. This explains why reaction rates depend on concentration. But what about some of the

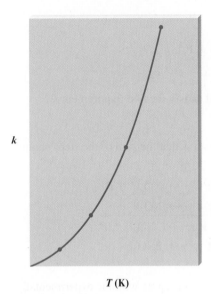

k

T (K)

Figure 12.9 | A plot showing the exponential dependence of the rate constant on absolute temperature. The exact temperature dependence of k is different for each reaction. This plot represents the behavior of a rate constant that doubles for every increase in temperature of 10 K.

The higher the activation energy, the slower the reaction at a given temperature.

other factors affecting reaction rates? For example, how does temperature affect the speed of a reaction?

We can answer this question qualitatively from our experience. We have refrigerators because food spoilage is retarded at low temperatures. The combustion of wood occurs at a measurable rate only at high temperatures. An egg cooks in boiling water much faster at sea level than in Leadville, Colorado (elevation 10,000 ft), where the boiling point of water is approximately 90°C. These observations and others lead us to conclude that *chemical reactions speed up when the temperature is increased.* Experiments have shown that virtually all rate constants show an exponential increase with absolute temperature, as represented in Fig. 12.9.

In this section we discuss a model used to account for the observed characteristics of reaction rates. This model, called the **collision model**, is built around the central idea that *molecules must collide to react.* We have already seen how this assumption explains the concentration dependence of reaction rates. Now we need to consider whether this model can account for the observed temperature dependence of reaction rates.

The kinetic molecular theory of gases predicts that an increase in temperature raises molecular velocities and so increases the frequency of collisions between molecules. This idea agrees with the observation that reaction rates are greater at higher temperatures. Thus there is qualitative agreement between the collision model and experimental observations. However, it is found that the rate of reaction is much smaller than the calculated collision frequency in a collection of gas particles. This must mean that *only a small fraction of the collisions produces a reaction.* Why?

This question was first addressed in the 1880s by Svante Arrhenius. He proposed the existence of a *threshold energy,* called the **activation energy**, that must be overcome to produce a chemical reaction. Such a proposal makes sense, as we can see by considering the decomposition of BrNO in the gas phase:

$$2BrNO(g) \longrightarrow 2NO(g) + Br_2(g)$$

In this reaction two Br—N bonds must be broken and one Br—Br bond must be formed. Breaking a Br—N bond requires considerable energy (243 kJ/mol), which must come from somewhere. The collision model postulates that the energy comes from the kinetic energies possessed by the reacting molecules before the collision. This kinetic energy is changed into potential energy as the molecules are distorted during a collision to break bonds and rearrange the atoms into the product molecules.

We can envision the reaction progress as shown in Fig. 12.10. The arrangement of atoms found at the top of the potential energy "hill," or barrier, is called the **activated complex**, or **transition state**. The conversion of BrNO to NO and Br_2 is exothermic, as indicated by the fact that the products have lower potential energy than the reactant. However, ΔE has no effect on the rate of the reaction. Rather, the rate depends on the size of the activation energy E_a.

The main point here is that a certain minimum energy is required for two BrNO molecules to "get over the hill" so that products can form. This energy is furnished by the energy of the collision. A collision between two BrNO molecules with small kinetic energies will not have enough energy to get over the barrier. At a given temperature only a certain fraction of the collisions possesses enough energy to be effective (to result in product formation).

We can be more precise by recalling from Chapter 5 that a distribution of velocities exists in a sample of gas molecules. Therefore, a distribution of collision energies also exists, as shown in Fig. 12.11 for two different temperatures. Figure 12.11 also shows the activation energy for the reaction in question. Only collisions with energy greater than the activation energy are able to react (get over the barrier). At the lower temperature, T_1, the fraction of effective collisions is quite small. However, as the temperature is increased to T_2, the fraction of collisions with the required activation energy increases dramatically. When the temperature is doubled, the fraction of

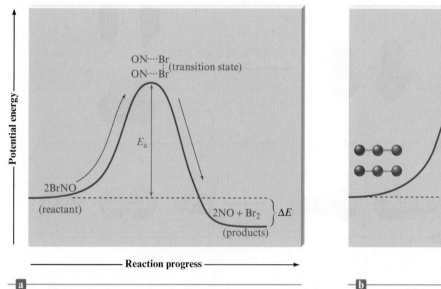

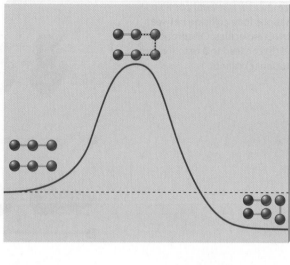

Figure 12.10 | (a) The change in potential energy as a function of reaction progress for the reaction $2BrNO \rightarrow 2NO + Br_2$. The activation energy E_a represents the energy needed to disrupt the BrNO molecules so that they can form products. The quantity ΔE represents the net change in energy in going from reactant to products. (b) A molecular representation of the reaction.

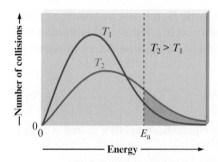

Figure 12.11 | Plot showing the number of collisions with a particular energy at T_1 and T_2, where $T_2 > T_1$.

effective collisions much more than doubles. In fact, the fraction of effective collisions increases *exponentially* with temperature. This is encouraging for our theory; remember that rates of reactions are observed to increase exponentially with temperature. Arrhenius postulated that the number of collisions having an energy greater than or equal to the activation energy is given by the expression:

$$\text{Number of collisions with the activation energy} = (\text{total number of collisions})e^{-E_a/RT}$$

where E_a is the activation energy, R is the universal gas constant, and T is the Kelvin temperature. The factor $e^{-E_a/RT}$ represents the fraction of collisions with energy E_a or greater at temperature T.

We have seen that not all molecular collisions are effective in producing chemical reactions because a minimum energy is required for the reaction to occur. There is, however, another complication. Experiments show that the *observed reaction rate is considerably smaller than the rate of collisions with enough energy to surmount the barrier.* This means that many collisions, even though they have the required energy, still do not produce a reaction. Why not?

The answer lies in the **molecular orientations** during collisions. We can illustrate this using the reaction between two BrNO molecules (Fig. 12.12). Some collision orientations can lead to reaction, and others cannot. Therefore, we must include a correction factor to allow for collisions with nonproductive molecular orientations.

To summarize, two requirements must be satisfied for reactants to collide successfully (to rearrange to form products):

1. The collision must involve enough energy to produce the reaction; that is, the collision energy must equal or exceed the activation energy.
2. The relative orientation of the reactants must allow formation of any new bonds necessary to produce products.

Taking these factors into account, we can represent the rate constant as

$$k = zpe^{-E_a/RT}$$

Figure 12.12 | Several possible orientations for a collision between two BrNO molecules. Orientations (a) and (b) can lead to a reaction, but orientation (c) cannot.

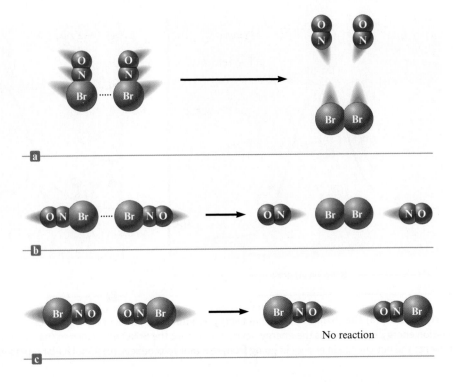

A snowy tree cricket. The frequency of a cricket's chirps depends on the temperature of the cricket.

Alvin E. Staffan/Photo Researchers, Inc.

where z is the collision frequency, p is called the **steric factor** (always less than 1) and reflects the fraction of collisions with effective orientations, and $e^{-E_a/RT}$ represents the fraction of collisions with sufficient energy to produce a reaction. This expression is most often written in form

$$k = Ae^{-E_a/RT} \tag{12.9}$$

which is called the **Arrhenius equation**. In this equation, A replaces zp and is called the **frequency factor** for the reaction.

Taking the natural logarithm of each side of the Arrhenius equation gives

$$\ln(k) = -\frac{E_a}{R}\left(\frac{1}{T}\right) + \ln(A) \tag{12.10}$$

Equation (12.10) is a linear equation of the type $y = mx + b$, where $y = \ln(k)$, $m = -E_a/R = \text{slope}$, $x = 1/T$, and $b = \ln(A) = \text{intercept}$. Thus for a reaction where the rate constant obeys the Arrhenius equation, a plot of $\ln(k)$ versus $1/T$ gives a straight line. The slope and intercept can be used to determine, respectively, the values of E_a and A characteristic of that reaction. The fact that most rate constants obey the Arrhenius equation to a good approximation indicates that the collision model for chemical reactions is physically reasonable.

Critical Thinking

There are many conditions that need to be met to result in a chemical reaction between molecules. What if all collisions between molecules resulted in a chemical reaction? How would life be different?

Example 12.7

Determining Activation Energy I

The reaction

$$2N_2O_5(g) \longrightarrow 4NO_2(g) + O_2(g)$$

was studied at several temperatures, and the following values of k were obtained:

k (s^{-1})	T (°C)
2.0×10^{-5}	20
7.3×10^{-5}	30
2.7×10^{-4}	40
9.1×10^{-4}	50
2.9×10^{-3}	60

Calculate the value of E_a for this reaction.

Solution

To obtain the value of E_a, we need to construct a plot of $\ln(k)$ versus $1/T$. First, we must calculate values of $\ln(k)$ and $1/T$, as shown below:

T (°C)	T (K)	$1/T$ (K)	k (s^{-1})	$\ln(k)$
20	293	3.41×10^{-3}	2.0×10^{-5}	-10.82
30	303	3.30×10^{-3}	7.3×10^{-5}	-9.53
40	313	3.19×10^{-3}	2.7×10^{-4}	-8.22
50	323	3.10×10^{-3}	9.1×10^{-4}	-7.00
60	333	3.00×10^{-3}	2.9×10^{-3}	-5.84

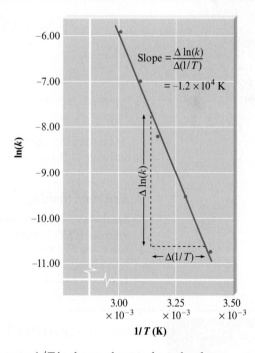

The plot of $\ln(k)$ versus $1/T$ is shown above, where the slope

$$\frac{\Delta \ln(k)}{\Delta\left(\dfrac{1}{T}\right)}$$

is found to be -1.2×10^4 K. The value of E_a can be determined by solving the following equation:

$$\text{Slope} = -\frac{E_a}{R}$$

$$E_a = -R(\text{slope}) = -(8.3145 \text{ J/K} \cdot \text{mol})(-1.2 \times 10^4 \text{ K})$$
$$= 1.0 \times 10^5 \text{ J/mol}$$

Thus the value of the activation energy for this reaction is 1.0×10^5 J/mol.

See Exercises 12.67 and 12.68

The most common procedure for finding E_a for a reaction involves measuring the rate constant k at several temperatures and then plotting $\ln(k)$ versus $1/T$, as shown in Example 12.7. However, E_a also can be calculated from the values of k at only two temperatures by using a formula that can be derived as follows from Equation (12.10).

At temperature T_1, where the rate constant is k_1,

$$\ln(k_1) = -\frac{E_a}{RT_1} + \ln(A)$$

At temperature T_2, where the rate constant is k_2,

$$\ln(k_2) = -\frac{E_a}{RT_2} + \ln(A)$$

Subtracting the first equation from the second gives

$$\ln(k_2) - \ln(k_1) = \left[-\frac{E_a}{RT_2} + \ln(A)\right] - \left[-\frac{E_a}{RT_1} + \ln(A)\right]$$
$$= -\frac{E_a}{RT_2} + \frac{E_a}{RT_1}$$

And

$$\ln\left(\frac{k_2}{k_1}\right) = \frac{E_a}{R}\left(\frac{1}{T_1} - \frac{1}{T_2}\right) \tag{12.11}$$

Therefore, the values of k_1 and k_2 measured at temperatures T_1 and T_2 can be used to calculate E_a, as shown in Example 12.8.

Critical Thinking

Most modern refrigerators have an internal temperature of 45°F. What if refrigerators were set at 55°F in the factory? How would this affect our lives?

Interactive Example 12.8

Sign in at http://login.cengagebrain.com to try this Interactive Example in **OWL**.

Determining Activation Energy II

The gas-phase reaction between methane and diatomic sulfur is given by the equation

$$CH_4(g) + 2S_2(g) \longrightarrow CS_2(g) + 2H_2S(g)$$

At 550°C the rate constant for this reaction is 1.1 L/mol · s, and at 625°C the rate constant is 6.4 L/mol · s. Using these values, calculate E_a for this reaction.

Solution

The relevant data are shown in the following table:

k (L/mol · s)	T (°C)	T (K)
$1.1 = k_1$	550	$823 = T_1$
$6.4 = k_2$	625	$898 = T_2$

Substituting these values into Equation (12.11) gives

$$\ln\left(\frac{6.4}{1.1}\right) = \frac{E_a}{8.3145 \text{ J/K} \cdot \text{mol}}\left(\frac{1}{823 \text{ K}} - \frac{1}{898 \text{ K}}\right)$$

Solving for E_a gives

$$E_a = \frac{(8.3145 \text{ J/K} \cdot \text{mol})\ln\left(\dfrac{6.4}{1.1}\right)}{\left(\dfrac{1}{823 \text{ K}} - \dfrac{1}{898 \text{ K}}\right)}$$

$$= 1.4 \times 10^5 \text{ J/mol}$$

See Exercises 12.69 through 12.72

12.7 | Catalysis

We have seen that the rate of a reaction increases dramatically with temperature. If a particular reaction does not occur fast enough at normal temperatures, we can speed it up by raising the temperature. However, sometimes this is not feasible. For example, living cells can survive only in a rather narrow temperature range, and the human body is designed to operate at an almost constant temperature of 98.6°F. But many of the complicated biochemical reactions keeping us alive would be much too slow at this temperature without intervention. We exist only because the body contains many substances called **enzymes**, which increase the rates of these reactions. In fact, almost every biologically important reaction is assisted by a specific enzyme.

Although it is possible to use higher temperatures to speed up commercially important reactions, such as the Haber process for synthesizing ammonia, this is very expensive. In a chemical plant an increase in temperature means significantly increased costs for energy. The use of an appropriate catalyst allows a reaction to proceed rapidly at a relatively low temperature and can therefore hold down production costs.

A **catalyst** is *a substance that speeds up a reaction without being consumed itself.* Just as virtually all vital biologic reactions are assisted by enzymes (biologic catalysts), almost all industrial processes also involve the use of catalysts. For example, the production of sulfuric acid uses vanadium(V) oxide, and the Haber process uses a mixture of iron and iron oxide.

How does a catalyst work? Remember that for each reaction a certain energy barrier must be surmounted. How can we make a reaction occur faster without raising the temperature to increase the molecular energies? The solution is to provide a new pathway for the reaction, one with a *lower activation energy.* This is what a catalyst does, as is shown in Fig. 12.13. Because the catalyst allows the reaction to occur with a lower activation energy, a much larger fraction of collisions is effective at a given temperature, and the reaction rate is increased. This effect is illustrated in Fig. 12.14. Note from this diagram that although a catalyst lowers the activation energy E_a for a reaction, it does not affect the energy difference ΔE between products and reactants.

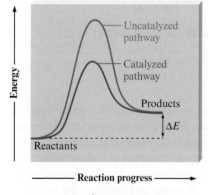

Figure 12.13 | Energy plots for a catalyzed and an uncatalyzed pathway for a given reaction.

Figure 12.14 │ Effect of a catalyst on the number of reaction-producing collisions. Because a catalyst provides a reaction pathway with a lower activation energy, a much greater fraction of the collisions is effective for the catalyzed pathway (b) than for the uncatalyzed pathway (a) (at a given temperature). This allows reactants to become products at a much higher rate, even though there is no temperature increase.

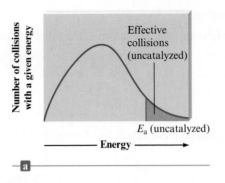

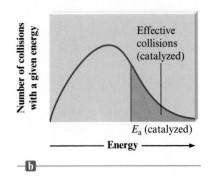

Catalysts are classified as homogeneous or heterogeneous. A **homogeneous catalyst** is one that is *present in the same phase as the reacting molecules.* A **heterogeneous catalyst** exists *in a different phase,* usually as a solid.

Heterogeneous Catalysis

Heterogeneous catalysis most often involves gaseous reactants being adsorbed on the surface of a solid catalyst. **Adsorption** refers to the collection of one substance on the surface of another substance; *absorption* refers to the penetration of one substance into another. Water is *absorbed* by a sponge.

An important example of heterogeneous catalysis occurs in the hydrogenation of unsaturated hydrocarbons, compounds composed mainly of carbon and hydrogen with some carbon–carbon double bonds. Hydrogenation is an important industrial process used to change unsaturated fats, occurring as oils, to saturated fats (solid shortenings such as Crisco) in which the C=C bonds have been converted to C—C bonds through addition of hydrogen.

A simple example of hydrogenation involves ethylene:

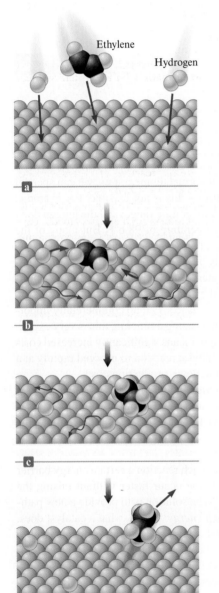

$$
\underset{\text{Ethylene}}{\begin{array}{c} \text{H} \\ \diagdown \\ \text{H} \end{array} \text{C}=\text{C} \begin{array}{c} \text{H} \\ \diagup \\ \text{H} \end{array}} (g) + \text{H}_2(g) \longrightarrow \underset{\text{Ethane}}{\text{H}-\overset{\displaystyle \text{H}}{\underset{\displaystyle \text{H}}{\text{C}}}-\overset{\displaystyle \text{H}}{\underset{\displaystyle \text{H}}{\text{C}}}-\text{H}(g)}
$$

This reaction is quite slow at normal temperatures, mainly because the strong bond in the hydrogen molecule results in a large activation energy for the reaction. However, the reaction rate can be greatly increased by using a solid catalyst of platinum, palladium, or nickel. The hydrogen and ethylene adsorb on the catalyst surface, where the reaction occurs. The main function of the catalyst apparently is to allow formation of metal–hydrogen interactions that weaken the H—H bonds and facilitate the reaction. The mechanism is illustrated in Fig. 12.15.

Figure 12.15 │ Heterogeneous catalysis of the hydrogenation of ethylene. (a) The reactants above the metal surface. (b) Hydrogen is adsorbed onto the metal surface, forming metal–hydrogen bonds and breaking the H—H bonds. The π bond in ethylene is broken and metal–hydrogen bonds are formed during adsorption. (c) The adsorbed molecules and atoms migrate toward each other on the metal surface, forming new C—H bonds. (d) The C atoms in ethane (C_2H_6) have completely saturated bonding capacities and so cannot bind strongly to the metal surfaces. The C_2H_6 molecule thus escapes.

Heterogeneous catalysis also occurs in the oxidation of gaseous sulfur dioxide to gaseous sulfur trioxide. This process is especially interesting because it illustrates both positive and negative consequences of chemical catalysis.

The negative side is the formation of damaging air pollutants. Recall that sulfur dioxide, a toxic gas with a choking odor, is formed whenever sulfur-containing fuels are burned. However, it is sulfur trioxide that causes most of the environmental damage, mainly through the production of acid rain. When sulfur trioxide combines with a droplet of water, sulfuric acid is formed:

$$H_2O(l) + SO_3(g) \longrightarrow H_2SO_4(aq)$$

This sulfuric acid can cause considerable damage to vegetation, buildings and statues, and fish populations.

Sulfur dioxide is *not* rapidly oxidized to sulfur trioxide in clean, dry air. Why, then, is there a problem? The answer is catalysis. Dust particles and water droplets catalyze the reaction between SO_2 and O_2 in the air.

On the positive side, the heterogeneous catalysis of the oxidation of SO_2 is used to advantage in the manufacture of sulfuric acid, where the reaction of O_2 and SO_2 to form SO_3 is catalyzed by a solid mixture of platinum and vanadium(V) oxide.

Heterogeneous catalysis is also utilized in the catalytic converters in automobile exhaust systems. The exhaust gases, containing compounds such as nitric oxide, carbon monoxide, and unburned hydrocarbons, are passed through a converter containing beads of solid catalyst (Fig. 12.16). The catalyst promotes the conversion of carbon monoxide to carbon dioxide, hydrocarbons to carbon dioxide and water, and nitric oxide to nitrogen gas to lessen the environmental impact of the exhaust gases. However, this beneficial catalysis can, unfortunately, be accompanied by the unwanted catalysis of the oxidation of SO_2 to SO_3, which reacts with the moisture present to form sulfuric acid.

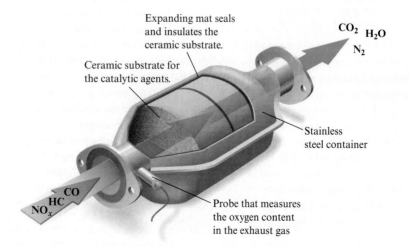

Figure 12.16 | The exhaust gases (HC, hydrocarbons; NO_x, nitrous oxides; and CO) from an automobile engine are passed through a catalytic converter to minimize environmental damage.

Chemical connections
Enzymes: Nature's Catalysts

The most impressive examples of homogeneous catalysis occur in nature, where the complex reactions necessary for plant and animal life are made possible by enzymes. Enzymes are large molecules specifically tailored to facilitate a given type of reaction. Usually enzymes are proteins, an important class of biomolecules constructed from α-amino acids that have the general structure

where R represents any one of 20 different substituents. These amino acid molecules can be "hooked together" to form a *polymer* (a word meaning "many parts") called a *protein*. The general structure of a protein can be represented as follows:

Many amino acid fragments | Fragment from an amino acid with substituent R | Fragment from an amino acid with substituent R' | Fragment from an amino acid with substituent R"

Since specific proteins are needed by the human body, the proteins in food must be broken into their constituent amino acids, which are then used to construct new proteins in the body's cells. The reaction in which a protein is broken down one amino acid at a time is shown in Fig. 12.17. Note that in this reaction a water molecule reacts with a protein molecule to produce an amino acid and a new protein containing one less amino acid.

Figure 12.17 | The removal of the end amino acid from a protein by reaction with a molecule of water. The products are an amino acid and a new, smaller protein.

Without the enzymes found in human cells, this reaction would be much too slow to be useful. One of these enzymes is *carboxypeptidase-A*, a zinc-containing protein (Fig. 12.18).

Carboxypeptidase-A captures the protein to be acted on (called the *substrate*) in a special groove and positions the substrate so that the end is in the active site, where the catalysis occurs (Fig. 12.19). Note that the Zn^{2+} ion bonds to the oxygen of the C=O (carbonyl) group. This polarizes the electron density in the carbonyl group, allowing the neighboring C—N bond to be broken much more easily. When the reaction is completed, the remaining portion of the substrate protein and the newly formed amino acid are released by enzyme.

The process just described for carboxypeptidase-A is characteristic of the behavior of other enzymes. Enzyme catalysis can be represented by the series of reactions shown below:

$$E + S \longrightarrow E \cdot S$$
$$E \cdot S \longrightarrow E + P$$

where E represents the enzyme, S represents the substrate, $E \cdot S$ represents the enzyme–substrate complex, and P represents the products. The enzyme and substrate form a complex, where the reaction occurs. The enzyme then releases the product and is ready to repeat the process. The most amazing thing about enzymes is their efficiency. Because an enzyme plays its catalytic role over and over and very rapidly, only a tiny amount of enzyme is required. This makes the isolation of enzymes for study quite difficult.

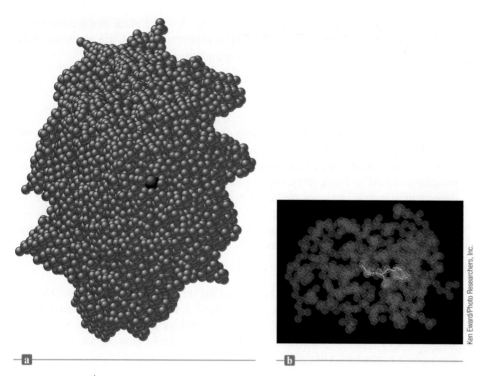

Ken Eward/Photo Researchers, Inc.

a   **b**

Figure 12.18 | (a) The structure of the enzyme carboxypeptidase-A, which contains 307 amino acids. The zinc ion is shown above as a black sphere in the center. (b) Carboxypeptidase-A with a substrate (pink) in place.

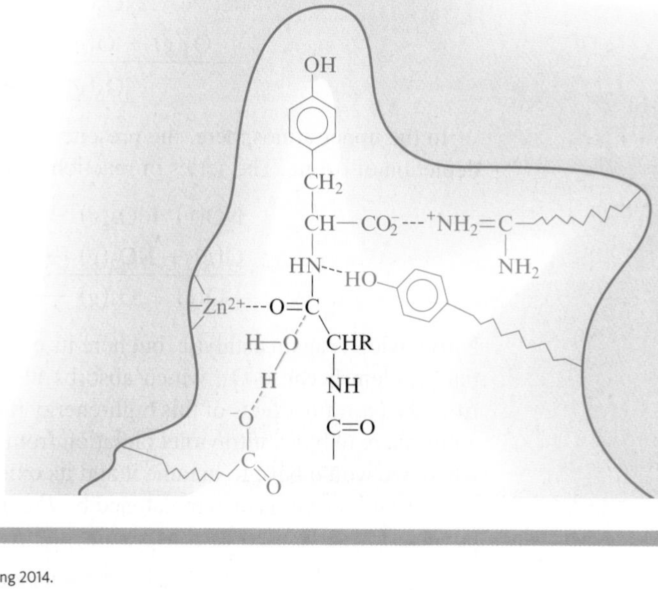

Figure 12.19 | Protein–substrate interaction. The substrate is shown in black and red, with the red representing the terminal amino acid. Blue indicates side chains from the enzyme that help bind the substrate.

Because of the complex nature of the reactions that take place in the converter, a mixture of catalysts is used. The most effective catalytic materials are transition metal oxides and noble metals such as palladium and platinum.

Homogeneous Catalysis

A homogeneous catalyst exists in the same phase as the reacting molecules. There are many examples in both the gas and liquid phases. One such example is the unusual catalytic behavior of nitric oxide toward ozone. In the troposphere, that part of the atmosphere closest to earth, nitric oxide catalyzes ozone production. However, in the upper atmosphere it catalyzes the decomposition of ozone. Both these effects are unfortunate environmentally.

In the lower atmosphere, NO is produced in any high-temperature combustion process where N_2 is present. The reaction

$$N_2(g) + O_2(g) \longrightarrow 2NO(g)$$

is very slow at normal temperatures because of the very strong $N\equiv N$ and $O\!=\!O$ bonds. However, at elevated temperatures, such as those found in the internal combustion engines of automobiles, significant quantities of NO form. Some of this NO is converted back to N_2 in the catalytic converter, but significant amounts escape into the atmosphere to react with oxygen:

$$2NO(g) + O_2(g) \longrightarrow 2NO_2(g)$$

In the atmosphere, NO_2 can absorb light and decompose as follows:

$$NO_2(g) \xrightarrow{\text{Light}} NO(g) + O(g)$$

The oxygen atom is very reactive and can combine with an oxygen molecule to form ozone:

$$O_2(g) + O(g) \longrightarrow O_3(g)$$

Ozone is a powerful oxidizing agent that can react with other air pollutants to form substances irritating to the eyes and lungs, and is itself very toxic.

In this series of reactions, nitric oxide is acting as a true catalyst because it assists the production of ozone without being consumed itself. This can be seen by summing the reactions:

$$
\begin{aligned}
NO(g) + \tfrac{1}{2}O_2(g) &\longrightarrow NO_2(g) \\
NO_2(g) &\xrightarrow{\text{Light}} NO(g) + O(g) \\
\underline{O_2(g) + O(g)} &\longrightarrow \underline{O_3(g)} \\
\tfrac{3}{2}O_2(g) &\longrightarrow O_3(g)
\end{aligned}
$$

In the margin: Although O_2 is represented here as the oxidizing agent for NO, the actual oxidizing agent is probably some type of peroxide compound produced by reaction of oxygen with pollutants. The direct reaction of NO and O_2 is very slow.

In the upper atmosphere, the presence of nitric oxide has the opposite effect—the depletion of ozone. The series of reactions involved is

$$
\begin{aligned}
NO(g) + O_3(g) &\longrightarrow NO_2(g) + O_2(g) \\
\underline{O(g) + NO_2(g)} &\longrightarrow \underline{NO(g) + O_2(g)} \\
O(g) + O_3(g) &\longrightarrow 2O_2(g)
\end{aligned}
$$

Nitric oxide is again catalytic, but here its effect is to change O_3 to O_2. This is a potential problem because O_3, which absorbs ultraviolet light, is necessary to protect us from the harmful effects of this high-energy radiation. That is, we want O_3 in the upper atmosphere to block ultraviolet radiation from the sun but not in the lower atmosphere, where we would have to breathe it and its oxidation products.

The ozone layer is also threatened by *Freons*, a group of stable, noncorrosive compounds, formerly used as refrigerants and as propellants in aerosol cans. The most

This graphic shows data from the Total Ozone Mapping Spectrometer (TOMS) Earth Probe.

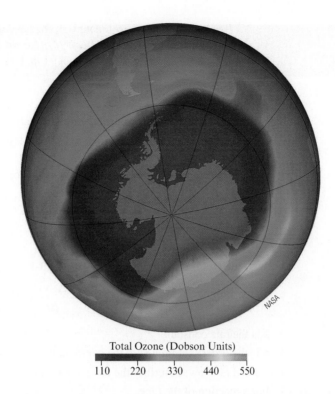

Total Ozone (Dobson Units)

110   220   330   440   550

Freon-12

commonly used substance of this type was Freon-12 (CCl_2F_2). The chemical inertness of Freons makes them valuable but also creates a problem, since they remain in the environment a long time. Eventually, they migrate into the upper atmosphere to be decomposed by high-energy light. Among the decomposition products are chlorine atoms:

$$CCl_2F_2(g) \xrightarrow{\text{Light}} CClF_2(g) + Cl(g)$$

These chlorine atoms can catalyze the decomposition of ozone:

$$\begin{array}{r} Cl(g) + O_3(g) \longrightarrow ClO(g) + O_2(g) \\ O(g) + ClO(g) \longrightarrow Cl(g) + O_2(g) \\ \hline O(g) + O_3(g) \longrightarrow 2O_2(g) \end{array}$$

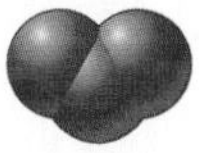

Ozone

The problem of Freons has been brought into strong focus by the discovery of a mysterious "hole" in the ozone layer in the stratosphere over Antarctica. Studies performed there to find the reason for the hole have found unusually high levels of chlorine monoxide (ClO). This strongly implicates the Freons in the atmosphere as being responsible for the ozone destruction.

Because they pose environmental problems, Freons have been banned by international agreement. Substitute compounds are now being used.

For review

Key terms

chemical kinetics

Section 12.1
reaction rate
instantaneous rate

Section 12.2
rate law
rate constant
order
(differential) rate law
integrated rate law

Section 12.3
method of initial rates
initial rate
overall reaction order

Section 12.4
first-order reaction
integrated first-order rate law
half-life of a reactant
integrated second-order rate
    law
zero-order reaction
integrated zero-order rate law
pseudo-first-order rate law

Section 12.5
reaction mechanism
intermediate
elementary step
molecularity
unimolecular step
bimolecular step
termolecular step
rate-determining step

Section 12.6
collision model
activation energy
activated complex (transition
    state)
molecular orientations
steric factor
Arrhenius equation
frequency factor

Section 12.7
enzyme
catalyst
homogeneous catalyst
heterogeneous catalyst
adsorption

Chemical kinetics

> The study of the factors that control the rate (speed) of a chemical reaction
>> Rate is defined in terms of the change in concentration of a given reaction component per unit time
>> Kinetic measurements are often made under conditions where the reverse reaction is insignificant
> The kinetic and thermodynamic properties of a reaction are not fundamentally related

Rate laws

> Differential rate law: describes the rate as a function of concentration

$$\text{Rate} = -\frac{\Delta[A]}{\Delta t} = k[A]^n$$

>> k is the rate constant
>> n is the order; not related to the coefficients in the balanced equation
> Integrated rate law: describes the concentration as a function of time
>> For a reaction of the type

$$aA \longrightarrow \text{products}$$

for which

$$\text{Rate} = k[A]^n$$

$n = 0$:
$$[A] = -kt + [A]_0$$
$$t_{1/2} = \frac{[A]_0}{2k}$$

$n = 1$:
$$\ln[A] = -kt + \ln[A]_0$$
$$t_{1/2} = \frac{0.693}{k}$$

$n = 2$:
$$\frac{1}{[A]} = kt + \frac{1}{[A]_0}$$
$$t_{1/2} = \frac{1}{k[A]_0}$$

> The value of k can be determined from the plot of the appropriate function of $[A]$ versus t

Reaction mechanism

> Series of elementary steps by which an overall reaction occurs
>> Elementary step: rate law for the step can be written from the molecularity of the reaction
> Two requirements for an acceptable mechanism:
>> The elementary steps sum to give the correct overall balanced equation
>> The mechanism agrees with the experimentally determined rate law
> Simple reactions can have an elementary step that is slower than all of the other steps; this is called the rate-determining step.

Kinetic models

> The simplest model to account for reaction kinetics is the collision model
>> Molecules must collide to react
>> The collision kinetic energy furnishes the potential energy needed to enable the reactants to rearrange to form products
>> A certain threshold energy called the activation energy (E_a) is necessary for a reaction to occur
>> The relative orientations of the colliding reactants are also a determining factor in the reaction rate
>> This model leads to the Arrhenius equation:

$$k = Ae^{-E_a/RT}$$

>> A depends on the collision frequency and relative orientation of the molecules
>> The value of E_a can be found by obtaining the values of k at several temperatures

Catalyst

> Speeds up a reaction without being consumed
> Works by providing a lower-energy pathway for the reaction
> Enzymes are biological catalysts
> Catalysts can be classified as homogeneous or heterogeneous
>> Homogeneous: exist in the same phase as the reactants
>> Heterogeneous: exist in a different phase than the reactants

Review questions   *Answers to the Review Questions can be found on the Student website (accessible from* **www.cengagebrain.com**).

1. Define *reaction rate*. Distinguish between the initial rate, average rate, and instantaneous rate of a chemical reaction. Which of these rates is usually fastest? The initial rate is the rate used by convention. Give a possible explanation as to why.

2. Distinguish between the differential rate law and the integrated rate law. Which of these is often called just the "rate law"? What is k in a rate law, and what are orders in a rate law? Explain.

3. One experimental procedure that can be used to determine the rate law of a reaction is the method of initial rates. What data are gathered in the method of initial rates, and how are these data manipulated to determine k and the orders of the species in the rate law? Are the units for k, the rate constant, the same for all rate laws? Explain. If a reaction is first order in A, what happens to the rate if [A] is tripled? If the initial rate for a reaction increases by a factor of 16 when [A] is quadrupled, what is the order of n? If a reaction is third order in A and [A] is doubled, what happens to the initial rate? If a reaction is zero order, what effect does [A] have on the initial rate of a reaction?

4. The initial rate for a reaction is equal to the slope of the tangent line at $t \approx 0$ in a plot of [A] versus time. From calculus, initial rate $= \dfrac{-d[A]}{dt}$. Therefore, the differen-tial rate law for a reaction is Rate $= \dfrac{-d[A]}{dt} = k[A]^n$. Assuming you have some calculus in your background, derive the zero-, first-, and second-order integrated rate laws using the differential rate law.

5. Consider the zero-, first-, and second-order integrated rate laws. If you have concentration versus time data for some species in a reaction, what plots would you make to "prove" a reaction is either zero, first, or second order? How would the rate constant, k, be determined from such a plot? What does the y-intercept equal in each plot? When a rate law contains the concentration of two or more species, how can plots be used to determine k and the orders of the species in the rate law?

6. Derive expressions for the half-life of zero-, first-, and second-order reactions using the integrated rate law for each order. How does each half-life depend on concentration? If the half-life for a reaction is 20. seconds, what would be the second half-life assuming the reaction is either zero, first, or second order?

7. Define each of the following.

   a. elementary step

   b. molecularity

   c. reaction mechanism

   d. intermediate

   e. rate-determining step

8. What two requirements must be met to call a mechanism plausible? Why say a "plausible" mechanism instead of the "correct" mechanism? Is it true that most reactions occur by a one-step mechanism? Explain.

9. What is the premise underlying the collision model? How is the rate affected by each of the following?

   a. activation energy

   b. temperature

   c. frequency of collisions

   d. orientation of collisions

   Sketch a potential energy versus reaction progress plot for an endothermic reaction and for an exothermic reaction. Show ΔE and E_a in both plots. When concentrations and temperatures are equal, would you expect the rate of the forward reaction to be equal to, greater than, or less than the rate of the reverse reaction if the reaction is exothermic? Endothermic?

10. Give the Arrhenius equation. Take the natural log of both sides and place this equation in the form of a straight-line equation ($y = mx + b$). What data would you need and how would you graph those data to get a linear relationship using the Arrhenius equation? What does the slope of the straight line equal? What does the y-intercept equal? What are the units of R in the Arrhenius equation? Explain how if you know the rate constant value at two different temperatures, you can determine the activation energy for the reaction.

11. Why does a catalyst increase the rate of a reaction? What is the difference between a homogeneous catalyst and a heterogeneous catalyst? Would a given reaction necessarily have the same rate law for both a catalyzed and an uncatalyzed pathway? Explain.

Active Learning Questions*

These questions are designed to be used by groups of students in class.

1. Define *stability* from both a kinetic and thermodynamic perspective. Give examples to show the differences in these concepts.

2. Describe at least two experiments you could perform to determine a rate law.

3. Make a graph of [A] versus time for zero-, first-, and second-order reactions. From these graphs, compare successive half-lives.

4. How does temperature affect k, the rate constant? Explain.

5. Consider the following statements: "In general, the rate of a chemical reaction increases a bit at first because it takes a while for the reaction to get 'warmed up.' After that, however, the rate of the reaction decreases because its rate is dependent on the concentrations of the reactants, and these are decreasing." Indicate everything that is correct in these statements, and indicate everything that is incorrect. Correct the incorrect statements and explain.

6. For the reaction $A + B \rightarrow C$, explain at least two ways in which the rate law could be zero order in chemical A.

7. A friend of yours states, "A balanced equation tells us how chemicals interact. Therefore, we can determine the rate law directly from the balanced equation." What do you tell your friend?

8. Provide a conceptual rationale for the differences in the half-lives of zero-, first-, and second-order reactions.

9. The rate constant (k) depends on which of the following (there may be more than one answer)?

   a. the concentration of the reactants

   b. the nature of the reactants

   c. the temperature

   d. the order of the reaction

   Explain.

A blue question or exercise number indicates that the answer to that question or exercise appears at the back of this book and a solution appears in the *Solutions Guide,* as found on PowerLecture.

Questions

10. Each of the statements given below is false. Explain why.

   a. The activation energy of a reaction depends on the overall energy change (ΔE) for the reaction.

   b. The rate law for a reaction can be deduced from examination of the overall balanced equation for the reaction.

   c. Most reactions occur by one-step mechanisms.

11. Define what is meant by unimolecular and bimolecular steps. Why are termolecular steps infrequently seen in chemical reactions?

12. The plot below shows the number of collisions with a particular energy for two different temperatures.

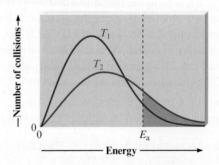

*In the Questions and the Exercises, the term *rate law* always refers to the differential rate law.

a. Which is greater, T_2 or T_1? How can you tell?

b. What does this plot tell us about the temperature of the rate of a chemical reaction? Explain your answer.

13. For the reaction

$$O_2(g) + 2NO(g) \longrightarrow 2NO_2(g)$$

the observed rate law is

$$\text{Rate} = k[NO]^2[O_2]$$

Which of the changes listed below would affect the value of the rate constant k?

a. increasing the partial pressure of oxygen gas

b. changing the temperature

c. using an appropriate catalyst

14. The rate law for a reaction can be determined only from experiment and not from the balanced equation. Two experimental procedures were outlined in Chapter 12. What are these two procedures? Explain how each method is used to determine rate laws.

15. Table 12.2 illustrates how the average rate of a reaction decreases with time. Why does the average rate decrease with time? How does the instantaneous rate of a reaction depend on time? Why are initial rates used by convention?

16. The type of rate law for a reaction, either the differential rate law or the integrated rate law, is usually determined by which data is easiest to collect. Explain.

17. The initial rate of a reaction doubles as the concentration of one of the reactants is quadrupled. What is the order of this reactant? If a reactant has a −1 order, what happens to the initial rate when the concentration of that reactant increases by a factor of two?

18. Hydrogen reacts explosively with oxygen. However, a mixture of H_2 and O_2 can exist indefinitely at room temperature. Explain why H_2 and O_2 do not react under these conditions.

19. The central idea of the collision model is that molecules must collide in order to react. Give two reasons why not all collisions of reactant molecules result in product formation.

20. Consider the following energy plots for a chemical reaction when answering the questions below.

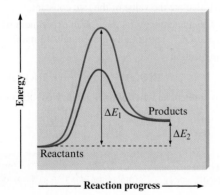

a. Which plot (purple or blue) is the catalyzed pathway? How do you know?

b. What does ΔE_1 represent?

c. What does ΔE_2 represent?

d. Is the reaction endothermic or exothermic?

21. Enzymes are kinetically important for many of the complex reactions necessary for plant and animal life to exist. However, only a tiny amount of any particular enzyme is required for these complex reactions to occur. Explain.

22. Would the slope of a $\ln(k)$ versus $1/T$ plot (with temperature in kelvin) for a catalyzed reaction be more or less negative than the slope of the $\ln(k)$ versus $1/T$ plot for the uncatalyzed reaction? Explain. Assume both rate laws are first-order overall.

Exercises

In this section similar exercises are paired.

Reaction Rates

23. Consider the reaction

$$4PH_3(g) \longrightarrow P_4(g) + 6H_2(g)$$

If, in a certain experiment, over a specific time period, 0.0048 mole of PH_3 is consumed in a 2.0-L container each second of reaction, what are the rates of production of P_4 and H_2 in this experiment?

24. In the Haber process for the production of ammonia,

$$N_2(g) + 3H_2(g) \longrightarrow 2NH_3(g)$$

what is the relationship between the rate of production of ammonia and the rate of consumption of hydrogen?

25. At 40°C, $H_2O_2(aq)$ will decompose according to the following reaction:

$$2H_2O_2(aq) \longrightarrow 2H_2O(l) + O_2(g)$$

The following data were collected for the concentration of H_2O_2 at various times.

Time (s)	[H_2O_2] (mol/L)
0	1.000
2.16×10^4	0.500
4.32×10^4	0.250

a. Calculate the average rate of decomposition of H_2O_2 between 0 and 2.16×10^4 s. Use this rate to calculate the average rate of production of $O_2(g)$ over the same time period.

b. What are these rates for the time period 2.16×10^4 s to 4.32×10^4 s?

26. Consider the general reaction

$$aA + bB \longrightarrow cC$$

and the following average rate data over some time period Δt:

$$-\frac{\Delta A}{\Delta t} = 0.0080 \text{ mol/L} \cdot \text{s}$$

$$-\frac{\Delta B}{\Delta t} = 0.0120 \text{ mol/L} \cdot \text{s}$$

$$\frac{\Delta C}{\Delta t} = 0.0160 \text{ mol/L} \cdot \text{s}$$

Determine a set of possible coefficients to balance this general reaction.

27. What are the units for each of the following if the concentrations are expressed in moles per liter and the time in seconds?

a. rate of a chemical reaction

b. rate constant for a zero-order rate law

c. rate constant for a first-order rate law

d. rate constant for a second-order rate law

e. rate constant for a third-order rate law

28. The rate law for the reaction

$$Cl_2(g) + CHCl_3(g) \longrightarrow HCl(g) + CCl_4(g)$$

is   Rate $= k[Cl_2]^{1/2}[CHCl_3]$

What are the units for k, assuming time in seconds and concentration in mol/L?

Rate Laws from Experimental Data: Initial Rates Method

29. The reaction

$$2NO(g) + Cl_2(g) \longrightarrow 2NOCl(g)$$

was studied at $-10°C$. The following results were obtained where

$$\text{Rate} = -\frac{\Delta[Cl_2]}{\Delta t}$$

$[NO]_0$ (mol/L)	$[Cl_2]_0$ (mol/L)	Initial Rate (mol/L · min)
0.10	0.10	0.18
0.10	0.20	0.36
0.20	0.20	1.45

a. What is the rate law?

b. What is the value of the rate constant?

30. The reaction

$$2I^-(aq) + S_2O_8^{2-}(aq) \longrightarrow I_2(aq) + 2SO_4^{2-}(aq)$$

was studied at $25°C$. The following results were obtained where

$$\text{Rate} = -\frac{\Delta[S_2O_8^{2-}]}{\Delta t}$$

$[I^-]_0$ (mol/L)	$[S_2O_8^{2-}]_0$ (mol/L)	Initial Rate (mol/L · s)
0.080	0.040	12.5×10^{-6}
0.040	0.040	6.25×10^{-6}
0.080	0.020	6.25×10^{-6}
0.032	0.040	5.00×10^{-6}
0.060	0.030	7.00×10^{-6}

a. Determine the rate law.

b. Calculate a value for the rate constant for each experiment and an average value for the rate constant.

31. The decomposition of nitrosyl chloride was studied:

$$2NOCl(g) \rightleftharpoons 2NO(g) + Cl_2(g)$$

The following data were obtained where

$$\text{Rate} = -\frac{\Delta[NOCl]}{\Delta t}$$

$[NOCl]_0$ (molecules/cm³)	Initial Rate (molecules/cm³ · s)
3.0×10^{16}	5.98×10^4
2.0×10^{16}	2.66×10^4
1.0×10^{16}	6.64×10^3
4.0×10^{16}	1.06×10^5

a. What is the rate law?

b. Calculate the value of the rate constant.

c. Calculate the value of the rate constant when concentrations are given in moles per liter.

32. The following data were obtained for the gas-phase decomposition of dinitrogen pentoxide,

$$2N_2O_5(g) \longrightarrow 4NO_2(g) + O_2(g)$$

$[N_2O_5]_0$ (mol/L)	Initial Rate (mol/L · s)
0.0750	8.90×10^{-4}
0.190	2.26×10^{-3}
0.275	3.26×10^{-3}
0.410	4.85×10^{-3}

Defining the rate as $-\Delta[N_2O_5]/\Delta t$, write the rate law and calculate the value of the rate constant.

33. The reaction

$$I^-(aq) + OCl^-(aq) \longrightarrow IO^-(aq) + Cl^-(aq)$$

was studied, and the following data were obtained:

$[I^-]_0$ (mol/L)	$[OCl^-]_0$ (mol/L)	Initial Rate (mol/L · s)
0.12	0.18	7.91×10^{-2}
0.060	0.18	3.95×10^{-2}
0.030	0.090	9.88×10^{-3}
0.24	0.090	7.91×10^{-2}

a. What is the rate law?

b. Calculate the value of the rate constant.

c. Calculate the initial rate for an experiment where both I^- and OCl^- are initially present at 0.15 mol/L.

34. The reaction

$$2NO(g) + O_2(g) \longrightarrow 2NO_2(g)$$

was studied, and the following data were obtained where

$$\text{Rate} = -\frac{\Delta[O_2]}{\Delta t}$$

$[NO]_0$ (molecules/cm³)	$[O_2]_0$ (molecules/cm³)	Initial Rate (molecules/cm³ · s)
1.00×10^{18}	1.00×10^{18}	2.00×10^{16}
3.00×10^{18}	1.00×10^{18}	1.80×10^{17}
2.50×10^{18}	2.50×10^{18}	3.13×10^{17}

What would be the initial rate for an experiment where $[NO]_0 = 6.21 \times 10^{18}$ molecules/cm³ and $[O_2]_0 = 7.36 \times 10^{18}$ molecules/cm³?

35. The rate of the reaction between hemoglobin (Hb) and carbon monoxide (CO) was studied at 20°C. The following data were collected with all concentration units in μmol/L. (A hemoglobin concentration of 2.21 μmol/L is equal to 2.21×10^{-6} mol/L.)

[Hb]$_0$ (μmol/L)	[CO]$_0$ (μmol/L)	Initial Rate (μmol/L · s)
2.21	1.00	0.619
4.42	1.00	1.24
4.42	3.00	3.71

a. Determine the orders of this reaction with respect to Hb and CO.

b. Determine the rate law.

c. Calculate the value of the rate constant.

d. What would be the initial rate for an experiment with [Hb]$_0$ = 3.36 μmol/L and [CO]$_0$ = 2.40 μmol/L?

36. The following data were obtained for the reaction

$$2ClO_2(aq) + 2OH^-(aq) \longrightarrow ClO_3^-(aq) + ClO_2^-(aq) + H_2O(l)$$

where

$$\text{Rate} = -\frac{\Delta[ClO_2]}{\Delta t}$$

[ClO$_2$]$_0$ (mol/L)	[OH$^-$]$_0$ (mol/L)	Initial Rate (mol/L · s)
0.0500	0.100	5.75×10^{-2}
0.100	0.100	2.30×10^{-1}
0.100	0.0500	1.15×10^{-1}

a. Determine the rate law and the value of the rate constant.

b. What would be the initial rate for an experiment with [ClO$_2$]$_0$ = 0.175 mol/L and [OH$^-$]$_0$ = 0.0844 mol/L?

Integrated Rate Laws

37. The decomposition of hydrogen peroxide was studied, and the following data were obtained at a particular temperature:

Time (s)	[H$_2$O$_2$] (mol/L)
0	1.00
120 ± 1	0.91
300 ± 1	0.78
600 ± 1	0.59
1200 ± 1	0.37
1800 ± 1	0.22
2400 ± 1	0.13
3000 ± 1	0.082
3600 ± 1	0.050

Assuming that

$$\text{Rate} = -\frac{\Delta[H_2O_2]}{\Delta t}$$

determine the rate law, the integrated rate law, and the value of the rate constant. Calculate [H$_2$O$_2$] at 4000. s after the start of the reaction.

38. A certain reaction has the following general form:

$$aA \longrightarrow bB$$

At a particular temperature and [A]$_0$ = 2.00×10^{-2} M, concentration versus time data were collected for this reaction, and a plot of ln[A] versus time resulted in a straight line with a slope value of -2.97×10^{-2} min^{-1}.

a. Determine the rate law, the integrated rate law, and the value of the rate constant for this reaction.

b. Calculate the half-life for this reaction.

c. How much time is required for the concentration of A to decrease to 2.50×10^{-3} M?

39. The rate of the reaction

$$NO_2(g) + CO(g) \longrightarrow NO(g) + CO_2(g)$$

depends only on the concentration of nitrogen dioxide below 225°C. At a temperature below 225°C, the following data were collected:

Time (s)	[NO$_2$] (mol/L)
0	0.500
1.20×10^3	0.444
3.00×10^3	0.381
4.50×10^3	0.340
9.00×10^3	0.250
1.80×10^4	0.174

Determine the rate law, the integrated rate law, and the value of the rate constant. Calculate [NO$_2$] at 2.70×10^4 s after the start of the reaction.

40. A certain reaction has the following general form:

$$aA \longrightarrow bB$$

At a particular temperature and [A]$_0$ = 2.80×10^{-3} M, concentration versus time data were collected for this reaction, and a plot of 1/[A] versus time resulted in a straight line with a slope value of $+3.60 \times 10^{-2}$ L/mol · s.

a. Determine the rate law, the integrated rate law, and the value of the rate constant for this reaction.

b. Calculate the half-life for this reaction.

c. How much time is required for the concentration of A to decrease to 7.00×10^{-4} M?

41. The decomposition of ethanol (C$_2$H$_5$OH) on an alumina (Al$_2$O$_3$) surface

$$C_2H_5OH(g) \longrightarrow C_2H_4(g) + H_2O(g)$$

was studied at 600 K. Concentration versus time data were collected for this reaction, and a plot of [A] versus time resulted in a straight line with a slope of -4.00×10^{-5} mol/L · s.

a. Determine the rate law, the integrated rate law, and the value of the rate constant for this reaction.

b. If the initial concentration of C$_2$H$_5$OH was 1.25×10^{-2} M, calculate the half-life for this reaction.

c. How much time is required for all the 1.25×10^{-2} M C$_2$H$_5$OH to decompose?

42. At 500 K in the presence of a copper surface, ethanol decomposes according to the equation

$$C_2H_5OH(g) \longrightarrow CH_3CHO(g) + H_2(g)$$

The pressure of C_2H_5OH was measured as a function of time and the following data were obtained:

Time (s)	$P_{C_2H_5OH}$ (torr)
0	250.
100.	237
200.	224
300.	211
400.	198
500.	185

Since the pressure of a gas is directly proportional to the concentration of gas, we can express the rate law for a gaseous reaction in terms of partial pressures. Using the above data, deduce the rate law, the integrated rate law, and the value of the rate constant, all in terms of pressure units in atm and time in seconds. Predict the pressure of C_2H_5OH after 900. s from the start of the reaction. (*Hint:* To determine the order of the reaction with respect to C_2H_5OH, compare how the pressure of C_2H_5OH decreases with each time listing.)

43. The dimerization of butadiene

$$2C_4H_6(g) \longrightarrow C_8H_{12}(g)$$

was studied at 500. K, and the following data were obtained:

Time (s)	$[C_4H_6]$ (mol/L)
195	1.6×10^{-2}
604	1.5×10^{-2}
1246	1.3×10^{-2}
2180	1.1×10^{-2}
6210	0.68×10^{-2}

Assuming that

$$\text{Rate} = -\frac{\Delta[C_4H_6]}{\Delta t}$$

determine the form of the rate law, the integrated rate law, and the value of the rate constant for this reaction.

44. The rate of the reaction

$$O(g) + NO_2(g) \longrightarrow NO(g) + O_2(g)$$

was studied at a certain temperature.

a. In one experiment, NO_2 was in large excess, at a concentration of 1.0×10^{13} molecules/cm^3 with the following data collected:

Time (s)	[O] (atoms/cm^3)
0	5.0×10^9
1.0×10^{-2}	1.9×10^9
2.0×10^{-2}	6.8×10^8
3.0×10^{-2}	2.5×10^8

What is the order of the reaction with respect to oxygen atoms?

b. The reaction is known to be first order with respect to NO_2. Determine the overall rate law and the value of the rate constant.

45. Experimental data for the reaction

$$A \longrightarrow 2B + C$$

have been plotted in the following three different ways (with concentration units in mol/L):

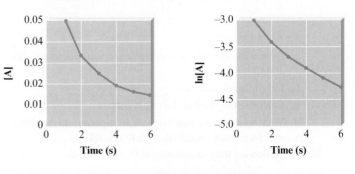

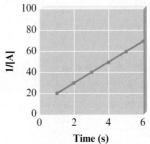

What is the order of the reaction with respect to A, and what is the initial concentration of A?

46. Consider the data plotted in Exercise 45 when answering the following questions.

a. What is the concentration of A after 9 s?

b. What are the first three half-lives for this experiment?

47. The reaction

$$A \longrightarrow B + C$$

is known to be zero order in A and to have a rate constant of 5.0×10^{-2} mol/L $\cdot$ s at 25°C. An experiment was run at 25°C where $[A]_0 = 1.0 \times 10^{-3}$ M.

a. Write the integrated rate law for this reaction.

b. Calculate the half-life for the reaction.

c. Calculate the concentration of B after 5.0×10^{-3} s has elapsed assuming $[B]_0 = 0$.

48. The decomposition of hydrogen iodide on finely divided gold at 150°C is zero order with respect to HI. The rate defined below is constant at 1.20×10^{-4} mol/L $\cdot$ s.

$$2HI(g) \xrightarrow{\text{Au}} H_2(g) + I_2(g)$$

$$\text{Rate} = -\frac{\Delta[HI]}{\Delta t} = k = 1.20 \times 10^{-4} \text{ mol/L} \cdot \text{s}$$

a. If the initial HI concentration was 0.250 mol/L, calculate the concentration of HI at 25 minutes after the start of the reaction.

b. How long will it take for all of the 0.250 M HI to decompose?

49. A certain first-order reaction is 45.0% complete in 65 s. What are the values of the rate constant and the half-life for this process?

50. A first-order reaction is 75.0% complete in 320. s.

   a. What are the first and second half-lives for this reaction?

   b. How long does it take for 90.0% completion?

51. The rate law for the decomposition of phosphine (PH_3) is

$$\text{Rate} = -\frac{\Delta[PH_3]}{\Delta t} = k[PH_3]$$

It takes 120. s for 1.00 M PH_3 to decrease to 0.250 M. How much time is required for 2.00 M PH_3 to decrease to a concentration of 0.350 M?

52. DDT (molar mass = 354.49 g/mol) was a widely used insecticide that was banned from use in the United States in 1973. This ban was brought about due to the persistence of DDT in many different ecosystems, leading to high accumulations of the substance in many birds of prey. The insecticide was shown to cause a thinning of egg shells, pushing many birds toward extinction. If a 20-L drum of DDT was spilled into a pond, resulting in a DDT concentration of 8.75×10^{-5} M, how long would it take for the levels of DDT to reach a concentration of 1.41×10^{-7} M (a level that is generally assumed safe in mammals)? Assume the decomposition of DDT is a first-order process with a half-life of 56.0 days.

53. Consider the following initial rate data for the decomposition of compound AB to give A and B:

$[AB]_0$ (mol/L)	Initial Rate (mol/L · s)
0.200	3.20×10^{-3}
0.400	1.28×10^{-2}
0.600	2.88×10^{-2}

Determine the half-life for the decomposition reaction initially having 1.00 M AB present.

54. The rate law for the reaction

$$2NOBr(g) \longrightarrow 2NO(g) + Br_2(g)$$

at some temperature is

$$\text{Rate} = -\frac{\Delta[NOBr]}{\Delta t} = k[NOBr]^2$$

   a. If the half-life for this reaction is 2.00 s when $[NOBr]_0 = 0.900$ M, calculate the value of k for this reaction.

   b. How much time is required for the concentration of NOBr to decrease to 0.100 M?

55. For the reaction A $\rightarrow$ products, successive half-lives are observed to be 10.0, 20.0, and 40.0 min for an experiment in which $[A]_0 = 0.10$ M. Calculate the concentration of A at the following times.

   a. 80.0 min

   b. 30.0 min

56. Theophylline is a pharmaceutical drug that is sometimes used to help with lung function. You observe a case where the initial lab results indicate that the concentration of theophylline in a patient's body decreased from 2.0×10^{-3} M to 1.0×10^{-3} M in 24 hours. In another 12 hours the drug concentration was found to be 5.0×10^{-4} M. What is the value of the rate constant for the metabolism of this drug in the body?

57. You and a coworker have developed a molecule that has shown potential as cobra antivenin (AV). This antivenin works by binding to the venom (V), thereby rendering it nontoxic. This reaction can be described by the rate law

$$\text{Rate} = k[AV]^1[V]^1$$

You have been given the following data from your coworker:

$$[V]_0 = 0.20 \ M$$
$$[AV]_0 = 1.0 \times 10^{-4} \ M$$

A plot of ln[AV] versus t (s) gives a straight line with a slope of -0.32 s^{-1}. What is the value of the rate constant (k) for this reaction?

58. Consider the hypothetical reaction

$$A + B + 2C \longrightarrow 2D + 3E$$

where the rate law is

$$\text{Rate} = -\frac{\Delta[A]}{\Delta t} = k[A][B]^2$$

An experiment is carried out where $[A]_0 = 1.0 \times 10^{-2}$ M, $[B]_0 = 3.0$ M, and $[C]_0 = 2.0$ M. The reaction is started, and after 8.0 seconds, the concentration of A is 3.8×10^{-3} M.

   a. Calculate the value of k for this reaction.

   b. Calculate the half-life for this experiment.

   c. Calculate the concentration of A after 13.0 seconds.

   d. Calculate the concentration of C after 13.0 seconds.

Reaction Mechanisms

59. Write the rate laws for the following elementary reactions.

   a. $CH_3NC(g) \rightarrow CH_3CN(g)$

   b. $O_3(g) + NO(g) \rightarrow O_2(g) + NO_2(g)$

   c. $O_3(g) \rightarrow O_2(g) + O(g)$

   d. $O_3(g) + O(g) \rightarrow 2O_2(g)$

60. A possible mechanism for the decomposition of hydrogen peroxide is

$$H_2O_2 \longrightarrow 2OH$$
$$H_2O_2 + OH \longrightarrow H_2O + HO_2$$
$$HO_2 + OH \longrightarrow H_2O + O_2$$

Using your results from Exercise 37, specify which step is the rate-determining step. What is the overall balanced equation for the reaction?

61. A proposed mechanism for a reaction is

$C_4H_9Br \longrightarrow C_4H_9^+ + Br^-$	Slow
$C_4H_9^+ + H_2O \longrightarrow C_4H_9OH_2^+$	Fast
$C_4H_9OH_2^+ + H_2O \longrightarrow C_4H_9OH + H_3O^+$	Fast

Write the rate law expected for this mechanism. What is the overall balanced equation for the reaction? What are the intermediates in the proposed mechanism?

62. The mechanism for the gas-phase reaction of nitrogen dioxide with carbon monoxide to form nitric oxide and carbon dioxide is thought to be

$$NO_2 + NO_2 \longrightarrow NO_3 + NO \qquad \text{Slow}$$
$$NO_3 + CO \longrightarrow NO_2 + CO_2 \qquad \text{Fast}$$

Write the rate law expected for this mechanism. What is the overall balanced equation for the reaction?

Temperature Dependence of Rate Constants and the Collision Model

63. For the following reaction profile, indicate

a. the positions of reactants and products.

b. the activation energy.

c. ΔE for the reaction.

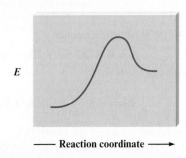

64. Draw a rough sketch of the energy profile for each of the following cases:

a. $\Delta E = +10$ kJ/mol, $E_a = 25$ kJ/mol

b. $\Delta E = -10$ kJ/mol, $E_a = 50$ kJ/mol

c. $\Delta E = -50$ kJ/mol, $E_a = 50$ kJ/mol

65. The activation energy for the reaction

$$NO_2(g) + CO(g) \longrightarrow NO(g) + CO_2(g)$$

is 125 kJ/mol, and ΔE for the reaction is -216 kJ/mol. What is the activation energy for the reverse reaction $[NO(g) + CO_2(g) \longrightarrow NO_2(g) + CO(g)]$?

66. The activation energy for some reaction

$$X_2(g) + Y_2(g) \longrightarrow 2XY(g)$$

is 167 kJ/mol, and ΔE for the reaction is $+28$ kJ/mol. What is the activation energy for the decomposition of XY?

67. The rate constant for the gas-phase decomposition of N_2O_5,

$$N_2O_5 \longrightarrow 2NO_2 + \tfrac{1}{2}O_2$$

has the following temperature dependence:

T (K)	k (s^{-1})
338	4.9×10^{-3}
318	5.0×10^{-4}
298	3.5×10^{-5}

Make the appropriate graph using these data, and determine the activation energy for this reaction.

68. The reaction

$$(CH_3)_3CBr + OH^- \longrightarrow (CH_3)_3COH + Br^-$$

in a certain solvent is first order with respect to $(CH_3)_3CBr$ and zero order with respect to OH^-. In several experiments, the rate constant k was determined at different temperatures. A plot of $\ln(k)$ versus $1/T$ was constructed resulting in a straight line with a slope value of -1.10×10^4 K and y-intercept of 33.5. Assume k has units of s^{-1}.

a. Determine the activation energy for this reaction.

b. Determine the value of the frequency factor A.

c. Calculate the value of k at 25°C.

69. The activation energy for the decomposition of $HI(g)$ to $H_2(g)$ and $I_2(g)$ is 186 kJ/mol. The rate constant at 555 K is 3.52×10^{-7} L/mol $\cdot$ s. What is the rate constant at 645 K?

70. A first-order reaction has rate constants of 4.6×10^{-2} s^{-1} and 8.1×10^{-2} s^{-1} at 0°C and 20.°C, respectively. What is the value of the activation energy?

71. A certain reaction has an activation energy of 54.0 kJ/mol. As the temperature is increased from 22°C to a higher temperature, the rate constant increases by a factor of 7.00. Calculate the higher temperature.

72. Chemists commonly use a rule of thumb that an increase of 10 K in temperature doubles the rate of a reaction. What must the activation energy be for this statement to be true for a temperature increase from 25 to 35°C?

73. Which of the following reactions would you expect to proceed at a faster rate at room temperature? Why? (*Hint:* Think about which reaction would have the lower activation energy.)

$$2Ce^{4+}(aq) + Hg_2^{2+}(aq) \longrightarrow 2Ce^{3+}(aq) + 2Hg^{2+}(aq)$$
$$H_3O^+(aq) + OH^-(aq) \longrightarrow 2H_2O(l)$$

74. One reason suggested for the instability of long chains of silicon atoms is that the decomposition involves the transition state shown below:

$$\begin{array}{ccc} H & H \\ | & | \\ H-Si-Si-H & \longrightarrow \\ | & | \\ H & H \end{array}$$

$$\left\{ \begin{array}{ccc} H & & H \\ | & & | \\ H-Si\text{-}\text{-}\text{-}\text{-}\text{-}\text{-}\text{-}Si\text{:} \\ | & & | \\ H & H & H \end{array} \right\} \longrightarrow SiH_4 + \text{:}SiH_2$$

The activation energy for such a process is 210 kJ/mol, which is less than either the Si—Si or the Si—H bond energy. Why would a similar mechanism not be expected to play a very important role in the decomposition of long chains of carbon atoms as seen in organic compounds?

Catalysts

75. One mechanism for the destruction of ozone in the upper atmosphere is

$$O_3(g) + NO(g) \longrightarrow NO_2(g) + O_2(g) \qquad \text{Slow}$$
$$NO_2(g) + O(g) \longrightarrow NO(g) + O_2(g) \qquad \text{Fast}$$

Overall reaction $O_3(g) + O(g) \longrightarrow 2O_2(g)$

  a. Which species is a catalyst?

  b. Which species is an intermediate?

  c. E_a for the uncatalyzed reaction

$$O_3(g) + O(g) \longrightarrow 2O_2(g)$$

  is 14.0 kJ. E_a for the same reaction when catalyzed is 11.9 kJ. What is the ratio of the rate constant for the catalyzed reaction to that for the uncatalyzed reaction at 25°C? Assume that the frequency factor A is the same for each reaction.

76. One of the concerns about the use of Freons is that they will migrate to the upper atmosphere, where chlorine atoms can be generated by the following reaction:

$$CCl_2F_2(g) \xrightarrow{h\nu} CF_2Cl(g) + Cl(g)$$
$$\text{Freon-12}$$

Chlorine atoms can act as a catalyst for the destruction of ozone. The activation energy for the reaction

$$Cl(g) + O_3(g) \longrightarrow ClO(g) + O_2(g)$$

is 2.1 kJ/mol. Which is the more effective catalyst for the destruction of ozone, Cl or NO? (See Exercise 75.)

77. Assuming that the mechanism for the hydrogenation of C_2H_4 given in Section 12.7 is correct, would you predict that the product of the reaction of C_2H_4 with D_2 would be CH_2D-CH_2D or CHD_2-CH_3? How could the reaction of C_2H_4 with D_2 be used to confirm the mechanism for the hydrogenation of C_2H_4 given in Section 12.7?

78. The decomposition of NH_3 to N_2 and H_2 was studied on two surfaces:

Surface	E_a (kJ/mol)
W	163
Os	197

Without a catalyst, the activation energy is 335 kJ/mol.

  a. Which surface is the better heterogeneous catalyst for the decomposition of NH_3? Why?

  b. How many times faster is the reaction at 298 K on the W surface compared with the reaction with no catalyst present? Assume that the frequency factor A is the same for each reaction.

  c. The decomposition reaction on the two surfaces obeys a rate law of the form

$$\text{Rate} = k \frac{[NH_3]}{[H_2]}$$

How can you explain the inverse dependence of the rate on the H_2 concentration?

79. The decomposition of many substances on the surface of a heterogeneous catalyst shows the following behavior:

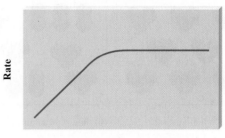

Concentration of reactant

How do you account for the rate law changing from first order to zero order in the concentration of reactant?

80. For enzyme-catalyzed reactions that follow the mechanism

$$E + S \rightleftharpoons E \cdot S$$
$$E \cdot S \rightleftharpoons E + P$$

a graph of the rate as a function of [S], the concentration of the substrate, has the following appearance:

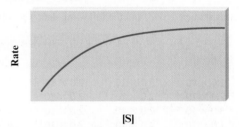

[S]

Note that at higher substrate concentrations the rate no longer changes with [S]. Suggest a reason for this.

81. A popular chemical demonstration is the "magic genie" procedure, in which hydrogen peroxide decomposes to water and oxygen gas with the aid of a catalyst. The activation energy of this (uncatalyzed) reaction is 70.0 kJ/mol. When the catalyst is added, the activation energy (at 20.°C) is 42.0 kJ/mol. Theoretically, to what temperature (°C) would one have to heat the hydrogen peroxide solution so that the rate of the uncatalyzed reaction is equal to the rate of the catalyzed reaction at 20.°C? Assume the frequency factor A is constant, and assume the initial concentrations are the same.

82. The activation energy for a reaction is changed from 184 kJ/mol to 59.0 kJ/mol at 600. K by the introduction of a catalyst. If the uncatalyzed reaction takes about 2400 years to occur, about how long will the catalyzed reaction take? Assume the frequency factor A is constant, and assume the initial concentrations are the same.

Additional Exercises

83. Consider the following representation of the reaction $2NO_2(g) \rightarrow 2NO(g) + O_2(g)$.

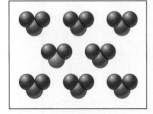 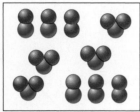 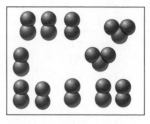

(a) time = 0 minutes   **(b)** time = 10 minutes   **(c)** time = ? minutes

Time →

Determine the time for the final representation above if the reaction is

a. first order

b. second order

c. zero order

84. The reaction

$$H_2SeO_3(aq) + 6I^-(aq) + 4H^+(aq)$$
$$\longrightarrow Se(s) + 2I_3^-(aq) + 3H_2O(l)$$

was studied at 0°C, and the following data were obtained:

$[H_2SeO_3]_0$ (mol/L)	$[H^+]_0$ (mol/L)	$[I^-]_0$ (mol/L)	Initial Rate (mol/L · s)
1.0×10^{-4}	2.0×10^{-2}	2.0×10^{-2}	1.66×10^{-7}
2.0×10^{-4}	2.0×10^{-2}	2.0×10^{-2}	3.33×10^{-7}
3.0×10^{-4}	2.0×10^{-2}	2.0×10^{-2}	4.99×10^{-7}
1.0×10^{-4}	4.0×10^{-2}	2.0×10^{-2}	6.66×10^{-7}
1.0×10^{-4}	1.0×10^{-2}	2.0×10^{-2}	0.42×10^{-7}
1.0×10^{-4}	2.0×10^{-2}	4.0×10^{-2}	13.2×10^{-7}
1.0×10^{-4}	1.0×10^{-2}	4.0×10^{-2}	3.36×10^{-7}

These relationships hold only if there is a very small amount of I_3^- present. What is the rate law and the value of the rate constant? $\left(\text{Assume that rate} = -\dfrac{\Delta[H_2SeO_3]}{\Delta t}. \right)$

85. Consider two reaction vessels, one containing A and the other containing B, with equal concentrations at $t = 0$. If both substances decompose by first-order kinetics, where

$$k_A = 4.50 \times 10^{-4} \, s^{-1}$$
$$k_B = 3.70 \times 10^{-3} \, s^{-1}$$

how much time must pass to reach a condition such that $[A] = 4.00[B]$?

86. Sulfuryl chloride (SO_2Cl_2) decomposes to sulfur dioxide (SO_2) and chlorine (Cl_2) by reaction in the gas phase. The following pressure data were obtained when a sample containing 5.00×10^{-2} mol sulfuryl chloride was heated to 600. K in a 5.00×10^{-1}-L container.

Time (hours):	0.00	1.00	2.00	4.00	8.00	16.00
$P_{SO_2Cl_2}$ (atm):	4.93	4.26	3.52	2.53	1.30	0.34

Defining the rate as $-\dfrac{\Delta[SO_2Cl_2]}{\Delta t}$,

a. determine the value of the rate constant for the decomposition of sulfuryl chloride at 600. K.

b. what is the half-life of the reaction?

c. what fraction of the sulfuryl chloride remains after 20.0 h?

87. For the reaction

$$2N_2O_5(g) \longrightarrow 4NO_2(g) + O_2(g)$$

the following data were collected, where

$$\text{Rate} = -\dfrac{\Delta[N_2O_5]}{\Delta t}$$

Time (s)	T = 338 K [N_2O_5]	T = 318 K [N_2O_5]
0	$1.00 \times 10^{-1} \, M$	$1.00 \times 10^{-1} \, M$
100.	$6.14 \times 10^{-2} \, M$	$9.54 \times 10^{-2} \, M$
300.	$2.33 \times 10^{-2} \, M$	$8.63 \times 10^{-2} \, M$
600.	$5.41 \times 10^{-3} \, M$	$7.43 \times 10^{-2} \, M$
900.	$1.26 \times 10^{-3} \, M$	$6.39 \times 10^{-2} \, M$

Calculate E_a for this reaction.

88. Experimental values for the temperature dependence of the rate constant for the gas-phase reaction

$$NO + O_3 \longrightarrow NO_2 + O_2$$

are as follows:

T (K)	k (L/mol · s)
195	1.08×10^9
230.	2.95×10^9
260.	5.42×10^9
298	12.0×10^9
369	35.5×10^9

Make the appropriate graph using these data, and determine the activation energy for this reaction.

89. Cobra venom helps the snake secure food by binding to acetylcholine receptors on the diaphragm of a bite victim, leading to the loss of function of the diaphragm muscle tissue and eventually death. In order to develop more potent antivenins, scientists have studied what happens to the toxin once it has bound the acetylcholine receptors. They have found that the toxin is released from the receptor in a process that can be described by the rate law

$$\text{Rate} = k[\text{acetylcholine receptor–toxin complex}]$$

If the activation energy of this reaction at 37.0°C is 26.2 kJ/mol and $A = 0.850 \text{ s}^{-1}$, what is the rate of reaction if you have a 0.200-M solution of receptor–toxin complex at 37.0°C?

90. Iodomethane (CH_3I) is a commonly used reagent in organic chemistry. When used properly, this reagent allows chemists to introduce methyl groups in many different useful applications. The chemical does pose a risk as a carcinogen, possibly owing to iodomethane's ability to react with portions of the DNA strand (if they were to come in contact). Consider the following hypothetical initial rates data:

$[\text{DNA}]_0$ (μmol/L)	$[CH_3I]_0$ (μmol/L)	Initial Rate (μmol/L · s)
0.100	0.100	3.20×10^{-4}
0.100	0.200	6.40×10^{-4}
0.200	0.200	1.28×10^{-3}

Which of the following could be a possible mechanism to explain the initial rate data?

Mechanism I $\text{DNA} + CH_3I \longrightarrow \text{DNA}-CH_3^+ + I^-$

Mechanism II $CH_3I \longrightarrow CH_3^+ + I^-$ Slow

$\text{DNA} + CH_3^+ \longrightarrow \text{DNA}-CH_3^+$ Fast

91. Experiments during a recent summer on a number of fireflies (small beetles, *Lampyridaes photinus*) showed that the average interval between flashes of individual insects was 16.3 s at 21.0°C and 13.0 s at 27.8°C.

   a. What is the apparent activation energy of the reaction that controls the flashing?

   b. What would be the average interval between flashes of an individual firefly at 30.0°C?

   c. Compare the observed intervals and the one you calculated in part b to the rule of thumb that the Celsius temperature is 54 minus twice the interval between flashes.

92. The activation energy of a certain uncatalyzed biochemical reaction is 50.0 kJ/mol. In the presence of a catalyst at 37°C, the rate constant for the reaction increases by a factor of 2.50×10^3 as compared with the uncatalyzed reaction. Assuming the frequency factor A is the same for both the catalyzed and uncatalyzed reactions, calculate the activation energy for the catalyzed reaction.

93. Consider the reaction

$$3A + B + C \longrightarrow D + E$$

where the rate law is defined as

$$-\frac{\Delta[A]}{\Delta t} = k[A]^2[B][C]$$

An experiment is carried out where $[B]_0 = [C]_0 = 1.00 \ M$ and $[A]_0 = 1.00 \times 10^{-4} \ M$.

   a. If after 3.00 min, $[A] = 3.26 \times 10^{-5} \ M$, calculate the value of k.

   b. Calculate the half-life for this experiment.

   c. Calculate the concentration of B and the concentration of A after 10.0 min.

ChemWork Problems

These multiconcept problems (and additional ones) are found interactively online with the same type of assistance a student would get from an instructor.

94. The thiosulfate ion ($S_2O_3^{2-}$) is oxidized by iodine as follows:

$$2S_2O_3^{2-}(aq) + I_2(aq) \longrightarrow S_4O_6^{2-}(aq) + 2I^-(aq)$$

In a certain experiment, 7.05×10^{-3} mol/L of $S_2O_3^{2-}$ is consumed in the first 11.0 seconds of the reaction. Calculate the rate of consumption of $S_2O_3^{2-}$. Calculate the rate of production of iodide ion.

95. The reaction $A(aq) + B(aq) \longrightarrow \text{products}(aq)$ was studied, and the following data were obtained:

$[A]_0$ (mol/L)	$[B]_0$ (mol/L)	Initial Rate (mol/L · s)
0.12	0.18	3.46×10^{-2}
0.060	0.12	1.15×10^{-2}
0.030	0.090	4.32×10^{-3}
0.24	0.090	3.46×10^{-2}

What is the order of the reaction with respect to A? What is the order of the reaction with respect to B? What is the value of the rate constant for the reaction?

96. A certain substance, initially present at 0.0800 M, decomposes by zero-order kinetics with a rate constant of 2.50×10^{-2} mol/L · s. Calculate the time (in seconds) required for the system to reach a concentration of 0.0210 M.

97. A reaction of the form

$$aA \longrightarrow \text{Products}$$

gives a plot of ln[A] versus time (in seconds), which is a straight line with a slope of -7.35×10^{-3}. Assuming $[A]_0 = 0.0100 \ M$, calculate the time (in seconds) required for the reaction to reach 22.9% completion.

98. A certain reaction has the form

$$aA \longrightarrow \text{Products}$$

At a particular temperature, concentration versus time data were collected. A plot of 1/[A] versus time (in seconds) gave a straight line with a slope of 6.90×10^{-2}. What is the differential rate law for this reaction? What is the integrated rate law for this reaction? What is the value of the rate constant for this reaction? If $[A]_0$ for this reaction is 0.100 M, what is the first half-life (in seconds)? If the original concentration (at $t = 0$) is 0.100 M, what is the second half-life (in seconds)?

99. Which of the following statement(s) is(are) *true*?

   a. The half-life for a zero-order reaction increases as the reaction proceeds.

   b. A catalyst does not change the value of the rate constant.

   c. The half-life for a reaction, $aA \longrightarrow$ products, that is first order in A increases with increasing $[A]_0$.

   d. The half-life for a second-order reaction increases as the reaction proceeds.

100. Consider the hypothetical reaction $A_2(g) + B_2(g) \longrightarrow 2AB(g)$, where the rate law is:

$$-\frac{\Delta[A_2]}{\Delta t} = k[A_2][B_2]$$

The value of the rate constant at 302°C is 2.45×10^{-4} L/mol · s, and at 508°C the rate constant is 0.891 L/mol · s. What is the activation energy for this reaction? What is the value of the rate constant for this reaction at 375°C?

101. Experiments have shown that the average frequency of chirping by a snowy tree cricket *(Oecanthus fultoni)* depends on temperature as shown in the table.

Chirping Rate (per min)	Temperature (°C)
178	25.0
126	20.3
100.	17.3

What is the apparent activation energy of the process that controls the chirping? What is the rate of chirping expected at a temperature of 7.5°C?

Challenge Problems

102. Consider a reaction of the type $aA \longrightarrow$ products, in which the rate law is found to be rate $= k[A]^3$ (termolecular reactions are improbable but possible). If the first half-life of the reaction is found to be 40. s, what is the time for the second half-life? *Hint:* Using your calculus knowledge, derive the integrated rate law from the differential rate law for a termolecular reaction:

$$\text{Rate} = \frac{-d[A]}{dt} = k[A]^3$$

103. A study was made of the effect of the hydroxide concentration on the rate of the reaction

$$I^-(aq) + OCl^-(aq) \longrightarrow IO^-(aq) + Cl^-(aq)$$

The following data were obtained:

$[I^-]_0$ (mol/L)	$[OCl^-]_0$ (mol/L)	$[OH^-]_0$ (mol/L)	Initial Rate (mol/L · s)
0.0013	0.012	0.10	9.4×10^{-3}
0.0026	0.012	0.10	18.7×10^{-3}
0.0013	0.0060	0.10	4.7×10^{-3}
0.0013	0.018	0.10	14.0×10^{-3}
0.0013	0.012	0.050	18.7×10^{-3}
0.0013	0.012	0.20	4.7×10^{-3}
0.0013	0.018	0.20	7.0×10^{-3}

Determine the rate law and the value of the rate constant for this reaction.

104. Two isomers (A and B) of a given compound dimerize as follows:

$$2A \xrightarrow{k_1} A_2$$
$$2B \xrightarrow{k_2} B_2$$

Both processes are known to be second order in reactant, and k_1 is known to be 0.250 L/mol · s at 25°C. In a particular experiment A and B were placed in separate containers at 25°C, where $[A]_0 = 1.00 \times 10^{-2}$ M and $[B]_0 = 2.50 \times 10^{-2}$ M. It was found that after each reaction had progressed for 3.00 min, $[A] = 3.00[B]$. In this case the rate laws are defined as

$$\text{Rate} = -\frac{\Delta[A]}{\Delta t} = k_1[A]^2$$

$$\text{Rate} = -\frac{\Delta[B]}{\Delta t} = k_2[B]^2$$

a. Calculate the concentration of A_2 after 3.00 min.

b. Calculate the value of k_2.

c. Calculate the half-life for the experiment involving A.

105. The reaction

$$NO(g) + O_3(g) \longrightarrow NO_2(g) + O_2(g)$$

was studied by performing two experiments. In the first experiment the rate of disappearance of NO was followed in the presence of a large excess of O_3. The results were as follows ($[O_3]$ remains effectively constant at 1.0×10^{14} molecules/cm³):

Time (ms)	[NO] (molecules/cm³)
0	6.0×10^8
100 ± 1	5.0×10^8
500 ± 1	2.4×10^8
700 ± 1	1.7×10^8
1000 ± 1	9.9×10^7

In the second experiment [NO] was held constant at 2.0×10^{14} molecules/cm³. The data for the disappearance of O_3 are as follows:

Time (ms)	[O_3] (molecules/cm³)
0	1.0×10^{10}
50 ± 1	8.4×10^9
100 ± 1	7.0×10^9
200 ± 1	4.9×10^9
300 ± 1	3.4×10^9

a. What is the order with respect to each reactant?

b. What is the overall rate law?

c. What is the value of the rate constant from each set of experiments?

$$\text{Rate} = k'[NO]^x \qquad \text{Rate} = k''[O_3]^y$$

d. What is the value of the rate constant for the overall rate law?

$$\text{Rate} = k[NO]^x[O_3]^y$$

106. Most reactions occur by a series of steps. The energy profile for a certain reaction that proceeds by a two-step mechanism is

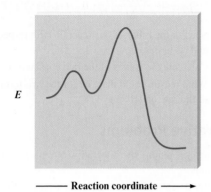

E

——— Reaction coordinate ———→

On the energy profile, indicate

a. the positions of reactants and products.

b. the activation energy for the overall reaction.

c. ΔE for the reaction.

d. Which point on the plot represents the energy of the intermediate in the two-step reaction?

e. Which step in the mechanism for this reaction is rate determining, the first or the second step? Explain.

107. You are studying the kinetics of the reaction $H_2(g) + F_2(g) \rightarrow 2HF(g)$ and you wish to determine a mechanism for the reaction. You run the reaction twice by keeping one reactant at a much higher pressure than the other reactant (this lower-pressure reactant begins at 1.000 atm). Unfortunately, you neglect to record which reactant was at the higher pressure, and you forget which it was later. Your data for the first experiment are:

Pressure of HF (atm)	Time (min)
0	0
0.300	30.0
0.600	65.8
0.900	110.4
1.200	169.1
1.500	255.9

When you ran the second experiment (in which the higher-pressure reactant was run at a much higher pressure), you determine the values of the apparent rate constants to be the same. It also turns out that you find data taken from another person in the lab. This individual found that the reaction proceeds 40.0 times faster at 55°C than at 35°C. You also know, from the energy-level diagram, that there are three steps to the mechanism, and the first step has the highest activation energy. You look up the bond energies of the species involved and they are (in kJ/mol): H—H (432), F—F (154), and H—F (565).

a. Sketch an energy-level diagram (qualitative) that is consistent with the one described previously. *Hint:* See Exercise 106.

b. Develop a reasonable mechanism for the reaction.

c. Which reactant was limiting in the experiments?

108. The decomposition of $NO_2(g)$ occurs by the following bimolecular elementary reaction:

$$2NO_2(g) \longrightarrow 2NO(g) + O_2(g)$$

The rate constant at 273 K is 2.3×10^{-12} L/mol · s, and the activation energy is 111 kJ/mol. How long will it take for the concentration of $NO_2(g)$ to decrease from an initial partial pressure of 2.5 atm to 1.5 atm at 500. K? Assume ideal gas behavior.

109. The following data were collected in two studies of the reaction

$$2A + B \longrightarrow C + D$$

Time (s)	Experiment 1 [A] (mol/L) × 10⁻²	Experiment 2 [A] (mol/L) × 10⁻²
0	10.0	10.0
20.	6.67	5.00
40.	5.00	3.33
60.	4.00	2.50
80.	3.33	2.00
100.	2.86	1.67
120.	2.50	1.43

In Experiment 1, $[B]_0 = 5.0\ M$.
In Experiment 2, $[B]_0 = 10.0\ M$.

$$\text{Rate} = \frac{-\Delta[A]}{\Delta t}$$

a. Why is [B] much greater than [A]?

b. Give the rate law and value for k for this reaction.

110. Consider the following hypothetical data collected in two studies of the reaction

$$2A + 2B \longrightarrow C + 2D$$

Time (s)	Experiment 1 [A] (mol/L)	Experiment 2 [A] (mol/L)
0	1.0×10^{-2}	1.0×10^{-2}
10.	8.4×10^{-3}	5.0×10^{-3}
20.	7.1×10^{-3}	2.5×10^{-3}
30.	?	1.3×10^{-3}
40.	5.0×10^{-3}	6.3×10^{-4}

In Experiment 1, $[B]_0 = 10.0\ M$.
In Experiment 2, $[B]_0 = 20.0\ M$.

$$\text{Rate} = \frac{-\Delta[A]}{\Delta t}$$

a. Use the concentration versus time data to determine the rate law for the reaction.

b. Solve for the value of the rate constant (k) for the reaction. Include units.

c. Calculate the concentration of A in Experiment 1 at $t = 30.$ s.

111. Consider the hypothetical reaction

$$A + B + 2C \longrightarrow 2D + 3E$$

In a study of this reaction three experiments were run at the same temperature. The rate is defined as $-\Delta[B]/\Delta t$.

Experiment 1:

$$[A]_0 = 2.0\,M \quad [B]_0 = 1.0 \times 10^{-3}\,M \quad [C]_0 = 1.0\,M$$

[B] (mol/L)	Time (s)
2.7×10^{-4}	1.0×10^5
1.6×10^{-4}	2.0×10^5
1.1×10^{-4}	3.0×10^5
8.5×10^{-5}	4.0×10^5
6.9×10^{-5}	5.0×10^5
5.8×10^{-5}	6.0×10^5

Experiment 2:

$$[A]_0 = 1.0 \times 10^{-2}\,M \quad [B]_0 = 3.0\,M \quad [C]_0 = 1.0\,M$$

[A] (mol/L)	Time (s)
8.9×10^{-3}	1.0
7.1×10^{-3}	3.0
5.5×10^{-3}	5.0
3.8×10^{-3}	8.0
2.9×10^{-3}	10.0
2.0×10^{-3}	13.0

Experiment 3:

$$[A]_0 = 10.0\,M \quad [B]_0 = 5.0\,M \quad [C]_0 = 5.0 \times 10^{-1}\,M$$

[C] (mol/L)	Time (s)
0.43	1.0×10^{-2}
0.36	2.0×10^{-2}
0.29	3.0×10^{-2}
0.22	4.0×10^{-2}
0.15	5.0×10^{-2}
0.08	6.0×10^{-2}

Write the rate law for this reaction, and calculate the value of the rate constant.

112. Hydrogen peroxide and the iodide ion react in acidic solution as follows:

$$H_2O_2(aq) + 3I^-(aq) + 2H^+(aq) \longrightarrow I_3^-(aq) + 2H_2O(l)$$

The kinetics of this reaction were studied by following the decay of the concentration of H_2O_2 and constructing plots of $\ln[H_2O_2]$ versus time. All the plots were linear and all solutions had $[H_2O_2]_0 = 8.0 \times 10^{-4}$ mol/L. The slopes of these straight lines depended on the initial concentrations of I^- and H^+. The results follow:

$[I^-]_0$ (mol/L)	$[H^+]_0$ (mol/L)	Slope (min^{-1})
0.1000	0.0400	-0.120
0.3000	0.0400	-0.360
0.4000	0.0400	-0.480
0.0750	0.0200	-0.0760
0.0750	0.0800	-0.118
0.0750	0.1600	-0.174

The rate law for this reaction has the form

$$\text{Rate} = \frac{-\Delta[H_2O_2]}{\Delta t} = (k_1 + k_2[H^+])[I^-]^m[H_2O_2]^n$$

a. Specify the order of this reaction with respect to $[H_2O_2]$ and $[I^-]$.

b. Calculate the values of the rate constants, k_1 and k_2.

c. What reason could there be for the two-term dependence of the rate on $[H^+]$?

Integrative Problems

These problems require the integration of multiple concepts to find the solutions.

113. Sulfuryl chloride undergoes first-order decomposition at 320.°C with a half-life of 8.75 h.

$$SO_2Cl_2(g) \longrightarrow SO_2(g) + Cl_2(g)$$

What is the value of the rate constant, k, in s^{-1}? If the initial pressure of SO_2Cl_2 is 791 torr and the decomposition occurs in a 1.25-L container, how many molecules of SO_2Cl_2 remain after 12.5 h?

114. Upon dissolving InCl(s) in HCl, $In^+(aq)$ undergoes a disproportionation reaction according to the following unbalanced equation:

$$In^+(aq) \longrightarrow In(s) + In^{3+}(aq)$$

This disproportionation follows first-order kinetics with a half-life of 667 s. What is the concentration of $In^+(aq)$ after 1.25 h if the initial solution of $In^+(aq)$ was prepared by dissolving 2.38 g InCl(s) in dilute HCl to make 5.00×10^2 mL of solution? What mass of In(s) is formed after 1.25 h?

115. The decomposition of iodoethane in the gas phase proceeds according to the following equation:

$$C_2H_5I(g) \longrightarrow C_2H_4(g) + HI(g)$$

At 660. K, $k = 7.2 \times 10^{-4}$ s^{-1}; at 720. K, $k = 1.7 \times 10^{-2}$ s^{-1}. What is the value of the rate constant for this first-order decomposition at 325°C? If the initial pressure of iodoethane is 894 torr at 245°C, what is the pressure of iodoethane after three half-lives?

Marathon Problem

This problem is designed to incorporate several concepts and techniques into one situation.

116. Consider the following reaction:

$$CH_3X + Y \longrightarrow CH_3Y + X$$

At 25°C, the following two experiments were run, yielding the following data:

Experiment 1: $[Y]_0 = 3.0\,M$

[CH$_3$X] (mol/L)	Time (h)
7.08×10^{-3}	1.0
4.52×10^{-3}	1.5
2.23×10^{-3}	2.3
4.76×10^{-4}	4.0
8.44×10^{-5}	5.7
2.75×10^{-5}	7.0

Experiment 2: $[Y]_0 = 4.5\ M$

$[CH_3X]$ (mol/L)	Time (h)
4.50×10^{-3}	0
1.70×10^{-3}	1.0
4.19×10^{-4}	2.5
1.11×10^{-4}	4.0
2.81×10^{-5}	5.5

Experiments also were run at 85°C. The value of the rate constant at 85°C was found to be 7.88×10^8 (with the time in units of hours), where $[CH_3X]_0 = 1.0 \times 10^{-2}\ M$ and $[Y]_0 = 3.0\ M$.

a. Determine the rate law and the value of k for this reaction at 25°C.

b. Determine the half-life at 85°C.

c. Determine E_a for the reaction.

d. Given that the C—X bond energy is known to be about 325 kJ/mol, suggest a mechanism that explains the results in parts a and c.

Chapter 13

Chemical Equilibrium

The equilibrium in a salt water aquarium must be carefully maintained to keep the sea life healthy. (Borissos/Dreamstime.com)

n doing stoichiometry calculations we assumed that reactions proceed to completion, that is, until one of the reactants runs out. Many reactions do proceed essentially to completion. For such reactions it can be assumed that the reactants are quantitatively converted to products and that the amount of limiting reactant that remains is negligible. On the other hand, there are many chemical reactions that stop far short of completion. An example is the dimerization of nitrogen dioxide:

$$NO_2(g) + NO_2(g) \longrightarrow N_2O_4(g)$$

The reactant, NO_2, is a dark brown gas, and the product, N_2O_4, is a colorless gas. When NO_2 is placed in an evacuated, sealed glass vessel at 25°C, the initial dark brown color decreases in intensity as it is converted to colorless N_2O_4. However, even over a long period of time, the contents of the reaction vessel do not become colorless. Instead, the intensity of the brown color eventually becomes constant, which means that the concentration of NO_2 is no longer changing. This is illustrated on the molecular level in Fig. 13.1. This observation is a clear indication that the reaction has stopped short of completion. In fact, the system has reached **chemical equilibrium**, *the state where the concentrations of all reactants and products remain constant with time.*

Any chemical reactions carried out in a closed vessel will reach equilibrium. For some reactions the equilibrium position so favors the products that the reaction appears to have gone to completion. We say that the equilibrium position for such reactions lies *far to the right* (in the direction of the products). For example, when gaseous hydrogen and oxygen are mixed in stoichiometric quantities and react to form water vapor, the reaction proceeds essentially to completion. The amounts of the reactants that remain when the system reaches equilibrium are so tiny as to be negligible. By contrast, some reactions occur only to a slight extent. For example, when solid CaO is placed in a closed vessel at 25°C, the decomposition to solid Ca and gaseous O_2 is virtually undetectable. In cases like this, the equilibrium position is said to lie *far to the left* (in the direction of the reactants).

In this chapter we will discuss how and why a chemical system comes to equilibrium and the characteristics of equilibrium. In particular, we will discuss how to calculate the concentrations of the reactants and products present for a given system at equilibrium.

13.1 | The Equilibrium Condition

Since no changes occur in the concentrations of reactants or products in a reaction system at equilibrium, it may appear that everything has stopped. However, this is not the case. On the molecular level, there is frantic activity. Equilibrium is not static but is a highly *dynamic* situation. The concept of chemical equilibrium is analogous to the flow of cars across a bridge connecting two island cities. Suppose the traffic flow on the bridge is the same in both directions. It is obvious that there is motion, since one can see the cars traveling back and forth across the bridge, but the number of cars in each city is not changing because equal numbers of cars are entering and leaving. The result is no *net* change in the car population.

Equilibrium is a dynamic situation.

To see how this concept applies to chemical reactions, consider the reaction between steam and carbon monoxide in a closed vessel at a high temperature where the reaction takes place rapidly:

$$H_2O(g) + CO(g) \rightleftharpoons H_2(g) + CO_2(g)$$

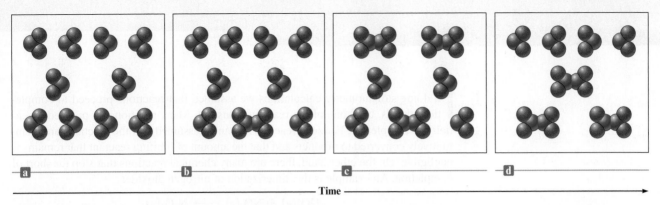

Figure 13.1 | A molecular representation of the reaction $2NO_2(g) \rightarrow N_2O_4(g)$ over time in a closed vessel. Note that the numbers of NO_2 and N_2O_4 in the container become constant (c and d) after sufficient time has passed.

Assume that the same number of moles of gaseous CO and gaseous H_2O are placed in a closed vessel and allowed to react. The plots of the concentrations of reactants and products versus time are shown in Fig. 13.2. Note that since CO and H_2O were originally present in equal molar quantities, and since they react in a 1:1 ratio, the concentrations of the two gases are always equal. Also, since H_2 and CO_2 are formed in equal amounts, they are always present in the same concentrations.

Figure 13.2 is a profile of the progress of the reaction. When CO and H_2O are mixed, they immediately begin to react to form H_2 and CO_2. This leads to a decrease in the concentrations of the reactants, but the concentrations of the products, which were initially at zero, are increasing. Beyond a certain time, indicated by the dashed line in Fig. 13.2, the concentrations of reactants and products no longer change—equilibrium has been reached. Unless the system is somehow disturbed, no further changes in concentrations will occur. Note that although the equilibrium position lies far to the right, the concentrations of reactants never go to zero; the reactants will always be present in small but constant concentrations. This is shown on the microscopic level in Fig. 13.3.

What would happen to the gaseous equilibrium mixture of reactants and products represented in Fig. 13.3, parts (c) and (d), if we injected some $H_2O(g)$ into the box? To answer this question, we need to be sure we understand the equilibrium condition: The concentrations of reactants and products remain constant at equilibrium because the forward and reverse reaction rates are equal. If we inject some H_2O molecules, what will happen to the forward reaction: $H_2O + CO \rightarrow H_2 + CO_2$? It will speed up because more H_2O molecules means more collisions between H_2O and CO molecules. This in turn will form more products and will cause the reverse reaction $H_2O + CO \leftarrow H_2 + CO_2$ to speed up. Thus the system will change until the forward and reverse reaction rates again become equal. Will this new equilibrium position contain more or fewer product molecules than are shown in Fig. 13.3(c) and (d)? Think about this carefully. If you are not sure of the answer now, keep reading. We will consider this type of situation in more detail later in this chapter.

Figure 13.2 | The changes in concentrations with time for the reaction $H_2O(g) + CO(g) \rightleftharpoons H_2(g) + CO_2(g)$ when equimolar quantities of $H_2O(g)$ and $CO(g)$ are mixed.

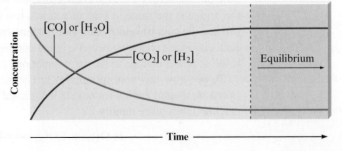

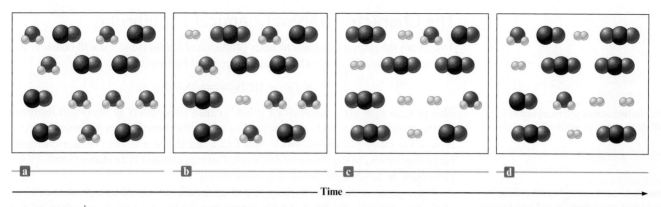

Figure 13.3 | (a) H_2O and CO are mixed in equal numbers and begin to react (b) to form CO_2 and H_2. After time has passed, equilibrium is reached (c) and the numbers of reactant and product molecules then remain constant over time (d).

Why does equilibrium occur? We saw in Chapter 12 that molecules react by colliding with one another, and the more collisions, the faster the reaction. This is why reaction rates depend on concentrations. In this case the concentrations of H_2O and CO are lowered by the forward reaction:

$$H_2O + CO \longrightarrow H_2 + CO_2$$

As the concentrations of the reactants decrease, the forward reaction slows down (Fig. 13.4). As in the bridge traffic analogy, there is also a reverse direction:

$$H_2O + CO \longleftarrow H_2 + CO_2$$

Initially in this experiment no H_2 and CO_2 were present, and this reverse reaction could not occur. However, as the forward reaction proceeds, the concentrations of H_2 and CO_2 build up, and the rate of the reverse reaction increases (Fig. 13.4) as the forward reaction slows down. Eventually, the concentrations reach levels where the rate of the forward reaction equals the rate of the reverse reaction. The system has reached equilibrium.

A double arrow ($\rightleftharpoons$) is used to show that a reaction can occur in either direction.

The relationship between equilibrium and thermodynamics is explored in Section 16.8.

The equilibrium position of a reaction—left, right, or somewhere in between—is determined by many factors: the initial concentrations, the relative energies of the reactants and products, and the relative degree of "organization" of the reactants and products. Energy and organization come into play because nature tries to achieve minimum energy and maximum disorder, as we will show in detail in Chapter 16. For now, we will simply view the equilibrium phenomenon in terms of the rates of opposing reactions.

Figure 13.4 | The changes with time in the rates of forward and reverse reactions for $H_2O(g) + CO(g) \rightleftharpoons H_2(g) + CO_2(g)$ when equimolar quantities of $H_2O(g)$ and $CO(g)$ are mixed. The rates do not change in the same way with time because the forward reaction has a much larger rate constant than the reverse reaction.

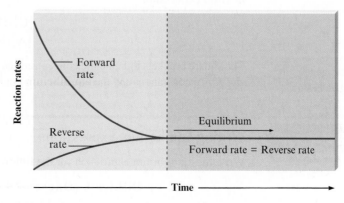

Reaction rates

Forward rate

Reverse rate

Equilibrium

Forward rate = Reverse rate

Time

The Characteristics of Chemical Equilibrium

To explore the important characteristics of chemical equilibrium, we will consider the synthesis of ammonia from elemental nitrogen and hydrogen:

$$N_2(g) + 3H_2(g) \rightleftharpoons 2NH_3(g)$$

This process is of great commercial value because ammonia is an important fertilizer for the growth of corn and other crops. Ironically, this beneficial process was discovered in Germany just before World War I in a search for ways to produce nitrogen-based explosives. In the course of this work, German chemist Fritz Haber (1868–1934) pioneered the large-scale production of ammonia.

When gaseous nitrogen, hydrogen, and ammonia are mixed in a closed vessel at 25°C, no apparent change in the concentrations occurs over time, regardless of the original amounts of the gases. Why? There are two possible reasons why the concentrations of the reactants and products of a given chemical reaction remain unchanged when mixed.

1. The system is at chemical equilibrium.
2. The forward and reverse reactions are so slow that the system moves toward equilibrium at a rate that cannot be detected.

The second reason applies to the nitrogen, hydrogen, and ammonia mixture at 25°C. As we saw in Chapters 8 and 9, the N_2 molecule has a very strong triple bond (941 kJ/mol) and thus is very unreactive. Also, the H_2 molecule has an unusually strong single bond (432 kJ/mol). Therefore, mixtures of N_2, H_2, and NH_3 at 25°C can exist with no apparent change over long periods of time, unless a catalyst is introduced to speed up the forward and reverse reactions. Under appropriate conditions, the system does reach equilibrium, as shown in Fig. 13.5. Note that because of the reaction stoichiometry, H_2 disappears three times as fast as N_2 does and NH_3 forms twice as fast as N_2 disappears.

The United States produces about 20 million tons of ammonia annually.

Molecules with strong bonds produce large activation energies and tend to react slowly at 25°C.

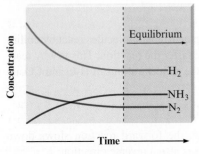

Figure 13.5 | A concentration profile for the reaction $N_2(g) + 3H_2(g) \rightleftharpoons 2NH_3(g)$ when only $N_2(g)$ and $H_2(g)$ are mixed initially.

13.2 | The Equilibrium Constant

Science is fundamentally empirical—it is based on experiment. The development of the equilibrium concept is typical. From their observations of many chemical reactions, two Norwegian chemists, Cato Maximilian Guldberg (1836–1902) and Peter Waage (1833–1900), proposed in 1864 the **law of mass action** as a general description of the equilibrium condition. Guldberg and Waage postulated that for a reaction of the type

$$jA + kB \rightleftharpoons lC + mD$$

where A, B, C, and D represent chemical species and j, k, l, and m are their coefficients in the balanced equation, the law of mass action is represented by the following **equilibrium expression**:

$$K = \frac{[C]^l[D]^m}{[A]^j[B]^k}$$

The square brackets indicate the concentrations of the chemical species *at equilibrium*, and K is a constant called the **equilibrium constant**.

The law of mass action is based on experimental observation.

Sign in at http://login.cengagebrain .com to try this Interactive Example in OWL.

Interactive Example 13.1

Writing Equilibrium Expressions

Write the equilibrium expression for the following reaction:

$$4NH_3(g) + 7O_2(g) \rightleftharpoons 4NO_2(g) + 6H_2O(g)$$

Solution

Applying the law of mass action gives

The square brackets indicate concentration in units of mol/L.

Coefficient of NO_2 Coefficient of H_2O

$$K = \frac{[NO_2]^4[H_2O]^6}{[NH_3]^4[O_2]^7}$$

Coefficient of O_2

Coefficient of NH_3

See Exercise 13.21

The value of the equilibrium constant at a given temperature can be calculated if we know the equilibrium concentrations of the reaction components, as illustrated in Example 13.2.

It is very important to note at this point that the equilibrium constants are customarily given without units. The reason for this is beyond the scope of this text, but it involves corrections for the nonideal behavior of the substances taking part in the reaction. When these corrections are made, the units cancel out and the corrected K has no units. Thus we will not use units for K in this text.

Interactive Example 13.2

Sign in at http://login.cengagebrain .com to try this Interactive Example in **OWL**.

Calculating the Values of K

The following equilibrium concentrations were observed for the Haber process for synthesis of ammonia at 127°C:

$$[NH_3] = 3.1 \times 10^{-2} \text{ mol/L}$$
$$[N_2] = 8.5 \times 10^{-1} \text{ mol/L}$$
$$[H_2] = 3.1 \times 10^{-3} \text{ mol/L}$$

a. Calculate the value of K at 127°C for this reaction.

b. Calculate the value of the equilibrium constant at 127°C for the reaction

$$2NH_3(g) \rightleftharpoons N_2(g) + 3H_2(g)$$

c. Calculate the value of the equilibrium constant at 127°C for the reaction given by the equation

$$\tfrac{1}{2}N_2(g) + \tfrac{3}{2}H_2(g) \rightleftharpoons NH_3(g)$$

Solution

a. What is the balanced equation for the Haber process?

$$N_2(g) + 3H_2(g) \rightleftharpoons 2NH_3(g)$$

Thus

$$K = \frac{[NH_3]^2}{[N_2][H_2]^3} = \frac{(3.1 \times 10^{-2})^2}{(8.5 \times 10^{-1})(3.1 \times 10^{-3})^3}$$

$$= 3.8 \times 10^4$$

Note that K is written without units.

b. What is the equilibrium expression? This reaction is written in the reverse order from the equation given in part a. This leads to the equilibrium expression

$$K' = \frac{[N_2][H_2]^3}{[NH_3]^2}$$

which is the reciprocal of the expression used in part a. Therefore,

$$K' = \frac{[N_2][H_2]^3}{[NH_3]^2} = \frac{1}{K} = \frac{1}{3.8 \times 10^4} = 2.6 \times 10^{-5}$$

c. What is the equilibrium constant? We use the law of mass action:

$$K'' = \frac{[NH_3]}{[N_2]^{1/2}[H_2]^{3/2}}$$

If we compare this expression to that obtained in part a, we see that since

$$\frac{[NH_3]}{[N_2]^{1/2}[H_2]^{3/2}} = \left(\frac{[NH_3]^2}{[N_2][H_2]^3}\right)^{1/2}$$

$$K'' = K^{1/2}$$

Thus

$$K'' = K^{1/2} = (3.8 \times 10^4)^{1/2} = 1.9 \times 10^2$$

See Exercises 13.23 and 13.25 through 13.27

We can draw some important conclusions from the results of Example 13.2. For a reaction of the form

$$jA + kB \rightleftharpoons lC + mD$$

the equilibrium expression is

$$K = \frac{[C]^l[D]^m}{[A]^j[B]^k}$$

If this reaction is reversed, then the new equilibrium expression is

$$K' = \frac{[A]^j[B]^k}{[C]^l[D]^m} = \frac{1}{K}$$

If the original reaction is multiplied by some factor n to give

$$njA + nkB \rightleftharpoons nlC + nmD$$

the equilibrium expression becomes

$$K'' = \frac{[C]^{nl}[D]^{nm}}{[A]^{nj}[B]^{nk}} = K^n$$

Let's Review | Conclusions About the Equilibrium Expression

❯ The equilibrium expression for a reaction is the reciprocal of that for the reaction written in reverse.

❯ When the balanced equation for a reaction is multiplied by a factor n, the equilibrium expression for the new reaction is the original expression raised to the nth power. Thus $K_{new} = (K_{original})^n$.

❯ K values are customarily written without units.

Table 13.1 | Results of Three Experiments for the Reaction $N_2(g) + 3H_2(g) \rightleftharpoons 2NH_3(g)$

Experiment	Initial Concentrations	Equilibrium Concentrations	$K = \dfrac{[NH_3]^2}{[N_2][H_2]^3}$
I	$[N_2]_0 = 1.000\ M$ $[H_2]_0 = 1.000\ M$ $[NH_3]_0 = 0$	$[N_2] = 0.921\ M$ $[H_2] = 0.763\ M$ $[NH_3] = 0.157\ M$	$K = 6.02 \times 10^{-2}$
II	$[N_2]_0 = 0$ $[H_2]_0 = 0$ $[NH_3]_0 = 1.000\ M$	$[N_2] = 0.399\ M$ $[H_2] = 1.197\ M$ $[NH_3] = 0.203\ M$	$K = 6.02 \times 10^{-2}$
III	$[N_2]_0 = 2.00\ M$ $[H_2]_0 = 1.00\ M$ $[NH_3]_0 = 3.00\ M$	$[N_2] = 2.59\ M$ $[H_2] = 2.77\ M$ $[NH_3] = 1.82\ M$	$K = 6.02 \times 10^{-2}$

The law of mass action applies to solution and gaseous equilibria.

Anhydrous ammonia is applied to soil to act as a fertilizer.

For a reaction at a given temperature, there are many equilibrium positions but only one value for K.

The law of mass action is widely applicable. It correctly describes the equilibrium behavior of an amazing variety of chemical systems in solution and in the gas phase. Although, as we will see later, corrections must be applied in certain cases, such as for concentrated aqueous solutions and for gases at high pressures, the law of mass action provides a remarkably accurate description of all types of chemical equilibria.

Consider again the ammonia synthesis reaction. The equilibrium constant K always has the same value at a given temperature. At 500°C the value of K is 6.0×10^{-2}. Whenever N_2, H_2, and NH_3 are mixed together at this temperature, the system will always come to an equilibrium position such that

$$\frac{[NH_3]^2}{[N_2][H_2]^3} = 6.0 \times 10^{-2}$$

This expression has the same value at 500°C, *regardless of the amounts of the gases that are mixed together initially.*

Although the special ratio of products to reactants defined by the equilibrium expression is constant for a given reaction system at a given temperature, the *equilibrium concentrations will not always be the same.* Table 13.1 gives three sets of data for the synthesis of ammonia, showing that even though the individual sets of equilibrium concentrations are quite different for the different situations, the *equilibrium constant, which depends on the ratio of the concentrations, remains the same* (within experimental error). Note that subscript zeros indicate initial concentrations.

Each *set of equilibrium concentrations* is called an **equilibrium position**. It is essential to distinguish between the equilibrium constant and the equilibrium positions for a given reaction system. There is only *one* equilibrium constant for a particular system at a particular temperature, but there are an *infinite* number of equilibrium positions. The specific equilibrium position adopted by a system depends on the initial concentrations, but the equilibrium constant does not.

Example 13.3

Equilibrium Positions

The following results were collected for two experiments involving the reaction at 600°C between gaseous sulfur dioxide and oxygen to form gaseous sulfur trioxide:

Experiment 1		Experiment 2	
Initial	Equilibrium	Initial	Equilibrium
$[SO_2]_0 = 2.00\ M$ $[O_2]_0 = 1.50\ M$ $[SO_3]_0 = 3.00\ M$	$[SO_2] = 1.50\ M$ $[O_2] = 1.25\ M$ $[SO_3] = 3.50\ M$	$[SO_2]_0 = 0.500\ M$ $[O_2]_0 = 0$ $[SO_3]_0 = 0.350\ M$	$[SO_2] = 0.590\ M$ $[O_2] = 0.0450\ M$ $[SO_3] = 0.260\ M$

Show that the equilibrium constant is the same in both cases.

Solution

The balanced equation for the reaction is

$$2SO_2(g) + O_2(g) \rightleftharpoons 2SO_3(g)$$

From the law of mass action,

$$K = \frac{[SO_3]^2}{[SO_2]^2[O_2]}$$

For Experiment 1,

$$K_1 = \frac{(3.50)^2}{(1.50)^2(1.25)} = 4.36$$

For Experiment 2,

$$K_2 = \frac{(0.260)^2}{(0.590)^2(0.0450)} = 4.32$$

The value of K is constant, within experimental error.

See Exercise 13.28

13.3 | Equilibrium Expressions Involving Pressures

So far we have been describing equilibria involving gases in terms of concentrations. Equilibria involving gases also can be described in terms of pressures. The relationship between the pressure and the concentration of a gas can be seen from the ideal gas equation:

The ideal gas equation was discussed in Section 5.3.

$$PV = nRT \qquad \text{or} \qquad P = \left(\frac{n}{V}\right)RT = CRT$$

where C equals n/V, or the number of moles n of gas per unit volume V. Thus C represents the *molar concentration of the gas*.

For the ammonia synthesis reaction, the equilibrium expression can be written in terms of concentrations, that is,

$$K = \frac{[NH_3]^2}{[N_2][H_2]^3} = \frac{C_{NH_3}^2}{(C_{N_2})(C_{H_2}^3)} = K_c$$

or in terms of the *equilibrium partial pressures of the gases,* that is,

$$K_p = \frac{P_{NH_3}^2}{(P_{N_2})(P_{H_2}^3)}$$

K involves concentrations; K_p involves pressures. In some books, the symbol K_c is used instead of K.

Sign in at http://login.cengagebrain .com to try this Interactive Example in OWL.

Both the symbols K and K_c are used commonly for an equilibrium constant in terms of concentrations. We will always use K in this book. The symbol K_p represents an equilibrium constant in terms of partial pressures.

Interactive Example 13.4

Calculating Values of K_p

The reaction for the formation of nitrosyl chloride

$$2NO(g) + Cl_2(g) \rightleftharpoons 2NOCl(g)$$

was studied at 25°C. The pressures at equilibrium were found to be

$$P_{NOCl} = 1.2 \text{ atm}$$
$$P_{NO} = 5.0 \times 10^{-2} \text{ atm}$$
$$P_{Cl_2} = 3.0 \times 10^{-1} \text{ atm}$$

Calculate the value of K_p for this reaction at 25°C.

Solution

For this reaction,

$$K_p = \frac{P_{NOCl}^2}{(P_{NO_2})^2(P_{Cl_2})} = \frac{(1.2)^2}{(5.0 \times 10^{-2})^2(3.0 \times 10^{-1})}$$
$$= 1.9 \times 10^3$$

See Exercises 13.29 and 13.30

The relationship between K and K_p for a particular reaction follows from the fact that for an ideal gas, $C = P/RT$. For example, for the ammonia synthesis reaction,

$P = CRT$ or $C = \dfrac{P}{RT}$

$$K = \frac{[NH_3]^2}{[N_2][H_2]^3} = \frac{C_{NH_3}^2}{(C_{N_2})(C_{H_2}^3)}$$

$$= \frac{\left(\dfrac{P_{NH_3}}{RT}\right)^2}{\left(\dfrac{P_{N_2}}{RT}\right)\left(\dfrac{P_{H_2}}{RT}\right)^3} = \frac{P_{NH_3}^2}{(P_{N_2})(P_{H_2}^3)} \times \frac{\left(\dfrac{1}{RT}\right)^2}{\left(\dfrac{1}{RT}\right)^4}$$

$$= \frac{P_{NH_3}^2}{(P_{N_2})(P_{H_2}^3)}(RT)^2$$

$$= K_p(RT)^2$$

However, for the synthesis of hydrogen fluoride from its elements,

$$H_2(g) + F_2(g) \rightleftharpoons 2HF(g)$$

the relationship between K and K_p is given by

$$K = \frac{[HF]^2}{[H_2][F_2]} = \frac{C_{HF}^2}{(C_{H_2})(C_{F_2})}$$

$$= \frac{\left(\dfrac{P_{HF}}{RT}\right)^2}{\left(\dfrac{P_{H_2}}{RT}\right)\left(\dfrac{P_{F_2}}{RT}\right)} = \frac{P_{HF}^2}{(P_{H_2})(P_{F_2})}$$

$$= K_p$$

Thus for this reaction, K is equal to K_p. This equality occurs because the sum of the coefficients on either side of the balanced equation is identical, so the terms in RT cancel out. In the equilibrium expression for the ammonia synthesis reaction, the sum of the powers in the numerator is different from that in the denominator, and K does not equal K_p.

For the general reaction

$$jA + kB \rightleftharpoons lC + mD$$

the relationship between K and K_p is

$$K_p = K(RT)^{\Delta n}$$

where Δn is the sum of the coefficients of the *gaseous* products minus the sum of the coefficients of the *gaseous* reactants. This equation is quite easy to derive from the definitions of K and K_p and the relationship between pressure and concentration. For the preceding general reaction,

$$K_p = \frac{(P_C^{\,l})(P_D^{\,m})}{(P_A^{\,j})(P_B^{\,k})} = \frac{(C_C \times RT)^l(C_D \times RT)^m}{(C_A \times RT)^j(C_B \times RT)^k}$$

$$= \frac{(C_C^{\,l})(C_D^{\,m})}{(C_A^{\,j})(C_B^{\,k})} \times \frac{(RT)^{l+m}}{(RT)^{j+k}} = K(RT)^{(l+m)-(j+k)}$$

$$= K(RT)^{\Delta n}$$

where $\Delta n = (l + m) - (j + k)$, the difference in the sums of the coefficients for the gaseous products and reactants.

Δn always involves products minus reactants.

Critical Thinking

The text gives an example reaction for which $K = K_p$. The text states this is true "because the sum of the coefficients on either side of the balanced equation is identical...." What if you are told that for a reaction, $K = K_p$, and the sum of the coefficients on either side of the balanced equation is not equal? How is this possible?

Interactive Example 13.5

Sign in at http://login.cengagebrain.com to try this Interactive Example in **OWL**.

Calculating K from K_p

Using the value of K_p obtained in Example 13.4, calculate the value of K at 25°C for the reaction

$$2NO(g) + Cl_2(g) \rightleftharpoons 2NOCl(g)$$

Solution

From the value of K_p, we can calculate K using

$$K_p = K(RT)^{\Delta n}$$

where $T = 25 + 273 = 298$ K and

$$\Delta n = 2 - (2 + 1) = -1$$

      ↗   ↖

Sum of product coefficients   Sum of reactant coefficients

Δn
↓

Thus

$$K_p = K(RT)^{-1} = \frac{K}{RT}$$

and

$$K = K_p(RT)$$

$$= (1.9 \times 10^3)(0.08206)(298)$$

$$= 4.6 \times 10^4$$

See Exercises 13.31 and 13.32

13.4 | Heterogeneous Equilibria

So far we have discussed equilibria only for systems in the gas phase, where all reactants and products are gases. These are **homogeneous equilibria**. However, many equilibria involve more than one phase and are called **heterogeneous equilibria**. For example, the thermal decomposition of calcium carbonate in the commercial preparation of lime occurs by a reaction involving both solid and gas phases:

$$CaCO_3(s) \rightleftharpoons CaO(s) + CO_2(g)$$
$$\underset{\text{Lime}}{\uparrow}$$

> Lime is among the top five chemicals manufactured in the United States in terms of the amount produced.

> The concentrations of pure liquids and solids are constant.

Straightforward application of the law of mass action leads to the equilibrium expression

$$K' = \frac{[CO_2][CaO]}{[CaCO_3]}$$

However, experimental results show that the *position of a heterogeneous equilibrium does not depend on the amounts of pure solids or liquids present* (Fig. 13.6). The fundamental reason for this behavior is that the concentrations of pure solids and liquids cannot change. Thus the equilibrium expression for the decomposition of solid calcium carbonate might be represented as

$$K' = \frac{[CO_2]C_1}{C_2}$$

where C_1 and C_2 are constants representing the concentrations of the solids CaO and $CaCO_3$, respectively. This expression can be rearranged to give

$$\frac{C_2 K'}{C_1} = K = [CO_2]$$

We can generalize from this result as follows: If pure solids or pure liquids are involved in a chemical reaction, their concentrations *are not included in the equilibrium expression* for the reaction. This simplification occurs *only* with pure solids or liquids, not with solutions or gases, since in these last two cases the concentrations can vary.

For example, in the decomposition of liquid water to gaseous hydrogen and oxygen,

$$2H_2O(l) \rightleftharpoons 2H_2(g) + O_2(g)$$

where

$$K = [H_2]^2[O_2] \quad \text{and} \quad K_p = (P_{H_2}^2)(P_{O_2})$$

The Seven Sisters chalk cliffs in East Sussex, England. The chalk is made up of compressed calcium carbonate skeletons of microscopic algae from the late Cretaceous Period.

David Callan/iStockphoto.com

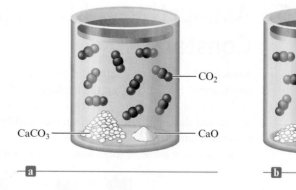

CaCO$_3$ CaO   CO$_2$

a   b

Figure 13.6 | The position of the equilibrium CaCO$_3$(s) $\rightleftharpoons$ CaO(s) + CO$_2$(g) does not depend on the amounts of CaCO$_3$(s) and CaO(s) present.

water is not included in either equilibrium expression because it is a pure liquid. However, if the reaction were carried out under conditions where the water is a gas rather than a liquid, that is,

$$2H_2O(g) \rightleftharpoons 2H_2(g) + O_2(g)$$

then

$$K = \frac{[H_2]^2[O_2]}{[H_2O]^2} \quad \text{and} \quad K_p = \frac{(P_{H_2}^2)(P_{O_2})}{P_{H_2O}^2}$$

because the concentration or pressure of water vapor can change.

Water applied to anhydrous copper(II) sulfate forms the dark blue hydrated compound.

Martyn F. Chillmaid/Photo Researchers, Inc.

Equilibrium Expressions for Heterogeneous Equilibria

Write the expressions for K and K_p for the following processes:

a. Solid phosphorus pentachloride decomposes to liquid phosphorus trichloride and chlorine gas.

b. Deep blue solid copper(II) sulfate pentahydrate is heated to drive off water vapor to form white solid copper(II) sulfate.

Solution

a. What is the balanced equation for the reaction?

$$PCl_5(s) \rightleftharpoons PCl_3(l) + Cl_2(g)$$

What are the equilibrium expressions?

$$K = [Cl_2] \quad \text{and} \quad K_p = P_{Cl_2}$$

In this case neither the pure solid PCl_5 nor the pure liquid PCl_3 is included in the equilibrium expressions.

b. What is the balanced equation for the reaction?

$$CuSO_4 \cdot 5H_2O(s) \rightleftharpoons CuSO_4(s) + 5H_2O(g)$$

What are the equilibrium expressions?

$$K = [H_2O]^5 \quad \text{and} \quad K_p = (P_{H_2O})^5$$

The solids are not included.

See Exercises 13.33 and 13.34

13.5 | Applications of the Equilibrium Constant

Knowing the equilibrium constant for a reaction allows us to predict several important features of the reaction: the tendency of the reaction to occur (but not the speed of the reaction), whether a given set of concentrations represents an equilibrium condition, and the equilibrium position that will be achieved from a given set of initial concentrations.

To introduce some of these ideas, we will first consider the reaction

where ● and ● represent two different types of atoms. Assume that this reaction has an equilibrium constant equal to 16.

In a given experiment, the two types of molecules are mixed together in the following amounts:

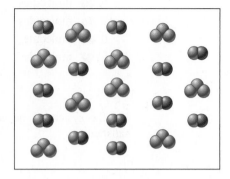

After the system reacts and comes to equilibrium, what will the system look like? We know that at equilibrium the ratio

$$\frac{(N_{●●})(N_{●●})}{(N_{●●})(N_{●●})} = 16$$

must be satisfied, where each N represents the number of molecules of each type. We originally have 9 ●● molecules and 12 ●● molecules. As a place to start, let's just assume that 5 ●● molecules disappear for the system to reach equilibrium. Since equal numbers of the ●● and ●● molecules react, this means that 5 ●● molecules also will disappear. This also means that 5 ●● molecules and 5 ●● molecules will be formed. We can summarize as follows:

Initial Conditions	*New Conditions*
9 ●● molecules	9 − 5 = 4 ●● molecules
12 ●● molecules	12 − 5 = 7 ●● molecules
0 ●● molecules	0 + 5 = 5 ●● molecules
0 ●● molecules	0 + 5 = 5 ●● molecules

Do the new conditions represent equilibrium for this reaction system? We can find out by taking the ratio of the numbers of molecules:

$$\frac{(N_{●●})(N_{●●})}{(N_{●●})(N_{●●})} = \frac{(5)(5)}{(4)(7)} = 0.9$$

Thus this is not an equilibrium position because the ratio is not 16, as required for equilibrium. In which direction must the system move to achieve equilibrium? Since the observed ratio is smaller than 16, we must increase the numerator and decrease the denominator: The system needs to move to the right (toward more products) to achieve equilibrium. That is, more than 5 of the original reactant molecules must disappear to reach equilibrium for this system. How can we find the correct number? Since we do not know the number of molecules that need to disappear to reach equilibrium, let's call this number x. Now we can set up a table similar to the one we used earlier:

Initial Conditions		*Equilibrium Conditions*
9 molecules	x disappear	$9 - x$ molecules
12 molecules	x disappear	$12 - x$ molecules
0 molecules	x form	x molecules
0 molecules	x form	x molecules

For the system to be at equilibrium, we know that the following ratio must be satisfied:

$$\frac{(N)(N)}{(N)(N)} = 16 = \frac{(x)(x)}{(9 - x)(12 - x)}$$

The easiest way to solve for x here is by trial and error. From our previous discussion, we know that x is greater than 5. Also, we know that it must be less than 9 because we have only 9 molecules to start. We can't use all of them or we will have a zero in the denominator, which causes the ratio to be infinitely large. By trial and error, we find that $x = 8$ because

$$\frac{(x)(x)}{(9 - x)(12 - x)} = \frac{(8)(8)}{(9 - 8)(12 - 8)} = \frac{64}{4} = 16$$

The equilibrium mixture can be pictured as follows:

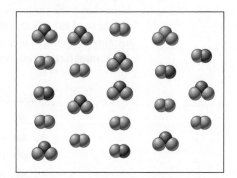

Note that it contains 8 molecules, 8 molecules, 1 molecule, and 4 molecules as required.

This pictorial example should help you understand the fundamental ideas of equilibrium. Now we will proceed to a more systematic, quantitative treatment of chemical equilibrium.

The Extent of a Reaction

The inherent tendency for a reaction to occur is indicated by the magnitude of the equilibrium constant. A value of K much larger than 1 means that at equilibrium the reaction system will consist of mostly products—the equilibrium lies to the right. Another way of saying this is that reactions with very large equilibrium constants go essentially to completion. On the other hand, a very small value of K means that the system at equilibrium will consist of mostly reactants—the equilibrium position is far to the left. The given reaction does not occur to any significant extent.

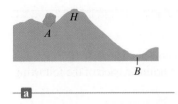

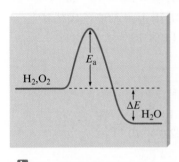

Figure 13.7 | (a) A physical analogy illustrating the difference between thermodynamic and kinetic stabilities. The boulder is thermodynamically more stable (lower potential energy) in position B than in position A but cannot get over the hump H. (b) The reactants H_2 and O_2 have a strong tendency to form H_2O. That is, H_2O has lower energy than H_2 and O_2. However, the large activation energy E_a prevents the reaction at 25°C. In other words, the magnitude of K for the reaction depends on ΔE, but the reaction rate depends on E_a.

It is important to understand that *the size of K and the time required to reach equilibrium are not directly related.* The time required to achieve equilibrium depends on the reaction rate, which is determined by the size of the activation energy. The size of K is determined by thermodynamic factors such as the difference in energy between products and reactants. This difference is represented in Fig. 13.7 and will be discussed in detail in Chapter 16.

Reaction Quotient

When the reactants and products of a given chemical reaction are mixed, it is useful to know whether the mixture is at equilibrium or, if not, the direction in which the system must shift to reach equilibrium. If the concentration of one of the reactants or products is zero, the system will shift in the direction that produces the missing component. However, if all the initial concentrations are nonzero, it is more difficult to determine the direction of the move toward equilibrium. To determine the shift in such cases, we use the **reaction quotient**, Q. The reaction quotient is obtained by applying the law of mass action using *initial concentrations* instead of equilibrium concentrations. For example, for the synthesis of ammonia

$$N_2(g) + 3H_2(g) \rightleftharpoons 2NH_3(g)$$

the expression for the reaction quotient is

$$Q = \frac{[NH_3]_0^2}{[N_2]_0[H_2]_0^3}$$

where the subscript zeros indicate initial concentrations.

To determine in which direction a system will shift to reach equilibrium, we compare the values of Q and K. There are three possible cases (Fig. 13.8):

1. *Q is equal to K.* The system is at equilibrium; no shift will occur.
2. *Q is greater than K.* In this case, the ratio of initial concentrations of products to initial concentrations of reactants is too large. To reach equilibrium, a net change of products to reactants must occur. The system *shifts to the left,* consuming products and forming reactants, until equilibrium is achieved.
3. *Q is less than K.* In this case, the ratio of initial concentrations of products to initial concentrations of reactants is too small. The *system must shift to the right,* consuming reactants and forming products, to attain equilibrium.

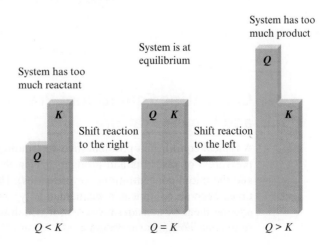

Figure 13.8 | The relationship between reaction quotient Q and the equilibrium constant K.

Using the Reaction Quotient

For the synthesis of ammonia at 500°C, the equilibrium constant is 6.0×10^{-2}. Predict the direction in which the system will shift to reach equilibrium in each of the following cases:

a. $[NH_3]_0 = 1.0 \times 10^{-3}\ M$; $[N_2]_0 = 1.0 \times 10^{-5}\ M$; $[H_2]_0 = 2.0 \times 10^{-3}\ M$

b. $[NH_3]_0 = 2.00 \times 10^{-4}\ M$; $[N_2]_0 = 1.50 \times 10^{-5}\ M$; $[H_2]_0 = 3.54 \times 10^{-1}\ M$

c. $[NH_3]_0 = 1.0 \times 10^{-4}\ M$; $[N_2]_0 = 5.0\ M$; $[H_2]_0 = 1.0 \times 10^{-2}\ M$

Solution

a. What is the value of Q?

$$Q = \frac{[NH_3]_0^2}{[N_2]_0[H_2]_0^3} = \frac{(1.0 \times 10^{-3})^2}{(1.0 \times 10^{-5})(2.0 \times 10^{-3})^3}$$

$$= 1.3 \times 10^7$$

Since $K = 6.0 \times 10^{-2}$, Q is much greater than K. To attain equilibrium, the concentrations of the products must be decreased and the concentrations of the reactants increased. The system will shift to the left:

$$N_2(g) + 3H_2(g) \longleftarrow 2NH_3(g)$$

b. What is the value of Q?

$$Q = \frac{[NH_3]_0^2}{[N_2]_0[H_2]_0^3} = \frac{(2.00 \times 10^{-4})^2}{(1.50 \times 10^{-5})(3.54 \times 10^{-1})^3}$$

$$= 6.01 \times 10^{-2}$$

In this case $Q = K$, so the system is at equilibrium. No shift will occur.

c. What is the value of Q?

$$Q = \frac{[NH_3]_0^2}{[N_2]_0[H_2]_0^3} = \frac{(1.0 \times 10^{-4})^2}{(5.0)(1.0 \times 10^{-2})^3}$$

$$= 2.0 \times 10^{-3}$$

Here Q is less than K, so the system will shift to the right to attain equilibrium by increasing the concentration of the product and decreasing the reactant concentrations:

$$N_2(g) + 3H_2(g) \longrightarrow 2NH_3(g)$$

See Exercises 13.39 through 13.42

Calculating Equilibrium Pressures and Concentrations

A typical equilibrium problem involves finding the equilibrium concentrations (or pressures) of reactants and products, given the value of the equilibrium constant and the initial concentrations (or pressures). However, since such problems sometimes become complicated mathematically, we will develop useful strategies for solving them by considering cases for which we know one or more of the equilibrium concentrations (or pressures).

Calculating Equilibrium Pressures I

Dinitrogen tetroxide in its liquid state was used as one of the fuels on the lunar lander for the NASA Apollo missions. In the gas phase, it decomposes to gaseous nitrogen dioxide:

$$N_2O_4(g) \rightleftharpoons 2NO_2(g)$$

Consider an experiment in which gaseous N_2O_4 was placed in a flask and allowed to reach equilibrium at a temperature where $K_p = 0.133$. At equilibrium, the pressure of N_2O_4 was found to be 2.71 atm. Calculate the equilibrium pressure of $NO_2(g)$.

Solution

We know that the equilibrium pressures of the gases NO_2 and N_2O_4 must satisfy the relationship

$$K_p = \frac{P_{NO_2}^2}{P_{N_2O_4}} = 0.133$$

Since we know $P_{N_2O_4}$, we can simply solve for P_{NO_2}:

$$P_{NO_2}^2 = K_p(P_{N_2O_4}) = (0.133)(2.71) = 0.360$$

Therefore,

$$P_{NO_2} = \sqrt{0.360} = 0.600$$

See Exercises 13.43 and 13.44

Apollo II *lunar landing module at Tranquility Base, 1969.*

Calculating Equilibrium Pressures II

At a certain temperature, a 1.00-L flask initially contained 0.298 mole of $PCl_3(g)$ and 8.70×10^{-3} mole of $PCl_5(g)$. After the system had reached equilibrium, 2.00×10^{-3} mole of $Cl_2(g)$ was found in the flask. Gaseous PCl_5 decomposes according to the reaction

$$PCl_5(g) \rightleftharpoons PCl_3(g) + Cl_2(g)$$

Calculate the equilibrium concentrations of all species and the value of K.

Solution

What is the equilibrium expression for this reaction?

$$K = \frac{[Cl_2][PCl_3]}{[PCl_5]}$$

To find the value of K, we must calculate the equilibrium concentrations of all species and then substitute these quantities into the equilibrium expression. The best method for finding the equilibrium concentrations is to begin with the initial concentrations, which we will define as the concentrations before any shift toward equilibrium has occurred. We will then modify these initial concentrations appropriately to find the equilibrium concentrations.

What are the initial concentrations?

$$[Cl_2]_0 = 0$$

$$[PCl_3]_0 = \frac{0.298 \text{ mol}}{1.00 \text{ L}} = 0.298 \ M$$

$$[PCl_5]_0 = \frac{8.70 \times 10^{-3} \text{ mol}}{1.00 \text{ L}} = 8.70 \times 10^{-3} \ M$$

What change is required to reach equilibrium?

Since no Cl_2 was initially present but $2.00 \times 10^{-3} \ M \ Cl_2$ is present at equilibrium, 2.00×10^{-3} mole of PCl_5 must have decomposed to form 2.00×10^{-3} mole of Cl_2 and 2.00×10^{-3} mole of PCl_3. In other words, to reach equilibrium, the reaction shifted to the right:

$$PCl_5(g) \longrightarrow PCl_3(g) + Cl_2(g)$$

$$2.00 \times 10^{-3} \text{ mol} \longrightarrow 2.00 \times 10^{-3} \text{ mol} + 2.00 \times 10^{-3} \text{ mol}$$

Net amount of PCl_5 decomposed   Net amounts of products formed

Now we apply this change to the initial concentrations.

What are the equilibrium concentrations?

> $[Cl_2] = 0 + \dfrac{2.00 \times 10^{-3} \text{ mol}}{1.00 \text{ L}} = 2.00 \times 10^{-3} \ M$

$[Cl_2]_0$

> $[PCl_3] = 0.298 \ M + \dfrac{2.00 \times 10^{-3} \text{ mol}}{1.00 \text{ L}} = 0.300 \ M$

$[PCl_3]_0$

> $[PCl_5] = 8.70 \times 10^{-3} \ M - \dfrac{2.00 \times 10^{-3} \text{ mol}}{1.00 \text{ L}} = 6.70 \times 10^{-3} \ M$

$[PCl_5]_0$

What is the value of K?

■ The equilibrium concentrations are substituted into the equilibrium expression:

$$K = \frac{[Cl_2][PCl_3]}{[PCl_5]} = \frac{(2.00 \times 10^{-3})(0.300)}{6.70 \times 10^{-3}}$$

$$= 8.96 \times 10^{-2}$$

See Exercises 13.45 through 13.48

Sometimes we are not given any of the equilibrium concentrations (or pressures), only the initial values. Then we must use the stoichiometry of the reaction to express concentrations (or pressures) at equilibrium in terms of the initial values. This is illustrated in Example 13.10.

Calculating Equilibrium Concentrations I

Carbon monoxide reacts with steam to produce carbon dioxide and hydrogen. At 700 K the equilibrium constant is 5.10. Calculate the equilibrium concentrations of all species if 1.000 mol of each component is mixed in a 1.000-L flask.

Solution

What is the balanced equation for the reaction?

$$CO(g) + H_2O(g) \rightleftharpoons CO_2(g) + H_2(g)$$

What is the equilibrium expression?

$$K = \frac{[CO_2][H_2]}{[CO][H_2O]} = 5.10$$

What are the initial concentrations?

$$[CO]_0 = [H_2O]_0 = [CO_2]_0 = [H_2]_0 = \frac{1.000\ \text{mol}}{1.000\ \text{L}} = 1.000\ M$$

Is the system at equilibrium, and if not, which way will it shift to reach the equilibrium position? These questions can be answered by calculating Q:

$$Q = \frac{[CO_2]_0[H_2]_0}{[CO]_0[H_2O]_0} = \frac{(1.000\ \text{mol/L})(1.000\ \text{mol/L})}{(1.000\ \text{mol/L})(1.000\ \text{mol/L})} = 1.000$$

Since Q is less than K, the system is not at equilibrium initially but must shift to the right.

What are the equilibrium concentrations?

As before, we start with the initial concentrations and modify them to obtain the equilibrium concentrations. We must ask this question: How much will the system shift to the right to attain the equilibrium condition? In Example 13.9 the change needed for the system to reach equilibrium was given. However, in this case we do not have this information.

Since the required change in concentrations is unknown at this point, we will define it in terms of x. Let's assume that x mol/L CO must react for the system to reach equilibrium. This means that the initial concentration of CO will decrease by x mol/L:

$$[CO] = [CO]_0 - x$$

Equilibrium   Initial   Change

Since each CO molecule reacts with one H_2O molecule, the concentration of water vapor also must decrease by x mol/L:

$$[H_2O] = [H_2O]_0 - x$$

As the reactant concentrations decrease, the product concentrations increase. Since all the coefficients are 1 in the balanced reaction, 1 mole of CO_2 reacting with 1 mole of H_2O will produce 1 mole of CO_2 and 1 mole of H_2. Or in the present case, to reach equilibrium, x mol/L CO will react with x mol/L H_2O to give an additional x mol/L CO_2 and x mol/L H_2:

$$x CO + x H_2O \longrightarrow x CO_2 + x H_2$$

Thus the initial concentrations of CO_2 and H_2 will increase by x mol/L:

$$[CO_2] = [CO_2]_0 + x$$
$$[H_2] = [H_2]_0 + x$$

Now we have all the equilibrium concentrations defined in terms of the initial concentrations and the change x:

Initial Concentration (mol/L)	Change (mol/L)	Equilibrium Concentration (mol/L)
$[CO]_0 = 1.000$	$-x$	$1.000 - x$
$[H_2O]_0 = 1.000$	$-x$	$1.000 - x$
$[CO_2]_0 = 1.000$	$+x$	$1.000 + x$
$[H_2]_0 = 1.000$	$+x$	$1.000 + x$

Note that the sign of x is determined by the direction of the shift. In this example, the system shifts to the right, so the product concentrations increase and the reactant concentrations decrease. Also note that because the coefficients in the balanced equation are all 1, the magnitude of the change is the same for all species.

Now since we know that the equilibrium concentrations must satisfy the equilibrium expression, we can find the value of x by substituting these concentrations into the expression

$$K = 5.10 = \frac{[CO_2][H_2]}{[CO][H_2O]} = \frac{(1.000 + x)(1.000 + x)}{(1.000 - x)(1.000 - x)} = \frac{(1.000 + x)^2}{(1.000 - x)^2}$$

Since the right side of the equation is a perfect square, the solution of the problem can be simplified by taking the square root of both sides:

$$\sqrt{5.10} = 2.26 = \frac{1.000 + x}{1.000 - x}$$

Multiplying and collecting terms gives

$$x = 0.387 \text{ mol/L}$$

Thus the system shifts to the right, consuming 0.387 mol/L CO and 0.387 mol/L H_2O and forming 0.387 mol/L CO_2 and 0.387 mol/L H_2.

Now the equilibrium concentrations can be calculated:

> $[CO] = [H_2O] = 1.000 - x = 1.000 - 0.387 = 0.613 \, M$

> $[CO_2] = [H_2] = 1.000 + x = 1.000 + 0.387 = 1.387 \, M$

Reality Check | These values can be checked by substituting them back into the equilibrium expression to make sure they give the correct value for K:

$$K = \frac{[CO_2][H_2]}{[CO][H_2O]} = \frac{(1.387)^2}{(0.613)^2} = 5.12$$

This result is the same as the given value of K (5.10) within round-off error, so the answer must be correct.

See Exercise 13.51

Interactive Example 13.11

Sign in at http://login.cengagebrain .com to try this Interactive Example in **OWL**.

Calculating Equilibrium Concentrations II

Assume that the reaction for the formation of gaseous hydrogen fluoride from hydrogen and fluorine has an equilibrium constant of 1.15×10^2 at a certain temperature. In a particular experiment, 3.000 moles of each component were added to a 1.500-L flask. Calculate the equilibrium concentrations of all species.

Solution

What is the balanced equation for the reaction?

$$H_2(g) + F_2(g) \rightleftharpoons 2HF(g)$$

What is the equilibrium expression?

$$K = 1.15 \times 10^2 = \frac{[HF]^2}{[H_2][F_2]}$$

What are the initial concentrations?

$$[HF]_0 = [H_2]_0 = [F_2]_0 = \frac{3.000 \text{ mol}}{1.500 \text{ L}} = 2.000 \text{ } M$$

What is the value of Q?

$$Q = \frac{[HF]_0^2}{[H_2]_0[F_2]_0} = \frac{(2.000)^2}{(2.000)(2.000)} = 1.000$$

Since Q is much less than K, the system must shift to the right to reach equilibrium.

What change in the concentrations is necessary?
Since this is presently unknown, we will define the change needed in terms of x. Let x equal the number of moles per liter of H_2 consumed to reach equilibrium. The stoichiometry of the reaction shows that x mol/L F_2 also will be consumed and $2x$ mol/L HF will be formed:

$$H_2(g) + F_2(g) \longrightarrow 2HF(g)$$
$$x \text{ mol/L} + x \text{ mol/L} \longrightarrow 2x \text{ mol/L}$$

Now the equilibrium concentrations can be expressed in terms of x:

Initial Concentration (mol/L)	Change (mol/L)	Equilibrium Concentration (mol/L)
$[H_2]_0 = 2.000$	$-x$	$[H_2] = 2.000 - x$
$[F_2]_0 = 2.000$	$-x$	$[F_2] = 2.000 - x$
$[HF]_0 = 2.000$	$+2x$	$[HF] = 2.000 + 2x$

These concentrations can be represented in a shorthand table as follows:

We often refer to this form as an **ICE** table (indicated by the first letters of Initial, Change, and Equilibrium).

	$H_2(g)$	$+$	$F_2(g)$	$\rightleftharpoons$	$2HF(g)$
Initial	2.000		2.000		2.000
Change	$-x$		$-x$		$+2x$
Equilibrium	$2.000 - x$		$2.000 - x$		$2.000 + 2x$

What is the value of x?
To solve for x, we substitute the equilibrium concentrations into the equilibrium expression:

$$K = 1.15 \times 10^2 = \frac{[HF]^2}{[H_2][F_2]} = \frac{(2.000 + 2x)^2}{(2.000 - x)^2}$$

The right side of this equation is a perfect square, so taking the square root of both sides gives

$$\sqrt{1.15 \times 10^2} = \frac{2.000 + 2x}{2.000 - x}$$

which yields $x = 1.528$.

What are the equilibrium concentrations?

> $[H_2] = [F_2] = 2.000\ M - x = 0.472\ M$

> $[HF] = 2.000\ M + 2x = 5.056\ M$

Reality Check | Checking these values by substituting them into the equilibrium expression gives

$$\frac{[HF]^2}{[H_2][F_2]} = \frac{(5.056)^2}{(0.472)^2} = 1.15 \times 10^2$$

which agrees with the given value of K.

See Exercise 13.52

13.6 | Solving Equilibrium Problems

We have already considered most of the strategies needed to solve equilibrium problems. The typical procedure for analyzing a chemical equilibrium problem can be summarized as follows:

Problem-Solving Strategy

Solving Equilibrium Problems

1. Write the balanced equation for the reaction.
2. Write the equilibrium expression using the law of mass action.
3. List the initial concentrations.
4. Calculate Q, and determine the direction of the shift to equilibrium.
5. Define the change needed to reach equilibrium, and define the equilibrium concentrations by applying the change to the initial concentrations.
6. Substitute the equilibrium concentrations into the equilibrium expression, and solve for the unknown.
7. Check your calculated equilibrium concentrations by making sure they give the correct value of K.

So far we have been careful to choose systems in which we can solve for the unknown by taking the square root of both sides of the equation. However, this type of system is not really very common, and we must now consider a more typical problem. Suppose for a synthesis of hydrogen fluoride from hydrogen and fluorine, 3.000 moles of H_2 and 6.000 moles of F_2 are mixed in a 3.000-L flask. Assume that the equilibrium constant for the synthesis reaction at this temperature is 1.15×10^2. We calculate the equilibrium concentration of each component as follows:

1. *What is the balanced equation for the reaction?*

$$H_2(g) + F_2(g) \rightleftharpoons 2HF(g)$$

2. *What is the equilibrium expression?*

$$K = 1.15 \times 10^2 = \frac{[HF]^2}{[H_2][F_2]}$$

3. *What are the initial concentrations?*

$$[H_2]_0 = \frac{3.000 \text{ mol}}{3.000 \text{ L}} = 1.000 \ M$$

$$[F_2]_0 = \frac{6.000 \text{ mol}}{3.000 \text{ L}} = 2.000 \ M$$

$$[HF]_0 = 0$$

4. *What is Q?*
There is no need to calculate Q because no HF is present initially, and we know that the system must shift to the right to reach equilibrium.

5. *What change is required to reach equilibrium?*
If we let x represent the number of moles per liter of H_2 consumed to reach equilibrium, we can represent the equilibrium concentrations as follows:

	$H_2(g)$	$+$	$F_2(g)$	$\rightleftharpoons$	$2HF(g)$
Initial	1.000		2.000		0
Change	$-x$		$-x$		$+2x$
Equilibrium	$1.000 - x$		$2.000 - x$		$2x$

6. *What is the value of K?*
Substituting the equilibrium concentrations into the equilibrium expression gives

$$K = 1.15 \times 10^2 = \frac{[HF]^2}{[H_2][F_2]} = \frac{(2x)^2}{(1.000 - x)(2.000 - x)}$$

Since the right side of this equation is not a perfect square, we cannot take the square root of both sides, but must use some other procedure.
First, do the indicated multiplication:

$$(1.000 - x)(2.000 - x)(1.15 \times 10^2) = (2x)^2$$

or $(1.15 \times 10^2)x^2 - 3.000(1.15 \times 10^2)x + 2.000(1.15 \times 10^2) = 4x^2$
and collect terms

$$(1.11 \times 10^2)x^2 - (3.45 \times 10^2)x + 2.30 \times 10^2 = 0$$

This is a quadratic equation of the general form

$$ax^2 + bx + c = 0$$

Use of the quadratic formula is explained in Appendix 1.4.

where the roots can be obtained from the quadratic formula:

$$x = \frac{-b \pm \sqrt{b^2 - 4ac}}{2a}$$

In this example, $a = 1.11 \times 10^2$, $b = -3.45 \times 10^2$, and $c = 2.30 \times 10^2$. Substituting these values into the quadratic formula gives two values for x:

$$x = 2.14 \text{ mol/L} \quad \text{and} \quad x = 0.968 \text{ mol/L}$$

Both of these results cannot be valid (since a *given* set of initial concentrations leads to only *one* equilibrium position). How can we choose between them? Since the expression for the equilibrium concentration of H_2 is

$$[H_2] = 1.000 \ M - x$$

the value of x cannot be 2.14 mol/L (because subtracting 2.14 M from 1.000 M gives a negative concentration of H_2, which is physically impossible). Thus the correct value for x is 0.968 mol/L, and the equilibrium concentrations are as follows:

> $[H_2] = 1.000\ M - 0.968\ M = 3.2 \times 10^{-2}\ M$

> $[F_2] = 2.000\ M - 0.968\ M = 1.032\ M$

> $[HF] = 2(0.968\ M) = 1.936\ M$

Reality Check

7. We can check these concentrations by substituting them into the equilibrium expression:

$$\frac{[HF]^2}{[H_2][F_2]} = \frac{(1.936)^2}{(3.2 \times 10^{-2})(1.032)} = 1.13 \times 10^2$$

This value is in close agreement with the given value for K (1.15×10^2), so the calculated equilibrium concentrations are correct.

This procedure is further illustrated for a problem involving pressures in Example 13.12.

Interactive Example 13.12

Sign in at http://login.cengagebrain .com to try this Interactive Example in **OWL**.

Calculating Equilibrium Pressures

Assume that gaseous hydrogen iodide is synthesized from hydrogen gas and iodine vapor at a temperature where the equilibrium constant is 1.00×10^2. Suppose HI at 5.000×10^{-1} atm, H_2 at 1.000×10^{-2} atm, and I_2 at 5.000×10^{-3} atm are mixed in a 5.000-L flask. Calculate the equilibrium pressures of all species.

Solution

1. *What is the balanced equation for this process?*

$$H_2(g) + I_2(g) \rightleftharpoons 2HI(g)$$

2. *What is the equilibrium expression in terms of pressure?*

$$K_p = \frac{P_{HI}^2}{(P_{H_2})(P_{I_2})} = 1.00 \times 10^2$$

3. *What are the given initial pressures?*

$$P_{HI}^0 = 5.000 \times 10^{-1}\ atm$$
$$P_{H_2}^0 = 1.000 \times 10^{-2}\ atm$$
$$P_{I_2}^0 = 5.000 \times 10^{-3}\ atm$$

4. *What is the value of Q for this system?*

$$Q = \frac{(P_{HI}^0)^2}{(P_{H_2}^0)(P_{I_2}^0)} = \frac{(5.000 \times 10^{-1}\ atm)^2}{(1.000 \times 10^{-2}\ atm)(5.000 \times 10^{-3}\ atm)} = 5.000 \times 10^3$$

Since Q is greater than K, the system will shift to the left to reach equilibrium.

So far we have used moles or concentrations in stoichiometric calculations. However, it is equally valid to use pressures for a gas-phase system at constant temperature and volume because in this case pressure is directly proportional to the number of moles:

$$P = n\left(\frac{RT}{V}\right) \longleftarrow \text{Constant if constant } T \text{ and } V$$

Thus we can represent the change needed to achieve equilibrium in terms of pressures.

5. *What change is required to reach equilibrium?*

Let x be the change in pressure (in atm) of H_2 as the system shifts left toward equilibrium. This leads to the following equilibrium pressures:

	$H_2(g)$	$+$	$I_2(g)$	$\rightleftharpoons$	$2HI(g)$
Initial	1.000×10^{-2}		5.000×10^{-3}		5.000×10^{-1}
Change	$+x$		$+x$		$-2x$
Equilibrium	$1.000 \times 10^{-2} + x$		$5.000 \times 10^{-3} + x$		$5.000 \times 10^{-1} - 2x$

6. *What is the value of K_p?*

Substitution into the equilibrium expression gives

$$K_p = \frac{(P_{HI})^2}{(P_{H_2})(P_{I_2})} = \frac{(5.000 \times 10^{-1} - 2x)^2}{(1.000 \times 10^{-2} + x)(5.000 \times 10^{-3} + x)}$$

Multiplying and collecting terms yield the quadratic equation where $a = 9.60 \times 10^1$, $b = 3.5$, and $c = -2.45 \times 10^{-1}$:

$$(9.60 \times 10^1)x^2 + 3.5x - (2.45 \times 10^{-1}) = 0$$

From the quadratic formula, the correct value for x is $x = 3.55 \times 10^{-2}$ atm.

What are the equilibrium pressures?

The equilibrium pressures can now be calculated from the expressions involving x:

> $P_{HI} = 5.000 \times 10^{-1}$ atm $- 2(3.55 \times 10^{-2})$ atm $= 4.29 \times 10^{-1}$ atm
> $P_{H_2} = 1.000 \times 10^{-2}$ atm $+ 3.55 \times 10^{-2}$ atm $= 4.55 \times 10^{-2}$ atm
> $P_{I_2} = 5.000 \times 10^{-3}$ atm $+ 3.55 \times 10^{-2}$ atm $= 4.05 \times 10^{-2}$ atm

Reality Check

7. $\dfrac{P_{HI}^2}{P_{H_2} \cdot P_{I_2}} = \dfrac{(4.29 \times 10^{-1})^2}{(4.55 \times 10^{-2})(4.05 \times 10^{-2})} = 99.9$

This agrees with the given value of K (1.00×10^2), so the calculated equilibrium concentrations are correct.

See Exercises 13.53 through 13.56

Treating Systems That Have Small Equilibrium Constants

We have seen that fairly complicated calculations are often necessary to solve equilibrium problems. However, under certain conditions, simplifications are possible that greatly reduce the mathematical difficulties. For example, gaseous NOCl decomposes to form the gases NO and Cl_2. At 35°C the equilibrium constant is 1.6×10^{-5}. In an experiment in which 1.0 mole of NOCl is placed in a 2.0-L flask, what are the equilibrium concentrations?

The balanced equation is

$$2NOCl(g) \rightleftharpoons 2NO(g) + Cl_2(g)$$

and

$$K = \frac{[NO]^2[Cl_2]}{[NOCl]^2} = 1.6 \times 10^{-5}$$

The initial concentrations are

$$[NOCl]_0 = \frac{1.0 \text{ mol}}{2.0 \text{ L}} = 0.50 \, M \qquad [NO]_0 = 0 \qquad [Cl_2]_0 = 0$$

Since there are no products initially, the system will move to the right to reach equilibrium. We will define x as the change in concentration of Cl_2 needed to reach equilibrium. The changes in the concentrations of NOCl and NO can then be obtained from the balanced equation:

$$2NOCl(g) \longrightarrow 2NO(g) + Cl_2(g)$$
$$2x \longrightarrow 2x + x$$

The concentrations can be summarized as follows:

	2NOCl(g)	$\rightleftharpoons$	2NO(g)	+	$Cl_2(g)$
Initial	0.50		0		0
Change	$-2x$		$+2x$		$+x$
Equilibrium	$0.50 - 2x$		$2x$		x

The equilibrium concentrations must satisfy the equilibrium expression

$$K = 1.6 \times 10^{-5} = \frac{[NO]^2[Cl_2]}{[NOCl]^2} = \frac{(2x)^2(x)}{(0.50 - 2x)^2}$$

Multiplying and collecting terms will give an equation with terms containing x^3, x^2, and x, which requires complicated methods to solve directly. However, we can avoid this situation by recognizing that since K is so small (1.6×10^{-5}), the system will not proceed far to the right to reach equilibrium. That is, x *represents a relatively small number.* The consequence of this fact is that the term $(0.50 - 2x)$ can be approximated by 0.50. That is, when x is small,

$$0.50 - 2x \approx 0.50$$

Making this approximation allows us to simplify the equilibrium expression:

$$1.6 \times 10^{-5} = \frac{(2x)^2(x)}{(0.50 - 2x)^2} \approx \frac{(2x)^2(x)}{(0.50)^2} = \frac{4x^3}{(0.50)^2}$$

Solving for x^3 gives

$$x^3 = \frac{(1.6 \times 10^{-5})(0.50)^2}{4} = 1.0 \times 10^{-6}$$

and $x = 1.0 \times 10^{-2}$.

How valid is this approximation? If $x = 1.0 \times 10^{-2}$, then

$$0.50 - 2x = 0.50 - 2(1.0 \times 10^{-2}) = 0.48$$

The difference between 0.50 and 0.48 is 0.02, or 4% of the initial concentration of NOCl, a relatively small discrepancy that will have little effect on the outcome. That is, since $2x$ is very small compared with 0.50, the value of x obtained in the approximate solution should be very close to the exact value. We use this approximate value of x to calculate the equilibrium concentrations:

> $[NOCl] = 0.50 - 2x \approx 0.50\ M$

> $[NO] = 2x = 2(1.0 \times 10^{-2}\ M) = 2.0 \times 10^{-2}\ M$

> $[Cl_2] = x = 1.0 \times 10^{-2}\ M$

Reality Check

$$\frac{[NO]^2[Cl_2]}{[NOCl]^2} = \frac{(2.0 \times 10^{-2})^2(1.0 \times 10^{-2})}{(0.50)^2} = 1.6 \times 10^{-5}$$

Since the given value of K is 1.6×10^{-5}, these calculated concentrations are correct.

This problem was much easier to solve than it appeared at first because the *small value of K and the resulting small shift to the right to reach equilibrium allowed simplification.*

Approximations can simplify complicated math, but their validity should be checked carefully.

Critical Thinking

You have learned how to treat systems that have small equilibrium constants by making approximations to simplify the math. What if the system has a very large equilibrium constant? What can you do to simplify the math for this case? Use the example from the text, but change the value of the equilibrium constant to 1.6×10^5 and rework the problem. Why can you not use approximations for the case in which $K = 1.6$?

13.7 | Le Châtelier's Principle

It is important to understand the factors that control the *position* of a chemical equilibrium. For example, when a chemical is manufactured, the chemists and chemical engineers in charge of production want to choose conditions that favor the desired product as much as possible. That is, they want the equilibrium to lie far to the right. When Fritz Haber was developing the process for the synthesis of ammonia, he did extensive studies on how temperature and pressure affect the equilibrium concentration of ammonia. Some of his results are given in Table 13.2. Note that the equilibrium amount of NH_3 increases with an increase in pressure but decreases as the temperature is increased. Thus the amount of NH_3 present at equilibrium is favored by conditions of low temperature and high pressure.

However, this is not the whole story. Carrying out the process at low temperatures is not feasible because then the reaction is too slow. Even though the equilibrium tends to shift to the right as the temperature is lowered, the attainment of equilibrium would be much too slow at low temperatures to be practical. This emphasizes once again that we must study both the thermodynamics and the kinetics of a reaction before we really understand the factors that control it.

We can qualitatively predict the effects of changes in concentration, pressure, and temperature on a system at equilibrium by using **Le Châtelier's principle**, which states that if a change is imposed on a system at equilibrium, the position of the equilibrium will shift in a direction that tends to reduce that change. Although this rule sometimes oversimplifies the situation, it works remarkably well.

The Effect of a Change in Concentration

To see how we can predict the effect of change in concentration on a system at equilibrium, we will consider the ammonia synthesis reaction. Suppose there is an equilibrium position described by these concentrations:

$$[N_2] = 0.399\ M \qquad [H_2] = 1.197\ M \qquad [NH_3] = 0.202\ M$$

Henri Louis Le Châtelier (1850–1936), the French physical chemist and metallurgist, seen here while a student at the École Polytechnique.

Science Photo Library/Photo Researchers, Inc.

Table 13.2 | The Percent by Mass of NH_3 at Equilibrium in a Mixture of N_2, H_2, and NH_3 as a Function of Temperature and Total Pressure*

| Temperature (°C) | Total Pressure | | |
	300 atm	400 atm	500 atm
400	48% NH_3	55% NH_3	61% NH_3
500	26% NH_3	32% NH_3	38% NH_3
600	13% NH_3	17% NH_3	21% NH_3

*Each experiment was begun with a 3:1 mixture of H_2 and N_2.

What will happen if 1.000 mol/L N_2 is suddenly injected into the system? We can answer this question by calculating the value of Q. The concentrations before the system adjusts are

$$[N_2]_0 = 0.399\ M + 1.000\ M = 1.399\ M$$

$$\uparrow$$
$$\text{Added } N_2$$

$$[H_2]_0 = 1.197\ M$$
$$[NH_3]_0 = 0.202\ M$$

Note we are labeling these as "initial concentrations" because the system is no longer at equilibrium. Then

$$Q = \frac{[NH_3]_0^2}{[N_2]_0[H_2]_0^3} = \frac{(0.202)^2}{(1.399)(1.197)^3} = 1.70 \times 10^{-2}$$

Since we are not given the value of K, we must calculate it from the first set of equilibrium concentrations:

$$K = \frac{[NH_3]^2}{[N_2][H_2]^3} = \frac{(0.202)^2}{(0.399)(1.197)^3} = 5.96 \times 10^{-2}$$

As expected, Q is less than K because the concentration of N_2 was increased.

The system will shift to the right to come to the new equilibrium position. Rather than do the calculations, we simply summarize the results:

Equilibrium Position I		Equilibrium Position II
$[N_2]$ = 0.399 M	$\xrightarrow[\text{of } N_2 \text{ added}]{1.000\ \text{mol/L}}$	$[N_2]$ = 1.348 M
$[H_2]$ = 1.197 M		$[H_2]$ = 1.044 M
$[NH_3]$ = 0.202 M		$[NH_3]$ = 0.304 M

Note from these data that the equilibrium position does in fact shift to the right: The concentration of H_2 decreases, the concentration of NH_3 increases, and, of course, since nitrogen is added, the concentration of N_2 shows an increase relative to the amount present in the original equilibrium position. (However, notice that the nitrogen showed a decrease relative to the amount present immediately after addition of the 1.000 mole of N_2.)

We can understand this shift by thinking about reaction rates. When we add N_2 molecules to the system, the number of collisions between N_2 and H_2 will increase, thus increasing the rate of the forward reaction and in turn increasing the rate of formation of NH_3 molecules. More NH_3 molecules will in turn lead to a higher rate for the reverse reaction. Eventually, the forward and reverse reaction rates will again become equal, and the system will reach its new equilibrium position.

We can predict this shift qualitatively by using Le Châtelier's principle. Since the change imposed is the addition of nitrogen, Le Châtelier's principle predicts that the system will shift in a direction that consumes nitrogen. This reduces the effect of the addition. Thus Le Châtelier's principle correctly predicts that adding nitrogen will cause the equilibrium to shift to the right (Fig. 13.9).

If ammonia had been added instead of nitrogen, the system would have shifted to the left to consume ammonia. So another way of stating Le Châtelier's principle is to say that if a component (reactant or product) is added to a reaction system at equilibrium (at constant T and P or constant T and V), the equilibrium position will shift in the direction that lowers the concentration of that component. If a component is removed, the opposite effect occurs.

The system shifts in the direction that compensates for the imposed change.

Figure 13.9 | (a) The initial equilibrium mixture of N_2, H_2, and NH_3. (b) Addition of N_2. (c) The new equilibrium position for the system containing more N_2 (due to addition of N_2), less H_2, and more NH_3 than the mixture in (a).

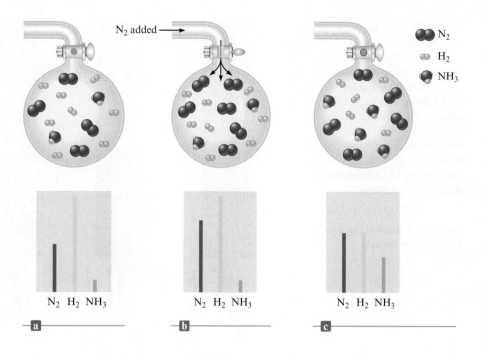

Using Le Châtelier's Principle I

Arsenic can be extracted from its ores by first reacting the ore with oxygen (called *roasting*) to form solid As_4O_6, which is then reduced using carbon:

$$As_4O_6(s) + 6C(s) \rightleftharpoons As_4(g) + 6CO(g)$$

Predict the direction of the shift of the equilibrium position in response to each of the following changes in conditions.

a. Addition of carbon monoxide

b. Addition or removal of carbon or tetraarsenic hexoxide (As_4O_6)

c. Removal of gaseous arsenic (As_4)

Solution

a. Le Châtelier's principle predicts that the shift will be away from the substance whose concentration is increased. The equilibrium position will shift to the left when carbon monoxide is added.

b. Since the amount of a pure solid has no effect on the equilibrium position, changing the amount of carbon or tetraarsenic hexoxide will have no effect.

c. If gaseous arsenic is removed, the equilibrium position will shift to the right to form more products. In industrial processes, the desired product is often continuously removed from the reaction system to increase the yield.

See Exercise 13.63

The Effect of a Change in Pressure

Basically, there are three ways to change the pressure of a reaction system involving gaseous components:

1. Add or remove a gaseous reactant or product.

2. Add an inert gas (one not involved in the reaction).

3. Change the volume of the container.

Figure 13.10 | (a) Brown $NO_2(g)$ and colorless $N_2O_4(g)$ in equilibrium in a syringe. (b) The volume is suddenly decreased, giving a greater concentration of both N_2O_4 and NO_2 (indicated by the darker brown color). (c) A few seconds after the sudden volume decrease, the color is much lighter brown as the equilibrium shifts the brown $NO_2(g)$ to colorless $N_2O_4(g)$ as predicted by Le Châtelier's principle, since in the equilibrium

$$2NO_2(g) \rightleftharpoons N_2O_4(g)$$

the product side has the smaller number of molecules.

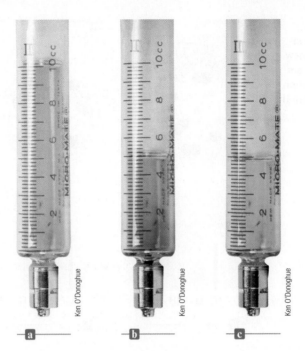

Ken O'Donoghue

We have already considered the addition or removal of a reactant or product. When an inert gas is added, there is no effect on the equilibrium position. The addition of an inert gas increases the total pressure but has no effect on the concentrations or partial pressures of the reactants or products. That is, in this case the added molecules do not participate in the reaction in any way and thus cannot affect the equilibrium in any way. Thus the system remains at the original equilibrium position.

When the volume of the container is changed, the concentrations (and thus the partial pressures) of both reactants and products are changed. We could calculate Q and predict the direction of the shift. However, for systems involving gaseous components, there is an easier way: We focus on the volume. The central idea is that when the volume of the container holding a gaseous system is reduced, the system responds by reducing its own volume. This is done by decreasing the total number of gaseous molecules in the system. This is illustrated by the NO_2/N_2O_4 system shown in Fig. 13.10.

To see that this is true, we can rearrange the ideal gas law to give

$$V = \left(\frac{RT}{P}\right)n$$

or at constant T and P,

$$V \propto n$$

That is, at constant temperature and pressure, the volume of a gas is directly proportional to the number of moles of gas present.

Suppose we have a mixture of the gases nitrogen, hydrogen, and ammonia at equilibrium (Fig. 13.11). If we suddenly reduce the volume, what will happen to the equilibrium position? The reaction system can reduce its volume by reducing the number of molecules present. This means that the reaction

$$N_2(g) + 3H_2(g) \rightleftharpoons 2NH_3(g)$$

will shift to the right, since in this direction four molecules (one of nitrogen and three of hydrogen) react to produce two molecules (of ammonia), thus *reducing the total number of gaseous molecules present*. The new equilibrium position will be farther to the right than the original one. That is, the equilibrium position will shift toward the

Figure 13.11 | (a) A mixture of $NH_3(g)$, $N_2(g)$, and $H_2(g)$ at equilibrium. (b) The volume is suddenly decreased. (c) The new equilibrium position for the system containing more NH_3 and less N_2 and H_2. The reaction $N_2(g) + 3H_2(g) \rightleftharpoons 2NH_3(g)$ shifts to the right (toward the side with fewer molecules) when the container volume is decreased.

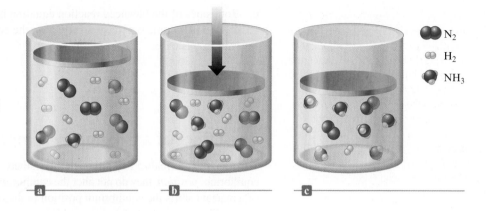

side of the reaction involving the smaller number of gaseous molecules in the balanced equation.

The opposite is also true. When the container volume is increased, the system will shift so as to increase its volume. An increase in volume in the ammonia synthesis system will produce a shift to the left to increase the total number of gaseous molecules present.

Critical Thinking

You and a friend are studying for a chemistry exam. What if your friend says, "Adding an inert gas to a system of gaseous components at equilibrium never changes the equilibrium position"? How do you explain to your friend that this holds true for a system at constant volume but is not necessarily true for a system at constant pressure? When would it hold true for a system at constant pressure?

Using Le Châtelier's Principle II

Predict the shift in equilibrium position that will occur for each of the following processes when the volume is reduced.

a. The preparation of liquid phosphorus trichloride by the reaction

$$P_4(s) + 6Cl_2(g) \rightleftharpoons 4PCl_3(l)$$

b. The preparation of gaseous phosphorus pentachloride according to the equation

$$PCl_3(g) + Cl_2(g) \rightleftharpoons PCl_5(g)$$

c. The reaction of phosphorus trichloride with ammonia:

$$PCl_3(g) + 3NH_3(g) \rightleftharpoons P(NH_2)_3(g) + 3HCl(g)$$

Solution

a. Since P_4 and PCl_3 are a pure solid and a pure liquid, respectively, we need to consider only the effect of the change in volume on Cl_2. The volume is decreased, so the position of the equilibrium will shift to the right, since the reactant side contains six gaseous molecules and the product side has none.

b. Decreasing the volume will shift the given reaction to the right, since the product side contains only one gaseous molecule while the reactant side has two.

c. Both sides of the balanced reaction equation have four gaseous molecules. A change in volume will have no effect on the equilibrium position. There is no shift in this case.

See Exercise 13.64

The Effect of a Change in Temperature

It is important to realize that although the changes we have just discussed may alter the equilibrium *position,* they do not alter the equilibrium *constant.* For example, the addition of a reactant shifts the equilibrium position to the right but has no effect on the value of the equilibrium constant; the new equilibrium concentrations satisfy the original equilibrium constant.

The effect of temperature on equilibrium is different, however, because *the value of K changes with temperature.* We can use Le Châtelier's principle to predict the direction of the change.

The synthesis of ammonia from nitrogen and hydrogen is exothermic. We can represent this by treating energy as a product:

$$N_2(g) + 3H_2(g) \rightleftharpoons 2NH_3(g) + 92 \text{ kJ}$$

Of course, energy is not a chemical product of this reaction, but thinking of it in this way makes it easy to apply Le Châtelier's principle.

If energy is added to this system at equilibrium by heating it, Le Châtelier's principle predicts that the shift will be in the direction that consumes energy, that is, to the left. Note that this shift decreases the concentration of NH_3 and increases the concentrations of N_2 and H_2, thus *decreasing the value of K.* The experimentally observed

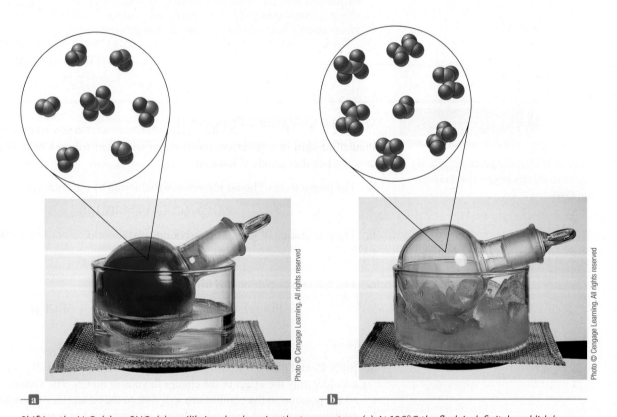

a ——————————————— **b** ———————————————

Shifting the $N_2O_4(g) \rightarrow 2NO_2(g)$ equilibrium by changing the temperature. (a) At 100°C the flask is definitely reddish brown due to a large amount of NO_2 present. (b) At 0°C the equilibrium is shifted toward colorless $N_2O_4(g)$.

Table 13.3 | Observed Value of *K* for the Ammonia Synthesis Reaction as a Function of Temperature*

Temperature (K)	*K*
500	90
600	3
700	0.3
800	0.04

*For this exothermic reaction, the value of *K* decreases as the temperature increases, as predicted by Le Châtelier's principle.

change in *K* with temperature for this reaction is indicated in Table 13.3. The value of *K* decreases with increased temperature, as predicted.

On the other hand, for an endothermic reaction, such as the decomposition of calcium carbonate,

$$556 \text{ kJ} + CaCO_3(s) \rightleftharpoons CaO(s) + CO_2(g)$$

an increase in temperature will cause the equilibrium to shift to the right and the value of *K* to increase.

In summary, to use Le Châtelier's principle to describe the effect of a temperature change on a system at equilibrium, treat energy as a reactant (in an endothermic process) or as a product (in an exothermic process), and predict the direction of the shift in the same way as when an actual reactant or product is added or removed. Although Le Châtelier's principle cannot predict the size of the change in *K*, it does correctly predict the direction of the change.

Interactive Example 13.15

Sign in at http://login.cengagebrain.com to try this Interactive Example in **OWL**.

Using Le Châtelier's Principle III

For each of the following reactions, predict how the value of *K* changes as the temperature is increased.

a. $N_2(g) + O_2(g) \rightleftharpoons 2NO(g)$ $\Delta H° = 181 \text{ kJ}$

b. $2SO_2(g) + O_2(g) \rightleftharpoons 2SO_3(g)$ $\Delta H° = -198 \text{ kJ}$

Solution

a. This is an endothermic reaction, as indicated by the positive value for $\Delta H°$. Energy can be viewed as a reactant, and *K* increases (the equilibrium shifts to the right) as the temperature is increased.

b. This is an exothermic reaction (energy can be regarded as a product). As the temperature is increased, the value of *K* decreases (the equilibrium shifts to the left).

See Exercises 13.69 and 13.70

We have seen how Le Châtelier's principle can be used to predict the effect of several types of changes on a system at equilibrium. To summarize these ideas, Table 13.4 shows how various changes affect the equilibrium position of the endothermic reaction

$$N_2O_4(g) \rightleftharpoons 2NO_2(g)   \Delta H° = 58 \text{ kJ}$$

Table 13.4 | Shifts in the Equilibrium Position for the Reaction $58 \text{ kJ} + N_2O_4(g) \rightleftharpoons 2NO_2(g)$

Change	Shift
Addition of $N_2O_4(g)$	Right
Addition of $NO_2(g)$	Left
Removal of $N_2O_4(g)$	Left
Removal of $NO_2(g)$	Right
Addition of He(*g*)	None
Decrease container volume	Left
Increase container volume	Right
Increase temperature	Right
Decrease temperature	Left

For review

Key terms

chemical equilibrium

Section 13.2
law of mass action
equilibrium expression
equilibrium constant
equilibrium position

Section 13.4
homogeneous equilibria
heterogeneous equilibria

Section 13.5
reaction quotient (Q)

Section 13.7
Le Châtelier's principle

Chemical equilibrium

> When a reaction takes place in a closed system, it reaches a condition where the concentrations of the reactants and products remain constant over time
> Dynamic state: reactants and products are interconverted continually
>> Forward rate = reverse rate
> The law of mass action: for the reaction

$$jA + kB \rightleftharpoons mC + nD$$

$$K = \frac{[C]^m[D]^n}{[A]^j[B]^k} = \text{equilibrium constant}$$

> A pure liquid or solid is never included in the equilibrium expression
> For a gas-phase reaction, the reactants and products can be described in terms of their partial pressures and the equilibrium constant is called K_p:

$$K_p = K(RT)^{\Delta n}$$

where Δn is the sum of the coefficients of the gaseous products minus the sum of the coefficients of the gaseous reactants

Equilibrium position

> A set of reactant and product concentrations that satisfies the equilibrium constant expression
>> There is one value of K for a given system at a given temperature
>> There are an infinite number of equilibrium positions at a given temperature depending on the initial concentrations
> A small value of K means the equilibrium lies to the left; a large value of K means the equilibrium lies to the right
>> The size of K has no relationship to the speed at which equilibrium is achieved
> Q, the reaction quotient, applies the law of mass action to initial concentrations rather than equilibrium concentrations
>> If $Q > K$, the system will shift to the left to achieve equilibrium
>> If $Q < K$, the system will shift to the right to achieve equilibrium
> Finding the concentrations that characterize a given equilibrium position:
>> Start with the given initial concentrations (pressures)
>> Define the change needed to reach equilibrium
>> Apply the change to the initial concentrations (pressures) and solve for the equilibrium concentrations (pressures)

Le Châtelier's principle

> Enables qualitative prediction of the effects of changes in concentration, pressure, and temperature on a system at equilibrium
> If a change in conditions is imposed on a system at equilibrium, the system will shift in a direction that compensates for the imposed change
>> In other words, when a stress is placed on a system at equilibrium, the system shifts in the direction that relieves the stress

Review questions
Answers to the Review Questions can be found on the Student website (accessible from **www.cengagebrain.com**).

1. Characterize a system at chemical equilibrium with respect to each of the following:

   a. the rates of the forward and reverse reactions

   b. the overall composition of the reaction mixture

   For a general reaction $3A(g) + B(g) \longrightarrow 2C(g)$, if one starts an experiment with only reactants present, show what the plot of concentrations of A, B, and C versus time would look like. Also sketch the plot illustrating the rate of the forward reaction and rate of the reverse reaction versus time.

2. What is the law of mass action? Is it true that the value of K depends on the amounts of reactants and products mixed together initially? Explain. Is it true that reactions with large equilibrium constant values are very fast? Explain. There is only one value of the equilibrium constant for a particular system at a particular temperature, but there is an infinite number of equilibrium positions. Explain.

3. Consider the following reactions at some temperature:

   $$2NOCl(g) \rightleftharpoons 2NO(g) + Cl_2(g) \qquad K = 1.6 \times 10^{-5}$$
   $$2NO(g) \rightleftharpoons N_2(g) + O_2(g) \qquad K = 1 \times 10^{31}$$

   For each reaction, assume some quantities of the reactants were placed in separate containers and allowed to come to equilibrium. Describe the relative amounts of reactants and products that would be present at equilibrium. At equilibrium, which is faster, the forward or reverse reaction in each case?

4. What is the difference between K and K_p? When does $K = K_p$ for a reaction? When does $K \neq K_p$ for a reaction? If the coefficients in a reaction equation are tripled, how is the new value of K related to the initial value of K? If an equation for a reaction is reversed, how is the value of K_p for the reversed equation related to the value of K_p for the initial equation?

5. What are homogeneous equilibria? Heterogeneous equilibria? What is the difference in writing K expressions for homogeneous versus heterogeneous reactions? Summarize which species are included in the K expression and which species are not included.

6. Distinguish between the terms *equilibrium constant* and *reaction quotient*. When $Q = K$, what does this say about a reaction? When $Q < K$, what does this say about a reaction? When $Q > K$, what does this say about a reaction?

7. Summarize the steps for solving equilibrium problems (see the beginning of Section 13.6). In general, when solving an equilibrium problem, you should always set up an ICE table. What is an ICE table?

8. A common type of reaction we will study is that having a very small K value ($K \ll 1$). Solving for equilibrium concentrations in an equilibrium problem usually requires many mathematical operations to be performed. However, the math involved when solving equilibrium problems for reactions having small K values ($K \ll 1$) is simplified. What assumption is made when solving the equilibrium concentrations for reactions with small K values? Whenever assumptions are made, they must be checked for validity. In general, the "5% rule" is used to check the validity of assuming x (or 2x, 3x, and so on) is very small compared to some number. When x (or 2x, 3x, and so on) is less than 5% of the number the assumption was made against, then the assumption is said to be valid. If the 5% rule fails, what do you do to solve for the equilibrium concentrations?

9. What is Le Châtelier's principle? Consider the reaction $2NOCl(g) \rightleftharpoons 2NO(g) + Cl_2(g)$. If this reaction is at equilibrium, what happens when the following changes occur?

   a. $NOCl(g)$ is added.

   b. $NO(g)$ is added.

   c. $NOCl(g)$ is removed.

   d. $Cl_2(g)$ is removed.

   e. The container volume is decreased.

   For each of these changes, what happens to the value of K for the reaction as equilibrium is reached again? Give an example of a reaction for which the addition or removal of one of the reactants or products has no effect on the equilibrium position.

   In general, how will the equilibrium position of a gas-phase reaction be affected if the volume of the reaction vessel changes? Are there reactions that will not have their equilibria shifted by a change in volume? Explain. Why does changing the pressure in a rigid container by adding an inert gas not shift the equilibrium position for a gas-phase reaction?

10. The only "stress" (change) that also changes the value of K is a change in temperature. For an exothermic reaction, how does the equilibrium position change as temperature increases, and what happens to the value of K? Answer the same questions for an endothermic reaction. If the value of K increases with a decrease in temperature, is the reaction exothermic or endothermic? Explain.

Active Learning Questions

These questions are designed to be used by groups of students in class.

1. Consider an equilibrium mixture of four chemicals (A, B, C, and D, all gases) reacting in a closed flask according to the equation:

$$A(g) + B(g) \rightleftharpoons C(g) + D(g)$$

a. You add more A to the flask. How does the concentration of each chemical compare to its original concentration after equilibrium is reestablished? Justify your answer.

b. You have the original setup at equilibrium, and you add more D to the flask. How does the concentration of each chemical compare to its original concentration after equilibrium is reestablished? Justify your answer.

2. The boxes shown below represent a set of initial conditions for the reaction:

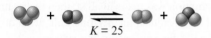

$$K = 25$$

Draw a quantitative molecular picture that shows what this system looks like after the reactants are mixed in one of the boxes and the system reaches equilibrium. Support your answer with calculations.

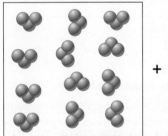 +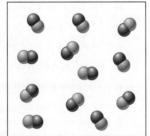

3. For the reaction $H_2(g) + I_2(g) \rightleftharpoons 2HI(g)$, consider two possibilities: (a) you mix 0.5 mole of each reactant, allow the system to come to equilibrium, and then add another mole of H_2 and allow the system to reach equilibrium again, or (b) you mix 1.5 moles of H_2 and 0.5 mole of I_2 and allow the system to reach equilibrium. Will the final equilibrium mixture be different for the two procedures? Explain.

4. Given the reaction $A(g) + B(g) \rightleftharpoons C(g) + D(g)$, consider the following situations:

   i. You have 1.3 M A and 0.8 M B initially.
   ii. You have 1.3 M A, 0.8 M B, and 0.2 M C initially.
   iii. You have 2.0 M A and 0.8 M B initially.

   Order the preceding situations in terms of increasing equilibrium concentration of D. Explain your order. Then give the order in terms of increasing equilibrium concentration of B and explain.

5. Consider the reaction $A(g) + 2B(g) \rightleftharpoons C(g) + D(g)$ in a 1.0-L rigid flask. Answer the following questions for each situation (a–d):

   i. Estimate a range (as small as possible) for the requested substance. For example, [A] could be between 95 M and 100 M.

   ii. Explain how you decided on the limits for the estimated range.

   iii. Indicate what other information would enable you to narrow your estimated range.

   iv. Compare the estimated concentrations for a through d, and explain any differences.

   a. If at equilibrium [A] = 1 M, and then 1 mole of C is added, estimate the value for [A] once equilibrium is reestablished.

   b. If at equilibrium [B] = 1 M, and then 1 mole of C is added, estimate the value for [B] once equilibrium is reestablished.

   c. If at equilibrium [C] = 1 M, and then 1 mole of C is added, estimate the value for [C] once equilibrium is reestablished.

   d. If at equilibrium [D] = 1 M, and then 1 mole of C is added, estimate the value for [D] once equilibrium is reestablished.

6. Consider the reaction $A(g) + B(g) \rightleftharpoons C(g) + D(g)$. A friend asks the following: "I know we have been told that if a mixture of A, B, C, and D is at equilibrium and more of A is added, more C and D will form. But how can more C and D form if we do not add more B?" What do you tell your friend?

7. Consider the following statements: "Consider the reaction $A(g) + B(g) \rightleftharpoons C(g)$, for which at equilibrium [A] = 2 M, [B] = 1 M, and [C] = 4 M. To a 1-L container of the system at equilibrium, you add 3 moles of B. A possible equilibrium condition is [A] = 1 M, [B] = 3 M, and [C] = 6 M because in both cases $K = 2$." Indicate everything that is correct in these statements and everything that is incorrect. Correct the incorrect statements, and explain.

8. Le Châtelier's principle is stated (Section 13.7) as follows: "If a change is imposed on a system at equilibrium, the position of the equilibrium will shift in a direction that tends to reduce that change." The system $N_2(g) + 3H_2(g) \rightleftharpoons 2NH_3(g)$ is used as an example in which the addition of nitrogen gas at equilibrium results in a decrease in H_2 concentration and an increase in NH_3 concentration. In the experiment the volume is assumed to be constant. On the other hand, if N_2 is added to the reaction system in a container with a piston so that the pressure can be held constant, the amount of NH_3 actually could decrease and the concentration of H_2 would increase as equilibrium is reestablished. Explain how this can happen. Also, if you consider this same system at equilibrium, the addition of an inert gas, holding the pressure constant, *does* affect the equilibrium position. Explain why the addition of an inert gas to this system in a rigid container does not affect the equilibrium position.

9. The value of the equilibrium constant K depends on which of the following (more than one answer may be correct)?

   a. the initial concentrations of the reactants

   b. the initial concentrations of the products

   c. the temperature of the system

   d. the nature of the reactants and products

   Explain.

A blue question or exercise number indicates that the answer to that question or exercise appears at the back of this book and a solution appears in the *Solutions Guide,* as found on PowerLecture.

Questions

10. Consider an initial mixture of N_2 and H_2 gases that can be represented as follows:

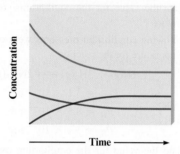

The gases react to form ammonia gas (NH_3) as represented by the following concentration profile:

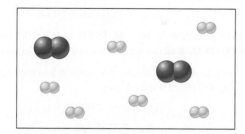

a. Label each plot on the graph as N_2, H_2, or NH_3, and explain your answers.

b. Explain the relative shapes of the plots.

c. When is equilibrium reached? How do you know?

11. Consider the following reaction:

$$H_2O(g) + CO(g) \rightleftharpoons H_2(g) + CO_2(g)$$

Amounts of H_2O, CO, H_2, and CO_2 are put into a flask so that the composition corresponds to an equilibrium position. If the CO placed in the flask is labeled with radioactive ^{14}C, will ^{14}C be found only in CO molecules for an indefinite period of time? Explain.

12. Consider the same reaction as in Question 11. In one experiment 1.0 mole of $H_2O(g)$ and 1.0 mole of $CO(g)$ are put into a flask and heated to 350°C. In a second experiment 1.0 mole of $H_2(g)$ and 1.0 mole of $CO_2(g)$ are put into another flask with the same volume as the first. This mixture is also heated to 350°C. After equilibrium is reached, will there be any difference in the composition of the mixtures in the two flasks?

13. Suppose a reaction has the equilibrium constant $K = 1.3 \times 10^8$. What does the magnitude of this constant tell you about the relative concentrations of products and reactants that will be present once equilibrium is reached? Is this reaction likely to be a good source of the products?

14. Suppose a reaction has the equilibrium constant $K = 1.7 \times 10^{-8}$ at a particular temperature. Will there be a large or small amount of unreacted starting material present when this reaction reaches equilibrium? Is this reaction likely to be a good source of products at this temperature?

15. Consider the following reaction at some temperature:

$$H_2O(g) + CO(g) \rightleftharpoons H_2(g) + CO_2(g) \qquad K = 2.0$$

Some molecules of H_2O and CO are placed in a 1.0-L container as shown below.

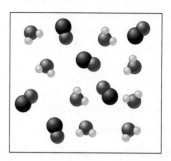

When equilibrium is reached, how many molecules of H_2O, CO, H_2, and CO_2 are present? Do this problem by trial and error—that is, if two molecules of CO react, is this equilibrium; if three molecules of CO react, is this equilibrium; and so on.

16. Consider the following generic reaction:

$$2A_2B(g) \rightleftharpoons 2A_2(g) + B_2(g)$$

Some molecules of A_2B are placed in a 1.0-L container. As time passes, several snapshots of the reaction mixture are taken as illustrated below.

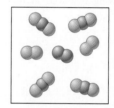

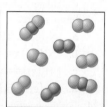

Which illustration is the first to represent an equilibrium mixture? Explain. How many molecules of A_2B reacted initially?

17. Explain the difference between K, K_p, and Q.

18. Consider the following reactions:

$$H_2(g) + I_2(g) \longrightarrow 2HI(g) \quad \text{and} \quad H_2(g) + I_2(s) \longrightarrow 2HI(g)$$

List two property differences between these two reactions that relate to equilibrium.

19. For a typical equilibrium problem, the value of K and the initial reaction conditions are given for a specific reaction, and you are asked to calculate the equilibrium concentrations. Many of these calculations involve solving a quadratic or cubic equation. What can you do to avoid solving a quadratic or cubic equation and still come up with reasonable equilibrium concentrations?

20. Which of the following statements is(are) *true*? Correct the false statement(s).

a. When a reactant is added to a system at equilibrium at a given temperature, the reaction will shift right to reestablish equilibrium.

b. When a product is added to a system at equilibrium at a given temperature, the value of K for the reaction will increase when equilibrium is reestablished.

c. When temperature is increased for a reaction at equilibrium, the value of K for the reaction will increase.

d. When the volume of a reaction container is increased for a system at equilibrium at a given temperature, the reaction will shift left to reestablish equilibrium.

e. Addition of a catalyst (a substance that increases the speed of the reaction) has no effect on the equilibrium position.

Exercises

In this section similar exercises are paired.

The Equilibrium Constant

21. Write the equilibrium expression (K) for each of the following gas-phase reactions.

a. $N_2(g) + O_2(g) \rightleftharpoons 2NO(g)$

b. $N_2O_4(g) \rightleftharpoons 2NO_2(g)$

c. $SiH_4(g) + 2Cl_2(g) \rightleftharpoons SiCl_4(g) + 2H_2(g)$

d. $2PBr_3(g) + 3Cl_2(g) \rightleftharpoons 2PCl_3(g) + 3Br_2(g)$

22. Write the equilibrium expression (K_p) for each reaction in Exercise 21.

23. At a given temperature, $K = 1.3 \times 10^{-2}$ for the reaction

$$N_2(g) + 3H_2(g) \rightleftharpoons 2NH_3(g)$$

Calculate values of K for the following reactions at this temperature.

a. $\frac{1}{2}N_2(g) + \frac{3}{2}H_2(g) \rightleftharpoons NH_3(g)$

b. $2NH_3(g) \rightleftharpoons N_2(g) + 3H_2(g)$

c. $NH_3(g) \rightleftharpoons \frac{1}{2}N_2(g) + \frac{3}{2}H_2(g)$

d. $2N_2(g) + 6H_2(g) \rightleftharpoons 4NH_3(g)$

24. For the reaction

$$H_2(g) + Br_2(g) \rightleftharpoons 2HBr(g)$$

$K_p = 3.5 \times 10^4$ at 1495 K. What is the value of K_p for the following reactions at 1495 K?

a. $HBr(g) \rightleftharpoons \frac{1}{2}H_2(g) + \frac{1}{2}Br_2(g)$

b. $2HBr(g) \rightleftharpoons H_2(g) + Br_2(g)$

c. $\frac{1}{2}H_2(g) + \frac{1}{2}Br_2(g) \rightleftharpoons HBr(g)$

25. For the reaction

$$2NO(g) + 2H_2(g) \rightleftharpoons N_2(g) + 2H_2O(g)$$

it is determined that, at equilibrium at a particular temperature, the concentrations are as follows: $[NO(g)] = 8.1 \times 10^{-3}$ M, $[H_2(g)] = 4.1 \times 10^{-5}$ M, $[N_2(g)] = 5.3 \times 10^{-2}$ M, and $[H_2O(g)] = 2.9 \times 10^{-3}$ M. Calculate the value of K for the reaction at this temperature.

26. At high temperatures, elemental nitrogen and oxygen react with each other to form nitrogen monoxide:

$$N_2(g) + O_2(g) \rightleftharpoons 2NO(g)$$

Suppose the system is analyzed at a particular temperature, and the equilibrium concentrations are found to be $[N_2] = 0.041$ M, $[O_2] = 0.0078$ M, and $[NO] = 4.7 \times 10^{-4}$ M. Calculate the value of K for the reaction.

27. At a particular temperature, a 3.0-L flask contains 2.4 moles of Cl_2, 1.0 mole of NOCl, and 4.5×10^{-3} mole of NO. Calculate K at this temperature for the following reaction:

$$2NOCl(g) \rightleftharpoons 2NO(g) + Cl_2(g)$$

28. At a particular temperature a 2.00-L flask at equilibrium contains 2.80×10^{-4} mole of N_2, 2.50×10^{-5} mole of O_2, and 2.00×10^{-2} mole of N_2O. Calculate K at this temperature for the reaction

$$2N_2(g) + O_2(g) \rightleftharpoons 2N_2O(g)$$

If $[N_2] = 2.00 \times 10^{-4}$ M, $[N_2O] = 0.200$ M, and $[O_2] = 0.00245$ M, does this represent a system at equilibrium?

29. The following equilibrium pressures at a certain temperature were observed for the reaction

$$2NO_2(g) \rightleftharpoons 2NO(g) + O_2(g)$$

$$P_{NO_2} = 0.55 \text{ atm}$$

$$P_{NO} = 6.5 \times 10^{-5} \text{ atm}$$

$$P_{O_2} = 4.5 \times 10^{-5} \text{ atm}$$

Calculate the value for the equilibrium constant K_p at this temperature.

30. The following equilibrium pressures were observed at a certain temperature for the reaction

$$N_2(g) + 3H_2(g) \rightleftharpoons 2NH_3(g)$$

$$P_{NH_3} = 3.1 \times 10^{-2} \text{ atm}$$

$$P_{N_2} = 8.5 \times 10^{-1} \text{ atm}$$

$$P_{H_2} = 3.1 \times 10^{-3} \text{ atm}$$

Calculate the value for the equilibrium constant K_p at this temperature.

If $P_{N_2} = 0.525$ atm, $P_{NH_3} = 0.0167$ atm, and $P_{H_2} = 0.00761$ atm, does this represent a system at equilibrium?

31. At 327°C, the equilibrium concentrations are $[CH_3OH] = 0.15$ M, $[CO] = 0.24$ M, and $[H_2] = 1.1$ M for the reaction

$$CH_3OH(g) \rightleftharpoons CO(g) + 2H_2(g)$$

Calculate K_p at this temperature.

32. At 1100 K, $K_p = 0.25$ for the reaction

$$2SO_2(g) + O_2(g) \rightleftharpoons 2SO_3(g)$$

What is the value of K at this temperature?

33. Write expressions for K and K_p for the following reactions.

a. $2NH_3(g) + CO_2(g) \rightleftharpoons N_2CH_4O(s) + H_2O(g)$

b. $2NBr_3(s) \rightleftharpoons N_2(g) + 3Br_2(g)$

c. $2KClO_3(s) \rightleftharpoons 2KCl(s) + 3O_2(g)$

d. $CuO(s) + H_2(g) \rightleftharpoons Cu(l) + H_2O(g)$

34. Write expressions for K_p for the following reactions.

a. $2Fe(s) + \frac{3}{2}O_2(g) \rightleftharpoons Fe_2O_3(s)$

b. $CO_2(g) + MgO(s) \rightleftharpoons MgCO_3(s)$

c. $C(s) + H_2O(g) \rightleftharpoons CO(g) + H_2(g)$

d. $4KO_2(s) + 2H_2O(g) \rightleftharpoons 4KOH(s) + 3O_2(g)$

35. For which reactions in Exercise 33 is K_p equal to K?

36. For which reactions in Exercise 34 is K_p equal to K?

37. Consider the following reaction at a certain temperature:

$$4Fe(s) + 3O_2(g) \rightleftharpoons 2Fe_2O_3(s)$$

An equilibrium mixture contains 1.0 mole of Fe, 1.0×10^{-3} mole of O_2, and 2.0 moles of Fe_2O_3 all in a 2.0-L container. Calculate the value of K for this reaction.

38. In a study of the reaction

$$3Fe(s) + 4H_2O(g) \rightleftharpoons Fe_3O_4(s) + 4H_2(g)$$

at 1200 K it was observed that when the equilibrium partial pressure of water vapor is 15.0 torr, the total pressure at equilibrium is 36.3 torr. Calculate the value of K_p for this reaction at 1200 K. (*Hint:* Apply Dalton's law of partial pressures.)

Equilibrium Calculations

39. The equilibrium constant is 0.0900 at 25°C for the reaction

$$H_2O(g) + Cl_2O(g) \rightleftharpoons 2HOCl(g)$$

For which of the following sets of conditions is the system at equilibrium? For those that are not at equilibrium, in which direction will the system shift?

a. A 1.0-L flask contains 1.0 mole of HOCl, 0.10 mole of Cl_2O, and 0.10 mole of H_2O.

b. A 2.0-L flask contains 0.084 mole of HOCl, 0.080 mole of Cl_2O, and 0.98 mole of H_2O.

c. A 3.0-L flask contains 0.25 mole of HOCl, 0.0010 mole of Cl_2O, and 0.56 mole of H_2O.

40. The equilibrium constant is 0.0900 at 25°C for the reaction

$$H_2O(g) + Cl_2O(g) \rightleftharpoons 2HOCl(g)$$

For which of the following sets of conditions is the system at equilibrium? For those that are not at equilibrium, in which direction will the system shift?

a. $P_{H_2O} = 1.00$ atm, $P_{Cl_2O} = 1.00$ atm, $P_{HOCl} = 1.00$ atm

b. $P_{H_2O} = 200.$ torr, $P_{Cl_2O} = 49.8$ torr, $P_{HOCl} = 21.0$ torr

c. $P_{H_2O} = 296$ torr, $P_{Cl_2O} = 15.0$ torr, $P_{HOCl} = 20.0$ torr

41. At 900°C, $K_p = 1.04$ for the reaction

$$CaCO_3(s) \rightleftharpoons CaO(s) + CO_2(g)$$

At a low temperature, dry ice (solid CO_2), calcium oxide, and calcium carbonate are introduced into a 50.0-L reaction chamber. The temperature is raised to 900°C, resulting in the dry ice converting to gaseous CO_2. For the following mixtures, will the initial amount of calcium oxide increase, decrease, or remain the same as the system moves toward equilibrium at 900°C?

a. 655 g $CaCO_3$, 95.0 g CaO, $P_{CO_2} = 2.55$ atm

b. 780 g $CaCO_3$, 1.00 g CaO, $P_{CO_2} = 1.04$ atm

c. 0.14 g $CaCO_3$, 5000 g CaO, $P_{CO_2} = 1.04$ atm

d. 715 g $CaCO_3$, 813 g CaO, $P_{CO_2} = 0.211$ atm

42. Ethyl acetate is synthesized in a nonreacting solvent (not water) according to the following reaction:

$$CH_3CO_2H + C_2H_5OH \rightleftharpoons CH_3CO_2C_2H_5 + H_2O \quad K = 2.2$$
Acetic acid   Ethanol   Ethyl acetate

For the following mixtures (a–d), will the concentration of H_2O increase, decrease, or remain the same as equilibrium is established?

a. $[CH_3CO_2C_2H_5] = 0.22$ M, $[H_2O] = 0.10$ M, $[CH_3CO_2H] = 0.010$ M, $[C_2H_5OH] = 0.010$ M

b. $[CH_3CO_2C_2H_5] = 0.22$ M, $[H_2O] = 0.0020$ M, $[CH_3CO_2H] = 0.0020$ M, $[C_2H_5OH] = 0.10$ M

c. $[CH_3CO_2C_2H_5] = 0.88$ M, $[H_2O] = 0.12$ M, $[CH_3CO_2H] = 0.044$ M, $[C_2H_5OH] = 6.0$ M

d. $[CH_3CO_2C_2H_5] = 4.4$ M, $[H_2O] = 4.4$ M, $[CH_3CO_2H] = 0.88$ M, $[C_2H_5OH] = 10.0$ M

e. What must the concentration of water be for a mixture with $[CH_3CO_2C_2H_5] = 2.0$ M, $[CH_3CO_2H] = 0.10$ M, and $[C_2H_5OH] = 5.0$ M to be at equilibrium?

f. Why is water included in the equilibrium expression for this reaction?

43. For the reaction

$$2H_2O(g) \rightleftharpoons 2H_2(g) + O_2(g)$$

$K = 2.4 \times 10^{-3}$ at a given temperature. At equilibrium in a 2.0-L container it is found that $[H_2O(g)] = 1.1 \times 10^{-1}$ M and $[H_2(g)] = 1.9 \times 10^{-2}$ M. Calculate the moles of $O_2(g)$ present under these conditions.

44. The reaction

$$2NO(g) + Br_2(g) \rightleftharpoons 2NOBr(g)$$

has $K_p = 109$ at 25°C. If the equilibrium partial pressure of Br_2 is 0.0159 atm and the equilibrium partial pressure of NOBr is 0.0768 atm, calculate the partial pressure of NO at equilibrium.

45. A 1.00-L flask was filled with 2.00 moles of gaseous SO_2 and 2.00 moles of gaseous NO_2 and heated. After equilibrium was reached, it was found that 1.30 moles of gaseous NO was present. Assume that the reaction

$$SO_2(g) + NO_2(g) \rightleftharpoons SO_3(g) + NO(g)$$

occurs under these conditions. Calculate the value of the equilibrium constant, K, for this reaction.

46. A sample of $S_8(g)$ is placed in an otherwise empty rigid container at 1325 K at an initial pressure of 1.00 atm, where it decomposes to $S_2(g)$ by the reaction

$$S_8(g) \rightleftharpoons 4S_2(g)$$

At equilibrium, the partial pressure of S_8 is 0.25 atm. Calculate K_p for this reaction at 1325 K.

47. At a particular temperature, 12.0 moles of SO_3 is placed into a 3.0-L rigid container, and the SO_3 dissociates by the reaction

$$2SO_3(g) \rightleftharpoons 2SO_2(g) + O_2(g)$$

At equilibrium, 3.0 moles of SO_2 is present. Calculate K for this reaction.

48. At a particular temperature, 8.0 moles of NO_2 is placed into a 1.0-L container and the NO_2 dissociates by the reaction

$$2NO_2(g) \rightleftharpoons 2NO(g) + O_2(g)$$

At equilibrium the concentration of NO(g) is 2.0 M. Calculate K for this reaction.

49. An initial mixture of nitrogen gas and hydrogen gas is reacted in a rigid container at a certain temperature by the reaction

$$3H_2(g) + N_2(g) \rightleftharpoons 2NH_3(g)$$

At equilibrium, the concentrations are $[H_2] = 5.0\ M$, $[N_2] = 8.0\ M$, and $[NH_3] = 4.0\ M$. What were the concentrations of nitrogen gas and hydrogen gas that were reacted initially?

50. Nitrogen gas (N_2) reacts with hydrogen gas (H_2) to form ammonia (NH_3). At 200°C in a closed container, 1.00 atm of nitrogen gas is mixed with 2.00 atm of hydrogen gas. At equilibrium, the total pressure is 2.00 atm. Calculate the partial pressure of hydrogen gas at equilibrium, and calculate the K_p value for this reaction.

51. At a particular temperature, $K = 3.75$ for the reaction

$$SO_2(g) + NO_2(g) \rightleftharpoons SO_3(g) + NO(g)$$

If all four gases had initial concentrations of 0.800 M, calculate the equilibrium concentrations of the gases.

52. At a particular temperature, $K = 1.00 \times 10^2$ for the reaction

$$H_2(g) + I_2(g) \rightleftharpoons 2HI(g)$$

In an experiment, 1.00 mole of H_2, 1.00 mole of I_2, and 1.00 mole of HI are introduced into a 1.00-L container. Calculate the concentrations of all species when equilibrium is reached.

53. At 2200°C, $K_p = 0.050$ for the reaction

$$N_2(g) + O_2(g) \rightleftharpoons 2NO(g)$$

What is the partial pressure of NO in equilibrium with N_2 and O_2 that were placed in a flask at initial pressures of 0.80 and 0.20 atm, respectively?

54. At 25°C, $K = 0.090$ for the reaction

$$H_2O(g) + Cl_2O(g) \rightleftharpoons 2HOCl(g)$$

Calculate the concentrations of all species at equilibrium for each of the following cases.

   a. 1.0 g H_2O and 2.0 g Cl_2O are mixed in a 1.0-L flask.

   b. 1.0 mole of pure HOCl is placed in a 2.0-L flask.

55. At 1100 K, $K_p = 0.25$ for the reaction

$$2SO_2(g) + O_2(g) \rightleftharpoons 2SO_3(g)$$

Calculate the equilibrium partial pressures of SO_2, O_2, and SO_3 produced from an initial mixture in which $P_{SO_2} = P_{O_2} = 0.50$ atm and $P_{SO_3} = 0$. (*Hint:* If you don't have a graphing calculator, then use the method of successive approximations to solve, as discussed in Appendix 1.4.)

56. At a particular temperature, $K_p = 0.25$ for the reaction

$$N_2O_4(g) \rightleftharpoons 2NO_2(g)$$

   a. A flask containing only N_2O_4 at an initial pressure of 4.5 atm is allowed to reach equilibrium. Calculate the equilibrium partial pressures of the gases.

   b. A flask containing only NO_2 at an initial pressure of 9.0 atm is allowed to reach equilibrium. Calculate the equilibrium partial pressures of the gases.

   c. From your answers to parts a and b, does it matter from which direction an equilibrium position is reached?

57. At 35°C, $K = 1.6 \times 10^{-5}$ for the reaction

$$2NOCl(g) \rightleftharpoons 2NO(g) + Cl_2(g)$$

Calculate the concentrations of all species at equilibrium for each of the following original mixtures.

   a. 2.0 moles of pure NOCl in a 2.0-L flask

   b. 1.0 mole of NOCl and 1.0 mole of NO in a 1.0-L flask

   c. 2.0 moles of NOCl and 1.0 mole of Cl_2 in a 1.0-L flask

58. At a particular temperature, $K = 4.0 \times 10^{-7}$ for the reaction

$$N_2O_4(g) \rightleftharpoons 2NO_2(g)$$

In an experiment, 1.0 mole of N_2O_4 is placed in a 10.0-L vessel. Calculate the concentrations of N_2O_4 and NO_2 when this reaction reaches equilibrium.

59. At a particular temperature, $K = 2.0 \times 10^{-6}$ for the reaction

$$2CO_2(g) \rightleftharpoons 2CO(g) + O_2(g)$$

If 2.0 moles of CO_2 is initially placed into a 5.0-L vessel, calculate the equilibrium concentrations of all species.

60. Lexan is a plastic used to make compact discs, eyeglass lenses, and bullet-proof glass. One of the compounds used to make Lexan is phosgene ($COCl_2$), an extremely poisonous gas. Phosgene decomposes by the reaction

$$COCl_2(g) \rightleftharpoons CO(g) + Cl_2(g)$$

for which $K_p = 6.8 \times 10^{-9}$ at 100°C. If pure phosgene at an initial pressure of 1.0 atm decomposes, calculate the equilibrium pressures of all species.

61. At 25°C, $K_p = 2.9 \times 10^{-3}$ for the reaction

$$NH_4OCONH_2(s) \rightleftharpoons 2NH_3(g) + CO_2(g)$$

In an experiment carried out at 25°C, a certain amount of NH_4OCONH_2 is placed in an evacuated rigid container and allowed to come to equilibrium. Calculate the total pressure in the container at equilibrium.

62. A sample of solid ammonium chloride was placed in an evacuated container and then heated so that it decomposed to ammonia gas and hydrogen chloride gas. After heating, the total pressure in the container was found to be 4.4 atm. Calculate K_p at this temperature for the decomposition reaction

$$NH_4Cl(s) \rightleftharpoons NH_3(g) + HCl(g)$$

Le Châtelier's Principle

63. Suppose the reaction system

$$UO_2(s) + 4HF(g) \rightleftharpoons UF_4(g) + 2H_2O(g)$$

has already reached equilibrium. Predict the effect that each of the following changes will have on the equilibrium position. Tell whether the equilibrium will shift to the right, will shift to the left, or will not be affected.

   a. Additional $UO_2(s)$ is added to the system.

   b. The reaction is performed in a glass reaction vessel; HF(g) attacks and reacts with glass.

   c. Water vapor is removed.

64. Predict the shift in the equilibrium position that will occur for each of the following reactions when the volume of the reaction container is increased.

   a. $N_2(g) + 3H_2(g) \rightleftharpoons 2NH_3(g)$

   b. $PCl_5(g) \rightleftharpoons PCl_3(g) + Cl_2(g)$

   c. $H_2(g) + F_2(g) \rightleftharpoons 2HF(g)$

   d. $COCl_2(g) \rightleftharpoons CO(g) + Cl_2(g)$

   e. $CaCO_3(s) \rightleftharpoons CaO(s) + CO_2(g)$

65. An important reaction in the commercial production of hydrogen is

$$CO(g) + H_2O(g) \rightleftharpoons H_2(g) + CO_2(g)$$

How will this system at equilibrium shift in each of the five following cases?

a. Gaseous carbon dioxide is removed.

b. Water vapor is added.

c. In a rigid reaction container, the pressure is increased by adding helium gas.

d. The temperature is increased (the reaction is exothermic).

e. The pressure is increased by decreasing the volume of the reaction container.

66. What will happen to the number of moles of SO_3 in equilibrium with SO_2 and O_2 in the reaction

$$2SO_3(g) \rightleftharpoons 2SO_2(g) + O_2(g)$$

in each of the following cases?

a. Oxygen gas is added.

b. The pressure is increased by decreasing the volume of the reaction container.

c. In a rigid reaction container, the pressure is increased by adding argon gas.

d. The temperature is decreased (the reaction is endothermic).

e. Gaseous sulfur dioxide is removed.

67. In which direction will the position of the equilibrium

$$2HI(g) \rightleftharpoons H_2(g) + I_2(g)$$

be shifted for each of the following changes?

a. $H_2(g)$ is added.

b. $I_2(g)$ is removed.

c. $HI(g)$ is removed.

d. In a rigid reaction container, some $Ar(g)$ is added.

e. The volume of the container is doubled.

f. The temperature is decreased (the reaction is exothermic).

68. Hydrogen for use in ammonia production is produced by the reaction

$$CH_4(g) + H_2O(g) \xrightarrow[750°C]{Ni\ catalyst} CO(g) + 3H_2(g)$$

What will happen to a reaction mixture at equilibrium if

a. $H_2O(g)$ is removed?

b. the temperature is increased (the reaction is endothermic)?

c. an inert gas is added to a rigid reaction container?

d. $CO(g)$ is removed?

e. the volume of the container is tripled?

69. Old-fashioned "smelling salts" consist of ammonium carbonate, $(NH_4)_2CO_3$. The reaction for the decomposition of ammonium carbonate

$$(NH_4)_2CO_3(s) \rightleftharpoons 2NH_3(g) + CO_2(g) + H_2O(g)$$

is endothermic. Would the smell of ammonia increase or decrease as the temperature is increased?

70. Ammonia is produced by the Haber process, in which nitrogen and hydrogen are reacted directly using an iron mesh impregnated with oxides as a catalyst. For the reaction

$$N_2(g) + 3H_2(g) \rightleftharpoons 2NH_3(g)$$

equilibrium constants (K_p values) as a function of temperature are

300°C, 4.34×10^{-3}
500°C, 1.45×10^{-5}
600°C, 2.25×10^{-6}

Is the reaction exothermic or endothermic?

Additional Exercises

71. Calculate a value for the equilibrium constant for the reaction

$$O_2(g) + O(g) \rightleftharpoons O_3(g)$$

given

$$NO_2(g) \xrightarrow{hv} NO(g) + O(g) \qquad K = 6.8 \times 10^{-49}$$

$$O_3(g) + NO(g) \rightleftharpoons NO_2(g) + O_2(g) \qquad K = 5.8 \times 10^{-34}$$

(*Hint:* When reactions are added together, the equilibrium expressions are multiplied.)

72. Given the following equilibrium constants at 427°C,

$$Na_2O(s) \rightleftharpoons 2Na(l) + \tfrac{1}{2}O_2(g) \qquad K_1 = 2 \times 10^{-25}$$

$$NaO(g) \rightleftharpoons Na(l) + \tfrac{1}{2}O_2(g) \qquad K_2 = 2 \times 10^{-5}$$

$$Na_2O_2(s) \rightleftharpoons 2Na(l) + O_2(g) \qquad K_3 = 5 \times 10^{-29}$$

$$NaO_2(s) \rightleftharpoons Na(l) + O_2(g) \qquad K_4 = 3 \times 10^{-14}$$

determine the values for the equilibrium constants for the following reactions:

a. $Na_2O(s) + \tfrac{1}{2}O_2(g) \rightleftharpoons Na_2O_2(s)$

b. $NaO(g) + Na_2O(s) \rightleftharpoons Na_2O_2(s) + Na(l)$

c. $2NaO(g) \rightleftharpoons Na_2O_2(s)$

(*Hint:* When reaction equations are added, the equilibrium expressions are multiplied.)

73. Consider the decomposition of the compound $C_5H_6O_3$ as follows:

$$C_5H_6O_3(g) \longrightarrow C_2H_6(g) + 3CO(g)$$

When a 5.63-g sample of pure $C_5H_6O_3(g)$ was sealed into an otherwise empty 2.50-L flask and heated to 200.°C, the pressure in the flask gradually rose to 1.63 atm and remained at that value. Calculate K for this reaction.

74. At 25°C, $K_p \approx 1 \times 10^{-31}$ for the reaction

$$N_2(g) + O_2(g) \rightleftharpoons 2NO(g)$$

a. Calculate the concentration of NO, in molecules/cm³, that can exist in equilibrium in air at 25°C. In air, $P_{N_2} = 0.8$ atm and $P_{O_2} = 0.2$ atm.

b. Typical concentrations of NO in relatively pristine environments range from 10^8 to 10^{10} molecules/cm³. Why is there a discrepancy between these values and your answer to part a?

75. The gas arsine, AsH_3, decomposes as follows:

$$2AsH_3(g) \rightleftharpoons 2As(s) + 3H_2(g)$$

In an experiment at a certain temperature, pure $AsH_3(g)$ was placed in an empty, rigid, sealed flask at a pressure of 392.0 torr. After 48 hours the pressure in the flask was observed to be constant at 488.0 torr.

a. Calculate the equilibrium pressure of $H_2(g)$.

b. Calculate K_p for this reaction.

76. At a certain temperature, $K = 9.1 \times 10^{-4}$ for the reaction

$$FeSCN^{2+}(aq) \rightleftharpoons Fe^{3+}(aq) + SCN^{-}(aq)$$

Calculate the concentrations of Fe^{3+}, SCN^{-}, and $FeSCN^{2+}$ in a solution that is initially 2.0 M $FeSCN^{2+}$.

77. At a certain temperature, $K = 1.1 \times 10^3$ for the reaction

$$Fe^{3+}(aq) + SCN^{-}(aq) \rightleftharpoons FeSCN^{2+}(aq)$$

Calculate the concentrations of Fe^{3+}, SCN^{-}, and $FeSCN^{2+}$ at equilibrium if 0.020 mole of $Fe(NO_3)_3$ is added to 1.0 L of 0.10 M KSCN. (Neglect any volume change.)

78. For the reaction

$$PCl_5(g) \rightleftharpoons PCl_3(g) + Cl_2(g)$$

at 600. K, the equilibrium constant, K_p, is 11.5. Suppose that 2.450 g PCl_5 is placed in an evacuated 500.-mL bulb, which is then heated to 600. K.

a. What would be the pressure of PCl_5 if it did not dissociate?

b. What is the partial pressure of PCl_5 at equilibrium?

c. What is the total pressure in the bulb at equilibrium?

d. What is the percent dissociation of PCl_5 at equilibrium?

79. At 25°C, gaseous SO_2Cl_2 decomposes to $SO_2(g)$ and $Cl_2(g)$ to the extent that 12.5% of the original SO_2Cl_2 (by moles) has decomposed to reach equilibrium. The total pressure (at equilibrium) is 0.900 atm. Calculate the value of K_p for this system.

80. For the following reaction at a certain temperature

$$H_2(g) + F_2(g) \rightleftharpoons 2HF(g)$$

it is found that the equilibrium concentrations in a 5.00-L rigid container are $[H_2] = 0.0500$ M, $[F_2] = 0.0100$ M, and $[HF] = 0.400$ M. If 0.200 mole of F_2 is added to this equilibrium mixture, calculate the concentrations of all gases once equilibrium is reestablished.

81. Novelty devices for predicting rain contain cobalt(II) chloride and are based on the following equilibrium:

$$CoCl_2(s) + 6H_2O(g) \rightleftharpoons CoCl_2 \cdot 6H_2O(s)$$
　　　Purple　　　　　　　　　　　　　　　Pink

What color will such an indicator be if rain is imminent?

82. Consider the reaction

$$Fe^{3+}(aq) + SCN^{-}(aq) \rightleftharpoons FeSCN^{2+}(aq)$$

How will the equilibrium position shift if

a. water is added, doubling the volume?

b. $AgNO_3(aq)$ is added? (AgSCN is insoluble.)

c. $NaOH(aq)$ is added? [$Fe(OH)_3$ is insoluble.]

d. $Fe(NO_3)_3(aq)$ is added?

83. Chromium(VI) forms two different oxyanions, the orange dichromate ion, $Cr_2O_7^{2-}$, and the yellow chromate ion, CrO_4^{2-}. (See the following photos.) The equilibrium reaction between the two ions is

$$Cr_2O_7^{2-}(aq) + H_2O(l) \rightleftharpoons 2CrO_4^{2-}(aq) + 2H^{+}(aq)$$

Explain why orange dichromate solutions turn yellow when sodium hydroxide is added.

84. The synthesis of ammonia gas from nitrogen gas and hydrogen gas represents a classic case in which a knowledge of kinetics and equilibrium was used to make a desired chemical reaction economically feasible. Explain how each of the following conditions helps to maximize the yield of ammonia.

a. running the reaction at an elevated temperature

b. removing the ammonia from the reaction mixture as it forms

c. using a catalyst

d. running the reaction at high pressure

85. Suppose $K = 4.5 \times 10^{-3}$ at a certain temperature for the reaction

$$PCl_5(g) \rightleftharpoons PCl_3(g) + Cl_2(g)$$

If it is found that the concentration of PCl_5 is twice the concentration of PCl_3, what must be the concentration of Cl_2 under these conditions?

86. For the reaction below, $K_p = 1.16$ at 800.°C.

$$CaCO_3(s) \rightleftharpoons CaO(s) + CO_2(g)$$

If a 20.0-g sample of $CaCO_3$ is put into a 10.0-L container and heated to 800.°C, what percentage by mass of the $CaCO_3$ will react to reach equilibrium?

87. Many sugars undergo a process called *mutarotation*, in which the sugar molecules interconvert between two isomeric forms, finally reaching an equilibrium between them. This is true for the simple sugar glucose, $C_6H_{12}O_6$, which exists in solution in isomeric forms called alpha-glucose and beta-glucose. If a solution of glucose at a certain temperature is analyzed, and it is found that the concentration of alpha-glucose is twice the concentration of beta-glucose, what is the value of K for the interconversion reaction?

88. Peptide decomposition is one of the key processes of digestion, where a peptide bond is broken into an acid group and an amine group. We can describe this reaction as follows:

$$Peptide(aq) + H_2O(l) \rightleftharpoons acid\ group(aq) + amine\ group(aq)$$

If we place 1.0 mole of peptide into 1.0 L water, what will be the equilibrium concentrations of all species in this reaction? Assume the K value for this reaction is 3.1×10^{-5}.

89. The creation of shells by mollusk species is a fascinating process. By utilizing the Ca^{2+} in their food and aqueous environment, as well as some complex equilibrium processes, a hard calcium carbonate shell can be produced. One important equilibrium reaction in this complex process is

$$HCO_3^-(aq) \rightleftharpoons H^+(aq) + CO_3^{2-}(aq) \qquad K = 5.6 \times 10^{-11}$$

If 0.16 mole of HCO_3^- is placed into 1.00 L of solution, what will be the equilibrium concentration of CO_3^{2-}?

90. Methanol, a common laboratory solvent, poses a threat of blindness or death if consumed in sufficient amounts. Once in the body, the substance is oxidized to produce formaldehyde (embalming fluid) and eventually formic acid. Both of these substances are also toxic in varying levels. The equilibrium between methanol and formaldehyde can be described as follows:

$$CH_3OH(aq) \rightleftharpoons H_2CO(aq) + H_2(aq)$$

Assuming the value of K for this reaction is 3.7×10^{-10}, what are the equilibrium concentrations of each species if you start with a 1.24 M solution of methanol? What will happen to the concentration of methanol as the formaldehyde is further converted to formic acid?

ChemWork Problems

These multiconcept problems (and additional ones) are found interactively online with the same type of assistance a student would get from an instructor.

91. For the reaction:

$$3O_2(g) \rightleftharpoons 2O_3(g)$$

$K = 1.8 \times 10^{-7}$ at a certain temperature. If at equilibrium $[O_2] = 0.062\ M$, calculate the equilibrium O_3 concentration.

92. An equilibrium mixture contains 0.60 g solid carbon and the gases carbon dioxide and carbon monoxide at partial pressures of 2.60 atm and 2.89 atm, respectively. Calculate the value of K_p for the reaction $C(s) + CO_2(g) \rightleftharpoons 2CO(g)$.

93. At a particular temperature, 8.1 moles of NO_2 gas is placed in a 3.0-L container. Over time the NO_2 decomposes to NO and O_2:

$$2NO_2(g) \rightleftharpoons 2NO(g) + O_2(g)$$

At equilibrium the concentration of $NO(g)$ was found to be 1.4 mol/L. Calculate the value of K for this reaction.

94. A sample of solid ammonium chloride was placed in an evacuated chamber and then heated, causing it to decompose according to the following reaction:

$$NH_4Cl(s) \rightleftharpoons NH_3(g) + HCl(g)$$

In a particular experiment, the equilibrium partial pressure of $NH_3(g)$ in the container was 2.9 atm. Calculate the value of K_p for the decomposition of $NH_4Cl(s)$ at this temperature.

95. In a given experiment, 5.2 moles of pure NOCl was placed in an otherwise empty 2.0-L container. Equilibrium was established by the following reaction:

$$2NOCl(g) \rightleftharpoons 2NO(g) + Cl_2(g) \qquad K = 1.6 \times 10^{-5}$$

a. Using numerical values for the concentrations in the Initial row and expressions containing the variable x in both the Change and Equilibrium rows, complete the following table summarizing what happens as this reaction reaches equilibrium. Let $x =$ the concentration of Cl_2 that is present at equilibrium.

	NOCl	NO	Cl_2
Initial			
Change			
Equilibrium	$2.6 - 2x$		

b. Calculate the equilibrium concentrations for all species.

96. For the reaction $N_2O_4(g) \rightleftharpoons 2NO_2(g)$, $K_p = 0.25$ at a certain temperature. If 0.040 atm of N_2O_4 is reacted initially, calculate the equilibrium partial pressures of $NO_2(g)$ and $N_2O_4(g)$.

97. Consider the following exothermic reaction at equilibrium:

$$N_2(g) + 3H_2(g) \rightleftharpoons 2NH_3(g)$$

Predict how the following changes affect the number of moles of each component of the system after equilibrium is reestablished by completing the table below. Complete the table with the terms *increase, decrease,* or *no change*.

	N_2	H_2	NH_3
Add $N_2(g)$			
Remove $H_2(g)$			
Add $NH_3(g)$			
Add $Ne(g)$ (constant V)			
Increase the temperature			
Decrease the volume (constant T)			
Add a catalyst			

98. For the following endothermic reaction at equilibrium:

$$2SO_3(g) \rightleftharpoons 2SO_2(g) + O_2(g)$$

which of the following changes will increase the value of K?
a. increasing the temperature
b. decreasing the temperature
c. removing $SO_3(g)$ (constant T)
d. decreasing the volume (constant T)
e. adding $Ne(g)$ (constant T)
f. adding $SO_2(g)$ (constant T)
g. adding a catalyst (constant T)

Challenge Problems

99. A 1.604-g sample of methane (CH_4) gas and 6.400 g oxygen gas are sealed into a 2.50-L vessel at 411°C and are allowed to reach equilibrium. Methane can react with oxygen to form gaseous carbon dioxide and water vapor, or methane can react with oxygen to form gaseous carbon monoxide and water vapor. At equilibrium, the pressure of oxygen is 0.326 atm, and the pressure of water vapor is 4.45 atm. Calculate the pressures of carbon monoxide and carbon dioxide present at equilibrium.

100. A 4.72-g sample of methanol (CH_3OH) was placed in an otherwise empty 1.00-L flask and heated to 250.°C to vaporize the methanol. Over time, the methanol vapor decomposed by the following reaction:

$$CH_3OH(g) \rightleftharpoons CO(g) + 2H_2(g)$$

After the system has reached equilibrium, a tiny hole is drilled in the side of the flask allowing gaseous compounds to effuse out of the flask. Measurements of the effusing gas show that it contains 33.0 times as much $H_2(g)$ as $CH_3OH(g)$. Calculate K for this reaction at 250.°C.

101. At 35°C, $K = 1.6 \times 10^{-5}$ for the reaction

$$2NOCl(g) \rightleftharpoons 2NO(g) + Cl_2(g)$$

If 2.0 moles of NO and 1.0 mole of Cl_2 are placed into a 1.0-L flask, calculate the equilibrium concentrations of all species.

102. Nitric oxide and bromine at initial partial pressures of 98.4 and 41.3 torr, respectively, were allowed to react at 300. K. At equilibrium the total pressure was 110.5 torr. The reaction is

$$2NO(g) + Br_2(g) \rightleftharpoons 2NOBr(g)$$

a. Calculate the value of K_p.

b. What would be the partial pressures of all species if NO and Br_2, both at an initial partial pressure of 0.30 atm, were allowed to come to equilibrium at this temperature?

103. At 25°C, $K_p = 5.3 \times 10^5$ for the reaction

$$N_2(g) + 3H_2(g) \rightleftharpoons 2NH_3(g)$$

When a certain partial pressure of $NH_3(g)$ is put into an otherwise empty rigid vessel at 25°C, equilibrium is reached when 50.0% of the original ammonia has decomposed. What was the original partial pressure of ammonia before any decomposition occurred?

104. Consider the reaction

$$P_4(g) \longrightarrow 2P_2(g)$$

where $K_p = 1.00 \times 10^{-1}$ at 1325 K. In an experiment where $P_4(g)$ is placed into a container at 1325 K, the equilibrium mixture of $P_4(g)$ and $P_2(g)$ has a total pressure of 1.00 atm. Calculate the equilibrium pressures of $P_4(g)$ and $P_2(g)$. Calculate the fraction (by moles) of $P_4(g)$ that has dissociated to reach equilibrium.

105. The partial pressures of an equilibrium mixture of $N_2O_4(g)$ and $NO_2(g)$ are $P_{N_2O_4} = 0.34$ atm and $P_{NO_2} = 1.20$ atm at a certain temperature. The volume of the container is doubled. Calculate the partial pressures of the two gases when a new equilibrium is established.

106. At 125°C, $K_p = 0.25$ for the reaction

$$2NaHCO_3(s) \rightleftharpoons Na_2CO_3(s) + CO_2(g) + H_2O(g)$$

A 1.00-L flask containing 10.0 g $NaHCO_3$ is evacuated and heated to 125°C.

a. Calculate the partial pressures of CO_2 and H_2O after equilibrium is established.

b. Calculate the masses of $NaHCO_3$ and Na_2CO_3 present at equilibrium.

c. Calculate the minimum container volume necessary for all of the $NaHCO_3$ to decompose.

107. A mixture of N_2, H_2, and NH_3 is at equilibrium [according to the equation $N_2(g) + 3H_2(g) \rightleftharpoons 2NH_3(g)$] as depicted below:

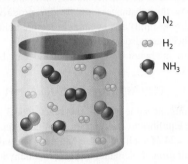

The volume is suddenly decreased (by increasing the external pressure) and a new equilibrium is established as depicted below:

a. If the volume of the final equilibrium mixture is 1.00 L, determine the value of the equilibrium constant, K, for the reaction. Assume temperature is constant.

b. Determine the volume of the initial equilibrium mixture assuming a final equilibrium volume of 1.00 L and assuming a constant temperature.

108. Consider the decomposition equilibrium for dinitrogen pentoxide:

$$2N_2O_5(g) \rightleftharpoons 4NO_2(g) + O_2(g)$$

At a certain temperature and a total pressure of 1.00 atm, the N_2O_5 is 0.50% decomposed (by moles) at equilibrium.

a. If the volume is increased by a factor of 10.0, will the mole percent of N_2O_5 decomposed at equilibrium be greater than, less than, or equal to 0.50%? Explain your answer.

b. Calculate the mole percent of N_2O_5 that will be decomposed at equilibrium if the volume is increased by a factor of 10.0.

109. An 8.00-g sample of SO_3 was placed in an evacuated container, where it decomposed at 600°C according to the following reaction:

$$SO_3(g) \rightleftharpoons SO_2(g) + \tfrac{1}{2}O_2(g)$$

At equilibrium the total pressure and the density of the gaseous mixture were 1.80 atm and 1.60 g/L, respectively. Calculate K_p for this reaction.

110. A sample of iron(II) sulfate was heated in an evacuated container to 920 K, where the following reactions occurred:

$$2FeSO_4(s) \rightleftharpoons Fe_2O_3(s) + SO_3(g) + SO_2(g)$$
$$SO_3(g) \rightleftharpoons SO_2(g) + \tfrac{1}{2}O_2(g)$$

After equilibrium was reached, the total pressure was 0.836 atm and the partial pressure of oxygen was 0.0275 atm. Calculate K_p for each of these reactions.

111. At 5000 K and 1.000 atm, 83.00% of the oxygen molecules in a sample have dissociated to atomic oxygen. At what pressure will 95.0% of the molecules dissociate at this temperature?

112. A sample of $N_2O_4(g)$ is placed in an empty cylinder at 25°C. After equilibrium is reached the total pressure is 1.5 atm and 16% (by moles) of the original $N_2O_4(g)$ has dissociated to $NO_2(g)$.
    a. Calculate the value of K_p for this dissociation reaction at 25°C.
    b. If the volume of the cylinder is increased until the total pressure is 1.0 atm (the temperature of the system remains constant), calculate the equilibrium pressure of $N_2O_4(g)$ and $NO_2(g)$.
    c. What percentage (by moles) of the original $N_2O_4(g)$ is dissociated at the new equilibrium position (total pressure = 1.00 atm)?

113. A sample of gaseous nitrosyl bromide (NOBr) was placed in a container fitted with a frictionless, massless piston, where it decomposed at 25°C according to the following equation:

$$2NOBr(g) \rightleftharpoons 2NO(g) + Br_2(g)$$

The initial density of the system was recorded as 4.495 g/L. After equilibrium was reached, the density was noted to be 4.086 g/L.
    a. Determine the value of the equilibrium constant K for the reaction.
    b. If $Ar(g)$ is added to the system at equilibrium at constant temperature, what will happen to the equilibrium position? What happens to the value of K? Explain each answer.

114. The equilibrium constant K_p for the reaction

$$CCl_4(g) \rightleftharpoons C(s) + 2Cl_2(g)$$

at 700°C is 0.76. Determine the initial pressure of carbon tetrachloride that will produce a total equilibrium pressure of 1.20 atm at 700°C.

Integrative Problems

These problems require the integration of multiple concepts to find the solutions.

115. For the reaction

$$NH_3(g) + H_2S(g) \rightleftharpoons NH_4HS(s)$$

$K = 400.$ at 35.0°C. If 2.00 moles each of NH_3, H_2S, and NH_4HS are placed in a 5.00-L vessel, what mass of NH_4HS will be present at equilibrium? What is the pressure of H_2S at equilibrium?

116. Given $K = 3.50$ at 45°C for the reaction

$$A(g) + B(g) \rightleftharpoons C(g)$$

and $K = 7.10$ at 45°C for the reaction

$$2A(g) + D(g) \rightleftharpoons C(g)$$

what is the value of K at the same temperature for the reaction

$$C(g) + D(g) \rightleftharpoons 2B(g)$$

What is the value of K_p at 45°C for the reaction? Starting with 1.50 atm partial pressures of both C and D, what is the mole fraction of B once equilibrium is reached?

117. In a solution with carbon tetrachloride as the solvent, the compound VCl_4 undergoes dimerization:

$$2VCl_4 \rightleftharpoons V_2Cl_8$$

When 6.6834 g VCl_4 is dissolved in 100.0 g carbon tetrachloride, the freezing point is lowered by 5.97°C. Calculate the value of the equilibrium constant for the dimerization of VCl_4 at this temperature. (The density of the equilibrium mixture is 1.696 g/cm³, and $K_f = 29.8$°C kg/mol for CCl_4.)

118. The hydrocarbon naphthalene was frequently used in mothballs until recently, when it was discovered that human inhalation of naphthalene vapors can lead to hemolytic anemia. Naphthalene is 93.71% carbon by mass, and a 0.256-mole sample of naphthalene has a mass of 32.8 g. What is the molecular formula of naphthalene? This compound works as a pesticide in mothballs by sublimation of the solid so that it fumigates enclosed spaces with its vapors according to the equation

$$Naphthalene(s) \rightleftharpoons naphthalene(g)$$
$$K = 4.29 \times 10^{-6} \text{ (at 298 K)}$$

If 3.00 g solid naphthalene is placed into an enclosed space with a volume of 5.00 L at 25°C, what percentage of the naphthalene will have sublimed once equilibrium has been established?

Marathon Problem

This problem is designed to incorporate several concepts and techniques into one situation.

119. A gaseous material $XY(g)$ dissociates to some extent to produce $X(g)$ and $Y(g)$:

$$XY(g) \rightleftharpoons X(g) + Y(g)$$

A 2.00-g sample of XY (molar mass = 165 g/mol) is placed in a container with a movable piston at 25°C. The pressure is held constant at 0.967 atm. As XY begins to dissociate, the piston moves until 35.0 mole percent of the original XY has dissociated and then remains at a constant position. Assuming ideal behavior, calculate the density of the gas in the container after the piston has stopped moving, and determine the value of K for this reaction of 25°C.

Chapter 14

Acids and Bases

The colors of hydrangea blossoms depend on the acidity of the soil. (Basie B/iStockphoto.com)

n this chapter we reencounter two very important classes of compounds, acids and bases. We will explore their interactions and apply the fundamentals of chemical equilibria discussed in Chapter 13 to systems involving proton-transfer reactions.

Acid–base chemistry is important in a wide variety of everyday applications. There are complex systems in our bodies that carefully control the acidity of our blood, since even small deviations may lead to serious illness and death. The same sensitivity exists in other life forms. If you have ever had tropical fish or goldfish, you know how important it is to monitor and control the acidity of the water in the aquarium.

Acids and bases are also important in industry. For example, the vast quantity of sulfuric acid manufactured in the United States each year is needed to produce fertilizers, polymers, steel, and many other materials.

The influence of acids on living things has assumed special importance in the United States, Canada, and Europe in recent years as a result of the phenomenon of acid rain. This problem is complex and has diplomatic and economic overtones that make it all the more difficult to solve.

14.1 | The Nature of Acids and Bases

Acids were first recognized as a class of substances that taste sour. Vinegar tastes sour because it is a dilute solution of acetic acid; citric acid is responsible for the sour taste of a lemon. Bases, sometimes called *alkalis*, are characterized by their bitter taste and slippery feel. Commercial preparations for unclogging drains are highly basic.

The first person to recognize the essential nature of acids and bases was Svante Arrhenius. Based on his experiments with electrolytes, Arrhenius postulated that

Don't taste chemicals!

Acids and bases were first discussed in Section 4.2.

Acidic fruits.

Liquids that contain acids and bases.

Household products that contain acids and bases.

acids produce hydrogen ions in aqueous solution, while bases produce hydroxide ions. At the time, the **Arrhenius concept** of acids and bases was a major step forward in quantifying acid–base chemistry, but this concept is limited because it applies only to aqueous solutions and allows for only one kind of base—the hydroxide ion. A more general definition of acids and bases was suggested by the Danish chemist Johannes Brønsted (1879–1947) and the English chemist Thomas Lowry (1874–1936). In terms of the **Brønsted–Lowry model**, *an acid is a proton (H^+) donor, and a base is a proton acceptor.* For example, when gaseous HCl dissolves in water, each HCl molecule donates a proton to a water molecule and so qualifies as a Brønsted–Lowry acid. The molecule that accepts the proton, in this case water, is a Brønsted–Lowry base.

To understand how water can act as a base, we need to remember that the oxygen of the water molecule has two unshared electron pairs, either of which can form a covalent bond with an H^+ ion. When gaseous HCl dissolves, the following reaction occurs:

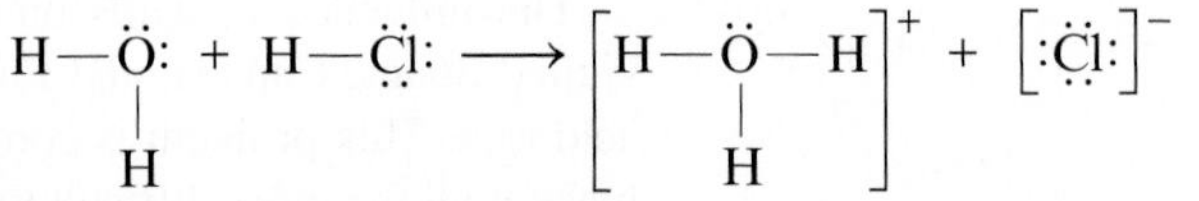

$$H-\ddot{O}: + H-\ddot{C}l: \longrightarrow \left[H-\overset{|}{\underset{H}{O}}-H\right]^+ + \left[:\ddot{C}l:\right]^-$$

Note that the proton is transferred from the HCl molecule to the water molecule to form H_3O^+, which is called the **hydronium ion*.** This reaction is represented in Fig. 14.1 using molecular models.

The general reaction that occurs when an acid is dissolved in water can best be represented as

$$HA(aq) + H_2O(l) \rightleftharpoons H_3O^+(aq) + A^-(aq) \qquad (14.1)$$

| Acid | Base | Conjugate acid | Conjugate base |

Recall that (*aq*) means the substance is hydrated.

This representation emphasizes the significant role of the polar water molecule in pulling the proton from the acid. Note that the **conjugate base** is everything that remains of the acid molecule after a proton is lost. The **conjugate acid** is formed when the proton is transferred to the base. A **conjugate acid–base pair** consists of two substances related to each other by the donating and accepting of a single proton. In Equation (14.1) there are two conjugate acid–base pairs: HA and A^- as well as H_2O and H_3O^+. This reaction is represented by molecular models in Fig. 14.2.

It is important to note that Equation (14.1) really represents *a competition for the proton between the two bases H_2O and A^-.* If H_2O is a much stronger base than A^-, that is, if H_2O has a much greater affinity for H^+ than does A^-, the equilibrium position will be far to the right; most of the acid dissolved will be in the ionized form. Conversely, if A^- is a much stronger base than H_2O, the equilibrium position will lie far to the left. In this case most of the acid dissolved will be present at equilibrium as HA.

The equilibrium expression for the reaction given in Equation (14.1) is

$$K_a = \frac{[H_3O^+][A^-]}{[HA]} = \frac{[H^+][A^-]}{[HA]} \qquad (14.2)$$

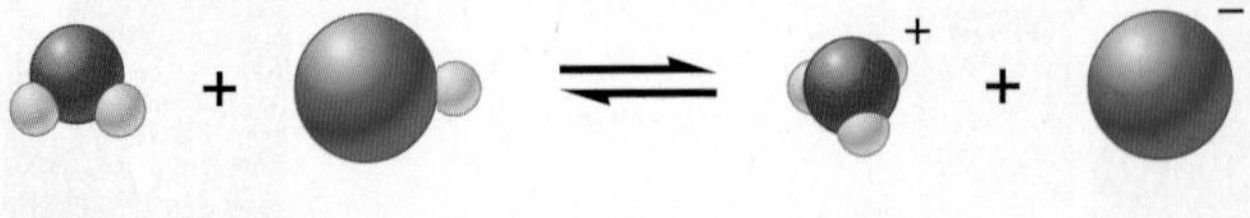

Figure 14.1 | The reaction of HCl and H_2O.

---

*The hydronium ion is actually a more complicated species than H_3O^+. See "The Solvated Proton Is NOT H_3O^+!" (*J. Chem. Ed.,* 2011, 88(7), p 875).

Figure 14.2 | The reaction of an acid HA with water to form H_3O^+ and a conjugate base A^-.

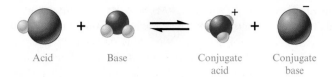

Acid   Base   Conjugate   Conjugate
                        acid   base

where K_a is called the **acid dissociation constant**. Both $H_3O^+(aq)$ and $H^+(aq)$ are commonly used to represent the hydrated proton. In this book we will often use simply H^+, but you should remember that it is hydrated in aqueous solutions.

In Chapter 13 we saw that the concentration of a pure solid or a pure liquid is always omitted from the equilibrium expression. In a dilute solution we can assume that the concentration of liquid water remains essentially constant when an acid is dissolved. Thus the term $[H_2O]$ is not included in Equation (14.2), and the equilibrium expression for K_a has the same form as that for the simple dissociation into ions:

In this chapter we will always represent an acid as simply dissociating. This does not mean we are using the Arrhenius model for acids. Since water does not affect the equilibrium position, it is simply easier to leave it out of the acid dissociation reaction.

$$HA(aq) \rightleftharpoons H^+(aq) + A^-(aq)$$

You should not forget, however, that water plays an important role in causing the acid to ionize.

Note that K_a is the equilibrium constant for the reaction in which a proton is removed from HA to form the conjugate base A^-. We use K_a to represent *only* this type of reaction. Knowing this, you can write the K_a expression for any acid, even one that is totally unfamiliar to you. As you do Example 14.1, focus on the definition of the reaction corresponding to K_a.

Interactive Example 14.1

Sign in at http://login.cengagebrain .com to try this Interactive Example in **OWL**.

Acid Dissociation (Ionization) Reactions

Write the simple dissociation (ionization) reaction (omitting water) for each of the following acids.

a.   hydrochloric acid (HCl)
b.   acetic acid ($HC_2H_3O_2$)
c.   the ammonium ion (NH_4^+)
d.   the anilinium ion ($C_6H_5NH_3^+$)
e.   the hydrated aluminum(III) ion $[Al(H_2O)_6]^{3+}$

Solution

a. $HCl(aq) \rightleftharpoons H^+(aq) + Cl^-(aq)$
b. $HC_2H_3O_2(aq) \rightleftharpoons H^+(aq) + C_2H_3O_2^-(aq)$
c. $NH_4^+(aq) \rightleftharpoons H^+(aq) + NH_3(aq)$
d. $C_6H_5NH_3^+(aq) \rightleftharpoons H^+(aq) + C_6H_5NH_2(aq)$
e.   Although this formula looks complicated, writing the reaction is simple if you concentrate on the meaning of K_a. Removing a proton, which can come only from one of the water molecules, leaves one OH^- and five H_2O molecules attached to the Al^{3+} ion. So the reaction is

$$Al(H_2O)_6^{3+}(aq) \rightleftharpoons H^+(aq) + Al(H_2O)_5OH^{2+}(aq)$$

See Exercises 14.35 and 14.36

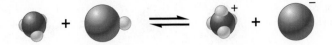

Figure 14.3 | The reaction of NH_3 with HCl to form NH_4^+ and Cl^-.

Ken O'Donoghue

When HCl(g) and NH_3(g) meet in a tube, a white ring of NH_4Cl(s) forms.

The Brønsted–Lowry model is not limited to aqueous solutions; it can be extended to reactions in the gas phase. For example, we discussed the reaction between gaseous hydrogen chloride and ammonia when we studied diffusion (see Section 5.7):

$$NH_3(g) + HCl(g) \rightleftharpoons NH_4Cl(s)$$

In this reaction, a proton is donated by the hydrogen chloride to the ammonia, as shown by these Lewis structures:

$$
\begin{array}{c}
\text{H} \\
| \\
\text{H—N:} + \text{H—\overset{..}{\underset{..}{Cl}}:} \rightleftharpoons
\end{array}
\left[
\begin{array}{c}
\text{H} \\
| \\
\text{H—N—H} \\
| \\
\text{H}
\end{array}
\right]^+
\left[:\overset{..}{\underset{..}{Cl}}: \right]^-
$$

Note that this is not considered an acid–base reaction according to the Arrhenius concept. Figure 14.3 shows a molecular representation of this reaction.

14.2 | Acid Strength

The strength of an acid is defined by the equilibrium position of its dissociation (ionization) reaction:

$$HA(aq) + H_2O(l) \rightleftharpoons H_3O^+(aq) + A^-(aq)$$

A strong acid has a weak conjugate base.

A **strong acid** is one for which *this equilibrium lies far to the right.* This means that almost all the original HA is dissociated (ionized) at equilibrium [Fig. 14.4(a)]. There is an important connection between the strength of an acid and that of its conjugate base. *A strong acid yields a weak conjugate base*—one that has a low affinity for a proton. A strong acid also can be described as an acid whose conjugate base is a much

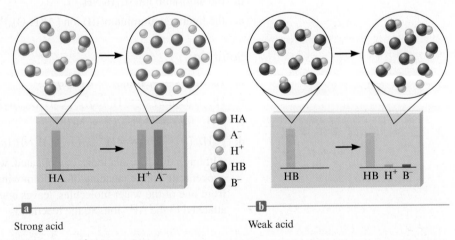

Strong acid   Weak acid

Figure 14.4 | (a) A strong acid HA is completely ionized in water. This is represented in terms of both molecules and a bar graph. (b) A weak acid HB consists of mostly undissociated HB molecules in water.

Table 14.1 | Various Ways to Describe Acid Strength

Property	Strong Acid	Weak Acid
K_a value	K_a is large	K_a is small
Position of the dissociation (ionization) equilibrium	Far to the right	Far to the left
Equilibrium concentration of [H$^+$] compared with original concentration of HA	$[H^+] \approx [HA]_0$	$[H^+] \ll [HA]_0$
Strength of conjugate base compared with that of water	A$^-$ much weaker base than H$_2$O	A$^-$ much stronger base than H$_2$O

≪ means much less than.
≫ means much greater than.

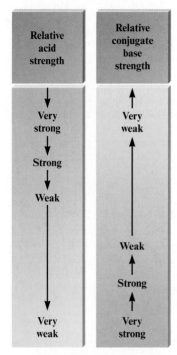

Figure 14.5 | The relationship of acid strength and conjugate base strength for the reaction

HA(*aq*) + H$_2$O(*l*) ⇌
Acid   H$_3$O$^+$(*aq*) + A$^-$(*aq*)
                              Conjugate base

Perchloric acid can explode if handled improperly.

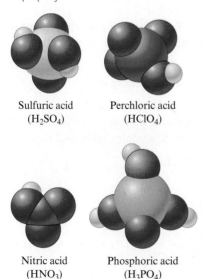

Sulfuric acid (H$_2$SO$_4$)   Perchloric acid (HClO$_4$)

Nitric acid (HNO$_3$)   Phosphoric acid (H$_3$PO$_4$)

weaker base than water (Fig. 14.5). In this case the water molecules win the competition for the H$^+$ ions.

Conversely, a **weak acid** is one for which *the equilibrium lies far to the left*. Most of the acid originally placed in the solution is still present as HA at equilibrium. That is, a weak acid dissociates only to a very small extent in aqueous solution [see Fig. 14.4(b)]. In contrast to a strong acid, a weak acid has a conjugate base that is a much stronger base than water. In this case a water molecule is not very successful in pulling an H$^+$ ion from the conjugate base. The weaker the acid, the stronger its conjugate base. The various ways of describing the strength of an acid are summarized in Table 14.1.

The common strong acids are sulfuric acid [H$_2$SO$_4$(*aq*)], hydrochloric acid [HCl(*aq*)], nitric acid [HNO$_3$(*aq*)], and perchloric acid [HClO$_4$(*aq*)]. Sulfuric acid is actually a **diprotic acid**, an acid having two acidic protons. The acid H$_2$SO$_4$ is a strong acid, virtually 100% dissociated (ionized) in water:

$$H_2SO_4(aq) \longrightarrow H^+(aq) + HSO_4^-(aq)$$

The HSO$_4^-$ ion, however, is a weak acid:

$$HSO_4^-(aq) \rightleftharpoons H^+(aq) + SO_4^{2-}(aq)$$

Most acids are **oxyacids**, in which the acidic proton is attached to an oxygen atom. The strong acids mentioned above, except hydrochloric acid, are typical examples. Many common weak acids, such as phosphoric acid (H$_3$PO$_4$), nitrous acid (HNO$_2$), and hypochlorous acid (HOCl), are also oxyacids. **Organic acids**, those with a carbon atom backbone, commonly contain the **carboxyl group**:

$$-C\overset{\displaystyle O}{\underset{\displaystyle O-H}{\Big\langle}}$$

Acids of this type are usually weak. Examples are acetic acid (CH$_3$COOH), often written HC$_2$H$_3$O$_2$, and benzoic acid (C$_6$H$_5$COOH). Note that the remainder of the hydrogens in these molecules are not acidic—they do not form H$^+$ in water.

There are some important acids in which the acidic proton is attached to an atom other than oxygen. The most significant of these are the hydrohalic acids HX, where X represents a halogen atom.

Table 14.2 contains a list of common **monoprotic acids** (those having *one* acidic proton) and their K_a values. Note that the strong acids are not listed. When a strong acid molecule such as HCl, for example, is placed in water, the position of the dissociation equilibrium

$$HCl(aq) \rightleftharpoons H^+(aq) + Cl^-(aq)$$

Relative acid strength

Very strong
↓
Strong
↓
Weak
↓
Very weak

Relative conjugate base strength

Very weak
↑
Weak
↑
Strong
↑
Very strong

Nitrous acid
(HNO$_2$)

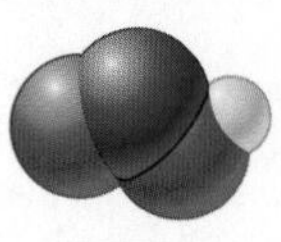

Hypochlorous acid
(HOCl)

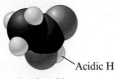

Acidic H

Acetic acid
(CH$_3$CO$_2$H)

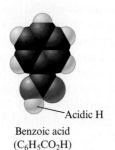

Acidic H

Benzoic acid
(C$_6$H$_5$CO$_2$H)

Table 14.2 │ Values of K_a for Some Common Monoprotic Acids

Formula	Name	Value of K_a*
HSO$_4^-$	Hydrogen sulfate ion	1.2×10^{-2}
HClO$_2$	Chlorous acid	1.2×10^{-2}
HC$_2$H$_2$ClO$_2$	Monochloracetic acid	1.35×10^{-3}
HF	Hydrofluoric acid	7.2×10^{-4}
HNO$_2$	Nitrous acid	4.0×10^{-4}
HC$_2$H$_3$O$_2$	Acetic acid	1.8×10^{-5}
[Al(H$_2$O)$_6$]$^{3+}$	Hydrated aluminum(III) ion	1.4×10^{-5}
HOCl	Hypochlorous acid	3.5×10^{-8}
HCN	Hydrocyanic acid	6.2×10^{-10}
NH$_4^+$	Ammonium ion	5.6×10^{-10}
HOC$_6$H$_5$	Phenol	1.6×10^{-10}

Increasing acid strength →

*The units of K_a are customarily omitted.

lies so far to the right that [HCl] cannot be measured accurately. This prevents an accurate calculation of K_a:

$$K_a = \frac{[\text{H}^+][\text{Cl}^-]}{[\text{HCl}]}$$

Very small and
highly uncertain

Critical Thinking

Vinegar contains acetic acid and is used in salad dressings. What if acetic acid was a strong acid instead of a weak acid? Would it be safe to use vinegar as a salad dressing?

Interactive Example 14.2

Sign in at http://login.cengagebrain
.com to try this Interactive Example
in **OWL**.

Relative Base Strength

Using Table 14.2, arrange the following species according to their strengths as bases: H$_2$O, F$^-$, Cl$^-$, NO$_2^-$, and CN$^-$.

Solution

Remember that water is a stronger base than the conjugate base of a strong acid but a weaker base than the conjugate base of a weak acid. This leads to the following order:

$$\text{Cl}^- < \text{H}_2\text{O} < \text{conjugate bases of weak acids}$$

Weakest bases ⟶ Strongest bases

We can order the remaining conjugate bases by recognizing that the strength of an acid is *inversely related* to the strength of its conjugate base. Since from Table 14.2 we have

$$K_a \text{ for HF} > K_a \text{ for HNO}_2 > K_a \text{ for HCN}$$

the base strengths increase as follows:

$$\text{F}^- < \text{NO}_2^- < \text{CN}^-$$

The combined order of increasing base strength is

$$\text{Cl}^- < \text{H}_2\text{O} < \text{F}^- < \text{NO}_2^- < \text{CN}^-$$

Appendix 5.1 contains a table of K_a values.

See Exercises 14.41 through 14.44

Water as an Acid and a Base

A substance is said to be **amphoteric** if it can behave either as an acid or as a base. Water is the most common **amphoteric substance**. We can see this clearly in the **autoionization** of water, which involves the transfer of a proton from one water molecule to another to produce a hydroxide ion and a hydronium ion:

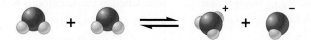

$H_2O \; + \; H_2O \rightleftharpoons H_3O^+ \; + \; OH^-$
acid(1)   base(1)   acid(2)   base(2)

In this reaction, one water molecule acts as an acid by furnishing a proton, and the other acts as a base by accepting the proton.

Autoionization can occur in other liquids besides water. For example, in liquid ammonia the autoionization reaction is

The autoionization reaction for water

$$2H_2O(l) \rightleftharpoons H_3O^+(aq) \; + \; OH^-(aq)$$

leads to the equilibrium expression

$$K_w = [H_3O^+][OH^-] = [H^+][OH^-]$$

where K_w, called the **ion-product constant** (or the **dissociation constant** for water), always refers to the autoionization of water.

Experiment shows that at 25°C in pure water,

$$[H^+] = [OH^-] = 1.0 \times 10^{-7} M$$

which means that at 25°C

$$K_w = [H^+][OH^-] = (1.0 \times 10^{-7})(1.0 \times 10^{-7})$$
$$= 1.0 \times 10^{-14}$$

$K_w = [H^+][OH^-]$
$= 1.0 \times 10^{-14}$

Let's Review | Using K_w

It is important to recognize the meaning of K_w. In any aqueous solution at 25°C, *no matter what it contains*, the product of $[H^+]$ and $[OH^-]$ must always equal 1.0×10^{-14}. There are three possible situations:

> a neutral solution, where $[H^+] = [OH^-]$
> an acidic solution, where $[H^+] > [OH^-]$
> a basic solution, where $[OH^-] > [H^+]$

In each case, however, at 25°C,

$$K_w = [H^+][OH^-] = 1.0 \times 10^{-14}$$

We normally consider the pH of solutions at 25°C. However, it is important to realize that K_w is temperature dependent. For example, at 37°C (normal body temperature) the value of $K_w = 2.42 \times 10^{-14}$. This means for a neutral solution at 37°C

$$[H^+][OH^-] = 2.42 \times 10^{-14}$$

and

$$[H^+] = [OH^-] = 1.55 \times 10^{-7}$$

So the pH for a neutral solution at 37°C is 6.81.

Interactive Example 14.3

Sign in at http://login.cengagebrain.com to try this Interactive Example in OWL.

Calculating $[H^+]$ and $[OH^-]$

Calculate $[H^+]$ or $[OH^-]$ as required for each of the following solutions at 25°C, and state whether the solution is neutral, acidic, or basic.

a. $1.0 \times 10^{-5}\ M\ OH^-$

b. $1.0 \times 10^{-7}\ M\ OH^-$

c. $10.0\ M\ H^+$

Solution

a. $K_w = [H^+][OH^-] = 1.0 \times 10^{-14}$. Since $[OH^-]$ is $1.0 \times 10^{-5}\ M$, solving for $[H^+]$ gives

$$[H^+] = \frac{1.0 \times 10^{-14}}{[OH^-]} = \frac{1.0 \times 10^{-14}}{1.0 \times 10^{-5}} = 1.0 \times 10^{-9}\ M$$

Since $[OH^-] > [H^+]$, the solution is basic.

b. As in part a, solving for $[H^+]$ gives

$$[H^+] = \frac{1.0 \times 10^{-14}}{[OH^-]} = \frac{1.0 \times 10^{-14}}{1.0 \times 10^{-7}} = 1.0 \times 10^{-7}\ M$$

Since $[H^+] = [OH^-]$, the solution is neutral.

c. Solving for $[OH^-]$ gives

$$[OH^-] = \frac{1.0 \times 10^{-14}}{[H^+]} = \frac{1.0 \times 10^{-14}}{10.0} = 1.0 \times 10^{-15}\ M$$

Since $[H^+] > [OH^-]$, the solution is acidic.

See Exercises 14.45 and 14.46

Since K_w is an equilibrium constant, it varies with temperature. The effect of temperature is considered in Example 14.4.

Example 14.4

Autoionization of Water

At 60°C, the value of K_w is 1×10^{-13}.

a. Using Le Châtelier's principle, predict whether the reaction

$$2H_2O(l) \rightleftharpoons H_3O^+(aq) + OH^-(aq)$$

is exothermic or endothermic.

b. Calculate $[H^+]$ and $[OH^-]$ in a neutral solution at 60°C.

Solution

a. K_w *increases* from 1×10^{-14} at 25°C to 1×10^{-13} at 60°C. Le Châtelier's principle states that if a system at equilibrium is heated, it will adjust to consume energy. Since the value of K_w increases with temperature, we must think of energy as a reactant, and so the process must be endothermic.

b. At 60°C,

$$[H^+][OH^-] = 1 \times 10^{-13}$$

For a neutral solution,

$$[H^+] = [OH^-] = \sqrt{1 \times 10^{-13}} = 3 \times 10^{-7} \, M$$

See Exercise 14.47

14.3 | The pH Scale

The pH scale is a compact way to represent solution acidity.

Appendix 1.2 has a review of logs.

Because $[H^+]$ in an aqueous solution is typically quite small, the **pH scale** provides a convenient way to represent solution acidity. The pH is a log scale based on 10, where

$$pH = -\log[H^+]$$

Thus for a solution where

$$[H^+] = 1.0 \times 10^{-7} \, M$$
$$pH = -(-7.00) = 7.00$$

At this point we need to discuss significant figures for logarithms. The number of decimal places in the log is equal to the number of significant figures in the original number. Thus

┌─ 2 significant figures
$$[H^+] = 1.0 \times 10^{-9} \, M$$
$$pH = 9.00$$
└─ 2 decimal places

Similar log scales are used for representing other quantities; for example,

$$pOH = -\log[OH^-]$$
$$pK = -\log K$$

Since pH is a log scale based on 10, *the pH changes by 1 for every power of 10 change in [H^+].* For example, a solution of pH 3 has an H^+ concentration 10 times that of a solution of pH 4 and 100 times that of a solution of pH 5. Also note that because pH is defined as $-\log[H^+]$, *the pH decreases as [H^+] increases.* The pH scale and the pH values for several common substances are shown in Fig. 14.6.

The pH of a solution is usually measured using a pH meter, an electronic device with a probe that can be inserted into a solution of unknown pH. The probe contains an acidic aqueous solution enclosed by a special glass membrane that allows migration of H^+ ions. If the unknown solution has a different pH from the solution in the probe, an electric potential results, which is registered on the meter (Fig. 14.7).

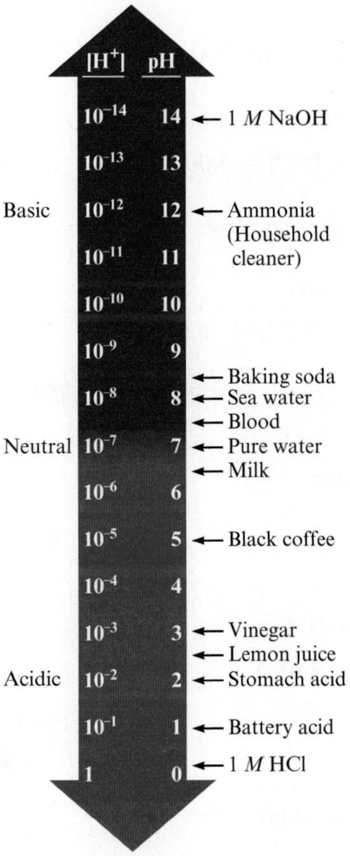

[H⁺]	pH	
10^{-14}	14	← 1 M NaOH
10^{-13}	13	
10^{-12}	12	← Ammonia (Household cleaner)
10^{-11}	11	
10^{-10}	10	
10^{-9}	9	
10^{-8}	8	← Baking soda / Sea water
10^{-7}	7	← Blood / Pure water / Milk
10^{-6}	6	
10^{-5}	5	← Black coffee
10^{-4}	4	
10^{-3}	3	← Vinegar / Lemon juice
10^{-2}	2	← Stomach acid
10^{-1}	1	← Battery acid
1	0	← 1 M HCl

Basic — Neutral — Acidic

Figure 14.6 | The pH scale and pH values of some common substances.

Critical Thinking

What if you lived on a planet identical to the earth but for which room temperature was 50°C? How would the pH scale be different?

Figure 14.7 | pH meters are used to measure acidity.

The pH meter is discussed more fully in Section 18.5.

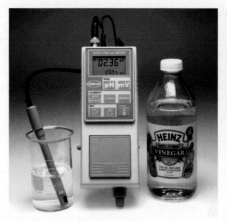

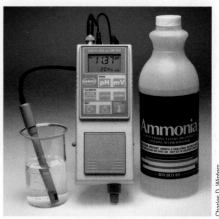

Charles D. Winters

Sign in at http://login.cengagebrain .com to try this Interactive Example in OWL.

Interactive Example 14.5

Calculating pH and pOH

Calculate pH and pOH for each of the following solutions at 25°C.

a. $1.0 \times 10^{-3} \, M \, OH^-$

b. $1.0 \, M \, OH^-$

Solution

a.
$$[H^+] = \frac{K_w}{[OH^-]} = \frac{1.0 \times 10^{-14}}{1.0 \times 10^{-3}} = 1.0 \times 10^{-11} \, M$$

$$pH = -\log[H^+] = -\log(1.0 \times 10^{-11}) = 11.00$$

$$pOH = -\log[OH^-] = -\log(1.0 \times 10^{-3}) = 3.00$$

b.
$$[OH^-] = \frac{K_w}{[H^+]} = \frac{1.0 \times 10^{-14}}{1.0} = 1.0 \times 10^{-14} \, M$$

$$pH = -\log[H^+] = -\log(1.0) = 0.00$$

$$pOH = -\log[OH^-] = -\log(1.0 \times 10^{-14}) = 14.00$$

See Exercise 14.49

It is useful to consider the log form of the expression

$$K_w = [H^+][OH^-]$$

That is,

$$\log K_w = \log[H^+] + \log[OH^-]$$

or

$$-\log K_w = -\log[H^+] - \log[OH^-]$$

Thus $$pK_w = pH + pOH \tag{14.3}$$

Since $K_w = 1.0 \times 10^{-14}$,

$$pK_w = -\log(1.0 \times 10^{-14}) = 14.00$$

Thus for *any* aqueous solution at 25°C, pH and pOH add up to 14.00:

$$pH + pOH = 14.00 \tag{14.4}$$

Chemical connections
Arnold Beckman, Man of Science

Arnold Beckman died at age 104 in May 2004. Beckman's leadership in science and business spans virtually the entire twentieth century. He was born in 1900 in Cullom, Illinois, a town of 500 people that had no electricity or telephones. Beckman says, "In Cullom we were forced to improvise. I think it was a good thing."

The son of a blacksmith, Beckman had his interest in science awakened at age nine. At that time, in the attic of his house he discovered J. Dorman Steele's *Fourteen Weeks in Chemistry*, a book containing instructions for doing chemistry experiments. Beckman became so fascinated with chemistry that his father built him a small "chemistry shed" in the backyard for his tenth birthday.

Beckman's interest in chemistry was fostered by his high school teachers, and he eventually attended the University of Illinois, Urbana–Champaign. He graduated with a bachelor's degree in chemical engineering in 1922 and stayed one more year to get a master's degree. He then went to Caltech, where he earned a Ph.D. and became a faculty member.

Beckman was always known for his inventiveness. As a youth he designed a pressurized fuel system for his Model T Ford to overcome problems with its normal gravity feed fuel system—you

had to *back* it up steep hills to keep it from starving for fuel. In 1927 he applied for his first patent: a buzzer to alert drivers that they were speeding.

In 1935 Beckman invented something that would cause a revolution in scientific instrumentation. A college friend who worked in a laboratory in the California citrus industry needed an accurate, convenient way to measure the acidity of orange juice. In response, Beckman invented the pH meter, which he initially called the acidimeter. This compact, sturdy device was an immediate hit. It signaled a new era in scientific instrumentation. In fact, business was so good that Beckman left Caltech to head his own company.

Over the years Beckman invented many other devices, including an improved potentiometer and an instrument for measuring the light absorbed by molecules. At age 65 he retired as president of Beckman Instruments (headquartered in Fullerton, California). After a merger the company became Beckman Coulter; it had sales of more than $2 billion in 2003.

After stepping down as president of Beckman Instruments, Beckman began a new career—donating his wealth for the improvement of science. In 1984 he and Mabel, his

Arnold Beckman.

wife of 58 years, donated $40 million to his alma mater—the University of Illinois—to fund the Beckman Institute. The Beckmans have also funded many other research institutes, including one at Caltech, and formed a foundation that currently gives $20 million each year to various scientific endeavors.

Arnold Beckman was a man known for his incredible creativity, but even more he was recognized as a man of absolute integrity. Beckman has important words for us: "Whatever you do, be enthusiastic about it."

Note: You can see Arnold Beckman's biography at the Chemical Heritage Foundation Web site (http://www.chemheritage.org).

Interactive Example 14.6

Sign in at http://login.cengagebrain.com to try this Interactive Example in **OWL**.

Calculations Using pH

The pH of a sample of human blood was measured to be 7.41 at 25°C. Calculate pOH, $[H^+]$, and $[OH^-]$ for the sample.

Solution

Since pH + pOH = 14.00,

$$\text{pOH} = 14.00 - \text{pH} = 14.00 - 7.41 = 6.59$$

To find $[H^+]$ we must go back to the definition of pH:

$$pH = -\log[H^+]$$

Thus

$$7.41 = -\log[H^+] \quad \text{or} \quad \log[H^+] = -7.41$$

We need to know the *antilog* of -7.41. As shown in Appendix 1.2, taking the antilog is the same as exponentiation; that is,

antilog(n) = log^{-1}(n)

$$\text{antilog}(n) = 10^n$$

Since $pH = -\log[H^+]$,

$$-pH = \log[H^+]$$

and $[H^+]$ can be calculated by taking the antilog of $-pH$:

$$[H^+] = \text{antilog}(-pH)$$

In the present case,

$$[H^+] = \text{antilog}(-pH) = \text{antilog}(-7.41) = 10^{-7.41} = 3.9 \times 10^{-8}\ M$$

Similarly, $[OH^-] = \text{antilog}(-pOH)$, and

$$[OH^-] = \text{antilog}(-6.59) = 10^{-6.59} = 2.6 \times 10^{-7}\ M$$

See Exercises 14.50 through 14.54

Now that we have considered all the fundamental definitions relevant to acid–base solutions, we can proceed to a quantitative description of the equilibria present in these solutions. The main reason that acid–base problems sometimes seem difficult is that a typical aqueous solution contains many components, so the problems tend to be complicated. However, you can deal with these problems successfully if you use the following general strategies:

Problem-Solving Strategy

Solving Acid–Base Problems

› *Think chemistry.* Focus on the solution components and their reactions. It will almost always be possible to choose one reaction that is the most important.

› *Be systematic.* Acid–base problems require a step-by-step approach.

› *Be flexible.* Although all acid–base problems are similar in many ways, important differences do occur. Treat each problem as a separate entity. Do not try to force a given problem into matching any you have solved before. Look for both the similarities and the differences.

› *Be patient.* The complete solution to a complicated problem cannot be seen immediately in all its detail. Pick the problem apart into its workable steps.

› *Be confident.* Look within the problem for the solution, and let the problem guide you. Assume that you can think it out. Do not rely on memorizing solutions to problems. In fact, memorizing solutions is usually detrimental because you tend to try to force a new problem to be the same as one you have seen before. *Understand and think; don't just memorize.*

14.4 | Calculating the pH of Strong Acid Solutions

When we deal with acid–base equilibria, *we must focus on the solution components and their chemistry.* For example, what species are present in a 1.0-*M* solution of HCl? Since hydrochloric acid is a strong acid, we assume that it is completely dissociated. Thus although the label on the bottle says 1.0 *M* HCl, the solution contains virtually no HCl molecules. Typically, container labels indicate the substance(s) used to make up the solution but do not necessarily describe the solution components after dissolution. Thus a 1.0-*M* HCl solution contains H$^+$ and Cl$^-$ ions rather than HCl molecules.

The next step in dealing with aqueous solutions is to determine which components are significant and which can be ignored. We need to focus on the **major species**, those solution components present in relatively large amounts. In 1.0 *M* HCl, for example, the major species are H$^+$, Cl$^-$, and H$_2$O. Since this is a very acidic solution, OH$^-$ is present only in tiny amounts and is classified as a minor species. In attacking acid–base problems, the importance of writing the major species in the solution as the first step cannot be overemphasized. This single step is the key to solving these problems successfully.

To illustrate the main ideas involved, let us calculate the pH of 1.0 *M* HCl. We first list the major species: H$^+$, Cl$^-$, and H$_2$O. Since we want to calculate the pH, we will focus on those major species that can furnish H$^+$. Obviously, we must consider H$^+$ from the dissociation of HCl. However, H$_2$O also furnishes H$^+$ by autoionization, which is often represented by the simple dissociation reaction

$$H_2O(l) \rightleftharpoons H^+(aq) + OH^-(aq)$$

However, is autoionization an important source of H$^+$ ions? In pure water at 25°C, [H$^+$] is 10^{-7} *M*. In 1.0 *M* HCl solution, the water will produce even less than 10^{-7} *M* H$^+$, since by Le Châtelier's principle the H$^+$ from the dissociated HCl will drive the position of the water equilibrium to the left. Thus the amount of H$^+$ contributed by water is negligible compared with the 1.0 *M* H$^+$ from the dissociation of HCl. Therefore, we can say that [H$^+$] in the solution is 1.0 *M*. The pH is then

$$pH = -\log[H^+] = -\log(1.0) = 0$$

Common Strong Acids

HCl(*aq*)
HNO$_3$(*aq*)
H$_2$SO$_4$(*aq*)
HClO$_4$(*aq*)

Always write the major species present in the solution.

The H$^+$ from the strong acid drives the equilibrium H$_2$O $\rightleftharpoons$ H$^+$ + OH$^-$ to the left.

---

pH of Strong Acids

In pure water, only 10^{-7} *M* H$^+$ is produced.

Major Species

 H$^+$

 NO$_3^-$

 H$_2$O

a. Calculate the pH of 0.10 *M* HNO$_3$.

b. Calculate the pH of 1.0×10^{-10} *M* HCl.

Solution

a. Since HNO$_3$ is a strong acid, the major species in solution are

$$H^+, \quad NO_3^-, \quad \text{and} \quad H_2O$$

The concentration of HNO$_3$ is virtually zero, since the acid completely dissociates in water. Also, [OH$^-$] will be very small because the H$^+$ ions from the acid will drive the equilibrium

$$H_2O(l) \rightleftharpoons H^+(aq) + OH^-(aq)$$

to the left. That is, this is an acidic solution where $[H^+] \gg [OH^-]$, so $[OH^-] \ll 10^{-7}\ M$. The sources of H^+ are

1. H^+ from HNO_3 (0.10 M)
2. H^+ from H_2O

The number of H^+ ions contributed by the autoionization of water will be very small compared with the 0.10 M contributed by the HNO_3 and can be neglected. Since the dissolved HNO_3 is the only important source of H^+ ions in this solution,

$$[H^+] = 0.10\ M \quad \text{and} \quad pH = -\log(0.10) = 1.00$$

b. Normally, in an aqueous solution of HCl the major species are H^+, Cl^-, and H_2O. However, in this case the amount of HCl in solution is so small that it has no effect; the only major species is H_2O. Thus the pH will be that of pure water, or pH = 7.00.

See Exercises 14.55, 14.57, and 14.58

14.5 | Calculating the pH of Weak Acid Solutions

Since a weak acid dissolved in water can be viewed as a prototype of almost any equilibrium occurring in aqueous solution, we will proceed carefully and systematically. Although some of the procedures we develop here may seem unnecessary, they will become essential as the problems become more complicated. We will develop the necessary strategies by calculating the pH of a 1.00-M solution of HF ($K_a = 7.2 \times 10^{-4}$).

First, *always* write the major species present in the solution.

The first step, as always, is to *write the major species in the solution.* From its small K_a value, we know that hydrofluoric acid is a weak acid and will be dissociated only to a slight extent. Thus when we write the major species, the hydrofluoric acid will be represented in its dominant form, as HF. The major species in solution are HF and H_2O.

Major Species

 HF

 H_2O

The next step (since this is a pH problem) is to decide which of the major species can furnish H^+ ions. Actually, both major species can do so:

$$HF(aq) \rightleftharpoons H^+(aq) + F^-(aq) \qquad K_a = 7.2 \times 10^{-4}$$
$$H_2O(l) \rightleftharpoons H^+(aq) + OH^-(aq) \qquad K_w = 1.0 \times 10^{-14}$$

In aqueous solutions, however, typically one source of H^+ can be singled out as dominant. By comparing K_a for HF with K_w for H_2O, we see that hydrofluoric acid, although weak, is still a much stronger acid than water. Thus we will assume that hydrofluoric acid will be the dominant source of H^+. We will ignore the tiny contribution by water.

Therefore, it is the dissociation of HF that will determine the equilibrium concentration of H^+ and hence the pH:

$$HF(aq) \rightleftharpoons H^+(aq) + F^-(aq)$$

The equilibrium expression is

$$K_a = 7.2 \times 10^{-4} = \frac{[H^+][F^-]}{[HF]}$$

To solve the equilibrium problem, we follow the procedures developed in Chapter 13 for gas-phase equilibria. First, we list the initial concentrations, the *concentrations before the reaction of interest has proceeded to equilibrium.* Before any HF dissociates, the concentrations of the species in the equilibrium are

$$[HF]_0 = 1.00\ M \quad [F^-]_0 = 0 \quad [H^+]_0 = 10^{-7}\ M \approx 0$$

(Note that the zero value for $[H^+]_0$ is an approximation, since we are neglecting the H^+ ions from the autoionization of water.)

The next step is to determine the change required to reach equilibrium. Since some HF will dissociate to come to equilibrium (but this amount is presently unknown), we let x be the change in the concentration of HF that is required to achieve equilibrium. That is, we assume that x mol/L HF will dissociate to produce x mol/L H^+ and x mol/L F^- as the system adjusts to its equilibrium position. Now the equilibrium concentrations can be defined in terms of x:

$$[HF] = [HF]_0 - x = 1.00 - x$$
$$[F^-] = [F^-]_0 + x = 0 + x = x$$
$$[H^+] = [H^+]_0 + x \approx 0 + x = x$$

Substituting these equilibrium concentrations into the equilibrium expression gives

$$K_a = 7.2 \times 10^{-4} = \frac{[H^+][F^-]}{[HF]} = \frac{(x)(x)}{1.00 - x}$$

This expression produces a quadratic equation that can be solved using the quadratic formula, as for the gas-phase systems in Chapter 13. However, since K_a for HF is so small, HF will dissociate only slightly, and x is expected to be small. This will allow us to simplify the calculation. If x is very small compared to 1.00, the term in the denominator can be approximated as follows:

$$1.00 - x \approx 1.00$$

The equilibrium expression then becomes

$$7.2 \times 10^{-4} = \frac{(x)(x)}{1.00 - x} \approx \frac{(x)(x)}{1.00}$$

which yields

$$x^2 \approx (7.2 \times 10^{-4})(1.00) = 7.2 \times 10^{-4}$$
$$x \approx \sqrt{7.2 \times 10^{-4}} = 2.7 \times 10^{-2}$$

The validity of an approximation should always be checked.

How valid is the approximation that $[HF] = 1.00\ M$? Because this question will arise often in connection with acid–base equilibrium calculations, we will consider it carefully. *The validity of the approximation depends on how much accuracy we demand for the calculated value of $[H^+]$.* Typically, the K_a values for acids are known to an accuracy of only about $\pm 5\%$. It is reasonable therefore to apply this figure when determining the validity of the approximation

$$[HA]_0 - x \approx [HA]_0$$

We will use the following test. First, we calculate the value of x by making the approximation

$$K_a = \frac{x^2}{[HA]_0 - x} \approx \frac{x^2}{[HA]_0}$$

where

$$x^2 \approx K_a[HA]_0 \quad \text{and} \quad x \approx \sqrt{K_a[HA]_0}$$

We then compare the sizes of x and $[HA]_0$. If the expression

$$\frac{x}{[HA]_0} \times 100\%$$

is less than or equal to 5%, the value of x is small enough that the approximation

$$[HA]_0 - x \approx [HA]_0$$

will be considered valid.

In our example,

$$x = 2.7 \times 10^{-2} \text{ mol/L}$$
$$[HA]_0 = [HF]_0 = 1.00 \text{ mol/L}$$

and

$$\frac{x}{[HA]_0} \times 100 = \frac{2.7 \times 10^{-2}}{1.00} \times 100\% = 2.7\%$$

The approximation we made is considered valid, and the value of x calculated using that approximation is acceptable. Thus

$$x = [H^+] = 2.7 \times 10^{-2} M \quad \text{and} \quad pH = -\log(2.7 \times 10^{-2}) = 1.57$$

This problem illustrates all the important steps for solving a typical equilibrium problem involving a weak acid. These steps are summarized as follows:

The K_a values for various weak acids are given in Table 14.2 and in Appendix 5.1.

Problem-Solving Strategy

Solving Weak Acid Equilibrium Problems

1. List the major species in the solution.
2. Choose the species that can produce H^+, and write balanced equations for the reactions producing H^+.
3. Using the values of the equilibrium constants for the reactions you have written, decide which equilibrium will dominate in producing H^+.
4. Write the equilibrium expression for the dominant equilibrium.
5. List the initial concentrations of the species participating in the dominant equilibrium.
6. Define the change needed to achieve equilibrium; that is, define x.
7. Write the equilibrium concentrations in terms of x.
8. Substitute the equilibrium concentrations into the equilibrium expression.
9. Solve for x the "easy" way, that is, by assuming that $[HA]_0 - x \approx [HA]_0$.
10. Use the 5% rule to verify whether the approximation is valid.
11. Calculate $[H^+]$ and pH.

Critical Thinking

Consider two aqueous solutions of different weak acids, HA and HB. What if all you know about the two acids is that the K_a value for HA is greater than that for HB? Can you tell which of the acids is stronger than the other? Can you tell which of the acid solutions has the lower pH? Defend your answers.

Major Species

HOCl

H$_2$O

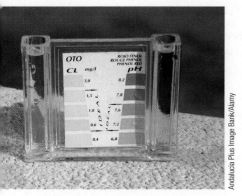

Swimming-pool water must be frequently tested for pH and chlorine content.

Andalucia Plus Image Bank/Alamy

The pH of Weak Acids

The hypochlorite ion (OCl$^-$) is a strong oxidizing agent often found in household bleaches and disinfectants. It is also the active ingredient that forms when swimming-pool water is treated with chlorine. In addition to its oxidizing abilities, the hypochlorite ion has a relatively high affinity for protons (it is a much stronger base than Cl$^-$, for example) and forms the weakly acidic hypochlorous acid (HOCl, $K_a = 3.5 \times 10^{-8}$). Calculate the pH of a 0.100-M aqueous solution of hypochlorous acid.

Solution

1. We list the major species. Since HOCl is a weak acid and remains mostly undissociated, the major species in a 0.100-M HOCl solution are

$$\text{HOCl} \quad \text{and} \quad \text{H}_2\text{O}$$

2. Both species can produce H$^+$:

$$\text{HOCl}(aq) \rightleftharpoons \text{H}^+(aq) + \text{OCl}^-(aq) \qquad K_a = 3.5 \times 10^{-8}$$
$$\text{H}_2\text{O}(l) \rightleftharpoons \text{H}^+(aq) + \text{OH}^-(aq) \qquad K_w = 1.0 \times 10^{-14}$$

3. Since HOCl is a significantly stronger acid than H$_2$O, it will dominate in the production of H$^+$.

4. We therefore use the following equilibrium expression:

$$K_a = 3.5 \times 10^{-8} = \frac{[\text{H}^+][\text{OCl}^-]}{[\text{HOCl}]}$$

5. The initial concentrations appropriate for this equilibrium are

$$[\text{HOCl}]_0 = 0.100 \; M$$
$$[\text{OCl}^-]_0 = 0$$
$$[\text{H}^+]_0 \approx 0 \qquad \text{(We neglect the contribution from H}_2\text{O.)}$$

6. Since the system will reach equilibrium by the dissociation of HOCl, let x be the amount of HOCl (in mol/L) that dissociates in reaching equilibrium.

7. The equilibrium concentrations in terms of x are

$$[\text{HOCl}] = [\text{HOCl}]_0 - x = 0.100 - x$$
$$[\text{OCl}^-] = [\text{OCl}^-]_0 + x = 0 + x = x$$
$$[\text{H}^+] = [\text{H}^+]_0 + x \approx 0 + x = x$$

8. Substituting these concentrations into the equilibrium expression gives

$$K_a = 3.5 \times 10^{-8} = \frac{(x)(x)}{0.100 - x}$$

9. Since K_a is so small, we can expect a small value for x. Thus we make the approximation $[\text{HA}]_0 - x \approx [\text{HA}]_0$, or $0.100 - x \approx 0.100$, which leads to the expression

$$K_a = 3.5 \times 10^{-8} = \frac{x^2}{0.100 - x} \approx \frac{x^2}{0.100}$$

Solving for x gives

$$x = 5.9 \times 10^{-5}$$

10. The approximation $0.100 - x \approx 0.100$ must be validated. To do this, we compare x to $[\text{HOCl}]_0$:

$$\frac{x}{[\text{HA}]_0} \times 100 = \frac{x}{[\text{HOCl}]_0} \times 100 = \frac{5.9 \times 10^{-5}}{0.100} \times 100 = 0.059\%$$

Since this value is much less than 5%, the approximation is considered valid.

11. We calculate $[\text{H}^+]$ and pH:

$$[\text{H}^+] = x = 5.9 \times 10^{-5}\, M \qquad \text{and} \qquad \text{pH} = 4.23$$

See Exercises 14.63 and 14.64

The pH of a Mixture of Weak Acids

The same systematic approach applies to all solution equilibria.

Sometimes a solution contains two weak acids of very different strengths. This case is considered in Example 14.9. Note that the steps are again followed (though not labeled).

The pH of Weak Acid Mixtures

Calculate the pH of a solution that contains 1.00 M HCN ($K_a = 6.2 \times 10^{-10}$) and 5.00 M HNO$_2$ ($K_a = 4.0 \times 10^{-4}$). Also calculate the concentration of cyanide ion (CN$^-$) in this solution at equilibrium.

Solution

Since HCN and HNO$_2$ are both weak acids and are largely undissociated, the major species in the solution are

$$\text{HCN}, \qquad \text{HNO}_2, \qquad \text{and} \qquad \text{H}_2\text{O}$$

Major Species

 HCN

 HNO$_2$

 H$_2$O

All three of these components produce H$^+$:

$$\text{HCN}(aq) \rightleftharpoons \text{H}^+(aq) + \text{CN}^-(aq) \qquad K_a = 6.2 \times 10^{-10}$$
$$\text{HNO}_2(aq) \rightleftharpoons \text{H}^+(aq) + \text{NO}_2^-(aq) \qquad K_a = 4.0 \times 10^{-4}$$
$$\text{H}_2\text{O}(l) \rightleftharpoons \text{H}^+(aq) + \text{OH}^-(aq) \qquad K_w = 1.0 \times 10^{-14}$$

A mixture of three acids might lead to a very complicated problem. However, the situation is greatly simplified by the fact that even though HNO$_2$ is a weak acid, it is much stronger than the other two acids present (as revealed by the K values). Thus HNO$_2$ can be assumed to be the dominant producer of H$^+$, and we will focus on the equilibrium expression

$$K_a = 4.0 \times 10^{-4} = \frac{[\text{H}^+][\text{NO}_2^-]}{[\text{HNO}_2]}$$

The initial concentrations, the definition of x, and the equilibrium concentrations are as follows:

Initial Concentration (mol/L)		Equilibrium Concentration (mol/L)
$[\text{HNO}_2]_0 = 5.00$		$[\text{HNO}_2] = 5.00 - x$
$[\text{NO}_2^-]_0 = 0$	$\xrightarrow{\substack{x\ \text{mol/L HNO}_2 \\ \text{dissociates}}}$	$[\text{NO}_2^-] = x$
$[\text{H}^+]_0 \approx 0$		$[\text{H}^+] = x$

It is convenient to represent these concentrations in the following shorthand form (called an ICE table):

	$HNO_2(aq)$	$\rightleftharpoons$	$H^+(aq)$	$+$	$NO_2^-(aq)$
Initial	5.00		0		0
Change	$-x$		$+x$		$+x$
Equilibrium	$5.00 - x$		x		x

To avoid clutter we do not show the units of concentration in the ICE tables. All terms have units of mol/L.

Substituting the equilibrium concentrations in the equilibrium expression and making the approximation that $5.00 - x = 5.00$ give

$$K_a = 4.0 \times 10^{-4} = \frac{(x)(x)}{5.00 - x} \approx \frac{x^2}{5.00}$$

We solve for x:

$$x = 4.5 \times 10^{-2}$$

Using the 5% rule, we show that the approximation is valid:

$$\frac{x}{[HNO_2]_0} \times 100 = \frac{4.5 \times 10^{-2}}{5.00} \times 100 = 0.90\%$$

Therefore,

$$[H^+] = x = 4.5 \times 10^{-2}\,M \quad \text{and} \quad pH = 1.35$$

We also want to calculate the equilibrium concentration of cyanide ion in this solution. The CN^- ions in this solution come from the dissociation of HCN:

$$HCN(aq) \rightleftharpoons H^+(aq) + CN^-(aq)$$

Although the position of this equilibrium lies far to the left and does not contribute *significantly* to $[H^+]$, HCN is the *only source* of CN^-. Thus we must consider the extent of the dissociation of HCN to calculate $[CN^-]$. The equilibrium expression for the preceding reaction is

$$K_a = 6.2 \times 10^{-10} = \frac{[H^+][CN^-]}{[HCN]}$$

We know $[H^+]$ for this solution from the results of the first part of the problem. It is important to understand that *there is only one kind of H^+ in this solution.* It does not matter from which acid the H^+ ions originate. The equilibrium $[H^+]$ we need to insert into the HCN equilibrium expression is $4.5 \times 10^{-2}\,M$, even though the H^+ was contributed almost entirely from the dissociation of HNO_2. What is $[HCN]$ at equilibrium? We know $[HCN]_0 = 1.00\,M$, and since K_a for HCN is so small, a negligible amount of HCN will dissociate. Thus

$$[HCN] = [HCN]_0 - \text{amount of HCN dissociated} \approx [HCN]_0 = 1.00\,M$$

Since $[H^+]$ and $[HCN]$ are known, we can find $[CN^-]$ from the equilibrium expression:

$$K_a = 6.2 \times 10^{-10} = \frac{[H^+][CN^-]}{[HCN]} = \frac{(4.5 \times 10^{-2})[CN^-]}{1.00}$$

$$[CN^-] = \frac{(6.2 \times 10^{-10})(1.00)}{4.5 \times 10^{-2}} = 1.4 \times 10^{-8}\,M$$

Note the significance of this result. Since $[CN^-] = 1.4 \times 10^{-8}\,M$ and HCN is the only source of CN^-, this means that only 1.4×10^{-8} mol/L of HCN dissociated. This is a very small amount compared with the initial concentration of HCN, which is exactly what we would expect from its very small K_a value, and $[HCN] = 1.00\,M$ as assumed.

See Exercise 14.71

Percent Dissociation

It is often useful to specify the amount of weak acid that has dissociated in achieving equilibrium in an aqueous solution. The **percent dissociation** is defined as follows:

$$\text{Percent dissociation} = \frac{\text{amount dissociated (mol/L)}}{\text{initial concentration (mol/L)}} \times 100\% \quad (14.5)$$

For example, we found earlier that in a $1.00\text{-}M$ solution of HF, $[H^+] = 2.7 \times 10^{-2}\,M$. To reach equilibrium, 2.7×10^{-2} mol/L of the original $1.00\,M$ HF dissociates, so

$$\text{Percent dissociation} = \frac{2.7 \times 10^{-2}\ \text{mol/L}}{1.00\ \text{mol/L}} \times 100\% = 2.7\%$$

For a given weak acid, the percent dissociation increases as the acid becomes more dilute. For example, the percent dissociation of acetic acid ($HC_2H_3O_2$, $K_a = 1.8 \times 10^{-5}$) is significantly greater in a $0.10\text{-}M$ solution than in a $1.0\text{-}M$ solution, as demonstrated in Example 14.10.

Calculating Percent Dissociation

Calculate the percent dissociation of acetic acid ($K_a = 1.8 \times 10^{-5}$) in each of the following solutions:

a. $1.00\,M$ $HC_2H_3O_2$

b. $0.100\,M$ $HC_2H_3O_2$

Major Species

 $HC_2H_3O_2$

 H_2O

Solution

a. Since acetic acid is a weak acid, the major species in this solution are $HC_2H_3O_2$ and H_2O. Both species are weak acids, but acetic acid is a much stronger acid than water. Thus the dominant equilibrium will be

$$HC_2H_3O_2(aq) \rightleftharpoons H^+(aq) + C_2H_3O_2{}^-(aq)$$

and the equilibrium expression is

$$K_a = 1.8 \times 10^{-5} = \frac{[H^+][C_2H_3O_2{}^-]}{[HC_2H_3O_2]}$$

The initial concentrations, definition of x, and equilibrium concentrations are:

	$HC_2H_3O_2(aq)$	$\rightleftharpoons$	$H^+(aq)$	$+$	$C_2H_3O_2{}^-(aq)$
Initial	1.00		0		0
Change	$-x$		x		x
Equilibrium	$1.00 - x$		x		x

Inserting the equilibrium concentrations into the equilibrium expression and making the usual approximation that x is small compared with $[HA]_0$ give

$$K_a = 1.8 \times 10^{-5} = \frac{[H^+][C_2H_3O_2{}^-]}{[HC_2H_3O_2]} = \frac{(x)(x)}{1.00 - x} \approx \frac{x^2}{1.00}$$

Thus

$$x^2 \approx 1.8 \times 10^{-5} \quad \text{and} \quad x \approx 4.2 \times 10^{-3}$$

The approximation $1.00 - x \approx 1.00$ is valid by the 5% rule (check this yourself), so

$$[H^+] = x = 4.2 \times 10^{-3}\,M$$

An acetic acid solution, which is a weak electrolyte, contains only a few ions and does not conduct as much current as a strong electrolyte. The bulb is only dimly lit.

The percent dissociation is

$$\frac{[H^+]}{[HC_2H_3O_2]_0} \times 100 = \frac{4.2 \times 10^{-3}}{1.00} \times 100\% = 0.42\%$$

b. This is a similar problem, except that in this case $[HC_2H_3O_2] = 0.100\ M$. Analysis of the problem leads to the expression

$$K_a = 1.8 \times 10^{-5} = \frac{[H^+][C_2H_3O_2^-]}{[HC_2H_3O_2]} = \frac{(x)(x)}{0.100 - x} \approx \frac{x^2}{0.100}$$

Thus

$$x = [H^+] = 1.3 \times 10^{-3}\ M$$

and

$$\text{Percent dissociation} = \frac{1.3 \times 10^{-3}}{0.10} \times 100\% = 1.3\%$$

See Exercises 14.73 and 14.74

The results in Example 14.10 show two important facts. The concentration of H^+ ion at equilibrium is smaller in the 0.10-M acetic acid solution than in the 1.0-M acetic acid solution, as we would expect. However, the percent dissociation is significantly greater in the 0.10-M solution than in the 1.0-M solution. This is a general result. *For solutions of any weak acid HA, $[H^+]$ decreases as $[HA]_0$ decreases, but the percent dissociation increases as $[HA]_0$ decreases.* This phenomenon can be explained as follows.

> The more dilute the weak acid solution, the greater is the percent dissociation.

Consider the weak acid HA with the initial concentration $[HA]_0$, where at equilibrium

$$[HA] = [HA]_0 - x \approx [HA]_0$$
$$[H^+] = [A^-] = x$$

Thus

$$K_a = \frac{[H^+][A^-]}{[HA]} \approx \frac{(x)(x)}{[HA]_0}$$

Now suppose enough water is added suddenly to dilute the solution by a factor of 10. The new concentrations before any adjustment occurs are

$$[A^-]_{new} = [H^+]_{new} = \frac{x}{10}$$

$$[HA]_{new} = \frac{[HA]_0}{10}$$

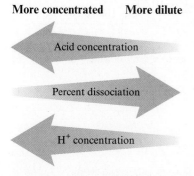

More concentrated   More dilute

Acid concentration

Percent dissociation

H^+ concentration

Figure 14.8 | The effect of dilution on the percent dissociation and $[H^+]$ of a weak acid solution.

and Q, the reaction quotient, is

$$Q = \frac{\left(\dfrac{x}{10}\right)\left(\dfrac{x}{10}\right)}{\dfrac{[HA]_0}{10}} = \frac{1}{10}\frac{(x)(x)}{[HA]_0} = \frac{1}{10}K_a$$

Since Q is less than K_a, the system must adjust to the right to reach the new equilibrium position. Thus the percent dissociation increases when the acid is diluted. This behavior is summarized in Fig. 14.8. In Example 14.11 we see how the percent dissociation can be used to calculate the K_a value for a weak acid.

Calculating K_a from Percent Dissociation

Lactic acid ($HC_3H_5O_3$) is a chemical that accumulates in muscle tissue during exertion. In a 0.100-M aqueous solution, lactic acid is 3.7% dissociated. Calculate the value of K_a for this acid.

Solution

From the small value for the percent dissociation, it is clear that $HC_3H_5O_3$ is a weak acid. Thus the major species in the solution are the undissociated acid and water:

$$HC_3H_5O_3 \quad \text{and} \quad H_2O$$

However, even though $HC_3H_5O_3$ is a weak acid, it is a much stronger acid than water and will be the dominant source of H^+ in the solution. The dissociation reaction is

$$HC_3H_5O_3(aq) \rightleftharpoons H^+(aq) + C_3H_5O_3^-(aq)$$

and the equilibrium expression is

$$K_a = \frac{[H^+][C_3H_5O_3^-]}{[HC_3H_5O_3]}$$

Major Species

$HC_3H_5O_3$

H_2O

The initial and equilibrium concentrations are as follows:

Initial Concentration (mol/L)		Equilibrium Concentration (mol/L)
$[HC_3H_5O_3]_0 = 0.10$	$\xrightarrow[\text{dissociates}]{\substack{x\ \text{mol/L} \\ HC_3H_5O_3}}$	$[HC_3H_5O_3] = 0.10 - x$
$[C_3H_5O_3^-]_0 = 0$		$[C_3H_5O_3^-] = x$
$[H^+]_0 \approx 0$		$[H^+] = x$

The change needed to reach equilibrium can be obtained from the percent dissociation and Equation (14.5). For this acid,

$$\text{Percent dissociation} = 3.7\% = \frac{x}{[HC_3H_5O_3]_0} \times 100\% = \frac{x}{0.10} \times 100\%$$

and

$$x = \frac{3.7}{100}(0.10) = 3.7 \times 10^{-3} \text{ mol/L}$$

Now we can calculate the equilibrium concentrations:

$$[HC_3H_5O_3] = 0.10 - x = 0.10\ M \qquad \text{(to the correct number of significant figures)}$$
$$[C_3H_5O_3^-] = [H^+] = x = 3.7 \times 10^{-3}\ M$$

These concentrations can now be used to calculate the value of K_a for lactic acid:

$$K_a = \frac{[H^+][C_3H_5O_3^-]}{[HC_3H_5O_3]} = \frac{(3.7 \times 10^{-3})(3.7 \times 10^{-3})}{0.10} = 1.4 \times 10^{-4}$$

See Exercises 14.75 and 14.76

Strenuous exercise causes a buildup of lactic acid in muscle tissues.

Bob Daemmrich

Problem-Solving Strategy

Solving Acid–Base Problems

1. List the major species in solution.
2. Look for reactions that can be assumed to go to completion—for example, a strong acid dissociating or H^+ reacting with OH^-.

3. For a reaction that can be assumed to go to completion:
   a. Determine the concentration of the products.
   b. Write down the major species in solution after the reaction.

4. Look at each major component of the solution and decide if it is an acid or a base.

5. Pick the equilibrium that will control the pH. Use known values of the dissociation constants for the various species to help decide on the dominant equilibrium.
   a. Write the equation for the reaction and the equilibrium expression.
   b. Compute the initial concentrations (assuming the dominant equilibrium has not yet occurred, that is, no acid dissociation, and so on).
   c. Define x.
   d. Compute the equilibrium concentrations in terms of x.
   e. Substitute the concentrations into the equilibrium expression, and solve for x.
   f. Check the validity of the approximation.
   g. Calculate the pH and other concentrations as required.

Although these steps may seem somewhat cumbersome, especially for simpler problems, they will become increasingly helpful as the aqueous solutions become more complicated. If you develop the habit of approaching acid–base problems systematically, the more complex cases will be much easier to manage.

14.6 | Bases

In a basic solution at 25°C, pH > 7.

According to the Arrhenius concept, a base is a substance that produces OH^- ions in aqueous solution. According to the Brønsted–Lowry model, a base is a proton acceptor. The bases sodium hydroxide (NaOH) and potassium hydroxide (KOH) fulfill both criteria. They contain OH^- ions in the solid lattice and, behaving as strong electrolytes, dissociate completely when dissolved in aqueous solution:

$$NaOH(s) \longrightarrow Na^+(aq) + OH^-(aq)$$

leaving virtually no undissociated NaOH. Thus a 1.0-M NaOH solution really contains 1.0 M Na^+ and 1.0 M OH^-. Because of their complete dissociation, NaOH and KOH are called **strong bases** in the same sense as we defined strong acids.

All the hydroxides of the Group 1A elements (LiOH, NaOH, KOH, RbOH, and CsOH) are strong bases, but only NaOH and KOH are common laboratory reagents, because the lithium, rubidium, and cesium compounds are expensive. The alkaline earth (Group 2A) hydroxides—$Ca(OH)_2$, $Ba(OH)_2$, and $Sr(OH)_2$—are also strong bases. For these compounds, 2 moles of hydroxide ion are produced for every mole of metal hydroxide dissolved in aqueous solution.

The alkaline earth hydroxides are not very soluble and are used only when the solubility factor is not important. In fact, the low solubility of these bases can sometimes be an advantage. For example, many antacids are suspensions of metal hydroxides, such as aluminum hydroxide and magnesium hydroxide. The low solubility of these compounds prevents a large hydroxide ion concentration that would harm the tissues of the mouth, esophagus, and stomach. Yet these suspensions furnish plenty of hydroxide ion to react with the stomach acid, since the salts dissolve as this reaction proceeds.

Calcium hydroxide, $Ca(OH)_2$, often called **slaked lime**, is widely used in industry because it is inexpensive and plentiful. For example, slaked lime is used in scrubbing stack gases to remove sulfur dioxide from the exhaust of power plants and factories. In

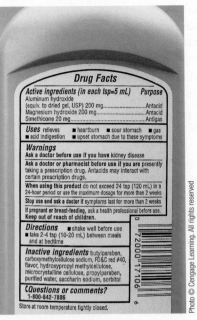

An antacid containing aluminum and magnesium hydroxides.

Calcium carbonate is also used in scrubbing, as discussed in Section 5.10.

the scrubbing process a suspension of slaked lime is sprayed into the stack gases to react with sulfur dioxide gas according to the following steps:

$$SO_2(g) + H_2O(l) \rightleftharpoons H_2SO_3(aq)$$
$$Ca(OH)_2(aq) + H_2SO_3(aq) \rightleftharpoons CaSO_3(s) + 2H_2O(l)$$

Slaked lime is also widely used in water treatment plants for softening hard water, which involves the removal of ions, such as Ca^{2+} and Mg^{2+}, that hamper the action of detergents. The softening method most often employed in water treatment plants is the **lime–soda process**, in which *lime* (CaO) and *soda ash* (Na_2CO_3) are added to the water. As we will see in more detail later in this chapter, the CO_3^{2-} ion reacts with water to produce the HCO_3^- ion. When the lime is added to the water, it forms slaked lime, that is,

$$CaO(s) + H_2O(l) \longrightarrow Ca(OH)_2(aq)$$

which then reacts with the HCO_3^- ion from the added soda ash and the Ca^{2+} ion in the hard water to produce calcium carbonate:

$$Ca(OH)_2(aq) + Ca^{2+}(aq) + 2HCO_3^-(aq) \longrightarrow 2CaCO_3(s) + 2H_2O(l)$$

From hard water

Thus, for every mole of $Ca(OH)_2$ consumed, 1 mole of Ca^{2+} is removed from the hard water, thereby softening it. Some hard water naturally contains bicarbonate ions. In this case, no soda ash is needed—simply adding the lime produces the softening.

Calculating the pH of a strong base solution is relatively simple, as illustrated in Example 14.12.

Interactive Example 14.12

Sign in at http://login.cengagebrain .com to try this Interactive Example in **OWL**.

Major Species

 Na^+

 OH^-

 H_2O

The pH of Strong Bases

Calculate the pH of a 5.0×10^{-2}-M NaOH solution.

Solution

The major species in this solution are

$$\underbrace{Na^+, \quad OH^-,}_{\text{From NaOH}} \quad \text{and} \quad H_2O$$

Although autoionization of water also produces OH^- ions, the pH will be dominated by the OH^- ions from the dissolved NaOH. Thus in the solution,

$$[OH^-] = 5.0 \times 10^{-2}\,M$$

and the concentration of H^+ can be calculated from K_w:

$$[H^+] = \frac{K_w}{[OH^-]} = \frac{1.0 \times 10^{-14}}{5.0 \times 10^{-2}} = 2.0 \times 10^{-13}\,M$$
$$pH = 12.70$$

Note that this is a basic solution for which

$$[OH^-] > [H^+] \quad \text{and} \quad pH > 7$$

The added OH^- from the salt has shifted the water autoionization equilibrium

$$H_2O(l) \rightleftharpoons H^+(aq) + OH^-(aq)$$

to the left, significantly lowering $[H^+]$ compared with that in pure water.

See Exercises 14.89 through 14.92

A base does not have to contain hydroxide ion.

Many types of proton acceptors (bases) do not contain the hydroxide ion. However, when dissolved in water, these substances increase the concentration of hydroxide ion because of their reaction with water. For example, ammonia reacts with water as follows:

$$NH_3(aq) + H_2O(l) \rightleftharpoons NH_4^+(aq) + OH^-(aq)$$

The ammonia molecule accepts a proton and thus functions as a base. Water is the acid in this reaction. Note that even though the base ammonia contains no hydroxide ion, it still increases the concentration of hydroxide ion to yield a basic solution.

Bases such as ammonia typically have at least one unshared pair of electrons that is capable of forming a bond with a proton. The reaction of an ammonia molecule with a water molecule can be represented as follows:

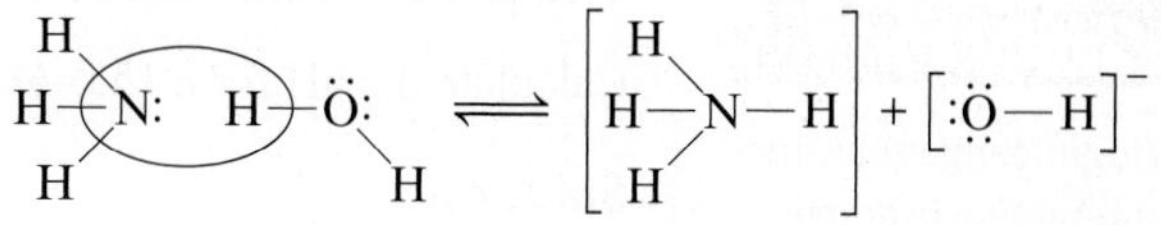

There are many bases like ammonia that produce hydroxide ion by reaction with water. In most of these bases, the lone pair is located on a nitrogen atom. Some examples are

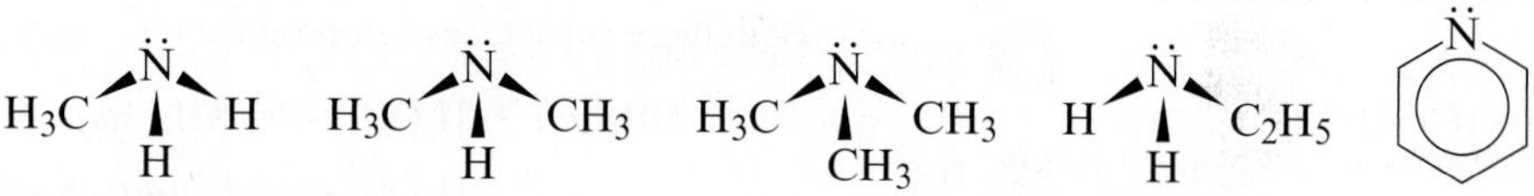

Methylamine   Dimethylamine   Trimethylamine   Ethylamine   Pyridine

Note that the first four bases can be thought of as substituted ammonia molecules with hydrogen atoms replaced by methyl (CH_3) or ethyl (C_2H_5) groups. The pyridine molecule is like benzene

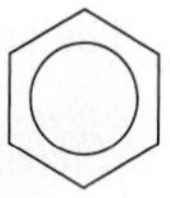

except that a nitrogen atom replaces one of the carbon atoms in the ring. The general reaction between a base B and water is given by

$$B(aq) + H_2O(l) \rightleftharpoons BH^+(aq) + OH^-(aq) \qquad (14.6)$$

Base   Acid   Conjugate acid   Conjugate base

The equilibrium constant for this general reaction is

Appendix 5.3 contains a table of K_b values.

$$K_b = \frac{[BH^+][OH^-]}{[B]}$$

where K_b always refers to the reaction of a base with water to form the conjugate acid and the hydroxide ion.

Bases of the type represented by B in Equation (14.6) compete with OH^-, a very strong base, for the H^+ ion. Thus their K_b values tend to be small (for example, for ammonia, $K_b = 1.8 \times 10^{-5}$), and they are called **weak bases**. The values of K_b for some common weak bases are listed in Table 14.3.

Refer to the steps for solving weak acid equilibrium problems. Use the same systematic approach for weak base equilibrium problems.

Typically, pH calculations for solutions of weak bases are very similar to those for weak acids, as illustrated by Examples 14.13 and 14.14.

Table 14.3 | Values of K_b for Some Common Weak Bases

Name	Formula	Conjugate Acid	K_b
Ammonia	NH_3	NH_4^+	1.8×10^{-5}
Methylamine	CH_3NH_2	$CH_3NH_3^+$	4.38×10^{-4}
Ethylamine	$C_2H_5NH_2$	$C_2H_5NH_3^+$	5.6×10^{-4}
Aniline	$C_6H_5NH_2$	$C_6H_5NH_3^+$	3.8×10^{-10}
Pyridine	C_5H_5N	$C_5H_5NH^+$	1.7×10^{-9}

Interactive Example 14.13

Sign in at http://login.cengagebrain.com to try this Interactive Example in **OWL**.

The pH of Weak Bases I

Calculate the pH for a 15.0-*M* solution of NH_3 ($K_b = 1.8 \times 10^{-5}$).

Solution

Since ammonia is a weak base, as can be seen from its small K_b value, most of the dissolved NH_3 will remain as NH_3. Thus the major species in solution are

$$NH_3 \quad \text{and} \quad H_2O$$

Major Species

NH_3

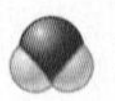

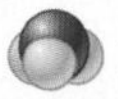

H_2O

Both these substances can produce OH^- according to the reactions

$$NH_3(aq) + H_2O(l) \rightleftharpoons NH_4^+(aq) + OH^-(aq) \qquad K_b = 1.8 \times 10^{-5}$$

$$H_2O(l) \rightleftharpoons H^+(aq) + OH^-(aq) \qquad K_w = 1.0 \times 10^{-14}$$

However, the contribution from water can be neglected, since $K_b \gg K_w$. The equilibrium for NH_3 will dominate, and the equilibrium expression to be used is

$$K_b = 1.8 \times 10^{-5} = \frac{[NH_4^+][OH^-]}{[NH_3]}$$

The appropriate concentrations are

Initial Concentration (mol/L)		Equilibrium Concentration (mol/L)
$[NH_3]_0 = 15.0$ $[NH_4^+]_0 = 0$ $[OH^-]_0 \approx 0$	x mol/L NH_3 reacts with H_2O to reach equilibrium →	$[NH_3] = 15.0 - x$ $[NH_4^+] = x$ $[OH^-] = x$

In terms of an ICE table, these concentrations are

	$NH_3(aq)$	+	$H_2O(l)$	$\rightleftharpoons$	$NH_4^+(aq)$	+	$OH^-(aq)$
Initial	15.0		—		0		0
Change	$-x$		—		$+x$		$+x$
Equilibrium	$15.0 - x$		—		x		x

Substituting the equilibrium concentrations into the equilibrium expression and making the usual approximation gives

$$K_b = 1.8 \times 10^{-5} = \frac{[NH_4^+][OH^-]}{[NH_3]} = \frac{(x)(x)}{15.0 - x} \approx \frac{x^2}{15.0}$$

Thus

$$x \approx 1.6 \times 10^{-2}$$

The 5% rule validates the approximation (check it yourself), so

$$[OH^-] = 1.6 \times 10^{-2}\, M$$

Chemical connections
Amines

We have seen that many bases have nitrogen atoms with one lone pair and can be viewed as substituted ammonia molecules, with the general formula $R_xNH_{(3-x)}$. Compounds of this type are called **amines**. Amines are widely distributed in animals and plants, and complex amines often serve as messengers or regulators. For example, in the human nervous system, there are two amine stimulants, *norepinephrine* and *adrenaline*.

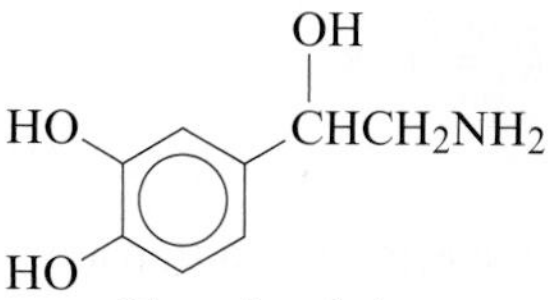

Norepinephrine

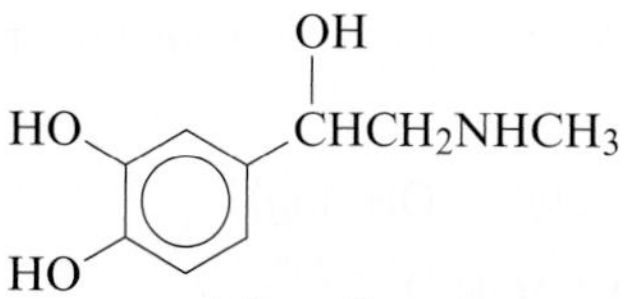

Adrenaline

Ephedrine, widely used as a decongestant, was a known drug in China over 2000 years ago. Indians in Mexico and the Southwest have used the hallucinogen *mescaline*, extracted from the peyote cactus, for centuries.

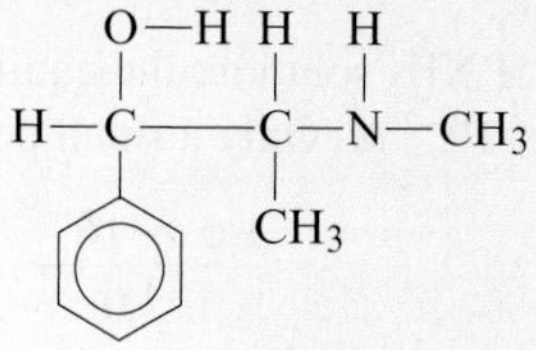

Ephedrine

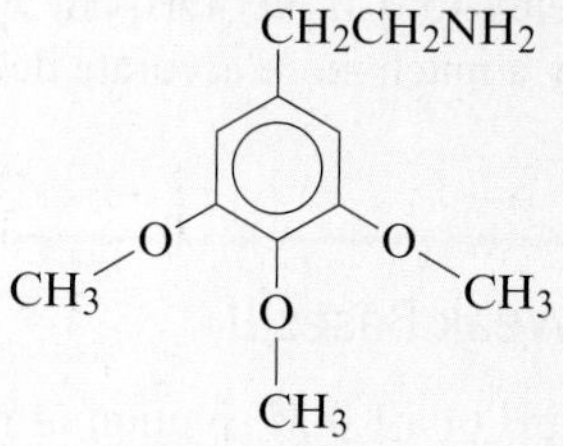

Mescaline

Many other drugs, such as codeine and quinine, are amines, but they are usually not used in their pure amine forms. Instead, they are treated with an acid to become acid salts. An example of an acid salt is ammonium chloride, obtained by the reaction

$$NH_3 + HCl \longrightarrow NH_4Cl$$

Amines also can be protonated in this way. The resulting acid salt, written as AHCl (where A represents the amine), contains AH^+ and Cl^-. In general, the acid salts are more stable and more soluble in water than the parent amines. For instance, the parent amine of the well-known local

Peyote cactus growing on a rock.

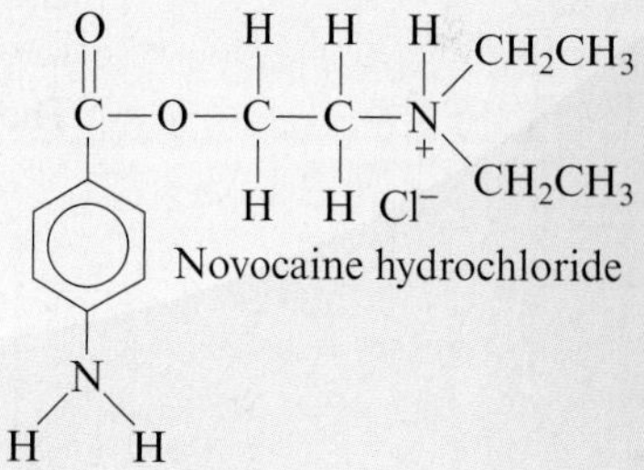

anesthetic *novocaine* is insoluble in water, whereas the acid salt is much more soluble.

Novocaine hydrochloride

Since we know that K_w must be satisfied for this solution, we can calculate $[H^+]$ as follows:

$$[H^+] = \frac{K_w}{[OH^-]} = \frac{1.0 \times 10^{-14}}{1.6 \times 10^{-2}} = 6.3 \times 10^{-13}\ M$$

Therefore, $pH = -\log(6.3 \times 10^{-13}) = 12.20$

See Exercises 14.95 and 14.96

A table of K_b values for bases is also given in Appendix 5.3.

Example 14.13 illustrates how a typical weak base equilibrium problem should be solved. Note two additional important points:

1. We calculated [H$^+$] from K_w and then calculated the pH, but another method is available. The pOH could have been calculated from [OH$^-$] and then used in Equation (14.3):

$$pK_w = 14.00 = pH + pOH$$
$$pH = 14.00 - pOH$$

2. In a 15.0-M NH$_3$ solution, the equilibrium concentrations of NH$_4^+$ and OH$^-$ are each 1.6×10^{-2} M. Only a small percentage,

$$\frac{1.6 \times 10^{-2}}{15.0} \times 100\% = 0.11\%$$

of the ammonia reacts with water. Bottles containing 15.0 M NH$_3$ solution are often labeled 15.0 M NH$_4$OH, but as you can see from these results, 15.0 M NH$_3$ is actually a much more accurate description of the solution contents.

Interactive Example 14.14

Sign in at http://login.cengagebrain .com to try this Interactive Example in **OWL**.

Major Species

 CH$_3$NH$_2$

 H$_2$O

The pH of Weak Bases II

Calculate the pH of a 1.0-M solution of methylamine ($K_b = 4.38 \times 10^{-4}$).

Solution

Since methylamine (CH$_3$NH$_2$) is a weak base, the major species in solution are

$$CH_3NH_2 \quad \text{and} \quad H_2O$$

Both are bases; however, water can be neglected as a source of OH$^-$, so the dominant equilibrium is

$$CH_3NH_2(aq) + H_2O(l) \rightleftharpoons CH_3NH_3^+(aq) + OH^-(aq)$$

and

$$K_b = 4.38 \times 10^{-4} = \frac{[CH_3NH_3^+][OH^-]}{[CH_3NH_2]}$$

The ICE table is:

	CH$_3$NH$_2$(aq)	+	H$_2$O(l)	$\rightleftharpoons$	CH$_3$NH$_3^+$(aq)	+	OH$^-$(aq)
Initial	1.0		—		0		0
Change	$-x$		—		$+x$		$+x$
Equilibrium	$1.0 - x$		—		x		x

Substituting the equilibrium concentrations in the equilibrium expression and making the usual approximation give

$$K_b = 4.38 \times 10^{-4} = \frac{[CH_3NH_3^+][OH^-]}{[CH_3NH_2]} = \frac{(x)(x)}{1.0 - x} \approx \frac{x^2}{1.0}$$

and

$$x \approx 2.1 \times 10^{-2}$$

The approximation is valid by the 5% rule, so

$$[OH^-] = x = 2.1 \times 10^{-2}\,M$$
$$pOH = 1.68$$
$$pH = 14.00 - 1.68 = 12.32$$

See Exercises 14.97 and 14.98

14.7 | Polyprotic Acids

Some important acids, such as sulfuric acid (H_2SO_4) and phosphoric acid (H_3PO_4), can furnish more than one proton and are called **polyprotic acids**. A polyprotic acid always dissociates in a *stepwise* manner, one proton at a time. For example, the diprotic (two-proton) acid carbonic acid (H_2CO_3), which is so important in maintaining a constant pH in human blood, dissociates in the following steps:

$$H_2CO_3(aq) \rightleftharpoons H^+(aq) + HCO_3^-(aq) \quad K_{a_1} = \frac{[H^+][HCO_3^-]}{[H_2CO_3]} = 4.3 \times 10^{-7}$$

$$HCO_3^-(aq) \rightleftharpoons H^+(aq) + CO_3^{2-}(aq) \quad K_{a_2} = \frac{[H^+][CO_3^{2-}]}{[HCO_3^-]} = 5.6 \times 10^{-11}$$

The successive K_a values for the dissociation equilibria are designated K_{a_1} and K_{a_2}. Note that the conjugate base HCO_3^- of the first dissociation equilibrium becomes the acid in the second step.

Carbonic acid is formed when carbon dioxide gas is dissolved in water. In fact, the first dissociation step for carbonic acid is best represented by the reaction

$$CO_2(aq) + H_2O(l) \rightleftharpoons H^+(aq) + HCO_3^-(aq)$$

since relatively little H_2CO_3 actually exists in solution. However, it is convenient to consider CO_2 in water as H_2CO_3 so that we can treat such solutions using the familiar dissociation reactions for weak acids.

Phosphoric acid is a **triprotic acid** (three protons) that dissociates in the following steps:

$$H_3PO_4(aq) \rightleftharpoons H^+(aq) + H_2PO_4^-(aq) \quad K_{a_1} = \frac{[H^+][H_2PO_4^-]}{[H_3PO_4]} = 7.5 \times 10^{-3}$$

$$H_2PO_4^-(aq) \rightleftharpoons H^+(aq) + HPO_4^{2-}(aq) \quad K_{a_2} = \frac{[H^+][HPO_4^{2-}]}{[H_2PO_4^-]} = 6.2 \times 10^{-8}$$

$$HPO_4^{2-}(aq) \rightleftharpoons H^+(aq) + PO_4^{3-}(aq) \quad K_{a_3} = \frac{[H^+][PO_4^{3-}]}{[HPO_4^{2-}]} = 4.8 \times 10^{-13}$$

For a typical weak polyprotic acid,

$$K_{a_1} > K_{a_2} > K_{a_3}$$

That is, the acid involved in each step of the dissociation is successively weaker, as shown by the stepwise dissociation constants given in Table 14.4. These values indicate that the loss of a second or third proton occurs less readily than loss of the first proton. This is not surprising; as the negative charge on the acid increases, it becomes more difficult to remove the positively charged proton.

A table of K_a values for polyprotic acids is also given in Appendix 5.2.

Table 14.4 | Stepwise Dissociation Constants for Several Common Polyprotic Acids

Name	Formula	K_{a_1}	K_{a_2}	K_{a_3}
Phosphoric acid	H_3PO_4	7.5×10^{-3}	6.2×10^{-8}	4.8×10^{-13}
Arsenic acid	H_3AsO_4	5.5×10^{-3}	1.7×10^{-7}	5.1×10^{-12}
Carbonic acid	H_2CO_3	4.3×10^{-7}	5.6×10^{-11}	
Sulfuric acid	H_2SO_4	Large	1.2×10^{-2}	
Sulfurous acid	H_2SO_3	1.5×10^{-2}	1.0×10^{-7}	
Hydrosulfuric acid*	H_2S	1.0×10^{-7}	$\sim10^{-19}$	
Oxalic acid	$H_2C_2O_4$	6.5×10^{-2}	6.1×10^{-5}	
Ascorbic acid (vitamin C)	$H_2C_6H_6O_6$	7.9×10^{-5}	1.6×10^{-12}	

*The K_{a_2} value for H_2S is very uncertain. Because it is so small, the K_{a_2} value is very difficult to measure accurately.

Although we might expect the pH calculations for solutions of polyprotic acids to be complicated, the most common cases are surprisingly straightforward. To illustrate, we will consider a typical case, phosphoric acid, and a unique case, sulfuric acid.

Phosphoric Acid

Phosphoric acid is typical of most polyprotic acids in that the successive K_a values are very different. For example, the ratios of successive K_a values (from Table 14.4) are

$$\frac{K_{a_1}}{K_{a_2}} = \frac{7.5 \times 10^{-3}}{6.2 \times 10^{-8}} = 1.2 \times 10^5$$

$$\frac{K_{a_2}}{K_{a_3}} = \frac{6.2 \times 10^{-8}}{4.8 \times 10^{-13}} = 1.3 \times 10^5$$

Thus the relative acid strengths are

$$H_3PO_4 \gg H_2PO_4^- \gg HPO_4^{2-}$$

For a typical polyprotic acid in water, only the first dissociation step is important in determining the pH.

This means that in a solution prepared by dissolving H_3PO_4 in water, *only the first dissociation step makes an important contribution to [H^+]*. This greatly simplifies the pH calculations for phosphoric acid solutions, as is illustrated in Example 14.15.

Interactive Example 14.15

Sign in at http://login.cengagebrain.com to try this Interactive Example in **OWL**.

The pH of a Polyprotic Acid

Calculate the pH of a 5.0-M H_3PO_4 solution and the equilibrium concentrations of the species H_3PO_4, $H_2PO_4^-$, HPO_4^{2-}, and PO_4^{3-}.

Solution

The major species in solution are

$$H_3PO_4 \quad \text{and} \quad H_2O$$

Major Species

H_3PO_4

H_2O

None of the dissociation products of H_3PO_4 is written, since the K_a values are all so small that they will be minor species. The dominant equilibrium is the dissociation of H_3PO_4:

$$H_3PO_4(aq) \rightleftharpoons H^+(aq) + H_2PO_4^-(aq)$$

where

$$K_{a_1} = 7.5 \times 10^{-3} = \frac{[H^+][H_2PO_4^-]}{[H_3PO_4]}$$

The ICE table is:

	$H_3PO_4(aq)$	$\rightleftharpoons$	$H^+(aq)$	$+$	$H_2PO_4^-(aq)$
Initial	5.0		0		0
Change	$-x$		$+x$		$+x$
Equilibrium	$5.0 - x$		x		x

Substituting the equilibrium concentrations into the expression for K_{a_1} and making the usual approximation give

$$K_{a_1} = 7.5 \times 10^{-3} = \frac{[H^+][H_2PO_4^-]}{[H_3PO_4]} = \frac{(x)(x)}{5.0 - x} \approx \frac{x^2}{5.0}$$

Thus

$$x \approx 1.9 \times 10^{-1}$$

Since 1.9×10^{-1} is less than 5% of 5.0, the approximation is acceptable, and

$$[H^+] = x = 0.19 \, M$$

$$pH = 0.72$$

So far we have determined that

$$[H^+] = [H_2PO_4^-] = 0.19 \ M$$

and

$$[H_3PO_4] = 5.0 - x = 4.8 \ M$$

The concentration of HPO_4^{2-} can be obtained by using the expression for K_{a_2}:

$$K_{a_2} = 6.2 \times 10^{-8} = \frac{[H^+][HPO_4^{2-}]}{[H_2PO_4^-]}$$

where

$$[H^+] = [H_2PO_4^-] = 0.19 \ M$$

Thus

$$[HPO_4^{2-}] = K_{a_2} = 6.2 \times 10^{-8} \ M$$

To calculate $[PO_4^{3-}]$, we use the expression for K_{a_3} and the values of $[H^+]$ and $[HPO_4^{2-}]$ calculated previously:

$$K_{a_3} = \frac{[H^+][PO_4^{3-}]}{[HPO_4^{2-}]} = 4.8 \times 10^{-13} = \frac{0.19[PO_4^{3-}]}{(6.2 \times 10^{-8})}$$

$$[PO_4^{3-}] = \frac{(4.8 \times 10^{-13})(6.2 \times 10^{-8})}{0.19} = 1.6 \times 10^{-19} \ M$$

These results show that the second and third dissociation steps do not make an important contribution to $[H^+]$. This is apparent from the fact that $[HPO_4^{2-}]$ is $6.2 \times 10^{-8} \ M$, which means that only 6.2×10^{-8} mol/L $H_2PO_4^-$ has dissociated. The value of $[PO_4^{3-}]$ shows that the dissociation of HPO_4^{2-} is even smaller. We must, however, use the second and third dissociation steps to calculate $[HPO_4^{2-}]$ and $[PO_4^{3-}]$, since these steps are the only sources of these ions.

See Exercises 14.107 and 14.108

Critical Thinking

What if the three values of K_a for phosphoric acid were closer to each other in value? Why would this complicate the calculation of the pH for an aqueous solution of phosphoric acid?

Sulfuric Acid

Sulfuric acid is unique among the common acids in that it is *a strong acid in its first dissociation step and a weak acid in its second step:*

$$H_2SO_4(aq) \longrightarrow H^+(aq) + HSO_4^-(aq) \quad K_{a_1} \text{ is very large}$$
$$HSO_4^-(aq) \rightleftharpoons H^+(aq) + SO_4^{2-}(aq) \quad K_{a_2} = 1.2 \times 10^{-2}$$

Example 14.16 illustrates how to calculate the pH for sulfuric acid solutions.

Interactive Example 14.16

Sign in at http://login.cengagebrain.com to try this Interactive Example in **OWL**.

The pH of Sulfuric Acid

Calculate the pH of a 1.0-M H_2SO_4 solution.

Solution

The major species in the solution are

$$H^+, \quad HSO_4^-, \quad \text{and} \quad H_2O$$

Major Species

H⁺

HSO₄⁻

H₂O

where the first two ions are produced by the complete first dissociation step of H_2SO_4. The concentration of H^+ in this solution will be at least 1.0 M, since this amount is produced by the first dissociation step of H_2SO_4. We must now answer this question: Does the HSO_4^- ion dissociate enough to produce a significant contribution to the concentration of H^+? This question can be answered by calculating the equilibrium concentrations for the dissociation reactions of HSO_4^-:

$$HSO_4^-(aq) \rightleftharpoons H^+(aq) + SO_4^{2-}(aq)$$

where

$$K_{a_2} = 1.2 \times 10^{-2} = \frac{[H^+][SO_4^{2-}]}{[HSO_4^-]}$$

The ICE table is:

	$HSO_4^-(aq)$	$\rightleftharpoons$	$H^+(aq)$	+	$SO_4^{2-}(aq)$
Initial	1.0		1.0		0
Change	$-x$		$+x$		$+x$
Equilibrium	$1.0 - x$		$1.0 + x$		x

Note that $[H^+]_0$ is not equal to zero, as it usually is for a weak acid, because the first dissociation step has already occurred. Substituting the equilibrium concentrations into the expression for K_{a_2} and making the usual approximation give

$$K_{a_2} = 1.2 \times 10^{-2} = \frac{[H^+][SO_4^{2-}]}{[HSO_4^-]} = \frac{(1.0 + x)(x)}{1.0 - x} \approx \frac{(1.0)(x)}{(1.0)}$$

Thus

$$x \approx 1.2 \times 10^{-2}$$

Since 1.2×10^{-2} is 1.2% of 1.0, the approximation is valid according to the 5% rule. Note that x is not equal to $[H^+]$ in this case. Instead,

$$[H^+] = 1.0\ M + x = 1.0\ M + (1.2 \times 10^{-2})\ M$$

$$= 1.0\ M \quad \text{(to the correct number of significant figures)}$$

Thus the dissociation of HSO_4^- does not make a significant contribution to the concentration of H^+, and

$$[H^+] = 1.0\ M \quad \text{and} \quad pH = 0.00$$

See Exercise 14.111

Only in dilute H_2SO_4 solutions does the second dissociation step contribute significantly to [H^+].

Example 14.16 illustrates the most common case for sulfuric acid in which only the first dissociation makes an important contribution to the concentration of H^+. In solutions more dilute than 1.0 M (for example, 0.10 M H_2SO_4), the dissociation of HSO_4^- is important, and solving the problem requires use of the quadratic formula, as shown in Example 14.17.

Example 14.17

The pH of Sulfuric Acid

Calculate the pH of a $1.00 \times 10^{-2}\ M$ H_2SO_4 solution.

Solution

The major species in solution are

$$H^+, \quad HSO_4^-, \quad \text{and} \quad H_2O$$

Major Species

H⁺

HSO₄⁻

H₂O

Proceeding as in Example 14.16, we consider the dissociation of HSO_4^-, which leads to the following ICE table:

	$HSO_4^-(aq)$	$\rightleftharpoons$	$H^+(aq)$	+	$SO_4^{2-}(aq)$
Initial	0.0100		0.0100		0
			From dissociation of H_2SO_4		
Change	$-x$		$+x$		$+x$
Equilibrium	$0.0100 - x$		$0.0100 + x$		x

Substituting the equilibrium concentrations into the expression for K_{a_2} gives

$$1.2 \times 10^{-2} = K_{a_2} = \frac{[H^+][SO_4^{2-}]}{[HSO_4^-]} = \frac{(0.0100 + x)(x)}{(0.0100 - x)}$$

If we make the usual approximation, then $0.0100 + x \approx 0.0100$ and $0.0100 - x \approx 0.0100$, and we have

$$1.2 \times 10^{-2} = \frac{(0.0100 + x)(x)}{(0.0100 - x)} \approx \frac{(0.0100)x}{(0.0100)}$$

The calculated value of x is

$$x = 1.2 \times 10^{-2} = 0.012$$

This value is larger than 0.010, clearly a ridiculous result. Thus we cannot make the usual approximation and must instead solve the quadratic equation. The expression

$$1.2 \times 10^{-2} = \frac{(0.0100 + x)(x)}{(0.0100 - x)}$$

leads to
$$(1.2 \times 10^{-2})(0.0100 - x) = (0.0100 + x)(x)$$
$$(1.2 \times 10^{-4}) - (1.2 \times 10^{-2})x = (1.0 \times 10^{-2})x + x^2$$
$$x^2 + (2.2 \times 10^{-2})x - (1.2 \times 10^{-4}) = 0$$

This equation can be solved using the quadratic formula

$$x = \frac{-b \pm \sqrt{b^2 - 4ac}}{2a}$$

where $a = 1$, $b = 2.2 \times 10^{-2}$, and $c = -1.2 \times 10^{-4}$. Use of the quadratic formula gives one negative root (which cannot be correct) and one positive root,

$$x = 4.5 \times 10^{-3}$$

Thus

$$[H^+] = 0.0100 + x = 0.0100 + 0.0045 = 0.0145$$

and

$$pH = 1.84$$

Note that in this case the second dissociation step produces about half as many H^+ ions as the initial step does.

This problem also can be solved by successive approximations, a method illustrated in Appendix 1.4.

See Exercise 14.112

14.8 | Acid–Base Properties of Salts

Salt is simply another name for *ionic compound*. When a salt dissolves in water, we assume that it breaks up into its ions, which move about independently, at least in dilute solutions. Under certain conditions, these ions can behave as acids or bases. In this section we explore such reactions.

Salts That Produce Neutral Solutions

The salt of a strong acid and a strong base gives a neutral solution.

Recall that the conjugate base of a strong acid has virtually no affinity for protons in water. This is why strong acids completely dissociate in aqueous solution. Thus, when anions such as Cl^- and NO_3^- are placed in water, they do not combine with H^+ and have no effect on the pH. Cations such as K^+ and Na^+ from strong bases have no affinity for H^+, nor can they produce H^+, so they too have no effect on the pH of an aqueous solution. Salts that consist of the cations of strong bases and the anions of strong acids have no effect on $[H^+]$ when dissolved in water. This means that aqueous solutions of salts such as KCl, NaCl, NaNO$_3$, and KNO$_3$ are neutral (have a pH of 7).

Salts That Produce Basic Solutions

Major Species

Na^+

$C_2H_3O_2^-$

H_2O

In an aqueous solution of sodium acetate (NaC$_2$H$_3$O$_2$), the major species are

$$Na^+, \quad C_2H_3O_2^-, \quad \text{and} \quad H_2O$$

What are the acid–base properties of each component? The Na^+ ion has neither acid nor base properties. The $C_2H_3O_2^-$ ion is the conjugate base of acetic acid, a weak acid. This means that $C_2H_3O_2^-$ has a significant affinity for a proton and is a base. Finally, water is a weakly amphoteric substance.

The pH of this solution will be determined by the $C_2H_3O_2^-$ ion. Since $C_2H_3O_2^-$ is a base, it will react with the best proton donor available. In this case, water is the *only* source of protons, and the reaction between the acetate ion and water is

$$C_2H_3O_2^-(aq) + H_2O(l) \rightleftharpoons HC_2H_3O_2(aq) + OH^-(aq) \qquad (14.7)$$

Note that this reaction, which yields a base solution, involves a *base reacting with water to produce hydroxide ion and a conjugate acid*. We have defined K_b as the equilibrium constant for such a reaction. In this case,

$$K_b = \frac{[HC_2H_3O_2][OH^-]}{[C_2H_3O_2^-]}$$

The value of K_a for acetic acid is well known (1.8×10^{-5}). But how can we obtain the K_b value for the acetate ion? The answer lies in the relationships among K_a, K_b, and

K_w. Note that when the expression for K_a for acetic acid is multiplied by the expression for K_b for the acetate ion, the result is K_w:

$$K_a \times K_b = \frac{[H^+][C_2H_3O_2^-]}{[HC_2H_3O_2]} \times \frac{[HC_2H_3O_2][OH^-]}{[C_2H_3O_2^-]} = [H^+][OH^-] = K_w$$

This is a very important result. For any weak acid and its conjugate base,

$$K_a \times K_b = K_w$$

We can obtain a different form of this equation by taking $-\log$ of both sides of the above equation. This gives the equation

$$pK_a + pK_b = pK_w$$

At 25°C, $K_w = 1.0 \times 10^{-14}$, so $pK_w = 14.00$. In solutions at 25°C, we can use the relationship

$$pK_a + pK_b = 14.00$$

Thus when either K_a or K_b is known, the other can be calculated. For the acetate ion, since $K_w = K_b \times K_a$ (acetic acid), the K_b for acetate is:

$$K_b = \frac{K_w}{K_a \text{ (for HC}_2\text{H}_3\text{O}_2)} = \frac{1.0 \times 10^{-14}}{1.8 \times 10^{-5}} = 5.6 \times 10^{-10}$$

This is the K_b value for the reaction described by Equation (14.7). Note that it is obtained from the K_a value of the parent weak acid, in this case acetic acid. The sodium acetate solution is an example of an important general case. For any salt whose cation has neutral properties (such as Na^+ or K^+) and whose anion is the conjugate base of a weak acid, the aqueous solution will be basic. The K_b value for the anion can be obtained from the relationship $K_b = K_w/K_a$. Equilibrium calculations of this type are illustrated in Example 14.18.

A basic solution is formed if the anion of the salt is the conjugate base of a weak acid.

Salts as Weak Bases

Calculate the pH of a 0.30-*M* NaF solution. The K_a value for HF is 7.2×10^{-4}.

Solution

The major species in solution are

$$Na^+, \quad F^-, \quad \text{and} \quad H_2O$$

Since HF is a weak acid, the F^- ion must have a significant affinity for protons, and the dominant reaction will be

$$F^-(aq) + H_2O(l) \rightleftharpoons HF(aq) + OH^-(aq)$$

which yields the K_b expression

$$K_b = \frac{[HF][OH^-]}{[F^-]}$$

The value of K_b can be calculated from K_w and the K_a value for HF:

$$K_b = \frac{K_w}{K_a \text{ (for HF)}} = \frac{1.0 \times 10^{-14}}{7.2 \times 10^{-4}} = 1.4 \times 10^{-11}$$

The corresponding ICE table is:

Major Species

 Na^+

 F^-

 H_2O

	$F^-(aq)$	+	$H_2O(l)$	$\rightleftharpoons$	$HF(aq)$	+	$OH^-(aq)$
Initial	0.30		—		0		≈0
Change	$-x$		—		$+x$		$+x$
Equilibrium	$0.30 - x$		—		x		x

Thus $\qquad K_b = 1.4 \times 10^{-11} = \dfrac{[HF][OH^-]}{[F^-]} = \dfrac{(x)(x)}{0.30 - x} \approx \dfrac{x^2}{0.30}$

and

$$x \approx 2.0 \times 10^{-6}$$

The approximation is valid by the 5% rule, so

$$[OH^-] = x = 2.0 \times 10^{-6}\,M$$
$$pOH = 5.69$$
$$pH = 14.00 - 5.69 = 8.31$$

As expected, the solution is basic.

See Exercise 14.117

Base Strength in Aqueous Solutions

To emphasize the concept of base strength, let us consider the basic properties of the cyanide ion. One relevant reaction is the dissociation of hydrocyanic acid in water:

$$HCN(aq) + H_2O(l) \rightleftharpoons H_3O^+(aq) + CN^-(aq) \qquad K_a = 6.2 \times 10^{-10}$$

Since HCN is such a weak acid, CN^- appears to be a *strong* base, showing a very high affinity for H^+ *compared to H_2O*, with which it is competing. However, we also need to look at the reaction in which cyanide ion reacts with water:

$$CN^-(aq) + H_2O(l) \rightleftharpoons HCN(aq) + OH^-(aq)$$

where $\qquad K_b = \dfrac{K_w}{K_a} = \dfrac{1.0 \times 10^{-14}}{6.2 \times 10^{-10}} = 1.6 \times 10^{-5}$

In this reaction CN^- appears to be a weak base; the K_b value is only 1.6×10^{-5}. What accounts for this apparent difference in base strength? The key idea is that in the reaction of CN^- with H_2O, *CN^- is competing with OH^- for H^+, instead of competing with H_2O,* as it does in the HCN dissociation reaction. These equilibria show the following relative base strengths:

$$OH^- > CN^- > H_2O$$

Similar arguments can be made for other "weak" bases, such as ammonia, the acetate ion, the fluoride ion, and so on.

Salts That Produce Acidic Solutions

Some salts produce acidic solutions when dissolved in water. For example, when solid NH_4Cl is dissolved in water, NH_4^+ and Cl^- ions are present, with NH_4^+ behaving as a weak acid:

$$NH_4^+(aq) \rightleftharpoons NH_3(aq) + H^+(aq)$$

The Cl^- ion, having virtually no affinity for H^+ in water, does not affect the pH of the solution.

In general, salts in which the anion is not a base and the cation is the conjugate acid of a weak base produce acidic solutions.

Major Species

Cl⁻

NH_4^+

H_2O

Salts as Weak Acids I

Calculate the pH of a 0.10-M NH_4Cl solution. The K_b value for NH_3 is 1.8×10^{-5}.

Solution

The major species in solution are

$$NH_4^+, \quad Cl^-, \quad \text{and} \quad H_2O$$

Note that both NH_4^+ and H_2O can produce H^+. The dissociation reaction for the NH_4^+ ion is

$$NH_4^+(aq) \rightleftharpoons NH_3(aq) + H^+(aq)$$

for which

$$K_a = \frac{[NH_3][H^+]}{[NH_4^+]}$$

Note that although the K_b value for NH_3 is given, the reaction corresponding to K_b is not appropriate here, since NH_3 is not a major species in the solution. Instead, the given value of K_b is used to calculate K_a for NH_4^+ from the relationship

$$K_a \times K_b = K_w$$

Thus K_a (for NH_4^+) $= \dfrac{K_w}{K_b \text{ (for } NH_3)} = \dfrac{1.0 \times 10^{-14}}{1.8 \times 10^{-5}} = 5.6 \times 10^{-10}$

Although NH_4^+ is a very weak acid, as indicated by its K_a value, it is stronger than H_2O and will dominate in the production of H^+. Thus we will focus on the dissociation reaction of NH_4^+ to calculate the pH in this solution.

We solve the weak acid problem in the usual way:

	$NH_4^+(aq)$	$\rightleftharpoons$	$H^+(aq)$	$+$	$NH_3(aq)$
Initial	0.10		≈ 0		0
Change	$-x$		$+x$		$+x$
Equilibrium	$0.10 - x$		x		x

Thus $5.6 \times 10^{-10} = K_a = \dfrac{[H^+][NH_3]}{[NH_4^+]} = \dfrac{(x)(x)}{0.10 - x} \approx \dfrac{x^2}{0.10}$

$$x \approx 7.5 \times 10^{-6}$$

The approximation is valid by the 5% rule, so

$$[H^+] = x = 7.5 \times 10^{-6} \, M \qquad \text{and} \qquad pH = 5.13$$

See Exercise 14.118

A second type of salt that produces an acidic solution is one that contains a *highly charged metal ion*. For example, when solid aluminum chloride ($AlCl_3$) is dissolved in water, the resulting solution is significantly acidic. Although the Al^{3+} ion is not itself a Brønsted–Lowry acid, the hydrated ion $Al(H_2O)_6^{3+}$ formed in water is a weak acid:

$$Al(H_2O)_6^{3+}(aq) \rightleftharpoons Al(OH)(H_2O)_5^{2+}(aq) + H^+(aq)$$

The high charge on the metal ion polarizes the O—H bonds in the attached water molecules, making the hydrogens in these water molecules more acidic than those in free water molecules. Typically, the higher the charge on the metal ion, the stronger the acidity of the hydrated ion.

Section 14.9 contains a further discussion
of the acidity of hydrated ions.

Major Species

Cl^-

$Al(H_2O)_6^{3+}$

H_2O

Salts as Weak Acids II

Calculate the pH of a 0.010-M $AlCl_3$ solution. The K_a value for $Al(H_2O)_6^{3+}$ is 1.4×10^{-5}.

Solution

The major species in solution are

$$Al(H_2O)_6^{3+}, \quad Cl^-, \quad \text{and} \quad H_2O$$

Since the $Al(H_2O)_6^{3+}$ ion is a stronger acid than water, the dominant equilibrium is

$$Al(H_2O)_6^{3+}(aq) \rightleftharpoons Al(OH)(H_2O)_5^{2+}(aq) + H^+(aq)$$

and

$$1.4 \times 10^{-5} = K_a = \frac{[Al(OH)(H_2O)_5^{2+}][H^+]}{[Al(H_2O)_6^{3+}]}$$

This is a typical weak acid problem, which we can solve with the usual procedure:

	$Al(H_2O)_6^{3+}(aq)$	$\rightleftharpoons$	$Al(OH)(H_2O)_5^{2+}(aq)$	+	$H^+(aq)$
Initial	0.010		0		≈ 0
Change	$-x$		$+x$		$+x$
Equilibrium	$0.010 - x$		x		x

Thus

$$1.4 \times 10^{-5} = K_a = \frac{[Al(OH)(H_2O)_5^{2+}][H^+]}{[Al(H_2O)_6^{3+}]} = \frac{(x)(x)}{0.010 - x} \approx \frac{x^2}{0.010}$$

$$x \approx 3.7 \times 10^{-4}$$

Since the approximation is valid by the 5% rule,

$$[H^+] = x = 3.7 \times 10^{-4}\,M \quad \text{and} \quad pH = 3.43$$

See Exercises 14.127 and 14.128

Table 14.5 | Qualitative Prediction of pH for Solutions of Salts for Which Both Cation and Anion Have Acidic or Basic Properties

$K_a > K_b$	pH < 7 (acidic)
$K_b > K_a$	pH > 7 (basic)
$K_a = K_b$	pH = 7 (neutral)

So far we have considered salts in which only one of the ions has acidic or basic properties. For many salts, such as ammonium acetate ($NH_4C_2H_3O_2$), both ions can affect the pH of the aqueous solution. Because the equilibrium calculations for these cases can be quite complicated, we will consider only the qualitative aspects of such problems. We can predict whether the solution will be basic, acidic, or neutral by comparing the K_a value for the acidic ion with the K_b value for the basic ion. If the K_a value for the acidic ion is larger than the K_b value for the basic ion, the solution will be acidic. If the K_b value is larger than the K_a value, the solution will be basic. Equal K_a and K_b values mean a neutral solution. These facts are summarized in Table 14.5.

The Acid–Base Properties of Salts

Predict whether an aqueous solution of each of the following salts will be acidic, basic, or neutral.

a. $NH_4C_2H_3O_2$ **b.** NH_4CN **c.** $Al_2(SO_4)_3$

Solution

a. The ions in solution are NH_4^+ and $C_2H_3O_2^-$. As we mentioned previously, K_a for NH_4^+ is 5.6×10^{-10} and K_b for $C_2H_3O_2^-$ is 5.6×10^{-10}. Thus K_a for NH_4^+ is equal to K_b for $C_2H_3O_2^-$, and the solution will be neutral (pH = 7).

b. The solution will contain NH_4^+ and CN^- ions. The K_a value for NH_4^+ is 5.6×10^{-10} and

$$K_b \text{ (for } CN^-) = \frac{K_w}{K_a \text{ (for HCN)}} = 1.6 \times 10^{-5}$$

Since K_b for CN^- is much larger than K_a for NH_4^+, CN^- is a much stronger base than NH_4^+ is an acid. This solution will be basic.

c. The solution will contain $Al(H_2O)_6^{3+}$ and SO_4^{2-} ions. The K_a value for $Al(H_2O)_6^{3+}$ is 1.4×10^{-5}, as given in Example 14.20. We must calculate K_b for SO_4^{2-}. The HSO_4^- ion is the conjugate acid of SO_4^{2-}, and its K_a value is K_{a_2} for sulfuric acid, or 1.2×10^{-2}. Therefore,

$$K_b \text{ (for } SO_4^{2-}) = \frac{K_w}{K_{a_2} \text{ (for sulfuric acid)}}$$

$$= \frac{1.0 \times 10^{-14}}{1.2 \times 10^{-2}} = 8.3 \times 10^{-13}$$

This solution will be acidic, since K_a for $Al(H_2O)_6^{3+}$ is much greater than K_b for SO_4^{2-}.

See Exercises 14.129 and 14.130

The acid–base properties of aqueous solutions of various salts are summarized in Table 14.6.

14.9 | The Effect of Structure on Acid–Base Properties

Further aspects of acid strengths are discussed in Section 20.13.

We have seen that when a substance is dissolved in water, it produces an acidic solution if it can donate protons and produces a basic solution if it can accept protons. What structural properties of a molecule cause it to behave as an acid or as a base?

Any molecule containing a hydrogen atom is potentially an acid. However, many such molecules show no acidic properties. For example, molecules containing C—H bonds, such as chloroform ($CHCl_3$) and nitromethane (CH_3NO_2), do not produce acidic aqueous solutions because a C—H bond is both strong and nonpolar and thus there is no tendency to donate protons. On the other hand, although the H—Cl bond in

Table 14.6 | Acid–Base Properties of Various Types of Salts

Type of Salt	Examples	Comment	pH of Solution
Cation is from strong base; anion is from strong acid	KCl, KNO_3, $NaCl$, $NaNO_3$	Neither acts as an acid or a base	Neutral
Cation is from strong base; anion is from weak acid	$NaC_2H_3O_2$, KCN, NaF	Anion acts as a base; cation has no effect on pH	Basic
Cation is conjugate acid of weak base; anion is from strong acid	NH_4Cl, NH_4NO_3	Cation acts as an acid; anion has no effect on pH	Acidic
Cation is conjugate acid of weak base; anion is conjugate base of weak acid	$NH_4C_2H_3O_2$, NH_4CN	Cation acts as an acid; anion acts as a base	Acidic if $K_a > K_b$, basic if $K_b > K_a$, neutral if $K_a = K_b$
Cation is highly charged metal ion; anion is from strong acid	$Al(NO_3)_3$, $FeCl_3$	Hydrated cation acts as an acid; anion has no effect on pH	Acidic

Table 14.7 | Bond Strengths and Acid Strengths for Hydrogen Halides

H—X Bond	Bond Strength (kJ/mol)	Acid Strength in Water
H—F	565	Weak
H—Cl	427	Strong
H—Br	363	Strong
H—I	295	Strong

gaseous hydrogen chloride is slightly stronger than a C—H bond, it is much more polar, and this molecule readily dissociates when dissolved in water.

Thus there are two main factors that determine whether a molecule containing an X—H bond will behave as a Brønsted–Lowry acid: the strength of the bond and the polarity of the bond.

To explore these factors let's consider the relative acid strengths of the hydrogen halides. The bond polarities vary as shown

$$H—F > H—Cl > H—Br > H—I$$

↑ Most polar	↑ Least polar

because electronegativity decreases going down the group. Based on the high polarity of the H—F bond, we might expect hydrogen fluoride to be a very strong acid. In fact, among HX molecules, HF is the only weak acid ($K_a = 7.2 \times 10^{-4}$) when dissolved in water. The H—F bond is unusually strong, as shown in Table 14.7, and thus is difficult to break. This contributes significantly to the reluctance of the HF molecules to dissociate in water.

Another important class of acids are the oxyacids, which as we saw in Section 14.2 characteristically contain the grouping H—O—X. Several series of oxyacids are listed with their K_a values in Table 14.8. Note from these data that for a given series the acid strength increases with an increase in the number of oxygen atoms attached to the central atom. For example, in the series containing chlorine and a varying number of oxygen atoms, HOCl is a weak acid, but the acid strength is successively greater as the number of oxygen atoms increases. This happens because the very electronegative oxygen atoms are able to draw electrons away from the chlorine atom and the O—H bond (Fig. 14.9). The net effect is to both polarize and weaken the O—H bond; this effect becomes more important as the number of attached oxygen atoms increases.

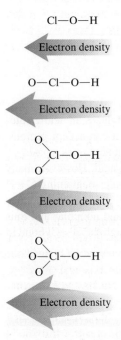

Figure 14.9 | The effect of the number of attached oxygens on the O—H bond in a series of chlorine oxyacids. As the number of oxygen atoms attached to the chlorine atom increases, they become more effective at withdrawing electron density from the O—H bond, thereby weakening and polarizing it. This increases the tendency for the molecule to produce a proton, and so its acid strength increases.

Table 14.8 | Several Series of Oxyacids and Their K_a Values

Oxyacid	Structure	K_a Value
$HClO_4$	H—O—Cl with O's (O, O, O)	Large (~10^7)
$HClO_3$	H—O—Cl with O's (O, O)	~1
$HClO_2$	H—O—Cl—O	1.2×10^{-2}
$HClO$	H—O—Cl	3.5×10^{-8}
H_2SO_4	H—O—S with O—H and O's	Large
H_2SO_3	H—O—S with O—H and O	1.5×10^{-2}
HNO_3	H—O—N with O's (O, O)	Large
HNO_2	H—O—N—O	4.0×10^{-4}

Table 14.9 | Comparison of Electronegativity of X and K_a Value for a Series of Oxyacids

Acid	X	Electronegativity of X	K_a for Acid
HOCl	Cl	3.0	4×10^{-8}
HOBr	Br	2.8	2×10^{-9}
HOI	I	2.5	2×10^{-11}
HOCH$_3$	CH$_3$	2.3 (for carbon in CH$_3$)	$\sim 10^{-15}$

This means that a proton is most readily produced by the molecule with the largest number of attached oxygen atoms ($HClO_4$).

This type of behavior is also observed for hydrated metal ions. Earlier in this chapter we saw that highly charged metal ions such as Al^{3+} produce acidic solutions. The acidity of the water molecules attached to the metal ion is increased by the attraction of electrons to the positive metal ion:

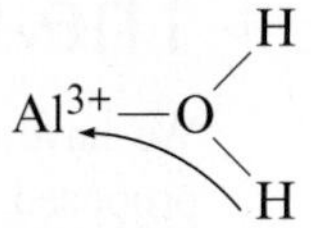

$$Al^{3+}{-}O\begin{smallmatrix}H\\\\H\end{smallmatrix}$$

The greater the charge on the metal ion, the more acidic the hydrated ion becomes.

For acids containing the H—O—X grouping, the greater the ability of X to draw electrons toward itself, the greater the acidity of the molecule. Since the electronegativity of X reflects its ability to attract the electrons involved in bonding, we might expect acid strength to depend on the electronegativity of X. In fact, there is an excellent correlation between the electronegativity of X and the acid strength for oxyacids, as shown in Table 14.9.

14.10 | Acid–Base Properties of Oxides

We have just seen that molecules containing the grouping H—O—X can behave as acids and that the acid strength depends on the electron-withdrawing ability of X. But substances with this grouping also can behave as bases if the hydroxide ion instead of a proton is produced. What determines which behavior will occur? The answer lies mainly in the nature of the O—X bond. If X has a relatively high electronegativity, the O—X bond will be covalent and strong. When the compound containing the H—O—X grouping is dissolved in water, the O—X bond will remain intact. It will be the polar and relatively weak H—O bond that will tend to break, releasing a proton. On the other hand, if X has a very low electronegativity, the O—X bond will be ionic and subject to being broken in polar water. Examples are the ionic substances NaOH and KOH that dissolve in water to give the metal cation and the hydroxide ion.

A compound containing the H—O—X group will produce an acidic solution in water if the O—X bond is strong and covalent. If the O—X bond is ionic, the compound will produce a basic solution in water.

We can use these principles to explain the acid–base behavior of oxides when they are dissolved in water. For example, when a covalent oxide such as sulfur trioxide is dissolved in water, an acidic solution results because sulfuric acid is formed:

$$SO_3(g) + H_2O(l) \longrightarrow H_2SO_4(aq)$$

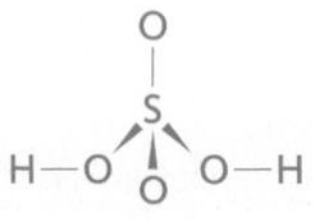

The structure of H_2SO_4 is shown in the margin. In this case, the strong, covalent O—S bonds remain intact and the H—O bonds break to produce protons. Other common covalent oxides that react with water to form acidic solutions are sulfur dioxide, carbon dioxide, and nitrogen dioxide, as shown by the following reactions:

$$SO_2(g) + H_2O(l) \longrightarrow H_2SO_3(aq)$$
$$CO_2(g) + H_2O(l) \longrightarrow H_2CO_3(aq)$$
$$2NO_2(g) + H_2O(l) \longrightarrow HNO_3(aq) + HNO_2(aq)$$

Thus when a covalent oxide dissolves in water, an acidic solution forms. These oxides are called **acidic oxides**.

On the other hand, when an ionic oxide dissolves in water, a basic solution results, as shown by the following reactions:

$$CaO(s) + H_2O(l) \longrightarrow Ca(OH)_2(aq)$$
$$K_2O(s) + H_2O(l) \longrightarrow 2KOH(aq)$$

These reactions can be explained by recognizing that the oxide ion has a high affinity for protons and reacts with water to produce hydroxide ions:

$$O^{2-}(aq) + H_2O(l) \longrightarrow 2OH^-(aq)$$

Thus the most ionic oxides, such as those of the Group 1A and 2A metals, produce basic solutions when they are dissolved in water. As a result, these oxides are called **basic oxides**.

14.11 | The Lewis Acid–Base Model

We have seen that the first successful conceptualization of acid–base behavior was proposed by Arrhenius. This useful but limited model was replaced by the more general Brønsted–Lowry model. An even more general model for acid–base behavior was suggested by G. N. Lewis in the early 1920s. A **Lewis acid** is an *electron-pair acceptor,* and a **Lewis base** is an *electron-pair donor.* Another way of saying this is that a Lewis acid has an empty atomic orbital that it can use to accept (share) an electron pair from a molecule that has a lone pair of electrons (Lewis base). The three models for acids and bases are summarized in Table 14.10.

Note that Brønsted–Lowry acid–base reactions (proton donor–proton acceptor reactions) are encompassed by the Lewis model. For example, the reaction between a proton and an ammonia molecule, that is,

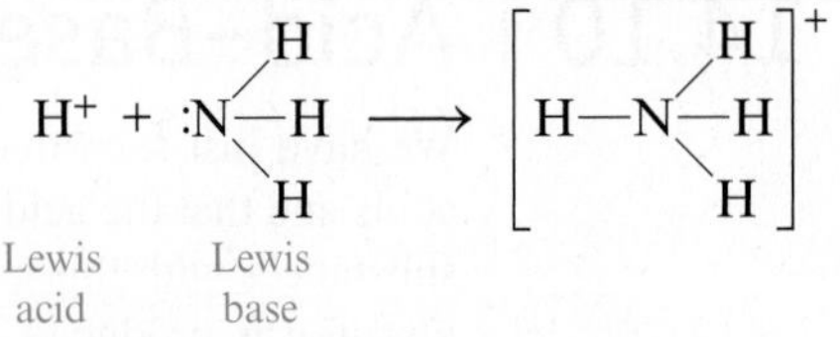

can be represented as a reaction between an electron-pair acceptor (H^+) and an electron-pair donor (NH_3). The same holds true for a reaction between a proton and a hydroxide ion:

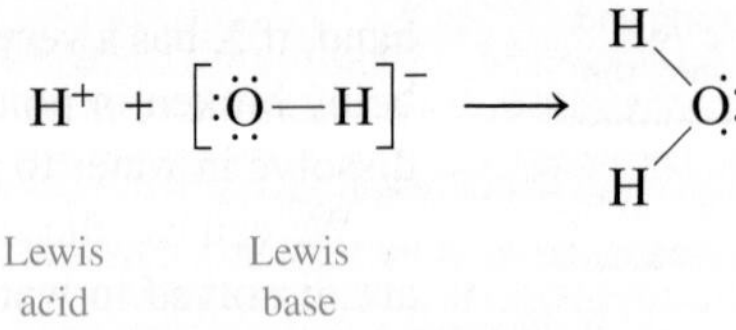

Table 14.10 | Three Models for Acids and Bases

Model	Definition of Acid	Definition of Base
Arrhenius	H^+ producer	OH^- producer
Brønsted–Lowry	H^+ donor	H^+ acceptor
Lewis	Electron-pair acceptor	Electron-pair donor

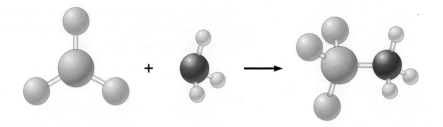

Figure 14.10 | Reaction of BF₃ with NH₃.

The Lewis model encompasses the Brønsted–Lowry model, but the reverse is not true.

The real value of the Lewis model for acids and bases is that it covers many reactions that do not involve Brønsted–Lowry acids. For example, consider the gas-phase reaction between boron trifluoride and ammonia:

$$\underset{\substack{\text{Lewis} \\ \text{acid}}}{\text{:F:}\!-\!\text{B}\!-\!\text{F:} \atop \text{:F:}} + \underset{\substack{\text{Lewis} \\ \text{base}}}{\text{:N}\!-\!\text{H} \atop \substack{\text{H} \\ \text{H}}} \longrightarrow \underset{}{\text{:F}\atop\text{:F}\!-\!\text{B}\!-\!\text{N}\!-\!\text{H} \atop \substack{\text{:F} \qquad \text{H}}}$$

Here the electron-deficient BF₃ molecule (there are only six electrons around the boron) completes its octet by reacting with NH₃, which has a lone pair of electrons (Fig. 14.10). In fact, as mentioned in Chapter 8, the electron deficiency of boron trifluoride makes it very reactive toward any electron-pair donor. That is, it is a strong Lewis acid.

The hydration of a metal ion, such as Al^{3+}, also can be viewed as a Lewis acid–base reaction:

$$Al^{3+} + 6 \underset{\substack{\text{Lewis} \\ \text{base}}}{\text{:O}} \begin{pmatrix} \text{H} \\ \text{H} \end{pmatrix} \longrightarrow \left[Al\!-\!\overset{..}{\text{O}} \begin{pmatrix} \text{H} \\ \text{H} \end{pmatrix}_6 \right]^{3+}$$

Here the Al^{3+} ion accepts one electron pair from each of six water molecules to form $Al(H_2O)_6^{3+}$ (Fig. 14.11).

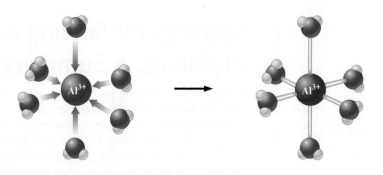

Figure 14.11 | The Al(H₂O)₆³⁺ ion.

In addition, the reaction between a covalent oxide and water to form a Brønsted–Lowry acid can be defined as a Lewis acid–base reaction. An example is the reaction between sulfur trioxide and water:

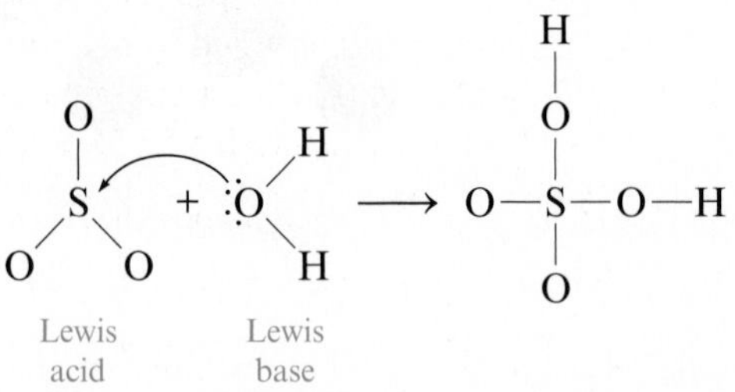

Lewis   Lewis
acid   base

Note that as the water molecule attaches to sulfur trioxide, a proton shift occurs to form sulfuric acid.

Example 14.22

Lewis Acids and Bases

For each reaction, identify the Lewis acid and base.

a. $Ni^{2+}(aq) + 6NH_3(aq) \longrightarrow Ni(NH_3)_6{}^{2+}(aq)$

b. $H^+(aq) + H_2O(aq) \rightleftharpoons H_3O^+(aq)$

Solution

a. Each NH_3 molecule donates an electron pair to the Ni^{2+} ion:

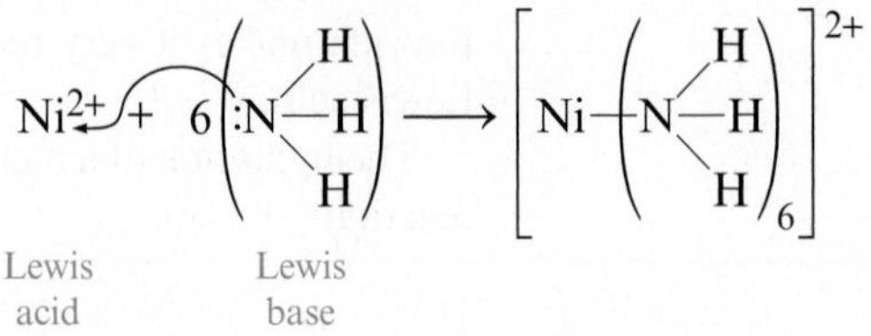

Lewis   Lewis
acid   base

The nickel(II) ion is the Lewis acid, and ammonia is the Lewis base.

b. The proton is the Lewis acid and the water molecule is the Lewis base:

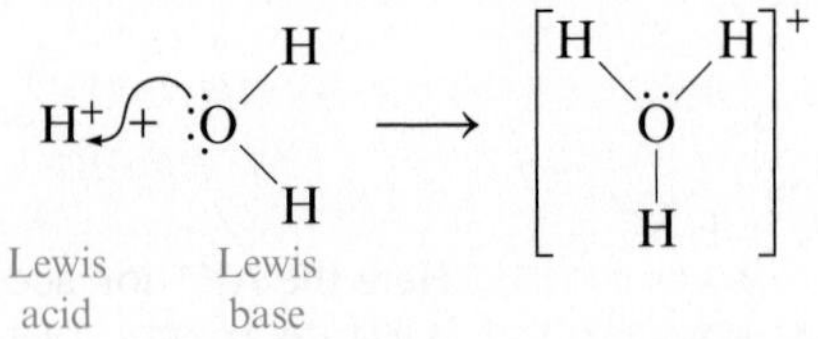

Lewis   Lewis
acid   base

See Exercises 14.137 and 14.138

14.12 | Strategy for Solving Acid–Base Problems: A Summary

In this chapter we have encountered many different situations involving aqueous solutions of acids and bases, and in the next chapter we will encounter still more. In solving for the equilibrium concentrations in these aqueous solutions, it is tempting to create a pigeonhole for each possible situation and to memorize the procedures necessary to deal with that particular case. This approach is just not practical and usually leads to frustration: Too many pigeonholes are required—there seems to be an infinite number of cases. But you can handle any case successfully by taking a systematic, patient, and thoughtful approach. When analyzing an acid–base equilibrium problem, do *not* ask yourself how a memorized solution can be used to solve the problem. Instead, ask this question: *What are the major species in the solution and what is their chemical behavior?*

The most important part of doing a complicated acid–base equilibrium problem is the analysis you do at the beginning of a problem.

What major species are present?

Does a reaction occur that can be assumed to go to completion?

What equilibrium dominates the solution?

Let the problem guide you. Be patient.

For review

Key terms

Section 14.1
Arrhenius concept
Brønsted–Lowry model
hydronium ion
conjugate base
conjugate acid
conjugate acid–base pair
acid dissociation constant

Section 14.2
strong acid
weak acid
diprotic acid
oxyacids
organic acids
carboxyl group
monoprotic acids
amphoteric substance
autoionization
ion-product (dissociation)
  constant

Section 14.3
pH scale

Section 14.4
major species

Section 14.5
percent dissociation

Section 14.6
strong bases
slaked lime
lime–soda process
weak bases
amine

Models for acids and bases

> Arrhenius model
> > Acids produce H^+ in solution
> > Bases produce OH^- in solution
> Brønsted–Lowry model
> > An acid is a proton donor
> > A base is a proton acceptor
> > In this model an acid molecule reacts with a water molecule, which behaves as a base:

$$HA(aq) + H_2O(l) \rightleftharpoons H_3O^+(aq) + A^-(aq)$$

   Acid   Base   Conjugate   Conjugate
                             acid   base

to form a new acid (conjugate acid) and a new base (conjugate base).

> Lewis model
> > A Lewis acid is an electron-pair acceptor
> > A Lewis base is an electron-pair donor

Acid–base equilibrium

> The equilibrium constant for an acid dissociating (ionizing) in water is called K_a
> The K_a expression is

$$K_a = \frac{[H_3O^+][A^-]}{[HA]}$$

which is often simplified as

$$K_a = \frac{[H^+][A^-]}{[HA]}$$

> $[H_2O]$ is never included because it is assumed to be constant

Key terms

Acid strength

> A strong acid has a very large K_a value
> > The acid completely dissociates (ionizes) in water
> > The dissociation (ionization) equilibrium position lies all the way to the right
> > Strong acids have very weak conjugate bases
> > The common strong acids are nitric acid [$HNO_3(aq)$], hydrochloric acid [$HCl(aq)$], sulfuric acid [$H_2SO(aq)$], and perchloric acid [$HClO_4(aq)$]

> A weak acid has a small K_a value
> > The acid dissociates (ionizes) to only a slight extent
> > The dissociation (ionization) equilibrium position lies far to the left
> > Weak acids have relatively strong conjugate bases
> > Percent dissociation of a weak acid

$$\% \text{ dissociation} = \frac{\text{amount dissociated (mol/L)}}{\text{initial concentration (mol/L)}} \times 100\%$$

> > The smaller the percent dissociation, the weaker the acid
> > Dilution of a weak acid increases its percent dissociation

Autoionization of water

> Water is an amphoteric substance: It behaves as both an acid and a base
> Water reacts with itself in an acid–base reaction

$$H_2O(l) + H_2O(l) \rightleftharpoons H_3O^+(aq) + OH^-(aq)$$

which leads to the equilibrium expression

$$K_w = [H_3O^+][OH^-] \quad \text{or} \quad [H^+][OH^-] = K_w$$

> > K_w is the ion-product constant for water
> > At 25°C in pure water $[H^+] = [OH^-] = 1.0 \times 10^{-7}$, so $K_w = 1.0 \times 10^{-14}$
> Acidic solution: $[H^+] > [OH^-]$
> Basic solution: $[OH^-] > [H^+]$
> Neutral solution: $[H^+] = [OH^-]$

The pH scale

> $pH = -\log[H^+]$
> Since pH is a log scale, the pH changes by 1 for every 10-fold change in $[H^+]$
> The log scale is also used for $[OH^-]$ and for K_a values

$$pOH = -\log[OH^-]$$
$$pK_a = -\log K_a$$

Bases

> Strong bases are hydroxide salts, such as NaOH and KOH
> Weak bases react with water to produce OH^-

$$B(aq) + H_2O(l) \rightleftharpoons BH^+(aq) + OH^-(aq)$$

> The equilibrium constant for this reaction is called K_b where

$$K_b = \frac{[BH^+][OH^-]}{[B]}$$

> In water a base B is always competing with OH^- for a proton (H^+), so K_b values tend to be very small, thus making B a weak base (compared to OH^-)

Polyprotic acids

> A polyprotic acid has more than one acidic proton
> Polyprotic acids dissociate one proton at a time
>> Each step has a characteristic K_a value
>> Typically for a weak polyprotic acid, $K_{a_1} > K_{a_2} > K_{a_3}$
> Sulfuric acid is unique:
>> It is a strong acid in the first dissociation step (K_{a_1} is very large)
>> It is a weak acid in the second step

Acid–base properties of salts

> Can produce acidic, basic, or neutral solutions
> Salts that contain
>> cations of strong bases and anions of strong acids produce neutral solutions
>> cations of strong bases and anions of weak acids produce basic solutions
>> cations of weak bases and anions of strong acids produce acidic solutions
> Acidic solutions are produced by salts containing a highly charged metal cation—for example, Al^{3+} and Fe^{3+}

Effect of structure on acid–base properties

> Many substances that function as acids or bases contain the H—O—X grouping
>> Molecules in which the O—X bond is strong and covalent tend to behave as acids
>> As X becomes more electronegative, the acid becomes stronger
>> When the O—X bond is ionic, the substance behaves as a base, releasing OH^- ions in water

Review questions *Answers to the Review Questions can be found on the Student website (accessible from* **www.cengagebrain.com**).

1. Define each of the following:
   a. Arrhenius acid
   b. Brønsted–Lowry acid
   c. Lewis acid

   Which of the definitions is most general? Write reactions to justify your answer.

2. Define or illustrate the meaning of the following terms:
   a. K_a reaction
   b. K_a equilibrium constant
   c. K_b reaction
   d. K_b equilibrium constant
   e. conjugate acid–base pair

3. Define or illustrate the meaning of the following terms:
   a. amphoteric
   b. K_w reaction
   c. K_w equilibrium constant
   d. pH
   e. pOH
   f. pK_w

   Give the conditions for a neutral aqueous solution at 25°C, in terms of [H^+], pH, and the relationship between [H^+] and [OH^-]. Do the same for an acidic solution and for a basic solution. As a solution becomes more acidic, what happens to pH, pOH, [H^+], and [OH^-]? As a solution becomes more basic, what happens to pH, pOH, [H^+], and [OH^-]?

4. How is acid strength related to the value of K_a? What is the difference between strong acids and weak acids (see Table 14.1)? As the strength of an acid increases, what happens to the strength of the conjugate base? How is base strength related to the value of K_b? As the strength of a base increases, what happens to the strength of the conjugate acid?

5. Two strategies are followed when solving for the pH of an acid in water. What is the strategy for calculating the pH of a strong acid in water? What major assumptions are made when solving strong acid problems? The best way to recognize strong acids is to memorize them. List the six common strong acids (the two not listed in the text are HBr and HI).

   Most acids, by contrast, are weak acids. When solving for the pH of a weak acid in water, you must have the K_a value. List two places in this text that provide K_a values for weak acids. You can utilize these tables to help you recognize weak acids. What is the strategy for calculating the pH of a weak acid in water? What assumptions are generally made? What is the 5% rule? If the 5% rule fails, how do you calculate the pH of a weak acid in water?

6. Two strategies are also followed when solving for the pH of a base in water. What is the strategy for calculating the pH of a strong base in water? List the strong bases mentioned in the text that should be committed to memory. Why is calculating the pH of $Ca(OH)_2$ solutions a little more difficult than calculating the pH of NaOH solutions?

   Most bases are weak bases. The presence of what element most commonly results in basic properties for an organic compound? What is present on this element in compounds that allows it to accept a proton?

   Table 14.3 and Appendix 5 of the text list K_b values for some weak bases. What strategy is used to solve for the pH of a weak base in water? What assumptions are made when solving for the pH of weak base solutions? If the 5% rule fails, how do you calculate the pH of a weak base in water?

7. Table 14.4 lists the stepwise K_a values for some polyprotic acids. What is the difference between a monoprotic acid, a diprotic acid, and a triprotic acid? Most polyprotic acids are weak acids; the major exception is H_2SO_4. To solve for the pH of a solution of H_2SO_4, you must generally solve a strong acid problem as well as a weak acid problem. Explain. Write out the reactions that refer to K_{a_1} and K_{a_2} for H_2SO_4.

   For H_3PO_4, $K_{a_1} = 7.5 \times 10^{-3}$, $K_{a_2} = 6.2 \times 10^{-8}$, and $K_{a_3} = 4.8 \times 10^{-13}$. Write out the reactions that refer to the K_{a_1}, K_{a_2}, and K_{a_3} equilibrium constants. What are the three acids in a solution of H_3PO_4? Which acid is strongest? What are the three conjugate bases in a solution of H_3PO_4? Which conjugate base is strongest? Summarize the strategy for calculating the pH of a polyprotic acid in water.

8. For conjugate acid–base pairs, how are K_a and K_b related? Consider the reaction of acetic acid in water

$$CH_3CO_2H(aq) + H_2O(l) \rightleftharpoons CH_3CO_2^-(aq) + H_3O^+(aq)$$

   where $K_a = 1.8 \times 10^{-5}$.

   a. Which two bases are competing for the proton?

   b. Which is the stronger base?

   c. In light of your answer to part b, why do we classify the acetate ion ($CH_3CO_2^-$) as a weak base? Use an appropriate reaction to justify your answer.

   In general, as base strength increases, conjugate acid strength decreases. Explain why the conjugate acid of the weak base NH_3 is a weak acid.

To summarize, the conjugate base of a weak acid is a weak base and the conjugate acid of a weak base is a weak acid (weak gives you weak). Assuming K_a for a monoprotic strong acid is 1×10^6, calculate K_b for the conjugate base of this strong acid. Why do conjugate bases of strong acids have no basic properties in water? List the conjugate bases of the six common strong acids. To tie it all together, some instructors have students think of Li^+, K^+, Rb^+, Cs^+, Ca^{2+}, Sr^{2+}, and Ba^{2+} as the conjugate acids of the strong bases LiOH, KOH, RbOH, CsOH, $Ca(OH)_2$, $Sr(OH)_2$, and $Ba(OH)_2$. Although not technically correct, the conjugate acid strength of these cations is similar to the conjugate base strength of the strong acids. That is, these cations have no acidic properties in water; similarly, the conjugate bases of strong acids have no basic properties (strong gives you worthless). Fill in the blanks with the correct response. The conjugate base of a weak acid is a _____ base. The conjugate acid of a weak base is a _____ acid. The conjugate base of a strong acid is a _____ base. The conjugate acid of a strong base is a _____ acid. (*Hint:* Weak gives you weak and strong gives you worthless.)

9. What is a salt? List some anions that behave as weak bases in water. List some anions that have no basic properties in water. List some cations that behave as weak acids in water. List some cations that have no acidic properties in water. Using these lists, give some formulas for salts that have only weak base properties in water. What strategy would you use to solve for the pH of these basic salt solutions? Identify some salts that have only weak acid properties in water. What strategy would you use to solve for the pH of these acidic salt solutions? Identify some salts that have no acidic or basic properties in water (produce neutral solutions). When a salt contains both a weak acid ion and a weak base ion, how do you predict whether the solution pH is acidic, basic, or neutral?

10. For oxyacids, how does acid strength depend on

   a. the strength of the bond to the acidic hydrogen atom?

   b. the electronegativity of the element bonded to the oxygen atom that bears the acidic hydrogen?

   c. the number of oxygen atoms?

   How does the strength of a conjugate base depend on these factors?

   What type of solution forms when a nonmetal oxide dissolves in water? Give an example of such an oxide. What type of solution forms when a metal oxide dissolves in water? Give an example of such an oxide.

Active Learning Questions

These questions are designed to be used by groups of students in class.

1. Consider two beakers of pure water at different temperatures. How do their pH values compare? Which is more acidic? more basic? Explain.

2. Differentiate between the terms *strength* and *concentration* as they apply to acids and bases. When is HCl strong? Weak? Concentrated? Dilute? Answer the same questions for ammonia. Is the conjugate base of a weak acid a strong base?

3. Sketch two graphs: (a) percent dissociation for weak acid HA versus the initial concentration of HA ($[HA]_0$) and (b) H^+ concentration versus $[HA]_0$. Explain both.

4. Consider a solution prepared by mixing a weak acid HA and HCl. What are the major species? Explain what is occurring in solution. How would you calculate the pH? What if you added NaA to this solution? Then added NaOH?

5. Explain why salts can be acidic, basic, or neutral, and show examples. Do this without specific numbers.

6. Consider two separate aqueous solutions: one of a weak acid HA and one of HCl. Assuming you started with 10 molecules of each:
   a. Draw a picture of what each solution looks like at equilibrium.
   b. What are the major species in each beaker?
   c. From your pictures, calculate the K_a values of each acid.
   d. Order the following from the strongest to the weakest base: H_2O, A^-, Cl^-. Explain your order.

7. You are asked to calculate the H^+ concentration in a solution of NaOH(aq). Because sodium hydroxide is a base, can we say there is no H^+, since having H^+ would imply that the solution is acidic?

8. Consider a solution prepared by mixing a weak acid HA, HCl, and NaA. Which of the following statements best describes what happens?
   a. The H^+ from the HCl reacts completely with the A^- from the NaA. Then the HA dissociates somewhat.
   b. The H^+ from the HCl reacts somewhat with the A^- from the NaA to make HA, while the HA is dissociating. Eventually you have equal amounts of everything.
   c. The H^+ from the HCl reacts somewhat with the A^- from the NaA to make HA while the HA is dissociating. Eventually all the reactions have equal rates.
   d. The H^+ from the HCl reacts completely with the A^- from the NaA. Then the HA dissociates somewhat until "too much" H^+ and A^- are formed, so the H^+ and A^- react to form HA, and so on. Eventually equilibrium is reached.

   Justify your choice, and for choices you did not pick, explain what is wrong with them.

9. Consider a solution formed by mixing 100.0 mL of 0.10 M HA ($K_a = 1.0 \times 10^{-6}$), 100.00 mL of 0.10 M NaA, and 100.0 mL of 0.10 M HCl. In calculating the pH for the final solution, you would make some assumptions about the order in which various reactions occur to simplify the calculations. State these assumptions. Does it matter whether the reactions actually occur in the assumed order? Explain.

10. A certain sodium compound is dissolved in water to liberate Na^+ ions and a certain negative ion. What evidence would you look for to determine whether the anion is behaving as an acid or a base? How could you tell whether the anion is a strong base? Explain how the anion could behave simultaneously as an acid and a base.

11. Acids and bases can be thought of as chemical opposites (acids are proton donors, and bases are proton acceptors). Therefore, one might think that $K_a = 1/K_b$. Why isn't this the case? What is the relationship between K_a and K_b? Prove it with a derivation.

12. Consider two solutions of the salts NaX(aq) and NaY(aq) at equal concentrations. What would you need to know to determine which solution has the higher pH? Explain how you would decide (perhaps even provide a sample calculation).

13. What is meant by pH? True or false: A strong acid solution always has a lower pH than a weak acid solution. Explain.

14. Why is the pH of water at 25°C equal to 7.00?

15. Can the pH of a solution be negative? Explain.

16. Is the conjugate base of a weak acid a strong base? Explain. Explain why Cl^- does not affect the pH of an aqueous solution.

17. Match the following pH values: 1, 2, 5, 6, 6.5, 8, 11, 11, and 13 with the following chemicals (of equal concentration): HBr, NaOH, NaF, NaCN, NH_4F, CH_3NH_3F, HF, HCN, and NH_3. Answer this question without performing calculations.

18. The salt BX, when dissolved in water, produces an acidic solution. Which of the following could be true? (There may be more than one correct answer.)
   a. The acid HX is a weak acid.
   b. The acid HX is a strong acid.
   c. The cation B^+ is a weak acid.
   Explain.

Questions

19. Anions containing hydrogen (for example, HCO_3^- and $H_2PO_4^-$) usually show amphoteric behavior. Write equations illustrating the amphoterism of these two anions.

20. Which of the following conditions indicate an *acidic* solution at 25°C?
   a. pH = 3.04
   b. $[H^+] > 1.0 \times 10^{-7} M$
   c. pOH = 4.51
   d. $[OH^-] = 3.21 \times 10^{-12} M$

21. Which of the following conditions indicate a *basic* solution at 25°C?
   a. pOH = 11.21
   b. pH = 9.42
   c. $[OH^-] > [H^+]$
   d. $[OH^-] > 1.0 \times 10^{-7} M$

22. Why is H_3O^+ the strongest acid and OH^- the strongest base that can exist in significant amounts in aqueous solutions?

23. How many significant figures are there in the following numbers: 10.78, 6.78, 0.78? If these were pH values, to how many significant figures can you express the $[H^+]$? Explain any discrepancies between your answers to the two questions.

24. In terms of orbitals and electron arrangements, what must be present for a molecule or an ion to act as a Lewis acid? What must be present for a molecule or an ion to act as a Lewis base?

25. Consider the autoionization of liquid ammonia:

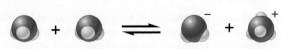

Label each of the species in the equation as an acid or a base and explain your answer.

26. The following are representations of acid–base reactions:

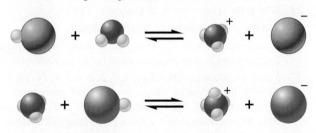

a. Label each of the species in both equations as an acid or a base and explain your answers.

b. For those species that are acids, which labels apply: Arrhenius acid, Brønsted–Lowry acid, Lewis acid? What about the bases?

27. Give three example solutions that fit each of the following descriptions.

a. a strong electrolyte solution that is very acidic

b. a strong electrolyte solution that is slightly acidic

c. a strong electrolyte solution that is very basic

d. a strong electrolyte solution that is slightly basic

e. a strong electrolyte solution that is neutral

28. Derive an expression for the relationship between pK_a and pK_b for a conjugate acid–base pair. ($pK = -\log K$.)

29. Consider the following statements. Write out an example reaction and K expression that is associated with each statement.

a. The autoionization of water.

b. An acid reacts with water to produce the conjugate base of the acid and the hydronium ion.

c. A base reacts with water to produce the conjugate acid of the base and the hydroxide ion.

30. Which of the following statements is(are) *true*? Correct the false statements.

a. When a base is dissolved in water, the lowest possible pH of the solution is 7.0.

b. When an acid is dissolved in water, the lowest possible pH is 0.

c. A strong acid solution will have a lower pH than a weak acid solution.

d. A 0.0010-M Ba(OH)$_2$ solution has a pOH that is twice the pOH value of a 0.0010-M KOH solution.

31. Consider the following mathematical expressions.

a. $[H^+] = [HA]_0$

b. $[H^+] = (K_a \times [HA]_0)^{1/2}$

c. $[OH^-] = 2[B]_0$

d. $[OH^-] = (K_b \times [B]_0)^{1/2}$

For each expression, give three solutions where the mathematical expression would give a good approximation for the $[H^+]$ or $[OH^-]$. $[HA]_0$ and $[B]_0$ represent initial concentrations of an acid or a base.

32. Consider a 0.10-M H$_2$CO$_3$ solution and a 0.10-M H$_2$SO$_4$ solution. Without doing any detailed calculations, choose one of the following statements that best describes the $[H^+]$ of each solution and explain your answer.

a. The $[H^+]$ is less than 0.10 M.

b. The $[H^+]$ is 0.10 M.

c. The $[H^+]$ is between 0.10 M and 0.20 M.

d. The $[H^+]$ is 0.20 M.

33. Of the hydrogen halides, only HF is a weak acid. Give a possible explanation.

34. Explain why the following are done, both of which are related to acid–base chemistry.

a. Power plants burning coal with high sulfur content use scrubbers to help eliminate sulfur emissions.

b. A gardener mixes lime (CaO) into the soil of his garden.

Exercises

In this section similar exercises are paired.

Nature of Acids and Bases

35. Write balanced equations that describe the following reactions.

a. the dissociation of perchloric acid in water

b. the dissociation of propanoic acid (CH$_3$CH$_2$CO$_2$H) in water

c. the dissociation of ammonium ion in water

36. Write the dissociation reaction and the corresponding K_a equilibrium expression for each of the following acids in water.

a. HCN

b. HOC$_6$H$_5$

c. C$_6$H$_5$NH$_3^+$

37. For each of the following aqueous reactions, identify the acid, the base, the conjugate base, and the conjugate acid.

a. $H_2O + H_2CO_3 \rightleftharpoons H_3O^+ + HCO_3^-$

b. $C_5H_5NH^+ + H_2O \rightleftharpoons C_5H_5N + H_3O^+$

c. $HCO_3^- + C_5H_5NH^+ \rightleftharpoons H_2CO_3 + C_5H_5N$

38. For each of the following aqueous reactions, identify the acid, the base, the conjugate base, and the conjugate acid.

a. $Al(H_2O)_6^{3+} + H_2O \rightleftharpoons H_3O^+ + Al(H_2O)_5(OH)^{2+}$

b. $H_2O + HONH_3^+ \rightleftharpoons HONH_2 + H_3O^+$

c. $HOCl + C_6H_5NH_2 \rightleftharpoons OCl^- + C_6H_5NH_3^+$

39. Classify each of the following as a strong acid or a weak acid.

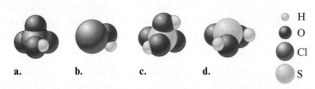

a.   b.   c.   d.

H
O
Cl
S

40. Consider the following illustrations:

H^+
A^-
B^-

Which beaker best illustrates what happens when the following acids are dissolved in water?

a. HNO_2 d. HF

b. HNO_3 e. $HC_2H_3O_2$

c. HCl

41. Use Table 14.2 to order the following from the strongest to the weakest acid.

$$HClO_2, \quad H_2O, \quad NH_4^+, \quad HClO_4$$

42. Use Table 14.2 to order the following from the strongest to the weakest base.

$$ClO_2^-, \quad H_2O, \quad NH_3, \quad ClO_4^-$$

43. You may need Table 14.2 to answer the following questions.

a. Which is the stronger acid, HCl or H_2O?

b. Which is the stronger acid, H_2O or HNO_2?

c. Which is the stronger acid, HCN or HOC_6H_5?

44. You may need Table 14.2 to answer the following questions.

a. Which is the stronger base, Cl^- or H_2O?

b. Which is the stronger base, H_2O or NO_2^-?

c. Which is the stronger base, CN^- or $OC_6H_5^-$?

Autoionization of Water and the pH Scale

45. Calculate the $[OH^-]$ of each of the following solutions at 25°C. Identify each solution as neutral, acidic, or basic.

a. $[H^+] = 1.0 \times 10^{-7} M$ c. $[H^+] = 12 M$

b. $[H^+] = 8.3 \times 10^{-16} M$ d. $[H^+] = 5.4 \times 10^{-5} M$

46. Calculate the $[H^+]$ of each of the following solutions at 25°C. Identify each solution as neutral, acidic, or basic.

a. $[OH^-] = 1.5 M$

b. $[OH^-] = 3.6 \times 10^{-15} M$

c. $[OH^-] = 1.0 \times 10^{-7} M$

d. $[OH^-] = 7.3 \times 10^{-4} M$

47. Values of K_w as a function of temperature are as follows:

Temperature (°C)	K_w
0	1.14×10^{-15}
25	1.00×10^{-14}
35	2.09×10^{-14}
40.	2.92×10^{-14}
50.	5.47×10^{-14}

a. Is the autoionization of water exothermic or endothermic?

b. Calculate $[H^+]$ and $[OH^-]$ in a neutral solution at 50.°C.

48. At 40.°C the value of K_w is 2.92×10^{-14}.

a. Calculate the $[H^+]$ and $[OH^-]$ in pure water at 40.°C.

b. What is the pH of pure water at 40.°C?

c. If the hydroxide ion concentration in a solution is 0.10 M, what is the pH at 40.°C?

49. Calculate the pH and pOH of the solutions in Exercises 45 and 46.

50. Calculate $[H^+]$ and $[OH^-]$ for each solution at 25°C. Identify each solution as neutral, acidic, or basic.

a. pH = 7.40 (the normal pH of blood)

b. pH = 15.3

c. pH = −1.0

d. pH = 3.20

e. pOH = 5.0

f. pOH = 9.60

51. Fill in the missing information in the following table.

	pH	pOH	$[H^+]$	$[OH^-]$	Acidic, Basic, or Neutral?
Solution a	6.88	___	___	___	___
Solution b	___	___	___	$8.4 \times 10^{-14}\ M$	___
Solution c	___	3.11	___	___	___
Solution d	___	___	$1.0 \times 10^{-7}\ M$	___	___

52. Fill in the missing information in the following table.

	pH	pOH	$[H^+]$	$[OH^-]$	Acidic, Basic, or Neutral?
Solution a	9.63	___	___	___	___
Solution b	___	___	___	$3.9 \times 10^{-6}\ M$	___
Solution c	___	___	0.027 M	___	___
Solution d	___	1.22	___	___	___

53. The pH of a sample of gastric juice in a person's stomach is 2.1. Calculate the pOH, $[H^+]$, and $[OH^-]$ for this sample. Is gastric juice acidic or basic?

54. The pOH of a sample of baking soda dissolved in water is 5.74 at 25°C. Calculate the pH, $[H^+]$, and $[OH^-]$ for this sample. Is the solution acidic or basic?

Solutions of Acids

55. What are the major species present in 0.250 M solutions of each of the following acids? Calculate the pH of each of these solutions.
   a. $HClO_4$
   b. HNO_3

56. A solution is prepared by adding 50.0 mL of 0.050 M HBr to 150.0 mL of 0.10 M HI. Calculate $[H^+]$ and the pH of this solution. HBr and HI are both considered strong acids.

57. Calculate the pH of each of the following solutions of a strong acid in water.
   a. 0.10 M HCl
   c. $1.0 \times 10^{-11}\ M$ HCl
   b. 5.0 M HCl

58. Calculate the pH of each of the following solutions containing a strong acid in water.
   a. $2.0 \times 10^{-2}\ M$ HNO_3
   c. $6.2 \times 10^{-12}\ M$ HNO_3
   b. 4.0 M HNO_3

59. Calculate the concentration of an aqueous HI solution that has pH = 2.50. HI is a strong acid.

60. Calculate the concentration of an aqueous HBr solution that has pH = 4.25. HBr is a strong acid.

61. How would you prepare 1600 mL of a pH = 1.50 solution using concentrated (12 M) HCl?

62. A solution is prepared by adding 50.0 mL concentrated hydrochloric acid and 20.0 mL concentrated nitric acid to 300 mL water. More water is added until the final volume is 1.00 L. Calculate $[H^+]$, $[OH^-]$, and the pH for this solution. [*Hint:* Concentrated HCl is 38% HCl (by mass) and has a density of 1.19 g/mL; concentrated HNO_3 is 70.% HNO_3 (by mass) and has a density of 1.42 g/mL.]

63. What are the major species present in 0.250 M solutions of each of the following acids? Calculate the pH of each of these solutions.
   a. HNO_2
   b. CH_3CO_2H ($HC_2H_3O_2$)

64. What are the major species present in 0.250 M solutions of each of the following acids? Calculate the pH of each of these solutions.
   a. HOC_6H_5
   b. HCN

65. Calculate the concentration of all species present and the pH of a 0.020-M HF solution.

66. Calculate the percent dissociation for a 0.22-M solution of chlorous acid ($HClO_2$, $K_a = 1.2 \times 10^{-2}$).

67. For propanoic acid ($HC_3H_5O_2$, $K_a = 1.3 \times 10^{-5}$), determine the concentration of all species present, the pH, and the percent dissociation of a 0.100-M solution.

68. A solution is prepared by dissolving 0.56 g benzoic acid ($C_6H_5CO_2H$, $K_a = 6.4 \times 10^{-5}$) in enough water to make 1.0 L of solution. Calculate $[C_6H_5CO_2H]$, $[C_6H_5CO_2^-]$, $[H^+]$, $[OH^-]$, and the pH of this solution.

69. Monochloroacetic acid, $HC_2H_2ClO_2$, is a skin irritant that is used in "chemical peels" intended to remove the top layer of dead skin from the face and ultimately improve the complexion. The value of K_a for monochloroacetic acid is 1.35×10^{-3}. Calculate the pH of a 0.10-M solution of monochloroacetic acid.

70. A typical aspirin tablet contains 325 mg acetylsalicylic acid ($HC_9H_7O_4$). Calculate the pH of a solution that is prepared by dissolving two aspirin tablets in enough water to make one cup (237 mL) of solution. Assume the aspirin tablets are pure acetylsalicylic acid, $K_a = 3.3 \times 10^{-4}$.

71. Calculate the pH of a solution that contains 1.0 M HF and 1.0 M HOC_6H_5. Also calculate the concentration of $OC_6H_5^-$ in this solution at equilibrium.

72. A solution is made by adding 50.0 mL of 0.200 M acetic acid ($K_a = 1.8 \times 10^{-5}$) to 50.0 mL of $1.00 \times 10^{-3}\ M$ HCl.
   a. Calculate the pH of the solution.
   b. Calculate the acetate ion concentration.

73. Calculate the percent dissociation of the acid in each of the following solutions.
   a. 0.50 M acetic acid
   b. 0.050 M acetic acid
   c. 0.0050 M acetic acid
   d. Use Le Châtelier's principle to explain why percent dissociation increases as the concentration of a weak acid decreases.
   e. Even though the percent dissociation increases from solutions a to c, the $[H^+]$ decreases. Explain.

74. Using the K_a values in Table 14.2, calculate the percent dissociation in a 0.20-M solution of each of the following acids.
   a. nitric acid (HNO_3)
   b. nitrous acid (HNO_2)
   c. phenol (HOC_6H_5)
   d. How is percent dissociation of an acid related to the K_a value for the acid (assuming equal initial concentrations of acids)?

75. A 0.15-M solution of a weak acid is 3.0% dissociated. Calculate K_a.

76. An acid HX is 25% dissociated in water. If the equilibrium concentration of HX is 0.30 M, calculate the K_a value for HX.

77. Trichloroacetic acid (CCl_3CO_2H) is a corrosive acid that is used to precipitate proteins. The pH of a 0.050-M solution of trichloroacetic acid is the same as the pH of a 0.040-M $HClO_4$ solution. Calculate K_a for trichloroacetic acid.

78. The pH of a 0.063-M solution of hypobromous acid (HOBr but usually written HBrO) is 4.95. Calculate K_a.

79. A solution of formic acid (HCOOH, $K_a = 1.8 \times 10^{-4}$) has a pH of 2.70. Calculate the initial concentration of formic acid in this solution.

80. A typical sample of vinegar has a pH of 3.0. Assuming that vinegar is only an aqueous solution of acetic acid ($K_a = 1.8 \times 10^{-5}$), calculate the concentration of acetic acid in vinegar.

81. One mole of a weak acid HA was dissolved in 2.0 L of solution. After the system had come to equilibrium, the concentration of HA was found to be 0.45 M. Calculate K_a for HA.

82. You have 100.0 g saccharin, a sugar substitute, and you want to prepare a pH = 5.75 solution. What volume of solution can be prepared? For saccharin, $HC_7H_4NSO_3$, $pK_a = 11.70$ ($pK_a = -\log K_a$).

Solutions of Bases

83. Write the reaction and the corresponding K_b equilibrium expression for each of the following substances acting as bases in water.

   a. NH_3 b. C_5H_5N

84. Write the reaction and the corresponding K_b equilibrium expression for each of the following substances acting as bases in water.

   a. aniline, $C_6H_5NH_2$ b. dimethylamine, $(CH_3)_2NH$

85. Use Table 14.3 to help order the following bases from strongest to weakest.

   $$NO_3^-, \quad H_2O, \quad NH_3, \quad C_5H_5N$$

86. Use Table 14.3 to help order the following acids from strongest to weakest.

   $$HNO_3, \quad H_2O, \quad NH_4^+, \quad C_5H_5NH^+$$

87. Use Table 14.3 to help answer the following questions.

   a. Which is the stronger base, ClO_4^- or $C_6H_5NH_2$?
   b. Which is the stronger base, H_2O or $C_6H_5NH_2$?
   c. Which is the stronger base, OH^- or $C_6H_5NH_2$?
   d. Which is the stronger base, $C_6H_5NH_2$ or CH_3NH_2?

88. Use Table 14.3 to help answer the following questions.

   a. Which is the stronger acid, $HClO_4$ or $C_6H_5NH_3^+$?
   b. Which is the stronger acid, H_2O or $C_6H_5NH_3^+$?
   c. Which is the stronger acid, $C_6H_5NH_3^+$ or $CH_3NH_3^+$?

89. Calculate the pH of the following solutions.

   a. 0.10 M NaOH
   b. 1.0×10^{-10} M NaOH
   c. 2.0 M NaOH

90. Calculate $[OH^-]$, pOH, and pH for each of the following.

   a. 0.00040 M $Ca(OH)_2$
   b. a solution containing 25 g KOH per liter
   c. a solution containing 150.0 g NaOH per liter

91. What are the major species present in 0.015 M solutions of each of the following bases?

   a. KOH   b. $Ba(OH)_2$

   What is $[OH^-]$ and the pH of each of these solutions?

92. What are the major species present in the following mixtures of bases?

   a. 0.050 M NaOH and 0.050 M LiOH
   b. 0.0010 M $Ca(OH)_2$ and 0.020 M RbOH

   What is $[OH^-]$ and the pH of each of these solutions?

93. What mass of KOH is necessary to prepare 800.0 mL of a solution having a pH = 11.56?

94. Calculate the concentration of an aqueous $Sr(OH)_2$ that has pH = 10.50.

95. What are the major species present in a 0.150-M NH_3 solution? Calculate the $[OH^-]$ and the pH of this solution.

96. For the reaction of hydrazine (N_2H_4) in water,

   $$H_2NNH_2(aq) + H_2O(l) \rightleftharpoons H_2NNH_3^+(aq) + OH^-(aq)$$

K_b is 3.0×10^{-6}. Calculate the concentrations of all species and the pH of a 2.0-M solution of hydrazine in water.

97. Calculate $[OH^-]$, $[H^+]$, and the pH of 0.20 M solutions of each of the following amines.

   a. triethylamine $[(C_2H_5)_3N, K_b = 4.0 \times 10^{-4}]$
   b. hydroxylamine ($HONH_2, K_b = 1.1 \times 10^{-8}$)

98. Calculate $[OH^-]$, $[H^+]$, and the pH of 0.40 M solutions of each of the following amines (the K_b values are found in Table 14.3).

   a. aniline   b. methylamine

99. Calculate the pH of a 0.20-M $C_2H_5NH_2$ solution ($K_b = 5.6 \times 10^{-4}$).

100. Calculate the pH of a 0.050-M $(C_2H_5)_2NH$ solution ($K_b = 1.3 \times 10^{-3}$).

101. What is the percent ionization in each of the following solutions?

   a. 0.10 M NH_3 c. 0.10 M CH_3NH_2
   b. 0.010 M NH_3

102. Calculate the percentage of pyridine (C_5H_5N) that forms pyridinium ion, $C_5H_5NH^+$, in a 0.10-M aqueous solution of pyridine ($K_b = 1.7 \times 10^{-9}$).

103. The pH of a 0.016-M aqueous solution of p-toluidine ($CH_3C_6H_4NH_2$) is 8.60. Calculate K_b.

104. Calculate the mass of $HONH_2$ required to dissolve in enough water to make 250.0 mL of solution having a pH of 10.00 ($K_b = 1.1 \times 10^{-8}$).

Polyprotic Acids

105. Write out the stepwise K_a reactions for the diprotic acid H_2SO_3.

106. Write out the stepwise K_a reactions for citric acid ($H_3C_6H_5O_7$), a triprotic acid.

107. A typical vitamin C tablet (containing pure ascorbic acid, $H_2C_6H_6O_6$) weighs 500. mg. One vitamin C tablet is dissolved in enough water to make 200.0 mL of solution. Calculate the pH of this solution. Ascorbic acid is a diprotic acid.

108. Arsenic acid (H_3AsO_4) is a triprotic acid with $K_{a_1} = 5.5 \times 10^{-3}$, $K_{a_2} = 1.7 \times 10^{-7}$, and $K_{a_3} = 5.1 \times 10^{-12}$. Calculate $[H^+]$, $[OH^-]$, $[H_3AsO_4]$, $[H_2AsO_4^-]$, $[HAsO_4^{2-}]$, and $[AsO_4^{3-}]$ in a 0.20-M arsenic acid solution.

109. Calculate the pH and $[S^{2-}]$ in a 0.10-M H_2S solution. Assume $K_{a_1} = 1.0 \times 10^{-7}$; $K_{a_2} = 1.0 \times 10^{-19}$.

110. Calculate $[CO_3^{2-}]$ in a 0.010-M solution of CO_2 in water (usually written as H_2CO_3). If all the CO_3^{2-} in this solution comes from the reaction

   $$HCO_3^-(aq) \rightleftharpoons H^+(aq) + CO_3^{2-}(aq)$$

what percentage of the H^+ ions in the solution is a result of the dissociation of HCO_3^-? When acid is added to a solution of sodium hydrogen carbonate ($NaHCO_3$), vigorous bubbling occurs. How is this reaction related to the existence of carbonic acid (H_2CO_3) molecules in aqueous solution?

111. Calculate the pH of a 2.0-M H_2SO_4 solution.

112. Calculate the pH of a 5.0×10^{-3}-M solution of H_2SO_4.

Acid–Base Properties of Salts

113. Arrange the following 0.10 M solutions in order of most acidic to most basic.

$$KOH, \quad KNO_3, \quad KCN, \quad NH_4Cl, \quad HCl$$

114. Arrange the following 0.10 M solutions in order from most acidic to most basic. See Appendix 5 for K_a and K_b values.

$$CaBr_2, \quad KNO_2, \quad HClO_4, \quad HNO_2, \quad HONH_3ClO_4$$

115. Given that the K_a value for acetic acid is 1.8×10^{-5} and the K_a value for hypochlorous acid is 3.5×10^{-8}, which is the stronger base, OCl^- or $C_2H_3O_2^-$?

116. The K_b values for ammonia and methylamine are 1.8×10^{-5} and 4.4×10^{-4}, respectively. Which is the stronger acid, NH_4^+ or $CH_3NH_3^+$?

117. Determine $[OH^-]$, $[H^+]$, and the pH of each of the following solutions.

    a. $1.0\ M$ KCl   **b.** $1.0\ M\ KC_2H_3O_2$

118. Calculate the concentrations of all species present in a 0.25-M solution of ethylammonium chloride ($C_2H_5NH_3Cl$).

119. Calculate the pH of each of the following solutions.

    a. $0.10\ M\ CH_3NH_3Cl$ **b.** $0.050\ M$ NaCN

120. Calculate the pH of each of the following solutions.

    a. $0.12\ M\ KNO_2$ **c.** $0.40\ M\ NH_4ClO_4$

    b. $0.45\ M$ NaOCl

121. Sodium azide (NaN_3) is sometimes added to water to kill bacteria. Calculate the concentration of all species in a 0.010-M solution of NaN_3. The K_a value for hydrazoic acid (HN_3) is 1.9×10^{-5}.

122. Papaverine hydrochloride (abbreviated papH$^+$Cl$^-$; molar mass = 378.85 g/mol) is a drug that belongs to a group of medicines called vasodilators, which cause blood vessels to expand, thereby increasing blood flow. This drug is the conjugate acid of the weak base papaverine (abbreviated pap; K_b = 8.33×10^{-9} at 35.0°C). Calculate the pH of a 30.0-mg/mL aqueous dose of papH$^+$Cl$^-$ prepared at 35.0°C. K_w at 35.0°C is 2.1×10^{-14}.

123. An unknown salt is either NaCN, $NaC_2H_3O_2$, NaF, NaCl, or NaOCl. When 0.100 mole of the salt is dissolved in 1.00 L of solution, the pH of the solution is 8.07. What is the identity of the salt?

124. Consider a solution of an unknown salt having the general formula BHCl, where B is one of the weak bases in Table 14.3. A 0.10-M solution of the unknown salt has a pH of 5.82. What is the actual formula of the salt?

125. A 0.050-M solution of the salt NaB has a pH of 9.00. Calculate the pH of a 0.010-M solution of HB.

126. A 0.20-M sodium chlorobenzoate ($NaC_7H_4ClO_2$) solution has a pH of 8.65. Calculate the pH of a 0.20-M chlorobenzoic acid ($HC_7H_4ClO_2$) solution.

127. Calculate the pH of a 0.050-M Al(NO₃)₃ solution. The K_a value for Al(H₂O)₆³⁺ is 1.4×10^{-5}.

128. Calculate the pH of a 0.10-M CoCl₃ solution. The K_a value for Co(H₂O)₆³⁺ is 1.0×10^{-5}.

129. Are solutions of the following salts acidic, basic, or neutral? For those that are not neutral, write balanced chemical equations for the reactions causing the solution to be acidic or basic. The relevant K_a and K_b values are found in Tables 14.2 and 14.3.

    a. $NaNO_3$ **c.** $C_5H_5NHClO_4$ **e.** KOCl

    b. $NaNO_2$ **d.** NH_4NO_2 **f.** NH_4OCl

130. Are solutions of the following salts acidic, basic, or neutral? For those that are not neutral, write balanced equations for the reactions causing the solution to be acidic or basic. The relevant K_a and K_b values are found in Tables 14.2 and 14.3.

    a. $Sr(NO_3)_2$ **c.** CH_3NH_3Cl **e.** NH_4F

    b. $NH_4C_2H_3O_2$ **d.** $C_6H_5NH_3ClO_2$ **f.** CH_3NH_3CN

Relationships Between Structure and Strengths of Acids and Bases

131. Place the species in each of the following groups in order of increasing acid strength. Explain the order you chose for each group.

    a. HIO_3, $HBrO_3$ **c.** HOCl, HOI

    b. HNO_2, HNO_3 **d.** H_3PO_4, H_3PO_3

132. Place the species in each of the following groups in order of increasing base strength. Give your reasoning in each case.

    a. IO_3^-, BrO_3^- **b.** NO_2^-, NO_3^- **c.** OCl^-, OI^-

133. Place the species in each of the following groups in order of increasing acid strength.

    a. H_2O, H_2S, H_2Se (bond energies: H—O, 467 kJ/mol; H—S, 363 kJ/mol; H—Se, 276 kJ/mol)

    b. CH_3CO_2H, FCH_2CO_2H, F_2CHCO_2H, F_3CCO_2H

    c. NH_4^+, $HONH_3^+$

    d. NH_4^+, PH_4^+ (bond energies: N—H, 391 kJ/mol; P—H, 322 kJ/mol)

Give reasons for the orders you chose.

134. Using your results from Exercise 133, place the species in each of the following groups in order of increasing base strength.

    a. OH^-, SH^-, SeH^- **b.** NH_3, PH_3 **c.** NH_3, $HONH_2$

135. Will the following oxides give acidic, basic, or neutral solutions when dissolved in water? Write reactions to justify your answers.

    a. CaO   **b.** SO_2 **c.** Cl_2O

136. Will the following oxides give acidic, basic, or neutral solutions when dissolved in water? Write reactions to justify your answers.

    a. Li_2O **b.** CO_2 **c.** SrO

Lewis Acids and Bases

137. Identify the Lewis acid and the Lewis base in each of the following reactions.

    a. $B(OH)_3(aq) + H_2O(l) \rightleftharpoons B(OH)_4^-(aq) + H^+(aq)$

    b. $Ag^+(aq) + 2NH_3(aq) \rightleftharpoons Ag(NH_3)_2^+(aq)$

    c. $BF_3(g) + F^-(aq) \rightleftharpoons BF_4^-(aq)$

138. Identify the Lewis acid and the Lewis base in each of the following reactions.

   a. $Fe^{3+}(aq) + 6H_2O(l) \rightleftharpoons Fe(H_2O)_6^{3+}(aq)$

   b. $H_2O(l) + CN^-(aq) \rightleftharpoons HCN(aq) + OH^-(aq)$

   c. $HgI_2(s) + 2I^-(aq) \rightleftharpoons HgI_4^{2-}(aq)$

139. Aluminum hydroxide is an amphoteric substance. It can act as either a Brønsted–Lowry base or a Lewis acid. Write a reaction showing $Al(OH)_3$ acting as a base toward H^+ and as an acid toward OH^-.

140. Zinc hydroxide is an amphoteric substance. Write equations that describe $Zn(OH)_2$ acting as a Brønsted–Lowry base toward H^+ and as a Lewis acid toward OH^-.

141. Would you expect Fe^{3+} or Fe^{2+} to be the stronger Lewis acid? Explain.

142. Use the Lewis acid–base model to explain the following reaction.

$$CO_2(g) + H_2O(l) \longrightarrow H_2CO_3(aq)$$

Additional Exercises

143. A 10.0-mL sample of an HCl solution has a pH of 2.000. What volume of water must be added to change the pH to 4.000?

144. Which of the following represent conjugate acid–base pairs? For those pairs that are not conjugates, write the correct conjugate acid or base for each species in the pair.

   a. H_2O, OH^- **c.** H_3PO_4, $H_2PO_4^-$

   b. H_2SO_4, SO_4^{2-} **d.** $HC_2H_3O_2$, $C_2H_3O_2^-$

145. A solution is tested for pH and conductivity as pictured below:

The solution contains one of the following substances: HCl, NaOH, NH_4Cl, HCN, NH_3, HF, or NaCN. If the solute concentration is about 1.0 M, what is the identity of the solute?

146. The pH of human blood is steady at a value of approximately 7.4 owing to the following equilibrium reactions:

$$CO_2(aq) + H_2O(l) \rightleftharpoons H_2CO_3(aq) \rightleftharpoons HCO_3^-(aq) + H^+(aq)$$

Acids formed during normal cellular respiration react with the HCO_3^- to form carbonic acid, which is in equilibrium with $CO_2(aq)$ and $H_2O(l)$. During vigorous exercise, a person's H_2CO_3 blood levels were 26.3 mM, whereas his CO_2 levels were 1.63 mM. On resting, the H_2CO_3 levels declined to 24.9 mM. What was the CO_2 blood level at rest?

147. Hemoglobin (abbreviated Hb) is a protein that is responsible for the transport of oxygen in the blood of mammals. Each hemoglobin molecule contains four iron atoms that are the binding sites for O_2 molecules. The oxygen binding is pH-dependent. The relevant equilibrium reaction is

$$HbH_4^{4+}(aq) + 4O_2(g) \rightleftharpoons Hb(O_2)_4(aq) + 4H^+(aq)$$

Use Le Châtelier's principle to answer the following.

   a. What form of hemoglobin, HbH_4^{4+} or $Hb(O_2)_4$, is favored in the lungs? What form is favored in the cells?

   b. When a person hyperventilates, the concentration of CO_2 in the blood is decreased. How does this affect the oxygen-binding equilibrium? How does breathing into a paper bag help to counteract this effect? (See Exercise 146.)

   c. When a person has suffered a cardiac arrest, injection of a sodium bicarbonate solution is given. Why is this necessary? (*Hint:* CO_2 blood levels increase during cardiac arrest.)

148. A 0.25-g sample of lime (CaO) is dissolved in enough water to make 1500 mL of solution. Calculate the pH of the solution.

149. At 25°C, a saturated solution of benzoic acid ($K_a = 6.4 \times 10^{-5}$) has a pH of 2.80. Calculate the water solubility of benzoic acid in moles per liter.

150. Calculate the pH of an aqueous solution containing 1.0×10^{-2} M HCl, 1.0×10^{-2} M H_2SO_4, and 1.0×10^{-2} M HCN.

151. Acrylic acid ($CH_2{=}CHCO_2H$) is a precursor for many important plastics. K_a for acrylic acid is 5.6×10^{-5}.

   a. Calculate the pH of a 0.10-M solution of acrylic acid.

   b. Calculate the percent dissociation of a 0.10-M solution of acrylic acid.

   c. Calculate the pH of a 0.050-M solution of sodium acrylate ($NaC_3H_3O_2$).

152. Classify each of the following as a strong acid, weak acid, strong base, or weak base in aqueous solution.

   a. HNO_2

   b. HNO_3

   c. CH_3NH_2

   d. NaOH

   e. NH_3

   f. HF

   g. $HC{-}OH$ (with O double-bonded to C, i.e. $\overset{\displaystyle O}{\overset{\|}{HC}}{-}OH$)

   h. $Ca(OH)_2$

   i. H_2SO_4

153. The following illustration displays the relative number of species when an acid, HA, is added to water.

a. Is HA a weak or strong acid? How can you tell?

b. Using the relative numbers given in the illustration, determine the value for K_a and the percent dissociation of the acid. Assume the initial acid concentration is 0.20 M.

154. Quinine ($C_{20}H_{24}N_2O_2$) is the most important alkaloid derived from cinchona bark. It is used as an antimalarial drug. For quinine, $pK_{b_1} = 5.1$ and $pK_{b_2} = 9.7$ ($pK_b = -\log K_b$). Only 1 g quinine will dissolve in 1900.0 mL of solution. Calculate the pH of a saturated aqueous solution of quinine. Consider only the reaction $Q + H_2O \rightleftharpoons QH^+ + OH^-$ described by pK_{b_1}, where Q = quinine.

155. Codeine ($C_{18}H_{21}NO_3$) is a derivative of morphine that is used as an analgesic, narcotic, or antitussive. It was once commonly used in cough syrups but is now available only by prescription because of its addictive properties. If the pH of a 1.7×10^{-3}-M solution of codeine is 9.59, calculate K_b.

156. A codeine-containing cough syrup lists codeine sulfate as a major ingredient instead of codeine. *The Merck Index* gives $C_{36}H_{44}N_2O_{10}S$ as the formula for codeine sulfate. Describe the composition of codeine sulfate. (See Exercise 155.) Why is codeine sulfate used instead of codeine?

157. The equilibrium constant K_a for the reaction

$$Fe(H_2O)_6^{3+}(aq) + H_2O(l) \rightleftharpoons$$
$$Fe(H_2O)_5(OH)^{2+}(aq) + H_3O^+(aq)$$

is 6.0×10^{-3}.

a. Calculate the pH of a 0.10-M solution of $Fe(H_2O)_6^{3+}$.

b. Will a 1.0-M solution of iron(II) nitrate have a higher or lower pH than a 1.0-M solution of iron(III) nitrate? Explain.

158. Rank the following 0.10 M solutions in order of increasing pH.

a. HI, HF, NaF, NaI

b. NH_4Br, HBr, KBr, NH_3

c. $C_6H_5NH_3NO_3$, $NaNO_3$, NaOH, HOC_6H_5, KOC_6H_5, $C_6H_5NH_2$, HNO_3

159. Is an aqueous solution of $NaHSO_4$ acidic, basic, or neutral? What reaction occurs with water? Calculate the pH of a 0.10-M solution of $NaHSO_4$.

160. Calculate the value for the equilibrium constant for each of the following aqueous reactions.

a. $NH_3 + H_3O^+ \rightleftharpoons NH_4^+ + H_2O$

b. $NO_2^- + H_3O^+ \rightleftharpoons HNO_2 + H_2O$

c. $NH_4^+ + OH^- \rightleftharpoons NH_3 + H_2O$

d. $HNO_2 + OH^- \rightleftharpoons H_2O + NO_2^-$

161. Students are often surprised to learn that organic acids, such as acetic acid, contain —OH groups. Actually, all oxyacids contain hydroxyl groups. Sulfuric acid, usually written as H_2SO_4, has the structural formula $SO_2(OH)_2$, where S is the central atom. Identify the acids whose structural formulas are shown below. Why do they behave as acids, while NaOH and KOH are bases?

a. $SO(OH)_2$ b. $ClO_2(OH)$ c. $HPO(OH)_2$

ChemWork Problems

These multiconcept problems (and additional ones) are found interactively online with the same type of assistance a student would get from an instructor.

162. For solutions of the same concentration, as acid strength increases, indicate what happens to each of the following (increases, decreases, or doesn't change).

a. $[H^+]$ d. pOH

b. pH   e. K_a

c. $[OH^-]$

163. Complete the table for each of the following solutions:

	$[H^+]$	pH	pOH	$[OH^-]$
0.0070 M HNO_3	_____	_____	_____	_____
3.0 M KOH	_____	_____	_____	_____

164. Consider a 0.60-M solution of $HC_3H_5O_3$, lactic acid ($K_a = 1.4 \times 10^{-4}$).

a. Which of the following are major species in the solution?

   i. $HC_3H_5O_3$

   ii. $C_3H_5O_3^-$

   iii. H^+

   iv. H_2O

   v. OH^-

b. Complete the following ICE table in terms of x, the amount (mol/L) of lactic acid that dissociates to reach equilibrium.

	$[HC_3H_5O_3]$	$[H^+]$	$[C_3H_5O_3^-]$
Initial	_____	_____	_____
Change	_____	_____	_____
Equilibrium	$0.60 - x$	_____	_____

c. What is the equilibrium concentration for $C_3H_5O_3^-$?

d. Calculate the pH of the solution.

165. Consider a 0.67-M solution of $C_2H_5NH_2$ ($K_b = 5.6 \times 10^{-4}$).

a. Which of the following are major species in the solution?

   i. $C_2H_5NH_2$

   ii. H^+

   iii. OH^-

   iv. H_2O

   v. $C_2H_5NH_3^+$

b. Calculate the pH of this solution.

166. Rank the following 0.10 M solutions in order of increasing pH.

a. NH_3 d. KCl

b. KOH   e. HCl

c. $HC_2H_3O_2$

167. Consider 0.25 M solutions of the following salts: NaCl, RbOCl, KI, $Ba(ClO_4)_2$, and NH_4NO_3. For each salt, indicate whether the solution is acidic, basic, or neutral.

168. Calculate the pH of the following solutions:

   a. 1.2 M $CaBr_2$

   b. 0.84 M $C_6H_5NH_3NO_3$ (K_b for $C_6H_5NH_2 = 3.8 \times 10^{-10}$)

   c. 0.57 M $KC_7H_5O_2$ (K_a for $HC_7H_5O_2 = 6.4 \times 10^{-5}$)

169. Consider 0.10 M solutions of the following compounds: $AlCl_3$, NaCN, KOH, $CsClO_4$, and NaF. Place these solutions in order of increasing pH.

Challenge Problems

170. The pH of 1.0×10^{-8} M hydrochloric acid is not 8.00. The correct pH can be calculated by considering the relationship between the molarities of the three principal ions in the solution (H^+, Cl^-, and OH^-). These molarities can be calculated from algebraic equations that can be derived from the considerations given below.

   a. The solution is electrically neutral.

   b. The hydrochloric acid can be assumed to be 100% ionized.

   c. The product of the molarities of the hydronium ions and the hydroxide ions must equal K_w.

   Calculate the pH of a 1.0×10^{-8}-M HCl solution.

171. Calculate the pH of a 1.0×10^{-7}-M solution of NaOH in water.

172. Calculate $[OH^-]$ in a 3.0×10^{-7}-M solution of $Ca(OH)_2$.

173. Consider 50.0 mL of a solution of weak acid HA ($K_a = 1.00 \times 10^{-6}$), which has a pH of 4.000. What volume of water must be added to make the pH = 5.000?

174. Making use of the assumptions we ordinarily make in calculating the pH of an aqueous solution of a weak acid, calculate the pH of a 1.0×10^{-6}-M solution of hypobromous acid (HBrO, $K_a = 2 \times 10^{-9}$). What is wrong with your answer? Why is it wrong? Without trying to solve the problem, explain what has to be included to solve the problem correctly.

175. Calculate the pH of a 0.200-M solution of C_5H_5NHF. *Hint:* C_5H_5NHF is a salt composed of $C_5H_5NH^+$ and F^- ions. The principal equilibrium in this solution is the best acid reacting with the best base; the reaction for the principal equilibrium is

$$C_5H_5NH^+(aq) + F^-(aq) \rightleftharpoons$$
$$C_5H_5N(aq) + HF(aq) \quad K = 8.2 \times 10^{-3}$$

176. Determine the pH of a 0.50-M solution of NH_4OCl. (See Exercise 175.)

177. Calculate $[OH^-]$ in a solution obtained by adding 0.0100 mol solid NaOH to 1.00 L of 15.0 M NH_3.

178. What mass of NaOH(s) must be added to 1.0 L of 0.050 M NH_3 to ensure that the percent ionization of NH_3 is no greater than 0.0010%? Assume no volume change on addition of NaOH.

179. Consider 1000. mL of a 1.00×10^{-4}-M solution of a certain acid HA that has a K_a value equal to 1.00×10^{-4}. How much water was added or removed (by evaporation) so that a solution remains in which 25.0% of HA is dissociated at equilibrium? Assume that HA is nonvolatile.

180. Calculate the mass of sodium hydroxide that must be added to 1.00 L of 1.00-M $HC_2H_3O_2$ to double the pH of the solution (assume that the added NaOH does not change the volume of the solution).

181. Consider the species PO_4^{3-}, HPO_4^{2-}, and $H_2PO_4^-$. Each ion can act as a base in water. Determine the K_b value for each of these species. Which species is the strongest base?

182. Calculate the pH of a 0.10-M solution of sodium phosphate. (See Exercise 181.)

183. Will 0.10 M solutions of the following salts be acidic, basic, or neutral? See Appendix 5 for K_a values.

   a. ammonium bicarbonate

   b. sodium dihydrogen phosphate

   c. sodium hydrogen phosphate

   d. ammonium dihydrogen phosphate

   e. ammonium formate

184. a. The principal equilibrium in a solution of $NaHCO_3$ is

$$HCO_3^-(aq) + HCO_3^-(aq) \rightleftharpoons H_2CO_3(aq) + CO_3^{2-}(aq)$$

   Calculate the value of the equilibrium constant for this reaction.

   b. At equilibrium, what is the relationship between $[H_2CO_3]$ and $[CO_3^{2-}]$?

   c. Using the equilibrium

$$H_2CO_3(aq) \rightleftharpoons 2H^+(aq) + CO_3^{2-}(aq)$$

   derive an expression for the pH of the solution in terms of K_{a_1} and K_{a_2} using the result from part b.

   d. What is the pH of a solution of $NaHCO_3$?

185. A 0.100-g sample of the weak acid HA (molar mass = 100.0 g/mol) is dissolved in 500.0 g water. The freezing point of the resulting solution is $-0.0056°C$. Calculate the value of K_a for this acid. Assume molality equals molarity in this solution.

186. A sample containing 0.0500 mole of $Fe_2(SO_4)_3$ is dissolved in enough water to make 1.00 L of solution. This solution contains hydrated SO_4^{2-} and Fe^{3+} ions. The latter behaves as an acid:

$$Fe(H_2O)_6^{3+}(aq) \rightleftharpoons Fe(H_2O)_5OH^{2+}(aq) + H^+(aq)$$

   a. Calculate the expected osmotic pressure of this solution at 25°C if the above dissociation is negligible.

   b. The actual osmotic pressure of the solution is 6.73 atm at 25°C. Calculate K_a for the dissociation reaction of $Fe(H_2O)_6^{3+}$. (To do this calculation, you must assume that none of the ions go through the semipermeable membrane. Actually, this is not a great assumption for the tiny H^+ ion.)

Integrative Problems

These problems require the integration of multiple concepts to find the solutions.

187. A 2.14-g sample of sodium hypoiodite is dissolved in water to make 1.25 L of solution. The solution pH is 11.32. What is K_b for the hypoiodite ion?

188. Isocyanic acid (HNCO) can be prepared by heating sodium cyanate in the presence of solid oxalic acid according to the equation

$$2NaOCN(s) + H_2C_2O_4(s) \longrightarrow 2HNCO(l) + Na_2C_2O_4(s)$$

Upon isolating pure HNCO(l), an aqueous solution of HNCO can be prepared by dissolving the liquid HNCO in water. What is the pH of a 100.-mL solution of HNCO prepared from the reaction of 10.0 g each of NaOCN and $H_2C_2O_4$, assuming all of the HNCO produced is dissolved in solution? (K_a of HNCO = 1.2×10^{-4}.)

189. A certain acid, HA, has a vapor density of 5.11 g/L when in the gas phase at a temperature of 25°C and a pressure of 1.00 atm. When 1.50 g of this acid is dissolved in enough water to make 100.0 mL of solution, the pH is found to be 1.80. Calculate K_a for the acid.

Marathon Problems

These problems are designed to incorporate several concepts and techniques into one situation.

190. An aqueous solution contains a mixture of 0.0500 M HCOOH ($K_a = 1.77 \times 10^{-4}$) and 0.150 M CH_3CH_2COOH ($K_a = 1.34 \times$ 10^{-5}). Calculate the pH of this solution. Because both acids are of comparable strength, the H^+ contribution from both acids must be considered.

191. For the following, mix equal volumes of one solution from Group I with one solution from Group II to achieve the indicated pH. Calculate the pH of each solution.

Group I: 0.20 M NH_4Cl, 0.20 M HCl, 0.20 M $C_6H_5NH_3Cl$, 0.20 M $(C_2H_5)_3NHCl$

Group II: 0.20 M KOI, 0.20 M NaCN, 0.20 M KOCl, 0.20 M $NaNO_2$

a. the solution with the lowest pH

b. the solution with the highest pH

c. the solution with the pH closest to 7.00

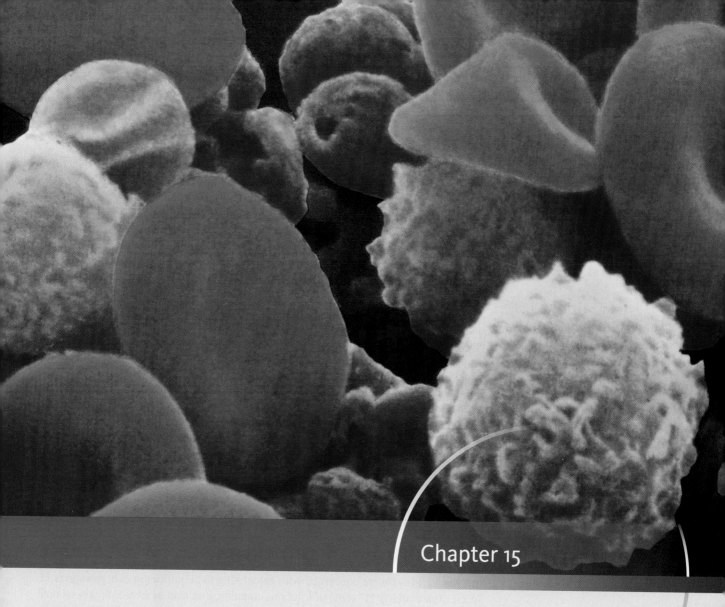

Chapter 15

Acid–Base Equilibria

Colored Scanning Electron Micrograph (SEM) of human blood, showing red and white cells and platelets. Blood has effective buffers to maintain the pH at a constant value. (National Cancer Institute/Photo Researchers Inc.)

Much important chemistry, including almost all the chemistry of the natural world, occurs in aqueous solution. We have already introduced one very significant class of aqueous reactions, those of acids and bases. In this chapter we will explore more applications of acid–base equilibria. In particular, we will examine buffered solutions, which contain components that enable the solution to be resistant to a change in pH. Buffered systems are especially important in living systems, which can survive only in a relatively narrow pH range. For example, although human blood contains many buffering systems, the most important of these consists of a mixture of carbonic acid (~0.0012 M) and bicarbonate ion (~0.024 M). These concentrations produce a pH of 7.4 for normal blood. Because our cells are so sensitive to pH, it is important that this pH value be maintained. So when reactions occur in our bodies, such as the formation of lactic acid ($HC_3H_5O_3$) when our muscles are exerted, the buffering systems must be capable of neutralizing the effects of this acid to maintain the pH at 7.4. We will see in this chapter how a buffered solution can deal with added acid without a significant change in pH.

In this chapter we will also study acid–base titrations to explore how the pH changes when a base is added to an acid and vice versa. This process is important because titrations are often used to determine the amount of acid or base present in an unknown sample. In addition, we will see how indicators can be used to mark the end point of an acid–base titration.

15.1 | Solutions of Acids or Bases Containing a Common Ion

In Chapter 14 we were concerned with calculating the equilibrium concentrations of species (particularly H^+ ions) in solutions containing an acid or a base. In this section we will discuss solutions that contain not only the weak acid HA but also its salt NaA. Although this appears to be a new type of problem, we will see that this case can be handled rather easily using the procedures developed in Chapter 14.

Suppose we have a solution containing the weak acid hydrofluoric acid (HF, $K_a = 7.2 \times 10^{-4}$) and its salt sodium fluoride (NaF). Recall that when a salt dissolves in water, it breaks up completely into its ions—it is a strong electrolyte:

$$NaF(s) \xrightarrow{\text{H}_2\text{O}(l)} Na^+(aq) + F^-(aq)$$

Since hydrofluoric acid is a weak acid and only slightly dissociated, the major species in the solution are HF, Na^+, F^-, and H_2O. The **common ion** in this solution is F^-, since it is produced by both hydrofluoric acid and sodium fluoride. What effect does the presence of the dissolved sodium fluoride have on the dissociation equilibrium of hydrofluoric acid?

To answer this question, we compare the extent of dissociation of hydrofluoric acid in two different solutions, the first containing 1.0 M HF and the second containing 1.0 M HF and 1.0 M NaF. By Le Châtelier's principle, we would expect the dissociation equilibrium for HF

$$HF(aq) \rightleftharpoons H^+(aq) + F^-(aq)$$

in the second solution to be *driven to the left by the presence of F⁻ ions from the NaF.* Thus the extent of dissociation of HF will be *less* in the presence of dissolved NaF:

$$HF(aq) \rightleftharpoons H^+(aq) + F^-(aq)$$

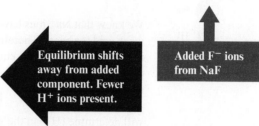

Equilibrium shifts away from added component. Fewer H⁺ ions present.

Added F⁻ ions from NaF

The common ion effect is an application of Le Châtelier's principle.

The shift in equilibrium position that occurs because of the addition of an ion already involved in the equilibrium reaction is called the **common ion effect**. This effect makes a solution of NaF and HF less acidic than a solution of HF alone.

The common ion effect is quite general. For example, solid NH_4Cl added to a 1.0-*M* NH_3 solution produces additional ammonium ions:

$$NH_4Cl(s) \xrightarrow{H_2O} NH_4^+(aq) + Cl^-(aq)$$

and this causes the position of the ammonia–water equilibrium to shift to the left:

$$NH_3(aq) + H_2O(l) \rightleftharpoons NH_4^+(aq) + OH^-(aq)$$

This reduces the equilibrium concentration of OH^- ions.

The common ion effect is also important in solutions of polyprotic acids. The production of protons by the first dissociation step greatly inhibits the succeeding dissociation steps, which, of course, also produce protons, the common ion in this case. We will see later in this chapter that the common ion effect is also important in dealing with the solubility of salts.

Equilibrium Calculations

The procedures for finding the pH of a solution containing a weak acid or base plus a common ion are very similar to the procedures, which we covered in Chapter 14, for solutions containing the acids or bases alone. For example, in the case of a weak acid, the only important difference is that the initial concentration of the anion A^- is not zero in a solution that also contains the salt NaA. Example 15.1 illustrates a typical example using the same general approach we developed in Chapter 14.

Interactive Example 15.1

Sign in at http://login.cengagebrain .com to try this Interactive Example in **OWL**.

Acidic Solutions Containing Common Ions

In Section 14.5 we found that the equilibrium concentration of H^+ in a 1.0-*M* HF solution is 2.7×10^{-2} *M*, and the percent dissociation of HF is 2.7%. Calculate $[H^+]$ and the percent dissociation of HF in a solution containing 1.0 *M* HF ($K_a = 7.2 \times 10^{-4}$) and 1.0 *M* NaF.

Solution

As the aqueous solutions we consider become more complex, it is more important than ever to be systematic and to *focus on the chemistry* occurring in the solution before think-

Major Species

 F⁻

 Na⁺

 HF

H₂O

ing about mathematical procedures. The way to do this is *always* to write the major species first and consider the chemical properties of each one.

In a solution containing 1.0 M HF and 1.0 M NaF, the major species are

$$HF, \quad F^-, \quad Na^+, \quad and \quad H_2O$$

We know that Na^+ ions have neither acidic nor basic properties and that water is a very weak acid (or base). Therefore, the important species are HF and F^-, which participate in the acid dissociation equilibrium that controls $[H^+]$ in this solution. That is, the position of the equilibrium

$$HF(aq) \rightleftharpoons H^+(aq) + F^-(aq)$$

will determine $[H^+]$ in the solution. The equilibrium expression is

$$K_a = \frac{[H^+][F^-]}{[HF]} = 7.2 \times 10^{-4}$$

The important concentrations are shown in the following table.

Initial Concentration (mol/L)		Equilibrium Concentration (mol/L)
$[HF]_0 = 1.0$ (from dissolved HF) $[F^-]_0 = 1.0$ (from dissolved NaF) $[H^+]_0 = 0$ (neglect contribution from H₂O)	$\xrightarrow[\text{dissociates}]{x \text{ mol/L HF}}$	$[HF] = 1.0 - x$ $[F^-] = 1.0 + x$ $[H^+] = x$

Note that $[F^-]_0 = 1.0$ M because of the dissolved sodium fluoride and that at equilibrium $[F^-] > 1.0$ M because when the acid dissociates it produces F^- as well as H^+. Then

$$K_a = 7.2 \times 10^{-4} = \frac{[H^+][F^-]}{[HF]} = \frac{(x)(1.0 + x)}{1.0 - x} \approx \frac{(x)(1.0)}{1.0}$$

(since x is expected to be small).

Solving for x gives

$$x = \frac{1.0}{1.0}(7.2 \times 10^{-4}) = 7.2 \times 10^{-4}$$

Noting that x is small compared to 1.0, we conclude that this result is acceptable. Thus

$$[H^+] = x = 7.2 \times 10^{-4} \ M \qquad \text{(The pH is 3.14.)}$$

The percent dissociation of HF in this solution is

$$\frac{[H^+]}{[HF]_0} \times 100 = \frac{7.2 \times 10^{-4} \ M}{1.0 \ M} \times 100 = 0.072\%$$

Compare these values for $[H^+]$ and the percent dissociation of HF with those for a 1.0-M HF solution, where $[H^+] = 2.7 \times 10^{-2}$ M and the percent dissociation is 2.7%. The large difference shows clearly that the presence of the F^- ions from the dissolved NaF greatly inhibits the dissociation of HF. The position of the acid dissociation equilibrium has been shifted to the left by the presence of F^- ions from NaF.

See Exercises 15.23 and 15.24

15.2 | Buffered Solutions

The most important application of acid–base solutions containing a common ion is for buffering. A **buffered solution** is one that *resists a change in its pH* when either hydroxide ions or protons are added. The most important practical example of a buffered solution is our blood, which can absorb the acids and bases produced in biologic reactions without changing its pH. A constant pH for blood is vital because cells can survive only in a very narrow pH range.

The most important buffering system in the blood involves HCO_3^- and H_2CO_3.

A buffered solution may contain a *weak* acid and its salt (for example, HF and NaF) or a *weak* base and its salt (for example, NH_3 and NH_4Cl). By choosing the appropriate components, a solution can be buffered at virtually any pH.

In treating buffered solutions in this chapter, we will start by considering the equilibrium calculations. We will then use these results to show how buffering works. That is, we will answer the question: How does a buffered solution resist changes in pH when an acid or a base is added?

The systematic approach developed in Chapter 14 for weak acids and bases applies to buffered solutions.

As you do the calculations associated with buffered solutions, keep in mind that these are merely solutions containing weak acids or bases, and the procedures required are the same ones we have already developed. Be sure to use the systematic approach introduced in Chapter 14.

Sign in at http://login.cengagebrain .com to try this Interactive Example in OWL.

The pH of a Buffered Solution I

A buffered solution contains 0.50 *M* acetic acid ($HC_2H_3O_2$, $K_a = 1.8 \times 10^{-5}$) and 0.50 *M* sodium acetate ($NaC_2H_3O_2$). Calculate the pH of this solution.

Solution

Major Species

  $HC_2H_3O_2$

  $C_2H_3O_2^-$

  Na^+

  H_2O

The major species in the solution are

$HC_2H_3O_2$,	Na^+,	$C_2H_3O_2^-$,	and	H_2O
↑	↑	↑		↑
Weak acid	Neither acid nor base	Base (conjugate base of $HC_2H_3O_2$)		Very weak acid or base

Examination of the solution components leads to the conclusion that the acetic acid dissociation equilibrium, which involves both $HC_2H_3O_2$ and $C_2H_3O_2^-$, will control the pH of the solution:

$$HC_2H_3O_2(aq) \rightleftharpoons H^+(aq) + C_2H_3O_2^-(aq)$$

$$K_a = 1.8 \times 10^{-5} = \frac{[H^+][C_2H_3O_2^-]}{[HC_2H_3O_2]}$$

The concentrations are as follows:

Initial Concentration (mol/L)		Equilibrium Concentration (mol/L)
$[HC_2H_3O_2]_0 = 0.50$ $[C_2H_3O_2^-]_0 = 0.50$ $[H^+]_0 \approx 0$	$\xrightarrow[\text{dissociates}]{\substack{x \text{ mol/L of} \\ HC_2H_3O_2}}$ to reach equilibrium	$[HC_2H_3O_2] = 0.50 - x$ $[C_2H_3O_2^-] = 0.50 + x$ $[H^+] = x$

Ken O'Donoghue © Cengage Learning

A digital pH meter shows the pH of the buffered solution to be 4.740.

The corresponding ICE table is

	$HC_2H_3O_2(aq)$	$\rightleftharpoons$	$H^+(aq)$	$+$	$C_2H_3O_2^-(aq)$
Initial	0.50		≈ 0		0.50
Change	$-x$		$+x$		$+x$
Equilibrium	$0.50 - x$		x		$0.50 + x$

Then

$$K_a = 1.8 \times 10^{-5} = \frac{[H^+][C_2H_3O_2^-]}{[HC_2H_3O_2]} = \frac{(x)(0.50 + x)}{0.50 - x} \approx \frac{(x)(0.50)}{0.50}$$

and

$$x \approx 1.8 \times 10^{-5}$$

The approximations are valid (by the 5% rule), so

$$[H^+] = x = 1.8 \times 10^{-5}\ M \quad \text{and} \quad pH = 4.74$$

See Exercises 15.31 and 15.32

Sign in at http://login.cengagebrain.com to try this Interactive Example in OWL.

Interactive Example 15.3

Major Species

 $HC_2H_3O_2$

 $C_2H_3O_2^-$

 Na^+

 OH^-

 H_2O

pH Changes in Buffered Solutions

Calculate the change in pH that occurs when 0.010 mole of solid NaOH is added to 1.0 L of the buffered solution described in Example 15.2. Compare this pH change with that which occurs when 0.010 mole of solid NaOH is added to 1.0 L water.

Solution

Since the added solid NaOH will completely dissociate, the major species in solution *before any reaction occurs* are $HC_2H_3O_2$, Na^+, $C_2H_3O_2^-$, OH^-, and H_2O. Note that the solution contains a relatively large amount of the very strong base hydroxide ion, which has a great affinity for protons. The best source of protons is the acetic acid, and the reaction that will occur is

$$OH^-(aq) + HC_2H_3O_2(aq) \longrightarrow H_2O(l) + C_2H_3O_2^-(aq)$$

Although acetic acid is a weak acid, the hydroxide ion is such a strong base that the reaction above will *proceed essentially to completion* (until the OH^- ions are consumed).

The best approach to this problem involves two distinct steps: (1) assume that the reaction goes to completion, and carry out the stoichiometric calculations, and then (2) carry out the equilibrium calculations.

1. *The stoichiometry problem.* The stoichiometry for the reaction is shown below.

Major Species

 $HC_2H_3O_2$

 $C_2H_3O_2^-$

 Na^+

 H_2O

	$HC_2H_3O_2(aq)$	$+$	$OH^-(aq)$	$\longrightarrow$	$C_2H_3O_2^-(aq)$	$+$	$H_2O(l)$
Before reaction	$1.0\ L \times 0.50\ M$ $= 0.50$ mol		0.010 mol		$1.0\ L \times 0.50\ M$ $= 0.50$ mol		
After reaction	$0.50 - 0.010$ $= 0.49$ mol		$0.010 - 0.010$ $= 0$ mol		$0.50 + 0.010$ $= 0.51$ mol		

Note that 0.010 mole of $HC_2H_3O_2$ has been converted to 0.010 mole of $C_2H_3O_2^-$ by the added OH^-.

2. *The equilibrium problem.* After the reaction between OH^- and $HC_2H_3O_2$ is complete, the major species in solution are

$$HC_2H_3O_2, \quad Na^+, \quad C_2H_3O_2^-, \quad \text{and} \quad H_2O$$

The dominant equilibrium involves the dissociation of acetic acid.

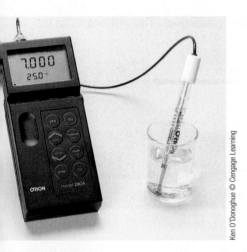

This problem is then very similar to that in Example 15.2. The only difference is that the addition of 0.010 mole of OH^- has consumed some $HC_2H_3O_2$ and produced some $C_2H_3O_2^-$, yielding the following ICE table:

	$HC_2H_3O_2(aq)$	$\rightleftharpoons$	$H^+(aq)$	$+$	$C_2H_3O_2^-(aq)$
Initial	0.49		0		0.51
Change	$-x$		$+x$		$+x$
Equilibrium	$0.49 - x$		x		$0.51 + x$

Note that the initial concentrations are defined after the reaction with OH^- is complete but before the system adjusts to equilibrium. Following the usual procedure gives

$$K_a = 1.8 \times 10^{-5} = \frac{[H^+][C_2H_3O_2^-]}{[HC_2H_3O_2]} = \frac{(x)(0.51 + x)}{0.49 - x} \approx \frac{(x)(0.51)}{0.49}$$

and

$$x \approx 1.7 \times 10^{-5}$$

The approximations are valid (by the 5% rule), so

$$[H^+] = x = 1.7 \times 10^{-5} \quad \text{and} \quad pH = 4.76$$

The change in pH produced by the addition of 0.01 mole of OH^- to this buffered solution is then

$$\underset{\substack{\uparrow \\ \text{New solution}}}{4.76} \quad - \quad \underset{\substack{\uparrow \\ \text{Original solution}}}{4.74} \quad = +0.02$$

The pH increased by 0.02 pH units.

Now compare this with what happens when 0.01 mole of solid NaOH is added to 1.0 L water to give 0.01 M NaOH. In this case $[OH^-] = 0.01\ M$ and

$$[H^+] = \frac{K_w}{[OH^-]} = \frac{1.0 \times 10^{-14}}{1.0 \times 10^{-2}} = 1.0 \times 10^{-12}$$

$$pH = 12.00$$

Thus the change in pH is

$$\underset{\substack{\uparrow \\ \text{New solution}}}{12.00} \quad - \quad \underset{\substack{\uparrow \\ \text{Pure water}}}{7.00} \quad = +5.00$$

The increase is 5.00 pH units. Note how well the buffered solution resists a change in pH as compared with pure water.

See Exercises 15.33 and 15.34

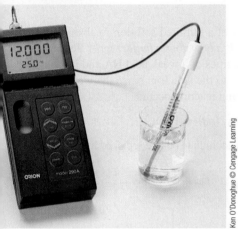

(top) Pure water at pH 7.000. (bottom) When 0.01 mole of NaOH is added to 1.0 L pure water, the pH jumps to 12.000.

Examples 15.2 and 15.3 represent typical buffer problems that involve all the concepts that you need to know to handle buffered solutions containing weak acids. Pay special attention to the following points:

1. Buffered solutions are simply solutions of weak acids or bases containing a common ion. The pH calculations on buffered solutions require exactly the same procedures introduced in Chapter 14. *This is not a new type of problem.*
2. When a strong acid or base is added to a buffered solution, it is best to deal with the stoichiometry of the resulting reaction first. After the stoichiometric calculations

are completed, then consider the equilibrium calculations. This procedure can be presented as follows:

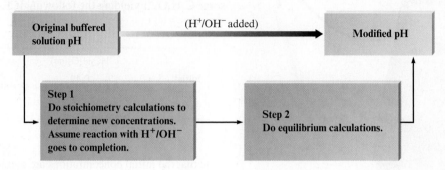

Buffering: How Does It Work?

Examples 15.2 and 15.3 demonstrate the ability of a buffered solution to absorb hydroxide ions without a significant change in pH. *But how does a buffer work?* Suppose a buffered solution contains relatively large quantities of a weak acid HA and its conjugate base A^-. When hydroxide ions are added to the solution, since the weak acid represents the best source of protons, the following reaction occurs:

$$OH^-(aq) + HA(aq) \longrightarrow A^-(aq) + H_2O(l)$$

The net result is that OH^- ions are not allowed to accumulate but are replaced by A^- ions.

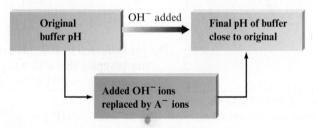

The stability of the pH under these conditions can be understood by examining the equilibrium expression for the dissociation of HA:

$$K_a = \frac{[H^+][A^-]}{[HA]}$$

or, rearranging,

$$[H^+] = K_a\frac{[HA]}{[A^-]}$$

In a buffered solution the pH is governed by the ratio [HA]/[A⁻].

In other words, the *equilibrium concentration of H^+, and thus the pH, is determined by the ratio [HA]/[A⁻]*. When OH^- ions are added, HA is converted to A^-, and the ratio [HA]/[A⁻] decreases. However, *if the amounts of HA and A^- originally present are very large compared with the amount of OH^- added*, the change in the [HA]/[A⁻] ratio will be small.

In Examples 15.2 and 15.3,

$$\frac{[HA]}{[A^-]} = \frac{0.50}{0.50} = 1.0 \quad \text{Initially}$$

$$\frac{[HA]}{[A^-]} = \frac{0.49}{0.51} = 0.96 \quad \text{After adding 0.01 mol/L } OH^-$$

The change in the ratio [HA]/[A⁻] is very small. Thus the [H⁺] and the pH remain essentially constant.

The essence of buffering, then, is that [HA] and [A⁻] are large compared with the amount of OH⁻ added. Thus when the OH⁻ is added, the concentrations of HA and A⁻ change, but only by small amounts. Under these conditions, the [HA]/[A⁻] ratio and thus the [H⁺] remain virtually constant.

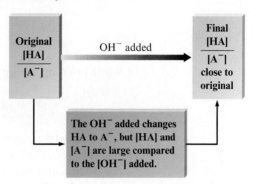

Similar reasoning applies when protons are added to a buffered solution of a weak acid and a salt of its conjugate base. Because the A⁻ ion has a high affinity for H⁺, the added H⁺ ions react with A⁻ to form the weak acid:

$$H^+(aq) + A^-(aq) \longrightarrow HA(aq)$$

and free H⁺ ions do not accumulate. In this case there will be a net change of A⁻ to HA. However, if [A⁻] and [HA] are large compared with the [H⁺] added, little change in the pH will occur.

Let's Review

$$HA(aq) \rightleftharpoons H^+(aq) + A^-(aq)$$

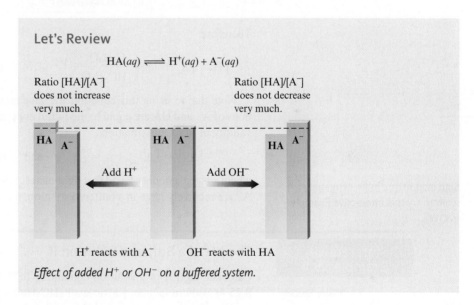

Effect of added H⁺ or OH⁻ on a buffered system.

The form of the acid dissociation equilibrium expression

$$[H^+] = K_a \frac{[HA]}{[A^-]} \tag{15.1}$$

is often useful for calculating [H⁺] in a buffered solution, since [HA] and [A⁻] are known. For example, to calculate [H⁺] in a buffered solution containing $0.10\ M$ HF ($K_a = 7.2 \times 10^{-4}$) and $0.30\ M$ NaF, we simply substitute into Equation (15.1):

$$[H^+] = (7.2 \times 10^{-4})\frac{0.10}{0.30} = 2.4 \times 10^{-4}\ M$$

Another useful form of Equation (15.1) can be obtained by taking the negative log of both sides:

$$-\log[H^+] = -\log(K_a) - \log\left(\frac{[HA]}{[A^-]}\right)$$

That is,

$$pH = pK_a - \log\left(\frac{[HA]}{[A^-]}\right)$$

or, where inverting the log term reverses the sign:

$$pH = pK_a + \log\left(\frac{[A^-]}{[HA]}\right) = pK_a + \log\left(\frac{[\text{base}]}{[\text{acid}]}\right) \qquad (15.2)$$

This log form of the expression for K_a is called the **Henderson–Hasselbalch equation** and is useful for calculating the pH of solutions when the ratio [HA]/[A⁻] is known.

For a particular buffering system (conjugate acid–base pair), all solutions that have the same ratio [A⁻]/[HA] will have the same pH. For example, a buffered solution containing 5.0 M $HC_2H_3O_2$ and 3.0 M $NaC_2H_3O_2$ will have the same pH as one containing 0.050 M $HC_2H_3O_2$ and 0.030 M $NaC_2H_3O_2$. This can be shown as follows:

System	[A⁻]/[HA]
5.0 M $HC_2H_3O_2$ and 3.0 M $NaC_2H_3O_2$	$\dfrac{3.0\ M}{5.0\ M} = 0.60$
0.050 M $HC_2H_3O_2$ and 0.030 M $NaC_2H_3O_2$	$\dfrac{0.030\ M}{0.050\ M} = 0.60$

Therefore,

$$pH = pK_a + \log\left(\frac{[C_2H_3O_2^-]}{[HC_2H_3O_2]}\right) = 4.74 + \log(0.60) = 4.74 - 0.22 = 4.52$$

Note that in using this equation we have assumed that the equilibrium concentrations of A⁻ and HA are equal to the initial concentrations. That is, we are assuming the validity of the approximations

$$[A^-] = [A^-]_0 + x \approx [A^-]_0 \quad \text{and} \quad [HA] = [HA]_0 - x \approx [HA]_0$$

where x is the amount of acid that dissociates. Since the initial concentrations of HA and A⁻ are relatively large in a buffered solution, this assumption is generally acceptable.

Sign in at http://login.cengagebrain.com to try this Interactive Example in OWL.

Interactive Example 15.4

The pH of a Buffered Solution II

Calculate the pH of a solution containing 0.75 M lactic acid ($K_a = 1.4 \times 10^{-4}$) and 0.25 M sodium lactate. Lactic acid ($HC_3H_5O_3$) is a common constituent of biologic systems. For example, it is found in milk and is present in human muscle tissue during exertion.

Solution

The major species in solution are

$$HC_3H_5O_3, \quad Na^+, \quad C_3H_5O_3^-, \quad \text{and} \quad H_2O$$

Since Na⁺ has no acid–base properties and H_2O is a weak acid or base, the pH will be controlled by the lactic acid dissociation equilibrium:

$$HC_3H_5O_3(aq) \rightleftharpoons H^+(aq) + C_3H_5O_3^-(aq)$$

$$K_a = \frac{[H^+][C_3H_5O_3^-]}{[HC_3H_5O_3]} = 1.4 \times 10^{-4}$$

Major Species

$HC_3H_5O_3$

$C_3H_5O_3^-$

Na^+

H_2O

Since $[HC_3H_5O_3]_0$ and $[C_3H_5O_3^-]_0$ are relatively large,

$$[HC_3H_5O_3] \approx [HC_3H_5O_3]_0 = 0.75\ M$$

and

$$[C_3H_5O_3^-] \approx [C_3H_5O_3^-]_0 = 0.25\ M$$

Thus using the rearranged K_a expression, we have

$$[H^+] = K_a \frac{[HC_3H_5O_3]}{[C_3H_5O_3^-]} = (1.4 \times 10^{-4}) \frac{(0.75\ M)}{(0.25\ M)} = 4.2 \times 10^{-4}\ M$$

and

$$pH = -\log(4.2 \times 10^{-4}) = 3.38$$

Alternatively, we could use the Henderson–Hasselbalch equation:

$$pH = pK_a + \log\left(\frac{[C_3H_5O_3^-]}{[HC_3H_5O_3]}\right) = 3.85 + \log\left(\frac{0.25\ M}{0.75\ M}\right) = 3.38$$

See Exercises 15.35 and 15.37

Buffered solutions also can be formed from a weak base and the corresponding conjugate acid. In these solutions, the weak base B reacts with any H^+ added:

$$B + H^+ \longrightarrow BH^+$$

and the conjugate acid BH^+ reacts with any added OH^-:

$$BH^+ + OH^- \longrightarrow B + H_2O$$

The approach needed to perform pH calculations for these systems is virtually identical to that used above. This makes sense because, as is true of all buffered solutions, a weak acid (BH^+) and a weak base (B) are present. A typical case is illustrated in Example 15.5.

Sign in at http://login.cengagebrain.com to try this Interactive Example in **OWL**.

Interactive Example 15.5

The pH of a Buffered Solution III

A buffered solution contains 0.25 M NH$_3$ ($K_b = 1.8 \times 10^{-5}$) and 0.40 M NH$_4$Cl. Calculate the pH of this solution.

Solution

The major species in solution are

$$NH_3, \quad \underbrace{NH_4^+, \quad Cl^-,}_{\text{From the dissolved NH}_4\text{Cl}} \quad \text{and} \quad H_2O$$

Since Cl^- is such a weak base and water is a weak acid or base, the important equilibrium is

$$NH_3(aq) + H_2O(l) \rightleftharpoons NH_4^+(aq) + OH^-(aq)$$

and

$$K_b = 1.8 \times 10^{-5} = \frac{[NH_4^+][OH^-]}{[NH_3]}$$

The appropriate ICE table is:

Major Species

 Cl^-

 NH_4^+

 NH_3

 H_2O

	NH$_3$(aq)	+	H$_2$O(l)	$\rightleftharpoons$	NH$_4^+$(aq)	+	OH$^-$(aq)
Initial	0.25	—	—		0.40		≈ 0
Change	$-x$	—	—		$+x$		$+x$
Equilibrium	$0.25 - x$	—	—		$0.40 + x$		x

Then

$$K_b = 1.8 \times 10^{-5} = \frac{[NH_4^+][OH^-]}{[NH_3]} = \frac{(0.40 + x)(x)}{0.25 - x} \approx \frac{(0.40)(x)}{0.25}$$

and

$$x \approx 1.1 \times 10^{-5}$$

The approximations are valid (by the 5% rule), so

$$[OH^-] = x = 1.1 \times 10^{-5}$$
$$pOH = 4.95$$
$$pH = 14.00 - 4.95 = 9.05$$

This case is typical of a buffered solution in that the initial and equilibrium concentrations of buffering materials are essentially the same.

Alternative Solution

There is another way of looking at this problem. Since the solution contains relatively large quantities of *both* NH_4^+ and NH_3, we can use the equilibrium

$$NH_3(aq) + H_2O(l) \rightleftharpoons NH_4^+(aq) + OH^-(aq)$$

to calculate $[OH^-]$ and then calculate $[H^+]$ from K_w as we have just done. Or we can use the dissociation equilibrium for NH_4^+, that is,

$$NH_4^+(aq) \rightleftharpoons NH_3(aq) + H^+(aq)$$

to calculate $[H^+]$ directly. *Either choice will give the same answer,* since the same equilibrium concentrations of NH_3 and NH_4^+ must satisfy both equilibria.

We can obtain the K_a value for NH_4^+ from the given K_b value for NH_3, since $K_a \times K_b = K_w$:

$$K_a = \frac{K_w}{K_b} = \frac{1.0 \times 10^{-14}}{1.8 \times 10^{-5}} = 5.6 \times 10^{-10}$$

Then, using the Henderson–Hasselbalch equation, we have

$$pH = pK_a + \log\left(\frac{[\text{base}]}{[\text{acid}]}\right)$$

$$= 9.25 + \log\left(\frac{0.25\ M}{0.40\ M}\right) = 9.25 - 0.20 = 9.05$$

See Exercises 15.36 and 15.38

Sign in at http://login.cengagebrain .com to try this Interactive Example in **OWL**.

Interactive Example 15.6

Major Species

Cl^-

H^+

NH_4^+

NH_3

H_2O

Adding Strong Acid to a Buffered Solution I

Calculate the pH of the solution that results when 0.10 mole of gaseous HCl is added to 1.0 L of the buffered solution from Example 15.5.

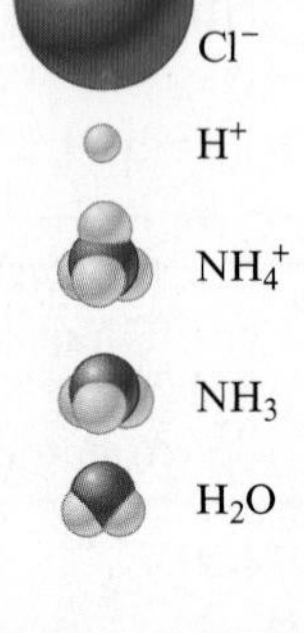

Solution

Before any reaction occurs, the solution contains the following major species:

$$NH_3, \quad NH_4^+, \quad \underbrace{Cl^-, \quad H^+,}_{\text{From added HCl}} \quad \text{and} \quad H_2O$$

What reaction can occur? We know that H^+ will not react with Cl^- to form HCl. In contrast to Cl^-, the NH_3 molecule has a great affinity for protons [this is demonstrated by the fact that NH_4^+ is such a weak acid ($K_a = 5.6 \times 10^{-10}$)]. Thus NH_3 will react with H^+ to form NH_4^+:

$$NH_3(aq) + H^+(aq) \longrightarrow NH_4^+(aq)$$

Since this reaction can be assumed to go essentially to completion to form the very weak acid NH_4^+, we will do the stoichiometry calculations before we consider the equilibrium calculations. That is, we will let the reaction run to completion and then consider the equilibrium.

The stoichiometry calculations for this process are shown below.

Remember: Think about the chemistry first. Ask yourself if a reaction will occur among the major species.

	NH_3	$+$	H^+	$\longrightarrow$	NH_4^+
Before reaction	(1.0 L)(0.25 M) = 0.25 mol		0.10 mol ↑ Limiting reactant		(1.0 L)(0.40 M) = 0.40 mol
After reaction	0.25 − 0.10 = 0.15 mol		0		0.40 + 0.10 = 0.50 mol

After the reaction goes to completion, the solution contains the major species

Major Species

 Cl^-

 NH_4^+

 NH_3

 H_2O

$$NH_3, \quad NH_4^+, \quad Cl^-, \quad \text{and} \quad H_2O$$

and

$$[NH_3]_0 = \frac{0.15 \text{ mol}}{1.0 \text{ L}} = 0.15 \ M$$

$$[NH_4^+]_0 = \frac{0.50 \text{ mol}}{1.0 \text{ L}} = 0.50 \ M$$

We can use the Henderson–Hasselbalch equation, where

$$[\text{Base}] = [NH_3] \approx [NH_3]_0 = 0.15 \ M$$
$$[\text{Acid}] = [NH_4^+] \approx [NH_4^+]_0 = 0.50 \ M$$

Then

$$pH = pK_a + \log\left(\frac{[NH_3]}{[NH_4^+]}\right)$$

$$= 9.25 + \log\left(\frac{0.15 \ M}{0.50 \ M}\right) = 9.25 - 0.52 = 8.73$$

Note that the addition of HCl only slightly decreases the pH, as we would expect in a buffered solution.

See Exercise 15.39

Let's Review | *Summary of the Most Important Characteristics of Buffered Solutions*

❯ Buffered solutions contain relatively large concentrations of a weak acid and the corresponding weak base. They can involve a weak acid HA and the conjugate base A^- or a weak base B and the conjugate acid BH^+.

❯ When H^+ is added to a buffered solution, it reacts essentially to completion with the weak base present:

$$H^+ + A^- \longrightarrow HA \quad \text{or} \quad H^+ + B \longrightarrow BH^+$$

(Box continues on following page)

15.3 | Buffering Capacity

The **buffering capacity** of a buffered solution represents the amount of protons or hydroxide ions the buffer can absorb without a significant change in pH. A buffer with a large capacity contains large concentrations of buffering components and so can absorb a relatively large amount of protons or hydroxide ions and show little pH change. *The pH of a buffered solution is determined by the ratio [A⁻]/[HA]. The capacity of a buffered solution is determined by the magnitudes of [HA] and [A⁻].*

A buffer with a large capacity contains large concentrations of the buffering components.

Example 15.7

Adding Strong Acid to a Buffered Solution II

Calculate the change in pH that occurs when 0.010 mole of gaseous HCl is added to 1.0 L of each of the following solutions:

Solution A: 5.00 M $HC_2H_3O_2$ and   5.00 M $NaC_2H_3O_2$

Solution B: 0.050 M $HC_2H_3O_2$ and   0.050 M $NaC_2H_3O_2$

For acetic acid, $K_a = 1.8 \times 10^{-5}$.

Solution

For both solutions the initial pH can be determined from the Henderson–Hasselbalch equation:

$$pH = pK_a + \log\left(\frac{[C_2H_3O_2^-]}{[HC_2H_3O_2]}\right)$$

Major Species

   Cl^-

   $HC_2H_3O_2$

   $C_2H_3O_2^-$

   Na^+

○ H^+

◍ H_2O

In each case, $[C_2H_3O_2^-] = [HC_2H_3O_2]$. Therefore, the initial pH for both A and B is

$$pH = pK_a + \log(1) = pK_a = -\log(1.8 \times 10^{-5}) = 4.74$$

After the addition of HCl to each of these solutions, the major species *before any reaction occurs* are

$$HC_2H_3O_2, \quad Na^+, \quad C_2H_3O_2^-, \quad \underbrace{H^+, \quad Cl^-,}_{\text{From the added HCl}} \quad \text{and} \quad H_2O$$

Will any reactions occur among these species? Note that we have a relatively large quantity of H^+, which will readily react with any effective base. We know that Cl^- will not react with H^+ to form HCl in water. However, $C_2H_3O_2^-$ will react with H^+ to form the weak acid $HC_2H_3O_2$:

$$H^+(aq) + C_2H_3O_2^-(aq) \longrightarrow HC_2H_3O_2(aq)$$

Because $HC_2H_3O_2$ is a weak acid, we assume that this reaction runs to completion; the 0.010 mole of added H^+ will convert 0.010 mole of $C_2H_3O_2^-$ to 0.010 mole of $HC_2H_3O_2$.

For solution A (since the solution volume is 1.0 L, the number of moles equals the molarity), the following calculations apply:

	H^+	+	$C_2H_3O_2^-$	$\longrightarrow$	$HC_2H_3O_2$
Before reaction	0.010 *M*		5.00 *M*		5.00 *M*
After reaction	0		4.99 *M*		5.01 *M*

The new pH can be obtained by substituting the new concentrations into the Henderson–Hasselbalch equation:

$$pH = pK_a + \log\left(\frac{[C_2H_3O_2^-]}{[HC_2H_3O_2]}\right)$$

$$= 4.74 + \log\left(\frac{4.99}{5.01}\right) = 4.74 - 0.0017 = 4.74$$

There is virtually no change in pH for solution A when 0.010 mole of gaseous HCl is added.

For solution B, the following calculations apply:

	H^+	+	$C_2H_3O_2^-$	$\longrightarrow$	$HC_2H_3O_2$
Before reaction	0.010 *M*		0.050 *M*		0.050 *M*
After reaction	0		0.040 *M*		0.060 *M*

The new pH is

$$pH = 4.74 + \log\left(\frac{0.040}{0.060}\right)$$

$$= 4.74 - 0.18 = 4.56$$

Although the pH change for solution B is small, a change did occur, which is in contrast to solution A.

These results show that solution A, which contains much larger quantities of buffering components, has a much higher buffering capacity than solution B.

See Exercises 15.39 and 15.40

We have seen that the pH of a buffered solution depends on the ratio of the concentrations of buffering components. When this ratio is least affected by added protons or hydroxide ions, the solution is the most resistant to a change in pH. To find the ratio that gives optimal buffering, let's suppose we have a buffered solution containing a large concentration of acetate ion and only a small concentration of acetic acid. Addition of protons to form acetic acid will produce a relatively large *percent* change in the concentration of acetic acid and so will produce a relatively large change in the ratio $[C_2H_3O_2^-]/[HC_2H_3O_2]$ (Table 15.1). Similarly, if hydroxide ions are added to remove some acetic acid, the percent change in the concentration of acetic acid is again large. The same effects are seen if the initial concentration of acetic acid is large and that of acetate ion is small.

Because large changes in the ratio $[A^-]/[HA]$ will produce large changes in pH, we want to avoid this situation for the most effective buffering. This type of reasoning leads us to the general conclusion that optimal buffering occurs when [HA] is equal to $[A^-]$. It is for this condition that the ratio $[A^-]/[HA]$ is most resistant to change when H^+ or OH^- is added to the buffered solution. This means that when choosing the

Original solution · New solution

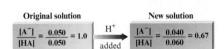

Original solution · New solution

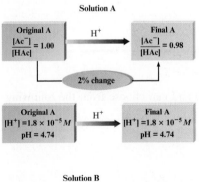

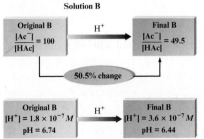

Table 15.1 | Change in $[C_2H_3O_2^-]/[HC_2H_3O_2]$ for Two Solutions When 0.01 mole of H^+ Is Added to 1.0 L of Each

Solution	$\left(\dfrac{[C_2H_3O_2^-]}{[HC_2H_3O_2]}\right)_{orig}$	$\left(\dfrac{[C_2H_3O_2^-]}{[HC_2H_3O_2]}\right)_{new}$	Change	Percent Change
A	$\dfrac{1.00\ M}{1.00\ M} = 1.00$	$\dfrac{0.99\ M}{1.01\ M} = 0.98$	$1.00 \rightarrow 0.98$	2.00%
B	$\dfrac{1.00\ M}{0.01\ M} = 100$	$\dfrac{0.99\ M}{0.02\ M} = 49.5$	$100 \rightarrow 49.5$	50.5%

buffering components for a specific application, we want $[A^-]/[HA]$ to equal 1. It follows that since

$$pH = pK_a + \log\left(\frac{[A^-]}{[HA]}\right) = pK_a + \log(1) = pK_a$$

the pK_a of the weak acid to be used in the buffer should be as close as possible to the desired pH. For example, suppose we need a buffered solution with a pH of 4.00. The most effective buffering will occur when $[HA]$ is equal to $[A^-]$. From the Henderson–Hasselbalch equation,

$$pH = pK_a + \log\left(\frac{[A^-]}{[HA]}\right)$$

4.00 is wanted

Ratio = 1 for most effective buffer

That is, $4.00 = pK_a + \log(1) = pK_a + 0$ and $pK_a = 4.00$

Thus the best choice of a weak acid is one that has $pK_a = 4.00$ or $K_a = 1.0 \times 10^{-4}$.

Critical Thinking

The text states that "the pK_a for a weak acid to be used in the buffer should be as close as possible to the desired pH." What is the problem with choosing a weak acid whose pK_a is not close to the desired pH when making a buffer?

Interactive Example 15.8

Sign in at http://login.cengagebrain .com to try this Interactive Example in **OWL**.

Preparing a Buffer

A chemist needs a solution buffered at pH 4.30 and can choose from the following acids (and their sodium salts):

a. chloroacetic acid ($K_a = 1.35 \times 10^{-3}$)

b. propanoic acid ($K_a = 1.3 \times 10^{-5}$)

c. benzoic acid ($K_a = 6.4 \times 10^{-5}$)

d. hypochlorous acid ($K_a = 3.5 \times 10^{-8}$)

Calculate the ratio $[HA]/[A^-]$ required for each system to yield a pH of 4.30. Which system will work best?

Solution

A pH of 4.30 corresponds to

$$[H^+] = 10^{-4.30} = \text{antilog}(-4.30) = 5.0 \times 10^{-5}\ M$$

Since K_a values rather than pK_a values are given for the various acids, we use Equation (15.1)

$$[H^+] = K_a \frac{[HA]}{[A^-]}$$

rather than the Henderson–Hasselbalch equation. We substitute the required $[H^+]$ and K_a for each acid into Equation (15.1) to calculate the ratio $[HA]/[A^-]$ needed in each case.

Acid	$[H^+] = K_a \dfrac{[HA]}{[A^-]}$	$\dfrac{[HA]}{[A^-]}$
a. Chloroacetic	$5.0 \times 10^{-5} = 1.35 \times 10^{-3}\left(\dfrac{[HA]}{[A^-]}\right)$	3.7×10^{-2}
b. Propanoic	$5.0 \times 10^{-5} = 1.3 \times 10^{-5}\left(\dfrac{[HA]}{[A^-]}\right)$	3.8
c. Benzoic	$5.0 \times 10^{-5} = 6.4 \times 10^{-5}\left(\dfrac{[HA]}{[A^-]}\right)$	0.78
d. Hypochlorous	$5.0 \times 10^{-5} = 3.5 \times 10^{-8}\left(\dfrac{[HA]}{[A^-]}\right)$	1.4×10^3

Since $[HA]/[A^-]$ for benzoic acid is closest to 1, the system of benzoic acid and its sodium salt will be the best choice among those given for buffering a solution at pH 4.30. This example demonstrates the principle that the optimal buffering system has a pK_a value close to the desired pH. The pK_a for benzoic acid is 4.19.

See Exercises 15.47 and 15.48

15.4 | Titrations and pH Curves

As we saw in Chapter 4, a titration is commonly used to determine the amount of acid or base in a solution. This process involves a solution of known concentration (the titrant) delivered from a buret into the unknown solution until the substance being analyzed is just consumed. The stoichiometric (equivalence) point is often signaled by the color change of an indicator. In this section we will discuss the pH changes that occur during an acid–base titration. We will use this information later to show how an appropriate indicator can be chosen for a particular titration.

The progress of an acid–base titration is often monitored by plotting the pH of the solution being analyzed as a function of the amount of titrant added. Such a plot is called a **pH curve** or **titration curve**.

Strong Acid–Strong Base Titrations

The net ionic reaction for a strong acid–strong base titration is

$$H^+(aq) + OH^-(aq) \longrightarrow H_2O(l)$$

To compute $[H^+]$ at a given point in the titration, we must determine the amount of H^+ that remains at that point and divide by the total volume of the solution. Before we proceed, we need to consider a new unit, which is especially convenient for titrations. Since titrations usually involve small quantities (burets are typically graduated in

A setup used to conduct the pH titration of an acid or a base.

$1\,\text{mmol} = 1 \times 10^{-3}\,\text{mol}$
$1\,\text{mL} = 1 \times 10^{-3}\,\text{L}$

$$\frac{\text{mmol}}{\text{mL}} = \frac{\text{mol}}{\text{L}} = M$$

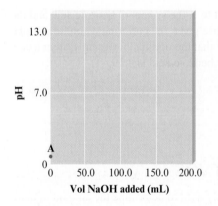

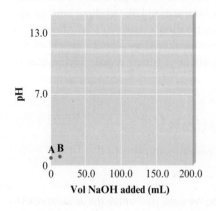

milliliters), the mole is inconveniently large. Therefore, we will use the **millimole** (abbreviated **mmol**), which, as the prefix indicates, is a thousandth of a mole:

$$1\,\text{mmol} = \frac{1\,\text{mol}}{1000} = 10^{-3}\,\text{mol}$$

So far we have defined molarity only in terms of moles per liter. We can now define it in terms of *millimoles per milliliter,* as shown below:

$$\text{Molarity} = \frac{\text{mol solute}}{\text{L solution}} = \frac{\dfrac{\text{mol solute}}{1000}}{\dfrac{\text{L solution}}{1000}} = \frac{\text{mmol solute}}{\text{mL solution}}$$

A 1.0-*M* solution thus contains 1.0 mole of solute per liter of solution or, *equivalently,* 1.0 mmol of solute per milliliter of solution. Just as we obtain the number of moles of solute from the product of the volume in liters and the molarity, we obtain the number of millimoles of solute from the product of the volume in milliliters and the molarity:

$$\text{Number of mmol} = \text{volume (in mL)} \times \text{molarity}$$

CASE STUDY │ Strong Acid–Strong Base Titration

We will illustrate the calculations involved in a strong acid–strong base titration by considering the titration of 50.0 mL of 0.200 *M* HNO$_3$ with 0.100 *M* NaOH. We will calculate the pH of the solution at selected points during the course of the titration, where specific volumes of 0.100 *M* NaOH have been added.

A. No NaOH has been added.
Since HNO$_3$ is a strong acid (is completely dissociated), the solution contains the major species

$$H^+, \quad NO_3^-, \quad \text{and} \quad H_2O$$

and the pH is determined by the H$^+$ from the nitric acid. Since 0.200 *M* HNO$_3$ contains 0.200 *M* H$^+$,

$$[H^+] = 0.200\,M \quad \text{and} \quad pH = 0.699$$

B. 10.0 mL of 0.100 *M* NaOH has been added.
In the mixed solution *before any reaction occurs,* the major species are

$$H^+, \quad NO_3^-, \quad Na^+, \quad OH^-, \quad \text{and} \quad H_2O$$

Note that large quantities of both H$^+$ and OH$^-$ are present. The 1.00 mmol (10.0 mL × 0.100 *M*) of added OH$^-$ will react with 1.00 mmol H$^+$ to form water:

	H$^+$	+	OH$^-$	⟶	H$_2$O
Before reaction	50.0 mL × 0.200 *M* = 10.0 mmol		10.0 mL × 0.100 *M* = 1.00 mmol		
After reaction	10.0 − 1.00 = 9.0 mmol		1.00 − 1.00 = 0		

After the reaction, the solution contains

$$H^+, \quad NO_3^-, \quad Na^+, \quad \text{and} \quad H_2O \text{ (the OH}^-\text{ ions have been consumed)}$$

and the pH will be determined by the H^+ remaining:

$$[H^+] = \frac{\text{mmol } H^+ \text{ left}}{\text{volume of solution (mL)}} = \frac{9.0 \text{ mmol}}{(50.0 + 10.0) \text{ mL}} = 0.15 \, M$$

The final solution volume is the sum of the original volume of HNO_3 and the volume of added NaOH.

Original volume of HNO_3 solution $\quad$ Volume of NaOH added

$$pH = -\log(0.15) = 0.82$$

C. 20.0 mL (total) of 0.100 M NaOH has been added.

We consider this point from the perspective that a total of 20.0 mL NaOH has been added to the *original* solution, rather than that 10.0 mL has been added to the solution from point B. It is best to go back to the original solution each time so that a mistake made at an earlier point does not show up in each succeeding calculation. As before, the added OH^- will react with H^+ to form water:

	H^+	$+$	OH^-	$\longrightarrow$	H_2O
Before reaction	50.0 mL × 0.200 M = 10.0 mmol		20.0 mL × 0.100 M = 2.00 mmol		
After reaction	10.0 − 2.00 = 8.00 mmol		2.00 − 2.00 = 0 mmol		

After the reaction

H^+ remaining

$$[H^+] = \frac{8.00 \text{ mmol}}{(50.0 + 20.0) \text{ mL}} = 0.11 \, M$$

$$pH = 0.942$$

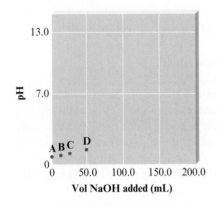

D. 50.0 mL (total) of 0.100 M NaOH has been added.

Proceeding exactly as for points B and C, the pH is found to be 1.301.

E. 100.0 mL (total) of 0.100 M NaOH has been added.

At this point the amount of NaOH that has been added is

$$100.0 \text{ mL} \times 0.100 \, M = 10.0 \text{ mmol}$$

The original amount of nitric acid was

$$50.0 \text{ mL} \times 0.200 \, M = 10.0 \text{ mmol}$$

Equivalence (stoichiometric) point: The point in the titration where an amount of base has been added to exactly react with all the acid originally present.

Enough OH^- has been added to react exactly with the H^+ from the nitric acid. This is the **stoichiometric point**, or **equivalence point**, of the titration. At this point the major species in solution are

$$Na^+, \quad NO_3^-, \quad \text{and} \quad H_2O$$

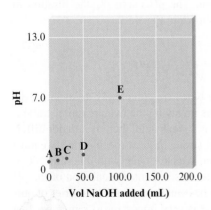

Since Na^+ has no acid or base properties and NO_3^- is the anion of the strong acid HNO_3 and is therefore a very weak base, neither NO_3^- nor Na^+ affects the pH, and the solution is neutral (the pH is 7.00).

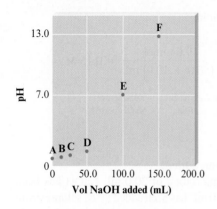

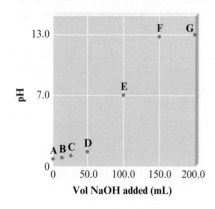

F. 150.0 mL (total) of 0.100 M NaOH has been added.

The stoichiometric calculations for the titration reaction are as follows:

	H^+	$+$	OH^-	$\longrightarrow$	H_2O
Before reaction	50.0 mL × 0.200 M = 10.0 mmol		150.0 mL × 0.100 M = 15.0 mmol		
After reaction	10.0 − 10.0 = 0 mmol		15.0 − 10.0 = 5.0 mmol		
			↑ Excess OH^- added		

Now OH^- is *in excess* and will determine the pH:

$$[OH^-] = \frac{\text{mmol } OH^- \text{ in excess}}{\text{volume (mL)}} = \frac{5.0 \text{ mmol}}{(50.0 + 150.0) \text{ mL}} = \frac{5.0 \text{ mmol}}{200.0 \text{ mL}} = 0.025 \ M$$

Since $[H^+][OH^-] = 1.0 \times 10^{-14}$,

$$[H^+] = \frac{1.0 \times 10^{-14}}{2.5 \times 10^{-2}} = 4.0 \times 10^{-13} \ M \quad \text{and} \quad \text{pH} = 12.40$$

G. 200.0 mL (total) of 0.100 M NaOH has been added.

Proceeding as for point F, the pH is found to be 12.60.

The results of these calculations are summarized by the pH curve shown in Fig. 15.1. Note that the pH changes very gradually until the titration is close to the equivalence point, where a dramatic change occurs. This behavior is due to the fact that early in the titration there is a relatively large amount of H^+ in the solution, and the addition of a given amount of OH^- thus produces a small change in pH. However, near the equivalence point $[H^+]$ is relatively small, and the addition of a small amount of OH^- produces a large change.

The pH curve in Fig. 15.1, typical of the titration of a strong acid with a strong base, has the following characteristics:

> Before the equivalence point, $[H^+]$ (and hence the pH) can be calculated by dividing the number of millimoles of H^+ remaining by the total volume of the solution in milliliters.

> At the equivalence point, the pH is 7.00.

> After the equivalence point, $[OH^-]$ can be calculated by dividing the number of millimoles of excess OH^- by the total volume of the solution. Then $[H^+]$ is obtained from K_w.

The titration of a strong base with a strong acid requires reasoning very similar to that used above, except, of course, that OH^- is in excess before the equivalence point and H^+ is in excess after the equivalence point. The pH curve for the titration of 100.0 mL of 0.50 M NaOH with 1.0 M HCl is shown in Fig. 15.2.

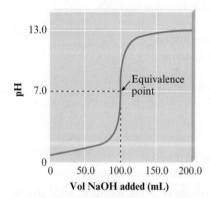

Figure 15.1 | The pH curve for the titration of 50.0 mL of 0.200 M HNO$_3$ with 0.100 M NaOH. Note that the equivalence point occurs at 100.0 mL NaOH added, the point where exactly enough OH^- has been added to react with all the H^+ originally present. The pH of 7 at the equivalence point is characteristic of a strong acid–strong base titration.

Titrations of Weak Acids with Strong Bases

We have seen that since strong acids and strong bases are completely dissociated, the calculations to obtain the pH curves for titrations involving the two are quite straightforward. However, when the acid being titrated is a weak acid, there is a major difference: To calculate $[H^+]$ after a certain amount of strong base has been added, we must deal with the weak acid dissociation equilibrium. We have dealt with this same situation earlier in this chapter when we treated buffered solutions. Calculation of the pH curve for a titration of a weak acid with a strong base really amounts to a series of buffer problems. In performing these calculations it is very important to remember that

Figure 15.2 | The pH curve for the titration of 100.0 mL of 0.50 *M* NaOH with 1.0 *M* HCl. The equivalence point occurs at 50.00 mL HCl added, since at this point 5.0 mmol H^+ has been added to react with the original 5.0 mmol OH^-.

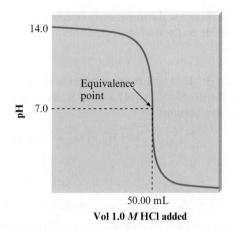

Vol 1.0 *M* HCl added

even though the acid is weak, it *reacts essentially to completion* with hydroxide ion, a very strong base.

This process always involves a two-step procedure.

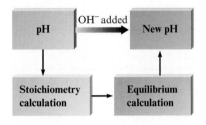

Treat the stoichiometry and equilibrium problems separately.

Problem-Solving Strategy

Calculating the pH Curve for a Weak Acid–Strong Base Titration

1. *A stoichiometry problem.* The reaction of hydroxide ion with the weak acid is assumed to run to completion, and the concentrations of the acid *remaining* and the conjugate base *formed* are determined.

2. *An equilibrium problem.* The position of the weak acid equilibrium is determined, and the pH is calculated.

It is *essential* to do these steps *separately*. Note that the procedures necessary to do these calculations have all been used before.

CASE STUDY | Weak Acid–Strong Base Titration

As an illustration, we will consider the titration of 50.0 mL of 0.10 *M* acetic acid ($HC_2H_3O_2$, $K_a = 1.8 \times 10^{-5}$) with 0.10 *M* NaOH. As before, we will calculate the pH at various points representing volumes of added NaOH.

A.   No NaOH has been added.
This is a typical weak acid calculation of the type introduced in Chapter 14. The pH is 2.87. (Check this yourself.)

B.   10.0 mL of 0.10 *M* NaOH has been added.
The major species in the mixed solution *before any reaction takes place* are

$$HC_2H_3O_2, \quad OH^-, \quad Na^+, \quad \text{and} \quad H_2O$$

The strong base OH^- will react with the strongest proton donor, which in this case is $HC_2H_3O_2$.

Stoichiometry Problem

	OH^-	+	$HC_2H_3O_2$	$\longrightarrow$	$C_2H_3O_2{}^-$	+	H_2O
Before reaction	10 mL × 0.10 *M* = 1.0 mmol		50.0 mL × 0.10 *M* = 5.0 mmol		0 mmol		
After reaction	1.0 − 1.0 = 0 mmol		5.0 − 1.0 = 4.0 mmol		1.0 mmol		
	↑ Limiting reactant				↑ Formed by the reaction		

Equilibrium Problem

We examine the major components left in the solution *after the reaction takes place* to decide on the dominant equilibrium. The major species are

$$HC_2H_3O_2, \quad C_2H_3O_2^-, \quad Na^+, \quad \text{and} \quad H_2O$$

Since $HC_2H_3O_2$ is a much stronger acid than H_2O, and since $C_2H_3O_2^-$ is the conjugate base of $HC_2H_3O_2$, the pH will be determined by the position of the acetic acid dissociation equilibrium:

$$HC_2H_3O_2(aq) \rightleftharpoons H^+(aq) + C_2H_3O_2^-(aq)$$

where
$$K_a = \frac{[H^+][C_2H_3O_2^-]}{[HC_2H_3O_2]}$$

We follow the usual steps to complete the equilibrium calculations:

The initial concentrations are defined after the reaction with OH^- has gone to completion but before any dissociation of $HC_2H_3O_2$ occurs.

Initial Concentration		Equilibrium Concentration
$[HC_2H_3O_2]_0 = \dfrac{4.0 \text{ mmol}}{(50.0 + 10.0) \text{ mL}} = \dfrac{4.0}{60.0}$		$[HC_2H_3O_2] = \dfrac{4.0}{60.0} - x$
$[C_2H_3O_2^-]_0 = \dfrac{1.0 \text{ mmol}}{(50.0 + 10.0) \text{ mL}} = \dfrac{1.0}{60.0}$	$\xrightarrow[\text{dissociates}]{\substack{x \text{ mmol/mL} \\ HC_2H_3O_2}}$	$[C_2H_3O_2^-] = \dfrac{1.0}{60.0} + x$
$[H^+]_0 \approx 0$		$[H^+] = x$

The appropriate ICE table is

	$HC_2H_3O_2(aq)$	$\rightleftharpoons$	$H^+(aq)$	$+$	$C_2H_3O_2^-(aq)$
Initial	$\dfrac{4.0}{60.0}$		≈ 0		$\dfrac{1.0}{60.0}$
Change	$-x$		$+x$		$+x$
Equilibrium	$\dfrac{4.0}{60.0} - x$		x		$\dfrac{1.0}{60.0} + x$

Therefore,

$$1.8 \times 10^{-5} = K_a = \frac{[H^+][C_2H_3O_2^-]}{[HC_2H_3O_2]} = \frac{x\left(\dfrac{1.0}{60.0} + x\right)}{\dfrac{4.0}{60.0} - x} \approx \frac{x\left(\dfrac{1.0}{60.0}\right)}{\dfrac{4.0}{60.0}} = \left(\dfrac{1.0}{4.0}\right)x$$

Note that the approximations made are well within the 5% rule.

$$x = \left(\frac{4.0}{1.0}\right)(1.8 \times 10^{-5}) = 7.2 \times 10^{-5} = [H^+] \quad \text{and} \quad pH = 4.14$$

C.   25.0 mL (total) of 0.10 *M* NaOH has been added.

The procedure here is very similar to that used at point B and will only be summarized briefly. The stoichiometry problem is summarized as follows:

Stoichiometry Problem

	OH^-	$+$	$HC_2H_3O_2$	$\longrightarrow$	$C_2H_3O_2^-$	$+$	H_2O
Before reaction	25.0 mL × 0.10 *M* = 2.5 mmol		50.0 mL × 0.10 *M* = 5.0 mmol		0 mmol		
After reaction	2.5 − 2.5 = 0		5.0 − 2.5 = 2.5 mmol		2.5 mmol		

Equilibrium Problem

After the reaction, the major species in solution are

$$HC_2H_3O_2, \quad C_2H_3O_2^-, \quad Na^+, \quad \text{and} \quad H_2O$$

The equilibrium that will control the pH is

$$HC_2H_3O_2(aq) \rightleftharpoons H^+(aq) + C_2H_3O_2^-(aq)$$

and the pertinent concentrations are as follows:

	Initial Concentration			Equilibrium Concentration
$[HC_2H_3O_2]_0 = \dfrac{2.5 \text{ mmol}}{(50.0 + 25.0) \text{ mL}}$		$\xrightarrow[\text{dissociates}]{x \text{ mmol/mL} \atop HC_2H_3O_2}$		$[HC_2H_3O_2] = \dfrac{2.5}{75.0} - x$
$[C_2H_3O_2{}^-]_0 = \dfrac{2.5 \text{ mmol}}{(50.0 + 25.0) \text{ mL}}$				$[C_2H_3O_2{}^-] = \dfrac{2.5}{75.0} + x$
$[H^+]_0 \approx 0$				$[H^+] = x$

The corresponding ICE table is

	$HC_2H_3O_2(aq)$	$\rightleftharpoons$	$H^+(aq)$	$+$	$C_2H_3O_2{}^-(aq)$
Initial	$\dfrac{2.5}{75.0}$		≈ 0		$\dfrac{2.5}{75.0}$
Change	$-x$		$+x$		$+x$
Equilibrium	$\dfrac{2.5}{75.0} - x$		x		$\dfrac{2.5}{75.0} + x$

Therefore,

$$1.8 \times 10^{-5} = K_a = \frac{[H^+][C_2H_3O_2{}^-]}{[HC_2H_3O_2]} = \frac{x\left(\dfrac{2.5}{75.0} + x\right)}{\dfrac{2.5}{75.0} - x} \approx \frac{x\left(\dfrac{2.5}{75.0}\right)}{\dfrac{2.5}{75.0}}$$

$$x = 1.8 \times 10^{-5} = [H^+] \quad \text{and} \quad pH = 4.74$$

This is a special point in the titration because it is *halfway to the equivalence point*. The original solution, 50.0 mL of 0.10 M $HC_2H_3O_2$, contained 5.0 mmol $HC_2H_3O_2$. Thus 5.0 mmol OH^- is required to reach the equivalence point. That is, 50 mL NaOH is required, since

$$(50.0 \text{ mL})(0.10 \text{ } M) = 5.0 \text{ mmol}$$

After 25.0 mL NaOH has been added, half the original $HC_2H_3O_2$ has been converted to $C_2H_3O_2{}^-$. At this point in the titration $[HC_2H_3O_2]_0$ is equal to $[C_2H_3O_2{}^-]_0$. We can neglect the effect of dissociation; that is,

At this point, half the acid has been used up, so

$$[HC_2H_3O_2] = [C_2H_3O_2{}^-]$$

$$[HC_2H_3O_2] = [HC_2H_3O_2]_0 - x \approx [HC_2H_3O_2]_0$$
$$[C_2H_3O_2{}^-] = [C_2H_3O_2{}^-]_0 + x \approx [C_2H_3O_2{}^-]_0$$

The expression for K_a at the halfway point is

$$K_a = \frac{[H^+][C_2H_3O_2{}^-]}{[HC_2H_3O_2]} = \frac{[H^+][C_2H_3O_2{}^-]_0}{[HC_2H_3O_2]_0} = [H^+]$$

Equal at the halfway point

Then, *at the halfway point* in the titration,

$$[H^+] = K_a \quad \text{and} \quad pH = pK_a$$

D. 40.0 mL (total) of 0.10 M NaOH has been added.
The procedures required here are the same as those used for points B and C. The pH is 5.35. (Check this yourself.)

Equilibrium Problem

E.　50.0 mL (total) of 0.10 M NaOH has been added.

This is the equivalence point of the titration; 5.0 mmol OH^- has been added, which will just react with the 5.0 mmol $HC_2H_3O_2$ originally present. At this point the solution contains the major species

$$Na^+, \quad C_2H_3O_2^-, \quad \text{and} \quad H_2O$$

Note that the solution contains $C_2H_3O_2^-$, which is a base. Remember that a base wants to combine with a proton, and the only source of protons in this solution is water. Thus the reaction will be

$$C_2H_3O_2^-(aq) + H_2O(l) \rightleftharpoons HC_2H_3O_2(aq) + OH^-(aq)$$

This is a *weak base* reaction characterized by K_b:

$$K_b = \frac{[HC_2H_3O_2][OH^-]}{[C_2H_3O_2^-]} = \frac{K_w}{K_a} = \frac{1.0 \times 10^{-14}}{1.8 \times 10^{-5}} = 5.6 \times 10^{-10}$$

The relevant concentrations are as follows:

Initial Concentration (before any $C_2H_3O_2^-$ reacts with H_2O)		Equilibrium Concentration
$[C_2H_3O_2^-]_0 = \dfrac{5.0 \text{ mmol}}{(50.0 + 50.0) \text{ mL}}$ $= 0.050 \ M$	$\xrightarrow[\text{with } H_2O]{\substack{x \text{ mmol/mL} \\ C_2H_3O_2^- \text{ reacts}}}$	$[C_2H_3O_2^-] = 0.050 - x$ $[OH^-] = x$ $[HC_2H_3O_2] = x$
$[OH^-]_0 \approx 0$		
$[HC_2H_3O_2]_0 = 0$		

The corresponding ICE table is

	$C_2H_3O_2^-(aq)$	+	$H_2O(l)$	$\rightleftharpoons$	$HC_2H_3O_2(aq)$	+	$OH^-(aq)$
Initial	0.050		—		0		≈ 0
Change	$-x$		—		$+x$		$+x$
Equilibrium	$0.050 - x$		—		x		x

Therefore,

$$5.6 \times 10^{-10} = K_b = \frac{[HC_2H_3O_2][OH^-]}{[C_2H_3O_2^-]} = \frac{(x)(x)}{0.050 - x} \approx \frac{x^2}{0.050}$$

$$x \approx 5.3 \times 10^{-6}$$

The approximation is valid (by the 5% rule), so

$$[OH^-] = 5.3 \times 10^{-6} \ M$$

and

$$[H^+][OH^-] = K_w = 1.0 \times 10^{-14}$$

$$[H^+] = 1.9 \times 10^{-9} \ M$$

$$pH = 8.72$$

This is another important result: The pH at the equivalence point of a titration of a weak acid with a strong base is always greater than 7. This is so because the anion of the acid, which remains in solution at the equivalence point, is a base. In contrast, for the titration of a strong acid with a strong base, the pH at the equivalence point is 7.0, because the anion remaining in this case is *not* an effective base.

The pH at the equivalence point of a titration of a weak acid with a strong base is always greater than 7.

F.   60.0 mL (total) of 0.10 *M* NaOH has been added.

At this point, excess OH^- has been added. The stoichiometric calculations are as follows:

Stoichiometry Problem

	OH^-	$+$	$HC_2H_3O_2$	$\longrightarrow$	$C_2H_3O_2^-$	$+$	H_2O
Before reaction	60.0 mL × 0.10 *M* = 6.0 mmol		50.0 mL × 0.10 *M* = 5.0 mmol		0 mmol		
After reaction	6.0 − 5.0 = 1.0 mmol in excess		5.0 − 5.0 = 0		5.0 mmol		

Equilibrium Problem

After the reaction is complete, the solution contains the major species

$$Na^+,\quad C_2H_3O_2^-,\quad OH^-,\quad \text{and}\quad H_2O$$

There are two bases in this solution, OH^- and $C_2H_3O_2^-$. However, $C_2H_3O_2^-$ is a weak base compared with OH^-. Therefore, the amount of OH^- produced by reaction of $C_2H_3O_2^-$ with H_2O will be small compared with the excess OH^- already in solution. You can verify this conclusion by looking at point E, where only 5.3×10^{-6} *M* OH^- was produced by $C_2H_3O_2^-$. The amount in this case will be even smaller, since the excess OH^- will push the K_b equilibrium to the left.

Thus the pH is determined by the excess OH^-:

$$[OH^-] = \frac{\text{mmol of } OH^- \text{ in excess}}{\text{volume (in mL)}} = \frac{1.0 \text{ mmol}}{(50.0 + 60.0) \text{ mL}}$$

$$= 9.1 \times 10^{-3} \, M$$

and

$$[H^+] = \frac{1.0 \times 10^{-14}}{9.1 \times 10^{-3}} = 1.1 \times 10^{-12} \, M$$

$$pH = 11.96$$

G.   75.0 mL (total) of 0.10 *M* NaOH has been added.

The procedure needed here is very similar to that for point F. The pH is 12.30. (Check this yourself.)

The pH curve for this titration is shown in Fig. 15.3. It is important to note the differences between this curve and that in Fig. 15.1. For example, the shapes of the plots are quite different before the equivalence point, although they are very similar after that point. (The shapes of the strong and weak acid curves are the same after the equivalence points because excess OH^- controls the pH in this region in both cases.) Near the beginning of the titration of the weak acid, the pH increases more rapidly than it does in the strong acid case. It levels off near the halfway point and then increases

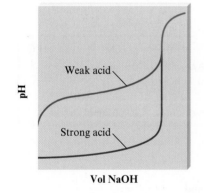

Figure 15.3 | The pH curve for the titration of 50.0 mL of 0.100 *M* $HC_2H_3O_2$ with 0.100 *M* NaOH. Note that the equivalence point occurs at 50.0 mL NaOH added, where the amount of added OH^- exactly equals the original amount of acid. The pH at the equivalence point is greater than 7.0 because the $C_2H_3O_2^-$ ion present at this point is a base and reacts with water to produce OH^-.

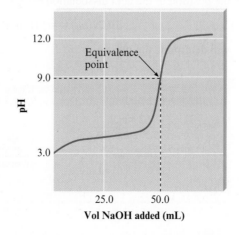

rapidly again. The leveling off near the halfway point is caused by buffering effects. Earlier in this chapter we saw that optimal buffering occurs when [HA] is equal to [A⁻]. This is exactly the case at the halfway point of the titration. As we can see from the curve, the pH changes least rapidly in this region of the titration.

The other notable difference between the curves for strong and weak acids is the value of the pH at the equivalence point. For the titration of a strong acid, the equivalence point occurs at pH 7. For the titration of a weak acid, the pH at the equivalence point is greater than 7 because of the basicity of the conjugate base of the weak acid.

It is important to understand that the equivalence point in an acid–base titration is *defined by the stoichiometry, not by the pH*. The equivalence point occurs when enough titrant has been added to react exactly with all the acid or base being titrated.

The equivalence point is defined by the stoichiometry, not by the pH.

Interactive Example 15.9

Sign in at http://login.cengagebrain .com to try this Interactive Example in OWL.

Titration of a Weak Acid

Hydrogen cyanide gas (HCN), a powerful respiratory inhibitor, is highly toxic. It is a very weak acid ($K_a = 6.2 \times 10^{-10}$) when dissolved in water. If a 50.0-mL sample of 0.100 M HCN is titrated with 0.100 M NaOH, calculate the pH of the solution

a. after 8.00 mL of 0.100 M NaOH has been added.

b. at the halfway point of the titration.

c. at the equivalence point of the titration.

Solution

Stoichiometry Problem

a. After 8.00 mL of 0.100 M NaOH has been added, the following calculations apply:

	HCN	+	OH⁻	⟶	CN⁻	+	H₂O
Before reaction	50.0 mL × 0.100 M = 5.00 mmol		8.00 mL × 0.100 M = 0.800 mmol		0 mmol		
After reaction	5.00 − 0.800 = 4.20 mmol		0.800 − 0.800 = 0		0.800 mmol		

Equilibrium Problem

Since the solution contains the major species

$$HCN, \quad CN^-, \quad Na^+, \quad \text{and} \quad H_2O$$

the position of the acid dissociation equilibrium

$$HCN(aq) \rightleftharpoons H^+(aq) + CN^-(aq)$$

will determine the pH.

Initial Concentration		Equilibrium Concentration
$[HCN]_0 = \dfrac{4.2 \text{ mmol}}{(50.0 + 8.0) \text{ mL}}$	$\xrightarrow[\text{dissociates}]{x \text{ mmol/mL}}$ HCN	$[HCN] = \dfrac{4.2}{58.0} - x$
$[CN^-]_0 = \dfrac{0.800 \text{ mmol}}{(50.0 + 8.0) \text{ mL}}$		$[CN^-] = \dfrac{0.80}{58.0} + x$
$[H^+]_0 \approx 0$		$[H^+] = x$

The corresponding ICE table is

	HCN(aq)	$\rightleftharpoons$	H$^+$(aq)	+	CN$^-$(aq)
Initial	$\dfrac{4.2}{58.0}$		≈ 0		$\dfrac{0.80}{58.0}$
Change	$-x$		$+x$		$+x$
Equilibrium	$\dfrac{4.2}{58.0} - x$		x		$\dfrac{0.80}{58.0} + x$

Substituting the equilibrium concentrations into the expression for K_a gives

The approximations made here are well within the 5% rule.

$$6.2 \times 10^{-10} = K_a = \frac{[\text{H}^+][\text{CN}^-]}{[\text{HCN}]} = \frac{x\left(\dfrac{0.80}{58.0} + x\right)}{\dfrac{4.2}{58.0} - x} \approx \frac{x\left(\dfrac{0.80}{58.0}\right)}{\left(\dfrac{4.2}{58.0}\right)} = x\left(\frac{0.80}{4.2}\right)$$

$$x = 3.3 \times 10^{-9}\, M = [\text{H}^+] \quad \text{and} \quad \text{pH} = 8.49$$

b. *At the halfway point of the titration.* The amount of HCN originally present can be obtained from the original volume and molarity:

$$50.0\ \text{mL} \times 0.100\ M = 5.00\ \text{mmol}$$

Thus the halfway point will occur when 2.50 mmol OH$^-$ has been added:

$$\text{Volume of NaOH (in mL)} \times 0.100\ M = 2.50\ \text{mmol OH}^-$$

or $\qquad\qquad$ Volume of NaOH = 25.0 mL

As was pointed out previously, at the halfway point [HCN] is equal to [CN$^-$] and pH is equal to pK_a. Thus, after 25.0 mL of 0.100 M NaOH has been added,

$$\text{pH} = pK_a = -\log(6.2 \times 10^{-10}) = 9.21$$

Equilibrium Problem

c. *At the equivalence point.* The equivalence point will occur when a total of 5.00 mmol OH$^-$ has been added. Since the NaOH solution is 0.100 M, the equivalence point occurs when 50.0 mL NaOH has been added. This amount will form 5.00 mmol CN$^-$. The major species in solution at the equivalence point are

$$\text{CN}^-, \quad \text{Na}^+, \quad \text{and} \quad \text{H}_2\text{O}$$

Thus the reaction that will control the pH involves the basic cyanide ion extracting a proton from water:

$$\text{CN}^-(aq) + \text{H}_2\text{O}(l) \rightleftharpoons \text{HCN}(aq) + \text{OH}^-(aq)$$

and $\qquad K_b = \dfrac{K_w}{K_a} = \dfrac{1.0 \times 10^{-14}}{6.2 \times 10^{-10}} = 1.6 \times 10^{-5} = \dfrac{[\text{HCN}][\text{OH}^-]}{[\text{CN}^-]}$

Initial Concentration		Equilibrium Concentration
$[\text{CN}^-]_0 = \dfrac{5.00\ \text{mmol}}{(50.0 + 50.0)\ \text{mL}}$ $= 5.00 \times 10^{-2}\ M$	$\xrightarrow[\text{with H}_2\text{O}]{\substack{x\ \text{mmol/mL of} \\ \text{CN}^-\ \text{reacts}}}$	$[\text{CN}^-] = (5.00 \times 10^{-2}) - x$
$[\text{HCN}]_0 = 0$		$[\text{HCN}] = x$
$[\text{OH}^-]_0 \approx 0$		$[\text{OH}^-] = x$

The corresponding ICE table is

	CN⁻(aq)	+	H₂O(l)	⇌	HCN(aq)	+	OH⁻(aq)
Initial	0.050		—		0		0
Change	$-x$		—		$+x$		$+x$
Equilibrium	$0.050 - x$		—		x		x

Substituting the equilibrium concentrations into the expression for K_b and solving in the usual way gives

$$[OH^-] = x = 8.9 \times 10^{-4}$$

Then, from K_w, we have

$$[H^+] = 1.1 \times 10^{-11}\,M \quad \text{and} \quad pH = 10.96$$

See Exercises 15.59, 15.61, and 15.62

The amount of acid present, not its strength, determines the equivalence point.

Two important conclusions can be drawn from a comparison of the titration of 50.0 mL of 0.1 M acetic acid covered earlier in this section and that of 50.0 mL of 0.1 M hydrocyanic acid analyzed in Example 15.9. First, the same amount of 0.1 M NaOH is required to reach the equivalence point in both cases. The fact that HCN is a much weaker acid than $HC_2H_3O_2$ has no bearing on the amount of base required. It is the *amount* of acid, not its strength, that determines the equivalence point. Second, the pH value at the equivalence point *is* affected by the acid strength. For the titration of acetic acid, the pH at the equivalence point is 8.72; for the titration of hydrocyanic acid, the pH at the equivalence point is 10.96. This difference occurs because the CN^- ion is a much stronger base than the $C_2H_3O_2^-$ ion. Also, the pH at the half-way point of the titration is much higher for HCN than for $HC_2H_3O_2$, again because of the greater base strength of the CN^- ion (or, equivalently, the smaller acid strength of HCN).

The strength of a weak acid has a significant effect on the shape of its pH curve. Figure 15.4 shows pH curves for 50-mL samples of 0.10 M solutions of various acids titrated with 0.10 M NaOH. Note that the equivalence point occurs in each case when the same volume of 0.10 M NaOH has been added but that the shapes of the curves are dramatically different. The weaker the acid, the greater the pH value at the equivalence point. In particular, note that the vertical region that surrounds the equivalence point becomes shorter as the acid being titrated becomes weaker. We will see in the next section that the choice of an indicator is more limited for such a titration.

Besides being used to analyze for the amount of acid or base in a solution, titrations can be used to determine the values of equilibrium constants, as shown in Example 15.10.

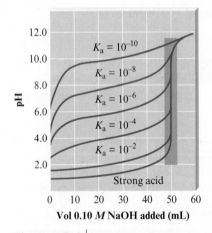

Figure 15.4 | The pH curves for the titrations of 50.0-mL samples of 0.10 M acids with various K_a values with 0.10 M NaOH.

Calculation of K_a

Calculating K_a

A chemist has synthesized a monoprotic weak acid and wants to determine its K_a value. To do so, the chemist dissolves 2.00 mmol of the solid acid in 100.0 mL water and titrates the resulting solution with 0.0500 M NaOH. After 20.0 mL NaOH has been added, the pH is 6.00. What is the K_a value for the acid?

Solution

Stoichiometry Problem

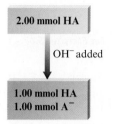

We represent the monoprotic acid as HA. The stoichiometry for the titration reaction is shown below.

	HA	+	OH⁻	⟶	A⁻	+	H₂O
Before reaction	2.00 mmol		20.0 mL × 0.0500 M = 1.00 mmol		0 mmol		
After reaction	2.00 − 1.00 = 1.00 mmol		1.00 − 1.00 = 0		1.00 mmol		

Equilibrium Problem

After the reaction the solution contains the major species

$$HA, \quad A^-, \quad Na^+, \quad and \quad H_2O$$

The pH will be determined by the equilibrium

$$HA(aq) \rightleftharpoons H^+(aq) + A^-(aq)$$

for which

$$K_a = \frac{[H^+][A^-]}{[HA]}$$

Initial Concentration		Equilibrium Concentration
$[HA]_0 = \dfrac{1.00 \text{ mmol}}{(100.0 + 20.0) \text{ mL}}$ $= 8.33 \times 10^{-3} M$		$[HA] = 8.33 \times 10^{-3} - x$
$[A^-]_0 = \dfrac{1.00 \text{ mmol}}{(100.0 + 20.0) \text{ mL}}$ $= 8.33 \times 10^{-3} M$	$\xrightarrow[\text{dissociates}]{\substack{x \text{ mmol/mL} \\ \text{HA}}}$	$[A^-] = 8.33 \times 10^{-3} + x$
$[H^+]_0 \approx 0$		$[H^+] = x$

The corresponding ICE table is

	HA(aq)	⇌	H⁺(aq)	+	A⁻(aq)
Initial	8.33×10^{-3}		≈ 0		8.33×10^{-3}
Change	$-x$		$+x$		$+x$
Equilibrium	$8.33 \times 10^{-3} - x$		x		$8.33 \times 10^{-3} + x$

Note that x is known here because the pH at this point is known to be 6.00. Thus

$$x = [H^+] = \text{antilog}(-pH) = 1.0 \times 10^{-6} M$$

Substituting the equilibrium concentrations into the expression for K_a allows calculation of the K_a value:

$$K_a = \frac{[H^+][A^-]}{[HA]} = \frac{x(8.33 \times 10^{-3} + x)}{(8.33 \times 10^{-3}) - x}$$

$$= \frac{(1.0 \times 10^{-6})(8.33 \times 10^{-3} + 1.0 \times 10^{-6})}{(8.33 \times 10^{-3}) - (1.0 \times 10^{-6})}$$

$$\approx \frac{(1.0 \times 10^{-6})(8.33 \times 10^{-3})}{8.33 \times 10^{-3}} = 1.0 \times 10^{-6}$$

There is an easier way to think about this problem. The original solution contained 2.00 mmol of HA, and since 20.0 mL of added 0.0500 M NaOH contains 1.0 mmol OH^-, this is the halfway point in the titration (where [HA] is equal to [A^-]). Thus

$$[H^+] = K_a = 1.0 \times 10^{-6}$$

See Exercise 15.67

Titrations of Weak Bases with Strong Acids

Titrations of weak bases with strong acids can be treated using the procedures we introduced previously. As always, you should *think first about the major species in solution* and decide whether a reaction occurs that runs essentially to completion. If such a reaction does occur, let it run to completion and do the stoichiometric calculations. Finally, choose the dominant equilibrium and calculate the pH.

CASE STUDY | Weak Base–Strong Acid Titration

The calculations involved for the titration of a weak base with a strong acid will be illustrated by the titration of 100.0 mL of 0.050 M NH_3 with 0.10 M HCl.

A. Before the addition of any HCl.
1. Major species:

$$NH_3 \quad \text{and} \quad H_2O$$

NH_3 is a base and will seek a source of protons. In this case H_2O is the only available source.

2. No reactions occur that go to completion, since NH_3 cannot readily take a proton from H_2O. This is evidenced by the small K_b value for NH_3.

3. The equilibrium that controls the pH involves the reaction of ammonia with water:

$$NH_3(aq) + H_2O(l) \rightleftharpoons NH_4^+(aq) + OH^-(aq)$$

Use K_b to calculate [OH^-]. Although NH_3 is a weak base (compared with OH^-), it produces much more OH^- in this reaction than is produced from the autoionization of H_2O.

B. Before the equivalence point.
1. Major species (before any reaction occurs):

$$NH_3, \quad \underbrace{H^+, \quad Cl^-,}_{\substack{\text{From added} \\ \text{HCl}}} \quad \text{and} \quad H_2O$$

2. The NH_3 will react with H^+ from the added HCl:

$$NH_3(aq) + H^+(aq) \rightleftharpoons NH_4^+(aq)$$

This reaction proceeds essentially to completion because the NH_3 readily reacts with a free proton. This case is much different from the previous case, where H_2O was the only source of protons. The stoichiometric calculations are then carried out using the known volume of 0.10 M HCl added.

3. After the reaction of NH_3 with H^+ is run to completion, the solution contains the following major species:

$$NH_3, \quad NH_4^+, \quad Cl^-, \quad \text{and} \quad H_2O$$

↑
Formed in
titration reaction

Note that the solution contains NH_3 and NH_4^+, and the equilibria involving these species will determine $[H^+]$. You can use either the dissociation reaction of NH_4^+

$$NH_4^+(aq) \rightleftharpoons NH_3(aq) + H^+(aq)$$

or the reaction of NH_3 with H_2O

$$NH_3(aq) + H_2O(l) \rightleftharpoons NH_4^+(aq) + OH^-(aq)$$

C. **At the equivalence point.**
1. By definition, the equivalence point occurs when all the original NH_3 is converted to NH_4^+. Thus the major species in solution are

$$NH_4^+, \quad Cl^-, \quad \text{and} \quad H_2O$$

2. No reactions occur that go to completion.

3. The dominant equilibrium (the one that controls the $[H^+]$) will be the dissociation of the weak acid NH_4^+, for which

$$K_a = \frac{K_w}{K_b(\text{for } NH_3)}$$

D. **Beyond the equivalence point.**
1. Excess HCl has been added, and the major species are

$$H^+, \quad NH_4^+, \quad Cl^-, \quad \text{and} \quad H_2O$$

2. No reaction occurs that goes to completion.

3. Although NH_4^+ will dissociate, it is such a weak acid that $[H^+]$ will be determined simply by the excess H^+:

$$[H^+] = \frac{\text{mmol } H^+ \text{ in excess}}{\text{mL solution}}$$

The results of these calculations are shown in Table 15.2. The pH curve is shown in Fig. 15.5.

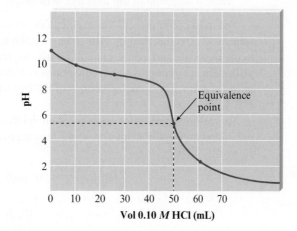

Figure 15.5 | The pH curve for the titration of 100.0 mL of 0.050 *M* NH_3 with 0.10 *M* HCl. Note the pH at the equivalence point is less than 7, since the solution contains the weak acid NH_4^+.

Table 15.2 | Summary of Results for the Titration of 100.0 mL 0.050 M NH₃ with 0.10 M HCl

Volume of 0.10 M HCl Added (mL)	$[NH_3]_0$	$[NH_4^+]_0$	$[H^+]$	pH
0	0.05 M	0	$1.1 \times 10^{-11}\ M$	10.96
10.0	$\dfrac{4.0\ \text{mmol}}{(100 + 10)\ \text{mL}}$	$\dfrac{1.0\ \text{mmol}}{(100 + 10)\ \text{mL}}$	$1.4 \times 10^{-10}\ M$	9.85
25.0*	$\dfrac{2.5\ \text{mmol}}{(100 + 25)\ \text{mL}}$	$\dfrac{2.5\ \text{mmol}}{(100 + 25)\ \text{mL}}$	$5.6 \times 10^{-10}\ M$	9.25
50.0†	0	$\dfrac{5.0\ \text{mmol}}{(100 + 50)\ \text{mL}}$	$4.3 \times 10^{-6}\ M$	5.36
60.0‡	0	$\dfrac{5.0\ \text{mmol}}{(100 + 60)\ \text{mL}}$	$\dfrac{1.0\ \text{mmol}}{160\ \text{mL}}$ $= 6.2 \times 10^{-3}\ M$	2.21

*Halfway point   †Equivalence point   ‡$[H^+]$ determined by the 1.0 mmol of excess H^+

Critical Thinking

You have read about titrations of strong acids with strong bases, weak acids with strong bases, and weak bases with strong acids. What if you titrated a weak acid with a weak base? Sketch a pH curve and defend its shape. Label the equivalence point and discuss the possibilities for the pH value at the equivalence point.

15.5 | Acid–Base Indicators

There are two common methods for determining the equivalence point of an acid–base titration:

1. Use a pH meter (see Fig. 14.7) to monitor the pH and then plot the titration curve. The center of the vertical region of the pH curve indicates the equivalence point (for example, see Figs. 15.1 through 15.5).
2. Use an **acid–base indicator**, which marks the end point of a titration by changing color. Although the *equivalence point of a titration, defined by the stoichiometry, is not necessarily the same as the end point* (where the indicator changes color), careful selection of the indicator will ensure that the error is negligible.

The most common acid–base indicators are complex molecules that are themselves weak acids (represented by HIn). They exhibit one color when the proton is attached to the molecule and a different color when the proton is absent. For example, **phenolphthalein**, a commonly used indicator, is colorless in its HIn form and pink in its In⁻, or basic, form. The actual structures of the two forms of phenolphthalein are shown in Fig. 15.6.

To see how molecules such as phenolphthalein function as indicators, consider the following equilibrium for some hypothetical indicator HIn, a weak acid with $K_a = 1.0 \times 10^{-8}$.

$$\underset{\text{Red}}{HIn(aq)} \rightleftharpoons H^+(aq) + \underset{\text{Blue}}{In^-(aq)}$$

$$K_a = \frac{[H^+][In^-]}{[HIn]}$$

The indicator phenolphthalein is colorless in acidic solution and pink in basic solution.

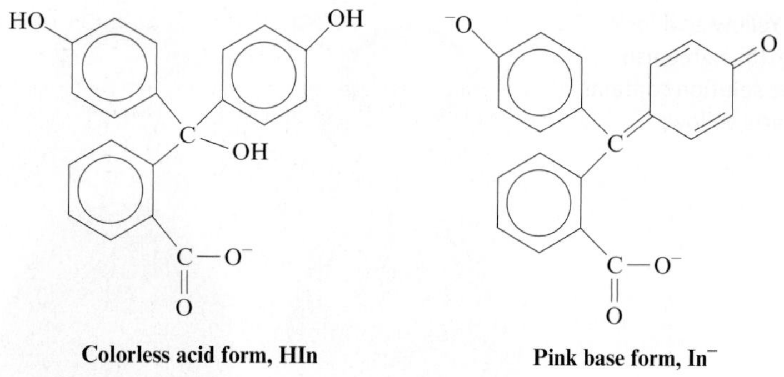

Figure 15.6 | The acid and base forms of the indicator phenolphthalein. In the acid form (HIn), the molecule is colorless. When a proton (plus H_2O) is removed to give the base form (In^-), the color changes to pink.

Colorless acid form, HIn

Pink base form, In^-

By rearranging, we get

$$\frac{K_a}{[H^+]} = \frac{[In^-]}{[HIn]}$$

Suppose we add a few drops of this indicator to an acidic solution whose pH is 1.0 ($[H^+] = 1.0 \times 10^{-1}$). Then

$$\frac{K_a}{[H^+]} = \frac{1.0 \times 10^{-8}}{1.0 \times 10^{-1}} = 10^{-7} = \frac{1}{10,000,000} = \frac{[In^-]}{[HIn]}$$

This ratio shows that the predominant form of the indicator is HIn, resulting in a red solution. As OH^- is added to this solution in a titration, $[H^+]$ decreases and the equilibrium shifts to the right, changing HIn to In^-. At some point in a titration, enough of the In^- form will be present in the solution so that a purple tint will be noticeable. That is, a color change from red to reddish purple will occur.

How much In^- must be present for the human eye to detect that the color is different from the original one? For most indicators, about a tenth of the initial form must be converted to the other form before a new color is apparent. We will assume, then, that in the titration of an acid with a base, the color change will occur at a pH where

The *end point* is defined by the change in color of the indicator. The *equivalence point* is defined by the reaction stoichiometry.

$$\frac{[In^-]}{[HIn]} = \frac{1}{10}$$

Example 15.11

Indicator Color Change

Bromthymol blue, an indicator with a K_a value of 1.0×10^{-7}, is yellow in its HIn form and blue in its In^- form. Suppose we put a few drops of this indicator in a strongly acidic solution. If the solution is then titrated with NaOH, at what pH will the indicator color change first be visible?

Solution

For bromthymol blue,

$$K_a = 1.0 \times 10^{-7} = \frac{[H^+][In^-]}{[HIn]}$$

We assume that the color change is visible when

$$\frac{[In^-]}{[HIn]} = \frac{1}{10}$$

Figure 15.7 | (a) Yellow acid form of bromthymol blue; (b) a greenish tint is seen when the solution contains 1 part blue and 10 parts yellow; (c) blue basic form.

Charles D. Winters   Charles D. Winters   Charles D. Winters

That is, we assume that we can see the first hint of a greenish tint (yellow plus a little blue) when the solution contains 1 part blue and 10 parts yellow (Fig. 15.7). Thus

$$K_a = 1.0 \times 10^{-7} = \frac{[H^+](1)}{10}$$

$$[H^+] = 1.0 \times 10^{-6} \quad \text{or} \quad pH = 6.00$$

The color change is first visible at pH 6.00.

See Exercises 15.69 through 15.72

The Henderson–Hasselbalch equation is very useful in determining the pH at which an indicator changes color. For example, application of Equation (15.2) to the K_a expression for the general indicator HIn yields

$$pH = pK_a + \log\left(\frac{[In^-]}{[HIn]}\right)$$

where K_a is the dissociation constant for the acid form of the indicator (HIn). Since we assume that the color change is visible when

$$\frac{[In^-]}{[HIn]} = \frac{1}{10}$$

we have the following equation for determining the pH at which the color change occurs:

$$pH = pK_a + \log(\tfrac{1}{10}) = pK_a - 1$$

For bromthymol blue ($K_a = 1 \times 10^{-7}$, or $pK_a = 7$), the pH at the color change is

$$pH = 7 - 1 = 6$$

as we calculated in Example 15.11.

When a basic solution is titrated, the indicator HIn will initially exist as In^- in solution, but as acid is added, more HIn will be formed. In this case the color change will be visible when there is a mixture of 10 parts In^- and 1 part HIn. That is, a color change from blue to blue-green will occur (see Fig. 15.7) due to the presence of some of the yellow HIn molecules. This color change will be first visible when

$$\frac{[In^-]}{[HIn]} = \frac{10}{1}$$

Note that this is the reciprocal of the ratio for the titration of an acid. Substituting this ratio into the Henderson–Hasselbalch equation gives

$$pH = pK_a + \log\left(\frac{10}{1}\right) = pK_a + 1$$

For bromthymol blue ($pK_a = 7$), we have a color change at

$$pH = 7 + 1 = 8$$

In summary, when bromthymol blue is used for the titration of an acid, the starting form will be HIn (yellow), and the color change occurs at a pH of about 6. When bromthymol blue is used for the titration of a base, the starting form is In⁻ (blue), and the color change occurs at a pH of about 8. Thus the useful pH range for bromthymol blue is

$$pK_a(\text{bromthymol blue}) \pm 1 = 7 \pm 1$$

or from 6 to 8. This is a general result. For a typical acid–base indicator with dissociation constant K_a, the color transition occurs over a range of pH values given by $pK_a \pm 1$. The useful pH ranges for several common indicators are shown in Fig. 15.8.

When we choose an indicator for a titration, we want the indicator end point (where the color changes) and the titration equivalence point to be as close as possible. Choosing an indicator is easier if there is a large change in pH near the equivalence point of the titration. The dramatic change in pH near the equivalence point in a strong acid–strong base titration (see Figs. 15.1 and 15.2) produces a sharp end point; that is, the complete color change (from the acid-to-base or base-to-acid colors) usually occurs over one drop of added titrant.

What indicator should we use for the titration of 100.00 mL of 0.100 M HCl with 0.100 M NaOH? We know that the equivalence point occurs at pH 7.00. In the initially acidic solution, the indicator will be predominantly in the HIn form. As OH⁻ ions are added, the pH increases rather slowly at first (see Fig. 15.1) and then rises rapidly at the equivalence point. This sharp change causes the indicator dissociation equilibrium

$$HIn \rightleftharpoons H^+ + In^-$$

to shift suddenly to the right, producing enough In⁻ ions to give a color change. Since we are titrating an acid, the indicator is predominantly in the acid form initially. Therefore, the first observable color change will occur at a pH where

$$\frac{[In^-]}{[HIn]} = \frac{1}{10}$$

Thus

$$pH = pK_a + \log\left(\frac{1}{10}\right) = pK_a - 1$$

If we want an indicator that changes color at pH 7, we can use this relationship to find the pK_a value for a suitable indicator:

$$pH = 7 = pK_a - 1 \quad \text{or} \quad pK_a = 7 + 1 = 8$$

Thus an indicator with a pK_a value of 8 ($K_a = 1 \times 10^{-8}$) changes color at about pH 7 and is ideal for marking the end point for a strong acid–strong base titration.

How crucial is it for a strong acid–strong base titration that the indicator change color exactly at pH 7? We can answer this question by examining the pH change near the equivalence point of the titration of 100 mL of 0.10 M HCl and 0.10 M NaOH. The data for a few points at or near the equivalence point are shown in Table 15.3. Note that in going from 99.99 to 100.01 mL of added NaOH solution (about half of a drop), the pH changes from 5.3 to 8.7—a very dramatic change. This behavior leads

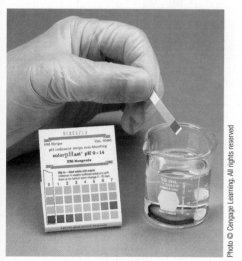

Universal indicator paper can be used to estimate the pH of a solution.

Table 15.3 | Selected pH Values Near the Equivalence Point in the Titration of 100.0 mL of 0.10 M HCl with 0.10 M NaOH

NaOH Added (mL)	pH
99.99	5.3
100.00	7.0
100.01	8.7

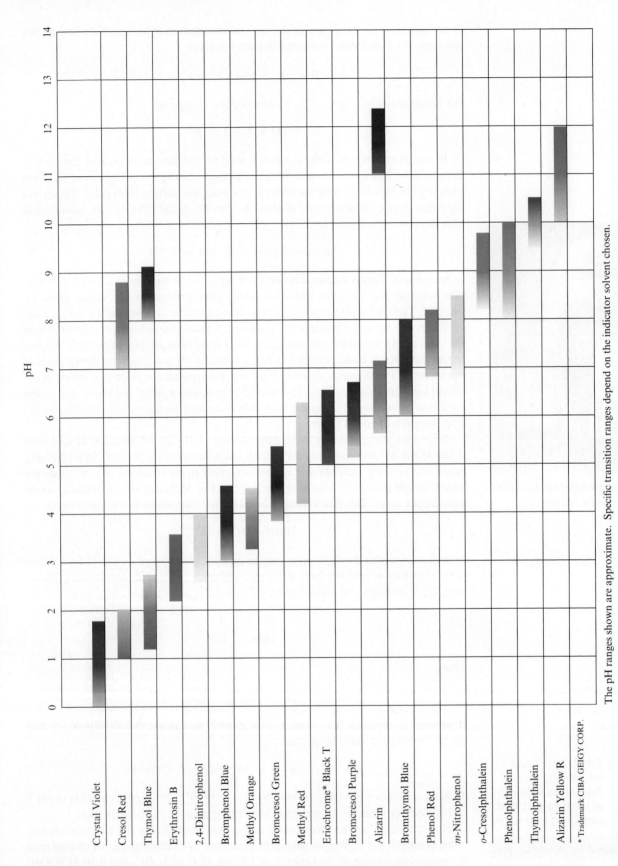

pH

Crystal Violet
Cresol Red
Thymol Blue
Erythrosin B
2,4-Dinitrophenol
Bromphenol Blue
Methyl Orange
Bromcresol Green
Methyl Red
Eriochrome* Black T
Bromcresol Purple
Alizarin
Bromthymol Blue
Phenol Red
m-Nitrophenol
o-Cresolphthalein
Phenolphthalein
Thymolphthalein
Alizarin Yellow R

* Trademark CIBA GEIGY CORP.

The pH ranges shown are approximate.   Specific transition ranges depend on the indicator solvent chosen.

Figure 15.8 | The useful pH ranges for several common indicators. Note that most indicators have a useful range of about two pH units, as predicted by the expression $pK_a \pm 1$.

Methyl red indicator is yellow in basic solution and red in acidic solution.

to the following general conclusions about indicators for a strong acid–strong base titration:

Indicator color changes will be sharp, occurring with the addition of a single drop of titrant.

There is a wide choice of suitable indicators. The results will agree within one drop of titrant, using indicators with end points as far apart as pH 5 and pH 9 (Fig. 15.9).

The titration of weak acids is somewhat different. Figure 15.4 shows that the weaker the acid being titrated, the smaller the vertical area around the equivalence point. This allows much less flexibility in choosing the indicator. We must choose an indicator whose useful pH range has a midpoint as close as possible to the pH at the equivalence point. For example, we saw earlier that in the titration of 0.1 M HC$_2$H$_3$O$_2$ with 0.1 M NaOH the pH at the equivalence point is 8.7 (see Fig. 15.3). A good indicator choice would be phenolphthalein, since its useful pH range is 8 to 10. Thymol blue (changes color, pH 8–9) also would be acceptable, but methyl red would not. The choice of an indicator is illustrated graphically in Fig. 15.10.

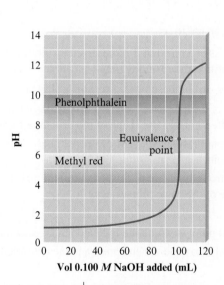

Figure 15.9 │ The pH curve for the titration of 100.0 mL of 0.10 M HCl with 0.10 M NaOH. Note that the end points of phenolphthalein and methyl red occur at virtually the same amounts of added NaOH.

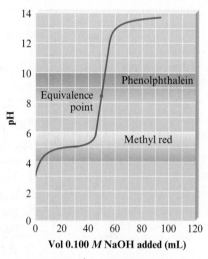

Figure 15.10 │ The pH curve for the titration of 50 mL of 0.1 M HC$_2$H$_3$O$_2$ with 0.1 M NaOH. Phenolphthalein will give an end point very close to the equivalence point of the titration. Methyl red would change color well before the equivalence point (so the end point would be very different from the equivalence point) and would not be a suitable indicator for this titration.

For review

Key terms

Section 15.1
common ion
common ion effect

Section 15.2
buffered solution
Henderson–Hasselbalch
equation

Section 15.3
buffering capacity

Section 15.4
pH curve (titration curve)
millimole (mmol)
equivalence point
(stoichiometric point)

Section 15.5
acid–base indicator
phenolphthalein

Buffered solutions

› Contains a weak acid (HA) and its salt (NaA) or a weak base (B) and its salt (BHCl)
› Resists a change in its pH when H^+ or OH^- is added
› For a buffered solution containing HA and A^-
  › The Henderson–Hasselbalch equation is useful:

$$pH = pK_a + \log\left(\frac{[A^-]}{[HA]}\right)$$

  › The capacity of the buffered solution depends on the amounts of HA and A^- present
› The most efficient buffering occurs when the $\frac{[A^-]}{[HA]}$ ratio is close to 1
› Buffering works because the amounts of HA (which reacts with added OH^-) and A^- (which reacts with added H^+) are large enough that the $\frac{[A^-]}{[HA]}$ ratio does not change significantly when strong acids or bases are added

Acid–base titrations

› The progress of a titration is represented by plotting the pH of the solution versus the volume of added titrant; the resulting graph is called a pH curve or titration curve
› Strong acid–strong base titrations show a sharp change in pH near the equivalence point
› The shape of the pH curve for a strong base–strong acid titration before the equivalence point is quite different from the shape of the pH curve for a strong base–weak acid titration
  › The strong base–weak acid pH curve shows the effects of buffering before the equivalence point
  › For a strong base–weak acid titration, the pH is greater than 7 at the equivalence point because of the basic properties of A^-
› Indicators are sometimes used to mark the equivalence point of an acid–base titration
  › The end point is where the indicator changes color
  › The goal is to have the end point and the equivalence point be as close as possible

Review questions *Answers to the Review Questions can be found on the Student website (accessible from* **www.cengagebrain.com**).

1. What is meant by the presence of a common ion? How does the presence of a common ion affect an equilibrium such as

$$HNO_2(aq) \rightleftharpoons H^+(aq) + NO_2^-(aq)$$

   What is an acid–base solution called that contains a common ion?

2. Define a buffer solution. What makes up a buffer solution? How do buffers absorb added H^+ or OH^- with little pH change?

   Is it necessary that the concentrations of the weak acid and the weak base in a buffered solution be equal? Explain. What is the pH of a buffer when the weak acid and conjugate base concentrations are equal?

   A buffer generally contains a weak acid and its weak conjugate base, or a weak base and its weak conjugate acid, in water. You can solve for the pH by setting up the equilibrium problem using the K_a reaction of the weak acid or the K_b reaction of the conjugate base. Both reactions give the same answer for the pH of the solution. Explain.

   A third method that can be used to solve for the pH of a buffer solution is the Henderson–Hasselbalch equation. What is the Henderson–Hasselbalch equation? What assumptions are made when using this equation?

3. One of the most challenging parts of solving acid–base problems is writing out the correct equation. When a

strong acid or a strong base is added to solutions, they are great at what they do and we always react them first. If a strong acid is added to a buffer, what reacts with the H^+ from the strong acid and what are the products? If a strong base is added to a buffer, what reacts with the OH^- from the strong base and what are the products? Problems involving the reaction of a strong acid or strong base are assumed to be stoichiometry problems and not equilibrium problems. What is assumed when a strong acid or strong base reacts to make it a stoichiometry problem?

4. A good buffer generally contains relatively equal concentrations of weak acid and conjugate base. If you wanted to buffer a solution at pH = 4.00 or pH = 10.00, how would you decide which weak acid–conjugate base or weak base–conjugate acid pair to use? The second characteristic of a good buffer is good buffering capacity. What is the *capacity* of a buffer? How do the following buffers differ in capacity? How do they differ in pH?

   0.01 M acetic acid/0.01 M sodium acetate

   0.1 M acetic acid/0.1 M sodium acetate

   1.0 M acetic acid/1.0 M sodium acetate

5. Draw the general titration curve for a strong acid titrated by a strong base. At the various points in the titration, list the major species present before any reaction takes place and the major species present after any reaction takes place. What reaction takes place in a strong acid–strong base titration? How do you calculate the pH at the various points along the curve? What is the pH at the equivalence point for a strong acid–strong base titration? Why?

6. Instead of the titration of a strong acid by a strong base considered in Question 5, consider the titration of a strong base by a strong acid. Compare and contrast a strong acid–strong base titration with a strong base–strong acid titration.

7. Sketch the titration curve for a weak acid titrated by a strong base. When performing calculations concerning weak acid–strong base titrations, the general two-step procedure is to solve a stoichiometry problem first, then to solve an equilibrium problem to determine the pH. What reaction takes place in the stoichiometry part of the problem? What is assumed about this reaction?

   At the various points in your titration curve, list the major species present after the strong base (NaOH, for example) reacts to completion with the weak acid, HA. What equilibrium problem would you solve at the various points in your titration curve to calculate the pH? Why is pH > 7.0 at the equivalence point of a weak acid–strong base titration? Does the pH at the halfway point to equivalence have to be less than 7.0? What does the pH at the halfway point equal? Compare and contrast the titration curves for a strong acid–strong base titration and a weak acid–strong base titration.

8. Sketch the titration curve for a weak base titrated by a strong acid. Weak base–strong acid titration problems also follow a two-step procedure. What reaction takes place in the stoichiometry part of the problem? What is assumed about this reaction? At the various points in your titration curve, list the major species present after the strong acid (HNO_3, for example) reacts to completion with the weak base, B. What equilibrium problem would you solve at the various points in your titration curve to calculate the pH? Why is pH < 7.0 at the equivalence point of a weak base–strong acid titration? If pH = 6.0 at the halfway point to equivalence, what is the K_b value for the weak base titrated? Compare and contrast the titration curves for a strong base–strong acid titration and a weak base–strong acid titration.

9. What is an acid–base indicator? Define the equivalence (stoichiometric) point and the end point of a titration. Why should you choose an indicator so that the two points coincide? Do the pH values of the two points have to be within ± 0.01 pH unit of each other? Explain.

10. Why does an indicator change from its acid color to its base color over a range of pH values? In general, when do color changes start to occur for indicators? Can the indicator thymol blue contain only a single $—CO_2H$ group and no other acidic or basic functional group? Explain.

Active Learning Questions

These questions are designed to be used by groups of students in class.

1. What are the major species in solution after $NaHSO_4$ is dissolved in water? What happens to the pH of the solution as more $NaHSO_4$ is added? Why? Would the results vary if baking soda ($NaHCO_3$) were used instead?

2. A friend asks the following: "Consider a buffered solution made up of the weak acid HA and its salt NaA. If a strong base like NaOH is added, the HA reacts with the OH^- to form A^-. Thus the amount of acid (HA) is decreased, and the amount of base (A^-) is increased. Analogously, adding HCl to the buffered solution forms more of the acid (HA) by reacting with the base (A^-). Thus how can we claim that a buffered solution resists changes in the pH of the solution?" How would you explain buffering to this friend?

3. Mixing together solutions of acetic acid and sodium hydroxide can make a buffered solution. Explain. How does the amount of each solution added change the effectiveness of the buffer?

4. Could a buffered solution be made by mixing aqueous solutions of HCl and NaOH? Explain. Why isn't a mixture of a strong acid and its conjugate base considered a buffered solution?

5. Sketch two pH curves, one for the titration of a weak acid with a strong base and one for a strong acid with a strong base. How are they similar? How are they different? Account for the similarities and the differences.

6. Sketch a pH curve for the titration of a weak acid (HA) with a strong base (NaOH). List the major species, and explain how you would go about calculating the pH of the solution at various points, including the halfway point and the equivalence point.

7. You have a solution of the weak acid HA and add some HCl to it. What are the major species in the solution? What do you need to know to calculate the pH of the solution, and how would you use this information? How does the pH of the solution of just the HA compare with that of the final mixture? Explain.

8. You have a solution of the weak acid HA and add some of the salt NaA to it. What are the major species in the solution? What do you need to know to calculate the pH of the solution, and how would you use this information? How does the pH of the solution of just the HA compare with that of the final mixture? Explain.

A blue question or exercise number indicates that the answer to that question or exercise appears at the back of this book and a solution appears in the *Solutions Guide*, as found on PowerLecture.

Questions

9. The common ion effect for weak acids is to significantly decrease the dissociation of the acid in water. Explain the common ion effect.

10. Consider a buffer solution where [weak acid] > [conjugate base]. How is the pH of the solution related to the pK_a value of the weak acid? If [conjugate base] > [weak acid], how is pH related to pK_a?

11. A best buffer has about equal quantities of weak acid and conjugate base present as well as having a large concentration of each species present. Explain.

12. Consider the following pH curves for 100.0 mL of two different acids with the same initial concentration each titrated by 0.10 M NaOH.

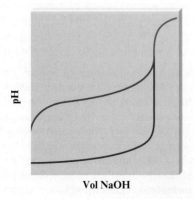

Vol NaOH

a. Which plot represents a pH curve of a weak acid, and which plot is for a strong acid? How can you tell? Cite three differences between the plots that help you decide.

b. In both cases the pH is relatively constant before the pH changes greatly. Does this mean that at some point in each titration each solution was a buffered solution?

c. True or false? The equivalence point volume for each titration is the same. Explain your answer.

d. True or false? The pH at the equivalence point for each titration is the same. Explain your answer.

13. An acid is titrated with NaOH. The following beakers are illustrations of the contents of the beaker at various times during the titration. These are presented out of order. *Note:* Counter-ions and water molecules have been omitted from the illustrations for clarity.

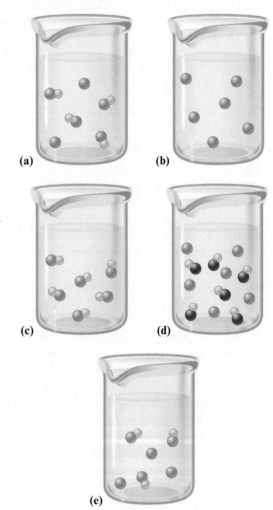

(a)   (b)

(c)   (d)

(e)

a. Is the acid a weak or strong acid? How can you tell?

b. Arrange the beakers in order of what the contents would look like as the titration progresses.

c. For which beaker would pH = pK_a? Explain your answer.

d. Which beaker represents the equivalence point of the titration? Explain your answer.

e. For which beaker would the K_a value for the acid not be necessary to determine the pH? Explain your answer.

14. Consider the following four titrations.

   i. 100.0 mL of 0.10 M HCl titrated by 0.10 M NaOH
  ii. 100.0 mL of 0.10 M NaOH titrated by 0.10 M HCl
 iii. 100.0 mL of 0.10 M CH$_3$NH$_2$ titrated by 0.10 M HCl
 iv. 100.0 mL of 0.10 M HF titrated by 0.10 M NaOH

Rank the titrations in order of:

a. increasing volume of titrant added to reach the equivalence point.

b. increasing pH initially before any titrant has been added.

c. increasing pH at the halfway point in equivalence.

d. increasing pH at the equivalence point.

How would the rankings change if C$_5$H$_5$N replaced CH$_3$NH$_2$ and if HOC$_6$H$_5$ replaced HF?

15. Figure 15.4 shows the pH curves for the titrations of six different acids by NaOH. Make a similar plot for the titration of three different bases by 0.10 M HCl. Assume 50.0 mL of 0.20 M of the bases and assume the three bases are a strong base (KOH), a weak base with $K_b = 1 \times 10^{-5}$, and another weak base with $K_b = 1 \times 10^{-10}$.

16. Acid–base indicators mark the end point of titrations by "magically" turning a different color. Explain the "magic" behind acid–base indicators.

Exercises

In this section similar exercises are paired.

Buffers

17. How many of the following are buffered solutions? Explain your answer. Note: Counter-ions and water molecules have been omitted from the illustrations for clarity.

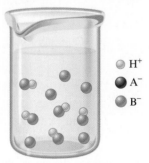

H$^+$
A$^-$
B$^-$

18. Which of the following can be classified as buffer solutions?

a. 0.25 M HBr + 0.25 M HOBr

b. 0.15 M HClO$_4$ + 0.20 M RbOH

c. 0.50 M HOCl + 0.35 M KOCl

d. 0.70 M KOH + 0.70 M HONH$_2$

e. 0.85 M H$_2$NNH$_2$ + 0.60 M H$_2$NNH$_3$NO$_3$

19. A certain buffer is made by dissolving NaHCO$_3$ and Na$_2$CO$_3$ in some water. Write equations to show how this buffer neutralizes added H$^+$ and OH$^-$.

20. A buffer is prepared by dissolving HONH$_2$ and HONH$_3$NO$_3$ in some water. Write equations to show how this buffer neutralizes added H$^+$ and OH$^-$.

21. Calculate the pH of each of the following solutions.

a. 0.100 M propanoic acid (HC$_3$H$_5$O$_2$, $K_a = 1.3 \times 10^{-5}$)

b. 0.100 M sodium propanoate (NaC$_3$H$_5$O$_2$)

c. pure H$_2$O

d. a mixture containing 0.100 M HC$_3$H$_5$O$_2$ and 0.100 M NaC$_3$H$_5$O$_2$

22. Calculate the pH of each of the following solutions.

a. 0.100 M HONH$_2$ ($K_b = 1.1 \times 10^{-8}$)

b. 0.100 M HONH$_3$Cl

c. pure H$_2$O

d. a mixture containing 0.100 M HONH$_2$ and 0.100 M HONH$_3$Cl

23. Compare the percent dissociation of the acid in Exercise 21a with the percent dissociation of the acid in Exercise 21d. Explain the large difference in percent dissociation of the acid.

24. Compare the percent ionization of the base in Exercise 22a with the percent ionization of the base in Exercise 22d. Explain any differences.

25. Calculate the pH after 0.020 mole of HCl is added to 1.00 L of each of the four solutions in Exercise 21.

26. Calculate the pH after 0.020 mole of HCl is added to 1.00 L of each of the four solutions in Exercise 22.

27. Calculate the pH after 0.020 mole of NaOH is added to 1.00 L of each of the four solutions in Exercise 21.

28. Calculate the pH after 0.020 mole of NaOH is added to 1.00 L of each of the solutions in Exercise 22.

29. Which of the solutions in Exercise 21 shows the least change in pH upon the addition of acid or base? Explain.

30. Which of the solutions in Exercise 22 is a buffered solution?

31. Calculate the pH of a solution that is 1.00 M HNO$_2$ and 1.00 M NaNO$_2$.

32. Calculate the pH of a solution that is 0.60 M HF and 1.00 M KF.

33. Calculate the pH after 0.10 mole of NaOH is added to 1.00 L of the solution in Exercise 31, and calculate the pH after 0.20 mole of HCl is added to 1.00 L of the solution in Exercise 31.

34. Calculate the pH after 0.10 mole of NaOH is added to 1.00 L of the solution in Exercise 32, and calculate the pH after 0.20 mole of HCl is added to 1.00 L of the solution in Exercise 32.

35. Calculate the pH of each of the following buffered solutions.

a. 0.10 M acetic acid/0.25 M sodium acetate

b. 0.25 M acetic acid/0.10 M sodium acetate

c. 0.080 M acetic acid/0.20 M sodium acetate

d. 0.20 M acetic acid/0.080 M sodium acetate

36. Calculate the pH of each of the following buffered solutions.

   a. 0.50 M $C_2H_5NH_2$/0.25 M $C_2H_5NH_3Cl$

   b. 0.25 M $C_2H_5NH_2$/0.50 M $C_2H_5NH_3Cl$

   c. 0.50 M $C_2H_5NH_2$/0.50 M $C_2H_5NH_3Cl$

37. Calculate the pH of a buffered solution prepared by dissolving 21.5 g benzoic acid ($HC_7H_5O_2$) and 37.7 g sodium benzoate in 200.0 mL of solution.

38. A buffered solution is made by adding 50.0 g NH_4Cl to 1.00 L of a 0.75-M solution of NH_3. Calculate the pH of the final solution. (Assume no volume change.)

39. Calculate the pH after 0.010 mole of gaseous HCl is added to 250.0 mL of each of the following buffered solutions.

   a. 0.050 M NH_3/0.15 M NH_4Cl

   b. 0.50 M NH_3/1.50 M NH_4Cl

   Do the two original buffered solutions differ in their pH or their capacity? What advantage is there in having a buffer with a greater capacity?

40. An aqueous solution contains dissolved $C_6H_5NH_3Cl$ and $C_6H_5NH_2$. The concentration of $C_6H_5NH_2$ is 0.50 M and pH is 4.20.

   a. Calculate the concentration of $C_6H_5NH_3^+$ in this buffer solution.

   b. Calculate the pH after 4.0 g NaOH(s) is added to 1.0 L of this solution. (Neglect any volume change.)

41. Calculate the mass of sodium acetate that must be added to 500.0 mL of 0.200 M acetic acid to form a pH = 5.00 buffer solution.

42. What volumes of 0.50 M HNO_2 and 0.50 M $NaNO_2$ must be mixed to prepare 1.00 L of a solution buffered at pH = 3.55?

43. Consider a solution that contains both C_5H_5N and $C_5H_5NHNO_3$. Calculate the ratio $[C_5H_5N]/[C_5H_5NH^+]$ if the solution has the following pH values:

   a. pH = 4.50   **c.** pH = 5.23

   b. pH = 5.00   **d.** pH = 5.50

44. Calculate the ratio $[NH_3]/[NH_4^+]$ in ammonia/ammonium chloride buffered solutions with the following pH values:

   a. pH = 9.00   **c.** pH = 10.00

   b. pH = 8.80   **d.** pH = 9.60

45. Carbonate buffers are important in regulating the pH of blood at 7.40. If the carbonic acid concentration in a sample of blood is 0.0012 M, determine the bicarbonate ion concentration required to buffer the pH of blood at pH = 7.40.

$$H_2CO_3(aq) \rightleftharpoons HCO_3^-(aq) + H^+(aq) \qquad K_a = 4.3 \times 10^{-7}$$

46. When a person exercises, muscle contractions produce lactic acid. Moderate increases in lactic acid can be handled by the blood buffers without decreasing the pH of blood. However, excessive amounts of lactic acid can overload the blood buffer system, resulting in a lowering of the blood pH. A condition called *acidosis* is diagnosed if the blood pH falls to 7.35 or lower. Assume the primary blood buffer system is the carbonate buffer system described in Exercise 45. Calculate what happens to the $[H_2CO_3]/[HCO_3^-]$ ratio in blood when the pH decreases from 7.40 to 7.35.

47. Consider the acids in Table 14.2. Which acid would be the best choice for preparing a pH = 7.00 buffer? Explain how to make 1.0 L of this buffer.

48. Consider the bases in Table 14.3. Which base would be the best choice for preparing a pH = 5.00 buffer? Explain how to make 1.0 L of this buffer.

49. Calculate the pH of a solution that is 0.40 M H_2NNH_2 and 0.80 M $H_2NNH_3NO_3$. In order for this buffer to have pH = pK_a, would you add HCl or NaOH? What quantity (moles) of which reagent would you add to 1.0 L of the original buffer so that the resulting solution has pH = pK_a?

50. Calculate the pH of a solution that is 0.20 M HOCl and 0.90 M KOCl. In order for this buffer to have pH = pK_a, would you add HCl or NaOH? What quantity (moles) of which reagent would you add to 1.0 L of the original buffer so that the resulting solution has pH = pK_a?

51. Which of the following mixtures would result in buffered solutions when 1.0 L of each of the two solutions are mixed?

   a. 0.1 M KOH and 0.1 M CH_3NH_3Cl

   b. 0.1 M KOH and 0.2 M CH_3NH_2

   c. 0.2 M KOH and 0.1 M CH_3NH_3Cl

   d. 0.1 M KOH and 0.2 M CH_3NH_3Cl

52. Which of the following mixtures would result in a buffered solution when 1.0 L of each of the two solutions are mixed?

   a. 0.2 M HNO_3 and 0.4 M $NaNO_3$

   b. 0.2 M HNO_3 and 0.4 M HF

   c. 0.2 M HNO_3 and 0.4 M NaF

   d. 0.2 M HNO_3 and 0.4 M NaOH

53. What quantity (moles) of NaOH must be added to 1.0 L of 2.0 M $HC_2H_3O_2$ to produce a solution buffered at each pH?

   a. pH = pK_a **b.** pH = 4.00   **c.** pH = 5.00

54. Calculate the number of moles of HCl(g) that must be added to 1.0 L of 1.0 M $NaC_2H_3O_2$ to produce a solution buffered at each pH.

   a. pH = pK_a **b.** pH = 4.20   **c.** pH = 5.00

Acid–Base Titrations

55. Consider the titration of a generic weak acid HA with a strong base that gives the following titration curve:

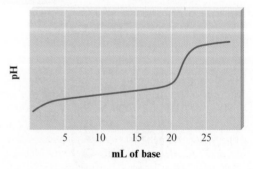

On the curve, indicate the points that correspond to the following:

   a. the stoichiometric (equivalence) point

   b. the region with maximum buffering

   c. pH = pK_a

d. pH depends only on [HA]

e. pH depends only on [A$^-$]

f. pH depends only on the amount of excess strong base added

56. Sketch the titration curve for the titration of a generic weak base B with a strong acid. The titration reaction is

$$B + H^+ \rightleftharpoons BH^+$$

On this curve, indicate the points that correspond to the following:

a. the stoichiometric (equivalence) point

b. the region with maximum buffering

c. pH = pK_a

d. pH depends only on [B]

e. pH depends only on [BH$^+$]

f. pH depends only on the amount of excess strong acid added

57. Consider the titration of 40.0 mL of 0.200 M HClO$_4$ by 0.100 M KOH. Calculate the pH of the resulting solution after the following volumes of KOH have been added.

a. 0.0 mL   **d.** 80.0 mL

b. 10.0 mL   **e.** 100.0 mL

c. 40.0 mL

58. Consider the titration of 80.0 mL of 0.100 M Ba(OH)$_2$ by 0.400 M HCl. Calculate the pH of the resulting solution after the following volumes of HCl have been added.

a. 0.0 mL   **d.** 40.0 mL

b. 20.0 mL   **e.** 80.0 mL

c. 30.0 mL

59. Consider the titration of 100.0 mL of 0.200 M acetic acid ($K_a = 1.8 \times 10^{-5}$) by 0.100 M KOH. Calculate the pH of the resulting solution after the following volumes of KOH have been added.

a. 0.0 mL   **d.** 150.0 mL

b. 50.0 mL   **e.** 200.0 mL

c. 100.0 mL   **f.** 250.0 mL

60. Consider the titration of 100.0 mL of 0.100 M H$_2$NNH$_2$ ($K_b = 3.0 \times 10^{-6}$) by 0.200 M HNO$_3$. Calculate the pH of the resulting solution after the following volumes of HNO$_3$ have been added.

a. 0.0 mL   **d.** 40.0 mL

b. 20.0 mL   **e.** 50.0 mL

c. 25.0 mL   **f.** 100.0 mL

61. Lactic acid is a common by-product of cellular respiration and is often said to cause the "burn" associated with strenuous activity. A 25.0-mL sample of 0.100 M lactic acid (HC$_3$H$_5$O$_3$, pK_a = 3.86) is titrated with 0.100 M NaOH solution. Calculate the pH after the addition of 0.0 mL, 4.0 mL, 8.0 mL, 12.5 mL, 20.0 mL, 24.0 mL, 24.5 mL, 24.9 mL, 25.0 mL, 25.1 mL, 26.0 mL, 28.0 mL, and 30.0 mL of the NaOH. Plot the results of your calculations as pH versus milliliters of NaOH added.

62. Repeat the procedure in Exercise 61, but for the titration of 25.0 mL of 0.100 M propanoic acid (HC$_3$H$_5$O$_2$, $K_a = 1.3 \times 10^{-5}$) with 0.100 M NaOH.

63. Repeat the procedure in Exercise 61, but for the titration of 25.0 mL of 0.100 M NH$_3$ ($K_b = 1.8 \times 10^{-5}$) with 0.100 M HCl.

64. Repeat the procedure in Exercise 61, but for the titration of 25.0 mL of 0.100 M pyridine with 0.100 M hydrochloric acid (K_b for pyridine is 1.7×10^{-9}). Do not calculate the points at 24.9 and 25.1 mL.

65. Calculate the pH at the halfway point and at the equivalence point for each of the following titrations.

a. 100.0 mL of 0.10 M HC$_7$H$_5$O$_2$ ($K_a = 6.4 \times 10^{-5}$) titrated by 0.10 M NaOH

b. 100.0 mL of 0.10 M C$_2$H$_5$NH$_2$ ($K_b = 5.6 \times 10^{-4}$) titrated by 0.20 M HNO$_3$

c. 100.0 mL of 0.50 M HCl titrated by 0.25 M NaOH

66. In the titration of 50.0 mL of 1.0 M methylamine, CH$_3$NH$_2$ ($K_b = 4.4 \times 10^{-4}$), with 0.50 M HCl, calculate the pH under the following conditions.

a. after 50.0 mL of 0.50 M HCl has been added

b. at the stoichiometric point

67. You have 75.0 mL of 0.10 M HA. After adding 30.0 mL of 0.10 M NaOH, the pH is 5.50. What is the K_a value of HA?

68. A student dissolves 0.0100 mol of an unknown weak base in 100.0 mL water and titrates the solution with 0.100 M HNO$_3$. After 40.0 mL of 0.100 M HNO$_3$ was added, the pH of the resulting solution was 8.00. Calculate the K_b value for the weak base.

Indicators

69. Two drops of indicator HIn ($K_a = 1.0 \times 10^{-9}$), where HIn is yellow and In$^-$ is blue, are placed in 100.0 mL of 0.10 M HCl.

a. What color is the solution initially?

b. The solution is titrated with 0.10 M NaOH. At what pH will the color change (yellow to greenish yellow) occur?

c. What color will the solution be after 200.0 mL NaOH has been added?

70. Methyl red has the following structure:

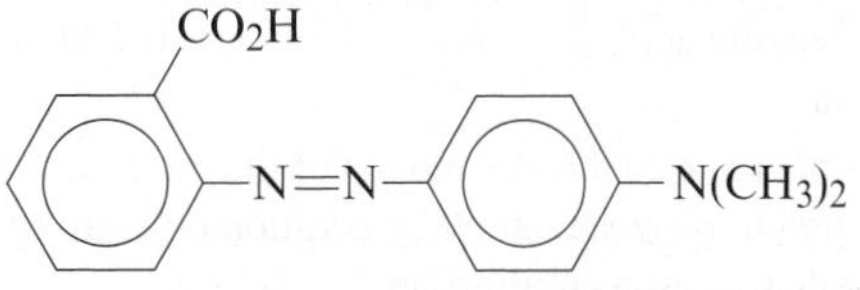

It undergoes a color change from red to yellow as a solution gets more basic. Calculate an approximate pH range for which methyl red is useful. What is the color change and the pH at the color change when a weak acid is titrated with a strong base using methyl red as an indicator? What is the color change and the pH at the color change when a weak base is titrated with a strong acid using methyl red as an indicator? For which of these two types of titrations is methyl red a possible indicator?

71. Potassium hydrogen phthalate, known as KHP (molar mass = 204.22 g/mol), can be obtained in high purity and is used to determine the concentration of solutions of strong bases by the reaction

$$HP^-(aq) + OH^-(aq) \longrightarrow H_2O(l) + P^{2-}(aq)$$

If a typical titration experiment begins with approximately 0.5 g KHP and has a final volume of about 100 mL, what is an appropriate indicator to use? The pK_a for HP^- is 5.51.

72. A certain indicator HIn has a pK_a of 3.00 and a color change becomes visible when 7.00% of the indicator has been converted to In^-. At what pH is this color change visible?

73. Which of the indicators in Fig. 15.8 could be used for the titrations in Exercises 57 and 59?

74. Which of the indicators in Fig. 15.8 could be used for the titrations in Exercises 58 and 60?

75. Which of the indicators in Fig. 15.8 could be used for the titrations in Exercises 61 and 63?

76. Which of the indicators in Fig. 15.8 could be used for the titrations in Exercises 62 and 64?

77. Estimate the pH of a solution in which bromcresol green is blue and thymol blue is yellow. (See Fig. 15.8.)

78. Estimate the pH of a solution in which crystal violet is yellow and methyl orange is red. (See Fig. 15.8.)

79. A solution has a pH of 7.0. What would be the color of the solution if each of the following indicators were added? (See Fig. 15.8.)

  a. thymol blue   **c.** methyl red

  b. bromthymol blue   **d.** crystal violet

80. A solution has a pH of 4.5. What would be the color of the solution if each of the following indicators were added? (See Fig. 15.8.)

  a. methyl orange   **c.** bromcresol green

  b. alizarin   **d.** phenolphthalein

Additional Exercises

81. Derive an equation analogous to the Henderson–Hasselbalch equation but relating pOH and pK_b of a buffered solution composed of a weak base and its conjugate acid, such as NH_3 and NH_4^+.

82. a. Calculate the pH of a buffered solution that is 0.100 M in $C_6H_5CO_2H$ (benzoic acid, $K_a = 6.4 \times 10^{-5}$) and 0.100 M in $C_6H_5CO_2Na$.

  b. Calculate the pH after 20.0% (by moles) of the benzoic acid is converted to benzoate anion by addition of a strong base. Use the dissociation equilibrium

$$C_6H_5CO_2H(aq) \rightleftharpoons C_6H_5CO_2^-(aq) + H^+(aq)$$

  to calculate the pH.

  c. Do the same as in part b, but use the following equilibrium to calculate the pH:

$$C_6H_5CO_2^-(aq) + H_2O(l) \rightleftharpoons C_6H_5CO_2H(aq) + OH^-(aq)$$

  d. Do your answers in parts b and c agree? Explain.

83. Tris(hydroxymethyl)aminomethane, commonly called TRIS or Trizma, is often used as a buffer in biochemical studies. Its buffering range is pH 7 to 9, and K_b is 1.19×10^{-6} for the aqueous reaction

$$(HOCH_2)_3CNH_2 + H_2O \rightleftharpoons (HOCH_2)_3CNH_3^+ + OH^-$$

          TRIS $TRISH^+$

  a. What is the optimal pH for TRIS buffers?

  b. Calculate the ratio $[TRIS]/[TRISH^+]$ at pH = 7.00 and at pH = 9.00.

  c. A buffer is prepared by diluting 50.0 g TRIS base and 65.0 g TRIS hydrochloride (written as TRISHCl) to a total volume of 2.0 L. What is the pH of this buffer? What is the pH after 0.50 mL of 12 M HCl is added to a 200.0-mL portion of the buffer?

84. You make 1.00 L of a buffered solution (pH = 4.00) by mixing acetic acid and sodium acetate. You have 1.00 M solutions of each component of the buffered solution. What volume of each solution do you mix to make such a buffered solution?

85. You have the following reagents on hand:

Solids (pK_a of Acid Form Is Given)	Solutions
Benzoic acid (4.19)	5.0 M HCl
Sodium acetate (4.74)	1.0 M acetic acid (4.74)
Potassium fluoride (3.14)	2.6 M NaOH
Ammonium chloride (9.26)	1.0 M HOCl (7.46)

What combinations of reagents would you use to prepare buffers at the following pH values?

  a. 3.0   **b.** 4.0   **c.** 5.0   **d.** 7.0   **e.** 9.0

86. Amino acids are the building blocks for all proteins in our bodies. A structure for the amino acid alanine is

$$\underset{\substack{\text{Amino} \\ \text{group}}}{H_2N} - \underset{\substack{| \\ H}}{\overset{\substack{CH_3 \\ |}}{C}} - \underset{\text{Carboxylic} \\ \text{acid group}}{\overset{O \\ ||}{C}} - OH$$

All amino acids have at least two functional groups with acidic or basic properties. In alanine, the carboxylic acid group has $K_a = 4.5 \times 10^{-3}$ and the amino group has $K_b = 7.4 \times 10^{-5}$. Because of the two groups with acidic or basic properties, three different charged ions of alanine are possible when alanine is dissolved in water. Which of these ions would predominate in a solution with $[H^+] = 1.0$ M? In a solution with $[OH^-] = 1.0$ M?

87. Phosphate buffers are important in regulating the pH of intracellular fluids at pH values generally between 7.1 and 7.2.

  a. What is the concentration ratio of $H_2PO_4^-$ to HPO_4^{2-} in intracellular fluid at pH = 7.15?

$$H_2PO_4^-(aq) \rightleftharpoons HPO_4^{2-}(aq) + H^+(aq) \qquad K_a = 6.2 \times 10^{-8}$$

  b. Why is a buffer composed of H_3PO_4 and $H_2PO_4^-$ ineffective in buffering the pH of intracellular fluid?

$$H_3PO_4(aq) \rightleftharpoons H_2PO_4^-(aq) + H^+(aq) \qquad K_a = 7.5 \times 10^{-3}$$

88. What quantity (moles) of HCl(g) must be added to 1.0 L of 2.0 M NaOH to achieve a pH of 0.00? (Neglect any volume changes.)

89. Calculate the value of the equilibrium constant for each of the following reactions in aqueous solution.

  a. $HC_2H_3O_2 + OH^- \rightleftharpoons C_2H_3O_2^- + H_2O$

  b. $C_2H_3O_2^- + H^+ \rightleftharpoons HC_2H_3O_2$

  c. $HCl + NaOH \rightleftharpoons NaCl + H_2O$

90. The following plot shows the pH curves for the titrations of various acids by 0.10 M NaOH (all of the acids were 50.0-mL samples of 0.10 M concentration).

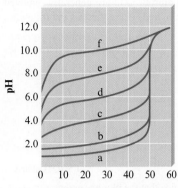

Vol 0.10 M NaOH added (mL)

a. Which pH curve corresponds to the weakest acid?

b. Which pH curve corresponds to the strongest acid? Which point on the pH curve would you examine to see if this acid is a strong acid or a weak acid (assuming you did not know the initial concentration of the acid)?

c. Which pH curve corresponds to an acid with $K_a \approx 1 \times 10^{-6}$?

91. Calculate the volume of 1.50×10^{-2} M NaOH that must be added to 500.0 mL of 0.200 M HCl to give a solution that has pH $= 2.15$.

92. Repeat the procedure in Exercise 61, but for the titration of 25.0 mL of 0.100 M HNO₃ with 0.100 M NaOH.

93. A certain acetic acid solution has pH $= 2.68$. Calculate the volume of 0.0975 M KOH required to reach the equivalence point in the titration of 25.0 mL of the acetic acid solution.

94. A 0.210-g sample of an acid (molar mass $= 192$ g/mol) is titrated with 30.5 mL of 0.108 M NaOH to a phenolphthalein end point. Is the acid monoprotic, diprotic, or triprotic?

95. The active ingredient in aspirin is acetylsalicylic acid. A 2.51-g sample of acetylsalicylic acid required 27.36 mL of 0.5106 M NaOH for complete reaction. Addition of 13.68 mL of 0.5106 M HCl to the flask containing the aspirin and the sodium hydroxide produced a mixture with pH $= 3.48$. Determine the molar mass of acetylsalicylic acid and its K_a value. State any assumptions you must make to reach your answer.

96. One method for determining the purity of aspirin ($C_9H_8O_4$) is to hydrolyze it with NaOH solution and then to titrate the remaining NaOH. The reaction of aspirin with NaOH is as follows:

$$C_9H_8O_4(s) + 2OH^-(aq)$$
Aspirin $\xrightarrow[\text{10 min}]{\text{Boil}}$ $C_7H_5O_3^-(aq) + C_2H_3O_2^-(aq) + H_2O(l)$
Salicylate ion   Acetate ion

A sample of aspirin with a mass of 1.427 g was boiled in 50.00 mL of 0.500 M NaOH. After the solution was cooled, it took 31.92 mL of 0.289 M HCl to titrate the excess NaOH. Calculate the purity of the aspirin. What indicator should be used for this titration? Why?

97. A student intends to titrate a solution of a weak monoprotic acid with a sodium hydroxide solution but reverses the two solutions and places the weak acid solution in the buret. After 23.75 mL of the weak acid solution has been added to 50.0 mL of the 0.100 M NaOH solution, the pH of the resulting solution is 10.50. Calculate the original concentration of the solution of weak acid.

98. A student titrates an unknown weak acid, HA, to a pale pink phenolphthalein end point with 25.0 mL of 0.100 M NaOH. The student then adds 13.0 mL of 0.100 M HCl. The pH of the resulting solution is 4.70. How is the value of pK_a for the unknown acid related to 4.70?

99. A sample of a certain monoprotic weak acid was dissolved in water and titrated with 0.125 M NaOH, requiring 16.00 mL to reach the equivalence point. During the titration, the pH after adding 2.00 mL NaOH was 6.912. Calculate K_a for the weak acid.

ChemWork Problems

These multiconcept problems (and additional ones) are found interactively online with the same type of assistance a student would get from an instructor.

100. Consider 1.0 L of a solution that is 0.85 M HOC₆H₅ and 0.80 M NaOC₆H₅. (K_a for HOC₆H₅ $= 1.6 \times 10^{-10}$.)

a. Calculate the pH of this solution.

b. Calculate the pH after 0.10 mole of HCl has been added to the original solution. Assume no volume change on addition of HCl.

c. Calculate the pH after 0.20 mole of NaOH has been added to the original buffer solution. Assume no volume change on addition of NaOH.

101. What concentration of NH₄Cl is necessary to buffer a 0.52-M NH₃ solution at pH $= 9.00$? (K_b for NH₃ $= 1.8 \times 10^{-5}$.)

102. Consider the following acids and bases:

HCO₂H	$K_a = 1.8 \times 10^{-4}$
HOBr	$K_a = 2.0 \times 10^{-9}$
(C₂H₅)₂NH	$K_b = 1.3 \times 10^{-3}$
HONH₂	$K_b = 1.1 \times 10^{-8}$

Choose substances from the following list that would be the best choice to prepare a pH $= 9.0$ buffer solution.

a. HCO₂H   **e.** (C₂H₅)₂NH

b. HOBr   **f.** (C₂H₅)₂NH₂Cl

c. KHCO₂   **g.** HONH₂

d. HONH₃NO₃   **h.** NaOBr

103. Consider a buffered solution containing CH₃NH₃Cl and CH₃NH₂. Which of the following statements concerning this solution is(are) *true*? (K_a for CH₃NH₃⁺ $= 2.3 \times 10^{-11}$.)

a. A solution consisting of 0.10 M CH₃NH₃Cl and 0.10 M CH₃NH₂ would have a higher buffering capacity than one containing 1.0 M CH₃NH₃Cl and 1.0 M CH₃NH₂.

b. If [CH₃NH₂] > [CH₃NH₃⁺], then the pH is larger than the pK_a value.

c. Adding more [CH₃NH₃Cl] to the initial buffer solution will decrease the pH.

d. If $[CH_3NH_2] < [CH_3NH_3^+]$, then pH < 3.36.

e. If $[CH_3NH_2] = [CH_3NH_3^+]$, then pH $= 10.64$.

104. Consider the titration of 150.0 mL of 0.100 M HI by 0.250 M NaOH.

   a. Calculate the pH after 20.0 mL of NaOH has been added.

   b. What volume of NaOH must be added so that the pH $= 7.00$?

105. Consider the titration of 100.0 mL of 0.100 M HCN by 0.100 M KOH at 25°C. (K_a for HCN $= 6.2 \times 10^{-10}$.)

   a. Calculate the pH after 0.0 mL of KOH has been added.

   b. Calculate the pH after 50.0 mL of KOH has been added.

   c. Calculate the pH after 75.0 mL of KOH has been added.

   d. Calculate the pH at the equivalence point.

   e. Calculate the pH after 125 mL of KOH has been added.

106. Consider the titration of 100.0 mL of 0.200 M HONH$_2$ by 0.100 M HCl. (K_b for HONH$_2$ $= 1.1 \times 10^{-8}$.)

   a. Calculate the pH after 0.0 mL of HCl has been added.

   b. Calculate the pH after 25.0 mL of HCl has been added.

   c. Calculate the pH after 70.0 mL of HCl has been added.

   d. Calculate the pH at the equivalence point.

   e. Calculate the pH after 300.0 mL of HCl has been added.

   f. At what volume of HCl added does the pH $= 6.04$?

107. Consider the following four titrations (i–iv):

   i. 150 mL of 0.2 M NH$_3$ ($K_b = 1.8 \times 10^{-5}$) by 0.2 M HCl

   ii. 150 mL of 0.2 M HCl by 0.2 M NaOH

   iii. 150 mL of 0.2 M HOCl ($K_a = 3.5 \times 10^{-8}$) by 0.2 M NaOH

   iv. 150 mL of 0.2 M HF ($K_a = 7.2 \times 10^{-4}$) by 0.2 M NaOH

   a. Rank the four titrations in order of increasing pH at the halfway point to equivalence (lowest to highest pH).

   b. Rank the four titrations in order of increasing pH at the equivalence point.

   c. Which titration requires the largest volume of titrant (HCl or NaOH) to reach the equivalence point?

Challenge Problems

108. Another way to treat data from a pH titration is to graph the absolute value of the change in pH per change in milliliters added versus milliliters added (ΔpH/ΔmL versus mL added). Make this graph using your results from Exercise 61. What advantage might this method have over the traditional method for treating titration data?

109. A buffer is made using 45.0 mL of 0.750 M HC$_3$H$_5$O$_2$ ($K_a = 1.3 \times 10^{-5}$) and 55.0 mL of 0.700 M NaC$_3$H$_5$O$_2$. What volume of 0.10 M NaOH must be added to change the pH of the original buffer solution by 2.5%?

110. A 0.400-M solution of ammonia was titrated with hydrochloric acid to the equivalence point, where the total volume was 1.50 times the original volume. At what pH does the equivalence point occur?

111. What volume of 0.0100 M NaOH must be added to 1.00 L of 0.0500 M HOCl to achieve a pH of 8.00?

112. Consider a solution formed by mixing 50.0 mL of 0.100 M H$_2$SO$_4$, 30.0 mL of 0.100 M HOCl, 25.0 mL of 0.200 M NaOH,

25.0 mL of 0.100 M Ba(OH)$_2$, and 10.0 mL of 0.150 M KOH. Calculate the pH of this solution.

113. When a diprotic acid, H$_2$A, is titrated with NaOH, the protons on the diprotic acid are generally removed one at a time, resulting in a pH curve that has the following generic shape:

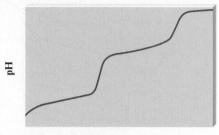

Vol NaOH added

   a. Notice that the plot has essentially two titration curves. If the first equivalence point occurs at 100.0 mL NaOH added, what volume of NaOH added corresponds to the second equivalence point?

   b. For the following volumes of NaOH added, list the major species present after the OH$^-$ reacts completely.

   i. 0 mL NaOH added

   ii. between 0 and 100.0 mL NaOH added

   iii. 100.0 mL NaOH added

   iv. between 100.0 and 200.0 mL NaOH added

   v. 200.0 mL NaOH added

   vi. after 200.0 mL NaOH added

   c. If the pH at 50.0 mL NaOH added is 4.0 and the pH at 150.0 mL NaOH added is 8.0, determine the values K_{a_1} and K_{a_2} for the diprotic acid.

114. Consider the following two acids:

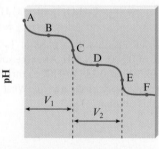

$pK_{a_1} = 2.98; pK_{a_2} = 13.40$

Salicylic acid

HO$_2$CCH$_2$CH$_2$CH$_2$CH$_2$CO$_2$H

Adipic acid $pK_{a_1} = 4.41; pK_{a_2} = 5.28$

In two separate experiments the pH was measured during the titration of 5.00 mmol of each acid with 0.200 M NaOH. Each experiment showed only one stoichiometric point when the data were plotted. In one experiment the stoichiometric point was at 25.00 mL added NaOH, and in the other experiment the stoichiometric point was at 50.00 mL NaOH. Explain these results. (See Exercise 113.)

115. The titration of Na$_2$CO$_3$ with HCl has the following qualitative profile:

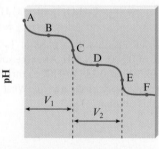

mL HCl

a. Identify the major species in solution at points A–F.

b. Calculate the pH at the halfway points to equivalence, B and D. (*Hint:* Refer to Exercise 113.)

116. Consider the titration curve in Exercise 115 for the titration of Na_2CO_3 with HCl.

a. If a mixture of $NaHCO_3$ and Na_2CO_3 was titrated, what would be the relative sizes of V_1 and V_2?

b. If a mixture of NaOH and Na_2CO_3 was titrated, what would be the relative sizes of V_1 and V_2?

c. A sample contains a mixture of $NaHCO_3$ and Na_2CO_3. When this sample was titrated with 0.100 *M* HCl, it took 18.9 mL to reach the first stoichiometric point and an additional 36.7 mL to reach the second stoichiometric point. What is the composition in mass percent of the sample?

117. A few drops of each of the indicators shown in the accompanying table were placed in separate portions of a 1.0-*M* solution of a weak acid, HX. The results are shown in the last column of the table. What is the approximate pH of the solution containing HX? Calculate the approximate value of K_a for HX.

Indicator	Color of HIn	Color of In⁻	pK_a of HIn	Color of 1.0 *M* HX
Bromphenol blue	Yellow	Blue	4.0	Blue
Bromcresol purple	Yellow	Purple	6.0	Yellow
Bromcresol green	Yellow	Blue	4.8	Green
Alizarin	Yellow	Red	6.5	Yellow

118. Malonic acid ($HO_2CCH_2CO_2H$) is a diprotic acid. In the titration of malonic acid with NaOH, stoichiometric points occur at pH = 3.9 and 8.8. A 25.00-mL sample of malonic acid of unknown concentration is titrated with 0.0984 *M* NaOH, requiring 31.50 mL of the NaOH solution to reach the phenolphthalein end point. Calculate the concentration of the initial malonic acid solution. (See Exercise 113.)

Integrative Problems

These problems require the integration of multiple concepts to find the solutions.

119. A buffer solution is prepared by mixing 75.0 mL of 0.275 *M* fluorobenzoic acid ($C_7H_5O_2F$) with 55.0 mL of 0.472 *M* sodium fluorobenzoate. The pK_a of this weak acid is 2.90. What is the pH of the buffer solution?

120. A 10.00-g sample of the ionic compound NaA, where A⁻ is the anion of a weak acid, was dissolved in enough water to make 100.0 mL of solution and was then titrated with 0.100 *M* HCl. After 500.0 mL HCl was added, the pH was 5.00. The experimenter found that 1.00 L of 0.100 *M* HCl was required to reach the stoichiometric point of the titration.

a. What is the molar mass of NaA?

b. Calculate the pH of the solution at the stoichiometric point of the titration.

121. Calculate the pH of a solution prepared by mixing 250. mL of 0.174 *m* aqueous HF (density = 1.10 g/mL) with 38.7 g of an aqueous solution that is 1.50% NaOH by mass (density = 1.02 g/mL). (K_a for HF = 7.2×10^{-4}.)

Marathon Problem

This problem is designed to incorporate several concepts and techniques into one situation.

122. Consider a solution prepared by mixing the following:

50.0 mL of 0.100 *M* Na_3PO_4
100.0 mL of 0.0500 *M* KOH
200.0 mL of 0.0750 *M* HCl
50.0 mL of 0.150 *M* NaCN

Determine the volume of 0.100 *M* HNO_3 that must be added to this mixture to achieve a final pH value of 7.21.

Solubility and Complex Ion Equilibria

Stalactites and stalagmites in the Drapery Room at Mammoth Cave in Kentucky. These formations are created when carbonate minerals dissolve in groundwater acidified by carbon dioxide and then solidify when the water evaporates. (Adam Jones/Danita Delimont)

Most of the chemistry of the natural world occurs in aqueous solution. We have already introduced one very significant class of aqueous equilibria, acid–base reactions. In this chapter we will consider more applications of aqueous equilibria, those involving the solubility of salts and those involving the formation of complex ions.

The interplay of acid–base, solubility, and complex ion equilibria is often important in natural processes, such as the weathering of minerals, the uptake of nutrients by plants, and tooth decay. For example, limestone ($CaCO_3$) will dissolve in water made acidic by dissolved carbon dioxide:

$$CO_2(aq) + H_2O(l) \rightleftharpoons H^+(aq) + HCO_3^-(aq)$$
$$H^+(aq) + CaCO_3(s) \rightleftharpoons Ca^{2+}(aq) + HCO_3^-(aq)$$

This two-step process and its reverse account for the formation of limestone caves and the stalactites and stalagmites found therein. In the forward direction of the process, the acidic water (containing carbon dioxide) dissolves the underground limestone deposits, thereby forming a cavern. The reverse process occurs as the water drips from the ceiling of the cave, and the carbon dioxide is lost to the air. This causes solid calcium carbonate to form, producing stalactites on the ceiling and stalagmites where the drops hit the cave floor.

In this chapter we will discuss the formation of solids from an aqueous solution and the resulting equilibria. We will also show how selective precipitation and the formation of complex ions can be used to do qualitative analysis.

16.1 | Solubility Equilibria and the Solubility Product

Solubility is a very important phenomenon. The fact that substances such as sugar and table salt dissolve in water allows us to flavor foods easily. The fact that calcium sulfate is less soluble in hot water than in cold water causes it to coat tubes in boilers, reducing thermal efficiency. Tooth decay involves solubility: When food lodges between the teeth, acids form that dissolve tooth enamel, which contains a mineral called *hydroxyapatite,* $Ca_5(PO_4)_3OH$. Tooth decay can be reduced by treating teeth with fluoride (see Chemical Connections, p. 763). Fluoride replaces the hydroxide in hydroxyapatite to produce the corresponding fluorapatite, $Ca_5(PO_4)_3F$, and calcium fluoride, CaF_2, both of which are less soluble in acids than the original enamel. Another important consequence of solubility involves the use of a suspension of barium sulfate to improve the clarity of X rays of the gastrointestinal tract. The very low solubility of barium sulfate, which contains the toxic ion Ba^{2+}, makes ingestion of the compound safe.

In this section we will consider the equilibria associated with solids dissolving to form aqueous solutions. We will assume that when a typical ionic solid dissolves in water, it dissociates completely into separate hydrated cations and anions. For example, calcium fluoride dissolves in water as follows:

$$CaF_2(s) \xrightarrow{H_2O} Ca^{2+}(aq) + 2F^-(aq)$$

When the solid salt is first added to the water, no Ca^{2+} and F^- ions are present. However, as the dissolution proceeds, the concentrations of Ca^{2+} and F^- increase, making

Adding F^- to drinking water is controversial. See Geoff Rayner-Canham, "Fluoride: Trying to Separate Fact from Fallacy," *Chem 13 News*, Sept. 2001, pp. 16–19.

For simplicity, we will ignore the effects of ion associations in these solutions.

it more and more likely that these ions will collide and re-form the solid phase. Thus two competing processes are occurring—the dissolution reaction and its reverse:

$$Ca^{2+}(aq) + 2F^-(aq) \longrightarrow CaF_2(s)$$

Ultimately, dynamic equilibrium is reached:

$$CaF_2(s) \rightleftharpoons Ca^{2+}(aq) + 2F^-(aq)$$

At this point no more solid dissolves (the solution is said to be *saturated*).

We can write an equilibrium expression for this process according to the law of mass action:

$$K_{sp} = [Ca^{2+}][F^-]^2$$

where $[Ca^{2+}]$ and $[F^-]$ are expressed in mol/L. The constant K_{sp} is called the **solubility product constant** or simply the **solubility product** for the equilibrium expression.

Since CaF_2 is a pure solid, it is not included in the equilibrium expression. The fact that the amount of excess solid present does not affect the position of the solubility equilibrium might seem strange at first; more solid means more surface area exposed to the solvent, which would seem to result in greater solubility. This is not the case, however. When the ions in solution re-form the solid, they do so on the surface of the solid. Thus doubling the surface area of the solid not only doubles the rate of dissolving, but also doubles the rate of re-formation of the solid. The amount of excess solid present therefore has no effect on the equilibrium position. Similarly, although either increasing the surface area by grinding up the solid or stirring the solution speeds up the attainment of equilibrium, neither procedure changes the amount of solid dissolved at equilibrium. Neither the amount of excess solid nor the size of the particles present will shift the *position* of the solubility equilibrium.

It is very important to distinguish between the *solubility* of a given solid and its *solubility product*. The solubility product is an *equilibrium constant* and has only *one* value for a given solid at a given temperature. Solubility, on the other hand, is an *equilibrium position*. In pure water at a specific temperature a given salt has a particular solubility. On the other hand, if a common ion is present in the solution, the solubility varies according to the concentration of the common ion. However, in all cases the product of the ion concentrations must satisfy the K_{sp} expression. The K_{sp} values at 25°C for many common ionic solids are listed in Table 16.1. The units are customarily omitted.

Solving solubility equilibria problems requires many of the same procedures we have used to deal with acid–base equilibria, as illustrated in Examples 16.1 and 16.2.

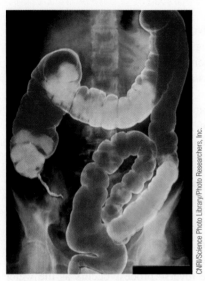

An X ray of the lower gastrointestinal tract using barium sulfate.

Pure liquids and pure solids are never included in an equilibrium expression (Section 13.4).

K_{sp} is an equilibrium constant; solubility is an equilibrium position.

Calculating K_{sp} from Solubility I

Copper(I) bromide has a measured solubility of 2.0×10^{-4} mol/L at 25°C. Calculate its K_{sp} value.

Solution

In this experiment the solid was placed in contact with water. Thus, before any reaction occurred, the system contained solid CuBr and H_2O. The process that occurs is the dissolving of CuBr to form the separated Cu^+ and Br^- ions:

$$CuBr(s) \rightleftharpoons Cu^+(aq) + Br^-(aq)$$

where

$$K_{sp} = [Cu^+][Br^-]$$

Initially, the solution contains no Cu^+ or Br^-, so the initial concentrations are

$$[Cu^+]_0 = [Br^-]_0 = 0$$

Table 16.1 | K_{sp} Values at 25°C for Common Ionic Solids

Ionic Solid	K_{sp} (at 25°C)	Ionic Solid	K_{sp} (at 25°C)	Ionic Solid	K_{sp} (at 25°C)
Fluorides		Hg_2CrO_4*	2×10^{-9}	$Co(OH)_2$	2.5×10^{-16}
BaF_2	2.4×10^{-5}	$BaCrO_4$	8.5×10^{-11}	$Ni(OH)_2$	1.6×10^{-16}
MgF_2	6.4×10^{-9}	Ag_2CrO_4	9.0×10^{-12}	$Zn(OH)_2$	4.5×10^{-17}
PbF_2	4×10^{-8}	$PbCrO_4$	2×10^{-16}	$Cu(OH)_2$	1.6×10^{-19}
SrF_2	7.9×10^{-10}			$Hg(OH)_2$	3×10^{-26}
CaF_2	4.0×10^{-11}	Carbonates		$Sn(OH)_2$	3×10^{-27}
		$NiCO_3$	1.4×10^{-7}	$Cr(OH)_3$	6.7×10^{-31}
Chlorides		$CaCO_3$	8.7×10^{-9}	$Al(OH)_3$	2×10^{-32}
$PbCl_2$	1.6×10^{-5}	$BaCO_3$	1.6×10^{-9}	$Fe(OH)_3$	4×10^{-38}
$AgCl$	1.6×10^{-10}	$SrCO_3$	7×10^{-10}	$Co(OH)_3$	2.5×10^{-43}
Hg_2Cl_2*	1.1×10^{-18}	$CuCO_3$	2.5×10^{-10}		
		$ZnCO_3$	2×10^{-10}	Sulfides	
Bromides		$MnCO_3$	8.8×10^{-11}	MnS	2.3×10^{-13}
$PbBr_2$	4.6×10^{-6}	$FeCO_3$	2.1×10^{-11}	FeS	3.7×10^{-19}
$AgBr$	5.0×10^{-13}	Ag_2CO_3	8.1×10^{-12}	NiS	3×10^{-21}
Hg_2Br_2*	1.3×10^{-22}	$CdCO_3$	5.2×10^{-12}	CoS	5×10^{-22}
		$PbCO_3$	1.5×10^{-15}	ZnS	2.5×10^{-22}
Iodides		$MgCO_3$	6.8×10^{-6}	SnS	1×10^{-26}
PbI_2	1.4×10^{-8}	Hg_2CO_3*	9.0×10^{-15}	CdS	1.0×10^{-28}
AgI	1.5×10^{-16}			PbS	7×10^{-29}
Hg_2I_2*	4.5×10^{-29}	Hydroxides		CuS	8.5×10^{-45}
		$Ba(OH)_2$	5.0×10^{-3}	Ag_2S	1.6×10^{-49}
Sulfates		$Sr(OH)_2$	3.2×10^{-4}	HgS	1.6×10^{-54}
$CaSO_4$	6.1×10^{-5}	$Ca(OH)_2$	1.3×10^{-6}		
Ag_2SO_4	1.2×10^{-5}	$AgOH$	2.0×10^{-8}	Phosphates	
$SrSO_4$	3.2×10^{-7}	$Mg(OH)_2$	8.9×10^{-12}	Ag_3PO_4	1.8×10^{-18}
$PbSO_4$	1.3×10^{-8}	$Mn(OH)_2$	2×10^{-13}	$Sr_3(PO_4)_2$	1×10^{-31}
$BaSO_4$	1.5×10^{-9}	$Cd(OH)_2$	5.9×10^{-15}	$Ca_3(PO_4)_2$	1.3×10^{-32}
		$Pb(OH)_2$	1.2×10^{-15}	$Ba_3(PO_4)_2$	6×10^{-39}
Chromates		$Fe(OH)_2$	1.8×10^{-15}	$Pb_3(PO_4)_2$	1×10^{-54}
$SrCrO_4$	3.6×10^{-5}				

*Contains Hg_2^{2+} ions. $K = [Hg_2^{2+}][X^-]^2$ for Hg_2X_2 salts, for example.

The equilibrium concentrations can be obtained from the measured solubility of CuBr, which is 2.0×10^{-4} mol/L. This means that 2.0×10^{-4} mole of solid CuBr dissolves per 1.0 L of solution to come to equilibrium with the excess solid. The reaction is

$$CuBr(s) \longrightarrow Cu^+(aq) + Br^-(aq)$$

Thus

2.0×10^{-4} mol/L $CuBr(s)$
$$\longrightarrow 2.0 \times 10^{-4} \text{ mol/L Cu}^+(aq) + 2.0 \times 10^{-4} \text{ mol/L Br}^-(aq)$$

We can now write the equilibrium concentrations:

$$[Cu^+] = [Cu^+]_0 + \text{change to reach equilibrium}$$
$$= 0 + 2.0 \times 10^{-4} \text{ mol/L}$$

and
$$[Br^-] = [Br^-]_0 + \text{change to reach equilibrium}$$
$$= 0 + 2.0 \times 10^{-4} \text{ mol/L}$$

These equilibrium concentrations allow us to calculate the value of K_{sp} for CuBr:

$$K_{sp} = [Cu^+][Br^-] = (2.0 \times 10^{-4}\ mol/L)(2.0 \times 10^{-4}\ mol/L)$$
$$= 4.0 \times 10^{-8}\ mol^2/L^2 = 4.0 \times 10^{-8}$$

The units for K_{sp} values are usually omitted.

See Exercises 16.21 and 16.22

Precipitation of bismuth sulfide.

Sulfide is a very basic anion and really exists in water as HS^-. We will not consider this complication.

Solubilities must be expressed in mol/L in K_{sp} calculations.

Calculating K_{sp} from Solubility II

Calculate the K_{sp} value for bismuth sulfide (Bi_2S_3), which has a solubility of 1.0×10^{-15} mol/L at 25°C.

Solution

The system initially contains H_2O and solid Bi_2S_3, which dissolves as follows:

$$Bi_2S_3(s) \rightleftharpoons 2Bi^{3+}(aq) + 3S^{2-}(aq)$$

Therefore,

$$K_{sp} = [Bi^{3+}]^2[S^{2-}]^3$$

Since no Bi^{3+} and S^{2-} ions were present in solution before the Bi_2S_3 dissolved,

$$[Bi^{3+}]_0 = [S^{2-}]_0 = 0$$

Thus the equilibrium concentrations of these ions will be determined by the amount of salt that dissolves to reach equilibrium, which in this case is 1.0×10^{-15} mol/L. Since each Bi_2S_3 unit contains $2Bi^{3+}$ and $3S^{2-}$ ions:

$$1.0 \times 10^{-15}\ mol/L\ Bi_2S_3(s)$$
$$\longrightarrow 2(1.0 \times 10^{-15}\ mol/L)\ Bi^{3+}(aq) + 3(1.0 \times 10^{-15}\ mol/L)\ S^{2-}(aq)$$

The equilibrium concentrations are

$$[Bi^{3+}] = [Bi^{3+}]_0 + change = 0 + 2.0 \times 10^{-15}\ mol/L$$
$$[S^{2-}] = [S^{2-}]_0 + change = 0 + 3.0 \times 10^{-15}\ mol/L$$

Then

$$K_{sp} = [Bi^{3+}]^2[S^{2-}]^3 = (2.0 \times 10^{-15})^2(3.0 \times 10^{-15})^3 = 1.1 \times 10^{-73}$$

See Exercises 16.23 through 16.26

We have seen that the experimentally determined solubility of an ionic solid can be used to calculate its K_{sp} value.* The reverse is also possible: The solubility of an ionic solid can be calculated if its K_{sp} value is known.

---

*This calculation assumes that all the dissolved solid is present as separated ions. In some cases, such as $CaSO_4$, large numbers of ion pairs exist in solution, so this method yields an incorrect value for K_{sp}.

Chemical connections
The Chemistry of Teeth

If dental chemistry continues to progress at the present rate, tooth decay may soon be a thing of the past. Cavities are holes that develop in tooth enamel, which is composed of the mineral hydroxyapatite, $Ca_5(PO_4)_3OH$. Recent research has shown that there is constant dissolving and re-forming of the tooth mineral in the saliva at the tooth's surface. Demineralization (dissolving of tooth enamel) is mainly caused by weak acids in the saliva created by bacteria as they metabolize carbohydrates in food. (The solubility of $Ca_5(PO_4)_3OH$ in acidic saliva should come as no surprise to you if you understand how pH affects the solubility of a salt with basic anions.)

In the first stages of tooth decay, parts of the tooth surface become porous and spongy and develop Swiss-cheese-like holes that, if untreated, eventually turn into cavities

(see photo). However, recent results indicate that if the affected tooth is bathed in a solution containing appropriate amounts of Ca^{2+}, PO_4^{3-}, and F^-, it remineralizes. Because the F^- replaces OH^- in the tooth mineral ($Ca_5(PO_4)_3OH$ is changed to $Ca_5(PO_4)_3F$), the remineralized area is more resistant to future decay, since fluoride is a weaker base than hydroxide ion. In addition, it has been shown that the presence of Sr^{2+} in the remineralizing fluid significantly increases resistance to decay.

If these results hold up under further study, the work of dentists will change dramatically. Dentists will be much more involved in preventing damage to teeth than in repairing

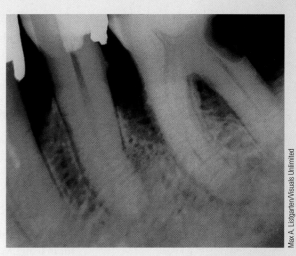

X-ray photo showing decay (dark area) on the molar (right).

damage that has already occurred. One can picture the routine use of a remineralization rinse that will repair problem areas before they become cavities. Dental drills could join leeches as a medical anachronism.

Interactive Example 16.3

Sign in at http://login.cengagebrain .com to try this Interactive Example in **OWL**.

Calculating Solubility from K_{sp}

The K_{sp} value for copper(II) iodate, $Cu(IO_3)_2$, is 1.4×10^{-7} at 25°C. Calculate its solubility at 25°C.

Solution

The system initially contains H_2O and solid $Cu(IO_3)_2$, which dissolves according to the following equilibrium:

$$Cu(IO_3)_2(s) \rightleftharpoons Cu^{2+}(aq) + 2IO_3^-(aq)$$

Therefore,

$$K_{sp} = [Cu^{2+}][IO_3^-]^2$$

To find the solubility of $Cu(IO_3)_2$, we must find the equilibrium concentrations of the Cu^{2+} and IO_3^- ions. We do this in the usual way by specifying the initial concentrations (before any solid has dissolved) and then defining the change required to reach equilibrium. Since in this case we do not know the solubility, we will assume that x mol/L of the solid dissolves to reach equilibrium. The 1:2 stoichiometry of the salt means that

$$x \text{ mol/L } Cu(IO_3)_2(s) \longrightarrow x \text{ mol/L } Cu^{2+}(aq) + 2x \text{ mol/L } IO_3^-(aq)$$

The concentrations are as follows:

Initial Concentration (mol/L) [before any $Cu(IO_3)_2$ dissolves]		Equilibrium Concentration (mol/L)
$[Cu^{2+}]_0 = 0$ $[IO_3^-]_0 = 0$	$\xrightarrow[\text{to reach equilibrium}]{\overset{\textstyle x \text{ mol/L}}{\text{dissolves}}}$	$[Cu^{2+}] = x$ $[IO_3^-] = 2x$

Substituting the equilibrium concentrations into the expression for K_{sp} gives

$$1.4 \times 10^{-7} = K_{sp} = [Cu^{2+}][IO_3^-]^2 = (x)(2x)^2 = 4x^3$$

Then

$$x = \sqrt[3]{3.5 \times 10^{-8}} = 3.3 \times 10^{-3} \text{ mol/L}$$

Thus the solubility of solid $Cu(IO_3)_2$ is 3.3×10^{-3} mol/L.

See Exercises 16.27 and 16.28

Relative Solubilities

A salt's K_{sp} value gives us information about its solubility. However, we must be careful in using K_{sp} values to predict the *relative* solubilities of a group of salts. There are two possible cases:

1. The salts being compared produce the same number of ions. For example, consider

$$
\begin{aligned}
\text{AgI}(s) & \quad K_{sp} = 1.5 \times 10^{-16} \\
\text{CuI}(s) & \quad K_{sp} = 5.0 \times 10^{-12} \\
\text{CaSO}_4(s) & \quad K_{sp} = 6.1 \times 10^{-5}
\end{aligned}
$$

Each of these solids dissolves to produce two ions:

$$\text{Salt} \rightleftharpoons \text{cation} + \text{anion}$$
$$K_{sp} = [\text{cation}][\text{anion}]$$

If x is the solubility in mol/L, then at equilibrium

$$
\begin{aligned}
[\text{Cation}] &= x \\
[\text{Anion}] &= x \\
K_{sp} &= [\text{cation}][\text{anion}] = x^2 \\
x &= \sqrt{K_{sp}} = \text{solubility}
\end{aligned}
$$

Therefore, in this case we can compare the solubilities for these solids by comparing the K_{sp} values:

$$
\underset{\substack{\text{Most soluble;}\\\text{largest } K_{sp}}}{\text{CaSO}_4(s)} > \text{CuI}(s) > \underset{\substack{\text{Least soluble;}\\\text{smallest } K_{sp}}}{\text{AgI}(s)}
$$

2. The salts being compared produce different numbers of ions. For example, consider

$$
\begin{aligned}
\text{CuS}(s) & \quad K_{sp} = 8.5 \times 10^{-45} \\
\text{Ag}_2\text{S}(s) & \quad K_{sp} = 1.6 \times 10^{-49} \\
\text{Bi}_2\text{S}_3(s) & \quad K_{sp} = 1.1 \times 10^{-73}
\end{aligned}
$$

Table 16.2 | Calculated Solubilities for CuS, Ag$_2$S, and Bi$_2$S$_3$ at 25°C

Salt	K_{sp}	Calculated Solubility (mol/L)
CuS	8.5×10^{-45}	9.2×10^{-23}
Ag$_2$S	1.6×10^{-49}	3.4×10^{-17}
Bi$_2$S$_3$	1.1×10^{-73}	1.0×10^{-15}

Because these salts produce different numbers of ions when they dissolve, the K_{sp} values cannot be compared *directly* to determine relative solubilities. In fact, if we calculate the solubilities (using the procedure in Example 16.3), we obtain the results summarized in Table 16.2. The order of solubilities is

$$\underset{\text{Most soluble}}{\text{Bi}_2\text{S}_3(s)} > \text{Ag}_2\text{S}(s) > \underset{\text{Least soluble}}{\text{CuS}(s)}$$

which is opposite to the order of the K_{sp} values.

Remember that relative solubilities can be predicted by comparing K_{sp} values *only* for salts that produce the same total number of ions.

Common Ion Effect

So far we have considered ionic solids dissolved in pure water. We will now see what happens when the water contains an ion in common with the dissolving salt. For example, consider the solubility of solid silver chromate (Ag$_2$CrO$_4$, $K_{sp} = 9.0 \times 10^{-12}$) in a 0.100-$M$ solution of AgNO$_3$. Before any Ag$_2$CrO$_4$ dissolves, the solution contains the major species Ag$^+$, NO$_3^-$, and H$_2$O, with solid Ag$_2$CrO$_4$ on the bottom of the container. Since NO$_3^-$ is not found in Ag$_2$CrO$_4$, we can ignore it. The relevant initial concentrations (before any Ag$_2$CrO$_4$ dissolves) are

$$[\text{Ag}^+]_0 = 0.100 \ M \ (\text{from the dissolved AgNO}_3)$$
$$[\text{CrO}_4{}^{2-}]_0 = 0$$

The system comes to equilibrium as the solid Ag$_2$CrO$_4$ dissolves according to the reaction

$$\text{Ag}_2\text{CrO}_4(s) \rightleftharpoons 2\text{Ag}^+(aq) + \text{CrO}_4{}^{2-}(aq)$$

for which

$$K_{sp} = [\text{Ag}^+]^2[\text{CrO}_4{}^{2-}] = 9.0 \times 10^{-12}$$

We assume that x mol/L of Ag$_2$CrO$_4$ dissolves to reach equilibrium, which means that

$$x \text{ mol/L Ag}_2\text{CrO}_4(s) \longrightarrow 2x \text{ mol/L Ag}^+(aq) + x \text{ mol/L CrO}_4{}^{2-}$$

Now we can specify the equilibrium concentrations in terms of x:

$$[\text{Ag}^+] = [\text{Ag}^+]_0 + \text{change} = 0.100 + 2x$$
$$[\text{CrO}_4{}^{2-}] = [\text{CrO}_4{}^{2-}]_0 + \text{change} = 0 + x = x$$

Substituting these concentrations into the expression for K_{sp} gives

$$9.0 \times 10^{-12} = [\text{Ag}^+]^2[\text{CrO}_4{}^{2-}] = (0.100 + 2x)^2(x)$$

The mathematics required here appear to be complicated, since the multiplication of terms on the right-hand side produces an expression that contains an x^3 term. However, as is usually the case, we can make simplifying assumptions. Since the K_{sp} value

A potassium chromate solution being added to aqueous silver nitrate, forming silver chromate.

Photo by Ken O'Donoghue © Cengage Learning

for Ag_2CrO_4 is small (the position of the equilibrium lies far to the left), x is expected to be small compared with $0.100\ M$. Therefore, $0.100 + 2x \approx 0.100$, which allows simplification of the expression:

$$9.0 \times 10^{-12} = (0.100 + 2x)^2(x) \approx (0.100)^2(x)$$

Then
$$x \approx \frac{9.0 \times 10^{-12}}{(0.100)^2} = 9.0 \times 10^{-10}\ \text{mol/L}$$

Since x is much less than $0.100\ M$, the approximation is valid (by the 5% rule). Thus

$$\text{Solubility of } Ag_2CrO_4 \text{ in } 0.100\ M\ AgNO_3 = x = 9.0 \times 10^{-10}\ \text{mol/L}$$

and the equilibrium concentrations are

$$[Ag^+] = 0.100 + 2x = 0.100 + 2(9.0 \times 10^{-10}) = 0.100\ M$$
$$[CrO_4^{2-}] = x = 9.0 \times 10^{-10}\ M$$

Now we compare the solubilities of Ag_2CrO_4 in pure water and in $0.100\ M\ AgNO_3$:

$$\text{Solubility of } Ag_2CrO_4 \text{ in pure water} = 1.3 \times 10^{-4}\ \text{mol/L}$$
$$\text{Solubility of } Ag_2CrO_4 \text{ in } 0.100\ M\ AgNO_3 = 9.0 \times 10^{-10}\ \text{mol/L}$$

Note that the solubility of Ag_2CrO_4 is much less in the presence of Ag^+ ions from $AgNO_3$. This is another example of the common ion effect. The solubility of a solid is lowered if the solution already contains ions common to the solid.

Critical Thinking

What if all you know about two salts is that the value of K_{sp} for salt A is greater than that of salt B? Why can we not compare relative solubilities of the salts? Use numbers to show how salt A could be more soluble than salt B, and how salt B could be more soluble than salt A.

Interactive Example 16.4

Sign in at http://login.cengagebrain .com to try this Interactive Example in **OWL**.

Solubility and Common Ions

Calculate the solubility of solid CaF_2 ($K_{sp} = 4.0 \times 10^{-11}$) in a 0.025-M NaF solution.

Solution

Before any CaF_2 dissolves, the solution contains the major species Na^+, F^-, and H_2O. The solubility equilibrium for CaF_2 is

$$CaF_2(s) \rightleftharpoons Ca^{2+}(aq) + 2F^-(aq)$$

and
$$K_{sp} = 4.0 \times 10^{-11} = [Ca^{2+}][F^-]^2$$

Initial Concentration (mol/L) (before any CaF₂ dissolves)		Equilibrium Concentration (mol/L)
$[Ca^{2+}]_0 = 0$ $[F^-]_0 = 0.025\ M$ ↗ From $0.025\ M$ NaF	$\xrightarrow[\text{to reach equilibrium}]{x \text{ mol/L } CaF_2 \text{ dissolves}}$	$[Ca^{2+}] = x$ $[F^-] = 0.025 + 2x$ ↗ ↗ From NaF From CaF₂

Substituting the equilibrium concentrations into the expression for K_{sp} gives

$$K_{sp} = 4.0 \times 10^{-11} = [Ca^{2+}][F^-]^2 = (x)(0.025 + 2x)^2$$

Assuming that $2x$ is negligible compared with 0.025 (since K_{sp} is small) gives

$$4.0 \times 10^{-11} \approx (x)(0.025)^2$$
$$x \approx 6.4 \times 10^{-8}$$

The approximation is valid (by the 5% rule), and

$$\text{Solubility} = x = 6.4 \times 10^{-8} \text{ mol/L}$$

Thus 6.4×10^{-8} mole of solid CaF_2 dissolves per liter of the 0.025-M NaF solution.

See Exercises 16.37 through 16.42

pH and Solubility

The pH of a solution can greatly affect a salt's solubility. For example, magnesium hydroxide dissolves according to the equilibrium

$$Mg(OH)_2(s) \rightleftharpoons Mg^{2+}(aq) + 2OH^-(aq)$$

Addition of OH^- ions (an increase in pH) will, by the common ion effect, force the equilibrium to the left, decreasing the solubility of $Mg(OH)_2$. On the other hand, an addition of H^+ ions (a decrease in pH) increases the solubility, because OH^- ions are removed from solution by reacting with the added H^+ ions. In response to the lower concentration of OH^-, the equilibrium position moves to the right. This is why a suspension of solid $Mg(OH)_2$, known as *milk of magnesia,* dissolves as required in the stomach to combat excess acidity.

This idea also applies to salts with other types of anions. For example, the solubility of silver phosphate (Ag_3PO_4) is greater in acid than in pure water because the PO_4^{3-} ion is a strong base that reacts with H^+ to form the HPO_4^{2-} ion. The reaction

$$H^+(aq) + PO_4^{3-}(aq) \longrightarrow HPO_4^{2-}(aq)$$

occurs in acidic solution, thus lowering the concentration of PO_4^{3-} and shifting the solubility equilibrium

$$Ag_3PO_4(s) \rightleftharpoons 3Ag^+(aq) + PO_4^{3-}(aq)$$

to the right. This, in turn, increases the solubility of silver phosphate.

Silver chloride (AgCl), however, has the same solubility in acid as in pure water. Why? Since the Cl^- ion is a very weak base (that is, HCl is a very strong acid), no HCl molecules are formed. Thus the addition of H^+ to a solution containing Cl^- does not affect $[Cl^-]$ and has no effect on the solubility of a chloride salt.

The general rule is that if the anion X^- is an effective base—that is, if HX is a weak acid—the salt MX will show increased solubility in an acidic solution. Examples of common anions that are effective bases are OH^-, S^{2-}, CO_3^{2-}, $C_2O_4^{2-}$, and CrO_4^{2-}. Salts containing these anions are much more soluble in an acidic solution than in pure water.

As mentioned at the beginning of this chapter, one practical result of the increased solubility of carbonates in acid is the formation of huge limestone caves such as Mammoth Cave in Kentucky and Carlsbad Caverns in New Mexico. Carbon dioxide dissolved in groundwater makes it acidic, increasing the solubility of calcium carbonate and eventually producing huge caverns. As the carbon dioxide escapes to the air, the pH of the dripping water goes up and the calcium carbonate precipitates, forming stalactites and stalagmites.

Stalactites and stalagmites in Luray Cavern in Virginia.

Critical Thinking

You and a friend are studying for a chemistry exam. What if your friend tells you that since acids are very reactive, all salts are more soluble in aqueous solutions of acids than in water? How would you explain to your friend that this is not true? Use a specific example to defend your answer.

16.2 | Precipitation and Qualitative Analysis

So far we have considered solids dissolving in solutions. Now we will consider the reverse process—the formation of a solid from solution. When solutions are mixed, various reactions can occur. We have already considered acid–base reactions in some detail. In this section we show how to predict whether a precipitate will form when two solutions are mixed. We will use the **ion product**, which is defined just like the expression for K_{sp} for a given solid except that *initial concentrations are used* instead of equilibrium concentrations. For solid CaF_2, the expression for the ion product Q is written

$$Q = [Ca^{2+}]_0[F^-]_0^2$$

If we add a solution containing Ca^{2+} ions to a solution containing F^- ions, a precipitate may or may not form, depending on the concentrations of these ions in the resulting mixed solution. To predict whether precipitation will occur, we consider the relationship between Q and K_{sp}.

Q is used here in a very similar way to the use of the reaction quotient in Chapter 13.

If Q is greater than K_{sp}, precipitation occurs and will continue until the concentrations are reduced to the point that they satisfy K_{sp}.

If Q is less than K_{sp}, no precipitation occurs.

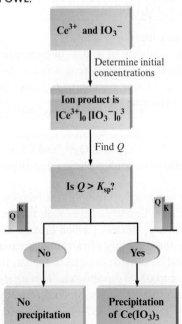

Interactive Example 16.5

Sign in at http://login.cengagebrain.com to try this Interactive Example in OWL.

Ce³⁺ and IO₃⁻

↓ Determine initial concentrations

Ion product is [Ce³⁺]₀ [IO₃⁻]₀³

↓ Find Q

Is Q > Ksp?

Q K   Q K

No   Yes

No precipitation

Precipitation of Ce(IO₃)₃

Determining Precipitation Conditions

A solution is prepared by adding 750.0 mL of 4.00×10^{-3} M $Ce(NO_3)_3$ to 300.0 mL of 2.00×10^{-2} M KIO_3. Will $Ce(IO_3)_3$ ($K_{sp} = 1.9 \times 10^{-10}$) precipitate from this solution?

Solution

First, we calculate $[Ce^{3+}]_0$ and $[IO_3^-]_0$ in the mixed solution before any reaction occurs:

$$[Ce^{3+}]_0 = \frac{(750.0 \text{ mL})(4.00 \times 10^{-3} \text{ mmol/mL})}{(750.0 + 300.0) \text{ mL}} = 2.86 \times 10^{-3} \text{ M}$$

$$[IO_3^-]_0 = \frac{(300.0 \text{ mL})(2.00 \times 10^{-2} \text{ mmol/mL})}{(750.0 + 300.0) \text{ mL}} = 5.71 \times 10^{-3} \text{ M}$$

The ion product for $Ce(IO_3)_3$ is

$$Q = [Ce^{3+}]_0[IO_3^-]_0^3 = (2.86 \times 10^{-3})(5.71 \times 10^{-3})^3 = 5.32 \times 10^{-10}$$

Since Q is greater than K_{sp}, $Ce(IO_3)_3$ will precipitate from the mixed solution.

See Exercises 16.49 through 16.52

Sometimes we want to do more than simply predict whether precipitation will occur; we may want to calculate the equilibrium concentrations in the solution after precipitation occurs. For example, let us calculate the equilibrium concentrations of Pb^{2+} and I^- ions in a solution formed by mixing 100.0 mL of 0.0500 M $Pb(NO_3)_2$ and

200.0 mL of 0.100 M NaI. First, we must determine whether solid PbI_2 ($K_{sp} = 1.4 \times 10^{-8}$) forms when the solutions are mixed. To do so, we need to calculate $[Pb^{2+}]_0$ and $[I^-]_0$ before any reaction occurs:

$$[Pb^{2+}]_0 = \frac{\text{mmol } Pb^{2+}}{\text{mL solution}} = \frac{(100.0 \text{ mL})(0.0500 \text{ mmol/mL})}{300.0 \text{ mL}} = 1.67 \times 10^{-2} \, M$$

$$[I^-]_0 = \frac{\text{mmol } I^-}{\text{mL solution}} = \frac{(200.0 \text{ mL})(0.100 \text{ mmol/mL})}{300.0 \text{ mL}} = 6.67 \times 10^{-2} \, M$$

The ion product for PbI_2 is

$$Q = [Pb^{2+}]_0[I^-]_0^2 = (1.67 \times 10^{-2})(6.67 \times 10^{-2})^2 = 7.43 \times 10^{-5}$$

Since Q is greater than K_{sp}, a precipitate of PbI_2 will form.

Since the K_{sp} for PbI_2 is quite small (1.4×10^{-8}), only very small quantities of Pb^{2+} and I^- can coexist in aqueous solution. In other words, when Pb^{2+} and I^- are mixed, most of these ions will precipitate out as PbI_2. That is, the reaction

$$Pb^{2+}(aq) + 2I^-(aq) \longrightarrow PbI_2(s)$$

The equilibrium constant for formation of solid PbI_2 is $1/K_{sp}$, or 7×10^7, so this equilibrium lies far to the right.

(which is the reverse of the dissolution reaction) goes essentially to completion.

If, when two solutions are mixed, a reaction occurs that goes virtually to completion, it is essential to do the stoichiometry calculations before considering the equilibrium calculations. Therefore, in this case we let the system go completely in the direction toward which it tends. Then we will let it adjust back to equilibrium. If we let Pb^{2+} and I^- react to completion, we have the following concentrations:

	Pb^{2+}	$+$	$2I^-$	$\longrightarrow$	PbI_2
Before reaction	(100.0 mL)(0.0500 M) = 5.00 mmol		(200.0 mL)(0.100 M) = 20.0 mmol		The amount of PbI_2 formed does
After reaction	0 mmol		20.0 − 2(5.00) = 10.0 mmol		not influence the equilibrium.

In this reaction, 10 mmol I^- is in excess.

Next we must allow the system to adjust to equilibrium. At equilibrium $[Pb^{2+}]$ is not actually zero because the reaction does not go quite to completion. The best way to think about this is that once the PbI_2 is formed, a very small amount redissolves to reach equilibrium. Since I^- is in excess, the PbI_2 is dissolving into a solution that contains 10.0 mmol I^- per 300.0 mL solution, or $3.33 \times 10^{-2} \, M \, I^-$.

We could state this problem as follows: What is the solubility of solid PbI_2 in a $3.33 \times 10^{-2} \, M$-NaI solution? The lead iodide dissolves according to the equation

$$PbI_2(s) \rightleftharpoons Pb^{2+}(aq) + 2I^-(aq)$$

The concentrations are as follows:

Initial Concentration (mol/L)		Equilibrium Concentration (mol/L)
$[Pb^{2+}]_0 = 0$ $[I^-]_0 = 3.33 \times 10^{-2}$	$\xrightarrow[\text{dissolves}]{\substack{x \text{ mol/L} \\ PbI_2(s)}}$	$[Pb^{2+}] = x$ $[I^-] = 3.33 \times 10^{-2} + 2x$

Substituting into the expression for K_{sp} gives

$$K_{sp} = 1.4 \times 10^{-8} = [Pb^{2+}][I^-]^2 = (x)(3.33 \times 10^{-2} + 2x)^2 \approx (x)(3.33 \times 10^{-2})^2$$

Then
$$[Pb^{2+}] = x = 1.3 \times 10^{-5} \, M$$
$$[I^-] = 3.33 \times 10^{-2} \, M$$

Note that $3.33 \times 10^{-2} \gg 2x$, so the approximation is valid. These Pb^{2+} and I^- concentrations thus represent the equilibrium concentrations present in a solution formed by mixing 100.0 mL of 0.0500 M $Pb(NO_3)_2$ and 200.0 mL of 0.100 M NaI.

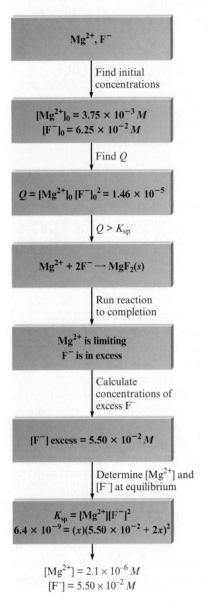

Precipitation

A solution is prepared by mixing 150.0 mL of $1.00 \times 10^{-2}\ M\ \mathrm{Mg(NO_3)_2}$ and 250.0 mL of $1.00 \times 10^{-1}\ M\ \mathrm{NaF}$. Calculate the concentrations of $\mathrm{Mg^{2+}}$ and $\mathrm{F^-}$ at equilibrium with solid $\mathrm{MgF_2}$ ($K_{sp} = 6.4 \times 10^{-9}$).

Solution

The first step is to determine whether solid $\mathrm{MgF_2}$ forms. To do this, we need to calculate the concentrations of $\mathrm{Mg^{2+}}$ and $\mathrm{F^-}$ in the mixed solution and find Q:

$$[\mathrm{Mg^{2+}}]_0 = \frac{\text{mmol } \mathrm{Mg^{2+}}}{\text{mL solution}} = \frac{(150.0\ \text{mL})(1.00 \times 10^{-2}\ M)}{400.0\ \text{mL}} = 3.75 \times 10^{-3}\ M$$

$$[\mathrm{F^-}]_0 = \frac{\text{mmol } \mathrm{F^-}}{\text{mL solution}} = \frac{(250.0\ \text{mL})(1.00 \times 10^{-1}\ M)}{400.0\ \text{mL}} = 6.25 \times 10^{-2}\ M$$

$$Q = [\mathrm{Mg^{2+}}]_0[\mathrm{F^-}]_0^2 = (3.75 \times 10^{-3})(6.25 \times 10^{-2})^2 = 1.46 \times 10^{-5}$$

Since Q is greater than K_{sp}, solid $\mathrm{MgF_2}$ will form.

The next step is to run the precipitation reaction to completion:

	$\mathrm{Mg^{2+}}$	+	$2\mathrm{F^-}$	$\longrightarrow$	$\mathrm{MgF_2}(s)$
Before reaction	$(150.0)(1.00 \times 10^{-2})$ $= 1.50$ mmol		$(250.0)(1.00 \times 10^{-1})$ $= 25.0$ mmol		
After reaction	$1.50 - 1.50 = 0$		$25.0 - 2(1.50)$ $= 22.0$ mmol		

Note that excess $\mathrm{F^-}$ remains after the precipitation reaction goes to completion. The concentration is

$$[\mathrm{F^-}]_{\text{excess}} = \frac{22.0\ \text{mmol}}{400.0\ \text{mL}} = 5.50 \times 10^{-2}\ M$$

Although we have assumed that the $\mathrm{Mg^{2+}}$ is completely consumed, we know that $[\mathrm{Mg^{2+}}]$ will not be zero at equilibrium. We can compute the equilibrium $[\mathrm{Mg^{2+}}]$ by letting $\mathrm{MgF_2}$ redissolve to satisfy the expression for K_{sp}. How much $\mathrm{MgF_2}$ will dissolve in a 5.50×10^{-2}-M NaF solution? We proceed as usual:

$$\mathrm{MgF_2}(s) \rightleftharpoons \mathrm{Mg^{2+}}(aq) + 2\mathrm{F^-}(aq)$$
$$K_{sp} = [\mathrm{Mg^{2+}}][\mathrm{F^-}]^2 = 6.4 \times 10^{-9}$$

Initial Concentration (mol/L)		Equilibrium Concentration (mol/L)
$[\mathrm{Mg^{2+}}]_0 = 0$ $[\mathrm{F^-}]_0 = 5.50 \times 10^{-2}$	$\xrightarrow[\text{dissolves}]{\begin{array}{c} x\ \text{mol/L} \\ \mathrm{MgF_2}(s) \end{array}}$	$[\mathrm{Mg^{2+}}] = x$ $[\mathrm{F^-}] = 5.50 \times 10^{-2} + 2x$

$$K_{sp} = 6.4 \times 10^{-9} = [\mathrm{Mg^{2+}}][\mathrm{F^-}]^2$$
$$= (x)(5.50 \times 10^{-2} + 2x)^2 \approx (x)(5.50 \times 10^{-2})^2$$
$$[\mathrm{Mg^{2+}}] = x = 2.1 \times 10^{-6}\ M$$
$$[\mathrm{F^-}] = 5.50 \times 10^{-2}\ M$$

See Exercises 16.53 through 16.56

Selective Precipitation

The approximations made here fall within the 5% rule.

Mixtures of metal ions in aqueous solution are often separated by **selective precipitation**, that is, by using a reagent whose anion forms a precipitate with only one or a few of the metal ions in the mixture. For example, suppose we have a solution containing both Ba^{2+} and Ag^+ ions. If NaCl is added to the solution, AgCl precipitates as a white solid, but since $BaCl_2$ is soluble, the Ba^{2+} ions remain in solution.

| Example 16.7 | Selective Precipitation |

A solution contains 1.0×10^{-4} M Cu^+ and 2.0×10^{-3} M Pb^{2+}. If a source of I^- is added gradually to this solution, will PbI_2 ($K_{sp} = 1.4 \times 10^{-8}$) or CuI ($K_{sp} = 5.3 \times 10^{-12}$) precipitate first? Specify the concentration of I^- necessary to begin precipitation of each salt.

Solution

For PbI_2, the K_{sp} expression is

$$1.4 \times 10^{-8} = K_{sp} = [Pb^{2+}][I^-]^2$$

Since $[Pb^{2+}]$ in this solution is known to be 2.0×10^{-3} M, the greatest concentration of I^- that can be present without causing precipitation of PbI_2 can be calculated from the K_{sp} expression:

$$1.4 \times 10^{-8} = [Pb^{2+}][I^-]^2 = (2.0 \times 10^{-3})[I^-]^2$$
$$[I^-] = 2.6 \times 10^{-3} M$$

Any I^- in excess of this concentration will cause solid PbI_2 to form.
  Similarly, for CuI, the K_{sp} expression is

$$5.3 \times 10^{-12} = K_{sp} = [Cu^+][I^-] = (1.0 \times 10^{-4})[I^-]$$

and
$$[I^-] = 5.3 \times 10^{-8} M$$

A concentration of I^- in excess of 5.3×10^{-8} M will cause formation of solid CuI.
  As I^- is added to the mixed solution, CuI will precipitate first, since the $[I^-]$ required is less. Therefore, Cu^+ would be separated from Pb^{2+} using this reagent.

See Exercises 16.57 through 16.60

We can compare K_{sp} values to find relative solubilities because FeS and MnS produce the same number of ions in solution.

Since metal sulfide salts differ dramatically in their solubilities, the sulfide ion is often used to separate metal ions by selective precipitation. For example, consider a solution containing a mixture of 10^{-3} M Fe^{2+} and 10^{-3} M Mn^{2+}. Since FeS ($K_{sp} = 3.7 \times 10^{-19}$) is much less soluble than MnS ($K_{sp} = 2.3 \times 10^{-13}$), careful addition of S^{2-} to the mixture will precipitate Fe^{2+} as FeS, leaving Mn^{2+} in solution.
  One real advantage of the sulfide ion as a precipitating reagent is that because it is basic, its concentration can be controlled by regulating the pH of the solution. H_2S is a diprotic acid that dissociates in two steps:

$$H_2S \rightleftharpoons H^+ + HS^- \qquad K_{a_1} = 1.0 \times 10^{-7}$$
$$HS^- \rightleftharpoons H^+ + S^{2-} \qquad K_{a_2} \approx 10^{-19}$$

Note from the small K_{a_2} value that S^{2-} ions have a high affinity for protons. In an acidic solution (large $[H^+]$), $[S^{2-}]$ will be relatively small, since under these conditions the dissociation equilibria will lie far to the left. On the other hand, in basic solutions $[S^{2-}]$ will be relatively large, since the very small value of $[H^+]$ will pull both equilibria to the right, producing S^{2-}.

Figure 16.1 | The separation of Cu^{2+} and Hg^{2+} from Ni^{2+} and Mn^{2+} using H_2S. At a low pH, $[S^{2-}]$ is relatively low and only the very insoluble HgS and CuS precipitate. When OH^- is added to lower $[H^+]$, the value of $[S^{2-}]$ increases, and MnS and NiS precipitate.

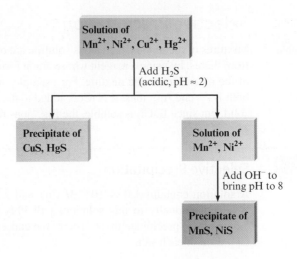

Flame test for potassium.

Flame test for sodium.

This means that the most insoluble sulfide salts, such as CuS ($K_{sp} = 8.5 \times 10^{-45}$) and HgS ($K_{sp} = 1.6 \times 10^{-54}$), can be precipitated from an acidic solution, leaving the more soluble ones, such as MnS ($K_{sp} = 2.3 \times 10^{-13}$) and NiS ($K_{sp} = 3 \times 10^{-21}$), still dissolved. The manganese and nickel sulfides can then be precipitated by making the solution slightly basic. This procedure is diagramed in Fig. 16.1.

Qualitative Analysis

The classic scheme for **qualitative analysis** of a mixture containing all the common cations (listed in Fig. 16.2) involves first separating them into five major groups based on solubilities. (These groups are not directly related to the groups of the periodic table.) Each group is then treated further to separate and identify the individual ions. We will be concerned here only with separation of the major groups.

Group I—Insoluble chlorides
When dilute aqueous HCl is added to a solution containing a mixture of the common cations, only Ag^+, Pb^{2+}, and Hg_2^{2+} will precipitate out as insoluble chlorides. All other chlorides are soluble and remain in solution. The Group I precipitate is removed, leaving the other ions in solution for treatment with sulfide ion.

Group II—Sulfides insoluble in acid solution
After the insoluble chlorides are removed, the solution is still acidic, since HCl was added. If H_2S is added to this solution, only the most insoluble sulfides (those of Hg^{2+}, Cd^{2+}, Bi^{3+}, Cu^{2+}, and Sn^{4+}) will precipitate, since $[S^{2-}]$ is relatively low because of the high concentration of H^+. The more soluble sulfides will remain dissolved under these conditions, and the precipitate of the insoluble salt is removed.

Group III—Sulfides insoluble in basic solution
The solution is made basic at this stage, and more H_2S is added. As we saw earlier, a basic solution produces a higher $[S^{2-}]$, which leads to precipitation of the more soluble sulfides. The cations precipitated as sulfides at this stage are Co^{2+}, Zn^{2+}, Mn^{2+}, Ni^{2+}, and Fe^{2+}. If any Cr^{3+} and Al^{3+} ions are present, they also will precipitate, but as insoluble hydroxides (remember the solution is now basic). The precipitate is separated from the solution containing the rest of the ions.

Group IV—Insoluble carbonates
At this point, all the cations have been precipitated except those from Groups 1A and 2A of the periodic table. The Group 2A cations form insoluble carbonates and can be precipitated by the addition of CO_3^{2-}. For example, Ba^{2+}, Ca^{2+}, and Mg^{2+} form solid carbonates and can be removed from the solution.

Figure 16.2 | A schematic diagram of the classic method for separating the common cations by selective precipitation.

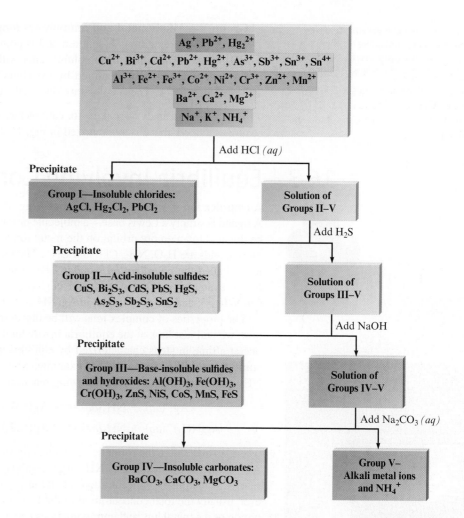

From left to right, cadmium sulfide, chromium(III) hydroxide, aluminum hydroxide, and nickel(II) hydroxide.

Group V—Alkali metal and ammonium ions

The only ions remaining in solution at this point are the Group 1A cations and the NH_4^+ ion, all of which form soluble salts with the common anions. The Group 1A cations are usually identified by the characteristic colors they produce when heated in a flame. These colors are due to the emission spectra of these ions.

The qualitative analysis scheme for cations based on the selective precipitation procedure described above is summarized in Fig. 16.2.

16.3 | Equilibria Involving Complex Ions

$CoCl_4^{2-}$

A **complex ion** is a charged species consisting of a metal ion surrounded by *ligands*. A ligand is simply a Lewis base—a molecule or ion having a lone electron pair that can be donated to an empty orbital on the metal ion to form a covalent bond. Some common ligands are H_2O, NH_3, Cl^-, and CN^-. The number of ligands attached to a metal ion is called the *coordination number*. The most common coordination numbers are 6, for example, in $Co(H_2O)_6^{2+}$ and $Ni(NH_3)_6^{2+}$; 4, for example, in $CoCl_4^{2-}$ and $Cu(NH_3)_4^{2+}$; and 2, for example, in $Ag(NH_3)_2^+$; but others are known.

The properties of complex ions will be discussed in more detail in Chapter 21. For now, we will just look at the equilibria involving these species. Metal ions add ligands one at a time in steps characterized by equilibrium constants called **formation constants** or **stability constants**. For example, when solutions containing Ag^+ ions and NH_3 molecules are mixed, the following reactions take place:

$$Ag^+(aq) + NH_3(aq) \rightleftharpoons Ag(NH_3)^+(aq) \qquad K_1 = 2.1 \times 10^3$$
$$Ag(NH_3)^+(aq) + NH_3(aq) \rightleftharpoons Ag(NH_3)_2^+(aq) \qquad K_2 = 8.2 \times 10^3$$

where K_1 and K_2 are the formation constants for the two steps. In a solution containing Ag^+ and NH_3, all the species NH_3, Ag^+, $Ag(NH_3)^+$, and $Ag(NH_3)_2^+$ exist at equilibrium. Calculating the concentrations of all these components can be complicated. However, usually the total concentration of the ligand is much larger than the total concentration of the metal ion, and approximations can greatly simplify the problems.

For example, consider a solution prepared by mixing 100.0 mL of 2.0 M NH_3 with 100.0 mL of 1.0×10^{-3} M $AgNO_3$. *Before any reaction occurs,* the mixed solution contains the major species Ag^+, NO_3^-, NH_3, and H_2O. What reaction or reactions will occur in this solution? From our discussions of acid–base chemistry, we know that one reaction is

$$NH_3(aq) + H_2O(l) \rightleftharpoons NH_4^+(aq) + OH^-(aq)$$

However, we are interested in the reaction between NH_3 and Ag^+ to form complex ions, and since the position of the preceding equilibrium lies far to the left (K_b for NH_3 is 1.8×10^{-5}), we can neglect the amount of NH_3 used up in the reaction with water. Therefore, before any complex ion formation, the concentrations in the mixed solution are

$$[Ag^+]_0 = \frac{(100.0 \text{ mL})(1.0 \times 10^{-3} \text{ M})}{(200.0 \text{ mL})} = 5.0 \times 10^{-4} \text{ M}$$

Total volume

$$[NH_3]_0 = \frac{(100.0 \text{ mL})(2.0 \text{ M})}{(200.0 \text{ mL})} = 1.0 \text{ M}$$

As mentioned already, the Ag^+ ion reacts with NH_3 in a stepwise fashion to form $AgNH_3^+$ and then $Ag(NH_3)_2^+$:

$$Ag^+(aq) + NH_3(aq) \rightleftharpoons Ag(NH_3)^+(aq) \qquad K_1 = 2.1 \times 10^3$$
$$Ag(NH_3)^+(aq) + NH_3(aq) \rightleftharpoons Ag(NH_3)_2^+(aq) \qquad K_2 = 8.2 \times 10^3$$

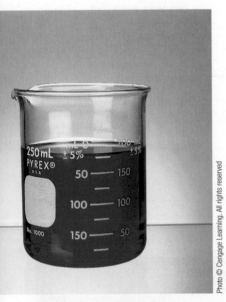

A solution containing the blue $CoCl_4^{2-}$ complex ion.

Since both K_1 and K_2 are large, and since there is a large excess of NH_3, *both reactions can be assumed to go essentially to completion.* This is equivalent to writing the net reaction in the solution as follows:

$$Ag^+ + 2NH_3 \longrightarrow Ag(NH_3)_2^+$$

The relevant stoichiometric calculations are as follows:

	Ag^+	+	$2NH_3$	$\longrightarrow$	$Ag(NH_3)_2^+$
Before reaction	$5.0 \times 10^{-4}\ M$		$1.0\ M$		0
After reaction	0		$1.0 - 2(5.0 \times 10^{-4}) \approx 1.0\ M$		$5.0 \times 10^{-4}\ M$

Twice as much NH_3 as
Ag^+ is required

Note that in this case we have used molarities when performing the stoichiometry calculations and we have assumed this reaction to be complete, using all the original Ag^+ to form $Ag(NH_3)_2^+$. In reality, a *very small amount* of the $Ag(NH_3)_2^+$ formed will dissociate to produce small amounts of $Ag(NH_3)^+$ and Ag^+. However, since the amount of $Ag(NH_3)_2^+$ dissociating will be so small, we can safely assume that $[Ag(NH_3)_2^+]$ is $5.0 \times 10^{-4}\ M$ at equilibrium. Also, we know that since so little NH_3 has been consumed, $[NH_3]$ is $1.0\ M$ at equilibrium. We can use these concentrations to calculate $[Ag^+]$ and $[Ag(NH_3)^+]$ using the K_1 and K_2 expressions.

To calculate the equilibrium concentration of $Ag(NH_3)^+$, we use

$$K_2 = 8.2 \times 10^3 = \frac{[Ag(NH_3)_2^+]}{[Ag(NH_3)^+][NH_3]}$$

since $[Ag(NH_3)_2^+]$ and $[NH_3]$ are known. Rearranging and solving for $[Ag(NH_3)^+]$ gives

$$[Ag(NH_3)^+] = \frac{[Ag(NH_3)_2^+]}{K_2[NH_3]} = \frac{5.0 \times 10^{-4}}{(8.2 \times 10^3)(1.0)} = 6.1 \times 10^{-8}\ M$$

Now the equilibrium concentration of Ag^+ can be calculated using K_1:

$$K_1 = 2.1 \times 10^3 = \frac{[Ag(NH_3)^+]}{[Ag^+][NH_3]} = \frac{6.1 \times 10^{-8}}{[Ag^+](1.0)}$$

$$[Ag^+] = \frac{6.1 \times 10^{-8}}{(2.1 \times 10^3)(1.0)} = 2.9 \times 10^{-11}\ M$$

So far we have assumed that $Ag(NH_3)_2^+$ is the dominant silver-containing species in solution. Is this a valid assumption? The calculated concentrations are

$$[Ag(NH_3)_2^+] = 5.0 \times 10^{-4}\ M$$
$$[Ag(NH_3)^+] = 6.1 \times 10^{-8}\ M$$
$$[Ag^+] = 2.9 \times 10^{-11}\ M$$

These values clearly support the conclusion that

Essentially all the Ag^+ ions originally present end up in $Ag(NH_3)_2^+$.

$$[Ag(NH_3)_2^+] \gg [Ag(NH_3)^+] \gg [Ag^+]$$

Thus the assumption that $Ag(NH_3)_2^+$ is the dominant Ag^+-containing species is valid, and the calculated concentrations are correct.

This analysis shows that although complex ion equilibria have many species present and look complicated, the calculations are actually quite straightforward, especially if the ligand is present in large excess.

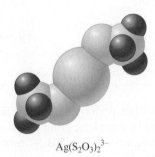

$Ag(S_2O_3)_2^{3-}$

Complex Ions

Calculate the concentrations of Ag^+, $Ag(S_2O_3)^-$, and $Ag(S_2O_3)_2^{3-}$ in a solution prepared by mixing 150.0 mL of $1.00 \times 10^{-3}\ M$ $AgNO_3$ with 200.0 mL of 5.00 M $Na_2S_2O_3$. The stepwise formation equilibria are

$$Ag^+ + S_2O_3^{2-} \rightleftharpoons Ag(S_2O_3)^- \qquad K_1 = 7.4 \times 10^8$$
$$Ag(S_2O_3)^- + S_2O_3^{2-} \rightleftharpoons Ag(S_2O_3)_2^{3-} \qquad K_2 = 3.9 \times 10^4$$

Solution

The concentrations of the ligand and metal ion in the mixed solution *before any reaction occurs* are

$$[Ag^+]_0 = \frac{(150.0\ \text{mL})(1.00 \times 10^{-3}\ M)}{(150.0\ \text{mL} + 200.0\ \text{mL})} = 4.29 \times 10^{-4}\ M$$

$$[S_2O_3^{2-}]_0 = \frac{(200.0\ \text{mL})(5.00\ M)}{(150.0\ \text{mL} + 200.0\ \text{mL})} = 2.86\ M$$

Since $[S_2O_3^{2-}]_0 \gg [Ag^+]_0$, and since K_1 and K_2 are large, both formation reactions can be assumed to go to completion, and the net reaction in the solution is as follows:

	Ag^+	+	$2S_2O_3^{2-}$	$\longrightarrow$	$Ag(S_2O_3)_2^{3-}$
Before reaction	$4.29 \times 10^{-4}\ M$		2.86 M		0
After reaction	~0		$2.86 - 2(4.29 \times 10^{-4})$ $\approx 2.86\ M$		$4.29 \times 10^{-4}\ M$

Note that Ag^+ is limiting and that the amount of $S_2O_3^{2-}$ consumed is negligible. Also note that since all these species are in the same solution, the molarities can be used to do the stoichiometry problem.

Of course, the concentration of Ag^+ is not zero at equilibrium, and there is some $Ag(S_2O_3)^-$ in the solution. To calculate the concentrations of these species, we must use the K_1 and K_2 expressions. We can calculate the concentration of $Ag(S_2O_3)^-$ from K_2:

$$3.9 \times 10^4 = K_2 = \frac{[Ag(S_2O_3)_2^{3-}]}{[Ag(S_2O_3)^-][S_2O_3^{2-}]} = \frac{4.29 \times 10^{-4}}{[Ag(S_2O_3)^-](2.86)}$$

$$[Ag(S_2O_3)^-] = 3.8 \times 10^{-9}\ M$$

We can calculate $[Ag^+]$ from K_1:

$$7.4 \times 10^8 = K_1 = \frac{[Ag(S_2O_3)^-]}{[Ag^+][S_2O_3^{2-}]} = \frac{3.8 \times 10^{-9}}{[Ag^+](2.86)}$$

$$[Ag^+] = 1.8 \times 10^{-18}\ M$$

These results show that $[Ag(S_2O_3)_2^{3-}] \gg [Ag(S_2O_3)^-] \gg [Ag^+]$

Thus the assumption is valid that essentially all the original Ag^+ is converted to $Ag(S_2O_3)_2^{3-}$ at equilibrium.

See Exercises 16.67 through 16.70

Complex Ions and Solubility

Often ionic solids that are very nearly water-insoluble must be dissolved somehow in aqueous solutions. For example, when the various qualitative analysis groups are precipitated out, the precipitates must be redissolved to separate the ions within each

(top) Aqueous ammonia is added to silver chloride (white). (bottom) Silver chloride, insoluble in water, dissolves to form Ag(NH₃)₂⁺(aq) and Cl⁻(aq).

When reactions are added, the equilibrium constant for the overall process is the product of the constants for the individual reactions.

group. Consider a solution of cations that contains Ag^+, Pb^{2+}, and Hg_2^{2+}, among others. When dilute aqueous HCl is added to this solution, the Group I ions will form the insoluble chlorides $AgCl$, $PbCl_2$, and Hg_2Cl_2. Once this mixed precipitate is separated from the solution, it must be redissolved to identify the cations individually. How can this be done? We know that some solids are more soluble in acidic than in neutral solutions. What about chloride salts? For example, can AgCl be dissolved by using a strong acid? The answer is no, because Cl^- ions have virtually no affinity for H^+ ions in aqueous solution. The position of the dissolution equilibrium

$$AgCl(s) \rightleftharpoons Ag^+(aq) + Cl^-(aq)$$

is not affected by the presence of H^+.

How can we pull the dissolution equilibrium to the right, even though Cl^- is an extremely weak base? The key is to lower the concentration of Ag^+ in solution by forming complex ions. For example, Ag^+ reacts with excess NH_3 to form the stable complex ion $Ag(NH_3)_2^+$. As a result, AgCl is quite soluble in concentrated ammonia solutions. The relevant reactions are

$$AgCl(s) \rightleftharpoons Ag^+(aq) + Cl^-(aq) \qquad K_{sp} = 1.6 \times 10^{-10}$$
$$Ag^+(aq) + NH_3(aq) \rightleftharpoons Ag(NH_3)^+(aq) \qquad K_1 = 2.1 \times 10^3$$
$$Ag(NH_3)^+(aq) + NH_3(aq) \rightleftharpoons Ag(NH_3)_2^+(aq) \qquad K_2 = 8.2 \times 10^3$$

The Ag^+ ion produced by dissolving solid AgCl combines with NH_3 to form $Ag(NH_3)_2^+$, which causes more AgCl to dissolve, until the point at which

$$[Ag^+][Cl^-] = K_{sp} = 1.6 \times 10^{-10}$$

Here $[Ag^+]$ refers only to the Ag^+ ion that is present as a separate species in solution. It is *not* the total silver content of the solution, which is

$$[Ag]_{\text{total dissolved}} = [Ag^+] + [Ag(NH_3)^+] + [Ag(NH_3)_2^+]$$

For reasons discussed in the previous section, virtually all the Ag^+ from the dissolved AgCl ends up in the complex ion $Ag(NH_3)_2^+$. Thus we can represent the dissolving of solid AgCl in excess NH_3 by the equation

$$AgCl(s) + 2NH_3(aq) \rightleftharpoons Ag(NH_3)_2^+(aq) + Cl^-(aq)$$

Since this equation is the *sum of the three stepwise reactions* given above, the equilibrium constant for the reaction is the product of the constants for the three reactions. (Demonstrate this to yourself by multiplying together the three expressions for K_{sp}, K_1, and K_2.) The equilibrium expression is

$$K = \frac{[Ag(NH_3)_2^+][Cl^-]}{[NH_3]^2}$$
$$= K_{sp} \times K_1 \times K_2 = (1.6 \times 10^{-10})(2.1 \times 10^3)(8.2 \times 10^3) = 2.8 \times 10^{-3}$$

Using this expression, we will now calculate the solubility of solid AgCl in a 10.0-M NH_3 solution. If we let x be the solubility (in mol/L) of AgCl in the solution, we can then write the following expressions for the equilibrium concentrations of the pertinent species:

$$[Cl^-] = x$$
$$[Ag(NH_3)_2^+] = x$$

x mol/L of AgCl dissolves to produce x mol/L Cl^- and x mol/L $Ag(NH_3)_2^+$

$$[NH_3] = 10.0 - 2x$$

Formation of x mol/L $Ag(NH_3)_2^+$ requires $2x$ mol/L NH_3, since each complex ion contains two NH_3 ligands

Substituting these concentrations into the equilibrium expression gives

$$K = 2.8 \times 10^{-3} = \frac{[Ag(NH_3)_2^+][Cl^-]}{[NH_3]^2} = \frac{(x)(x)}{(10.0 - 2x)^2} = \frac{x^2}{(10.0 - 2x)^2}$$

No approximations are necessary here. Taking the square root of both sides of the equation gives

$$\sqrt{2.8 \times 10^{-3}} = \frac{x}{10.0 - 2x}$$

$$x = 0.48 \text{ mol/L} = \text{solubility of AgCl}(s) \text{ in } 10.0 \text{ M NH}_3$$

Thus the solubility of AgCl in 10.0 M NH$_3$ is much greater than its solubility in pure water, which is

$$\sqrt{K_{sp}} = 1.3 \times 10^{-5} \text{ mol/L}$$

In this chapter we have considered two strategies for dissolving a water-insoluble ionic solid. If the *anion* of the solid is a good base, the solubility is greatly increased by acidifying the solution. In cases where the anion is not sufficiently basic, the ionic solid often can be dissolved in a solution containing a ligand that forms stable complex ions with its *cation*.

Sometimes solids are so insoluble that combinations of reactions are needed to dissolve them. For example, to dissolve the extremely insoluble HgS ($K_{sp} = 10^{-54}$), it is necessary to use a mixture of concentrated HCl and concentrated HNO$_3$, called *aqua regia*. The H$^+$ ions in the aqua regia react with the S^{2-} ions to form H$_2$S, and Cl$^-$ reacts with Hg^{2+} to form various complex ions, including HgCl$_4^{2-}$. In addition, NO$_3^-$ oxidizes S^{2-} to elemental sulfur. These processes lower the concentrations of Hg^{2+} and S^{2-} and thus promote the solubility of HgS.

Since the solubility of many salts increases with temperature, simple heating is sometimes enough to make a salt sufficiently soluble. For example, earlier in this

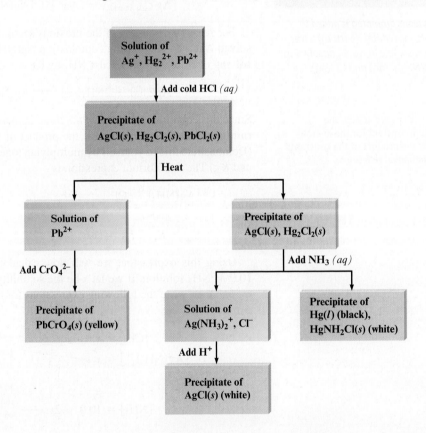

Figure 16.3 | The separation of the Group I ions in the classic scheme of qualitative analysis.

section we considered the mixed chloride precipitates of the Group I ions—$PbCl_2$, AgCl, and Hg_2Cl_2. The effect of temperature on the solubility of $PbCl_2$ is such that we can precipitate $PbCl_2$ with cold aqueous HCl and then redissolve it by heating the solution to near boiling. The silver and mercury(I) chlorides remain precipitated, since they are not significantly soluble in hot water. However, solid AgCl can be dissolved using aqueous ammonia. The solid Hg_2Cl_2 reacts with NH_3 to form a mixture of elemental mercury and $HgNH_2Cl$:

$$Hg_2Cl_2(s) + 2NH_3(aq) \longrightarrow \underset{\text{White}}{HgNH_2Cl(s)} + \underset{\text{Black}}{Hg(l)} + NH_4^+(aq) + Cl^-(aq)$$

The mixed precipitate appears gray. This is an oxidation–reduction reaction in which one mercury(I) ion in Hg_2Cl_2 is oxidized to Hg^{2+} in $HgNH_2Cl$ and the other mercury(I) ion is reduced to Hg, or elemental mercury.

The treatment of the Group I ions is summarized in Fig. 16.3. Note that the presence of Pb^{2+} is confirmed by adding CrO_4^{2-}, which forms bright yellow lead(II) chromate ($PbCrO_4$). Also note that H^+ added to a solution containing $Ag(NH_3)_2^+$ reacts with the NH_3 to form NH_4^+, destroying the $Ag(NH_3)_2^+$ complex. Silver chloride then re-forms:

$$2H^+(aq) + Ag(NH_3)_2^+(aq) + Cl^-(aq) \longrightarrow 2NH_4^+(aq) + AgCl(s)$$

Note that the qualitative analysis of cations by selective precipitation involves all the types of reactions we have discussed and represents an excellent application of the principles of chemical equilibrium.

For review

Key terms

Section 16.1
solubility product constant (solubility product)

Section 16.2
ion product
selective precipitation
qualitative analysis

Section 16.3
complex ion
formation (stability) constants

Solids dissolving in water

› For a slightly soluble salt, an equilibrium is set up between the excess solid (MX) and the ions in solution:

$$MX(s) \rightleftharpoons M^+(aq) + X^-(aq)$$

› The corresponding constant is called K_{sp}:

$$K_{sp} = [M^+][X^-]$$

› The solubility of $MX(s)$ is decreased by the presence of another source of either M^+ or X^-; this is called the common ion effect

› Predicting whether precipitation will occur when two solutions are mixed involves calculating Q for the initial concentrations:
  › If $Q > K_{sp}$, precipitation occurs
  › If $Q \leq K_{sp}$, no precipitation occurs

Qualitative analysis

› A mixture of ions can be separated by selective precipitation
  › The ions are first separated into groups by adding $HCl(aq)$, then $H_2S(aq)$, then $NaOH(aq)$, and finally $Na_2CO_3(aq)$
  › The ions in the groups are separated and identified by further selective dissolution and precipitation

Complex ions

> Complex ions consist of a metal ion surrounded by attached ligands
>> A ligand is a Lewis base
>> The number of ligands is called the coordination number, which is commonly 2, 4, or 6
> Complex ion equilibria in solution are described by formation (stability) constants
> The formation of complex ions can be used to selectively dissolve solids in the qualitative analysis scheme

Review questions   *Answers to the Review Questions can be found on the Student website (accessible from* **www.cengagebrain.com**).

1. To what reaction does the solubility product constant, K_{sp}, refer? Table 16.1 lists K_{sp} values for several ionic solids. For any of these ionic compounds, you should be able to calculate the solubility. What is the solubility of a salt, and what procedures do you follow to calculate the solubility of a salt? How would you calculate the K_{sp} value for a salt given the solubility?

2. Under what circumstances can you compare the relative solubilities of two salts directly by comparing the values of their solubility products? When can relative solubilities not be compared based on K_{sp} values?

3. What is a common ion and how does its presence affect the solubility?

4. List some salts whose solubility increases as the pH becomes more acidic. What is true about the anions in these salts? List some salts whose solubility remains unaffected by the solution pH. What is true about the anions in these salts?

5. What is the difference between the ion product, Q, and the solubility product, K_{sp}? What happens when $Q > K_{sp}$? $Q < K_{sp}$? $Q = K_{sp}$?

6. Mixtures of metal ions in aqueous solution can sometimes be separated by selective precipitation. What is selective precipitation? If a solution contained 0.10 M Mg^{2+}, 0.10 M Ca^{2+}, and 0.10 M Ba^{2+}, how could addition of NaF be used to separate the cations out of

solution—that is, what would precipitate first, then second, then third? How could addition of K_3PO_4 be used to separate out the cations in a solution that is 1.0 M Ag^+, 1.0 M Pb^{2+}, and 1.0 M Sr^{2+}?

7. Figure 16.2 summarizes the classic method for separating a mixture of common cations by selective precipitation. Explain the chemistry involved with each of the four steps in the diagram.

8. What is a complex ion? The stepwise formation constants for the complex ion $Cu(NH_3)_4^{2+}$ are $K_1 \approx 1 \times 10^3$, $K_2 \approx 1 \times 10^4$, $K_3 \approx 1 \times 10^3$, and $K_4 \approx 1 \times 10^3$. Write the reactions that refer to each of these formation constants. Given that the values of the formation constants are large, what can you deduce about the equilibrium concentration of $Cu(NH_3)_4^{2+}$ versus the equilibrium concentration of Cu^{2+}?

9. When 5 M ammonia is added to a solution containing $Cu(OH)_2(s)$, the precipitate will eventually dissolve in solution. Why? If 5 M HNO_3 is then added, the $Cu(OH)_2$ precipitate re-forms. Why? In general, what effect does the ability of a cation to form a complex ion have on the solubility of salts containing that cation?

10. Figure 16.3 outlines the classic scheme for separating a mixture of insoluble chloride salts from one another. Explain the chemistry involved in the various steps of the figure.

Active Learning Questions

These questions are designed to be used by groups of students in class.

1. Which of the following will affect the total amount of solute that can dissolve in a given amount of solvent?
   a. The solution is stirred.
   b. The solute is ground to fine particles before dissolving.
   c. The temperature changes.

2. Devise as many ways as you can to experimentally determine the K_{sp} value of a solid. Explain why each of these would work.

3. You are browsing through the *Handbook of Hypothetical Chemistry* when you come across a solid that is reported to have a K_{sp} value of zero in water at 25°C. What does this mean?

4. A friend tells you: "The constant K_{sp} of a salt is called the solubility product constant and is calculated from the concentrations of ions in the solution. Thus, if salt A dissolves to a greater extent than salt B, salt A must have a higher K_{sp} than salt B." Do you agree with your friend? Explain.

5. Explain the following phenomenon: You have a test tube with an aqueous solution of silver nitrate as shown in test tube 1 on the following page. A few drops of aqueous sodium chromate solution was added with the end result shown in test tube 2. A

few drops of aqueous sodium chloride solution was then added with the end result shown in test tube 3.

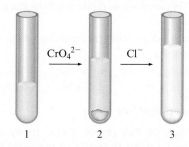

Use the K_{sp} values in the book to support your explanation, and include the balanced equations. Also, list the ions that are present in solution in each test tube.

6. What happens to the K_{sp} value of a solid as the temperature of the solution changes? Consider both increasing and decreasing temperatures, and explain your answer.

7. Which is more likely to dissolve in an acidic solution, silver sulfide or silver chloride? Why?

A blue question or exercise number indicates that the answer to that question or exercise appears at the back of this book and a solution appears in the *Solutions Guide*, as found on PowerLecture.

Questions

8. For which of the following is the K_{sp} value of the ionic compound the largest? The smallest? Explain your answer.

9. $Ag_2S(s)$ has a larger molar solubility than CuS even though Ag_2S has the smaller K_{sp} value. Explain how this is possible.

10. Solubility is an equilibrium position, whereas K_{sp} is an equilibrium constant. Explain the difference.

11. The salts in Table 16.1, with the possible exception of the hydroxide salts, generally have one of the following mathematical relationships between the K_{sp} value and the molar solubility s.

  i. $K_{sp} = s^2$ **iii.** $K_{sp} = 27s^4$

  ii. $K_{sp} = 4s^3$ **iv.** $K_{sp} = 108s^5$

For each mathematical relationship, give an example of a salt in Table 16.1 that exhibits that relationship.

12. When $Na_3PO_4(aq)$ is added to a solution containing a metal ion and a precipitate forms, the precipitate generally could be one of two possibilities. What are the two possibilities?

13. The common ion effect for ionic solids (salts) is to significantly decrease the solubility of the ionic compound in water. Explain the common ion effect.

14. Sulfide precipitates are generally grouped as sulfides insoluble in acidic solution and sulfides insoluble in basic solution. Explain why there is a difference between the two groups of sulfide precipitates.

15. List some ways one can increase the solubility of a salt in water.

16. The stepwise formation constants for a complex ion usually have values much greater than 1. What is the significance of this?

17. Silver chloride dissolves readily in 2 *M* NH_3 but is quite insoluble in 2 *M* NH_4NO_3. Explain.

18. If a solution contains either $Pb^{2+}(aq)$ or $Ag^+(aq)$, how can temperature be manipulated to help identify the ion in solution?

Exercises

In this section similar exercises are paired.

Solubility Equilibria

19. Write balanced equations for the dissolution reactions and the corresponding solubility product expressions for each of the following solids.

  a. $AgC_2H_3O_2$ **b.** $Al(OH)_3$ **c.** $Ca_3(PO_4)_2$

20. Write balanced equations for the dissolution reactions and the corresponding solubility product expressions for each of the following solids.

  a. Ag_2CO_3 **b.** $Ce(IO_3)_3$ **c.** BaF_2

21. Use the following data to calculate the K_{sp} value for each solid.

  a. The solubility of CaC_2O_4 is 4.8×10^{-5} mol/L.

  b. The solubility of BiI_3 is 1.32×10^{-5} mol/L.

22. Use the following data to calculate the K_{sp} value for each solid.

  a. The solubility of $Pb_3(PO_4)_2$ is 6.2×10^{-12} mol/L.

  b. The solubility of Li_2CO_3 is 7.4×10^{-2} mol/L.

23. Approximately 0.14 g nickel(II) hydroxide, $Ni(OH)_2(s)$, dissolves per liter of water at 20°C. Calculate K_{sp} for $Ni(OH)_2(s)$ at this temperature.

24. The solubility of the ionic compound M_2X_3, having a molar mass of 288 g/mol, is 3.60×10^{-7} g/L. Calculate the K_{sp} of the compound.

25. The concentration of Pb^{2+} in a solution saturated with $PbBr_2(s)$ is 2.14×10^{-2} *M*. Calculate K_{sp} for $PbBr_2$.

26. The concentration of Ag^+ in a solution saturated with $Ag_2C_2O_4(s)$ is 2.2×10^{-4} *M*. Calculate K_{sp} for $Ag_2C_2O_4$.

27. Calculate the solubility of each of the following compounds in moles per liter. Ignore any acid–base properties.

  a. Ag_3PO_4, $K_{sp} = 1.8 \times 10^{-18}$

  b. $CaCO_3$, $K_{sp} = 8.7 \times 10^{-9}$

  c. Hg_2Cl_2, $K_{sp} = 1.1 \times 10^{-18}$ (Hg_2^{2+} is the cation in solution.)

28. Calculate the solubility of each of the following compounds in moles per liter. Ignore any acid–base properties.

  a. PbI_2, $K_{sp} = 1.4 \times 10^{-8}$

  b. $CdCO_3$, $K_{sp} = 5.2 \times 10^{-12}$

  c. $Sr_3(PO_4)_2$, $K_{sp} = 1 \times 10^{-31}$

29. Cream of tartar, a common ingredient in cooking, is the common name for potassium bitartrate (abbreviated KBT, molar mass = 188.2 g/mol). Historically, KBT was a crystalline solid that formed on the casks of wine barrels during the fermentation process. Calculate the maximum mass of KBT that can dissolve in 250.0 mL of solution to make a saturated solution. The K_{sp} value for KBT is 3.8×10^{-4}.

30. Barium sulfate is a contrast agent for X-ray scans that are most often associated with the gastrointestinal tract. Calculate the mass of $BaSO_4$ that can dissolve in 100.0 mL of solution. The K_{sp} value for $BaSO_4$ is 1.5×10^{-9}.

31. Calculate the molar solubility of $Mg(OH)_2$, $K_{sp} = 8.9 \times 10^{-12}$.

32. Calculate the molar solubility of $Cd(OH)_2$, $K_{sp} = 5.9 \times 10^{-11}$.

33. Calculate the molar solubility of $Al(OH)_3$, $K_{sp} = 2 \times 10^{-32}$.

34. Calculate the molar solubility of $Co(OH)_3$, $K_{sp} = 2.5 \times 10^{-43}$.

35. For each of the following pairs of solids, determine which solid has the smallest molar solubility.
    a. $CaF_2(s)$, $K_{sp} = 4.0 \times 10^{-11}$, or $BaF_2(s)$, $K_{sp} = 2.4 \times 10^{-5}$
    b. $Ca_3(PO_4)_2(s)$, $K_{sp} = 1.3 \times 10^{-32}$, or $FePO_4(s)$, $K_{sp} = 1.0 \times 10^{-22}$

36. For each of the following pairs of solids, determine which solid has the smallest molar solubility.
    a. FeC_2O_4, $K_{sp} = 2.1 \times 10^{-7}$, or $Cu(IO_4)_2$, $K_{sp} = 1.4 \times 10^{-7}$
    b. Ag_2CO_3, $K_{sp} = 8.1 \times 10^{-12}$, or $Mn(OH)_2$, $K_{sp} = 2 \times 10^{-13}$

37. Calculate the solubility (in moles per liter) of $Fe(OH)_3$ ($K_{sp} = 4 \times 10^{-38}$) in each of the following.
    a. water
    b. a solution buffered at pH = 5.0
    c. a solution buffered at pH = 11.0

38. Calculate the solubility of $Co(OH)_2(s)$ ($K_{sp} = 2.5 \times 10^{-16}$) in a buffered solution with a pH of 11.00.

39. The K_{sp} for silver sulfate (Ag_2SO_4) is 1.2×10^{-5}. Calculate the solubility of silver sulfate in each of the following.
    a. water
    b. 0.10 M $AgNO_3$
    c. 0.20 M K_2SO_4

40. The K_{sp} for lead iodide (PbI_2) is 1.4×10^{-8}. Calculate the solubility of lead iodide in each of the following.
    a. water
    b. 0.10 M $Pb(NO_3)_2$
    c. 0.010 M NaI

41. Calculate the solubility of solid $Ca_3(PO_4)_2$ ($K_{sp} = 1.3 \times 10^{-32}$) in a 0.20-M Na_3PO_4 solution.

42. Calculate the solubility of solid $Pb_3(PO_4)_2$ ($K_{sp} = 1 \times 10^{-54}$) in a 0.10-M $Pb(NO_3)_2$ solution.

43. The solubility of $Ce(IO_3)_3$ in a 0.20-M KIO_3 solution is 4.4×10^{-8} mol/L. Calculate K_{sp} for $Ce(IO_3)_3$.

44. The solubility of $Pb(IO_3)_2(s)$ in a 0.10-M KIO_3 solution is 2.6×10^{-11} mol/L. Calculate K_{sp} for $Pb(IO_3)_2(s)$.

45. Which of the substances in Exercises 27 and 28 show increased solubility as the pH of the solution becomes more acidic? Write equations for the reactions that occur to increase the solubility.

46. For which salt in each of the following groups will the solubility depend on pH?
    a. AgF, AgCl, AgBr
    c. $Sr(NO_3)_2$, $Sr(NO_2)_2$
    b. $Pb(OH)_2$, $PbCl_2$
    d. $Ni(NO_3)_2$, $Ni(CN)_2$

Precipitation Conditions

47. What mass of ZnS ($K_{sp} = 2.5 \times 10^{-22}$) will dissolve in 300.0 mL of 0.050 M $Zn(NO_3)_2$? Ignore the basic properties of S^{2-}.

48. The concentration of Mg^{2+} in seawater is 0.052 M. At what pH will 99% of the Mg^{2+} be precipitated as the hydroxide salt? [K_{sp} for $Mg(OH)_2 = 8.9 \times 10^{-12}$.]

49. Will a precipitate form when 100.0 mL of 4.0×10^{-4} M $Mg(NO_3)_2$ is added to 100.0 mL of 2.0×10^{-4} M NaOH?

50. A solution contains 1.0×10^{-5} M Ag^+ and 2.0×10^{-6} M CN^-. Will AgCN(s) precipitate? (K_{sp} for AgCN(s) is 2.2×10^{-12}.)

51. A solution is prepared by mixing 100.0 mL of 1.0×10^{-2} M $Pb(NO_3)_2$ and 100.0 mL of 1.0×10^{-3} M NaF. Will $PbF_2(s)$ ($K_{sp} = 4 \times 10^{-8}$) precipitate?

52. A solution contains 2.0×10^{-3} M Ce^{3+} and 1.0×10^{-2} M IO_3^-. Will $Ce(IO_3)_3(s)$ precipitate? [K_{sp} for $Ce(IO_3)_3$ is 3.2×10^{-10}.]

53. Calculate the final concentrations of $K^+(aq)$, $C_2O_4^{2-}(aq)$, $Ba^{2+}(aq)$, and $Br^-(aq)$ in a solution prepared by adding 0.100 L of 0.200 M $K_2C_2O_4$ to 0.150 L of 0.250 M $BaBr_2$. (For BaC_2O_4, $K_{sp} = 2.3 \times 10^{-8}$.)

54. A solution is prepared by mixing 75.0 mL of 0.020 M $BaCl_2$ and 125 mL of 0.040 M K_2SO_4. What are the concentrations of barium and sulfate ions in this solution? Assume only SO_4^{2-} ions (no HSO_4^-) are present.

55. A 50.0-mL sample of 0.00200 M $AgNO_3$ is added to 50.0 mL of 0.0100 M $NaIO_3$. What is the equilibrium concentration of Ag^+ in solution? (K_{sp} for $AgIO_3$ is 3.0×10^{-8}.)

56. A solution is prepared by mixing 50.0 mL of 0.10 M $Pb(NO_3)_2$ with 50.0 mL of 1.0 M KCl. Calculate the concentrations of Pb^{2+} and Cl^- at equilibrium. [K_{sp} for $PbCl_2(s)$ is 1.6×10^{-5}.]

57. A solution contains 1.0×10^{-5} M Na_3PO_4. What is the minimum concentration of $AgNO_3$ that would cause precipitation of solid Ag_3PO_4 ($K_{sp} = 1.8 \times 10^{-18}$)?

58. The K_{sp} of $Al(OH)_3$ is 2×10^{-32}. At what pH will a 0.2-M Al^{3+} solution begin to show precipitation of $Al(OH)_3$?

59. A solution is 1×10^{-4} M in NaF, Na_2S, and Na_3PO_4. What would be the order of precipitation as a source of Pb^{2+} is added gradually to the solution? The relevant K_{sp} values are $K_{sp}(PbF_2) = 4 \times 10^{-8}$, $K_{sp}(PbS) = 7 \times 10^{-29}$, and $K_{sp}[Pb_3(PO_4)_2] = 1 \times 10^{-54}$.

60. A solution contains 0.25 M $Ni(NO_3)_2$ and 0.25 M $Cu(NO_3)_2$. Can the metal ions be separated by slowly adding Na_2CO_3? Assume that for successful separation 99% of the metal ion must be precipitated before the other metal ion begins to precipitate, and assume no volume change on addition of Na_2CO_3.

Complex Ion Equilibria

61. Write equations for the stepwise formation of each of the following complex ions.

a. $Ni(CN)_4^{2-}$

b. $V(C_2O_4)_3^{3-}$

62. Write equations for the stepwise formation of each of the following complex ions.

a. CoF_6^{3-}

b. $Zn(NH_3)_4^{2+}$

63. In the presence of CN^-, Fe^{3+} forms the complex ion $Fe(CN)_6^{3-}$. The equilibrium concentrations of Fe^{3+} and $Fe(CN)_6^{3-}$ are 8.5×10^{-40} M and 1.5×10^{-3} M, respectively, in a 0.11-M KCN solution. Calculate the value for the overall formation constant of $Fe(CN)_6^{3-}$.

$$Fe^{3+}(aq) + 6CN^-(aq) \rightleftharpoons Fe(CN)_6^{3-}(aq) \qquad K_{overall} = ?$$

64. In the presence of NH_3, Cu^{2+} forms the complex ion $Cu(NH_3)_4^{2+}$. If the equilibrium concentrations of Cu^{2+} and $Cu(NH_3)_4^{2+}$ are 1.8×10^{-17} M and 1.0×10^{-3} M, respectively, in a 1.5-M NH_3 solution, calculate the value for the overall formation constant of $Cu(NH_3)_4^{2+}$.

$$Cu^{2+}(aq) + 4NH_3(aq) \rightleftharpoons Cu(NH_3)_4^{2+}(aq) \qquad K_{overall} = ?$$

65. When aqueous KI is added gradually to mercury(II) nitrate, an orange precipitate forms. Continued addition of KI causes the precipitate to dissolve. Write balanced equations to explain these observations. (*Hint:* Hg^{2+} reacts with I^- to form HgI_4^{2-}.)

66. As sodium chloride solution is added to a solution of silver nitrate, a white precipitate forms. Ammonia is added to the mixture and the precipitate dissolves. When potassium bromide solution is then added, a pale yellow precipitate appears. When a solution of sodium thiosulfate is added, the yellow precipitate dissolves. Finally, potassium iodide is added to the solution and a yellow precipitate forms. Write equations for all the changes mentioned above. What conclusions can you draw concerning the sizes of the K_{sp} values for AgCl, AgBr, and AgI?

67. The overall formation constant for HgI_4^{2-} is 1.0×10^{30}. That is,

$$1.0 \times 10^{30} = \frac{[HgI_4^{2-}]}{[Hg^{2+}][I^-]^4}$$

What is the concentration of Hg^{2+} in 500.0 mL of a solution that was originally 0.010 M Hg^{2+} and 0.78 M I^-? The reaction is

$$Hg^{2+}(aq) + 4I^-(aq) \rightleftharpoons HgI_4^{2-}(aq)$$

68. A solution is prepared by adding 0.10 mole of $Ni(NH_3)_6Cl_2$ to 0.50 L of 3.0 M NH_3. Calculate $[Ni(NH_3)_6^{2+}]$ and $[Ni^{2+}]$ in this solution. $K_{overall}$ for $Ni(NH_3)_6^{2+}$ is 5.5×10^8. That is,

$$5.5 \times 10^8 = \frac{[Ni(NH_3)_6^{2+}]}{[Ni^{2+}][NH_3]^6}$$

for the overall reaction

$$Ni^{2+}(aq) + 6NH_3(aq) \rightleftharpoons Ni(NH_3)_6^{2+}(aq)$$

69. A solution is formed by mixing 50.0 mL of 10.0 M NaX with 50.0 mL of 2.0×10^{-3} M $CuNO_3$. Assume that Cu^+ forms complex ions with X^- as follows:

$$Cu^+(aq) + X^-(aq) \rightleftharpoons CuX(aq) \qquad K_1 = 1.0 \times 10^2$$
$$CuX(aq) + X^-(aq) \rightleftharpoons CuX_2^-(aq) \qquad K_2 = 1.0 \times 10^4$$
$$CuX_2^-(aq) + X^-(aq) \rightleftharpoons CuX_3^{2-}(aq) \qquad K_3 = 1.0 \times 10^3$$

with an overall reaction

$$Cu^+(aq) + 3X^-(aq) \rightleftharpoons CuX_3^{2-}(aq) \qquad K = 1.0 \times 10^9$$

Calculate the following concentrations at equilibrium.

a. CuX_3^{2-} **b.** CuX_2^- **c.** Cu^+

70. A solution is prepared by mixing 100.0 mL of 1.0×10^{-4} M $Be(NO_3)_2$ and 100.0 mL of 8.0 M NaF.

$$Be^{2+}(aq) + F^-(aq) \rightleftharpoons BeF^+(aq) \qquad K_1 = 7.9 \times 10^4$$
$$BeF^+(aq) + F^-(aq) \rightleftharpoons BeF_2(aq) \qquad K_2 = 5.8 \times 10^3$$
$$BeF_2(aq) + F^-(aq) \rightleftharpoons BeF_3^-(aq) \qquad K_3 = 6.1 \times 10^2$$
$$BeF_3^-(aq) + F^-(aq) \rightleftharpoons BeF_4^{2-}(aq) \qquad K_4 = 2.7 \times 10^1$$

Calculate the equilibrium concentrations of F^-, Be^{2+}, BeF^+, BeF_2, BeF_3^-, and BeF_4^{2-} in this solution.

71. a. Calculate the molar solubility of AgI in pure water. K_{sp} for AgI is 1.5×10^{-16}.

b. Calculate the molar solubility of AgI in 3.0 M NH_3. The overall formation constant for $Ag(NH_3)_2^+$ is 1.7×10^7.

c. Compare the calculated solubilities from parts a and b. Explain any differences.

72. Solutions of sodium thiosulfate are used to dissolve unexposed AgBr ($K_{sp} = 5.0 \times 10^{-13}$) in the developing process for black-and-white film. What mass of AgBr can dissolve in 1.00 L of 0.500 M $Na_2S_2O_3$? Ag^+ reacts with $S_2O_3^{2-}$ to form a complex ion:

$$Ag^+(aq) + 2S_2O_3^{2-}(aq) \rightleftharpoons Ag(S_2O_3)_2^{3-}(aq)$$
$$K = 2.9 \times 10^{13}$$

73. K_f for the complex ion $Ag(NH_3)_2^+$ is 1.7×10^7. K_{sp} for AgCl is 1.6×10^{-10}. Calculate the molar solubility of AgCl in 1.0 M NH_3.

74. The copper(I) ion forms a chloride salt that has $K_{sp} = 1.2 \times 10^{-6}$. Copper(I) also forms a complex ion with Cl^-:

$$Cu^+(aq) + 2Cl^-(aq) \rightleftharpoons CuCl_2^-(aq) \qquad K = 8.7 \times 10^4$$

a. Calculate the solubility of copper(I) chloride in pure water. (Ignore $CuCl_2^-$ formation for part a.)

b. Calculate the solubility of copper(I) chloride in 0.10 M NaCl.

75. A series of chemicals were added to some $AgNO_3(aq)$. $NaCl(aq)$ was added first to the silver nitrate solution with the end result shown below in test tube 1, $NH_3(aq)$ was then added with the end result shown in test tube 2, and $HNO_3(aq)$ was added last with the end result shown in test tube 3.

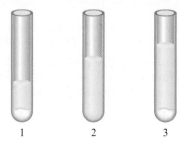

1 2 3

Explain the results shown in each test tube. Include a balanced equation for the reaction(s) taking place.

76. The solubility of copper(II) hydroxide in water can be increased by adding either the base NH_3 or the acid HNO_3. Explain. Would added NH_3 or HNO_3 have the same effect on the solubility of silver acetate or silver chloride? Explain.

Additional Exercises

77. A solution contains 0.018 mole each of I^-, Br^-, and Cl^-. When the solution is mixed with 200. mL of 0.24 M $AgNO_3$, what mass of $AgCl(s)$ precipitates out, and what is $[Ag^+]$? Assume no volume change.

$$AgI: K_{sp} = 1.5 \times 10^{-16}$$
$$AgBr: K_{sp} = 5.0 \times 10^{-13}$$
$$AgCl: K_{sp} = 1.6 \times 10^{-10}$$

78. You have two salts, AgX and AgY, with very similar K_{sp} values. You know that K_a for HX is much greater than K_a for HY. Which salt is more soluble in acidic solution? Explain.

79. Tooth enamel is composed of the mineral hydroxyapatite. The K_{sp} of hydroxyapatite, $Ca_5(PO_4)_3OH$, is 6.8×10^{-37}. Calculate the solubility of hydroxyapatite in pure water in moles per liter. How is the solubility of hydroxyapatite affected by adding acid? When hydroxyapatite is treated with fluoride, the mineral fluorapatite, $Ca_5(PO_4)_3F$, forms. The K_{sp} of this substance is 1×10^{-60}. Calculate the solubility of fluorapatite in water. How do these calculations provide a rationale for the fluoridation of drinking water?

80. The U.S. Public Health Service recommends the fluoridation of water as a means for preventing tooth decay. The recommended concentration is 1 mg F^- per liter. The presence of calcium ions in hard water can precipitate the added fluoride. What is the maximum molarity of calcium ions in hard water if the fluoride concentration is at the USPHS recommended level? (K_{sp} for $CaF_2 = 4.0 \times 10^{-11}$)

81. What mass of $Ca(NO_3)_2$ must be added to 1.0 L of a 1.0-M HF solution to begin precipitation of $CaF_2(s)$? For CaF_2, $K_{sp} = 4.0 \times 10^{-11}$ and K_a for HF $= 7.2 \times 10^{-4}$. Assume no volume change on addition of $Ca(NO_3)_2(s)$.

82. Calculate the mass of manganese hydroxide present in 1300 mL of a saturated manganese hydroxide solution. For $Mn(OH)_2$, $K_{sp} = 2.0 \times 10^{-13}$.

83. On a hot day, a 200.0-mL sample of a saturated solution of PbI_2 was allowed to evaporate until dry. If 240 mg of solid PbI_2 was collected after evaporation was complete, calculate the K_{sp} value for PbI_2 on this hot day.

84. The active ingredient of Pepto-Bismol is the compound bismuth subsalicylate, which undergoes the following dissociation when added to water:

$$C_7H_5BiO_4(s) + H_2O(l) \rightleftharpoons C_7H_4O_3{}^{2-}(aq)$$
$$+ Bi^{3+}(aq) + OH^-(aq) \qquad K = ?$$

If the maximum amount of bismuth subsalicylate that reacts by this reaction is 3.2×10^{-19} mol/L, calculate the equilibrium constant for the preceding reaction.

85. Nanotechnology has become an important field, with applications ranging from high-density data storage to the design of "nano machines." One common building block of nanostructured architectures is manganese oxide nanoparticles. The particles can be formed from manganese oxalate nanorods, the formation of which can be described as follows:

$$Mn^{2+}(aq) + C_2O_4{}^{2-}(aq) \rightleftharpoons MnC_2O_4(aq) \qquad K_1 = 7.9 \times 10^3$$
$$MnC_2O_4(aq) + C_2O_4{}^{2-}(aq) \rightleftharpoons Mn(C_2O_4)_2{}^{2-}(aq)$$
$$K_2 = 7.9 \times 10^1$$

Calculate the value for the overall formation constant for $Mn(C_2O_4)_2{}^{2-}$:

$$K = \frac{[Mn(C_2O_4)_2{}^{2-}]}{[Mn^{2+}][C_2O_4{}^{2-}]^2}$$

86. The equilibrium constant for the following reaction is 1.0×10^{23}:

$$Cr^{3+}(aq) + H_2EDTA^{2-}(aq) \rightleftharpoons CrEDTA^-(aq) + 2H^+(aq)$$

$$EDTA^{4-} = \begin{array}{c} {}^-O_2C-CH_2 \\ {}^-O_2C-CH_2 \end{array} \!\!\!\! N-CH_2-CH_2-N \!\!\!\! \begin{array}{c} CH_2-CO_2{}^- \\ CH_2-CO_2{}^- \end{array}$$

Ethylenediaminetetraacetate

EDTA is used as a complexing agent in chemical analysis. Solutions of EDTA, usually containing the disodium salt Na_2H_2EDTA, are used to treat heavy metal poisoning. Calculate $[Cr^{3+}]$ at equilibrium in a solution originally 0.0010 M in Cr^{3+} and 0.050 M in H_2EDTA^{2-} and buffered at pH $= 6.00$.

87. Calculate the concentration of Pb^{2+} in each of the following.
  a. a saturated solution of $Pb(OH)_2$, $K_{sp} = 1.2 \times 10^{-15}$
  b. a saturated solution of $Pb(OH)_2$ buffered at pH $= 13.00$
  c. Ethylenediaminetetraacetate ($EDTA^{4-}$) is used as a complexing agent in chemical analysis and has the following structure:

$$\begin{array}{c} {}^-O_2C-CH_2 \\ {}^-O_2C-CH_2 \end{array} \!\!\!\! N-CH_2-CH_2-N \!\!\!\! \begin{array}{c} CH_2-CO_2{}^- \\ CH_2-CO_2{}^- \end{array}$$

Ethylenediaminetetraacetate

Solutions of $EDTA^{4-}$ are used to treat heavy metal poisoning by removing the heavy metal in the form of a soluble complex ion. The reaction of $EDTA^{4-}$ with Pb^{2+} is

$$Pb^{2+}(aq) + EDTA^{4-}(aq) \rightleftharpoons PbEDTA^{2-}(aq)$$
$$K = 1.1 \times 10^{18}$$

Consider a solution with 0.010 mole of $Pb(NO_3)_2$ added to 1.0 L of an aqueous solution buffered at pH $= 13.00$ and containing 0.050 M Na_4EDTA. Does $Pb(OH)_2$ precipitate from this solution?

88. Will a precipitate of $Cd(OH)_2$ form if 1.0 mL of 1.0 M $Cd(NO_3)_2$ is added to 1.0 L of 5.0 M NH_3?

$$Cd^{2+}(aq) + 4NH_3(aq) \rightleftharpoons Cd(NH_3)_4{}^{2+}(aq)$$
$$K = 1.0 \times 10^7$$
$$Cd(OH)_2(s) \rightleftharpoons Cd^{2+}(aq) + 2OH^-(aq)$$
$$K_{sp} = 5.9 \times 10^{-15}$$

89. a. Using the K_{sp} value for $Cu(OH)_2$ (1.6×10^{-19}) and the overall formation constant for $Cu(NH_3)_4^{2+}$ (1.0×10^{13}), calculate the value for the equilibrium constant for the following reaction:

$$Cu(OH)_2(s) + 4NH_3(aq) \rightleftharpoons Cu(NH_3)_4^{2+}(aq) + 2OH^-(aq)$$

b. Use the value of the equilibrium constant you calculated in part a to calculate the solubility (in mol/L) of $Cu(OH)_2$ in 5.0 M NH_3. In 5.0 M NH_3 the concentration of OH^- is 0.0095 M.

90. Describe how you could separate the ions in each of the following groups by selective precipitation.

a. Ag^+, Mg^{2+}, Cu^{2+} **c.** Pb^{2+}, Bi^{3+}

b. Pb^{2+}, Ca^{2+}, Fe^{2+}

91. The solubility rules outlined in Chapter 4 say that $Ba(OH)_2$, $Sr(OH)_2$, and $Ca(OH)_2$ are marginally soluble hydroxides. Calculate the pH of a saturated solution of each of these marginally soluble hydroxides.

92. In the chapter discussion of precipitate formation, we ran the precipitation reaction to completion and then let some of the precipitate redissolve to get back to equilibrium. To see why, redo Example 16.6, where

Initial Concentration (mol/L)	Equilibrium Concentration (mol/L)
$[Mg^{2+}]_0 = 3.75 \times 10^{-3}$	$[Mg^{2+}] = 3.75 \times 10^{-3} - y$
$[F^-]_0 = 6.25 \times 10^{-2}$	$[F^-] = 6.25 \times 10^{-2} - 2y$
$\xrightarrow[\text{reacts to}]{y \text{ mol/Mg}^{2+}}$ form MgF_2	

ChemWork Problems

These multiconcept problems (and additional ones) are found interactively online with the same type of assistance a student would get from an instructor.

93. Assuming that the solubility of $Ca_3(PO_4)_2(s)$ is 1.6×10^{-7} mol/L at 25°C, calculate the K_{sp} for this salt. Ignore any potential reactions of the ions with water.

94. Order the following solids (a–d) from least soluble to most soluble. Ignore any potential reactions of the ions with water.

a. $AgCl$ $K_{sp} = 1.6 \times 10^{-10}$

b. Ag_2S $K_{sp} = 1.6 \times 10^{-49}$

c. CaF_2 $K_{sp} = 4.0 \times 10^{-11}$

d. CuS $K_{sp} = 8.5 \times 10^{-45}$

95. The K_{sp} for $PbI_2(s)$ is 1.4×10^{-8}. Calculate the solubility of $PbI_2(s)$ in 0.048 M NaI.

96. The solubility of $Pb(IO_3)_2(s)$ in a 7.2×10^{-2}-M KIO_3 solution is 6.0×10^{-9} mol/L. Calculate the K_{sp} value for $Pb(IO_3)_2(s)$.

97. A 50.0-mL sample of 0.0413 M $AgNO_3(aq)$ is added to 50.0 mL of 0.100 M $NaIO_3(aq)$. Calculate the $[Ag^+]$ at equilibrium in the resulting solution. [K_{sp} for $AgIO_3(s) = 3.17 \times 10^{-8}$.]

98. The Hg^{2+} ion forms complex ions with I^- as follows:

$$Hg^{2+}(aq) + I^-(aq) \rightleftharpoons HgI^+(aq) \qquad K_1 = 1.0 \times 10^8$$
$$HgI^+(aq) + I^-(aq) \rightleftharpoons HgI_2(aq) \qquad K_2 = 1.0 \times 10^5$$
$$HgI_2(aq) + I^-(aq) \rightleftharpoons HgI_3^-(aq) \qquad K_3 = 1.0 \times 10^9$$
$$HgI_3^-(aq) + I^-(aq) \rightleftharpoons HgI_4^{2-}(aq) \qquad K_4 = 1.0 \times 10^8$$

A solution is prepared by dissolving 0.088 mole of $Hg(NO_3)_2$ and 5.00 mole of NaI in enough water to make 1.0 L of solution.

a. Calculate the equilibrium concentration of $[HgI_4^{2-}]$.

b. Calculate the equilibrium concentration of $[I^-]$.

c. Calculate the equilibrium concentration of $[Hg^{2+}]$.

Challenge Problems

99. The copper(I) ion forms a complex ion with CN^- according to the following equation:

$$Cu^+(aq) + 3CN^-(aq) \rightleftharpoons Cu(CN)_3^{2-}(aq) \qquad K = 1.0 \times 10^{11}$$

a. Calculate the solubility of $CuBr(s)$ ($K_{sp} = 1.0 \times 10^{-5}$) in 1.0 L of 1.0 M $NaCN$.

b. Calculate the concentration of Br^- at equilibrium.

c. Calculate the concentration of CN^- at equilibrium.

100. Consider a solution made by mixing 500.0 mL of 4.0 M NH_3 and 500.0 mL of 0.40 M $AgNO_3$. Ag^+ reacts with NH_3 to form $AgNH_3^+$ and $Ag(NH_3)_2^+$:

$$Ag^+(aq) + NH_3(aq) \rightleftharpoons AgNH_3^+(aq) \qquad K_1 = 2.1 \times 10^3$$
$$AgNH_3^+(aq) + NH_3(aq) \rightleftharpoons Ag(NH_3)_2^+(aq) \qquad K_2 = 8.2 \times 10^3$$

Determine the concentration of all species in solution.

101. a. Calculate the molar solubility of $AgBr$ in pure water. K_{sp} for $AgBr$ is 5.0×10^{-13}.

b. Calculate the molar solubility of $AgBr$ in 3.0 M NH_3. The overall formation constant for $Ag(NH_3)_2^+$ is 1.7×10^7, that is,

$$Ag^+(aq) + 2NH_3(aq) \longrightarrow Ag(NH_3)_2^+(aq) \qquad K = 1.7 \times 10^7.$$

c. Compare the calculated solubilities from parts a and b. Explain any differences.

d. What mass of $AgBr$ will dissolve in 250.0 mL of 3.0 M NH_3?

e. What effect does adding HNO_3 have on the solubilities calculated in parts a and b?

102. Calculate the equilibrium concentrations of NH_3, Cu^{2+}, $Cu(NH_3)^{2+}$, $Cu(NH_3)_2^{2+}$, $Cu(NH_3)_3^{2+}$, and $Cu(NH_3)_4^{2+}$ in a solution prepared by mixing 500.0 mL of 3.00 M NH_3 with 500.0 mL of 2.00×10^{-3} M $Cu(NO_3)_2$. The stepwise equilibria are

$$Cu^{2+}(aq) + NH_3(aq) \rightleftharpoons CuNH_3^{2+}(aq)$$
$$K_1 = 1.86 \times 10^4$$

$$CuNH_3^{2+}(aq) + NH_3(aq) \rightleftharpoons Cu(NH_3)_2^{2+}(aq)$$
$$K_2 = 3.88 \times 10^3$$

$$Cu(NH_3)_2^{2+}(aq) + NH_3(aq) \rightleftharpoons Cu(NH_3)_3^{2+}(aq)$$
$$K_3 = 1.00 \times 10^3$$

$$Cu(NH_3)_3^{2+}(aq) + NH_3(aq) \rightleftharpoons Cu(NH_3)_4^{2+}(aq)$$
$$K_4 = 1.55 \times 10^2$$

103. Calculate the solubility of $AgCN(s)$ ($K_{sp} = 2.2 \times 10^{-12}$) in a solution containing $1.0\ M\ H^+$. (K_a for HCN is 6.2×10^{-10}.)

104. Calcium oxalate (CaC_2O_4) is relatively insoluble in water ($K_{sp} = 2 \times 10^{-9}$). However, calcium oxalate is more soluble in acidic solution. How much more soluble is calcium oxalate in $0.10\ M\ H^+$ than in pure water? In pure water, ignore the basic properties of $C_2O_4^{2-}$.

105. What is the maximum possible concentration of Ni^{2+} ion in water at 25°C that is saturated with $0.10\ M\ H_2S$ and maintained at pH 3.0 with HCl?

106. A mixture contains $1.0 \times 10^{-3}\ M\ Cu^{2+}$ and $1.0 \times 10^{-3}\ M\ Mn^{2+}$ and is saturated with $0.10\ M\ H_2S$. Determine a pH where CuS precipitates but MnS does not precipitate. K_{sp} for CuS $= 8.5 \times 10^{-45}$ and K_{sp} for MnS $= 2.3 \times 10^{-13}$.

107. Sodium tripolyphosphate ($Na_5P_3O_{10}$) is used in many synthetic detergents. Its major effect is to soften the water by complexing Mg^{2+} and Ca^{2+} ions. It also increases the efficiency of surfactants, or wetting agents that lower a liquid's surface tension. The K value for the formation of $MgP_3O_{10}^{3-}$ is 4.0×10^8. The reaction is $Mg^{2+}(aq) + P_3O_{10}^{5-}(aq) \rightleftharpoons MgP_3O_{10}^{3-}(aq)$. Calculate the concentration of Mg^{2+} in a solution that was originally 50. ppm Mg^{2+} (50. mg/L of solution) after 40. g $Na_5P_3O_{10}$ is added to 1.0 L of the solution.

108. You add an excess of solid MX in 250 g water. You measure the freezing point and find it to be $-0.028°C$. What is the K_{sp} of the solid? Assume the density of the solution is 1.0 g/cm³.

109. **a.** Calculate the molar solubility of SrF_2 in water, ignoring the basic properties of F^-. (For SrF_2, $K_{sp} = 7.9 \times 10^{-10}$.)

  b. Would the measured molar solubility of SrF_2 be greater than or less than the value calculated in part a? Explain.

  c. Calculate the molar solubility of SrF_2 in a solution buffered at pH $= 2.00$. (K_a for HF is 7.2×10^{-4}.)

Integrative Problems

These problems require the integration of multiple concepts to find the solutions.

110. A solution saturated with a salt of the type M_3X_2 has an osmotic pressure of 2.64×10^{-2} atm at 25°C. Calculate the K_{sp} value for the salt, assuming ideal behavior.

111. Consider 1.0 L of an aqueous solution that contains 0.10 M sulfuric acid to which 0.30 mole of barium nitrate is added. Assuming no change in volume of the solution, determine the pH, the concentration of barium ions in the final solution, and the mass of solid formed.

112. The K_{sp} for Q, a slightly soluble ionic compound composed of M_2^{2+} and X^- ions, is 4.5×10^{-29}. The electron configuration of M^+ is $[Xe]6s^14f^{14}5d^{10}$. The X^- anion has 54 electrons. What is the molar solubility of Q in a solution of NaX prepared by dissolving 1.98 g NaX in 150. mL solution?

Marathon Problem

This problem is designed to incorporate several concepts and techniques into one situation.

113. Aluminum ions react with the hydroxide ion to form the precipitate $Al(OH)_3(s)$, but can also react to form the soluble complex ion $Al(OH)_4^-$. In terms of solubility, $Al(OH)_3(s)$ will be more soluble in very acidic solutions as well as more soluble in very basic solutions.

  a. Write equations for the reactions that occur to increase the solubility of $Al(OH)_3(s)$ in very acidic solutions and in very basic solutions.

  b. Let's study the pH dependence of the solubility of $Al(OH)_3(s)$ in more detail. Show that the solubility of $Al(OH)_3$, as a function of $[H^+]$, obeys the equation

$$S = [H^+]^3 K_{sp}/K_w^3 + KK_w/[H^+]$$

  where S = solubility = $[Al^{3+}] + [Al(OH)_4^-]$ and K is the equilibrium constant for

$$Al(OH)_3(s) + OH^-(aq) \rightleftharpoons Al(OH)_4^-(aq)$$

  c. The value of K is 40.0 and K_{sp} for $Al(OH)_3$ is 2×10^{-32}. Plot the solubility of $Al(OH)_3$ in the pH range 4–12.

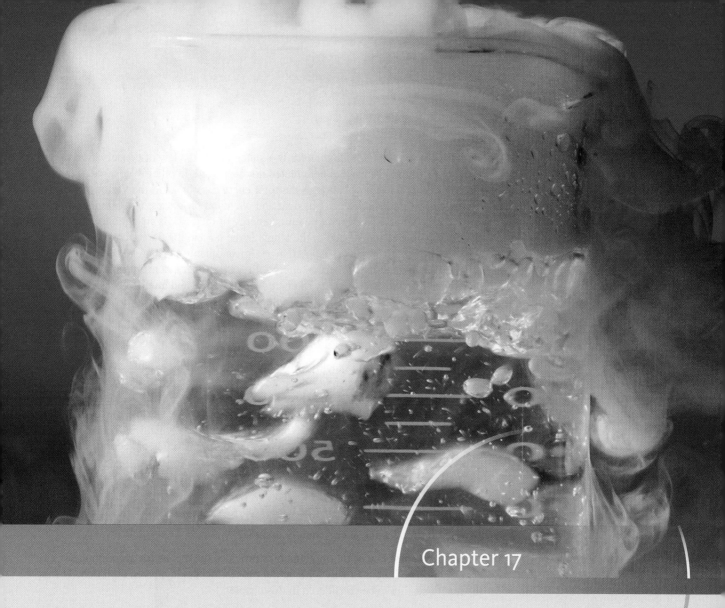

Chapter 17

Spontaneity, Entropy, and Free Energy

Solid carbon dioxide (dry ice), when placed in water, causes violent bubbling as gaseous CO_2 is released. The "fog" is moisture condensed from the cold air. (Phil Degginger/Alamy)

The *first law of thermodynamics* is a statement of the law of conservation of energy: Energy can be neither created nor destroyed. In other words, *the energy of the universe is constant.* Although the total energy is constant, the various forms of energy can be interchanged in physical and chemical processes. For example, if you drop a book, some of the initial potential energy of the book is changed to kinetic energy, which is then transferred to the atoms in the air and the floor as random motion. The net effect of this process is to change a given quantity of potential energy to exactly the same quantity of thermal energy. Energy has been converted from one form to another, but the same quantity of energy exists before and after the process.

Now let's consider a chemical example. When methane is burned in excess oxygen, the major reaction is

$$CH_4(g) + 2O_2(g) \longrightarrow CO_2(g) + 2H_2O(g) + \text{energy}$$

This reaction produces a quantity of energy, which is released as heat. This energy flow results from the lowering of the potential energy stored in the bonds of CH_4 and O_2 as they react to form CO_2 and H_2O. This is illustrated in Fig. 17.1. Potential energy has been converted to thermal energy, but the energy content of the universe has remained constant in accordance with the first law of thermodynamics.

The first law of thermodynamics is used mainly for energy bookkeeping, that is, to answer such questions as

How much energy is involved in the change?

Does energy flow into or out of the system?

What form does the energy finally assume?

> The first law of thermodynamics: The energy of the universe is constant.

Although the first law of thermodynamics provides the means for accounting for energy, it gives no hint as to why a particular process occurs in a given direction. This is the main question to be considered in this chapter.

17.1 | Spontaneous Processes and Entropy

A process is said to be *spontaneous* if it *occurs without outside intervention.* **Spontaneous processes** may be fast or slow. As we will see in this chapter, thermodynamics can tell us the *direction* in which a process will occur but can say nothing about the *speed* of the process. As we saw in Chapter 12, the rate of a reaction depends on many factors, such as activation energy, temperature, concentration, and catalysts, and we were able to explain these effects using a simple collision model. In describing a chemical reaction, the discipline of chemical kinetics focuses on the pathway between reactants and products; thermodynamics considers only the initial and final states and does not require knowledge of the pathway between reactants and products (Fig. 17.2).

In summary, thermodynamics lets us predict whether a process will occur but gives no information about the amount of time required for the process. For example,

> *Spontaneous* does not mean fast.

> An alternative way to describe a spontaneous process is to say that it is "thermodynamically favored."

Figure 17.1 | When methane and oxygen react to form carbon dioxide and water, the products have lower potential energy than the reactants. This change in potential energy results in energy flow (heat) to the surroundings.

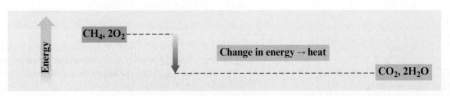

Figure 17.2 | The rate of a reaction depends on the pathway from reactants to products; this is the domain of kinetics. Thermodynamics tells us whether a reaction is spontaneous based only on the properties of the reactants and products. The predictions of thermodynamics do not require knowledge of the pathway between reactants and products.

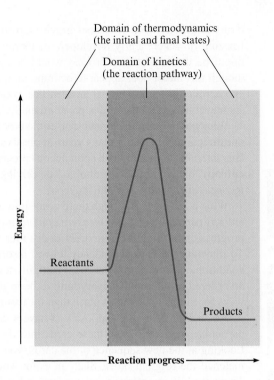

Domain of thermodynamics
(the initial and final states)

Domain of kinetics
(the reaction pathway)

Energy

Reactants

Products

◄── **Reaction progress** ──►

The process of diamond turning to graphite, while thermodynamically favored, is said to be under kinetic control; that is, it does not proceed at a noticeable rate.

according to the principles of thermodynamics, a diamond should change spontaneously to graphite. The fact that we do not observe this process does not mean the prediction is wrong; it simply means the process is very slow. Thus we need both thermodynamics and kinetics to describe reactions fully.

To explore the idea of spontaneity, consider the following physical and chemical processes:

A ball rolls down a hill but never spontaneously rolls back up the hill.

If exposed to air and moisture, steel rusts spontaneously. However, the iron oxide in rust does not spontaneously change back to iron metal and oxygen gas.

A gas fills its container uniformly. It never spontaneously collects at one end of the container.

Heat flow always occurs from a hot object to a cooler one. The reverse process never occurs spontaneously.

Wood burns spontaneously in an exothermic reaction to form carbon dioxide and water, but wood is not formed when carbon dioxide and water are heated together.

At temperatures below 0°C, water spontaneously freezes, and at temperatures above 0°C, ice spontaneously melts.

What thermodynamic principle will provide an explanation of why, under a given set of conditions, each of these diverse processes occurs in one direction and never in the reverse? In searching for an answer, we could explain the behavior of a ball on a hill in

Iron spontaneously rusts when it comes in contact with water.

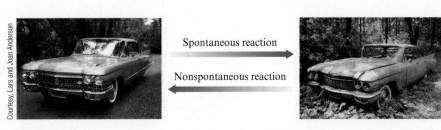

Spontaneous reaction

Nonspontaneous reaction

terms of gravity. But what does gravity have to do with the rusting of a nail or the freezing of water? Early developers of thermodynamics thought that exothermicity might be the key—that a process would be spontaneous if it were exothermic. Although this factor does appear to be important, since many spontaneous processes are exothermic, it is not the total answer. For example, the melting of ice, which occurs spontaneously at temperatures greater than 0°C, is an endothermic process.

What common characteristic causes the processes listed above to be spontaneous in one direction only? After many years of observation, scientists have concluded that the characteristic common to all spontaneous processes is an increase in a property called **entropy**, denoted by the symbol S. The driving force for a spontaneous process is an increase in the entropy of the universe.

What is entropy? Although there is no simple definition that is completely accurate, *entropy can be viewed as a measure of molecular randomness or disorder.* The natural progression of things is from order to disorder, from lower entropy to higher entropy. To illustrate the natural tendency toward disorder, you only have to think about the condition of your room. Your room naturally tends to get messy (disordered), because an ordered room requires everything to be in its place. There are simply many more ways for things to be out of place than for them to be in their places.

As another example, suppose you have a deck of playing cards ordered in some particular way. You throw these cards into the air and pick them all up at random. Looking at the new sequence of the cards, you would be very surprised to find that it matched the original order. Such an event would be possible, but *very improbable.* There are billions of ways for the deck to be disordered, but only one way to be ordered according to your definition. Thus the chances of picking the cards up out of order are much greater than the chance of picking them up in order. It is natural for disorder to increase.

Entropy is a thermodynamic function that describes the *number of arrangements* (positions and/or energy levels) that are *available to a system* existing in a given state. Entropy is closely associated with probability. The key concept is that the more ways a particular state can be achieved, the greater is the likelihood (probability) of finding that state. In other words, *nature spontaneously proceeds toward the states that have the highest probabilities of existing.* This conclusion is not surprising at all. The difficulty comes in connecting this concept to real-life processes. For example, what does the spontaneous rusting of steel have to do with probability? Understanding the connection between entropy and spontaneity will allow us to answer such questions. We will begin to explore this connection by considering a very simple process, the expansion of an ideal gas into a vacuum, as represented in Fig. 17.3. Why is this process

A disordered pile of playing cards.

Probability refers to likelihood.

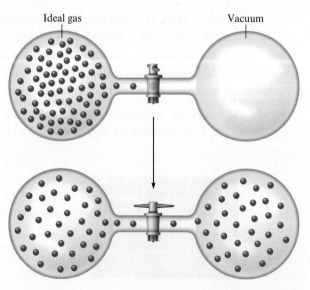

Figure 17.3 | The expansion of an ideal gas into an evacuated bulb.

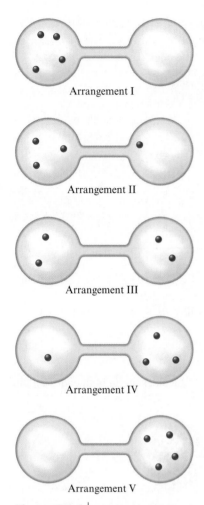

Figure 17.4 | Possible arrangements (states) of four molecules in a two-bulbed flask.

Table 17.1 | The Microstates That Give a Particular Arrangement (State)

Arrangement	Microstates	Number of Microstates
I		1
II		4
III		6
IV		4
V		1

spontaneous? The driving force is probability. Because there are more ways of having the gas evenly spread throughout the container than there are ways for it to be in any other possible state, the gas spontaneously attains the uniform distribution.

To understand this conclusion, we will greatly simplify the system and consider the possible arrangements of only four gas molecules in the two-bulbed container (Fig. 17.4). How many ways can each arrangement (state) be achieved? Arrangements I and V can be achieved in only one way—all the molecules must be in one end. Arrangements II and IV can be achieved in four ways, as shown in Table 17.1. Each configuration that gives a particular arrangement is called a *microstate*. Arrangement I has one microstate, and arrangement II has four microstates. Arrangement III can be achieved in six ways (six microstates), as shown in Table 17.1. *Which arrangement is most likely to occur?* The one that can be achieved in the greatest number of ways. Thus arrangement III is most probable. The relative probabilities of arrangements III, II, and I are $6:4:1$. We have discovered an important principle: The probability of occurrence of a particular arrangement (state) depends on the number of ways (microstates) in which that arrangement can be achieved.

The consequences of this principle are dramatic for large numbers of molecules. One gas molecule in the flask in Fig. 17.4 has one chance in two of being in the left bulb. We say that the probability of finding the molecule in the left bulb is $\frac{1}{2}$. For two molecules in the flask, there is one chance in two of finding each molecule in the left bulb, so there is one chance in four ($\frac{1}{2} \times \frac{1}{2} = \frac{1}{4}$) that *both* molecules will be in the left bulb. As the number of molecules increases, the relative probability of finding all of them in the left bulb decreases, as shown in Table 17.2. For 1 mole of gas, the probability of finding all the molecules in the left bulb is so small that this arrangement would "never" occur.

Thus a gas placed in one end of a container will spontaneously expand to fill the entire vessel evenly because, for a large number of gas molecules, there is a huge

For two molecules in the flask, there are four possible microstates:

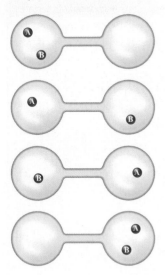

Thus there is one chance in four of finding

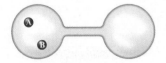

Table 17.2 | Probability of Finding All the Molecules in the Left Bulb as a Function of the Total Number of Molecules

Number of Molecules	Relative Probability of Finding All Molecules in the Left Bulb
1	$\dfrac{1}{2}$
2	$\dfrac{1}{2} \times \dfrac{1}{2} = \dfrac{1}{2^2} = \dfrac{1}{4}$
3	$\dfrac{1}{2} \times \dfrac{1}{2} \times \dfrac{1}{2} = \dfrac{1}{2^3} = \dfrac{1}{8}$
5	$\dfrac{1}{2} \times \dfrac{1}{2} \times \dfrac{1}{2} \times \dfrac{1}{2} \times \dfrac{1}{2} = \dfrac{1}{2^5} = \dfrac{1}{32}$
10	$\dfrac{1}{2^{10}} = \dfrac{1}{1024}$
n	$\dfrac{1}{2^n} = \left(\dfrac{1}{2}\right)^n$
6×10^{23} (1 mole)	$\left(\dfrac{1}{2}\right)^{6 \times 10^{23}} \approx 10^{-(2 \times 10^{23})}$

number of microstates in which equal numbers of molecules are in both ends. On the other hand, the opposite process,

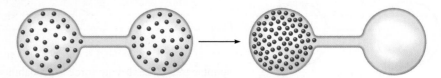

although not impossible, is *highly* improbable, since only one microstate leads to this arrangement. Therefore, this process does not occur spontaneously.

The type of probability we have been considering in this example is called **positional probability** because it depends on the number of configurations in space (positional microstates) that yield a particular state. A gas expands into a vacuum to give a uniform distribution because the expanded state has the highest positional probability, that is, the largest entropy, of the states available to the system.

Solid, liquid, and gaseous states were compared in Chapter 10.

Positional probability is also illustrated by changes of state. In general, positional entropy increases in going from solid to liquid to gas. A mole of a substance has a much smaller volume in the solid state than it does in the gaseous state. In the solid state, the molecules are close together, with relatively few positions available to them; in the gaseous state, the molecules are far apart, with many more positions available to them. The liquid state is closer to the solid state than it is to the gaseous state in these terms. We can summarize these comparisons as follows:

Solids are more ordered than liquids or gases and thus have lower entropy.

$$S_{\text{solid}} \quad < \quad S_{\text{liquid}} \quad \ll \quad S_{\text{gas}}$$

The tendency to mix is due to the increased volume available to the particles of each component of the mixture. For example, when two liquids are mixed, the molecules of each liquid have more available volume and thus more available positions.

Positional entropy is also very important in the formation of solutions. In Chapter 11 we saw that solution formation is favored by the natural tendency for substances to mix. We can now be more precise. The entropy change associated with the mixing of two pure substances is expected to be positive. An increase in entropy is expected because there are many more microstates for the mixed condition than for the separated condition. This effect is due principally to the increased volume available to a given "particle" after mixing occurs. For example, when two liquids are mixed to form a solution, the molecules of each liquid have more available volume and thus more available positions. Therefore, the increase in positional entropy associated with mixing favors the formation of solutions.

Interactive Example 17.1

Sign in at http://login.cengagebrain.com to try this Interactive Example in **OWL**.

Positional Entropy

For each of the following pairs, choose the substance with the higher positional entropy (per mole) at a given temperature.

a. Solid CO_2 and gaseous CO_2

b. N_2 gas at 1 atm and N_2 gas at 1.0×10^{-2} atm

Solution

a. Since a mole of gaseous CO_2 has the greater volume by far, the molecules have many more available positions than in a mole of solid CO_2. Thus gaseous CO_2 has the higher positional entropy.

b. A mole of N_2 gas at 1×10^{-2} atm has a volume 100 times that (at a given temperature) of a mole of N_2 gas at 1 atm. Thus N_2 gas at 1×10^{-2} atm has the higher positional entropy.

See Exercise 17.31

Interactive Example 17.2

Sign in at http://login.cengagebrain.com to try this Interactive Example in **OWL**.

Predicting Entropy Changes

Predict the sign of the entropy change for each of the following processes.

a. Solid sugar is added to water to form a solution.

b. Iodine vapor condenses on a cold surface to form crystals.

Solution

a. The sugar molecules become randomly dispersed in the water when the solution forms and thus have access to a larger volume and a larger number of possible positions. The positional disorder is increased, and there will be an increase in entropy. ΔS is positive, since the final state has a larger entropy than the initial state, and $\Delta S = S_{\text{final}} - S_{\text{initial}}$.

b. Gaseous iodine is forming a solid. This process involves a change from a relatively large volume to a much smaller volume, which results in lower positional disorder. For this process ΔS is negative (the entropy decreases).

See Exercise 17.32

Chemical connections
Entropy: An Organizing Force?

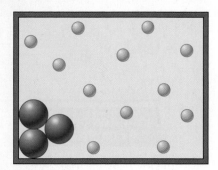

In this text we have emphasized the meaning of the second law of thermodynamics—that the entropy of the universe is always increasing. Although the results of all our experiments support this conclusion, this does not mean that order cannot appear spontaneously in a given part of the universe. The best example of this phenomenon involves the assembly of cells in living organisms. Of course, when a process that creates an ordered system is examined in detail, it is found that other parts of the process involve an increase in disorder such that the sum of all the entropy changes is positive. In fact, scientists are now finding that the search for maximum entropy in one part of a system can be a powerful force for organization in another part of the system.

To understand how entropy can be an organizing force, look at the accompanying figure. In a system containing large and small "balls" as shown in the figure, the small balls can "herd" the large balls into clumps in the corners and near the walls. This clears out the maximum space for the small balls so that they can move more freely, thus maximizing the entropy of the system, as demanded by the second law of thermodynamics.

In essence, the ability to maximize entropy by sorting different-sized objects creates a kind of attractive force, called a *depletion*, or *excluded-volume, force*. These "entropic forces" operate for objects in the size range of approximately 10^{-8} to approximately 10^{-6} m. For entropy-induced ordering to occur, the particles must be constantly jostling each other and must be constantly agitated by solvent molecules, thus making gravity unimportant.

There is increasing evidence that entropic ordering is important in many biological systems. For example, this phenomenon seems to be responsible for the clumping of sickle-cell hemoglobin in the presence of much smaller proteins that act as the "smaller balls." Entropic forces also have been linked to the clustering of DNA in cells without nuclei, and Allen Minton of the National Institutes of Health in Bethesda, Maryland, is studying the role of entropic forces in the binding of proteins to cell membranes.

Entropic ordering also appears in nonbiological settings, especially in the ways polymer molecules clump together. For example, polymers added to paint to improve the flow characteristics of the paint actually caused it to coagulate because of depletion forces.

Thus, as you probably have concluded already, entropy is a complex issue. As entropy drives the universe to its ultimate death of maximum chaos, it provides some order along the way.

17.2 | Entropy and the Second Law of Thermodynamics

The total energy of the universe is constant, but the entropy is increasing.

We have seen that processes are spontaneous when they result in an increase in disorder. Nature always moves toward the most probable state available to it. We can state this principle in terms of entropy: In any spontaneous process there is always an increase in the entropy of the universe. This is the **second law of thermodynamics**. Contrast this with the first law of thermodynamics, which tells us that the energy of the universe is constant. Energy is conserved in the universe, but entropy is not. In fact, the second law can be paraphrased as follows: *The entropy of the universe is increasing.*

As in Chapter 6, we find it convenient to divide the universe into a system and the surroundings. Thus we can represent the change in the entropy of the universe as

$$\Delta S_{\text{univ}} = \Delta S_{\text{sys}} + \Delta S_{\text{surr}}$$

where ΔS_{sys} and ΔS_{surr} represent the changes in entropy that occur in the system and surroundings, respectively.

To predict whether a given process will be spontaneous, we must know the sign of ΔS_{univ}. If ΔS_{univ} is positive, the entropy of the universe increases, and the process is spontaneous in the direction written. If ΔS_{univ} is negative, the process is spontaneous in the *opposite* direction. If ΔS_{univ} is zero, the process has no tendency to occur, and the system is at equilibrium. To predict whether a process is spontaneous, we must consider the entropy changes that occur both in the system and in the surroundings and then take their sum.

Critical Thinking

What if ΔS_{univ} was a state function? How would the world be different?

Example 17.3

The Second Law

In a living cell, large molecules are assembled from simple ones. Is this process consistent with the second law of thermodynamics?

Solution

To reconcile the operation of an order-producing cell with the second law of thermodynamics, we must remember that ΔS_{univ}, not ΔS_{sys}, must be positive for a process to be spontaneous. A process for which ΔS_{sys} is negative can be spontaneous if the associated ΔS_{surr} is both larger and positive. The operation of a cell is such a process.

See Questions 17.11 and 17.12

17.3 | The Effect of Temperature on Spontaneity

To explore the interplay of ΔS_{sys} and ΔS_{surr} in determining the sign of ΔS_{univ}, we will first discuss the change of state for 1 mole of water from liquid to gas,

$$H_2O(l) \longrightarrow H_2O(g)$$

considering the water to be the system and everything else the surroundings.

What happens to the entropy of water in this process? A mole of liquid water (18 g) has a volume of approximately 18 mL. A mole of gaseous water at 1 atmosphere and 100°C occupies a volume of approximately 31 L. Clearly, there are many more positions available to the water molecules in a volume of 31 L than in 18 mL, and the vaporization of water is favored by this increase in positional probability. That is, for this process the entropy of the system increases; ΔS_{sys} has a positive sign.

What about the entropy change in the surroundings? Although we will not prove it here, entropy changes in the surroundings are determined primarily by the flow of energy into or out of the system as heat. To understand this, suppose an exothermic process transfers 50 J of energy as heat to the surroundings, where it becomes thermal energy, that is, kinetic energy associated with the random motions of atoms. Thus this flow of energy into the surroundings increases the random motions of atoms there and thereby increases the entropy of the surroundings. The sign of ΔS_{surr} is positive. When an endothermic process occurs in the system, it produces the opposite effect. Heat flows from the surroundings to the system, and the random motions of the atoms in the

Boiling water to form steam increases its volume and thus its entropy.

Can Balcioglu/Shutterstock

surroundings decrease, decreasing the entropy of the surroundings. The vaporization of water is an endothermic process. Thus, for this change of state, ΔS_{surr} is negative.

Remember it is the sign of ΔS_{univ} that tells us whether the vaporization of water is spontaneous. We have seen that ΔS_{sys} is positive and favors the process and that ΔS_{surr} is negative and unfavorable. Thus the components of ΔS_{univ} are in opposition. Which one controls the situation? The answer *depends on the temperature.* We know that at a pressure of 1 atmosphere, water changes spontaneously from liquid to gas at all temperatures above 100°C. Below 100°C, the opposite process (condensation) is spontaneous.

Since ΔS_{sys} and ΔS_{surr} are in opposition for the vaporization of water, the temperature must have an effect on the relative importance of these two terms. To understand why this is so, we must discuss in more detail the factors that control the entropy changes in the surroundings. The central idea is that the entropy changes in the surroundings are primarily determined by heat flow. An exothermic process in the system increases the entropy of the surroundings, because the resulting energy flow increases the random motions in the surroundings. This means that exothermicity is an important driving force for spontaneity. In earlier chapters we have seen that a system tends to undergo changes that lower its energy. We now understand the reason for this tendency. When a system at constant temperature moves to a lower energy, the energy it gives up is transferred to the surroundings, leading to an increase in entropy there.

The significance of exothermicity as a driving force *depends on the temperature at which the process occurs.* That is, the magnitude of ΔS_{surr} depends on the temperature at which the heat is transferred. We will not attempt to prove this fact here. Instead, we offer an analogy. Suppose that you have $50 to give away. Giving it to a millionaire would not create much of an impression—a millionaire has money to spare. However, to a poor college student, $50 would represent a significant sum and would be received with considerable joy. The same principle can be applied to energy transfer via the flow of heat. If 50 J of energy is transferred to the surroundings, the impact of that event depends greatly on the temperature. If the temperature of the surroundings is very high, the atoms there are in rapid motion. The 50 J of energy will not make a large percent change in these motions. On the other hand, if 50 J of energy is transferred to the surroundings at a very low temperature, where atomic motion is slow, the energy will cause a large percent change in these motions. The impact of the transfer of a given quantity of energy as heat to or from the surroundings will be greater at lower temperatures.

For our purposes, there are two important characteristics of the entropy changes that occur in the surroundings:

1. *The sign of ΔS_{surr} depends on the direction of the heat flow.* At constant temperature, an exothermic process in the system causes heat to flow into the surroundings, increasing the random motions and thus the entropy of the surroundings. For this case, ΔS_{surr} is positive. The opposite is true for an endothermic process in a system at constant temperature. Note that although the driving force described here really results from the change in entropy, it is often described in terms of energy: Nature tends to seek the lowest possible energy.

2. *The magnitude of ΔS_{surr} depends on the temperature.* The transfer of a given quantity of energy as heat produces a much greater percent change in the randomness of the surroundings at a low temperature than it does at a high temperature. Thus ΔS_{surr} depends directly on the quantity of heat transferred and inversely on temperature. In other words, the tendency for the system to lower its energy becomes a more important driving force at lower temperatures.

> In an endothermic process, heat flows from the surroundings into the system. In an exothermic process, heat flows into the surroundings from the system.

> In a process occurring at constant temperature, the tendency for the system to lower its energy results from the positive value of ΔS_{surr}.

$$\begin{array}{c} \text{Driving force} \\ \text{provided by} \\ \text{the energy flow} \\ \text{(heat)} \end{array} = \begin{array}{c} \text{magnitude of the} \\ \text{entropy change of} \\ \text{the surroundings} \end{array} = \frac{\text{quantity of heat (J)}}{\text{temperature (K)}}$$

These ideas are summarized as follows:

Exothermic process:
ΔS_{surr} = positive

Endothermic process:
ΔS_{surr} = negative

Exothermic process:	$\Delta S_{surr} = +\dfrac{\text{quantity of heat (J)}}{\text{temperature (K)}}$
Endothermic process:	$\Delta S_{surr} = -\dfrac{\text{quantity of heat (J)}}{\text{temperature (K)}}$

We can express ΔS_{surr} in terms of the change in enthalpy ΔH for a process occurring at constant pressure, since

$$\text{Heat flow (constant } P) = \text{change in enthalpy} = \Delta H$$

When no subscript is present, the quantity (for example, ΔH) refers to the system.

Recall that ΔH consists of two parts: a sign and a number. The *sign* indicates the direction of flow, where a plus sign means into the system (endothermic) and a minus sign means out of the system (exothermic). The *number* indicates the quantity of energy.

Combining all these concepts produces the following definition of ΔS_{surr} for a reaction that takes place under conditions of constant temperature (in kelvins) and pressure:

$$\Delta S_{surr} = -\frac{\Delta H}{T}$$

The minus sign changes the point of view from the system to the surroundings.

The minus sign is necessary because the sign of ΔH is determined with respect to the reaction system, and this equation expresses a property of the surroundings. This means that if the reaction is exothermic, ΔH has a negative sign, but since heat flows into the surroundings, ΔS_{surr} is positive.

Interactive Example 17.4

Sign in at http://login.cengagebrain .com to try this Interactive Example in OWL.

The mineral stibnite contains Sb_2S_3.

Determining ΔS_{surr}

In the metallurgy of antimony, the pure metal is recovered via different reactions, depending on the composition of the ore. For example, iron is used to reduce antimony in sulfide ores:

$$Sb_2S_3(s) + 3Fe(s) \longrightarrow 2Sb(s) + 3FeS(s) \qquad \Delta H = -125 \text{ kJ}$$

Carbon is used as the reducing agent for oxide ores:

$$Sb_4O_6(s) + 6C(s) \longrightarrow 4Sb(s) + 6CO(g) \qquad \Delta H = 778 \text{ kJ}$$

Calculate ΔS_{surr} for each of these reactions at 25°C and 1 atm.

Solution

We use

$$\Delta S_{surr} = -\frac{\Delta H}{T}$$

where $\qquad\qquad T = 25 + 273 = 298 \text{ K}$

For the sulfide ore reaction,

$$\Delta S_{surr} = -\frac{-125 \text{ kJ}}{298 \text{ K}} = 0.419 \text{ kJ/K} = 419 \text{ J/K}$$

Note that ΔS_{surr} is positive, as it should be, since this reaction is exothermic and heat flow occurs to the surroundings, increasing the randomness of the surroundings.

Table 17.3 | Interplay of ΔS_{sys} and ΔS_{surr} in Determining the Sign of ΔS_{univ}

Signs of Entropy Changes			
ΔS_{sys}	ΔS_{surr}	ΔS_{univ}	Process Spontaneous?
+	+	+	Yes
−	−	−	No (reaction will occur in opposite direction)
+	−	?	Yes, if ΔS_{sys} has a larger magnitude than ΔS_{surr}
−	+	?	Yes, if ΔS_{surr} has a larger magnitude than ΔS_{sys}

For the oxide ore reaction,

$$\Delta S_{surr} = -\frac{778 \text{ kJ}}{298} = -2.61 \text{ kJ/K} = -2.61 \times 10^3 \text{ J/K}$$

In this case ΔS_{surr} is negative because heat flow occurs from the surroundings to the system.

See Exercises 17.33 and 17.34

We have seen that the spontaneity of a process is determined by the entropy change it produces in the universe. We also have seen that ΔS_{univ} has two components, ΔS_{sys} and ΔS_{surr}. If for some process both ΔS_{sys} and ΔS_{surr} are positive, then ΔS_{univ} is positive, and the process is spontaneous. If, on the other hand, both ΔS_{sys} and ΔS_{surr} are negative, the process does not occur in the direction indicated but is spontaneous in the opposite direction. Finally, if ΔS_{sys} and ΔS_{surr} have opposite signs, the spontaneity of the process depends on the sizes of the opposing terms. These cases are summarized in Table 17.3.

We can now understand why spontaneity is often dependent on temperature and thus why water spontaneously freezes below 0°C and melts above 0°C. The term ΔS_{surr} is temperature-dependent. Since

$$\Delta S_{surr} = -\frac{\Delta H}{T}$$

at constant pressure, the value of ΔS_{surr} changes markedly with temperature. The magnitude of ΔS_{surr} will be very small at high temperatures and will increase as the temperature decreases. That is, exothermicity is most important as a driving force at low temperatures.

17.4 | Free Energy

The symbol G for free energy honors Josiah Willard Gibbs (1839–1903), who was professor of mathematical physics at Yale University from 1871 to 1903. He laid the foundations of many areas of thermodynamics, particularly as they apply to chemistry.

So far we have used ΔS_{univ} to predict the spontaneity of a process. However, another thermodynamic function is also related to spontaneity and is especially useful in dealing with the temperature dependence of spontaneity. This function is called the **free energy**, which is symbolized by G and defined by the relationship

$$G = H - TS$$

where H is the enthalpy, T is the Kelvin temperature, and S is the entropy.

For a process that occurs at constant temperature, the change in free energy (ΔG) is given by the equation

$$\Delta G = \Delta H - T\Delta S$$

Note that all quantities here refer to the system. From this point on we will follow the usual convention that when no subscript is included, the quantity refers to the system.

To see how this equation relates to spontaneity, we divide both sides of the equation by $-T$ to produce

$$-\frac{\Delta G}{T} = -\frac{\Delta H}{T} + \Delta S$$

Remember that at constant temperature and pressure

$$\Delta S_{surr} = -\frac{\Delta H}{T}$$

So we can write

$$-\frac{\Delta G}{T} = -\frac{\Delta H}{T} + \Delta S = \Delta S_{surr} + \Delta S = \Delta S_{univ}$$

We have shown that

$$\Delta S_{univ} = -\frac{\Delta G}{T} \qquad \text{at constant } T \text{ and } P$$

This result is very important. It means that a process carried out at constant temperature and pressure will be spontaneous only if ΔG is negative. A process (at constant T and P) is spontaneous in the direction in which the free energy decreases ($-\Delta G$ means $+\Delta S_{univ}$).

Now we have two functions that can be used to predict spontaneity: the entropy of the universe, which applies to all processes, and free energy, which can be used for processes carried out at constant temperature and pressure. Since so many chemical reactions occur under the latter conditions, free energy is the more useful to chemists.

Let's use the free energy equation to predict the spontaneity of the melting of ice:

$$H_2O(s) \longrightarrow H_2O(l)$$

for which $\Delta H° = 6.03 \times 10^3$ J/mol and $\Delta S° = 22.1$ J/K · mol

The superscript degree symbol (°) indicates all substances are in their standard states.

Results of the calculations of ΔS_{univ} and $\Delta G°$ at $-10°C$, $0°C$, and $10°C$ are shown in Table 17.4. These data predict that the process is spontaneous at $10°C$; that is, ice melts at this temperature because ΔS_{univ} is positive and $\Delta G°$ is negative. The opposite is true at $-10°C$, where water freezes spontaneously.

To review the definitions of standard states, see Section 6.4.

Why is this so? The answer lies in the fact that ΔS_{sys} ($\Delta S°$) and ΔS_{surr} oppose each other. The term $\Delta S°$ favors the melting of ice because of the increase in positional entropy, and ΔS_{surr} favors the freezing of water because it is an exothermic process. At temperatures below $0°C$, the change of state occurs in the exothermic direction because ΔS_{surr} is larger in magnitude than ΔS_{sys}. But above $0°C$ the change occurs in the direction in which ΔS_{sys} is favorable, since in this case ΔS_{sys} is larger in magnitude than ΔS_{surr}. At $0°C$ the *opposing tendencies just balance,* and the two states coexist;

Table 17.4 | Results of the Calculation of ΔS_{univ} and $\Delta G°$ for the Process $H_2O(s) \rightarrow H_2O(l)$ at $-10°C$, $0°C$, and $10°C$*

T (°C)	T (K)	$\Delta H°$ (J/mol)	$\Delta S°$ (J/K · mol)	$\Delta S_{surr} = -\dfrac{\Delta H°}{T}$ (J/K · mol)	$\Delta S_{univ} = \Delta S° + \Delta S_{surr}$ (J/K · mol)	$T\Delta S°$ (J/mol)	$\Delta G° = \Delta H° - T\Delta S°$ (J/mol)
−10	263	6.03×10^3	22.1	−22.9	−0.8	5.81×10^3	$+2.2 \times 10^2$
0	273	6.03×10^3	22.1	−22.1	0	6.03×10^3	0
10	283	6.03×10^3	22.1	−21.3	+0.8	6.25×10^3	-2.2×10^2

*Note that at $10°C$, $\Delta S°$ (ΔS_{sys}) controls, and the process occurs even though it is endothermic. At $-10°C$, the magnitude of ΔS_{surr} is larger than that of $\Delta S°$, so the process is spontaneous in the opposite (exothermic) direction.

Note that although ΔH and ΔS are somewhat temperature-dependent, it is a good approximation to assume they are constant over a relatively small temperature range.

Table 17.5 | Various Possible Combinations of ΔH and ΔS for a Process and the Resulting Dependence of Spontaneity on Temperature

Case	Result
ΔS positive, ΔH negative	Spontaneous at all temperatures
ΔS positive, ΔH positive	Spontaneous at high temperatures (where exothermicity is relatively unimportant)
ΔS negative, ΔH negative	Spontaneous at low temperatures (where exothermicity is dominant)
ΔS negative, ΔH positive	Process not spontaneous at *any* temperature (reverse process is spontaneous at *all* temperatures)

there is no driving force in either direction. An equilibrium exists between the two states of water. Note that ΔS_{univ} is equal to 0 at 0°C.

We can reach the same conclusions by examining $\Delta G°$. At $-10°$, $\Delta G°$ is positive because the $\Delta H°$ term is larger than the $T\Delta S°$ term. The opposite is true at 10°C. At 0°C, $\Delta H°$ is equal to $T\Delta S°$ and $\Delta G°$ is equal to 0. This means that solid H_2O and liquid H_2O have the same free energy at 0°C ($\Delta G° = G_{liquid} - G_{solid}$), and the system is at equilibrium.

We can understand the temperature dependence of spontaneity by examining the behavior of ΔG. For a process occurring at constant temperature and pressure,

$$\Delta G = \Delta H - T\Delta S$$

If ΔH and ΔS favor opposite processes, spontaneity will depend on temperature in such a way that the exothermic direction will be favored at low temperatures. For example, for the process

$$H_2O(s) \longrightarrow H_2O(l)$$

ΔH is positive and ΔS is positive. The natural tendency for this system to lower its energy is in opposition to its natural tendency to increase its positional randomness. At low temperatures, ΔH dominates, and at high temperatures, ΔS dominates. The various possible cases are summarized in Table 17.5.

Critical Thinking

Consider an ideal gas in a container fitted with a frictionless, massless piston. What if weight is added to the top of the piston? We would expect the gas to be compressed at constant temperature. For this to be true, ΔS would be negative (since the gas is compressed) and ΔH would be zero (since the process is at constant temperature). This would make ΔG positive. Does this mean the isothermal compression of the gas is not spontaneous? Defend your answer.

Interactive Example 17.5

Sign in at http://login.cengagebrain .com to try this Interactive Example in **OWL**.

Free Energy and Spontaneity

At what temperatures is the following process spontaneous at 1 atm?

$$Br_2(l) \longrightarrow Br_2(g)$$
$$\Delta H° = 31.0 \text{ kJ/mol} \quad \text{and} \quad \Delta S° = 93.0 \text{ J/K} \cdot \text{mol}$$

What is the normal boiling point of liquid Br_2?

Solution

The vaporization process will be spontaneous at all temperatures where $\Delta G°$ is negative. Note that $\Delta S°$ favors the vaporization process because of the increase in positional entropy, and $\Delta H°$ favors the *opposite* process, which is exothermic. These opposite tendencies will exactly balance at the boiling point of liquid Br_2, since at this temperature liquid and gaseous Br_2 are in equilibrium ($\Delta G° = 0$). We can find this temperature by setting $\Delta G° = 0$ in the equation

$$\Delta G° = \Delta H° - T\Delta S°$$
$$0 = \Delta H° - T\Delta S°$$
$$\Delta H° = T\Delta S°$$

Then
$$T = \frac{\Delta H°}{\Delta S°} = \frac{3.10 \times 10^4 \text{ J/mol}}{93.0 \text{ J/K} \cdot \text{mol}} = 333 \text{ K}$$

At temperatures above 333 K, $T\Delta S°$ has a larger magnitude than $\Delta H°$, and $\Delta G°$ (or $\Delta H° - T\Delta S°$) is negative. Above 333 K, the vaporization process is spontaneous; the opposite process occurs spontaneously below this temperature. At 333 K, liquid and gaseous Br_2 coexist in equilibrium. These observations can be summarized as follows (the pressure is 1 atm in each case):

1. $T > 333$ K. The term $\Delta S°$ controls. The increase in entropy when liquid Br_2 is vaporized is dominant.
2. $T < 333$ K. The process is spontaneous in the direction in which it is exothermic. The term $\Delta H°$ controls.
3. $T = 333$ K. The opposing driving forces are just balanced ($\Delta G° = 0$), and the liquid and gaseous phases of bromine coexist. This is the normal boiling point.

See Exercises 17.37 through 17.39

17.5 | Entropy Changes in Chemical Reactions

The second law of thermodynamics tells us that a process will be spontaneous if the entropy of the universe increases when the process occurs. We saw in Section 17.4 that for a process at constant temperature and pressure, we can use the change in free energy of the system to predict the sign of ΔS_{univ} and thus the direction in which it is spontaneous. So far we have applied these ideas only to physical processes, such as changes of state and the formation of solutions. However, the main business of chemistry is studying chemical reactions, and, therefore, we want to apply the second law to reactions.

First, we will consider the entropy changes accompanying chemical reactions that occur under conditions of constant temperature and pressure. As for the other types of processes we have considered, the entropy changes in the *surroundings* are determined by the heat flow that occurs as the reaction takes place. However, the entropy changes in the *system* (the reactants and products of the reaction) are determined by positional probability.

For example, in the ammonia synthesis reaction

$$N_2(g) + 3H_2(g) \longrightarrow 2NH_3(g)$$

four reactant molecules become two product molecules, lowering the number of independent units in the system, which leads to less positional disorder.

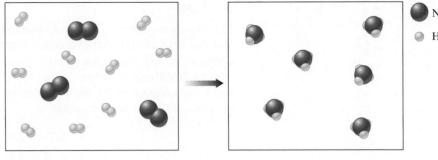

| Greater entropy | Less entropy |

Fewer molecules mean fewer possible configurations. To help clarify this idea, consider a special container with a million compartments, each large enough to hold a hydrogen molecule. Thus there are a million ways one H_2 molecule can be placed in this container. But suppose we break the H—H bond and place the two independent H atoms in the same container. A little thought will convince you that there are *many* more than a million ways to place the two separate atoms. The number of arrangements possible for the two independent atoms is much greater than the number for the molecule. Thus for the process

$$H_2 \longrightarrow 2H$$

positional entropy increases.

Does positional entropy increase or decrease when the following reaction takes place?

$$4NH_3(g) + 5O_2(g) \longrightarrow 4NO(g) + 6H_2O(g)$$

In this case 9 gaseous molecules are changed to 10 gaseous molecules, and the positional entropy increases. There are more independent units as products than as reactants. In general, when a reaction involves gaseous molecules, the change in positional entropy is dominated by the relative numbers of molecules of gaseous reactants and products. If the number of molecules of the gaseous products is greater than the number of molecules of the gaseous reactants, positional entropy typically increases, and ΔS will be positive for the reaction.

Interactive Example 17.6

Sign in at http://login.cengagebrain .com to try this Interactive Example in **OWL**.

Predicting the Sign of $\Delta S°$

Predict the sign of $\Delta S°$ for each of the following reactions.

a. The thermal decomposition of solid calcium carbonate:

$$CaCO_3(s) \longrightarrow CaO(s) + CO_2(g)$$

b. The oxidation of SO_2 in air:

$$2SO_2(g) + O_2(g) \longrightarrow 2SO_3(g)$$

Solution

a. Since in this reaction a gas is produced from a solid reactant, the positional entropy increases, and $\Delta S°$ is positive.

b. Here three molecules of gaseous reactants become two molecules of gaseous products. Since the number of gas molecules decreases, positional entropy decreases, and $\Delta S°$ is negative.

See Exercises 17.41 and 17.42

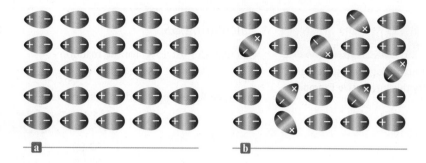

Figure 17.5 | (a) An idealized perfect crystal of hydrogen chloride at 0 K; the dipolar HCl molecules are represented by ⊕ ⊖. The entropy is zero ($S = 0$) for this crystal at 0 K. (b) As the temperature rises above 0 K, lattice vibrations allow some dipoles to change their orientations, producing some disorder and an increase in entropy ($S > 0$).

A perfect crystal at 0 K is an unattainable ideal, taken as a standard but never actually observed.

In thermodynamics it is the *change* in a certain function that usually is important. The change in enthalpy determines if a reaction is exothermic or endothermic at constant pressure. The change in free energy determines if a process is spontaneous at constant temperature and pressure. It is fortunate that changes in thermodynamic functions are sufficient for most purposes, since absolute values for many thermodynamic characteristics of a system, such as enthalpy or free energy, cannot be determined.

However, we can assign absolute entropy values. Consider a solid at 0 K, where molecular motion virtually ceases. If the substance is a perfect crystal, its internal arrangement is absolutely regular [Fig. 17.5(a)]. There is only *one way* to achieve this perfect order: Every particle must be in its place. For example, with N coins there is only one way to achieve the state of all heads. Thus a perfect crystal represents the lowest possible entropy; that is, *the entropy of a perfect crystal at 0 K is zero*. This is a statement of the **third law of thermodynamics**.

As the temperature of a perfect crystal is increased, the random vibrational motions increase, and disorder increases within the crystal [Fig. 17.5(b)]. Thus the entropy of a substance increases with temperature. Since S is zero for a perfect crystal at 0 K, the entropy value for a substance at a particular temperature can be calculated by knowing the temperature dependence of entropy. (We will not show such calculations here.)

The standard entropy values represent the increase in entropy that occurs when a substance is heated from 0 K to 298 K at 1 atm pressure.

The *standard entropy values* $(S°)$ of many common substances at 298 K and 1 atm are listed in Appendix 4. From these values you will see that the entropy of a substance does indeed increase in going from solid to liquid to gas. One especially interesting feature of this table is the very low $S°$ value for diamond. The structure of diamond is highly ordered, with each carbon strongly bound to a tetrahedral arrangement of four other carbon atoms (see Section 10.5, Fig. 10.22). This type of structure allows very little disorder and has a very low entropy, even at 298 K. Graphite has a slightly higher entropy because its layered structure allows for a little more disorder.

Because *entropy is a state function of the system* (it is not pathway-dependent), the entropy change for a given chemical reaction can be calculated by taking the difference between the standard entropy values of products and those of the reactants:

$$\Delta S°_{\text{reaction}} = \Sigma n_{\text{p}} S°_{\text{products}} - \Sigma n_{\text{r}} S°_{\text{reactants}}$$

where, as usual, Σ represents the sum of the terms. It is important to note that entropy is an extensive property (it depends on the amount of substance present). This means that the number of moles of a given reactant (n_{r}) or product (n_{p}) must be taken into account.

Interactive Example 17.7

Sign in at http://login.cengagebrain.com to try this Interactive Example in OWL.

Calculating $\Delta S°$

Calculate $\Delta S°$ at 25°C for the reaction

$$2NiS(s) + 3O_2(g) \longrightarrow 2SO_2(g) + 2NiO(s)$$

given the following standard entropy values:

Substance	$S°$ (J/K · mol)
$SO_2(g)$	248
$NiO(s)$	38
$O_2(g)$	205
$NiS(s)$	53

Solution

Since

$$\begin{aligned}\Delta S° &= \Sigma n_p S°_{products} - \Sigma n_r S°_{reactants} \\ &= 2S°_{SO_2(g)} + 2S°_{NiO(s)} - 2S°_{NiS(s)} - 3S°_{O_2(g)} \\ &= 2\text{ mol}\left(248 \frac{J}{K \cdot mol}\right) + 2\text{ mol}\left(38 \frac{J}{K \cdot mol}\right) \\ &\quad -2\text{ mol}\left(53 \frac{J}{K \cdot mol}\right) - 3\text{ mol}\left(205 \frac{J}{K \cdot mol}\right) \\ &= 496\text{ J/K} + 76\text{ J/K} - 106\text{ J/K} - 615\text{ J/K} \\ &= -149\text{ J/K}\end{aligned}$$

We would expect $\Delta S°$ to be negative because the number of gaseous molecules decreases in this reaction.

See Exercise 17.45

Interactive Example 17.8

Sign in at http://login.cengagebrain.com to try this Interactive Example in OWL.

Calculating $\Delta S°$

Calculate $\Delta S°$ for the reduction of aluminum oxide by hydrogen gas:

$$Al_2O_3(s) + 3H_2(g) \longrightarrow 2Al(s) + 3H_2O(g)$$

Use the following standard entropy values:

Substance	$S°$ (J/K · mol)
$Al_2O_3(s)$	51
$H_2(g)$	131
$Al(s)$	28
$H_2O(g)$	189

Solution

$$\begin{aligned}\Delta S° &= \Sigma n_p S°_{products} - \Sigma n_r S°_{reactants} \\ &= 2S°_{Al(s)} + 3S°_{H_2O(g)} - 3S°_{H_2(g)} - S°_{Al_2O_3(s)} \\ &= 2\text{ mol}\left(28 \frac{J}{K \cdot mol}\right) + 3\text{ mol}\left(189 \frac{J}{K \cdot mol}\right) \\ &\quad -3\text{ mol}\left(131 \frac{J}{K \cdot mol}\right) - 1\text{ mol}\left(51 \frac{J}{K \cdot mol}\right) \\ &= 56\text{ J/K} + 567\text{ J/K} - 393\text{ J/K} - 51\text{ J/K} \\ &= 179\text{ J/K}\end{aligned}$$

See Exercises 17.46 through 17.48

Figure 17.6 | The H_2O molecule can vibrate and rotate in several ways, some of which are shown here. This freedom of motion leads to a higher entropy for water than for a substance like hydrogen, a simple diatomic molecule with fewer possible motions.

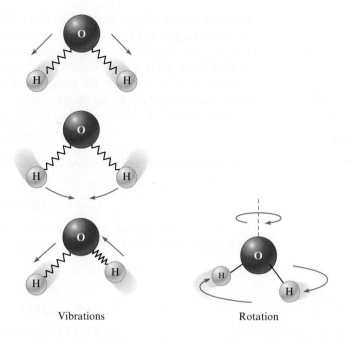

Vibrations   Rotation

The reaction considered in Example 17.8 involves 3 moles of hydrogen gas on the reactant side and 3 moles of water vapor on the product side. Would you expect ΔS to be large or small for such a case? We have assumed that ΔS depends on the relative numbers of molecules of gaseous reactants and products. Based on this assumption, ΔS should be near zero for this reaction. However, ΔS is large and positive. Why is this so? The large value for ΔS results from the difference in the entropy values for hydrogen gas and water vapor. The reason for this difference can be traced to the difference in molecular structure. Because it is a nonlinear, triatomic molecule, H_2O has more rotational and vibrational motions (Fig. 17.6) than does the diatomic H_2 molecule. Thus the standard entropy value for $H_2O(g)$ is greater than that for $H_2(g)$. Generally, *the more complex the molecule, the higher the standard entropy value.*

17.6 | Free Energy and Chemical Reactions

For chemical reactions we are often interested in the **standard free energy change** ($\Delta G°$), *the change in free energy that will occur if the reactants in their standard states are converted to the products in their standard states.* For example, for the ammonia synthesis reaction at 25°C,

$$N_2(g) + 3H_2(g) \rightleftharpoons 2NH_3(g) \qquad \Delta G° = -33.3 \text{ kJ} \qquad (17.1)$$

This $\Delta G°$ value represents the change in free energy when 1 mole of nitrogen gas at 1 atm reacts with 3 moles of hydrogen gas at 1 atm to produce 2 moles of gaseous NH_3 at 1 atm.

It is important to recognize that the standard free energy change for a reaction is not measured directly. For example, we can measure heat flow in a calorimeter to determine $\Delta H°$, but we cannot measure $\Delta G°$ this way. The value of $\Delta G°$ for the ammonia synthesis in Equation (17.1) was *not* obtained by mixing 1 mole of N_2 and 3 moles of H_2 in a flask and measuring the change in free energy as 2 moles of NH_3 formed. For one thing, if we mixed 1 mole of N_2 and 3 moles of H_2 in a flask, the system would go to equilibrium rather than to completion. Also, we have no instrument that measures free energy. However, while we cannot directly measure $\Delta G°$ for a reaction, we can calculate it from other measured quantities, as we will see later in this section.

Why is it useful to know $\Delta G°$ for a reaction? As we will see in more detail later in this chapter, knowing the $\Delta G°$ values for several reactions allows us to compare the relative tendency of these reactions to occur. The more negative the value of $\Delta G°$, the further a reaction will go to the right to reach equilibrium. We must use standard-state free energies to make this comparison because free energy varies with pressure or concentration. Thus, to get an accurate comparison of reaction tendencies, we must compare all reactions under the same pressure or concentration conditions. We will have more to say about the significance of $\Delta G°$ later.

The value of $\Delta G°$ tells us nothing about the rate of a reaction, only its eventual equilibrium position.

There are several ways to calculate $\Delta G°$. One common method uses the equation

$$\Delta G° = \Delta H° - T\Delta S°$$

which applies to a reaction carried out at constant temperature. For example, for the reaction

$$C(s) + O_2(g) \longrightarrow CO_2(g)$$

the values of $\Delta H°$ and $\Delta S°$ are known to be -393.5 kJ and 3.05 J/K, respectively, and $\Delta G°$ can be calculated at 298 K as follows:

$$\Delta G° = \Delta H° - T\Delta S°$$
$$= -3.935 \times 10^5 \text{ J} - (298 \text{ K})(3.05 \text{ J/K})$$
$$= -3.944 \times 10^5 \text{ J}$$
$$= -394.4 \text{ kJ (per mole of CO}_2\text{)}$$

Interactive Example 17.9

Sign in at http://login.cengagebrain.com to try this Interactive Example in OWL.

Calculating $\Delta H°$, $\Delta S°$, and $\Delta G°$

Consider the reaction

$$2SO_2(g) + O_2(g) \longrightarrow 2SO_3(g)$$

carried out at 25°C and 1 atm. Calculate $\Delta H°$, $\Delta S°$, and $\Delta G°$ using the following data:

Substance	$H_f°$ (kJ/mol)	$S°$ (J/K · mol)
$SO_2(g)$	-297	248
$SO_3(g)$	-396	257
$O_2(g)$	0	205

Solution

The value of $\Delta H°$ can be calculated from the enthalpies of formation using the equation we discussed in Section 6.4:

$$\Delta H° = \Sigma n_p \Delta H_{f\,(products)}° - \Sigma n_r \Delta H_{f\,(reactants)}°$$

Then
$$\Delta H° = 2\Delta H_{f(SO_3(g))}° - 2\Delta H_{f(SO_2(g))}° - \Delta H_{f(O_2(g))}°$$
$$= 2 \text{ mol}(-396 \text{ kJ/mol}) - 2 \text{ mol}(-297 \text{ kJ/mol}) - 0$$
$$= -792 \text{ kJ} + 594 \text{ kJ}$$
$$= -198 \text{ kJ}$$

The value of $\Delta S°$ can be calculated using the standard entropy values and the equation discussed in Section 17.5:

$$\Delta S° = \Sigma n_p S_{products}° - \Sigma n_r S_{reactants}°$$

Thus
$$\Delta S° = 2S_{SO_3(g)}° - 2S_{SO_2(g)}° - S_{O_2(g)}°$$
$$= 2 \text{ mol}(257 \text{ J/K} \cdot \text{mol}) - 2 \text{ mol}(248 \text{ J/K} \cdot \text{mol}) - 1 \text{ mol}(205 \text{ J/K} \cdot \text{mol})$$
$$= 514 \text{ J/K} - 496 \text{ J/K} - 205 \text{ J/K}$$
$$= -187 \text{ J/K}$$

We would expect $\Delta S°$ to be negative because three molecules of gaseous reactants give two molecules of gaseous products.

The value of $\Delta G°$ can now be calculated from the equation

$$\Delta G° = \Delta H° - T\Delta S°$$

$$= -198 \text{ kJ} - (298 \text{ K})\left(-187 \frac{\text{J}}{\text{K}}\right)\left(\frac{1 \text{ kJ}}{1000 \text{ J}}\right)$$

$$= -198 \text{ kJ} + 55.7 \text{ kJ} = -142 \text{ kJ}$$

See Exercise 17.53

A second method for calculating ΔG for a reaction takes advantage of the fact that, like enthalpy, *free energy is a state function.* Therefore, we can use procedures for finding ΔG that are similar to those for finding ΔH using Hess's law.

To illustrate this method for calculating the free energy change, we will obtain $\Delta G°$ for the reaction

$$2CO(g) + O_2(g) \longrightarrow 2CO_2(g) \tag{17.2}$$

from the following data:

$$2CH_4(g) + 3O_2(g) \longrightarrow 2CO(g) + 4H_2O(g) \qquad \Delta G° = -1088 \text{ kJ} \tag{17.3}$$
$$CH_4(g) + 2O_2(g) \longrightarrow CO_2(g) + 2H_2O(g) \qquad \Delta G° = -801 \text{ kJ} \tag{17.4}$$

Note that $CO(g)$ is a reactant in Equation (17.2). This means that Equation (17.3) must be reversed, since $CO(g)$ is a product in that reaction as written. When a reaction is reversed, the sign of $\Delta G°$ is also reversed. In Equation (17.4), $CO_2(g)$ is a product, as it is in Equation (17.2), but only one molecule of CO_2 is formed. Thus Equation (17.4) must be multiplied by 2, which means the $\Delta G°$ value for Equation (17.4) also must be multiplied by 2. Free energy is an extensive property, since it is defined by two extensive properties, H and S.

Reversed Equation (17.3)
$$2CO(g) + 4H_2O(g) \longrightarrow 2CH_4(g) + 3O_2(g) \qquad \Delta G° = -(-1088 \text{ kJ})$$

2 × Equation (17.4)
$$\underline{2CH_4(g) + 4O_2(g) \longrightarrow 2CO_2(g) + 4H_2O(g)} \qquad \underline{\Delta G° = 2(-801 \text{ kJ})}$$
$$2CO(g) + O_2(g) \longrightarrow 2CO_2(g) \qquad \Delta G° = -(-1088 \text{ kJ})$$
$$+ 2(-801 \text{ kJ})$$
$$= -514 \text{ kJ}$$

This example shows that the ΔG values for reactions are manipulated in exactly the same way as the ΔH values.

Interactive Example 17.10

Sign in at http://login.cengagebrain.com to try this Interactive Example in **OWL**.

Calculating $\Delta G°$

Using the following data (at 25°C)

$$C_{\text{diamond}}(s) + O_2(g) \longrightarrow CO_2(g) \qquad \Delta G° = -397 \text{ kJ} \tag{17.5}$$
$$C_{\text{graphite}}(s) + O_2(g) \longrightarrow CO_2(g) \qquad \Delta G° = -394 \text{ kJ} \tag{17.6}$$

calculate $\Delta G°$ for the reaction

$$C_{\text{diamond}}(s) \longrightarrow C_{\text{graphite}}(s)$$

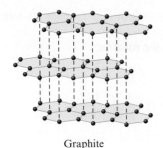

Graphite

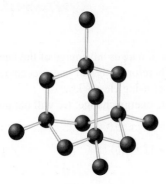

Diamond

The standard state of an element is its most stable state of 25°C and 1 atm.

Calculating $\Delta G°$ is very similar to calculating $\Delta H°$, as shown in Section 6.4.

Solution

We reverse Equation (17.6) to make graphite a product, as required, and then add the new equation to Equation (17.5):

$$C_{diamond}(s) + O_2(g) \longrightarrow CO_2(g) \qquad \Delta G° = -397 \text{ kJ}$$

Reversed Equation (17.6)

$$\frac{CO_2(g) \longrightarrow C_{graphite}(s) + O_2(g) \qquad \Delta G° = -(-394 \text{ kJ})}{C_{diamond}(s) \longrightarrow C_{graphite}(s) \qquad \Delta G° = -397 \text{ kJ} + 394 \text{ kJ}}$$

$$= -3 \text{ kJ}$$

Since $\Delta G°$ is negative for this process, diamond should spontaneously change to graphite at 25°C and 1 atm. However, the reaction is so slow under these conditions that we do not observe the process. This is another example of kinetic rather than thermodynamic control of a reaction. We can say that diamond is kinetically stable with respect to graphite even though it is thermodynamically unstable.

See Exercises 17.57 and 17.58

In Example 17.10 we saw that the process

$$C_{diamond}(s) \longrightarrow C_{graphite}(s)$$

is spontaneous but very slow at 25°C and 1 atm. The reverse process can be made to occur at high temperatures and pressures. Diamond has a more compact structure and thus a higher density than graphite, so exerting very high pressure causes it to become thermodynamically favored. If high temperatures are also used to make the process fast enough to be feasible, diamonds can be made from graphite. The conditions typically used involve temperatures greater than 1000°C and pressures of about 10^5 atm. About half of all industrial diamonds are made this way.

A third method for calculating the free energy change for a reaction uses standard free energies of formation. The **standard free energy of formation** $(\Delta G_f°)$ of a substance is defined as the *change in free energy that accompanies the formation of 1 mole of that substance from its constituent elements with all reactants and products in their standard states.* For the formation of glucose $(C_6H_{12}O_6)$, the appropriate reaction is

$$6C(s) + 6H_2(g) + 3O_2(g) \longrightarrow C_6H_{12}O_6(s)$$

The standard free energy associated with this process is called the *free energy of formation of glucose.* Values of the standard free energy of formation are useful in calculating $\Delta G°$ for specific chemical reactions using the equation

$$\Delta G° = \Sigma n_p \Delta G_f°_{(products)} - \Sigma n_r \Delta G_f°_{(reactants)}$$

Values of $\Delta G_f°$ for many common substances are listed in Appendix 4. Note that, analogous to the enthalpy of formation, *the standard free energy of formation of an element in its standard state is zero.* Also note that the number of moles of each reactant (n_r) and product (n_p) must be used when calculating $\Delta G°$ for a reaction.

Interactive Example 17.11

Sign in at http://login.cengagebrain.com to try this Interactive Example in OWL.

Calculating $\Delta G°$

Methanol is a high-octane fuel used in high-performance racing engines. Calculate $\Delta G°$ for the reaction

$$2CH_3OH(g) + 3O_2(g) \longrightarrow 2CO_2(g) + 4H_2O(g)$$

given the following free energies of formation:

Substance	ΔG_f° (kJ/mol)
$CH_3OH(g)$	-163
$O_2(g)$	0
$CO_2(g)$	-394
$H_2O(g)$	-229

Solution

We use the equation

$$\Delta G^\circ = \Sigma n_p \Delta G_{f(\text{products})}^\circ - \Sigma n_r \Delta G_{f(\text{reactants})}^\circ$$
$$= 2\Delta G_{f(CO_2(g))}^\circ + 4\Delta G_{f(H_2O(g))}^\circ - 3\Delta G_{f(O_2(g))}^\circ - 2\Delta G_{f(CH_3OH(g))}^\circ$$
$$= 2\ \text{mol}(-394\ \text{kJ/mol}) + 4\ \text{mol}(-229\ \text{kJ/mol}) - 3(0) - 2\ \text{mol}(-163\ \text{kJ/mol})$$
$$= -1378\ \text{kJ}$$

The large magnitude and the negative sign of ΔG° indicate that this reaction is very favorable thermodynamically.

See Exercises 17.59 and 17.60

Interactive Example 17.12

Sign in at http://login.cengagebrain .com to try this Interactive Example in **OWL**.

Ethylene

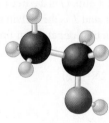

Ethanol

Free Energy and Spontaneity

A chemical engineer wants to determine the feasibility of making ethanol (C_2H_5OH) by reacting water with ethylene (C_2H_4) according to the equation

$$C_2H_4(g) + H_2O(l) \longrightarrow C_2H_5OH(l)$$

Is this reaction spontaneous under standard conditions?

Solution

To determine the spontaneity of this reaction under standard conditions, we must determine ΔG° for the reaction. We can do this using standard free energies of formation at 25°C from Appendix 4:

$$\Delta G_{f(C_2H_5OH(l))}^\circ = -175\ \text{kJ/mol}$$
$$\Delta G_{f(H_2O(l))}^\circ = -237\ \text{kJ/mol}$$
$$\Delta G_{f(C_2H_4(g))}^\circ = 68\ \text{kJ/mol}$$

Then
$$\Delta G^\circ = \Delta G_{f(C_2H_5OH(l))}^\circ - \Delta G_{f(H_2O(l))}^\circ - \Delta G_{f(C_2H_4(g))}^\circ$$
$$= -175\ \text{kJ} - (-237\ \text{kJ}) - 68\ \text{kJ}$$
$$= -6\ \text{kJ}$$

Thus the process is spontaneous under standard conditions at 25°C.

See Exercises 17.61 and 17.62

Although the reaction considered in Example 17.12 is spontaneous, other features of the reaction must be studied to see if the process is feasible. For example, the chemical engineer will need to study the kinetics of the reaction to determine whether it is fast enough to be useful and, if it is not, whether a catalyst can be found to enhance the rate. In doing these studies, the engineer must remember that ΔG° depends on temperature:

$$\Delta G^\circ = \Delta H^\circ - T\Delta S^\circ$$

Thus, if the process must be carried out at high temperatures to be fast enough to be feasible, $\Delta G°$ must be recalculated at that temperature from the $\Delta H°$ and $\Delta S°$ values for the reaction.

17.7 | The Dependence of Free Energy on Pressure

In this chapter we have seen that a system at constant temperature and pressure will proceed spontaneously in the direction that lowers its free energy. This is why reactions proceed until they reach equilibrium. The equilibrium position represents the lowest free energy value available to a particular reaction system. The free energy of a reaction system changes as the reaction proceeds, because free energy is dependent on the pressure of a gas or on the concentration of species in solution. We will deal only with the pressure dependence of the free energy of an ideal gas. The dependence of free energy on concentration can be developed using similar reasoning.

To understand the pressure dependence of free energy, we need to know how pressure affects the thermodynamic functions that comprise free energy, that is, enthalpy and entropy (recall that $G = H - TS$). For an ideal gas, enthalpy is not pressure-dependent. However, entropy *does* depend on pressure because of its dependence on volume. Consider 1 mole of an ideal gas at a given temperature. At a volume of 10.0 L, the gas has many more positions available for its molecules than if its volume is 1.0 L. The positional entropy is greater in the larger volume. In summary, at a given temperature for 1 mole of ideal gas

$$S_{\text{large volume}} > S_{\text{small volume}}$$

or, since pressure and volume are inversely related,

$$S_{\text{low pressure}} > S_{\text{high pressure}}$$

We have shown qualitatively that the entropy and therefore the free energy of an ideal gas depend on its pressure. Using a more detailed argument, which we will not consider here, it can be shown that

$$G = G° + RT \ln(P)$$

where $G°$ is the free energy of the gas at a pressure of 1 atm, G is the free energy of the gas at a pressure of P atm, R is the universal gas constant, and T is the Kelvin temperature.

To see how the change in free energy for a reaction depends on pressure, we will consider the ammonia synthesis reaction

$$N_2(g) + 3H_2(g) \longrightarrow 2NH_3(g)$$

In general,
$$\Delta G = \Sigma n_p G_{\text{products}} - \Sigma n_r G_{\text{reactants}}$$

For this reaction
$$\Delta G = 2G_{NH_3} - G_{N_2} - 3G_{H_2}$$

See Appendix 1.2 to review logarithms.

where
$$G_{NH_3} = G°_{NH_3} + RT \ln(P_{NH_3})$$
$$G_{N_2} = G°_{N_2} + RT \ln(P_{N_2})$$
$$G_{H_2} = G°_{H_2} + RT \ln(P_{H_2})$$

Substituting these values into the equation gives

$$\Delta G = 2[G°_{NH_3} + RT \ln(P_{NH_3})] - [G°_{N_2} + RT \ln(P_{N_2})] - 3[G°_{H_2} + RT \ln(P_{H_2})]$$
$$= 2G°_{NH_3} - G°_{N_2} - 3G°_{H_2} + 2RT \ln(P_{NH_3}) - RT \ln(P_{N_2}) - 3RT \ln(P_{H_2})$$
$$= \underbrace{(2G°_{NH_3} - G°_{N_2} - 3G°_{H_2})}_{\Delta G° \text{ reaction}} + RT[2 \ln(P_{NH_3}) - \ln(P_{N_2}) - 3 \ln(P_{H_2})]$$

The first term (in parentheses) is ΔG° for the reaction. Thus we have

$$\Delta G = \Delta G^\circ_{\text{reaction}} + RT[2\ln(P_{NH_3}) - \ln(P_{N_2}) - 3\ln(P_{H_2})]$$

and since

$$2\ln(P_{NH_3}) = \ln(P_{NH_3}{}^2)$$
$$-\ln(P_{N_2}) = \ln\left(\frac{1}{P_{N_2}}\right)$$
$$-3\ln(P_{H_2}) = \ln\left(\frac{1}{P_{H_2}{}^3}\right)$$

the equation becomes

$$\Delta G = \Delta G^\circ + RT\ln\left(\frac{P_{NH_3}{}^2}{(P_{N_2})(P_{H_2}{}^3)}\right)$$

But the term $\dfrac{P_{NH_3}{}^2}{(P_{N_2})(P_{H_2}{}^3)}$

is the reaction quotient Q discussed in Section 13.5. Therefore, we have

$$\Delta G = \Delta G^\circ + RT\ln(Q)$$

where Q is the reaction quotient (from the law of mass action), T is the temperature (K), R is the gas law constant and is equal to 8.3145 J/K · mol, ΔG° is the free energy change for the reaction with all reactants and products at a pressure of 1 atm, and ΔG is the free energy change for the reaction for the specified pressures of reactants and products.

Interactive Example 17.13

Sign in at http://login.cengagebrain.com to try this Interactive Example in OWL.

Calculating ΔG°

One method for synthesizing methanol (CH_3OH) involves reacting carbon monoxide and hydrogen gases:

$$CO(g) + 2H_2(g) \longrightarrow CH_3OH(l)$$

Calculate ΔG at 25°C for this reaction where carbon monoxide gas at 5.0 atm and hydrogen gas at 3.0 atm are converted to liquid methanol.

Solution

To calculate ΔG for this process, we use the equation

$$\Delta G = \Delta G^\circ + RT\ln(Q)$$

We must first compute ΔG° from standard free energies of formation (see Appendix 4). Since

$$\Delta G^\circ_{f\,(CH_3OH(l))} = -166\text{ kJ}$$
$$\Delta G^\circ_{f\,(H_2(g))} = 0$$
$$\Delta G^\circ_{f\,(CO(g))} = -137\text{ kJ}$$
$$\Delta G^\circ = -166\text{ kJ} - (-137\text{ kJ}) - 0 = -29\text{ kJ} = -2.9 \times 10^4\text{ J}$$

Note that this is the value of ΔG° for the reaction of 1 mole of CO with 2 moles of H_2 to produce 1 mole of CH_3OH. We might call this the value of ΔG° for one "round" of the reaction or for 1 mole of the reaction. Thus the ΔG° value might better be written as -2.9×10^4 J/mol of reaction, or -2.9×10^4 J/mol rxn.

We can now calculate ΔG using

$$\Delta G^\circ = -2.9 \times 10^4 \text{ J/mol rxn}$$

$$R = 8.3145 \text{ J/K} \cdot \text{mol}$$

$$T = 273 + 25 = 298 \text{ K}$$

$$Q = \frac{1}{(P_{CO})(P_{H_2}^2)} = \frac{1}{(5.0)(3.0)^2} = 2.2 \times 10^{-2}$$

Note in this case that ΔG is defined for "1 mole of the reaction," that is, for 1 mole of $CO(g)$ reacting with 2 moles of $H_2(g)$ to form 1 mole of $CH_3OH(l)$. Thus ΔG, ΔG°, and $RT \ln(Q)$ all have units of J/mol of reaction. In this case the units of R are actually J/K · mol of reaction, although they are usually not written this way.

Note that the pure liquid methanol is not included in the calculation of Q. Then

$$
\begin{aligned}
\Delta G &= \Delta G^\circ + RT \ln(Q) \\
&= (-2.9 \times 10^4 \text{ J/mol rxn}) + (8.3145 \text{ J/K} \cdot \text{mol rxn})(298 \text{ K}) \ln (2.2 \times 10^{-2}) \\
&= (-2.9 \times 10^4 \text{ J/mol rxn}) - (9.4 \times 10^3 \text{ J/mol rxn}) = -3.8 \times 10^4 \text{ J/mol rxn} \\
&= -38 \text{ kJ/mol rxn}
\end{aligned}
$$

Note that ΔG is significantly more negative than ΔG°, implying that the reaction is more spontaneous at reactant pressures greater than 1 atm. We might expect this result from Le Châtelier's principle.

See Exercises 17.65 and 17.66

The Meaning of ΔG for a Chemical Reaction

In this section we have learned to calculate ΔG for chemical reactions under various conditions. For example, in Example 17.13 the calculations show that the formation of $CH_3OH(l)$ from $CO(g)$ at 5.0 atm reacting with $H_2(g)$ at 3.0 atm is spontaneous. What does this result mean? Does it mean that if we mixed 1.0 mole of $CO(g)$ and 2.0 moles of $H_2(g)$ together at pressures of 5.0 and 3.0 atm, respectively, that 1.0 mole of $CH_3OH(l)$ would form in the reaction flask? The answer is no. This answer may surprise you in view of what has been said in this section. It is true that 1.0 mole of $CH_3OH(l)$ has a lower free energy than 1.0 mole of $CO(g)$ at 5.0 atm plus 2.0 moles of $H_2(g)$ at 3.0 atm. However, when $CO(g)$ and $H_2(g)$ are mixed under these conditions, there is *an even lower free energy available to this system than 1.0 mole of pure $CH_3OH(l)$.* For reasons we will discuss shortly, *the system can achieve the lowest possible free energy by going to equilibrium, not by going to completion.* At the equilibrium position, some of the $CO(g)$ and $H_2(g)$ will remain in the reaction flask. So even though 1.0 mole of pure $CH_3OH(l)$ is at a lower free energy than 1.0 mole of $CO(g)$ and 2.0 moles of $H_2(g)$ at 5.0 and 3.0 atm, respectively, the reaction system will stop short of forming 1.0 mole of $CH_3OH(l)$. The reaction stops short of completion because the equilibrium mixture of $CH_3OH(l)$, $CO(g)$, and $H_2(g)$ exists at the lowest possible free energy available to the system.

To illustrate this point, we will explore a mechanical example. Consider balls rolling down the two hills shown in Fig. 17.7. Note that in both cases point B has a lower potential energy than point A.

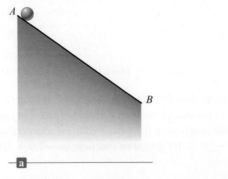

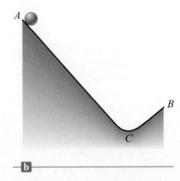

Figure 17.7 | Schematic representations of balls rolling down two types of hills.

In Fig. 17.7(a) the ball will roll to point *B*. This diagram is analogous to a phase change. For example, at 25°C ice will spontaneously change completely to liquid water, because the latter has the lowest free energy. In this case liquid water is the only choice. There is no intermediate mixture of ice and water with lower free energy.

The situation is different for a chemical reaction system, as illustrated in Fig. 17.7(b). In Fig. 17.7(b) the ball will not get to point *B* because there is a lower potential energy at point *C*. Like the ball, a chemical system will seek the *lowest possible* free energy, which, for reasons we will discuss below, is the equilibrium position.

Therefore, although the value of ΔG for a given reaction system tells us whether the products or reactants are favored under a given set of conditions, it does not mean that the system will proceed to pure products (if ΔG is negative) or remain at pure reactants (if ΔG is positive). Instead, the system will spontaneously go to the equilibrium position, the lowest possible free energy available to it. In the next section we will see that the value of $\Delta G°$ for a particular reaction tells us exactly where this position will be.

17.8 | Free Energy and Equilibrium

When the components of a given chemical reaction are mixed, they will proceed, rapidly or slowly depending on the kinetics of the process, to the equilibrium position. In Chapter 13 we defined the equilibrium position as the point at which the forward and reverse reaction rates are equal. In this chapter we look at equilibrium from a thermodynamic point of view, and we find that the **equilibrium point** *occurs at the lowest value of free energy available to the reaction system*. As it turns out, the two definitions give the same equilibrium state, which must be the case for both the kinetic and thermodynamic models to be valid.

To understand the relationship of free energy to equilibrium, let's consider the following simple hypothetical reaction:

$$A(g) \rightleftharpoons B(g)$$

where 1.0 mole of gaseous A is initially placed in a reaction vessel at a pressure of 2.0 atm. The free energies for A and B are diagramed as shown in Fig. 17.8(a). As A reacts to form B, the total free energy of the system changes, yielding the following results:

$$\text{Free energy of A} = G_A = G_A° + RT \ln(P_A)$$
$$\text{Free energy of B} = G_B = G_B° + RT \ln(P_B)$$
$$\text{Total free energy of system} = G = G_A + G_B$$

As A changes to B, G_A will decrease because P_A is decreasing [Fig. 17.8(b)]. In contrast, G_B will increase because P_B is increasing. The reaction will proceed to the right as long as the total free energy of the system decreases (as long as G_B is less than G_A). At some point the pressures of A and B reach the values P_A^e and P_B^e that make G_A equal to G_B. *The system has reached equilibrium* [Fig. 17.8(c)]. Since A at pressure P_A^e and B at pressure P_B^e have the same free energy (G_A equals G_B), ΔG is zero for A at pressure P_A^e changing to B at pressure P_B^e. *The system has reached minimum free energy.* There is no longer any driving force to change A to B or B to A, so the system remains at this position (the pressures of A and B remain constant).

Figure 17.8 | (a) The initial free energies of A and B. (b) As A(*g*) changes to B(*g*), the free energy of A decreases and that of B increases. (c) Eventually, pressures of A and B are achieved such that $G_A = G_B$, the equilibrium position.

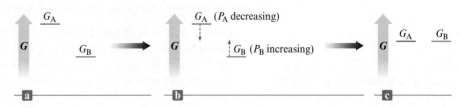

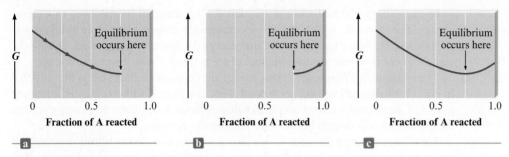

Figure 17.9 | (a) The change in free energy to reach equilibrium, beginning with 1.0 mole of A(g) at P_A = 2.0 atm. (b) The change in free energy to reach equilibrium, beginning with 1.0 mole of B(g) at P_B = 2.0 atm. (c) The free energy profile for A(g) $\rightleftharpoons$ B(g) in a system containing 1.0 mole of gas (A plus B) at P_{TOTAL} = 2.0 atm. Each point on the curve corresponds to the total free energy of the system for a given combination of A and B.

Suppose that for the experiment described above, the plot of free energy versus the mole fraction of A reacted is defined as shown in Fig. 17.9(a). In this experiment, minimum free energy is reached when 75% of A has been changed to B. At this point, the pressure of A is 0.25 times the original pressure, or

$$(0.25)(2.0 \text{ atm}) = 0.50 \text{ atm}$$

The pressure of B is

$$(0.75)(2.0 \text{ atm}) = 1.5 \text{ atm}$$

Since this is the equilibrium position, we can use the equilibrium pressures to calculate a value for K for the reaction in which A is converted to B at this temperature:

$$K = \frac{P_B^e}{P_A^e} = \frac{1.5 \text{ atm}}{0.50 \text{ atm}} = 3.0$$

For the reaction A(g) $\rightleftharpoons$ B(g), the pressure is constant during the reaction, since the same number of gas molecules is always present.

Exactly the same equilibrium point would be achieved if we placed 1.0 mole of pure B(g) in the flask at a pressure of 2.0 atm. In this case B would change to A until equilibrium ($G_B = G_A$) is reached. This is shown in Fig. 17.9(b).

The overall free energy curve for this system is shown in Fig. 17.9(c). Note that any mixture of A(g) and B(g) containing 1.0 mole of gas (A plus B) at a total pressure of 2.0 atm will react until it reaches the minimum in the curve.

In summary, when substances undergo a chemical reaction, the reaction proceeds to the minimum free energy (equilibrium), which corresponds to the point where

$$G_{\text{products}} = G_{\text{reactants}} \quad \text{or} \quad \Delta G = G_{\text{products}} - G_{\text{reactants}} = 0$$

We can now establish a quantitative relationship between free energy and the value of the equilibrium constant. We have seen that

$$\Delta G = \Delta G° + RT \ln(Q)$$

and at equilibrium ΔG equals 0 and Q equals K.

So
$$\Delta G = 0 = \Delta G° + RT \ln(K)$$

or
$$\Delta G° = -RT \ln(K)$$

We must note the following characteristics of this very important equation.

Case 1: $\Delta G° = 0$. When $\Delta G°$ equals zero for a particular reaction, the free energies of the reactants and products are equal when all components are in the standard states (1 atm for gases). The system is at equilibrium when the pressures of all reactants and products are 1 atm, which means that K equals 1.

Table 17.6 | Qualitative Relationship Between the Change in Standard Free Energy and the Equilibrium Constant for a Given Reaction

$\Delta G°$	K
$\Delta G° = 0$	$K = 1$
$\Delta G° < 0$	$K > 1$
$\Delta G° > 0$	$K < 1$

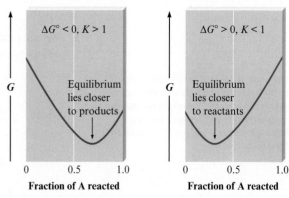

Figure 17.10 | The relationship of $\Delta G°$ for a reaction to its eventual equilibrium position.

Case 2: $\Delta G° < 0$. In this case $\Delta G°$ ($G°_{\text{products}} - G°_{\text{reactants}}$) is negative, which means that

$$G°_{\text{products}} < G°_{\text{reactants}}$$

If a flask contains the reactants and products, all at 1 atm, the system will *not* be at equilibrium. Since $G°_{\text{products}}$ is less than $G°_{\text{reactants}}$, the system will adjust to the right to reach equilibrium. In this case K will be *greater than 1,* since the pressures of the products at equilibrium will be greater than 1 atm and the pressures of the reactants at equilibrium will be less than 1 atm.

Case 3: $\Delta G° > 0$. Since $\Delta G°$ ($G°_{\text{products}} - G°_{\text{reactants}}$) is positive,

$$G°_{\text{reactants}} < G°_{\text{products}}$$

If a flask contains the reactants and products, all at 1 atm, the system will *not* be at equilibrium. In this case the system will adjust to the left (toward the reactants, which have a lower free energy) to reach equilibrium. The value of K will be *less than 1,* since at equilibrium the pressures of the reactants will be greater than 1 atm and the pressures of the products will be less than 1 atm.

These results are summarized in Table 17.6 and Figure 17.10. The value of K for a specific reaction can be calculated from the equation

$$\Delta G° = -RT \ln(K)$$

as is shown in Examples 17.14 and 17.15.

Interactive Example 17.14

Sign in at http://login.cengagebrain.com to try this Interactive Example in **OWL**.

Free Energy and Equilibrium I

Consider the ammonia synthesis reaction

$$N_2(g) + 3H_2(g) \rightleftharpoons 2NH_3(g)$$

where $\Delta G° = -33.3$ kJ per mole of N_2 consumed at 25°C. For each of the following mixtures of reactants and products at 25°C, predict the direction in which the system will shift to reach equilibrium.

a. $P_{NH_3} = 1.00$ atm, $P_{N_2} = 1.47$ atm, $P_{H_2} = 1.00 \times 10^{-2}$ atm

b. $P_{NH_3} = 1.00$ atm, $P_{N_2} = 1.00$ atm, $P_{H_2} = 1.00$ atm

Solution

a. We can predict the direction of reaction to equilibrium by calculating the value of ΔG using the equation

$$\Delta G = \Delta G° + RT \ln(Q)$$

The units of ΔG, $\Delta G°$, and $RT \ln(Q)$ all refer to the balanced reaction with all amounts expressed in moles. We might say that the units are joules per "mole of reaction," although only the "per mole" is indicated for R (as is customary).

where
$$Q = \frac{P_{NH_3}{}^2}{(P_{N_2})(P_{H_2}{}^3)} = \frac{(1.00)^2}{(1.47)(1.00 \times 10^{-2})^3} = 6.80 \times 10^5$$

$$T = 25 + 273 = 298 \text{ K}$$
$$R = 8.3145 \text{ J/K} \cdot \text{mol}$$

and

$$\Delta G° = -33.3 \text{ kJ/mol} = -3.33 \times 10^4 \text{ J/mol}$$

Then

$$\Delta G = (-3.33 \times 10^4 \text{ J/mol}) + (8.3145 \text{ J/K} \cdot \text{mol})(298 \text{ K}) \ln(6.8 \times 10^5)$$
$$= (-3.33 \times 10^4 \text{ J/mol}) + (3.33 \times 10^4 \text{ J/mol}) = 0$$

Since $\Delta G = 0$, the reactants and products have the same free energies at these partial pressures. The system is already at equilibrium, and no shift will occur.

b. The partial pressures given here are all 1.00 atm, which means that the system is in the standard state. That is,

$$\Delta G = \Delta G° + RT \ln(Q) = \Delta G° + RT \ln\frac{(1.00)^2}{(1.00)(1.00)^3}$$
$$= \Delta G° + RT \ln(1.00) = \Delta G° + 0 = \Delta G°$$

For this reaction at 25°C,

$$\Delta G° = -33.3 \text{ kJ/mol}$$

The negative value for $\Delta G°$ means that in their standard states the products have a lower free energy than the reactants. Thus the system will move to the right to reach equilibrium. That is, K is greater than 1.

See Exercise 17.67

Interactive
Example 17.15

Sign in at http://login.cengagebrain
.com to try this Interactive Example
in OWL.

Free Energy and Equilibrium II

The overall reaction for the corrosion (rusting) of iron by oxygen is

$$4Fe(s) + 3O_2(g) \rightleftharpoons 2Fe_2O_3(s)$$

Using the following data, calculate the equilibrium constant for this reaction at 25°C.

Substance	$\Delta H_f°$ (kJ/mol)	$S°$ (J/K · mol)
$Fe_2O_3(s)$	−826	90
$Fe(s)$	0	27
$O_2(g)$	0	205

Solution

To calculate K for this reaction, we will use the equation

$$\Delta G° = -RT \ln(K)$$

We must first calculate $\Delta G°$ from

$$\Delta G° = \Delta H° - T\Delta S°$$

where

$$\Delta H° = 2\Delta H°_{f\,(Fe_2O_3(s))} - 3\Delta H°_{f\,(O_2(g))} - 4\Delta H°_{f\,(Fe(s))}$$
$$= 2\text{ mol}(-826\text{ kJ/mol}) - 0 - 0$$
$$= -1652\text{ kJ} = -1.652 \times 10^6\text{ J}$$

$$\Delta S° = 2S°_{Fe_2O_3} - 3S°_{O_2} - 4S°_{Fe}$$
$$= 2\text{ mol}(90\text{ J/K} \cdot \text{mol}) - 3\text{ mol}(205\text{ J/K} \cdot \text{mol}) - 4\text{ mol}(27\text{ J/K} \cdot \text{mol})$$
$$= -543\text{ J/K}$$

and

$$T = 273 + 25 = 298\text{ K}$$

Then

$$\Delta G° = \Delta H° - T\Delta S° = (-1.652 \times 10^6\text{ J}) - (298\text{ K})(-543\text{ J/K})$$
$$= -1.490 \times 10^6\text{ J}$$

and

$$\Delta G° = -RT\ln(K) = -1.490 \times 10^6\text{ J} = -(8.3145\text{ J/K} \cdot \text{mol})(298\text{ K})\ln(K)$$

Thus

$$\ln(K) = \frac{1.490 \times 10^6}{2.48 \times 10^3} = 601$$

and

$$K = e^{601}$$

This is a very large equilibrium constant. The rusting of iron is clearly very favorable from a thermodynamic point of view.

See Exercises 17.69 and 17.70

Formation of rust on bare steel is a spontaneous process.

The Temperature Dependence of K

In Chapter 13 we used Le Châtelier's principle to predict qualitatively how the value of K for a given reaction would change with a change in temperature. Now we can specify the quantitative dependence of the equilibrium constant on temperature from the relationship

$$\Delta G° = -RT\ln(K) = \Delta H° - T\Delta S°$$

We can rearrange this equation to give

$$\ln(K) = -\frac{\Delta H°}{RT} + \frac{\Delta S°}{R} = -\frac{\Delta H°}{R}\left(\frac{1}{T}\right) + \frac{\Delta S°}{R}$$

Note that this is a linear equation of the form $y = mx + b$, where $y = \ln(K)$, $m = -\Delta H°/R = $ slope, $x = 1/T$, and $b = \Delta S°/R = $ intercept. This means that if values of K for a given reaction are determined at various temperatures, a plot of $\ln(K)$ versus $1/T$ will be linear, with slope $-\Delta H°/R$ and intercept $\Delta S°/R$. This result assumes that both $\Delta H°$ and $\Delta S°$ are independent of temperature over the temperature range considered. This assumption is a good approximation over a relatively small temperature range.

17.9 | Free Energy and Work

One of the main reasons we are interested in physical and chemical processes is that we want to use them to do work for us, and we want this work done as efficiently and economically as possible. We have already seen that at constant temperature and

Thinkstock/Jupiter images

pressure, the sign of the change in free energy tells us whether a given process is spontaneous. This is very useful information because it prevents us from wasting effort on a process that has no inherent tendency to occur. Although a thermodynamically favorable chemical reaction may not occur to any appreciable extent because it is too slow, it makes sense in this case to try to find a catalyst to speed up the reaction. On the other hand, if the reaction is prevented from occurring by its thermodynamic characteristics, we would be wasting our time looking for a catalyst.

In addition to its qualitative usefulness (telling us whether a process is spontaneous), the change in free energy is important quantitatively because it can tell us how much work can be done with a given process. The maximum possible useful work obtainable from a process at constant temperature and pressure is equal to the change in free energy:

$$w_{max} = \Delta G$$

Note that "PV work" is not counted as useful work here.

This relationship explains why this function is called the *free* energy. Under certain conditions, ΔG for a spontaneous process represents the energy that is *free to do useful work*. On the other hand, for a process that is not spontaneous, the value of ΔG tells us the minimum amount of work that must be *expended* to make the process occur.

Knowing the value of ΔG for a process thus gives us valuable information about how close the process is to 100% efficiency. For example, when gasoline is burned in a car's engine, the work produced is about 20% of the maximum work available.

For reasons we will only briefly introduce in this book, the amount of work we actually obtain from a spontaneous process is *always* less than the maximum possible amount.

To explore this idea more fully, let's consider an electric current flowing through the starter motor of a car. The current is generated from a chemical change in a battery, and we can calculate ΔG for the battery reaction and so determine the energy available to do work. Can we use all this energy to do work? No, because a current flowing through a wire causes frictional heating, and the greater the current, the greater the heat. This heat represents wasted energy—it is not useful for running the starter motor. We can minimize this energy waste by running very low currents through the motor circuit. However, zero current flow would be necessary to eliminate frictional heating entirely, and we cannot derive any work from the motor if no current flows. This represents the difficulty in which nature places us. Using a process to do work requires that some of the energy be wasted, and usually the faster we run the process, the more energy we waste.

Achieving the maximum work available from a spontaneous process can occur only via a hypothetical pathway. Any real pathway wastes energy. If we could discharge the battery infinitely slowly by an infinitesimal current flow, we would achieve the maximum useful work. Also, if we could then recharge the battery using an infinitesimally small current, exactly the same amount of energy would be used to return the battery to its original state. After we cycle the battery in this way, the universe (the system and surroundings) is exactly the same as it was before the cyclic process. This is a **reversible process** (Fig. 17.11).

However, if the battery is discharged to run the starter motor and then recharged using a *finite* current flow, as is the case in reality, *more* work will always be required to recharge the battery than the battery produces as it discharges. This means that even though the battery (the system) has returned to its original state, the surroundings have not, because the surroundings had to furnish a net amount of work as the battery was cycled. The *universe is different* after this cyclic process is performed, and this function is called an **irreversible process**. *All real processes are irreversible.*

In general, after any real cyclic process is carried out in a system, the surroundings have less ability to do work and contain more thermal energy. In any real cyclic process in the system, work is changed to heat in the surroundings, and the entropy of the universe increases. This is another way of stating the second law of thermodynamics.

Figure 17.11 | A battery can do work by sending current to a starter motor. The battery can then be recharged by forcing current through it in the opposite direction. If the current flow in both processes is infinitesimally small, $w_1 = w_2$. This is a *reversible process*. But if the current flow is finite, as it would be in any real case, $w_2 > w_1$. This is an *irreversible process* (the *universe is different* after the cyclic process occurs). All real processes are irreversible.

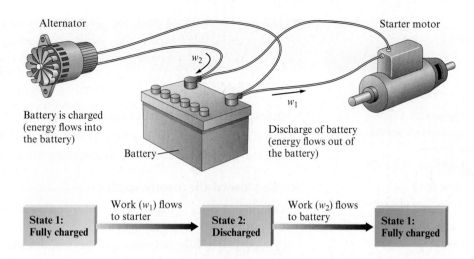

Alternator

Starter motor

w_2

Battery is charged
(energy flows into
the battery)

w_1

Discharge of battery
(energy flows out of
the battery)

Battery

| State 1:
Fully charged | Work (w_1) flows
to starter → | State 2:
Discharged | Work (w_2) flows
to battery → | State 1:
Fully charged |

When energy is used to do work, it becomes less organized and less concentrated and thus less useful.

Thus thermodynamics tells us the work potential of a process and then tells us that we can never achieve this potential. In this spirit, thermodynamicist Henry Bent has paraphrased the first two laws of thermodynamics as follows:

First law: You can't win, you can only break even.

Second law: You can't break even.

The ideas we have discussed in this section are applicable to the energy crisis that will probably increase in severity over the next 25 years. The crisis is obviously not one of supply; the first law tells us that the universe contains a constant supply of energy. The problem is the availability of *useful* energy. As we use energy, we degrade its usefulness. For example, when gasoline reacts with oxygen in the combustion reaction, the change in potential energy results in heat flow. Thus the energy concentrated in the bonds of the gasoline and oxygen molecules ends up *spread* over the surroundings as thermal energy, where it is much more difficult to harness for useful work. This is a way in which the entropy of the universe increases: Concentrated energy becomes spread out—more disordered and less useful. Thus the crux of the energy problem is that we are rapidly consuming the concentrated energy found in fossil fuels. It took millions of years to concentrate the sun's energy in these fuels, and we will consume these same fuels in a few hundred years. Thus we must use these energy sources as wisely as possible.

Critical Thinking

What if the first law of thermodynamics was true but the second law was not? How would the world be different?

For review

Key terms

First law of thermodynamics

> States that the energy of the universe is constant
> Provides a way to keep track of energy as it changes form
> Gives no information about why a particular process occurs in a given direction

Second law of thermodynamics

> States that for any spontaneous process there is always an increase in the entropy of the universe
> Entropy(S) is a thermodynamic function that describes the number of arrangements (positions and/or energy levels) available to a system existing in a given state
>> Nature spontaneously proceeds toward states that have the highest probability of occurring
>> Using entropy, thermodynamics can predict the direction in which a process will occur spontaneously:

$$\Delta S_{univ} = \Delta S_{sys} + \Delta S_{surr}$$

> For a spontaneous process, ΔS_{univ} must be positive
> For a process at constant temperature and pressure:
>> ΔS_{sys} is dominated by "positional" entropy
>> For a chemical reaction, ΔS_{sys} is dominated by changes in the number of gaseous molecules
>> ΔS_{surr} is determined by heat:

$$\Delta S_{surr} = -\frac{\Delta H}{T}$$

>> ΔS_{surr} is positive for an exothermic process (ΔH is negative)
>> Because ΔS_{surr} depends inversely on T, exothermicity becomes a more important driving force at low temperatures
> Thermodynamics cannot predict the rate at which a system will spontaneously change; the principles of kinetics are necessary to do this

Third law of thermodynamics

> States that the entropy of a perfect crystal at 0 K is zero

Free energy (G)

> Free energy is a state function:

$$G = H - TS$$

> A process occurring at constant temperature and pressure is spontaneous in the direction in which its free energy decreases ($\Delta G < 0$)
> For a reaction the standard free energy change (ΔG°) is the change in free energy that occurs when reactants in their standard states are converted to products in their standard states
> The standard free energy change for a reaction can be determined from the standard free energies of formation (ΔG_f°) of the reactants and products:

$$\Delta G^\circ = \Sigma n_p \Delta G_f^\circ(\text{products}) - \Sigma n_r \Delta G_f^\circ(\text{reactants})$$

> Free energy depends on temperature and pressure:

$$G = G^\circ + RT \ln P$$

> This relationship can be used to derive the relationship between $\Delta G°$ for a reaction and the value of its equilibrium constant K:

$$\Delta G° = -RT \ln K$$

> For $\Delta G° = 0$, $K = 1$
> For $\Delta G° < 0$, $K > 1$
> For $\Delta G° > 0$, $K < 1$

> The maximum possible useful work obtainable from a process at constant temperature and pressure is equal to the change in free energy:

$$w_{max} = \Delta G$$

> In any real process, $w < w_{max}$
> When energy is used to do work in a real process, the energy of the universe remains constant but the usefulness of the energy decreases
>> Concentrated energy is spread out in the surroundings as thermal energy

Review questions   *Answers to the Review Questions can be found on the Student website (accessible from* **www.cengagebrain.com**)*.*

1. Define the following:
   a. spontaneous process
   b. entropy
   c. positional probability
   d. system
   e. surroundings
   f. universe

2. What is the second law of thermodynamics? For any process, there are four possible sign combinations for ΔS_{sys} and ΔS_{surr}. Which sign combination(s) always give a spontaneous process? Which sign combination(s) always give a nonspontaneous process? Which sign combination(s) may or may not give a spontaneous process?

3. What determines ΔS_{surr} for a process? To calculate ΔS_{surr} at constant pressure and temperature, we use the following equation: $\Delta S_{surr} = -\Delta H/T$. Why does a minus sign appear in the equation, and why is ΔS_{surr} inversely proportional to temperature?

4. The free energy change, ΔG, for a process at constant temperature and pressure is related to ΔS_{univ} and reflects the spontaneity of the process. How is ΔG related to ΔS_{univ}? When is a process spontaneous? Nonspontaneous? At equilibrium? ΔG is a composite term composed of ΔH, T, and ΔS. What is the ΔG equation? Give the four possible sign combinations for ΔH and ΔS. What temperatures are required for each sign combination to yield a spontaneous process? If ΔG is positive, what does it say about the reverse process? How does the $\Delta G = \Delta H - T\Delta S$ equation reduce when at the melting-point temperature of a solid-to-liquid phase change or at the boiling-point temperature of a liquid-to-gas phase change? What is the sign of ΔG for the solid-to-liquid phase change at temperatures above the freezing point?

What is the sign of ΔG for the liquid-to-gas phase change at temperatures below the boiling point?

5. What is the third law of thermodynamics? What are standard entropy values, $S°$, and how are these $S°$ values (listed in Appendix 4) used to calculate $\Delta S°$ for a reaction? How would you use Hess's law to calculate $\Delta S°$ for a reaction? What does the superscript $°$ indicate?

   Predicting the sign of $\Delta S°$ for a reaction is an important skill to master. For a gas-phase reaction, what do you concentrate on to predict the sign of $\Delta S°$? For a phase change, what do you concentrate on to predict the sign of $\Delta S°$? That is, how are $S°_{solid}$, $S°_{liquid}$, and $S°_{gas}$ related to one another? When a solute dissolves in water, what is usually the sign of $\Delta S°$ for this process?

6. What is the standard free energy change, $\Delta G°$, for a reaction? What is the standard free energy of formation, $\Delta G°_f$, for a substance? How are $\Delta G°_f$ values used to calculate $\Delta G°_{rxn}$? How can you use Hess's law to calculate $\Delta G°_{rxn}$? How can you use $\Delta H°$ and $\Delta S°$ values to calculate $\Delta G°_{rxn}$? Of the functions $\Delta H°$, $\Delta S°$, and $\Delta G°$, which depends most strongly on temperature? When $\Delta G°$ is calculated at temperatures other than 25°C, what assumptions are generally made concerning $\Delta H°$ and $\Delta S°$?

7. If you calculate a value for $\Delta G°$ for a reaction using the values of $\Delta G°_f$ in Appendix 4 and get a negative number, is it correct to say that the reaction is always spontaneous? Why or why not? Free energy changes also depend on concentration. For gases, how is G related to the pressure of the gas? What are standard pressures for gases and standard concentrations for solutes? How do you calculate ΔG for a reaction at nonstandard conditions? The equation to determine ΔG at nonstandard conditions has Q in it: What is Q? A reaction is spontaneous as long as ΔG is negative; that is, reactions

always proceed as long as the products have a lower free energy than the reactants. What is so special about equilibrium? Why don't reactions move away from equilibrium?

8. Consider the equation $\Delta G = \Delta G° + RT \ln(Q)$. What is the value of ΔG for a reaction at equilibrium? What does Q equal at equilibrium? At equilibrium, the previous equation reduces to $\Delta G° = -RT \ln(K)$. When $\Delta G° > 0$, what does it indicate about K? When $\Delta G° < 0$, what does it indicate about K? When $\Delta G° = 0$, what does it indicate about K? ΔG predicts spontaneity for a reaction, whereas $\Delta G°$ predicts the

equilibrium position. Explain what this statement means. Under what conditions can you use $\Delta G°$ to determine the spontaneity of a reaction?

9. Even if ΔG is negative, the reaction may not occur. Explain the interplay between the thermodynamics and the kinetics of a reaction. High temperatures are favorable to a reaction kinetically but may be unfavorable to a reaction thermodynamically. Explain.

10. Discuss the relationship between w_{max} and the magnitude and sign of the free energy change for a reaction. Also discuss w_{max} for real processes. What is a reversible process?

Active Learning Questions

These questions are designed to be used by groups of students in class.

1. For the process $A(l) \longrightarrow A(g)$, which direction is favored by changes in energy probability? Positional probability? Explain your answers. If you wanted to favor the process as written, would you raise or lower the temperature of the system? Explain.

2. For a liquid, which would you expect to be larger, ΔS_{fusion} or $\Delta S_{evaporation}$? Why?

3. Gas A_2 reacts with gas B_2 to form gas AB at a constant temperature. The bond energy of AB is much greater than that of either reactant. What can be said about the sign of ΔH? ΔS_{surr}? ΔS? Explain how potential energy changes for this process. Explain how random kinetic energy changes during the process.

4. What types of experiments can be carried out to determine whether a reaction is spontaneous? Does spontaneity have any relationship to the final equilibrium position of a reaction? Explain.

5. A friend tells you, "Free energy G and pressure P are related by the equation $G = G° + RT \ln(P)$. Also, G is related to the equilibrium constant K in that when $G_{products} = G_{reactants}$, the system is at equilibrium. Therefore, it must be true that a system is at equilibrium when all the pressures are equal." Do you agree with this friend? Explain.

6. You remember that $\Delta G°$ is related to $RT \ln(K)$ but cannot remember if it's $RT \ln(K)$ or $-RT \ln(K)$. Realizing what $\Delta G°$ and K mean, how can you figure out the correct sign?

7. Predict the sign of ΔS for each of the following and explain.
   a. the evaporation of alcohol
   b. the freezing of water
   c. compressing an ideal gas at constant temperature
   d. dissolving NaCl in water

8. Is ΔS_{surr} favorable or unfavorable for exothermic reactions? Endothermic reactions? Explain.

9. At 1 atm, liquid water is heated above 100°C. For this process, which of the following choices (i–iv) is correct for ΔS_{surr}? ΔS? ΔS_{univ}? Explain each answer.
   i. greater than zero
   ii. less than zero

   iii. equal to zero
   iv. cannot be determined

10. When (if ever) are high temperatures unfavorable to a reaction thermodynamically?

A blue question or exercise number indicates that the answer to that question or exercise appears at the back of this book and a solution appears in the *Solutions Guide*, as found on PowerLecture.

Questions

11. The synthesis of glucose directly from CO_2 and H_2O and the synthesis of proteins directly from amino acids are both nonspontaneous processes under standard conditions. Yet it is necessary for these to occur for life to exist. In light of the second law of thermodynamics, how can life exist?

12. When the environment is contaminated by a toxic or potentially toxic substance (for example, from a chemical spill or the use of insecticides), the substance tends to disperse. How is this consistent with the second law of thermodynamics? In terms of the second law, which requires the least work: cleaning the environment after it has been contaminated or trying to prevent the contamination before it occurs? Explain.

13. Entropy has been described as "time's arrow." Interpret this view of entropy.

14. Human DNA contains almost twice as much information as is needed to code for all the substances produced in the body. Likewise, the digital data sent from *Voyager II* contained one redundant bit out of every two bits of information. The Hubble space telescope transmits three redundant bits for every bit of information. How is entropy related to the transmission of information? What do you think is accomplished by having so many redundant bits of information in both DNA and the space probes?

15. A mixture of hydrogen gas and chlorine gas remains unreacted until it is exposed to ultraviolet light from a burning magnesium strip. Then the following reaction occurs very rapidly:

$$H_2(g) + Cl_2(g) \longrightarrow 2HCl(g)$$

Explain.

16. Consider the following potential energy plots:

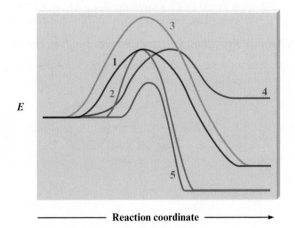

⎯⎯⎯⎯ **Reaction coordinate** ⎯⎯⎯⎯→

a. Rank the reactions from fastest to slowest and explain your answer. If any reactions have equal rates, explain why.

b. Label the reactions as endothermic or exothermic, and support your answer.

c. Rank the exothermic reactions from greatest to least change in potential energy, and support your answer.

17. ΔS_{surr} is sometimes called the energy disorder term. Explain.

18. Given the following illustration, what can be said about the sign of ΔS for the process of solid NaCl dissolving in water? What can be said about ΔH for this process?

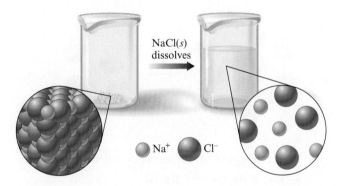

NaCl(s) dissolves

Na⁺   Cl⁻

19. The third law of thermodynamics states that the entropy of a perfect crystal at 0 K is zero. In Appendix 4, $F^-(aq)$, $OH^-(aq)$, and $S^{2-}(aq)$ all have negative standard entropy values. How can $S°$ values be less than zero?

20. The deciding factor on why HF is a weak acid and not a strong acid like the other hydrogen halides is entropy. What occurs when HF dissociates in water as compared to the other hydrogen halides?

21. List three different ways to calculate the standard free energy change, $\Delta G°$, for a reaction at 25°C. How is $\Delta G°$ estimated at temperatures other than 25°C? What assumptions are made?

22. What information can be determined from ΔG for a reaction? Does one get the same information from $\Delta G°$, the standard free energy change? $\Delta G°$ allows determination of the equilibrium constant K for a reaction. How? How can one estimate the value of K at temperatures other than 25°C for a reaction? How can one estimate the temperature where $K = 1$ for a reaction? Do all reactions have a specific temperature where $K = 1$?

23. Monochloroethane (C_2H_5Cl) can be produced by the direct reaction of ethane gas (C_2H_6) with chlorine gas or by the reaction of ethylene gas (C_2H_4) with hydrogen chloride gas. The second reaction gives almost a 100% yield of pure C_2H_5Cl at a rapid rate without catalysis. The first method requires light as an energy source or the reaction would not occur. Yet $\Delta G°$ for the first reaction is considerably more negative than $\Delta G°$ for the second reaction. Explain how this can be so.

24. At 1500 K, the process

$$I_2(g) \longrightarrow 2I(g)$$
$$\text{10 atm} \qquad \text{10 atm}$$

is not spontaneous. However, the process

$$I_2(g) \longrightarrow 2I(g)$$
$$\text{0.10 atm} \qquad \text{0.10 atm}$$

is spontaneous at 1500 K. Explain.

Exercises

In this section similar exercises are paired.

Spontaneity, Entropy, and the Second Law of Thermodynamics: Free Energy

25. Which of the following processes are spontaneous?

a. Salt dissolves in H_2O.

b. A clear solution becomes a uniform color after a few drops of dye are added.

c. Iron rusts.

d. You clean your bedroom.

26. Which of the following processes are spontaneous?

a. A house is built.

b. A satellite is launched into orbit.

c. A satellite falls back to earth.

d. The kitchen gets cluttered.

27. Table 17.1 shows the possible arrangements of four molecules in a two-bulbed flask. What are the possible arrangements if there is one molecule in this two-bulbed flask or two molecules or three molecules? For each, what arrangement is most likely?

28. Consider the following illustration of six molecules of gas in a two-bulbed flask.

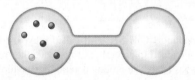

a. What is the most likely arrangement of molecules? How many microstates are there for this arrangement?

b. Determine the probability of finding the gas in its most likely arrangement.

29. Consider the following energy levels, each capable of holding two particles:

$E = 2$ kJ _____

E $E = 1$ kJ _____

$E = 0$ _XX_

Draw all the possible arrangements of the two identical particles (represented by X) in the three energy levels. What total energy is most likely, that is, occurs the greatest number of times? Assume that the particles are indistinguishable from each other.

30. Redo Exercise 29 with two particles A and B, which can be distinguished from each other.

---

31. Choose the substance with the larger positional probability in each case.

  a. 1 mole of H_2 (at STP) or 1 mole of H_2 (at 100°C, 0.5 atm)

  b. 1 mole of N_2 (at STP) or 1 mole of N_2 (at 100 K, 2.0 atm)

  c. 1 mole of $H_2O(s)$ (at 0°C) or 1 mole of $H_2O(l)$ (at 20°C)

32. Which of the following involve an increase in the entropy of the system?

  a. melting of a solid

  b. sublimation

  c. freezing

  d. mixing

  e. separation

  f. boiling

---

33. Predict the sign of ΔS_{surr} for the following processes.

  a. $H_2O(l) \longrightarrow H_2O(g)$

  b. $I_2(g) \longrightarrow I_2(s)$

34. Calculate ΔS_{surr} for the following reactions at 25°C and 1 atm.

  a. $C_3H_8(g) + 5O_2(g) \longrightarrow 3CO_2(g) + 4H_2O(l)$
  $$\Delta H° = -2221 \text{ kJ}$$

  b. $2NO_2(g) \longrightarrow 2NO(g) + O_2(g)$ $\Delta H° = 112 \text{ kJ}$

---

35. Given the values of ΔH and ΔS, which of the following changes will be spontaneous at constant T and P?

  a. $\Delta H = +25 \text{ kJ}, \Delta S = +5.0 \text{ J/K}, T = 300. \text{ K}$

  b. $\Delta H = +25 \text{ kJ}, \Delta S = +100. \text{ J/K}, T = 300. \text{ K}$

  c. $\Delta H = -10. \text{ kJ}, \Delta S = +5.0 \text{ J/K}, T = 298 \text{ K}$

  d. $\Delta H = -10. \text{ kJ}, \Delta S = -40. \text{ J/K}, T = 200. \text{ K}$

36. At what temperatures will the following processes be spontaneous?

  a. $\Delta H = -18 \text{ kJ and } \Delta S = -60. \text{ J/K}$

  b. $\Delta H = +18 \text{ kJ and } \Delta S = +60. \text{ J/K}$

  c. $\Delta H = +18 \text{ kJ and } \Delta S = -60. \text{ J/K}$

  d. $\Delta H = -18 \text{ kJ and } \Delta S = +60. \text{ J/K}$

---

37. Ethanethiol (C_2H_5SH; also called ethyl mercaptan) is commonly added to natural gas to provide the "rotten egg" smell of a gas leak. The boiling point of ethanethiol is 35°C and its heat of vaporization is 27.5 kJ/mol. What is the entropy of vaporization for this substance?

38. For mercury, the enthalpy of vaporization is 58.51 kJ/mol and the entropy of vaporization is 92.92 J/K · mol. What is the normal boiling point of mercury?

---

39. For ammonia (NH_3), the enthalpy of fusion is 5.65 kJ/mol and the entropy of fusion is 28.9 J/K · mol.

  a. Will $NH_3(s)$ spontaneously melt at 200. K?

  b. What is the approximate melting point of ammonia?

40. The enthalpy of vaporization of ethanol is 38.7 kJ/mol at its boiling point (78°C). Determine ΔS_{sys}, ΔS_{surr}, and ΔS_{univ} when 1.00 mole of ethanol is vaporized at 78°C and 1.00 atm.

Chemical Reactions: Entropy Changes and Free Energy

41. Predict the sign of $\Delta S°$ for each of the following changes. Assume all equations are balanced.

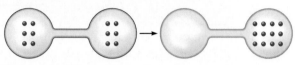

a.

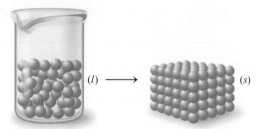

b.

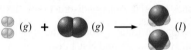

c.

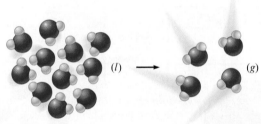

d.

42. Predict the sign of $\Delta S°$ for each of the following changes.

  a. $K(s) + \frac{1}{2}Br_2(g) \longrightarrow KBr(s)$

  b. $N_2(g) + 3H_2(g) \longrightarrow 2NH_3(g)$

  c. $KBr(s) \longrightarrow K^+(aq) + Br^-(aq)$

  d. $KBr(s) \longrightarrow KBr(l)$

---

43. For each of the following pairs of substances, which substance has the greater value of $S°$?

  a. $C_{graphite}(s)$ or $C_{diamond}(s)$

  b. $C_2H_5OH(l)$ or $C_2H_5OH(g)$

  c. $CO_2(s)$ or $CO_2(g)$

44. For each of the following pairs, which substance has the greater value of S?

  a. N_2O (at 0 K) or He (at 10 K)

  b. $N_2O(g)$ (at 1 atm, 25°C) or He(g) (at 1 atm, 25°C)

  c. $NH_3(s)$ (at 196 K) $\longrightarrow$ $NH_3(l)$ (at 196 K)

45. Predict the sign of $\Delta S°$ and then calculate $\Delta S°$ for each of the following reactions.

   a. $2H_2S(g) + SO_2(g) \longrightarrow 3S_{rhombic}(s) + 2H_2O(g)$
   b. $2SO_3(g) \longrightarrow 2SO_2(g) + O_2(g)$
   c. $Fe_2O_3(s) + 3H_2(g) \longrightarrow 2Fe(s) + 3H_2O(g)$

46. Predict the sign of $\Delta S°$ and then calculate $\Delta S°$ for each of the following reactions.

   a. $H_2(g) + \frac{1}{2}O_2(g) \longrightarrow H_2O(l)$
   b. $2CH_3OH(g) + 3O_2(g) \longrightarrow 2CO_2(g) + 4H_2O(g)$
   c. $HCl(g) \longrightarrow H^+(aq) + Cl^-(aq)$

47. For the reaction

$$C_2H_2(g) + 4F_2(g) \longrightarrow 2CF_4(g) + H_2(g)$$

   $\Delta S°$ is equal to -358 J/K. Use this value and data from Appendix 4 to calculate the value of $S°$ for $CF_4(g)$.

48. For the reaction

$$CS_2(g) + 3O_2(g) \longrightarrow CO_2(g) + 2SO_2(g)$$

   $\Delta S°$ is equal to -143 J/K. Use this value and data from Appendix 4 to calculate the value of $S°$ for $CS_2(g)$.

49. It is quite common for a solid to change from one structure to another at a temperature below its melting point. For example, sulfur undergoes a phase change from the rhombic crystal structure to the monoclinic crystal form at temperatures above 95°C.

   a. Predict the signs of ΔH and ΔS for the process
   $S_{rhombic}(s) \longrightarrow S_{monoclinic}(s)$.
   b. Which form of sulfur has the more ordered crystalline structure (has the smaller positional probability)?

50. Two crystalline forms of white phosphorus are known. Both forms contain P_4 molecules, but the molecules are packed together in different ways. The α form is always obtained when the liquid freezes. However, below $-76.9°C$, the α form spontaneously converts to the β form:

$$P_4(s, \alpha) \longrightarrow P_4(s, \beta)$$

   a. Predict the signs of ΔH and ΔS for this process.
   b. Predict which form of phosphorus has the more ordered crystalline structure (has the smaller positional probability).

51. Consider the reaction

$$2O(g) \longrightarrow O_2(g)$$

   a. Predict the signs of ΔH and ΔS.
   b. Would the reaction be more spontaneous at high or low temperatures?

52. Hydrogen cyanide is produced industrially by the following exothermic reaction:

$$2NH_3(g) + 3O_2(g) + 2CH_4(g) \xrightarrow[\text{Pt-Rh}]{100°C} 2HCN(g) + 6H_2O(g)$$

   Is the high temperature needed for thermodynamic or kinetic reasons?

53. From data in Appendix 4, calculate $\Delta H°$, $\Delta S°$, and $\Delta G°$ for each of the following reactions at 25°C.

   a. $CH_4(g) + 2O_2(g) \longrightarrow CO_2(g) + 2H_2O(g)$
   b. $6CO_2(g) + 6H_2O(l) \longrightarrow C_6H_{12}O_6(s) + 6O_2(g)$
   Glucose

   c. $P_4O_{10}(s) + 6H_2O(l) \longrightarrow 4H_3PO_4(s)$
   d. $HCl(g) + NH_3(g) \longrightarrow NH_4Cl(s)$

54. The major industrial use of hydrogen is in the production of ammonia by the Haber process:

$$3H_2(g) + N_2(g) \longrightarrow 2NH_3(g)$$

   a. Using data from Appendix 4, calculate $\Delta H°$, $\Delta S°$, and $\Delta G°$ for the Haber process reaction.
   b. Is the reaction spontaneous at standard conditions?
   c. At what temperatures is the reaction spontaneous at standard conditions? Assume $\Delta H°$ and $\Delta S°$ do not depend on temperature.

55. For the reaction at 298 K,

$$2NO_2(g) \rightleftharpoons N_2O_4(g)$$

   the values of $\Delta H°$ and $\Delta S°$ are -58.03 kJ and -176.6 J/K, respectively. What is the value of $\Delta G°$ at 298 K? Assuming that $\Delta H°$ and $\Delta S°$ do not depend on temperature, at what temperature is $\Delta G° = 0$? Is $\Delta G°$ negative above or below this temperature?

56. At 100.°C and 1.00 atm, $\Delta H° = 40.6$ kJ/mol for the vaporization of water. Estimate $\Delta G°$ for the vaporization of water at 90.°C and 110.°C. Assume $\Delta H°$ and $\Delta S°$ at 100.°C and 1.00 atm do not depend on temperature.

57. Given the following data:

$$\begin{array}{ll} 2H_2(g) + C(s) \longrightarrow CH_4(g) & \Delta G° = -51 \text{ kJ} \\ 2H_2(g) + O_2(g) \longrightarrow 2H_2O(l) & \Delta G° = -474 \text{ kJ} \\ C(s) + O_2(g) \longrightarrow CO_2(g) & \Delta G° = -394 \text{ kJ} \end{array}$$

   Calculate $\Delta G°$ for $CH_4(g) + 2O_2(g) \rightarrow CO_2(g) + 2H_2O(l)$.

58. Given the following data:

$$\begin{array}{ll} 2C_6H_6(l) + 15O_2(g) \longrightarrow 12CO_2(g) + 6H_2O(l) & \\ & \Delta G° = -6399 \text{ kJ} \\ C(s) + O_2(g) \longrightarrow CO_2(g) & \Delta G° = -394 \text{ kJ} \\ H_2(g) + \frac{1}{2}O_2(g) \longrightarrow H_2O(l) & \Delta G° = -237 \text{ kJ} \end{array}$$

   calculate $\Delta G°$ for the reaction

$$6C(s) + 3H_2(g) \longrightarrow C_6H_6(l)$$

59. For the reaction

$$SF_4(g) + F_2(g) \longrightarrow SF_6(g)$$

   the value of $\Delta G°$ is -374 kJ. Use this value and data from Appendix 4 to calculate the value of $\Delta G_f°$ for $SF_4(g)$.

60. The value of $\Delta G°$ for the reaction

$$2C_4H_{10}(g) + 13O_2(g) \longrightarrow 8CO_2(g) + 10H_2O(l)$$

   is $-5490.$ kJ. Use this value and data from Appendix 4 to calculate the standard free energy of formation for $C_4H_{10}(g)$.

61. Consider the reaction

$$Fe_2O_3(s) + 3H_2(g) \longrightarrow 2Fe(s) + 3H_2O(g)$$

   a. Use $\Delta G_f°$ values in Appendix 4 to calculate $\Delta G°$ for this reaction.
   b. Is this reaction spontaneous under standard conditions at 298 K?

c. The value of $\Delta H°$ for this reaction is 100. kJ. At what temperatures is this reaction spontaneous at standard conditions? Assume that $\Delta H°$ and $\Delta S°$ do not depend on temperature.

62. Consider the reaction

$$2POCl_3(g) \longrightarrow 2PCl_3(g) + O_2(g)$$

a. Calculate $\Delta G°$ for this reaction. The $\Delta G_f°$ values for $POCl_3(g)$ and $PCl_3(g)$ are -502 kJ/mol and $-270.$ kJ/mol, respectively.

b. Is this reaction spontaneous under standard conditions at 298 K?

c. The value of $\Delta S°$ for this reaction is 179 J/K · mol. At what temperatures is this reaction spontaneous at standard conditions? Assume that $\Delta H°$ and $\Delta S°$ do not depend on temperature.

63. Using data from Appendix 4, calculate $\Delta H°$, $\Delta S°$, and $\Delta G°$ for the following reactions that produce acetic acid:

$$CH_4(g) + CO_2(g) \longrightarrow CH_3\overset{\overset{\displaystyle O}{\|}}{C}—OH(l)$$

$$CH_3OH(g) + CO(g) \longrightarrow CH_3\overset{\overset{\displaystyle O}{\|}}{C}—OH(l)$$

Which reaction would you choose as a commercial method for producing acetic acid (CH_3CO_2H) at standard conditions? What temperature conditions would you choose for the reaction? Assume $\Delta H°$ and $\Delta S°$ do not depend on temperature.

64. Consider two reactions for the production of ethanol:

$$C_2H_4(g) + H_2O(g) \longrightarrow CH_3CH_2OH(l)$$
$$C_2H_6(g) + H_2O(g) \longrightarrow CH_3CH_2OH(l) + H_2(g)$$

Which would be the more thermodynamically feasible at standard conditions? Why?

Free Energy: Pressure Dependence and Equilibrium

65. Using data from Appendix 4, calculate ΔG for the reaction

$$NO(g) + O_3(g) \longrightarrow NO_2(g) + O_2(g)$$

for these conditions:

$$T = 298 \text{ K}$$
$$P_{NO} = 1.00 \times 10^{-6} \text{ atm}, P_{O_3} = 2.00 \times 10^{-6} \text{ atm}$$
$$P_{NO_2} = 1.00 \times 10^{-7} \text{ atm}, P_{O_2} = 1.00 \times 10^{-3} \text{ atm}$$

66. Using data from Appendix 4, calculate ΔG for the reaction

$$2H_2S(g) + SO_2(g) \rightleftharpoons 3S_{rhombic}(s) + 2H_2O(g)$$

for the following conditions at 25°C:

$$P_{H_2S} = 1.0 \times 10^{-4} \text{ atm}$$
$$P_{SO_2} = 1.0 \times 10^{-2} \text{ atm}$$
$$P_{H_2O} = 3.0 \times 10^{-2} \text{ atm}$$

67. Consider the reaction

$$2NO_2(g) \rightleftharpoons N_2O_4(g)$$

For each of the following mixtures of reactants and products at 25°C, predict the direction in which the reaction will shift to reach equilibrium.

a. $P_{NO_2} = P_{N_2O_4} = 1.0$ atm
b. $P_{NO_2} = 0.21$ atm, $P_{N_2O_4} = 0.50$ atm
c. $P_{NO_2} = 0.29$ atm, $P_{N_2O_4} = 1.6$ atm

68. Consider the following reaction:

$$N_2(g) + 3H_2(g) \rightleftharpoons 2NH_3(g)$$

Calculate ΔG for this reaction under the following conditions (assume an uncertainty of ± 1 in all quantities):

a. $T = 298$ K, $P_{N_2} = P_{H_2} = 200$ atm, $P_{NH_3} = 50$ atm
b. $T = 298$ K, $P_{N_2} = 200$ atm, $P_{H_2} = 600$ atm, $P_{NH_3} = 200$ atm

69. One of the reactions that destroys ozone in the upper atmosphere is

$$NO(g) + O_3(g) \rightleftharpoons NO_2(g) + O_2(g)$$

Using data from Appendix 4, calculate $\Delta G°$ and K (at 298 K) for this reaction.

70. Hydrogen sulfide can be removed from natural gas by the reaction

$$2H_2S(g) + SO_2(g) \rightleftharpoons 3S(s) + 2H_2O(g)$$

Calculate $\Delta G°$ and K (at 298 K) for this reaction. Would this reaction be favored at a high or low temperature?

71. Consider the following reaction at 25.0°C:

$$2NO_2(g) \rightleftharpoons N_2O_4(g)$$

The values of $\Delta H°$ and $\Delta S°$ are -58.03 kJ/mol and -176.6 J/K · mol, respectively. Calculate the value of K at 25.0°C. Assuming $\Delta H°$ and $\Delta S°$ are temperature independent, estimate the value of K at 100.0°C.

72. The standard free energies of formation and the standard enthalpies of formation at 298 K for difluoroacetylene (C_2F_2) and hexafluorobenzene (C_6F_6) are

	$\Delta G_f°$ (kJ/mol)	$\Delta H_f°$ (kJ/mol)
$C_2F_2(g)$	191.2	241.3
$C_6F_6(g)$	78.2	132.8

For the following reaction:

$$C_6F_6(g) \rightleftharpoons 3C_2F_2(g)$$

a. calculate $\Delta S°$ at 298 K.
b. calculate K at 298 K.
c. estimate K at 3000. K, assuming $\Delta H°$ and $\Delta S°$ do not depend on temperature.

73. Calculate $\Delta G°$ for $H_2O(g) + \frac{1}{2}O_2(g) \rightleftharpoons H_2O_2(g)$ at 600. K, using the following data:

$$H_2(g) + O_2(g) \rightleftharpoons H_2O_2(g) \quad K = 2.3 \times 10^6 \text{ at 600. K}$$
$$2H_2(g) + O_2(g) \rightleftharpoons 2H_2O(g) \quad K = 1.8 \times 10^{37} \text{ at 600. K}$$

74. The Ostwald process for the commercial production of nitric acid involves three steps:

$$4NH_3(g) + 5O_2(g) \xrightarrow[825°C]{Pt} 4NO(g) + 6H_2O(g)$$
$$2NO(g) + O_2(g) \longrightarrow 2NO_2(g)$$
$$3NO_2(g) + H_2O(l) \longrightarrow 2HNO_3(l) + NO(g)$$

a. Calculate $\Delta H°$, $\Delta S°$, $\Delta G°$, and K (at 298 K) for each of the three steps in the Ostwald process (see Appendix 4).

b. Calculate the equilibrium constant for the first step at 825°C, assuming $\Delta H°$ and $\Delta S°$ do not depend on temperature.

c. Is there a thermodynamic reason for the high temperature in the first step, assuming standard conditions?

75. Cells use the hydrolysis of adenosine triphosphate, abbreviated as ATP, as a source of energy. Symbolically, this reaction can be written as

$$\text{ATP}(aq) + \text{H}_2\text{O}(l) \longrightarrow \text{ADP}(aq) + \text{H}_2\text{PO}_4^{-}(aq)$$

where ADP represents adenosine diphosphate. For this reaction, $\Delta G° = -30.5$ kJ/mol.

a. Calculate K at 25°C.

b. If all the free energy from the metabolism of glucose

$$\text{C}_6\text{H}_{12}\text{O}_6(s) + 6\text{O}_2(g) \longrightarrow 6\text{CO}_2(g) + 6\text{H}_2\text{O}(l)$$

goes into forming ATP from ADP, how many ATP molecules can be produced for every molecule of glucose?

76. One reaction that occurs in human metabolism is

$$\text{HO}_2\text{CCH}_2\text{CH}_2\text{CHCO}_2\text{H}(aq) + \text{NH}_3(aq) \Longrightarrow$$
$$\underset{|}{}$$
$$\text{NH}_2$$

Glutamic acid

$$\overset{\text{O}}{\overset{||}{\text{H}_2\text{NCCH}_2\text{CH}_2\text{CHCO}_2\text{H}}}(aq) + \text{H}_2\text{O}(l)$$
$$\underset{|}{}$$
$$\text{NH}_2$$

Glutamine

For this reaction $\Delta G° = 14$ kJ at 25°C.

a. Calculate K for this reaction at 25°C.

b. In a living cell this reaction is coupled with the hydrolysis of ATP. (See Exercise 75.) Calculate $\Delta G°$ and K at 25°C for the following reaction:

$$\text{Glutamic acid}(aq) + \text{ATP}(aq) + \text{NH}_3(aq) \Longrightarrow$$
$$\text{Glutamine}(aq) + \text{ADP}(aq) + \text{H}_2\text{PO}_4^{-}(aq)$$

77. Consider the following reaction at 800. K:

$$\text{N}_2(g) + 3\text{F}_2(g) \longrightarrow 2\text{NF}_3(g)$$

An equilibrium mixture contains the following partial pressures: $P_{\text{N}_2} = 0.021$ atm, $P_{\text{F}_2} = 0.063$ atm, $P_{\text{NF}_3} = 0.48$ atm. Calculate $\Delta G°$ for the reaction at 800. K.

78. Consider the following reaction at 298 K:

$$2\text{SO}_2(g) + \text{O}_2(g) \longrightarrow 2\text{SO}_3(g)$$

An equilibrium mixture contains $\text{O}_2(g)$ and $\text{SO}_3(g)$ at partial pressures of 0.50 atm and 2.0 atm, respectively. Using data from Appendix 4, determine the equilibrium partial pressure of SO_2 in the mixture. Will this reaction be most favored at a high or a low temperature, assuming standard conditions?

79. Consider the relationship

$$\ln(K) = \frac{-\Delta H°}{RT} + \frac{\Delta S°}{R}$$

The equilibrium constant for some hypothetical process was determined as a function of temperature (Kelvin) with the results plotted below.

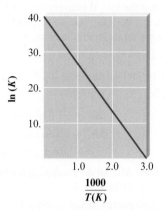

From the plot, determine the values of $\Delta H°$ and $\Delta S°$ for this process. What would be the major difference in the $\ln(K)$ versus $1/T$ plot for an endothermic process as compared to an exothermic process?

80. The equilibrium constant K for the reaction

$$2\text{Cl}(g) \Longrightarrow \text{Cl}_2(g)$$

was measured as a function of temperature (Kelvin). A graph of $\ln(K)$ versus $1/T$ for this reaction gives a straight line with a slope of 1.352×10^4 K and a y-intercept of -14.51. Determine the values of $\Delta H°$ and $\Delta S°$ for this reaction. See Exercise 79.

Additional Exercises

81. Using Appendix 4 and the following data, determine $S°$ for $\text{Fe(CO)}_5(g)$.

$$\text{Fe}(s) + 5\text{CO}(g) \longrightarrow \text{Fe(CO)}_5(g) \qquad \Delta S° = ?$$
$$\text{Fe(CO)}_5(l) \longrightarrow \text{Fe(CO)}_5(g) \qquad \Delta S° = 107 \text{ J/K}$$
$$\text{Fe}(s) + 5\text{CO}(g) \longrightarrow \text{Fe(CO)}_5(l) \qquad \Delta S° = -677 \text{ J/K}$$

82. Some water is placed in a coffee-cup calorimeter. When 1.0 g of an ionic solid is added, the temperature of the solution increases from 21.5°C to 24.2°C as the solid dissolves. For the dissolving process, what are the signs for ΔS_{sys}, ΔS_{surr}, and ΔS_{univ}?

83. Consider the following system at equilibrium at 25°C:

$$\text{PCl}_3(g) + \text{Cl}_2(g) \Longrightarrow \text{PCl}_5(g) \qquad \Delta G° = -92.50 \text{ kJ}$$

What will happen to the ratio of partial pressure of PCl_5 to partial pressure of PCl_3 if the temperature is raised? Explain completely.

84. Calculate the entropy change for the vaporization of liquid methane and liquid hexane using the following data.

	Boiling Point (1 atm)	ΔH_{vap}
Methane	112 K	8.20 kJ/mol
Hexane	342 K	28.9 kJ/mol

Compare the molar volume of gaseous methane at 112 K with that of gaseous hexane at 342 K. How do the differences in molar volume affect the values of ΔS_{vap} for these liquids?

85. As $O_2(l)$ is cooled at 1 atm, it freezes at 54.5 K to form solid I. At a lower temperature, solid I rearranges to solid II, which has a different crystal structure. Thermal measurements show that ΔH for the I → II phase transition is -743.1 J/mol, and ΔS for the same transition is -17.0 J/K · mol. At what temperature are solids I and II in equilibrium?

86. Consider the following reaction:

$$H_2O(g) + Cl_2O(g) \rightleftharpoons 2HOCl(g) \qquad K_{298} = 0.090$$

For $Cl_2O(g)$,

$$\Delta G_f^\circ = 97.9 \text{ kJ/mol}$$
$$\Delta H_f^\circ = 80.3 \text{ kJ/mol}$$
$$S^\circ = 266.1 \text{ J/K} \cdot \text{mol}$$

a. Calculate ΔG° for the reaction using the equation $\Delta G^\circ = -RT \ln(K)$.

b. Use bond energy values (Table 8.4) to estimate ΔH° for the reaction.

c. Use the results from parts a and b to estimate ΔS° for the reaction.

d. Estimate ΔH_f° and S° for $HOCl(g)$.

e. Estimate the value of K at 500. K.

f. Calculate ΔG at 25°C when $P_{H_2O} = 18$ torr, $P_{Cl_2O} = 2.0$ torr, and $P_{HOCl} = 0.10$ torr.

87. Using the following data, calculate the value of K_{sp} for $Ba(NO_3)_2$, one of the *least* soluble of the common nitrate salts.

Species	ΔG_f°
$Ba^{2+}(aq)$	-561 kJ/mol
$NO_3^-(aq)$	-109 kJ/mol
$Ba(NO_3)_2(s)$	-797 kJ/mol

88. Many biochemical reactions that occur in cells require relatively high concentrations of potassium ion (K^+). The concentration of K^+ in muscle cells is about 0.15 M. The concentration of K^+ in blood plasma is about 0.0050 M. The high internal concentration in cells is maintained by pumping K^+ from the plasma. How much work must be done to transport 1.0 mole of K^+ from the blood to the inside of a muscle cell at 37°C, normal body temperature? When 1.0 mole of K^+ is transferred from blood to the cells, do any other ions have to be transported? Why or why not?

89. Carbon monoxide is toxic because it bonds much more strongly to the iron in hemoglobin (Hgb) than does O_2. Consider the following reactions and approximate standard free energy changes:

$$Hgb + O_2 \longrightarrow HgbO_2 \qquad \Delta G^\circ = -70 \text{ kJ}$$
$$Hgb + CO \longrightarrow HgbCO \qquad \Delta G^\circ = -80 \text{ kJ}$$

Using these data, estimate the equilibrium constant value at 25°C for the following reaction:

$$HgbO_2 + CO \rightleftharpoons HgbCO + O_2$$

90. In the text, the equation

$$\Delta G = \Delta G^\circ + RT \ln(Q)$$

was derived for gaseous reactions where the quantities in Q were expressed in units of pressure. We also can use units of mol/L for the quantities in Q, specifically for aqueous reactions. With this in mind, consider the reaction

$$HF(aq) \rightleftharpoons H^+(aq) + F^-(aq)$$

for which $K_a = 7.2 \times 10^{-4}$ at 25°C. Calculate ΔG for the reaction under the following conditions at 25°C.

a. $[HF] = [H^+] = [F^-] = 1.0$ M
b. $[HF] = 0.98$ M, $[H^+] = [F^-] = 2.7 \times 10^{-2}$ M
c. $[HF] = [H^+] = [F^-] = 1.0 \times 10^{-5}$ M
d. $[HF] = [F^-] = 0.27$ M, $[H^+] = 7.2 \times 10^{-4}$ M
e. $[HF] = 0.52$ M, $[F^-] = 0.67$ M, $[H^+] = 1.0 \times 10^{-3}$ M

Based on the calculated ΔG values, in what direction will the reaction shift to reach equilibrium for each of the five sets of conditions?

91. Consider the reactions

$$Ni^{2+}(aq) + 6NH_3(aq) \longrightarrow Ni(NH_3)_6^{2+}(aq) \qquad (1)$$
$$Ni^{2+}(aq) + 3en(aq) \longrightarrow Ni(en)_3^{2+}(aq) \qquad (2)$$

where

$$en = H_2N-CH_2-CH_2-NH_2$$

The ΔH values for the two reactions are quite similar, yet $K_{reaction\ 2} > K_{reaction\ 1}$. Explain.

92. Use the equation in Exercise 79 to determine ΔH° and ΔS° for the autoionization of water:

$$H_2O(l) \rightleftharpoons H^+(aq) + OH^-(aq)$$

$T(°C)$	K_w
0	1.14×10^{-15}
25	1.00×10^{-14}
35	2.09×10^{-14}
40.	2.92×10^{-14}
50.	5.47×10^{-14}

93. Consider the reaction

$$Fe_2O_3(s) + 3H_2(g) \longrightarrow 2Fe(s) + 3H_2O(g)$$

Assuming ΔH° and ΔS° do not depend on temperature, calculate the temperature where $K = 1.00$ for this reaction.

94. Consider the following diagram of free energy (G) versus fraction of A reacted in terms of moles for the reaction $2A(g) \rightarrow B(g)$.

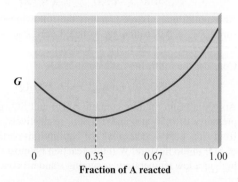

Before any A has reacted, $P_A = 3.0$ atm and $P_B = 0$. Determine the sign of ΔG° and the value of K_p for this reaction.

ChemWork Problems

These multiconcept problems (and additional ones) are found inter-actively online with the same type of assistance a student would get from an instructor.

95. Which of the following reactions (or processes) are expected to have a negative value for $\Delta S°$?

  a. $SiF_6(aq) + H_2(g) \longrightarrow 2HF(g) + SiF_4(g)$

  b. $4Al(s) + 3O_2(g) \longrightarrow 2Al_2O_3(s)$

  c. $CO(g) + Cl_2(g) \longrightarrow COCl_2(g)$

  d. $C_2H_4(g) + H_2O(l) \longrightarrow C_2H_5OH(l)$

  e. $H_2O(s) \longrightarrow H_2O(l)$

96. For rubidium $\Delta H°_{vap} = 69.0$ kJ/mol at 686°C, its boiling point. Calculate $\Delta S°$, q, w, and ΔE for the vaporization of 1.00 mole of rubidium at 686°C and 1.00 atm pressure.

97. Given the thermodynamic data below, calculate ΔS and ΔS_{surr} for the following reaction at 25°C and 1 atm:

$$XeF_6(g) \longrightarrow XeF_4(s) + F_2(g)$$

	$\Delta H°_f$ (kJ/mol)	$S°$ (J/K · mol)
$XeF_6(g)$	−294	300.
$XeF_4(s)$	−251	146
$F_2(g)$	0	203

98. Consider the reaction:

$$H_2S(g) + SO_2(g) \longrightarrow 3S(g) + 2H_2O(l)$$

for which ΔH is −233 kJ and ΔS is −424 J/K.

  a. Calculate the free energy change for the reaction (ΔG) at 393 K.

  b. Assuming ΔH and ΔS do not depend on temperature, at what temperatures is this reaction spontaneous?

99. The following reaction occurs in pure water:

$$H_2O(l) + H_2O(l) \longrightarrow H_3O^+(aq) + OH^-(aq)$$

which is often abbreviated as

$$H_2O(l) \longrightarrow H^+(aq) + OH^-(aq)$$

For this reaction, $\Delta G° = 79.9$ kJ/mol at 25°C. Calculate the value of ΔG for this reaction at 25°C when $[OH^-] = 0.15\ M$ and $[H^+] = 0.71\ M$.

100. Consider the dissociation of a weak acid HA ($K_a = 4.5 \times 10^{-3}$) in water:

$$HA(aq) \rightleftharpoons H^+(aq) + A^-(aq)$$

Calculate $\Delta G°$ for this reaction at 25°C.

101. Consider the reaction:

$$PCl_3(g) + Cl_2(g) \rightleftharpoons PCl_5(g)$$

At 25°C, $\Delta H° = -92.50$ kJ.

Which of the following statements is(are) *true*?

  a. This is an endothermic reaction.

  b. $\Delta S°$ for this reaction is negative.

  c. If the temperature is increased, the ratio $\dfrac{PCl_5}{PCl_3}$ will increase.

  d. $\Delta G°$ for this reaction has to be negative at all temperatures.

  e. When $\Delta G°$ for this reaction is negative, then K_p is greater than 1.00.

102. The equilibrium constant for a certain reaction increases by a factor of 6.67 when the temperature is increased from 300.0 K to 350.0 K. Calculate the standard change in enthalpy ($\Delta H°$) for this reaction (assuming $\Delta H°$ is temperature-independent).

Challenge Problems

103. Consider two perfectly insulated vessels. Vessel 1 initially contains an ice cube at 0°C and water at 0°C. Vessel 2 initially contains an ice cube at 0°C and a saltwater solution at 0°C. Consider the process $H_2O(s) \rightarrow H_2O(l)$.

  a. Determine the sign of ΔS, ΔS_{surr}, and ΔS_{univ} for the process in vessel 1.

  b. Determine the sign of ΔS, ΔS_{surr}, and ΔS_{univ} for the process in vessel 2.

  (*Hint:* Think about the effect that a salt has on the freezing point of a solvent.)

104. Liquid water at 25°C is introduced into an evacuated, insulated vessel. Identify the signs of the following thermodynamic functions for the process that occurs: ΔH, ΔS, ΔT_{water}, ΔS_{surr}, ΔS_{univ}.

105. Using data from Appendix 4, calculate $\Delta H°$, $\Delta G°$, and K (at 298 K) for the production of ozone from oxygen:

$$3O_2(g) \rightleftharpoons 2O_3(g)$$

At 30 km above the surface of the earth, the temperature is about 230. K and the partial pressure of oxygen is about 1.0×10^{-3} atm. Estimate the partial pressure of ozone in equilibrium with oxygen at 30 km above the earth's surface. Is it reasonable to assume that the equilibrium between oxygen and ozone is maintained under these conditions? Explain.

106. Entropy can be calculated by a relationship proposed by Ludwig Boltzmann:

$$S = k \ln(W)$$

where $k = 1.38 \times 10^{-23}$ J/K and W is the number of ways a particular state can be obtained. (This equation is engraved on Boltzmann's tombstone.) Calculate S for the five arrangements of particles in Table 17.1.

107. a. Using the free energy profile for a simple one-step reaction, show that at equilibrium $K = k_f/k_r$, where k_f and k_r are the rate constants for the forward and reverse reactions. *Hint:* Use the relationship $\Delta G° = -RT \ln(K)$ and represent k_f and k_r using the Arrhenius equation ($k = Ae^{-E_a/RT}$).

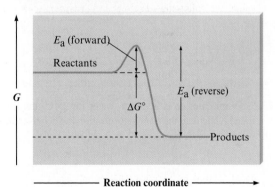

b. Why is the following statement false? "A catalyst can increase the rate of a forward reaction but not the rate of the reverse reaction."

108. Consider the reaction

$$H_2(g) + Br_2(g) \rightleftharpoons 2HBr(g)$$

where $\Delta H° = -103.8$ kJ/mol. In a particular experiment, equal moles of $H_2(g)$ at 1.00 atm and $Br_2(g)$ at 1.00 atm were mixed in a 1.00-L flask at 25°C and allowed to reach equilibrium. Then the molecules of H_2 at equilibrium were counted using a very sensitive technique, and 1.10×10^{13} molecules were found. For this reaction, calculate the values of K, $\Delta G°$, and $\Delta S°$.

109. Consider the system

$$A(g) \longrightarrow B(g)$$

at 25°C.

a. Assuming that $G_A° = 8996$ J/mol and $G_B° = 11,718$ J/mol, calculate the value of the equilibrium constant for this reaction.

b. Calculate the equilibrium pressures that result if 1.00 mole of $A(g)$ at 1.00 atm and 1.00 mole of $B(g)$ at 1.00 atm are mixed at 25°C.

c. Show by calculations that $\Delta G = 0$ at equilibrium.

110. The equilibrium constant for a certain reaction decreases from 8.84 to 3.25×10^{-2} when the temperature increases from 25°C to 75°C. Estimate the temperature where $K = 1.00$ for this reaction. Estimate the value of $\Delta S°$ for this reaction. (*Hint:* Manipulate the equation in Exercise 79.)

111. If wet silver carbonate is dried in a stream of hot air, the air must have a certain concentration level of carbon dioxide to prevent silver carbonate from decomposing by the reaction

$$Ag_2CO_3(s) \rightleftharpoons Ag_2O(s) + CO_2(g)$$

$\Delta H°$ for this reaction is 79.14 kJ/mol in the temperature range of 25 to 125°C. Given that the partial pressure of carbon dioxide in equilibrium with pure solid silver carbonate is 6.23×10^{-3} torr at 25°C, calculate the partial pressure of CO_2 necessary to prevent decomposition of Ag_2CO_3 at 110.°C. (*Hint:* Manipulate the equation in Exercise 79.)

112. Carbon tetrachloride (CCl_4) and benzene (C_6H_6) form ideal solutions. Consider an equimolar solution of CCl_4 and C_6H_6 at 25°C. The vapor above the solution is collected and condensed. Using the following data, determine the composition in mole fraction of the condensed vapor.

Substance	$\Delta G_f°$
$C_6H_6(l)$	124.50 kJ/mol
$C_6H_6(g)$	129.66 kJ/mol
$CCl_4(l)$	−65.21 kJ/mol
$CCl_4(g)$	−60.59 kJ/mol

113. Sodium chloride is added to water (at 25°C) until it is saturated. Calculate the Cl^- concentration in such a solution.

Species	$\Delta G°$ (kJ/mol)
$NaCl(s)$	−384
$Na^+(aq)$	−262
$Cl^-(aq)$	−131

114. You have a 1.00-L sample of hot water (90.0°C) sitting open in a 25.0°C room. Eventually the water cools to 25.0°C while the temperature of the room remains unchanged. Calculate ΔS_{surr} for this process. Assume the density of water is 1.00 g/cm^3 over this temperature range, and the heat capacity of water is constant over this temperature range and equal to 75.4 J/K · mol.

115. Consider a weak acid, HX. If a 0.10-M solution of HX has a pH of 5.83 at 25°C, what is $\Delta G°$ for the acid's dissociation reaction at 25°C?

Integrative Problems

These problems require the integration of multiple concepts to find the solutions.

116. Some nonelectrolyte solute (molar mass = 142 g/mol) was dissolved in 150. mL of a solvent (density = 0.879 g/cm^3). The elevated boiling point of the solution was 355.4 K. What mass of solute was dissolved in the solvent? For the solvent, the enthalpy of vaporization is 33.90 kJ/mol, the entropy of vaporization is 95.95 J/K · mol, and the boiling-point elevation constant is 2.5 K · kg/mol.

117. For the equilibrium

$$A(g) + 2B(g) \rightleftharpoons C(g)$$

the initial concentrations are [A] = [B] = [C] = 0.100 atm. Once equilibrium has been established, it is found that [C] = 0.040 atm. What is $\Delta G°$ for this reaction at 25°C?

118. What is the pH of a 0.125-M solution of the weak base B if $\Delta H° = -28.0$ kJ and $\Delta S° = -175$ J/K for the following equilibrium reaction at 25°C?

$$B(aq) + H_2O(l) \rightleftharpoons BH^+(aq) + OH^-(aq)$$

Marathon Problem

This problem is designed to incorporate several concepts and techniques into one situation.

119. Impure nickel, refined by smelting sulfide ores in a blast furnace, can be converted into metal from 99.90% to 99.99% purity by the Mond process. The primary reaction involved in the Mond process is

$$Ni(s) + 4CO(g) \rightleftharpoons Ni(CO)_4(g)$$

a. Without referring to Appendix 4, predict the sign of $\Delta S°$ for the above reaction. Explain.

b. The spontaneity of the above reaction is temperature-dependent. Predict the sign of ΔS_{surr} for this reaction. Explain.

c. For $Ni(CO)_4(g)$, $\Delta H_f° = -607$ kJ/mol and $S° = 417$ J/K · mol at 298 K. Using these values and data in Appendix 4, calculate $\Delta H°$ and $\Delta S°$ for the above reaction.

d. Calculate the temperature at which $\Delta G° = 0$ ($K = 1$) for the above reaction, assuming that $\Delta H°$ and $\Delta S°$ do not depend on temperature.

e. The first step of the Mond process involves equilibrating impure nickel with $CO(g)$ and $Ni(CO)_4(g)$ at about 50°C. The purpose of this step is to convert as much nickel as possible into the gas phase. Calculate the equilibrium constant for the above reaction at 50.°C.

f. In the second step of the Mond process, the gaseous $Ni(CO)_4$ is isolated and heated to 227°C. The purpose of this step is to deposit as much nickel as possible as pure solid (the reverse of the preceding reaction). Calculate the equilibrium constant for the preceding reaction at 227°C.

g. Why is temperature increased for the second step of the Mond process?

h. The Mond process relies on the volatility of $Ni(CO)_4$ for its success. Only pressures and temperatures at which $Ni(CO)_4$ is a gas are useful. A recently developed variation of the Mond process carries out the first step at higher pressures and a temperature of 152°C. Estimate the maximum pressure of $Ni(CO)_4(g)$ that can be attained before the gas will liquefy at 152°C. The boiling point for $Ni(CO)_4$ is 42°C and the enthalpy of vaporization is 29.0 kJ/mol.

[*Hint:* The phase change reaction and the corresponding equilibrium expression are

$$Ni(CO)_4(l) \rightleftharpoons Ni(CO)_4(g) \qquad K = P_{Ni(CO)_4}$$

$Ni(CO)_4(g)$ will liquefy when the pressure of $Ni(CO)_4$ is greater than the K value.]

Chapter 18

Electrochemistry

The Hiriko (meaning urban in Basque) is an electric car that folds up to park in a 1.5-meter space. The car was designed at MIT and is produced in Spain by a consortium of small businesses. (© Kika/ZumaPress/Newscom.com)

lectrochemistry constitutes one of the most important interfaces between chemistry and everyday life. Every time you start your car, turn on your calculator, use your smartphone, or listen to a radio at the beach, you are depending on electrochemical reactions. Our society sometimes seems to run almost entirely on batteries. Certainly the advent of small, dependable batteries along with silicon-chip technology has made possible the tiny calculators, portable audio players, and cell phones that we take for granted.

Electrochemistry is important in other less obvious ways. For example, the corrosion of iron, which has tremendous economic implications, is an electrochemical process. In addition, many important industrial materials such as aluminum, chlorine, and sodium hydroxide are prepared by electrolytic processes. In analytical chemistry, electrochemical techniques use electrodes that are specific for a given molecule or ion, such as H^+ (pH meters), F^-, Cl^-, and many others. These increasingly important methods are used to analyze for trace pollutants in natural waters or for the tiny quantities of chemicals in human blood that may signal the development of a specific disease.

Electrochemistry is best defined as *the study of the interchange of chemical and electrical energy.* It is primarily concerned with two processes that involve oxidation–reduction reactions: the generation of an electric current from a spontaneous chemical reaction and, the opposite process, the use of a current to produce chemical change.

18.1 | Balancing Oxidation–Reduction Equations

Oxidation–reduction reactions in aqueous solutions are often complicated, which means that it can be difficult to balance their equations by simple inspection. In this section we will discuss a special technique for balancing the equations of redox reactions that occur in aqueous solutions. It is called the *half-reaction method.*

The Half-Reaction Method for Balancing Oxidation–Reduction Reactions in Aqueous Solutions

For oxidation–reduction reactions that occur in aqueous solution, it is useful to separate the reaction into two **half-reactions**: one involving oxidation and the other involving reduction. For example, consider the unbalanced equation for the oxidation–reduction reaction between cerium(IV) ion and tin(II) ion:

$$Ce^{4+}(aq) + Sn^{2+}(aq) \longrightarrow Ce^{3+}(aq) + Sn^{4+}(aq)$$

This reaction can be separated into a half-reaction involving the substance being *reduced,*

$$Ce^{4+}(aq) \longrightarrow Ce^{3+}(aq)$$

and one involving the substance being *oxidized,*

$$Sn^{2+}(aq) \longrightarrow Sn^{4+}(aq)$$

The general procedure is to balance the equations for the half-reactions separately and then to add them to obtain the overall balanced equation. The half-reaction method for balancing oxidation–reduction equations differs slightly depending on whether the reaction takes place in acidic or basic solution.

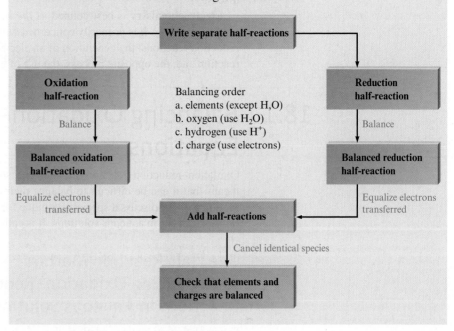

Problem-Solving Strategy

The Half-Reaction Method for Balancing Equations for Oxidation–Reduction Reactions Occurring in Acidic Solution

1. Write separate equations for the oxidation and reduction half-reactions.
2. For each half-reaction,
   a. balance all the elements except hydrogen and oxygen.
   b. balance oxygen using H_2O.
   c. balance hydrogen using H^+.
   d. balance the charge using electrons.
3. If necessary, multiply one or both balanced half-reactions by an integer to equalize the number of electrons transferred in the two half-reactions.
4. Add the half-reactions, and cancel identical species.
5. Check that the elements and charges are balanced.

We will illustrate this method by balancing the equation for the reaction between permanganate and iron(II) ions in acidic solution:

$$MnO_4^-(aq) + Fe^{2+}(aq) \xrightarrow{\text{Acid}} Fe^{3+}(aq) + Mn^{2+}(aq)$$

This reaction can be used to analyze iron ore for its iron content.

1. *Identify and write equations for the half-reactions.* The oxidation states for the half-reaction involving the permanganate ion show that manganese is reduced:

$$MnO_4^- \longrightarrow Mn^{2+}$$

$$\underset{+7}{} \quad \underset{-2 \text{ (each O)}}{} \quad \underset{+2}{}$$

This is the *reduction half-reaction.* The other half-reaction involves the oxidation of iron(II) to iron(III) ion and is the *oxidation half-reaction:*

$$Fe^{2+} \longrightarrow Fe^{3+}$$
$$\uparrow \qquad \uparrow$$
$$+2 \qquad +3$$

2. *Balance each half-reaction.* For the reduction reaction, we have

$$MnO_4^-(aq) \longrightarrow Mn^{2+}(aq)$$

a. The manganese is balanced.

b. We balance oxygen by adding $4H_2O$ to the right side of the equation:

$$MnO_4^-(aq) \longrightarrow Mn^{2+}(aq) + 4H_2O(l)$$

c. Next, we balance hydrogen by adding $8H^+$ to the left side:

$$8H^+(aq) + MnO_4^-(aq) \longrightarrow Mn^{2+}(aq) + 4H_2O(l)$$

d. All the elements have been balanced, but we need to balance the charge using electrons. At this point we have the following overall charges for reactants and products in the reduction half-reaction:

$$\underbrace{8H^+(aq) + MnO_4^-(aq)}_{\underbrace{8+ \qquad + \qquad 1-}_{7+}} \longrightarrow \underbrace{Mn^{2+}(aq) + 4H_2O(l)}_{\underbrace{2+ \qquad + \qquad 0}_{2+}}$$

We can equalize the charges by adding five electrons to the left side:

$$\underbrace{5e^- + 8H^+(aq) + MnO_4^-(aq)}_{2+} \longrightarrow \underbrace{Mn^{2+}(aq) + 4H_2O(l)}_{2+}$$

Both the *elements* and the *charges* are now balanced, so this represents the balanced reduction half-reaction. The fact that five electrons appear on the reactant side of the equation makes sense, since five electrons are required to reduce MnO_4^- (Mn has an oxidation state of $+7$) to Mn^{2+} (Mn has an oxidation state of $+2$).

For the oxidation reaction

$$Fe^{2+}(aq) \longrightarrow Fe^{3+}(aq)$$

the elements are balanced, and we must simply balance the charge:

$$\underbrace{Fe^{2+}(aq)}_{2+} \longrightarrow \underbrace{Fe^{3+}(aq)}_{3+}$$

One electron is needed on the right side to give a net $2+$ charge on both sides:

$$\underbrace{Fe^{2+}(aq)}_{2+} \longrightarrow \underbrace{Fe^{3+}(aq) + e^-}_{2+}$$

The number of electrons gained in the reduction half-reaction must equal the number of electrons lost in the oxidation half-reaction.

3. *Equalize the electron transfer in the two half-reactions.* Since the reduction half-reaction involves a transfer of five electrons and the oxidation half-reaction involves a transfer of only one electron, the oxidation half-reaction must be multiplied by 5:

$$5Fe^{2+}(aq) \longrightarrow 5Fe^{3+}(aq) + 5e^-$$

4. *Add the half-reactions.* The half-reactions are added to give

$$5e^- + 5Fe^{2+}(aq) + MnO_4^-(aq) + 8H^+(aq) \longrightarrow$$
$$5Fe^{3+}(aq) + Mn^{2+}(aq) + 4H_2O(l) + 5e^-$$

Note that the electrons cancel (as they must) to give the final balanced equation:

$$5Fe^{2+}(aq) + MnO_4^-(aq) + 8H^+(aq) \longrightarrow 5Fe^{3+}(aq) + Mn^{2+}(aq) + 4H_2O(l)$$

5. *Check that elements and charges are balanced.*

Elements balance: 5Fe, 1Mn, 4O, 8H $\longrightarrow$ 5Fe, 1Mn, 4O, 8H

Charges balance: $5(2+) + (1-) + 8(1+) = 17+ \longrightarrow$

$$5(3+) + (2+) + 0 = 17+$$

The equation is balanced.

Interactive Example 18.1

Sign in at http://login.cengagebrain.com to try this Interactive Example in **OWL**.

Balancing Oxidation–Reduction Reactions (Acidic)

Potassium dichromate ($K_2Cr_2O_7$) is a bright orange compound that can be reduced to a blue-violet solution of Cr^{3+} ions. Under certain conditions, $K_2Cr_2O_7$ reacts with ethanol (C_2H_5OH) as follows:

$$H^+(aq) + Cr_2O_7^{2-}(aq) + C_2H_5OH(l) \longrightarrow Cr^{3+}(aq) + CO_2(g) + H_2O(l)$$

Balance this equation using the half-reaction method.

Solution

1. The reduction half-reaction is

$$Cr_2O_7^{2-}(aq) \longrightarrow Cr^{3+}(aq)$$

Chromium is reduced from an oxidation state of +6 in $Cr_2O_7^{2-}$ to one of +3 in Cr^{3+}.

The oxidation half-reaction is

$$C_2H_5OH(l) \longrightarrow CO_2(g)$$

Carbon is oxidized from an oxidation state of −2 in C_2H_5OH to +4 in CO_2.

2. Balancing all elements except hydrogen and oxygen in the first half-reaction, we have

$$Cr_2O_7^{2-}(aq) \longrightarrow 2Cr^{3+}(aq)$$

Balancing oxygen using H_2O, we have

$$Cr_2O_7^{2-}(aq) \longrightarrow 2Cr^{3+}(aq) + 7H_2O(l)$$

Balancing hydrogen using H^+, we have

$$14H^+(aq) + Cr_2O_7^{2-}(aq) \longrightarrow 2Cr^{3+}(aq) + 7H_2O(l)$$

Balancing the charge using electrons, we have

$$6e^- + 14H^+(aq) + Cr_2O_7^{2-}(aq) \longrightarrow 2Cr^{3+}(aq) + 7H_2O(l)$$

When potassium dichromate reacts with ethanol, a blue-violet solution containing Cr^{3+} is formed.

Next, we turn to the oxidation half-reaction

$$C_2H_5OH(l) \longrightarrow CO_2(g)$$

Balancing carbon, we have

$$C_2H_5OH(l) \longrightarrow 2CO_2(g)$$

Balancing oxygen using H_2O, we have

$$C_2H_5OH(l) + 3H_2O(l) \longrightarrow 2CO_2(g)$$

Balancing hydrogen using H^+, we have

$$C_2H_5OH(l) + 3H_2O(l) \longrightarrow 2CO_2(g) + 12H^+(aq)$$

We then balance the charge by adding $12e^-$ to the right side:

$$C_2H_5OH(l) + 3H_2O(l) \longrightarrow 2CO_2(g) + 12H^+(aq) + 12e^-$$

3. In the reduction half-reaction, there are 6 electrons on the left-hand side, and there are 12 electrons on the right-hand side of the oxidation half-reaction. Thus we multiply the reduction half-reaction by 2 to give

$$12e^- + 28H^+(aq) + 2Cr_2O_7{}^{2-}(aq) \longrightarrow 4Cr^{3+}(aq) + 14H_2O(l)$$

4. Adding the half-reactions and canceling identical species, we have

Reduction Half-Reaction: $12e^- + 28H^+(aq) + 2Cr_2O_7{}^{2-}(aq) \longrightarrow 4Cr^{3+}(aq) + 14H_2O(l)$

Oxidation Half-Reaction: $C_2H_5OH(l) + 3H_2O(l) \longrightarrow 2CO_2(g) + 12H^+(aq) + 12e^-$

Complete Reaction: $16H^+(aq) + 2Cr_2O_7{}^{2-}(aq) + C_2H_5OH(l) \longrightarrow 4Cr^{3+} + 11H_2O(l) + 2CO_2(g)$

5. Check that elements and charges are balanced.

Elements balance: 22H, 4Cr, 15O, 2C $\longrightarrow$ 22H, 4Cr, 15O, 2C

Charges balance: $+16 + 2(-2) + 0 = +12 \longrightarrow 4(+3) + 0 + 0 = +12$

See Exercises 18.29 and 18.30

Oxidation–reduction reactions can occur in basic solutions (the reactions involve OH^- ions) as well as in acidic solutions (the reactions involve H^+ ions). The half-reaction method for balancing equations is slightly different for the two cases.

Problem-Solving Strategy

The Half-Reaction Method for Balancing Equations for Oxidation–Reduction Reactions Occurring in Basic Solution

1. Use the half-reaction method as specified for acidic solutions to obtain the final balanced equation *as if H^+ ions were present.*

2. To both sides of the equation obtained above, add a number of OH^- ions that is equal to the number of H^+ ions. (We want to eliminate H^+ by forming H_2O.)

3. Form H_2O on the side containing both H^+ and OH^- ions, and eliminate the number of H_2O molecules that appear on both sides of the equation.

4. Check that elements and charges are balanced.

(continued)

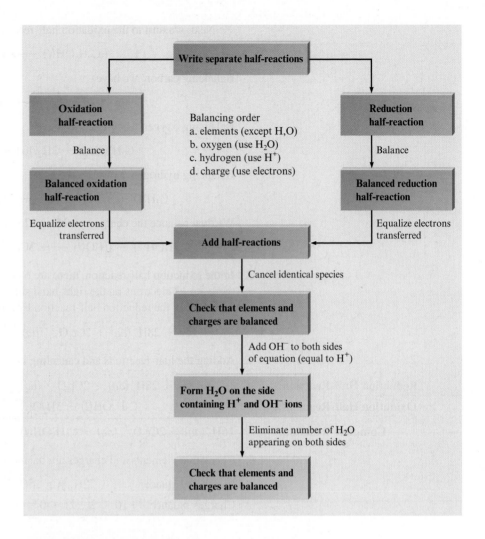

Critical Thinking

When balancing redox reactions occurring in basic solutions, the text instructs you to first use the half-reaction method as specified for acidic solutions. What if you started by adding OH^- first instead of H^+? What potential problem could there be with this approach?

Balancing Oxidation–Reduction Reactions (Basic)

Silver is sometimes found in nature as large nuggets; more often it is found mixed with other metals and their ores. An aqueous solution containing cyanide ion is often used to extract the silver using the following reaction that occurs in basic solution:

$$Ag(s) + CN^-(aq) + O_2(g) \xrightarrow{\text{Base}} Ag(CN)_2^-(aq)$$

Balance this equation using the half-reaction method.

Solution

1. Balance the equation as if H^+ ions were present. Balance the oxidation half-reaction:

$$CN^-(aq) + Ag(s) \longrightarrow Ag(CN)_2^-(aq)$$

Balance carbon and nitrogen:

$$2CN^-(aq) + Ag(s) \longrightarrow Ag(CN)_2^-(aq)$$

Balance the charge:

$$2CN^-(aq) + Ag(s) \longrightarrow Ag(CN)_2^-(aq) + e^-$$

Balance the reduction half-reaction:

$$O_2(g) \longrightarrow$$

Balance oxygen:

$$O_2(g) \longrightarrow 2H_2O(l)$$

Balance hydrogen:

$$O_2(g) + 4H^+(aq) \longrightarrow 2H_2O(l)$$

Balance the charge:

$$4e^- + O_2(g) + 4H^+(aq) \longrightarrow 2H_2O(l)$$

Multiply the balanced oxidation half-reaction by 4:

$$8CN^-(aq) + 4Ag(s) \longrightarrow 4Ag(CN)_2^-(aq) + 4e^-$$

Add the half-reactions, and cancel identical species:

Oxidation Half-Reaction: $\quad 8CN^-(aq) + 4Ag(s) \longrightarrow 4Ag(CN)_2^-(aq) + 4e^-$

Reduction Half-Reaction: $\quad 4e^- + O_2(g) + 4H^+(aq) \longrightarrow 2H_2O(l)$

Complete Reaction: $\quad 8CN^-(aq) + 4Ag(s) + O_2(g) + 4H^+(aq) \longrightarrow 4Ag(CN)_2^-(aq) + 2H_2O(l)$

2. Add OH^- ions to both sides of the balanced equation to eliminate the H^+ ions. We need to add $4OH^-$ to each side:

$$8CN^-(aq) + 4Ag(s) + O_2(g) + \underbrace{4H^+(aq) + 4OH^-(aq)}_{4H_2O(l)} \longrightarrow$$
$$4Ag(CN)_2^-(aq) + 2H_2O(l) + 4OH^-(aq)$$

3. Eliminate as many H_2O molecules as possible:

$$8CN^-(aq) + 4Ag(s) + O_2(g) + 2H_2O(l) \longrightarrow 4Ag(CN)_2^-(aq) + 4OH^-(aq)$$

4. Check that elements and charges are balanced.

Elements balance: $8C, 8N, 4Ag, 4O, 4H \longrightarrow 8C, 8N, 4Ag, 4O, 4H$

Charges balance: $8(1-) + 0 + 0 + 0 = 8- \longrightarrow 4(1-) + 4(1-) = 8-$

See Exercises 18.31 and 18.32

18.2 | Galvanic Cells

To understand how a redox reaction can be used to generate a current, let's consider the reaction between MnO_4^- and Fe^{2+}:

$$8H^+(aq) + MnO_4^-(aq) + 5Fe^{2+}(aq) \longrightarrow Mn^{2+}(aq) + 5Fe^{3+}(aq) + 4H_2O(l)$$

In this reaction, Fe^{2+} is oxidized and MnO_4^- is reduced; electrons are transferred from Fe^{2+} (the reducing agent) to MnO_4^- (the oxidizing agent).

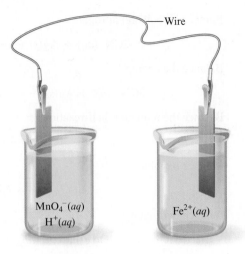

It is useful to break a redox reaction into half-reactions, one involving oxidation and one involving reduction. For the previous reaction, the half-reactions are

$$\text{Reduction:} \quad 8H^+ + MnO_4^- + 5e^- \longrightarrow Mn^{2+} + 4H_2O$$
$$\text{Oxidation:} \quad 5(Fe^{2+} \longrightarrow Fe^{3+} + e^-)$$

The multiplication of the second half-reaction by 5 indicates that this reaction must occur five times for each time the first reaction occurs. The balanced overall reaction is the sum of the half-reactions.

When MnO_4^- and Fe^{2+} are present in the same solution, the electrons are transferred directly when the reactants collide. Under these conditions, no useful work is obtained from the chemical energy involved in the reaction, which instead is released as heat. How can we harness this energy? The key is to separate the oxidizing agent from the reducing agent, thus requiring the electron transfer to occur through a wire. The current produced in the wire by the electron flow can then be directed through a device, such as an electric motor, to provide useful work.

For example, consider the system illustrated in Fig. 18.1. If our reasoning has been correct, electrons should flow through the wire from Fe^{2+} to MnO_4^-. However, when we construct the apparatus as shown, no flow of electrons is apparent. Why? Careful observation shows that when we connect the wires from the two compartments, current flows for an instant and then ceases. The current stops flowing because of charge buildups in the two compartments. If electrons flowed from the right to the left compartment in the apparatus as shown, the left compartment (receiving electrons) would become negatively charged, and the right compartment (losing electrons) would become positively charged. Creating a charge separation of this type requires a large amount of energy. Thus sustained electron flow cannot occur under these conditions.

However, we can solve this problem very simply. The solutions must be connected so that ions can flow to keep the net charge in each compartment zero. This connection might involve a **salt bridge** (a U-tube filled with an electrolyte) or a **porous disk** in a tube connecting the two solutions (Fig. 18.2). Either of these devices allows ions to flow without extensive mixing of the solutions. When we make the provision for ion flow, the circuit is complete. Electrons flow through the wire from reducing agent to oxidizing agent, and ions flow from one compartment to the other to keep the net charge zero.

We now have covered all the essential characteristics of a **galvanic cell**, *a device in which chemical energy is changed to electrical energy.* (The opposite process is called *electrolysis* and will be considered in Section 18.8.)

The reaction in an electrochemical cell occurs at the interface between the electrode and the solution where the electron transfer occurs. The electrode compartment in

A galvanic cell uses a spontaneous redox reaction to produce a current that can be used to do work.

Figure 18.2 | Galvanic cells can contain a salt bridge as in (a) or a porous-disk connection as in (b). A salt bridge contains a strong electrolyte held in a Jello-like matrix. A porous disk contains tiny passages that allow hindered flow of ions.

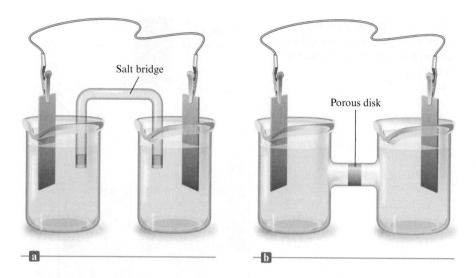

Oxidation occurs at the anode. Reduction occurs at the cathode.

which *oxidation* occurs is called the **anode**; the electrode compartment in which *reduction* occurs is called the **cathode** (Fig. 18.3).

Cell Potential

A galvanic cell consists of an oxidizing agent in one compartment that pulls electrons through a wire from a reducing agent in the other compartment. The "pull," or driving force, on the electrons is called the **cell potential** ($\mathscr{E}_{cell}$), or the **electromotive force** (emf) of the cell. The unit of electrical potential is the **volt** (abbreviated V), which is defined as 1 joule of work per coulomb of charge transferred.

A volt is 1 joule of work per coulomb of charge transferred: $1\ V = 1\ J/C$.

How can we measure the cell potential? One possible instrument is a crude **voltmeter**, which works by drawing current through a known resistance. However, when current flows through a wire, the frictional heating that occurs wastes some of the potentially useful energy of the cell. A traditional voltmeter will therefore measure a potential that is less than the maximum cell potential. The key to determining the maximum potential is to do the measurement under conditions of zero current so that no energy is wasted. Traditionally, this has been accomplished by inserting a variable-voltage device (powered from an external source) in *opposition* to the cell potential.

Figure 18.3 | An electrochemical process involves electron transfer at the interface between the electrode and the solution. (a) The species in the solution acting as the reducing agent supplies electrons to the anode. (b) The species in the solution acting as the oxidizing agent receives electrons from the cathode.

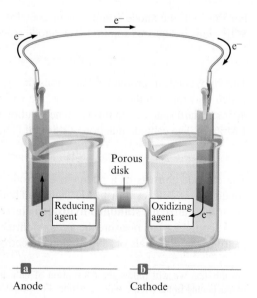

Figure 18.4 | Digital voltmeters draw only a negligible current and are convenient to use to measure the cell potential.

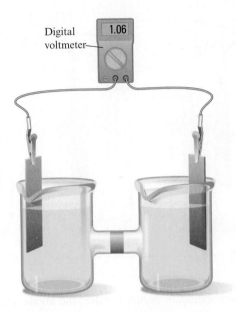

Digital voltmeter

The voltage on this instrument (called a **potentiometer**) is adjusted until no current flows in the cell circuit. Under such conditions, the cell potential is equal in magnitude and opposite in sign to the voltage setting of the potentiometer. This value represents the *maximum* cell potential, since no energy is wasted heating the wire. More recently, advances in electronic technology have allowed the design of *digital voltmeters* that draw only a negligible amount of current (Fig. 18.4). Since these instruments are more convenient to use, they have replaced potentiometers in the modern laboratory.

18.3 | Standard Reduction Potentials

The name *galvanic cell* honors Luigi Galvani (1737–1798), an Italian scientist generally credited with the discovery of electricity. These cells are sometimes called *voltaic cells* after Alessandro Volta (1745–1827), another Italian, who first constructed cells of this type around 1800.

The reaction in a galvanic cell is always an oxidation–reduction reaction that can be broken down into two half-reactions. It would be convenient to assign a potential to *each* half-reaction so that when we construct a cell from a given pair of half-reactions we can obtain the cell potential by summing the half-cell potentials. For example, the observed potential for the cell shown in Fig. 18.5(a) is 0.76 V, and the cell reaction* is

$$2H^+(aq) + Zn(s) \longrightarrow Zn^{2+}(aq) + H_2(g)$$

For this cell, the anode compartment contains a zinc metal electrode with Zn^{2+} and SO_4^{2-} ions in aqueous solution. The anode reaction is the oxidation half-reaction:

$$Zn \longrightarrow Zn^{2+} + 2e^-$$

The zinc metal, in producing Zn^{2+} ions that go into solution, is giving up electrons, which flow through the wire. For now, we will assume that all cell components are in their standard states, so in this case the solution in the anode compartment will contain $1\ M\ Zn^{2+}$. The cathode reaction of this cell is the reduction half-reaction:

$$2H^+ + 2e^- \longrightarrow H_2$$

The cathode consists of a platinum electrode (used because it is a chemically inert conductor) in contact with $1\ M\ H^+$ ions and bathed by hydrogen gas at 1 atm. Such an electrode, called the **standard hydrogen electrode**, is shown in Fig. 18.5(b).

Although we can measure the *total* potential of this cell (0.76 V), there is no way to measure the potentials of the individual electrode processes. Thus, if we want poten-

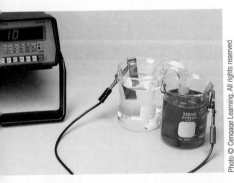

An electrochemical cell with a measured potential of 1.10 V.

---

*In this text we will follow the convention of indicating the physical states of the reactants and products only in the overall redox reaction. For simplicity, half-reactions will *not* include the physical states.

Figure 18.5 | (a) A galvanic cell involving the reactions Zn → Zn^{2+} + $2e^-$ (at the anode) and $2H^+ + 2e^- \rightarrow H_2$ (at the cathode) has a potential of 0.76 V. (b) The standard hydrogen electrode where $H_2(g)$ at 1 atm is passed over a platinum electrode in contact with 1 M H^+ ions. This electrode process (assuming ideal behavior) is arbitrarily assigned a value of exactly zero volts.

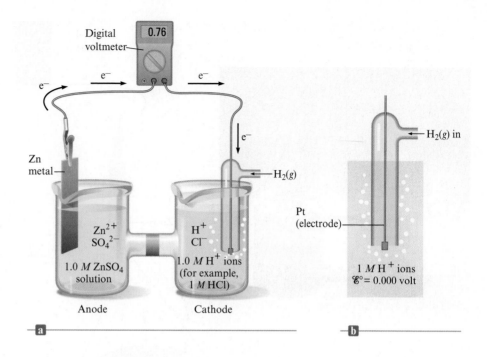

tials for the half-reactions (half-cells), we must arbitrarily divide the total cell potential. For example, if we assign the reaction

$$2H^+ + 2e^- \longrightarrow H_2$$

where $\qquad [H^+] = 1\ M \quad$ and $\quad P_{H_2} = 1$ atm

a potential of exactly zero volts, then the reaction

$$Zn \longrightarrow Zn^{2+} + 2e^-$$

will have a potential of 0.76 V because

$$\mathscr{E}^\circ_{cell} = \mathscr{E}^\circ_{H^+ \rightarrow H_2} + \mathscr{E}^\circ_{Zn \rightarrow Zn^{2+}}$$
$$\uparrow \qquad\qquad \uparrow \qquad\qquad \uparrow$$
$$0.76\ V \qquad 0\ V \qquad 0.76\ V$$

where the superscript ° indicates that *standard states* are used. In fact, by setting the standard potential for the half-reaction $2H^+ + 2e^- \rightarrow H_2$ equal to zero, we can assign values to all other half-reactions.

Standard states were discussed in Section 6.4.

For example, the measured potential for the cell shown in Fig. 18.6 is 1.10 V. The cell reaction is

$$Zn(s) + Cu^{2+}(aq) \longrightarrow Zn^{2+}(aq) + Cu(s)$$

which can be divided into the half-reactions

$$\text{Anode:} \quad Zn \longrightarrow Zn^{2+} + 2e^-$$
$$\text{Cathode:} \quad Cu^{2+} + 2e^- \longrightarrow Cu$$

Then

$$\mathscr{E}^\circ_{cell} = \mathscr{E}^\circ_{Zn \longrightarrow Zn^{2+}} + \mathscr{E}^\circ_{Cu^{2+} \longrightarrow Cu}$$

Since $\mathscr{E}^\circ_{Zn \rightarrow Zn^{2+}}$ was earlier assigned a value of 0.76 V, the value of $\mathscr{E}^\circ_{Cu^{2+} \rightarrow Cu}$ must be 0.34 V because

$$1.10\ V = 0.76\ V + 0.34\ V$$

The standard hydrogen potential is the reference potential against which all half-reaction potentials are assigned.

The scientific community has universally accepted the half-reaction potentials based on the assignment of 0 V to the process $2H^+ + 2e^- \rightarrow H_2$ (under standard

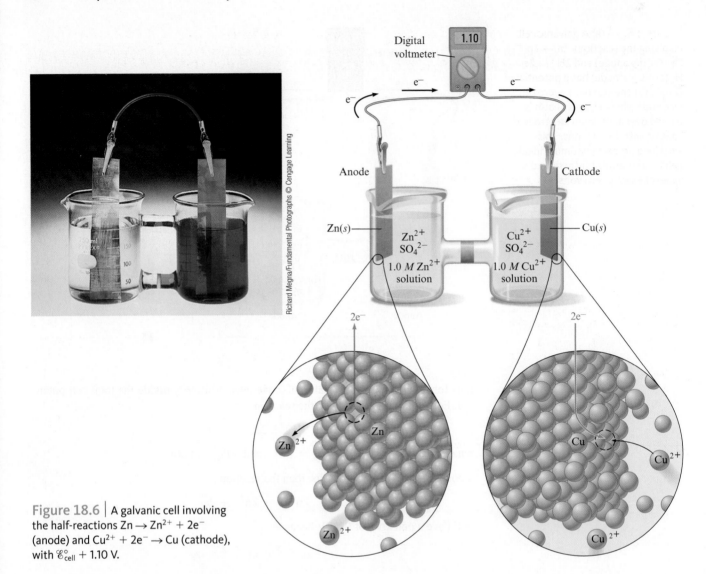

Richard Megna/Fundamental Photographs © Cengage Learning

Figure 18.6 | A galvanic cell involving the half-reactions Zn → Zn²⁺ + 2e⁻ (anode) and Cu²⁺ + 2e⁻ → Cu (cathode), with $\mathscr{E}^\circ_{cell}$ + 1.10 V.

conditions where ideal behavior is assumed). However, before we can use these values to calculate cell potentials, we need to understand several essential characteristics of half-cell potentials.

The accepted convention is to give the potentials of half-reactions as *reduction* processes. For example:

$$2H^+ + 2e^- \longrightarrow H_2$$
$$Cu^{2+} + 2e^- \longrightarrow Cu$$
$$Zn^{2+} + 2e^- \longrightarrow Zn$$

All half-reactions are given as reduction processes in standard tables.

The $\mathscr{E}^\circ$ values corresponding to reduction half-reactions with all solutes at 1 *M* and all gases at 1 atm are called **standard reduction potentials**. Standard reduction potentials for the most common half-reactions are given in Table 18.1 and Appendix 5.5.

Combining two half-reactions to obtain a balanced oxidation–reduction reaction often requires two manipulations:

1. One of the reduction half-reactions must be reversed (since redox reactions must involve a substance being oxidized and a substance being reduced). The half-reaction with the largest positive potential will run as written (as a reduction), and the other half-reaction will be forced to run in reverse (will be the oxidation

When a half-reaction is reversed, the sign of $\mathscr{E}^\circ$ is reversed.

Table 18.1 | Standard Reduction Potentials at 25°C (298 K) for Many Common Half-Reactions

Half-Reaction	$\mathcal{E}°$ (V)	Half-Reaction	$\mathcal{E}°$ (V)
$F_2 + 2e^- \rightarrow 2F^-$	2.87	$O_2 + 2H_2O + 4e^- \rightarrow 4OH^-$	0.40
$Ag^{2+} + e^- \rightarrow Ag^+$	1.99	$Cu^{2+} + 2e^- \rightarrow Cu$	0.34
$Co^{3+} + e^- \rightarrow Co^{2+}$	1.82	$Hg_2Cl_2 + 2e^- \rightarrow 2Hg + 2Cl^-$	0.27
$H_2O_2 + 2H^+ + 2e^- \rightarrow 2H_2O$	1.78	$AgCl + e^- \rightarrow Ag + Cl^-$	0.22
$Ce^{4+} + e^- \rightarrow Ce^{3+}$	1.70	$SO_4^{2-} + 4H^+ + 2e^- \rightarrow H_2SO_3 + H_2O$	0.20
$PbO_2 + 4H^+ + SO_4^{2-} + 2e^- \rightarrow PbSO_4 + 2H_2O$	1.69	$Cu^{2+} + e^- \rightarrow Cu^+$	0.16
$MnO_4^- + 4H^+ + 3e^- \rightarrow MnO_2 + 2H_2O$	1.68	$2H^+ + 2e^- \rightarrow H_2$	0.00
$2e^- + 2H^+ + IO_4^- \rightarrow IO_3^- + H_2O$	1.60	$Fe^{3+} + 3e^- \rightarrow Fe$	-0.036
$MnO_4^- + 8H^+ + 5e^- \rightarrow Mn^{2+} + 4H_2O$	1.51	$Pb^{2+} + 2e^- \rightarrow Pb$	-0.13
$Au^{3+} + 3e^- \rightarrow Au$	1.50	$Sn^{2+} + 2e^- \rightarrow Sn$	-0.14
$PbO_2 + 4H^+ + 2e^- \rightarrow Pb^{2+} + 2H_2O$	1.46	$Ni^{2+} + 2e^- \rightarrow Ni$	-0.23
$Cl_2 + 2e^- \rightarrow 2Cl^-$	1.36	$PbSO_4 + 2e^- \rightarrow Pb + SO_4^{2-}$	-0.35
$Cr_2O_7^{2-} + 14H^+ + 6e^- \rightarrow 2Cr^{3+} + 7H_2O$	1.33	$Cd^{2+} + 2e^- \rightarrow Cd$	-0.40
$O_2 + 4H^+ + 4e^- \rightarrow 2H_2O$	1.23	$Fe^{2+} + 2e^- \rightarrow Fe$	-0.44
$MnO_2 + 4H^+ + 2e^- \rightarrow Mn^{2+} + 2H_2O$	1.21	$Cr^{3+} + e^- \rightarrow Cr^{2+}$	-0.50
$IO_3^- + 6H^+ + 5e^- \rightarrow \frac{1}{2}I_2 + 3H_2O$	1.20	$Cr^{3+} + 3e^- \rightarrow Cr$	-0.73
$Br_2 + 2e^- \rightarrow 2Br^-$	1.09	$Zn^{2+} + 2e^- \rightarrow Zn$	-0.76
$VO_2^+ + 2H^+ + e^- \rightarrow VO^{2+} + H_2O$	1.00	$2H_2O + 2e^- \rightarrow H_2 + 2OH^-$	-0.83
$AuCl_4^- + 3e^- \rightarrow Au + 4Cl^-$	0.99	$Mn^{2+} + 2e^- \rightarrow Mn$	-1.18
$NO_3^- + 4H^+ + 3e^- \rightarrow NO + 2H_2O$	0.96	$Al^{3+} + 3e^- \rightarrow Al$	-1.66
$ClO_2 + e^- \rightarrow ClO_2^-$	0.954	$H_2 + 2e^- \rightarrow 2H^-$	-2.23
$2Hg^{2+} + 2e^- \rightarrow Hg_2^{2+}$	0.91	$Mg^{2+} + 2e^- \rightarrow Mg$	-2.37
$Ag^+ + e^- \rightarrow Ag$	0.80	$La^{3+} + 3e^- \rightarrow La$	-2.37
$Hg_2^{2+} + 2e^- \rightarrow 2Hg$	0.80	$Na^+ + e^- \rightarrow Na$	-2.71
$Fe^{3+} + e^- \rightarrow Fe^{2+}$	0.77	$Ca^{2+} + 2e^- \rightarrow Ca$	-2.76
$O_2 + 2H^+ + 2e^- \rightarrow H_2O_2$	0.68	$Ba^{2+} + 2e^- \rightarrow Ba$	-2.90
$MnO_4^- + e^- \rightarrow MnO_4^{2-}$	0.56	$K^+ + e^- \rightarrow K$	-2.92
$I_2 + 2e^- \rightarrow 2I^-$	0.54	$Li^+ + e^- \rightarrow Li$	-3.05
$Cu^+ + e^- \rightarrow Cu$	0.52		

reaction). The net potential of the cell will be the *difference* between the two. Since the reduction process occurs at the cathode and the oxidation process occurs at the anode, we can write

$$\mathcal{E}°_{cell} = \mathcal{E}° \text{ (cathode)} - \mathcal{E}° \text{ (anode)}$$

Because subtraction means "change the sign and add," in the examples done here we will change the sign of the oxidation (anode) reaction when we reverse it and add it to the reduction (cathode) reaction.

2. Since the number of electrons lost must equal the number gained, the half-reactions must be multiplied by integers as necessary to achieve the balanced equation. However, the *value of $\mathcal{E}°$ is not changed* when a half-reaction is multiplied by an integer. Since a standard reduction potential is an *intensive property* (it does not depend on how many times the reaction occurs), the potential is *not* multiplied by the integer required to balance the cell reaction.

When a half-reaction is multiplied by an integer, $\mathcal{E}°$ remains the same.

Consider a galvanic cell based on the redox reaction

$$\text{Fe}^{3+}(aq) + \text{Cu}(s) \longrightarrow \text{Cu}^{2+}(aq) + \text{Fe}^{2+}(aq)$$

The pertinent half-reactions are

$$\text{Fe}^{3+} + e^- \longrightarrow \text{Fe}^{2+} \qquad \mathcal{E}° = 0.77 \text{ V} \tag{1}$$

$$\text{Cu}^{2+} + 2e^- \longrightarrow \text{Cu} \qquad \mathcal{E}° = 0.34 \text{ V} \tag{2}$$

To balance the cell reaction and calculate the standard cell potential, reaction (2) must be reversed:

$$Cu \longrightarrow Cu^{2+} + 2e^- \qquad -\mathscr{E}° = -0.34 \text{ V}$$

Note the change in sign for the $\mathscr{E}°$ value. Now, since each Cu atom produces two electrons but each Fe^{3+} ion accepts only one electron, reaction (1) must be multiplied by 2:

$$2Fe^{3+} + 2e^- \longrightarrow 2Fe^{2+} \qquad \mathscr{E}° = 0.77 \text{ V}$$

Note that $\mathscr{E}°$ is not changed in this case.

Now we can obtain the balanced cell reaction by summing the appropriately modified half-reactions:

$$2Fe^{3+} + 2e^- \longrightarrow 2Fe^{2+} \qquad\qquad \mathscr{E} \text{ (cathode)} = -0.77 \text{ V}$$
$$Cu \longrightarrow Cu^{2+} + 2e^- \qquad\qquad -\mathscr{E} \text{ (anode)} = -0.34 \text{ V}$$

Cell reaction: $Cu(s) + 2Fe^{3+}(aq) \longrightarrow Cu^{2+}(aq) + 2Fe^{2+}(aq)$ $\qquad \mathscr{E}°_{cell} = \mathscr{E}° \text{ (cathode)} - \mathscr{E}° \text{ (anode)}$
$$= 0.77 \text{ V} - 0.34 \text{ V} = 0.43 \text{ V}$$

Critical Thinking

What if you want to "plate out" copper metal from an aqueous Cu^{2+} solution? Use Table 18.1 to determine several metals you can place in the solution to plate copper metal from the solution. Defend your choices. Why can Zn not be plated out from an aqueous solution of Zn^{2+} using the choices in Table 18.1?

Interactive Example 18.3

Sign in at http://login.cengagebrain .com to try this Interactive Example in OWL.

Galvanic Cells

a. Consider a galvanic cell based on the reaction

$$Al^{3+}(aq) + Mg(s) \longrightarrow Al(s) + Mg^{2+}(aq)$$

The half-reactions are

$$Al^{3+} + 3e^- \longrightarrow Al \qquad \mathscr{E}° = -1.66 \text{ V} \qquad (1)$$
$$Mg^{2+} + 2e^- \longrightarrow Mg \qquad \mathscr{E}° = -2.37 \text{ V} \qquad (2)$$

Give the balanced cell reaction, and calculate $\mathscr{E}°$ for the cell.

b. A galvanic cell is based on the reaction

$$MnO_4^-(aq) + H^+(aq) + ClO_3^-(aq) \longrightarrow ClO_4^-(aq) + Mn^{2+}(aq) + H_2O(l)$$

The half-reactions are

$$MnO_4^- + 5e^- + 8H^+ \longrightarrow Mn^{2+} + 4H_2O \qquad \mathscr{E}° = 1.51 \text{ V} \qquad (1)$$
$$ClO_4^- + 2H^+ + 2e^- \longrightarrow ClO_3^- + H_2O \qquad \mathscr{E}° = 1.19 \text{ V} \qquad (2)$$

Give the balanced cell reaction, and calculate $\mathscr{E}°$ for the cell.

Solution

a. The half-reaction involving magnesium must be reversed and since this is the oxidation process, it is the anode:

$$Mg \longrightarrow Mg^{2+} + 2e^- \qquad -\mathscr{E}° \text{ (anode)} = -(-2.37 \text{ V}) = 2.37 \text{ V}$$

Also, since the two half-reactions involve different numbers of electrons, they must be multiplied by integers as follows:

$$2(Al^{3+} + 3e^- \longrightarrow Al) \qquad\qquad \mathscr{E}° \text{ (cathode)} = -1.66 \text{ V}$$
$$3(Mg \longrightarrow Mg^{2+} + 2e^-) \qquad\qquad -\mathscr{E}° \text{ (anode)} = 2.37 \text{ V}$$

$$2Al^{3+}(aq) + 3Mg(s) \longrightarrow 2Al(s) + 3Mg^{2+}(aq) \qquad \mathscr{E}°_{\text{cell}} = \mathscr{E}° \text{ (cathode)} - \mathscr{E}° \text{ (anode)}$$
$$= -1.66 \text{ V} + 2.37 \text{ V} = 0.71 \text{ V}$$

b. Half-reaction (2) must be reversed (it is the anode), and both half-reactions must be multiplied by integers to make the number of electrons equal:

$$2(MnO_4^- + 5e^- + 8H^+ \longrightarrow Mn^{2+} + 4H_2O) \qquad \mathscr{E}° \text{ (cathode)} = 1.51 \text{ V}$$
$$5(ClO_3^- + H_2O \longrightarrow ClO_4^- + 2H^+ + 2e^-) \qquad -\mathscr{E}° \text{ (anode)} = -1.19 \text{ V}$$

$$2MnO_4^-(aq) + 6H^+(aq) + 5ClO_3^-(aq) \longrightarrow \qquad\qquad \mathscr{E}°_{\text{cell}} = \mathscr{E}° \text{ (cathode)} - \mathscr{E}° \text{ (anode)}$$
$$2Mn^{2+}(aq) + 3H_2O(l) + 5ClO_4^-(aq) \qquad\qquad\qquad = 1.51 \text{ V} - 1.19 \text{ V} = 0.32 \text{ V}$$

See Exercises 18.39 and 18.40

Line Notation

We now will introduce a handy line notation used to describe electrochemical cells. In this notation the anode components are listed on the left and the cathode components are listed on the right, separated by double vertical lines (indicating the salt bridge or porous disk). For example, the line notation for the cell described in Example 18.3(a) is

$$Mg(s) | Mg^{2+}(aq) || Al^{3+}(aq) | Al(s)$$

In this notation a phase difference (boundary) is indicated by a single vertical line. Thus, in this case, vertical lines occur between the solid Mg metal and the Mg^{2+} in aqueous solution and between solid Al and Al^{3+} in aqueous solution. Also note that the substance constituting the anode is listed at the far left and the substance constituting the cathode is listed at the far right.

For the cell described in Example 18.3(b), all the components involved in the oxidation–reduction reaction are ions. Since none of these dissolved ions can serve as an electrode, a nonreacting (inert) conductor must be used. The usual choice is platinum. Thus, for the cell described in Example 18.3(b), the line notation is

$$Pt(s) | ClO_3^-(aq), ClO_4^-(aq), H^+(aq) || H^+(aq), MnO_4^-(aq), Mn^{2+}(aq) | Pt(s)$$

Complete Description of a Galvanic Cell

Next we want to consider how to describe a galvanic cell fully, given just its half-reactions. This description will include the cell reaction, the cell potential, and the physical setup of the cell. Let's consider a galvanic cell based on the following half-reactions:

$$Fe^{2+} + 2e^- \longrightarrow Fe \qquad\qquad \mathscr{E}° = -0.44 \text{ V}$$
$$MnO_4^- + 5e^- + 8H^+ \longrightarrow Mn^{2+} + 4H_2O \qquad \mathscr{E}° = 1.51 \text{ V}$$

In a working galvanic cell, one of these reactions must run in reverse. Which one?

A galvanic cell runs spontaneously in the direction that gives a positive value for $\mathscr{E}_{\text{cell}}$.

We can answer this question by considering the sign of the potential of a working cell: *A cell will always run spontaneously in the direction that produces a positive cell potential.* Thus, in the present case, it is clear that the half-reaction involving iron must be reversed, since this choice leads to a positive cell potential:

$$Fe \longrightarrow Fe^{2+} + 2e^- \qquad -\mathscr{E}° = 0.44 \text{ V} \qquad \text{Anode reaction}$$
$$MnO_4^- + 5e^- + 8H^+ \longrightarrow Mn^{2+} + 4H_2O \qquad \mathscr{E}° = 1.51 \text{ V} \qquad \text{Cathode reaction}$$

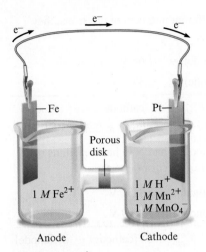

Figure 18.7 | The schematic of a galvanic cell based on the half-reactions:

$$Fe \longrightarrow Fe^{2+} + 2e^-$$
$$MnO_4^- + 5e^- + 8H^+ \longrightarrow Mn^{2+} + 4H_2O$$

where $\mathscr{E}^{\circ}_{cell} = \mathscr{E}^{\circ}(\text{cathode}) - \mathscr{E}^{\circ}(\text{anode}) = 1.51\ V + 0.44\ V = 1.95\ V$

The balanced cell reaction is obtained as follows:

$$5(Fe \longrightarrow Fe^{2+} + 2e^-)$$
$$2(MnO_4^- + 5e^- + 8H^+ \longrightarrow Mn^{2+} + 4H_2O)$$

$$\overline{2MnO_4^-(aq) + 5Fe(s) + 16H^+(aq) \longrightarrow 5Fe^{2+}(aq) + 2Mn^{2+}(aq) + 8H_2O(l)}$$

Now consider the physical setup of the cell, shown schematically in Fig. 18.7. In the left compartment the active components in their standard states are pure metallic iron (Fe) and 1.0 M Fe^{2+}. The anion present depends on the iron salt used. In this compartment the anion does not participate in the reaction but simply balances the charge. The half-reaction that takes place at this electrode is

$$Fe \longrightarrow Fe^{2+} + 2e^-$$

which is an oxidation reaction, so this compartment is the anode. The electrode consists of pure iron metal.

In the right compartment the active components in their standard states are 1.0 M MnO$_4^-$, 1.0 M H$^+$, and 1.0 M Mn^{2+}, with appropriate unreacting ions (often called *counterions*) to balance the charge. The half-reaction in this compartment is

$$MnO_4^- + 5e^- + 8H^+ \longrightarrow Mn^{2+} + 4H_2O$$

which is a reduction reaction, so this compartment is the cathode. Since neither MnO$_4^-$ nor Mn^{2+} ions can serve as the electrode, a nonreacting conductor such as platinum must be used.

The next step is to determine the direction of electron flow. In the left compartment the half-reaction involves the oxidation of iron:

$$Fe \longrightarrow Fe^{2+} + 2e^-$$

In the right compartment the half-reaction is the reduction of MnO$_4^-$:

$$MnO_4^- + 5e^- + 8H^+ \longrightarrow Mn^{2+} + 4H_2O$$

Thus the electrons flow from Fe to MnO$_4^-$ in this cell, or from the anode to the cathode, as is always the case. The line notation for this cell is

$$Fe(s)\,|\,Fe^{2+}(aq)\,\|\,MnO_4^-(aq),\ Mn^{2+}(aq)\,|\,Pt(s)$$

Let's Review | *Description of a Galvanic Cell*

A complete description of a galvanic cell usually includes four items:

> The cell potential (always positive for a galvanic cell where $\mathscr{E}^{\circ}_{cell} = \mathscr{E}^{\circ}(\text{cathode}) - \mathscr{E}^{\circ}(\text{anode})$) and the balanced cell reaction.

> The direction of electron flow, obtained by inspecting the half-reactions and using the direction that gives a positive $\mathscr{E}_{cell}$.

> Designation of the anode and cathode.

> The nature of each electrode and the ions present in each compartment. A chemically inert conductor is required if none of the substances participating in the half-reaction is a conducting solid.

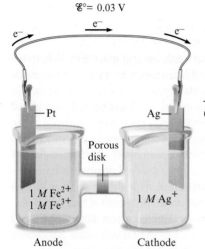

Figure 18.8 | Schematic diagram for the galvanic cell based on the half-reactions:

$$Ag^+ + e^- \longrightarrow Ag$$
$$Fe^{2+} \longrightarrow Fe^{3+} + e^-$$

| Example 18.4 | Description of a Galvanic Cell |

Describe completely the galvanic cell based on the following half-reactions under standard conditions:

$$Ag^+ + e^- \longrightarrow Ag \qquad \mathscr{E}^\circ = 0.80 \text{ V} \qquad (1)$$
$$Fe^{3+} + e^- \longrightarrow Fe^{2+} \qquad \mathscr{E}^\circ = 0.77 \text{ V} \qquad (2)$$

Solution

> Since a positive $\mathscr{E}^\circ_{cell}$ value is required, reaction (2) must run in reverse:

$$Ag^+ + e^- \longrightarrow Ag \qquad \mathscr{E}^\circ \text{ (cathode)} = 0.80 \text{ V}$$
$$Fe^{2+} \longrightarrow Fe^{3+} + e^- \qquad -\mathscr{E}^\circ \text{ (anode)} = -0.77 \text{ V}$$

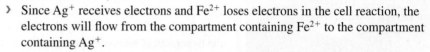

Cell reaction: $Ag^+(aq) + Fe^{2+}(aq) \longrightarrow Fe^{3+}(aq) + Ag(s) \qquad \mathscr{E}^\circ_{cell} = 0.03 \text{ V}$

> Since Ag^+ receives electrons and Fe^{2+} loses electrons in the cell reaction, the electrons will flow from the compartment containing Fe^{2+} to the compartment containing Ag^+.

> Oxidation occurs in the compartment containing Fe^{2+} (electrons flow from Fe^{2+} to Ag^+). Hence this compartment functions as the anode. Reduction occurs in the compartment containing Ag^+, so this compartment functions as the cathode.

> The electrode in the Ag/Ag^+ compartment is silver metal, and an inert conductor, such as platinum, must be used in the Fe^{2+}/Fe^{3+} compartment. Appropriate counterions are assumed to be present. The diagram for this cell is shown in Fig. 18.8. The line notation for this cell is

$$Pt(s)\,|\,Fe^{2+}(aq),\,Fe^{3+}(aq)\,||\,Ag^+(aq)\,|\,Ag(s)$$

See Exercises 18.41 and 18.42

18.4 | Cell Potential, Electrical Work, and Free Energy

So far we have considered electrochemical cells in a very practical fashion without much theoretical background. The next step will be to explore the relationship between thermodynamics and electrochemistry.

The work that can be accomplished when electrons are transferred through a wire depends on the "push" (the thermodynamic driving force) behind the electrons. This driving force (the emf) is defined in terms of a *potential difference* (in volts) between two points in the circuit. Recall that a volt represents a joule of work per coulomb of charge transferred:

$$\text{emf} = \text{potential difference (V)} = \frac{\text{work (J)}}{\text{charge (C)}}$$

Thus 1 joule of work is produced or required (depending on the direction) when 1 coulomb of charge is transferred between two points in the circuit that differ by a potential of 1 volt.

In this book, *work is viewed from the point of view of the system.* Thus work flowing out of the system is indicated by a minus sign. When a cell produces a current, the cell potential is positive, and the current can be used to do work—to run a motor, for instance. Thus the cell potential $\mathcal{E}$ and the work w have opposite signs:

$$\mathcal{E} = \frac{-w \leftarrow \text{Work}}{q \leftarrow \text{Charge}}$$

Therefore,

$$-w = q\mathcal{E}$$

From this equation it can be seen that the maximum work in a cell would be obtained at the maximum cell potential:

$$-w_{max} = q\mathcal{E}_{max} \quad \text{or} \quad w_{max} = -q\mathcal{E}_{max}$$

However, there is a problem. To obtain electrical work, current must flow. When current flows, some energy is inevitably wasted through frictional heating, and the maximum work is not obtained. This reflects the important general principle introduced in Section 17.9: In any real, spontaneous process some energy is always wasted—the actual work realized is always less than the calculated maximum. This is a consequence of the fact that the entropy of the universe must increase in any spontaneous process. Recall from Section 17.9 that the only process from which maximum work could be realized is the hypothetical reversible process. For a galvanic cell this would involve an infinitesimally small current flow and thus an infinite amount of time to do the work. Even though we can never achieve the maximum work through the actual discharge of a galvanic cell, we can measure the maximum potential. There is negligible current flow when a cell potential is measured with a potentiometer or an efficient digital voltmeter. No current flow implies no waste of energy, so the potential measured is the maximum.

Although we can never actually realize the maximum work from a cell reaction, the value for it is still useful in evaluating the efficiency of a real process based on the cell reaction. For example, suppose a certain galvanic cell has a maximum potential (at zero current) of 2.50 V. In a particular experiment 1.33 moles of electrons were passed through this cell at an average actual potential of 2.10 V. The actual work done is

$$w = -q\mathcal{E}$$

where $\mathcal{E}$ represents the actual potential difference at which the current flowed (2.10 V or 2.10 J/C) and q is the quantity of charge in coulombs transferred. The charge on 1 mole of electrons is a constant called the **faraday** (abbreviated F), which has the value *96,485 coulombs of charge per mole of electrons*. Thus q equals the number of moles of electrons times the charge per mole of electrons:

$$q = nF = 1.33 \text{ mol e}^- \times 96{,}485 \text{ C/mol e}^-$$

Then, for the preceding experiment, the actual work is

$$w = -q\mathcal{E} = -(1.33 \text{ mol e}^- \times 96{,}485 \text{ C/mol e}^-)(2.10 \text{ J/C})$$
$$= -2.69 \times 10^5 \text{ J}$$

For the maximum possible work, the calculation is similar, except that the maximum potential is used:

$$w_{max} = -q\mathcal{E}$$
$$= -\left(1.33 \text{ mol e}^- \times 96{,}485 \frac{\text{C}}{\text{mol e}^-}\right)\left(2.50 \frac{\text{J}}{\text{C}}\right)$$
$$= -3.21 \times 10^5 \text{ J}$$

A workman using a battery-powered drill on a construction site.

Work is never the maximum possible if any current is flowing.

Michael Faraday lecturing at the Royal Institution before Prince Albert and others (1855). The faraday was named in honor of Michael Faraday (1791–1867), an Englishman who may have been the greatest experimental scientist of the nineteenth century. Among his many achievements were the invention of the electric motor and generator and the development of the principles of electrolysis.

Thus, in its actual operation, the efficiency of this cell is

$$\frac{w}{w_{max}} \times 100\% = \frac{-2.69 \times 10^5 \text{ J}}{-3.21 \times 10^5 \text{ J}} \times 100\% = 83.8\%$$

Next we want to relate the potential of a galvanic cell to free energy. In Section 17.9 we saw that for a process carried out at constant temperature and pressure, the change in free energy equals the maximum useful work obtainable from that process:

$$w_{max} = \Delta G$$

For a galvanic cell,

$$w_{max} = -q\mathscr{E}_{max} = \Delta G$$

Since

$$q = nF$$

we have

$$\Delta G = -q\mathscr{E}_{max} = -nF\mathscr{E}_{max}$$

From now on the subscript on $\mathscr{E}_{max}$ will be deleted, with the understanding that any potential given in this book is the maximum potential. Thus

$$\Delta G = -nF\mathscr{E}$$

For standard conditions,

$$\Delta G° = -nF\mathscr{E}°$$

This equation states that the maximum cell potential is directly related to the free energy difference between the reactants and the products in the cell. This relationship is important because it provides an experimental means to obtain ΔG for a reaction. It also confirms that a galvanic cell will run in the direction that gives a positive value for $\mathscr{E}_{cell}$; a positive $\mathscr{E}_{cell}$ value corresponds to a negative ΔG value, which is the condition for spontaneity.

Calculating $\Delta G°$ for a Cell Reaction

Using the data in Table 18.1, calculate $\Delta G°$ for the reaction

$$Cu^{2+}(aq) + Fe(s) \longrightarrow Cu(s) + Fe^{2+}(aq)$$

Is this reaction spontaneous?

Solution

The half-reactions are

$$
\begin{array}{ll}
Cu^{2+} + 2e^- \longrightarrow Cu & \mathscr{E}° \text{ (cathode)} = 0.34 \text{ V} \\
\underline{\hspace{0.5em} Fe \longrightarrow Fe^{2+} + 2e^-} & \underline{-\mathscr{E}° \text{ (anode)} = 0.44 \text{ V}} \\
Cu^{2+} + Fe \longrightarrow Fe^{2+} + Cu & \mathscr{E}°_{cell} = 0.78 \text{ V}
\end{array}
$$

We can calculate $\Delta G°$ from the equation

$$\Delta G° = -nF\mathscr{E}°$$

Since two electrons are transferred per atom in the reaction, 2 moles of electrons are required per mole of reactants and products. Thus $n = 2$ mol e$^-$, $F = 96{,}485$ C/mol e$^-$, and $\mathscr{E}° = 0.78$ V $= 0.78$ J/C. Therefore,

$$\Delta G° = -(2 \text{ mol e}^-)\left(96{,}485 \, \frac{C}{\text{mol e}^-}\right)\left(0.78 \, \frac{J}{C}\right)$$

$$= -1.5 \times 10^5 \text{ J}$$

The process is spontaneous, as indicated by both the negative sign of $\Delta G°$ and the positive sign of $\mathscr{E}°_{cell}$.

This reaction is used industrially to deposit copper metal from solutions resulting from the dissolving of copper ores.

See Exercises 18.49 and 18.50

Example 18.6

A gold ring does not dissolve in nitric acid.

Predicting Spontaneity

Using the data from Table 18.1, predict whether 1 M HNO$_3$ will dissolve gold metal to form a 1-M Au^{3+} solution.

Solution

The half-reaction for HNO$_3$ acting as an oxidizing agent is

$$NO_3^- + 4H^+ + 3e^- \longrightarrow NO + 2H_2O \qquad \mathscr{E}° \text{ (cathode)} = 0.96 \text{ V}$$

The reaction for the oxidation of solid gold to Au^{3+} ions is

$$Au \longrightarrow Au^{3+} + 3e^- \qquad -\mathscr{E}° \text{ (anode)} = -1.50 \text{ V}$$

The sum of these half-reactions gives the required reaction:

$$Au(s) + NO_3^-(aq) + 4H^+(aq) \longrightarrow Au^{3+}(aq) + NO(g) + 2H_2O(l)$$

and $\mathscr{E}°_{cell} = \mathscr{E}° \text{ (cathode)} - \mathscr{E}° \text{ (anode)} = 0.96 \text{ V} - 1.50 \text{ V} = -0.54 \text{ V}$

Since the $\mathscr{E}°$ value is negative, the process will *not* occur under standard conditions. That is, gold will not dissolve in 1 M HNO$_3$ to give 1 M Au^{3+}. In fact, a mixture (1:3 by volume) of concentrated nitric and hydrochloric acids, called *aqua regia,* is required to dissolve gold.

See Exercises 18.59 and 18.60

18.5 | Dependence of Cell Potential on Concentration

So far we have described cells under standard conditions. In this section we consider the dependence of the cell potential on concentration. Under standard conditions (all concentrations 1 M), the cell with the reaction

$$Cu(s) + 2Ce^{4+}(aq) \longrightarrow Cu^{2+}(aq) + 2Ce^{3+}(aq)$$

has a potential of 1.36 V. What will the cell potential be if [Ce^{4+}] is greater than 1.0 M? This question can be answered qualitatively in terms of Le Châtelier's principle. An increase in the concentration of Ce^{4+} will favor the forward reaction and thus increase the driving force on the electrons. The cell potential will increase. On the other hand, an increase in the concentration of a product (Cu^{2+} or Ce^{3+}) will oppose the forward reaction, thus decreasing the cell potential.

These ideas are illustrated in Example 18.7.

The Effects of Concentration on $\mathscr{E}$

For the cell reaction

$$2Al(s) + 3Mn^{2+}(aq) \longrightarrow 2Al^{3+}(aq) + 3Mn(s) \qquad \mathscr{E}^{\circ}_{cell} = 0.48 \text{ V}$$

predict whether $\mathscr{E}_{cell}$ is larger or smaller than $\mathscr{E}^{\circ}_{cell}$ for the following cases.

a. $[Al^{3+}] = 2.0 \ M$, $[Mn^{2+}] = 1.0 \ M$

b. $[Al^{3+}] = 1.0 \ M$, $[Mn^{2+}] = 3.0 \ M$

Solution

a. A product concentration has been raised above $1.0 \ M$. This will oppose the cell reaction and will cause $\mathscr{E}_{cell}$ to be less than $\mathscr{E}^{\circ}_{cell}$ ($\mathscr{E}_{cell} < 0.48$ V).

b. A reactant concentration has been increased above $1.0 \ M$, and $\mathscr{E}_{cell}$ will be greater than $\mathscr{E}^{\circ}_{cell}$ ($\mathscr{E}_{cell} > 0.48$ V).

See Exercise 18.67

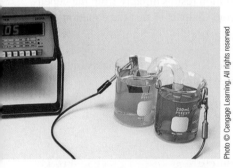

A concentration cell with 1.0 M Cu^{2+} on the right and 0.010 M Cu^{2+} on the left.

Concentration Cells

Because cell potentials depend on concentration, we can construct galvanic cells where both compartments contain the same components but at different concentrations. For example, in the cell in Fig. 18.9, both compartments contain aqueous $AgNO_3$, but with different molarities. Let's consider the potential of this cell and the direction of electron flow. The half-reaction relevant to both compartments of this cell is

$$Ag^+ + e^- \longrightarrow Ag \qquad \mathscr{E}^{\circ} = 0.80 \text{ V}$$

If the cell had $1 \ M$ Ag^+ in both compartments,

$$\mathscr{E}^{\circ}_{cell} = 0.80 \text{ V} - 0.80 \text{ V} = 0 \text{ V}$$

However, in the cell described here, the concentrations of Ag^+ in the two compartments are $1 \ M$ and $0.1 \ M$. Because the concentrations of Ag^+ are unequal, the half-cell potentials will not be identical, and the cell will exhibit a positive voltage. In which direction will the electrons flow in this cell? The best way to think about this question is to recognize that nature will try to equalize the concentrations of Ag^+ in the two compartments. This can be done by transferring electrons from the compartment containing $0.1 \ M$ Ag^+ to the one containing $1 \ M$ Ag^+ (left to right in Fig. 18.9). This electron transfer will produce more Ag^+ in the left compartment and consume Ag^+ (to form Ag) in the right compartment.

A cell in which both compartments have the same components but at different concentrations is called a **concentration cell**. The difference in concentration is the only factor that produces a cell potential in this case, and the voltages are typically small.

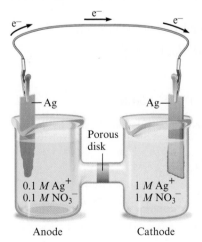

Anode   Cathode

Figure 18.9 | A concentration cell that contains a silver electrode and aqueous silver nitrate in both compartments. Because the right compartment contains $1 \ M$ Ag^+ and the left compartment contains $0.1 \ M$ Ag^+, there will be a driving force to transfer electrons from left to right. Silver will be deposited on the right electrode, thus lowering the concentration of Ag^+ in the right compartment. In the left compartment the silver electrode dissolves (producing Ag^+ ions) to raise the concentration of Ag^+ in solution.

Example 18.8

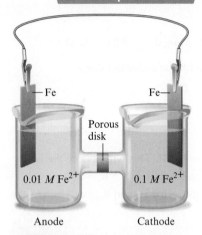

0.01 M Fe^{2+} 0.1 M Fe^{2+}

Anode   Cathode

Figure 18.10 | A concentration cell containing iron electrodes and different concentrations of Fe^{2+} ion in the two compartments.

Concentration Cells

Determine the direction of electron flow, and designate the anode and cathode for the cell represented in Fig. 18.10.

Solution

The concentrations of Fe^{2+} ion in the two compartments can (eventually) be equalized by transferring electrons from the left compartment to the right. This will cause Fe^{2+} to be formed in the left compartment, and iron metal will be deposited (by reducing Fe^{2+} ions to Fe) on the right electrode. Since electron flow is from left to right, oxidation occurs in the left compartment (the anode) and reduction occurs in the right (the cathode).

See Exercise 18.68

The Nernst Equation

Nernst was one of the pioneers in the development of electrochemical theory and is generally given credit for first stating the third law of thermodynamics. He won the Nobel Prize in chemistry in 1920.

The dependence of the cell potential on concentration results directly from the dependence of free energy on concentration. Recall from Chapter 17 that the equation

$$\Delta G = \Delta G^\circ + RT \ln(Q)$$

where Q is the reaction quotient, was used to calculate the effect of concentration on ΔG. Since $\Delta G = -nF\mathscr{E}$ and $\Delta G^\circ = -nF\mathscr{E}^\circ$, the equation becomes

$$-nF\mathscr{E} = -nF\mathscr{E}^\circ + RT \ln(Q)$$

Dividing each side of the equation by $-nF$ gives

$$\mathscr{E} = \mathscr{E}^\circ - \frac{RT}{nF} \ln(Q) \qquad (18.1)$$

Equation (18.1), which gives the relationship between the cell potential and the concentrations of the cell components, is commonly called the **Nernst equation**, after the German chemist Walther Hermann Nernst (1864–1941).

The Nernst equation is often given in a form that is valid at 25°C:

$$\mathscr{E} = \mathscr{E}^\circ - \frac{0.0591}{n} \log(Q)$$

Using this relationship, we can calculate the potential of a cell in which some or all of the components are not in their standard states.

For example, $\mathscr{E}^\circ_{\text{cell}}$ is 0.48 V for the galvanic cell based on the reaction

$$2\text{Al}(s) + 3\text{Mn}^{2+}(aq) \longrightarrow 2\text{Al}^{3+}(aq) + 3\text{Mn}(s)$$

Consider a cell in which

$$[\text{Mn}^{2+}] = 0.50 \, M \quad \text{and} \quad [\text{Al}^{3+}] = 1.50 \, M$$

The cell potential at 25°C for these concentrations can be calculated using the Nernst equation:

$$\mathscr{E}_{\text{cell}} = \mathscr{E}^\circ_{\text{cell}} - \frac{0.0591}{n} \log(Q)$$

We know that

$$\mathscr{E}^\circ_{\text{cell}} = 0.48 \text{ V}$$

and
$$Q = \frac{[Al^{3+}]^2}{[Mn^{2+}]^3} = \frac{(1.50)^2}{(0.50)^3} = 18$$

Since the half-reactions are

Oxidation: $\quad 2Al \longrightarrow 2Al^{3+} + 6e^-$

Reduction: $\quad 3Mn^{2+} + 6e^- \longrightarrow 3Mn$

we know that

$$n = 6$$

Thus
$$\mathscr{E}_{cell} = 0.48 - \frac{0.0591}{6} \log(18)$$

$$= 0.48 - \frac{0.0591}{6} (1.26) = 0.48 - 0.01 = 0.47 \text{ V}$$

Note that the cell voltage decreases slightly because of the nonstandard concentrations. This change is consistent with the predictions of Le Châtelier's principle (see Example 18.7). In this case, since the reactant concentration is lower than 1.0 M and the product concentration is higher than 1.0 M, $\mathscr{E}_{cell}$ is less than $\mathscr{E}_{cell}^\circ$.

The potential calculated from the Nernst equation is the maximum potential before any current flow has occurred. As the cell discharges and current flows from anode to cathode, the concentrations will change, and as a result, $\mathscr{E}_{cell}$ will change. In fact, *the cell will spontaneously discharge until it reaches equilibrium,* at which point

$$Q = K \text{ (the equilibrium constant)} \quad \text{and} \quad \mathscr{E}_{cell} = 0$$

A "dead" battery is one in which the cell reaction has reached equilibrium, and there is no longer any chemical driving force to push electrons through the wire. In other words, *at equilibrium, the components in the two cell compartments have the same free energy,* and $\Delta G = 0$ for the cell reaction at the equilibrium concentrations. The cell no longer has the ability to do work.

Critical Thinking

What if you are told that $\mathscr{E}^\circ = 0$ for an electrolytic cell? Does this mean the cell is "dead"? What if $\mathscr{E} = 0$? Explain your answer in each case.

Example 18.9

The Nernst Equation

Describe the cell based on the following half-reactions:

$$VO_2^+ + 2H^+ + e^- \longrightarrow VO^{2+} + H_2O \qquad \mathscr{E}^\circ = 1.00 \text{ V} \qquad (1)$$
$$Zn^{2+} + 2e^- \longrightarrow Zn \qquad \mathscr{E}^\circ = -0.76 \text{ V} \qquad (2)$$

where
$$T = 25°C$$
$$[VO_2^+] = 2.0 \ M$$
$$[H^+] = 0.50 \ M$$
$$[VO^{2+}] = 1.0 \times 10^{-2} \ M$$
$$[Zn^{2+}] = 1.0 \times 10^{-1} \ M$$

Solution

The balanced cell reaction is obtained by reversing reaction (2) and multiplying reaction (1) by 2:

2 × reaction (1)	$2VO_2^+ + 4H^+ + 2e^- \longrightarrow 2VO^{2+} + 2H_2O$	$\mathscr{E}°$ (cathode) = 1.00 V
Reaction (2) reversed	$Zn \longrightarrow Zn^{2+} + 2e^-$	$-\mathscr{E}°$ (anode) = 0.76 V
Cell reaction:	$2VO_2^+(aq) + 4H^+(aq) + Zn(s) \longrightarrow 2VO^{2+}(aq) + 2H_2O(l) + Zn^{2+}(aq)$	$\mathscr{E}°_{cell}$ = 1.76 V

Since the cell contains components at concentrations other than 1 M, we must use the Nernst equation, where $n = 2$ (since two electrons are transferred), to calculate the cell potential. At 25°C we can use the equation

$$\mathscr{E} = \mathscr{E}°_{cell} - \frac{0.0591}{n} \log(Q)$$

$$= 1.76 - \frac{0.0591}{2} \log\left(\frac{[Zn^{2+}][VO^{2+}]^2}{[VO_2^+]^2[H^+]^4}\right)$$

$$= 1.76 - \frac{0.0591}{2} \log\left(\frac{(1.0 \times 10^{-1})(1.0 \times 10^{-2})^2}{(2.0)^2(0.50)^4}\right)$$

$$= 1.76 - \frac{0.0591}{2} \log(4 \times 10^{-5}) = 1.76 + 0.13 = 1.89 \text{ V}$$

The cell diagram is given in Fig. 18.11.

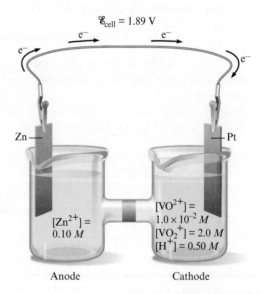

Figure 18.11 | Schematic diagram of the cell described in Example 18.9.

See Exercises 18.71 through 18.74

Ion-Selective Electrodes

Because the cell potential is sensitive to the concentrations of the reactants and products involved in the cell reaction, measured potentials can be used to determine the concentration of an ion. A pH meter (see Fig. 14.7) is a familiar example of an instrument that measures concentration using an observed potential. The pH meter has three main components: a standard electrode of known potential, a special **glass electrode** that changes potential depending on the concentration of H^+ ions in the solution into which it is dipped, and a potentiometer that measures the potential between the

Figure 18.12 | A glass electrode contains a reference solution of dilute hydrochloric acid in contact with a thin glass membrane in which a silver wire coated with silver chloride has been embedded. When the electrode is dipped into a solution containing H^+ ions, the electrode potential is determined by the difference in $[H^+]$ between the two solutions.

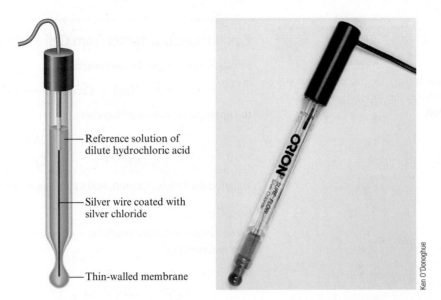

— Reference solution of dilute hydrochloric acid

— Silver wire coated with silver chloride

— Thin-walled membrane

Table 18.2 | Some Ions Whose Concentrations Can Be Detected by Ion-Selective Electrodes

Cations	Anions
H^+	Br^-
Cd^{2+}	Cl^-
Ca^{2+}	CN^-
Cu^{2+}	F^-
K^+	NO_3^-
Ag^+	S^{2-}
Na^+	

electrodes. The potentiometer reading is automatically converted electronically to a direct reading of the pH of the solution being tested.

The glass electrode (Fig. 18.12) contains a reference solution of dilute hydrochloric acid in contact with a thin glass membrane. The electrical potential of the glass electrode depends on the difference in $[H^+]$ between the reference solution and the solution into which the electrode is dipped. Thus the electrical potential varies with the pH of the solution being tested.

Electrodes that are sensitive to the concentration of a particular ion are called **ion-selective electrodes**, of which the glass electrode for pH measurement is just one example. Glass electrodes can be made sensitive to such ions as Na^+, K^+, or NH_4^+ by changing the composition of the glass. Other ions can be detected if an appropriate crystal replaces the glass membrane. For example, a crystal of lanthanum(III) fluoride (LaF_3) can be used in an electrode to measure $[F^-]$. Solid silver sulfide (Ag_2S) can be used to measure $[Ag^+]$ and $[S^{2-}]$. Some of the ions that can be detected by ion-selective electrodes are listed in Table 18.2.

Calculation of Equilibrium Constants for Redox Reactions

The quantitative relationship between $\mathscr{E}°$ and $\Delta G°$ allows calculation of equilibrium constants for redox reactions. For a cell at equilibrium,

$$\mathscr{E}_{cell} = 0 \quad \text{and} \quad Q = K$$

Applying these conditions to the Nernst equation valid at 25°C,

$$\mathscr{E} = \mathscr{E}° - \frac{0.0591}{n} \log(Q)$$

gives

$$0 = \mathscr{E}° - \frac{0.0591}{n} \log(K)$$

or

$$\log(K) = \frac{n\mathscr{E}°}{0.0591} \quad \text{at 25°C}$$

**Interactive
Example 18.10**

Sign in at http://login.cengagebrain
.com to try this Interactive Example
in **OWL**.

Equilibrium Constants from Cell Potentials

For the oxidation–reduction reaction

$$S_4O_6^{2-}(aq) + Cr^{2+}(aq) \longrightarrow Cr^{3+}(aq) + S_2O_3^{2-}(aq)$$

the appropriate half-reactions are

$$S_4O_6^{2-} + 2e^- \longrightarrow 2S_2O_3^{2-} \qquad \mathscr{E}° = 0.17 \text{ V} \qquad (1)$$
$$Cr^{3+} + e^- \longrightarrow Cr^{2+} \qquad \mathscr{E}° = -0.50 \text{ V} \qquad (2)$$

Balance the redox reaction, and calculate $\mathscr{E}°$ and K (at 25°C).

Solution

To obtain the balanced reaction, we must reverse reaction (2), multiply it by 2, and add it to reaction (1):

Reaction (1)	$S_4O_6^{2-} + 2e^- \longrightarrow 2S_2O_3^{2-}$	$\mathscr{E}°\text{ (cathode)} = 0.17 \text{ V}$
$2 \times$ reaction (2) reversed	$2(Cr^{2+} \longrightarrow Cr^{3+} + e^-)$	$-\mathscr{E}°\text{ (anode)} = -(-0.50) \text{ V}$
Cell reaction:	$2Cr^{2+}(aq) + S_4O_6^{2-}(aq) \longrightarrow 2Cr^{3+}(aq) + 2S_2O_3^{2-}(aq)$	$\mathscr{E}° = 0.67 \text{ V}$

In this reaction, 2 moles of electrons are transferred for every unit of reaction, that is, for every 2 moles of Cr^{2+} reacting with 1 mole of $S_4O_6^{2-}$ to form 2 moles of Cr^{3+} and 2 moles of $S_2O_3^{2-}$. Thus $n = 2$. Then

$$\log(K) = \frac{n\mathscr{E}°}{0.0591} = \frac{2(0.67)}{0.0591} = 22.6$$

The value of K is found by taking the antilog of 22.6:

$$K = 10^{22.6} = 4 \times 10^{22}$$

This very large equilibrium constant is not unusual for a redox reaction.

The blue solution contains Cr^{2+} ions, and the green solution contains Cr^{3+} ions.

See Exercises 18.75 through 18.78

18.6 | Batteries

A **battery** is a galvanic cell or, more commonly, a group of galvanic cells connected in series, where the potentials of the individual cells add to give the total battery potential. Batteries are a source of direct current and have become an essential source of portable power in our society. In this section we examine the most common types of batteries. Some new batteries currently being developed are described at the end of the chapter.

Lead Storage Battery

Since about 1915 when self-starters were first used in automobiles, the **lead storage battery** has been a major factor in making the automobile a practical means of transportation. This type of battery can function for several years under temperature extremes from −30°F to 120°F and under incessant punishment from rough roads.

In this battery, lead serves as the anode, and lead coated with lead dioxide serves as the cathode. Both electrodes dip into an electrolyte solution of sulfuric acid. The electrode reactions are

Anode reaction: $\qquad\qquad\qquad\qquad\qquad\qquad Pb + HSO_4^- \longrightarrow PbSO_4 + H^+ + 2e^-$

Cathode reaction: $\qquad\qquad\qquad PbO_2 + HSO_4^- + 3H^+ + 2e^- \longrightarrow PbSO_4 + 2H_2O$

Cell reaction: $\quad Pb(s) + PbO_2(s) + 2H^+(aq) + 2HSO_4^-(aq) \longrightarrow 2PbSO_4(s) + 2H_2O(l)$

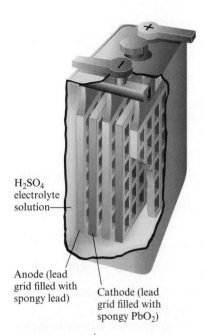

H$_2$SO$_4$
electrolyte
solution

Anode (lead
grid filled with
spongy lead)

Cathode (lead
grid filled with
spongy PbO$_2$)

Figure 18.13 | One of the six cells in a 12-V lead storage battery. The anode consists of a lead grid filled with spongy lead, and the cathode is a lead grid filled with lead dioxide. The cell also contains 38% (by mass) sulfuric acid.

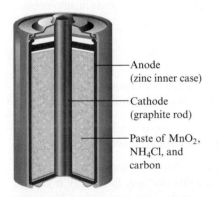

Anode
(zinc inner case)

Cathode
(graphite rod)

Paste of MnO$_2$,
NH$_4$Cl, and
carbon

Figure 18.14 | A common dry cell battery.

Sorinus/Dreamstime.com

Batteries for electronic watches are, by necessity, very tiny.

The typical automobile lead storage battery has six cells connected in series. Each cell contains multiple electrodes in the form of grids (Fig. 18.13) and produces approximately 2 V, to give a total battery potential of about 12 V. Note from the cell reaction that sulfuric acid is consumed as the battery discharges. This lowers the density of the electrolyte solution from its initial value of about 1.28 g/cm^3 in the fully charged battery. As a result, the condition of the battery can be monitored by measuring the density of the sulfuric acid solution. The solid lead sulfate formed in the cell reaction during discharge adheres to the grid surfaces of the electrodes. The battery is recharged by forcing current through it in the opposite direction to reverse the cell reaction. A car's battery is continuously charged by an alternator driven by the automobile engine.

An automobile with a dead battery can be "jump-started" by connecting its battery to the battery in a running automobile. This process can be dangerous, however, because the resulting flow of current causes electrolysis of water in the dead battery, producing hydrogen and oxygen gases (see Section 18.8 for details). Disconnecting the jumper cables after the disabled car starts causes an arc that can ignite the gaseous mixture. If this happens, the battery may explode, ejecting corrosive sulfuric acid. This problem can be avoided by connecting the ground jumper cable to a part of the engine remote from the battery. Any arc produced when this cable is disconnected will then be harmless.

Traditional types of storage batteries require periodic "topping off" because the water in the electrolyte solution is depleted by the electrolysis that accompanies the charging process. Recent types of batteries have electrodes made of an alloy of calcium and lead that inhibits the electrolysis of water. These batteries can be sealed, since they require no addition of water.

It is rather amazing that in the 100 years in which lead storage batteries have been used, no better system has been found. Although a lead storage battery does provide excellent service, it has a useful lifetime of 3 to 5 years in an automobile. While it might seem that the battery could undergo an indefinite number of discharge/charge cycles, physical damage from road shock and chemical side-reactions eventually cause the battery to fail.

Other Batteries

The calculators, electronic games, digital watches, and portable audio players that are so familiar to us are all powered by small, efficient batteries. The common **dry cell battery** was invented more than 100 years ago by Georges Leclanché (1839–1882), a French chemist. In its *acid version,* the dry cell battery contains a zinc inner case that acts as the anode and a carbon rod in contact with a moist paste of solid MnO$_2$, solid NH$_4$Cl, and carbon that acts as the cathode (Fig. 18.14). The half-reactions are complex but can be approximated as follows:

Anode reaction: $\quad\quad\quad\quad\quad\quad\quad\quad\quad$ Zn $\longrightarrow$ Zn^{2+} + 2e$^-$

Cathode reaction: $\quad$ 2NH$_4^+$ + 2MnO$_2$ + 2e$^-$ $\longrightarrow$ Mn$_2$O$_3$ + 2NH$_3$ + H$_2$O

This cell produces a potential of about 1.5 V.

In the *alkaline version* of the dry cell battery, the solid NH$_4$Cl is replaced with KOH or NaOH. In this case the half-reactions can be approximated as follows:

Anode reaction: $\quad\quad\quad\quad\quad\quad$ Zn + 2OH$^-$ $\longrightarrow$ ZnO + H$_2$O + 2e$^-$

Cathode reaction: $\quad$ 2MnO$_2$ + H$_2$O + 2e$^-$ $\longrightarrow$ Mn$_2$O$_3$ + 2OH$^-$

The alkaline dry cell lasts longer mainly because the zinc anode corrodes less rapidly under basic conditions than under acidic conditions.

Other types of useful batteries include the *silver cell,* which has a Zn anode and a cathode that uses Ag$_2$O as the oxidizing agent in a basic environment. *Mercury cells,*

Figure 18.15 | A mercury battery of the type used in calculators.

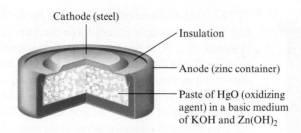

Figure 18.15 | A mercury battery of the type used in calculators.

Cathode (steel)

Insulation

Anode (zinc container)

Paste of HgO (oxidizing agent) in a basic medium of KOH and $Zn(OH)_2$

often used in calculators, have a Zn anode and a cathode involving HgO as the oxidizing agent in a basic medium (Fig. 18.15).

An especially important type of battery is the *nickel–cadmium battery,* in which the electrode reactions are

$$\text{Anode reaction:} \qquad Cd + 2OH^- \longrightarrow Cd(OH)_2 + 2e^-$$

$$\text{Cathode reaction:} \quad NiO_2 + 2H_2O + 2e^- \longrightarrow Ni(OH)_2 + 2OH^-$$

As in the lead storage battery, the products adhere to the electrodes. Therefore, a nickel–cadmium battery can be recharged an indefinite number of times.

Lithium-ion batteries involve the migration of Li^+ ions from the cathode to the anode, where they intercalate (enter the interior) as the battery is charged. At the same time, charge-balancing electrons travel to the anode through the external circuit in the charger. On discharge, the opposite process occurs. The cathode of the first successful lithium-ion batteries originally contained $LiCoO_2$ and a lithium-intercalated carbon (LiC_6) anode. More recently manufacturers have included transition metals such as nickel and manganese in the cathode in addition to cobalt. The mixed-metal cathodes have greater charge capacity and power output and shorter recharge times.

Lithium-ion batteries are used in a wide variety of applications, including cell phones, laptop computers, power tools, and even electric drive systems in automobiles and motorcycles.

Fuel Cells

A **fuel cell** is *a galvanic cell for which the reactants are continuously supplied.* To illustrate the principles of fuel cells, let's consider the exothermic redox reaction of methane with oxygen:

$$CH_4(g) + 2O_2(g) \longrightarrow CO_2(g) + 2H_2O(g) + \text{energy}$$

Usually the energy from this reaction is released as heat to warm homes and to run machines. However, in a fuel cell designed to use this reaction, the energy is used to produce an electric current: The electrons flow from the reducing agent (CH_4) to the oxidizing agent (O_2) through a conductor.

The U.S. space program has supported extensive research to develop fuel cells. The space shuttle uses a fuel cell based on the reaction of hydrogen and oxygen to form water:

$$2H_2(g) + O_2(g) \longrightarrow 2H_2O(l)$$

A schematic of a fuel cell that uses this reaction is shown in Fig. 18.16. The half-reactions are

$$\text{Anode reaction:} \qquad 2H_2 + 4OH^- \longrightarrow 4H_2O + 4e^-$$

$$\text{Cathode reaction:} \quad 4e^- + O_2 + 2H_2O \longrightarrow 4OH^-$$

A cell of this type weighing about 500 pounds has been designed for space vehicles, but this fuel cell is not practical enough for general use as a source of portable power. However, current research on portable electrochemical power is now proceeding at a rapid pace. In fact, cars powered by fuel cells are now being tested on the streets.

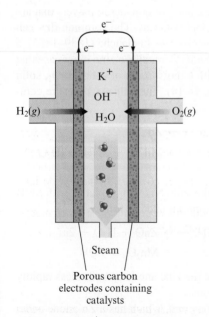

e^-

e^- e^-

K^+

OH^-

$H_2(g)$ H_2O $O_2(g)$

Steam

Porous carbon electrodes containing catalysts

Figure 18.16 | Schematic of the hydrogen–oxygen fuel cell.

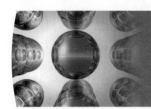

Chemical connections
Fuel Cells—Portable Energy

The promise of an energy-efficient, environmentally sound source of electrical power has spurred an intense interest in fuel cells in recent years. Although fuel cells have long been used in the U.S. space program, no practical fuel cell for powering automobiles has been developed. However, we are now on the verge of practical fuel-cell–powered cars. For example, DaimlerChrysler's NECAR 5 was driven across the United States, a 3000-mile trip that took 16 days. NECAR 5 is powered by a H_2/O_2 fuel cell that generates its H_2 from decomposition of methanol (CH_3OH).

General Motors, which has also been experimenting with H_2/O_2 fuel cells, in 2005 introduced the Chevrolet Sequel, which can go from 0 to 60 mph in 10 seconds with a range of 300 miles. The car has composite tanks for storage of 8 kg of liquid hydrogen.

In reality, fuel cells have a long way to go before they can be economically viable in automobiles. The main problem is the membrane that separates the hydrogen electrode from the oxygen electrode. This membrane must prevent H_2 molecules from passing through and still allow ions to pass between the electrodes. Current membranes cost over $3000 for an automobile-size fuel cell. The hope is to reduce this by a factor of 10 in the next few years.

Besides providing power for automobiles, fuel cells are being considered for powering small electronic devices such as cameras, cell phones, and laptop computers. Many of these micro fuel cells currently use methanol as the fuel (reducing agent) rather than H_2. However, these direct-methanol fuel cells are rife with problems. A major difficulty is water management. Water is needed at the anode to react with the methanol and is produced at the cathode. Water is also needed to moisten the electrolyte to promote charge migration.

Although the direct-methanol fuel cell is currently the leader among micro fuel-cell designs, its drawbacks have encouraged the development of other designs. For example, Richard Masel at the University of Illinois at Urbana–Champaign has designed a micro fuel cell that uses formic acid as the fuel. Masel and others are also experimenting with mini hot chambers external to the fuel cell that break down hydrogen-rich fuels into hydrogen gas, which is then fed into the tiny fuel cells.

To replace batteries, fuel cells must be demonstrated to be economically feasible, safe, and dependable. Today, rapid progress is being made to overcome the current problems. A recent estimate indicates that by late in this decade annual sales of the little power plants may reach 200 million units per year. It appears that after years of hype about the virtues of fuel cells, we are finally going to realize their potential.

Part of the fleet of 16 hydrogen fuel cars recently bought by the US Army.

Fuel cells are also finding use as permanent power sources. For example, a power plant built in New York City contains stacks of hydrogen–oxygen fuel cells, which can be rapidly put on-line in response to fluctuating power demands. The hydrogen gas is obtained by decomposing the methane in natural gas. A plant of this type also has been constructed in Tokyo.

In addition, new fuel cells are under development that can use fuels such as methane and diesel directly without having to produce hydrogen first.

18.7 | Corrosion

Corrosion can be viewed as the process of returning metals to their natural state—the ores from which they were originally obtained. Corrosion involves oxidation of the metal. Since corroded metal often loses its structural integrity and attractiveness, this spontaneous process has great economic impact. Approximately one-fifth of the iron and steel produced annually is used to replace rusted metal.

Metals corrode because they oxidize easily. Table 18.1 shows that, with the exception of gold, those metals commonly used for structural and decorative purposes all have standard reduction potentials less positive than that of oxygen gas. When any of these half-reactions is reversed (to show oxidation of the metal) and combined with the reduction half-reaction for oxygen, the result is a positive $\mathscr{E}°$ value. Thus the oxidation of most metals by oxygen is spontaneous (although we cannot tell from the potential how fast it will occur).

In view of the large difference in reduction potentials between oxygen and most metals, it is surprising that the problem of corrosion does not completely prevent the use of metals in air. However, most metals develop a thin oxide coating, which tends to protect their internal atoms against further oxidation. The metal that best demonstrates this phenomenon is aluminum. With a reduction potential of -1.7 V, aluminum should be easily oxidized by O_2. According to the apparent thermodynamics of the reaction, an aluminum airplane could dissolve in a rainstorm. The fact that this very active metal can be used as a structural material is due to the formation of a thin, adherent layer of aluminum oxide (Al_2O_3), more properly represented as $Al_2(OH)_6$, which greatly inhibits further corrosion. The potential of the "passive," oxide-coated aluminum is -0.6 V, a value that causes it to behave much like a noble metal.

Iron also can form a protective oxide coating. This coating is not an infallible shield against corrosion, however; when steel is exposed to oxygen in moist air, the oxide that forms tends to scale off and expose new metal surfaces to corrosion.

The corrosion products of noble metals such as copper and silver are complex and affect the use of these metals as decorative materials. Under normal atmospheric conditions, copper forms an external layer of greenish copper carbonate called *patina*. *Silver tarnish* is silver sulfide (Ag_2S), which in thin layers gives the silver surface a richer appearance. Gold, with a positive standard reduction potential of 1.50 V, significantly larger than that for oxygen (1.23 V), shows no appreciable corrosion in air.

Corrosion of Iron

Since steel is the main structural material for bridges, buildings, and automobiles, controlling its corrosion is extremely important. To do this, we must understand the corrosion mechanism. Instead of being a direct oxidation process as we might expect, the corrosion of iron is an electrochemical reaction, as shown in Fig. 18.17.

Steel has a nonuniform surface because the chemical composition is not completely homogeneous. Also, physical strains leave stress points in the metal. These nonuniformities cause areas where the iron is more easily oxidized (*anodic regions*) than it is at others (*cathodic regions*). In the anodic regions each iron atom gives up two electrons to form the Fe^{2+} ion:

$$Fe \longrightarrow Fe^{2+} + 2e^-$$

Some metals, such as copper, gold, silver, and platinum, are relatively difficult to oxidize. These are often called *noble metals*.

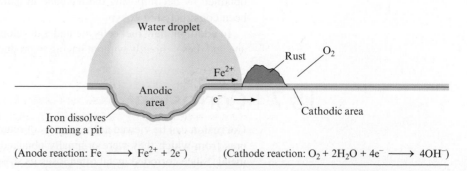

Figure 18.17 | The electrochemical corrosion of iron.

(Anode reaction: Fe $\longrightarrow$ Fe^{2+} + 2e$^-$)   (Cathode reaction: O$_2$ + 2H$_2$O + 4e$^-$ $\longrightarrow$ 4OH$^-$)

The electrons that are released flow through the steel, as they do through the wire of a galvanic cell, to a cathodic region, where they react with oxygen:

$$O_2 + 2H_2O + 4e^- \longrightarrow 4OH^-$$

The Fe^{2+} ions formed in the anodic regions travel to the cathodic regions through the moisture on the surface of the steel, just as ions travel through a salt bridge in a galvanic cell. In the cathodic regions Fe^{2+} ions react with oxygen to form rust, which is hydrated iron(III) oxide of variable composition:

$$4Fe^{2+}(aq) + O_2(g) + (4 + 2n)H_2O(l) \longrightarrow \underset{\text{Rust}}{2Fe_2O_3 \cdot nH_2O(s)} + 8H^+(aq)$$

Because of the migration of ions and electrons, rust often forms at sites that are remote from those where the iron dissolved to form pits in the steel. The degree of hydration of the iron oxide affects the color of the rust, which may vary from black to yellow to the familiar reddish brown.

The electrochemical nature of the rusting of iron explains the importance of moisture in the corrosion process. Moisture must be present to act as a kind of salt bridge between anodic and cathodic regions. Steel does not rust in dry air, a fact that explains why cars last much longer in the arid Southwest than in the relatively humid Midwest. Salt also accelerates rusting, a fact all too easily recognized by car owners in the colder parts of the United States, where salt is used on roads to melt snow and ice. The severity of rusting is greatly increased because the dissolved salt on the moist steel surface increases the conductivity of the aqueous solution formed there and thus accelerates the electrochemical corrosion process. Chloride ions also form very stable complex ions with Fe^{3+}, and this factor tends to encourage the dissolving of the iron, again accelerating the corrosion.

Prevention of Corrosion

Prevention of corrosion is an important way of conserving our natural resources of energy and metals. The primary means of protection is the application of a coating, most commonly paint or metal plating, to protect the metal from oxygen and moisture. Chromium and tin are often used to plate steel (see Section 18.9) because they oxidize to form a durable, effective oxide coating. Zinc, also used to coat steel in a process called **galvanizing**, forms a mixed oxide–carbonate coating. Since zinc is a more active metal than iron, as the potentials for the oxidation half-reactions show,

$$Fe \longrightarrow Fe^{2+} + 2e^- \qquad -\mathscr{E}° = 0.44 \text{ V}$$
$$Zn \longrightarrow Zn^{2+} + 2e^- \qquad -\mathscr{E}° = 0.76 \text{ V}$$

any oxidation that occurs dissolves zinc rather than iron. Recall that the reaction with the most positive standard potential has the greatest thermodynamic tendency to occur. Thus zinc acts as a "sacrificial" coating on steel.

Alloying is also used to prevent corrosion. *Stainless steel* contains chromium and nickel, both of which form oxide coatings that change steel's reduction potential to one characteristic of the noble metals. In addition, a new technology is now being developed to create surface alloys. That is, instead of forming a metal alloy such as stainless steel, which has the same composition throughout, a cheaper carbon steel is treated by ion bombardment to produce a thin layer of stainless steel or other desirable alloy on the surface. In this process, a "plasma" or "ion gas" of the alloying ions is formed at high temperatures and is then directed onto the surface of the metal.

Cathodic protection is a method most often used to protect steel in buried fuel tanks and pipelines. An active metal, such as magnesium, is connected by a wire to the pipeline or tank to be protected (Fig. 18.18). Because the magnesium is a better reducing agent than iron, electrons are furnished by the magnesium rather than by the iron, keeping the iron from being oxidized. As oxidation occurs, the magnesium anode dissolves, and so it must be replaced periodically. Ships' hulls are protected in a similar way by

Figure 18.18 | Cathodic protection of an underground pipe.

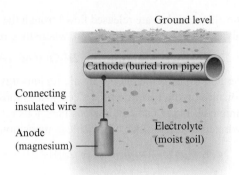

attaching bars of titanium metal to the steel hull. In salt water the titanium acts as the anode and is oxidized instead of the steel hull (the cathode).

18.8 | Electrolysis

An electrolytic cell uses electrical energy to produce a chemical change that would otherwise not occur spontaneously.

A galvanic cell produces current when an oxidation–reduction reaction proceeds spontaneously. A similar apparatus, an **electrolytic cell**, uses electrical energy to produce chemical change. The process of **electrolysis** involves *forcing a current through a cell to produce a chemical change for which the cell potential is negative;* that is, electrical work causes an otherwise nonspontaneous chemical reaction to occur. Electrolysis has great practical importance; for example, charging a battery, producing aluminum metal, and chrome plating an object are all done electrolytically.

To illustrate the difference between a galvanic cell and an electrolytic cell, consider the cell shown in Fig. 18.19(a) as it runs spontaneously to produce 1.10 V. In this *galvanic* cell, the reaction at the anode is

$$Zn \longrightarrow Zn^{2+} + 2e^-$$

whereas at the cathode the reaction is

$$Cu^{2+} + 2e^- \longrightarrow Cu$$

Figure 18.19(b) shows an external power source forcing electrons through the cell in the *opposite* direction to that in (a). This requires an external potential greater than 1.10 V, which must be applied in opposition to the natural cell potential. This device is an *electrolytic cell.* Notice that since electron flow is opposite in the two cases, the anode and cathode are reversed between (a) and (b). Also, ion flow through the salt bridge is opposite in the two cells.

Now we will consider the stoichiometry of electrolytic processes, that is, *how much chemical change occurs with the flow of a given current for a specified time.* Suppose we wish to determine the mass of copper that is plated out when a current of 10.0 amps (an **ampere** [amp], abbreviated A, is *1 coulomb of charge per second*) is passed for 30.0 minutes through a solution containing Cu^{2+}. *Plating* means depositing the neutral metal on the electrode by reducing the metal ions in solution. In this case each Cu^{2+} ion requires two electrons to become an atom of copper metal:

1 A = 1 C/s

$$Cu^{2+}(aq) + 2e^- \longrightarrow Cu(s)$$

This reduction process will occur at the cathode of the electrolytic cell.

To solve this stoichiometry problem, we need the following steps:

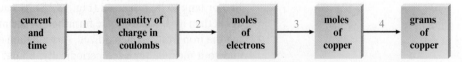

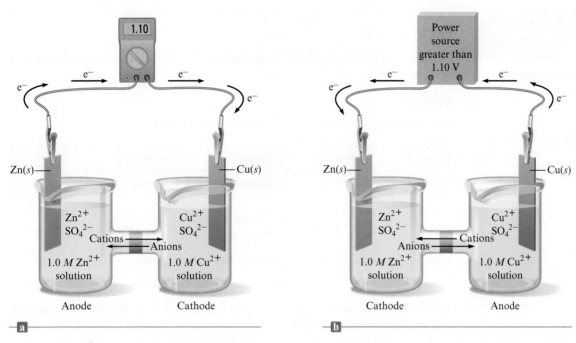

Figure 18.19 | (a) A standard galvanic cell based on the spontaneous reaction

$$Zn + Cu^{2+} \longrightarrow Zn^{2+} + Cu$$

(b) A standard electrolytic cell. A power source forces the opposite reaction

$$Cu + Zn^{2+} \longrightarrow Cu^{2+} + Zn$$

1. Since an amp is a coulomb of charge per second, we multiply the current by the time in seconds to obtain the total coulombs of charge passed into the Cu^{2+} solution at the cathode:

$$\text{Coulombs of charge} = \text{amps} \times \text{seconds} = \frac{C}{s} \times s$$

$$= 10.0 \, \frac{C}{s} \times 30.0 \, \text{min} \times 60.0 \, \frac{s}{\text{min}}$$

$$= 1.80 \times 10^4 \, C$$

2. Since 1 mole of electrons carries a charge of 1 faraday, or 96,485 coulombs, we can calculate the number of moles of electrons required to carry 1.80×10^4 coulombs of charge:

$$1.80 \times 10^4 \, C \times \frac{1 \, \text{mol e}^-}{96{,}485 \, C} = 1.87 \times 10^{-1} \, \text{mol e}^-$$

This means that 0.187 mole of electrons flowed into the Cu^{2+} solution.

3. Each Cu^{2+} ion requires two electrons to become a copper atom. Thus each mole of electrons produces $\frac{1}{2}$ mole of copper metal:

$$1.87 \times 10^{-1} \, \text{mol e}^- \times \frac{1 \, \text{mol Cu}}{2 \, \text{mol e}^-} = 9.35 \times 10^{-2} \, \text{mol Cu}$$

4. We now know the moles of copper metal plated onto the cathode, and we can calculate the mass of copper formed:

$$9.35 \times 10^{-2} \, \text{mol Cu} \times \frac{63.546 \, \text{g}}{\text{mol Cu}} = 5.94 \, \text{g Cu}$$

Electroplating

How long must a current of 5.00 A be applied to a solution of Ag^+ to produce 10.5 g silver metal?

Solution

In this case, we must use the steps given earlier in reverse:

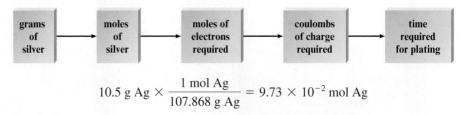

$$10.5 \text{ g Ag} \times \frac{1 \text{ mol Ag}}{107.868 \text{ g Ag}} = 9.73 \times 10^{-2} \text{ mol Ag}$$

Each Ag^+ ion requires one electron to become a silver atom:

$$Ag^+ + e^- \longrightarrow Ag$$

Thus 9.73×10^{-2} mole of electrons is required, and we can calculate the quantity of charge carried by these electrons:

$$9.73 \times 10^{-2} \text{ mol } e^- \times \frac{96,485 \text{ C}}{\text{mol } e^-} = 9.39 \times 10^3 \text{ C}$$

The 5.00 A (5.00 C/s) of current must produce 9.39×10^3 C of charge. Thus

$$\left(5.00 \frac{\text{C}}{\text{s}}\right) \times (\text{time, in s}) = 9.39 \times 10^3 \text{ C}$$

$$\text{Time} = \frac{9.39 \times 10^3}{5.00} \text{ s} = 1.88 \times 10^3 \text{ s} = 31.3 \text{ min}$$

See Exercises 18.93 through 18.96

Electrolysis of Water

Figure 18.20 | The electrolysis of water produces hydrogen gas at the cathode (on the right) and oxygen gas at the anode (on the left). Note that twice as much hydrogen is produced as oxygen.

We have seen that hydrogen and oxygen combine spontaneously to form water and that the accompanying decrease in free energy can be used to run a fuel cell to produce electricity. The reverse process, which is of course nonspontaneous, can be forced by electrolysis:

Anode reaction:	$2H_2O \longrightarrow O_2 + 4H^+ + 4e^-$	$-\mathscr{E}° = -1.23$ V
Cathode reaction:	$4H_2O + 4e^- \longrightarrow 2H_2 + 4OH^-$	$\mathscr{E}° = -0.83$ V
Net reaction:	$6H_2O \longrightarrow 2H_2 + O_2 + \underbrace{4(H^+ + OH^-)}_{4H_2O}$	$\mathscr{E}° = -2.06$ V

or

$$2H_2O \longrightarrow 2H_2 + O_2$$

Note that these potentials assume an anode chamber with 1 M H^+ and a cathode chamber with 1 M OH^-. In pure water, where $[H^+] = [OH^-] = 10^{-7}$ M, the potential for the overall process is -1.23 V.

In practice, however, if platinum electrodes connected to a 6-V battery are dipped into pure water, no reaction is observed because pure water contains so few ions that only a negligible current can flow. However, addition of even a small amount of a soluble salt causes an immediate evolution of bubbles of hydrogen and oxygen, as illustrated in Fig. 18.20.

Electrolysis of Mixtures of Ions

Suppose a solution in an electrolytic cell contains the ions Cu^{2+}, Ag^+, and Zn^{2+}. If the voltage is initially very low and is gradually turned up, in which order will the metals be plated out onto the cathode? This question can be answered by looking at the standard reduction potentials of these ions:

$$Ag^+ + e^- \longrightarrow Ag \qquad \mathscr{E}° = 0.80 \text{ V}$$
$$Cu^{2+} + 2e^- \longrightarrow Cu \qquad \mathscr{E}° = 0.34 \text{ V}$$
$$Zn^{2+} + 2e^- \longrightarrow Zn \qquad \mathscr{E}° = -0.76 \text{ V}$$

Remember that the more *positive* the $\mathscr{E}°$ value, the more the reaction has a tendency to proceed in the direction indicated. Of the three reactions listed, the reduction of Ag^+ occurs most easily, and the order of oxidizing ability is

$$Ag^+ > Cu^{2+} > Zn^{2+}$$

This means that silver will plate out first as the potential is increased, followed by copper, and finally zinc.

Interactive Example 18.12

Sign in at http://login.cengagebrain.com to try this Interactive Example in **OWL**.

Relative Oxidizing Abilities

An acidic solution contains the ions Ce^{4+}, VO_2^+, and Fe^{3+}. Using the $\mathscr{E}°$ values listed in Table 18.1, give the order of oxidizing ability of these species, and predict which one will be reduced at the cathode of an electrolytic cell at the lowest voltage.

Solution

The half-reactions and $\mathscr{E}°$ values are

$$Ce^{4+} + e^- \longrightarrow Ce^{3+} \qquad \mathscr{E}° = 1.70 \text{ V}$$
$$VO_2^+ + 2H^+ + e^- \longrightarrow VO^{2+} + H_2O \qquad \mathscr{E}° = 1.00 \text{ V}$$
$$Fe^{3+} + e^- \longrightarrow Fe^{2+} \qquad \mathscr{E}° = 0.77 \text{ V}$$

The order of oxidizing ability is therefore

$$Ce^{4+} > VO_2^+ > Fe^{3+}$$

The Ce^{4+} ion will be reduced at the lowest voltage in an electrolytic cell.

See Exercise 18.105

The principle described in this section is very useful, but it must be applied with some caution. For example, in the electrolysis of an aqueous solution of sodium chloride, we should be able to use $\mathscr{E}°$ values to predict the products. Of the major species in the solution (Na^+, Cl^-, and H_2O), only Cl^- and H_2O can be readily oxidized. The half-reactions (written as oxidization processes) are

$$2Cl^- \longrightarrow Cl_2 + 2e^- \qquad -\mathscr{E}° = -1.36 \text{ V}$$
$$2H_2O \longrightarrow O_2 + 4H^+ + 4e^- \qquad -\mathscr{E}° = -1.23 \text{ V}$$

Since water has the more positive potential, we would expect to see O_2 produced at the anode because it is easier (thermodynamically) to oxidize H_2O than Cl^-. Actually, this does not happen. As the voltage is increased in the cell, the Cl^- ion is the first to be oxidized. A much higher potential than expected is required to oxidize water. The voltage required in excess of the expected value (called the *overvoltage*) is much greater for the production of O_2 than for Cl_2, which explains why chlorine is produced first.

Chemical connections
The Chemistry of Sunken Treasure

When the galleon *Atocha* was destroyed on a reef by a hurricane in 1622, it was bound for Spain carrying approximately 47 tons of copper, gold, and silver from the New World. The bulk of the treasure was silver bars and coins packed in wooden chests. When treasure hunter Mel Fisher salvaged the silver in 1985, corrosion and marine growth had transformed the shiny metal into something that looked like coral. Restoring the silver to its original condition required an understanding of the chemical changes that had occurred in 350 years of being submerged in the ocean. Much of this chemistry we have already considered at various places in this text.

As the wooden chests containing the silver decayed, the oxygen supply was depleted, favoring the growth of certain bacteria that use the sulfate ion rather than oxygen as an oxidizing agent to generate energy. As these bacteria consume sulfate ions, they release hydrogen sulfide gas that reacts with silver to form black silver sulfide:

$$2Ag(s) + H_2S(aq) \longrightarrow Ag_2S(s) + H_2(g)$$

Thus, over the years, the surface of the silver became covered with a tightly adhering layer of corrosion, which fortunately protected the silver underneath and thus prevented total conversion of the silver to silver sulfide.

Another change that took place as the wood decomposed was the formation of carbon dioxide. This shifted the equilibrium that is present in the ocean,

$$CO_2(aq) + H_2O(l) \rightleftharpoons HCO_3^-(aq) + H^+(aq)$$

to the right, producing higher concentrations of HCO_3^-. In turn, the HCO_3^- reacted with Ca^{2+} ions present in the seawater to form calcium carbonate:

$$Ca^{2+}(aq) + HCO_3^-(aq) \rightleftharpoons CaCO_3(s) + H^+(aq)$$

Calcium carbonate is the main component of limestone. Thus, over time, the corroded silver coins and bars became encased in limestone.

The causes of overvoltage are very complex. Basically, the phenomenon is caused by difficulties in transferring electrons from the species in the solution to the atoms on the electrode across the electrode–solution interface. Because of this situation, $\mathcal{E}°$ values must be used cautiously in predicting the actual order of oxidation or reduction of species in an electrolytic cell.

18.9 | Commercial Electrolytic Processes

The chemistry of metals is characterized by their ability to donate electrons to form ions. Because metals are typically such good reducing agents, most are found in nature in *ores,* mixtures of ionic compounds often containing oxide, sulfide, and silicate anions. The noble metals, such as gold, silver, and platinum, are more difficult to oxidize and are often found as pure metals.

Production of Aluminum

Aluminum is one of the most abundant elements on earth, ranking third behind oxygen and silicon. Since aluminum is a very active metal, it is found in nature as its oxide in an ore called *bauxite* (named after Les Baux, France, where it was discovered in 1821). Production of aluminum metal from its ore proved to be more difficult than production of most other metals. In 1782 Lavoisier recognized aluminum to be a metal "whose affinity for oxygen is so strong that it cannot be overcome by any known reducing agent." As a result, pure aluminum metal remained unknown. Finally, in 1854 a process was found for producing metallic aluminum using sodium, but aluminum remained a very expensive rarity. In fact, it is said that Napoleon III served his most

Silver coins and tankards salvaged from the wreck of the Atocha.

Courtesy, Mel Fisher's Motivation, Inc.

Both the limestone formation and the corrosion had to be dealt with. Since $CaCO_3$ contains the basic anion CO_3^{2-}, acid dissolves limestone:

$$2H^+(aq) + CaCO_3(s) \longrightarrow Ca^{2+}(aq) + CO_2(g) + H_2O(l)$$

Soaking the mass of coins in a buffered acidic bath for several hours allowed the individual pieces to be separated, and the black Ag_2S on the surfaces was revealed. An abrasive could not be used to remove this corrosion; it would have destroyed the details of the engraving—a very valuable feature of the coins to a historian or a collector—and it would have washed away some of the silver. Instead, the corrosion reaction was reversed through electrolytic reduction. The coins were connected to the cathode of an electrolytic cell in a dilute sodium hydroxide solution as represented in the figure.

As electrons flow, the Ag^+ ions in the silver sulfide are reduced to silver metal:

$$Ag_2S + 2e^- \longrightarrow Ag + S^{2-}$$

As a by-product, bubbles of hydrogen gas from the reduction of water form on the surface of the coins:

$$2H_2O + 2e^- \longrightarrow H_2(g) + 2OH^-$$

The agitation caused by the bubbles loosens the flakes of metal sulfide and helps clean the coins.

These procedures have made it possible to restore the treasure to very nearly its condition when the *Atocha* sailed many years ago.

honored guests with aluminum forks and spoons, while the others had to settle for gold and silver utensils.

The breakthrough came in 1886 when two men, Charles M. Hall in the United States and Paul Heroult in France, almost simultaneously discovered a practical electrolytic process for producing aluminum (Fig. 18.21). The key factor in the *Hall–Heroult process* is the use of molten cryolite (Na_3AlF_6) as the solvent for the aluminum oxide.

Electrolysis is possible only if ions can move to the electrodes. A common method for producing ion mobility is dissolving the substance to be electrolyzed in water. This is not possible in the case of aluminum because water is more easily reduced than Al^{3+}, as the following standard reduction potentials show:

$$Al^{3+} + 3e^- \longrightarrow Al \qquad \mathscr{E}° = -1.66 \text{ V}$$
$$2H_2O + 2e^- \longrightarrow H_2 + 2OH^- \qquad \mathscr{E}° = -0.83 \text{ V}$$

Thus aluminum metal cannot be plated out of an aqueous solution of Al^{3+}.

Courtesy, Oberlin College Archives, Oberlin, Ohio

Figure 18.21 | Charles Martin Hall (1863–1914) was a student at Oberlin College in Ohio when he first became interested in aluminum. One of his professors commented that anyone who could manufacture aluminum cheaply would make a fortune, and Hall decided to give it a try. The 21-year-old Hall worked in a wooden shed near his house with an iron frying pan as a container, a blacksmith's forge as a heat source, and galvanic cells constructed from fruit jars. Using these crude galvanic cells, Hall found that he could produce aluminum by passing a current through a molten Al_2O_3/Na_3AlF_6 mixture. By a strange coincidence, Paul Heroult, a Frenchman who was born and died in the same years as Hall, made the same discovery at about the same time.

Table 18.3 | The Price of Aluminum over the Past Century

Date	Price of Aluminum ($/lb)*
1855	100,000
1885	100
1890	2
1895	0.50
1970	0.30
1980	0.80
1990	0.74

*Note the precipitous drop in price after the discovery of the Hall–Heroult process.

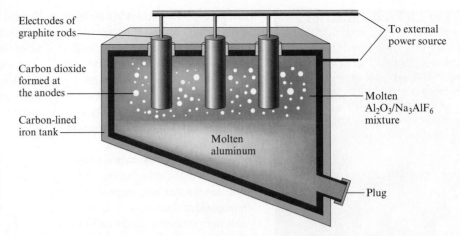

Figure 18.22 | A schematic diagram of an electrolytic cell for producing aluminum by the Hall–Heroult process. Because molten aluminum is more dense than the mixture of molten cryolite and alumina, it settles to the bottom of the cell and is drawn off periodically. The graphite electrodes are gradually eaten away and must be replaced from time to time. The cell operates at a current flow of up to 250,000 A.

Ion mobility also can be produced by melting the salt. But the melting point of solid Al_2O_3 is much too high (2050°C) to allow practical electrolysis of the molten oxide. A mixture of Al_2O_3 and Na_3AlF_6, however, has a melting point of 1000°C, and the resulting molten mixture can be used to obtain aluminum metal electrolytically. Because of this discovery by Hall and Heroult, the price of aluminum plunged (Table 18.3), and its use became economically feasible.

Bauxite is not pure aluminum oxide (called *alumina*); it also contains the oxides of iron, silicon, and titanium, and various silicate materials. To obtain the pure hydrated alumina ($Al_2O_3 \cdot nH_2O$), the crude bauxite is treated with aqueous sodium hydroxide. Being amphoteric, alumina dissolves in the basic solution:

$$Al_2O_3(s) + 2OH^-(aq) \longrightarrow 2AlO_2^-(aq) + H_2O(l)$$

The other metal oxides, which are basic, remain as solids. The solution containing the aluminate ion (AlO_2^-) is separated from the sludge of the other oxides and is acidified with carbon dioxide gas, causing the hydrated alumina to reprecipitate:

$$2CO_2(g) + 2AlO_2^-(aq) + (n + 1)H_2O(l) \longrightarrow 2HCO_3^-(aq) + Al_2O_3 \cdot nH_2O(s)$$

The purified alumina is then mixed with cryolite and melted, and the aluminum ion is reduced to aluminum metal in an electrolytic cell of the type shown in Fig. 18.22. Because the electrolyte solution contains a large number of aluminum-containing ions, the chemistry is not completely clear. However, the alumina probably reacts with the cryolite anion as follows:

$$Al_2O_3 + 4AlF_6^{3-} \longrightarrow 3Al_2OF_6^{2-} + 6F^-$$

The electrode reactions are thought to be

Cathode reaction: $\quad AlF_6^{3-} + 3e^- \longrightarrow Al + 6F^-$

Anode reaction: $\quad 2Al_2OF_6^{2-} + 12F^- + C \longrightarrow 4AlF_6^{3-} + CO_2 + 4e^-$

The overall cell reaction can be written as

$$2Al_2O_3 + 3C \longrightarrow 4Al + 3CO_2$$

The aluminum produced in this electrolytic process is 99.5% pure. To be useful as a structural material, aluminum is alloyed with metals such as zinc (used for trailer and aircraft construction) and manganese (used for cooking utensils, storage tanks, and

Figure 18.23 | Ultrapure copper sheets that serve as the cathodes are lowered between slabs of impure copper that serve as the anodes into a tank containing an aqueous solution of copper sulfate ($CuSO_4$). It takes about 4 weeks for the anodes to dissolve and for the pure copper to be deposited on the cathodes.

highway signs). The production of aluminum consumes about 5% of all the electricity used in the United States.

Electrorefining of Metals

Purification of metals is another important application of electrolysis. For example, impure copper from the chemical reduction of copper ore is cast into large slabs that serve as the anodes for electrolytic cells. Aqueous copper sulfate is the electrolyte, and thin sheets of ultrapure copper function as the cathodes (Fig. 18.23).

The main reaction at the anode is

$$Cu \longrightarrow Cu^{2+} + 2e^-$$

Other metals such as iron and zinc are also oxidized from the impure anode:

$$Zn \longrightarrow Zn^{2+} + 2e^-$$
$$Fe \longrightarrow Fe^{2+} + 2e^-$$

Noble metal impurities in the anode are not oxidized at the voltage used; they fall to the bottom of the cell to form a sludge, which is processed to remove the valuable silver, gold, and platinum.

The Cu^{2+} ions from the solution are deposited onto the cathode

$$Cu^{2+} + 2e^- \longrightarrow Cu$$

producing copper that is 99.95% pure.

Metal Plating

Metals that readily corrode can often be protected by the application of a thin coating of a metal that resists corrosion. Examples are "tin" cans, which are actually steel cans with a thin coating of tin, and chrome-plated steel bumpers for automobiles.

An object can be plated by making it the cathode in a tank containing ions of the plating metal. The silver plating of a spoon is shown schematically in Fig. 18.24(b). In an actual plating process, the solution also contains ligands that form complexes with the silver ion. By lowering the concentration of Ag^+ in this way, a smooth, even coating of silver is obtained.

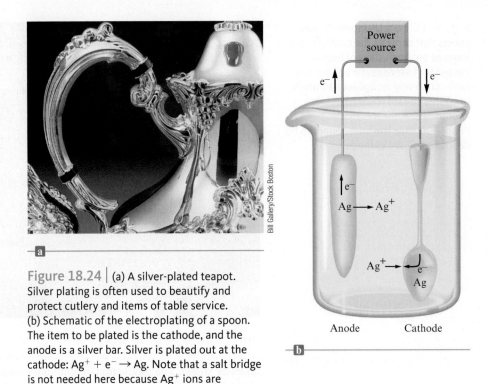

Bill Gallery/Stock Boston

a

Figure 18.24 | (a) A silver-plated teapot. Silver plating is often used to beautify and protect cutlery and items of table service. (b) Schematic of the electroplating of a spoon. The item to be plated is the cathode, and the anode is a silver bar. Silver is plated out at the cathode: $Ag^+ + e^- \rightarrow Ag$. Note that a salt bridge is not needed here because Ag^+ ions are involved at both electrodes.

Electrolysis of Sodium Chloride

Addition of a nonvolatile solute lowers the melting point of the solvent, molten NaCl in this case.

Sodium metal is mainly produced by the electrolysis of molten sodium chloride. Because solid NaCl has a rather high melting point (800°C), it is usually mixed with solid $CaCl_2$ to lower the melting point to about (600°C). The mixture is then electrolyzed in a **Downs cell**, as illustrated in Fig. 18.25, where the reactions are

$$\text{Anode reaction:} \qquad 2Cl^- \longrightarrow Cl_2 + 2e^-$$

$$\text{Cathode reaction:} \quad Na^+ + e^- \longrightarrow Na$$

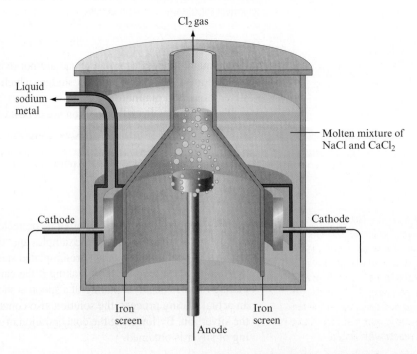

Figure 18.25 | The Downs cell for the electrolysis of molten sodium chloride. The cell is designed so that the sodium and chlorine produced cannot come into contact with each other to re-form NaCl.

At the temperatures in the Downs cell, the sodium is liquid and is drained off, then cooled, and cast into blocks. Because it is so reactive, sodium must be stored in an inert solvent, such as mineral oil, to prevent its oxidation.

Electrolysis of aqueous sodium chloride (brine) is an important industrial process for the production of chlorine and sodium hydroxide. In fact, this process is the second largest consumer of electricity in the United States, after the production of aluminum. Sodium is not produced in this process under normal circumstances because H_2O is more easily reduced than Na^+, as the standard reduction potentials show:

$$Na^+ + e^- \longrightarrow Na \qquad\qquad \mathscr{E}° = -2.71 \text{ V}$$
$$2H_2O + 2e^- \longrightarrow H_2 + 2OH^- \qquad \mathscr{E}° = -0.83 \text{ V}$$

Hydrogen, not sodium, is produced at the cathode.

For the reasons we discussed in Section 18.8, chlorine gas is produced at the anode. Thus the electrolysis of brine produces hydrogen and chlorine:

Anode reaction: $\qquad\qquad 2Cl^- \longrightarrow Cl_2 + 2e^-$

Cathode reaction: $\qquad 2H_2O + 2e^- \longrightarrow H_2 + 2OH^-$

It leaves a solution containing dissolved NaOH and NaCl.

The contamination of the sodium hydroxide by NaCl can be virtually eliminated using a special **mercury cell** for electrolyzing brine (Fig. 18.26). In this cell, mercury is the conductor at the cathode, and because hydrogen gas has an extremely high overvoltage with a mercury electrode, Na^+ is reduced instead of H_2O. The resulting sodium metal dissolves in the mercury, forming a liquid alloy, which is then pumped to a chamber where the dissolved sodium reacts with water to produce hydrogen:

$$2Na(s) + 2H_2O(l) \longrightarrow 2Na^+(aq) + 2OH^-(aq) + H_2(g)$$

Relatively pure solid NaOH can be recovered from the aqueous solution, and the regenerated mercury is then pumped back to the electrolysis cell. This process, called the **chlor–alkali process**, was the main method for producing chlorine and sodium hydroxide in the United States for many years. However, because of the environmental problems associated with the mercury cell, it has been largely displaced in the chlor–alkali industry by other technologies. In the United States, nearly 75% of the chlor–alkali production is now carried out in diaphragm cells. In a diaphragm cell the cathode and anode are separated by a diaphragm that allows passage of H_2O molecules, Na^+ ions, and, to a limited extent, Cl^- ions. The diaphragm does not allow OH^- ions

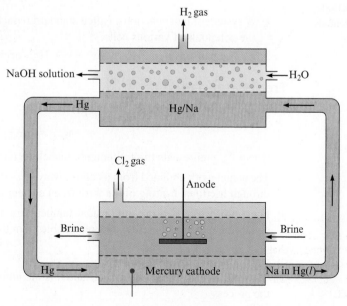

Figure 18.26 | The mercury cell for production of chlorine and sodium hydroxide. The large overvoltage required to produce hydrogen at a mercury electrode means that Na^+ ions are reduced rather than water. The sodium formed dissolves in the liquid mercury and is pumped to a chamber, where it reacts with water.

to pass through it. Thus the H_2 and OH^- formed at the cathode are kept separate from the Cl_2 formed at the anode. The major disadvantage of this process is that the aqueous effluent pumped from the cathode compartment contains a mixture of sodium hydroxide and unreacted sodium chloride, which must be separated if pure sodium hydroxide is a desired product.

In the past 30 years, a new process has been developed in the chlor–alkali industry that uses a membrane to separate the anode and cathode compartments in brine electrolysis cells. The membrane is superior to a diaphragm because the membrane is impermeable to anions. Only cations can flow through the membrane. Because neither Cl^- nor OH^- ions can pass through the membrane separating the anode and cathode compartments, NaCl contamination of the NaOH formed at the cathode does not occur. Although membrane technology is now just becoming prominent in the United States, it is the dominant method for chlor–alkali production in Japan.

For review

Key terms

electrochemistry

Section 18.1
half-reactions

Section 18.2
salt bridge
porous disk
galvanic cell
anode
cathode
cell potential (electromotive
    force)
volt
voltmeter
potentiometer

Section 18.3
standard hydrogen electrode
standard reduction potentials

Section 18.4
faraday

Section 18.5
concentration cell
Nernst equation
glass electrode
ion-selective electrode

Section 18.6
battery
lead storage battery
dry cell battery
fuel cell

Electrochemistry

> The study of the interchange of chemical and electrical energy
> Uses oxidation–reduction reactions
> Galvanic cell: chemical energy is transformed into electrical energy by separating the oxidizing and reducing agents and forcing the electrons to travel through a wire
> Electrolytic cell: electrical energy is used to produce a chemical change

Galvanic cell

> Anode: the electrode where oxidation occurs
> Cathode: the electrode where reduction occurs
> The driving force behind the electron transfer is called the cell potential ($\mathscr{E}_{cell}$)
> > The potential is measured in units of volts (V), defined as a joule of work per coulomb of charge:

$$\mathscr{E}(V) = \frac{-\text{work (J)}}{\text{charge (C)}} = -\frac{w}{q}$$

> A system of half-reactions, called standard reduction potentials, can be used to calculate the potentials of various cells
> > The half-reaction $2H^+ + 2e^- \longrightarrow H_2$ is arbitrarily assigned a potential of 0 V

Free energy and work

> The maximum work that a cell can perform is

$$-w_{max} = q\mathscr{E}_{max}$$

where $\mathscr{E}_{max}$ represents the cell potential when no current is flowing

> The actual work obtained from a cell is always less than the maximum because energy is lost through frictional heating of the wire when current flows
> For a process carried out at constant temperature and pressure, the change in free energy equals the maximum useful work obtainable from that process:

$$\Delta G = w_{max} = -q\mathscr{E}_{max} = -nF\mathscr{E}$$

where F (faraday) equals 96,485 C and n is the number of moles of electrons transferred in the process

Key terms

Concentration cell

> A galvanic cell in which both compartments have the same components but at different concentrations

> The electrons flow in the direction that tends to equalize the concentrations

Nernst equation

> Shows how the cell potential depends on the concentrations of the cell components:

$$\mathscr{E} = \mathscr{E}_0 - \frac{0.0591}{n}\log Q \qquad \text{at } 25°C$$

> When a galvanic cell is at equilibrium, $\mathscr{E} = 0$ and $Q = K$

Batteries

> A battery consists of a galvanic cell or group of cells connected in series that serve as a source of direct current.

> Lead storage battery
>> Anode: lead
>> Cathode: lead coated with PbO_2
>> Electrolyte: $H_2SO_4(aq)$

> Dry cell battery
>> Contains a moist paste instead of a liquid electrolyte
>> Anode: usually Zn
>> Cathode: carbon rod in contact with an oxidizing agent (which varies depending on the application)

Fuel cells

> Galvanic cells in which the reactants are continuously supplied

> The H_2/O_2 fuel cell is based on the reaction between H_2 and O_2 to form water

Corrosion

> Involves the oxidation of metals to form mainly oxides and sulfides

> Some metals, such as aluminum and chromium, form a thin, protective oxide coating that prevents further corrosion

> The corrosion of iron to form rust is an electrochemical process
>> The Fe^{2+} ions formed at anodic areas of the surface migrate through the moisture layer to cathodic regions, where they react with oxygen from the air
>> Iron can be protected from corrosion by coating it with paint or with a thin layer of metal such as chromium, tin, or zinc; by alloying; and by cathodic protection

Electrolysis

> Used to place a thin coating of metal onto steel

> Used to produce pure metals such as aluminum and copper

Review questions
Answers to the Review Questions can be found on the Student website (accessible from **www.cengagebrain.com**)*.*

1. What is a half-reaction? Why must the number of electrons lost in the oxidation half-reaction equal the number of electrons gained in the reduction half-reaction? Summarize briefly the steps in the half-reaction method for balancing redox reactions. What two items must be balanced in a redox reaction (or any reaction)?

2. Galvanic cells harness spontaneous oxidation–reduction reactions to produce work by producing a current. They

do so by controlling the flow of electrons from the species oxidized to the species reduced. How is a galvanic cell designed? What is in the cathode compartment? The anode compartment? What purpose do electrodes serve? Which way do electrons always flow in the wire connecting the two electrodes in a galvanic cell? Why is it necessary to use a salt bridge or a porous disk in a galvanic cell? Which way do cations flow in the salt bridge? Which way do the anions flow? What is a cell potential and what is a volt?

3. Table 18.1 lists common half-reactions along with the standard reduction potential associated with each half-reaction. These standard reduction potentials are all relative to some standard. What is the standard (zero point)? If $\mathscr{E}°$ is positive for a half-reaction, what does it mean? If $\mathscr{E}°$ is negative for a half-reaction, what does it mean? Which species in Table 18.1 is most easily reduced? Least easily reduced? The reverse of the half-reactions in Table 18.1 are the oxidation half-reactions. How are standard oxidation potentials determined? In Table 18.1, which species is the best reducing agent? The worst reducing agent?

   To determine the standard cell potential for a redox reaction, the standard reduction potential is added to the standard oxidation potential. What must be true about this sum if the cell is to be spontaneous (produce a galvanic cell)? Standard reduction and oxidation potentials are intensive. What does this mean? Summarize how line notation is used to describe galvanic cells.

4. Consider the equation $\Delta G° = -nF\mathscr{E}°$. What are the four terms in this equation? Why does a minus sign appear in the equation? What does the superscript $°$ indicate?

5. The Nernst equation allows determination of the cell potential for a galvanic cell at nonstandard conditions. Write out the Nernst equation. What are nonstandard conditions? What do $\mathscr{E}$, $\mathscr{E}°$, n, and Q stand for in the Nernst equation? What does the Nernst equation reduce to when a redox reaction is at equilibrium? What are the signs of $\Delta G°$ and $\mathscr{E}°$ when $K < 1$? When $K > 1$? When $K = 1$? Explain the following statement: $\mathscr{E}$ determines spontaneity, while $\mathscr{E}°$ determines the equilibrium position. Under what conditions can you use $\mathscr{E}°$ to predict spontaneity?

6. What are concentration cells? What is $\mathscr{E}°$ in a concentration cell? What is the driving force for a concentration cell to produce a voltage? Is the higher or the lower ion concentration solution present at the anode? When the anode ion concentration is decreased and/or the cathode ion concentration is increased, both give rise to larger cell potentials. Why? Concentration cells are commonly used to calculate the value of

equilibrium constants for various reactions. For example, the silver concentration cell illustrated in Fig. 18.9 can be used to determine the K_{sp} value for $AgCl(s)$. To do so, NaCl is added to the anode compartment until no more precipitate forms. The [Cl$^-$] in solution is then determined somehow. What happens to $\mathscr{E}_{cell}$ when NaCl is added to the anode compartment? To calculate the K_{sp} value, [Ag$^+$] must be calculated. Given the value of $\mathscr{E}_{cell}$, how is [Ag$^+$] determined at the anode?

7. Batteries are galvanic cells. What happens to $\mathscr{E}_{cell}$ as a battery discharges? Does a battery represent a system at equilibrium? Explain. What is $\mathscr{E}_{cell}$ when a battery reaches equilibrium? How are batteries and fuel cells alike? How are they different? The U.S. space program utilizes hydrogen–oxygen fuel cells to produce power for its spacecraft. What is a hydrogen–oxygen fuel cell?

8. Not all spontaneous redox reactions produce wonderful results. Corrosion is an example of a spontaneous redox process that has negative effects. What happens in the corrosion of a metal such as iron? What must be present for the corrosion of iron to take place? How can moisture and salt increase the severity of corrosion? Explain how the following protect metals from corrosion:

   a. paint

   b. durable oxide coatings

   c. galvanizing

   d. sacrificial metal

   e. alloying

   f. cathodic protection

9. What characterizes an electrolytic cell? What is an ampere? When the current applied to an electrolytic cell is multiplied by the time in seconds, what quantity is determined? How is this quantity converted to moles of electrons required? How are moles of electrons required converted to moles of metal plated out? What does plating mean? How do you predict the cathode and the anode half-reactions in an electrolytic cell? Why is the electrolysis of molten salts much easier to predict in terms of what occurs at the anode and cathode than the electrolysis of aqueous dissolved salts? What is overvoltage?

10. Electrolysis has many important industrial applications. What are some of these applications? The electrolysis of molten NaCl is the major process by which sodium metal is produced. However, the electrolysis of aqueous NaCl does not produce sodium metal under normal circumstances. Why? What is purification of a metal by electrolysis?

Active Learning Questions

These questions are designed to be used by groups of students in class.

1. Sketch a galvanic cell, and explain how it works. Look at Figs. 18.1 and 18.2. Explain what is occurring in each container and why the cell in Fig. 18.2 "works" but the one in Fig. 18.1 does not.

2. In making a specific galvanic cell, explain how one decides on the electrodes and the solutions to use in the cell.

3. You want to "plate out" nickel metal from a nickel nitrate solution onto a piece of metal inserted into the solution. Should you use copper or zinc? Explain.

4. A copper penny can be dissolved in nitric acid but not in hydrochloric acid. Using reduction potentials from the book, show why this is so. What are the products of the reaction? Newer pennies contain a mixture of zinc and copper. What happens to the zinc in the penny when the coin is placed in nitric acid? Hydrochloric acid? Support your explanations with data from the book, and include balanced equations for all reactions.

5. Sketch a cell that forms iron metal from iron(II) while changing chromium metal to chromium(III). Calculate the voltage, show the electron flow, label the anode and cathode, and balance the overall cell equation.

6. Which of the following is the best reducing agent: F_2, H_2, Na, Na^+, F^-? Explain. Order as many of these species as possible from the best to the worst oxidizing agent. Why can't you order all of them? From Table 18.1 choose the species that is the best oxidizing agent. Choose the best reducing agent. Explain.

7. You are told that metal A is a better reducing agent than metal B. What, if anything, can be said about A^+ and B^+? Explain.

8. Explain the following relationships: ΔG and w, cell potential and w, cell potential and ΔG, cell potential and Q. Using these relationships, explain how you could make a cell in which both electrodes are the same metal and both solutions contain the same compound, but at different concentrations. Why does such a cell run spontaneously?

9. Explain why cell potentials are not multiplied by the coefficients in the balanced redox equation. (Use the relationship between ΔG and cell potential to do this.)

10. What is the difference between $\mathscr{E}$ and $\mathscr{E}°$? When is $\mathscr{E}$ equal to zero? When is $\mathscr{E}°$ equal to zero? (Consider "regular" galvanic cells as well as concentration cells.)

11. Consider the following galvanic cell:

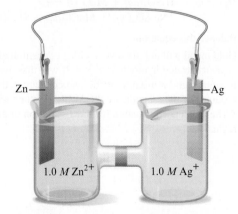

Zn | 1.0 M Zn^{2+} ‖ 1.0 M Ag^+ | Ag

What happens to $\mathscr{E}$ as the concentration of Zn^{2+} is increased? As the concentration of Ag^+ is increased? What happens to $\mathscr{E}°$ in these cases?

12. Look up the reduction potential for Fe^{3+} to Fe^{2+}. Look up the reduction potential for Fe^{2+} to Fe. Finally, look up the reduction potential for Fe^{3+} to Fe. You should notice that adding the reduction potentials for the first two does not give the potential for the third. Why not? Show how you can use the first two potentials to calculate the third potential.

13. If the cell potential is proportional to work and the standard reduction potential for the hydrogen ion is zero, does this mean that the reduction of the hydrogen ion requires no work?

14. Is the following statement true or false? Concentration cells work because standard reduction potentials are dependent on concentration. Explain.

A blue question or exercise number indicates that the answer to that question or exercise appears at the back of this book and a solution appears in the *Solutions Guide*, as found on PowerLecture.

Review of Oxidation–Reduction Reactions

If you have trouble with these exercises, you should review Section 4.9.

15. Define *oxidation* and *reduction* in terms of both change in oxidation number and electron loss or gain.

16. Assign oxidation numbers to all the atoms in each of the following:

    a. HNO_3 e. $C_6H_{12}O_6$ i. $Na_2C_2O_4$
    b. $CuCl_2$ f. Ag   j. CO_2
    c. O_2 g. $PbSO_4$ k. $(NH_4)_2Ce(SO_4)_3$
    d. H_2O_2 h. PbO_2 l. Cr_2O_3

17. Specify which of the following equations represent oxidation–reduction reactions, and indicate the oxidizing agent, the reducing agent, the species being oxidized, and the species being reduced.

    a. $CH_4(g) + H_2O(g) \rightarrow CO(g) + 3H_2(g)$
    b. $2AgNO_3(aq) + Cu(s) \rightarrow Cu(NO_3)_2(aq) + 2Ag(s)$
    c. $Zn(s) + 2HCl(aq) \rightarrow ZnCl_2(aq) + H_2(g)$
    d. $2H^+(aq) + 2CrO_4^{2-}(aq) \rightarrow Cr_2O_7^{2-}(aq) + H_2O(l)$

18. The Ostwald process for the commercial production of nitric acid involves the following three steps:

    $$4NH_3(g) + 5O_2(g) \longrightarrow 4NO(g) + 6H_2O(g)$$
    $$2NO(g) + O_2(g) \longrightarrow 2NO_2(g)$$
    $$3NO_2(g) + H_2O(l) \longrightarrow 2HNO_3(aq) + NO(g)$$

    a. Which reactions in the Ostwald process are oxidation–reduction reactions?
    b. Identify each oxidizing agent and reducing agent.

Questions

19. What is electrochemistry? What are redox reactions? Explain the difference between a galvanic and an electrolytic cell.

20. When balancing equations in Chapter 3, we did not mention that reactions must be charge balanced as well as mass

balanced. What do *charge balanced* and *mass balanced* mean? How are redox equations charge balanced?

21. When magnesium metal is added to a beaker of HCl(*aq*), a gas is produced. Knowing that magnesium is oxidized and that hydrogen is reduced, write the balanced equation for the reaction. How many electrons are transferred in the balanced equation? What quantity of useful work can be obtained when Mg is added directly to the beaker of HCl? How can you harness this reaction to do useful work?

22. How can one construct a galvanic cell from two substances, each having a negative standard reduction potential?

23. The free energy change for a reaction, ΔG, is an extensive property. What is an extensive property? Surprisingly, one can calculate ΔG from the cell potential, $\mathscr{E}$, for the reaction. This is surprising because $\mathscr{E}$ is an intensive property. How can the extensive property ΔG be calculated from the intensive property $\mathscr{E}$?

24. What is wrong with the following statement: The best concentration cell will consist of the substance having the most positive standard reduction potential. What drives a concentration cell to produce a large voltage?

25. When jump-starting a car with a dead battery, the ground jumper should be attached to a remote part of the engine block. Why?

26. In theory, most metals should easily corrode in air. Why? A group of metals called the noble metals are relatively difficult to corrode in air. Some noble metals include gold, platinum, and silver. Reference Table 18.1 to come up with a possible reason why the noble metals are relatively difficult to corrode.

27. Consider the electrolysis of a molten salt of some metal. What information must you know to calculate the mass of metal plated out in the electrolytic cell?

28. Consider the following electrochemical cell:

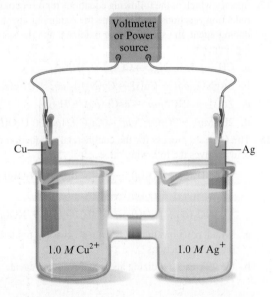

a. If silver metal is a product of the reaction, is the cell a galvanic cell or electrolytic cell? Label the cathode and anode, and describe the direction of the electron flow.

b. If copper metal is a product of the reaction, is the cell a galvanic cell or electrolytic cell? Label the cathode and anode, and describe the direction of the electron flow.

c. If the above cell is a galvanic cell, determine the standard cell potential.

d. If the above cell is an electrolytic cell, determine the minimum external potential that must be applied to cause the reaction to occur.

Exercises

In this section similar exercises are paired.

Balancing Oxidation–Reduction Equations

29. Balance the following oxidation–reduction reactions that occur in acidic solution using the half-reaction method.

   a. $I^-(aq) + ClO^-(aq) \rightarrow I_3^-(aq) + Cl^-(aq)$

   b. $As_2O_3(s) + NO_3^-(aq) \rightarrow H_3AsO_4(aq) + NO(g)$

   c. $Br^-(aq) + MnO_4^-(aq) \rightarrow Br_2(l) + Mn^{2+}(aq)$

   d. $CH_3OH(aq) + Cr_2O_7^{2-}(aq) \rightarrow CH_2O(aq) + Cr^{3+}(aq)$

30. Balance the following oxidation–reduction reactions that occur in acidic solution using the half-reaction method.

   a. $Cu(s) + NO_3^-(aq) \rightarrow Cu^{2+}(aq) + NO(g)$

   b. $Cr_2O_7^{2-}(aq) + Cl^-(aq) \rightarrow Cr^{3+}(aq) + Cl_2(g)$

   c. $Pb(s) + PbO_2(s) + H_2SO_4(aq) \rightarrow PbSO_4(s)$

   d. $Mn^{2+}(aq) + NaBiO_3(s) \rightarrow Bi^{3+}(aq) + MnO_4^-(aq)$

   e. $H_3AsO_4(aq) + Zn(s) \rightarrow AsH_3(g) + Zn^{2+}(aq)$

31. Balance the following oxidation–reduction reactions that occur in basic solution.

   a. $Al(s) + MnO_4^-(aq) \rightarrow MnO_2(s) + Al(OH)_4^-(aq)$

   b. $Cl_2(g) \rightarrow Cl^-(aq) + OCl^-(aq)$

   c. $NO_2^-(aq) + Al(s) \rightarrow NH_3(g) + AlO_2^-(aq)$

32. Balance the following oxidation–reduction reactions that occur in basic solution.

   a. $Cr(s) + CrO_4^{2-}(aq) \rightarrow Cr(OH)_3(s)$

   b. $MnO_4^-(aq) + S^{2-}(aq) \rightarrow MnS(s) + S(s)$

   c. $CN^-(aq) + MnO_4^-(aq) \rightarrow CNO^-(aq) + MnO_2(s)$

33. Chlorine gas was first prepared in 1774 by C. W. Scheele by oxidizing sodium chloride with manganese(IV) oxide. The reaction is

$$NaCl(aq) + H_2SO_4(aq) + MnO_2(s) \longrightarrow$$
$$Na_2SO_4(aq) + MnCl_2(aq) + H_2O(l) + Cl_2(g)$$

Balance this equation.

34. Gold metal will not dissolve in either concentrated nitric acid or concentrated hydrochloric acid. It will dissolve, however, in aqua regia, a mixture of the two concentrated acids. The products of the reaction are the $AuCl_4^-$ ion and gaseous NO. Write a balanced equation for the dissolution of gold in aqua regia.

Galvanic Cells, Cell Potentials, Standard Reduction Potentials, and Free Energy

35. Consider the following galvanic cell:

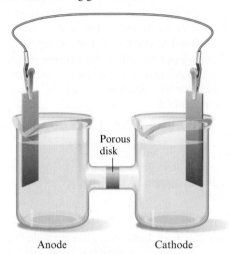

Porous disk

Anode   Cathode

Label the reducing agent and the oxidizing agent, and describe the direction of the electron flow.

36. Consider the following galvanic cell:

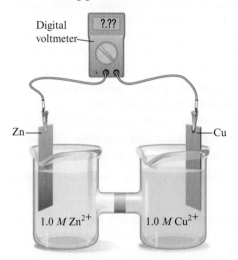

Digital voltmeter — ?.??

Zn ——   —— Cu

1.0 M Zn^{2+} 1.0 M Cu^{2+}

a. Label the reducing agent and the oxidizing agent, and describe the direction of the electron flow.
b. Determine the standard cell potential.
c. Which electrode increases in mass as the reaction proceeds, and which electrode decreases in mass?

37. Sketch the galvanic cells based on the following overall reactions. Show the direction of electron flow, and identify the cathode and anode. Give the overall balanced equation. Assume that all concentrations are 1.0 M and that all partial pressures are 1.0 atm.

a. $Cr^{3+}(aq) + Cl_2(g) \rightleftharpoons Cr_2O_7^{2-}(aq) + Cl^-(aq)$
b. $Cu^{2+}(aq) + Mg(s) \rightleftharpoons Mg^{2+}(aq) + Cu(s)$

38. Sketch the galvanic cells based on the following overall reactions. Show the direction of electron flow, the direction of ion migration through the salt bridge, and identify the cathode and anode. Give the overall balanced equation. Assume that all concentrations are 1.0 M and that all partial pressures are 1.0 atm.

a. $IO_3^-(aq) + Fe^{2+}(aq) \rightleftharpoons Fe^{3+}(aq) + I_2(aq)$
b. $Zn(s) + Ag^+(aq) \rightleftharpoons Zn^{2+}(aq) + Ag(s)$

39. Calculate $\mathscr{E}°$ values for the galvanic cells in Exercise 37.
40. Calculate $\mathscr{E}°$ values for the galvanic cells in Exercise 38.

41. Sketch the galvanic cells based on the following half-reactions. Show the direction of electron flow, show the direction of ion migration through the salt bridge, and identify the cathode and anode. Give the overall balanced equation, and determine $\mathscr{E}°$ for the galvanic cells. Assume that all concentrations are 1.0 M and that all partial pressures are 1.0 atm.

a. $Cl_2 + 2e^- \rightarrow 2Cl^-$ $\mathscr{E}° = 1.36$ V
    $Br_2 + 2e^- \rightarrow 2Br^-$ $\mathscr{E}° = 1.09$ V
b. $MnO_4^- + 8H^+ + 5e^- \rightarrow Mn^{2+} + 4H_2O$ $\mathscr{E}° = 1.51$ V
    $IO_4^- + 2H^+ + 2e^- \rightarrow IO_3^- + H_2O$ $\mathscr{E}° = 1.60$ V

42. Sketch the galvanic cells based on the following half-reactions. Show the direction of electron flow, show the direction of ion migration through the salt bridge, and identify the cathode and anode. Give the overall balanced equation, and determine $\mathscr{E}°$ for the galvanic cells. Assume that all concentrations are 1.0 M and that all partial pressures are 1.0 atm.

a. $H_2O_2 + 2H^+ + 2e^- \rightarrow 2H_2O$ $\mathscr{E}° = 1.78$ V
    $O_2 + 2H^+ + 2e^- \rightarrow H_2O_2$ $\mathscr{E}° = 0.68$ V
b. $Mn^{2+} + 2e^- \rightarrow Mn$ $\mathscr{E}° = -1.18$ V
    $Fe^{3+} + 3e^- \rightarrow Fe$ $\mathscr{E}° = -0.036$ V

43. Give the standard line notation for each cell in Exercises 37 and 41.

44. Give the standard line notation for each cell in Exercises 38 and 42.

45. Consider the following galvanic cells:

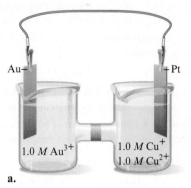

Au —   — Pt

1.0 M Au^{3+} 1.0 M Cu$^+$
                     1.0 M Cu^{2+}

a.

Cd —   — Pt

1.0 M Cd^{2+} 1.0 M VO$_2^+$
                     1.0 M H$^+$
                     1.0 M VO^{2+}

b.

For each galvanic cell, give the balanced cell equation and determine $\mathscr{E}°$. Standard reduction potentials are found in Table 18.1.

46. Give the balanced cell equation and determine $\mathscr{E}°$ for the galvanic cells based on the following half-reactions. Standard reduction potentials are found in Table 18.1.

a. $Cr_2O_7^{2-} + 14H^+ + 6e^- \rightarrow 2Cr^{3+} + 7H_2O$
   $H_2O_2 + 2H^+ + 2e^- \rightarrow 2H_2O$

b. $2H^+ + 2e^- \rightarrow H_2$
   $Al^{3+} + 3e^- \rightarrow Al$

47. Calculate $\mathscr{E}°$ values for the following cells. Which reactions are spontaneous as written (under standard conditions)? Balance the equations. Standard reduction potentials are found in Table 18.1.

a. $MnO_4^-(aq) + I^-(aq) \longrightarrow I_2(aq) + Mn^{2+}(aq)$

b. $MnO_4^-(aq) + F^-(aq) \longrightarrow F_2(g) + Mn^{2+}(aq)$

48. Calculate $\mathscr{E}°$ values for the following cells. Which reactions are spontaneous as written (under standard conditions)? Balance the equations that are not already balanced. Standard reduction potentials are found in Table 18.1.

a. $H_2(g) \longrightarrow H^+(aq) + H^-(aq)$

b. $Au^{3+}(aq) + Ag(s) \longrightarrow Ag^+(aq) + Au(s)$

49. Chlorine dioxide (ClO_2), which is produced by the reaction

$$2NaClO_2(aq) + Cl_2(g) \longrightarrow 2ClO_2(g) + 2NaCl(aq)$$

has been tested as a disinfectant for municipal water treatment. Using data from Table 18.1, calculate $\mathscr{E}°$ and $\Delta G°$ at 25°C for the production of ClO_2.

50. The amount of manganese in steel is determined by changing it to permanganate ion. The steel is first dissolved in nitric acid, producing Mn^{2+} ions. These ions are then oxidized to the deeply colored MnO_4^- ions by periodate ion (IO_4^-) in acid solution.

a. Complete and balance an equation describing each of the above reactions.

b. Calculate $\mathscr{E}°$ and $\Delta G°$ at 25°C for each reaction.

51. Calculate the maximum amount of work that can be obtained from the galvanic cells at standard conditions in Exercise 45.

52. Calculate the maximum amount of work that can be obtained from the galvanic cells at standard conditions in Exercise 46.

53. Estimate $\mathscr{E}°$ for the half-reaction

$$2H_2O + 2e^- \longrightarrow H_2 + 2OH^-$$

given the following values of $\Delta G_f°$:

$$H_2O(l) = -237 \text{ kJ/mol}$$
$$H_2(g) = 0.0$$
$$OH^-(aq) = -157 \text{ kJ/mol}$$
$$e^- = 0.0$$

Compare this value of $\mathscr{E}°$ with the value of $\mathscr{E}°$ given in Table 18.1.

54. The equation $\Delta G° = -nF\mathscr{E}°$ also can be applied to half-reactions. Use standard reduction potentials to estimate $\Delta G_f°$ for $Fe^{2+}(aq)$ and $Fe^{3+}(aq)$. ($\Delta G_f°$ for $e^- = 0.$)

55. Glucose is the major fuel for most living cells. The oxidative breakdown of glucose by our body to produce energy is called

respiration. The reaction for the complete combustion of glucose is

$$C_6H_{12}O_6(s) + 6O_2(g) \longrightarrow 6CO_2(g) + 6H_2O(l)$$

If this combustion reaction could be harnessed as a fuel cell, calculate the theoretical voltage that could be produced at standard conditions. (*Hint:* Use $\Delta G_f°$ values from Appendix 4.)

56. Direct methanol fuel cells (DMFCs) have shown some promise as a viable option for providing "green" energy to small electrical devices. Calculate $\mathscr{E}°$ for the reaction that takes place in DMFCs:

$$CH_3OH(l) + 3/2O_2(g) \longrightarrow CO_2(g) + 2H_2O(l)$$

Use values of $\Delta G_f°$ from Appendix 4.

57. Using data from Table 18.1, place the following in order of increasing strength as oxidizing agents (all under standard conditions).

$$Cd^{2+}, \quad IO_3^-, \quad K^+, \quad H_2O, \quad AuCl_4^-, \quad I_2$$

58. Using data from Table 18.1, place the following in order of increasing strength as reducing agents (all under standard conditions).

$$Cu^+, \quad F^-, \quad H^-, \quad H_2O, \quad I_2, \quad K$$

59. Answer the following questions using data from Table 18.1 (all under standard conditions).

a. Is $H^+(aq)$ capable of oxidizing $Cu(s)$ to $Cu^{2+}(aq)$?

b. Is $Fe^{3+}(aq)$ capable of oxidizing $I^-(aq)$?

c. Is $H_2(g)$ capable of reducing $Ag^+(aq)$?

60. Answer the following questions using data from Table 18.1 (all under standard conditions).

a. Is $H_2(g)$ capable of reducing $Ni^{2+}(aq)$?

b. Is $Fe^{2+}(aq)$ capable of reducing $VO_2^+(aq)$?

c. Is $Fe^{2+}(aq)$ capable of reducing $Cr^{3+}(aq)$ to $Cr^{2+}(aq)$?

61. Consider only the species (at standard conditions)

$$Na^+, \quad Cl^-, \quad Ag^+, \quad Ag, \quad Zn^{2+}, \quad Zn, \quad Pb$$

in answering the following questions. Give reasons for your answers. (Use data from Table 18.1.)

a. Which is the strongest oxidizing agent?

b. Which is the strongest reducing agent?

c. Which species can be oxidized by $SO_4^{2-}(aq)$ in acid?

d. Which species can be reduced by $Al(s)$?

62. Consider only the species (at standard conditions)

$$Br^-, \quad Br_2, \quad H^+, \quad H_2, \quad La^{3+}, \quad Ca, \quad Cd$$

in answering the following questions. Give reasons for your answers.

a. Which is the strongest oxidizing agent?

b. Which is the strongest reducing agent?

c. Which species can be oxidized by MnO_4^- in acid?

d. Which species can be reduced by $Zn(s)$?

63. Use the table of standard reduction potentials (Table 18.1) to pick a reagent that is capable of each of the following oxidations (under standard conditions in acidic solution).

a. oxidize Br^- to Br_2 but not oxidize Cl^- to Cl_2

b. oxidize Mn to Mn^{2+} but not oxidize Ni to Ni^{2+}

64. Use the table of standard reduction potentials (Table 18.1) to pick a reagent that is capable of each of the following reductions (under standard conditions in acidic solution).

    a. reduce Cu^{2+} to Cu but not reduce Cu^{2+} to Cu^+

    b. reduce Br_2 to Br^- but not reduce I_2 to I^-

65. Hydrazine is somewhat toxic. Use the half-reactions shown below to explain why household bleach (a highly alkaline solution of sodium hypochlorite) should not be mixed with household ammonia or glass cleansers that contain ammonia.

$$ClO^- + H_2O + 2e^- \longrightarrow 2OH^- + Cl^- \qquad \mathscr{E}° = 0.90 \text{ V}$$
$$N_2H_4 + 2H_2O + 2e^- \longrightarrow 2NH_3 + 2OH^- \qquad \mathscr{E}° = -0.10 \text{ V}$$

66. The compound with the formula TlI_3 is a black solid. Given the following standard reduction potentials,

$$Tl^{3+} + 2e^- \longrightarrow Tl^+ \qquad \mathscr{E}° = 1.25 \text{ V}$$
$$I_3^- + 2e^- \longrightarrow 3I^- \qquad \mathscr{E}° = 0.55 \text{ V}$$

would you formulate this compound as thallium(III) iodide or thallium(I) triiodide?

The Nernst Equation

67. A galvanic cell is based on the following half-reactions at 25°C:

$$Ag^+ + e^- \longrightarrow Ag$$
$$H_2O_2 + 2H^+ + 2e^- \longrightarrow 2H_2O$$

Predict whether $\mathscr{E}_{cell}$ is larger or smaller than $\mathscr{E}°_{cell}$ for the following cases.

    a. $[Ag^+] = 1.0 \ M$, $[H_2O_2] = 2.0 \ M$, $[H^+] = 2.0 \ M$

    b. $[Ag^+] = 2.0 \ M$, $[H_2O_2] = 1.0 \ M$, $[H^+] = 1.0 \times 10^{-7} \ M$

68. Consider the concentration cell in Fig. 18.10. If the Fe^{2+} concentration in the right compartment is changed from 0.1 M to $1 \times 10^{-7} \ M \ Fe^{2+}$, predict the direction of electron flow, and designate the anode and cathode compartments.

69. Consider the concentration cell shown below. Calculate the cell potential at 25°C when the concentration of Ag^+ in the compartment on the right is the following.

    a. 1.0 M

    b. 2.0 M

    c. 0.10 M

    d. $4.0 \times 10^{-5} \ M$

    e. Calculate the potential when both solutions are 0.10 M in Ag^+.

For each case, also identify the cathode, the anode, and the direction in which electrons flow.

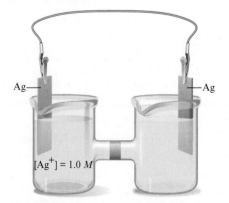

$[Ag^+] = 1.0 \ M$

70. Consider a concentration cell similar to the one shown in Exercise 69, except that both electrodes are made of Ni and in the left-hand compartment $[Ni^{2+}] = 1.0 \ M$. Calculate the cell potential at 25°C when the concentration of Ni^{2+} in the compartment on the right has each of the following values.

    a. 1.0 M

    b. 2.0 M

    c. 0.10 M

    d. $4.0 \times 10^{-5} \ M$

    e. Calculate the potential when both solutions are 2.5 M in Ni^{2+}.

For each case, also identify the cathode, anode, and the direction in which electrons flow.

71. The overall reaction in the lead storage battery is

$$Pb(s) + PbO_2(s) + 2H^+(aq) + 2HSO_4^-(aq) \longrightarrow$$
$$2PbSO_4(s) + 2H_2O(l)$$

Calculate $\mathscr{E}$ at 25°C for this battery when $[H_2SO_4] = 4.5 \ M$, that is, $[H^+] = [HSO_4^-] = 4.5 \ M$. At 25°C, $\mathscr{E}° = 2.04$ V for the lead storage battery.

72. Calculate the pH of the cathode compartment for the following reaction given $\mathscr{E}_{cell} = 3.01$ V when $[Cr^{3+}] = 0.15 \ M$, $[Al^{3+}] = 0.30 \ M$, and $[Cr_2O_7^{2-}] = 0.55 \ M$.

$$2Al(s) + Cr_2O_7^{2-}(aq) + 14H^+(aq) \longrightarrow$$
$$2Al^{3+}(aq) + 2Cr^{3+}(aq) + 7H_2O(l)$$

73. Consider the cell described below:

$$Zn\,|\,Zn^{2+}(1.00 \ M)\,||\,Cu^{2+}(1.00 \ M)\,|\,Cu$$

Calculate the cell potential after the reaction has operated long enough for the $[Zn^{2+}]$ to have changed by 0.20 mol/L. (Assume $T = 25°C$.)

74. Consider the cell described below:

$$Al\,|\,Al^{3+}(1.00 \ M)\,||\,Pb^{2+}(1.00 \ M)\,|\,Pb$$

Calculate the cell potential after the reaction has operated long enough for the $[Al^{3+}]$ to have changed by 0.60 mol/L. (Assume $T = 25°C$.)

75. Calculate $\Delta G°$ and K at 25°C for the reactions in Exercises 37 and 41.

76. Calculate $\Delta G°$ and K at 25°C for the reactions in Exercises 38 and 42.

77. Consider the galvanic cell based on the following half-reactions:

$$Zn^{2+} + 2e^- \longrightarrow Zn \qquad \mathscr{E}° = -0.76 \text{ V}$$
$$Fe^{2+} + 2e^- \longrightarrow Fe \qquad \mathscr{E}° = -0.44 \text{ V}$$

    a. Determine the overall cell reaction and calculate $\mathscr{E}°_{cell}$.

    b. Calculate $\Delta G°$ and K for the cell reaction at 25°C.

    c. Calculate $\mathscr{E}_{cell}$ at 25°C when $[Zn^{2+}] = 0.10 \ M$ and $[Fe^{2+}] = 1.0 \times 10^{-5} \ M$.

78. Consider the galvanic cell based on the following half-reactions:

$$Au^{3+} + 3e^- \longrightarrow Au \qquad \mathscr{E}° = 1.50 \text{ V}$$
$$Tl^+ + e^- \longrightarrow Tl \qquad \mathscr{E}° = -0.34 \text{ V}$$

a. Determine the overall cell reaction and calculate $\mathscr{E}_{cell}^{\circ}$.

b. Calculate ΔG° and K for the cell reaction at 25°C.

c. Calculate $\mathscr{E}_{cell}$ at 25°C when $[Au^{3+}] = 1.0 \times 10^{-2}\ M$ and $[Tl^{+}] = 1.0 \times 10^{-4}\ M$.

79. An electrochemical cell consists of a standard hydrogen electrode and a copper metal electrode.

a. What is the potential of the cell at 25°C if the copper electrode is placed in a solution in which $[Cu^{2+}] = 2.5 \times 10^{-4}\ M$?

b. The copper electrode is placed in a solution of unknown $[Cu^{2+}]$. The measured potential at 25°C is 0.195 V. What is $[Cu^{2+}]$? (Assume Cu^{2+} is reduced.)

80. An electrochemical cell consists of a nickel metal electrode immersed in a solution with $[Ni^{2+}] = 1.0\ M$ separated by a porous disk from an aluminum metal electrode.

a. What is the potential of this cell at 25°C if the aluminum electrode is placed in a solution in which $[Al^{3+}] = 7.2 \times 10^{-3}\ M$?

b. When the aluminum electrode is placed in a certain solution in which $[Al^{3+}]$ is unknown, the measured cell potential at 25°C is 1.62 V. Calculate $[Al^{3+}]$ in the unknown solution. (Assume Al is oxidized.)

81. An electrochemical cell consists of a standard hydrogen electrode and a copper metal electrode. If the copper electrode is placed in a solution of 0.10 M NaOH that is saturated with $Cu(OH)_2$, what is the cell potential at 25°C? [For $Cu(OH)_2$, $K_{sp} = 1.6 \times 10^{-19}$.]

82. An electrochemical cell consists of a nickel metal electrode immersed in a solution with $[Ni^{2+}] = 1.0\ M$ separated by a porous disk from an aluminum metal electrode immersed in a solution with $[Al^{3+}] = 1.0\ M$. Sodium hydroxide is added to the aluminum compartment, causing $Al(OH)_3(s)$ to precipitate. After precipitation of $Al(OH)_3$ has ceased, the concentration of OH^- is $1.0 \times 10^{-4}\ M$ and the measured cell potential is 1.82 V. Calculate the K_{sp} value for $Al(OH)_3$.

$$Al(OH)_3(s) \rightleftharpoons Al^{3+}(aq) + 3OH^-(aq) \quad K_{sp} = ?$$

83. Consider a concentration cell that has both electrodes made of some metal M. Solution A in one compartment of the cell contains $1.0\ M\ M^{2+}$. Solution B in the other cell compartment has a volume of 1.00 L. At the beginning of the experiment 0.0100 mole of $M(NO_3)_2$ and 0.0100 mole of Na_2SO_4 are dissolved in solution B (ignore volume changes), where the reaction

$$M^{2+}(aq) + SO_4^{2-}(aq) \rightleftharpoons MSO_4(s)$$

occurs. For this reaction equilibrium is rapidly established, whereupon the cell potential is found to be 0.44 V at 25°C. Assume that the process

$$M^{2+} + 2e^- \longrightarrow M$$

has a standard reduction potential of -0.31 V and that no other redox process occurs in the cell. Calculate the value of K_{sp} for $MSO_4(s)$ at 25°C.

84. You have a concentration cell in which the cathode has a silver electrode with 0.10 M Ag^+. The anode also has a silver electrode with $Ag^+(aq)$, 0.050 M $S_2O_3^{2-}$, and $1.0 \times 10^{-3}\ M$ $Ag(S_2O_3)_2^{3-}$. You read the voltage to be 0.76 V.

a. Calculate the concentration of Ag^+ at the anode.

b. Determine the value of the equilibrium constant for the formation of $Ag(S_2O_3)_2^{3-}$.

$$Ag^+(aq) + 2S_2O_3^{2-}(aq) \rightleftharpoons Ag(S_2O_3)_2^{3-}(aq) \quad K = ?$$

85. Under standard conditions, what reaction occurs, if any, when each of the following operations is performed?

a. Crystals of I_2 are added to a solution of NaCl.

b. Cl_2 gas is bubbled into a solution of NaI.

c. A silver wire is placed in a solution of $CuCl_2$.

d. An acidic solution of $FeSO_4$ is exposed to air.

For the reactions that occur, write a balanced equation and calculate $\mathscr{E}^{\circ}$, ΔG°, and K at 25°C.

86. A disproportionation reaction involves a substance that acts as both an oxidizing and a reducing agent, producing higher and lower oxidation states of the same element in the products. Which of the following disproportionation reactions are spontaneous under standard conditions? Calculate ΔG° and K at 25°C for those reactions that are spontaneous under standard conditions.

a. $2Cu^+(aq) \rightarrow Cu^{2+}(aq) + Cu(s)$

b. $3Fe^{2+}(aq) \rightarrow 2Fe^{3+}(aq) + Fe(s)$

c. $HClO_2(aq) \rightarrow ClO_3^-(aq) + HClO(aq)$ (unbalanced)

Use the half-reactions:

$ClO_3^- + 3H^+ + 2e^- \longrightarrow HClO_2 + H_2O \qquad \mathscr{E}^{\circ} = 1.21\ V$

$HClO_2 + 2H^+ + 2e^- \longrightarrow HClO + H_2O \qquad \mathscr{E}^{\circ} = 1.65\ V$

87. Consider the following galvanic cell at 25°C:

$$Pt\,|\,Cr^{2+}(0.30\ M),\ Cr^{3+}(2.0\ M)\,\|\,Co^{2+}(0.20\ M)\,|\,Co$$

The overall reaction and equilibrium constant value are

$$2Cr^{2+}(aq) + Co^{2+}(aq) \longrightarrow$$
$$2Cr^{3+}(aq) + Co(s) \quad K = 2.79 \times 10^7$$

Calculate the cell potential, $\mathscr{E}$, for this galvanic cell and ΔG for the cell reaction at these conditions.

88. An electrochemical cell consists of a silver metal electrode immersed in a solution with $[Ag^+] = 1.0\ M$ separated by a porous disk from a copper metal electrode. If the copper electrode is placed in a solution of 5.0 M NH_3 that is also 0.010 M in $Cu(NH_3)_4^{2+}$, what is the cell potential at 25°C?

$$Cu^{2+}(aq) + 4NH_3(aq) \rightleftharpoons Cu(NH_3)_4^{2+}(aq)$$
$$K = 1.0 \times 10^{13}$$

89. Calculate K_{sp} for iron(II) sulfide given the following data:

$FeS(s) + 2e^- \longrightarrow Fe(s) + S^{2-}(aq) \qquad \mathscr{E}^{\circ} = -1.01\ V$

$Fe^{2+}(aq) + 2e^- \longrightarrow Fe(s) \qquad \mathscr{E}^{\circ} = -0.44\ V$

90. For the following half-reaction, $\mathscr{E}^{\circ} = -2.07\ V$:

$$AlF_6^{3-}(aq) + 3e^- \longrightarrow Al(s) + 6F^-(aq)$$

Using data from Table 18.1, calculate the equilibrium constant at 25°C for the reaction

$$Al^{3+}(aq) + 6F^-(aq) \rightleftharpoons AlF_6^{3-}(aq) \quad K = ?$$

91. Calculate $\mathscr{E}°$ for the following half-reaction:

$$AgI(s) + e^- \longrightarrow Ag(s) + I^-(aq)$$

(*Hint:* Reference the K_{sp} value for AgI and the standard reduction potential for Ag^+.)

92. The solubility product for $CuI(s)$ is 1.1×10^{-12}. Calculate the value of $\mathscr{E}°$ for the half-reaction

$$CuI(s) + e^- \longrightarrow Cu(s) + I^-(aq)$$

Electrolysis

93. How long will it take to plate out each of the following with a current of 100.0 A?

  a. 1.0 kg Al from aqueous Al^{3+}

  b. 1.0 g Ni from aqueous Ni^{2+}

  c. 5.0 moles of Ag from aqueous Ag^+

94. The electrolysis of BiO^+ produces pure bismuth. How long would it take to produce 10.0 g Bi by the electrolysis of a BiO^+ solution using a current of 25.0 A?

95. What mass of each of the following substances can be produced in 1.0 h with a current of 15 A?

  a. Co from aqueous Co^{2+}

  b. Hf from aqueous Hf^{4+}

  c. I_2 from aqueous KI

  d. Cr from molten CrO_3

96. Aluminum is produced commercially by the electrolysis of Al_2O_3 in the presence of a molten salt. If a plant has a continuous capacity of 1.00 million A, what mass of aluminum can be produced in 2.00 h?

97. An unknown metal M is electrolyzed. It took 74.1 s for a current of 2.00 A to plate out 0.107 g of the metal from a solution containing $M(NO_3)_3$. Identify the metal.

98. Electrolysis of an alkaline earth metal chloride using a current of 5.00 A for 748 s deposits 0.471 g of metal at the cathode. What is the identity of the alkaline earth metal chloride?

99. What volume of F_2 gas, at 25°C and 1.00 atm, is produced when molten KF is electrolyzed by a current of 10.0 A for 2.00 h? What mass of potassium metal is produced? At which electrode does each reaction occur?

100. What volumes of $H_2(g)$ and $O_2(g)$ at STP are produced from the electrolysis of water by a current of 2.50 A in 15.0 min?

101. A single Hall–Heroult cell (as shown in Fig. 18.22) produces about 1 ton of aluminum in 24 h. What current must be used to accomplish this?

102. A factory wants to produce 1.00×10^3 kg barium from the electrolysis of molten barium chloride. What current must be applied for 4.00 h to accomplish this?

103. It took 2.30 min using a current of 2.00 A to plate out all the silver from 0.250 L of a solution containing Ag^+. What was the original concentration of Ag^+ in the solution?

104. A solution containing Pt^{4+} is electrolyzed with a current of 4.00 A. How long will it take to plate out 99% of the platinum in 0.50 L of a 0.010-M solution of Pt^{4+}?

105. A solution at 25°C contains 1.0 M Cd^{2+}, 1.0 M Ag^+, 1.0 M Au^{3+}, and 1.0 M Ni^{2+} in the cathode compartment of an electrolytic cell. Predict the order in which the metals will plate out as the voltage is gradually increased.

106. Consider the following half-reactions:

$$IrCl_6^{3-} + 3e^- \longrightarrow Ir + 6Cl^- \qquad \mathscr{E}° = 0.77 \text{ V}$$
$$PtCl_4^{2-} + 2e^- \longrightarrow Pt + 4Cl^- \qquad \mathscr{E}° = 0.73 \text{ V}$$
$$PdCl_4^{2-} + 2e^- \longrightarrow Pd + 4Cl^- \qquad \mathscr{E}° = 0.62 \text{ V}$$

A hydrochloric acid solution contains platinum, palladium, and iridium as chloro-complex ions. The solution is a constant 1.0 M in chloride ion and 0.020 M in each complex ion. Is it feasible to separate the three metals from this solution by electrolysis? (Assume that 99% of a metal must be plated out before another metal begins to plate out.)

107. In the electrolysis of an aqueous solution of Na_2SO_4, what reactions occur at the anode and the cathode (assuming standard conditions)?

	$\mathscr{E}°$
$S_2O_8^{2-} + 2e^- \longrightarrow 2SO_4^{2-}$	2.01 V
$O_2 + 4H^+ + 4e^- \longrightarrow 2H_2O$	1.23 V
$2H_2O + 2e^- \longrightarrow H_2 + 2OH^-$	-0.83 V
$Na^+ + e^- \longrightarrow Na$	-2.71 V

108. Copper can be plated onto a spoon by placing the spoon in an acidic solution of $CuSO_4(aq)$ and connecting it to a copper strip via a power source as illustrated below:

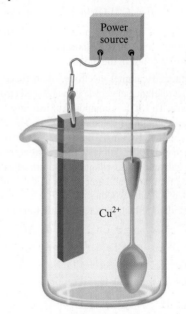

  a. Label the anode and cathode, and describe the direction of the electron flow.

  b. Write out the chemical equations for the reactions that occur at each electrode.

109. What reactions take place at the cathode and the anode when each of the following is electrolyzed?

  a. molten $NiBr_2$ **b.** molten AlF_3 **c.** molten MnI_2

110. What reaction will take place at the cathode and the anode when each of the following is electrolyzed?

  a. molten KF   **b.** molten $CuCl_2$ **c.** molten MgI_2

111. What reactions take place at the cathode and the anode when each of the following is electrolyzed? (Assume standard conditions.)

    a. 1.0 M $NiBr_2$ solution

    b. 1.0 M AlF_3 solution

    c. 1.0 M MnI_2 solution

112. What reaction will take place at the cathode and the anode when each of the following is electrolyzed? (Assume standard conditions.)

    a. 1.0 M KF solution

    b. 1.0 M $CuCl_2$ solution

    c. 1.0 M MgI_2 solution

Additional Exercises

113. Gold is produced electrochemically from an aqueous solution of $Au(CN)_2^-$ containing an excess of CN^-. Gold metal and oxygen gas are produced at the electrodes. What amount (moles) of O_2 will be produced during the production of 1.00 mole of gold?

114. The blood alcohol (C_2H_5OH) level can be determined by titrating a sample of blood plasma with an acidic potassium dichromate solution, resulting in the production of $Cr^{3+}(aq)$ and carbon dioxide. The reaction can be monitored because the dichromate ion ($Cr_2O_7^{2-}$) is orange in solution, and the Cr^{3+} ion is green. The unbalanced redox equation is

    $$Cr_2O_7^{2-}(aq) + C_2H_5OH(aq) \longrightarrow Cr^{3+}(aq) + CO_2(g)$$

    If 31.05 mL of 0.0600 M potassium dichromate solution is required to titrate 30.0 g blood plasma, determine the mass percent of alcohol in the blood.

115. The saturated calomel electrode, abbreviated SCE, is often used as a reference electrode in making electrochemical measurements. The SCE is composed of mercury in contact with a saturated solution of calomel (Hg_2Cl_2). The electrolyte solution is saturated KCl. $\mathscr{E}_{SCE}$ is +0.242 V relative to the standard hydrogen electrode. Calculate the potential for each of the following galvanic cells containing a saturated calomel electrode and the given half-cell components at standard conditions. In each case, indicate whether the SCE is the cathode or the anode. Standard reduction potentials are found in Table 18.1.

    a. $Cu^{2+} + 2e^- \longrightarrow Cu$ d. $Al^{3+} + 3e^- \longrightarrow Al$

    b. $Fe^{3+} + e^- \longrightarrow Fe^{2+}$ e. $Ni^{2+} + 2e^- \longrightarrow Ni$

    c. $AgCl + e^- \longrightarrow Ag + Cl^-$

116. Consider the following half-reactions:

    $$Pt^{2+} + 2e^- \longrightarrow Pt \qquad \mathscr{E}° = 1.188 \text{ V}$$
    $$PtCl_4^{2-} + 2e^- \longrightarrow Pt + 4Cl^- \qquad \mathscr{E}° = 0.755 \text{ V}$$
    $$NO_3^- + 4H^+ + 3e^- \longrightarrow NO + 2H_2O \qquad \mathscr{E}° = 0.96 \text{ V}$$

    Explain why platinum metal will dissolve in aqua regia (a mixture of hydrochloric and nitric acids) but not in either concentrated nitric or concentrated hydrochloric acid individually.

117. Consider the standard galvanic cell based on the following half-reactions:

    $$Cu^{2+} + 2e^- \longrightarrow Cu$$
    $$Ag^+ + e^- \longrightarrow Ag$$

The electrodes in this cell are Ag(s) and Cu(s). Does the cell potential increase, decrease, or remain the same when the following changes occur to the standard cell?

    a. $CuSO_4(s)$ is added to the copper half-cell compartment (assume no volume change).

    b. $NH_3(aq)$ is added to the copper half-cell compartment. [*Hint:* Cu^{2+} reacts with NH_3 to form $Cu(NH_3)_4^{2+}(aq)$.]

    c. NaCl(s) is added to the silver half-cell compartment. [*Hint:* Ag^+ reacts with Cl^- to form AgCl(s).]

    d. Water is added to both half-cell compartments until the volume of solution is doubled.

    e. The silver electrode is replaced with a platinum electrode.

    $$Pt^{2+} + 2e^- \longrightarrow Pt \qquad \mathscr{E}° = 1.19 \text{ V}$$

118. A standard galvanic cell is constructed so that the overall cell reaction is

    $$2Al^{3+}(aq) + 3M(s) \longrightarrow 3M^{2+}(aq) + 2Al(s)$$

    where M is an unknown metal. If $\Delta G° = -411$ kJ for the overall cell reaction, identify the metal used to construct the standard cell.

119. The black silver sulfide discoloration of silverware can be removed by heating the silver article in a sodium carbonate solution in an aluminum pan. The reaction is

    $$3Ag_2S(s) + 2Al(s) \rightleftharpoons 6Ag(s) + 3S^{2-}(aq) + 2Al^{3+}(aq)$$

    a. Using data in Appendix 4, calculate $\Delta G°$, K, and $\mathscr{E}°$ for the above reaction at 25°C. [For $Al^{3+}(aq)$, $\Delta G_f° = -480.$ kJ/mol.]

    b. Calculate the value of the standard reduction potential for the following half-reaction:

    $$2e^- + Ag_2S(s) \longrightarrow 2Ag(s) + S^{2-}(aq)$$

120. In 1973 the wreckage of the Civil War ironclad USS *Monitor* was discovered near Cape Hatteras, North Carolina. [The *Monitor* and the CSS *Virginia* (formerly the USS *Merrimack*) fought the first battle between iron-armored ships.] In 1987 investigations were begun to see if the ship could be salvaged. It was reported in *Time* (June 22, 1987) that scientists were considering adding sacrificial anodes of zinc to the rapidly corroding metal hull of the *Monitor*. Describe how attaching zinc to the hull would protect the *Monitor* from further corrosion.

121. When aluminum foil is placed in hydrochloric acid, nothing happens for the first 30 seconds or so. This is followed by vigorous bubbling and the eventual disappearance of the foil. Explain these observations.

122. Which of the following statements concerning corrosion is(are) *true*? For the false statements, correct them.

    a. Corrosion is an example of an electrolytic process.

    b. Corrosion of steel involves the reduction of iron coupled with the oxidation of oxygen.

    c. Steel rusts more easily in the dry (arid) Southwest states than in the humid Midwest states.

    d. Salting roads in the winter has the added benefit of hindering the corrosion of steel.

    e. The key to cathodic protection is to connect via a wire a metal more easily oxidized than iron to the steel surface to be protected.

123. A fuel cell designed to react grain alcohol with oxygen has the following net reaction:

$$C_2H_5OH(l) + 3O_2(g) \longrightarrow 2CO_2(g) + 3H_2O(l)$$

The maximum work that 1 mole of alcohol can do is 1.32×10^3 kJ. What is the theoretical maximum voltage this cell can achieve at 25°C?

124. The overall reaction and equilibrium constant value for a hydrogen–oxygen fuel cell at 298 K is

$$2H_2(g) + O_2(g) \longrightarrow 2H_2O(l) \qquad K = 1.28 \times 10^{83}$$

a. Calculate $\mathscr{E}°$ and $\Delta G°$ at 298 K for the fuel cell reaction.

b. Predict the signs of $\Delta H°$ and $\Delta S°$ for the fuel cell reaction.

c. As temperature increases, does the maximum amount of work obtained from the fuel cell reaction increase, decrease, or remain the same? Explain.

125. What is the maximum work that can be obtained from a hydrogen–oxygen fuel cell at standard conditions that produces 1.00 kg water at 25°C? Why do we say that this is the maximum work that can be obtained? What are the advantages and disadvantages in using fuel cells rather than the corresponding combustion reactions to produce electricity?

126. The overall reaction and standard cell potential at 25°C for the rechargeable nickel–cadmium alkaline battery is

$$Cd(s) + NiO_2(s) + 2H_2O(l) \longrightarrow$$
$$Ni(OH)_2(s) + Cd(OH)_2(s) \qquad \mathscr{E}° = 1.10 \text{ V}$$

For every mole of Cd consumed in the cell, what is the maximum useful work that can be obtained at standard conditions?

127. An experimental fuel cell has been designed that uses carbon monoxide as fuel. The overall reaction is

$$2CO(g) + O_2(g) \longrightarrow 2CO_2(g)$$

The two half-cell reactions are

$$CO + O^{2-} \longrightarrow CO_2 + 2e^-$$
$$O_2 + 4e^- \longrightarrow 2O^{2-}$$

The two half-reactions are carried out in separate compartments connected with a solid mixture of CeO_2 and Gd_2O_3. Oxide ions can move through this solid at high temperatures (about 800°C). ΔG for the overall reaction at 800°C under certain concentration conditions is -380 kJ. Calculate the cell potential for this fuel cell at the same temperature and concentration conditions.

128. The ultimate electron acceptor in the respiration process is molecular oxygen. Electron transfer through the respiratory chain takes place through a complex series of oxidation–reduction reactions. Some of the electron transport steps use iron-containing proteins called *cytochromes*. All cytochromes transport electrons by converting the iron in the cytochromes from the $+3$ to the $+2$ oxidation state. Consider the following reduction potentials for three different cytochromes used in the transfer process of electrons to oxygen (the potentials have been corrected for pH and for temperature):

$$\text{cytochrome a}(Fe^{3+}) + e^- \longrightarrow \text{cytochrome a}(Fe^{2+})$$
$$\mathscr{E} = 0.385 \text{ V}$$

$$\text{cytochrome b}(Fe^{3+}) + e^- \longrightarrow \text{cytochrome b}(Fe^{2+})$$
$$\mathscr{E} = 0.030 \text{ V}$$

$$\text{cytochrome c}(Fe^{3+}) + e^- \longrightarrow \text{cytochrome c}(Fe^{2+})$$
$$\mathscr{E} = 0.254 \text{ V}$$

In the electron transfer series, electrons are transferred from one cytochrome to another. Using this information, determine the cytochrome order necessary for spontaneous transport of electrons from one cytochrome to another, which eventually will lead to electron transfer to O_2.

129. One of the few industrial-scale processes that produce organic compounds electrochemically is used by the Monsanto Company to produce 1,4-dicyanobutane. The reduction reaction is

$$2CH_2{=}CHCN + 2H^+ + 2e^- \longrightarrow NC{-}(CH_2)_4{-}CN$$

The $NC{-}(CH_2)_4{-}CN$ is then chemically reduced using hydrogen gas to $H_2N{-}(CH_2)_6{-}NH_2$, which is used in the production of nylon. What current must be used to produce 150. kg $NC{-}(CH_2)_4{-}CN$ per hour?

130. It took 150. s for a current of 1.25 A to plate out 0.109 g of a metal from a solution containing its cations. Show that it is not possible for the cations to have a charge of $1+$.

131. It takes 15 kWh (kilowatt-hours) of electrical energy to produce 1.0 kg aluminum metal from aluminum oxide by the Hall–Heroult process. Compare this to the amount of energy necessary to melt 1.0 kg aluminum metal. Why is it economically feasible to recycle aluminum cans? [The enthalpy of fusion for aluminum metal is 10.7 kJ/mol (1 watt = 1 J/s).]

132. In the electrolysis of a sodium chloride solution, what volume of $H_2(g)$ is produced in the same time it takes to produce 257 L $Cl_2(g)$, with both volumes measured at 50.°C and 2.50 atm?

133. An aqueous solution of an unknown salt of ruthenium is electrolyzed by a current of 2.50 A passing for 50.0 min. If 2.618 g Ru is produced at the cathode, what is the charge on the ruthenium ions in solution?

ChemWork Problems

These multiconcept problems (and additional ones) are found interactively online with the same type of assistance a student would get from an instructor.

134. Which of the following statement(s) is/are *true*?

a. Copper metal can be oxidized by Ag^+ (at standard conditions).

b. In a galvanic cell the oxidizing agent in the cell reaction is present at the anode.

c. In a cell using the half reactions $Al^{3+} + 3e^- \longrightarrow Al$ and $Mg^{2+} + 2e^- \longrightarrow Mg$, aluminum functions as the anode.

d. In a concentration cell electrons always flow from the compartment with the lower ion concentration to the compartment with the higher ion concentration.

e. In a galvanic cell the negative ions in the salt bridge flow in the same direction as the electrons.

135. Consider a galvanic cell based on the following half-reactions:

	$\mathscr{E}°$ (V)
$La^{3+} + 3e^- \longrightarrow La$	-2.37
$Fe^{2+} + 2e^- \longrightarrow Fe$	-0.44

a. What is the expected cell potential with all components in their standard states?

b. What is the oxidizing agent in the overall cell reaction?

c. What substances make up the anode compartment?

d. In the standard cell, in which direction do the electrons flow?

e. How many electrons are transferred per unit of cell reaction?

f. If this cell is set up at 25°C with $[Fe^{2+}] = 2.00 \times 10^{-4}\ M$ and $[La^{3+}] = 3.00 \times 10^{-3}\ M$, what is the expected cell potential?

136. Consider a galvanic cell based on the following theoretical half-reactions:

	$\mathscr{E}°$ (V)
$M^{4+} + 4e^- \longrightarrow M$	0.66
$N^{3+} + 3e^- \longrightarrow N$	0.39

What is the value of $\Delta G°$ and K for this cell?

137. Consider a galvanic cell based on the following half-reactions:

	$\mathscr{E}°$ (V)
$Au^{3+} + 3e^- \longrightarrow Au$	1.50
$Mg^{2+} + 2e^- \longrightarrow Mg$	−2.37

a. What is the standard potential for this cell?

b. A nonstandard cell is set up at 25°C with $[Mg^{2+}] = 1.00 \times 10^{-5}\ M$. The cell potential is observed to be 4.01 V. Calculate $[Au^{3+}]$ in this cell.

138. An electrochemical cell consists of a silver metal electrode immersed in a solution with $[Ag^+] = 1.00\ M$ separated by a porous disk from a compartment with a copper metal electrode immersed in a solution of 10.00 M NH_3 that also contains $2.4 \times 10^{-3}\ M$ $Cu(NH_3)_4^{2+}$. The equilibrium between Cu^{2+} and NH_3 is:

$$Cu^{2+}(aq) + 4NH_3(aq) \rightleftharpoons Cu(NH_3)_4^{2+}(aq) \qquad K = 1.0 \times 10^{13}$$

and the two cell half-reactions are:

$$Ag^+ + e^- \longrightarrow Ag \qquad \mathscr{E}° = 0.80\ V$$
$$Cu^{2+} + 2e^- \longrightarrow Cu \qquad \mathscr{E}° = 0.34\ V$$

Assuming Ag^+ is reduced, what is the cell potential at 25°C?

139. An aqueous solution of $PdCl_2$ is electrolyzed for 48.6 seconds, and during this time 0.1064 g of Pd is deposited on the cathode. What is the average current used in the electrolysis?

Challenge Problems

140. Balance the following equations by the half-reaction method.

a. $Fe(s) + HCl(aq) \longrightarrow HFeCl_4(aq) + H_2(g)$

b. $IO_3^-(aq) + I^-(aq) \xrightarrow{\text{Acid}} I_3^-(aq)$

c. $Cr(NCS)_6^{4-}(aq) + Ce^{4+}(aq) \xrightarrow{\text{Acid}}$
$Cr^{3+}(aq) + Ce^{3+}(aq) + NO_3^-(aq) + CO_2(g) + SO_4^{2-}(aq)$

d. $CrI_3(s) + Cl_2(g) \xrightarrow{\text{Base}}$
$CrO_4^{2-}(aq) + IO_4^-(aq) + Cl^-(aq)$

e. $Fe(CN)_6^{4-}(aq) + Ce^{4+}(aq) \xrightarrow{\text{Base}}$
$Ce(OH)_3(s) + Fe(OH)_3(s) + CO_3^{2-}(aq) + NO_3^-(aq)$

141. Combine the equations

$$\Delta G° = -nF\mathscr{E}° \quad \text{and} \quad \Delta G° = \Delta H° - T\Delta S°$$

to derive an expression for $\mathscr{E}°$ as a function of temperature. Describe how one can graphically determine $\Delta H°$ and $\Delta S°$ from measurements of $\mathscr{E}°$ at different temperatures, assuming that $\Delta H°$ and $\Delta S°$ do not depend on temperature. What property would you look for in designing a reference half-cell that would produce a potential relatively stable with respect to temperature?

142. The overall reaction in the lead storage battery is

$$Pb(s) + PbO_2(s) + 2H^+(aq) + 2HSO_4^-(aq) \longrightarrow$$
$$2PbSO_4(s) + 2H_2O(l)$$

a. For the cell reaction $\Delta H° = -315.9$ kJ and $\Delta S° = 263.5$ J/K. Calculate $\mathscr{E}°$ at −20.°C. Assume $\Delta H°$ and $\Delta S°$ do not depend on temperature.

b. Calculate $\mathscr{E}$ at −20.°C when $[HSO_4^-] = [H^+] = 4.5\ M$.

c. Consider your answer to Exercise 71. Why does it seem that batteries fail more often on cold days than on warm days?

143. Consider the following galvanic cell:

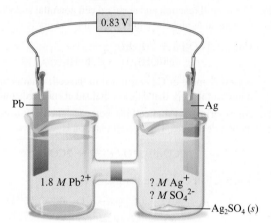

Calculate the K_{sp} value for $Ag_2SO_4(s)$. Note that to obtain silver ions in the right compartment (the cathode compartment), excess solid Ag_2SO_4 was added and some of the salt dissolved.

144. A zinc–copper battery is constructed as follows at 25°C:

$$Zn\,|\,Zn^{2+}(0.10\ M)\,||\,Cu^{2+}(2.50\ M)\,|\,Cu$$

The mass of each electrode is 200. g.

a. Calculate the cell potential when this battery is first connected.

b. Calculate the cell potential after 10.0 A of current has flowed for 10.0 h. (Assume each half-cell contains 1.00 L of solution.)

c. Calculate the mass of each electrode after 10.0 h.

d. How long can this battery deliver a current of 10.0 A before it goes dead?

145. A galvanic cell is based on the following half-reactions:

$$Fe^{2+} + 2e^- \longrightarrow Fe(s) \qquad \mathscr{E}° = -0.440\ V$$
$$2H^+ + 2e^- \longrightarrow H_2(g) \qquad \mathscr{E}° = 0.000\ V$$

where the iron compartment contains an iron electrode and $[Fe^{2+}] = 1.00 \times 10^{-3}\ M$ and the hydrogen compartment contains a platinum electrode, $P_{H_2} = 1.00$ atm, and a weak acid, HA, at an initial concentration of 1.00 M. If the observed cell potential is 0.333 V at 25°C, calculate the K_a value for the weak acid HA.

146. Consider a cell based on the following half-reactions:

$$Au^{3+} + 3e^- \longrightarrow Au \qquad \mathscr{E}° = 1.50 \text{ V}$$
$$Fe^{3+} + e^- \longrightarrow Fe^{2+} \qquad \mathscr{E}° = 0.77 \text{ V}$$

a. Draw this cell under standard conditions, labeling the anode, the cathode, the direction of electron flow, and the concentrations, as appropriate.

b. When enough NaCl(s) is added to the compartment containing gold to make the $[Cl^-] = 0.10$ M, the cell potential is observed to be 0.31 V. Assume that Au^{3+} is reduced and assume that the reaction in the compartment containing gold is

$$Au^{3+}(aq) + 4Cl^-(aq) \rightleftharpoons AuCl_4^-(aq)$$

Calculate the value of K for this reaction at 25°C.

147. The measurement of pH using a glass electrode obeys the Nernst equation. The typical response of a pH meter at 25.00°C is given by the equation

$$\mathscr{E}_{meas} = \mathscr{E}_{ref} + 0.05916 \text{ pH}$$

where $\mathscr{E}_{ref}$ contains the potential of the reference electrode and all other potentials that arise in the cell that are not related to the hydrogen ion concentration. Assume that $\mathscr{E}_{ref} = 0.250$ V and that $\mathscr{E}_{meas} = 0.480$ V.

a. What is the uncertainty in the values of pH and $[H^+]$ if the uncertainty in the measured potential is ± 1 mV (± 0.001 V)?

b. To what precision must the potential be measured for the uncertainty in pH to be ± 0.02 pH unit?

148. You have a concentration cell with Cu electrodes and $[Cu^{2+}]$ = 1.00 M (right side) and 1.0×10^{-4} M (left side).

a. Calculate the potential for this cell at 25°C.

b. The Cu^{2+} ion reacts with NH_3 to form $Cu(NH_3)_4^{2+}$ by the following equation:

$$Cu^{2+}(aq) + 4NH_3(aq) \rightleftharpoons Cu(NH_3)_4^{2+}(aq)$$
$$K = 1.0 \times 10^{13}$$

Calculate the new cell potential after enough NH_3 is added to the left cell compartment such that at equilibrium $[NH_3] = 2.0$ M.

149. A galvanic cell is based on the following half-reactions:

$$Ag^+ + e^- \longrightarrow Ag(s) \qquad \mathscr{E}° = 0.80 \text{ V}$$
$$Cu^{2+} + 2e^- \longrightarrow Cu(s) \qquad \mathscr{E}° = 0.34 \text{ V}$$

In this cell, the silver compartment contains a silver electrode and excess AgCl(s) ($K_{sp} = 1.6 \times 10^{-10}$), and the copper compartment contains a copper electrode and $[Cu^{2+}] = 2.0$ M.

a. Calculate the potential for this cell at 25°C.

b. Assuming 1.0 L of 2.0 M Cu^{2+} in the copper compartment, calculate the moles of NH_3 that would have to be added to give a cell potential of 0.52 V at 25°C (assume no volume change on addition of NH_3).

$$Cu^{2+}(aq) + 4NH_3(aq) \rightleftharpoons$$
$$Cu(NH_3)_4^{2+}(aq) \qquad K = 1.0 \times 10^{13}$$

150. Given the following two standard reduction potentials,

$$M^{3+} + 3e^- \longrightarrow M \qquad \mathscr{E}° = -0.10 \text{ V}$$
$$M^{2+} + 2e^- \longrightarrow M \qquad \mathscr{E}° = -0.50 \text{ V}$$

solve for the standard reduction potential of the half-reaction

$$M^{3+} + e^- \longrightarrow M^{2+}$$

(*Hint:* You must use the extensive property $\Delta G°$ to determine the standard reduction potential.)

151. Consider the following galvanic cell:

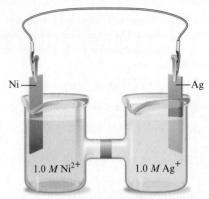

1.0 M Ni^{2+} | 1.0 M Ag^+

Calculate the concentrations of $Ag^+(aq)$ and $Ni^{2+}(aq)$ once the cell is "dead."

152. A chemist wishes to determine the concentration of CrO_4^{2-} electrochemically. A cell is constructed consisting of a saturated calomel electrode (SCE; see Exercise 115) and a silver wire coated with Ag_2CrO_4. The $\mathscr{E}°$ value for the following half-reaction is 0.446 V relative to the standard hydrogen electrode:

$$Ag_2CrO_4 + 2e^- \longrightarrow 2Ag + CrO_4^{2-}$$

a. Calculate $\mathscr{E}_{cell}$ and ΔG at 25°C for the cell reaction when $[CrO_4^{2-}] = 1.00$ mol/L.

b. Write the Nernst equation for the cell. Assume that the SCE concentrations are constant.

c. If the coated silver wire is placed in a solution (at 25°C) in which $[CrO_4^{2-}] = 1.00 \times 10^{-5}$ M, what is the expected cell potential?

d. The measured cell potential at 25°C is 0.504 V when the coated wire is dipped into a solution of unknown $[CrO_4^{2-}]$. What is $[CrO_4^{2-}]$ for this solution?

e. Using data from this problem and from Table 18.1, calculate the solubility product (K_{sp}) for Ag_2CrO_4.

153. Consider the following galvanic cell:

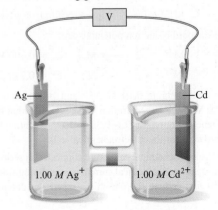

1.00 M Ag^+ | 1.00 M Cd^{2+}

A 15.0-mole sample of NH_3 is added to the Ag compartment (assume 1.00 L of total solution after the addition). The silver ion reacts with ammonia to form complex ions as shown:

$$Ag^+(aq) + NH_3(aq) \rightleftharpoons AgNH_3^+(aq)$$
$$K_1 = 2.1 \times 10^3$$

$$AgNH_3^+(aq) + NH_3(aq) \rightleftharpoons Ag(NH_3)_2^+(aq)$$
$$K_2 = 8.2 \times 10^3$$

Calculate the cell potential after the addition of 15.0 moles of NH_3.

154. When copper reacts with nitric acid, a mixture of $NO(g)$ and $NO_2(g)$ is evolved. The volume ratio of the two product gases depends on the concentration of the nitric acid according to the equilibrium

$$2H^+(aq) + 2NO_3^-(aq) + NO(g) \rightleftharpoons 3NO_2(g) + H_2O(l)$$

Consider the following standard reduction potentials at 25°C:

$$3e^- + 4H^+(aq) + NO_3^-(aq) \longrightarrow NO(g) + 2H_2O(l)$$
$$\mathscr{E}° = 0.957 \text{ V}$$

$$e^- + 2H^+(aq) + NO_3^-(aq) \longrightarrow NO_2(g) + 2H_2O(l)$$
$$\mathscr{E}° = 0.775 \text{ V}$$

a. Calculate the equilibrium constant for the above reaction.

b. What concentration of nitric acid will produce a NO and NO_2 mixture with only 0.20% NO_2 (by moles) at 25°C and 1.00 atm? Assume that no other gases are present and that the change in acid concentration can be neglected.

Integrative Problems

These problems require the integration of multiple concepts to find the solutions.

155. The following standard reduction potentials have been determined for the aqueous chemistry of indium:

$$In^{3+}(aq) + 2e^- \longrightarrow In^+(aq) \quad \mathscr{E}° = -0.444 \text{ V}$$
$$In^+(aq) + e^- \longrightarrow In(s) \quad \mathscr{E}° = -0.126 \text{ V}$$

a. What is the equilibrium constant for the disproportionation reaction, where a species is both oxidized and reduced, shown below?

$$3In^+(aq) \longrightarrow 2In(s) + In^{3+}(aq)$$

b. What is $\Delta G_f°$ for $In^+(aq)$ if $\Delta G_f° = -97.9$ kJ/mol for $In^{3+}(aq)$?

156. An electrochemical cell is set up using the following unbalanced reaction:

$$M^{a+}(aq) + N(s) \longrightarrow N^{2+}(aq) + M(s)$$

The standard reduction potentials are:

$$M^{a+} + ae^- \longrightarrow M \quad \mathscr{E}° = 0.400 \text{ V}$$
$$N^{2+} + 2e^- \longrightarrow N \quad \mathscr{E}° = 0.240 \text{ V}$$

The cell contains 0.10 M N^{2+} and produces a voltage of 0.180 V. If the concentration of M^{a+} is such that the value of the reaction quotient Q is 9.32×10^{-3}, calculate $[M^{a+}]$. Calculate w_{max} for this electrochemical cell.

157. Three electrochemical cells were connected in series so that the same quantity of electrical current passes through all three cells. In the first cell, 1.15 g chromium metal was deposited from a chromium(III) nitrate solution. In the second cell, 3.15 g osmium was deposited from a solution made of Os^{n+} and nitrate ions. What is the name of the salt? In the third cell, the electrical charge passed through a solution containing X^{2+} ions caused deposition of 2.11 g metallic X. What is the electron configuration of X?

158. A silver concentration cell is set up at 25°C as shown below:

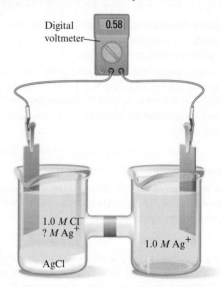

The AgCl(s) is in excess in the left compartment.

a. Label the anode and cathode, and describe the direction of the electron flow.

b. Determine the value of K_{sp} for AgCl at 25°C.

Marathon Problems

These problems are designed to incorporate several concepts and techniques into one situation.

159. A galvanic cell is based on the following half-reactions:

$$Cu^{2+}(aq) + 2e^- \longrightarrow Cu(s) \quad \mathscr{E}° = 0.34 \text{ V}$$
$$V^{2+}(aq) + 2e^- \longrightarrow V(s) \quad \mathscr{E}° = -1.20 \text{ V}$$

In this cell, the copper compartment contains a copper electrode and $[Cu^{2+}] = 1.00$ M, and the vanadium compartment contains a vanadium electrode and V^{2+} at an unknown concentration. The compartment containing the vanadium (1.00 L of solution) was titrated with 0.0800 M H_2EDTA^{2-}, resulting in the reaction

$$H_2EDTA^{2-}(aq) + V^{2+}(aq) \rightleftharpoons VEDTA^{2-}(aq) + 2H^+(aq)$$
$$K = ?$$

The potential of the cell was monitored to determine the stoichiometric point for the process, which occurred at a volume of 500.0 mL H_2EDTA^{2-} solution added. At the stoichiometric point, $\mathscr{E}_{cell}$ was observed to be 1.98 V. The solution was buffered at a pH of 10.00.

a. Calculate $\mathscr{E}_{cell}$ before the titration was carried out.

b. Calculate the value of the equilibrium constant, K, for the titration reaction.

c. Calculate $\mathscr{E}_{cell}$ at the halfway point in the titration.

160. The table below lists the cell potentials for the 10 possible galvanic cells assembled from the metals A, B, C, D, and E, and their respective 1.00 M 2+ ions in solution. Using the data in the table, establish a standard reduction potential table similar to Table 18.1 in the text. Assign a reduction potential of 0.00 V to the half-reaction that falls in the middle of the series. You should get two different tables. Explain why, and discuss what you could do to determine which table is correct.

	A(s) in $A^{2+}(aq)$	B(s) in $B^{2+}(aq)$	C(s) in $C^{2+}(aq)$	D(s) in $D^{2+}(aq)$
E(s) in $E^{2+}(aq)$	0.28 V	0.81 V	0.13 V	1.00 V
D(s) in $D^{2+}(aq)$	0.72 V	0.19 V	1.13 V	—
C(s) in $C^{2+}(aq)$	0.41 V	0.94 V	—	—
B(s) in $B^{2+}(aq)$	0.53 V	—	—	—

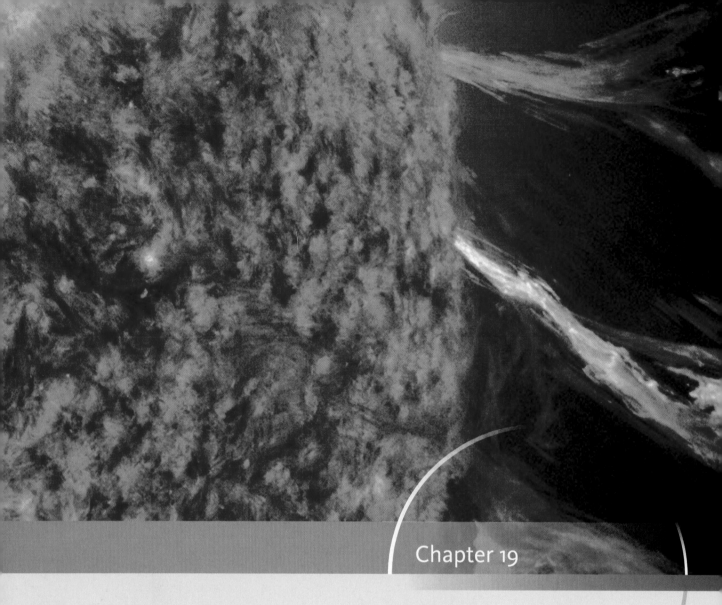

The Nucleus: A Chemist's View

The sun is powered by nuclear fusion reactions. (NASA/SDO/AIA)

Since the chemistry of an atom is determined by the number and arrangement of its electrons, the properties of the nucleus are not of primary importance to chemists. In the simplest view, the nucleus provides the positive charge to bind the electrons in atoms and molecules. However, a quick reading of any daily newspaper will show you that the nucleus and its properties have an important impact on our society. This chapter considers those aspects of the nucleus about which everyone should have some knowledge.

Several aspects of the nucleus are immediately impressive: its very small size, its very large density, and the magnitude of the energy that holds it together. The radius of a typical nucleus appears to be about 10^{-13} cm. This can be compared to the radius of a typical atom, which is on the order of 10^{-8} cm. A visualization will help you appreciate the small size of the nucleus: If the nucleus of the hydrogen atom were the size of a Ping-Pong ball, the electron in the $1s$ orbital would be, on average, 0.5 kilometer (0.3 mile) away. The density of the nucleus is equally impressive—approximately 1.6×10^{14} g/cm³. A sphere of nuclear material the size of a Ping-Pong ball would have a mass of 2.5 *billion tons!* In addition, the energies involved in nuclear processes are typically millions of times larger than those associated with normal chemical reactions. This fact makes nuclear processes very attractive for feeding the voracious energy appetite of our civilization.

Atomos, the Greek root of the word *atom,* means "indivisible." It was originally believed that the atom was the ultimate indivisible particle of which all matter was composed. However, as we discussed in Chapter 2, Lord Rutherford showed in 1911 that the atom is not homogeneous, but rather has a dense, positively charged center surrounded by electrons. Subsequently, scientists have learned that the nucleus of the atom can be subdivided into particles called **neutrons** and **protons**. In fact, in the past two decades it has become apparent that even the protons and neutrons are composed of smaller particles called *quarks.*

For most purposes, the nucleus can be regarded as a collection of **nucleons** (neutrons and protons), and the internal structures of these particles can be ignored. As we discussed in Chapter 2, the number of protons in a particular nucleus is called the **atomic number** (Z), and the sum of the neutrons and protons is the **mass number** (A). Atoms that have identical atomic numbers but different mass number values are called **isotopes**. However, we usually do not use the singular form *isotope* to refer to a particular member of a group of isotopes. Rather, we use the term *nuclide*. A **nuclide** is a unique atom, represented by the symbol

$$_{Z}^{A}X$$

where X represents the symbol for a particular element. For example, the following nuclides constitute the isotopes of carbon: carbon-12 ($_{6}^{12}C$), carbon-13 ($_{6}^{13}C$), and carbon-14 ($_{6}^{14}C$).

The atomic number Z is the number of protons in a nucleus; the mass number A is the sum of protons and neutrons in a nucleus.

The term *isotopes* refers to a group of nuclides with the same atomic number. Each individual atom is properly called a *nuclide*, not an isotope.

19.1 | Nuclear Stability and Radioactive Decay

Nuclear stability is the central topic of this chapter and forms the basis for all the important applications related to nuclear processes. Nuclear stability can be considered from both a kinetic and a thermodynamic point of view. **Thermodynamic stability**, as we will use the term here, refers to the potential energy of a particular nucleus as compared with the sum of the potential energies of its component protons and neutrons. We will use the term **kinetic stability** to describe the probability that a nucleus will undergo decomposition to form a different nucleus—a process called **radioactive decay**. We will consider radioactivity in this section.

Many nuclei are radioactive; that is, they decompose, forming another nucleus and producing one or more particles. An example is carbon-14, which decays as follows:

$$^{14}_{6}C \longrightarrow {}^{14}_{7}N + {}^{0}_{-1}e$$

where $^{0}_{-1}e$ represents an electron, which is called a **beta particle**, or **β particle**, in nuclear terminology. This equation is typical of those representing radioactive decay in that both A and Z must be conserved. That is, the Z values must give the same sum on both sides of the equation ($6 = 7 - 1$), as must the A values ($14 = 14 + 0$).

Of the approximately 2000 known nuclides, only 279 are stable with respect to radioactive decay. Tin has the largest number of stable isotopes—10.

It is instructive to examine how the numbers of neutrons and protons in a nucleus are related to its stability with respect to radioactive decay. Figure 19.1 shows a plot of the positions of the stable nuclei as a function of the number of protons (Z) and the number of neutrons ($A - Z$). The stable nuclides are said to reside in the **zone of stability**.

The following are some important observations concerning radioactive decay:

- All nuclides with 84 or more protons are unstable with respect to radioactive decay.

- Light nuclides are stable when Z equals $A - Z$, that is, when the neutron-to-proton ratio is 1. However, for heavier elements the neutron-to-proton ratio required for stability is greater than 1 and increases with Z.

- Certain combinations of protons and neutrons seem to confer special stability. For example, nuclides with even numbers of protons and neutrons are more often stable than those with odd numbers, as shown by the data in Table 19.1.

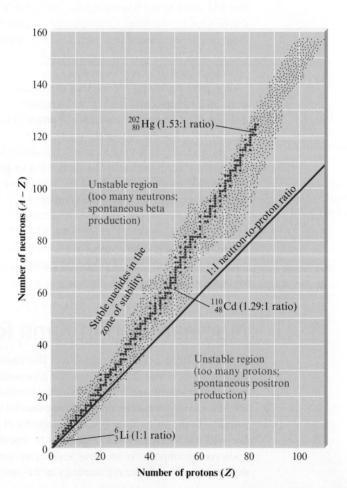

Figure 19.1 | The zone of stability. The red dots indicate the nuclides that *do not* undergo radioactive decay. Note that as the number of protons in a nuclide increases, the neutron-to-proton ratio required for stability also increases.

Table 19.1 | Number of Stable Nuclides Related to Numbers of Protons and Neutrons

Number of Protons	Number of Neutrons	Number of Stable Nuclides	Examples
Even	Even	168	$^{12}_{6}C$, $^{16}_{8}O$
Even	Odd	57	$^{13}_{6}C$, $^{47}_{22}Ti$
Odd	Even	50	$^{19}_{9}F$, $^{23}_{11}Na$
Odd	Odd	4	$^{2}_{1}H$, $^{6}_{3}Li$

Note: Even numbers of protons and neutrons seem to favor stability.

- There are also certain specific numbers of protons or neutrons that produce especially stable nuclides. These *magic numbers* are 2, 8, 20, 28, 50, 82, and 126. This behavior parallels that for atoms in which certain numbers of electrons (2, 10, 18, 36, 54, and 86) produce special chemical stability (the noble gases).

Types of Radioactive Decay

Radioactive nuclei can undergo decomposition in various ways. These decay processes fall into two categories: those that involve a change in the mass number of the decaying nucleus and those that do not. We will consider the former type of process first.

An **alpha particle**, or **α particle**, is a helium nucleus ($^{4}_{2}He$). **Alpha-particle production** is a very common mode of decay for heavy radioactive nuclides. For example, $^{238}_{92}U$, the predominant (99.3%) isotope of natural uranium, decays by α-particle production:

α-particle production involves a change in A for the decaying nucleus; β-particle production has no effect on A.

$$^{238}_{92}U \longrightarrow {}^{4}_{2}He + {}^{234}_{90}Th$$

Another α-particle producer is $^{230}_{90}Th$:

$$^{230}_{90}Th \longrightarrow {}^{4}_{2}He + {}^{226}_{88}Ra$$

Another decay process in which the mass number of the decaying nucleus changes is **spontaneous fission**, the splitting of a heavy nuclide into two lighter nuclides with similar mass numbers. Although this process occurs at an extremely slow rate for most nuclides, it is important in some cases, such as for $^{254}_{98}Cf$, where spontaneous fission is the predominant mode of decay.

The most common decay process in which the mass number of the decaying nucleus remains constant is **β-particle production**. For example, the thorium-234 nuclide produces a β particle and is converted to protactinium-234:

$$^{234}_{90}Th \longrightarrow {}^{234}_{91}Pa + {}^{0}_{-1}e$$

Iodine-131 is also a β-particle producer:

$$^{131}_{53}I \longrightarrow {}^{0}_{-1}e + {}^{131}_{54}Xe$$

The β particle is assigned the mass number 0, since its mass is tiny compared with that of a proton or neutron. Because the value of Z is -1 for the β particle, the atomic number for the new nuclide is greater by 1 than for the original nuclide. Thus *the net effect of β-particle production is to change a neutron to a proton*. We therefore expect nuclides that lie above the zone of stability (those nuclides whose neutron/proton ratios are too high) to be β-particle producers.

It should be pointed out that although the β particle is an electron, the emitting nucleus does not contain electrons. As we shall see later in this chapter, a given quantity of energy (which is best regarded as a form of matter) can become a particle (another form of matter) under certain circumstances. The unstable nuclide creates an electron as it releases energy in the decay process. The electron thus results from the decay process rather than being present before the decay occurs. Think of this as somewhat like talking: Words are not stored inside us but are formed as we speak. Later in this chapter we will discuss in more detail this very interesting phenomenon where matter in the form of particles and matter in the form of energy can interchange.

A **gamma ray**, or **γ ray**, refers to a high-energy photon. Frequently, γ-ray production accompanies nuclear decays and particle reactions, such as in the α-particle decay of $^{238}_{92}U$:

$$^{238}_{92}U \longrightarrow {}^{4}_{2}He + {}^{234}_{90}Th + 2\,{}^{0}_{0}\gamma$$

where two γ rays of different energies are produced in addition to the α particle. The emission of γ rays is one way a nucleus with excess energy (in an excited nuclear state) can relax to its ground state.

Positron production occurs for nuclides that are below the zone of stability (those nuclides whose neutron/proton ratios are too small). The positron is a particle with the same mass as the electron but opposite charge. An example of a nuclide that decays by positron production is sodium-22:

$$^{22}_{11}Na \longrightarrow {}^{0}_{1}e + {}^{22}_{10}Ne$$

Note that *the net effect is to change a proton to a neutron,* causing the product nuclide to have a higher neutron/proton ratio than the original nuclide.

Besides being oppositely charged, the positron shows an even more fundamental difference from the electron: It is the *antiparticle* of the electron. When a positron collides with an electron, the particulate matter is changed to electromagnetic radiation in the form of high-energy photons:

$$^{0}_{-1}e + {}^{0}_{1}e \longrightarrow 2\,{}^{0}_{0}\gamma$$

This process, which is characteristic of matter–antimatter collisions, is called *annihilation* and is another example of the interchange of the forms of matter.

Electron capture is a process in which one of the inner-orbital electrons is captured by the nucleus, as illustrated by the process

$$^{201}_{80}Hg + \underbrace{{}^{0}_{-1}e}_{\text{Inner-orbital electron}} \longrightarrow {}^{201}_{79}Au + {}^{0}_{0}\gamma$$

This reaction would have been of great interest to the alchemists, but unfortunately it does not occur at a rate that would make it a practical means for changing mercury to gold. Gamma rays are always produced along with electron capture to release excess energy. The various types of radioactive decay are summarized in Table 19.2.

Critical Thinking

What if a nuclide were to undergo two successive decays such that it became the original nuclide? Which decays could account for this? Provide an example.

Table 19.2 | Various Types of Radioactive Processes Showing the Changes That Take Place in the Nuclides

Process	Change in A	Change in Z	Change in Neutron/Proton Ratio	Example
β-particle (electron) production	0	+1	Decrease	$^{227}_{89}Ac \longrightarrow {}^{227}_{90}Th + {}^{0}_{-1}e$
Positron production	0	−1	Increase	$^{13}_{7}N \longrightarrow {}^{13}_{6}C + {}^{0}_{1}e$
Electron capture	0	−1	Increase	$^{73}_{33}As + {}^{0}_{-1}e \longrightarrow {}^{73}_{32}Ge$
α-particle production	−4	−2	Increase	$^{210}_{84}Po \longrightarrow {}^{206}_{82}Pb + {}^{4}_{2}He$
γ-ray production	0	0	—	Excited nucleus $\longrightarrow$ ground-state nucleus $+ {}^{0}_{0}\gamma$
Spontaneous fission	—	—	—	$^{254}_{98}Cf \longrightarrow$ lighter nuclides $+$ neutrons

Nuclear Equations I

Write balanced equations for each of the following processes.

a. $^{11}_{6}C$ produces a positron.

b. $^{214}_{83}Bi$ produces a β particle.

c. $^{237}_{93}Np$ produces an α particle.

Solution

a. We must find the product nuclide represented by $^{A}_{Z}X$ in the following equation:

$$^{11}_{6}C \longrightarrow \underset{\underset{\text{Positron}}{\uparrow}}{^{0}_{1}e} + ^{A}_{Z}X$$

We can find the identity of $^{A}_{Z}X$ by recognizing that the total of the Z and A values must be the same on both sides of the equation. Thus for X, Z must be $6 - 1 = 5$ and A must be $11 - 0 = 11$. Therefore, $^{A}_{Z}X$ is $^{11}_{5}B$. (The fact that Z is 5 tells us that the nuclide is boron.) Thus the balanced equation is

$$^{11}_{6}C \longrightarrow ^{0}_{1}e + ^{11}_{5}B$$

b. Knowing that a β particle is represented by $_{-1}^{0}e$ and that Z and A are conserved, we can write

$$^{214}_{83}Bi \longrightarrow _{-1}^{0}e + ^{214}_{84}X$$

so $^{A}_{Z}X$ must be $^{214}_{84}Po$.

c. Since an α particle is represented by $^{4}_{2}He$, the balanced equation must be

$$^{237}_{93}Np \longrightarrow ^{4}_{2}He + ^{233}_{91}Pa$$

See Exercises 19.11, 19.14, and 19.15

Nuclear Equations II

In each of the following nuclear reactions, supply the missing particle.

a. $^{195}_{79}Au + ? \rightarrow ^{195}_{78}Pt$

b. $^{38}_{19}K \rightarrow ^{38}_{18}Ar + ?$

Solution

a. Since A does not change and Z decreases by 1, the missing particle must be an electron:

$$^{195}_{79}Au + _{-1}^{0}e \longrightarrow ^{195}_{78}Pt$$

This is an example of electron capture.

b. To conserve Z and A, the missing particle must be a positron:

$$^{38}_{19}K \longrightarrow ^{38}_{18}Ar + ^{0}_{1}e$$

Thus potassium-38 decays by positron production.

See Exercises 19.12, 19.13, and 19.16

Often a radioactive nucleus cannot reach a stable state through a single decay process. In such a case, a **decay series** occurs until a stable nuclide is formed. A

Figure 19.2 | The decay series from $^{238}_{92}$U to $^{206}_{82}$Pb. Each nuclide in the series except $^{206}_{82}$Pb is radioactive, and the successive transformations (shown by the arrows) continue until $^{238}_{82}$Pb is finally formed. The horizontal red arrows indicate β-particle production (Z increases by 1 and A is unchanged). The diagonal blue arrows signify α-particle production (both A and Z decrease).

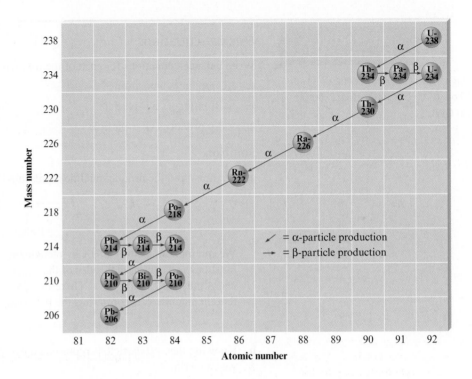

well-known example is the decay series that starts with $^{238}_{92}$U and ends with $^{206}_{82}$Pb, as shown in Fig. 19.2. Similar series exist for $^{235}_{92}$U:

$$^{235}_{92}\text{U} \xrightarrow[\text{decays}]{\text{Series of}} {}^{207}_{82}\text{Pb}$$

and for $^{232}_{90}$Th:

$$^{232}_{90}\text{Th} \xrightarrow[\text{decays}]{\text{Series of}} {}^{208}_{82}\text{Pb}$$

19.2 | The Kinetics of Radioactive Decay

Rates of reaction are discussed in Chapter 12.

In a sample containing radioactive nuclides of a given type, each nuclide has a certain probability of undergoing decay. Suppose that a sample of 1000 atoms of a certain nuclide produces 10 decay events per hour. This means that over the span of an hour, 1 out of every 100 nuclides will decay. Given that this probability of decay is characteristic for this type of nuclide, we could predict that a 2000-atom sample would give 20 decay events per hour. Thus, for radioactive nuclides, the **rate of decay**, which is the negative of the change in the number of nuclides per unit time

$$\left(-\frac{\Delta N}{\Delta t}\right)$$

is directly proportional to the number of nuclides N in a given sample:

$$\text{Rate} = -\frac{\Delta N}{\Delta t} \propto N$$

The negative sign is included because the number of nuclides is decreasing. We now insert a proportionality constant k to give

$$\text{Rate} = -\frac{\Delta N}{\Delta t} = kN$$

This is the rate law for a first-order process, as we saw in Chapter 12. As shown in Section 12.4, the integrated first-order rate law is

$$\ln\left(\frac{N}{N_0}\right) = -kt$$

where N_0 represents the original number of nuclides (at $t = 0$) and N represents the number *remaining* at time t.

Half-Life

The **half-life** ($t_{1/2}$) of a radioactive sample is defined as the time required for the number of nuclides to reach half the original value ($N_0/2$). We can use this definition in connection with the integrated first-order rate law (as we did in Section 12.4) to produce the following expression for $t_{1/2}$:

$$t_{1/2} = \frac{\ln(2)}{k} = \frac{0.693}{k}$$

Thus, if the half-life of a radioactive nuclide is known, the rate constant can be easily calculated, and vice versa.

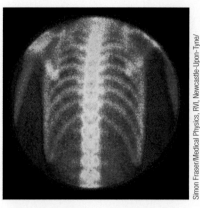

The image of a bone scan of a normal chest (posterior view). Radioactive technetium-99m is injected into the patient and is then concentrated in bones, allowing a physician to look for abnormalities such as might be caused by cancer.

Simon Fraser/Medical Physics, RVI, Newcastle-Upon-Tyne/ Photo Researchers, Inc.

Kinetics of Nuclear Decay I

Technetium-99m is used to form pictures of internal organs in the body and is often used to assess heart damage. The *m* for this nuclide indicates an excited nuclear state that decays to the ground state by gamma emission. The rate constant for decay of $^{99m}_{43}\text{Tc}$ is known to be $1.16 \times 10^{-1}/\text{h}$. What is the half-life of this nuclide?

Solution

The half-life can be calculated from the expression

$$t_{1/2} = \frac{0.693}{k} = \frac{0.693}{1.16 \times 10^{-1}/\text{h}}$$
$$= 5.98 \text{ h}$$

Thus it will take 5.98 h for a given sample of technetium-99m to decrease to half the original number of nuclides.

See Exercise 19.25

As we saw in Section 12.4, the half-life for a first-order process is constant. This is shown for the β-particle decay of strontium-90 in Fig. 19.3; it takes 28.9 years for each halving of the amount of $^{90}_{38}\text{Sr}$. Contamination of the environment with $^{90}_{38}\text{Sr}$ poses serious health hazards because of the similar chemistry of strontium and calcium (both are in Group 2A). Strontium-90 in grass and hay is incorporated into cow's milk along with calcium and is then passed on to humans, where it lodges in the bones. Because of its relatively long half-life, it persists for years in humans, causing radiation damage that may lead to cancer.

The harmful effects of radiation will be discussed in Section 19.7.

Figure 19.3 │ The decay of a 10.0-g sample of strontium-90 over time. Note that the half-life is a constant 28.9 years.

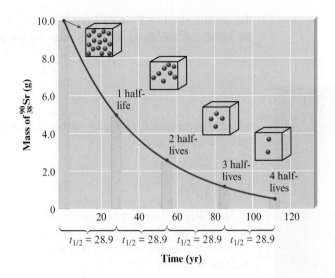

$t_{1/2} = 28.9$ $t_{1/2} = 28.9$ $t_{1/2} = 28.9$ $t_{1/2} = 28.9$

Time (yr)

Example 19.4

^{99}Mo

β decay,
$t_{1/2} = 66$ h

^{99m}Tc

Kinetics of Nuclear Decay II

The half-life of molybdenum-99 is 66.0 h. How much of a 1.000-mg sample of $^{99}_{42}$Mo is left after 330 h?

Solution

The easiest way to solve this problem is to recognize that 330 h represents five half-lives for $^{99}_{42}$Mo:

$$330 = 5 \times 66.0$$

We can sketch the change that occurs, as is shown in Fig. 19.4. Thus, after 330 h, 0.031 mg $^{99}_{42}$Mo remains.

See Exercise 19.27

The half-lives of radioactive nuclides vary over a tremendous range. For example, $^{144}_{60}$Nd has a half-life of 2.3×10^{15} years, while $^{214}_{84}$Po has a half-life of 2×10^{-4} second. To give you some perspective on this, the half-lives of the nuclides in the $^{238}_{92}$U decay series are given in Table 19.3.

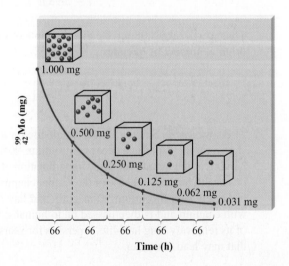

Figure 19.4 │ The change in the amount of $^{99}_{42}$Mo with time ($t_{1/2} = 66$ h).

Table 19.3 | The Half-Lives of Nuclides in the $^{238}_{92}$U Decay Series

Nuclide	Particle Produced	Half-Life
Uranium-238 ($^{238}_{92}$U)	α	4.47×10^9 years
↓		
Thorium-234 ($^{234}_{90}$Th)	β	24.1 days
↓		
Protactinium-234 ($^{234}_{91}$Pa)	β	6.7 hours
↓		
Uranium-234 ($^{234}_{92}$U)	α	2.46×10^5 years
↓		
Thorium-230 ($^{230}_{90}$Th)	α	7.5×10^4 years
↓		
Radium-226 ($^{226}_{88}$Ra)	α	1.60×10^3 years
↓		
Radon-222 ($^{222}_{86}$Rn)	α	3.82 days
↓		
Polonium-218 ($^{218}_{84}$Po)	α	3.1 minutes
↓		
Lead-214 ($^{214}_{82}$Pb)	β	26.8 minutes
↓		
Bismuth-214 ($^{214}_{83}$Bi)	β	19.9 minutes
↓		
Polonium-214 ($^{214}_{84}$Po)	α	1.6×10^{-4} second
↓		
Lead-210 ($^{210}_{82}$Pb)	β	22.2 years
↓		
Bismuth-210 ($^{210}_{83}$Bi)	β	5.0 days
↓		
Polonium-210 ($^{210}_{84}$Po)	α	138.4 days
↓		
Lead-206 ($^{206}_{82}$Pb)	—	Stable

19.3 | Nuclear Transformations

In 1919 Lord Rutherford observed the first **nuclear transformation**, *the change of one element into another*. He found that by bombarding $^{14}_{7}$N with α particles, the nuclide $^{17}_{8}$O could be produced:

$$^{14}_{7}\text{N} + {}^{4}_{2}\text{He} \longrightarrow {}^{17}_{8}\text{O} + {}^{1}_{1}\text{H}$$

Fourteen years later, Irene Curie and her husband Frederick Joliot observed a similar transformation from aluminum to phosphorus:

$$^{27}_{13}\text{Al} + {}^{4}_{2}\text{He} \longrightarrow {}^{30}_{15}\text{P} + {}^{1}_{0}\text{n}$$

where $^{1}_{0}$n represents a neutron.

Over the years, many other nuclear transformations have been achieved, mostly using **particle accelerators**, which, as the name reveals, are devices used to give particles very high velocities. Because of the electrostatic repulsion between the target nucleus and a positive ion, accelerators are needed when positive ions are used as bombarding particles. The particle, accelerated to a very high velocity, can overcome the repulsion and penetrate the target nucleus, thus effecting the transformation. A schematic diagram of one type of particle accelerator, the **cyclotron**, is shown in Fig. 19.5. The ion is introduced at the center of the cyclotron and is accelerated in an expanding spiral path by use of alternating electric fields in the presence of a magnetic field. The **linear accelerator** illustrated in Fig. 19.6 uses changing electric fields to achieve high velocities on a linear pathway.

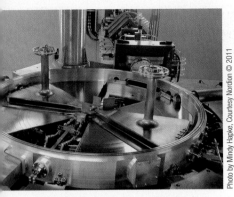

A cyclotron at TRIUMF, Canada's national laboratory of particle and nuclear physics.

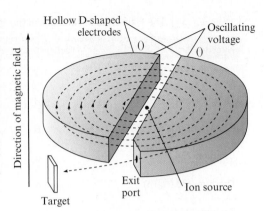

Hollow D-shaped electrodes

Oscillating voltage

Direction of magnetic field

Target

Exit port

Ion source

Figure 19.5 | A schematic diagram of a cyclotron. The ion is introduced in the center and is pulled back and forth between the hollow D-shaped electrodes by constant reversals of the electric field. Magnets above and below these electrodes produce a spiral path that expands as the particle velocity increases. When the particle has sufficient speed, it exits the accelerator and is directed at the target nucleus.

Figure 19.6 | Schematic diagram of a linear accelerator, which uses a changing electric field to accelerate a positive ion along a linear path. As the ion leaves the source, the odd-numbered tubes are negatively charged, and the even-numbered tubes are positively charged. The positive ion is thus attracted into tube 1. As the ion leaves tube 1, the tube polarities are reversed. Now tube 1 is positive, repelling the positive ion, and tube 2 is negative, attracting the positive ion. This process continues, eventually producing high particle velocity.

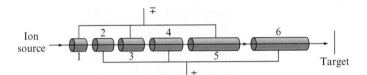

Ion source

Target

In addition to positive ions, neutrons are often used as bombarding particles to effect nuclear transformations. Because neutrons are uncharged and thus not repelled electrostatically by a target nucleus, they are readily absorbed by many nuclei, leading to new nuclides. The most common source of neutrons for this purpose is a fission reactor (see Section 19.6).

By using neutron and positive-ion bombardment, scientists have been able to extend the periodic table. Prior to 1940, the heaviest known element was uranium ($Z = 92$), but in 1940, neptunium ($Z = 93$) was produced by neutron bombardment of $^{238}_{92}\text{U}$. The process initially gives $^{239}_{92}\text{U}$, which decays to $^{239}_{93}\text{Np}$ by β-particle production:

$$^{238}_{92}\text{U} + {}^{1}_{0}\text{n} \longrightarrow {}^{239}_{92}\text{U} \xrightarrow[t_{1/2} = 23 \text{ min}]{} {}^{238}_{92}\text{Np} + {}^{0}_{-1}\text{e}$$

In the years since 1940, the elements with atomic numbers greater than 92, called the **transuranium elements**,* have been synthesized. Many of these elements have very short half-lives, as shown in Table 19.4. As a result, only a few atoms of some have ever been formed. This, of course, makes the chemical characterization of these elements extremely difficult.

Model of the large hadron collider at CERN #81594945

---

*For more information, see G. B. Kauffman, "Beyond uranium," *Chem. Eng. News* (Nov. 19, 1990): 18.

Chemical connections
Element 117

The discovery of element 117 (Uus) provides an excellent illustration of the importance of teamwork in modern scientific activities. Although element 117 was prepared in Dubna, Russia, the target nuclides were prepared at Oak Ridge National Laboratory (ORNL) in Tennessee, and data analysis on the discovery was carried out at Lawrence Livermore Laboratory in California.

The birth of element 117 started at ORNL where a 250-day irradiation experiment produced 22 mg of berkelium-249, which has a 320-day half-life. This process was followed by a 90-day effort to separate and purify the berkelium, after which the berkelium was sent to the Joint Institute for Nuclear Research (JINR) in Dubna, Russia. JINR has an accelerator beam that enabled calcium-48 to be directed at the berkelium-249 target. This 150-day process resulted in the production of six atoms of element 117 by the following nuclear reactions:

$$^{48}_{20}\text{Ca} + {}^{249}_{97}\text{Bk} \rightarrow {}^{297}_{117}\text{Uus}^* \rightarrow {}^{294}_{117}\text{Uus} + 3{}^{1}_{0}\text{n}$$
(1 atom produced)

$$^{48}_{20}\text{Ca} + {}^{249}_{97}\text{Bk} \rightarrow {}^{297}_{117}\text{Uus}^* \rightarrow {}^{293}_{117}\text{Uus} + 4{}^{1}_{0}\text{n}$$
(5 atoms produced)

The periodic table shows Uus as the heaviest element in the halogen family. However, because only six atoms of Uus were produced and they existed for only about 0.01 second, no evidence is available at present of the chemical behavior of Uus.

																	Noble gases
1 1A	Alkaline earth metals **2** 2A												Halogens			**17** 7A	**18** 8A
1 H	2 2A											**13** 3A	**14** 4A	**15** 5A	**16** 6A	**17** 7A	2 He
3 Li	4 Be											5 B	6 C	7 N	8 O	9 F	10 Ne
11 Na	12 Mg	**3**	**4**	**5**	**6** Transition metals	**7**	**8**	**9**	**10**	**11**	**12**	13 Al	14 Si	15 P	16 S	17 Cl	18 Ar
19 K	20 Ca	21 Sc	22 Ti	23 V	24 Cr	25 Mn	26 Fe	27 Co	28 Ni	29 Cu	30 Zn	31 Ga	32 Ge	33 As	34 Se	35 Br	36 Kr
37 Rb	38 Sr	39 Y	40 Zr	41 Nb	42 Mo	43 Tc	44 Ru	45 Rh	46 Pd	47 Ag	48 Cd	49 In	50 Sn	51 Sb	52 Te	53 I	54 Xe
55 Cs	56 Ba	57 La*	72 Hf	73 Ta	74 W	75 Re	76 Os	77 Ir	78 Pt	79 Au	80 Hg	81 Tl	82 Pb	83 Bi	84 Po	85 At	86 Rn
87 Fr	88 Ra	89 Ac†	104 Rf	105 Db	106 Sg	107 Bh	108 Hs	109 Mt	110 Ds	111 Rg	112 Cn	113 Uut	114 Fl	115 Uup	116 Lv	117 Uus	118 Uuo

Alkali metals (label on left side of periodic table)

*Lanthanides	58 Ce	59 Pr	60 Nd	61 Pm	62 Sm	63 Eu	64 Gd	65 Tb	66 Dy	67 Ho	68 Er	69 Tm	70 Yb	71 Lu
†Actinides	90 Th	91 Pa	92 U	93 Np	94 Pu	95 Am	96 Cm	97 Bk	98 Cf	99 Es	100 Fm	101 Md	102 No	103 Lr

Table 19.4 | Syntheses of Some of the Transuranium Elements

Element	Neutron Bombardment	Half-Life
Neptunium (Z = 93)	$^{238}_{92}U + ^{1}_{0}n \longrightarrow ^{239}_{93}Np + ^{0}_{-1}e$	2.36 days ($^{239}_{93}$Np)
Plutonium (Z = 94)	$^{239}_{93}Np \longrightarrow ^{239}_{94}Pu + ^{0}_{-1}e$	24,110 years ($^{239}_{94}$Pu)
Americium (Z = 95)	$^{239}_{94}Pu + 2^{1}_{0}n \longrightarrow ^{241}_{94}Pu \longrightarrow ^{241}_{95}Am + ^{0}_{-1}e$	433 years ($^{241}_{95}$Am)

Element	Positive-Ion Bombardment	Half-Life
Curium (Z = 96)	$^{239}_{94}Pu + ^{4}_{2}He \longrightarrow ^{242}_{96}Cm + ^{1}_{0}n$	163 days ($^{242}_{96}$Cm)
Californium (Z = 98)	$^{242}_{96}Cm + ^{4}_{2}He \longrightarrow ^{245}_{98}Cf + ^{1}_{0}n$ or $^{238}_{92}U + ^{12}_{6}C \longrightarrow ^{246}_{98}Cf + 4^{1}_{0}n$	45 minutes ($^{245}_{98}$Cf)
Rutherfordium (Z = 104)	$^{249}_{98}Cf + ^{12}_{6}C \longrightarrow ^{257}_{104}Rf + 4^{1}_{0}n$	
Dubnium (Z = 105)	$^{249}_{98}Cf + ^{15}_{7}N \longrightarrow ^{260}_{105}Db + 4^{1}_{0}n$	
Seaborgium (Z = 106)	$^{249}_{98}Cf + ^{18}_{8}O \longrightarrow ^{263}_{106}Sg + 4^{1}_{0}n$	

19.4 | Detection and Uses of Radioactivity

Geiger counters are often called survey meters in the industry.

Although various instruments measure radioactivity levels, the most familiar of them is the **Geiger–Müller counter**, or **Geiger counter** (Fig. 19.7). This instrument takes advantage of the fact that the high-energy particles from radioactive decay processes produce ions when they travel through matter. The probe of the Geiger counter is filled with argon gas, which can be ionized by a rapidly moving particle. This reaction is demonstrated by the equation:

$$Ar(g) \xrightarrow[\text{particle}]{\text{High-energy}} Ar^{+}(g) + e^{-}$$

Normally, a sample of argon gas will not conduct a current when an electrical potential is applied. However, the formation of ions and electrons produced by the passage of the high-energy particle allows a momentary current to flow. Electronic devices detect this current flow, and the number of these events can be counted. Thus the decay rate of the radioactive sample can be determined.

Figure 19.7 | A schematic representation of a Geiger–Müller counter. The high-energy radioactive particle enters the window and ionizes argon atoms along its path. The resulting ions and electrons produce a momentary current pulse, which is amplified and counted.

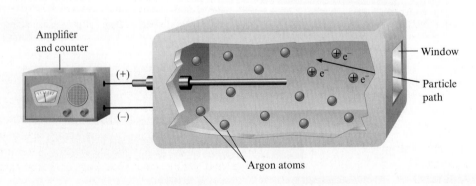

Brigham Young researcher Scott Woodward taking a bone sample for carbon-14 dating at an archaeological site in Egypt.

Radioactive nuclides are often called *radionuclides*. Carbon dating is based on the radionuclide $^{14}_{6}C$.

The $^{14}_{6}C/^{12}_{6}C$ ratio is the basis for carbon-14 dating.

Another instrument often used to detect levels of radioactivity is a **scintillation counter**, which takes advantage of the fact that certain substances, such as zinc sulfide, give off light when they are struck by high-energy radiation. A photocell senses the flashes of light that occur as the radiation strikes and thus measures the number of decay events per unit of time.

Dating by Radioactivity

Archaeologists, geologists, and others involved in reconstructing the ancient history of the earth rely heavily on radioactivity to provide accurate dates for artifacts and rocks. A method that has been very important for dating ancient articles made from wood or cloth is **radiocarbon dating**, or **carbon-14 dating**, a technique originated in the 1940s by Willard Libby, an American chemist who received a Nobel Prize for his efforts in this field.

Radiocarbon dating is based on the radioactivity of the nuclide $^{14}_{6}C$, which decays via β-particle production:

$$^{14}_{6}C \longrightarrow {}^{0}_{-1}e + {}^{14}_{7}N$$

Carbon-14 is continuously produced in the atmosphere when high-energy neutrons from space collide with nitrogen-14:

$$^{14}_{7}N + {}^{1}_{0}n \longrightarrow {}^{14}_{6}C + {}^{1}_{1}H$$

Thus carbon-14 is continuously produced by this process, and it continuously decomposes through β-particle production. Over the years, the rates for these two processes have become equal, and like a participant in a chemical reaction at equilibrium, the amount of $^{14}_{6}C$ that is present in the atmosphere remains approximately constant.

Carbon-14 can be used to date wood and cloth artifacts because the $^{14}_{6}C$, along with the other carbon isotopes in the atmosphere, reacts with oxygen to form carbon dioxide. A living plant consumes carbon dioxide in the photosynthesis process and incorporates the carbon, including $^{14}_{6}C$, into its molecules. As long as the plant lives, the $^{14}_{6}C/^{12}_{6}C$ ratio in its molecules remains the same as in the atmosphere because of the continuous uptake of carbon. However, as soon as a tree is cut to make a wooden bowl or a flax plant is harvested to make linen, the $^{14}_{6}C/^{12}_{6}C$ ratio begins to decrease because of the radioactive decay of $^{14}_{6}C$ (the $^{12}_{6}C$ nuclide is stable). Since the half-life of $^{14}_{6}C$ is 5730 years, a wooden bowl found in an archaeological dig showing a $^{14}_{6}C/^{12}_{6}C$ ratio that is half that found in currently living trees is approximately 5730 years old. This reasoning assumes that the current $^{14}_{6}C/^{12}_{6}C$ ratio is the same as that found in ancient times.

Dendrochronologists, scientists who date trees from annual growth rings, have used data collected from long-lived species of trees, such as bristlecone pines and sequoias, to show that the $^{14}_{6}C$ content of the atmosphere has changed significantly over the ages. These data have been used to derive correction factors that allow very accurate dates to be determined from the observed $^{14}_{6}C/^{12}_{6}C$ ratio in an artifact, especially for artifacts 10,000 years old or younger. Recent measurements of uranium-to-thorium ratios in ancient coral indicate that dates in the 20,000- to 30,000-year range may have errors as large as 3000 years. As a result, efforts are now being made to recalibrate the $^{14}_{6}C$ dates over this period.

Dr. Thomas Swetnam, a dendrochronologist at the University of Arizona in Tucson.

Interactive Example 19.5

Sign in at http://login.cengagebrain.com to try this Interactive Example in **OWL**.

^{14}C Dating

The remnants of an ancient fire in a cave in Africa showed a $^{14}_{6}C$ decay rate of 3.1 counts per minute per gram of carbon. Assuming that the decay rate of $^{14}_{6}C$ in freshly cut wood (corrected for changes in the $^{14}_{6}C$ content of the atmosphere) is 13.6 counts per minute per gram of carbon, calculate the age of the remnants. The half-life of $^{14}_{6}C$ is 5730 years.

Solution

The key to solving this problem is to realize that the decay rates given are directly proportional to the number of $^{14}_{6}C$ nuclides present. Radioactive decay follows first-order kinetics:

$$\text{Rate} = kN$$

Thus

$$\frac{3.1 \text{ counts/min} \cdot \text{g}}{13.6 \text{ counts/min} \cdot \text{g}} = \frac{\text{rate at time } t}{\text{rate at time } 0} = \frac{kN}{kN_0}$$

Number of nuclides
present at time t

Number of nuclides
present at time 0

$$= \frac{N}{N_0} = 0.23$$

We can now use the integrated first-order rate law:

$$\ln\left(\frac{N}{N_0}\right) = -kt$$

where

$$k = \frac{0.693}{t_{1/2}} = \frac{0.693}{5730 \text{ years}}$$

to solve for t, the time elapsed since the campfire:

$$\ln\left(\frac{N}{N_0}\right) = \ln(0.23) = -\left(\frac{0.693}{5730 \text{ years}}\right)t$$

Solving this equation gives $t = 12{,}000$ years; the campfire in the cave occurred about 12,000 years ago.

See Exercises 19.37 and 19.38

One drawback of radiocarbon dating is that a fairly large piece of the object (from a half to several grams) must be burned to form carbon dioxide, which is then analyzed for radioactivity. Another method for counting $^{14}_{6}C$ nuclides avoids destruction of a significant portion of a valuable artifact. This technique, requiring only about 10^{-3} g, uses a mass spectrometer (see Chapter 3), in which the carbon atoms are ionized and accelerated through a magnetic field that deflects their path. Because of their different masses, the various ions are deflected by different amounts and can be counted separately. This allows a very accurate determination of the $^{14}_{6}C/^{12}_{6}C$ ratio in the sample.

In their attempts to establish the geologic history of the earth, geologists have made extensive use of radioactivity. For example, since $^{238}_{92}U$ decays to the stable $^{206}_{82}Pb$ nuclide, the ratio of $^{206}_{82}Pb$ to $^{238}_{92}U$ in a rock can, under favorable circumstances, be used to estimate the age of the rock. The radioactive nuclide $^{176}_{71}Lu$, which decays to $^{176}_{72}Hf$, has a half-life of 37 billion years (only 186 nuclides out of 10 trillion decay each year!). Thus this nuclide can be used to date very old rocks. With this technique, scientists have estimated that the earth's crust formed 4.3 billion years ago.

Interactive Example 19.6

Sign in at http://login.cengagebrain .com to try this Interactive Example in **OWL**.

Dating by Radioactivity

A rock containing $^{238}_{92}U$ and $^{206}_{82}Pb$ was examined to determine its approximate age. Analysis showed the ratio of $^{206}_{82}Pb$ atoms to $^{238}_{92}U$ atoms to be 0.115. Assuming that no lead was originally present, that all the $^{206}_{82}Pb$ formed over the years has remained in the rock, and that the number of nuclides in intermediate stages of decay between

$^{238}_{92}U$ and $^{206}_{82}Pb$ is negligible, calculate the age of the rock. The half-life of $^{238}_{92}U$ is 4.5×10^9 years.

Solution

This problem can be solved using the integrated first-order rate law:

$$\ln\left(\frac{N}{N_0}\right) = -kt = -\left(\frac{0.693}{4.5 \times 10^9 \text{ years}}\right)t$$

where N/N_0 represents the ratio of $^{238}_{92}U$ atoms now found in the rock to the number present when the rock was formed. We are assuming that each $^{206}_{82}Pb$ nuclide present must have come from decay of a $^{238}_{92}U$ atom:

$$^{238}_{92}U \longrightarrow {}^{206}_{82}Pb$$

Thus

Number of $^{238}_{92}U$ atoms originally present	=	number of $^{206}_{82}Pb$ atoms now present	+	number of $^{238}_{92}U$ atoms now present

$$\frac{\text{Atoms of } ^{206}_{82}Pb \text{ now present}}{\text{Atoms of } ^{238}_{92}U \text{ now present}} = 0.115 = \frac{0.115}{1.000} = \frac{115}{1000}$$

Think carefully about what this means. For every 1115 $^{238}_{92}U$ atoms originally present in the rock, 115 have been changed to $^{206}_{82}Pb$ and 1000 remain as $^{238}_{92}U$. Thus

$$\frac{N}{N_0} = \frac{\overset{\text{Now present}}{^{238}_{92}U}}{\underset{^{238}_{92}U \text{ originally present}}{\underbrace{^{206}_{82}Pb + {}^{238}_{92}U}}} = \frac{1000}{1115} = 0.8969$$

$$\ln\left(\frac{N}{N_0}\right) = \ln(0.8969) = -\left(\frac{0.693}{4.5 \times 10^9 \text{ years}}\right)t$$

$$t = 7.1 \times 10^8 \text{ years}$$

This is the approximate age of the rock. It was formed sometime in the Cambrian period.

See Exercises 19.39 and 19.40

Medical Applications of Radioactivity

Although the rapid advances of the medical sciences in recent decades are due to many causes, one of the most important has been the discovery and use of **radiotracers**, radioactive nuclides that can be introduced into organisms in food or drugs and whose pathways can be *traced* by monitoring their radioactivity. For example, the incorporation of nuclides such as $^{14}_{6}C$ and $^{32}_{15}P$ into nutrients has produced important information about metabolic pathways.

Iodine-131 has proved very useful in the diagnosis and treatment of illnesses of the thyroid gland. Patients drink a solution containing small amounts of $Na^{131}I$, and the uptake of the iodine by the thyroid gland is monitored with a scanner (Fig. 19.8).

Thallium-201 can be used to assess the damage to the heart muscle in a person who has suffered a heart attack, because thallium is concentrated in healthy muscle tissue. Technetium-99m is also taken up by normal heart tissue and is used for damage assessment in a similar way.

Radiotracers provide sensitive and noninvasive methods for learning about biological systems, for detection of disease, for monitoring the action and effectiveness of

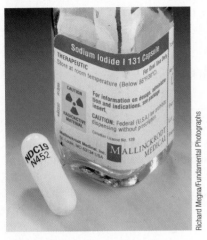

A pellet containing radioactive ^{131}I.

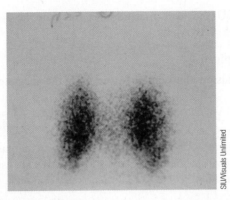

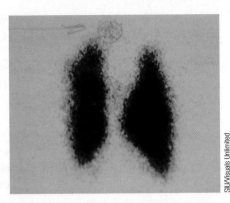

Figure 19.8 | After consumption of Na^{131}I, the patient's thyroid is scanned for radioactivity levels to determine the efficiency of iodine absorption. (left) A normal thyroid. (right) An enlarged thyroid.

Table 19.5 | Some Radioactive Nuclides, with Half-Lives and Medical Applications as Radiotracers

Nuclide	Half-Life	Area of the Body Studied
^{131}I	8.0 days	Thyroid
^{59}Fe	44.5 days	Red blood cells
^{99}Mo	66 hours	Metabolism
^{32}P	14.3 days	Eyes, liver, tumors
^{51}Cr	27.7 days	Red blood cells
^{87}Sr	2.8 hours	Bones
^{99m}Tc	6.0 hours	Heart, bones, liver, and lungs
^{133}Xe	5.2 days	Lungs
^{24}Na	15.0 hours	Circulatory system

drugs, and for early detection of pregnancy, and their usefulness should continue to grow. Some useful radiotracers are listed in Table 19.5.

19.5 | Thermodynamic Stability of the Nucleus

We can determine the thermodynamic stability of a nucleus by calculating the change in potential energy that would occur if that nucleus were formed from its constituent protons and neutrons. For example, let's consider the hypothetical process of forming a $^{16}_{8}O$ nucleus from eight neutrons and eight protons:

$$8\,^{1}_{0}n + 8\,^{1}_{1}H \longrightarrow\ ^{16}_{8}O$$

The energy change associated with this process can be calculated by comparing the sum of the masses of eight protons and eight neutrons with that of the oxygen nucleus:

$$\text{Mass of } (8\,^{1}_{0}n + 8\,^{1}_{1}H) = \underset{\underset{\text{Mass of }^{1}_{0}n}{\uparrow}}{8(1.67493 \times 10^{-24}\text{ g})} + \underset{\underset{\text{Mass of }^{1}_{1}H}{\uparrow}}{8(1.67262 \times 10^{-24}\text{ g})}$$

$$= 2.67804 \times 10^{-23}\text{ g}$$

$$\text{Mass of }^{16}_{8}O \text{ nucleus} = 2.65535 \times 10^{-23}\text{ g}$$

The difference in mass for one nucleus is

$$\text{Mass of } {}^{16}_{8}\text{O} - \text{mass of } (8\,{}^{1}_{0}\text{n} + 8\,{}^{1}_{1}\text{H}) = -2.269 \times 10^{-25}\text{ g}$$

The difference in mass for formation of 1 mole of ${}^{16}_{8}$O nuclei is therefore

$$(-2.269 \times 10^{-25}\text{ g/nucleus})(6.022 \times 10^{23}\text{ nuclei/mol}) = -0.1366\text{ g/mol}$$

Thus 0.1366 g of mass would be lost if 1 mole of oxygen-16 were formed from protons and neutrons. What is the reason for this difference in mass, and how can this information be used to calculate the energy change that accompanies this process?

The answers to these questions can be found in the work of Albert Einstein. As we discussed in Section 7.2, Einstein's theory of relativity showed that energy should be considered a form of matter. His famous equation

$$E = mc^2$$

Energy is a form of matter.

where c is the speed of light, gives the relationship between a quantity of energy and its mass. When a system gains or loses energy, it also gains or loses a quantity of mass, given by E/c^2. Thus the mass of a nucleus is less than that of its component nucleons because the process is so exothermic.

The energy changes associated with normal chemical reactions are small enough that the corresponding mass changes are not detectable.

Einstein's equation in the form

$$\text{Energy change} = \Delta E = \Delta mc^2$$

where Δm is the change in mass, or the **mass defect**, can be used to calculate ΔE for the hypothetical formation of a nucleus from its component nucleons.

Nuclear Binding Energy I

Calculate the change in energy if 1 mole of ${}^{16}_{8}$O nuclei was formed from neutrons and protons.

Solution

We have already calculated that 0.1366 g of mass would be lost in the hypothetical process of assembling 1 mole of ${}^{16}_{8}$O nuclei from the component nucleons. We can calculate the change in energy for this process from

$$\Delta E = \Delta mc^2$$

where

$$c = 3.00 \times 10^8\text{ m/s} \quad \text{and} \quad \Delta m = -0.1366\text{ g/mol} = -1.366 \times 10^{-4}\text{ kg/mol}$$

Thus

$$\Delta E = (-1.366 \times 10^{-4}\text{ kg/mol})(3.00 \times 10^8\text{ m/s})^2 = -1.23 \times 10^{13}\text{ J/mol}$$

The negative sign for the ΔE value indicates that the process is exothermic. Energy, and thus mass, is lost from the system.

See Exercises 19.41 through 19.43

The energy changes observed for nuclear processes are extremely large compared with those observed for chemical and physical changes. Thus nuclear processes constitute a potentially valuable energy resource.

The thermodynamic stability of a particular nucleus is normally represented as energy released per nucleon. To illustrate how this quantity is obtained, we will continue

Figure 19.9 | The binding energy per nucleon as a function of mass number. The most stable nuclei are at the top of the curve. The most stable nucleus is $^{56}_{26}$Fe.

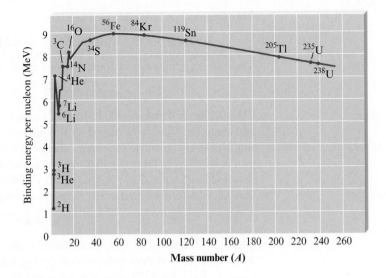

to consider $^{16}_{8}$O. First, we calculate ΔE per nucleus by dividing the molar value from Example 19.7 by Avogadro's number:

$$\Delta E \text{ per } {}^{16}_{8}\text{O nucleus} = \frac{-1.23 \times 10^{13} \text{ J/mol}}{6.022 \times 10^{23} \text{ nuclei/mol}} = -2.04 \times 10^{-11} \text{ J/nucleus}$$

In terms of a more convenient energy unit, a million electronvolts (MeV), where

$$1 \text{ MeV} = 1.60 \times 10^{-13} \text{ J}$$

$$\Delta E \text{ per } {}^{16}_{8}\text{O nucleus} = (-2.04 \times 10^{-11} \text{ J/nucleus}) \left(\frac{1 \text{ MeV}}{1.60 \times 10^{-13} \text{ J}}\right)$$

$$= -1.28 \times 10^{2} \text{ MeV/nucleus}$$

Next, we can calculate the value of ΔE per nucleon by dividing by A, the sum of neutrons and protons:

$$\Delta E \text{ per nucleon for } {}^{16}_{8}\text{O} = \frac{-1.28 \times 10^{2} \text{ MeV/nucleus}}{16 \text{ nucleons/nucleus}}$$

$$= -7.98 \text{ MeV/nucleon}$$

This means that 7.98 MeV of energy per nucleon would be *released* if $^{16}_{8}$O were formed from neutrons and protons. The energy required to *decompose* this nucleus into its components has the same numeric value but a positive sign (since energy is required). This is called the **binding energy** per nucleon for $^{16}_{8}$O.

The values of the binding energy per nucleon for the various nuclides are shown in Fig. 19.9. Note that the most stable nuclei (those requiring the largest energy per nucleon to decompose the nucleus) occur at the top of the curve. The most stable nucleus known is $^{56}_{26}$Fe, which has a binding energy per nucleon of 8.79 MeV.

Sign in at http://login.cengagebrain.com to try this Interactive Example in OWL.

| Interactive Example 19.8 | Nuclear Binding Energy II |

Calculate the binding energy per nucleon for the $^{4}_{2}$He nucleus (atomic masses: $^{4}_{2}$He = 4.0026 amu; $^{1}_{1}$H = 1.0078 amu).

Solution

First, we must calculate the mass defect (Δm) for ^4_2He. Since atomic masses (which include the electrons) are given, we must decide how to account for the electron mass:

$$4.0026 = \text{mass of } ^4_2\text{He atom} = \text{mass of } ^4_2\text{He nucleus} + 2m_e$$

Electron mass

$$1.0078 = \text{mass of } ^1_1\text{H atom} = \text{mass of } ^1_1\text{H nucleus} + m_e$$

> Since atomic masses include the masses of the electrons, to obtain the mass of a given atomic nucleus from its atomic mass, we must subtract the mass of the electrons.

Thus, since a ^4_2He nucleus is "synthesized" from two protons and two neutrons, we see that

$$\Delta m = \underbrace{(4.0026 - 2m_e)}_{\substack{\text{Mass of } ^4_2\text{He} \\ \text{nucleus}}} - [2\underbrace{(1.0078 - m_e)}_{\substack{\text{Mass of } ^1_1\text{H} \\ \text{nucleus (proton)}}} + 2\underbrace{(1.0087)}_{\substack{\text{Mass of} \\ \text{neutron}}}]$$

$$= 4.0026 - 2m_e - 2(1.0078) + 2m_e - 2(1.0087)$$

$$= 4.0026 - 2(1.0078) - 2(1.0087)$$

$$= -0.0304 \text{ u}$$

Note that in this case the electron mass cancels out in taking the difference. This will always happen in this type of calculation if the atomic masses are used both for the nuclide of interest and for ^1_1H. Thus 0.0304 of mass is *lost* per ^4_2He nucleus formed.

The corresponding energy change can be calculated from

$$\Delta E = \Delta mc^2$$

where

$$\Delta m = -0.0304 \frac{\text{u}}{\text{nucleus}} = \left(-0.0304 \frac{\text{u}}{\text{nucleus}}\right)\left(1.66 \times 10^{-27} \frac{\text{kg}}{\text{u}}\right)$$

$$= -5.04 \times 10^{-29} \frac{\text{kg}}{\text{nucleus}}$$

and

$$c = 3.00 \times 10^8 \text{ m/s}$$

Thus

$$\Delta E = \left(-5.04 \times 10^{-29} \frac{\text{kg}}{\text{nucleus}}\right)\left(3.00 \times 10^8 \frac{\text{m}}{\text{s}}\right)^2$$

$$= -4.54 \times 10^{-12} \text{ J/nucleus}$$

This means that 4.54×10^{-12} J of energy is *released* per nucleus formed and that 4.54×10^{-12} J would be required to decompose the nucleus into the constituent neutrons and protons. Thus the binding energy (BE) per nucleon is

$$\text{BE per nucleon} = \frac{4.54 \times 10^{-12} \text{ J/nucleus}}{4 \text{ nucleons/nucleus}}$$

$$= 1.14 \times 10^{-12} \text{ J/nucleon}$$

$$= \left(1.14 \times 10^{-12} \frac{\text{J}}{\text{nucleon}}\right)\left(\frac{1 \text{ MeV}}{1.60 \times 10^{-13} \text{ J}}\right)$$

$$= 7.13 \text{ MeV/nucleon}$$

See Exercises 19.44 through 19.46

Figure 19.10 | Both fission and fusion produce more stable nuclides and are thus exothermic.

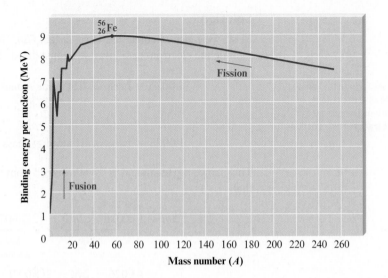

19.6 | Nuclear Fission and Nuclear Fusion

The graph shown in Fig. 19.9 has very important implications for the use of nuclear processes as sources of energy. Recall that energy is released, that is, ΔE is negative, when a process goes from a less stable to a more stable state. The higher a nuclide is on the curve, the more stable it is. This means that two types of nuclear processes will be exothermic (Fig. 19.10):

1. Combining two light nuclei to form a heavier, more stable nucleus. This process is called **fusion**.
2. Splitting a heavy nucleus into two nuclei with smaller mass numbers. This process is called **fission**.

Because of the large binding energies involved in holding the nucleus together, both these processes involve energy changes more than a million times larger than those associated with chemical reactions.

Nuclear Fission

Nuclear fission was discovered in the late 1930s when $^{235}_{92}U$ nuclides bombarded with neutrons were observed to split into two lighter elements:

$$^{1}_{0}n + {}^{235}_{92}U \longrightarrow {}^{141}_{56}Ba + {}^{92}_{36}Kr + 3\,{}^{1}_{0}n$$

This process, shown schematically in Fig. 19.11, releases 3.5×10^{-11} J of energy per event, which translates to 2.1×10^{13} J per mole of $^{235}_{92}U$. Compare this figure with that for the combustion of methane, which releases only 8.0×10^5 J of energy per mole. The fission of $^{235}_{92}U$ produces about 26 million times more energy than the combustion of methane.

The process shown in Fig. 19.11 is only one of the many fission reactions that $^{235}_{92}U$ can undergo. Another is

$$^{1}_{0}n + {}^{235}_{92}U \longrightarrow {}^{137}_{52}Te + {}^{97}_{40}Zr + 2\,{}^{1}_{0}n$$

In fact, over 200 different isotopes of 35 different elements have been observed among the fission products of $^{235}_{92}U$.

In addition to the product nuclides, neutrons are produced in the fission reactions of $^{235}_{92}U$. This makes it possible to have a self-sustaining fission process—a **chain reaction** (Fig. 19.12). For the fission process to be self-sustaining, at least one neutron from each fission event must go on to split another nucleus. If, on average, *less than one* neutron causes another fission event, the process dies out and the reaction is said

Figure 19.11 | On capturing a neutron, the $_{92}^{235}U$ nucleus undergoes fission to produce two lighter nuclides, free neutrons (typically three), and a large amount of energy.

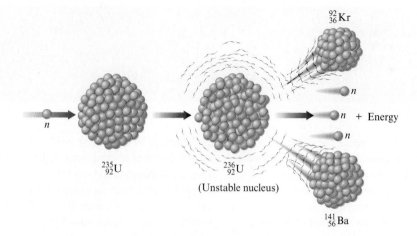

$_{36}^{92}Kr$

n

n + Energy

n

$_{92}^{235}U$

$_{92}^{236}U$
(Unstable nucleus)

$_{56}^{141}Ba$

Figure 19.12 | Representation of a fission process in which each event produces two neutrons, which can go on to split other nuclei, leading to a self-sustaining chain reaction.

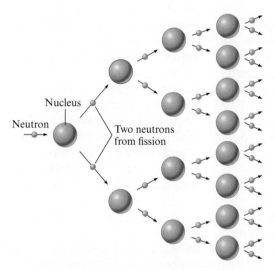

Nucleus

Neutron

Two neutrons
from fission

to be **subcritical**. If *exactly one* neutron from each fission event causes another fission event, the process sustains itself at the same level and is said to be **critical**. If *more than one* neutron from each fission event causes another fission event, the process rapidly escalates and the heat buildup causes a violent explosion. This situation is described as **supercritical**.

To achieve the critical state, a certain mass of fissionable material, called the **critical mass**, is needed. If the sample is too small, too many neutrons escape before they have a chance to cause a fission event, and the process stops. This is illustrated in Fig. 19.13.

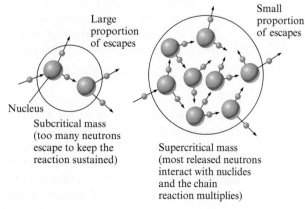

Large
proportion
of escapes

Small
proportion
of escapes

Nucleus

Subcritical mass
(too many neutrons
escape to keep the
reaction sustained)

Supercritical mass
(most released neutrons
interact with nuclides
and the chain
reaction multiplies)

Figure 19.13 | If the mass of fissionable material is too small, most of the neutrons escape before causing another fission event, and the process dies out.

Chemical connections
Future Nuclear Power

Energy—a crucial commodity in today's world—will become even more important as the pace of world development increases. Because the energy content of the universe is constant, the challenge of energy is not its quantity but rather its quality. We must find economical and environmentally friendly ways to change the energy available in the universe to forms useful to humanity. This process always involves tradeoffs.

Currently, about 65% of the world's energy consumption involves combustion of fossil fuels (coal, 39%; natural gas, 19%; oil, 7%). The use of these fuels causes significant pollution and contributes huge amounts of greenhouse gases (mostly CO_2) to the atmosphere.

One of the most abundant sources of energy is the energy that binds the atomic nucleus together. We can derive

useful energy by assembling small nuclei (fusion) or splitting large nuclei (fission). Although fusion reactors are being studied, a practical fusion reactor appears to be decades away. By contrast, fission reactors have been used since the 1950s. In fact, the production of electricity via fission reactors is widespread. At present, more than 400 nuclear reactors operate in 31 countries, producing over 355 billion watts of electrical power (see accompanying table). More than 30 reactors are currently under construction, and at least 100 more are in the planning stages. The 103 reactors currently operating in the United States produce almost 100 billion watts of electricity—about 20% of the country's electrical demands.

Forecasts indicate that the United States will need an additional 355 billion watts of generating capacity in the next 20 years. Where will this energy come from? A significant amount will be derived from coal-fired power plants with their inherent environmental problems.

Another potential source of power is solar energy. It should be an excellent pollution-free energy source, but significant technical problems remain to be solved before it sees widespread use. Wind power is also being developed but promises to make only a limited contribution to overall energy use.

The most important power source is nuclear energy. To provide all of the 355 billion watts from nuclear energy

would require hundreds of new reactors. However, nuclear power generation is very controversial because of safety, waste disposal, and cost issues. On the other hand, nuclear energy produces no greenhouse gases, has a much lower volume of waste products than the combustion of fossil fuels, has an almost unlimited supply of fuel, and has an excellent safety record. Research is now under way to improve existing reactor designs and to find new types of reactors that will be safer, more efficient, and generate much less waste by finding ways to reprocess the reactor fuel.

Actually, the United States has demonstrated an ability to deal successfully with the long-term storage of nuclear wastes. The Waste Isolation Pilot Plant (WIPP) in New Mexico has been receiving nuclear wastes since 1999 with no accidents in either transporting or storing the wastes. WIPP uses tunnels carved into the salt beds of an ancient ocean. Once a repository room becomes full, the salt will collapse around the waste, encapsulating it forever.

There is no doubt that nuclear energy will be important to the United States and to the world. An excellent source of information about all aspects of nuclear energy use is the book *Power to Save the World—The Truth About Nuclear Energy*, by Gwyneth Cravens (Alfred A. Knopf, New York, 2007). This book is a thorough but very readable treatment of the subject.

Top Ten Countries Producing Electricity by Nuclear Power (in order of total nuclear output)

Country	Percentage of Country's Total Power Production
United States	21.9
France	77.4
Japan	34.0
Germany	30.3
Russia	13.1
Canada	16.0
Ukraine	43.8
United Kingdom	26.0
Sweden	52.4
South Korea	35.8

During World War II, an intense research effort called the Manhattan Project was carried out by the United States to build a bomb based on the principles of nuclear fission. This program produced the fission bombs that were used with devastating effects on the cities of Hiroshima and Nagasaki in 1945. Basically, a fission bomb operates by suddenly combining subcritical masses of fissionable material to form a supercritical mass, thereby producing an explosion of incredible intensity.

Nuclear Reactors

Because of the tremendous energies involved, it seemed desirable to develop the fission process as an energy source to produce electricity. To accomplish this, reactors were designed in which controlled fission can occur. The resulting energy is used to heat water to produce steam to run turbine generators, in much the same way that a coal-burning power plant generates energy. A schematic diagram of a nuclear power plant is shown in Fig. 19.14.

In the **reactor core**, shown in Fig. 19.15, uranium that has been enriched to approximately 3% $^{235}_{92}U$ (natural uranium contains only 0.7% $^{235}_{92}U$) is housed in cylinders. A **moderator** surrounds the cylinders to slow down the neutrons so that the uranium fuel can capture them more efficiently. **Control rods**, composed of substances that absorb neutrons, are used to regulate the power level of the reactor. The reactor is designed so that should a malfunction occur, the control rods are automatically inserted into the core to stop the reaction. A liquid (usually water) is circulated through the core to extract the heat generated by the energy of fission; the energy can then be passed on via a heat exchanger to water in the turbine system.

Although the concentration of $^{235}_{92}U$ in the fuel elements is not great enough to allow a supercritical mass to develop in the core, a failure of the cooling system can lead to temperatures high enough to melt the core. As a result, the building housing the core must be designed to contain the core even if meltdown occurs. A great deal of controversy now exists about the efficiency of the safety systems in nuclear power plants. Accidents such as the one at the Three Mile Island facility in Pennsylvania in 1979 and in Chernobyl,* Ukraine, in 1986 have led to questions about the wisdom of continuing to build fission-based power plants.

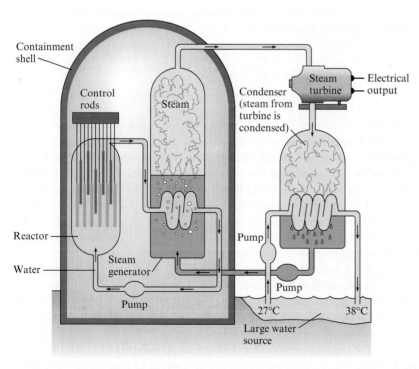

Figure 19.14 | A schematic diagram of a nuclear power plant.

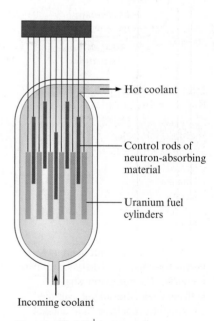

Figure 19.15 | A schematic of a reactor core. The position of the control rods determines the level of energy production by regulating the amount of fission taking place.

*See C. A. Atwood, "Chernobyl—What happened?" *J. Chem. Ed.* 65 (1988): 1037.

Uranium oxide (refined uranium).

Martin Lazarus/Photo Researchers, Inc.

Breeder Reactors

One potential problem facing the nuclear power industry is the supply of $^{235}_{92}U$. Some scientists have suggested that we have nearly depleted those uranium deposits rich enough in $^{235}_{92}U$ to make production of fissionable fuel economically feasible. Because of this possibility, **breeder reactors** have been developed, in which fissionable fuel is actually produced while the reactor runs. In the breeder reactors now being studied, the major component of natural uranium, nonfissionable $^{238}_{92}U$, is changed to fissionable $^{239}_{94}Pu$. The reaction involves absorption of a neutron, followed by production of two β particles:

$$^{1}_{0}n + {}^{238}_{92}U \longrightarrow {}^{239}_{92}U$$

$$^{239}_{92}U \longrightarrow {}^{239}_{93}Np + {}^{0}_{-1}e$$

$$^{239}_{93}Np \longrightarrow {}^{239}_{94}Pu + {}^{0}_{-1}e$$

As the reactor runs and $^{235}_{92}U$ is split, some of the excess neutrons are absorbed by $^{238}_{92}U$ to produce $^{239}_{94}Pu$. The $^{239}_{94}Pu$ is then separated out and used to fuel another reactor. Such a reactor thus "breeds" nuclear fuel as it operates.

Although breeder reactors are now used in France, the United States is proceeding slowly with their development because of their controversial nature. One problem involves the hazards in handling plutonium, which flames on contact with air and is very toxic.

Fusion

Large quantities of energy are also produced by the fusion of two light nuclei. In fact, stars produce their energy through nuclear fusion. Our sun, which presently consists of 73% hydrogen, 26% helium, and 1% other elements, gives off vast quantities of energy from the fusion of protons to form helium:

$$^{1}_{1}H + {}^{1}_{1}H \longrightarrow {}^{2}_{1}H + {}^{0}_{1}e$$

$$^{1}_{1}H + {}^{2}_{1}H \longrightarrow {}^{3}_{2}He$$

$$^{3}_{2}He + {}^{3}_{2}He \longrightarrow {}^{4}_{2}He + 2\,{}^{1}_{1}H$$

$$^{3}_{2}He + {}^{1}_{1}H \longrightarrow {}^{4}_{2}He + {}^{0}_{1}e$$

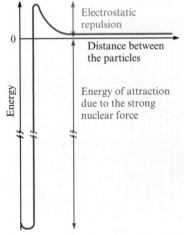

Figure 19.16 | A plot of energy versus the separation distance for two $^{2}_{1}H$ nuclei. The nuclei must have sufficient velocities to get over the electrostatic repulsion "hill" and get close enough for the nuclear binding forces to become effective, thus "fusing" the particles into a new nucleus and releasing large quantities of energy. The binding force is at least 100 times the electrostatic repulsion.

Intense research is under way to develop a feasible fusion process because of the ready availability of many light nuclides (deuterium, $^{2}_{1}H$, in seawater, for example) that can serve as fuel in fusion reactors. The major stumbling block is that high temperatures are required to initiate fusion. The forces that bind nucleons together to form a nucleus are effective only at *very small* distances ($\sim 10^{-13}$ cm). Thus, for two protons to bind together and thereby release energy, they must get very close together. But protons, because they are identically charged, repel each other electrostatically. This means that to get two protons (or two deuterons) close enough to bind together (the nuclear binding force is *not* electrostatic), they must be "shot" at each other at speeds high enough to overcome the electrostatic repulsion.

The electrostatic repulsion forces between two $^{2}_{1}H$ nuclei are so great that a temperature of 4×10^{7} K is required to give them velocities large enough to cause them to collide with sufficient energy that the nuclear forces can bind the particles together and thus release the binding energy. This situation is represented in Fig. 19.16.

Currently, scientists are studying two types of systems to produce the extremely high temperatures required: high-powered lasers and heating by electric currents. At present, many technical problems remain to be solved, and it is not clear which method will prove more useful or when fusion might become a practical energy source. However, there is still hope that fusion will be a major energy source sometime in the future.

19.7 | Effects of Radiation

Everyone knows that being hit by a train is very serious. The problem is the energy transfer involved. In fact, any source of energy is potentially harmful to organisms. Energy transferred to cells can break chemical bonds and cause malfunctioning of the cell systems. This fact is behind the concern about the ozone layer in the earth's upper atmosphere, which screens out high-energy ultraviolet radiation from the sun. Radioactive elements, which are sources of high-energy particles, are also potentially hazardous, although the effects are usually quite subtle. The reason for the subtlety of radiation damage is that even though high-energy particles are involved, the quantity of energy actually deposited in tissues *per event* is quite small. However, the resulting damage is no less real, although the effects may not be apparent for years.

The ozone layer is discussed in Section 20.11.

Radiation damage to organisms can be classified as somatic or genetic damage. **Somatic damage** is damage to the organism itself, resulting in sickness or death. The effects may appear almost immediately if a massive dose of radiation is received; for smaller doses, damage may appear years later, usually in the form of cancer. **Genetic damage** is damage to the genetic machinery, which produces malfunctions in the offspring of the organism.

The biological effects of a particular source of radiation depend on several factors:

1. *The energy of the radiation.* The higher the energy content of the radiation, the more damage it can cause. Radiation doses are measured in **rads** (which is short for *r*adiation *a*bsorbed *do*se), where 1 rad corresponds to 10^{-2} J of energy deposited per kilogram of tissue.

2. *The penetrating ability of the radiation.* The particles and rays produced in radioactive processes vary in their abilities to penetrate human tissue: γ rays are highly penetrating, β particles can penetrate approximately 1 cm, and α particles are stopped by the skin.

3. *The ionizing ability of the radiation.* Extraction of electrons from biomolecules to form ions is particularly detrimental to their functions. The ionizing ability of radiation varies dramatically. For example, γ rays penetrate very deeply but cause only occasional ionization. On the other hand, α particles, although not very penetrating, are very effective at causing ionization and produce a dense trail of damage. Thus ingestion of an α-particle producer, such as plutonium, is particularly damaging.

4. *The chemical properties of the radiation source.* When a radioactive nuclide is ingested into the body, its effectiveness in causing damage depends on its residence time. For example, $^{85}_{36}Kr$ and $^{90}_{38}Sr$ are both β-particle producers. However, since krypton is chemically inert, it passes through the body quickly and does not have much time to do damage. Strontium, being chemically similar to calcium, can collect in bones, where it may cause leukemia and bone cancer.

Because of the differences in the behavior of the particles and rays produced by radioactive decay, both the energy dose of the radiation and its effectiveness in causing biological damage must be taken into account. The **rem** (which is short for *r*oentgen *e*quivalent for *m*an) is defined as follows:

$$\text{Number of rems} = (\text{number of rads}) \times \text{RBE}$$

where RBE represents the relative effectiveness of the radiation in causing biological damage.

Table 19.6 | Effects of Short-Term Exposures to Radiation

Dose (rem)	Clinical Effect
0–25	Nondetectable
25–50	Temporary decrease in white blood cell counts
100–200	Strong decrease in white blood cell counts
500	Death of half the exposed population within 30 days after exposure

Table 19.7 | Typical Radiation Exposures for a Person Living in the United States (1 millirem = 10^{-3} rem)

	Exposure (millirems/year)
Cosmic radiation	50
From the earth	47
From building materials	3
In human tissues	21
Inhalation of air	5
Total from natural sources	126
X-ray diagnosis	50
Radiotherapy	10
Internal diagnosis/therapy	1
Nuclear power industry	0.2
TV tubes, industrial wastes, etc.	2
Radioactive fallout	4
Total from human activities	67
Total	193

Table 19.6 shows the physical effects of short-term exposure to various doses of radiation, and Table 19.7 gives the sources and amounts of radiation exposure for a typical person in the United States. Note that natural sources contribute about twice as much as human activities to the total exposure. However, although the nuclear industry contributes only a small percentage of the total exposure, the major controversy associated with nuclear power plants is the *potential* for radiation hazards. These arise mainly from two sources: accidents allowing the release of radioactive materials and improper disposal of the radioactive products in spent fuel elements. The radioactive products of the fission of $^{235}_{92}U$, although only a small percentage of the total products, have half-lives of several hundred years and remain dangerous for a long time. Various schemes have been advanced for the disposal of these wastes. The one that seems to hold the most promise is the incorporation of the wastes into ceramic blocks and the burial of these blocks in geologically stable formations. At present, however, no disposal method has been accepted, and nuclear wastes continue to accumulate in temporary storage facilities.

Even if a satisfactory method for permanent disposal of nuclear wastes is found, there will continue to be concern about the effects of exposure to low levels of radiation. Exposure is inevitable from natural sources such as cosmic rays and radioactive minerals, and many people are also exposed to low levels of radiation from reactors, radioactive tracers, or diagnostic X rays. Currently, we have little reliable information on the long-term effects of low-level exposure to radiation.

Two models of radiation damage, illustrated in Fig. 19.17, have been proposed: the *linear model* and the *threshold model*. The linear model postulates that damage from radiation is proportional to the dose, even at low levels of exposure. Thus any exposure is dangerous. The threshold model, on the other hand, assumes that no significant damage occurs below a certain exposure, called the *threshold exposure*. Note that if the linear model is correct, radiation exposure should be limited to a bare minimum (ideally at the natural levels). If the threshold model is correct, a certain level of radiation exposure beyond natural levels can be tolerated. Most scientists feel that since there is little evidence available to evaluate these models, it is safest to assume that the linear hypothesis is correct and to minimize radiation exposure.

Figure 19.17 | The two models for radiation damage. In the linear model, even a small dosage causes a proportional risk. In the threshold model, risk begins only after a certain dosage.

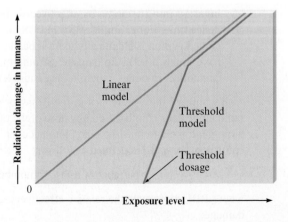

For review

Key terms

neutron
proton
nucleon
atomic number
mass number
isotopes
nuclide

Section 19.1
thermodynamic stability
kinetic stability
radioactive decay
beta (β) particle
zone of stability
alpha (α) particle
α-particle production
spontaneous fission
β-particle production
gamma (γ) ray
positron production
electron capture
decay series

Section 19.2
rate of decay
half-life

Section 19.3
nuclear transformation
particle accelerator
cyclotron
linear accelerator
transuranium elements

Section 19.4
Geiger–Müller counter (Geiger counter)
scintillation counter
radiocarbon dating (carbon-14 dating)
radiotracers

Section 19.5
mass defect
binding energy

Section 19.6
fusion
fission
chain reaction
subcritical reaction
critical reaction
supercritical reaction
critical mass

Radioactivity

> Certain nuclei decay spontaneously into more stable nuclei
> Types of radioactive decay:
>> α-particle ($^4_2$He) production
>> β-particle ($_{-1}^{\;0}$e) production
>> Positron ($^0_1$e) production
>> γ rays are usually produced in a radioactive decay event
> A decay series involves several radioactive decays to finally reach a stable nuclide
> Radioactive decay follows first-order kinetics
>> Half-life of a radioactive sample: the time required for half of the nuclides to decay
> The transuranium elements (those beyond uranium in the periodic table) can be synthesized by particle bombardment of uranium or heavier elements
> Radiocarbon dating uses the $^{14}_6$C/$^{12}_6$C ratio in an object to establish its date of origin

Thermodynamic stability of a nucleus

> Compares the mass of a nucleus to the sum of the masses of its component nucleons
> When a system gains or loses energy, it also gains or loses mass as described by the relationship $E = mc^2$
> The difference between the sum of the masses of the component nucleons and the actual mass of a nucleus (called the mass defect) can be used to calculate the nuclear binding energy

Nuclear energy production

> Fusion: the process of combining two light nuclei to form a heavier, more stable nucleus
> Fission: the process of splitting a heavy nucleus into two lighter, more stable nuclei
>> Current nuclear power reactors use controlled fission to produce energy

Radiation damage

> Radiation can cause direct (somatic) damage to a living organism or genetic damage to the organism's offspring
> The biological effects of radiation depend on the energy, the penetrating ability, the ionizing ability of the radiation, and the chemical properties of the nuclide producing the radiation

Key terms

reactor core
moderator
control rods
breeder reactor

Section 19.7
somatic damage
genetic damage
rad
rem

Review questions *Answers to the Review Questions can be found on the Student website (accessible from* **www.cengagebrain.com***).*

1. Define or illustrate the following terms:

   a. thermodynamic stability

   b. kinetic stability

   c. radioactive decay

   d. beta-particle production

   e. alpha-particle production

   f. positron production

   g. electron capture

   h. gamma-ray emissions

   In radioactive decay processes, A and Z are conserved. What does this mean?

2. Figure 19.1 illustrates the zone of stability. What is the zone of stability? Stable light nuclides have about equal numbers of neutrons and protons. What happens to the neutron-to-proton ratio for stable nuclides as the number of protons increases? Nuclides that are not already in the zone of stability undergo radioactive processes to get to the zone of stability. If a nuclide has too many neutrons, which process(es) can the nuclide undergo to become more stable? Answer the same question for a nuclide having too many protons.

3. All radioactive decay processes follow first-order kinetics. What does this mean? What happens to the rate of radioactive decay as the number of nuclides is halved? Write the first-order rate law and the integrated first-order rate law. Define the terms in each equation. What is the half-life equation for radioactive decay processes? How does the half-life depend on how many nuclides are present? Are the half-life and rate constant k directly related or inversely related?

4. What is a nuclear transformation? How do you balance nuclear transformation reactions? Particle accelerators are used to perform nuclear transformations. What is a particle accelerator?

5. What is a Geiger counter, and how does it work? What is a scintillation counter, and how does it work? Radiotracers are used in the medical sciences to learn about metabolic pathways. What are radiotracers? Explain why ^{14}C and ^{32}P radioactive nuclides would be

very helpful in learning about metabolic pathways. Why is iodine-131 useful for diagnosis of diseases of the thyroid? How could you use a radioactive nuclide to demonstrate that chemical equilibrium is a dynamic process?

6. Explain the theory behind carbon-14 dating. What assumptions must be made and what problems arise when using carbon-14 dating?

   The decay of uranium-238 to lead-206 is also used to estimate the age of objects. Specifically, $^{206}Pb/^{238}U$ ratios allow dating of rocks. Why is the ^{238}U decay to ^{206}Pb useful for dating rocks but useless for dating objects 10,000 years old or younger? Similarly, why is carbon-14 dating useful for dating objects 10,000 years old or younger but useless for dating rocks?

7. Define *mass defect* and *binding energy*. How do you determine the mass defect for a nuclide? How do you convert the mass defect into the binding energy for a nuclide? Iron-56 has the largest binding energy per nucleon among all known nuclides. Is this good or bad for iron-56? Explain.

8. Define *fission* and *fusion*. How does the energy associated with fission or fusion processes compare to the energy changes associated with chemical reactions? Fusion processes are more likely to occur for lighter elements, whereas fission processes are more likely to occur for heavier elements. Why? (*Hint:* Refer to Fig. 19.10.) The major stumbling block for turning fusion reactions into a feasible source of power is the high temperature required to initiate a fusion reaction. Why are elevated temperatures necessary to initiate fusion reactions but not fission reactions?

9. The fission of uranium-235 is used exclusively in nuclear power plants located in the United States. There are many different fission reactions of uranium-235, but all the fission reactions are self-sustaining chain reactions. Explain. Differentiate between the terms *critical, subcritical,* and *supercritical*. What is the critical mass? How does a nuclear power plant produce electricity? What are the purposes of the moderator and the control rods in a fission reactor? What are some

problems associated with nuclear reactors? What are breeder reactors? What are some problems associated with breeder reactors?

10. The biological effects of a particular source of radiation depend on several factors. List some of these factors. Even though ^{85}Kr and ^{90}Sr are both β-particle emitters, the dangers associated with the decay of ^{90}Sr are much greater than those linked to ^{85}Kr. Why? Although γ rays are far more penetrating than α particles, the latter are more likely to cause damage to an organism. Why? Which type of radiation is more effective at promoting the ionization of biomolecules?

A blue question or exercise number indicates that the answer to that question or exercise appears at the back of this book and a solution appears in the *Solutions Guide*, as found on PowerLecture.

Questions

1. When nuclei undergo nuclear transformations, γ rays of characteristic frequencies are observed. How does this fact, along with other information in the chapter on nuclear stability, suggest that a quantum mechanical model may apply to the nucleus?

2. There is a trend in the United States toward using coal-fired power plants to generate electricity rather than building new nuclear fission power plants. Is the use of coal-fired power plants without risk? Make a list of the risks to society from the use of each type of power plant.

3. Which type of radioactive decay has the net effect of changing a neutron into a proton? Which type of decay has the net effect of turning a proton into a neutron?

4. Consider the following graph of binding energy per nucleon as a function of mass number.

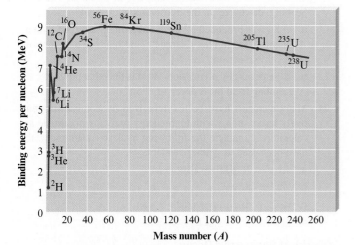

a. What does this graph tell us about the relative half-lives of the nuclides? Explain your answer.

b. Which nuclide shown is the most thermodynamically stable? Which is the least thermodynamically stable?

c. What does this graph tell us about which nuclides undergo fusion and which undergo fission to become more stable? Support your answer.

5. What are transuranium elements and how are they synthesized?

6. Scientists have estimated that the earth's crust was formed 4.3 billion years ago. The radioactive nuclide ^{176}Lu, which decays to ^{176}Hf, was used to estimate this age. The half-life of ^{176}Lu is 37 billion years. How are ratios of ^{176}Lu to ^{176}Hf utilized to date very old rocks?

7. Why are the observed energy changes for nuclear processes so much larger than the energy changes for chemical and physical processes?

8. Natural uranium is mostly nonfissionable ^{238}U; it contains only about 0.7% of fissionable ^{235}U. For uranium to be useful as a nuclear fuel, the relative amount of ^{235}U must be increased to about 3%. This is accomplished through a gas diffusion process. In the diffusion process, natural uranium reacts with fluorine to form a mixture of ^{238}UF$_6$(g) and ^{235}UF$_6$(g). The fluoride mixture is then enriched through a multistage diffusion process to produce a 3% ^{235}U nuclear fuel. The diffusion process utilizes Graham's law of effusion (see Chapter 5, Section 5.7). Explain how Graham's law of effusion allows natural uranium to be enriched by the gaseous diffusion process.

9. Much of the research on controlled fusion focuses on the problem of how to contain the reacting material. Magnetic fields appear to be the most promising mode of containment. Why is containment such a problem? Why must one resort to magnetic fields for containment?

10. A recent study concluded that any amount of radiation exposure can cause biological damage. Explain the differences between the two models of radiation damage, the linear model and the threshold model.

Exercises

In this section similar exercises are paired.

Radioactive Decay and Nuclear Transformations

11. Write an equation describing the radioactive decay of each of the following nuclides. (The particle produced is shown in parentheses, except for electron capture, where an electron is a reactant.)

   a. $^3_1$H (β)

   b. $^8_3$Li (β followed by α)

   c. $^7_4$Be (electron capture)

   d. $^8_5$B (positron)

12. In each of the following radioactive decay processes, supply the missing particle.

   a. ^{60}Co → ^{60}Ni + ?

   b. ^{97}Tc + ? → ^{97}Mo

   c. ^{99}Tc → ^{99}Ru + ?

   d. ^{239}Pu → ^{235}U + ?

13. Supply the missing particle, and state the type of decay for each of the following nuclear processes.

    a.

    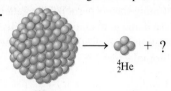

    $^{238}_{92}U$

    b.

    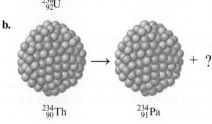

    $^{234}_{90}Th$ $^{234}_{91}Pa$

14. Write balanced equations for each of the processes described below.

    a. Chromium-51, which targets the spleen and is used as a tracer in studies of red blood cells, decays by electron capture.

    b. Iodine-131, used to treat hyperactive thyroid glands, decays by producing a β particle.

    c. Phosphorus-32, which accumulates in the liver, decays by β-particle production.

15. Write an equation describing the radioactive decay of each of the following nuclides. (The particle produced is shown in parentheses, except for electron capture, where an electron is a reactant.)

    a. ^{68}Ga (electron capture)   c. ^{212}Fr (α)
    b. ^{62}Cu (positron)   d. ^{129}Sb (β)

16. In each of the following radioactive decay processes, supply the missing particle.

    a. $^{73}Ga \rightarrow {}^{73}Ge + ?$
    b. $^{192}Pt \rightarrow {}^{188}Os + ?$
    c. $^{205}Bi \rightarrow {}^{205}Pb + ?$
    d. $^{241}Cm + ? \rightarrow {}^{241}Am$

17. Uranium-235 undergoes a series of α-particle and β-particle productions to end up as lead-207. How many α particles and β particles are produced in the complete decay series?

18. The radioactive isotope ^{247}Bk decays by a series of α-particle and β-particle productions, taking ^{247}Bk through many transformations to end up as ^{207}Pb. In the complete decay series, how many α particles and β particles are produced?

19. One type of commercial smoke detector contains a minute amount of radioactive americium-241 (^{241}Am), which decays by α-particle production. The α particles ionize molecules in the air, allowing it to conduct an electric current. When smoke particles enter, the conductivity of the air is changed and the alarm buzzes.

    a. Write the equation for the decay of $^{241}_{95}Am$ by α-particle production.

    b. The complete decay of ^{241}Am involves successively α, α, β, α, α, β, α, α, α, β, α, and β production. What is the final stable nucleus produced in this decay series?

    c. Identify the 11 intermediate nuclides.

20. Thorium-232 is known to undergo a progressive decay series until it reaches stability at lead-208. For each step of the series indicated in the table below, which nuclear particle is emitted?

Parent Nuclide	Particle Emitted
Th-232	_____
Ra-228	_____
Ac-228	_____
Th-228	_____
Ra-224	_____
Rn-220	_____
Po-216	_____
Pb-212	_____
Bi-212	_____
Po-212	_____
Pb-208	_____

21. There are four stable isotopes of iron with mass numbers 54, 56, 57, and 58. There are also two radioactive isotopes: iron-53 and iron-59. Predict modes of decay for these two isotopes. (See Table 19.2.)

22. The only stable isotope of fluorine is fluorine-19. Predict possible modes of decay for fluorine-21, fluorine-18, and fluorine-17.

23. In 1994 it was proposed (and eventually accepted) that element 106 be named seaborgium, Sg, in honor of Glenn T. Seaborg, discoverer of the transuranium elements.

    a. ^{263}Sg was produced by the bombardment of ^{249}Cf with a beam of ^{18}O nuclei. Complete and balance an equation for this reaction.

    b. ^{263}Sg decays by α emission. What is the other product resulting from the α decay of ^{263}Sg?

24. Many elements have been synthesized by bombarding relatively heavy atoms with high-energy particles in particle accelerators. Complete the following nuclear equations, which have been used to synthesize elements.

    a. $\_\_\_\_\_ + {}^{4}_{2}He \rightarrow {}^{243}_{97}Bk + {}^{1}_{0}n$
    b. $^{238}_{92}U + {}^{12}_{6}C \rightarrow \_\_\_\_\_ + 6 {}^{1}_{0}n$
    c. $^{249}_{98}Cf + \_\_\_\_\_ \rightarrow {}^{260}_{105}Db + 4 {}^{1}_{0}n$
    d. $^{249}_{98}Cf + {}^{10}_{5}B \rightarrow {}^{257}_{103}Lr + \_\_\_\_\_$

Kinetics of Radioactive Decay

25. The rate constant for a certain radioactive nuclide is 1.0×10^{-3} h^{-1}. What is the half-life of this nuclide?

26. Americium-241 is widely used in smoke detectors. The radiation released by this element ionizes particles that are then detected by a charged-particle collector. The half-life of ^{241}Am is 433 years, and it decays by emitting α particles. How many α particles are emitted each second by a 5.00-g sample of ^{241}Am?

27. Krypton consists of several radioactive isotopes, some of which are listed in the following table.

	Half-Life
^{73}Kr	27 s
^{74}Kr	11.5 min
^{76}Kr	14.8 h
^{81}Kr	2.1×10^5 yr

Which of these isotopes is most stable, and which isotope is "hottest"? How long does it take for 87.5% of each isotope to decay?

28. Radioactive copper-64 decays with a half-life of 12.8 days.

   a. What is the value of k in s^{-1}?

   b. A sample contains 28.0 mg ^{64}Cu. How many decay events will be produced in the first second? Assume the atomic mass of ^{64}Cu is 64.0 u.

   c. A chemist obtains a fresh sample of ^{64}Cu and measures its radioactivity. She then determines that to do an experiment, the radioactivity cannot fall below 25% of the initial measured value. How long does she have to do the experiment?

29. A chemist wishing to do an experiment requiring ^{47}Ca^{2+} (half-life = 4.5 days) needs 5.0 μg of the nuclide. What mass of ^{47}CaCO$_3$ must be ordered if it takes 48 h for delivery from the supplier? Assume that the atomic mass of ^{47}Ca is 47.0 u.

30. The curie (Ci) is a commonly used unit for measuring nuclear radioactivity: 1 curie of radiation is equal to 3.7×10^{10} decay events per second (the number of decay events from 1 g radium in 1 s).

   a. What mass of Na$_2$^{38}SO$_4$ has an activity of 10.0 mCi? Sulfur-38 has an atomic mass of 38.0 u and a half-life of 2.87 h.

   b. How long does it take for 99.99% of a sample of sulfur-38 to decay?

31. The first atomic explosion was detonated in the desert north of Alamogordo, New Mexico, on July 16, 1945. What percentage of the strontium-90 ($t_{1/2}$ = 28.9 years) originally produced by that explosion still remains as of July 16, 2013?

32. Iodine-131 is used in the diagnosis and treatment of thyroid disease and has a half-life of 8.0 days. If a patient with thyroid disease consumes a sample of Na^{131}I containing 10. μg ^{131}I, how long will it take for the amount of ^{131}I to decrease to 1/100 of the original amount?

33. Technetium-99 has been used as a radiographic agent in bone scans ($^{99}_{43}$Tc is absorbed by bones). If $^{99}_{43}$Tc has a half-life of 6.0 hours, what fraction of an administered dose of 100. μg $^{99}_{43}$Tc remains in a patient's body after 2.0 days?

34. Phosphorus-32 is a commonly used radioactive nuclide in biochemical research, particularly in studies of nucleic acids. The half-life of phosphorus-32 is 14.3 days. What mass of phosphorus-32 is left from an original sample of 175 mg Na$_3$^{32}PO$_4$ after 35.0 days? Assume the atomic mass of ^{32}P is 32.0 u.

35. The bromine-82 nucleus has a half-life of 1.0×10^3 min. If you wanted 1.0 g ^{82}Br and the delivery time was 3.0 days, what mass of NaBr should you order (assuming all of the Br in the NaBr was ^{82}Br)?

36. Fresh rainwater or surface water contains enough tritium ($^3_1$H) to show 5.5 decay events per minute per 100. g water. Tritium has a half-life of 12.3 years. You are asked to check a vintage wine that is claimed to have been produced in 1946. How many decay events per minute should you expect to observe in 100. g of that wine?

37. A living plant contains approximately the same fraction of carbon-14 as in atmospheric carbon dioxide. Assuming that the observed rate of decay of carbon-14 from a living plant is 13.6 counts per minute per gram of carbon, how many counts per minute per gram of carbon will be measured from a 15,000-year-old sample? Will radiocarbon dating work well for small samples of 10 mg or less? (For ^{14}C, $t_{1/2}$ = 5730 years.)

38. Assume a constant ^{14}C/^{12}C ratio of 13.6 counts per minute per gram of living matter. A sample of a petrified tree was found to give 1.2 counts per minute per gram. How old is the tree? (For ^{14}C, $t_{1/2}$ = 5730 years.)

39. A rock contains 0.688 mg ^{206}Pb for every 1.000 mg ^{238}U present. Assuming that no lead was originally present, that all the ^{206}Pb formed over the years has remained in the rock, and that the number of nuclides in intermediate stages of decay between ^{238}U and ^{206}Pb is negligible, calculate the age of the rock. (For ^{238}U, $t_{1/2}$ = 4.5×10^9 years.)

40. The mass ratios of ^{40}Ar to ^{40}K also can be used to date geologic materials. Potassium-40 decays by two processes:

$$^{40}_{19}\text{K} + ^{0}_{-1}\text{e} \longrightarrow ^{40}_{18}\text{Ar} \ (10.7\%) \qquad t_{1/2} = 1.27 \times 10^9 \text{ years}$$

$$^{40}_{19}\text{K} \longrightarrow ^{40}_{20}\text{Ca} + ^{0}_{-1}\text{e} \ (89.3\%)$$

   a. Why are ^{40}Ar/^{40}K ratios used to date materials rather than ^{40}Ca/^{40}K ratios?

   b. What assumptions must be made using this technique?

   c. A sedimentary rock has an ^{40}Ar/^{40}K ratio of 0.95. Calculate the age of the rock.

   d. How will the measured age of a rock compare to the actual age if some ^{40}Ar escaped from the sample?

Energy Changes in Nuclear Reactions

41. The sun radiates 3.9×10^{23} J of energy into space every second. What is the rate at which mass is lost from the sun?

42. The earth receives 1.8×10^{14} kJ/s of solar energy. What mass of solar material is converted to energy over a 24-h period to provide the daily amount of solar energy to the earth? What mass of coal would have to be burned to provide the same amount of energy? (Coal releases 32 kJ of energy per gram when burned.)

43. Many transuranium elements, such as plutonium-232, have very short half-lives. (For ^{232}Pu, the half-life is 36 minutes.) However, some, like protactinium-231 (half-life = 3.34×10^4 years), have relatively long half-lives. Use the masses given in the following table to calculate the change in energy when 1 mole of ^{232}Pu nuclei and 1 mole of ^{231}Pa nuclei are each formed from their respective number of protons and neutrons.

Atom or Particle	Atomic Mass
Neutron	1.67493×10^{-24} g
Proton	1.67262×10^{-24} g
Electron	9.10939×10^{-28} g
^{232}Pu	3.85285×10^{-22} g
^{231}Pa	3.83616×10^{-22} g

(Since the masses of ^{232}Pu and ^{231}Pa are atomic masses, they each include the mass of the electrons present. The mass of the nucleus will be the atomic mass minus the mass of the electrons.)

44. The most stable nucleus in terms of binding energy per nucleon is ^{56}Fe. If the atomic mass of ^{56}Fe is 55.9349 u, calculate the binding energy per nucleon for ^{56}Fe.

45. Calculate the binding energy in J/nucleon for carbon-12 (atomic mass = 12.0000 u) and uranium-235 (atomic mass = 235.0439 u). The atomic mass of $^1_1$H is 1.00782 u and the mass of a neutron is 1.00866 u. The most stable nucleus known is ^{56}Fe (see Exercise 44). Would the binding energy per nucleon for ^{56}Fe be larger or smaller than that of ^{12}C or ^{235}U? Explain.

46. Calculate the binding energy per nucleon for $^2_1$H and $^3_1$H. The atomic masses are $^2_1$H, 2.01410 u; and $^3_1$H, 3.01605 u.

47. The mass defect for a lithium-6 nucleus is -0.03434 g/mol. Calculate the atomic mass of ^{6}Li.

48. The binding energy per nucleon for magnesium-27 is 1.326×10^{-12} J/nucleon. Calculate the atomic mass of ^{27}Mg.

49. Calculate the amount of energy released per gram of hydrogen nuclei reacted for the following reaction. The atomic masses are $^1_1$H, 1.00782 u; $^2_1$H, 2.01410 u; and an electron, 5.4858×10^{-4} u. (*Hint:* Think carefully about how to account for the electron mass.)

$$^1_1H + ^1_1H \longrightarrow ^2_1H + ^0_{+1}e$$

50. The easiest fusion reaction to initiate is

$$^2_1H + ^3_1H \longrightarrow ^4_2He + ^1_0n$$

Calculate the energy released per $^4_2$He nucleus produced and per mole of $^4_2$He produced. The atomic masses are $^2_1$H, 2.01410 u; $^3_1$H, 3.01605 u; and $^4_2$He, 4.00260 u. The masses of the electron and neutron are 5.4858×10^{-4} u and 1.00866 u, respectively.

Detection, Uses, and Health Effects of Radiation

51. The typical response of a Geiger–Müller tube is shown below. Explain the shape of this curve.

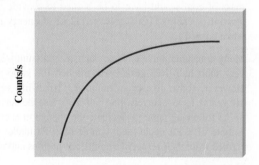

Disintegrations/s from sample

52. When using a Geiger–Müller counter to measure radioactivity, it is necessary to maintain the same geometrical orientation between the sample and the Geiger–Müller tube to compare different measurements. Why?

53. Consider the following reaction to produce methyl acetate:

$$CH_3OH + CH_3\overset{O}{\overset{\|}{C}}OH \longrightarrow CH_3\overset{O}{\overset{\|}{C}}OCH_3 + H_2O$$
Methyl acetate

When this reaction is carried out with CH_3OH containing oxygen-18, the water produced does not contain oxygen-18. Explain.

54. A chemist studied the reaction mechanism for the reaction

$$2NO(g) + O_2(g) \longrightarrow 2NO_2(g)$$

by reacting $N^{16}O$ with $^{18}O_2$. If the reaction mechanism is

$$NO + O_2 \rightleftharpoons NO_3 \text{ (fast equilibrium)}$$
$$NO_3 + NO \longrightarrow 2NO_2 \text{ (slow)}$$

what distribution of ^{18}O would you expect in the NO_2? Assume that N is the central atom in NO_3, assume only $N^{16}O^{18}O_2$ forms, and assume stoichiometric amounts of reactants are combined.

55. Uranium-235 undergoes many different fission reactions. For one such reaction, when ^{235}U is struck with a neutron, ^{144}Ce and ^{90}Sr are produced along with some neutrons and electrons. How many neutrons and β-particles are produced in this fission reaction?

56. Breeder reactors are used to convert the nonfissionable nuclide $^{238}_{92}$U to a fissionable product. Neutron capture of the $^{238}_{92}$U is followed by two successive beta decays. What is the final fissionable product?

57. Which do you think would be the greater health hazard: the release of a radioactive nuclide of Sr or a radioactive nuclide of Xe into the environment? Assume the amount of radioactivity is the same in each case. Explain your answer on the basis of the chemical properties of Sr and Xe. Why are the chemical properties of a radioactive substance important in assessing its potential health hazards?

58. Consider the following information:
  i. The layer of dead skin on our bodies is sufficient to protect us from most α-particle radiation.
  ii. Plutonium is an α-particle producer.
  iii. The chemistry of Pu^{4+} is similar to that of Fe^{3+}.
  iv. Pu oxidizes readily to Pu^{4+}.

Why is plutonium one of the most toxic substances known?

Additional Exercises

59. Predict whether each of the following nuclides is stable or unstable (radioactive). If the nuclide is unstable, predict the type of radioactivity you would expect it to exhibit.
  a. $^{45}_{19}$K   **b.** $^{56}_{26}$Fe   **c.** $^{20}_{11}$Na   **d.** $^{194}_{81}$Tl

60. Each of the following isotopes has been used medically for the purpose indicated. Suggest reasons why the particular element might have been chosen for this purpose.

a. cobalt-57, for study of the body's use of vitamin B_{12}

b. calcium-47, for study of bone metabolism

c. iron-59, for study of red blood cell function

61. The mass percent of carbon in a typical human is 18%, and the mass percent of ^{14}C in natural carbon is $1.6 \times 10^{-10}\%$. Assuming a 180-lb person, how many decay events per second occur in this person due exclusively to the β-particle decay of ^{14}C (for ^{14}C, $t_{1/2} = 5730$ years)?

62. At a flea market, you've found a very interesting painting done in the style of Rembrandt's "dark period" (1642–1672). You suspect that you really do not have a genuine Rembrandt, but you take it to the local university for testing. Living wood shows a carbon-14 activity of 15.3 counts per minute per gram. Your painting showed a carbon-14 activity of 15.1 counts per minute per gram. Could it be a genuine Rembrandt?

63. Define "third-life" in a similar way to "half-life", and determine the "third-life" for a nuclide that has a half-life of 31.4 years.

64. A proposed system for storing nuclear wastes involves storing the radioactive material in caves or deep mine shafts. One of the most toxic nuclides that must be disposed of is plutonium-239, which is produced in breeder reactors and has a half-life of 24,100 years. A suitable storage place must be geologically stable long enough for the activity of plutonium-239 to decrease to 0.1% of its original value. How long is this for plutonium-239?

65. During World War II, tritium (^3H) was a component of fluorescent watch dials and hands. Assume you have such a watch that was made in January 1944. If 17% or more of the original tritium was needed to read the dial in dark places, until what year could you read the time at night? (For ^3H, $t_{1/2} = 12.3$ yr.)

66. A positron and an electron can annihilate each other on colliding, producing energy as photons:

$$_{-1}^{0}e + _{+1}^{0}e \longrightarrow 2\,_{0}^{0}\gamma$$

Assuming that both γ rays have the same energy, calculate the wavelength of the electromagnetic radiation produced.

67. A small atomic bomb releases energy equivalent to the detonation of 20,000 tons of TNT; a ton of TNT releases 4×10^9 J of energy when exploded. Using 2×10^{13} J/mol as the energy released by fission of ^{235}U, approximately what mass of ^{235}U undergoes fission in this atomic bomb?

68. During the research that led to production of the two atomic bombs used against Japan in World War II, different mechanisms for obtaining a supercritical mass of fissionable material were investigated. In one type of bomb, a "gun" shot one piece of fissionable material into a cavity containing another piece of fissionable material. In the second type of bomb, the fissionable material was surrounded with a high explosive that, when detonated, compressed the fissionable material into a smaller volume. Discuss what is meant by critical mass, and explain why the ability to achieve a critical mass is essential to sustaining a nuclear reaction.

69. Using the kinetic molecular theory (see Section 5.6), calculate the root mean square velocity and the average kinetic energy of $_1^2H$ nuclei at a temperature of 4×10^7 K. (See Exercise 50 for the appropriate mass values.)

70. Consider the following reaction, which can take place in particle accelerators:

$$_1^1H + _0^1n \longrightarrow 2\,_1^1H + _0^1n + _{-1}^1H$$

Calculate the energy change for this reaction. Is energy released or absorbed? What is a possible source for this energy?

71. Photosynthesis in plants can be represented by the following overall equation:

$$6CO_2(g) + 6H_2O(l) \xrightarrow{\text{Light}} C_6H_{12}O_6(s) + 6O_2(g)$$

Algae grown in water containing some ^{18}O (in $H_2^{18}O$) evolve oxygen gas with the same isotopic composition as the oxygen in the water. When algae growing in water containing only ^{16}O were furnished carbon dioxide containing ^{18}O, no ^{18}O was found to be evolved from the oxygen gas produced. What conclusions about photosynthesis can be drawn from these experiments?

72. Strontium-90 and radon-222 both pose serious health risks. ^{90}Sr decays by β-particle production and has a relatively long half-life (28.9 years). Radon-222 decays by α-particle production and has a relatively short half-life (3.82 days). Explain why each decay process poses health risks.

ChemWork Problems

These multiconcept problems (and additional ones) are found interactively online with the same type of assistance a student would get from an instructor.

73. Complete the following table with the nuclear particle that is produced in each nuclear reaction.

Initial Nuclide	Product Nuclide	Particle Produced
$_{94}^{239}Pu$	$_{92}^{235}U$	_____
$_{82}^{214}Pb$	$_{83}^{214}Bi$	_____
$_{27}^{60}Co$	$_{28}^{60}Ni$	_____
$_{43}^{99}Tc$	$_{44}^{99}Ru$	_____
$_{93}^{239}Np$	$_{94}^{239}Pu$	_____

74. A certain radioactive nuclide has a half-life of 3.00 hours.

a. Calculate the rate constant in s^{-1} for this nuclide.

b. Calculate the decay rate in decays/s for 1.000 mole of this nuclide.

75. Iodine-131 has a half-life of 8.0 days. How many days will it take for 174 g of ^{131}I to decay to 83 g of ^{131}I?

76. Rubidium-87 decays by β-particle production to strontium-87 with a half-life of 4.7×10^{10} years. What is the age of a rock sample that contains 109.7 μg of ^{87}Rb and 3.1 μg of ^{87}Sr? Assume that no ^{87}Sr was present when the rock was formed. The atomic masses for ^{87}Rb and ^{87}Sr are 86.90919 u and 86.90888 u, respectively.

77. Given the following information:

Mass of proton = 1.00728 u
Mass of neutron = 1.00866 u
Mass of electron = 5.486×10^{-4} u
Speed of light = 2.9979×10^8 m/s

Calculate the nuclear binding energy of $_{12}^{24}Mg$, which has an atomic mass of 23.9850 u.

78. Which of the following statement(s) is(are) *true*?

  a. A radioactive nuclide that decays from 1.00×10^{10} atoms to 2.5×10^9 atoms in 10 minutes has a half-life of 5.0 minutes.

  b. Nuclides with large Z values are observed to be α-particle producers.

  c. As Z increases, nuclides need a greater proton-to-neutron ratio for stability.

  d. Those "light" nuclides that have twice as many neutrons as protons are expected to be stable.

Challenge Problems

79. Naturally occurring uranium is composed mostly of ^{238}U and ^{235}U, with relative abundances of 99.28% and 0.72%, respectively. The half-life for ^{238}U is 4.5×10^9 years, and the half-life for ^{235}U is 7.1×10^8 years. Assuming that the earth was formed 4.5 billion years ago, calculate the relative abundances of the ^{238}U and ^{235}U isotopes when the earth was formed.

80. The curie (Ci) is a commonly used unit for measuring nuclear radioactivity: 1 curie of radiation is equal to 3.7×10^{10} decay events per second (the number of decay events from 1 g radium in 1 s). A 1.7-mL sample of water containing tritium was injected into a 150-lb person. The total activity of radiation injected was 86.5 mCi. After some time to allow the tritium activity to equally distribute throughout the body, a sample of blood plasma containing 2.0 mL water at an activity of 3.6 μCi was removed. From these data, calculate the mass percent of water in this 150-lb person.

81. A 0.10-cm^3 sample of a solution containing a radioactive nuclide (5.0×10^3 counts per minute per milliliter) is injected into a rat. Several minutes later 1.0 cm^3 blood is removed. The blood shows 48 counts per minute of radioactivity. Calculate the volume of blood in the rat. What assumptions must be made in performing this calculation?

82. Zirconium is one of the few metals that retains its structural integrity upon exposure to radiation. The fuel rods in most nuclear reactors therefore are often made of zirconium. Answer the following questions about the redox properties of zirconium based on the half-reaction

$$\text{ZrO}_2 \cdot \text{H}_2\text{O} + \text{H}_2\text{O} + 4e^- \longrightarrow \text{Zr} + 4\text{OH}^- \quad \mathscr{E}^\circ = -2.36 \text{ V}$$

  a. Is zirconium metal capable of reducing water to form hydrogen gas at standard conditions?

  b. Write a balanced equation for the reduction of water by zirconium.

  c. Calculate $\mathscr{E}^\circ$, ΔG°, and K for the reduction of water by zirconium metal.

  d. The reduction of water by zirconium occurred during the accidents at Three Mile Island in 1979. The hydrogen produced was successfully vented and no chemical explosion occurred. If 1.00×10^3 kg Zr reacts, what mass of H$_2$ is produced? What volume of H$_2$ at 1.0 atm and 1000.°C is produced?

  e. At Chernobyl in 1986, hydrogen was produced by the reaction of superheated steam with the graphite reactor core:

$$\text{C}(s) + \text{H}_2\text{O}(g) \longrightarrow \text{CO}(g) + \text{H}_2(g)$$

It was not possible to prevent a chemical explosion at Chernobyl. In light of this, do you think it was a correct decision to vent the hydrogen and other radioactive gases into the atmosphere at Three Mile Island? Explain.

83. In addition to the process described in the text, a second process called the *carbon–nitrogen cycle* occurs in the sun:

$$^1_1\text{H} + {}^{12}_6\text{C} \longrightarrow {}^{13}_7\text{N} + {}^0_0\gamma$$
$$^{13}_7\text{N} \longrightarrow {}^{13}_6\text{C} + {}^0_{+1}\text{e}$$
$$^1_1\text{H} + {}^{13}_6\text{C} \longrightarrow {}^{14}_7\text{N} + {}^0_0\gamma$$
$$^1_1\text{H} + {}^{14}_7\text{N} \longrightarrow {}^{15}_8\text{O} + {}^0_0\gamma$$
$$^{15}_8\text{O} \longrightarrow {}^{15}_7\text{N} + {}^0_{+1}\text{e}$$
$$^1_1\text{H} + {}^{15}_7\text{N} \longrightarrow {}^{12}_6\text{C} + {}^4_2\text{He} + {}^0_0\gamma$$

Overall reaction: $\quad 4 {}^1_1\text{H} \longrightarrow {}^4_2\text{He} + 2 {}^0_{+1}\text{e}$

  a. What is the catalyst in this process?

  b. What nucleons are intermediates?

  c. How much energy is released per mole of hydrogen nuclei in the overall reaction? (The atomic masses of ${}^1_1$H and ${}^4_2$He are 1.00782 u and 4.00260 u, respectively.)

84. The most significant source of natural radiation is radon-222. ^{222}Rn, a decay product of ^{238}U, is continuously generated in the earth's crust, allowing gaseous Rn to seep into the basements of buildings. Because ^{222}Rn is an α-particle producer with a relatively short half-life of 3.82 days, it can cause biological damage when inhaled.

  a. How many α particles and β particles are produced when ^{238}U decays to ^{222}Rn? What nuclei are produced when ^{222}Rn decays?

  b. Radon is a noble gas so one would expect it to pass through the body quickly. Why is there a concern over inhaling ^{222}Rn?

  c. Another problem associated with ^{222}Rn is that the decay of ^{222}Rn produces a more potent α-particle producer ($t_{1/2} = 3.11$ min) that is a solid. What is the identity of the solid? Give the balanced equation of this species decaying by α-particle production. Why is the solid a more potent α-particle producer?

  d. The U.S. Environmental Protection Agency (EPA) recommends that ^{222}Rn levels not exceed 4 pCi per liter of air (1 Ci = 1 curie = 3.7×10^{10} decay events per second; 1 pCi = 1×10^{-12} Ci). Convert 4.0 pCi per liter of air into concentrations units of ^{222}Rn atoms per liter of air and moles of ^{222}Rn per liter of air.

85. To determine the K_{sp} value of Hg$_2$I$_2$, a chemist obtained a solid sample of Hg$_2$I$_2$ in which some of the iodine is present as radioactive ^{131}I. The count rate of the Hg$_2$I$_2$ sample is 5.0×10^{11} counts per minute per mole of I. An excess amount of Hg$_2$I$_2$(s) is placed into some water, and the solid is allowed to come to equilibrium with its respective ions. A 150.0-mL sample of the saturated solution is withdrawn and the radioactivity measured at 33 counts per minute. From this information, calculate the K_{sp} value for Hg$_2$I$_2$.

$$\text{Hg}_2\text{I}_2(s) \rightleftharpoons \text{Hg}_2{}^{2+}(aq) + 2\text{I}^-(aq) \quad K_{sp} = [\text{Hg}_2{}^{2+}][\text{I}^-]^2$$

86. Estimate the temperature needed to achieve the fusion of deuterium to make an α particle. The energy required can be estimated from Coulomb's law [use the form $E = 9.0 \times 10^9$ (Q_1Q_2/r), using $Q = 1.6 \times 10^{-19}$ C for a proton, and $r = 2 \times 10^{-15}$ m for the helium nucleus; the unit for the proportionality constant in Coloumb's law is J · m/C^2].

Integrative Problems

These problems require the integration of multiple concepts to find the solutions.

87. A reported synthesis of the transuranium element bohrium (Bh) involved the bombardment of berkelium-249 with neon-22 to produce bohrium-267. Write a nuclear reaction for this synthesis. The half-life of bohrium-267 is 15.0 seconds. If 199 atoms of bohrium-267 could be synthesized, how much time would elapse before only 11 atoms of bohrium-267 remain? What is the expected electron configuration of elemental bohrium?

88. Radioactive cobalt-60 is used to study defects in vitamin B$_{12}$ absorption because cobalt is the metallic atom at the center of the vitamin B$_{12}$ molecule. The nuclear synthesis of this cobalt isotope involves a three-step process. The overall reaction is iron-58 reacting with two neutrons to produce cobalt-60 along with the emission of another particle. What particle is emitted in this nuclear synthesis? What is the binding energy in J per nucleon for the cobalt-60 nucleus (atomic masses: ^{60}Co = 59.9338 u; ^{1}H = 1.00782 u)? What is the de Broglie wavelength of the emitted particle if it has a velocity equal to $0.90c$, where c is the speed of light?

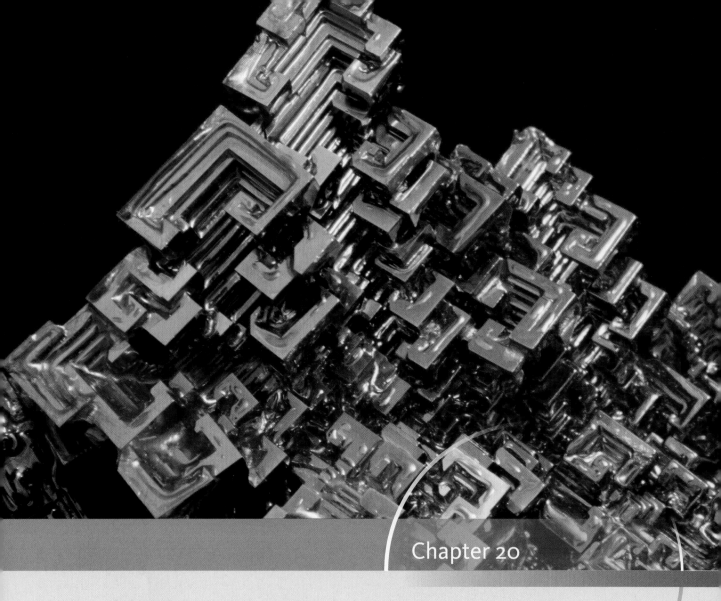

Chapter 20

The Representative Elements

A magnified crystal of bismuth. (© Walter Geiersperger/Corbis)

So far in this book we have covered the major principles and explored the most important models of chemistry. In particular, we have seen that the chemical properties of the elements can be explained very successfully by the quantum mechanical model of the atom. In fact, the most convincing evidence of that model's validity is its ability to relate the observed periodic properties of the elements to the number of valence electrons in their atoms.

We have learned many properties of the elements and their compounds, but we have not discussed extensively the relationship between the chemical properties of a specific element and its position on the periodic table. In this chapter we will explore the chemical similarities and differences among the elements in the several groups of the periodic table and will try to interpret these data using the wave mechanical model of the atom. In the process we will illustrate a great variety of chemical properties and further demonstrate the practical importance of chemistry.

20.1 | A Survey of the Representative Elements

The traditional form of the periodic table is shown in Fig. 20.1. Recall that the **representative elements**, whose chemical properties are determined by the valence-level s and p electrons, are designated Groups 1A through 8A. The **transition metals**, in the center of the table, result from the filling of d orbitals. The elements that correspond to the filling of the $4f$ and $5f$ orbitals are listed separately as the **lanthanides** and **actinides**, respectively.

The heavy black line in Fig. 20.1 separates the metals from the nonmetals. Some elements just on either side of this line, such as silicon and germanium, exhibit both metallic and nonmetallic properties. These elements are often called **metalloids**, or **semimetals**. The fundamental chemical difference between metals and nonmetals is that metals tend to lose their valence electrons to form *cations,* which usually have the valence electron configuration of the noble gas from the preceding period. On the other hand, nonmetals tend to gain electrons to form *anions* that exhibit the electron configuration of the noble gas in the same period. Metallic character is observed to increase in going down a given group, which is consistent with the trends in ionization energy, electron affinity, and electronegativity discussed earlier (see Sections 7.12 and 8.2).

Metallic character increases going down a group in the periodic table.

Atomic Size and Group Anomalies

Although the chemical properties of the members of a group have many similarities, there are also important differences. The most dramatic differences usually occur between the first and second member. For example, hydrogen in Group 1A is a nonmetal, whereas lithium is a very active metal. This extreme difference results primarily from the very large difference in the atomic radii of hydrogen and lithium, as shown in Fig. 20.2. Since the small hydrogen atom has a much greater attraction for electrons than do the larger members of Group 1A, it forms covalent bonds with nonmetals. In contrast, the other members of Group 1A lose their valence electrons to nonmetals to form 1+ cations in ionic compounds.

The effect of size is also evident in other groups. For example, the oxides of the metals in Group 2A are all quite basic except for the first member of the series; beryllium oxide (BeO) is amphoteric. The basicity of an oxide depends on its ionic

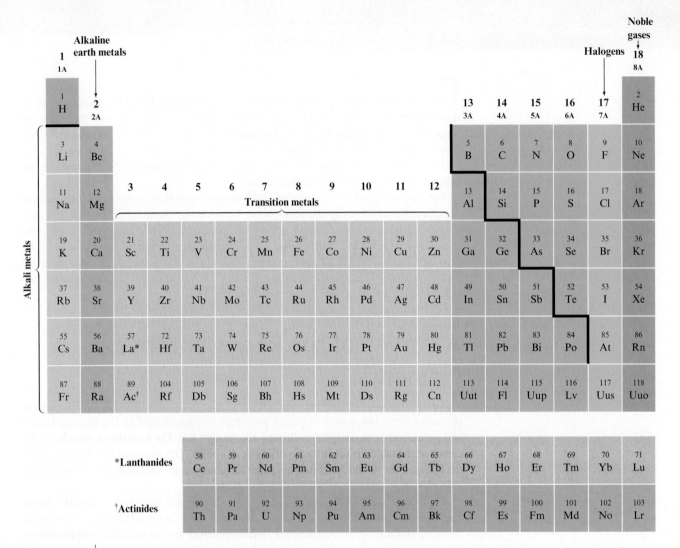

Figure 20.1 | The periodic table. The elements in the A groups are the representative elements. The elements shown in yellow are called transition metals. The heavy black line approximately separates the nonmetals from the metals.

character. Ionic oxides contain the O^{2-} ion, which reacts with water to form two OH^- ions. All the oxides of the Group 2A metals are highly ionic except for beryllium oxide, which has considerable covalent character. The small Be^{2+} ion can effectively polarize the electron "cloud" of the O^{2-} ion, thereby producing significant electron sharing. We see the same pattern in Group 3A, where only the small boron atom behaves as a nonmetal, or sometimes as a semimetal, whereas aluminum and the other members are active metals.

In Group 4A the effect of size is reflected in the dramatic differences between the chemical properties of carbon and silicon. The chemistry of carbon is dominated by molecules containing chains of C—C bonds, but silicon compounds mainly contain Si—O bonds rather than Si—Si bonds. Silicon does form compounds with chains of Si—Si bonds, but these compounds are much more reactive than the corresponding carbon compounds. The reasons for the difference in reactivity between the carbon and silicon compounds are quite complex but are likely related to the differences in the sizes of the carbon and silicon atoms.

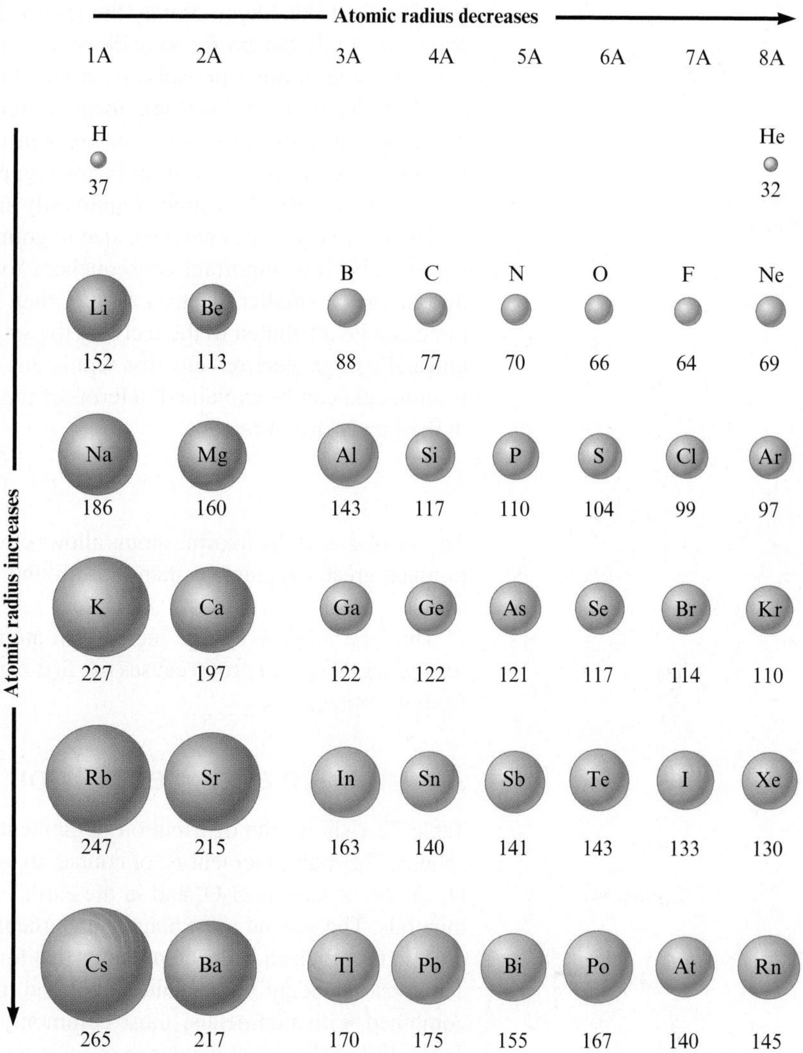

Figure 20.2 | Some atomic radii (in picometers).

Atomic radius decreases →

	1A	2A	3A	4A	5A	6A	7A	8A
	H 37							He 32
	Li 152	Be 113	B 88	C 77	N 70	O 66	F 64	Ne 69
	Na 186	Mg 160	Al 143	Si 117	P 110	S 104	Cl 99	Ar 97
	K 227	Ca 197	Ga 122	Ge 122	As 121	Se 117	Br 114	Kr 110
	Rb 247	Sr 215	In 163	Sn 140	Sb 141	Te 143	I 133	Xe 130
	Cs 265	Ba 217	Tl 170	Pb 175	Bi 155	Po 167	At 140	Rn 145

Atomic radius increases ↓

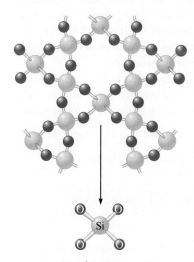

Figure 20.3 | The structure of quartz, which has the empirical formula SiO_2. Note that the structure is based on interlocking SiO_4 tetrahedra (shown below the arrow), in which each oxygen atom is shared by two silicon atoms.

Carbon and silicon also differ markedly in their abilities to form π bonds. As we discussed in Section 9.1, carbon dioxide is composed of discrete CO_2 molecules with the Lewis structure

$$\ddot{O}=C=\ddot{O}$$

where the carbon and oxygen atoms achieve the [Ne] configuration by forming π bonds. In contrast, the structure of silica (empirical formula SiO_2) is based on SiO_4 tetrahedra with Si—O—Si bridges, as shown in Fig. 20.3. The silicon $3p$ valence orbitals do not overlap very effectively with the smaller oxygen $2p$ orbitals to form π bonds; therefore, discrete SiO_2 molecules with the Lewis structure

$$\ddot{O}=Si=\ddot{O}$$

are not stable. Instead, the silicon atoms achieve a noble gas configuration by forming four Si—O single bonds.

The importance of π bonding for the relatively small elements of the second period also explains the different elemental forms of the members of Groups 5A and 6A. For example, elemental nitrogen exists as very stable N_2 molecules with the Lewis structure :N≡N:. Elemental phosphorus forms larger aggregates of atoms, the simplest being the tetrahedral P_4 molecules found in white phosphorus (see

Fig. 20.18). Like silicon atoms, the relatively large phosphorus atoms do not form strong π bonds but prefer to achieve a noble gas configuration by forming single bonds to several other phosphorus atoms. In contrast, its very strong π bonds make the N_2 molecule the most stable form of elemental nitrogen. Similarly, in Group 6A the most stable form of elemental oxygen is the O_2 molecule with a double bond. However, the larger sulfur atom forms bigger aggregates, such as the cyclic S_8 molecule (see Fig. 20.22), which contain only single bonds.

The relatively large change in size in going from the first to the second member of a group also has important consequences for the Group 7A elements. For example, fluorine has a smaller electron affinity than chlorine. This violation of the expected trend can be attributed to the fact that the small size of the fluorine $2p$ orbitals causes unusually large electron–electron repulsions. The relative weakness of the bond in the F_2 molecule can be explained in terms of the repulsions among the lone pairs, shown in the Lewis structure:

$$\ddot{\text{F}} - \ddot{\text{F}}\text{:}$$

The small size of the fluorine atoms allows close approach of the lone pairs, which leads to much greater repulsions than those found in the Cl_2 molecule with its much larger atoms.

Thus the relatively large increase in atomic radius in going from the first to the second member of a group causes the first element to exhibit properties quite different from the others.

Abundance and Preparation

Table 20.1 shows the distribution of elements in the earth's crust, oceans, and atmosphere. The major element is, of course, oxygen, which is found in the atmosphere as O_2, in the oceans as H_2O, and in the earth's crust primarily in silicate and carbonate minerals. The second most abundant element, silicon, is found throughout the earth's crust in the silica and silicate minerals that form the basis of most sand, rocks, and soil. The most abundant metals, aluminum and iron, are found in ores, in which they are combined with nonmetals, most commonly oxygen. One notable fact revealed by Table 20.1 is the small incidence of most transition metals. Since many of these relatively rare elements are assuming increasing importance in our high-technology society, it is possible that the control of transition metal ores may ultimately have more significance in world politics than will control of petroleum supplies.

The distribution of elements in living materials is very different from that found in the earth's crust. Table 20.2 shows the distribution of elements in the human body. Oxygen, carbon, hydrogen, and nitrogen form the basis for all biologically important

Table 20.1 | Distribution (Mass Percent) of the 18 Most Abundant Elements in the Earth's Crust, Oceans, and Atmosphere

Element	Mass Percent	Element	Mass Percent
Oxygen	49.2	Chlorine	0.19
Silicon	25.7	Phosphorus	0.11
Aluminum	7.50	Manganese	0.09
Iron	4.71	Carbon	0.08
Calcium	3.39	Sulfur	0.06
Sodium	2.63	Barium	0.04
Potassium	2.40	Nitrogen	0.03
Magnesium	1.93	Fluorine	0.03
Hydrogen	0.87	All others	0.49
Titanium	0.58		

Table 20.2 | Abundance of Elements in the Human Body

Major Elements	Mass Percent	Trace Elements (in alphabetical order)
Oxygen	65.0	Arsenic
Carbon	18.0	Chromium
Hydrogen	10.0	Cobalt
Nitrogen	3.0	Copper
Calcium	1.4	Fluorine
Phosphorus	1.0	Iodine
Magnesium	0.50	Manganese
Potassium	0.34	Molybdenum
Sulfur	0.26	Nickel
Sodium	0.14	Selenium
Chlorine	0.14	Silicon
Iron	0.004	Vanadium
Zinc	0.003	

molecules. The other elements, even though they are found in relatively small amounts, are often crucial for life. For example, zinc is found in over 150 different biomolecules in the human body.

Carbon is the cheapest and most readily available industrial reducing agent for metallic ions.

Only about one-fourth of the elements occur naturally in the free state. Most are found in a combined state. The *process of obtaining a metal from its ore* is called **metallurgy**. Since the metals in ores are found in the form of cations, the chemistry of *metallurgy always involves reduction of the ions to the elemental metal (with an oxidation state of zero)*. A variety of reducing agents can be used, but carbon is the usual choice because of its wide availability and relatively low cost.

Electrolysis is often used to reduce the most active metals. In Chapter 18 we considered the electrolytic production of aluminum metal. The alkali metals are also produced by electrolysis, usually of their molten halide salts.

The preparation of nonmetals varies widely. Elemental nitrogen and oxygen are usually obtained from the **liquefaction** of air, which is based on the principle that a gas cools as it expands. After each expansion, part of the cooler gas is compressed, whereas the rest is used to carry away the heat of the compression. The compressed gas is then allowed to expand again. This cycle is repeated many times. Eventually, the remaining gas becomes cold enough to form the liquid state. Because liquid nitrogen and liquid oxygen have different boiling points, they can be separated by the distillation of liquid

Sand, such as that found in the massive sand dunes bordering the desert plain near Namib, Namibia, is composed of silicon and oxygen.

Frank Krahmer/Radius Images/Masterfile

air. Both substances are important industrial chemicals, with nitrogen ranking second in terms of amount manufactured in the United States (approximately 60 billion pounds per year) and oxygen ranking third (over 40 billion pounds per year). Hydrogen can be obtained from the electrolysis of water, but more commonly it is obtained from the decomposition of the methane in natural gas. Sulfur is found underground in its elemental form and is recovered by the Frasch process (see Section 20.12). The halogens are obtained by oxidation of the anions from halide salts (see Section 20.13).

The preparation of sulfur and the halogens is discussed later in this chapter.

20.2 | The Group 1A Elements

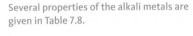

The Group 1A elements with their ns^1 valence electron configurations are all very active metals (they lose their valence electrons very readily), except for hydrogen, which behaves as a nonmetal. We will discuss the chemistry of hydrogen in the next section. Many of the properties of the **alkali metals** have been given previously (see Section 7.13). The sources and methods of preparation of pure alkali metals are given in Table 20.3. The ionization energies, standard reduction potentials, ionic radii, and melting points for the alkali metals are listed in Table 20.4.

In Section 7.13 we saw that the alkali metals all react vigorously with water to release hydrogen gas:

$$2M(s) + 2H_2O(l) \longrightarrow 2M^+(aq) + 2OH^-(aq) + H_2(g)$$

Several properties of the alkali metals are given in Table 7.8.

We will reconsider this process briefly because it illustrates several important concepts. From the ionization energies, we might expect lithium to be the weakest of the alkali metals as a reducing agent in water. However, the standard reduction potentials indicate that it is the strongest. This reversal results mainly from the very large energy of hydration of the small Li^+ ion. Because of its relatively high charge density, the Li^+ ion very effectively attracts water molecules. A large quantity of energy is released in the process, favoring the formation of the Li^+ ion and making lithium a strong reducing agent in aqueous solution.

We also saw in Section 7.13 that lithium, although it is the strongest reducing agent, reacts more slowly with water than sodium or potassium. From the discussions in Chapters 12 and 17, we know that the *equilibrium position* for a reaction (in this case indicated by the $\mathscr{E}°$ values) is controlled by thermodynamic factors but that the *rate* of a reaction is controlled by kinetic factors. There is *no* direct connection between these factors. Lithium reacts more slowly with water than sodium or potassium because as a solid lithium has a higher melting point than either of the other elements. Since lithium does not become molten from the heat of reaction with water as sodium and potassium do, it has a smaller area of contact with the water.

Figure 20.4 | Lepidolite is composed of mainly lithium, aluminum, silicon, and oxygen, but it also contains significant amounts of rubidium and cesium.

Kevin Webb/NHM Image Resources/SPL/Photo Researchers, Inc.

Table 20.3 | Sources and Methods of Preparation of the Pure Alkali Metals

Element	Source	Method of Preparation
Lithium	Silicate minerals such as spodumene, $LiAl(Si_2O_6)$	Electrolysis of molten LiCl
Sodium	NaCl	Electrolysis of molten NaCl
Potassium	KCl	Electrolysis of molten KCl
Rubidium	Impurity in lepidolite, $Li_2(F,OH)_2Al_2(SiO_3)_3$	Reduction of RbOH with Mg and H_2
Cesium	Pollucite ($Cs_4Al_4Si_9O_{26} \cdot H_2O$) and an impurity in lepidolite (Fig. 20.4)	Reduction of CsOH with Mg and H_2

Sodium reacts violently with water.

Table 20.4 | Selected Physical Properties of the Alkali Metals

Element	Ionization Energy (kJ/mol)	Standard Reduction Potential (V) for $M^+ + e^- \rightarrow M$	Radius of M^+ (pm)	Melting Point (°C)
Lithium	520	−3.05	60	180
Sodium	495	−2.71	95	98
Potassium	419	−2.92	133	64
Rubidium	409	−2.99	148	39
Cesium	382	−3.02	169	29

Table 20.5 summarizes some important reactions of the alkali metals.

The alkali metal ions are very important for the proper functioning of biological systems such as nerves and muscles; Na^+ and K^+ ions are present in all body cells and fluids. In human blood plasma, the concentrations are

$$[Na^+] \approx 0.15\ M \quad \text{and} \quad [K^+] \approx 0.005\ M$$

In the fluids *inside* the cells, the concentrations are reversed:

$$[Na^+] \approx 0.005\ M \quad \text{and} \quad [K^+] \approx 0.16\ M$$

Since the concentrations are so different inside and outside the cells, an elaborate mechanism involving selective ligands is needed to transport Na^+ and K^+ ions through the cell membranes.

Table 20.5 | Selected Reactions of the Alkali Metals

Reaction	Comment
$2M + X_2 \longrightarrow 2MX$	X_2 = any halogen molecule
$4Li + O_2 \longrightarrow 2Li_2O$	Excess oxygen
$2Na + O_2 \longrightarrow Na_2O_2$	
$M + O_2 \longrightarrow MO_2$	M = K, Rb, or Cs
$2M + S \longrightarrow M_2S$	
$6Li + N_2 \longrightarrow 2Li_3N$	Li only
$12M + P_4 \longrightarrow 4M_3P$	
$2M + H_2 \longrightarrow 2MH$	
$2M + 2H_2O \longrightarrow 2MOH + H_2$	
$2M + 2H^+ \longrightarrow 2M^+ + H_2$	Violent reaction!

20.3 | The Chemistry of Hydrogen

Under ordinary conditions of temperature and pressure, hydrogen is a colorless, odorless gas composed of H_2 molecules. Because of its low molar mass and nonpolarity, hydrogen has a very low boiling point (−253°C) and melting point (−260°C). Hydrogen gas is highly flammable; mixtures of air containing from 18% to 60% hydrogen by volume are explosive. In a common lecture demonstration, hydrogen and oxygen gases are bubbled into soapy water. The resulting bubbles are then ignited with a candle on a long stick, producing a loud explosion.

The major industrial source of hydrogen gas is the reaction of methane with water at high temperatures (800–1000°C) and pressures (10–50 atm) in the presence of a metallic catalyst (often nickel):

$$CH_4(g) + H_2O(g) \xrightarrow[\text{Catalyst}]{\text{Heat, pressure}} CO(g) + 3H_2(g)$$

(left) Hydrogen gas being used to blow soap bubbles. (right) As the bubbles float upward, they are lighted by using a candle on a long stick. The orange flame results from the heat of the reaction between hydrogen and oxygen, which excites sodium atoms in the soap bubbles.

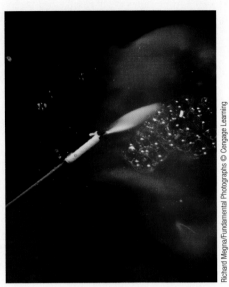

Large quantities of hydrogen are also formed as a by-product of gasoline production, when hydrocarbons with high molecular masses are broken down (or *cracked*) to produce smaller molecules more suitable for use as a motor fuel.

Very pure hydrogen can be produced by the electrolysis of water (see Section 18.7), but this method currently is not economically feasible for large-scale production because of the relatively high cost of electricity.

The major industrial use of hydrogen is in the production of ammonia by the Haber process. Large quantities of hydrogen are also used for hydrogenating unsaturated vegetable oils (those containing carbon–carbon double bonds) to produce solid shortenings that are saturated (containing carbon–carbon single bonds):

$$
\begin{array}{ccc}
\text{H} \quad \text{H} & & \text{H} \quad \text{H} \\
| \quad\ | & & | \quad\ | \\
\sim\!\!\sim\!\!\sim\,\text{C}=\text{C}\,\sim\!\!\sim\!\!\sim & \xrightarrow{\ \text{H}_2\ } & \sim\!\!\sim\!\!\sim\,\text{C}-\text{C}\,\sim\!\!\sim\!\!\sim \\
& & | \quad\ | \\
& & \text{H} \quad \text{H}
\end{array}
$$

The catalysis of this process was discussed in Section 12.8.

Chemically, hydrogen behaves as a typical nonmetal, forming covalent compounds with other nonmetals and forming salts with very active metals. Binary compounds containing hydrogen are called **hydrides**, of which there are three classes. The **ionic** (or **saltlike**) **hydrides** are formed when hydrogen combines with the most active metals, those from Groups 1A and 2A. Examples are LiH and CaH_2, which can best be characterized as containing hydride ions (H^-) and metal cations. Because the presence of two electrons in the small $1s$ orbital produces large electron–electron repulsions and because the nucleus has only a $1+$ charge, the hydride ion is a strong reducing agent. For example, when ionic hydrides are placed in water, a violent reaction takes place. This reaction results in the formation of hydrogen gas, as seen in the equation

$$\text{LiH}(s) + \text{H}_2\text{O}(l) \longrightarrow \text{H}_2(g) + \text{Li}^+(aq) + \text{OH}^-(aq)$$

Covalent hydrides are formed when hydrogen combines with other nonmetals. We have encountered many of these compounds already: HCl, CH_4, NH_3, H_2O, and so on. The most important covalent hydride is water. The polarity of the H_2O molecule leads to many of water's unusual properties. Water has a much higher boiling point than is expected from its molar mass. It has a large heat of vaporization and a large heat capacity, both of which make it a very useful coolant. Water has a higher density as a liquid than as a solid because of the open structure of ice, which results from maximizing the

Figure 20.5 | The structure of ice, showing the hydrogen bonding.

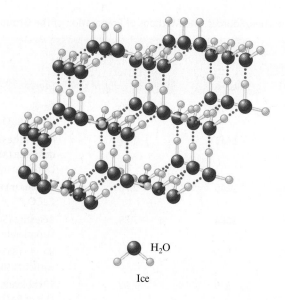

H_2O

Ice

hydrogen bonding (Fig. 20.5). Because water is an excellent solvent for ionic and polar substances, it provides an effective medium for life processes. In fact, water is one of the few covalent hydrides that is nontoxic to organisms.

The third class of hydrides is the **metallic**, or **interstitial**, **hydrides**, which are formed when transition metal crystals are treated with hydrogen gas. The hydrogen molecules dissociate at the metal's surface, and the small hydrogen atoms migrate into the crystal structure to occupy holes, or *interstices*. These metal–hydrogen mixtures are more like solid solutions than true compounds. Palladium can absorb about *900 times* its own volume of hydrogen gas. In fact, hydrogen can be purified by placing it under slight pressure in a vessel containing a thin wall of palladium. The hydrogen diffuses into and through the metal wall, leaving the impurities behind.

Although hydrogen can react with transition metals to form compounds such as UH_3 and FeH_6, most of the interstitial hydrides have variable compositions (often called *nonstoichiometric* compositions) with formulas such as $LaH_{2.76}$ and $VH_{0.56}$. The compositions of the nonstoichiometric hydrides vary with the length of exposure of the metal to hydrogen gas.

When interstitial hydrides are heated, much of the absorbed hydrogen is lost as hydrogen gas. Because of this behavior, these materials offer possibilities for storing hydrogen for use as a portable fuel. The internal combustion engines in current automobiles can burn hydrogen gas with little modification, but storage of enough hydrogen to provide an acceptable mileage range remains a problem. One possible solution might be to use a fuel tank containing a porous solid that includes a transition metal. The hydrogen gas could be pumped into the solid to form the interstitial hydride. The hydrogen gas could then be released when the engine requires additional energy. This system is now being tested by several automobile companies.

Boiling points of covalent hydrides were discussed in Section 10.1.

See Section 6.6 for a discussion of the feasibility of using hydrogen gas as a fuel.

20.4 | The Group 2A Elements

The Group 2A elements (with the valence electron configuration ns^2) are very reactive, losing their two valence electrons to form ionic compounds that contain M^{2+} cations. These elements are commonly called the **alkaline earth metals** because of the basicity of their oxides:

$$MO(s) + H_2O(l) \longrightarrow M^{2+}(aq) + 2OH^-(aq)$$

Table 20.6 | Selected Physical Properties, Sources, and Methods of Preparation of the Group 2A Elements

Element	Radius of M^{2+} (pm)	Ionization Energy (kJ/mol)		$\mathscr{E}°$ (V) for $M^{2+} + 2e^- \longrightarrow M$	Source	Method of Preparation
		First	Second			
Beryllium	≈ 30	900	1760	-1.70	Beryl $(Be_3Al_2Si_6O_{18})$	Electrolysis of molten $BeCl_2$
Magnesium	65	735	1445	-2.37	Magnesite $(MgCO_3)$, dolomite $(MgCO_3 \cdot CaCO_3)$, carnallite $(MgCl_2 \cdot KCl \cdot 6H_2O)$	Electrolysis of molten $MgCl_2$
Calcium	99	590	1146	-2.76	Various minerals containing $CaCO_3$	Electrolysis of molten $CaCl_2$
Strontium	113	549	1064	-2.89	Celestite $(SrSO_4)$, strontianite $(SrCO_3)$	Electrolysis of molten $SrCl_2$
Barium	135	503	965	-2.90	Baryte $(BaSO_4)$, witherite $(BaCO_3)$	Electrolysis of molten $BaCl_2$
Radium	140	509	979	-2.92	Pitchblende (1 g of Ra/7 tons of ore)	Electrolysis of molten $RaCl_2$

2A

| Be |
| Mg |
| Ca |
| Sr |
| Ba |
| Ra |

An amphoteric oxide displays both acidic and basic properties.

Only the amphoteric beryllium oxide (BeO) also shows some acidic properties, such as dissolving in aqueous solutions containing hydroxide ions:

$$BeO(s) + 2OH^-(aq) + H_2O(l) \longrightarrow Be(OH)_4^{2-}(aq)$$

The more active alkaline earth metals react with water as the alkali metals do, producing hydrogen gas:

$$M(s) + 2H_2O(l) \longrightarrow M^{2+}(aq) + 2OH^-(aq) + H_2(g)$$

Calcium, strontium, and barium react vigorously at 25°C. The less easily oxidized beryllium and magnesium show no observable reaction with water at 25°C, although magnesium reacts with boiling water. Table 20.6 summarizes various properties, sources, and methods of preparation of the alkaline earth metals.

The alkaline earth metals have great practical importance. Calcium and magnesium ions are essential for human life. Calcium is found primarily in the structural minerals composing bones and teeth. Magnesium (as the Mg^{2+} ion) plays a vital role in metabolism and in muscle functions. Because magnesium metal has a relatively low density and displays moderate strength, it is a useful structural material, especially if alloyed with aluminum.

Table 20.7 summarizes some important reactions involving the alkaline earth metals.

Calcium metal reacting with water to form bubbles of hydrogen gas.

Table 20.7 | Selected Reactions of the Group 2A Elements

Reaction	Comment
$M + X_2 \longrightarrow MX_2$	X_2 = any halogen molecule
$2M + O_2 \longrightarrow 2MO$	Ba gives BaO_2 as well
$M + S \longrightarrow MS$	
$3M + N_2 \longrightarrow M_3N_2$	High temperatures
$6M + P_4 \longrightarrow 2M_3P_2$	High temperatures
$M + H_2 \longrightarrow MH_2$	M = Ca, Sr, or Ba; high temperatures; Mg at high pressure
$M + 2H_2O \longrightarrow M(OH)_2 + H_2$	M = Ca, Sr, or Ba
$M + 2H^+ \longrightarrow M^{2+} + H_2$	
$Be + 2OH^- + 2H_2O \longrightarrow Be(OH)_4^{2-} + H_2$	

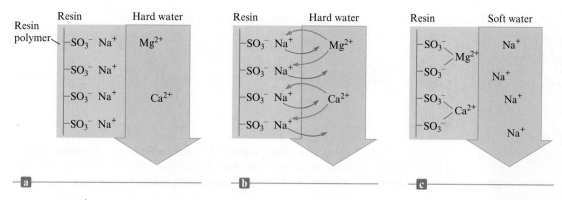

Figure 20.6 | (a) A schematic representation of a typical cation-exchange resin. (b) and (c) When hard water is passed over the cation-exchange resin, the Ca^{2+} and Mg^{2+} bind to the resin.

Relatively large concentrations of Ca^{2+} and Mg^{2+} ions are often found in natural water supplies. These ions in this so-called **hard water** interfere with the action of detergents and form precipitates with soap. In Section 7.6 we saw that Ca^{2+} is often removed by precipitation as $CaCO_3$ in large-scale water softening. In individual homes Ca^{2+}, Mg^{2+}, and other cations are removed by **ion exchange**. An **ion-exchange resin** consists of large molecules (polymers) that have many ionic sites. A cation-exchange resin is represented schematically in Fig. 20.6(a), showing Na^+ ions bound ionically to the SO_3^- groups that are covalently attached to the resin polymer. When hard water is passed over the resin, Ca^{2+} and Mg^{2+} bind to the resin in place of Na^+, which is released into the solution [Fig. 20.6(b)]. Replacing Mg^{2+} and Ca^{2+} by Na^+ [Fig. 20.6(c)] "softens" the water because the sodium salts of soap are soluble.

20.5 | The Group 3A Elements

3A

B
Al
Ga
In
Tl

The Group 3A elements (valence electron configuration ns^2np^1) generally show the increase in metallic character in going down the group that is characteristic of the representative elements. Some physical properties, sources, and methods of preparation of the Group 3A elements are summarized in Table 20.8.

Boron is a typical nonmetal, and most of its compounds are covalent. The most interesting compounds of boron are the covalent hydrides called **boranes**. We might expect BH_3 to be the simplest hydride, since boron has three valence electrons to share with three hydrogen atoms. However, this compound is unstable, and the simplest known member of the series is diborane (B_2H_6), with the structure shown in Fig. 20.7(a). In this

Table 20.8 | Selected Physical Properties, Sources, and Methods of Preparation of the Group 3A Elements

Element	Radius of M^{3+} (pm)	Ionization Energy (kJ/mol)	$\mathscr{E}°$ (V) for $M^{3+} + 3e^- \longrightarrow M$	Source	Method of Preparation
Boron	20	798	—	Kernite, a form of borax ($Na_2B_4O_7 \cdot 4H_2O$)	Reduction by Mg or H_2
Aluminum	50	581	−1.66	Bauxite (Al_2O_3)	Electrolysis of Al_2O_3 in molten Na_3AlF_6
Gallium	62	577	−0.53	Traces in various minerals	Reduction with H_2 or electrolysis
Indium	81	556	−0.34	Traces in various minerals	Reduction with H_2 or electrolysis
Thallium	95	589	0.72	Traces in various minerals	Electrolysis

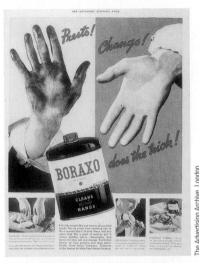

An old ad from The Saturday Evening Post *for Boraxo, a hand-cleaning product containing sodium tetraborate (NaB_4O_7). Extensive natural deposits of borax ($Na_2B_4O_7 \cdot 10H_2O$) found in saline lakes near Death Valley, California, were hauled to a factory in wagons pulled by teams of 20 mules—hence the name 20 Mule Team Borax.*

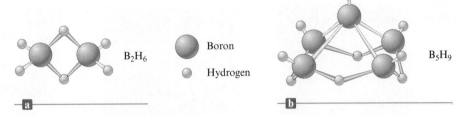

Figure 20.7 | (a) The structure of B_2H_6 with its two three-center B—H—B bridging bonds and four "normal" B—H bonds. (b) The structure of B_5H_9. There are five "normal" B—H bonds to terminal hydrogens and four three-center bridging bonds around the base.

molecule the terminal B—H bonds are normal covalent bonds, each involving one electron pair. The bridging bonds are three-center bonds similar to those in solid BeH_2. Another interesting borane contains the square pyramidal B_5H_9 molecule [Fig. 20.7(b)], which has four three-center bonds situated around the base of the pyramid. Because the boranes are extremely electron-deficient, they are highly reactive. The boranes react very exothermically with oxygen and were once evaluated as potential fuels for rockets in the U.S. space program.

Aluminum, the most abundant metal on earth, has metallic physical properties, such as high thermal and electrical conductivities and a lustrous appearance; however, its bonds to nonmetals are significantly covalent. This covalency is responsible for the amphoteric nature of Al_2O_3, which dissolves in acidic or basic solution, and for the acidity of $Al(H_2O)_6^{3+}$ (see Section 14.8):

$$Al(H_2O)_6^{3+}(aq) \rightleftharpoons Al(OH)(H_2O)_5^{2+}(aq) + H^+(aq)$$

One especially interesting property of *gallium* is its unusually low melting point of 29.8°C, which is in contrast to the 660°C melting point of aluminum. Also, since gallium's boiling point is about 2400°C, it has the largest liquid range of any metal. This makes it useful for thermometers, especially to measure high temperatures. Gallium, like water, expands when it freezes. The chemistry of gallium is quite similar to that of aluminum. For example, Ga_2O_3 is amphoteric.

The chemistry of *indium* is similar to that of aluminum and gallium except that compounds containing the 1+ ion are known, such as InCl and In_2O, in addition to those with the more common 3+ ion. The chemistry of *thallium* is completely metallic.

Table 20.9 summarizes some important reactions of the Group 3A elements.

Gallium melts in the hand.

Table 20.9 | Selected Reactions of the Group 3A Elements

Reaction	Comment
$2M + 3X_2 \longrightarrow 2MX_3$	X_2 = any halogen molecule; Tl gives TlX as well, but no TlI_3
$4M + 3O_2 \longrightarrow 2M_2O_3$	High temperatures; Tl gives Tl_2O as well
$2M + 3S \longrightarrow M_2S_3$	High temperatures; Tl gives Tl_2S as well
$2M + N_2 \longrightarrow 2MN$	M = Al only
$2M + 6H^+ \longrightarrow 2M^{3+} + 3H_2$	M = Al, Ga, or In; Tl gives Tl^+
$2M + 2OH^- + 6H_2O \longrightarrow 2M(OH)_4^- + 3H_2$	M = Al or Ga

Table 20.10 | Selected Physical Properties, Sources, and Methods of Preparation of the Group 4A Elements

Element	Electronegativity	Melting Point (°C)	Boiling Point (°C)	Source	Method of Preparation
Carbon	2.6	3727 (sublimes)	—	Graphite, diamond, petroleum, coal	—
Silicon	1.9	1410	2355	Silicate minerals, silica	Reduction of K_2SiF_6 with Al, or reduction of SiO_2 with Mg
Germanium	2.0	937	2830	Germinate (mixture of copper, iron, and germanium sulfides)	Reduction of GeO_2 with H_2 or C
Tin	2.0	232	2270	Cassiterite (SnO_2)	Reduction of SnO_2 with C
Lead	2.3	327	1740	Galena (PbS)	Roasting of PbS with O_2 to form PbO_2 and then reduction with C

20.6 | The Group 4A Elements

4A

C

Si

Ge

Sn

Pb

Group 4A (with the valence electron configuration ns^2np^2) contains two of the most important elements on earth: carbon, the fundamental constituent of the molecules necessary for life, and silicon, which forms the basis of the geological world. The change from nonmetallic to metallic properties seen in Group 3A is also apparent in going down Group 4A from carbon, a typical nonmetal, to silicon and germanium, usually considered semimetals, to the metals tin and lead. Table 20.10 summarizes some physical properties, sources, and methods of preparation of the elements in this group.

All the Group 4A elements can form four covalent bonds to nonmetals—for example, CH_4, SiF_4, $GeBr_4$, $SnCl_4$, and $PbCl_4$. In each of these tetrahedral molecules, the central atom is described as sp^3 hybridized by the localized electron model.

We have seen that carbon also differs markedly from the other members of Group 4A in its ability to form π bonds. This accounts for the completely different structures and properties of CO_2 and SiO_2. Note from Table 20.11 that C—C bonds and Si—O bonds are stronger than Si—Si bonds. This partly explains why the chemistry of carbon is dominated by C—C bonds, whereas that of silicon is dominated by Si—O bonds.

Carbon occurs in the allotropic forms graphite, diamond, and fullerenes, whose structures were given in Section 10.5. The most important chemistry of carbon is organic chemistry, which is described in detail in Chapter 22.

Silicon, the second most abundant element in the earth's crust, is a semimetal found widely distributed in silica and silicates (see Section 10.5). About 85% of the earth's crust is composed of these substances. Although silicon is found in some steel and aluminum alloys, its major use is in semiconductors for electronic devices (see Chapter 10).

Germanium, a relatively rare element, is a semimetal used mainly in the manufacture of semiconductors for transistors and similar electronic devices.

Tin is a soft, silvery metal that can be rolled into thin sheets (tin foil) and has been used for centuries in various alloys such as bronze (20% Sn and 80% Cu), solder (33% Sn and 67% Pb), and pewter (85% Sn, 7% Cu, 6% Bi, and 2% Sb). Tin exists as three allotropes: *white tin,* stable at normal temperatures; *gray tin,* stable at temperatures below 13.2°C; and *brittle tin,* found at temperatures above 161°C. When tin is exposed to low temperatures, it gradually changes to powdery gray tin and crumbles away; this is known as *tin disease.*

Currently, tin is used mainly as a protective coating for steel, especially for cans used as food containers. The thin layer of tin, applied electrolytically, forms a protective oxide coating that prevents further corrosion.

Lead is easily obtained from its ore, galena (PbS). Because lead melts at such a low temperature, it may have been the first pure metal obtained from its ore. We know that

A new form of elemental carbon, the fullerenes, was discussed in Chapter 2.

Table 20.11 | Strengths of C—C, Si—Si, and Si—O Bonds

Bond	Bond Energy (kJ/mol)
C—C	347
Si—Si	340
Si—O	368

Chemical connections

Beethoven: Hair Is the Story

Ludwig van Beethoven, arguably the greatest composer who ever lived, led a troubled life fraught with sickness, deafness, and personality aberrations. Now we may know the source of these difficulties: lead poisoning. Scientists have recently reached this conclusion through analysis of Beethoven's hair. When Beethoven died in 1827 at age 56, many mourners took samples of the great man's hair. In fact, it was said at the time that he was practically bald by the time he was buried. The hair that was recently analyzed consisted of 582 strands—3 to 6 inches long—bought for the Center of Beethoven Studies for $7300 in 1994 from Sotheby's auction house in London.

According to William Walsh of the Health Research Institute (HRI) in suburban Chicago, Beethoven's hair showed a lead concentration 100 times the normal levels. The scientists concluded that Beethoven's exposure to lead came as an adult, possibly from the mineral water he drank and swam in when he visited spas.

The lead poisoning may well explain Beethoven's volatile temper—the composer was subject to towering rages and sometimes had the look of a wild animal. In rare cases lead poisoning has been known to cause deafness, but the researchers remain unsure if this problem led to Beethoven's hearing loss.

According to Walsh, the scientists at HRI were originally looking for mercury, a common treatment for syphilis in the early nineteenth century, in Beethoven's hair. The absence of mercury supports the consensus of scholars that Beethoven did not have this disease. Not surprisingly, Beethoven himself wanted to know what made him so ill. In a letter to his brothers in 1802, he asked them to have doctors find the cause of his frequent abdominal pain after his death.

Portrait of Beethoven by Josef Karl Stieler.

The Granger Collection, New York

Roman baths such as these in Bath, England, used lead pipes for water.

Michael Holford

Table 20.12 | Selected Reactions of the Group 4A Elements

Reaction	Comment
$M + 2X_2 \longrightarrow MX_4$	X_2 = any halogen molecule; M = Ge or Sn; Pb gives PbX_2
$M + O_2 \longrightarrow MO_2$	M = Ge or Sn; high temperatures; Pb gives PbO or Pb_3O_4
$M + 2H^+ \longrightarrow M^{2+} + H_2$	M = Sn or Pb

lead was used as early as 3000 B.C. by the Egyptians. It was later used by the Romans to make eating utensils, glazes on pottery, and even intricate plumbing systems. The Romans also prepared a sweetener called *sapa* by boiling down grape juice in lead-lined vessels. The sweetness of this syrup was partly caused by the formation of lead(II) acetate (formerly called sugar of lead), a very sweet-tasting compound. The problem with these practices is that lead is very toxic. In fact, the Romans had so much contact with lead that it may have contributed to the demise of their civilization. Analysis of bones from that era shows significant levels of lead.

Although lead poisoning has been known since at least the second century B.C., lead continues to be a problem. For example, many children have been exposed to lead by eating chips of lead-based paint. Because of this problem, lead-based paints are no longer used for children's furniture, and many states have banned lead-based paint for interior use. Lead poisoning can also occur when acidic foods and drinks leach the lead from lead-glazed pottery dishes that were improperly fired and when liquor is stored in leaded crystal decanters, producing toxic levels of lead in the drink in a relatively short time. In addition, the widespread use of tetraethyl lead [$(C_2H_5)_4Pb$] as an antiknock agent in gasoline has increased the lead levels in our environment. Concern about the effects of this lead pollution has caused the U.S. government to require the gradual replacement of the lead in gasoline with other antiknock agents. The largest commercial use of lead (over one million tons annually) is for electrodes in the lead storage batteries used in automobiles (see Section 18.5).

Table 20.12 summarizes some important reactions of the Group 4A elements.

20.7 | The Group 5A Elements

5A

| N |
| P |
| As |
| Sb |
| Bi |

The Group 5A elements (with the valence electron configuration ns^2np^3), which are prepared as shown in Table 20.13, exhibit remarkably varied chemical properties. As usual, metallic character increases going down the group, as is apparent from the electronegativity values (Table 20.13). Nitrogen and phosphorus are nonmetals that can gain three electrons to form $3-$ anions in salts with active metals; examples are magnesium nitride (Mg_3N_2) and beryllium phosphide (Be_3P_2). The chemistry of these two important elements is discussed in the next two sections.

Bismuth and *antimony* tend to be metallic, readily losing electrons to form cations. Although these elements have five valence electrons, so much energy is required to remove all five that no ionic compounds containing Bi^{5+} or Sb^{5+} ions are known.

The Group 5A elements can form molecules or ions that involve three, five, or six covalent bonds to the Group 5A atom. Examples involving three single bonds are NH_3, PH_3, NF_3, and $AsCl_3$. Each of these molecules has a lone pair of electrons (and thus can behave as a Lewis base) and a pyramidal shape as predicted by the VSEPR model (see Fig. 20.8).

All the Group 5A elements except nitrogen can form molecules with five covalent bonds (of general formula MX_5). Nitrogen cannot form such molecules because of its small size. The MX_5 molecules have a trigonal bipyramidal shape (see Fig. 20.9)

Figure 20.8 | The pyramidal shape of the Group 5A MX_3 molecules.

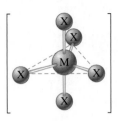

Figure 20.9 | The trigonal bipyramidal shape of the MX_5 molecules.

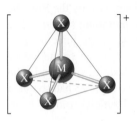

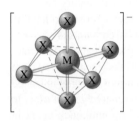

Figure 20.10 | The structures of the tetrahedral MX_4^+ and the octahedral MX_6^- ions.

Table 20.13 | Selected Physical Properties, Sources, and Methods of Preparation of the Group 5A Elements

Element	Electronegativity	Source	Method of Preparation
Nitrogen	3.0	Air	Liquefaction of air
Phosphorus	2.2	Phosphate rock [$Ca_3(PO_4)_2$], fluorapatite [$Ca_5(PO_4)_3F$]	$2Ca_3(PO_4)_2 + 6SiO_2 \longrightarrow 6CaSiO_3 + P_4O_{10}$ $P_4O_{10} + 10C \longrightarrow 4P + 10CO$
Arsenic	2.2	Arsenopyrite (Fe_3As_2, FeS)	Heating arsenopyrite in the absence of air
Antimony	2.1	Stibnite (Sb_2S_3)	Roasting Sb_2S_3 in air to form Sb_2O_3 and then reduction with carbon
Bismuth	2.0	Bismite (Bi_2O_3), bismuth glance (Bi_2S_3)	Roasting Bi_2S_3 in air to form Bi_2O_3 and then reduction with carbon

as predicted by the VSEPR model, and the central atom can be described as dsp^3 hybridized.

Although the MX_5 molecules have a trigonal bipyramidal structure in the gas phase, the solids of many of these compounds contain a 1:1 mixture of the ions MX_4^+ and MX_6^- (Fig. 20.10). The MX_4^+ cation is tetrahedral (the atom represented by M is sp^3 hybridized), and the MX_6^- anion is octahedral (the atom represented by M is d^2sp^3 hybridized). Examples are PCl_5 (which in the solid state contains PCl_4^+ and PCl_6^-) and AsF_3Cl_2 (which in the solid state contains $AsCl_4^+$ and AsF_6^-).

As discussed in Section 20.1, the ability of the Group 5A elements to form π bonds decreases dramatically after nitrogen. This explains why elemental nitrogen exists as N_2 molecules containing two π bonds, whereas the other elements in the group exist as larger aggregates containing single bonds. For example, in the gas phase the elements phosphorus, arsenic, and antimony consist of P_4, As_4, and Sb_4 molecules, respectively.

20.8 | The Chemistry of Nitrogen

At the earth's surface virtually all elemental nitrogen exists as the N_2 molecule with its very strong triple bond (941 kJ/mol). Because of this large bond strength, the N_2 molecule is so unreactive that it can coexist with most other elements under normal conditions without undergoing any appreciable reaction. This property makes nitrogen gas very useful as a medium for experiments involving substances that react with oxygen or water. Such experiments can be done using an inert atmosphere box of the type shown in Fig. 20.11.

The strength of the triple bond in the N_2 molecule is important both thermodynamically and kinetically. Thermodynamically, the great stability of the $N \equiv N$ bond means that most binary compounds containing nitrogen decompose exothermically to the elements, for example:

$$N_2O(g) \longrightarrow N_2(g) + \tfrac{1}{2}O_2(g) \qquad \Delta H° = -82 \text{ kJ}$$
$$NO(g) \longrightarrow \tfrac{1}{2}N_2(g) + \tfrac{1}{2}O_2(g) \qquad \Delta H° = -90 \text{ kJ}$$
$$NO_2(g) \longrightarrow \tfrac{1}{2}N_2(g) + O_2(g) \qquad \Delta H° = -34 \text{ kJ}$$
$$N_2H_4(g) \longrightarrow N_2(g) + 2H_2(g) \qquad \Delta H° = -95 \text{ kJ}$$
$$NH_3(g) \longrightarrow \tfrac{1}{2}N_2(g) + \tfrac{3}{2}H_2(g) \qquad \Delta H° = +46 \text{ kJ}$$

Figure 20.11 | An inert atmosphere box used when working with oxygen- or water-sensitive materials. The box is filled with an inert gas such as nitrogen, and work is done through the ports fitted with large rubber gloves.

Paul Ridgway/Lawrence Livermore National Laboratory

Of these compounds, only ammonia is thermodynamically more stable than its component elements. That is, only for ammonia is energy required to decompose the molecule to its elements. For the remaining molecules, energy is released when decomposition to the elements occurs, as a result of the great stability of N_2.

The importance of the thermodynamic stability of N_2 can be clearly seen in the power of nitrogen-based explosives, such as nitroglycerin ($C_3H_5N_3O_9$), which has the following skeletal structure:

$$\text{(skeletal structure of nitroglycerin)}$$

When ignited or subjected to sudden impact, nitroglycerin decomposes very rapidly and exothermically:

$$4C_3H_5N_3O_9(l) \longrightarrow 6N_2(g) + 12CO_2(g) + 10H_2O(g) + O_2(g) + \text{energy}$$

An explosion occurs; that is, large volumes of gas are produced in a fast, highly exothermic reaction. Note that 4 moles of liquid nitroglycerin produce 29 (6 + 12 + 10 + 1) moles of gaseous products. This alone produces a large increase in volume. However, also note that the products, which include N_2, are very stable molecules with strong bonds. Their formation is therefore accompanied by the release of large quantities of energy as heat, which increases the gaseous volume. The hot, rapidly expanding gases produce a pressure surge and damaging shock wave.

Most high explosives are organic compounds that, like nitroglycerin, contain nitro (—NO_2) groups and produce nitrogen and other gases as products. Another example is *trinitrotoluene* (TNT), a solid at normal temperatures, which decomposes as follows:

$$2C_7H_5N_3O_6(s) \longrightarrow 12CO(g) + 5H_2(g) + 3N_2(g) + 2C(s) + \text{energy}$$

Note that 2 moles of solid TNT produce 20 moles of gaseous products plus energy.

The effect of bond strength on the kinetics of reactions involving the N_2 molecule is illustrated by the synthesis of ammonia from nitrogen and hydrogen, a reaction we

TNT

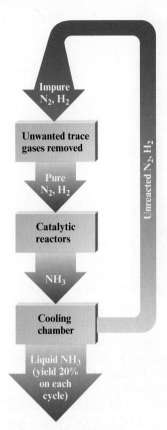

Impure
N_2, H_2

Unwanted trace
gases removed

Pure
N_2, H_2

Catalytic
reactors

NH_3

Cooling
chamber

Liquid NH_3
(yield 20%
on each
cycle)

Unreacted N_2, H_2

Figure 20.12 | A schematic diagram of the Haber process for the manufacture of ammonia.

Nodules on the roots of pea plants contain nitrogen-fixing bacteria.

Hugh Spencer/Photo Researchers, Inc.

have discussed many times before. Because a large quantity of energy is required to disrupt the $N\equiv N$ bond, the ammonia synthesis reaction occurs at a negligible rate at room temperature, even though the equilibrium constant is very large ($K \approx 10^8$). Of course, the most direct way to increase the rate of a reaction is to raise the temperature. However, since this reaction is very exothermic,

$$N_2(g) + 3H_2(g) \longrightarrow 2NH_3(g) \qquad \Delta H^\circ = -92 \text{ kJ}$$

the value of K decreases significantly with a temperature increase (at 500°C, $K \approx 10^{-2}$).

Obviously, the kinetics and the thermodynamics of this reaction are in opposition. A compromise must be reached, involving high pressure to force the equilibrium to the right and high temperature to produce a reasonable rate. The **Haber process** for manufacturing ammonia represents such a compromise (Fig. 20.12). The process is carried out at a pressure of about 250 atm and a temperature of approximately 400°C. Even higher temperatures would be required if a catalyst consisting of a solid iron oxide mixed with small amounts of potassium oxide and aluminum oxide were not used to facilitate the reaction.

Nitrogen is essential to living systems. The problem with nitrogen is not one of supply—we are surrounded by it—but rather one of changing it from the inert N_2 molecule to a form usable by plants and animals. The process of transforming N_2 to other nitrogen-containing compounds is called **nitrogen fixation**. The Haber process is one example of nitrogen fixation. The ammonia produced can be applied to the soil as a fertilizer, since plants can readily use the nitrogen in ammonia to make the nitrogen-containing biomolecules essential for their growth.

Nitrogen fixation also results from the high-temperature combustion process in automobile engines. The nitrogen in the air is drawn into the engine and reacts at a significant rate with oxygen to form nitric oxide (NO), which further reacts with oxygen from the air to form nitrogen dioxide (NO_2). This nitrogen dioxide, which contributes to photochemical smog in many urban areas (see Section 12.8), reacts with moisture in the air and eventually reaches the soil to form nitrate salts, which are plant nutrients.

Nitrogen fixation also occurs naturally. For example, lightning provides the energy to disrupt N_2 and O_2 molecules in the air, producing highly reactive nitrogen and oxygen atoms. These atoms in turn attack other N_2 and O_2 molecules to form nitrogen oxides that eventually become nitrates. Although lightning has traditionally been credited with forming about 10% of the total fixed nitrogen, recent studies indicate that lightning may account for as much as half of the fixed nitrogen available on earth. Another natural nitrogen fixation process involves bacteria that reside in the root nodules of plants such as beans, peas, and alfalfa. These **nitrogen-fixing bacteria** readily allow the conversion of nitrogen to ammonia and to other nitrogen-containing compounds useful to plants. The efficiency of these bacteria is intriguing: They produce ammonia at soil temperatures and 1 atm pressure, whereas the Haber process requires severe conditions of 400°C and 250 atm. For obvious reasons, researchers are studying these bacteria intensively.

When plants and animals die and decompose, the elements they consist of are returned to the environment. In the case of nitrogen, the return of the element to the atmosphere as nitrogen gas, called **denitrification**, is carried out by bacteria that change nitrates to nitrogen. The complex **nitrogen cycle** is summarized in Fig. 20.13. It has been estimated that as much as 10 million tons more nitrogen per year is currently being fixed by natural and human processes than is being returned to the atmosphere. This fixed nitrogen is accumulating in soil, lakes, rivers, and oceans, where it promotes the growth of algae and other undesirable organisms.

Nitrogen Hydrides

By far the most important hydride of nitrogen is **ammonia**. A toxic, colorless gas with a pungent odor, ammonia is manufactured in huge quantities (approximately 40 billion pounds per year), mainly for use in fertilizers.

Figure 20.13 | The nitrogen cycle. To be used by plants and animals, nitrogen must be converted from N_2 to nitrogen-containing compounds, such as nitrates, ammonia, and proteins. The nitrogen is returned to the atmosphere by natural decay processes.

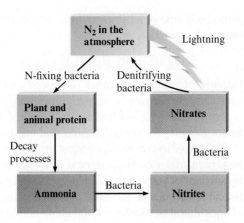

The pyramidal ammonia molecule has a lone pair of electrons on its nitrogen atom (see Fig. 20.8) and polar N—H bonds. This structure leads to a high degree of intermolecular interaction by hydrogen bonding in the liquid state, thereby producing an unusually high boiling point ($-33.4°C$) for a substance with such a low molar mass. Note, however, that the hydrogen bonding in liquid ammonia is clearly not as important as that in liquid water, which has about the same molar mass but a much higher boiling point. The water molecule has two polar bonds involving hydrogen and two lone pairs—the right combination for optimum hydrogen bonding—in contrast to the one lone pair and three polar bonds of the ammonia molecule.

As we saw in Chapter 14, ammonia behaves as a base, reacting with acids to produce ammonium salts. For example,

$$NH_3(g) + HCl(g) \longrightarrow NH_4Cl(s)$$

A second nitrogen hydride of major importance is **hydrazine** (N_2H_4). The Lewis structure of hydrazine

indicates that each nitrogen atom should be sp^3 hybridized with bond angles close to 109.5° (the tetrahedral angle), since the nitrogen atom is surrounded by four electron pairs. The observed structure with bond angles of 112° (Fig. 20.14) agrees reasonably well with these predictions. Hydrazine, a colorless liquid with an ammoniacal odor, freezes at 2°C and boils at 113.5°C. This boiling point is quite high for a compound with a molar mass of 32; this suggests that considerable hydrogen bonding occurs among the polar hydrazine molecules.

Hydrazine is a powerful reducing agent and has been widely used as a rocket propellant. For example, its reaction with oxygen is highly exothermic:

$$N_2H_4(l) + O_2(g) \longrightarrow N_2(g) + 2H_2O(g) \qquad \Delta H° = -622 \text{ kJ}$$

Since hydrazine also reacts vigorously with the halogens, fluorine is often used instead of oxygen as the oxidizer in rocket engines. Substituted hydrazines, where one or more of the hydrogen atoms are replaced by other groups, are also useful rocket fuels. For example, monomethylhydrazine,

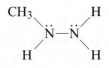

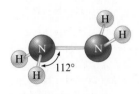

Figure 20.14 | The molecular structure of hydrazine (N_2H_4). This arrangement minimizes the repulsion between the lone pairs on the nitrogen atoms by placing them on opposite sides.

Blowing agents—such as hydrazine, which forms nitrogen gas on decomposition—are used to produce porous plastics like these polystyrene products.

is used with the oxidizer dinitrogen tetroxide (N_2O_4) to power the U.S. space shuttle orbiter. The reaction is

$$5N_2O_4(l) + 4N_2H_3(CH_3)(l) \longrightarrow 12H_2O(g) + 9N_2(g) + 4CO_2(g)$$

Because of the large number of gaseous molecules produced and the exothermic nature of this reaction, a very high thrust per mass of fuel is achieved. The reaction is also self-starting—it begins immediately when the fuels are mixed—which is a useful property for rocket engines that must be started and stopped frequently.

The use of hydrazine as a rocket propellant is a rather specialized application. The main industrial use of hydrazine is as a "blowing" agent in the manufacture of plastics. Hydrazine decomposes to form nitrogen gas, which causes foaming in the liquid plastic and results in a porous texture. Another major use of hydrazine is in the production of agricultural pesticides. Of the many hundreds of hydrazine derivatives (substituted hydrazines) that have been tested, 40 are used as fungicides, herbicides, insecticides, or plant growth regulators.

Nitrogen Oxides

Nitrogen forms a series of oxides in which its oxidation state ranges from +1 to +5, as shown in Table 20.14.

Dinitrogen monoxide (N_2O), more commonly called *nitrous oxide* or laughing gas, has an inebriating effect and has been used as a mild anesthetic by dentists. Because of its high solubility in fats, nitrous oxide is widely used as a propellant in aerosol cans of whipped cream. It is dissolved in the liquid inside the can at high pressure and forms bubbles that produce foaming as the liquid is released from the can. A significant

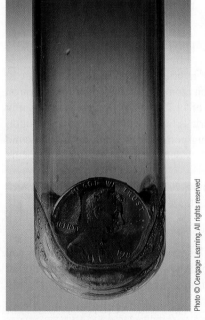

A copper penny reacts with nitric acid to produce NO gas, which is immediately oxidized in air to give reddish brown NO_2.

Table 20.14 | Some Common Nitrogen Compounds

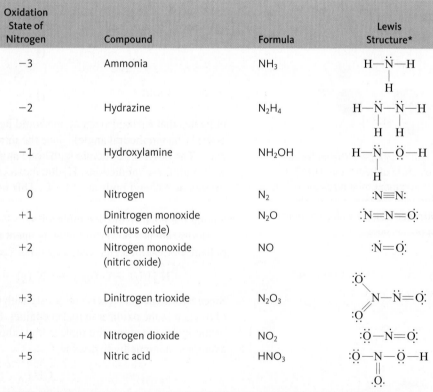

Oxidation State of Nitrogen	Compound	Formula	Lewis Structure*
−3	Ammonia	NH_3	
−2	Hydrazine	N_2H_4	
−1	Hydroxylamine	NH_2OH	
0	Nitrogen	N_2	
+1	Dinitrogen monoxide (nitrous oxide)	N_2O	
+2	Nitrogen monoxide (nitric oxide)	NO	
+3	Dinitrogen trioxide	N_2O_3	
+4	Nitrogen dioxide	NO_2	
+5	Nitric acid	HNO_3	

*In some cases additional resonance structures are needed to fully describe the electron distribution.

Figure 20.15 | The molecular orbital energy-level diagram for nitric oxide (NO). The bond order is 2.5, or $(8 - 3)/2$.

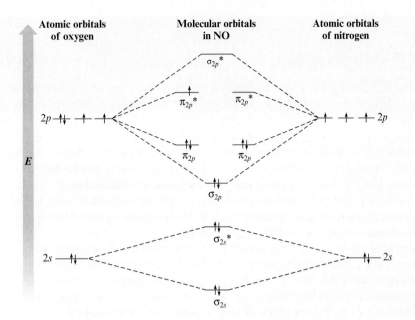

amount of N_2O exists in the atmosphere, mostly produced by soil microorganisms, and its concentration appears to be gradually increasing. Because it can strongly absorb infrared radiation, nitrous oxide plays a small but probably significant role in controlling the earth's temperature in the same way that atmospheric carbon dioxide and water vapor do (see the discussion of the greenhouse effect in Section 6.5). Some scientists fear that the rapid decrease of tropical rain forests resulting from the development of countries such as Brazil will significantly affect the rate of production of N_2O by soil organisms and thus will have important effects on the earth's temperature.

Nitrogen monoxide (NO), commonly called *nitric oxide*, has been found to be an important regulator in biological systems. Nitric oxide is a colorless gas under normal conditions and can be produced in the laboratory by reacting $6\ M$ nitric acid with copper metal:

$$8H^+(aq) + 2NO_3^-(aq) + 3Cu(s) \longrightarrow 3Cu^{2+}(aq) + 4H_2O(l) + 2NO(g)$$

When this reaction is carried out in the air, the nitric oxide is immediately oxidized by O_2 to reddish brown nitrogen dioxide (NO_2).

Since the NO molecule has an odd number of electrons, it is most conveniently described in terms of the molecular orbital model. The molecular orbital energy-level diagram is shown in Fig. 20.15. Note that the NO molecule should be paramagnetic and have a bond order of 2.5, predictions that are supported by experimental observations. Since the NO molecule has one high-energy electron, it is not surprising that it can be rather easily oxidized to form NO^+, the *nitrosyl ion*. Because an antibonding electron is removed in going from NO to NO^+, the resulting ion should have a stronger bond (the predicted bond order is 3) than the molecule. This is borne out by experiment. The bond lengths and bond energies for nitric oxide and the nitrosyl ion are shown in Table 20.15.

Nitric oxide is thermodynamically unstable and decomposes to nitrous oxide and nitrogen dioxide:

$$3NO(g) \longrightarrow N_2O(g) + NO_2(g)$$

Nitrogen dioxide (NO_2), which is also an odd-electron molecule, has a V-shaped structure. The reddish brown, paramagnetic NO_2 molecule readily dimerizes to form dinitrogen tetroxide,

$$2NO_2(g) \rightleftharpoons N_2O_4(g)$$

Table 20.15 | Comparison of the Bond Lengths and Bond Energies for Nitric Oxide and the Nitrosyl Ion

	NO	NO$^+$
Bond length (Å)	1.15	1.09
Bond energy (kJ/mol)	630	1020
Bond order (predicted by MO model)	2.5	3

Chemical connections
Nitrous Oxide: Laughing Gas That Propels Whipped Cream and Cars

Nitrous oxide (N_2O), more properly called dinitrogen monoxide, is a compound with many interesting uses. It was discovered in 1772 by Joseph Priestley (who is also given credit for discovering oxygen gas), and its intoxicating effects were noted almost immediately. In 1798, the 20-year-old Humphry Davy became director of the Pneumatic Institute, which was set up to investigate the medical effects of various gases. Davy tested the effects of N_2O on himself, reporting that after inhaling 16 quarts of the gas in 7 minutes, he became "absolutely intoxicated."

Over the next century "laughing gas," as nitrous oxide became known,

was developed as an anesthetic, particularly for dental procedures. Nitrous oxide is still used as an anesthetic, although it has been primarily replaced by more modern drugs.

One major use of nitrous oxide today is as the propellant in cans of "instant" whipped cream. The high solubility of N_2O in the whipped cream mixture makes it an excellent candidate for pressurizing the cans of whipping cream.

Another current use of nitrous oxide is to produce "instant horse-power" for street racers. Because the reaction of N_2O with O_2 to form NO actually absorbs heat, this reaction has

a cooling effect when placed in the fuel mixture in an automobile engine. This cooling effect lowers combustion temperatures, thus allowing the fuel–air mixture to be significantly more dense (the density of a gas is inversely proportional to temperature). The effect can produce a burst of additional power in excess of 200 horsepower. Because engines are not designed to run steadily at such high power levels, the nitrous oxide is injected from a tank when extra power is desired.

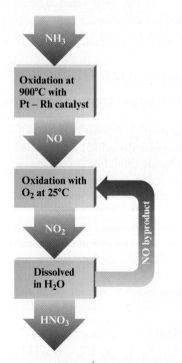

Figure 20.16 | The Ostwald process.

which is diamagnetic and colorless. The value of the equilibrium constant is approximately 1 for this process at 55°C, and since the dimerization is exothermic, K decreases as the temperature increases.

The least common of the nitrogen oxides are *dinitrogen trioxide* (N_2O_3), a blue liquid that readily dissociates to gaseous nitric oxide and nitrogen dioxide, and *dinitrogen pentoxide* (N_2O_5), which under normal conditions is a solid that is best viewed as a mixture of NO_2^+ and NO_3^- ions. Although N_2O_5 molecules can exist in the gas phase, they readily dissociate to nitrogen dioxide and oxygen:

$$2N_2O_5(g) \rightleftharpoons 4NO_2(g) + O_2(g)$$

This reaction follows first-order kinetics, as discussed in Section 12.4.

Oxyacids of Nitrogen

Nitric acid is an important industrial chemical (approximately 8 million tons produced annually) used in the manufacture of many products, such as nitrogen-based explosives and ammonium nitrate for use as a fertilizer.

Nitric acid is produced commercially by the oxidation of ammonia in the **Ostwald process** (Fig. 20.16). In the first step of this process, ammonia is oxidized to nitric oxide:

$$4NH_3(g) + 5O_2(g) \longrightarrow 4NO(g) + 6H_2O(g) \qquad \Delta H° = -905 \text{ kJ}$$

Although this reaction is highly exothermic, it is very slow at 25°C. A side reaction occurs between nitric oxide and ammonia:

$$4NH_3(g) + 6NO(g) \longrightarrow 5N_2(g) + 6H_2O(g)$$

Figure 20.17 | (a) The molecular structure of HNO₃. (b) The resonance structures of HNO₃.

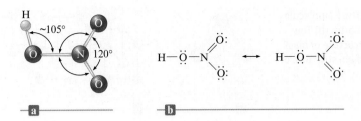

which is particularly undesirable because it traps the nitrogen in the very unreactive N₂ molecules. The desired reaction can be accelerated and the effects of the competing reaction can be minimized if the ammonia oxidation is carried out by using a catalyst of a platinum–rhodium alloy heated to 900°C. Under these conditions, there is a 97% conversion of the ammonia to nitric oxide.

In the second step, nitric oxide is reacted with oxygen to produce nitrogen dioxide:

$$2NO(g) + O_2(g) \longrightarrow 2NO_2(g) \qquad \Delta H° = -113 \text{ kJ}$$

This oxidation reaction has a rate that *decreases* with increasing temperature. Because of this very unusual behavior, the reaction is carried out at approximately 25°C and is kept at this temperature by cooling with water.

The third step in the Ostwald process is the absorption of nitrogen dioxide by water:

$$3NO_2(g) + H_2O(l) \longrightarrow 2HNO_3(aq) + NO(g) \qquad \Delta H° = -139 \text{ kJ}$$

The gaseous NO produced in the reaction is recycled so that it can be oxidized to NO₂. The aqueous nitric acid from this process is about 50% HNO₃ by mass, which can be increased to 68% by distillation to remove some of the water. The maximum concentration attainable by this method is 68% because nitric acid and water form an *azeotrope* at this concentration. The solution can be further concentrated to 95% HNO₃ by treatment with concentrated sulfuric acid, which strongly absorbs water; H₂SO₄ is often used as a *dehydrating (water-removing) agent*.

Nitric acid is a colorless, fuming liquid (bp = 83°C) with a pungent odor; it decomposes in sunlight by the following reaction:

$$4HNO_3(l) \xrightarrow{h\nu} 4NO_2(g) + 2H_2O(l) + O_2(g)$$

As a result, nitric acid turns yellow as it ages because of the dissolved nitrogen dioxide. The common laboratory reagent called *concentrated nitric acid* is 15.9 *M* HNO₃ (70.4% HNO₃ by mass) and is a very strong oxidizing agent. The resonance structures and molecular structure of HNO₃ are shown in Fig. 20.17. Note that the hydrogen is bound to an oxygen atom rather than to nitrogen as the formula might suggest.

Nitrous acid (HNO₂) is a weak acid,

$$HNO_2(aq) \rightleftharpoons H^+(aq) + NO_2^-(aq) \qquad K_a = 4.0 \times 10^{-4}$$

that forms pale yellow nitrite (NO₂⁻) salts. In contrast to nitrates, which are often used as explosives, nitrites are quite stable even at high temperatures.

An *azeotrope* is a solution that, like a pure liquid, distills at a constant temperature without a change in composition.

20.9 | The Chemistry of Phosphorus

Although phosphorus lies directly below nitrogen in Group 5A of the periodic table, its chemical properties are significantly different from those of nitrogen. The differences arise mainly from four factors: nitrogen's ability to form much stronger π bonds, the greater electronegativity of nitrogen, the larger size of phosphorus atoms, and the potential availability of empty valence *d* orbitals on phosphorus.

Figure 20.18 | (a) The P$_4$ molecule found in white phosphorus. (b) The crystalline network structure of black phosphorus. (c) The chain structure of red phosphorus.

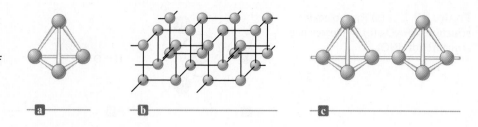

White phosphorus reacts vigorously with the oxygen in air and must be stored under water. Red phosphorus is stable in air.

Charles D. Winters

The chemical differences between nitrogen and phosphorus are apparent in their elemental forms. In contrast to the diatomic form of elemental nitrogen, which is stabilized by strong π bonds, there are several solid forms of phosphorus that all contain aggregates of atoms. *White phosphorus,* which contains discrete tetrahedral P$_4$ molecules [Fig. 20.18(a)], is very reactive; it bursts into flames on contact with air (it is said to be *pyrophoric*). Consequently, white phosphorus is commonly stored under water. White phosphorus is quite toxic; the P$_4$ molecules are very damaging to tissue, particularly the cartilage and bones of the nose and jaw. The much less reactive forms, called *black phosphorus* and *red phosphorus,* are network solids (see Section 10.5). Black phosphorus has a regular crystalline structure [Fig. 20.18(b)], but red phosphorus is amorphous and is thought to consist of chains of P$_4$ units [Fig. 20.18(c)]. Red phosphorus can be obtained by heating white phosphorus in the absence of air at 1 atm. Black phosphorus is obtained from either white or red phosphorus by heating at high pressures.

Even though phosphorus has a lower electronegativity than nitrogen, it will form phosphides (ionic substances containing the P^{3-} anion) such as Na$_3$P and Ca$_3$P$_2$. Phosphide salts react vigorously with water to produce *phosphine* (PH$_3$), a toxic, colorless gas:

$$2Na_3P(s) + 6H_2O(l) \longrightarrow 2PH_3(g) + 6Na^+(aq) + 6OH^-(aq)$$

Phosphine is analogous to ammonia, although it is a much weaker base ($K_b \approx 10^{-26}$) and is much less soluble in water.

Phosphine has the Lewis structure

$$\text{H}-\overset{\displaystyle ..}{\text{P}}-\text{H}$$
$$\underset{\text{H}}{|}$$

and a pyramidal molecular structure, as we would predict from the VSEPR model. However, it has bond angles of 94° rather than 107°, as found in the ammonia molecule. The reasons for this are complex; therefore, we will simply regard phosphine as an exception to the simple version of the VSEPR model that we use.

Phosphorus Oxides and Oxyacids

The terminal oxygens are the nonbridging oxygen atoms.

Phosphorus reacts with oxygen to form oxides in which its oxidation states are +5 and +3. The oxide P$_4$O$_6$ is formed when elemental phosphorus is burned in a limited supply of oxygen, and P$_4$O$_{10}$ is produced when the oxygen is in excess. Picture these oxides (shown in Fig. 20.19) as being constructed by adding oxygen atoms to the fundamental P$_4$ structure. The intermediate states, P$_4$O$_7$, P$_4$O$_8$, and P$_4$O$_9$, which contain one, two, and three terminal oxygen atoms, respectively, are also known.

Tetraphosphorus decoxide (P$_4$O$_{10}$), which was formerly represented as P$_2$O$_5$ and called phosphorus pentoxide, has a great affinity for water and thus is a powerful dehydrating agent. For example, it can be used to convert HNO$_3$ and H$_2$SO$_4$ to their parent oxides, N$_2$O$_5$ and SO$_3$, respectively.

Figure 20.19 | The structures of P_4O_6 and P_4O_{10}.

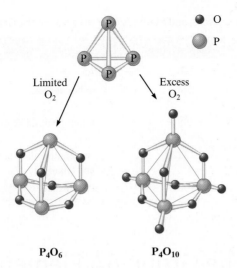

When tetraphosphorus decoxide dissolves in water, **phosphoric acid** (H_3PO_4), also called **orthophosphoric acid**, is produced:

$$P_4O_{10}(s) + 6H_2O(l) \longrightarrow 4H_3PO_4(aq)$$

Pure phosphoric acid is a white solid that melts at 42°C. Aqueous phosphoric acid is a much weaker acid ($K_{a_1} \approx 10^{-2}$) than nitric acid or sulfuric acid and is a poor oxidizing agent.

When the oxide P_4O_6 is placed in water, **phosphorous acid** (H_3PO_3) is formed [Fig. 20.20(a)]. Although the formula suggests a triprotic acid, phosphorous acid is a *diprotic* acid. The hydrogen atom bonded directly to the phosphorus atom is not acidic in aqueous solution; only those hydrogen atoms bonded to the oxygen atoms in H_3PO_3 can be released as protons.

A third oxyacid of phosphorus is *hypophosphorous acid* (H_3PO_2) [Fig. 20.20(b)], which is a monoprotic acid.

Phosphorus in Fertilizers

Phosphorus is essential for plant growth. Although most soil contains large amounts of phosphorus, it is often present in insoluble minerals, making it inaccessible to the plants. Soluble phosphate fertilizers are manufactured by treating phosphate rock with sulfuric acid to make **superphosphate of lime**, a mixture of $CaSO_4 \cdot 2H_2O$ and $Ca(H_2PO_4)_2 \cdot H_2O$. If phosphate rock is treated with phosphoric acid, $Ca(H_2PO_4)_2$, known as *triple phosphate,* is produced. The reaction of ammonia with phosphoric acid gives *ammonium dihydrogen phosphate* ($NH_4H_2PO_4$), a very efficient fertilizer that furnishes both phosphorus and nitrogen.

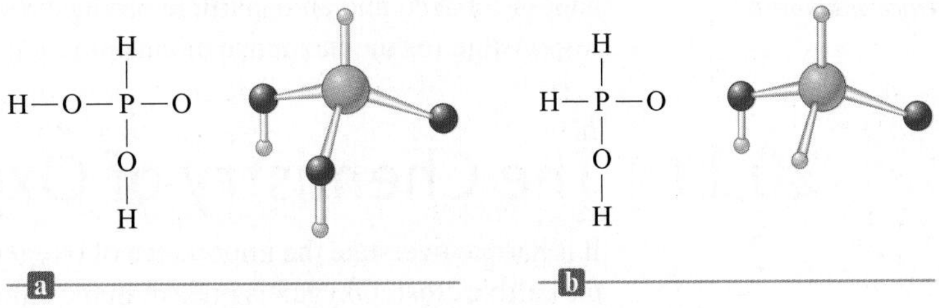

Figure 20.20 | (a) The structure of phosphorous acid (H_3PO_3). (b) The structure of hypophosphorous acid (H_3PO_2).

Table 20.16 | Selected Physical Properties, Sources, and Methods of Preparation of the Group 6A Elements

Element	Electronegativity	Radius of X^{2-} (pm)	Source	Method of Preparation
Oxygen	3.4	140	Air	Distillation from liquid air
Sulfur	2.6	184	Sulfur deposits	Melted with hot water and pumped to the surface
Selenium	2.6	198	Impurity in sulfide ores	Reduction of H_2SeO_4 with SO_2
Tellurium	2.1	221	Nagyagite (mixed sulfide and telluride)	Reduction of ore with SO_2
Polonium	2.0	230	Pitchblende	

20.10 | The Group 6A Elements

6A

O

S

Se

Te

Po

Although in Group 6A (Table 20.16) there is the usual tendency for metallic properties to increase going down the group, none of the Group 6A elements behaves as a typical metal. The most common chemical behavior of a Group 6A element involves reacting with a metal to achieve a noble gas electron configuration by adding two electrons to become a 2− anion in ionic compounds. In fact, for most metals, the oxides and sulfides constitute the most common minerals.

The Group 6A elements can form covalent bonds with other nonmetals. For example, they combine with hydrogen to form a series of covalent hydrides of the general formula H_2X. Those members of the group that have valence d orbitals available (all except oxygen) commonly form molecules in which they are surrounded by more than eight electrons. Examples are SF_4, SF_6, TeI_4, and $SeBr_4$.

The two heaviest members of Group 6A can lose electrons to form cations. Although they do not lose all six valence electrons because of the high energies that would be required, tellurium and polonium appear to exhibit some chemistry involving their 4+ cations. However, the chemistry of these Group 6A cations is much more limited than that of the Group 5A elements bismuth and antimony.

In recent years there has been a growing interest in the chemistry of selenium, an element found throughout the environment in trace amounts. Selenium's toxicity has long been known, but some medical studies have shown an *inverse* relationship between the incidence of cancer and the selenium levels in soil. It has been suggested that the greater dietary intake of selenium by people living in areas with relatively high selenium levels somehow furnishes protection from cancer. These studies are only preliminary, but selenium is definitely known to be physiologically important (it is involved in the activity of vitamin E and certain enzymes). Selenium (as well as tellurium) is also a semiconductor and therefore finds some application in the electronics industry.

Walnuts contain trace amounts of selenium.

Polonium was discovered in 1898 by Marie and Pierre Curie in their search for the sources of radioactivity in pitchblende. Polonium has 27 isotopes and is highly toxic and very radioactive. It has been suggested that the isotope ^{210}Po, a natural contaminant of tobacco and an α-particle producer (see Section 19.1), might be at least partly responsible for the incidence of cancer in smokers.

20.11 | The Chemistry of Oxygen

It is hard to overstate the importance of oxygen, the most abundant element in and near the earth's crust. Oxygen is present in the atmosphere as oxygen gas and ozone; in soil and rocks in oxide, silicate, and carbonate minerals; in the oceans in water; and in our bodies in water and a myriad of other molecules. In addition, most of the energy we need to live and run our civilization comes from the exothermic reactions of oxygen with carbon-containing molecules.

The most common elemental form of oxygen (O_2) constitutes 21% of the volume of the earth's atmosphere. Since nitrogen has a lower boiling point than oxygen, nitrogen can be boiled away from liquid air, leaving oxygen and small amounts of argon, another component of air. Liquid oxygen is a pale blue liquid that freezes at $-219°C$ and boils at $-183°C$. The paramagnetism of the O_2 molecule can be demonstrated by pouring liquid oxygen between the poles of a strong magnet, where it "sticks" until it boils away (see Fig. 9.39). The paramagnetism of the O_2 molecule can be accounted for by the molecular orbital model (Fig. 9.38), which also explains its bond strength.

The other form of elemental oxygen is **ozone** (O_3), a molecule that can be represented by the resonance structures

$$\ddot{O}{=}\ddot{O}{-}\ddot{\underset{\cdot\cdot}{O}}\colon \longleftrightarrow \colon\ddot{\underset{\cdot\cdot}{O}}{-}\ddot{O}{=}\ddot{O}$$

The bond angle in the O_3 molecule is 117°, in reasonable agreement with the prediction of the VSEPR model (three effective pairs require a trigonal planar arrangement). That the bond angle is slightly less than 120° can be explained by concluding that more space is required for the lone pair than for the bonding pairs.

Ozone can be prepared by passing an electrical discharge through pure oxygen gas. The electrical energy disrupts the bonds in some O_2 molecules, thereby producing oxygen atoms, which react with other O_2 molecules to form O_3. Ozone is much less stable than oxygen at 25°C and 1 atm. For example, $K \approx 10^{-56}$ for the equilibrium

$$3O_2(g) \rightleftharpoons 2O_3(g)$$

A pale blue, highly toxic gas, ozone is a much more powerful oxidizing agent than oxygen. The strong oxidizing power of ozone makes it useful for killing bacteria in swimming pools, hot tubs, and aquariums. It is also increasingly being used in municipal water treatment and for washing produce after it comes out of the fields. One of the main advantages of using ozone for water purification is that it does not leave potentially toxic residues behind. On the other hand, chlorine, which is widely used for water purification, leaves residues of chloro compounds, such as chloroform ($CHCl_3$), which may cause cancer after long-term exposure. Although ozone effectively kills the bacteria in water, one problem with **ozonolysis** is that the water supply is not protected against recontamination, since virtually no ozone remains after the initial treatment. In contrast, for chlorination, significant residual chlorine remains after treatment.

The oxidizing ability of ozone can be detrimental, especially when it is present in the pollution from automobile exhausts (see Section 5.10).

Ozone exists naturally in the upper atmosphere of the earth. The *ozone layer* is especially important because it absorbs ultraviolet light and thus acts as a screen to prevent this radiation, which can cause skin cancer, from penetrating to the earth's surface. When an ozone molecule absorbs this energy, it splits into an oxygen molecule and an oxygen atom:

$$O_3 \xrightarrow{\ h\nu\ } O_2 + O$$

If the oxygen molecule and atom collide, they will not stay together as ozone unless a "third body," such as a nitrogen molecule, is present to help absorb the energy released by bond formation. The third body absorbs this energy as kinetic energy; its temperature is increased. Therefore, the energy originally absorbed as ultraviolet radiation is eventually changed to thermal energy. Thus the ozone prevents the harmful high-energy ultraviolet light from reaching the earth.

A U.S. Navy test pilot in his F/A-18F Super Hornet wearing an oxygen mask.

U.S. Navy photo by Cmdr. Ian C. Anderson

Scientists have become concerned that Freons and nitrogen dioxide are promoting the destruction of the ozone layer (see Section 12.8).

Figure 20.21 | The Frasch method for recovering sulfur from underground deposits.

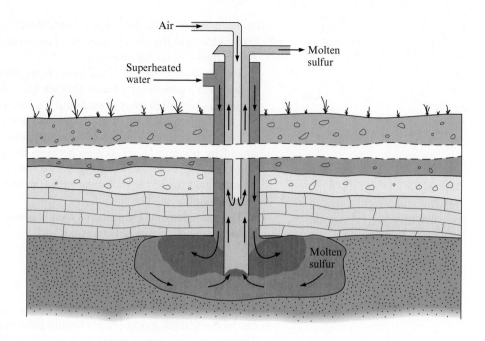

Air

Molten sulfur

Superheated water

Molten sulfur

The mineral cinnabar

Charles D. Winters

20.12 | The Chemistry of Sulfur

Sulfur is found in nature both in large deposits of the free element and in widely distributed ores, such as galena (PbS), cinnabar (HgS), pyrite (FeS_2), gypsum ($CaSO_4 \cdot 2H_2O$), epsomite ($MgSO_4 \cdot 7H_2O$), and glauberite ($Na_2SO_4 \cdot CaSO_4$).

About 60% of the sulfur produced in the United States comes from the underground deposits of elemental sulfur found in Texas and Louisiana. This sulfur is recovered by using the **Frasch process** developed by Herman Frasch in the 1890s. Superheated water is pumped into the deposit to melt the sulfur (mp = 113°C), which is then forced to the surface by air pressure (Fig. 20.21). The remaining 40% of sulfur produced in the United States either is a by-product of the purification of fossil fuels before combustion to prevent pollution or comes from the sulfur dioxide (SO_2) scrubbed from the exhaust gases when sulfur-containing fuels are burned.

In contrast to oxygen, elemental sulfur exists as S_2 molecules only in the gas phase at high temperatures. Because sulfur atoms form much stronger σ bonds than π bonds, S_2 is less stable at 25°C than larger aggregates such as S_6 and S_8 rings and S_n chains (Fig. 20.22). The most stable form of sulfur at 25°C and 1 atm is called *rhombic sulfur* [Fig. 20.23(a)], which contains stacked S_8 rings. If rhombic sulfur is melted and heated to 120°C, it forms *monoclinic sulfur* as it slowly cools [Fig. 20.23(b)]. The monoclinic form also contains S_8 rings, but the rings are stacked differently than in rhombic sulfur.

Sulfur obtained from underground deposits by the Frasch process.

Kevin Burke/Corbis

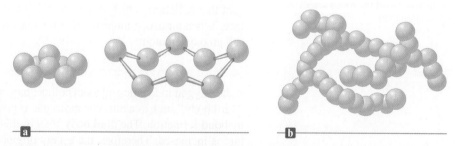

Figure 20.22 | (a) The S_8 molecule. (b) Chains of sulfur atoms in viscous liquid sulfur. The chains may contain as many as 10,000 atoms.

Figure 20.23 | (a) Crystals of rhombic sulfur. (b) Crystals of monoclinic sulfur.

Sulfur Oxides

The scrubbing of sulfur dioxide from exhaust gases was discussed in Section 5.10.

From its position below oxygen in the periodic table, we might expect the simplest stable oxide of sulfur to have the formula SO. However, *sulfur monoxide,* which can be produced in small amounts when gaseous sulfur dioxide (SO_2) is subjected to an electrical discharge, is very unstable. The difference in the stabilities of the O_2 and SO molecules probably reflects the much stronger π bonding between oxygen atoms than between a sulfur and an oxygen atom.

Sulfur burns in air with a bright blue flame to give *sulfur dioxide* (SO_2), a colorless gas with a pungent odor, which condenses to a liquid at $-10°C$ and 1 atm. Sulfur dioxide is a V-shaped molecule, which is a very effective antibacterial agent often used to preserve stored fruit.

Sulfur dioxide reacts with oxygen to produce *sulfur trioxide* (SO_3):

$$2SO_2(g) + O_2(g) \longrightarrow 2SO_3(g)$$

However, this reaction is very slow in the absence of a catalyst. One of the mysteries during early research on air pollution was how the sulfur dioxide produced from the combustion of sulfur-containing fuels is so rapidly converted to sulfur trioxide in the atmosphere. It is now known that dust and other particles can act as heterogeneous catalysts for this process (see Section 12.8).

Oxyacids of Sulfur

Sulfur dioxide dissolves in water to form an acidic solution. The reaction is often represented as

$$SO_2(g) + H_2O(l) \longrightarrow H_2SO_3(aq)$$

where H_2SO_3 is called *sulfurous acid.* However, very little H_2SO_3 actually exists in the solution. The major form of sulfur dioxide in water is SO_2, and the acid dissociation equilibria are best represented as

$$SO_2(aq) + H_2O(l) \rightleftharpoons H^+(aq) + HSO_3^-(aq) \qquad K_{a_1} = 1.5 \times 10^{-2}$$
$$HSO_3^-(aq) \rightleftharpoons H^+(aq) + SO_3^{2-}(aq) \qquad K_{a_2} = 1.0 \times 10^{-7}$$

This situation is analogous to the behavior of carbon dioxide in water (see Section 14.7). Although H_2SO_3 cannot be isolated, salts of SO_3^{2-} (*sulfites*) and HSO_3^- (*hydrogen sulfites*) are well known.

Figure 20.24 | The reaction of H_2SO_4 with sucrose (left) to produce a blackened column of carbon (right).

Sulfur trioxide reacts violently with water to produce the diprotic acid **sulfuric acid**:

$$SO_3(g) + H_2O(l) \longrightarrow H_2SO_4(aq)$$

Manufactured in greater amounts than any other chemical, sulfuric acid is usually produced by the *contact process*. About 60% of the sulfuric acid manufactured in the United States is used to produce fertilizers from phosphate rock. The other 40% is used in lead storage batteries, in petroleum refining, in steel manufacturing, and for various other purposes in the chemical industry.

Because sulfuric acid has a high affinity for water, it is often used as a dehydrating agent. Gases that do not react with sulfuric acid, such as oxygen, nitrogen, and carbon dioxide, are often dried by bubbling them through concentrated solutions of the acid. Sulfuric acid is such a powerful dehydrating agent that it will remove hydrogen and oxygen from a substance in a 2:1 ratio even when the substance contains no molecular water. For example, concentrated sulfuric acid reacts vigorously with common table sugar (sucrose), leaving a charred mass of carbon (Fig. 20.24):

$$\underset{\text{Sucrose}}{C_{12}H_{22}O_{11}(s)} + 11H_2SO_4(conc) \longrightarrow 12C(s) + 11H_2SO_4 \cdot H_2O(l)$$

20.13 | The Group 7A Elements

7A
F
Cl
Br
I
At

In our coverage of the representative elements we have progressed from the groups of metallic elements (Groups 1A and 2A), through groups in which the lighter members are nonmetals and the heavier members are metals (Groups 3A, 4A, and 5A), to a group containing all nonmetals (Group 6A—although some might prefer to call polonium a metal). The Group 7A elements, the **halogens** (with the valence electron configuration ns^2np^5), are all nonmetals whose properties generally vary smoothly going down the group. The only notable exceptions are the unexpectedly low value for the electron affinity of fluorine and the unexpectedly small bond energy of the F_2 molecule (see Section 20.1). Table 20.17 summarizes the trends in some physical properties of the halogens.

Because of their high reactivities, the halogens are not found as free elements in nature. Instead, they are found as halide ions (X^-) in various minerals and in seawater (Table 20.18).

Although astatine is a member of Group 7A, its chemistry is of no practical importance because all its known isotopes are radioactive. The longest-lived isotope, ^{210}At, has a half-life of only 8.3 hours.

The halogens, particularly fluorine, have very high electronegativity values (see Table 20.17). They tend to form polar covalent bonds with other nonmetals and ionic bonds with metals in their lower oxidation states. When a metal ion is in a higher oxidation state, such as +3 or +4, the metal–halogen bonds are polar and covalent. For example, $TiCl_4$ and $SnCl_4$ are both covalent compounds that are liquids under normal conditions.

Samples of chlorine gas, liquid bromine, and solid iodine.

Table 20.17 | Trends in Selected Physical Properties of the Group 7A Elements

Element	Electronegativity	Radius of X^- (pm)	$\mathscr{E}°$ (V) for $X_2 + 2e \rightarrow 2X^-$	Bond Energy of X_2 (kJ/mol)
Fluorine	4.0	136	2.87	154
Chlorine	3.2	181	1.36	239
Bromine	3.0	195	1.09	193
Iodine	2.7	216	0.54	149
Astatine	2.2	—	—	—

Table 20.18 | Some Physical Properties, Sources, and Methods of Preparation of the Group 7A Elements

Element	Color and State	Percentage of Earth's Crust	Melting Point (°C)	Boiling Point (°C)	Source	Method of Preparation
Fluorine	Pale yellow gas	0.07	−220	−188	Fluorospar (CaF_2), cryolite (Na_3AlF_6), fluorapatite [$Ca_5(PO_4)_3F$]	Electrolysis of molten KHF_2
Chlorine	Yellow-green gas	0.14	−101	−34	Rock salt (NaCl), halite (NaCl), sylvite (KCl)	Electrolysis of aqueous NaCl
Bromine	Red-brown liquid	2.5×10^{-4}	−7.3	59	Seawater, brine wells	Oxidation of Br^- by Cl_2
Iodine	Violet-black solid	3×10^{-5}	113	184	Seaweed, brine wells	Oxidation of I^- by electrolysis or MnO_2

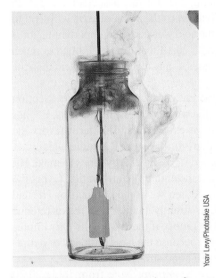

A candle burning in an atmosphere of $Cl_2(g)$. The exothermic reaction, which involves breaking C—C and C—H bonds in the wax and forming C—Cl bonds in their places, produces enough heat to make the gases in the region incandescent (a flame results).

Yoav Levy/Phototake USA

Hydrogen Halides

The hydrogen halides can be prepared by a reaction of the elements

$$H_2(g) + X_2(g) \longrightarrow 2HX(g)$$

This reaction occurs with explosive vigor when fluorine and hydrogen are mixed. On the other hand, hydrogen and chlorine can coexist with little apparent reaction for relatively long periods in the dark. However, ultraviolet light causes an explosively fast reaction, and this is the basis of a popular lecture demonstration, the "hydrogen–chlorine cannon." Bromine and iodine also react with hydrogen, but more slowly.

Some physical properties of the hydrogen halides are listed in Table 20.19. Note the very high boiling point for hydrogen fluoride, which results from extensive hydrogen bonding among the very polar HF molecules (Fig. 20.25). Fluoride ion has such a high affinity for protons that in concentrated aqueous solutions of hydrogen fluoride, the ion [F---H---F]$^-$ exists, in which an H$^+$ ion is centered between two F$^-$ ions.

When dissolved in water, the hydrogen halides behave as acids, and all except hydrogen fluoride are completely dissociated. Because water is a much stronger base than the Cl$^-$, Br$^-$, or I$^-$ ion, the acid strengths of HCl, HBr, and HI cannot be differentiated in water. However, in a less basic solvent, such as glacial (pure) acetic acid, the acids show different strengths:

$$H—I > H—Br > H—Cl \gg H—F$$

Strongest acid   Weakest acid

Figure 20.25 | The hydrogen bonding among HF molecules in liquid hydrogen fluoride.

Table 20.19 | Some Physical Properties of the Hydrogen Halides

HX	Melting Point (°C)	Boiling Point (°C)	H—X Bond Energy (kJ/mol)
HF	−83	20	565
HCl	−114	−85	427
HBr	−87	−67	363
HI	−51	−35	295

To see why hydrogen fluoride is the only weak acid in water among the HX molecules, let's consider the dissociation equilibrium,

$$HX(aq) \rightleftharpoons H^+(aq) + X^-(aq) \qquad \text{where} \qquad K_a = \frac{[H^+][X^-]}{[HX]}$$

from a thermodynamic point of view. Recall that acid strength is reflected by the magnitude of K_a—a small K_a value means a weak acid. Also recall that the value of an equilibrium constant is related to the standard free energy change for the reaction,

$$\Delta G° = -RT \ln(K)$$

As $\Delta G°$ becomes more negative, K becomes larger; a *decrease* in free energy favors a given reaction. As we saw in Chapter 17, free energy depends on enthalpy, entropy, and temperature. For a process at constant temperature,

$$\Delta G° = \Delta H° - T\Delta S°$$

Thus, to explain the various acid strengths of the hydrogen halides, we must focus on the factors that determine $\Delta H°$ and $\Delta S°$ for the acid dissociation reaction.

What energy terms are important in determining $\Delta H°$ for the dissociation of HX in water? (Keep in mind that large, positive contributions to the value of $\Delta H°$ will tend to make $\Delta G°$ more highly positive, K_a smaller, and the acid weaker.) One important factor is certainly the H—X bond strength. Note from Table 20.19 that the H—F bond is much stronger than the other H—X bonds. This factor tends to make HF a weaker acid than the others.

Another important contribution to $\Delta H°$ is the enthalpy of hydration (see Section 11.2) of X^- (Table 20.20). As we would expect, the smallest of the halide ions, F^-, has the most negative value—its hydration is the most exothermic. This term favors the dissociation of HF into its ions more so than it does for the other HX molecules.

So far we have two conflicting factors: The large HF bond energy tends to make HF a weaker acid than the other hydrogen halides, but the enthalpy of hydration favors the dissociation of HF more than that of the others. When we compare data for HF and HCl, the difference in bond energy (138 kJ/mol) is slightly smaller than the difference in the enthalpies of hydration for the anions (144 kJ/mol). If these were the *only* important factors, HF should be a stronger acid than HCl because the large enthalpy of hydration of F^- more than compensates for the large HF bond strength.

As it turns out, the *deciding factor appears to be entropy*. Note from Table 20.20 that the entropy of hydration for F^- is much more negative than the entropy of hydration for the other halides because of the high degree of ordering that occurs as the water molecules associate with the small F^- ion. Remember that a negative change in entropy is unfavorable. Thus, although the enthalpy of hydration favors dissociation of HF, the *entropy* of hydration strongly opposes it.

When all these factors are taken into account, $\Delta G°$ for the dissociation of HF in water is positive; that is, K_a is small. In contrast, $\Delta G°$ for dissociation of the other HX molecules in water is negative (K_a is large). This example illustrates the complexity of the processes that occur in aqueous solutions and the importance of entropy effects in that medium.

In practical terms, **hydrochloric acid** is the most important of the **hydrohalic acids**, the aqueous solutions of the hydrogen halides. About 3 million tons of hydrochloric acid are produced annually for use in cleaning steel before galvanizing and in the manufacture of many other chemicals.

Hydrofluoric acid is used to etch glass by reacting with the silica in glass to form the volatile gas SiF_4:

$$SiO_2(s) + 4HF(aq) \longrightarrow SiF_4(g) + 2H_2O(l)$$

Table 20.20 | The Enthalpies and Entropies of Hydration for the Halide Ions

$X^-(g) \xrightarrow{H_2O} X^-(aq)$		
X^-	$\Delta H°$ (kJ/mol)	$\Delta S°$ (J/K mol)
F^-	−510	−159
Cl^-	−366	−96
Br^-	−334	−81
I^-	−291	−64

Hydration becomes more exothermic as the charge density of an ion increases. Thus, for ions of a given charge, the smallest is most strongly hydrated.

When H_2O molecules cluster around an ion, an ordering effect occurs; thus $\Delta S°_{hyd}$ is negative.

Stomach acid is 0.1 M HCl.

Table 20.21 | The Known Oxyacids of the Halogens

Oxidation State of Halogen	Fluorine	Chlorine	Bromine	Iodine*	General Name of Acids	General Name of Salts
+1	HOF†	HOCl	HOBr	HOI	Hypohalous acid	Hypohalites, MOX
+3	‡	HOClO	‡	‡	Halous acid	Halites, MXO$_2$
+5	‡	HOClO$_2$	HOBrO$_2$	HOIO$_2$	Halic acid	Halates, MXO$_3$
+7	‡	HOClO$_3$	HOBrO$_3$	HOIO$_3$	Perhalic acid	Perhalates, MXO$_4$

*Iodine also forms $H_4I_2O_9$ (mesodiperiodic acid) and H_5IO_6 (paraperiodic acid).
†HOF oxidation state is best represented as −1.
‡Compound is unknown.

Oxyacids and Oxyanions

All the halogens except fluorine combine with various numbers of oxygen atoms to form a series of oxyacids, as shown in Table 20.21. The strengths of these acids vary in direct proportion to the number of oxygen atoms attached to the halogen, with the acid strength increasing as more oxygens are added.

The only member of the chlorine series that has been obtained in the pure state is *perchloric acid* ($HOClO_3$), a strong acid and a powerful oxidizing agent. Because perchloric acid reacts explosively with many organic materials, it must be handled with great caution. The other oxyacids of chlorine are known only in solution, although salts containing their anions are well known (Fig. 20.26).

Hypochlorous acid (HOCl) is formed when chlorine gas is dissolved in cold water:

$$Cl_2(aq) + H_2O(l) \rightleftharpoons HOCl(aq) + H^+(aq) + Cl^-(aq)$$

Note that in this reaction chlorine is both oxidized (from 0 in Cl_2 to +1 in HOCl) and reduced (from 0 in Cl_2 to −1 in Cl^-). Such a reaction, *in which a given element is both oxidized and reduced,* is called a **disproportionation reaction**. Hypochlorous acid and its salts are strong oxidizing agents; their solutions are widely used as household bleaches and disinfectants.

Chlorate salts, such as $KClO_3$, are also strong oxidizing agents and are used as weed killers and as oxidizers in fireworks (see Chapter 7) and explosives.

Fluorine forms only one oxyacid, hypofluorous acid (HOF), but it forms at least two oxides. When fluorine gas is bubbled into a dilute solution of sodium hydroxide, the compound *oxygen difluoride* (OF_2) is formed:

$$4F_2(g) + 3H_2O(l) \longrightarrow 6HF(aq) + OF_2(g) + O_2(g)$$

The name for OF_2 is oxygen difluoride rather than difluorine oxide because fluorine has a higher electronegativity than oxygen and thus is named as the anion.

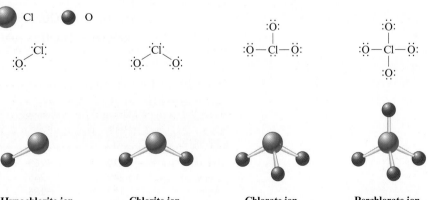

Figure 20.26 | The structures of the oxychloro anions.

Hypochlorite ion, OCl⁻   Chlorite ion, ClO$_2^-$ Chlorate ion, ClO$_3^-$ Perchlorate ion, ClO$_4^-$

Oxygen difluoride is a pale yellow gas (bp $= -145°C$) that is a strong oxidizing agent. The oxide *dioxygen difluoride* (O_2F_2) is an orange solid that can be prepared by an electric discharge in an equimolar mixture of fluorine and oxygen gases:

$$F_2(g) + O_2(g) \xrightarrow[\text{discharge}]{\text{Electric}} O_2F_2(s)$$

20.14 | The Group 8A Elements

8A

He

Ne

Ar

Kr

Xe

Rn

The Group 8A elements, the **noble gases**, are characterized by filled *s* and *p* valence orbitals (electron configurations of $2s^2$ for helium and ns^2np^6 for the others). Because of their completed valence shells, these elements are very unreactive. In fact, no noble gas compounds were known prior to 1962. Selected properties of the Group 8A elements are summarized in Table 20.22.

Helium was identified by its characteristic emission spectrum as a component of the sun before it was found on earth. The major sources of helium on earth are natural gas deposits, where helium was formed from the α-particle decay of radioactive elements. The α particle is a helium nucleus that can easily pick up electrons from the environment to form a helium atom. Although helium forms no compounds, it is an important substance that is used as a coolant, as a pressurizing gas for rocket fuels, as a diluent in the gases used for deep-sea diving and spaceship atmospheres, and as the gas in lighter-than-air airships (blimps).

Like helium, *neon* forms no compounds, but it is a very useful element. For example, neon is widely used in luminescent lighting (neon signs). *Argon,* which recently has been shown to form chemical bonds under special circumstances, is used to provide the noncorrosive atmosphere in incandescent light bulbs, which prolongs the life of the tungsten filament.

Krypton and *xenon* have been observed to form many stable chemical compounds. The first of these was prepared in 1962 by Neil Bartlett (1932-2008), an English chemist who made an ionic compound that he thought had the formula $XePtF_6$. Subsequent studies indicated that the compound might be better represented as $XeFPtF_6$ and contains the XeF^+ and PtF_6^- ions.

Less than a year after Bartlett's report, a group at Argonne National Laboratory near Chicago prepared xenon tetrafluoride by reacting xenon and fluorine gases in a nickel reaction vessel at $400°C$ and 6 atm:

$$Xe(g) + 2F_2(g) \longrightarrow XeF_4(s)$$

Xenon tetrafluoride forms stable colorless crystals. Two other xenon fluorides, XeF_2 and XeF_6, were synthesized by the group at Argonne, and a highly explosive xenon oxide (XeO_3) was also found. The xenon fluorides react with water to form hydrogen fluoride and oxycompounds. For example:

$$XeF_6(s) + 3H_2O(l) \longrightarrow XeO_3(aq) + 6HF(aq)$$
$$XeF_6(s) + H_2O(l) \longrightarrow XeOF_4(aq) + 2HF(aq)$$

AP Photo/Donna McWilliam

Neon signmaker and artist Jess Baird shows off a few of the items he has made in his Weatherford, Texas, shop.

Table 20.22 | Selected Properties of Group 8A Elements

Element	Melting Point (°C)	Boiling Point (°C)	Atmospheric Abundance (% by volume)	Examples of Compounds
Helium	-270	-269	5×10^{-4}	None
Neon	-249	-246	1×10^{-3}	None
Argon	-189	-186	9×10^{-1}	HArF
Krypton	-157	-153	1×10^{-4}	KrF_2
Xenon	-112	-107	9×10^{-6}	XeF_4, XeO_3, XeF_6

Figure 20.27 | The structures of several known xenon compounds.

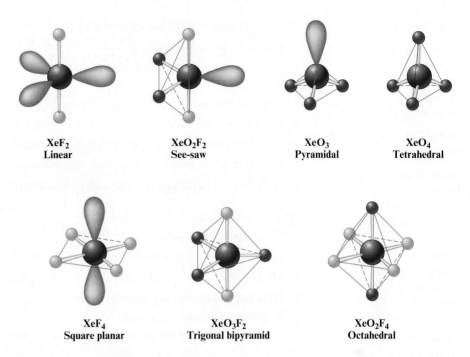

XeF$_2$
Linear

XeO$_2$F$_2$
See-saw

XeO$_3$
Pyramidal

XeO$_4$
Tetrahedral

XeF$_4$
Square planar

XeO$_3$F$_2$
Trigonal bipyramid

XeO$_2$F$_4$
Octahedral

In the past 35 years, other xenon compounds have been prepared. Examples are XeO$_4$ (explosive), XeOF$_4$, XeOF$_2$, and XeO$_3$F$_2$. These compounds contain discrete molecules with covalent bonds between the xenon atom and the other atoms. A few compounds of krypton, such as KrF$_2$ and KrF$_4$, have also been observed. The structures of several known xenon compounds are shown in Fig. 20.27. Radon also has been shown to form compounds similar to those of xenon and krypton.

For review

Key terms

Section 20.1
representative elements
transition metals
lanthanides
actinides
metalloids (semimetals)
metallurgy
liquefaction

Section 20.2
alkali metals

Section 20.3
hydride
ionic (saltlike) hydride
covalent hydride
metallic (interstitial) hydride

Representative elements

› Chemical properties are determined by their s and p valence-electron configurations
› Metallic character increases going down the group
› The properties of the first element in a group usually differ most from the properties of the other elements in the group due to a significant difference in size
  › In Group 1A, hydrogen is a nonmetal and the other members of the group are active metals
  › The first member of a group forms the strongest π bonds, causing nitrogen and oxygen to exist as N$_2$ and O$_2$ molecules

Elemental abundances on earth

› Oxygen is the most abundant element, followed by silicon
› The most abundant metals are aluminum and iron, which are found as ores

Key terms

Section 20.4
alkaline earth metals
hard water
ion exchange
ion-exchange resin

Section 20.5
boranes

Section 20.8
Haber process
nitrogen fixation
nitrogen-fixing bacteria
denitrification
nitrogen cycle
ammonia
hydrazine
nitric acid
Ostwald process

Section 20.9
phosphoric (orthophosphoric) acid
phosphorous acid
superphosphate of lime

Section 20.11
ozone
ozonolysis

Section 20.12
Frasch process
sulfuric acid

Section 20.13
halogens
hydrochloric acid
hydrohalic acids
disproportionation reaction

Section 20.14
noble gases

Group 1A elements (alkali metals)

- Have valence configuration ns^1
- Except for hydrogen, readily lose one electron to form M^+ ions in their compounds with nonmetals
- React vigorously with water to form M^+ and OH^- ions and hydrogen gas
- Form a series of oxides of the types M_2O (oxide), M_2O_2 (peroxide), and MO_2 (superoxide)
  - Not all metals form all types of oxide compounds
- Hydrogen forms covalent compounds with nonmetals
- With very active metals, hydrogen forms hydrides that contain the H^- ion

Group 2A (alkaline earth metals)

- Have valence configuration ns^2
- React less violently with water than alkali metals
- The heavier alkaline earth metals form nitrides and hydrides
- Hard water contains Ca^{2+} and Mg^{2+} ions
  - Form precipitates with soap
  - Usually removed by ion-exchange resins that replace the Ca^{2+} and Mg^{2+} ions with Na^+

Group 3A

- Have valence configuration ns^2np^1
- Show increasing metallic character going down the group
- Boron is a nonmetal that forms many types of covalent compounds, including boranes, which are highly electron-deficient and thus are very reactive
- The metals aluminum, gallium, and indium show some covalent tendencies

Group 4A

- Have valence configuration ns^2np^2
- Lighter members are nonmetals; heavier members are metals
  - All group members can form covalent bonds to nonmetals
- Carbon forms a huge variety of compounds, most of which are classified as organic compounds

Group 5A

- Elements show a wide variety of chemical properties
  - Nitrogen and phosphorus are nonmetals
  - Antimony and bismuth tend to be metallic, although no ionic compounds containing Sb^{5+} and Bi^{5+} are known; the compounds containing Sb(V) and Bi(V) are molecular rather than ionic
  - All group members except N form molecules with five covalent bonds
  - The ability to form π bonds decreases dramatically after N
- Chemistry of nitrogen
  - Most nitrogen-containing compounds decompose exothermically, forming the very stable N_2 molecule, which explains the power of nitrogen-based explosives
  - The nitrogen cycle, which consists of a series of steps, shows how nitrogen is cycled in the natural environment

> Nitrogen fixation changes the N_2 in air into compounds useful to plants
  > The Haber process is a synthetic method of nitrogen fixation
  > In the natural world, nitrogen fixation occurs through nitrogen-fixing bacteria in the root nodules of certain plants and through lightning in the atmosphere
> Ammonia is the most important hydride of nitrogen
  > Contains pyramidal NH_3 molecules
  > Widely used as a fertilizer
> Hydrazine (N_2H_4) is a powerful reducing agent
> Nitrogen forms a series of oxides including N_2O, NO, NO_2, and N_2O_5
> Nitric acid (HNO_3) is a very important strong acid manufactured by the Ostwald process

> Chemistry of phosphorus
  > Exists in three elemental forms: white (contains P_4 molecules), red, and black
  > Phosphine (PH_3) has bond angles close to 90 degrees
  > Phosphorus forms oxides including P_4O_6 and P_4O_{10} (which dissolves in water to form phosphoric acid, H_3PO_4)

Group 6A

> Metallic character increases going down the group but no element behaves as a typical metal
> The lighter members tend to gain two electrons to form X^{2-} ions in compounds with metals
> Chemistry of oxygen
  > Elemental forms are O_2 and O_3
  > Oxygen forms a wide variety of oxides
  > O_2 and especially O_3 are powerful oxidizing agents
> Chemistry of sulfur
  > The elemental forms are called rhombic and monoclinic sulfur, both of which contain S_8 molecules
  > The most important oxides are SO_2 (which forms H_2SO_3 in water) and SO_3 (which forms H_2SO_4 in water)
  > Sulfur forms a wide variety of compounds in which it shows the oxidation states $+6$, $+4$, $+2$, 0, and -2

Group 7A (halogens)

> All nonmetals
> Form hydrides of the type HX that behave as strong acids in water except for HF, which is a weak acid
> The oxyacids of the halogens become stronger as more oxygen atoms are present
> The interhalogens contain two or more different halogens

Group 8A (noble gases)

> All elements are monatomic gases and are generally very unreactive
> The heavier elements form compounds with electronegative elements such as fluorine and oxygen

Review questions   *Answers to the Review Questions can be found on the Student website (accessible from* **www.cengagebrain.com***).*

1. What are the two most abundant elements by mass in the earth's crust, oceans, and atmosphere? Does this make sense? Why? What are the four most abundant elements by mass in the human body? Does this make sense? Why?

2. What evidence supports putting hydrogen in Group 1A of the periodic table? In some periodic tables hydrogen is listed separately from any of the groups. In what ways is hydrogen unlike a typical Group 1A element? What is the valence electron configuration for the alkali

metals? List some common properties of alkali metals. How are the pure metals prepared? Predict the formulas of the compounds formed when an alkali metal reacts with F_2, S, P_4, H_2, and H_2O.

3. What is the valence electron configuration for alkaline earth metals? List some common properties of alkaline earth metals. How are alkaline earth metals prepared? Predict the formulas of the compounds formed when an alkaline earth metal reacts with F_2, O_2, S, N_2, H_2, and H_2O.

4. What is the valence electron configuration for the Group 3A elements? How does metallic character change as one goes down this group? How are boron and aluminum different? Predict the formulas of the compounds formed when aluminum reacts with F_2, O_2, S, and N_2.

5. What is the valence electron configuration for Group 4A elements? Group 4A contains two of the most important elements on earth. What are they, and why are they so important? How does metallic character change as one goes down Group 4A? What are the three allotropic forms of carbon? List some properties of germanium, tin, and lead. Predict the formulas of the compounds formed when Ge reacts with F_2 and O_2.

6. What is the valence electron configuration for Group 5A elements? Metallic character increases when going down a group. Give some examples illustrating how Bi and Sb have metallic characteristics not associated with N, P, and As. Elemental nitrogen exists as N_2, whereas in the gas phase the elements phosphorus, arsenic, and antimony consist of P_4, As_4, and Sb_4 molecules, respectively. Give a possible reason for this difference between N_2 and the other Group 5A elements. White phosphorus is much more reactive than black or red phosphorus. Explain.

7. Table 20.14 lists some common nitrogen compounds having oxidation states ranging from -3 to $+5$. Rationalize this spread in oxidation states. For each substance listed in Table 20.14, list some of its special properties. Ammonia forms hydrogen-bonding intermolecular forces resulting in an unusually high boiling point for a substance with the small size of NH_3. Can hydrazine (N_2H_4) also form hydrogen-bonding interactions? How is phosphine's (PH_3) structure different from that of ammonia?

8. What is the valence electron configuration of Group 6A elements? What are some property differences between oxygen and polonium? What are the Lewis structures for the two allotropic forms of oxygen? What is the molecular structure and the bond angle in ozone? The most stable form of solid sulfur is the rhombic form; however, a solid form called monoclinic sulfur can also form. What is the difference between rhombic and monoclinic sulfur? Explain why O_2 is much more stable than S_2 or SO. When $SO_2(g)$ or $SO_3(g)$ reacts with water, an acidic solution forms. Explain. What are the molecular structures and bond angles in SO_2 and SO_3? H_2SO_4 is a powerful dehydrating agent: What does this mean?

9. What is the valence electron configuration of the halogens? Why do the boiling points and melting points of the halogens increase steadily from F_2 to I_2? Give two reasons why F_2 is the most reactive of the halogens. Explain why the boiling point of HF is much higher than the boiling points of HCl, HBr, and HI. In nature, the halogens are generally found as halide ions in various minerals and seawater. What is a halide ion, and why are halide salts so stable? The oxidation states of the halogens vary from -1 to $+7$. Identify compounds of chlorine that have -1, $+1$, $+3$, $+5$, and $+7$ oxidation states.

10. What special property of the noble gases makes them unreactive? The boiling points and melting points of the noble gases increase steadily from He to Xe. Explain. The noble gases were among the last elements discovered; their existence was not predicted by Mendeleev when he published his first periodic table. Explain. In chemistry textbooks written before 1962, the noble gases were referred to as the inert gases. Why do we no longer use this term? For the structures of the xenon compounds in Fig. 20.27, give the bond angles exhibited and the hybridization of the central atom in each compound.

A blue question or exercise number indicates that the answer to that question or exercise appears at the back of this book and a solution appears in the *Solutions Guide*, as found on PowerLecture.

Questions

1. Although the earth was formed from the same interstellar material as the sun, there is little elemental hydrogen (H_2) in the earth's atmosphere. Explain.

2. List two major industrial uses of hydrogen.

3. How do the acidities of the aqueous solutions of the alkaline earth metal ions (M^{2+}) change in going down the group?

4. Diagonal relationships in the periodic table exist as well as the vertical relationships. For example, Be and Al are similar in some of their properties as are B and Si. Rationalize why these diagonal relationships hold for properties such as size, ionization energy, and electron affinity.

5. Atomic size seems to play an important role in explaining some of the differences between the first element in a group and the subsequent group elements. Explain.

6. Silicon carbide (SiC) is an extremely hard substance. Propose a structure for SiC.

7. In most compounds, the solid phase is denser than the liquid phase. Why isn't this true for water?

8. What is nitrogen fixation? Give some examples of nitrogen fixation.

9. All the Group 1A and 2A metals are produced by electrolysis of molten salts. Why?

10. Why are the tin(IV) halides more volatile than the tin(II) halides?

Exercises

In this section similar exercises are paired.

Group 1A Elements

11. Hydrogen is produced commercially by the reaction of methane with steam:

$$CH_4(g) + H_2O(g) \rightleftharpoons CO(g) + 3H_2(g)$$

  a. Calculate $\Delta H°$ and $\Delta S°$ for this reaction (use the data in Appendix 4).
  b. What temperatures will favor product formation at standard conditions? Assume $\Delta H°$ and $\Delta S°$ do not depend on temperature.

12. The major industrial use of hydrogen is in the production of ammonia by the Haber process:

$$3H_2(g) + N_2(g) \longrightarrow 2NH_3(g)$$

  a. Using data from Appendix 4, calculate $\Delta H°$, $\Delta S°$, and $\Delta G°$ for the Haber process reaction.
  b. Is the reaction spontaneous at standard conditions?
  c. At what temperatures is the reaction spontaneous at standard conditions? Assume $\Delta H°$ and $\Delta S°$ do not depend on temperature.

13. Write balanced equations describing the reaction of lithium metal with each of the following: O_2, S, Cl_2, P_4, H_2, H_2O, and HCl.

14. The electrolysis of aqueous sodium chloride (brine) is an important industrial process for the production of chlorine and sodium hydroxide. In fact, this process is the second largest consumer of electricity in the United States, after the production of aluminum. Write a balanced equation for the electrolysis of aqueous sodium chloride (hydrogen gas is also produced).

15. Refer to Table 20.5 and give examples of the three types of alkali metal oxides that form. How do they differ?

16. Label the following hydrides as ionic, covalent, or interstitial and support your answer. *Note:* The light blue atoms are hydrogen atoms.

17. Many lithium salts are hygroscopic (absorb water), but the corresponding salts of the other alkali metals are not. Why are lithium salts different from the others?

18. What will be the atomic number of the next alkali metal to be discovered? How would you expect the physical properties of the next alkali metal to compare with the properties of the other alkali metals summarized in Table 20.4?

Group 2A Elements

19. One harmful effect of acid rain is the deterioration of structures and statues made of marble or limestone, both of which are essentially calcium carbonate. The reaction of calcium carbonate with sulfuric acid yields carbon dioxide, water, and calcium sulfate. Because calcium sulfate is marginally soluble in water, part of the object is washed away by the rain. Write a balanced chemical equation for the reaction of sulfuric acid with calcium carbonate.

20. Write balanced equations describing the reaction of Sr with each of the following: O_2, S, Cl_2, P_4, H_2, H_2O, and HCl.

21. What mass of barium is produced when molten $BaCl_2$ is electrolyzed by a current of 2.50×10^5 A for 6.00 h?

22. How long will it take to produce 1.00×10^3 kg of magnesium metal by the electrolysis of molten magnesium chloride using a current of 5.00×10^4 A?

23. Beryllium shows some covalent characteristics in some of its compounds, unlike the other alkaline earth compounds. Give a possible explanation for this phenomenon.

24. What ions are found in hard water? What happens when water is "softened"?

Group 3A Elements

25. Consider element 113. What is the expected electron configuration for element 113? What oxidation states would be exhibited by element 113 in its compounds?

26. Thallium and indium form +1 and +3 oxidation states when in compounds. Predict the formulas of the possible compounds between thallium and oxygen and between indium and chlorine. Name the compounds.

27. Boron hydrides were once evaluated for possible use as rocket fuels. Complete and balance the following equation for the combustion of diborane.

$$B_2H_6(g) + O_2(g) \longrightarrow B(OH)_3(s)$$

28. Elemental boron is produced by reduction of boron oxide with magnesium to give boron and magnesium oxide. Write a balanced equation for this reaction.

29. Write equations describing the reactions of Ga with each of the following: F_2, O_2, S, and HCl.

30. Write a balanced equation describing the reaction of aluminum metal with concentrated aqueous sodium hydroxide.

31. Al_2O_3 is amphoteric. What does this mean?

32. What are three-centered bonds?

Group 4A Elements

33. Discuss the importance of the C—C and Si—Si bond strengths and of π bonding to the properties of carbon and silicon.

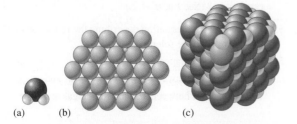

(a)   (b)   (c)

34. Besides the central atom, what are the differences between CO_2 and SiO_2?

35. The following illustration shows the orbitals used to form the bonds in carbon dioxide.

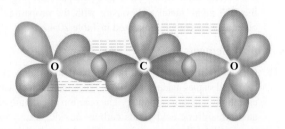

Each color represents a different orbital. Label each orbital, draw the Lewis structure for carbon dioxide, and explain how the localized electron model describes the bonding in CO_2.

36. In addition to CO_2, two additional stable oxides of carbon form. The space-filling models for CO_2 and the other two stable oxides are:

What are the formulas for the two additional stable oxides of carbon? Explain the bonding in each of these two forms using the localized electron model.

37. Silicon is produced for the chemical and electronics industries by the following reactions. Give the balanced equation for each reaction.

a. $SiO_2(s) + C(s) \longrightarrow Si(s) + CO(g)$

b. Silicon tetrachloride is reacted with very pure magnesium, producing silicon and magnesium chloride.

c. $Na_2SiF_6(s) + Na(s) \longrightarrow Si(s) + NaF(s)$

38. Write equations describing the reactions of Sn with each of the following: Cl_2, O_2, and HCl.

39. The compound Pb_3O_4 (red lead) contains a mixture of lead(II) and lead(IV) oxidation states. What is the mole ratio of lead(II) to lead(IV) in Pb_3O_4?

40. Tin forms compounds in the $+2$ and $+4$ oxidation states. Therefore, when tin reacts with fluorine, two products are possible. Write balanced equations for the production of the two tin halide compounds and name them.

Group 5A Elements

41. The oxyanion of nitrogen in which it has the highest oxidation state is the nitrate ion (NO_3^-). The corresponding oxyanion of phosphorus is PO_4^{3-}. The NO_4^{3-} ion is known but not very stable. The PO_3^- ion is not known. Account for these differences in terms of the bonding in the four anions.

42. In each of the following pairs of substances, one is stable and known, and the other is unstable. For each pair, choose the stable substance, and explain why the other is unstable.

a. NF_5 or PF_5 **b.** AsF_5 or AsI_5 **c.** NF_3 or NBr_3

43. Write balanced equations for the reactions described in Table 20.13 for the production of Bi and Sb.

44. Arsenic reacts with oxygen to form oxides that react with water in a manner analogous to that of the phosphorus oxides.

Write balanced chemical equations describing the reaction of arsenic with oxygen and the reaction of the resulting oxide with water.

45. The Group 5A elements can form molecules or ions that involve three, five, or six covalent bonds; NH_3, $AsCl_5$, and PF_6^- are examples. Draw the Lewis structure for each of these substances, and predict the molecular structure and hybridization for each. Why doesn't NF_5 or NCl_6^- form?

46. Compare the Lewis structures with the molecular orbital view of the bonding in NO, NO^+, and NO^-. Account for any discrepancies between the two models.

47. Many oxides of nitrogen have positive values for the standard free energy of formation. Using NO as an example, explain why this is the case.

48. Using data from Appendix 4 calculate $\Delta H°$, $\Delta S°$, and $\Delta G°$ for the reaction

$$N_2(g) + O_2(g) \longrightarrow 2NO(g)$$

Why does NO form in an automobile engine but then does not readily decompose back to N_2 and O_2 in the atmosphere?

49. In many natural waters, nitrogen and phosphorus are the least abundant nutrients available for plant life. Some waters that become polluted from agricultural runoff or municipal sewage become infested with algae. The algae flourish, and fish life dies off as a result. Describe how these events are chemically related.

50. Phosphate buffers are important in regulating the pH of intracellular fluids. If the concentration ratio of $H_2PO_4^-/HPO_4^{2-}$ in a sample of intracellular fluid is 1.1:1, what is the pH of this sample of intracullular fluid?

$$H_2PO_4^-(aq) \rightleftharpoons HPO_4^{2-}(aq) + H^+(aq) \quad K_a = 6.2 \times 10^{-8}$$

51. Phosphoric acid (H_3PO_4) is a triprotic acid, phosphorous acid (H_3PO_3) is a diprotic acid, and hypophosphorous acid (H_3PO_2) is a monoprotic acid. Explain this phenomenon.

52. Trisodium phosphate (TSP) is an effective grease remover. Like many cleaners, TSP acts as a base in water. Write a balanced equation to account for this basic behavior.

Group 6A Elements

53. Use bond energies to estimate the maximum wavelength of light that will cause the reaction

$$O_3 \xrightarrow{h\nu} O_2 + O$$

54. The xerographic (dry writing) process was invented in 1938 by C. Carlson. In xerography, an image is produced on a photoconductor by exposing it to light. Selenium is commonly used, since its conductivity increases three orders of magnitude upon exposure to light in the range from 400 to 500 nm. What color light should be used to cause selenium to become conductive? (See Figure 7.2.)

55. Write a balanced equation describing the reduction of H_2SeO_4 by SO_2 to produce selenium.

56. Complete and balance each of the following reactions.

a. the reaction between sulfur dioxide gas and oxygen gas

b. the reaction between sulfur trioxide gas and water

c. the reaction between concentrated sulfuric acid and sucrose ($C_{12}H_{22}O_{11}$)

57. Ozone is desirable in the upper atmosphere but undesirable in the lower atmosphere. A dictionary states that ozone has the scent of a spring thunderstorm. How can these seemingly conflicting statements be reconciled in terms of the chemical properties of ozone?

58. Ozone is a possible replacement for chlorine in municipal water purification. Unlike chlorine, virtually no ozone remains after treatment. This has good and bad consequences. Explain.

59. How can the paramagnetism of O_2 be explained using the molecular orbital model?

60. Describe the bonding in SO_2 and SO_3 using the localized electron model (hybrid orbital theory). How would the molecular orbital model describe the π bonding in these two compounds?

Group 7A Elements

61. Write the Lewis structure for O_2F_2. Predict the bond angles and hybridization of the two central oxygen atoms. Assign oxidation states and formal charges to the atoms in O_2F_2. The compound O_2F_2 is a vigorous and potent oxidizing and fluorinating agent. Are oxidation states or formal charges more useful in accounting for these properties of O_2F_2?

62. Give the Lewis structure, molecular structure, and hybridization of the oxygen atom for OF_2. Would you expect OF_2 to be a strong oxidizing agent like O_2F_2 discussed in Exercise 61?

63. Fluorine reacts with sulfur to form several different covalent compounds. Three of these compounds are SF_2, SF_4, and SF_6. Draw the Lewis structures for these compounds, and predict the molecular structures (including bond angles). Would you expect OF_4 to be a stable compound?

64. Predict some possible compounds that could form between chlorine and selenium. (*Hint:* See Exercise 63.)

65. How does the oxyacid strength of the halogens vary as the number of oxygens in the formula increases?

66. Explain why HF is a weak acid, whereas HCl, HBr, and HI are all strong acids.

Group 8A Elements

67. The xenon halides and oxides are isoelectronic with many other compounds and ions containing halogens. Give a molecule or ion in which iodine is the central atom that is isoelectronic with each of the following.
    a. xenon tetroxide   d. xenon tetrafluoride
    b. xenon trioxide   e. xenon hexafluoride
    c. xenon difluoride

68. For each of the following, write the Lewis structure(s), predict the molecular structure (including bond angles), and give the expected hybridization of the central atom.
    a. KrF_2 b. KrF_4 c. XeO_2F_2 d. XeO_2F_4

69. Although He is the second most abundant element in the universe, it is very rare on the earth. Why?

70. The noble gas with the largest atmospheric abundance is argon. Using the data in Table 20.22, calculate the mass of argon at 25°C and 1.0 atm in a room 10.0 m × 10.0 m × 10.0 m.

How many Ar atoms are in this room? How many Ar atoms do you inhale in one breath (approximately 2 L) of air at 25°C and 1.0 atm? Argon gas is inert, so it poses no serious health risks. However, if significant amounts of radon are inhaled into the lungs, lung cancer is a possible result. Explain the health risk differences between argon gas and radon gas.

71. There is evidence that radon reacts with fluorine to form compounds similar to those formed by xenon and fluorine. Predict the formulas of these RnF_x compounds. Why is the chemistry of radon difficult to study?

72. For the RnF_x compounds you predicted in Exercise 71, give the molecular structure (including bond angles).

Additional Exercises

73. Hydrazine (N_2H_4) is used as a fuel in liquid-fueled rockets. When hydrazine reacts with oxygen gas, nitrogen gas and water vapor are produced. Write a balanced equation and use bond energies from Table 8.4 to estimate ΔH for this reaction.

74. The inert-pair effect is sometimes used to explain the tendency of heavier members of Group 3A to exhibit +1 and +3 oxidation states. What does the inert-pair effect reference? (*Hint:* Consider the valence electron configuration for Group 3A elements.)

75. How could you determine experimentally whether the compound Ga_2Cl_4 contains two gallium(II) ions or one gallium(I) and one gallium(III) ion? (*Hint:* Consider the electron configurations of the three possible ions.)

76. The resistivity (a measure of electrical resistance) of graphite is $(0.4 \text{ to } 5.0) \times 10^{-4}$ ohm · cm in the basal plane. (The basal plane is the plane of the six-membered rings of carbon atoms.) The resistivity is 0.2 to 1.0 ohm · cm along the axis perpendicular to the plane. The resistivity of diamond is 10^{14} to 10^{16} ohm · cm and is independent of direction. How can you account for this behavior in terms of the structures of graphite and diamond?

77. Slaked lime, $Ca(OH)_2$, is used to soften hard water by removing calcium ions from hard water through the reaction

$$Ca(OH)_2(aq) + Ca^{2+}(aq) + 2HCO_3^-(aq) \rightarrow 2CaCO_3(s) + 2H_2O(l)$$

Although $CaCO_3(s)$ is considered insoluble, some of it does dissolve in aqueous solutions. Calculate the molar solubility of $CaCO_3$ in water ($K_{sp} = 8.7 \times 10^{-9}$).

78. EDTA is used as a complexing agent in chemical analysis. Solutions of EDTA, usually containing the disodium salt Na_2H_2EDTA, are also used to treat heavy metal poisoning. The equilibrium constant for the following reaction is 6.7×10^{21}:

$$Pb^{2+}(aq) = + H_2EDTA^{2-}(aq) \rightleftharpoons PbEDTA^{2-}(aq) + 2H^+(aq)$$

$$EDTA^{4-} = \begin{array}{c} {}^-O_2C-CH_2 \\ {}^-O_2C-CH_2 \end{array} N-CH_2-CH_2-N \begin{array}{c} CH_2-CO_2^- \\ CH_2-CO_2^- \end{array}$$

Ethylenediaminetetraacetate

Calculate $[Pb^{2+}]$ at equilibrium in a solution originally 0.0050 *M* in Pb^{2+}, 0.075 *M* in H_2EDTA^{2-}, and buffered at pH = 7.00.

79. Photogray lenses contain small embedded crystals of solid silver chloride. Silver chloride is light-sensitive because of the reaction

$$AgCl(s) \xrightarrow{hv} Ag(s) + Cl$$

Small particles of metallic silver cause the lenses to darken. In the lenses this process is reversible. When the light is removed, the reverse reaction occurs. However, when pure white silver chloride is exposed to sunlight it darkens; the reverse reaction does not occur in the dark.

a. How do you explain this difference?

b. Photogray lenses do become permanently dark in time. How do you account for this?

80. Draw Lewis structures for the $AsCl_4^+$ and $AsCl_6^-$ ions. What type of reaction (acid–base, oxidation–reduction, or the like) is the following?

$$2AsCl_5(g) \longrightarrow AsCl_4AsCl_6(s)$$

81. Provide a reasonable estimate for the number of atoms in a 150-lb adult human. Use the information given in Table 20.2.

82. In large doses, selenium is toxic. However, in moderate intake, selenium is a physiologically important element. How is selenium physiologically important?

83. In the 1950s and 1960s, several nations conducted tests of nuclear warheads in the atmosphere. It was customary, following each test, to monitor the concentration of strontium-90 (a radioactive isotope of strontium) in milk. Why would strontium-90 tend to accumulate in milk?

84. What is a disproportionation reaction? Use the following reduction potentials

$$ClO_3^- + 3H^+ + 2e^- \longrightarrow HClO_2 + H_2O \qquad \mathscr{E}° = 1.21 \text{ V}$$
$$HClO_2 + 2H^+ + 2e^- \longrightarrow HClO + H_2O \qquad \mathscr{E}° = 1.65 \text{ V}$$

to predict whether $HClO_2$ will disproportionate.

85. Sulfur forms a wide variety of compounds in which it has +6, +4, +2, 0, and −2 oxidation states. Give examples of sulfur compounds having each of these oxidation states.

86. Halogens form a variety of covalent compounds with each other. For example, chlorine and fluorine form the compounds ClF, ClF_3, and ClF_5. Predict the molecular structure (including bond angles) for each of these three compounds. Would you expect FCl_3 to be a stable compound? Explain.

ChemWork Problems

These multiconcept problems (and additional ones) are found interactively online with the same type of assistance a student would get from an instructor.

87. Hydrogen gas is being considered as a fuel for automobiles. There are many chemical means for producing hydrogen gas from water. One of these reactions is

$$C(s) + H_2O(g) \longrightarrow CO(g) + H_2(g)$$

In this case the form of carbon used is graphite.

a. Calculate $\Delta H°$ and $\Delta S°$ for this reaction using data from Appendix 4.

b. At what temperature is $\Delta G° = $ zero for this reaction? Assume $\Delta H°$ and $\Delta S°$ do not depend on temperature.

88. Molten $CaCl_2$ is electrolyzed for 8.00 h to produce Ca(s) and $Cl_2(g)$.

a. What current is needed to produce 5.52 kg of calcium metal?

b. If 5.52 kg calcium metal is produced, what mass (in kg) of Cl_2 is produced?

89. Calculate the solubility of $Mg(OH)_2$ ($K_{sp} = 8.9 \times 10^{-12}$) in an aqueous solution buffered at pH = 9.42.

90. Which of the following statement(s) is(are) *true*?

a. The alkali metals are found in the earth's crust in the form of pure elements.

b. Gallium has one of the highest melting points known for metals.

c. When calcium metal reacts with water, one of the products is $H_2(g)$.

d. When $AlCl_3$ is dissolved in water, it produces an acidic solution.

e. Lithium reacts in the presence of excess oxygen gas to form lithium superoxide.

91. What is the hybridization of the underlined nitrogen atom in each of the following molecules or ions?

a. $\underline{N}O^+$

b. N_2O_3 ($O_2N\underline{N}O$)

c. $\underline{N}O_2^-$

d. $\underline{N}_2$

92. Nitrous oxide (N_2O) can be produced by thermal decomposition of ammonium nitrate:

$$NH_4NO_3(s) \xrightarrow{heat} N_2O(g) + 2H_2O(l)$$

What volume of $N_2O(g)$ collected over water at a total pressure of 94.0 kPa and 22°C can be produced from thermal decomposition of 8.68 g NH_4NO_3? The vapor pressure of water at 22°C is 21 torr.

93. What is the hybridization of the central atom in each of the following molecules?

a. SF_6

b. ClF_3

c. $GeCl_4$

d. XeF_4

94. What is the molecular structure for each of the following molecules or ions?

a. OCl_2

b. ClO_4^-

c. ICl_5

d. PF_6^-

95. The atmosphere contains $9.0 \times 10^{-6}\%$ Xe by volume at 1.0 atm and 25°C.

a. Calculate the mass of Xe in a room 7.26 m by 8.80 m by 5.67 m.

b. A typical person takes in about 2 L of air during a breath. How many Xe atoms are inhaled in each breath?

96. Which of following statement(s) is/are *true*?

a. Phosphoric acid is a stronger acid than nitric acid.

b. The noble gas with the lowest boiling point is helium.

c. Sulfur is found as the free element in the earth's crust.

d. One of the atoms in Teflon is fluorine.

e. The P_4 molecule has a square planar structure.

Challenge Problems

97. Suppose 10.00 g of an alkaline earth metal reacts with 10.0 L water to produce 6.10 L hydrogen gas at 1.00 atm and 25°C. Identify the metal and determine the pH of the solution.

98. From the information on the temperature stability of white and gray tin given in this chapter, which form would you expect to have the more ordered structure (have the smaller positional probability)?

99. Lead forms compounds in the $+2$ and $+4$ oxidation states. All lead(II) halides are known (and are known to be ionic). Only PbF_4 and $PbCl_4$ are known among the possible lead(IV) halides. Presumably lead(IV) oxidizes bromide and iodide ions, producing the lead(II) halide and the free halogen:

$$PbX_4 \longrightarrow PbX_2 + X_2$$

Suppose 25.00 g of a lead(IV) halide reacts to form 16.12 g of a lead(II) halide and the free halogen. Identify the halogen.

100. Many structures of phosphorus-containing compounds are drawn with some P=O bonds. These bonds are not the typical π bonds we've considered, which involve the overlap of two p orbitals. Instead, they result from the overlap of a d orbital on the phosphorus atom with a p orbital on oxygen. This type of π bonding is sometimes used as an explanation for why H_3PO_3 has the first structure below rather than the second:

$$
\begin{array}{ccc}
\text{O} & & \text{OH} \\
\parallel & & | \\
\text{H}-\text{P}-\text{OH} & & \text{HO}-\text{P}: \\
| & & | \\
\text{OH} & & \text{OH}
\end{array}
$$

Draw a picture showing how a d orbital and a p orbital overlap to form a π bond.

101. Use bond energies (see Table 8.4) to show that the preferred products for the decomposition of N_2O_3 are NO_2 and NO rather than O_2 and N_2O. (The N—O single bond energy is 201 kJ/mol.) (*Hint:* Consider the reaction kinetics.)

102. A proposed two-step mechanism for the destruction of ozone in the upper atmosphere is

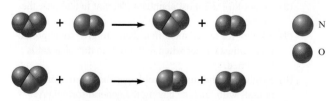

a. What is the overall balanced equation for the ozone destruction reaction?

b. Which species is a catalyst?

c. Which species is an intermediate?

d. What is the rate law derived from this mechanism if the first step in the mechanism is slow and the second step is fast?

e. One of the concerns about the use of Freons is that they will migrate to the upper atmosphere, where chlorine atoms can be generated by the reaction

$$CCl_2F_2 \xrightarrow{h\nu} CF_2Cl + Cl$$
Freon-12

Chlorine atoms also can act as a catalyst for the destruction of ozone. The first step of a proposed mechanism for chlorine-catalyzed ozone destruction is

$$Cl(g) + O_3(g) \longrightarrow ClO(g) + O_2(g) \quad \text{Slow}$$

Assuming a two-step mechanism, propose the second step in the mechanism and give the overall balanced equation.

103. You travel to a distant, cold planet where the ammonia flows like water. In fact, the inhabitants of this planet use ammonia (an abundant liquid on their planet) much as earthlings use water. Ammonia is also similar to water in that it is amphoteric and undergoes autoionization. The K value for the autoionization of ammonia is 1.8×10^{-12} at the standard temperature of the planet. What is the pH of ammonia at this temperature?

104. Nitrogen gas reacts with hydrogen gas to form ammonia gas (NH_3). Consider the following illustration representing the original reaction mixture in a 15.0 L container (the numbers of each molecule shown are relative numbers):

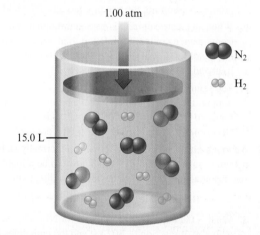

Assume this reaction mixture goes to completion. The piston apparatus allows the container volume to change in order to keep the pressure constant at 1.00 atm. Assume ideal behavior and constant temperature.

a. What is the partial pressure of ammonia in the container when the reaction is complete?

b. What is the mole fraction of ammonia in the container when the reaction is complete?

c. What is the volume of the container when the reaction is complete?

105. A cylinder fitted with a movable piston initially contains 2.00 moles of $O_2(g)$ and an unknown amount of $SO_2(g)$. The oxygen is known to be in excess. The density of the mixture is 0.8000 g/L at some T and P. After the reaction has gone to completion, forming $SO_3(g)$, the density of the resulting gaseous mixture is 0.8471 g/L at the same T and P. Calculate the mass of SO_3 formed in the reaction.

Integrative Problems

These problems require the integration of multiple concepts to find the solutions.

106. The heaviest member of the alkaline earth metals is radium (Ra), a naturally radioactive element discovered by Pierre and Marie Curie in 1898. Radium was initially isolated from the uranium ore pitchblende, in which it is present as approximately 1.0 g per 7.0 metric tons of pitchblende. How many atoms of radium can be isolated from 1.75×10^8 g pitchblende (1 metric ton = 1000 kg)? One of the early uses of radium was as an additive to paint so that watch dials coated with this paint would glow in the dark. The longest-lived isotope of radium has a half-life of 1.60×10^3 years. If an antique watch, manufactured in 1925, contains 15.0 mg radium, how many atoms of radium will remain in 2025?

107. Indium(III) phosphide is a semiconducting material that has been frequently used in lasers, light-emitting diodes (LED), and fiber-optic devices. This material can be synthesized at 900. K according to the following reaction:

$$In(CH_3)_3(g) + PH_3(g) \longrightarrow InP(s) + 3CH_4(g)$$

a. If 2.56 L $In(CH_3)_3$ at 2.00 atm is allowed to react with 1.38 L PH_3 at 3.00 atm, what mass of $InP(s)$ will be produced assuming the reaction has an 87% yield?

b. When an electric current is passed through an optoelectronic device containing InP, the light emitted has an energy of 2.03×10^{-19} J. What is the wavelength of this light and is it visible to the human eye?

c. The semiconducting properties of InP can be altered by doping. If a small number of phosphorus atoms are replaced by atoms with an electron configuration of $[Kr]5s^24d^{10}5p^4$, is this n-type or p-type doping?

108. Although nitrogen trifluoride (NF_3) is a thermally stable compound, nitrogen triiodide (NI_3) is known to be a highly explosive material. NI_3 can be synthesized according to the equation

$$BN(s) + 3IF(g) \longrightarrow BF_3(g) + NI_3(g)$$

a. What is the enthalpy of formation for $NI_3(s)$ given the enthalpy of reaction (-307 kJ) and the enthalpies of formation for $BN(s)$ (-254 kJ/mol), $IF(g)$ (-96 kJ/mol), and $BF_3(g)$ (-1136 kJ/mol)?

b. It is reported that when the synthesis of NI_3 is conducted using 4 moles of IF for every 1 mole of BN, one of the by-products isolated is $[IF_2]^+[BF_4]^-$. What are the molecular geometries of the species in this by-product? What are the hybridizations of the central atoms in each species in the by-product?

109. While selenic acid has the formula H_2SeO_4 and thus is directly related to sulfuric acid, telluric acid is best visualized as H_6TeO_6 or $Te(OH)_6$.

a. What is the oxidation state of tellurium in $Te(OH)_6$?

b. Despite its structural differences with sulfuric and selenic acid, telluric acid is a diprotic acid with $pK_{a_1} = 7.68$ and $pK_{a_2} = 11.29$. Telluric acid can be prepared by hydrolysis of tellurium hexafluoride according to the equation

$$TeF_6(g) + 6H_2O(l) \longrightarrow Te(OH)_6(aq) + 6HF(aq)$$

Tellurium hexafluoride can be prepared by the reaction of elemental tellurium with fluorine gas:

$$Te(s) + 3F_2(g) \longrightarrow TeF_6(g)$$

If a cubic block of tellurium (density = 6.240 g/cm³) measuring 0.545 cm on edge is allowed to react with 2.34 L fluorine gas at 1.06 atm and 25°C, what is the pH of a solution of $Te(OH)_6$ formed by dissolving the isolated $TeF_6(g)$ in 115 mL solution? Assume 100% yield in all reactions.

Marathon Problems

These problems are designed to incorporate several concepts and techniques into one situation.

110. Captain Kirk has set a trap for the Klingons who are threatening an innocent planet. He has sent small groups of fighter rockets to sites that are invisible to Klingon radar and put a decoy in the open. He calls this the "fishhook" strategy. Mr. Spock has sent a coded message to the chemists on the fighters to tell the ships what to do next. The outline of the message is

```
___   ___      ___         ___      ___      ___
(1)   (2)      (3)         (4)      (5)      (6)

___      ___  ___      ___  ___  ___  ___  ___
(7)      (8)  (9)      (10) (11) (12) (10) (11)
```

Fill in the blanks of the message using the following clues.

(1) Symbol of the halogen whose hydride has the second highest boiling point in the series of HX compounds that are hydrogen halides.

(2) Symbol of the halogen that is the only hydrogen halide, HX, that is a weak acid in aqueous solution.

(3) Symbol of the element whose existence on the sun was known before its existence on earth was discovered.

(4) The Group 5A element in Table 20.13 that should have the most metallic character.

(5) Symbol of the Group 6A element that, like selenium, is a semiconductor.

(6) Symbol for the element known in rhombic and monoclinic forms.

(7) Symbol for the element that exists as diatomic molecules in a yellow-green gas when not combined with another element.

(8) Symbol for the most abundant element in and near the earth's crust.

(9) Symbol for the element that seems to give some protection against cancer when a diet rich in this element is consumed.

(10) Symbol for the smallest noble gas that forms compounds with fluorine having the general formula AF_2 and AF_4 (reverse the symbol and split the letters as shown).

(11) Symbol for the toxic element that, like phosphorus and antimony, forms tetrameric molecules when uncombined with other elements (split the letters of the symbol as shown).

(12) Symbol for the element that occurs as an inert component of air but is a very prominent part of fertilizers and explosives.

111. Use the symbols of the elements described in the following clues to fill in the blanks that spell out the name of a famous American scientist. Although this scientist was better known as a physicist than as a chemist, the Philadelphia institute that bears his name does include a biochemistry research facility.

___ ___ ___   ___ ___ ___ ___ ___ ___ ___
(1)   (2)   (3)   (4)   (5)   (6)   (7)

(1) The oxide of this alkaline earth metal is amphoteric.
(2) The element that makes up approximately 3.0% by mass of the human body.
(3) The element having a $7s^1$ valence electron configuration.
(4) This element is the alkali metal with the least negative standard reduction potential. Write its symbol in reverse order.
(5) The alkali metal whose ion is more concentrated in intracellular fluids as compared with blood plasma.
(6) This is the only alkali metal that reacts directly with nitrogen to make a binary compound with formula M_3N.
(7) This element is the first in Group 3A for which the +1 oxidation state is exhibited in stable compounds. Use only the second letter of its symbol.

Chapter 21

Transition Metals and Coordination Chemistry

The brilliant color of rubies results from trace concentrations of Cr^{3+} ions. (Chip Clark/Smithsonian Institution)

T ransition metals have many uses in our society. Iron is used for steel; copper for electrical wiring and water pipes; titanium for paint; silver for photographic paper; manganese, chromium, vanadium, and cobalt as additives to steel; platinum for industrial and automotive catalysts; and so on.

One indication of the importance of transition metals is the great concern shown by the U.S. government for continuing the supply of these elements. In recent years the United States has been a net importer of about 60 "strategic and critical" minerals, including cobalt, manganese, platinum, palladium, and chromium. All these metals play a vital role in the U.S. economy and defense, and approximately 90% of the required amounts must be imported (Table 21.1).

In addition to being important in industry, transition metal ions play a vital role in living organisms. For example, complexes of iron provide for the transport and storage of oxygen, molybdenum and iron compounds are catalysts in nitrogen fixation, zinc is found in more than 150 biomolecules in humans, copper and iron play a crucial role in the respiratory cycle, and cobalt is found in essential biomolecules such as vitamin B_{12}.

In this chapter we will explore the general properties of transition metals, paying particular attention to the bonding, structure, and properties of the complex ions of these metals.

21.1 | The Transition Metals: A Survey

General Properties

One striking characteristic of the representative elements is that their chemistry changes markedly across a given period as the number of valence electrons changes. The chemical similarities occur mainly within the vertical groups. In contrast, *the transition metals show great similarities within a given period as well as within a given vertical group.* This difference occurs because the last electrons added for transition metals are inner electrons: *d* electrons for the *d*-block transition metals and *f* electrons for the lanthanides and actinides. These inner *d* and *f* electrons cannot participate as easily in bonding as can the valence *s* and *p* electrons. Thus the chemistry of the transition elements is not affected as greatly by the gradual change in the number of electrons as is the chemistry of the representative elements.

Table 21.1 | Some Transition Metals Important to the U.S. Economy and Defense

Metal	Uses	Percentage Imported
Chromium	Stainless steel (especially for parts exposed to corrosive gases and high temperatures)	~91%
Cobalt	High-temperature alloys in jet engines, magnets, catalysts, drill bits	~93%
Manganese	Steelmaking	~97%
Platinum and palladium	Catalysts	~87%

Figure 21.1 | The position of the transition elements on the periodic table. The *d*-block elements correspond to filling the 3*d*, 4*d*, 5*d*, or 6*d* orbitals. The inner transition metals correspond to filling the 4*f* (lanthanides) or 5*f* (actinides) orbitals.

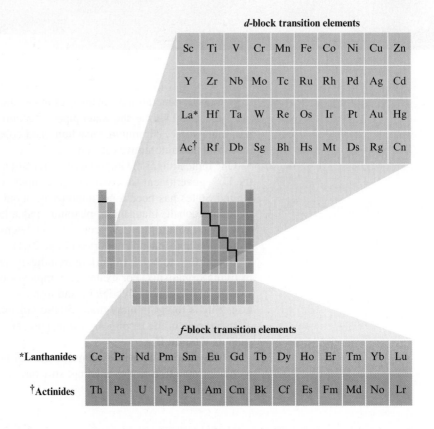

d-block transition elements

Sc	Ti	V	Cr	Mn	Fe	Co	Ni	Cu	Zn
Y	Zr	Nb	Mo	Tc	Ru	Rh	Pd	Ag	Cd
La*	Hf	Ta	W	Re	Os	Ir	Pt	Au	Hg
Ac†	Rf	Db	Sg	Bh	Hs	Mt	Ds	Rg	Cn

f-block transition elements

*Lanthanides	Ce	Pr	Nd	Pm	Sm	Eu	Gd	Tb	Dy	Ho	Er	Tm	Yb	Lu
†Actinides	Th	Pa	U	Np	Pu	Am	Cm	Bk	Cf	Es	Fm	Md	No	Lr

Sport trophies are often made from silver.

Group designations are traditionally given on the periodic table for the *d*-block transition metals (Fig. 21.1). However, these designations do not relate as directly to the chemical behavior of these elements as they do for the representative elements (the A groups), so we will not use them.

As a class, the transition metals behave as typical metals, possessing metallic luster and relatively high electrical and thermal conductivities. Silver is the best conductor of heat and electric current. However, copper is a close second, which explains copper's wide use in the electrical systems of homes and factories.

Despite their many similarities, the transition metals do vary considerably in certain properties. For example, tungsten has a melting point of 3400°C and is used for filaments in light bulbs; mercury is a liquid at 25°C. Some transition metals such as iron and titanium are hard and strong and make very useful structural materials; others such as copper, gold, and silver are relatively soft. The chemical reactivity of the transition metals also varies significantly. Some react readily with oxygen to form oxides. Of these metals, some, such as chromium, nickel, and cobalt, form oxides that adhere tightly to the metallic surface, protecting the metal from further oxidation. Others, such as iron, form oxides that scale off, constantly exposing new metal to the corrosion process. On the other hand, the noble metals—primarily gold, silver, platinum, and palladium—do not readily form oxides.

In forming ionic compounds with nonmetals, the transition metals exhibit several typical characteristics:

More than one oxidation state is often found. For example, iron combines with chlorine to form $FeCl_2$ and $FeCl_3$.

The cations are often **complex ions**, species where *the transition metal ion is surrounded by a certain number of ligands* (molecules or ions that behave as Lewis bases). For example, the compound $[Co(NH_3)_6]Cl_3$ contains the $Co(NH_3)_6^{3+}$ cation and Cl^- anions.

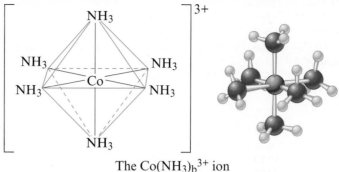

The $Co(NH_3)_b^{3+}$ ion

Most compounds are colored, because the transition metal ion in the complex ion can absorb visible light of specific wavelengths.

Many compounds are paramagnetic (they contain unpaired electrons).

In this chapter we will concentrate on the **first-row transition metals** (scandium through zinc) because they are representative of the other transition series and because they have great practical significance. Some important properties of these elements are summarized in Table 21.2 and are discussed in the next section.

Electron Configurations

The electron configurations of the first-row transition metals were discussed in Section 7.11. The $3d$ orbitals begin to fill after the $4s$ orbital is complete, that is, after calcium ($[Ar]4s^2$). The first transition metal, *scandium,* has one electron in the $3d$ orbitals; the

(clockwise from upper left) Calcite stalactites colored by traces of iron. Quartz is often colored by the presence of transition metals such as Mn, Fe, and Ni. Wulfenite contains PbMoO₄. Rhodochrosite is a mineral containing MnCO₃.

Table 21.2 | Selected Properties of the First-Row Transition Metals

	Scandium	Titanium	Vanadium	Chromium	Manganese	Iron	Cobalt	Nickel	Copper	Zinc
Atomic number	21	22	23	24	25	26	27	28	29	30
Electron configuration*	$4s^2 3d^1$	$4s^2 3d^2$	$4s^2 3d^3$	$4s^1 3d^5$	$4s^2 3d^5$	$4s^2 3d^6$	$4s^2 3d^7$	$4s^2 3d^8$	$4s^1 3d^{10}$	$4s^2 3d^{10}$
Atomic radius (pm)	162	147	134	130	135	126	125	124	128	138
Ionization energies (eV/atom)										
First	6.54	6.82	6.74	6.77	7.44	7.87	7.86	7.64	7.73	9.39
Second	12.80	13.58	14.65	16.50	15.64	16.18	17.06	18.17	20.29	17.96
Third	24.76	27.49	29.31	30.96	33.67	30.65	33.50	35.17	36.83	39.72
Reduction potential† (V)	−2.08	−1.63	−1.2	−0.91	−1.18	−0.44	−0.28	−0.23	+0.34	−0.76
Common oxidation states	+3	+2,+3, +4	+2,+3, +4,+5	+2,+3, +6	+2,+3, +4,+7	+2,+3	+2,+3	+2	+1,+2	+2
Melting point (°C)	1397	1672	1710	1900	1244	1530	1495	1455	1083	419
Density (g/cm³)	2.99	4.49	5.96	7.20	7.43	7.86	8.9	8.90	8.92	7.14
Electrical conductivity‡	—	2	3	10	2	17	24	24	97	27

*Each atom has an argon inner-core configuration.
†For the reduction process $M^{2+} + 2e^- \rightarrow M$ (except for scandium, where the ion is Sc^{3+}).
‡Compared with an arbitrarily assigned value of 100 for silver.

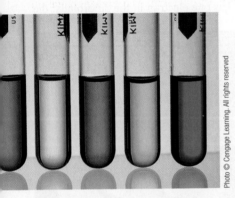

(from left to right) Aqueous solutions containing the metal ions Co^{2+}, Mn^{2+}, Cr^{3+}, Fe^{3+}, and Ni^{2+}.

Chromium has the electron configuration $[Ar]4s^1 3d^5$.

A set of orbitals with the same energy is said to be *degenerate*.

Copper has the electron configuration $[Ar]4s^1 3d^{10}$.

In transition metal *ions*, the 3d orbitals are lower in energy than the 4s orbitals.

second, *titanium,* has two; and the third, *vanadium,* has three. We would expect *chromium,* the fourth transition metal, to have the electron configuration $[Ar]4s^2 3d^4$. However, the actual configuration is $[Ar]4s^1 3d^5$, which shows a half-filled 4s orbital and a half-filled set of 3d orbitals (one electron in each of the five 3d orbitals). It is tempting to say that the configuration results because half-filled "shells" are especially stable. Although there are some reasons to think that this explanation might be valid, it is an oversimplification. For instance, tungsten, which is in the same vertical group as chromium, has the configuration $[Xe]6s^2 4f^{14} 5d^4$, where half-filled s and d shells are not found. There are several similar cases.

Basically, the chromium configuration occurs because the energies of the 3d and 4s orbitals are very similar for the first-row transition elements. We saw in Section 7.11 that when electrons are placed in a set of degenerate orbitals, they first occupy each orbital singly to minimize electron repulsions. Since the 4s and 3d orbitals are virtually degenerate in the chromium atom, we would expect the configuration

$$4s \; \uparrow \qquad 3d \; \uparrow \; \uparrow \; \uparrow \; \uparrow \; \uparrow$$

rather than

$$4s \; \uparrow\downarrow \qquad 3d \; \uparrow \; \uparrow \; \uparrow \; \uparrow \; \_\_$$

since the second arrangement has greater electron–electron repulsions and thus a higher energy.

The only other unexpected configuration among the first-row transition metals is that of copper, which is $[Ar]4s^1 3d^{10}$ rather than the expected $[Ar]4s^2 3d^9$.

In contrast to the neutral transition metals, where the 3d and 4s orbitals have very similar energies, the *energy of the 3d orbitals in transition metal ions is significantly less than that of the 4s orbital.* This means that the electrons remaining after the ion is formed occupy the 3d orbitals, since they are lower in energy. *First-row transition metal ions do not have 4s electrons.* For example, manganese has the configuration

Figure 21.2 | Plots of the first (red dots) and third (blue dots) ionization energies for the first-row transition metals.

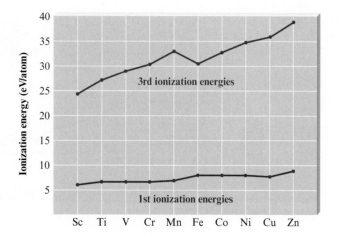

Figure 21.2 | Plots of the first (red dots) and third (blue dots) ionization energies for the first-row transition metals.

$[Ar]4s^2 3d^5$, while that of Mn^{2+} is $[Ar]3d^5$. The neutral titanium atom has the configuration $[Ar]4s^2 3d^2$, while that of Ti^{3+} is $[Ar]3d^1$.

Oxidation States and Ionization Energies

The transition metals can form a variety of ions by losing one or more electrons. The common oxidation states of these elements are shown in Table 21.2. Note that for the first five metals the maximum possible oxidation state corresponds to the loss of all the $4s$ and $3d$ electrons. For example, the maximum oxidation state of chromium ($[Ar]4s^1 3d^5$) is +6. Toward the right end of the period, the maximum oxidation states are not observed; in fact, the 2+ ions are the most common. The higher oxidation states are not seen for these metals because the $3d$ orbitals become lower in energy as the nuclear charge increases, and the electrons become increasingly difficult to remove. From Table 21.2 we see that ionization energy increases gradually going from left to right across the period. However, the third ionization energy (when an electron is removed from a $3d$ orbital) increases faster than the first ionization energy, clear evidence of the significant decrease in the energy of the $3d$ orbitals going across the period (Fig. 21.2).

Standard Reduction Potentials

When a metal acts as a *reducing agent,* the half-reaction is

$$M \longrightarrow M^{n+} + ne^-$$

This is the reverse of the conventional listing for half-reactions in tables. Thus, to rank the transition metals in order of reducing ability, it is most convenient to reverse the reactions and the signs given in Table 21.2. The metal with the most positive potential is then the best reducing agent. The transition metals are listed in order of reducing ability in Table 21.3.

Since $\mathscr{E}°$ is zero for the process

$$2H^+ + 2e^- \longrightarrow H_2$$

all the metals except copper can reduce H^+ ions to hydrogen gas in 1 M aqueous solutions of strong acid:

$$M(s) + 2H^+(aq) \longrightarrow H_2(g) + M^{2+}(aq)$$

As Table 21.3 shows, the reducing abilities of the first-row transition metals generally decrease going from left to right across the period. Only chromium and zinc do not follow this trend.

Table 21.3 | Relative Reducing Abilities of the First-Row Transition Metals in Aqueous Solution

Reaction	Potential (V)
$Sc \rightarrow Sc^{3+} + 3e^-$	2.08
$Ti \rightarrow Ti^{2+} + 2e^-$	1.63
$V \rightarrow V^{2+} + 2e^-$	1.2
$Mn \rightarrow Mn^{2+} + 2e^-$	1.18
$Cr \rightarrow Cr^{2+} + 2e^-$	0.91
$Zn \rightarrow Zn^{2+} + 2e^-$	0.76
$Fe \rightarrow Fe^{2+} + 2e^-$	0.44
$Co \rightarrow Co^{2+} + 2e^-$	0.28
$Ni \rightarrow Ni^{2+} + 2e^-$	0.23
$Cu \rightarrow Cu^{2+} + 2e^-$	−0.34

Reducing ability →

Figure 21.3 | Atomic radii of the 3*d*, 4*d*, and 5*d* transition series.

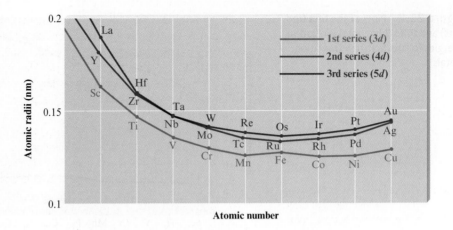

The 4*d* and 5*d* Transition Series

In comparing the 3*d*, 4*d*, and 5*d* transition series, it is instructive to consider the atomic radii of these elements (Fig. 21.3). Note that there is a general, although not regular, decrease in size going from left to right for each of the series. Also note that although there is a significant increase in radius in going from the 3*d* to the 4*d* metals, the 4*d* and 5*d* metals are remarkably similar in size. This latter phenomenon is the result of the **lanthanide contraction**. In the **lanthanide series**, consisting of the elements between lanthanum and hafnium (see Fig. 21.1), electrons are filling the 4*f* orbitals. Since the 4*f* orbitals are buried in the interior of these atoms, the additional electrons do not add to the atomic size. In fact, the increasing nuclear charge (remember that a proton is added to the nucleus for each electron) causes the radii of the lanthanide elements to decrease significantly going from left to right. This lanthanide contraction just offsets the normal increase in size due to going from one principal quantum level to another. Thus the 5*d* elements, instead of being significantly larger than the 4*d* elements, are almost identical to them in size. This leads to a great similarity in the chemistry of the 4*d* and 5*d* elements in a given vertical group. For example, the chemical properties of hafnium and zirconium are remarkably similar, and they always occur together in nature. Their separation, which is probably more difficult than the separation of any other pair of elements, often requires fractional distillation of their compounds.

In general, the differences between the 4*d* and 5*d* elements in a group increase gradually going from left to right. For example, niobium and tantalum are also quite similar, but less so than zirconium and hafnium.

Although generally less well known than the 3*d* elements, the 4*d* and 5*d* transition metals have certain very useful properties. For example, zirconium and zirconium oxide (ZrO_2) have great resistance to high temperatures and are used, along with niobium and molybdenum alloys, for space vehicle parts that are exposed to high temperatures during re-entry into the earth's atmosphere. Niobium and molybdenum are also important alloying materials for certain types of steel. Tantalum, which has a high resistance to attack by body fluids, is often used for replacement of bones. The *platinum group metals*—ruthenium, osmium, rhodium, iridium, palladium, and platinum—are all quite similar and are widely used as catalysts for many types of industrial processes.

Niobium was originally called columbium *and is still occasionally referred to by that name.*

21.2 | The First-Row Transition Metals

We have seen that the transition metals are similar in many ways but also show important differences. We will now explore some of the specific properties of each of the 3*d* transition metals.

Scandium is a rare element that exists in compounds mainly in the +3 oxidation state—for example, in $ScCl_3$, Sc_2O_3, and $Sc_2(SO_4)_3$. The chemistry of scandium

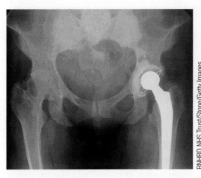

An X ray of a patient who has had a hip replacement. The normal hip joint is on the left; the hip joint constructed from the transition metal tantalum is on the right.

$Ti(H_2O)_6^{3+}$ is purple in solution.

The manufacture of sulfuric acid was discussed at the end of Chapter 3.

The most common oxidation state for vanadium is +5.

strongly resembles that of the lanthanides, with most of its compounds being colorless and diamagnetic. This is not surprising; as we will see in Section 21.6, the color and magnetism of transition metal compounds usually arise from the d electrons on the metal ion, and Sc^{3+} has no d electrons. Scandium metal, which can be prepared by electrolysis of molten $ScCl_3$, is not widely used because of its rarity, but it is found in some electronic devices, such as high-intensity lamps.

Titanium is widely distributed in the earth's crust (0.6% by mass). Because of its relatively low density and high strength, titanium is an excellent structural material, especially in jet engines, where light weight and stability at high temperatures are required. Nearly 5000 kg of titanium alloys is used in each engine of a Boeing 747 jetliner. In addition, the resistance of titanium to chemical attack makes it a useful material for pipes, pumps, and reaction vessels in the chemical industry.

The most familiar compound of titanium is no doubt responsible for the white color of this paper. Titanium dioxide, or more correctly, *titanium(IV) oxide* (TiO_2), is a highly opaque substance used as the white pigment in paper, paint, linoleum, plastics, synthetic fibers, whitewall tires, and cosmetics (sunscreens, for example). Over one million tons is used annually in these and other products. Titanium(IV) oxide is widely dispersed in nature, but the main ores are rutile (impure TiO_2) and ilmenite ($FeTiO_3$). Rutile is processed by treatment with chlorine to form volatile $TiCl_4$, which is separated from the impurities and burned to form TiO_2:

$$TiCl_4(g) + O_2(g) \longrightarrow TiO_2(s) + 2Cl_2(g)$$

Ilmenite is treated with sulfuric acid to form a soluble sulfate:

$$FeTiO_3(s) + 2H_2SO_4(aq) \longrightarrow Fe^{2+}(aq) + TiO^{2+}(aq) + 2SO_4^{2-}(aq) + 2H_2O(l)$$

When this aqueous mixture is allowed to stand, under vacuum, solid $FeSO_4 \cdot 7H_2O$ forms first and is removed. The mixture is then heated, and the insoluble titanium(IV) oxide hydrate ($TiO_2 \cdot H_2O$) forms. The water of hydration is driven off by heating to form pure TiO_2:

$$TiO_2 \cdot H_2O(s) \xrightarrow{\text{Heat}} TiO_2(s) + H_2O(g)$$

In its compounds, titanium is most often found in the +4 oxidation state. Examples are TiO_2 and $TiCl_4$, the latter a colorless liquid (bp = 137°C) that fumes in moist air to produce TiO_2:

$$TiCl_4(l) + 2H_2O(l) \longrightarrow TiO_2(s) + 4HCl(g)$$

Titanium(III) compounds can be produced by reduction of the +4 state. In aqueous solution, Ti^{3+} exists as the purple $Ti(H_2O)_6^{3+}$ ion, which is slowly oxidized to titanium(IV) by air. Titanium(II) is not stable in aqueous solution but does exist in the solid state in compounds such as TiO and the dihalides of general formula TiX_2.

Vanadium is widely spread throughout the earth's crust (0.02% by mass). It is used mostly in alloys with other metals such as iron (80% of vanadium is used in steel) and titanium. Vanadium(V) oxide (V_2O_5) is used as an industrial catalyst in the production of materials such as sulfuric acid.

Pure vanadium can be obtained from the electrolytic reduction of fused salts, such as VCl_2, to produce a metal similar to titanium that is steel gray, hard, and corrosion resistant. Often the pure element is not required for alloying. For example, *ferrovanadium,* produced by the reduction of a mixture of V_2O_5 and Fe_2O_3 with aluminum, is added to iron to form *vanadium steel,* a hard steel used for engine parts and axles.

The principal oxidation state of vanadium is +5, found in compounds such as the orange V_2O_5 (mp = 650°C) and the colorless VF_5 (mp = 19.5°C). The oxidation states from +5 to +2 all exist in aqueous solution (Table 21.4). The higher oxidation states, +5 and +4, do not exist as hydrated ions of the type $V^{n+}(aq)$ because the highly charged ion causes the attached water molecules to be very acidic. The H^+ ions are lost to give the oxycations VO_2^+ and VO^{2+}. The hydrated V^{3+} and V^{2+} ions are easily oxidized and thus can function as reducing agents in aqueous solution.

Chemical connections
Titanium Dioxide—Miracle Coating

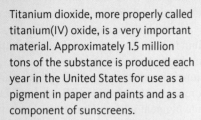

Titanium dioxide, more properly called titanium(IV) oxide, is a very important material. Approximately 1.5 million tons of the substance is produced each year in the United States for use as a pigment in paper and paints and as a component of sunscreens.

In recent years, however, scientists have found a new use for TiO_2. When surfaces are coated with titanium dioxide, they become resistant to dirt and bacteria. For example, the Pilkington Glass Company is now making glass coated with TiO_2 that cleans itself. All the glass needs is sun and rain to keep itself clean. The self-cleaning action arises from two effects. First, the coating of TiO_2 acts as a catalyst in the presence of ultraviolet (UV) light to break down

carbon-based pollutants to carbon dioxide and water. Second, because TiO_2 reduces surface tension, rain-water "sheets" instead of forming droplets on the glass, thereby washing away the grime on the surface of the glass. Although this self-cleaning glass is bad news for window washers, it could save millions of dollars in maintenance costs for owners of commercial buildings.

Because the TiO_2-treated glass requires UV light for its action, it does not work well for interior surfaces where UV light is present only in small amounts. However, a team of Japanese researchers has found that if the TiO_2 coating is doped with nitrogen atoms, it will catalyze the breakdown of dirt in the presence of visible light as well as

UV light. Studies also show that this N-doped TiO_2 surface coating kills many types of bacteria in the presence of visible or ultraviolet light. This discovery could lead to products such as self-sterilizing bathroom tiles, counters, and toilets. In addition, because the TiO_2 on the surface of glass has such a strong attraction for water molecules (greatly lowering the surface tension), water does not bead up to form droplets. Just as this effect produces sheeting action on exterior glass, so it prevents interior windows and mirrors from "fogging up."

Titanium dioxide, a cheap and plentiful material, may prove to be worth its weight in gold as a surface coating.

Table 21.4 | Oxidation States and Species for Vanadium in Aqueous Solution

Oxidation State of Vanadium	Species in Aqueous Solution
+5	VO_2^+ (yellow)
+4	VO^{2+} (blue)
+3	$V^{3+}(aq)$ (blue-green)
+2	$V^{2+}(aq)$ (violet)

Table 21.5 | Typical Chromium Compounds

Oxidation State of Chromium	Examples of Compounds (X = halogen)
+2	CrX_2
+3	CrX_3
	Cr_2O_3 (green)
	$Cr(OH)_3$ (blue-green)
+6	$K_2Cr_2O_7$ (orange)
	Na_2CrO_4 (yellow)
	CrO_3 (red)

Although *chromium* is relatively rare, it is a very important industrial material. The chief ore of chromium is chromite ($FeCr_2O_4$), which can be reduced by carbon to give *ferrochrome*,

$$FeCr_2O_4(s) + 4C(s) \longrightarrow \underbrace{Fe(s) + 2Cr(s)}_{\text{Ferrochrome}} + 4CO(g)$$

which can be added directly to iron in the steelmaking process. Chromium metal, which is often used to plate steel, is hard and brittle and maintains a bright surface by developing a tough invisible oxide coating.

Chromium commonly forms compounds in which it has the oxidation state +2, +3, or +6, as shown in Table 21.5. The Cr^{2+} (chromous) ion is a powerful reducing agent in aqueous solution. In fact, traces of O_2 in other gases can be removed by bubbling through a Cr^{2+} solution:

$$4Cr^{2+}(aq) + O_2(g) + 4H^+(aq) \longrightarrow 4Cr^{3+}(aq) + 2H_2O(l)$$

The chromium(VI) species are excellent oxidizing agents, especially in acidic solution, where chromium(VI) as the dichromate ion ($Cr_2O_7^{2-}$) is reduced to the Cr^{3+} ion:

$$Cr_2O_7^{2-}(aq) + 14H^+(aq) + 6e^- \longrightarrow 2Cr^{3+}(aq) + 7H_2O(l) \qquad \mathscr{E}° = 1.33 \text{ V}$$

The oxidizing ability of the dichromate ion is strongly pH-dependent, increasing as $[H^+]$ increases, as predicted by Le Châtelier's principle. In basic solution, chromium(VI) exists as the chromate ion, a much less powerful oxidizing agent:

$$CrO_4^{2-}(aq) + 4H_2O(l) + 3e^- \longrightarrow Cr(OH)_3(s) + 5OH^-(aq) \qquad \mathscr{E}° = -0.13 \text{ V}$$

The structures of the $Cr_2O_7^{2-}$ and CrO_4^{2-} ions are shown in Fig. 21.4.

Figure 21.4 | The structures of the chromium(VI) anions: (a) $Cr_2O_7^{2-}$, which exists in acidic solution, and (b) CrO_4^{2-}, which exists in basic solution.

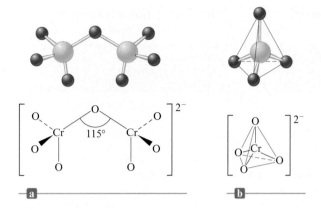

Red chromium(VI) oxide (CrO_3) dissolves in water to give a strongly acidic, red-orange solution:

$$2CrO_3(s) + H_2O(l) \longrightarrow 2H^+(aq) + Cr_2O_7^{2-}(aq)$$

It is possible to precipitate bright orange dichromate salts, such as $K_2Cr_2O_7$, from these solutions. When made basic, the solution turns yellow, and chromate salts such as Na_2CrO_4 can be obtained. A mixture of chromium(VI) oxide and concentrated sulfuric acid, commonly called *cleaning solution,* is a powerful oxidizing medium that can remove organic materials from analytical glassware, yielding a very clean surface.

Manganese is relatively abundant (0.1% of the earth's crust), although no significant sources are found in the United States. The most common use of manganese is in the production of an especially hard steel used for rock crushers, bank vaults, and armor plate. One interesting source of manganese is from *manganese nodules* found on the ocean floor. These roughly spherical "rocks" contain mixtures of manganese and iron oxides as well as smaller amounts of other metals such as cobalt, nickel, and copper. Apparently, the nodules were formed at least partly by the action of marine organisms. Because of the abundance of these nodules, there is much interest in developing economical methods for their recovery and processing.

Manganese can exist in all oxidation states from +2 to +7, although +2 and +7 are the most common. Manganese(II) forms an extensive series of salts with all the common anions. In aqueous solution Mn^{2+} forms $Mn(H_2O)_6^{2+}$, which has a light pink color. Manganese(VII) is found in the intensely purple permanganate ion (MnO_4^-). Widely used as an analytical reagent in acidic solution, the MnO_4^- ion behaves as a strong oxidizing agent, with the manganese becoming Mn^{2+}:

$$MnO_4^-(aq) + 8H^+(aq) + 5e^- \longrightarrow Mn^{2+}(aq) + 4H_2O(l) \qquad \mathscr{E}° = 1.51 \text{ V}$$

Several typical compounds of manganese are shown in Table 21.6.

Iron is the most abundant heavy metal (4.7% of the earth's crust) and the most important to our civilization. It is a white, lustrous, not particularly hard metal that is very reactive toward oxidizing agents. For example, in moist air it is rapidly oxidized by oxygen to form rust, a mixture of iron oxides.

The chemistry of iron involves mainly its +2 and +3 oxidation states. Typical compounds are listed in Table 21.7. In aqueous solutions iron(II) salts are generally light green because of the presence of $Fe(H_2O)_6^{2+}$. Although the $Fe(H_2O)_6^{3+}$ ion is colorless, aqueous solutions of iron(III) salts are usually yellow to brown in color due to the presence of $Fe(OH)(H_2O)_5^{2+}$, which results from the acidity of $Fe(H_2O)_6^{3+}$ ($K_a = 6 \times 10^{-3}$):

$$Fe(H_2O)_6^{3+}(aq) \rightleftharpoons Fe(OH)(H_2O)_5^{2+}(aq) + H^+(aq)$$

Although *cobalt* is relatively rare, it is found in ores such as smaltite ($CoAs_2$) and cobaltite (CoAsS) in large enough concentrations to make its production economically

Table 21.6 | Some Compounds of Manganese in Its Most Common Oxidation States

Oxidation State of Manganese	Examples of Compounds
+2	$Mn(OH)_2$ (pink) MnS (salmon) $MnSO_4$ (reddish) $MnCl_2$ (pink)
+4	MnO_2 (dark brown)
+7	$KMnO_4$ (purple)

Table 21.7 | Typical Compounds of Iron

Oxidation State of Iron	Examples of Compounds
+2	FeO (black) FeS (brownish black) $FeSO_4 \cdot 7H_2O$ (green) $K_4Fe(CN)_6$ (yellow)
+3	$FeCl_3$ (brownish black) Fe_2O_3 (reddish brown) $K_3Fe(CN)_6$ (red) $Fe(SCN)_3$ (red)
+2, +3 (mixture)	Fe_3O_4 (black) $KFe[Fe(CN)_6]$ (deep blue, "Prussian blue")

Table 21.8 | Typical Compounds of Cobalt

Oxidation State of Cobalt	Examples of Compounds
+2	$CoSO_4$ (dark blue) $[Co(H_2O)_6]Cl_2$ (pink) $[Co(H_2O)_6](NO_3)_2$ (red) CoS (black) CoO (greenish brown)
+3	CoF_3 (brown) Co_2O_3 (charcoal) $K_3[Co(CN)_6]$ (yellow) $[Co(NH_3)_6]Cl_3$ (yellow)

Table 21.9 | Typical Compounds of Nickel

Oxidation State of Nickel	Examples of Compounds
+2	$NiCl_2$ (yellow) $[Ni(H_2O)_6]Cl_2$ (green) NiO (greenish black) NiS (black) $[Ni(H_2O)_6]SO_4$ (green) $[Ni(NH_3)_6](NO_3)_2$ (blue)

feasible. Cobalt is a hard, bluish white metal mainly used in alloys such as stainless steel and stellite, an alloy of iron, copper, and tungsten that is used in surgical instruments.

The chemistry of cobalt involves mainly its +2 and +3 oxidation states, although compounds containing cobalt in the 0, +1, or +4 oxidation state are known. Aqueous solutions of cobalt(II) salts contain the $Co(H_2O)_6^{2+}$ ion, which has a characteristic rose color. Cobalt forms a wide variety of coordination compounds, many of which will be discussed in later sections of this chapter. Some typical cobalt compounds are listed in Table 21.8.

Nickel, which ranks twenty-fourth in elemental abundance in the earth's crust, is found in ores, where it is combined mainly with arsenic, antimony, and sulfur. Nickel metal, a silvery white substance with high electrical and thermal conductivities, is quite resistant to corrosion and is often used for plating more active metals. Nickel is also widely used in the production of alloys such as steel.

Nickel in compounds is almost exclusively in the +2 oxidation state. Aqueous solutions of nickel(II) salts contain the $Ni(H_2O)_6^{2+}$ ion, which has a characteristic emerald green color. Coordination compounds of nickel(II) will be discussed later in this chapter. Some typical nickel compounds are listed in Table 21.9.

Copper, widely distributed in nature in ores containing sulfides, arsenides, chlorides, and carbonates, is valued for its high electrical conductivity and its resistance to corrosion. It is widely used for plumbing, and 50% of all copper produced annually is used for electrical applications. Copper is a major constituent in several well-known alloys (Table 21.10).

Although copper is not highly reactive (it will not reduce H^+ to H_2, for example), the reddish metal does slowly corrode in air, producing the characteristic green *patina* consisting of basic copper sulfate

$$3Cu(s) + 2H_2O(l) + SO_2(g) + 2O_2(g) \longrightarrow Cu_3(OH)_4SO_4(s)$$

Basic copper sulfate

and other similar compounds.

Copper roofs and bronze statues, such as the Statue of Liberty, turn green in air because $Cu_3(OH)_4SO_4$ and $Cu_4(OH)_6SO_4$ form.

An aqueous solution containing the Ni^{2+} ion.

Table 21.10 | Alloys Containing Copper

Alloy	Composition (% by mass)
Brass	Cu (20–97), Zn (2–80), Sn (0–14), Pb (0–12), Mn (0–25)
Bronze	Cu (50–98), Sn (0–35), Zn (0–29), Pb (0–50), P (0–3)
Sterling silver	Cu (7.5), Ag (92.5)
Gold (18-karat)	Cu (5–15), Au (75), Ag (10–20)
Gold (14-karat)	Cu (12–28), Au (58), Ag (4–30)

Table 21.11 | Typical Compounds of Copper

Oxidation State of Copper	Examples of Compounds
+1	Cu_2O (red)
	Cu_2S (black)
	CuCl (white)
+2	CuO (black)
	$CuSO_4 \cdot 5H_2O$ (blue)
	$CuCl_2 \cdot 2H_2O$ (green)
	$[Cu(H_2O)_6](NO_3)_2$ (blue)

The chemistry of copper principally involves the $+2$ oxidation state, but many compounds containing copper(I) are also known. Aqueous solutions of copper(II) salts are a characteristic bright blue color due to the presence of the $Cu(H_2O)_6^{2+}$ ion. Table 21.11 lists some typical copper compounds.

Although trace amounts of copper are essential for life, copper in large amounts is quite toxic; copper salts are used to kill bacteria, fungi, and algae. For example, paints containing copper are used on ship hulls to prevent fouling by marine organisms.

Widely dispersed in the earth's crust, *zinc* is refined mainly from sphalerite (ZnS), which often occurs with galena (PbS). Zinc is a white, lustrous, very active metal that behaves as an excellent reducing agent and tarnishes rapidly. About 90% of the zinc produced is used for galvanizing steel. Zinc forms colorless salts in the $+2$ oxidation state.

21.3 | Coordination Compounds

Transition metal ions characteristically form coordination compounds, which are usually colored and often paramagnetic. A **coordination compound** typically consists of a *complex ion,* a transition metal ion with its attached ligands (see Section 15.8), and **counterions**, anions or cations as needed to produce a compound with no net charge. The substance $[Co(NH_3)_5Cl]Cl_2$ is a typical coordination compound. The brackets indicate the composition of the complex ion, in this case $Co(NH_3)_5Cl^{2+}$, and the two Cl^- counterions are shown outside the brackets. Note that in this compound one Cl^- acts as a ligand along with the five NH_3 molecules. In the solid state this compound consists of the large $Co(NH_3)_5Cl^{2+}$ cations and twice as many Cl^- anions, all packed together as efficiently as possible. When dissolved in water, the solid behaves like any ionic solid; the cations and anions are assumed to separate and move about independently:

$$[Co(NH_3)_5Cl]Cl_2(s) \xrightarrow{H_2O} Co(NH_3)_5Cl^{2+}(aq) + 2Cl^-(aq)$$

Coordination compounds have been known since about 1700, but their true nature was not understood until the 1890s when a young Swiss chemist named Alfred Werner (1866–1919) proposed that transition metal ions have two types of valence (combining ability). One type of valence, which Werner called the *secondary valence,* refers to the ability of a metal ion to bind to Lewis bases (ligands) to form complex ions. The other type, the *primary valence,* refers to the ability of the metal ion to form ionic bonds with oppositely charged ions. Thus Werner explained that the compound, originally written as $CoCl_3 \cdot 5NH_3$, was really $[Co(NH_3)_5Cl]Cl_2$, where the Co^{3+} ion has a primary valence of 3, satisfied by the three Cl^- ions, and a secondary valence of 6, satisfied by the six ligands (five NH_3 and one Cl^-). We now call the primary valence the **oxidation state** and the secondary valence the **coordination number**, which reflects the number of bonds formed between the metal ion and the ligands in the complex ion.

Coordination Number

The number of bonds formed by metal ions to ligands in complex ions varies from two to eight depending on the size, charge, and electron configuration of the transition metal ion. As shown in Table 21.12, 6 is the most common coordination number, followed closely by 4, with a few metal ions showing a coordination number of 2. Many metal ions show more than one coordination number, and there is really no simple way to predict what the coordination number will be in a particular case. The typical geometries for the various common coordination numbers are shown in Fig. 21.5. Note that six ligands produce an octahedral arrangement around the metal ion. Four ligands can form either a tetrahedral or a square planar arrangement, and two ligands give a linear structure.

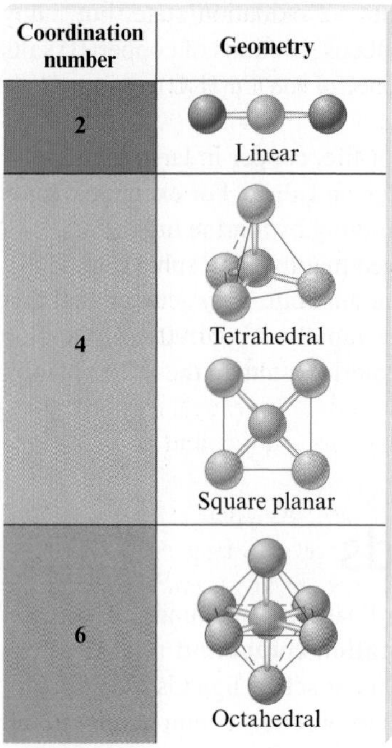

Coordination number	Geometry
2	Linear
4	Tetrahedral
4	Square planar
6	Octahedral

Figure 21.5 | The ligand arrangements for coordination numbers 2, 4, and 6.

Table 21.12 | Typical Coordination Numbers for Some Common Metal Ions

M	Coordination Numbers	M^{2+}	Coordination Numbers	M^{3+}	Coordination Numbers
Cu^+	2, 4	Mn^{2+}	4, 6	Sc^{3+}	6
Ag^+	2	Fe^{2+}	6	Cr^{3+}	6
Au^+	2, 4	Co^{2+}	4, 6	Co^{3+}	6
		Ni^{2+}	4, 6		
		Cu^{2+}	4, 6	Au^{3+}	4
		Zn^{2+}	4, 6		

Ligands

A **ligand** is a *neutral molecule or ion having a lone electron pair that can be used to form a bond to a metal ion.* The formation of a metal–ligand bond therefore can be described as the interaction between a Lewis base (the ligand) and a Lewis acid (the metal ion). The resulting bond is often called a **coordinate covalent bond**.

A *ligand that can form one bond to a metal ion* is called a **monodentate ligand**, or a **unidentate ligand** (from root words meaning "one tooth"). Examples of unidentate ligands are shown in Table 21.13.

Some ligands have more than one atom with a lone electron pair that can be used to bond to a metal ion. Such ligands are said to be **chelating ligands**, or **chelates** (from the Greek word *chele* for "claw"). A ligand that can form two bonds to a metal ion is called a **bidentate ligand**. A very common bidentate ligand is ethylenediamine (abbreviated en), which is shown coordinating to a metal ion in Fig. 21.6(a). Note the re-

Table 21.13 | Some Common Ligands

Type	Examples
Unidentate/ monodentate	H_2O CN^- SCN^- (thiocyanate) X^- (halides) NH_3 NO_2^- (nitrite) OH^-
Bidentate	Oxalate 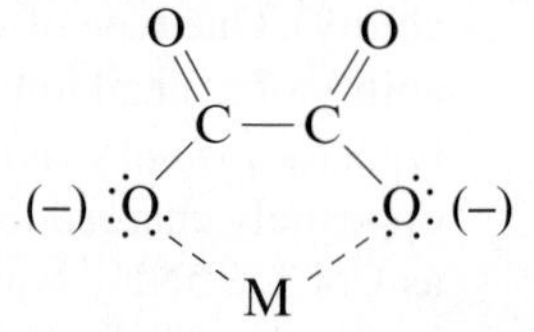 Ethylenediamine (en) 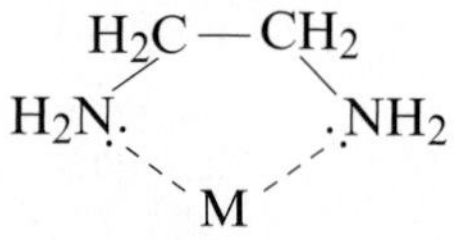
Polydentate	Diethylenetriamine (dien) $H_2\ddot{N}-(CH_2)_2-\ddot{N}H-(CH_2)_2-\ddot{N}H_2$ Three coordinating atoms Ethylenediaminetetraacetate (EDTA) Six coordinating atoms

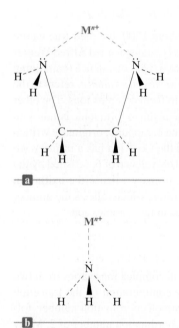

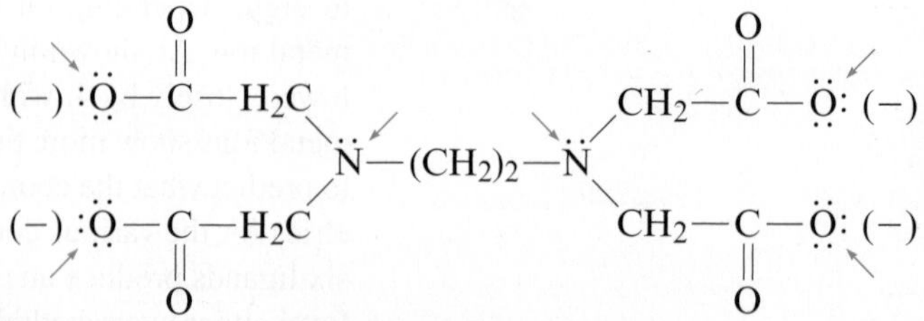

Figure 21.6 | (a) The bidentate ligand ethylenediamine can bond to the metal ion through the lone pair on each nitrogen atom, thus forming two coordinate covalent bonds. (b) Ammonia is a monodentate ligand.

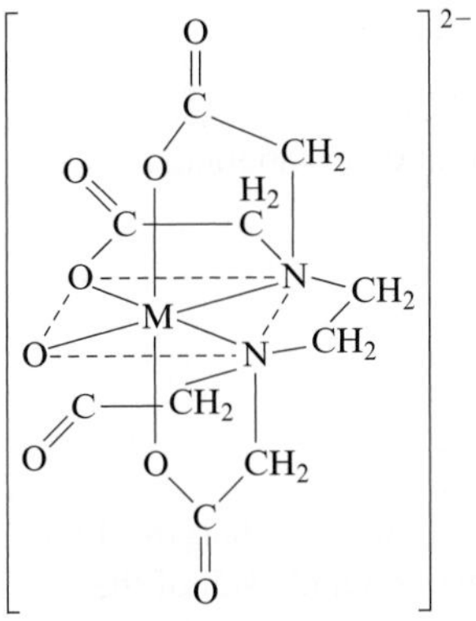

Figure 21.7 | The coordination of EDTA with a 2+ metal ion.

Table 21.14 | Names of Some Common Unidentate Ligands

Neutral Molecules	
Aqua	H_2O
Ammine	NH_3
Methylamine	CH_3NH_2
Carbonyl	CO
Nitrosyl	NO
Anions	
Fluoro	F^-
Chloro	Cl^-
Bromo	Br^-
Iodo	I^-
Hydroxo	OH^-
Cyano	CN^-

Table 21.15 | Latin Names Used for Some Metal Ions in Anionic Complex Ions

Metal	Name in an Anionic Complex
Iron	Ferrate
Copper	Cuprate
Lead	Plumbate
Silver	Argentate
Gold	Aurate
Tin	Stannate

lationship between this ligand and the unidentate ligand ammonia [Fig. 21.6(b)]. Oxalate, another typical bidentate ligand, is shown in Table 21.13.

Ligands that can form more than two bonds to a metal ion are called *polydentate ligands*. Some ligands can form as many as six bonds to a metal ion. The best-known example is ethylenediaminetetraacetate (abbreviated EDTA), which is shown in Table 21.13. This ligand virtually surrounds the metal ion (Fig. 21.7), coordinating through six atoms (a *hexadentate ligand*). As might be expected from the large number of coordination sites, EDTA forms very stable complex ions with most metal ions and is used as a "scavenger" to remove toxic heavy metals such as lead from the human body. It is also used as a reagent to analyze solutions for their metal ion content. EDTA is found in countless consumer products, such as soda, beer, salad dressings, bar soaps, and most cleaners. In these products EDTA ties up trace metal ions that would otherwise catalyze decomposition and produce unwanted precipitates.

Some even more complicated ligands are found in biological systems, where metal ions play crucial roles in catalyzing reactions, transferring electrons, and transporting and storing oxygen. A discussion of these complex ligands will follow in Section 21.7.

Nomenclature

In Werner's lifetime, no system was used to name coordination compounds. Names of the compounds were commonly based on colors and names of discoverers. As the field expanded and more coordination compounds were identified, an orderly system of nomenclature became necessary. A simplified version of this system is summarized by the following rules.

Rules for Naming Coordination Compounds

> As with any ionic compound, *the cation is named before the anion.*

> In naming a complex ion, *the ligands are named before the metal ion.*

> In naming ligands, *an o is added to the root name of an anion.* For example, the halides as ligands are called *fluoro, chloro, bromo,* and *iodo;* hydroxide is *hydroxo;* cyanide is *cyano;* and so on. *For a neutral ligand, the name of the molecule is used,* with the exception of H_2O, NH_3, CO, and NO, as illustrated in Table 21.14.

> *The prefixes mono-, di-, tri-, tetra-, penta-, and hexa- are used to denote the number of simple ligands.* The prefixes *bis-, tris-, tetrakis-,* and so on are also used, especially for more complicated ligands or ones that already contain *di-, tri-,* and so on.

> *The oxidation state of the central metal ion is designated by a Roman numeral in parentheses.*

> *When more than one type of ligand is present, they are named alphabetically.** Prefixes do not affect the order.

> *If the complex ion has a negative charge, the suffix -ate is added to the name of the metal.* Sometimes the Latin name is used to identify the metal (Table 21.15).

The application of these rules is shown in Example 21.1.

---

*In an older system the negatively charged ligands were named first, then neutral ligands, with positively charged ligands named last. We will follow the newer convention in this text.

Naming Coordination Compounds I

Give the systematic name for each of the following coordination compounds.

a. $[Co(NH_3)_5Cl]Cl_2$

b. $K_3Fe(CN)_6$

c. $[Fe(en)_2(NO_2)_2]_2SO_4$

Solution

a. To determine the oxidation state of the metal ion, we examine the charges of all ligands and counterions. The ammonia molecules are neutral and each of the chloride ions has a $1-$ charge, so the cobalt ion must have a $3+$ charge to produce a neutral compound. Thus cobalt has the oxidation state $+3$, and we use cobalt(III) in the name.

  The ligands include one Cl^- ion and five NH_3 molecules. The chloride ion is designated as *chloro,* and each ammonia molecule is designated as *ammine.* The prefix *penta-* indicates that there are five NH_3 ligands present. The name of the complex cation is therefore pentaamminechlorocobalt(III). Note that the ligands are named alphabetically, disregarding the prefix. Since the counterions are chloride ions, the compound is named as a chloride salt:

$$\underbrace{\text{Pentaamminechlorocobalt(III)}}_{\text{Cation}} \underbrace{\text{chloride}}_{\text{Anion}}$$

An aqueous solution of $[Co(NH_3)_5Cl]Cl_2$.

b. First, we determine the oxidation state of the iron by considering the other charged species. The compound contains three K^+ ions and six CN^- ions. Therefore, the iron must carry a charge of $3+$, giving a total of six positive charges to balance the six negative charges. The complex ion present is thus $Fe(CN)_6^{3-}$. The cyanide ligands are each designated *cyano,* and the prefix *hexa-* indicates that six are present. Since the complex ion is an anion, we use the Latin name *ferrate.* The oxidation state is indicated by (III) at the end of the name. The anion name is therefore hexacyanoferrate(III). The cations are K^+ ions, which are simply named potassium. Putting this together gives the name

$$\underbrace{\text{Potassium}}_{\text{Cation}} \underbrace{\text{hexacyanoferrate(III)}}_{\text{Anion}}$$

(The common name of this compound is potassium ferricyanide.)

c. We first determine the oxidation state of the iron by looking at the other charged species: four NO_2^- ions and one SO_4^{2-} ion. The ethylenediamine is neutral. Thus the two iron ions must carry a total positive charge of 6 to balance the six negative charges. This means that each iron has a $+3$ oxidation state and is designated as iron(III).

  Since the name ethylenediamine already contains *di,* we use *bis-* instead of *di-* to indicate the two en ligands. The name for NO_2^- as a ligand is *nitro,* and the prefix *di-* indicates the presence of two NO_2^- ligands. Since the anion is sulfate, the compound's name is

$$\underbrace{\text{Bis(ethylenediamine)dinitroiron(III)}}_{\text{Cation}} \underbrace{\text{sulfate}}_{\text{Anion}}$$

Because the complex ion is a cation, the Latin name for iron is not used.

Solid $K_3Fe(CN)_6$.

See Exercises 21.34 through 21.36

Sign in at http://login.cengagebrain .com to try this Interactive Example in **OWL**.

Interactive Example 21.2

Naming Coordination Compounds II

Given the following systematic names, give the formula of each coordination compound.

a. Triamminebromoplatinum(II) chloride

b. Potassium hexafluorocobaltate(III)

Solution

a. *Triammine* signifies three ammonia ligands, and *bromo* indicates one bromide ion as a ligand. The oxidation state of platinum is +2, as indicated by the Roman numeral II. Thus the complex ion is $[Pt(NH_3)_3Br]^+$. One chloride ion is needed to balance the 1+ charge of this cation. The formula of the compound is $[Pt(NH_3)_3Br]Cl$. Note that brackets enclose the complex ion.

b. The complex ion contains six fluoride ligands attached to a Co^{3+} ion to give $CoF_6{}^{3-}$. Note that the *-ate* ending indicates that the complex ion is an anion. The cations are K^+ ions, and three are required to balance the 3− charge on the complex ion. Thus the formula is $K_3[CoF_6]$.

See Exercises 21.37 and 21.38

21.4 | Isomerism

When two or more species have the same formula but different properties, they are said to be **isomers**. Although isomers contain exactly the same types and numbers of atoms, the arrangements of the atoms differ, and this leads to different properties. We will consider two main types of isomerism: **structural isomerism**, where the isomers contain the same atoms but one or more bonds differ, and **stereoisomerism**, where all the bonds in the isomers are the same but the spatial arrangements of the atoms are different. Each of these classes also has subclasses (Fig. 21.8), which we will now consider.

Structural Isomerism

The first type of structural isomerism we will consider is **coordination isomerism**, in which the composition of the complex ion varies. For example, $[Cr(NH_3)_5SO_4]Br$ and $[Cr(NH_3)_5Br]SO_4$ are coordination isomers. In the first case, $SO_4{}^{2-}$ is coordinated to Cr^{3+}, and Br^- is the counterion; in the second case, the roles of these ions are reversed.

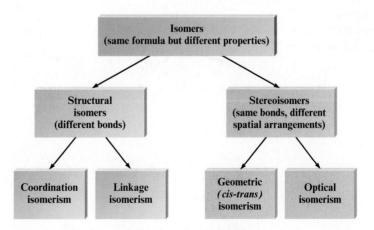

Figure 21.8 | Some classes of isomers.

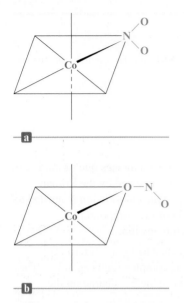

Figure 21.9 | As a ligand, NO_2^- can bond to a metal ion (a) through a lone pair on the nitrogen atom or (b) through a lone pair on one of the oxygen atoms.

Another example of coordination isomerism is the $[Co(en)_3][Cr(ox)_3]$ and $[Cr(en)_3]$ $[Co(ox)_3]$ pair, where ox represents the oxalate ion, a bidentate ligand shown in Table 21.13.

In a second type of structural isomerism, **linkage isomerism**, the composition of the complex ion is the same, but the point of attachment of at least one of the ligands differs. Two ligands that can attach to metal ions in different ways are thiocyanate (SCN^-), which can bond through lone electron pairs on the nitrogen or the sulfur atom, and the nitrite ion (NO_2^-), which can bond through lone electron pairs on the nitrogen or the oxygen atom. For example, the following two compounds are linkage isomers:

$$[Co(NH_3)_4(NO_2)Cl]Cl$$
Tetraamminechloronitrocobalt(III) chloride
(yellow)

$$[Co(NH_3)_4(ONO)Cl]Cl$$
Tetraamminechloronitritocobalt(III) chloride
(red)

In the first case, the NO_2^- ligand is called *nitro* and is attached to Co^{3+} through the nitrogen atom; in the second case, the NO_2^- ligand is called *nitrito* and is attached to Co^{3+} through an oxygen atom (Fig. 21.9).

Stereoisomerism

Stereoisomers have the same bonds but different spatial arrangements of the atoms. One type, **geometrical isomerism**, or ***cis–trans* isomerism**, occurs when atoms or groups of atoms can assume different positions around a rigid ring or bond. An important example is the compound $Pt(NH_3)_2Cl_2$, which has a square planar structure. The two possible arrangements of the ligands are shown in Fig. 21.10. In the ***trans* isomer**, the ammonia molecules are across (*trans*) from each other. In the ***cis* isomer**, the ammonia molecules are next (*cis*) to each other.

Geometrical isomerism also occurs in octahedral complex ions. For example, the compound $[Co(NH_3)_4Cl_2]Cl$ has *cis* and *trans* isomers (Fig. 21.11).

A second type of stereoisomerism is called **optical isomerism** because the isomers have opposite effects on plane-polarized light. When light is emitted from a source such as a glowing filament, the oscillating electric fields of the photons in the beam are oriented randomly, as shown in Fig. 21.12. If this light is passed through a polarizer, only the photons with electric fields oscillating in a single plane remain, constituting *plane-polarized light.*

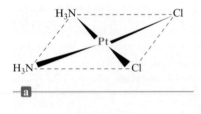

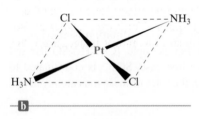

Figure 21.10 | (a) The *cis* isomer of $Pt(NH_3)_2Cl_2$ (yellow). (b) The *trans* isomer of $Pt(NH_3)_2Cl_2$ (pale yellow).

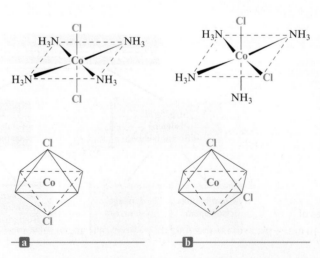

Figure 21.11 |
(a) The *trans* isomer of $[Co(NH_3)_4Cl_2]^+$. The chloride ligands are directly across from each other. (b) The *cis* isomer of $[Co(NH_3)_4Cl_2]^+$. The chloride ligands in this case share an edge of the octahedron. Because of their different structures, the *trans* isomer of $[Co(NH_3)_4Cl_2]Cl$ is green and the *cis* isomer is violet.

Figure 21.12 | Unpolarized light consists of waves vibrating in many different planes (indicated by the arrows). The polarizing filter blocks all waves except those vibrating in a given plane.

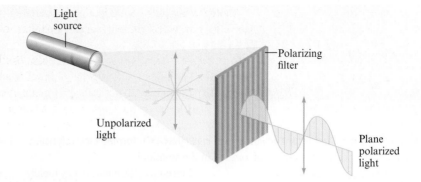

In 1815, a French physicist, Jean Biot (1774–1862), showed that certain crystals could rotate the plane of polarization of light. Later it was found that solutions of certain compounds could do the same thing (Fig. 21.13). Louis Pasteur (1822–1895) was the first to understand this behavior. In 1848 he noted that solid sodium ammonium tartrate ($NaNH_4C_4H_4O_4$) existed as a mixture of two types of crystals, which he painstakingly separated with tweezers. Separate solutions of these two types of crystals rotated plane-polarized light in exactly opposite directions. This led to a connection between optical activity and molecular structure.

We now realize that optical activity is exhibited by molecules that have *nonsuperimposable mirror images.* Your hands are nonsuperimposable mirror images (Fig. 21.15). The two hands are related like an object and its mirror image; one hand cannot be turned to make it identical to the other. Many molecules show this same feature, such as the complex ion $[Co(en)_3]^{3+}$ shown in Fig. 21.16. Objects that have nonsuperimposable mirror images are said to be **chiral** (from the Greek word *cheir,* meaning "hand").

The isomers of $[Co(en)_3]^{3+}$ (Fig. 21.17) are nonsuperimposable mirror images called **enantiomers**, which rotate plane-polarized light in opposite directions and are thus optical isomers. The isomer that rotates the plane of light to the right (when viewed down the beam of oncoming light) is said to be *dextrorotatory,* designated by *d.* The isomer that rotates the plane of light to the left is *levorotatory* (*l*). An equal mixture of the *d* and *l* forms in solution, called a *racemic mixture,* does not rotate the plane of the polarized light at all because the two opposite effects cancel each other.

Geometrical isomers are not necessarily optical isomers. For instance, the *trans* isomer of $[Co(en)_2Cl_2]^+$ shown in Fig. 21.17 is identical to its mirror image. Since this

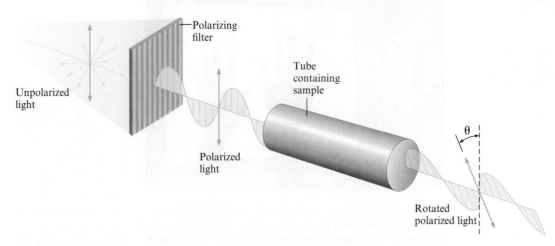

Figure 21.13 | The rotation of the plane of polarized light by an optically active substance. The angle of rotation is called theta (θ).

Chemical connections
The Importance of Being cis

Some of the most important advancements of science are the results of accidental discoveries—for example, penicillin, Teflon, and the sugar substitutes cyclamate and aspartame. Another important chance discovery occurred in 1964, when a group of scientists using platinum electrodes to apply an electric field to a colony of *E. coli* bacteria noticed that the bacteria failed to divide but continued to grow, forming long, fibrous cells. Further study revealed that cell division was inhibited by small concentrations of cis-Pt(NH$_3$)$_2$Cl$_2$ and cis-Pt(NH$_3$)$_2$Cl$_4$ formed electrolytically in the solution.

Cancerous cells multiply very rapidly because cell division is uncontrolled. Thus these and similar platinum complexes were evaluated as *antitumor agents*, which inhibit the division of cancer cells. The results showed that cis-Pt(NH$_3$)$_2$Cl$_2$ was active against a wide variety of tumors, including testicular and ovarian tumors, which are very resistant to treatment by more traditional methods. However, although the *cis* complex showed significant antitumor activity, the corresponding trans complex had no effect on tumors. This shows the importance of isomerism in biological systems. When drugs are synthesized, great care must be taken to obtain the correct isomer.

Although cis-Pt(NH$_3$)$_2$Cl$_2$ has proven to be a valuable drug, unfortunately it has some troublesome side effects, the most serious being kidney damage. As a result, the search continues for even more effective antitumor agents. Promising candidates are shown in Fig. 21.14. Note that they are all cis complexes.

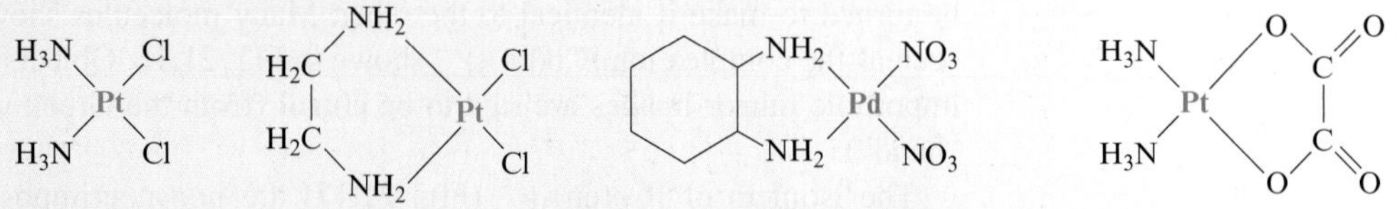

Figure 21.14 | Some *cis* complexes of platinum and palladium that show significant antitumor activity. It is thought that the *cis* complexes work by losing two adjacent ligands and forming coordinate covalent bonds to adjacent bases on a DNA molecule.

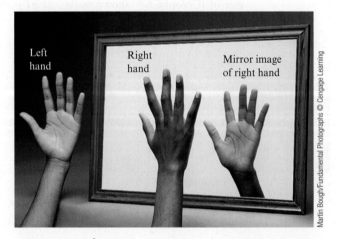

Figure 21.15 | A human hand exhibits a nonsuperimposable mirror image. Note that the mirror image of the right hand (while identical to the left hand) cannot be turned in any way to make it identical to (superimposable on) the actual right hand.

Figure 21.16 │ Isomers I and II of $Co(en)_3^{3+}$ are mirror images (the image of I is identical to II) that cannot be superimposed. That is, there is no way that I can be turned in space so that it is the same as II.

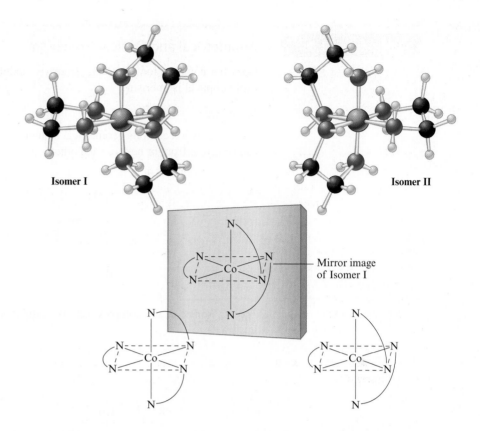

Isomer I

Isomer II

Mirror image of Isomer I

isomer is superimposable on its mirror image, it does not exhibit optical isomerism and is not chiral. On the other hand, cis-$[Co(en)_2Cl_2]^+$ is *not* superimposable on its mirror image; a pair of enantiomers exists for this complex ion (the *cis* isomer is chiral).

Most important biomolecules are chiral, and their reactions are highly structure dependent. For example, a drug can have a particular effect because its molecules can bind to chiral molecules in the body. To bind correctly, however, the correct optical isomer of the drug must be administered. Just as the right hand of one person requires the right hand of another to perform a handshake, a given isomer in the body requires a specific isomer of the drug to bind together. Because of this, the syntheses of drugs, which are usually very complicated molecules, must be carried out in a way that produces the correct "handedness," a requirement that greatly adds to the synthetic difficulties.

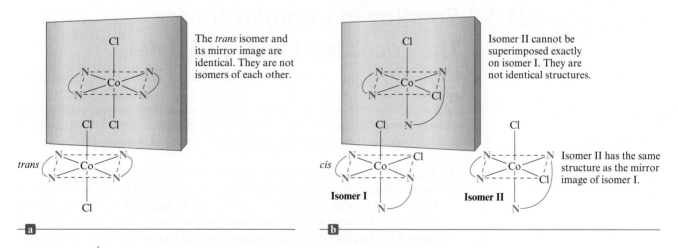

The *trans* isomer and its mirror image are identical. They are not isomers of each other.

trans

Isomer II cannot be superimposed exactly on isomer I. They are not identical structures.

cis

Isomer I

Isomer II

Isomer II has the same structure as the mirror image of isomer I.

ⓐ ⓑ

Figure 21.17 │ (a) The *trans* isomer of $Co(en)_2Cl_2^+$ and its mirror image are identical (superimposable). (b) The *cis* isomer of $Co(en)_2Cl_2^+$ and its mirror image are not superimposable and are thus a pair of optical isomers.

Example 21.3

Geometrical and Optical Isomerism

Does the complex ion $[Co(NH_3)Br(en)_2]^{2+}$ exhibit geometrical isomerism? Does it exhibit optical isomerism?

Solution

The complex ion exhibits geometrical isomerism because the ethylenediamine ligands can be across from or next to each other:

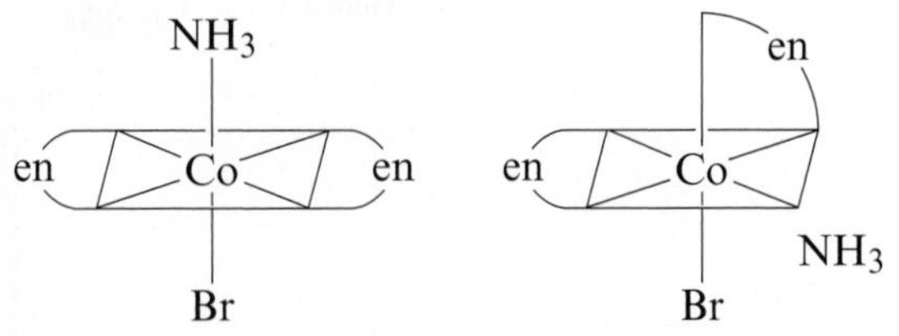

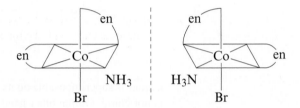

The *cis* isomer of the complex ion also exhibits optical isomerism because its mirror images

cannot be turned in any way to make them superimposable. Thus these mirror-image isomers of the *cis* complex are shown to be enantiomers that will rotate plane-polarized light in opposite directions.

See Exercises 21.47 and 21.48

21.5 | Bonding in Complex Ions: The Localized Electron Model

In Chapters 8 and 9 we considered the localized electron model, a very useful model for describing the bonding in molecules. Recall that a central feature of this model is the formation of hybrid atomic orbitals that are used to share electron pairs to form σ bonds between atoms. This same model can be used to account for the bonding in complex ions, but there are two important points to keep in mind:

1. The VSEPR model for predicting structure generally *does not work for complex ions*. However, we can safely assume that a complex ion with a coordination number of 6 will have an octahedral arrangement of ligands, and complexes with two ligands will be linear. On the other hand, complex ions with a coordination number of 4 can be either tetrahedral or square planar, and there is no completely reliable way to predict which will occur in a particular case.

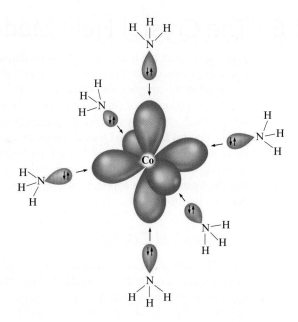

Figure 21.18 | A set of six d^2sp^3 hybrid orbitals on Co^{3+} can accept an electron pair from each of six NH_3 ligands to form the $Co(NH_3)_6^{3+}$ ion.

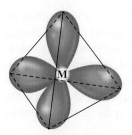

Tetrahedral ligand arrangement; sp^3 hybridization

Square planar ligand arrangement; dsp^2 hybridization

Linear ligand arrangement; sp hybridization

Figure 21.19 | The hybrid orbitals required for tetrahedral, square planar, and linear complex ions. The metal ion hybrid orbitals are empty, and the metal ion bonds to the ligands by accepting lone pairs.

2. The interaction between a metal ion and a ligand can be viewed as a Lewis acid–base reaction with the ligand donating a lone pair of electrons to an *empty* orbital of the metal ion to form a coordinate covalent bond:

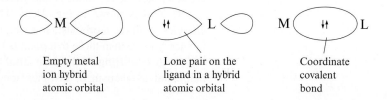

| Empty metal ion hybrid atomic orbital | Lone pair on the ligand in a hybrid atomic orbital | Coordinate covalent bond |

The hybrid orbitals used by the metal ion depend on the number and arrangement of the ligands. For example, accommodating the lone pairs from the six ammonia molecules in the octahedral $Co(NH_3)_6^{3+}$ ion requires a set of six empty hybrid atomic orbitals in an octahedral arrangement. As we discussed in Section 9.1, an octahedral set of orbitals is formed by the hybridization of two d, one s, and three p orbitals to give a set of six d^2sp^3 orbitals (Fig. 21.18).

The hybrid orbitals required on a metal ion in a four-coordinate complex depend on whether the structure is tetrahedral or square planar. For a tetrahedral arrangement of ligands, an sp^3 hybrid set is required (Fig. 21.19). For example, in the tetrahedral $CoCl_4^{2-}$ ion, the Co^{2+} can be described as sp^3 hybridized. A square planar arrangement of ligands requires a dsp^2 hybrid orbital set on the metal ion (Fig. 21.19). For example, in the square planar $Ni(CN)_4^{2-}$ ion, the Ni^{2+} is described as dsp^2 hybridized.

A linear complex requires two hybrid orbitals 180 degrees from each other. This arrangement is given by an sp hybrid set (see Fig. 21.19). Thus, in the linear $Ag(NH_3)_2^+$ ion, the Ag^+ can be described as sp hybridized.

Although the localized electron model can account in a general way for metal–ligand bonds, it is rarely used today because it cannot readily account for important properties of complex ions, such as magnetism and color. Thus we will not pursue the model any further.

21.6 | The Crystal Field Model

The main reason the localized electron model cannot fully account for the properties of complex ions is that it gives no information about how the energies of the *d* orbitals are affected by complex ion formation. This is critical because, as we will see, the color and magnetism of complex ions result from changes in the energies of the metal ion *d* orbitals caused by the metal–ligand interactions.

The **crystal field model** focuses on the energies of the *d* orbitals. In fact, this model is not so much a bonding model as it is an attempt to account for the colors and magnetic properties of complex ions. In its simplest form, the crystal field model assumes that the ligands can be approximated by *negative point charges* and that metal–ligand bonding is *entirely ionic*.

Octahedral Complexes

We will illustrate the fundamental principles of the crystal field model by applying it to an octahedral complex. Figure 21.20 shows the orientation of the 3*d* orbitals relative to an octahedral arrangement of point-charge ligands. The important thing to note is that two of the orbitals, d_{z^2} and $d_{x^2-y^2}$, point their lobes *directly at* the point-charge ligands and three of the orbitals, d_{xz}, d_{yz}, and d_{xy}, point their lobes *between* the point charges.

To understand the effect of this difference, we need to consider which type of orbital is lower in energy. Because the negative point-charge ligands repel negatively charged electrons, the electrons will first fill the *d* orbitals farthest from the ligands to minimize repulsions. In other words, the d_{xz}, d_{yz}, and d_{xy} orbitals (known as the t_{2g} set) are at a *lower energy* in the octahedral complex than are the d_{z^2} and $d_{x^2-y^2}$ orbitals (the e_g set). This is shown in Fig. 21.21. The negative point-charge ligands increase the energies of all the *d* orbitals. However, the orbitals that point at the ligands are raised in energy more than those that point between the ligands.

It is this **splitting of the 3*d* orbital energies** (symbolized by Δ) that explains the color and magnetism of complex ions of the first-row transition metal ions. For

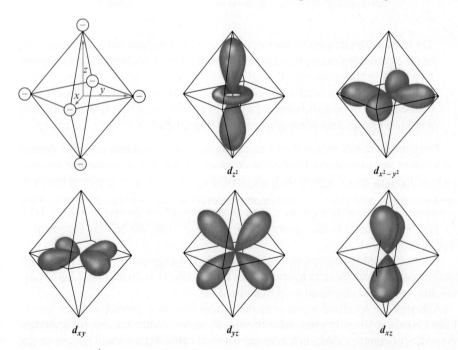

Figure 21.20 | An octahedral arrangement of point-charge ligands and the orientation of the 3*d* orbitals.

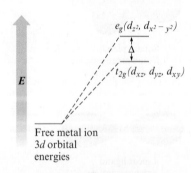

Figure 21.21 | The energies of the 3*d* orbitals for a metal ion in an octahedral complex. The 3*d* orbitals are degenerate (all have the same energy) in the free metal ion. In the octahedral complex the orbitals are split into two sets as shown. The difference in energy between the two sets is designated as Δ (delta).

Figure 21.22 | Possible electron arrangements in the split $3d$ orbitals in an octahedral complex of Co^{3+} (electron configuration $3d^6$). (a) In a strong field (large Δ value), the electrons fill the t_{2g} set first, giving a diamagnetic complex. (b) In a weak field (small Δ value), the electrons occupy all five orbitals before any pairing occurs.

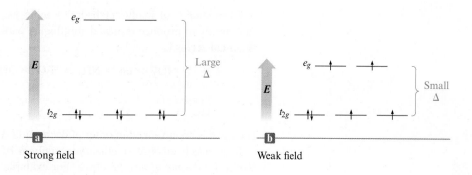

example, in an octahedral complex of Co^{3+} (a metal ion with six $3d$ electrons), there are two possible ways to place the electrons in the split $3d$ orbitals (Fig. 21.22). If the splitting produced by the ligands is very large, a situation called the **strong-field case**, the electrons will pair in the lower-energy t_{2g} orbitals. This gives a *diamagnetic* complex in which all the electrons are paired. On the other hand, if the splitting is small (the **weak-field case**), the electrons will occupy all five orbitals before pairing occurs. In this case the complex has four unpaired electrons and is *paramagnetic*.

The crystal field model allows us to account for the differences in the magnetic properties of $Co(NH_3)_6^{3+}$ and CoF_6^{3-}. The $Co(NH_3)_6^{3+}$ ion is known to be diamagnetic and thus corresponds to the strong-field case, also called the **low-spin case**, since it yields the *minimum* number of unpaired electrons. In contrast, the CoF_6^{3-} ion, which is known to have four unpaired electrons, corresponds to the weak-field case, also known as the **high-spin case**, since it gives the *maximum* number of unpaired electrons.

Critical Thinking

What if you are told the number of unpaired electrons for a coordinate covalent ion and are asked to tell if the ligand produced a strong or weak field? Give an example of a coordinate covalent ion for which you could decide if it produced a strong or weak field and one for which you couldn't, and explain your answers.

Interactive Example 21.4

Sign in at http://login.cengagebrain.com to try this Interactive Example in **OWL**.

Crystal Field Model I

The $Fe(CN)_6^{3-}$ ion is known to have one unpaired electron. Does the CN^- ligand produce a strong or weak field?

Solution

Since the ligand is CN^- and the overall complex ion charge is $3-$, the metal ion must be Fe^{3+}, which has a $3d^5$ electron configuration. The two possible arrangements of the five electrons in the d orbitals split by the octahedrally arranged ligands are

The strong-field case gives one unpaired electron, which agrees with the experimental observation. The CN^- ion is a strong-field ligand toward the Fe^{3+} ion.

See Exercises 21.53 and 21.54

From studies of many octahedral complexes, we can arrange ligands in order of their ability to produce *d*-orbital splitting. A partial listing of ligands in this **spectrochemical series** is

$$CN^- > NO_2^- > en > NH_3 > H_2O > OH^- > F^- > Cl^- > Br^- > I^-$$

Strong-field
ligands
(large Δ)

Weak-field
ligands
(small Δ)

The ligands are arranged in order of decreasing Δ values toward a given metal ion.

It also has been observed that *the magnitude of Δ for a given ligand increases as the charge on the metal ion increases.* For example, NH_3 is a weak-field ligand toward Co^{2+} but acts as a strong-field ligand toward Co^{3+}. This makes sense; as the metal ion charge increases, the ligands will be drawn closer to the metal ion because of the increased charge density. As the ligands move closer, they cause greater splitting of the *d* orbitals and produce a larger Δ value.

Interactive Example 21.5

Sign in at http://login.cengagebrain.com to try this Interactive Example in **OWL**.

Crystal Field Model II

Predict the number of unpaired electrons in the complex ion $Cr(CN)_6^{4-}$.

Solution

The net charge of $4-$ means that the metal ion present must be Cr^{2+} ($-6 + 2 = -4$), which has a $3d^4$ electron configuration. Since CN^- is a strong-field ligand (see the spectrochemical series), the correct crystal field diagram for $Cr(CN)_6^{4-}$ is

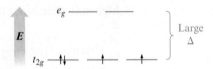

The complex ion will have two unpaired electrons. Note that the CN^- ligand produces such a large splitting that all four electrons will occupy the t_{2g} set even though two of the electrons must be paired in the same orbital.

See Exercises 21.55 and 21.56

We have seen how the crystal field model can account for the magnetic properties of octahedral complexes. The same model also can explain the colors of these complex ions. For example, $Ti(H_2O)_6^{3+}$, an octahedral complex of Ti^{3+}, which has a $3d^1$ electron configuration, is violet because it absorbs light in the middle of the visible region of the spectrum (Fig. 21.23). When a substance absorbs certain wavelengths of light in the visible region, the color of the substance is determined by the wavelengths of visible light that remain. We say that the substance exhibits the color *complementary* to those absorbed. The $Ti(H_2O)_6^{3+}$ ion is violet because it absorbs light in the

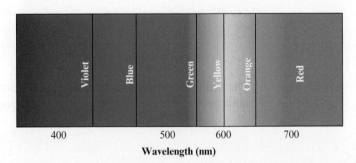

Figure 21.23 | The visible spectrum.

Chemical connections
Transition Metal Ions Lend Color to Gems

The beautiful pure color of gems, so valued by cultures everywhere, arises from trace transition metal ion impurities in minerals that would otherwise be colorless. For example, the stunning red of a ruby, the most valuable of all gemstones, is caused by Cr^{3+} ions, which replace about 1% of the Al^{3+} ions in the mineral corundum, which is a form of aluminum oxide (Al_2O_3) that is nearly as hard as diamond. In the corundum structure the Cr^{3+} ions are surrounded by six oxide ions at the vertices of an octahedron. This leads to the characteristic octahedral splitting of chromium's $3d$ orbitals, such that the Cr^{3+} ions absorb strongly in the blue-violet and yellow-green regions of the visible spectrum but transmit red light to give the characteristic ruby color. (On the other hand, if some of the Al^{3+} ions in the corundum are replaced by a mixture of Fe^{2+}, Fe^{3+}, and Ti^{4+} ions, the gem is a sapphire with its brilliant blue color; or if some of the Al^{3+} ions are replaced by Fe^{3+} ions, the stone is a yellow topaz.)

Emeralds are derived from the mineral beryl, a beryllium aluminum silicate (empirical formula $3BeO \cdot Al_2O_3 \cdot 6SiO_2$). When some of the Al^{3+} ions in beryl are replaced by Cr^{3+} ions, the characteristic green color of emerald results. In this environment the splitting of the Cr^{3+} $3d$ orbitals causes it to strongly absorb yellow and blue-violet light and to transmit green light.

A gem closely related to ruby and emerald is alexandrite, named after Alexander II of Russia. This gem is based on the mineral chrysoberyl, a beryllium aluminate with the empirical formula $BeO \cdot Al_2O_3$ in which approximately 1% of the Al^{3+} ions are replaced by Cr^{3+} ions. In the chrysoberyl environment Cr^{3+} absorbs strongly in the yellow region of the spectrum. Alexandrite has the interesting property of changing colors depending on the light source. When the first alexandrite stone was discovered deep in a mine in the Russian Ural Mountains in 1831, it appeared to be a deep red color in the firelight of the miners' lamps. However, when the stone was brought to the surface, its color was blue. This seemingly magical color change occurs because the firelight of a miner's helmet is rich in the yellow and red wavelengths of the visible spectrum but does not contain much blue. Absorption of the yellow by the stone produces a reddish color. However, daylight has much more intensity in the blue region than firelight. Thus the extra blue in the light transmitted by the stone gives it bluish color in daylight.

Once the structure of a natural gem is known, it is usually not very difficult to make the gem artificially. For example, rubies and sapphires are made on a large scale by fusing $Al(OH)_3$ with the appropriate transition metal salts at approximately 1200°C to make the "doped" corundum. With these techniques gems of astonishing size can be manufactured: Rubies as large as 10 lb and sapphires up to 100 lb have been synthesized. Smaller synthetic stones produced for jewelry are virtually identical to the corresponding natural stones, and it takes great skill for a gemologist to tell the difference.

Alexandrite, a gem closely related to ruby and emerald.

yellow-green region, thus letting red light and blue light pass, which gives the observed violet color. This is shown schematically in Fig. 21.24. Table 21.16 shows the general relationship between the wavelengths of visible light absorbed and the approximate color observed.

The reason that the $Ti(H_2O)_6^{3+}$ ion absorbs a specific wavelength of visible light can be traced to the transfer of the lone d electron between the split d orbitals, as shown in Fig. 21.25. A given photon of light can be absorbed by a molecule only if the wavelength of the light provides exactly the energy needed by the molecule. In other words, the wavelength absorbed is determined by the relationship

$$\Delta E = \frac{hc}{\lambda}$$

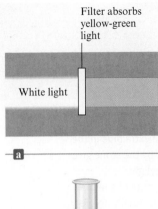

White light

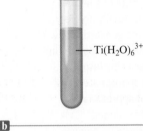

—Ti(H$_2$O)$_6$$^{3+}$

Figure 21.24 | (a) When white light shines on a filter that absorbs in the yellow-green region, the emerging light is violet. (b) Because the complex ion Ti(H$_2$O)$_6$$^{3+}$ absorbs yellow-green light, a solution of it is violet.

Table 21.16 | Approximate Relationship of Wavelength of Visible Light Absorbed to Color Observed

Absorbed Wavelength in nm (Color)	Observed Color
400 (violet)	Greenish yellow
450 (blue)	Yellow
490 (blue-green)	Red
570 (yellow-green)	Violet
580 (yellow)	Dark blue
600 (orange)	Blue
650 (red)	Green

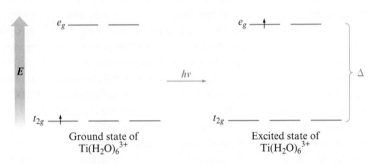

Figure 21.25 | The complex ion Ti(H$_2$O)$_6$$^{3+}$ can absorb visible light in the yellow-green region to transfer the lone d electron from the t_{2g} to the e_g set.

Table 21.17 | Several Octahedral Complexes of Cr^{3+} and Their Colors

Isomer	Color
[Cr(H$_2$O)$_6$]Cl$_3$	Violet
[Cr(H$_2$O)$_5$Cl]Cl$_2$	Blue-green
[Cr(H$_2$O)$_4$Cl$_2$]Cl	Green
[Cr(NH$_3$)$_6$]Cl$_3$	Yellow
[Cr(NH$_3$)$_5$Cl]Cl$_2$	Purple
[Cr(NH$_3$)$_4$Cl$_2$]Cl	Violet

Solutions of [Cr(NH$_3$)$_6$]Cl$_3$ (left) and [Cr(NH$_3$)$_5$Cl]Cl$_2$ (right).

where ΔE represents the energy spacing in the molecule (we have used simply Δ in this chapter) and λ represents the wavelength of light needed. Because the d-orbital splitting in most octahedral complexes corresponds to the energies of photons in the visible region, octahedral complex ions are usually colored.

Since the ligands coordinated to a given metal ion determine the size of the d-orbital splitting, the color changes as the ligands are changed. This occurs because a change in Δ means a change in the wavelength of light needed to transfer electrons between the t_{2g} and e_g orbitals. Several octahedral complexes of Cr^{3+} and their colors are listed in Table 21.17.

Other Coordination Geometries

Using the same principles developed for octahedral complexes, we will now consider complexes with other geometries. For example, Fig. 21.26 shows a tetrahedral arrangement of point charges in relation to the $3d$ orbitals of a metal ion. There are two important facts to note:

1. None of the $3d$ orbitals "point at the ligands" in the tetrahedral arrangement, as the $d_{x^2-y^2}$ and d_{z^2} orbitals do in the octahedral case. Thus the tetrahedrally arranged ligands do not differentiate the d orbitals as much in the tetrahedral case as in the octahedral case. That is, the difference in energy between the split d orbitals will be significantly less in tetrahedral complexes. Although we will not derive it here, the tetrahedral splitting is $\frac{4}{9}$ that of the octahedral splitting for a given ligand and metal ion:

$$\Delta_{\text{tet}} = \tfrac{4}{9}\Delta_{\text{oct}}$$

2. Although not exactly pointing at the ligands, the d_{xy}, d_{xz}, and d_{yz} orbitals are closer to the point charges than are the d_{z^2} and $d_{x^2-y^2}$ orbitals. This means that the

Figure 21.26 | (a) Tetrahedral and octahedral arrangements of ligands shown inscribed in cubes. Note that in the two types of arrangements, the point charges occupy opposite parts of the cube; the octahedral point charges are at the centers of the cube faces, and the tetrahedral point charges occupy opposite corners of the cube. (b) The orientations of the $3d$ orbitals relative to the tetrahedral set of point charges.

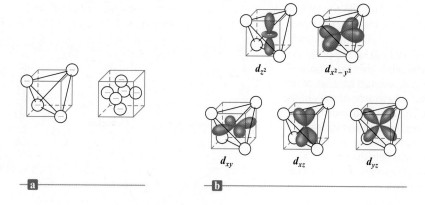

tetrahedral d-orbital splitting will be opposite to that for the octahedral arrangement. The two arrangements are contrasted in Fig. 21.27. Because the d-orbital splitting is relatively small for the tetrahedral case, the weak-field case (high-spin case) *always* applies. There are no known ligands powerful enough to produce the strong-field case in a tetrahedral complex.

Interactive Example 21.6

Sign in at http://login.cengagebrain .com to try this Interactive Example in **OWL**.

Crystal Field Model III

Give the crystal field diagram for the tetrahedral complex ion $CoCl_4{}^{2-}$.

Solution

The complex ion contains Co^{2+}, which has a $3d^7$ electron configuration. The splitting of the d orbitals will be small, since this is a tetrahedral complex, giving the high-spin case with three unpaired electrons.

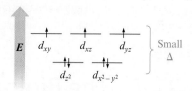

See Exercises 21.61 and 21.63

The crystal field model also applies to square planar and linear complexes. The crystal field diagrams for these cases are shown in Fig. 21.28. The ranking of orbitals in these diagrams can be explained by considering the relative orientations of the point charges and the orbitals. The diagram in Fig. 21.27 for the octahedral arrangement can be used to obtain these orientations. We can obtain the square planar complex by

Figure 21.27 | The crystal field diagrams for octahedral and tetrahedral complexes. The relative energies of the sets of d orbitals are reversed. For a given type of ligand, the splitting is much larger for the octahedral complex ($\Delta_{oct} > \Delta_{tet}$) because in this arrangement the d_{z^2} and $d_{x^2-y^2}$ orbitals point their lobes directly at the point charges and are thus relatively high in energy.

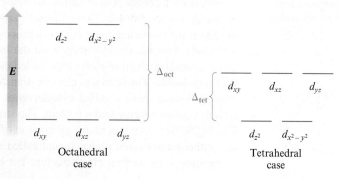

Figure 21.28 | (a) The crystal field diagram for a square planar complex oriented in the xy plane with ligands along the x and y axes. The position of the d_{z^2} orbital is higher than those of the d_{xz} and d_{yz} orbitals because of the "doughnut" of electron density in the xy plane. The actual position of d_{z^2} is somewhat uncertain and varies in different square planar complexes. (b) The crystal field diagram for a linear complex where the ligands lie along the z axis.

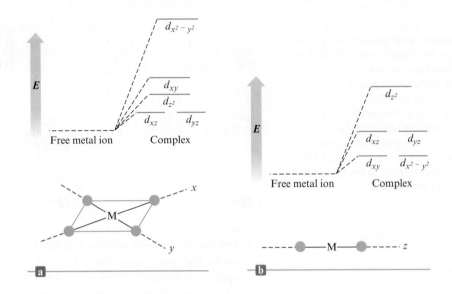

removing the two point charges along the z axis. This will greatly lower the energy of d_{z^2}, leaving only $d_{x^2-y^2}$, which points at the four remaining ligands as the highest-energy orbital. We can obtain the linear complex from the octahedral arrangement by leaving the two ligands along the z axis and removing the four in the xy plane. This means that only the d_{z^2} points at the ligands and is highest in energy.

Critical Thinking

Figure 21.28(a) shows a crystal field diagram for a square planar complex oriented in the xy plane. What if you oriented the complex in the xz plane? Sketch the crystal field diagram and contrast it with Fig. 21.28(a).

21.7 | The Biological Importance of Coordination Complexes

The ability of metal ions to coordinate with and release ligands and to easily undergo oxidation and reduction makes them ideal for use in biological systems. For example, metal ion complexes are used in humans for the transport and storage of oxygen, as electron-transfer agents, as catalysts, and as drugs. Most of the first-row transition metals are essential for human health, as summarized in Table 21.18. We will concentrate on iron's role in biological systems, since several of its coordination complexes have been studied extensively.

Iron plays a central role in almost all living cells. In mammals, the principal source of energy comes from the oxidation of carbohydrates, proteins, and fats. Although oxygen is the oxidizing agent for these processes, it does not react directly with these molecules. Instead, the electrons from the breakdown of these nutrients are passed along a complex chain of molecules, called the *respiratory chain,* eventually reaching the O_2 molecule. The principal electron-transfer molecules in the respiratory chain are iron-containing species called **cytochromes**, consisting of two main parts: an iron complex called a **heme** and a protein. The structure of the heme complex is shown in Fig. 21.29. Note that it contains an iron ion (it can be either Fe^{2+} or Fe^{3+}) coordinated to a rather complicated planar ligand called a **porphyrin**. As a class, porphyrins all contain the same central ring structure but have different substituent groups at the

A protein is a large molecule assembled from amino acids, which have the general structure in which R represents various groups.

$$H_2N-\underset{\underset{H}{|}}{\overset{\overset{R}{|}}{C}}-COOH$$

Table 21.18 | The First-Row Transition Metals and Their Biological Significance

First-Row Transition Metal	Biological Function(s)
Scandium	None known.
Titanium	None known.
Vanadium	None known in humans.
Chromium	Assists insulin in the control of blood sugar; may also be involved in the control of cholesterol.
Manganese	Necessary for a number of enzymatic reactions.
Iron	Component of hemoglobin and myoglobin; involved in the electron-transport chain.
Cobalt	Component of vitamin B_{12}, which is essential for the metabolism of carbohydrates, fats, and proteins.
Nickel	Component of the enzymes urease and hydrogenase.
Copper	Component of several enzymes; assists in iron storage; involved in the production of color pigments of hair, skin, and eyes.
Zinc	Component of insulin and many enzymes.

edges of the rings. The various porphyrin molecules act as tetradentate ligands for many metal ions, including iron, cobalt, and magnesium. In fact, *chlorophyll,* a substance essential to the process of photosynthesis, is a magnesium–porphyrin complex of the type shown in Fig. 21.30.

In addition to participating in the transfer of electrons from nutrients to oxygen, iron plays a principal role in the transport and storage of oxygen in mammalian blood and tissues. Oxygen is stored in a molecule called **myoglobin**, which consists of a heme complex and a protein in a structure very similar to that of the cytochromes. In myoglobin, the Fe^{2+} ion is coordinated to four nitrogen atoms of the porphyrin ring

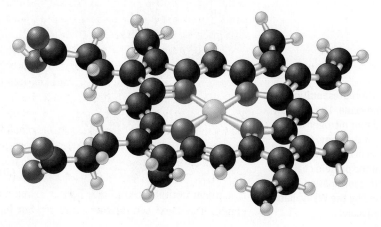

Figure 21.29 | The heme complex, in which an Fe^{2+} ion is coordinated to four nitrogen atoms of a planar porphyrin ligand.

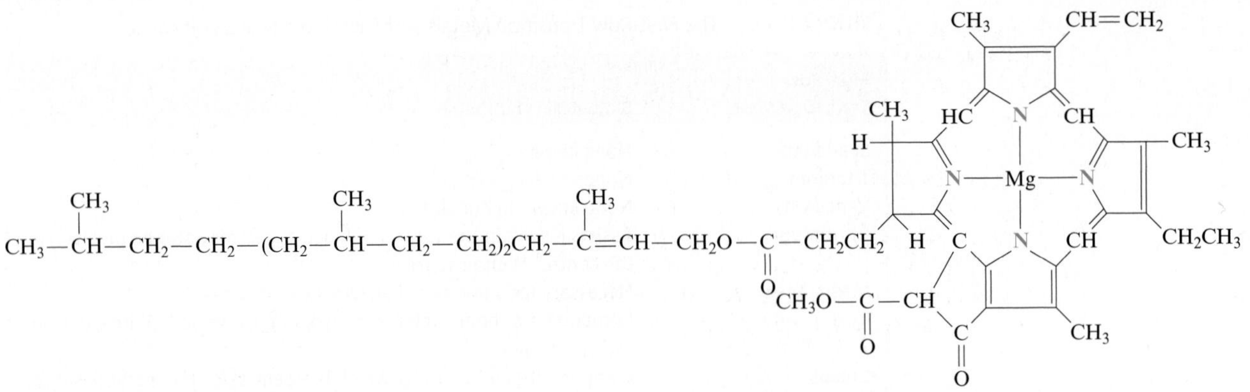

Figure 21.30 | Chlorophyll is a porphyrin complex of Mg^{2+}. There are two similar forms of chlorophyll, one of which is shown here.

and to a nitrogen atom of the protein chain, as shown in Fig. 21.31. Since Fe^{2+} is normally six-coordinate, this leaves one position open for attachment of an O_2 molecule.

One especially interesting feature of myoglobin is that it involves an O_2 molecule attaching directly to Fe^{2+}. However, if gaseous O_2 is bubbled into an aqueous solution containing heme, the Fe^{2+} is immediately oxidized to Fe^{3+}. This oxidation of the Fe^{2+} in heme does not happen in myoglobin. This fact is of crucial importance because Fe^{3+} does not form a coordinate covalent bond with O_2, and myoglobin would not function if the bound Fe^{2+} could be oxidized. Since the Fe^{2+} in the "bare" heme complex can be oxidized, it must be the protein that somehow prevents the oxidation. How? Based on much research, the answer seems to be that the oxidation of Fe^{2+} to Fe^{3+} involves an oxygen bridge between two iron ions (the circles indicate the ligands):

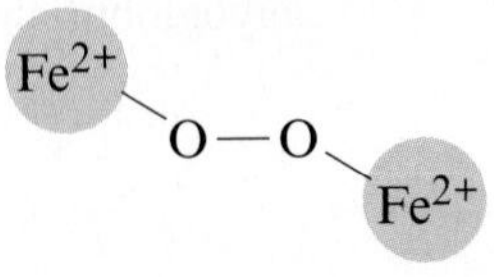

$$Fe^{2+} \quad O\!-\!O \quad Fe^{2+}$$

The bulky protein around the heme group in myoglobin prevents two molecules from getting close enough to form the oxygen bridge, and so oxidation of the Fe^{2+} is prevented.

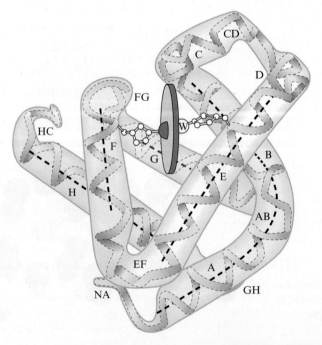

Figure 21.31 | A representation of the myoglobin molecule. The Fe^{2+} ion is coordinated to four nitrogen atoms in the porphyrin of the heme (represented by the disk in the figure) and on nitrogen from the protein chain. This leaves a sixth coordination position (indicated by the W) available for an oxygen molecule.

Figure 21.32 | A representation of the hemoglobin structure. There are two slightly different types of protein chains (α and β). Each hemoglobin has two α chains and two β chains, each with a heme complex near the center. Thus each hemoglobin molecule can complex with four O_2 molecules.

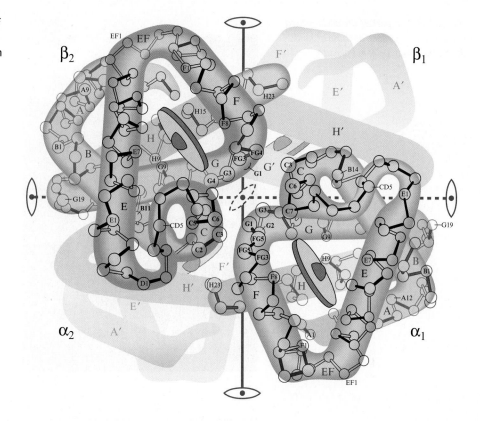

The transport of O_2 in the blood is carried out by **hemoglobin**, a molecule consisting of four myoglobin-like units, as shown in Fig. 21.32. Each hemoglobin can therefore bind four O_2 molecules to form a bright red diamagnetic complex. The diamagnetism occurs because oxygen is a strong-field ligand toward Fe^{2+}, which has a $3d^6$ electron configuration. When the oxygen molecule is released, water molecules occupy the sixth coordination position around each Fe^{2+}, giving a bluish paramagnetic complex (H_2O is a weak-field ligand toward Fe^{2+}) that gives venous blood its characteristic bluish tint.

Hemoglobin dramatically demonstrates how sensitive the function of a biomolecule is to its structure. In certain people, in the synthesis of the proteins needed for hemoglobin, an improper amino acid is inserted into the protein in two places. This may not seem very serious, since there are several hundred amino acids present. However, because the incorrect amino acid has a nonpolar substituent instead of the polar one found on the proper amino acid, the hemoglobin drastically changes its shape. The red blood cells are then sickle-shaped rather than disk-shaped, as shown in Fig. 21.33. The misshapen cells can aggregate, causing clogging of tiny capillaries. This condition, known as *sickle cell anemia,* is the subject of intense research.

Our knowledge of the workings of hemoglobin allows us to understand the effects of high altitudes on humans. The reaction between hemoglobin and oxygen can be represented by the following equilibrium:

$$\text{Hb}(aq) + 4O_2(g) \rightleftharpoons \text{Hb}(O_2)_4(aq)$$

Hemoglobin   Oxyhemoglobin

At high altitudes, where the oxygen content of the air is lower, the position of this equilibrium will shift to the left, according to Le Châtelier's principle. Because less oxyhemoglobin is formed, fatigue, dizziness, and even a serious illness called *high-altitude sickness* can result. One way to combat this problem is to use supplemental oxygen, as most high-altitude mountain climbers do. However, this is impractical for people who live at high elevations. In fact, the human body adapts to the lower oxygen concentrations by making more hemoglobin, causing the equilibrium to shift back to

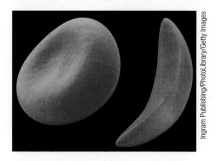

Figure 21.33 | A normal red blood cell (right) and a sickle cell (left), both magnified 18,000 times.

Sherpa and Balti porters are acclimatized to high elevations such as those around the K2 mountain peak in Pakistan.

Gopal Chitrakar/Reuters

the right. Someone moving from Chicago to Boulder, Colorado (5300 feet), would notice the effects of the new altitude for a couple of weeks, but as the hemoglobin level increased, the effects would disappear. This change is called *high-altitude acclimatization,* which explains why athletes who want to compete at high elevations should practice there for several weeks prior to the event.

Our understanding of the biological role of iron also allows us to explain the toxicities of substances such as carbon monoxide and the cyanide ion. Both CO and CN$^-$ are very good ligands toward iron and so can interfere with the normal workings of the iron complexes in the body. For example, carbon monoxide has about 200 times the affinity for the Fe^{2+} in hemoglobin as oxygen does. The resulting stable complex, **carboxyhemoglobin,** prevents the normal uptake of O$_2$, thus depriving the body of needed oxygen. Asphyxiation can result if enough carbon monoxide is present in the air. The mechanism for the toxicity of the cyanide ion is somewhat different. Cyanide coordinates strongly to cytochrome oxidase, an iron-containing cytochrome enzyme that catalyzes the oxidation–reduction reactions of certain cytochromes. The coordinated cyanide thus prevents the electron-transfer process and rapid death results. Because of its behavior, cyanide is called a *respiratory inhibitor.*

21.8 | Metallurgy and Iron and Steel Production

In the preceding section we saw the importance of iron in biological systems. Of course, iron is also very important in many other ways in our world. In this section we will discuss the isolation of metals from their natural sources and the formulation of metals into useful materials, with special emphasis on the role of iron.

Metals are very important for structural applications, electrical wires, cooking utensils, tools, decorative items, and many other purposes. However, because the main chemical characteristic of a metal is its ability to give up electrons, almost all metals in nature are found in ores, combined with nonmetals such as oxygen, sulfur, and the

Pouring molten metal in a steel mill.

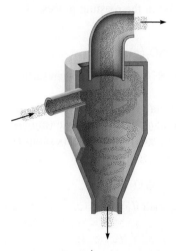

Figure 21.34 | A schematic diagram of a cyclone separator. The ore is pulverized and blown into the separator. The more dense mineral particles are thrown toward the walls by centrifugal force and fall down the funnel. The lighter particles (gangue) tend to stay closer to the center and are drawn out through the top by the stream of air.

halogens. To recover and use these metals, we must separate them from their ores and reduce the metal ions. Then, because most metals are unsuitable for use in the pure state, we must form alloys that have the desired properties. The process of separating a metal from its ore and preparing it for use is known as **metallurgy**. The steps in this process are typically

1. Mining
2. Pretreatment of the ore
3. Reduction to the free metal
4. Purification of the metal (refining)
5. Alloying

An ore can be viewed as a mixture containing **minerals** (relatively pure metal compounds) and **gangue** (sand, clay, and rock). Some typical minerals are listed in Table 21.19. Although silicate minerals are the most common in the earth's crust, they are typically very hard and difficult to process, making metal extraction relatively expensive. Therefore, other ores are used when available.

After mining, an ore must be treated to remove the gangue and to concentrate the mineral. The ore is first pulverized and then processed in a variety of devices, including cyclone separators (Fig. 21.34), inclined vibrating tables, and flotation tanks.

In the **flotation process**, the crushed ore is fed into a tank containing a water–oil–detergent mixture. Because of the difference in the surface characteristics of the mineral particles and the silicate rock particles, the oil wets the mineral particles. A stream of air blown through the mixture causes tiny bubbles to form on the oil-covered pieces, which then float to the surface, where they can be skimmed off.

After the mineral has been concentrated, it is often chemically altered in preparation for the reduction step. For example, nonoxide minerals are often converted to oxides before reduction. Carbonates and hydroxides can be converted by simple heating:

$$CaCO_3(s) \xrightarrow{\text{Heat}} CaO(s) + CO_2(g)$$

$$Mg(OH)_2(s) \xrightarrow{\text{Heat}} MgO(s) + H_2O(g)$$

Table 21.19 | Common Minerals Found in Ores

Anion	Examples
None (free metal)	Au, Ag, Pt, Pd, Rh, Ir, Ru
Oxide	Fe_2O_3 (hematite)
	Fe_3O_4 (magnetite)
	Al_2O_3 (bauxite)
	SnO_2 (cassiterite)
Sulfide	PbS (galena)
	ZnS (sphalerite)
	FeS_2 (pyrite)
	HgS (cinnabar)
	Cu_2S (chalcocite)
Chloride	NaCl (rock salt)
	KCl (sylvite)
	$KCl \cdot MgCl_2$ (carnalite)
Carbonate	$FeCO_3$ (siderite)
	$CaCO_3$ (limestone)
	$MgCO_3$ (magnesite)
	$MgCO_3 \cdot CaCO_3$ (dolomite)
Sulfate	$CaSO_4 \cdot 2H_2O$ (gypsum)
	$BaSO_4$ (barite)
Silicate	$Be_3Al_2Si_6O_{18}$ (beryl)
	$Al_2(Si_2O_8)(OH)_4$ (kaolinite)
	$LiAl(SiO_3)_2$ (spodumene)

sdigzps/iStockphoto.com

Figure 21.35 | A schematic representation of zone refining.

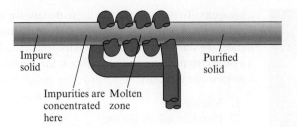

Impure solid

Purified solid

Impurities are concentrated here Molten zone

Sulfide minerals can be converted to oxides by heating in air at temperatures below their melting points, a process called **roasting**:

$$2ZnS(s) + 3O_2(g) \xrightarrow{\text{Heat}} 2ZnO(s) + 2SO_2(g)$$

As we have seen earlier, sulfur dioxide causes severe problems if released into the atmosphere, and modern roasting operations collect this gas and use it in the manufacture of sulfuric acid.

The method chosen to reduce the metal ion to the free metal, a process called **smelting**, depends on the affinity of the metal ion for electrons. Some metals are good enough oxidizing agents that the free metal is produced in the roasting process. For example, the roasting reaction for cinnabar is

$$HgS(s) + O_2(g) \xrightarrow{\text{Heat}} Hg(l) + SO_2(g)$$

where the Hg^{2+} is reduced by electrons donated by the S^{2-} ion, which is then further oxidized by O_2 to form SO_2.

The roasting of a more active metal produces the metal oxide, which must be reduced to obtain the free metal. The most common reducing agents are coke (impure carbon), carbon monoxide, and hydrogen. The following are some common examples of the reduction process:

$$Fe_2O_3(s) + 3CO(g) \xrightarrow{\text{Heat}} 2Fe(l) + 3CO_2(g)$$

$$WO_3(s) + 3H_2(g) \xrightarrow{\text{Heat}} W(l) + 3H_2O(g)$$

$$ZnO(s) + C(s) \xrightarrow{\text{Heat}} Zn(l) + CO(g)$$

The most active metals, such as aluminum and the alkali metals, must be reduced electrolytically, usually from molten salts (see Section 18.9).

The metal obtained in the reduction step is invariably impure and must be refined. The methods of refining include electrolytic refining (see Section 18.9), oxidation of impurities (as for iron, see below), and distillation of low-boiling metals such as mercury and zinc. One process used when highly pure metals are needed is **zone refining**. In this process a bar of the impure metal travels through a heater (Fig. 21.35), which causes melting and recrystallizing of the metal as the bar cools. Purification of the metal occurs because as the crystal re-forms, the metal ions are likely to fit much better in the crystal lattice than are the atoms of impurities. Thus the impurities tend to be excluded and carried to the end of the bar. Several repetitions of this process give a very pure metal bar.

Hydrometallurgy

The metallurgical processes we have considered so far are usually called **pyrometallurgy** (*pyro* means "at high temperatures"). These traditional methods require large quantities of energy and have two other serious problems: atmospheric pollution (mainly by sulfur dioxide) and relatively high costs that make treatment of low-grade ores economically unfeasible.

In the past hundred years, a different process, **hydrometallurgy** (*hydro* means "water"), has been used to extract metals from ores by use of aqueous chemical solutions, a process called **leaching**. The first two uses of hydrometallurgy were for the extraction of gold from low-grade ores and for the production of aluminum oxide, or alumina, from bauxite, an aluminum-bearing ore.

Gold is sometimes found in ores in the elemental state, but it usually occurs in relatively small concentrations. A process called **cyanidation** treats the crushed ore with an aqueous cyanide solution in the presence of air to dissolve the gold by forming the complex ion $Au(CN)_2^-$:

$$4Au(s) + 8CN^-(aq) + O_2(g) + 2H_2O(l) \longrightarrow 4Au(CN)_2^-(aq) + 4OH^-(aq)$$

Pure gold is then recovered by reaction of the solution of $Au(CN)_2^-$ with zinc powder to reduce Au^+ to Au:

$$2Au(CN)_2^-(aq) + Zn(s) \longrightarrow 2Au(s) + Zn(CN)_4^{2-}(aq)$$

The extraction of alumina from bauxite (the Bayer process) leaches the ore with sodium hydroxide at high temperatures and pressures to dissolve the amphoteric aluminum oxide:

$$Al_2O_3(s) + 2OH^-(aq) \longrightarrow 2AlO_2^-(aq) + H_2O(l)$$

This process leaves behind solid impurities such as SiO_2, Fe_2O_3, and TiO_2, which are not appreciably soluble in basic solution. After the solid impurities are removed, the pH of the solution is lowered, causing the pure aluminum oxide to re-form. It is then electrolyzed to produce aluminum metal (see Section 18.9).

As illustrated by these processes, hydrometallurgy involves two distinct steps: *selective leaching* of a given metal ion from the ore and recovery of the metal ion from the solution by *selective precipitation* as an ionic compound.

Precipitation reactions are discussed in Section 16.2.

The leaching agent can simply be water if the metal-containing compound is a water-soluble chloride or sulfate. However, most commonly, the metal is present in a water-insoluble substance that must somehow be dissolved. The leaching agents used in such cases are usually aqueous solutions containing acids, bases, oxidizing agents, salts, or some combination of these. Often the dissolving process involves the formation of complex ions. For example, when an ore containing water-insoluble lead sulfate is treated with an aqueous sodium chloride solution, the soluble complex ion $PbCl_4^{2-}$ is formed:

$$PbSO_4(s) + 4Na^+(aq) + 4Cl^-(aq) \longrightarrow 4Na^+(aq) + PbCl_4^{2-}(aq) + SO_4^{2-}(aq)$$

Formation of a complex ion also occurs in the cyanidation process for the recovery of gold. However, since the gold is present in the ore as particles of metal, it must first be oxidized by oxygen to produce Au^+, which then reacts with CN^- to form the soluble $Au(CN)_2^-$ species. Thus, in this case, the leaching process involves a combination of oxidation and complexation.

Sometimes just oxidation is used. For example, insoluble zinc sulfide can be converted to soluble zinc sulfate by pulverizing the ore and suspending it in water to form a slurry through which oxygen is bubbled:

$$ZnS(s) + 2O_2(g) \longrightarrow Zn^{2+}(aq) + SO_4^{2-}(aq)$$

One advantage of hydrometallurgy over the traditional processes is that sometimes the leaching agent can be pumped directly into the ore deposits in the earth. For example, aqueous sodium carbonate (Na_2CO_3) can be injected into uranium-bearing ores to form water-soluble complex carbonate ions.

Recovering the metal ions from the leaching solution involves forming an insoluble solid containing the metal ion to be recovered. This step may involve addition of an anion to form an insoluble salt, reduction to the solid metal, or a combination of reduction and precipitation of a salt. Examples of these processes are shown in Table 21.20. Because of its suitability for treating low-grade ores economically and without significant pollution, hydrometallurgy is becoming more popular for recovering many important metals such as copper, nickel, zinc, and uranium.

Table 21.20 | Examples of Methods for Recovery of Metal Ions from Leaching Solutions

Method		Examples
Precipitation of a salt		$Cu^{2+}(aq) + S^{2-}(aq) \longrightarrow CuS(s)$
		$Cu^+(aq) + HCN(aq) \longrightarrow CuCN(s) + H^+(aq)$
	Chemical	$\begin{cases} Au^+(aq) + Fe^{2+}(aq) \longrightarrow Au(s) + Fe^{3+}(aq) \\ Cu^{2+}(aq) + Fe(s) \longrightarrow Cu(s) + Fe^{2+}(aq) \\ Ni^{2+}(aq) + H_2(g) \longrightarrow Ni(s) + 2H^+(aq) \end{cases}$
Reduction	Electrolytic	$\begin{cases} Cu^{2+}(aq) + 2e^- \longrightarrow Cu(s) \\ Al^{3+}(aq) + 3e^- \longrightarrow Al(s) \end{cases}$
Reduction plus precipitation		$2Cu^{2+}(aq) + 2Cl^-(aq) + H_2SO_3(aq) + H_2O(l) \longrightarrow$ $2CuCl(s) + 3H^+(aq) + HSO_4^-(aq)$

The Metallurgy of Iron

Iron is present in the earth's crust in many types of minerals. *Iron pyrite* (FeS_2) is widely distributed but is not suitable for production of metallic iron and steel because it is almost impossible to remove the last traces of sulfur. The presence of sulfur makes the resulting steel too brittle to be useful. *Siderite* ($FeCO_3$) is a valuable iron mineral that can be converted to iron oxide by heating. The iron oxide minerals are *hematite* (Fe_2O_3), the more abundant, and *magnetite* (Fe_3O_4, really $FeO \cdot Fe_2O_3$). *Taconite* ores contain iron oxides mixed with silicates and are more difficult to process than the others. However, taconite ores are being increasingly used as the more desirable ores are consumed.

To concentrate the iron in iron ores, advantage is taken of the natural magnetism of Fe_3O_4 (hence its name, *magnetite*). The Fe_3O_4 particles can be separated from the gangue by magnets. The ores that are not magnetic are often converted to Fe_3O_4; hematite is partially reduced to magnetite, while siderite is first converted to FeO thermally, then oxidized to Fe_2O_3, and then reduced to Fe_3O_4:

$$FeCO_3(s) \xrightarrow{\text{Heat}} FeO(s) + CO_2(g)$$

$$4FeO(s) + O_2(g) \longrightarrow 2Fe_2O_3(s)$$

$$3Fe_2O_3(s) + C(s) \longrightarrow 2Fe_3O_4(s) + CO(g)$$

Sometimes the nonmagnetic ores are concentrated by flotation processes.

The most commonly used reduction process for iron takes place in the **blast furnace** (Fig. 21.36). The raw materials required are concentrated iron ore, coke, and limestone (which serves as a *flux* to trap impurities). The furnace, which is approximately 25 feet in diameter, is charged from the top with a mixture of iron ore, coke, and limestone. A very strong blast (~350 mi/h) of hot air is injected at the bottom, where the oxygen reacts with the carbon in the coke to form carbon monoxide, the reducing agent for the iron. The temperature of the charge increases as it travels down the furnace, with reduction of the iron to iron metal occurring in steps:

$$3Fe_2O_3 + CO \longrightarrow 2Fe_3O_4 + CO_2$$

$$Fe_3O_4 + CO \longrightarrow 3FeO + CO_2$$

$$FeO + CO \longrightarrow Fe + CO_2$$

Iron can reduce carbon dioxide,

$$Fe + CO_2 \longrightarrow FeO + CO$$

so complete reduction of the iron occurs only if the carbon dioxide is destroyed by adding excess coke:

$$CO_2 + C \longrightarrow 2CO$$

Figure 21.36 | The blast furnace used in the production of iron.

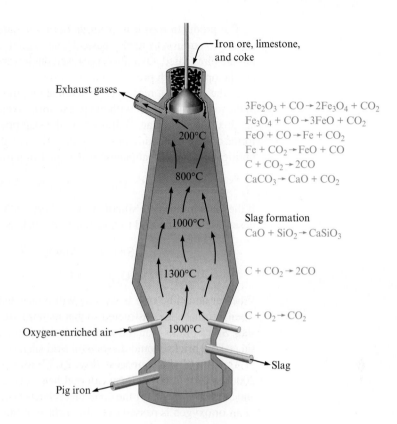

The limestone ($CaCO_3$) in the charge loses carbon dioxide, or *calcines*, in the hot furnace and combines with silica and other impurities to form **slag**, which is mostly molten calcium silicate, $CaSiO_3$,

$$CaO + SiO_2 \longrightarrow CaSiO_3$$

and alumina (Al_2O_3). The slag floats on the molten iron and is skimmed off. The gas that escapes from the top of the furnace contains carbon monoxide, which is combined with air to form carbon dioxide. The energy released in this exothermic reaction is collected in a heat exchanger and used in heating the furnace.

The iron collected from the blast furnace, called **pig iron**, is quite impure. It contains ~90% iron, ~5% carbon, ~2% manganese, ~1% silicon, ~0.3% phosphorus, and ~0.04% sulfur (from impurities in the coke). The production of 1 ton of pig iron requires approximately 1.7 tons of iron ore, 0.5 ton of coke, 0.25 ton of limestone, and 2 tons of air.

Iron oxide also can be reduced in a **direct reduction furnace**, which operates at much lower temperatures (1300–2000°F) than a blast furnace and produces a solid "sponge iron" rather than molten iron. Because of the milder reaction conditions, the direct reduction furnace requires a higher grade of iron ore (with fewer impurities) than that used in a blast furnace. The iron from the direct reduction furnace is called *DRI (directly reduced iron)* and contains ~95% iron, with the balance mainly silica and alumina.

Production of Steel

Steel is an alloy and can be classified as **carbon steel**, which contains up to about 1.5% carbon, or **alloy steel**, which contains carbon plus other metals such as Cr, Co, Mn, and Mo. The wide range of mechanical properties associated with steel is determined by its chemical composition and by the heat treatment of the final product.

The production of iron from its ore is fundamentally a reduction process, but the conversion of iron to steel is basically an oxidation process in which unwanted impurities are eliminated. Oxidation is carried out in various ways, but the two most common are the open hearth process and the basic oxygen process.

In the oxidation reactions of steelmaking, the manganese, phosphorus, and silicon in the impure iron react with oxygen to form oxides, which in turn react with appropriate fluxes to form slag. Sulfur enters the slag primarily as sulfides, and excess carbon forms carbon monoxide or carbon dioxide. The flux chosen depends on the major impurities present. If manganese is the chief impurity, an acidic flux of silica is used:

$$MnO(s) + SiO_2(s) \xrightarrow{\text{Heat}} MnSiO_3(l)$$

If the main impurity is silicon or phosphorus, a basic flux, usually lime (CaO) or magnesia (MgO), is needed to give reactions such as

$$SiO_2(s) + MgO(s) \xrightarrow{\text{Heat}} MgSiO_3(l)$$
$$P_4O_{10}(s) + 6CaO(s) \xrightarrow{\text{Heat}} 2Ca_3(PO_4)_2(l)$$

Whether an acidic or a basic slag will be needed is a factor that must be considered when a furnace is constructed so that its refractory linings will be compatible with the slag. Silica bricks would deteriorate quickly in the presence of basic slag, and magnesia or lime bricks would dissolve in acid slag.

The **open hearth process** (Fig. 21.37) uses a dishlike container that holds 100 to 200 tons of molten iron. An external heat source is required to keep the iron molten, and a concave roof over the container reflects heat back toward the iron surface. A blast of air or oxygen is passed over the surface of the iron to react with impurities. Silicon and manganese are oxidized first and enter the slag, followed by oxidation of carbon to carbon monoxide, which causes agitation and foaming of the molten bath. The exothermic oxidation of carbon raises the temperature of the bath, causing the limestone flux to calcine:

$$CaCO_3 \xrightarrow{\text{Heat}} CaO + CO_2$$

The resulting lime floats to the top of the molten mixture (an event called the *lime boil*), where it combines with phosphates, sulfates, and silicates. Next comes the refining process, which involves continued oxidation of carbon and other impurities. Because the melting point increases as the carbon content decreases, the bath temperatures must be increased during this phase of the operation. If the carbon content falls below that desired in the final product, coke or pig iron may be added.

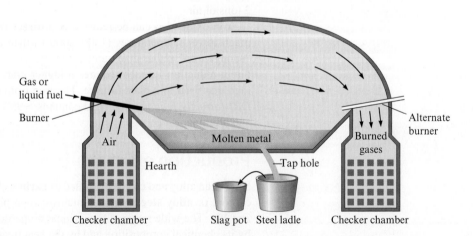

Figure 21.37 | A schematic diagram of the open hearth process for steelmaking. The checker chambers contain bricks that absorb heat from gases passing over the molten charge. The flow of air and gases is reversed periodically.

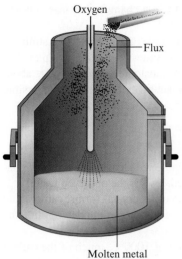

Figure 21.38 | The basic oxygen process for steelmaking.

The final composition of the steel is "fine-tuned" after the charge is poured. For example, aluminum is sometimes added at this stage to remove trace amounts of oxygen via the reaction

$$4Al + 3O_2 \longrightarrow 2Al_2O_3$$

Alloying metals such as vanadium, chromium, titanium, manganese, and nickel are also added to give the steel the properties needed for a specific application.

The processing of a batch of steel by the open hearth process is quite slow, taking up to 8 hours. The **basic oxygen process** is much faster. Molten pig iron and scrap iron are placed in a barrel-shaped container (Fig. 21.38) that can hold as much as 300 tons of material. A high-pressure blast of oxygen is directed at the surface of the molten iron, oxidizing impurities in a manner very similar to that used in the open hearth process. Fluxes are added after the oxygen blow begins. One advantage of this process is that the exothermic oxidation reactions proceed so rapidly that they produce enough heat to raise the temperature nearly to the boiling point of iron without an external heat source. Also, at these high temperatures only about an hour is needed to complete the oxidation processes.

The **electric arc method**, which was once used only for small batches of specialty steels, is being utilized more and more in the steel industry. In this process an electric arc between carbon electrodes is used to melt the charge. This means that no fuel-borne impurities are added to the steel, since no fuel is needed. Also, higher temperatures are possible than in the open hearth or basic oxygen processes, and this leads to more effective removal of sulfur and phosphorus impurities. Oxygen is added in this process so that the oxide impurities in the steel can be controlled effectively.

Heat Treatment of Steel

One way of producing the desired physical properties in steel is by controlling the chemical composition (Table 21.21). Another method for tailoring the properties of steel involves heat treatment. Pure iron exists in two different crystalline forms, depending on the temperature. At any temperature below 912°C, iron has a body-centered cubic structure and is called *α-iron*. Between 912°C and 1394°C, iron has a face-centered

Refer to Section 10.3 for a review of packing and crystal lattices.

Table 21.21 | Percent Composition and Uses of Various Types of Steel

Type of Steel	% Carbon	% Manganese	% Phosphorus	% Sulfur	% Silicon	% Nickel	% Chromium	% Other	Uses
Plain carbon	≤1.35	≤1.65	≤0.04	≤0.05	≤0.60	—	—	—	Sheet steel, tools
High-strength (low-alloy)	≤0.25	≤1.65	≤0.04	≤0.05	0.15–0.9	0.4–1	0.3–1.3	Cu (0.2–0.6) Sb (0.01–0.08) V (0.01–0.08)	Transportation equipment, structural beams
Alloy	≤1.00	≤3.50	≤0.04	≤0.05	0.15–2.0	0.25–10.0	0.25–4.0	Mo (0.08–4.0) V (0–0.2) W (0–18) Co (0–5)	Automobile and aircraft engine parts
Stainless	0.03–1.2	1.0–10	0.04–0.06	≤0.03	1–3	1–22	4.0–27	—	Engine parts, steam turbine parts, kitchen utensils
Silicon	—	—	—	—	0.5–5.0	—	—	—	Electric motors and transformers

cubic structure called *austentite,* or γ-*iron*. At 1394°C, iron changes to δ-*iron*, a body-centered cubic structure identical to α-*iron*.

When iron is alloyed with carbon, which fits into holes among the iron atoms to form the interstitial alloy *carbon steel,* the situation becomes even more complex. For example, the temperature at which α-iron changes to austentite is lowered by about 200°C. Also, at high temperatures iron and carbon react by an endothermic reversible reaction to form an iron carbide called *cementite*:

$$3Fe + C + \underset{\text{(Heat)}}{\text{energy}} \rightleftharpoons \underset{\text{Cementite}}{Fe_3C}$$

By Le Châtelier's principle, we can predict that cementite will become more stable relative to iron and carbon as the temperature is increased. This is the observed result.

Thus steel is really a mixture of iron metal in one of its crystal forms, carbon, and cementite. The proportions of these components are very important in determining the physical properties of steel.

When steel is heated to temperatures in the region of 1000°C, much of the carbon is converted to cementite. If the steel is then allowed to cool slowly, the equilibrium shown above shifts to the left, and small crystals of carbon precipitate, giving a steel that is relatively ductile. If the cooling is very rapid, the equilibrium does not have time to adjust. The cementite is trapped, and the steel has a high cementite content, making it quite brittle. The proportions of carbon crystals and cementite can be "fine-tuned" to give the desired properties by heating to intermediate temperatures followed by rapid cooling, a process called **tempering**. The rate of heating and cooling determines not only the amounts of cementite present but also the size of its crystals and the form of crystalline iron present.

For review

Key terms

Section 21.1
complex ion
first-row transition metals
lanthanide contraction
lanthanide series

Section 21.3
coordination compound
counterion
oxidation state
coordination number
ligand
coordinate covalent bond
monodentate (unidentate)
 ligand
chelating ligand (chelate)
bidentate ligand

Section 21.4
isomers
structural isomerism
stereoisomerism
coordination isomerism
linkage isomerism

First-row transition metals (scandium–zinc)

> All have one or more electrons in the 4s orbital and various numbers of 3d electrons
> All exhibit metallic properties
>> A particular element often shows more than one oxidation state in its compounds
> Most compounds are colored, and many are paramagnetic
> Most commonly form coordination compounds containing a complex ion involving ligands (Lewis bases) attached to a central transition metal ion
>> The number of attached ligands (called the coordination number) can vary from 2 to 8, with 4 and 6 being most common
> Many transition metal ions have major biological importance in molecules such as enzymes and those that transport and store oxygen
>> Chelating ligands form more than one bond to the transition metal ion

Isomerism

> Isomers: two or more compounds with the same formula but different properties
>> Coordination isomerism: the composition of the coordination sphere varies
>> Linkage isomerism: the point of attachment of one or more ligands varies
>> Stereoisomerism: isomers have identical bonds but different spatial arrangements
>>> Geometric isomerism: ligands assume different relative positions in the coordination sphere; examples are *cis* and *trans* isomers
>>> Optical isomerism: molecules with nonsuperimposable mirror images rotate plane-polarized light in opposite directions

Key terms

geometrical (*cis–trans*)
   isomerism
trans isomer
cis isomer
optical isomerism
chiral
enantiomers

Section 21.6
crystal field model
d-orbital splitting
strong-field (low-spin) case
weak-field (high-spin) case
spectrochemical series

Section 21.7
cytochromes
heme
porphyrin
myoglobin
hemoglobin
carboxyhemoglobin

Section 21.8
metallurgy
minerals
gangue
flotation process
roasting
smelting
zone refining
pyrometallurgy
hydrometallurgy
leaching
cyanidation
blast furnace
slag
pig iron
direct reduction furnace
carbon steel
alloy steel
open hearth process
basic oxygen process
electric arc method
tempering

Spectral and magnetic properties

> Usually explained in terms of the crystal field model
> Model assumes the ligands are point charges that split the energies of the $3d$ orbitals
> Color and magnetism are explained in terms of how the $3d$ electrons occupy the split $3d$ energy levels
>> Strong-field case: relatively large orbital splitting
>> Weak-field case: relatively small orbital splitting

Metallurgy

> The processes connected with separating a metal from its ore
>> The minerals in ores are often converted to oxides (roasting) before being reduced to the metal (smelting)
> The metallurgy of iron: most common method for reduction uses a blast furnace; process involves iron ore, coke, and limestone
>> Impure product (~90% iron) is called pig iron
> Steel is manufactured by oxidizing the impurities in pig iron

Review questions   *Answers to the Review Questions can be found on the Student website (accessible from* www.cengagebrain.com)*.*

1. What two first-row transition metals have unexpected electron configurations? A statement in the text says that first-row transition metal ions do not have 4*s* electrons. Why not? Why do transition metal ions often have several oxidation states, whereas representative metals generally have only one?

2. Define each of the following terms:
   a. coordination compound
   b. complex ion
   c. counterions
   d. coordination number

e. ligand

f. chelate

g. bidentate

How would transition metal ions be classified using the Lewis definition of acids and bases? What must a ligand have to bond to a metal? What do we mean when we say that a bond is a "coordinate covalent bond"?

3. When a metal ion has a coordination number of 2, 4, or 6, what are the observed geometries and associated bond angles? For each of the following, give the correct formulas for the complex ions.

a. linear Ag^+ complex ions having CN^- ligands

b. tetrahedral Cu^+ complex ions having H_2O ligands

c. tetrahedral Mn^{2+} complex ions having oxalate ligands

d. square planar Pt^{2+} complex ions having NH_3 ligands

e. octahedral Fe^{3+} complex ions having EDTA ligands

f. octahedral Co^{2+} complex ions having Cl^- ligands

g. octahedral Cr^{3+} complex ions having ethylenediamine ligands

What is the electron configuration for the metal ion in each of the complex ions in a–g?

4. What is wrong with the following formula–name combinations? Give the correct names for each.

a. $[Cu(NH_3)_4]Cl_2$
   copperammine chloride

b. $[Ni(en)_2]SO_4$
   bis(ethylenediamine)nickel(IV) sulfate

c. $K[Cr(H_2O)_2Cl_4]$
   potassium tetrachlorodiaquachromium(III)

d. $Na_4[Co(CN)_4C_2O_4]$
   tetrasodium tetracyanooxalatocobaltate(II)

5. Define each of the following and give examples of each.

a. isomer

b. structural isomer

c. stereoisomer

d. coordination isomer

e. linkage isomer

f. geometrical isomer

g. optical isomer

Consider the *cis* and *trans* forms of the octahedral complex $Cr(en)_2Cl_2$. Are both of these isomers optically active? Explain.

Another way to determine whether a substance is optically active is to look for a plane of symmetry in the molecule. If a substance has a plane of symmetry, then it will not exhibit optical activity (the mirror image will be superimposable). Show the plane of symmetry in the *trans* isomer and prove to yourself that the *cis* isomer does not have a plane of symmetry.

6. What is the major focus of the crystal field model? Why are the *d* orbitals split into two sets for an octahedral complex? What are the two sets of orbitals?

Define each of the following.

a. weak-field ligand

b. strong-field ligand

c. low-spin complex

d. high-spin complex

Why is $Co(NH_3)_6^{3+}$ diamagnetic whereas CoF_6^{3-} is paramagnetic? Some octahedral complex ions have the same *d*-orbital splitting diagrams whether they are high-spin or low-spin. For which of the following is this true?

a. V^{3+}

b. Ni^{2+}

c. Ru^{2+}

7. The crystal field model predicts magnetic properties of complex ions and explains the colors of these complex ions. How? Solutions of $[Cr(NH_3)_6]Cl_3$ are yellow, but $Cr(NH_3)_6^{3+}$ does not absorb yellow light. Why? What color light is absorbed by $Cr(NH_3)_6^{3+}$? What is the spectrochemical series, and how can the study of light absorbed by various complex ions be used to develop this series? Would you expect $Co(NH_3)_6^{2+}$ to absorb light of a longer or shorter wavelength than $Co(NH_3)_6^{3+}$? Explain.

8. Why do tetrahedral complex ions have a different crystal field diagram than octahedral complex ions? What is the tetrahedral crystal field diagram? Why are virtually all tetrahedral complex ions "high spin"?

Explain the crystal field diagram for square planar complex ions and for linear complex ions.

9. Review Table 21.18, which lists some important biological functions associated with different first-row transition metals. The transport of O_2 in the blood is carried out by hemoglobin. Briefly explain how hemoglobin transports O_2 in the blood.

10. Define and give an example of each of the following.

a. roasting

b. smelting

c. flotation

d. leaching

e. gangue

What are the advantages and disadvantages of hydrometallurgy? Describe the process by which a metal is purified by zone refining.

Active Learning Questions

These questions are designed to be used by groups of students in class.

1. You isolate a compound with the formula $PtCl_4 \cdot 2KCl$. From electrical conductance tests of an aqueous solution of the compound, you find that three ions per formula unit are present, and you also notice that addition of $AgNO_3$ does not cause a precipitate. Give the formula for this compound that shows the complex ion present. Explain your findings. Name this compound.

2. Both $Ni(NH_3)_4^{2+}$ and $Ni(SCN)_4^{2-}$ have four ligands. The first is paramagnetic, and the second is diamagnetic. Are the complex ions tetrahedral or square planar? Explain.

3. Which is more likely to be paramagnetic, $Fe(CN)_6^{4-}$ or $Fe(H_2O)_6^{2+}$? Explain.

4. A metal ion in a high-spin octahedral complex has two more unpaired electrons than the same ion does in a low-spin octahedral complex. Name some possible metal ions for which this would be true.

A blue question or exercise number indicates that the answer to that question or exercise appears at the back of this book and a solution appears in the *Solutions Guide*, as found on PowerLecture.

Questions

5. What is the lanthanide contraction? How does the lanthanide contraction affect the properties of the $4d$ and $5d$ transition metals?

6. Four different octahedral chromium coordination compounds exist that all have the same oxidation state for chromium and have H_2O and Cl^- as the ligands and counterions. When 1 mole of each of the four compounds is dissolved in water, how many moles of silver chloride will precipitate upon addition of excess $AgNO_3$?

7. Figure 21.17 shows that the *cis* isomer of $Co(en)_2Cl_2^+$ is optically active while the *trans* isomer is not optically active. Is the same true for $Co(NH_3)_4Cl_2^+$? Explain.

8. A certain first-row transition metal ion forms many different colored solutions. When four coordination compounds of this metal, each having the same coordination number, are dissolved in water, the colors of the solutions are red, yellow, green, and blue. Further experiments reveal that two of the complex ions are paramagnetic with four unpaired electrons and the other two are diamagnetic. What can be deduced from this information about the four coordination compounds?

9. Oxalic acid is often used to remove rust stains. What properties of oxalic acid allow it to do this?

10. For the following crystal field diagrams, label each as low spin, high spin, or cannot tell. Explain your answers.

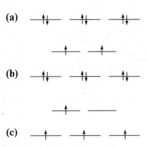

11. $CoCl_4^{2-}$ forms a tetrahedral complex ion and $Co(CN)_6^{3-}$ forms an octahedral complex ion. What is wrong about the following statements concerning each complex ion and the d orbital splitting diagrams?

   a. $CoCl_4^{2-}$ is an example of a strong-field case having two unpaired electrons.

   b. Because CN^- is a weak-field ligand, $Co(CN)_6^{3-}$ will be a low-spin case having four unpaired electrons.

12. The following statements discuss some coordination compounds. For each coordination compound, give the complex ion and the counterions, the electron configuration of the transition metal, and the geometry of the complex ion.

   a. $CoCl_2 \cdot 6H_2O$ is a compound used in novelty devices that predict rain.

   b. During the developing process of black-and-white film, silver bromide is removed from photographic film by the fixer. The major component of the fixer is sodium thiosulfate. The equation for the reaction is:

   $$AgBr(s) + 2Na_2S_2O_3(aq) \longrightarrow$$
   $$Na_3[Ag(S_2O_3)_2](aq) + NaBr(aq)$$

   c. In the production of printed circuit boards for the electronics industry, a thin layer of copper is laminated onto an insulating plastic board. Next, a circuit pattern made of a chemically resistant polymer is printed on the board. The unwanted copper is removed by chemical etching, and the protective polymer is finally removed by solvents. One etching reaction is:

   $$Cu(NH_3)_4Cl_2(aq) + 4NH_3(aq) + Cu(s) \longrightarrow$$
   $$2Cu(NH_3)_4Cl(aq)$$

   Assume these copper complex ions have tetrahedral geometry.

13. When concentrated hydrochloric acid is added to a red solution containing the $Co(H_2O)_6^{2+}$ complex ion, the solution turns blue as the tetrahedral $CoCl_4^{2-}$ complex ion forms. Explain this color change.

14. Tetrahedral complexes of Co^{2+} are quite common. Use a d-orbital splitting diagram to rationalize the stability of Co^{2+} tetrahedral complex ions.

15. Which of the following ligands are capable of linkage isomerism? Explain your answer.

   $$SCN^-, N_3^-, NO_2^-, NH_2CH_2CH_2NH_2, OCN^-, I^-$$

16. Compounds of copper(II) are generally colored, but compounds of copper(I) are not. Explain. Would you expect $Cd(NH_3)_4Cl_2$ to be colored? Explain.

17. Compounds of Sc^{3+} are not colored, but those of Ti^{3+} and V^{3+} are. Why?

18. Almost all metals in nature are found as ionic compounds in ores instead of being in the pure state. Why? What must be done to a sample of ore to obtain a metal substance that has desirable properties?

19. What causes high-altitude sickness, and what is high-altitude acclimatization?

20. Why are CN^- and CO toxic to humans?

Exercises

In this section similar exercises are paired.

Transition Metals and Coordination Compounds

21. Write electron configurations for the following metals.
   a. Ni   **b.** Cd   **c.** Zr   **d.** Os

22. Write electron configurations for the following ions.
   a. Ni^{2+} **c.** Zr^{3+} and Zr^{4+}
   b. Cd^{2+} **d.** Os^{2+} and Os^{3+}

23. Write electron configurations for each of the following.
   a. Ti, Ti^{2+}, Ti^{4+} **b.** Re, Re^{2+}, Re^{3+} **c.** Ir, Ir^{2+}, Ir^{3+}

24. Write electron configurations for each of the following.
   a. Cr, Cr^{2+}, Cr^{3+} **b.** Cu, Cu^+, Cu^{2+} **c.** V, V^{2+}, V^{3+}

25. What is the electron configuration for the transition metal ion in each of the following compounds?
   a. $K_3[Fe(CN)_6]$
   b. $[Ag(NH_3)_2]Cl$
   c. $[Ni(H_2O)_6]Br_2$
   d. $[Cr(H_2O)_4(NO_2)_2]I$

26. What is the electron configuration for the transition metal ion(s) in each of the following compounds?
   a. $(NH_4)_2[Fe(H_2O)_2Cl_4]$
   b. $[Co(NH_3)_2(NH_2CH_2CH_2NH_2)_2]I_2$
   c. $Na_2[TaF_7]$
   d. $[Pt(NH_3)_4I_2][PtI_4]$
   Pt forms +2 and +4 oxidation states in compounds.

27. Molybdenum is obtained as a by-product of copper mining or is mined directly (primary deposits are in the Rocky Mountains in Colorado). In both cases it is obtained as MoS_2, which is then converted to MoO_3. The MoO_3 can be used directly in the production of stainless steel for high-speed tools (which accounts for about 85% of the molybdenum used). Molybdenum can be purified by dissolving MoO_3 in aqueous ammonia and crystallizing ammonium molybdate. Depending on conditions, either $(NH_4)_2Mo_2O_7$ or $(NH_4)_6Mo_7O_{24} \cdot 4H_2O$ is obtained.
   a. Give names for MoS_2 and MoO_3.
   b. What is the oxidation state of Mo in each of the compounds mentioned above?

28. Titanium dioxide, the most widely used white pigment, occurs naturally but is often colored by the presence of impurities. The chloride process is often used in purifying rutile, a mineral form of titanium dioxide.
   a. Show that the unit cell for rutile, shown below, conforms to the formula TiO_2. (*Hint:* Recall the discussion in Sections 10.4 and 10.7.)

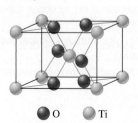

   ● O   ○ Ti

b. The reactions for the chloride process are

$$2TiO_2(s) + 3C(s) + 4Cl_2(g)$$
$$\xrightarrow{950°C} 2TiCl_4(g) + CO_2(g) + 2CO(g)$$
$$TiCl_4(g) + O_2(g) \xrightarrow{1000-1400°C} TiO_2(s) + 2Cl_2(g)$$

Assign oxidation states to the elements in both reactions. Which elements are being reduced, and which are being oxidized? Identify the oxidizing agent and the reducing agent in each reaction.

29. When 6 *M* ammonia is added gradually to aqueous copper(II) nitrate, a white precipitate forms. The precipitate dissolves as more 6 *M* ammonia is added. Write balanced equations to explain these observations. [*Hint:* Cu^{2+} reacts with NH_3 to form $Cu(NH_3)_4^{2+}$.]

30. When an aqueous solution of KCN is added to a solution containing Ni^{2+} ions, a precipitate forms, which redissolves on addition of more KCN solution. Write reactions describing what happens in this solution. [*Hint:* CN^- is a Brønsted–Lowry base $(K_b \approx 10^{-5})$ and a Lewis base.]

31. Consider aqueous solutions of the following coordination compounds: $Co(NH_3)_6I_3$, $Pt(NH_3)_4I_4$, Na_2PtI_6, and $Cr(NH_3)_4I_3$. If aqueous $AgNO_3$ is added to separate beakers containing solutions of each coordination compound, how many moles of AgI will precipitate per mole of transition metal present? Assume that each transition metal ion forms an octahedral complex.

32. A coordination compound of cobalt(III) contains four ammonia molecules, one sulfate ion, and one chloride ion. Addition of aqueous $BaCl_2$ solution to an aqueous solution of the compound gives no precipitate. Addition of aqueous $AgNO_3$ to an aqueous solution of the compound produces a white precipitate. Propose a structure for this coordination compound.

33. Name the following complex ions:

a.

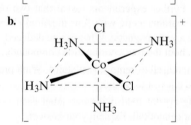

b.

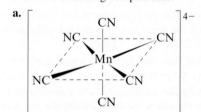

c.

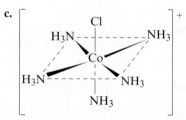

34. Name the following complex ions.

  a. $Ru(NH_3)_5Cl^{2+}$

  b. $Fe(CN)_6^{4-}$

  c. $Mn(NH_2CH_2CH_2NH_2)_3^{2+}$

  d. $Co(NH_3)_5NO_2^{2+}$

  e. $Ni(CN)_4^{2-}$

  f. $Cr(NH_3)_4Cl_2^{+}$

  g. $Fe(C_2O_4)_3^{3-}$

  h. $Co(SCN)_2(H_2O)_4^{+}$

35. Name the following coordination compounds.

  a. $[Co(NH_3)_6]Cl_2$

  b. $[Co(H_2O)_6]I_3$

  c. $K_2[PtCl_4]$

  d. $K_4[PtCl_6]$

  e. $[Co(NH_3)_5Cl]Cl_2$

  f. $[Co(NH_3)_3(NO_2)_3]$

36. Name the following coordination compounds.

  a. $[Cr(H_2O)_5Br]Br_2$

  b. $Na_3[Co(CN)_6]$

  c. $[Fe(NH_2CH_2CH_2NH_2)_2(NO_2)_2]Cl$

  d. $[Pt(NH_3)_4I_2][PtI_4]$

37. Give formulas for the following.

  a. potassium tetrachlorocobaltate(II)

  b. aquatricarbonylplatinum(II) bromide

  c. sodium dicyanobis(oxalato)ferrate(III)

  d. triamminechloroethylenediaminechromium(III) iodide

38. Give formulas for the following complex ions.

  a. tetrachloroferrate(III) ion

  b. pentaammineaquaruthenium(III) ion

  c. tetracarbonyldihydroxochromium(III) ion

  d. amminetrichloroplatinate(II) ion

39. Draw geometrical isomers of each of the following complex ions.

  a. $Co(C_2O_4)_2(H_2O)_2^{-}$

  b. $Pt(NH_3)_4I_2^{2+}$

  c. $Ir(NH_3)_3Cl_3$

  d. $Cr(en)(NH_3)_2I_2^{+}$

40. Draw structures of each of the following.

  a. *cis*-dichloroethylenediamineplatinum(II)

  b. *trans*-dichlorobis(ethylenediamine)cobalt(II)

  c. *cis*-tetraamminechloronitrocobalt(III) ion

  d. *trans*-tetraamminechloronitritocobalt(III) ion

  e. *trans*-diaquabis(ethylenediamine)copper(II) ion

41. The carbonate ion (CO_3^{2-}) can act as either a monodentate or a bidentate ligand. Draw a picture of CO_3^{2-} coordinating to a metal ion as a monodentate and as a bidentate ligand. The carbonate ion can also act as a bridge between two metal ions. Draw a picture of a CO_3^{2-} ion bridging between two metal ions.

42. Amino acids can act as ligands toward transition metal ions. The simplest amino acid is glycine $(NH_2CH_2CO_2H)$. Draw a structure of the glycinate anion $(NH_2CH_2CO_2^{-})$ acting as a bidentate ligand. Draw the structural isomers of the square planar complex $Cu(NH_2CH_2CO_2)_2$.

43. How many bonds could each of the following chelating ligands form with a metal ion?

  a. acetylacetone (acacH), a common ligand in organometallic catalysts:

$$CH_3-\overset{\overset{\displaystyle O}{\|}}{C}-CH_2-\overset{\overset{\displaystyle O}{\|}}{C}-CH_3$$

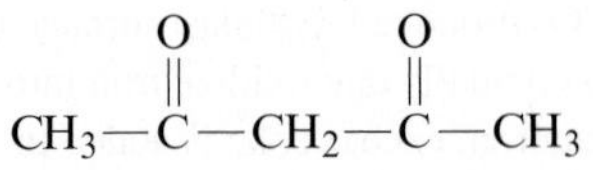

  b. diethylenetriamine, used in a variety of industrial processes:

$$NH_2-CH_2-CH_2-NH-CH_2-CH_2-NH_2$$

  c. salen, a common ligand for chiral organometallic catalysts:

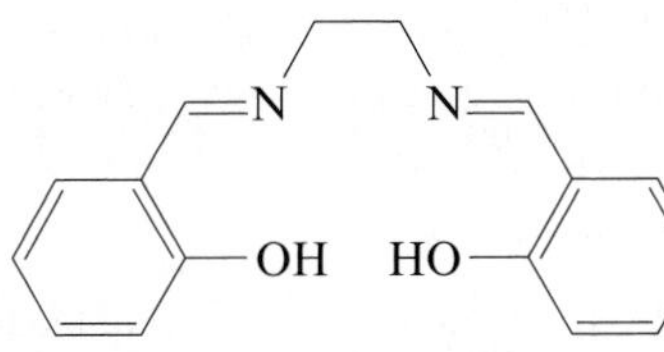

  d. porphine, often used in supermolecular chemistry as well as catalysis; biologically, porphine is the basis for many different types of porphyrin-containing proteins, including heme proteins:

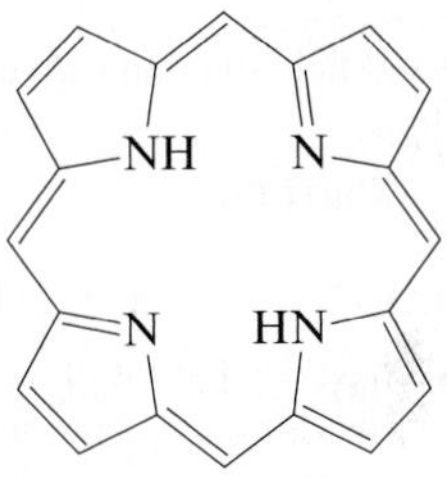

44. BAL is a chelating agent used in treating heavy metal poisoning. It acts as a bidentate ligand. What type of linkage isomers are possible when BAL coordinates to a metal ion?

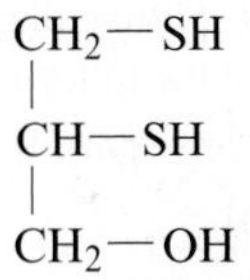

$$\begin{array}{c} CH_2-SH \\ | \\ CH-SH \\ | \\ CH_2-OH \end{array}$$

BAL

45. Draw all geometrical and linkage isomers of $Co(NH_3)_4(NO_2)_2$.

46. Draw all geometrical and linkage isomers of square planar $Pt(NH_3)_2(SCN)_2$.

47. Acetylacetone, abbreviated acacH, is a bidentate ligand. It loses a proton and coordinates as acac$^-$, as shown below, where M is a transition metal:

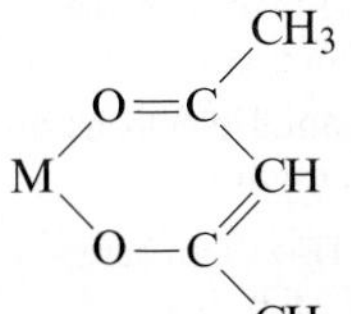

Which of the following complexes are optically active: *cis*-$Cr(acac)_2(H_2O)_2$, *trans*-$Cr(acac)_2(H_2O)_2$, and $Cr(acac)_3$?

48. Draw all geometrical isomers of $Pt(CN)_2Br_2(H_2O)_2$. Which of these isomers has an optical isomer? Draw the various optical isomers.

Bonding, Color, and Magnetism in Coordination Compounds

49. Match the crystal field diagrams given below with the following complex ions.

$Cr(NH_3)_5Cl^{2+}$ $Co(NH_3)_4Br_2^+$ $Fe(H_2O)_6^{3+}$
            (assume strong field)   (assume weak field)

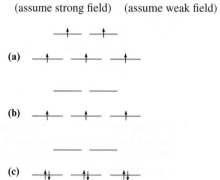

50. Match the crystal field diagrams given below with the following complex ions.

$Fe(CN)_6^{3-}$ $Mn(H_2O)_6^{2+}$

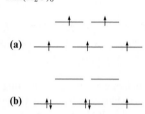

51. Draw the *d*-orbital splitting diagrams for the octahedral complex ions of each of the following.
  a. Fe^{2+} (high and low spin)
  b. Fe^{3+} (high spin)
  c. Ni^{2+}

52. Draw the *d*-orbital splitting diagrams for the octahedral complex ions of each of the following.
  a. Zn^{2+}
  b. Co^{2+} (high and low spin)
  c. Ti^{3+}

53. The CrF_6^{4-} ion is known to have four unpaired electrons. Does the F^- ligand produce a strong or weak field?

54. The $Co(NH_3)_6^{3+}$ ion is diamagnetic, but $Fe(H_2O)_6^{2+}$ is paramagnetic. Explain.

55. How many unpaired electrons are in the following complex ions?
  a. $Ru(NH_3)_6^{2+}$ (low-spin case)
  b. $Ni(H_2O)_6^{2+}$
  c. $V(en)_3^{3+}$

56. The complex ion $Fe(CN)_6^{3-}$ is paramagnetic with one unpaired electron. The complex ion $Fe(SCN)_6^{3-}$ has five unpaired electrons. Where does SCN^- lie in the spectrochemical series relative to CN^-?

57. Rank the following complex ions in order of increasing wavelength of light absorbed.

$Co(H_2O)_6^{3+}, Co(CN)_6^{3-}, CoI_6^{3-}, Co(en)_3^{3+}$

58. The complex ion $Cu(H_2O)_6^{2+}$ has an absorption maximum at around 800 nm. When four ammonias replace water, $Cu(NH_3)_4(H_2O)_2^{2+}$, the absorption maximum shifts to around 600 nm. What do these results signify in terms of the relative field splittings of NH_3 and H_2O? Explain.

59. The following test tubes each contain a different chromium complex ion.

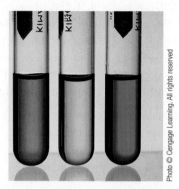

For each complex ion, predict the predominant color of light absorbed. If the complex ions are $Cr(NH_3)_6^{3+}$, $Cr(H_2O)_6^{3+}$, and $Cr(H_2O)_4Cl_2^+$, what is the identity of the complex ion in each test tube? (*Hint:* Reference the spectrochemical series.)

60. Consider the complex ions $Co(NH_3)_6^{3+}$, $Co(CN)_6^{3-}$, and CoF_6^{3-}. The wavelengths of absorbed electromagnetic radiation for these compounds (in no specific order) are 770 nm, 440 nm, and 290 nm. Match the complex ion to the wavelength of absorbed electromagnetic radiation.

61. The wavelength of absorbed electromagnetic radiation for $CoBr_4^{2-}$ is 3.4×10^{-6} m. Will the complex ion $CoBr_6^{4-}$ absorb electromagnetic radiation having a wavelength longer or shorter than 3.4×10^{-6} m? Explain.

62. The complex ion $NiCl_4^{2-}$ has two unpaired electrons, whereas $Ni(CN)_4^{2-}$ is diamagnetic. Propose structures for these two complex ions.

63. How many unpaired electrons are present in the tetrahedral ion $FeCl_4^-$?

64. The complex ion $PdCl_4^{2-}$ is diamagnetic. Propose a structure for $PdCl_4^{2-}$.

Metallurgy

65. A blast furnace is used to reduce iron oxides to elemental iron. The reducing agent for this reduction process is carbon monoxide.
  a. Given the following data:

$$Fe_2O_3(s) + 3CO(g) \longrightarrow 2Fe(s) + 3CO_2(g) \qquad \Delta H° = -23 \text{ kJ}$$
$$3Fe_2O_3(s) + CO(g) \longrightarrow 2Fe_3O_4(s) + CO_2(g) \qquad \Delta H° = -39 \text{ kJ}$$
$$Fe_3O_4(s) + CO(g) \longrightarrow 3FeO(s) + CO_2(g) \qquad \Delta H° = 18 \text{ kJ}$$

    determine $\Delta H°$ for the reaction

$$FeO(s) + CO(g) \longrightarrow Fe(s) + CO_2(g)$$

  b. The CO_2 produced in a blast furnace during the reduction process actually can oxidize iron into FeO. To eliminate this reaction, excess coke is added to convert CO_2 into CO by the reaction

$$CO_2(g) + C(s) \longrightarrow 2CO(g)$$

Using data from Appendix 4, determine $\Delta H°$ and $\Delta S°$ for this reaction. Assuming $\Delta H°$ and $\Delta S°$ do not depend on temperature, at what temperature is the conversion reaction of CO_2 into CO spontaneous at standard conditions?

66. Use the data in Appendix 4 for the following.

   a. Calculate $\Delta H°$ and $\Delta S°$ for the reaction

   $$3Fe_2O_3(s) + CO(g) \longrightarrow 2Fe_3O_4(s) + CO_2(g)$$

   that occurs in a blast furnace.

   b. Assume that $\Delta H°$ and $\Delta S°$ are independent of temperature. Calculate $\Delta G°$ at 800.°C for this reaction.

67. Iron is present in the earth's crust in many types of minerals. The iron oxide minerals are hematite (Fe_2O_3) and magnetite (Fe_3O_4). What is the oxidation state of iron in each mineral? The iron ions in magnetite are a mixture of Fe^{2+} and Fe^{3+} ions. What is the ratio of Fe^{3+} to Fe^{2+} ions in magnetite? The formula for magnetite is often written as $FeO \cdot Fe_2O_3$. Does this make sense? Explain.

68. What roles do kinetics and thermodynamics play in the effect that the following reaction has on the properties of steel?

   $$3Fe + C \rightleftharpoons Fe_3C$$

69. Silver is sometimes found in nature as large nuggets; more often it is found mixed with other metals and their ores. Cyanide ion is often used to extract the silver by the following reaction that occurs in basic solution:

   $$Ag(s) + CN^-(aq) + O_2(g) \xrightarrow{\text{Basic}} Ag(CN)_2^-(aq)$$

   Balance this equation by using the half-reaction method.

70. One of the classic methods for the determination of the manganese content in steel involves converting all the manganese to the deeply colored permanganate ion and then measuring the absorption of light. The steel is first dissolved in nitric acid, producing the manganese(II) ion and nitrogen dioxide gas. This solution is then reacted with an acidic solution containing periodate ion; the products are the permanganate and iodate ions. Write balanced chemical equations for both of these steps.

Additional Exercises

71. Acetylacetone (see Exercise 43, part a), abbreviated acacH, is a bidentate ligand. It loses a proton and coordinates as acac$^-$, as shown below:

Acetylacetone reacts with an ethanol solution containing a salt of europium to give a compound that is 40.1% C and 4.71% H by mass. Combustion of 0.286 g of the compound gives 0.112 g Eu_2O_3. Assuming the compound contains only C, H, O, and Eu, determine the formula of the compound formed from the reaction of acetylacetone and the europium salt. (Assume that the compound contains one europium ion.)

72. The compound cisplatin, $Pt(NH_3)_2Cl_2$, has been studied extensively as an antitumor agent. The reaction for the synthesis of cisplatin is:

   $$K_2PtCl_4(aq) + 2NH_3(aq) \longrightarrow Pt(NH_3)_2Cl_2(s) + 2KCl(aq)$$

   Write the electron configuration for platinum ion in cisplatin. Most d^8 transition metal ions exhibit square planar geometry. With this and the name in mind, draw the structure of cisplatin.

73. Use standard reduction potentials to calculate $\mathscr{E}°$, $\Delta G°$, and K (at 298 K) for the reaction that is used in production of gold:

   $$2Au(CN)_2^-(aq) + Zn(s) \longrightarrow 2Au(s) + Zn(CN)_4^{2-}(aq)$$

   The relevant half-reactions are

   $$Au(CN)_2^- + e^- \longrightarrow Au + 2CN^- \quad \mathscr{E}° = -0.60 \text{ V}$$
   $$Zn(CN)_4^{2-} + 2e^- \longrightarrow Zn + 4CN^- \quad \mathscr{E}° = -1.26 \text{ V}$$

74. Until the discoveries of Alfred Werner, it was thought that carbon had to be present in a compound for it to be optically active. Werner prepared the following compound containing OH$^-$ ions as bridging groups and separated the optical isomers.

   a. Draw structures of the two optically active isomers of this compound.

   b. What are the oxidation states of the cobalt ions?

   c. How many unpaired electrons are present if the complex is the low-spin case?

75. Draw all the geometrical isomers of $Cr(en)(NH_3)_2BrCl^+$. Which of these isomers also have an optical isomer? Draw the various isomers.

76. A compound related to acetylacetone is 1,1,1-trifluoroacetylacetone (abbreviated Htfa):

   Htfa forms complexes in a manner similar to acetylacetone. (See Exercise 47.) Both Be^{2+} and Cu^{2+} form complexes with tfa$^-$ having the formula $M(tfa)_2$. Two isomers are formed for each metal complex.

   a. The Be^{2+} complexes are tetrahedral. Draw the two isomers of $Be(tfa)_2$. What type of isomerism is exhibited by $Be(tfa)_2$?

   b. The Cu^{2+} complexes are square planar. Draw the two isomers of $Cu(tfa)_2$. What type of isomerism is exhibited by $Cu(tfa)_2$?

77. Would it be better to use octahedral Ni^{2+} complexes or octahedral Cr^{2+} complexes to determine whether a given ligand is a strong-field or weak-field ligand by measuring the number of unpaired electrons? How else could the relative ligand field strengths be determined?

78. Name the following coordination compounds.

   a. $Na_4[Ni(C_2O_4)_3]$

   b. $K_2[CoCl_4]$

   c. $[Cu(NH_3)_4]SO_4$

   d. $[Co(en)_2(SCN)Cl]Cl$

79. Give formulas for the following.

   a. hexakis(pyridine)cobalt(III) chloride

   b. pentaammineiodochromium(III) iodide

   c. tris(ethylenediamine)nickel(II) bromide

   d. potassium tetracyanonickelate(II)

   e. tetraamminedichloroplatinum(IV) tetrachloroplatinate(II)

80. The complex ion $Ru(phen)_3^{2+}$ has been used as a probe for the structure of DNA. (Phen is a bidentate ligand.)

   a. What type of isomerism is found in $Ru(phen)_3^{2+}$?

   b. $Ru(phen)_3^{2+}$ is diamagnetic (as are all complex ions of Ru^{2+}). Draw the crystal field diagram for the d orbitals in this complex ion.

Phen = 1,10-phenanthroline =

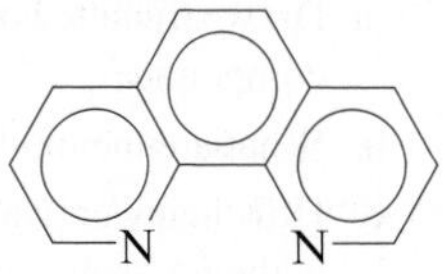

81. Carbon monoxide is toxic because it binds more strongly to iron in hemoglobin (Hb) than does O_2. Consider the following reactions and approximate standard free energy changes:

$$Hb + O_2 \longrightarrow HbO_2 \qquad \Delta G° = -70 \text{ kJ}$$
$$Hb + CO \longrightarrow HbCO \qquad \Delta G° = -80 \text{ kJ}$$

Using these data, estimate the equilibrium constant value at 25°C for the following reaction:

$$HbO_2 + CO \rightleftharpoons HbCO + O_2$$

82. For the process

$$Co(NH_3)_5Cl^{2+} + Cl^- \longrightarrow Co(NH_3)_4Cl_2^+ + NH_3$$

what would be the expected ratio of *cis* to *trans* isomers in the product?

ChemWork Problems

These multiconcept problems (and additional ones) are found interactively online with the same type of assistance a student would get from an instructor.

83. In which of the following is(are) the electron configuration(s) correct for the species indicated?

   a. Cu $[Ar]4s^2 3d^9$

   b. Fe^{3+} $[Ar]3d^5$

   c. Co $[Ar]4s^2 3d^7$

   d. La $[Ar]6s^2 4f^1$

   e. Pt^{2+} $[Xe]4f^{14} 5d^8$

84. Which of the following molecules exhibit(s) optical isomerism?

   a. *cis*-$Pt(NH_3)_2Cl_2$

   b. *trans*-$Ni(en)_2Br_2$ (en is ethylenediamine)

   c. *cis*-$Ni(en)_2Br_2$ (en is ethylenediamine)

   d.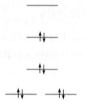

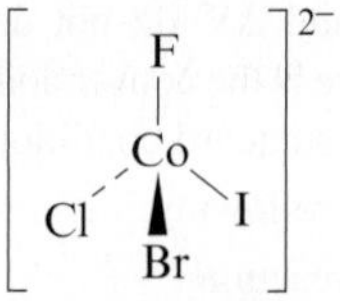

85. Which of the following ions is(are) expected to form colored octahedral aqueous complex ions?

   a. Zn^{2+}

   b. Cu^{2+}

   c. Mn^{3+}

   d. Ti^{4+}

86. The following table indicates the number of unpaired electrons in the crystal field diagrams for some complexes. Complete the table by classifying each species as weak field, strong field, or insufficient information.

Species	Unpaired Electrons	Classification
$Fe(CNS)_6^{4-}$	0	_____
$CoCl_4^{2-}$	3	_____
$Fe(H_2O)_6^{3+}$	5	_____
$Fe(CN)_6^{4-}$	0	_____

87. Which of the following crystal field diagram(s) is(are) correct for the complex given?

   a. $Zn(NH_3)_4^{2+}$ (tetrahedral)

   b. $Mn(CN)_6^{3-}$ (strong field)

   c. $Ni(CN)_4^{2-}$ (square planar, diamagnetic)

88. Which of the following statement(s) is(are) *true*?

   a. The coordination number of a metal ion in an octahedral complex ion is 8.

   b. All tetrahedral complex ions are low-spin.

   c. The formula for triaquatriamminechromium(III) sulfate is $[Cr(H_2O)_3(NH_3)_3]_2(SO_4)_3$.

   d. The electron configuration of Hf^{2+} is $[Xe]4f^{12}6s^2$.

   e. Hemoglobin contains Fe^{3+}.

Challenge Problems

89. Consider the following complex ion, where A and B represent ligands.

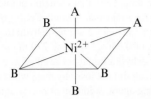

The complex is known to be diamagnetic. Do A and B produce very similar or very different crystal fields? Explain.

90. Consider the pseudo-octahedral complex ion of Cr^{3+}, where A and B represent ligands.

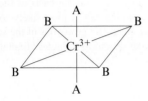

Ligand A produces a stronger crystal field than ligand B. Draw an appropriate crystal field diagram for this complex ion (assume the A ligands are on the z-axis).

91. Consider the following data:

$$Co^{3+} + e^- \longrightarrow Co^{2+} \qquad \mathscr{E}° = 1.82 \text{ V}$$
$$Co^{2+} + 3en \longrightarrow Co(en)_3^{2+} \qquad K = 1.5 \times 10^{12}$$
$$Co^{3+} + 3en \longrightarrow Co(en)_3^{3+} \qquad K = 2.0 \times 10^{47}$$

where en = ethylenediamine.

a. Calculate $\mathscr{E}°$ for the half-reaction

$$Co(en)_3^{3+} + e^- \longrightarrow Co(en)_3^{2+}$$

b. Based on your answer to part a, which is the stronger oxidizing agent, Co^{3+} or $Co(en)_3^{3+}$?

c. Use the crystal field model to rationalize the result in part b.

92. Henry Taube, 1983 Nobel Prize winner in chemistry, has studied the mechanisms of the oxidation–reduction reactions of transition metal complexes. In one experiment he and his students studied the following reaction:

$$Cr(H_2O)_6^{2+}(aq) + Co(NH_3)_5Cl^{2+}(aq)$$
$$\longrightarrow Cr(III) \text{ complexes} + Co(II) \text{ complexes}$$

Chromium(III) and cobalt(III) complexes are substitutionally inert (no exchange of ligands) under conditions of the experiment. Chromium(II) and cobalt(II) complexes can exchange ligands very rapidly. One of the products of the reaction is $Cr(H_2O)_5Cl^{2+}$. Is this consistent with the reaction proceeding through formation of $(H_2O)_5Cr—Cl—Co(NH_3)_5$ as an intermediate? Explain.

93. Chelating ligands often form more stable complex ions than the corresponding monodentate ligands with the same donor atoms. For example,

$$Ni^{2+}(aq) + 6NH_3(aq) \rightleftharpoons Ni(NH_3)_6^{2+}(aq) \qquad K = 3.2 \times 10^8$$
$$Ni^{2+}(aq) + 3en(aq) \rightleftharpoons Ni(en)_3^{2+}(aq) \qquad K = 1.6 \times 10^{18}$$
$$Ni^{2+}(aq) + penten(aq) \rightleftharpoons Ni(penten)^{2+}(aq) \qquad K = 2.0 \times 10^{19}$$

where en is ethylenediamine and penten is

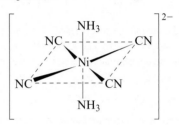

This increased stability is called the *chelate effect.* Based on bond energies, would you expect the enthalpy changes for the above reactions to be very different? What is the order (from least favorable to most favorable) of the entropy changes for the above reactions? How do the values of the formation constants correlate with $\Delta S°$? How can this be used to explain the chelate effect?

94. Qualitatively draw the crystal field splitting of the d orbitals in a trigonal planar complex ion. (Let the z axis be perpendicular to the plane of the complex.)

95. Qualitatively draw the crystal field splitting for a trigonal bipyramidal complex ion. (Let the z axis be perpendicular to the trigonal plane.)

96. Sketch a d-orbital energy diagram for the following.

a. a linear complex ion with ligands on the x axis

b. a linear complex ion with ligands on the y axis

97. Sketch and explain the most likely crystal field diagram for the following complex ion:

$$\left[\begin{array}{c} NH_3 \\ NC{-}\!\!-\!\!-\!\!-\!\!-CN \\ Ni \\ NC{-}\!\!-\!\!-\!\!-\!\!-CN \\ NH_3 \end{array} \right]^{2-}$$

Note: The CN^- ligand produces a *much* stronger crystal field than NH_3. Assume the NH_3 ligands lie on the z axis.

Integrative Problems

These problems require the integration of multiple concepts to find the solutions.

98. The ferrate ion, FeO_4^{2-}, is such a powerful oxidizing agent that in acidic solution, aqueous ammonia is reduced to elemental nitrogen along with the formation of the iron(III) ion.

a. What is the oxidation state of iron in FeO_4^{2-}, and what is the electron configuration of iron in this polyatomic ion?

b. If 25.0 mL of a 0.243 M FeO_4^{2-} solution is allowed to react with 55.0 mL of 1.45 M aqueous ammonia, what volume of nitrogen gas can form at 25°C and 1.50 atm?

99. Ammonia and potassium iodide solutions are added to an aqueous solution of $Cr(NO_3)_3$. A solid is isolated (compound A), and the following data are collected:

i. When 0.105 g of compound A was strongly heated in excess O_2, 0.0203 g CrO_3 was formed.

ii. In a second experiment it took 32.93 mL of 0.100 M HCl to titrate completely the NH_3 present in 0.341 g compound A.

iii. Compound A was found to contain 73.53% iodine by mass.

iv. The freezing point of water was lowered by 0.64°C when 0.601 g compound A was dissolved in 10.00 g H_2O ($K_f = 1.86°C \cdot$ kg/mol).

What is the formula of the compound? What is the structure of the complex ion present? (*Hints:* Cr^{3+} is expected to be six-coordinate, with NH_3 and possibly I^- as ligands. The I^- ions will be the counterions if needed.)

100. **a.** In the absorption spectrum of the complex ion $Cr(NCS)_6^{3-}$, there is a band corresponding to the absorption of a photon of light with an energy of 1.75×10^4 cm^{-1}. Given 1 cm^{-1} = 1.986×10^{-23} J, what is the wavelength of this photon?

   b. The Cr—N—C bond angle in $Cr(NCS)_6^{3-}$ is predicted to be 180°. What is the hybridization of the N atom in the NCS^- ligand when a Lewis acid–base reaction occurs between Cr^{3+} and NCS^- that would give a 180° Cr—N—C bond angle? $Cr(NCS)_6^{3-}$ undergoes substitution by ethylenediamine (en) according to the equation

   $$Cr(NCS)_6^{3-} + 2en \longrightarrow Cr(NCS)_2(en)_2^+ + 4NCS^-$$

   Does $Cr(NCS)_2(en)_2^+$ exhibit geometric isomerism? Does $Cr(NCS)_2(en)_2^+$ exhibit optical isomerism?

Marathon Problem

This problem is designed to incorporate several concepts and techniques into one situation.

101. There are three salts that contain complex ions of chromium and have the molecular formula $CrCl_3 \cdot 6H_2O$. Treating 0.27 g of the first salt with a strong dehydrating agent resulted in a mass loss of 0.036 g. Treating 270 mg of the second salt with the same dehydrating agent resulted in a mass loss of 18 mg. The third salt did not lose any mass when treated with the same dehydrating agent. Addition of excess aqueous silver nitrate to 100.0-mL portions of 0.100 *M* solutions of each salt resulted in the formation of different masses of silver chloride; one solution yielded 1430 mg AgCl; another, 2870 mg AgCl; the third, 4300 mg AgCl. Two of the salts are green and one is violet.

   Suggest probable structural formulas for these salts, defending your answer on the basis of the preceding observations. State which salt is most likely to be violet. Would a study of the magnetic properties of the salts be helpful in determining the structural formulas? Explain.

Chapter 22

Organic and Biological Molecules

This complex web was spun by an orb weaver spider in Costa Rica. It is composed of a natural polymer.
(© Michael & Patricia Fogden/Corbis)

Two Group 4A elements, carbon and silicon, form the basis of most natural substances. Silicon, with its great affinity for oxygen, forms chains and rings containing Si—O—Si bridges to produce the silica and silicates that form the basis for most rocks, sands, and soils. What silicon is to the geological world, carbon is to the biological world. Carbon has the unusual ability of bonding strongly to itself to form long chains or rings of carbon atoms. In addition, carbon forms strong bonds to other nonmetals such as hydrogen, nitrogen, oxygen, sulfur, and the halogens. Because of these bonding properties, there are a myriad of carbon compounds; several million are now known, and the number continues to grow rapidly. Among these many compounds are the **biomolecules**, those responsible for maintaining and reproducing life.

The study of carbon-containing compounds and their properties is called **organic chemistry**. Although a few compounds involving carbon, such as its oxides and carbonates, are considered to be inorganic substances, the vast majority are organic compounds that typically contain chains or rings of carbon atoms.

Originally, the distinction between inorganic and organic substances was based on whether a compound was produced by living systems. For example, until the early nineteenth century it was believed that organic compounds had some sort of "life force" and could be synthesized only by living organisms. This view was dispelled in 1828 when the German chemist Friedrich Wöhler (1800–1882) prepared urea from the inorganic salt ammonium cyanate by simple heating:

$$\text{NH}_4\text{OCN} \xrightarrow{\text{Heat}} \text{H}_2\text{N}-\overset{\displaystyle \underset{\|}{}}{\text{C}}-\text{NH}_2$$
$$\text{O}$$

Ammonium cyanate   Urea

Urea is a component of urine, so it is clearly an organic material; yet here was clear evidence that it could be produced in the laboratory as well as by living things.

Organic chemistry plays a vital role in our quest to understand living systems. Beyond that, the synthetic fibers, plastics, artificial sweeteners, and drugs that are such an accepted part of modern life are products of industrial organic chemistry. In addition, the energy on which we rely so heavily to power our civilization is based mostly on the organic materials found in coal and petroleum.

Because organic chemistry is such a vast subject, we can provide only a brief introduction to it in this book. We will begin with the simplest class of organic compounds, the hydrocarbons, and then show how most other organic compounds can be considered to be derivatives of hydrocarbons.

22.1 | Alkanes: Saturated Hydrocarbons

As the name indicates, **hydrocarbons** are compounds composed of carbon and hydrogen. Those compounds whose carbon–carbon bonds are all single bonds are said to be **saturated**, because each carbon is bound to four atoms, the maximum number. Hydrocarbons containing carbon–carbon multiple bonds are described as being **unsaturated**, since the carbon atoms involved in a multiple bond can react with additional atoms, as shown by the *addition* of hydrogen to ethylene:

Unsaturated   Saturated

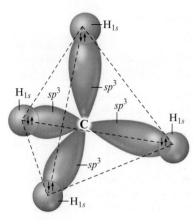

Figure 22.1 | The C—H bonds in methane.

Note that each carbon in ethylene is bonded to three atoms (one carbon and two hydrogens) but that each can bond to one additional atom if one bond of the carbon–carbon double bond is broken.

The simplest member of the saturated hydrocarbons, which are also called the **alkanes**, is *methane* (CH_4). As discussed in Section 9.1, methane has a tetrahedral structure and can be described in terms of a carbon atom using an sp^3 hybrid set of orbitals to bond to the four hydrogen atoms (Fig. 22.1). The next alkane, the one containing two carbon atoms, is *ethane* (C_2H_6), as shown in Fig. 22.2. Each carbon in ethane is surrounded by four atoms and thus adopts a tetrahedral arrangement and sp^3 hybridization, as predicted by the localized electron model.

The next two members of the series are *propane* (C_3H_8) and *butane* (C_4H_{10}), shown in Fig. 22.3. Again, each carbon is bonded to four atoms and is described as sp^3 hybridized.

Alkanes in which the carbon atoms form long "strings" or chains are called **normal**, **straight-chain**, or **unbranched hydrocarbons**. As can be seen from Fig. 22.3, the chains in normal alkanes are not really straight but zig-zag, since the tetrahedral C—C—C angle is 109.5°. The normal alkanes can be represented by the structure

$$H-\overset{\displaystyle H}{\underset{\displaystyle H}{C}}-\left(\overset{\displaystyle H}{\underset{\displaystyle H}{C}}\right)_{\!n}-\overset{\displaystyle H}{\underset{\displaystyle H}{C}}-H$$

where n is an integer. Note that each member is obtained from the previous one by inserting a *methylene* (CH_2) group. We can condense the structural formulas by omitting some of the C—H bonds. For example, the general formula for normal alkanes shown above can be condensed to

$$CH_3-(CH_2)_n-CH_3$$

The first ten normal alkanes and some of their properties are listed in Table 22.1. Note that all alkanes can be represented by the general formula C_nH_{2n+2}. For example, nonane, which has nine carbon atoms, is represented by $C_9H_{(2 \times 9) + 2}$, or C_9H_{20}. Also note from Table 22.1 that the melting points and boiling points increase as the molar masses increase, as we would expect.

Isomerism in Alkanes

Butane and all succeeding members of the alkanes exhibit **structural isomerism**. Recall from Section 21.4 that structural isomerism occurs when two molecules have the same atoms but different bonds. For example, butane can exist as a straight-chain molecule (normal butane, or *n*-butane) or with a branched-chain structure (called isobutane), as shown in Fig. 22.4. Because of their different structures, these molecules exhibit different properties. For example, the boiling point of *n*-butane is −0.5°C, whereas that of isobutane is −12°C.

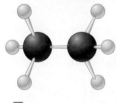

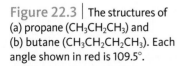

Figure 22.2 | (a) The Lewis structure of ethane (C_2H_6). (b) The molecular structure of ethane represented by space-filling and ball-and-stick models.

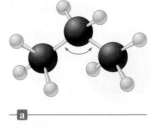

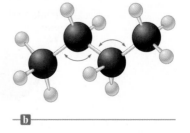

Figure 22.3 | The structures of (a) propane ($CH_3CH_2CH_3$) and (b) butane ($CH_3CH_2CH_2CH_3$). Each angle shown in red is 109.5°.

Table 22.1 | Selected Properties of the First Ten Normal Alkanes

Name	Formula	Molar Mass	Melting Point (°C)	Boiling Point (°C)	Number of Structural Isomers
Methane	CH_4	16	−182	−162	1
Ethane	C_2H_6	30	−183	−89	1
Propane	C_3H_8	44	−187	−42	1
Butane	C_4H_{10}	58	−138	0	2
Pentane	C_5H_{12}	72	−130	36	3
Hexane	C_6H_{14}	86	−95	68	5
Heptane	C_7H_{16}	100	−91	98	9
Octane	C_8H_{18}	114	−57	126	18
Nonane	C_9H_{20}	128	−54	151	35
Decane	$C_{10}H_{22}$	142	−30	174	75

Figure 22.4 | (a) Normal butane (abbreviated *n*-butane). (b) The branched isomer of butane (called isobutane).

Example 22.1

Structural Isomerism

Draw the isomers of pentane.

Solution

Pentane (C_5H_{12}) has the following isomeric structures:

1. 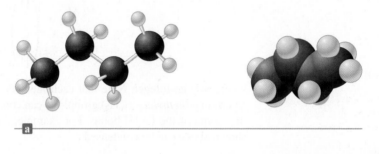 $CH_3—CH_2—CH_2—CH_2—CH_3$

n-Pentane

2.

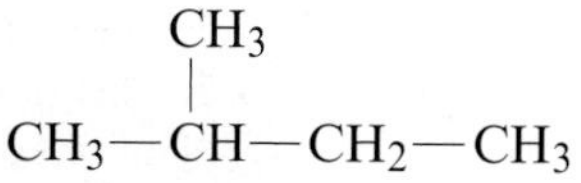

$$CH_3-\overset{\overset{\displaystyle CH_3}{|}}{CH}-CH_2-CH_3$$

Isopentane

3.

$$CH_3-\overset{\overset{\displaystyle CH_3}{|}}{\underset{\underset{\displaystyle CH_3}{|}}{C}}-CH_3$$

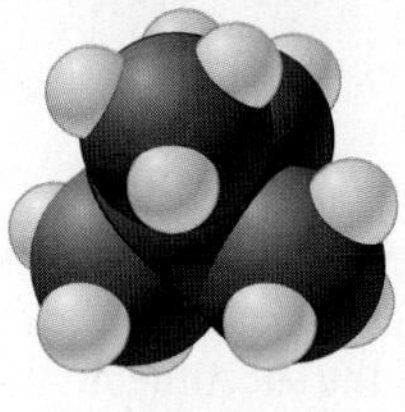

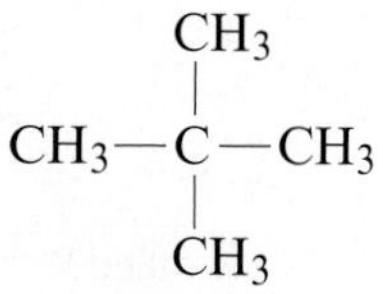

Neopentane

Note that the structures

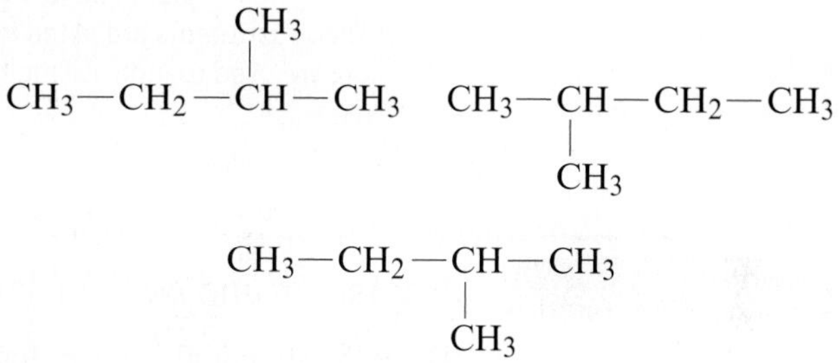

$$CH_3-CH_2-\overset{\overset{\displaystyle CH_3}{|}}{CH}-CH_3 \qquad CH_3-\overset{\overset{\displaystyle CH_3}{|}}{\underset{\underset{\displaystyle CH_3}{|}}{CH}}-CH_2-CH_3$$

$$CH_3-CH_2-\overset{\overset{\displaystyle CH_3}{|}}{\underset{\underset{\displaystyle CH_3}{|}}{CH}}-CH_3$$

which might appear to be other isomers, are actually identical to structure 2.

See Exercise 22.13

Nomenclature

Because there are literally millions of organic compounds, it would be impossible to remember common names for all of them. We must have a systematic method for naming them. The following rules are used in naming alkanes.

Rules for Naming Alkanes

1. The names of the alkanes beyond butane are obtained by adding the suffix *-ane* to the Greek root for the number of carbon atoms (*pent-* for five, *hex-* for six, and so on). For a branched hydrocarbon, the longest continuous chain of carbon atoms determines the root name for the hydrocarbon. For example, in the alkane

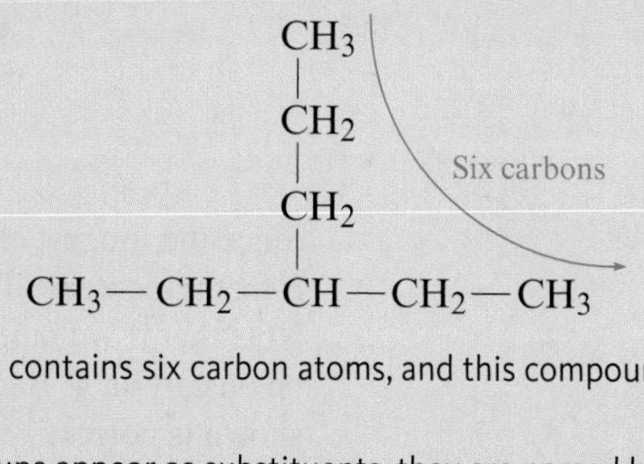

$$\begin{array}{c} CH_3 \\ | \\ CH_2 \\ | \\ CH_2 \\ | \\ CH_3-CH_2-CH-CH_2-CH_3 \end{array}$$

Six carbons

the longest chain contains six carbon atoms, and this compound is named as a hexane.

2. When alkane groups appear as substituents, they are named by dropping the *-ane* and adding *-yl*. For example, $-CH_3$ is obtained by removing a hydrogen from methane and is called *methyl*, $-C_2H_5$ is called *ethyl*, $-C_3H_7$ is called *propyl*, and so on. The compound above is therefore an ethylhexane (Table 22.2).

continued

Table 22.2 | The Most Common Alkyl Substituents and Their Names

Structure*	Name†
$-CH_3$	Methyl
$-CH_2CH_3$	Ethyl
$-CH_2CH_2CH_3$	Propyl
$CH_3\overset{|}{C}HCH_3$	Isopropyl
$-CH_2CH_2CH_2CH_3$	Butyl
$CH_3\overset{|}{C}HCH_2CH_3$	sec-Butyl
$-CH_2-\overset{\overset{\displaystyle H}{|}}{\underset{\underset{\displaystyle CH_3}{|}}{C}}-CH_3$	Isobutyl
$-\overset{\overset{\displaystyle CH_3}{|}}{\underset{\underset{\displaystyle CH_3}{|}}{C}}-CH_3$	tert-Butyl

*The bond with one end open shows the point of attachment of the substituent to the carbon chain.

†For the butyl groups, *sec-* indicates attachment to the chain through a secondary carbon, a carbon atom attached to *two* other carbon atoms. The designation *tert-* signifies attachment through a tertiary carbon, a carbon attached to *three* other carbon atoms.

3. The positions of substituent groups are specified by numbering the longest chain of carbon atoms sequentially, starting at the end closest to the branching. For example, the compound

$$CH_3$$
$$|$$
$$CH_3-CH_2-CH-CH_2-CH_2-CH_3$$

| 1 | 2 | 3 | 4 | 5 | 6 | Correct numbering |
| 6 | 5 | 4 | 3 | 2 | 1 | Incorrect numbering |

is called 3-methylhexane. Note that the top set of numbers is correct since the left end of the molecule is closest to the branching, and this gives the smallest number for the position of the substituent. Also, note that a hyphen is written between the number and the substituent name.

4. The location and name of each substituent are followed by the root alkane name. The substituents are listed in alphabetical order, and the prefixes *di-*, *tri-*, and so on, are used to indicate multiple, identical substituents.

Example 22.2

Isomerism and Nomenclature

Draw the structural isomers for the alkane C_6H_{14} and give the systematic name for each one.

Solution

We will proceed systematically, starting with the longest chain and then rearranging the carbons to form the shorter, branched chains.

1. $CH_3CH_2CH_2CH_2CH_2CH_3$ Hexane

Note that although a structure such as

$$CH_3$$
$$|$$
$$CH_2CH_2CH_2CH_2$$
$$|$$
Six carbon atoms CH_3

may look different it is still hexane, since the longest carbon chain has six atoms.
2. We now take one carbon out of the chain and make it a methyl substituent.

$$\overset{1}{C}H_3\overset{2}{C}H\overset{3}{C}H_2\overset{4}{C}H_2\overset{5}{C}H_3 \quad \text{2-Methylpentane}$$
$$|$$
$$CH_3$$

Since the longest chain consists of five carbons, this is a substituted pentane: 2-methylpentane. The 2 indicates the position of the methyl group on the chain. Note that if we numbered the chain from the right end, the methyl group would be on carbon 4. Because we want the smallest possible number, the numbering shown is correct.
3. The methyl substituent can also be on carbon 3 to give

$$\overset{1}{C}H_3\overset{2}{C}H_2\overset{3}{C}H\overset{4}{C}H_2\overset{5}{C}H_3 \quad \text{3-Methylpentane}$$
$$|$$
$$CH_3$$

Note that we have now exhausted all possibilities for placing a single methyl group on pentane.

4. Next, we can take two carbons out of the original six-member chain:

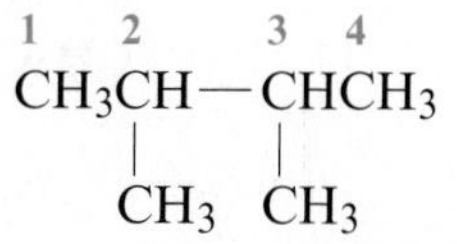

$$\underset{1}{CH_3}\underset{2}{CH}-\underset{3}{CH}\underset{4}{CH_3} \qquad \textbf{2,3-Dimethylbutane}$$
$$\quad\quad\;\; | \quad\;\; |$$
$$\quad\quad CH_3 \;\; CH_3$$

Since the longest chain now has four carbons, the root name is butane. Since there are two methyl groups, we use the prefix *di-*. The numbers denote that the two methyl groups are positioned on the second and third carbons in the butane chain. Note that when two or more numbers are used, they are separated by a comma.

5. The two methyl groups can also be attached to the same carbon atom as shown here:

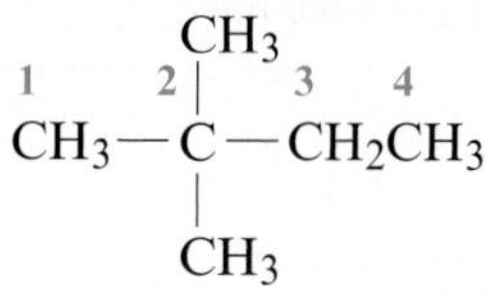

$$\quad\quad\quad\quad CH_3$$
$$\underset{1}{CH_3}-\underset{2}{\overset{|}{C}}-\underset{3}{CH_2}\underset{4}{CH_3} \qquad \textbf{2,2-Dimethylbutane}$$
$$\quad\quad\quad | $$
$$\quad\quad\; CH_3$$

We might also try ethyl-substituted butanes, such as

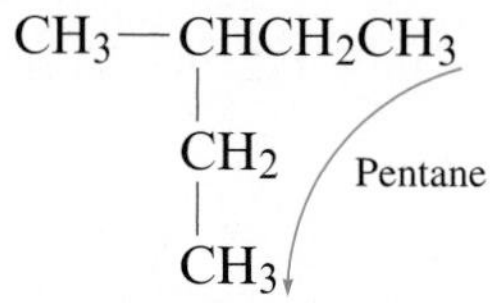

$$CH_3-CHCH_2CH_3$$
$$\quad\quad\quad |$$
$$\quad\quad\quad CH_2 \quad \text{Pentane}$$
$$\quad\quad\quad |$$
$$\quad\quad\quad CH_3$$

However, note that this is instead a pentane (3-methylpentane), since the longest chain has five carbon atoms. Thus it is not a new isomer. Trying to reduce the chain to three atoms provides no further isomers either. For example, the structure

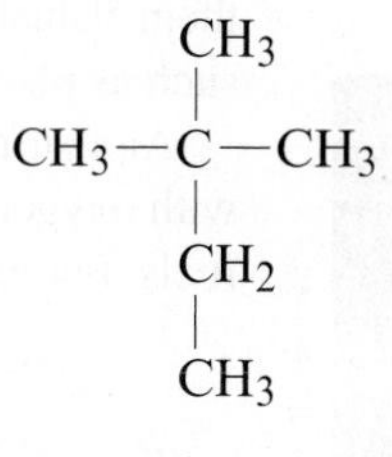

$$\quad\quad\quad\quad CH_3$$
$$\quad\quad\quad\quad |$$
$$CH_3-\overset{|}{\underset{|}{C}}-CH_3$$
$$\quad\quad\quad\quad CH_2$$
$$\quad\quad\quad\quad |$$
$$\quad\quad\quad\quad CH_3$$

is actually 2,2-dimethylbutane.

Thus there are only five distinct structural isomers of C_6H_{14}: hexane, 2-methylpentane, 3-methylpentane, 2,3-dimethylbutane, and 2,2-dimethylbutane.

See Exercises 22.15 and 22.16

Structures from Names

Determine the structure for each of the following compounds.

a. 4-ethyl-3,5-dimethylnonane

b. 4-*tert*-butylheptane

Solution

a. The root name *nonane* signifies a nine-carbon chain. Thus we have

$$\underset{1}{CH_3}\underset{2}{CH_2}\underset{3}{CH}-\underset{4}{CH}-\underset{5}{CH}\underset{6}{CH_2}\underset{7}{CH_2}\underset{8}{CH_2}\underset{9}{CH_3}$$

with CH_3, CH_2, CH_3 substituents and CH_3 below.

b. Heptane signifies a seven-carbon chain, and the *tert*-butyl group is

$$H_3C-\underset{\underset{CH_3}{|}}{\overset{\overset{|}{}}{C}}-CH_3$$

Thus we have

$$\underset{1}{CH_3}\underset{2}{CH_2}\underset{3}{CH_2}\underset{4}{CH}\underset{5}{CH_2}\underset{6}{CH_2}\underset{7}{CH_3}$$

$$H_3C-\underset{\underset{CH_3}{|}}{\overset{\overset{|}{}}{C}}-CH_3$$

See Exercises 22.19 and 22.20

Reactions of Alkanes

Because they are saturated compounds and because the C—C and C—H bonds are relatively strong, the alkanes are fairly unreactive. For example, at 25°C they do not react with acids, bases, or strong oxidizing agents. This chemical inertness makes them valuable as lubricating materials and as the backbone for structural materials such as plastics.

At a sufficiently high temperature alkanes do react vigorously and exothermically with oxygen, and these **combustion reactions** are the basis for their widespread use as fuels. For example, the reaction of butane with oxygen is

$$2C_4H_{10}(g) + 13O_2(g) \longrightarrow 8CO_2(g) + 10H_2O(g)$$

The alkanes can also undergo **substitution reactions**, primarily where halogen atoms replace hydrogen atoms. For example, methane can be successively chlorinated as follows:

$$CH_4 + Cl_2 \xrightarrow{hv} \underset{\text{Chloromethane}}{CH_3Cl} + HCl$$

$$CH_3Cl + Cl_2 \xrightarrow{hv} \underset{\text{Dichloromethane}}{CH_2Cl_2} + HCl$$

$$CH_2Cl_2 + Cl_2 \xrightarrow{hv} \underset{\substack{\text{Trichloromethane}\\\text{(chloroform)}}}{CHCl_3} + HCl$$

$$CHCl_3 + Cl_2 \xrightarrow{hv} \underset{\substack{\text{Tetrachloromethane}\\\text{(carbon tetrachloride)}}}{CCl_4} + HCl$$

Note that the products of the last two reactions have two names; the systematic name is given first, followed by the common name in parentheses. (This format will be used

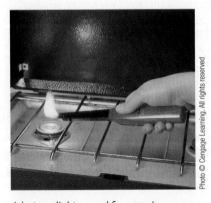

A butane lighter used for camping.

The *hv* above the arrow represents ultraviolet light.

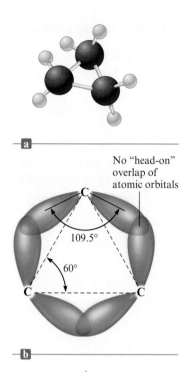

No "head-on" overlap of atomic orbitals

109.5°

60°

Figure 22.5 | (a) The molecular structure of cyclopropane (C_3H_6). (b) The overlap of the sp^3 orbitals that form the C—C bonds in cyclopropane.

throughout this chapter for compounds that have common names.) Also, note that ultraviolet light (hv) furnishes the energy to break the Cl—Cl bond to produce chlorine atoms:

$$Cl_2 \longrightarrow Cl\cdot + Cl\cdot$$

A chlorine atom has an unpaired electron, as indicated by the dot, which makes it very reactive and able to attack the C—H bond.

Substituted methanes with the general formula CF_xCl_{4-x} containing both chlorine and fluorine as substituents are called chlorofluorocarbons (CFCs) and are also known as *Freons*. These substances are very unreactive and have been extensively used as coolant fluids in refrigerators and air conditioners. Unfortunately, their chemical inertness allows Freons to remain in the atmosphere so long that they eventually reach altitudes where they are a threat to the protective ozone layer (see Section 12.8), and the use of these compounds is being rapidly phased out.

Alkanes can also undergo **dehydrogenation reactions** in which hydrogen atoms are removed and the product is an unsaturated hydrocarbon. For example, in the presence of chromium(III) oxide at high temperatures, ethane can be dehydrogenated, yielding ethylene:

$$CH_3CH_3 \xrightarrow[500°C]{Cr_2O_3} CH_2{=}CH_2 + H_2$$

Ethylene

Cyclic Alkanes

Besides forming chains, carbon atoms also form rings. The simplest of the **cyclic alkanes** (general formula C_nH_{2n}) is cyclopropane (C_3H_6), shown in Fig. 22.5(a). Since the carbon atoms in cyclopropane form an equilateral triangle with 60° bond angles, their sp^3 hybrid orbitals do not overlap head-on as in normal alkanes [Fig. 22.5(b)]. This results in unusually weak, or *strained*, C—C bonds; thus the cyclopropane molecule is much more reactive than straight-chain propane. The carbon atoms in cyclobutane (C_4H_8) form a square with 88° bond angles, and cyclobutane is also quite reactive.

The next two members of the series, cyclopentane (C_5H_{10}) and cyclohexane (C_6H_{12}), are quite stable, because their rings have bond angles very close to the tetrahedral angles, which allows the sp^3 hybrid orbitals on adjacent carbon atoms to overlap head-on and form normal C—C bonds, which are quite strong. To attain the tetrahedral angles, the cyclohexane ring must "pucker"—that is, become nonplanar. Cyclohexane can exist in two forms, the *chair* and the *boat* forms, as shown in Fig. 22.6. The two hydrogen atoms above the ring in the boat form are quite close to each other, and the resulting repulsion between these atoms causes the chair form to be preferred. At 25°C more than 99% of cyclohexane exists in the chair form.

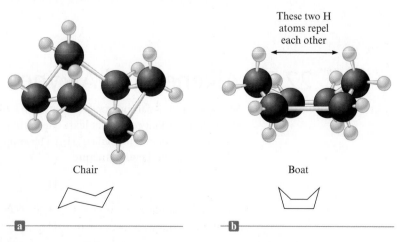

These two H atoms repel each other

Chair

Boat

Figure 22.6 | The (a) chair and (b) boat forms of cyclohexane.

For simplicity, the cyclic alkanes are often represented by the following structures:

Thus the structure CH_3 represents methylcyclopropane.

The nomenclature for cycloalkanes follows the same rules as for the other alkanes except that the root name is preceded by the prefix *cyclo-*. The ring is numbered to yield the smallest substituent numbers possible.

Naming Cyclic Alkanes

Name each of the following cyclic alkanes.

a. 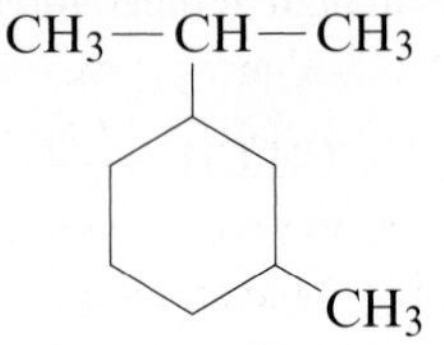 $CH_3 - CH - CH_3$

(attached to a cyclohexane ring with CH_3 substituent)

b. 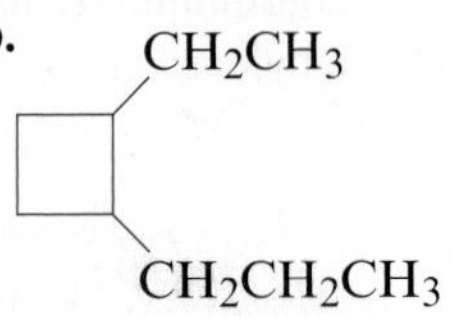 cyclobutane ring with CH_2CH_3 and $CH_2CH_2CH_3$ substituents

Solution

a. The six-carbon cyclohexane ring is numbered as follows:

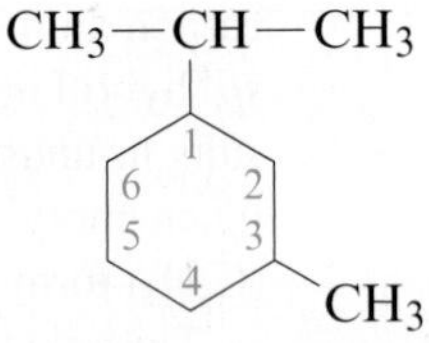

$$CH_3 - CH - CH_3$$

(cyclohexane ring numbered 1–6 with CH_3 at carbon 3)

There is an isopropyl group at carbon 1 and a methyl group at carbon 3. The name is 1-isopropyl-3-methylcyclohexane, since the alkyl groups are named in alphabetical order.

b. This is a cyclobutane ring, which is numbered as follows:

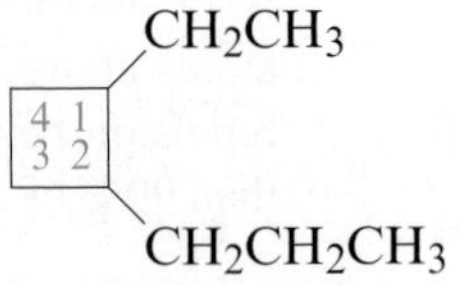

$$CH_2CH_3$$

(cyclobutane ring numbered 1–4 with CH_2CH_3 and $CH_2CH_2CH_3$ substituents)

The name is 1-ethyl-2-propylcyclobutane.

See Exercise 22.22

22.2 | Alkenes and Alkynes

Multiple carbon–carbon bonds result when hydrogen atoms are removed from alkanes. Hydrocarbons that contain at least one carbon–carbon double bond are called **alkenes** and have the general formula C_nH_{2n}. The simplest alkene (C_2H_4), commonly known as *ethylene,* has the Lewis structure

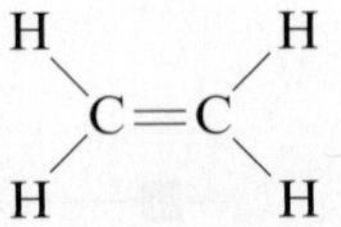

$$\begin{array}{ccc} H & & H \\ & C=C & \\ H & & H \end{array}$$

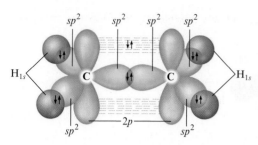

Figure 22.7 | The bonding in ethylene.

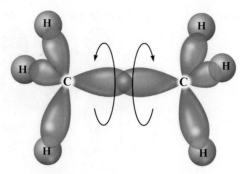

Figure 22.8 | The bonding in ethane.

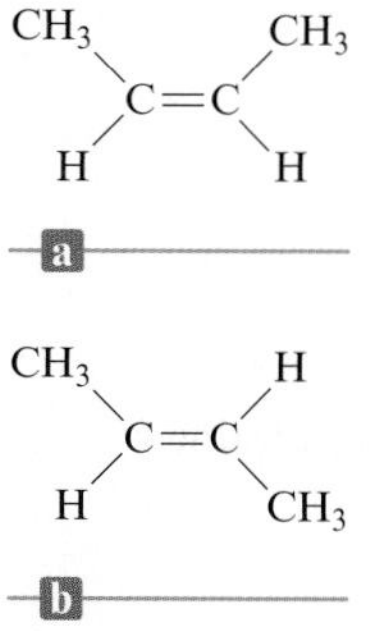

As discussed in Section 9.1, each carbon in ethylene can be described as sp^2 hybridized. The C—C σ bond is formed by sharing an electron pair between sp^2 orbitals, and the π bond is formed by sharing a pair of electrons between p orbitals (Fig. 22.7).

The systematic nomenclature for alkenes is quite similar to that for alkanes.

1. The root hydrocarbon name ends in *-ene* rather than *-ane*. Thus the systematic name for C_2H_4 is *ethene* and the name for C_3H_6 is *propene*.
2. In alkenes containing more than three carbon atoms, the location of the double bond is indicated by the lowest-numbered carbon atom involved in the bond. Thus $CH_2\!=\!CHCH_2CH_3$ is called 1-butene, and $CH_3CH\!=\!CHCH_3$ is called 2-butene.

Note from Fig. 22.7 that the p orbitals on the two carbon atoms in ethylene must be lined up (parallel) to allow formation of the π bond. This prevents rotation of the two CH_2 groups relative to each other at ordinary temperatures, in contrast to alkanes, where free rotation is possible (Fig. 22.8). The restricted rotation around doubly bonded carbon atoms means that alkenes exhibit **cis–trans isomerism**. For example, there are two stereoisomers of 2-butene (Fig. 22.9). Identical substituents on the same side of the double bond are designated *cis* and those on opposite sides are labeled *trans*.

Alkynes are unsaturated hydrocarbons containing at least one triple carbon–carbon bond. The simplest alkyne is C_2H_2 (commonly called *acetylene*), which has the systematic name *ethyne*. As discussed in Section 9.1, the triple bond in acetylene can be described as one σ bond between two sp hybrid orbitals on the two carbon atoms and two π bonds involving two $2p$ orbitals on each carbon atom (Fig. 22.10).

The nomenclature for alkynes involves the use of *-yne* as a suffix to replace the *-ane* of the parent alkane. Thus the molecule $CH_3CH_2C\!\equiv\!CCH_3$ has the name 2-pentyne.

Like alkanes, unsaturated hydrocarbons can exist as ringed structures, for example,

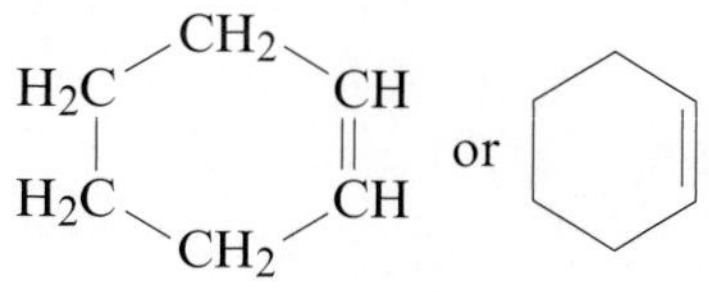

Figure 22.9 | The two stereoisomers of 2-butene: (a) *cis*-2-butene and (b) *trans*-2-butene.

For cyclic alkenes, number through the double bond toward the substituent.

Cyclohexene

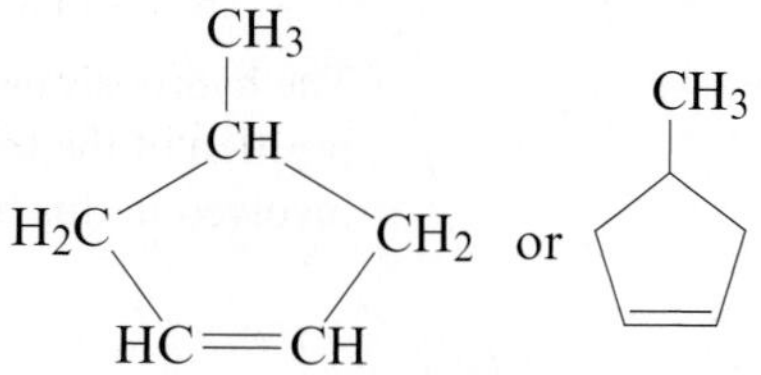

4-Methyl-cyclopentene

Figure 22.10 | The bonding in acetylene.

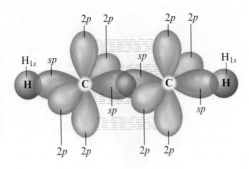

Naming Alkenes and Alkynes

Name each of the following molecules.

a.

$$\begin{array}{ccc} H & & CH_3 \\ & C{=}C & \\ CH_3CH_2CH & & H \\ & | & \\ & CH_3 & \end{array}$$

b. $CH_3CH_2C{\equiv}CCHCH_2CH_3$
         $\quad\quad\quad\quad\quad\quad | $
         $\quad\quad\quad\quad\quad\quad CH_2$
         $\quad\quad\quad\quad\quad\quad | $
         $\quad\quad\quad\quad\quad\quad CH_3$

A worker using an oxyacetylene torch.

Solution

a. The longest chain, which contains six carbon atoms, is numbered as follows:

$$\begin{array}{ccccccc} & & & & & & 1 \\ & & H & & & & CH_3 \\ & & & \underset{3}{} & & \underset{2}{} & \\ 6 & 5 & 4 & & C{=}C & & \\ CH_3 & CH_2 & CH & & & & H \\ & & | & & & & \\ & & CH_3 & & & & \end{array}$$

Thus the hydrocarbon is a 2-hexene. Since the hydrogen atoms are located on opposite sides of the double bond, this molecule corresponds to the *trans* isomer. The name is *trans*-4-methyl-2-hexene.

b. The longest chain, consisting of seven carbon atoms, is numbered as shown (giving the triple bond the lowest possible number):

$$\underset{1}{CH_3}\underset{2}{CH_2}\underset{3}{C}{\equiv}\underset{4}{C}\underset{5}{CH}\underset{6}{CH_2}\underset{7}{CH_3}$$
$$| $$
$$CH_2$$
$$| $$
$$CH_3$$

The hydrocarbon is a 3-heptyne. The full name is 5-ethyl-3-heptyne, where the position of the triple bond is indicated by the lower-numbered carbon atom involved in this bond.

See Exercises 22.25, 22.26, and 22.44

Reactions of Alkenes and Alkynes

Because alkenes and alkynes are unsaturated, their most important reactions are **addition reactions**. In these reactions π bonds, which are weaker than the C—C σ bonds, are broken, and new σ bonds are formed to the atoms being added. For example, **hydrogenation reactions** involve the addition of hydrogen atoms:

$$CH_2{=}CHCH_3 + H_2 \xrightarrow{\text{Catalyst}} CH_3CH_2CH_3$$
$$\text{1-Propene} \qquad\qquad\qquad \text{Propane}$$

For this reaction to proceed rapidly at normal temperatures, a catalyst of platinum, palladium, or nickel is used. The catalyst serves to help break the relatively strong H—H bond, as was discussed in Section 12.7. Hydrogenation of alkenes is an important industrial process, particularly in the manufacture of solid shortenings where unsaturated fats (fats containing double bonds), which are generally liquid, are converted to solid saturated fats.

Halogenation of unsaturated hydrocarbons involves addition of halogen atoms. For example,

$$CH_2{=}CHCH_2CH_2CH_3 + Br_2 \longrightarrow CH_2BrCHBrCH_2CH_2CH_3$$
$$\text{1-Pentene} \qquad\qquad\qquad\qquad \text{1,2-Dibromopentane}$$

Another important reaction involving certain unsaturated hydrocarbons is **polymerization**, a process in which many small molecules are joined together to form a large molecule. Polymerization will be discussed in Section 22.5.

22.3 | Aromatic Hydrocarbons

A special class of cyclic unsaturated hydrocarbons is known as the **aromatic hydrocarbons**. The simplest of these is benzene (C_6H_6), which has a planar ring structure, as shown in Fig. 22.11(a). In the localized electron model of the bonding in benzene, resonance structures of the type shown in Fig. 22.11(b) are used to account for the known equivalence of all the carbon–carbon bonds. But as we discussed in Section 9.5, the best description of the benzene molecule assumes that sp^2 hybrid orbitals on each carbon are used to form the C—C and C—H σ bonds, while the remaining $2p$ orbital on each carbon is used to form π molecular orbitals. The delocalization of these π electrons is usually indicated by a circle inside the ring [Fig. 22.11(c)].

The delocalization of the π electrons makes the benzene ring behave quite differently from a typical unsaturated hydrocarbon. As we have seen previously, unsaturated hydrocarbons generally undergo rapid addition reactions. However, benzene does not. Instead, it undergoes substitution reactions in which *hydrogen atoms are replaced by other atoms*. For example,

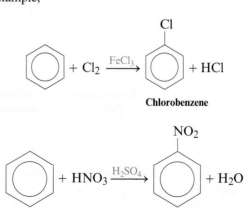

Chlorobenzene

Nitrobenzene

Figure 22.11 | (a) The structure of benzene, a planar ring system in which all bond angles are 120°. (b) Two of the resonance structures of benzene. (c) The usual representation of benzene. The circle represents the electrons in the delocalized π system. All C—C bonds in benzene are equivalent.

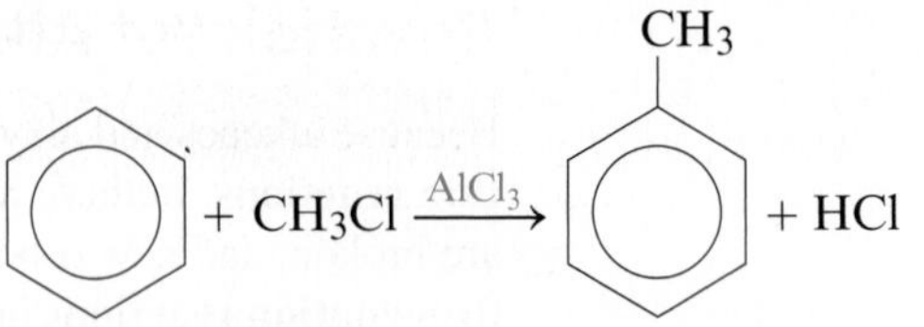

$$\text{benzene} + CH_3Cl \xrightarrow{AlCl_3} \text{toluene} + HCl$$

Toluene

In each case the substance shown over the arrow is needed to catalyze these substitution reactions.

Substitution reactions are characteristic of saturated hydrocarbons, and addition reactions are characteristic of unsaturated ones. The fact that benzene reacts more like a saturated hydrocarbon indicates the great stability of the delocalized π electron system.

The nomenclature of benzene derivatives is similar to the nomenclature for saturated ring systems. If there is more than one substituent present, numbers are used to indicate substituent positions. For example, the compound

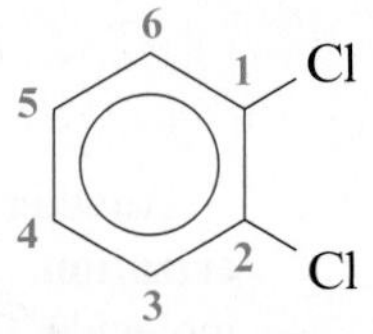

is named 1,2-dichlorobenzene. Another nomenclature system uses the prefix *ortho-* (*o-*) for two adjacent substituents, *meta-* (*m-*) for two substituents with one carbon between them, and *para-* (*p-*) for two substituents opposite each other. When benzene is used as a substituent, it is called the **phenyl group**. Examples of some aromatic compounds are shown in Fig. 22.12.

Benzene is the simplest aromatic molecule. More complex aromatic systems can be viewed as consisting of a number of "fused" benzene rings. Some examples are given in Table 22.3.

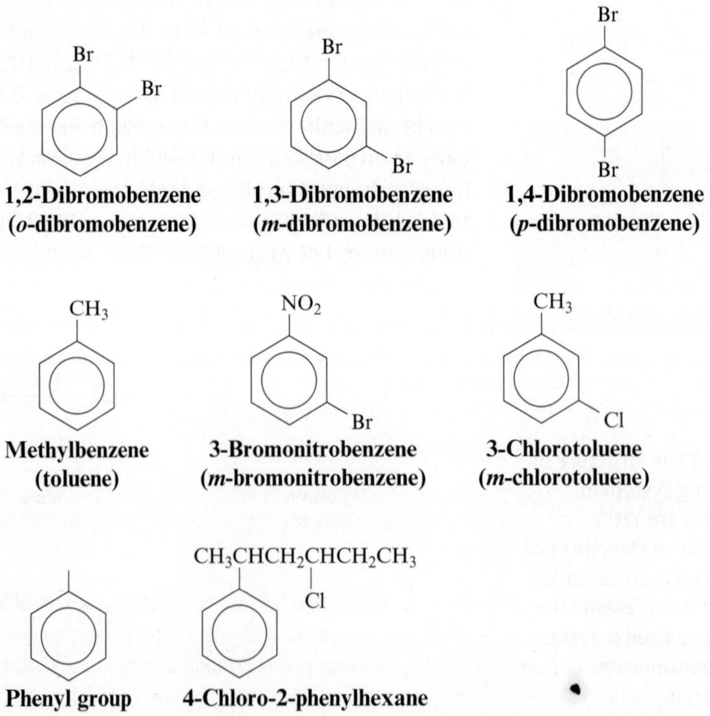

1,2-Dibromobenzene
(*o*-dibromobenzene)

1,3-Dibromobenzene
(*m*-dibromobenzene)

1,4-Dibromobenzene
(*p*-dibromobenzene)

Methylbenzene
(toluene)

3-Bromonitrobenzene
(*m*-bromonitrobenzene)

3-Chlorotoluene
(*m*-chlorotoluene)

Phenyl group

4-Chloro-2-phenylhexane

Figure 22.12 | Some selected substituted benzenes and their names. Common names are given in parentheses.

Table 22.3 | More Complex Aromatic Systems

Structural Formula	Name	Use of Effect
	Naphthalene	Formerly used in mothballs
	Anthracene	Dyes
	Phenanthrene	Dyes, explosives, and synthesis of drugs
	3,4-Benzpyrene	Active carcinogen found in smoke and smog

22.4 | Hydrocarbon Derivatives

The vast majority of organic molecules contain elements in addition to carbon and hydrogen. However, most of these substances can be viewed as **hydrocarbon derivatives**, molecules that are fundamentally hydrocarbons but that have additional atoms or groups of atoms called **functional groups**. The common functional groups are listed in Table 22.4. Because each functional group exhibits characteristic chemistry, we will consider the groups separately.

Alcohols

Alcohols are characterized by the presence of the hydroxyl group (—OH). Some common alcohols are listed in Table 22.5. The systematic name for an alcohol is obtained by replacing the final -*e* of the parent hydrocarbon with -*ol*. The position of the —OH group is specified by a number (where necessary) chosen so that it is the smallest of the substituent numbers. Alcohols are classified according to the number of hydrocarbon fragments bonded to the carbon where the —OH group is attached (see below), where R, R′, and R″ (which may be the same or different) represent hydrocarbon fragments.

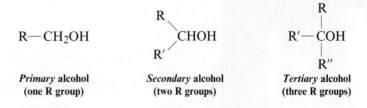

R—CH₂OH	R∖ CHOH ∕R′	R ∣ R′—COH ∣ R″
***Primary* alcohol** (one R group)	***Secondary* alcohol** (two R groups)	***Tertiary* alcohol** (three R groups)

Alcohols usually have much higher boiling points than might be expected from their molar masses. For example, both methanol and ethane have a molar mass of 30, but the boiling point for methanol is 65°C while that for ethane is −89°C. This difference can be understood if we consider the types of intermolecular attractions that

Compounds containing aromatic rings are often used in dyes, such as these for sale in a market in Nepal.

Digital Vision/Getty Images

Table 22.4 | The Common Functional Groups

Class	Functional Group	General Formula*	Example
Halohydrocarbons	—X (F, Cl, Br, I)	R—X	CH_3I Iodomethane (methyl iodide)
Alcohols	—OH	R—OH	CH_3OH Methanol (methyl alcohol)
Ethers	—O—	R—O—R′	CH_3OCH_3 Dimethyl ether
Aldehydes	$\overset{\displaystyle O}{\overset{\displaystyle \|}{-C-H}}$	$\overset{\displaystyle O}{\overset{\displaystyle \|}{R-C-H}}$	CH_2O Methanal (formaldehyde)
Ketones	$\overset{\displaystyle O}{\overset{\displaystyle \|}{-C-}}$	$\overset{\displaystyle O}{\overset{\displaystyle \|}{R-C-R'}}$	CH_3COCH_3 Propanone (dimethyl ketone or acetone)
Carboxylic acids	$\overset{\displaystyle O}{\overset{\displaystyle \|}{-C-OH}}$	$\overset{\displaystyle O}{\overset{\displaystyle \|}{R-C-OH}}$	CH_3COOH Ethanoic acid (acetic acid)
Esters	$\overset{\displaystyle O}{\overset{\displaystyle \|}{-C-O-}}$	$\overset{\displaystyle O}{\overset{\displaystyle \|}{R-C-O-R'}}$	$CH_3COOCH_2CH_3$ Ethyl acetate
Amines	$-NH_2$	$R-NH_2$	CH_3NH_2 Aminomethane (methylamine)

*R and R′ represent hydrocarbon fragments.

occur in these liquids. Ethane molecules are nonpolar and exhibit only weak London dispersion interactions. However, the polar —OH group of methanol produces extensive hydrogen bonding similar to that found in water (see Section 10.1), which results in the relatively high boiling point.

Although there are many important alcohols, the simplest ones, methanol and ethanol, have the greatest commercial value. Methanol, also known as *wood alcohol* because it was formerly obtained by heating wood in the absence of air, is prepared industrially (approximately 4 million tons annually in the United States) by the hydrogenation of carbon monoxide:

$$CO + 2H_2 \xrightarrow[\text{ZnO/Cr}_2\text{O}_3]{400°C} CH_3OH$$

A winemaker draws off a glass of wine in a modern wine cellar.

Ian Shaw/Stone/Getty Images

Table 22.5 | Some Common Alcohols

Formula	Systematic Name	Common Name
CH_3OH	Methanol	Methyl alcohol
CH_3CH_2OH	Ethanol	Ethyl alcohol
$CH_3CH_2CH_2OH$	1-Propanol	n-Propyl alcohol
$CH_3\underset{\underset{OH}{\|}}{C}HCH_3$	2-Propanol	Isopropyl alcohol

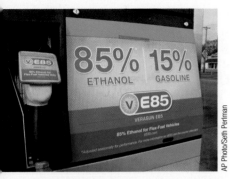

AP Photo/Seth Perlman

An E85 fuel pump.

Methanol is used as a starting material for the synthesis of acetic acid and for many types of adhesives, fibers, and plastics. It is also used (and such use may increase) as a motor fuel. Methanol is highly toxic to humans and can lead to blindness and death if ingested.

Ethanol is the alcohol found in beverages such as beer, wine, and whiskey; it is produced by the fermentation of glucose in corn, barley, grapes, and so on:

$$\underset{\text{Glucose}}{C_6H_{12}O_6} \xrightarrow{\text{Yeast}} \underset{\text{Ethanol}}{2CH_3CH_2OH} + 2CO_2$$

The reaction is catalyzed by the enzymes found in yeast. This reaction can proceed only until the alcohol content reaches about 13% (the percentage found in most wines), at which point the yeast can no longer survive. Beverages with higher alcohol content are made by distilling the fermentation mixture.

Ethanol, like methanol, can be burned in the internal combustion engines of automobiles and is now commonly added to gasoline to form gasohol (see Section 6.6). It is also used in industry as a solvent and for the preparation of acetic acid. One common method for the production of ethanol in the chemical industry is by reaction of water with ethylene:

$$CH_2{=}CH_2 + H_2O \xrightarrow[\text{Catalyst}]{\text{Acid}} CH_3CH_2OH$$

Many polyhydroxyl (more than one —OH group) alcohols are known, the most important being *1,2-ethanediol* (ethylene glycol),

```
H2C—OH
 |
H2C—OH
```

a toxic substance that is the major constituent of most automobile antifreeze solutions.

The simplest aromatic compound with an attached —OH group is

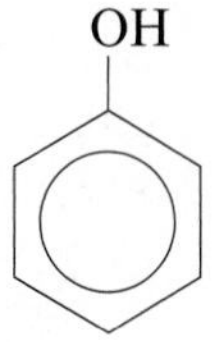

which is commonly called **phenol**. While this compound looks like an alcohol, its properties are very different from alcohols. Most of the 1 million tons of phenol produced annually in the United States is used to make polymers for adhesives and plastics.

Interactive Example 22.6

Sign in at http://login.cengagebrain.com to try this Interactive Example in **OWL**.

Naming and Classifying Alcohols

For each of the following alcohols, give the systematic name and specify whether the alcohol is primary, secondary, or tertiary.

a.
```
CH3CHCH2CH3
   |
   OH
```

b. $ClCH_2CH_2CH_2OH$

c.
```
    CH3
    |
CH3CCH2CH2CH2CH2Br
    |
    OH
```

Solution

a. The chain is numbered as follows:

$$\overset{1}{C}H_3\overset{2}{C}H\overset{3}{C}H_2\overset{4}{C}H_3$$
$$\vert$$
$$OH$$

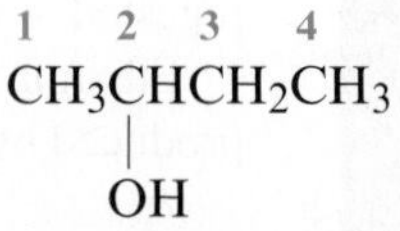

The compound is called 2-butanol, since the —OH group is located at the number 2 position of a four-carbon chain. Note that the carbon to which the —OH is attached also has —CH₃ and —CH₂CH₃ groups attached:

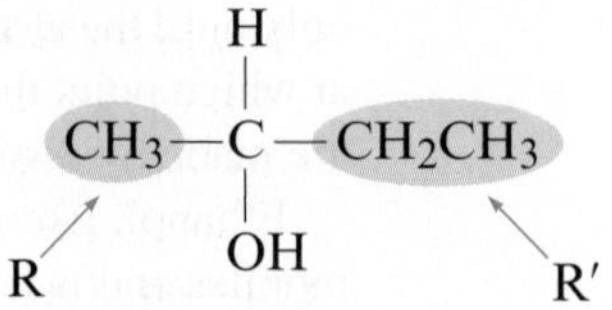

$$\begin{array}{c} H \\ \vert \\ CH_3 - C - CH_2CH_3 \\ \vert \\ OH \end{array}$$

R   OH   R′

Therefore, this is a *secondary* alcohol.

b. The chain is numbered as follows:

$$Cl - \overset{3}{C}H_2 - \overset{2}{C}H_2 - \overset{1}{C}H_2 - OH$$

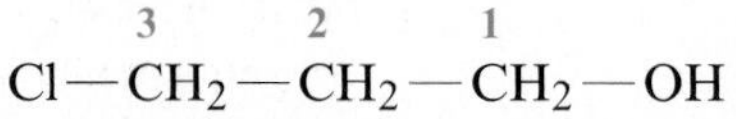

The name is 3-chloro-1-propanol. This is a *primary* alcohol:

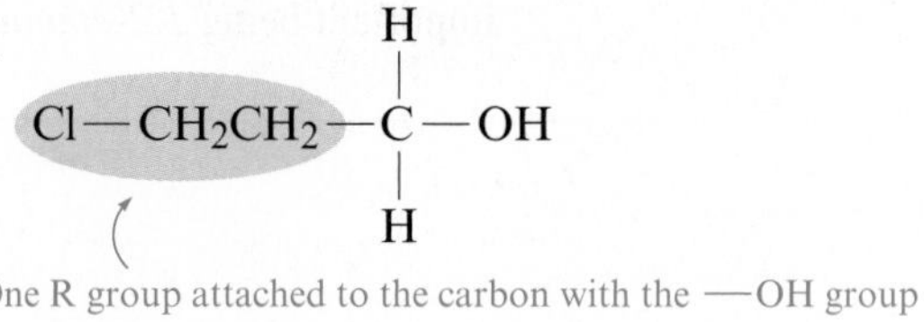

$$\begin{array}{c} H \\ \vert \\ Cl - CH_2CH_2 - C - OH \\ \vert \\ H \end{array}$$

One R group attached to the carbon with the —OH group

c. The chain is numbered as follows:

$$\begin{array}{c} CH_3 \\ \vert \\ \overset{1}{C}H_3 - \overset{2}{C} - \overset{3}{C}H_2 - \overset{4}{C}H_2 - \overset{5}{C}H_2 - \overset{6}{C}H_2Br \\ \vert \\ OH \end{array}$$

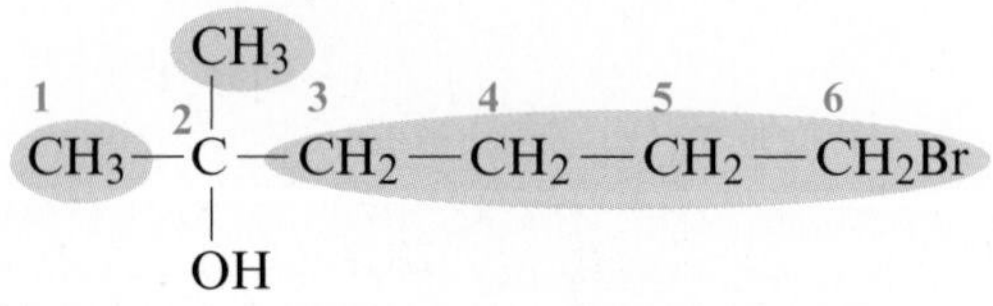

The name is 6-bromo-2-methyl-2-hexanol. This is a *tertiary* alcohol since the carbon where the —OH is attached also has three other R groups attached.

See Exercise 22.51

Cinnamaldehyde produces the characteristic odor of cinnamon.

JL Varga/kStockphoto.com

Aldehydes and Ketones

Aldehydes and ketones contain the **carbonyl group**,

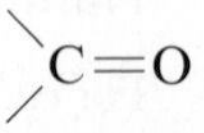

$$\overset{\diagdown}{\underset{\diagup}{}}C=O$$

In **ketones** this group is bonded to two carbon atoms, as in acetone,

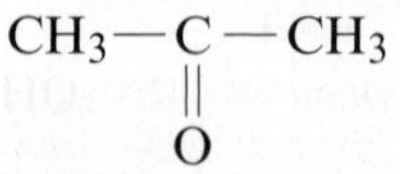

$$\begin{array}{c} CH_3 - C - CH_3 \\ \Vert \\ O \end{array}$$

In **aldehydes** the carbonyl group is bonded to at least one hydrogen atom, as in formaldehyde,

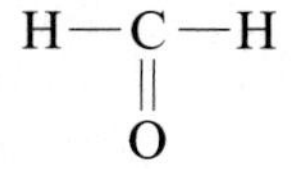

$$H-\overset{\underset{\|}{O}}{C}-H$$

or acetaldehyde,

$$CH_3-\overset{\underset{\|}{O}}{C}-H$$

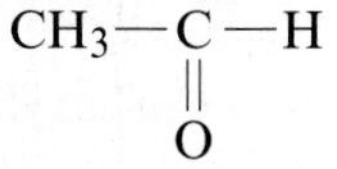

The systematic name for an aldehyde is obtained from the parent alkane by removing the final -e and adding -al. For ketones the final -e is replaced by -one, and a number indicates the position of the carbonyl group where necessary. Examples of common aldehydes and ketones are shown in Fig. 22.13. Note that since the aldehyde functional group always occurs at the end of the carbon chain, the aldehyde carbon is assigned the number 1 when substituent positions are listed in the name.

Ketones often have useful solvent properties (acetone is found in nail polish remover, for example) and are frequently used in industry for this purpose. Aldehydes typically have strong odors. Vanillin is responsible for the pleasant odor in vanilla beans; cinnamaldehyde produces the characteristic odor of cinnamon. On the other hand, the unpleasant odor in rancid butter arises from the presence of butyraldehyde.

Aldehydes and ketones are most often produced commercially by the oxidation of alcohols. For example, oxidation of a *primary* alcohol yields the corresponding aldehyde:

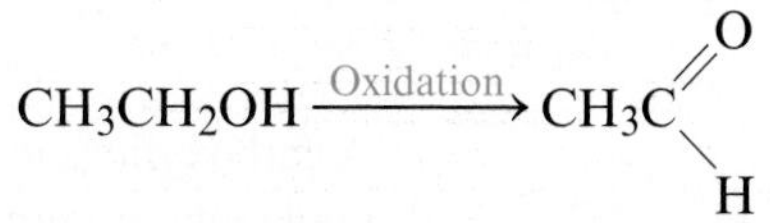

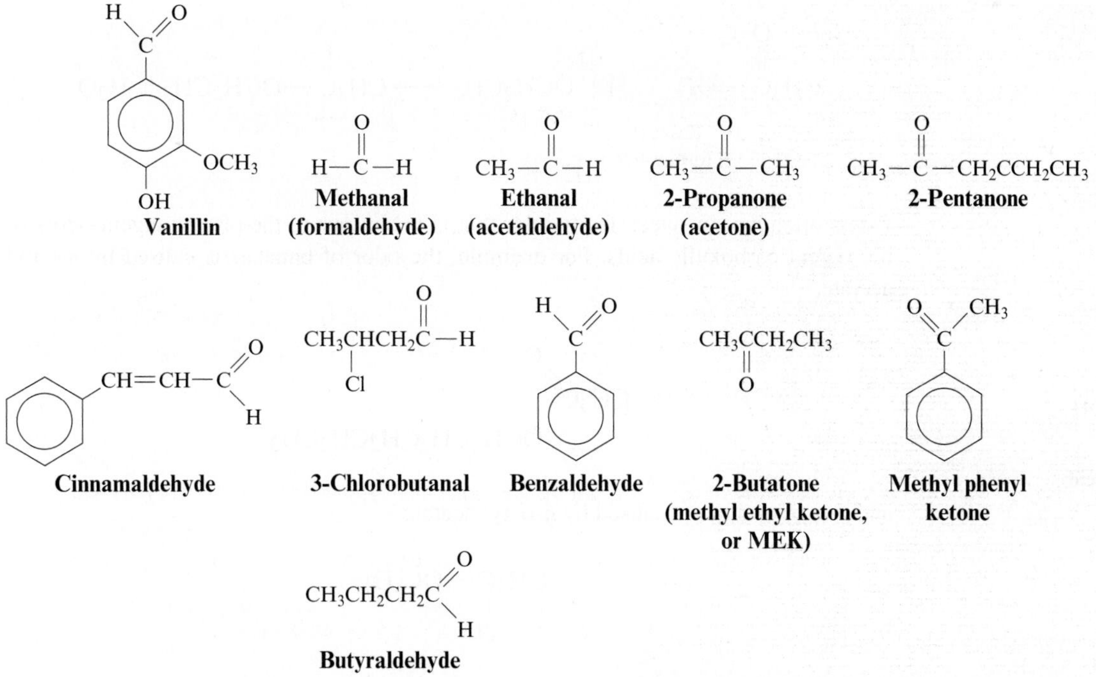

$$CH_3CH_2OH \xrightarrow{\text{Oxidation}} CH_3C\overset{\overset{O}{\diagup}}{\underset{\diagdown}{H}}$$

Figure 22.13 | Some common ketones and aldehydes. Note that since the aldehyde functional group always appears at the end of a carbon chain, the aldehyde carbon is assigned the number 1 when the compound is named.

Oxidation of a *secondary* alcohol results in a ketone:

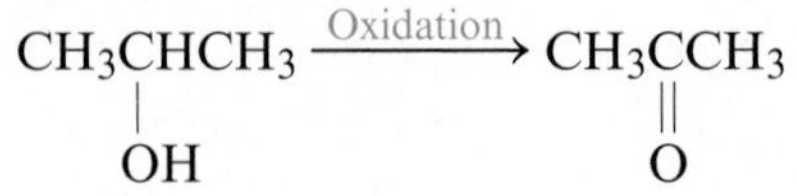

$$CH_3CHCH_3 \xrightarrow{\text{Oxidation}} CH_3CCH_3$$

(with OH below the first structure and O (double bond) below the product)

Carboxylic Acids and Esters

Carboxylic acids are characterized by the presence of the **carboxyl group**

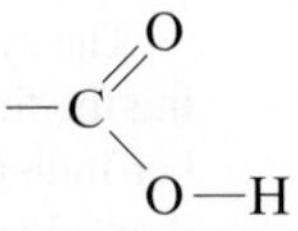

(structure of carboxyl group shown)

that gives an acid of the general formula RCOOH. Typically, these molecules are weak acids in aqueous solution (see Section 14.5). Organic acids are named from the parent alkane by dropping the final *-e* and adding *-oic*. Thus CH_3COOH, commonly called acetic acid, has the systematic name ethanoic acid, since the parent alkane is ethane. Other examples of carboxylic acids are shown in Fig. 22.14.

Many carboxylic acids are synthesized by oxidizing primary alcohols with a strong oxidizing agent. For example, ethanol can be oxidized to acetic acid by using potassium permanganate:

$$CH_3CH_2OH \xrightarrow{\text{KMnO}_4(aq)} CH_3COOH$$

A carboxylic acid reacts with an alcohol to form an **ester** and a water molecule. For example, the reaction of acetic acid with ethanol produces ethyl acetate and water:

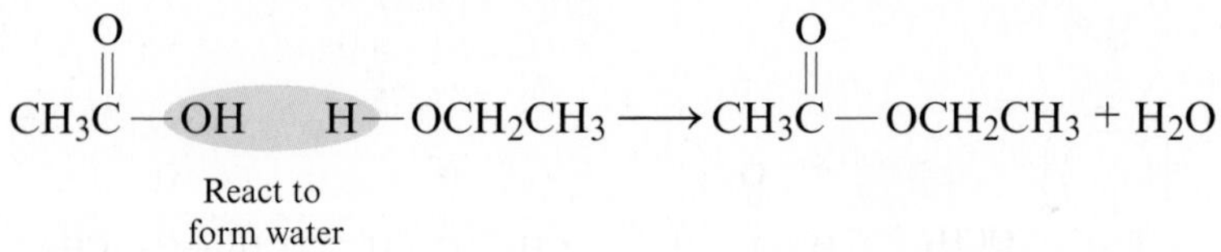

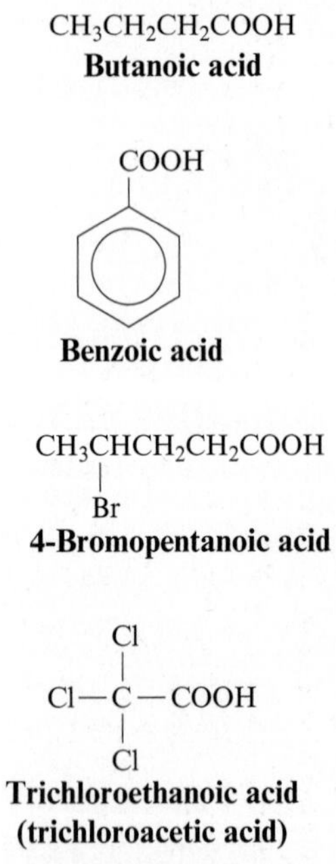

$$CH_3C-OH \quad H-OCH_2CH_3 \longrightarrow CH_3C-OCH_2CH_3 + H_2O$$

React to form water

Esters often have a sweet, fruity odor that is in contrast to the often pungent odors of the parent carboxylic acids. For example, the odor of bananas is caused by *n*-amyl acetate,

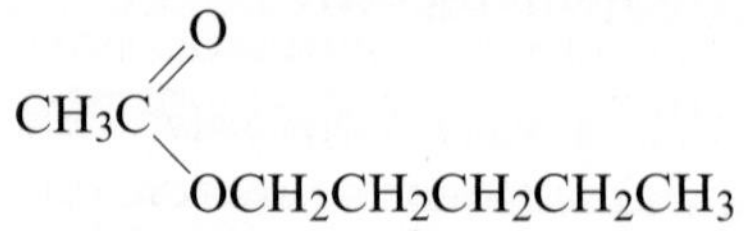

$$CH_3C\!\!\overset{O}{\underset{OCH_2CH_2CH_2CH_2CH_3}{\diagup}}$$

and that of oranges is caused by *n*-octyl acetate,

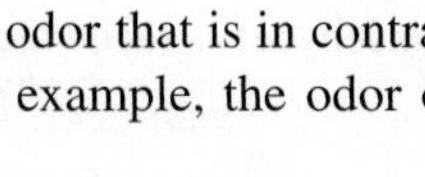

$$CH_3C-OC_8H_{17}$$

(with O double bond below)

The systematic name for an ester is formed by changing the *-oic* ending of the parent acid to *-oate*. The parent alcohol chain is named first with a *-yl* ending. For example, the systematic name for *n*-octyl acetate is *n*-octylethanoate (from ethanoic acid).

$CH_3CH_2CH_2COOH$
Butanoic acid

COOH
(benzene ring structure)

Benzoic acid

$CH_3CHCH_2CH_2COOH$
|
Br
4-Bromopentanoic acid

Cl
|
Cl—C—COOH
|
Cl
Trichloroethanoic acid
(trichloroacetic acid)

Figure 22.14 | Some carboxylic acids.

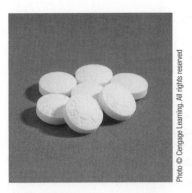

Aspirin tablets.

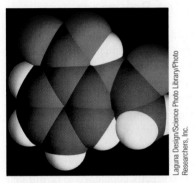

Computer-generated space-filling model of acetylsalicylic acid (aspirin).

A very important ester is formed from the reaction of salicylic acid and acetic acid:

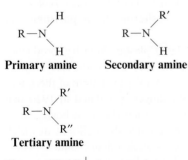

Salicylic acid   **Acetic acid**   **Acetylsalicylic acid**

The product is acetylsalicylic acid, commonly known as *aspirin,* which is used in huge quantities as an analgesic (painkiller).

Amines

Amines are probably best viewed as derivatives of ammonia in which one or more N—H bonds are replaced by N—C bonds. The resulting amines are classified as *primary* if one N—C bond is present, *secondary* if two N—C bonds are present, and *tertiary* if all three N—H bonds in NH_3 have been replaced by N—C bonds (Fig. 22.15). Examples of some common amines are given in Table 22.6.

Common names are often used for simple amines; the systematic nomenclature for more complex molecules uses the name *amino-* for the —NH_2 functional group. For example, the molecule

$$CH_3CHCH_2CH_3$$
$$|$$
$$NH_2$$

is named 2-aminobutane.

Many amines have unpleasant "fishlike" odors. For example, the odors associated with decaying animal and human tissues are caused by amines such as putrescine ($H_2NCH_2CH_2CH_2NH_2$) and cadaverine ($H_2NCH_2CH_2CH_2CH_2CH_2NH_2$).

Figure 22.15 | The general formulas for primary, secondary, and tertiary amines. R, R′, and R″ represent carbon-containing substituents.

Table 22.6 | Some Common Amines

Formula	Common Name	Type
CH_3NH_2	Methylamine	Primary
$CH_3CH_2NH_2$	Ethylamine	Primary
$(CH_3)_2NH$	Dimethylamine	Secondary
$(CH_3)_3N$	Trimethylamine	Tertiary
NH_2 (aniline)	Aniline	Primary
Diphenylamine	Diphenylamine	Secondary

Aromatic amines are used primarily to make dyes. Since many of them are carcinogenic, they must be handled with great care.

22.5 | Polymers

Polymers are large, usually chainlike molecules that are built from small molecules called *monomers*. Polymers form the basis for synthetic fibers, rubbers, and plastics and have played a leading role in the revolution that has been brought about in daily life by chemistry. It has been estimated that about 50% of the industrial chemists in the United States work in some area of polymer chemistry, a fact that illustrates just how important polymers are to our economy and standard of living.

The Development and Properties of Polymers

The development of the polymer industry provides a striking example of the importance of serendipity in the progress of science. Many discoveries in polymer chemistry arose from accidental observations that scientists followed up.

The age of plastics might be traced to a day in 1846 when Christian Schoenbein, a chemistry professor at the University of Basel in Switzerland, spilled a flask containing nitric and sulfuric acids. In his hurry to clean up the spill, he grabbed his wife's cotton apron, which he then rinsed out and hung up in front of a hot stove to dry. Instead of drying, the apron flared and burned.

Very interested in this event, Schoenbein repeated the reaction under more controlled conditions and found that the new material, which he correctly concluded to be nitrated cellulose, had some surprising properties. As he had experienced, the nitrated cellulose is extremely flammable and, under certain circumstances, highly explosive. In addition, he found that it could be molded at moderate temperatures to give objects that were, upon cooling, tough but elastic. Predictably, the explosive nature of the substance was initially of more interest than its other properties, and cellulose nitrate rapidly became the basis for smokeless gunpowder. Although Schoenbein's discovery cannot be described as a truly synthetic polymer (because he simply found a way to modify the natural polymer cellulose), it formed the basis for a large number of industries that grew up to produce photographic films, artificial fibers, and molded objects of all types.

The first synthetic polymers were produced as by-products of various organic reactions and were regarded as unwanted contaminants. Thus the first preparations of many of the polymers now regarded as essential to our modern lifestyle were thrown away in disgust. One chemist who refused to be defeated by the "tarry" products obtained when he reacted phenol with formaldehyde was the Belgian-American chemist Leo H. Baekeland (1863–1944). Baekeland's work resulted in the first completely synthetic plastic (called Bakelite), a substance that when molded to a certain shape under high pressure and temperature cannot be softened again or dissolved. Bakelite is a **thermoset polymer**. In contrast, cellulose nitrate is a **thermoplastic polymer**; that is, it can be remelted after it has been molded.

The discovery of Bakelite in 1907 spawned a large plastics industry, producing telephones, billiard balls, and insulators for electrical devices. During the early days of polymer chemistry, there was a great deal of controversy over the nature of these materials. Although the German chemist Hermann Staudinger speculated in 1920 that polymers were very large molecules held together by strong chemical bonds, most chemists of the time assumed that these materials were much like colloids, in which small molecules are aggregated into large units by forces weaker than chemical bonds.

One chemist who contributed greatly to the understanding of polymers as giant molecules was Wallace H. Carothers of the DuPont Chemical Company. Among his accomplishments was the preparation of nylon. The nylon story further illustrates the importance of serendipity in scientific research. When nylon is first prepared, the resulting product is a sticky material with little structural integrity. Because of this, it

A radio from the 1930s made of Bakelite.

Phil Nelson

Chemical connections
Wallace Hume Carothers

Wallace H. Carothers, a brilliant organic chemist who was principally responsible for the development of nylon and the first synthetic rubber (Neoprene), was born in 1896 in Burlington, Iowa. As a youth, Carothers was fascinated by tools and mechanical devices and spent many hours experimenting. In 1915 he entered Tarkio College in Missouri. Carothers so excelled in chemistry that even before his graduation, he was made a chemistry instructor.

Carothers eventually moved to the University of Illinois at Urbana–Champaign, where he was appointed to the faculty when he completed his Ph.D. in organic chemistry in 1924. He moved to Harvard University in 1926, and then to DuPont in 1928 to participate in a new program in fundamental research. At DuPont,

Carothers headed the organic chemistry division, and during his ten years there played a prominent role in laying the foundations of polymer chemistry.

By the age of 33, Carothers had become a world-famous chemist whose advice was sought by almost everyone working in polymers. He was the first industrial chemist to be elected to the prestigious National Academy of Sciences.

Carothers was an avid reader of poetry and a lover of classical music. Unfortunately, he also suffered from severe bouts of depression that finally led to his suicide in 1937 in a Philadelphia hotel room, where he drank a cyanide solution. He was 41 years old. Despite the brevity of his career, Carothers was truly one of the finest American chemists of all time. His

Wallace H. Carothers.

great intellect, his love of chemistry, and his insistence on perfection produced his special genius.

Nylon netting magnified 62 times.

was initially put aside as having no apparently useful characteristics. However, Julian Hill, a chemist in the Carothers research group, one day put a small ball of this nylon on the end of a stirring rod and drew it away from the remaining sticky mass, forming a string. He noticed the silky appearance and strength of this thread and realized that nylon could be drawn into useful fibers.

The reason for this behavior of nylon is now understood. When nylon is first formed, the individual polymer chains are oriented randomly, like cooked spaghetti, and the substance is highly amorphous. However, when drawn out into a thread, the chains tend to line up (the nylon becomes more crystalline), which leads to increased hydrogen bonding between adjacent chains. This increase in crystallinity, along with the resulting increase in hydrogen-bonding interactions, leads to strong fibers and thus to a highly useful material. Commercially, nylon is produced by forcing the raw material through a *spinneret,* a plate containing small holes, which forces the polymer chains to line up.

Another property that adds strength to polymers is **crosslinking**, the existence of covalent bonds between adjacent chains. The structure of Bakelite is highly crosslinked, which accounts for the strength and toughness of this polymer. Another example of crosslinking occurs in the manufacture of rubber. Raw natural rubber consists of chains of the type

$$\text{~~~}CH_2-CH_2-CH=C-CH_2-CH_2-CH=C-CH_2\text{~~~}$$
$$\qquad\qquad\qquad\quad | \qquad\qquad\qquad\qquad\quad |$$
$$\qquad\qquad\qquad CH_3 \qquad\qquad\qquad\qquad\; CH_3$$

Chemical connections

Super-Slippery Slope

One of the most amazing characteristics of insects is their ability to cling to almost any surface, whether it is vertical or upside down. However, the walls of the *Nepenthes* pitcher plant are so slippery that any insect that lands on those walls slips to its death in the digestive juices at the bottom of the "pitcher." One can envision this type of slippery surface being useful in human activities, such as in pipes for handling biomedical fluids and fuels. Slippery surfaces also could be useful

in repelling ice inside freezers or on ship hulls used in polar regions.

Joanna Aizenberg and her colleagues at Harvard University have designed a synthetic system that mimics the slippery pitcher plant surface. Their omniphobic (repels virtually all liquids) surface, which they call SLIPS, is prepared from a porous network of Teflon nanofibers that is infused with a special oil-and-water fluid. It is the layer of fluid on the surface of the Teflon "sponge" that

makes SLIPS so slippery. Another advantage of SLIPS is its self-healing property. If any damage occurs to the surface, more fluid flows from the interior to heal cracks or dents. The surface can be maintained indefinitely because a reservoir of fluid sustains the fluid.

SLIPS is a great example of how useful it can be to mimic nature's problem-solving skills.

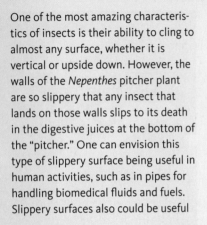

Charles Goodyear tried for many years to change natural rubber into a useful product. In 1839 he accidentally dropped some rubber containing sulfur on a hot stove. Noting that the rubber did not melt as expected, Goodyear pursued this lead and developed vulcanization.

and is a soft, sticky material unsuitable for tires. However, in 1839 Charles Goodyear (1800–1860), an American chemist, accidentally found that if sulfur is added to rubber and the resulting mixture is heated (a process called **vulcanization**), the resulting rubber is still elastic (reversibly stretchable) but is much stronger. This change in character occurs because sulfur atoms become bonded between carbon atoms on different chains. These sulfur atoms form bridges between the polymer chains, thus linking the chains together.

Types of Polymers

The simplest and one of the best-known polymers is *polyethylene,* which is constructed from ethylene monomers:

$$n\text{CH}_2{=}\text{CH}_2 \xrightarrow{\text{Catalyst}} \left(\begin{array}{cc} \text{H} & \text{H} \\ | & | \\ \text{C} - \text{C} \\ | & | \\ \text{H} & \text{H} \end{array} \right)_n$$

where n represents a large number (usually several thousand). Polyethylene is a tough, flexible plastic used for piping, bottles, electrical insulation, packaging films, garbage bags, and many other purposes. Its properties can be varied by using substituted ethylene monomers. For example, when tetrafluoroethylene is the monomer, the polymer Teflon is obtained:

$$n \left(\begin{array}{cc} \text{F} & \text{F} \\ & \diagdown \diagup \\ & \text{C}{=}\text{C} \\ & \diagup \diagdown \\ \text{F} & \text{F} \end{array} \right) \longrightarrow \left(\begin{array}{cc} \text{F} & \text{F} \\ | & | \\ \text{C} - \text{C} \\ | & | \\ \text{F} & \text{F} \end{array} \right)_n$$

Tetrafluoroethylene   **Teflon**

Ashiga/Shutterstock.com

Crosslinking gives the rubber in these tires strength and toughness.

The discovery of Teflon, a very important substituted polyethylene, is another illustration of the role of chance in chemical research. In 1938 a DuPont chemist named

Roy Plunkett was studying the chemistry of gaseous tetrafluoroethylene. He synthesized about 100 pounds of the chemical and stored it in steel cylinders. When one of the cylinders failed to produce tetrafluoroethylene gas when the valve was opened, the cylinder was cut open to reveal a white powder. This powder turned out to be a polymer of tetrafluoroethylene, which was eventually developed into Teflon. Because of the resistance of the strong C—F bonds to chemical attack, Teflon is an inert, tough, and nonflammable material widely used for electrical insulation, nonstick coatings on cooking utensils, and bearings for low-temperature applications.

Other polyethylene-type polymers are made from monomers containing chloro, methyl, cyano, and phenyl substituents, as summarized in Table 22.7. In each case the double carbon–carbon bond in the substituted ethylene monomer becomes a single bond in the polymer. The different substituents lead to a wide variety of properties.

The polyethylene polymers illustrate one of the major types of polymerization reactions, called **addition polymerization**, in which the monomers simply "add together" to produce the polymer. No other products are formed. The polymerization process is initiated by a **free radical** (a species with an unpaired electron) such as the hydroxyl

Table 22.7 | Some Common Synthetic Polymers and Their Monomers and Applications

Monomer		Polymer		
Name	Formula	Name	Formula	Uses
Ethylene	$H_2C{=}CH_2$	Polyethylene	$-(CH_2-CH_2)_n-$	Plastic piping, bottles, electrical insulation, toys
Propylene	$H_2C{=}C$ with H and CH_3	Polypropylene	$-(CH-CH_2-CH-CH_2)_n-$ with CH_3 and CH_3	Film for packaging, carpets, lab wares, toys
Vinyl chloride	$H_2C{=}C$ with H and Cl	Polyvinyl chloride (PVC)	$-(CH_2-CH)_n-$ with Cl	Piping, siding, floor tile, clothing, toys
Acrylonitrile	$H_2C{=}C$ with H and CN	Polyacrylonitrile (PAN)	$-(CH_2-CH)_n-$ with CN	Carpets, fabrics
Tetrafluoro-ethylene	$F_2C{=}CF_2$	Teflon	$-(CF_2-CF_2)_n-$	Cooking utensils, electrical insulation, bearings
Styrene	$H_2C{=}C$ with H and phenyl ring	Polystyrene	$-(CH_2CH)_n-$ with phenyl ring	Containers, thermal insulation, toys
Butadiene	$H_2C{=}C-C{=}CH_2$ with H and H	Polybutadiene	$-(CH_2CH{=}CHCH_2)_n-$	Tire tread, coating resin
Butadiene and styrene	(See above.)	Styrene-butadiene rubber	$-(CH-CH_2-CH_2-CH{=}CH-CH_2)_n-$ with phenyl ring	Synthetic rubber

radical (HO·). The free radical attacks and breaks the π bond of an ethylene molecule to form a new free radical,

which is then available to attack another ethylene molecule:

Repetition of this process thousands of times creates a long-chain polymer. Termination of the growth of the chain occurs when *two radicals* react to form a bond, a process that consumes two radicals without producing any others.

Another common type of polymerization is **condensation polymerization**, in which a small molecule, such as water, is formed for each extension of the polymer chain. The most familiar polymer produced by condensation is *nylon*. Nylon is a **copolymer**, since two different types of monomers combine to form the chain; a **homopolymer** is the result of polymerizing a single type of monomer. One common form of nylon is produced when hexamethylenediamine and adipic acid react by splitting out a water molecule to form a C—N bond:

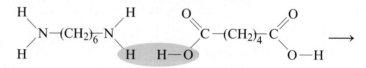

Hexamethylenediamine   **Adipic acid**

The molecule formed, called a **dimer** (two monomers joined), can undergo further condensation reactions since it has an amino group at one end and a carboxyl group at the other. Thus both ends are free to react with another monomer. Repetition of this process leads to a long chain of the type

which is the basic structure of nylon. The reaction to form nylon occurs quite readily and is often used as a lecture demonstration (Fig. 22.16). The properties of nylon can be varied by changing the number of carbon atoms in the chain of the acid or amine monomer.

Charles D. Winters/Photo Researchers, Inc.

Figure 22.16 | The reaction to form nylon can be carried out at the interface of two immiscible liquid layers in a beaker. The bottom layer contains adipoyl chloride,

$$Cl—C—(CH_2)_4—C—Cl$$
$$\quad\ \|\qquad\qquad\ \|$$
$$\quad\ O\qquad\qquad O$$

dissolved in CCl₄, and the top layer contains hexamethylenediamine,

$$H_2N—(CH_2)_6—NH_2$$

dissolved in water. A molecule of HCl is formed as each C—N bond forms.

More than 1 million tons of nylon is produced annually in the United States for use in clothing, carpets, rope, and so on. Many other types of condensation polymers are also produced. For example, Dacron is a copolymer formed from the condensation reaction of ethylene glycol (a dialcohol) and *p*-terephthalic acid (a dicarboxylic acid):

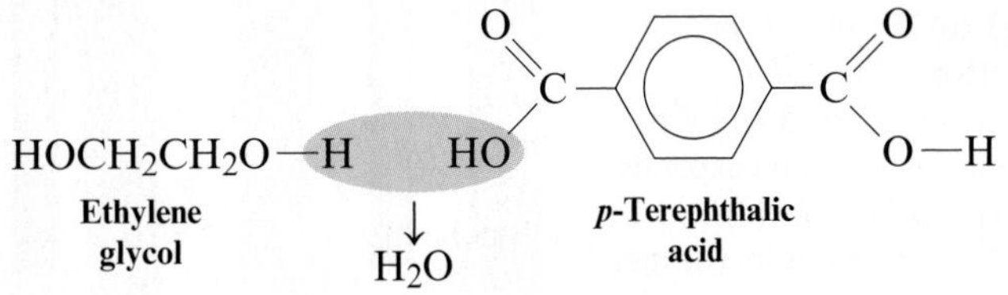

The repeating unit of Dacron is

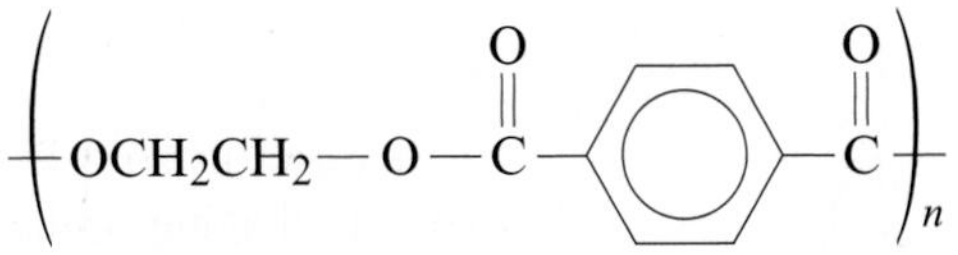

Note that this polymerization involves a carboxylic acid and an alcohol forming an ester group:

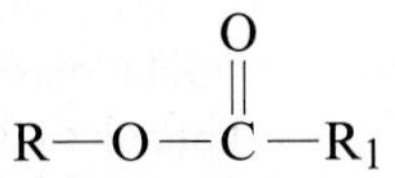

Thus Dacron is called a **polyester**. By itself or blended with cotton, Dacron is widely used in fibers for the manufacture of clothing.

Polymers Based on Ethylene

A large section of the polymer industry involves the production of macromolecules from ethylene or substituted ethylenes. As discussed previously, ethylene molecules polymerize by addition after the double bond has been broken by some initiator:

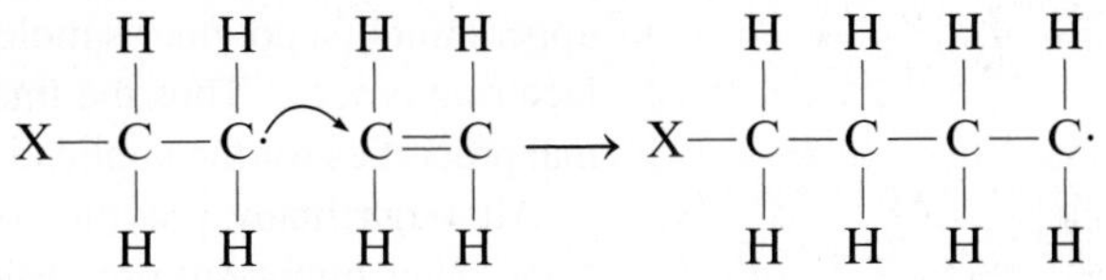

This process continues by adding new ethylene molecules to eventually give polyethylene, a thermoplastic material.

There are two forms of polyethylene: low-density polyethylene (LDPE) and high-density polyethylene (HDPE). The chains in LDPE contain many branches and thus do not pack as tightly as those in HDPE, which consist of mostly straight-chain molecules.

Traditionally, LDPE has been manufactured under conditions of high pressure ($\approx$20,000 psi) and high temperature (500°C). These severe reaction conditions require specially designed equipment, and for safety reasons the reaction usually has been run behind a reinforced concrete barrier. More recently, lower reaction pressures and temperatures have become possible through the use of catalysts. One catalytic system using triethylaluminum, $Al(C_2H_5)_3$, and titanium(IV) chloride was developed by Karl Ziegler in Germany and Giulio Natta in Italy. Although this catalyst is very efficient, it catches fire on contact with air and must be handled very carefully. A safer catalytic system was developed at Phillips Petroleum Company. It uses a chromium(III) oxide (Cr_2O_3) and aluminosilicate catalyst and has mainly taken over in the United States. The product of the catalyzed reaction is highly linear (unbranched) and is often called *linear low-density polyethylene*. It is very similar to HDPE.

psi is the abbreviation for pounds per square inch: 15 psi $\approx$ 1 atm.

Figure 22.17 | A major use of HDPE is for blow-molded objects such as bottles for soft drinks, shampoos, bleaches, and so on. (a) A tube composed of HDPE is inserted into the mold (die). (b) The die closes, sealing the bottom of the tube. (c) Compressed air is forced into the warm HDPE tube, which then expands to take the shape of the die. (d) The molded bottle is removed from the die.

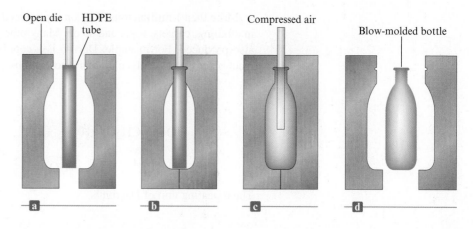

Molecular weight (not molar mass) is the common terminology in the polymer industry.

The major use of LDPE is in the manufacture of the tough, transparent film that is used in packaging so many consumer goods. Two-thirds of the approximately 10 billion pounds of LDPE produced annually in the United States are used for this purpose. The major use of HDPE is for blow-molded products, such as bottles for consumer products (Fig. 22.17).

The useful properties of polyethylene are due primarily to its high molecular weight (molar mass). Although the strengths of the interactions between specific points on the nonpolar chains are quite small, the chains are so long that these small attractions accumulate to a very significant value, so that the chains stick together very tenaciously. There is also a great deal of physical tangling of the lengthy chains. The combination of these interactions gives the polymer strength and toughness. However, a material like polyethylene can be melted and formed into a new shape (thermoplastic behavior), because in the melted state the molecules can readily flow past one another.

Since a high molecular weight gives a polymer useful properties, one might think that the goal would be to produce polymers with chains as long as possible. However, this is not the case—polymers become much more difficult to process as the molecular weights increase. Most industrial operations require that the polymer flow through pipes as it is processed. But as the chain lengths increase, viscosity also increases. In practice, the upper limit of a polymer's molecular weight is set by the flow requirements of the manufacturing process. Thus the final product often reflects a compromise between the optimal properties for the application and those needed for ease of processing.

Although many polymer properties are greatly influenced by molecular weight, some other important properties are not. For example, chain length does not affect a polymer's resistance to chemical attack. Physical properties such as color, refractive index, hardness, density, and electrical conductivity are also not greatly influenced by molecular weight.

We have already seen that one way of altering the strength of a polymeric material is to vary the chain length. Another method for modifying polymer behavior involves varying the substituents. For example, if we use a monomer of the type

$$\begin{matrix} H & & H \\ & C{=}C & \\ H & & X \end{matrix}$$

the properties of the resulting polymer depend on the identity of X. The simplest example is polypropylene, whose monomer is

$$\begin{matrix} H & & H \\ & C{=}C & \\ H & & CH_3 \end{matrix}$$

and that has the form

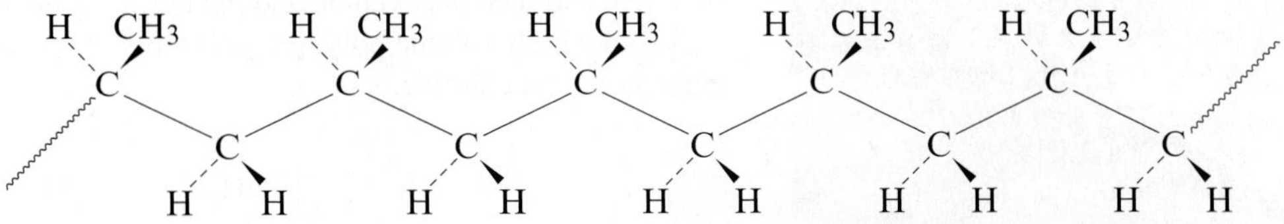

The CH_3 groups can be arranged on the same side of the chain (called an **isotactic chain**) as shown above, can alternate (called a **syndiotactic chain**) as shown below,

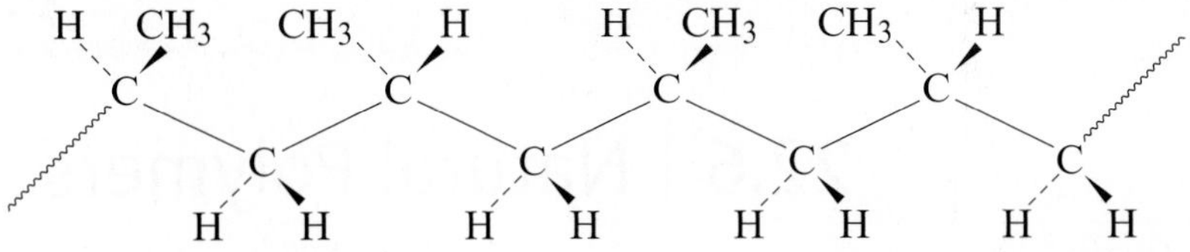

or can be randomly distributed (called an **atactic chain**).

The chain arrangement has a significant effect on the polymer's properties. Most polypropylene is made using the Ziegler-Natta catalyst, $Al(C_2H_5)_3 \cdot TiCl_4$, which produces highly isotactic chains that pack together quite closely. As a result, polypropylene is more crystalline, and therefore stronger and harder, than polyethylene. The major uses of polypropylene are for molded parts (40%), fibers (35%), and packaging films (10%). Polypropylene fibers are especially useful for athletic wear because they do not absorb water from perspiration, as cotton does. Rather, the moisture is drawn away from the skin to the surface of the polypropylene garment, where it can evaporate. The annual U.S. production of polypropylene is about 7 billion pounds.

Another related polymer, **polystyrene**, is constructed from the monomer styrene,

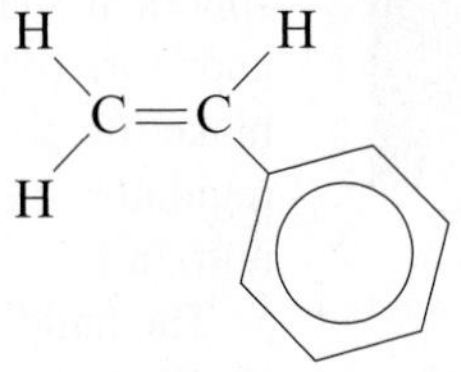

Pure polystyrene is too brittle for many uses, so most polystyrene-based polymers are actually *copolymers* of styrene and butadiene,

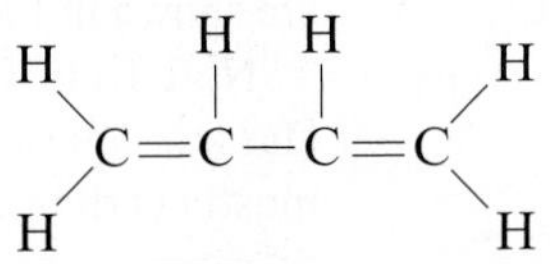

thus incorporating bits of butadiene rubber into the polystyrene matrix. The resulting polymer is very tough and is often used as a substitute for wood in furniture.

Another polystyrene-based product is acrylonitrile-butadiene-styrene (ABS), a tough, hard, and chemically resistant plastic used for pipes and for items such as radio housings, telephone cases, and golf club heads, for which shock resistance is an essential property. Originally, ABS was produced by copolymerization of the three monomers:

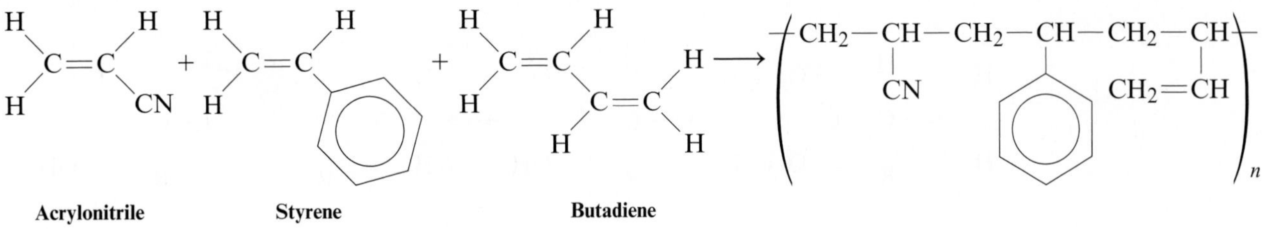

Acrylonitrile   **Styrene**   **Butadiene**

PVC pipe is widely used in industry.

It is now prepared by a special process called *grafting,* in which butadiene is polymerized first, and then the cyanide and phenyl substituents are added chemically.

Another high-volume polymer, **polyvinyl chloride (PVC)**, is constructed from the monomer vinyl chloride,

$$\underset{H}{\overset{H}{\text{C}}}=\underset{Cl}{\overset{H}{\text{C}}}$$

22.6 | Natural Polymers

Proteins

The protein in muscles enables them to contract.

We have seen that many useful synthetic materials are polymers. Thus it should not be surprising that a great many natural materials are also polymers: starch, hair, silicate chains in soil and rocks, silk and cotton fibers, and the cellulose in woody plants, to name only a few.

In this section we will consider a class of natural polymers, the **proteins**, which make up about 15% of our bodies and have molecular weights (molar masses) that range from about 6000 to over 1,000,000 grams per mole. Proteins perform many functions in the human body. **Fibrous proteins** provide structural integrity and strength for many types of tissue and are the main components of muscle, hair, and cartilage. Other proteins, usually called **globular proteins** because of their roughly spherical shape, are the "worker" molecules of the body. These proteins transport and store oxygen and nutrients, act as catalysts for the thousands of reactions that make life possible, fight invasion by foreign objects, participate in the body's many regulatory systems, and transport electrons in the complex process of metabolizing nutrients.

The building blocks of all proteins are the **α-amino acids**, where R may represent H, CH_3, or a more complex substituent. These molecules are called α-amino acids because the amino group (—NH_2) is always attached to the α-carbon, the one next to the carboxyl group (—CO_2H). The 20 amino acids most commonly found in proteins are shown in Fig. 22.18.

Note from Fig. 22.18 that the amino acids are grouped into polar and nonpolar classes, determined by the R groups, or **side chains**. Nonpolar side chains contain mostly carbon and hydrogen atoms, whereas polar side chains contain large numbers of nitrogen and oxygen atoms. This difference is important, because polar side chains are *hydrophilic* (water-loving), but nonpolar side chains are *hydrophobic* (water-fearing), and this characteristic greatly affects the three-dimensional structure of the resulting protein.

The protein polymer is built by condensation reactions between amino acids. For example,

α-Carbon

$$\underset{NH_2}{\overset{H}{\underset{|}{R-C-C}}}\overset{O}{\underset{OH}{\diagup}}$$

At the pH in biological fluids, the amino acids shown in Fig. 22.18 exist in a different form, with the proton of the —COOH group transferred to the —NH_2 group. For example, glycine would be in the form $H_3{}^+NCH_2COO^-$.

$$\underset{H}{\overset{H}{N}}-\underset{R}{\overset{H}{\underset{|}{C}}}-\overset{O}{\underset{OH}{\diagup}}\;+\;\underset{H}{\overset{H}{N}}-\underset{R'}{\overset{H}{\underset{|}{C}}}-\overset{O}{\underset{OH}{\diagup}}\;\longrightarrow\;\underset{H}{\overset{H}{N}}-\underset{R}{\overset{H}{\underset{|}{C}}}-\overset{O}{\overset{\parallel}{C}}-\underset{H}{\overset{H}{N}}-\underset{R'}{\overset{H}{\underset{|}{C}}}-\overset{O}{\underset{OH}{\diagup}}\;+\;H_2O$$

Peptide linkage

The product shown before is called a **dipeptide**. This name is used because the structure

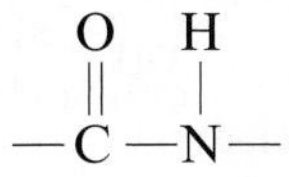

The peptide linkage is also found in nylon (see Section 22.5).

$$\begin{array}{cc} O & H \\ \parallel & \mid \\ -C-N- \end{array}$$

is called a **peptide linkage** by biochemists. (The same grouping is called an amide by organic chemists.) Additional condensation reactions lengthen the chain to produce a **polypeptide**, eventually yielding a protein.

You can imagine that with 20 amino acids, which can be assembled in any order, there is essentially an infinite variety possible in the construction of proteins. This flexibility allows an organism to tailor proteins for the many types of functions that must be carried out.

The order, or sequence, of amino acids in the protein chain is called the **primary structure**, conveniently indicated by using three-letter codes for the amino acids (see Fig. 22.18), where it is understood that the terminal carboxyl group is on the right and the terminal amino group is on the left. For example, one possible sequence for a tri-peptide containing the amino acids lysine, alanine, and leucine is

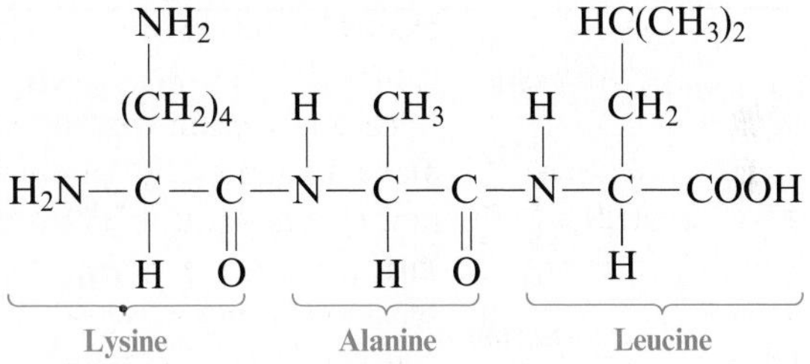

$$\begin{array}{c} NH_2 \qquad\qquad\qquad HC(CH_3)_2 \\ \mid \qquad\qquad\qquad\qquad \mid \\ (CH_2)_4 \quad H \quad CH_3 \quad\quad H \quad CH_2 \\ \mid \quad\quad \mid \quad\ \mid \quad\quad\quad \mid \quad\ \mid \\ H_2N-C-C-N-C-C-N-C-COOH \\ \mid \quad \parallel \qquad \mid \quad \parallel \qquad \mid \\ H \quad O \qquad H \quad O \qquad H \end{array}$$

$$\underbrace{\qquad\qquad}_{\text{Lysine}} \quad \underbrace{\qquad\qquad}_{\text{Alanine}} \quad \underbrace{\qquad\qquad}_{\text{Leucine}}$$

which is represented in the shorthand notation by

$$\text{lys-ala-leu}$$

Note from Example 22.7 that there are six sequences possible for a polypeptide with three given amino acids. There are three possibilities for the first amino acid (any one of the three given amino acids), there are two possibilities for the second amino acid (one has already been accounted for), but there is only one possibility left for the third amino acid. Thus the number of sequences is $3 \times 2 \times 1 = 6$. The product $3 \times 2 \times 1$ is often written 3! (and is called 3 factorial). Similar reasoning shows that for a polypeptide with four amino acids, there are 4!, or $4 \times 3 \times 2 \times 1 = 24$, possible sequences.

Interactive Example 22.7

Sign in at http://login.cengagebrain .com to try this Interactive Example in **OWL**.

Tripeptide Sequences

Write the sequences of all possible tripeptides composed of the amino acids tyrosine, histidine, and cysteine.

Solution

There are six possible sequences:

tyr-his-cys	his-tyr-cys	cys-tyr-his
tyr-cys-his	his-cys-tyr	cys-his-tyr

See Exercise 22.89

Nonpolar R groups

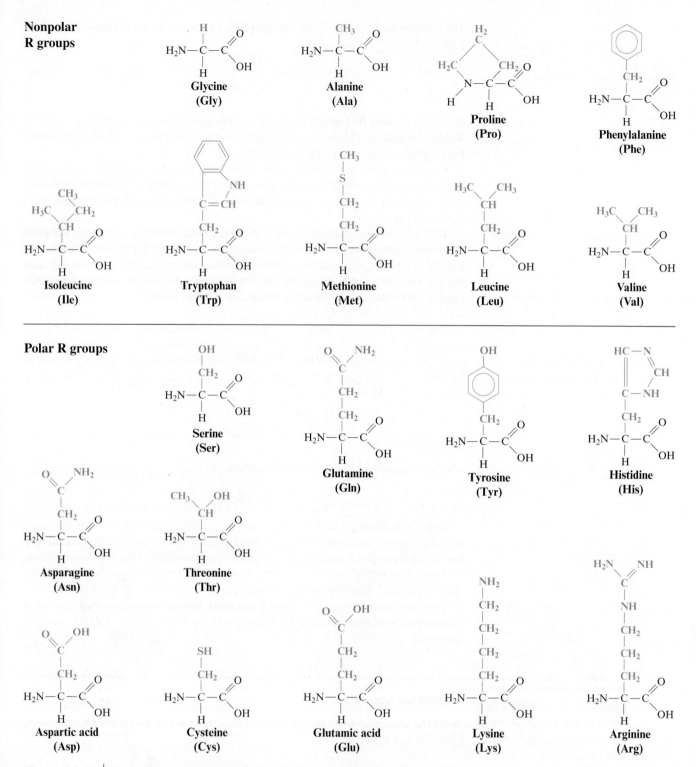

Polar R groups

Figure 22.18 | The 20 α-amino acids found in most proteins. The R group is shown in color.

Polypeptide Sequences

What number of possible sequences exists for a polypeptide composed of 20 different
amino acids?

Solution
The answer is 20!, or

$$20 \times 19 \times 18 \times 17 \times 16 \times \cdots \times 5 \times 4 \times 3 \times 2 \times 1 = 2.43 \times 10^{18}$$

See Exercise 22.90

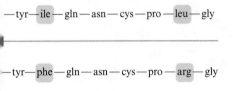

—tyr—ile—gln—asn—cys—pro—leu—gly

—tyr—phe—gln—asn—cys—pro—arg—gly

Figure 22.19 | The amino acid
sequences in (a) oxytocin and
(b) vasopressin. The differing amino
acids are boxed.

A striking example of the importance of the primary structure of polypeptides can
be seen in the differences between *oxytocin* and *vasopressin*. Both of these molecules
are nine-unit polypeptides that differ by only two amino acids (Fig. 22.19), yet they
perform completely different functions in the human body. Oxytocin is a hormone that
triggers contraction of the uterus and milk secretion. Vasopressin raises blood pressure
levels and regulates kidney function.

A second level of structure in proteins, beyond the sequence of amino acids, is the
arrangement of the chain of the long molecule. The **secondary structure** is determined to a large extent by hydrogen bonding between lone pairs on an oxygen atom in
the carbonyl group of an amino acid and a hydrogen atom attached to a nitrogen of
another amino acid:

$$C{=}\ddot{O}{:}\text{---}H{-}N$$
$$\delta^- \qquad \delta^+$$

Such interactions can occur *within* the chain coils to form a spiral structure called
an **α-helix**, as shown in Figs. 22.20 and 22.21. This type of secondary structure gives
the protein elasticity (springiness) and is found in the fibrous proteins in wool, hair,

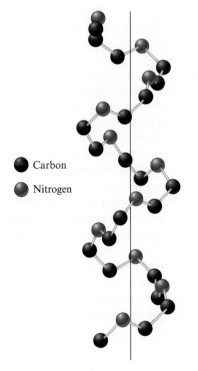

- Carbon
- Nitrogen

Figure 22.20 | Hydrogen bonding
within a protein chain causes it to
form a stable helical structure called
the α-helix. Only the main atoms in
the helical backbone are shown here.
The hydrogen bonds are not shown.

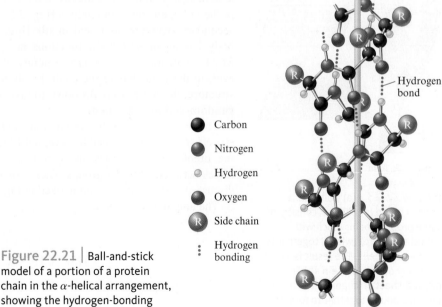

Imaginary axis
of α-helix

Hydrogen
bond

- Carbon
- Nitrogen
- Hydrogen
- Oxygen
- R Side chain
- : Hydrogen bonding

Figure 22.21 | Ball-and-stick
model of a portion of a protein
chain in the α-helical arrangement,
showing the hydrogen-bonding
interactions.

Figure 22.22 | When hydrogen bonding occurs between protein chains rather than within them, a stable structure (the pleated sheet) results. This structure contains many protein chains and is found in natural fibers, such as silk, and in muscles.

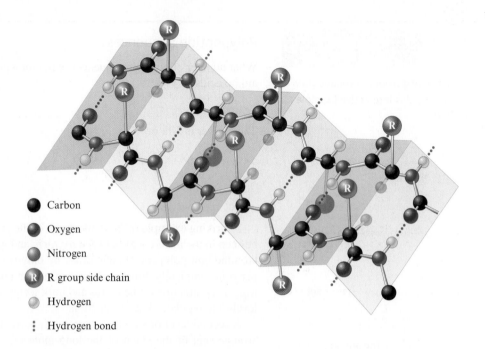

- ● Carbon
- ● Oxygen
- ● Nitrogen
- ⊛ R group side chain
- ○ Hydrogen
- ⋮ Hydrogen bond

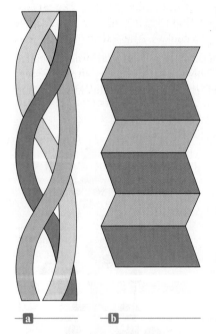

Figure 22.23 | (a) Collagen, a protein found in tendons, consists of three protein chains (each with a helical structure) twisted together to form a superhelix. The result is a long, relatively narrow protein. (b) The pleated-sheet arrangement of many proteins bound together to form the elongated protein found in silk fibers.

and tendons. Hydrogen bonding can also occur *between different* protein chains, joining them together in an arrangement called a **pleated sheet**, as shown in Fig. 22.22. Silk contains this arrangement of proteins, making its fibers flexible yet very strong and resistant to stretching. The pleated sheet is also found in muscle fibers. The hydrogen bonds in the α-helical protein are called *intrachain* (within a given protein chain), and those in the pleated sheet are said to be *interchain* (between protein chains).

As you might imagine, a molecule as large as a protein has a great deal of flexibility and can assume a variety of overall shapes. The specific shape that a protein assumes depends on its function. For long, thin structures, such as hair, wool and silk fibers, and tendons, an elongated shape is required. This may involve an α-helical secondary structure, as found in the protein α-keratin in hair and wool or in the collagen found in tendons [Fig. 22.23(a)], or it may involve a pleated-sheet secondary structure, as found in silk [Fig. 22.23(b)]. Many of the proteins in the body having nonstructural functions are globular, such as myoglobin (see Fig. 21.31). Note that the secondary structure of myoglobin is basically α-helical. However, in the areas where the chain bends to give the protein its compact globular structure, the α-helix breaks down to give a secondary configuration known as the **random-coil arrangement**.

The overall shape of the protein, long and narrow or globular, is called its **tertiary structure** and is maintained by several different types of interactions: hydrogen bonding, dipole–dipole interactions, ionic bonds, covalent bonds, and London dispersion forces between nonpolar groups. These bonds, which represent all the bonding types discussed in this text, are summarized in Fig. 22.24.

The amino acid *cysteine*

$$\underset{\underset{\displaystyle H}{|}}{\overset{\displaystyle HS-CH_2-\overset{\overset{\displaystyle H}{|}}{C}-\overset{\overset{\displaystyle O}{\|}}{C}-OH}{N}}$$

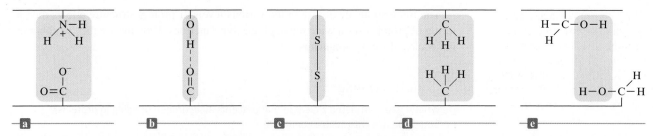

Figure 22.24 | Summary of the various types of interactions that stabilize the tertiary structure of a protein: (a) ionic, (b) hydrogen bonding, (c) covalent, (d) London dispersion, and (e) dipole–dipole.

Natural cysteine linkages in hair

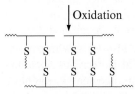

Figure 22.25 | The permanent waving of hair.

plays a special role in stabilizing the tertiary structure of many proteins because the —SH groups on two cysteines can react in the presence of an oxidizing agent to form a S—S bond called a **disulfide linkage**:

$$C-CH_2-S-H + H-S-CH_2-C \longrightarrow C-CH_2-S-S-CH_2-C$$

A practical application of the chemistry of disulfide bonds is permanent waving of hair, as summarized in Fig. 22.25. The S—S linkages in the protein of hair are broken by treatment with a reducing agent. The hair is then set in curlers to change the tertiary protein structure to the desired shape. Then treatment with an oxidizing agent causes new S—S bonds to form, which allow the hair protein to retain the new structure.

The three-dimensional structure of a protein is crucial to its function. The process of breaking down this structure is called **denaturation** (Fig. 22.26). For example, the denaturation of egg proteins occurs when an egg is cooked. Any source of energy can cause denaturation of proteins and is thus potentially dangerous to living organisms. For example, ultraviolet and X-ray radiation or nuclear radioactivity can disrupt protein structure, which may lead to cancer or genetic damage. Protein damage is also caused by chemicals like benzene, trichloroethane, and 1,2-dibromoethane. The metals lead and mercury, which have a very high affinity for sulfur, cause protein denaturation by disrupting disulfide bonds between protein chains.

Energy

Figure 22.26 | A schematic representation of the thermal denaturation of a protein.

The tremendous flexibility in the various levels of protein structure allows the tailoring of proteins for a wide range of specific functions. Proteins are the "workhorse" molecules of living organisms.

> ### Critical Thinking
>
> What if you contracted a disease that prevents all hydrogen bonding in proteins? Could you live with such a condition?

Carbohydrates

Carbohydrates form another class of biologically important molecules. They serve as a food source for most organisms and as a structural material for plants. Because many carbohydrates have the empirical formula CH_2O, it was originally believed that these substances were hydrates of carbon, thus accounting for the name.

Most important carbohydrates, such as starch and cellulose, are polymers composed of monomers called **monosaccharides**, or **simple sugars**. The monosaccharides are polyhydroxy ketones and aldehydes. The most important contain five carbon atoms (**pentoses**) or six carbon atoms (**hexoses**). One important hexose is *fructose,* a sugar found in honey and fruit. Its structure is

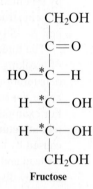

Fructose

where the asterisks indicate chiral carbon atoms. In Section 21.4 we saw that molecules with nonsuperimposable mirror images exhibit optical isomerism. A carbon atom with four *different* groups bonded to it in a tetrahedral arrangement *always* has a nonsuperimposable mirror image (Fig. 22.27), which gives rise to a pair of optical isomers. For example, the simplest sugar, glyceraldehyde,

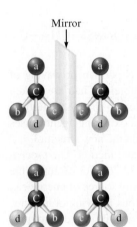

Figure 22.27 | When a tetrahedral carbon atom has four different substituents, there is no way that its mirror image can be superimposed. The lower two forms show other possible orientations of the molecule. Compare these with the mirror image and note that they cannot be superimposed.

which has one chiral carbon, has two optical isomers, as shown in Fig. 22.28.

In fructose each of the three chiral carbon atoms satisfies the requirement of being surrounded by four different groups. This leads to a total of 2^3, or 8, isomers that differ in their ability to rotate polarized light. The particular isomer whose structure is shown in Table 22.8 is called D-fructose. Generally, monosaccharides have one isomer that is more common in nature than the others. The most important pentoses and hexoses are shown in Table 22.8.

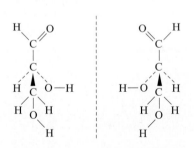

Figure 22.28 | The mirror image optical isomers of glyceraldehyde. Note that these mirror images cannot be superimposed.

Chemical connections

Tanning in the Shade

Among today's best-selling cosmetics are self-tanning lotions. Many light-skinned people want to look like they have just spent a vacation in the Caribbean, but they recognize the dangers of too much sun—it causes premature aging and may lead to skin cancer. Chemistry has come to the rescue in the form of lotions that produce an authentic-looking tan. All of these lotions have the same active ingredient: dihydroxyacetone (DHA). DHA, which has the structure

$$\begin{array}{ccccc} & H & & O & & H \\ & | & & \| & & | \\ H-O- & C & - & C & - & C & -OH \\ & | & & & & | \\ & H & & & & H \end{array}$$

is a nontoxic, simple sugar that occurs as an intermediate in carbohydrate metabolism in higher-order plants and animals. The DHA used in self-tanners is prepared by bacterial fermentation of glycerine,

$$\begin{array}{ccccccc} & & & H & & & \\ & & & | & & & \\ & & & O & & & \\ & H & & | & & H & \\ & | & & | & & | & \\ H-O- & C & - & C & - & C & -O-H \\ & | & & | & & | & \\ & H & & H & & H & \end{array}$$

The tanning effects of DHA were discovered by accident in the 1950s at Children's Hospital at the University of Cincinnati, where DHA was being used to treat children with glycogen storage disease. When the DHA was accidentally spilled on the skin, it produced brown spots.

The mechanism of the browning process involves the Maillard reaction, which was discovered by Louis-Camille Maillard in 1912. In this process amino acids react with sugars to create brown or golden brown products. The same reaction is responsible for much of the browning that occurs during the manufacture and storage of foods. It is also the reason that beer is golden brown.

The browning of skin occurs in the stratum corneum—the outermost, dead layer—where the DHA reacts with free amino ($-NH_2$) groups of the proteins found there.

DHA is present in most tanning lotions at concentrations between 2% and 5%, although some products designed to give a deeper tan are more concentrated. Because the lotions themselves turn brown above pH 7, the tanning lotions are buffered at pH 5.

Thanks to these new products, tanning is now both safe and easy.

Ingredients: Water, Aloe Vera Gel, Glycerin, Propylene Glycol, Polysorbate 20, Fragrance, Tocopheryl Acetate (Vitamin E Acetate), Imidazolidinyl Urea, Dihydroxyacetone.

© 1995, 1998 Distributed by Schering-Plough HealthCare Products, Inc., Memphis, TN 38151, USA. All rights Reserved. Made in USA.

www.coppertone.com

21621-C

Self-tanning products and a close-up of a label showing the contents.

General Name of Sugar	Number of Carbon Atoms
Triose	3
Tetrose	4
Pentose	5
Hexose	6
Heptose	7
Octose	8
Nonose	9

Table 22.8 | Some Important Monosaccharides

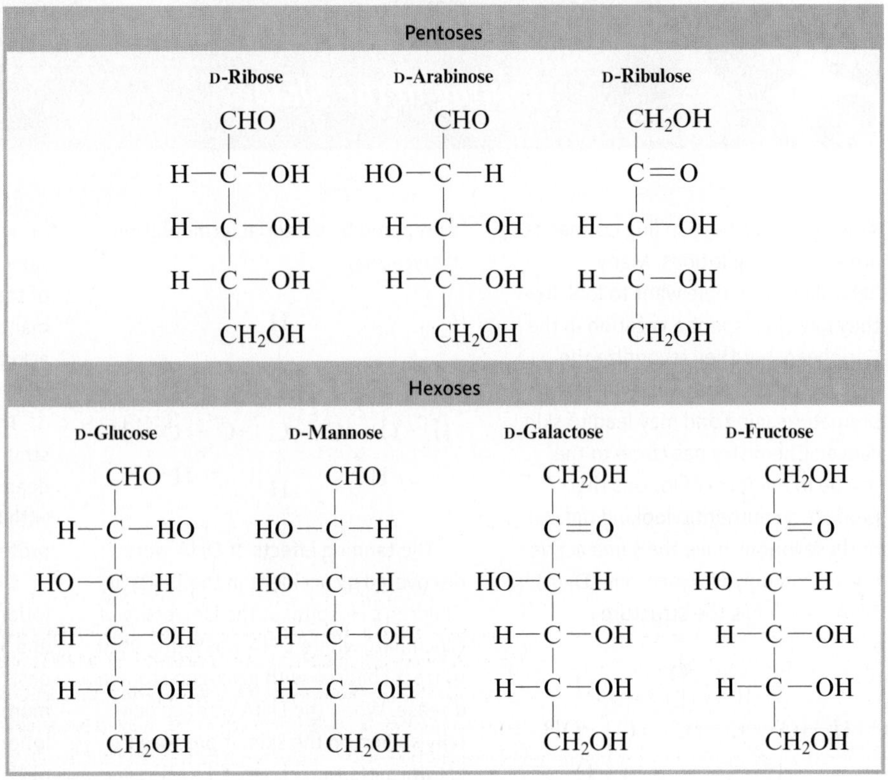

Pentoses

D-Ribose

$$\begin{array}{c} CHO \\ | \\ H-C-OH \\ | \\ H-C-OH \\ | \\ H-C-OH \\ | \\ CH_2OH \end{array}$$

D-Arabinose

$$\begin{array}{c} CHO \\ | \\ HO-C-H \\ | \\ H-C-OH \\ | \\ H-C-OH \\ | \\ CH_2OH \end{array}$$

D-Ribulose

$$\begin{array}{c} CH_2OH \\ | \\ C=O \\ | \\ H-C-OH \\ | \\ H-C-OH \\ | \\ CH_2OH \end{array}$$

Hexoses

D-Glucose

$$\begin{array}{c} CHO \\ | \\ H-C-HO \\ | \\ HO-C-H \\ | \\ H-C-OH \\ | \\ H-C-OH \\ | \\ CH_2OH \end{array}$$

D-Mannose

$$\begin{array}{c} CHO \\ | \\ HO-C-H \\ | \\ HO-C-H \\ | \\ H-C-OH \\ | \\ H-C-OH \\ | \\ CH_2OH \end{array}$$

D-Galactose

$$\begin{array}{c} CH_2OH \\ | \\ C=O \\ | \\ HO-C-H \\ | \\ H-C-OH \\ | \\ H-C-OH \\ | \\ CH_2OH \end{array}$$

D-Fructose

$$\begin{array}{c} CH_2OH \\ | \\ C=O \\ | \\ HO-C-H \\ | \\ H-C-OH \\ | \\ H-C-OH \\ | \\ CH_2OH \end{array}$$

Interactive Example 22.9

Sign in at http://login.cengagebrain .com to try this Interactive Example in OWL.

Chiral Carbons in Carbohydrates

Determine the number of chiral carbon atoms in the following pentose:

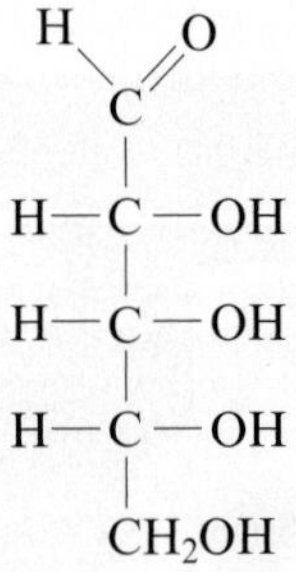

$$\begin{array}{c} H \quad O \\ \diagdown\diagup \\ C \\ | \\ H-C-OH \\ | \\ H-C-OH \\ | \\ H-C-OH \\ | \\ CH_2OH \end{array}$$

Solution

We must look for carbon atoms that have four different substituents. The top carbon has only three substituents and thus cannot be chiral. The three carbon atoms shown in blue each have four different groups attached to them:

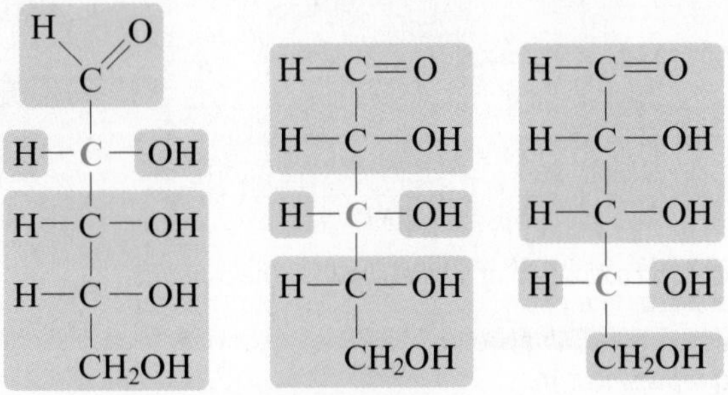

Since the fifth carbon atom has only three types of substituents (it has two hydrogen atoms), it is not chiral.

Thus the three chiral carbon atoms in this pentose are those shown in blue:

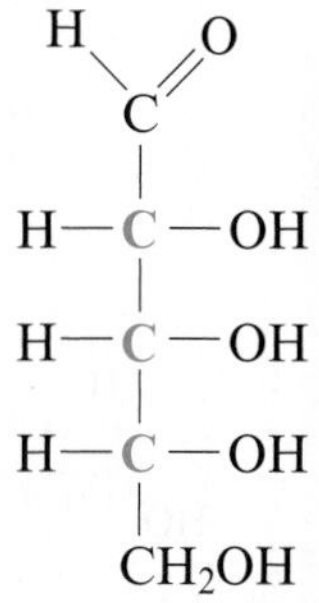

$$
\begin{array}{c}
\text{H} \quad\quad \text{O} \\
\diagdown\text{C}\diagup \\
| \\
\text{H---C---OH} \\
| \\
\text{H---C---OH} \\
| \\
\text{H---C---OH} \\
| \\
\text{CH}_2\text{OH}
\end{array}
$$

Note that D-ribose and D-arabinose, shown in Table 22.8, are two of the eight isomers of this pentose.

See Exercises 22.96 and 22.101 through 22.104

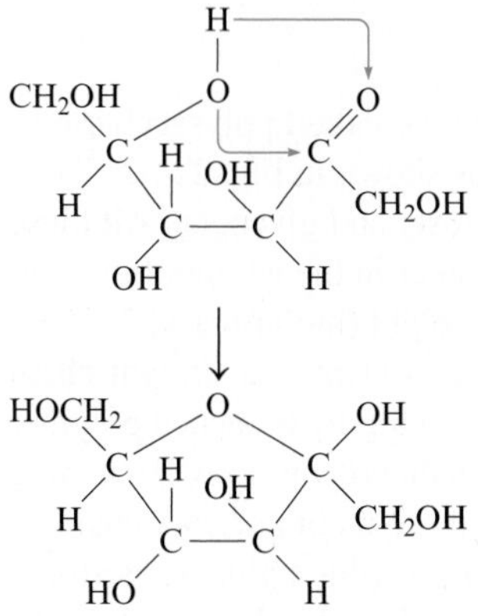

Although we have so far represented the monosaccharides as straight-chain molecules, they usually cyclize, or form a ring structure, in aqueous solution. Figure 22.29 shows this reaction for fructose. Note that a new bond is formed between the oxygen of the terminal hydroxyl group and the carbon of the ketone group. In the cyclic form fructose is a five-membered ring containing a C—O—C bond. The same type of reaction can occur between a hydroxyl group and an aldehyde group, as shown for D-glucose in Fig. 22.30. In this case a six-membered ring is formed.

More complex carbohydrates are formed by combining monosaccharides. For example, **sucrose**, common table sugar, is a **disaccharide** formed from glucose and fructose by elimination of water to form a C—O—C bond between the rings, which is called a **glycoside linkage** (Fig. 22.31). When sucrose is consumed in food, the above reaction is reversed. An enzyme in saliva catalyzes the breakdown of this disaccharide.

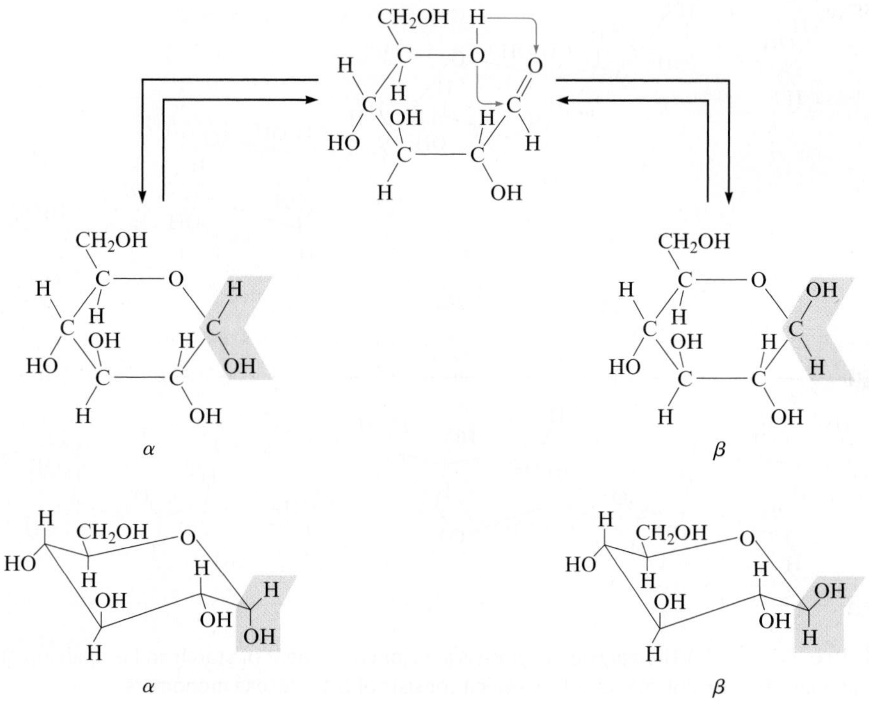

Figure 22.29 | The cyclization of D-fructose.

Figure 22.30 | The cyclization of glucose. Two different rings are possible; they differ in the orientation of the hydroxy group and hydrogen on one carbon, as indicated. The two forms are designated α and β and are shown here in two representations.

Figure 22.31 | Sucrose is a disaccharide formed from α-D-glucose and fructose.

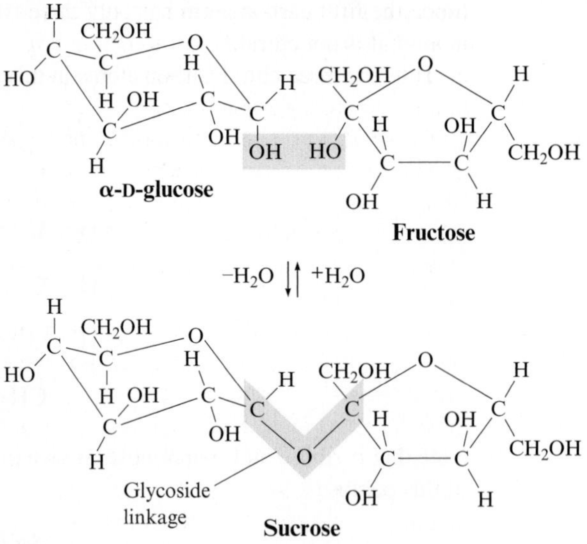

Bowl of sugar cubes.

α-D-glucose

Fructose

$-H_2O \updownarrow +H_2O$

Glycoside linkage

Sucrose

Large polymers consisting of many monosaccharide units, called polysaccharides, can form when each ring forms two glycoside linkages, as shown in Fig. 22.32. Three of the most important of these polymers are starch, cellulose, and glycogen. All these substances are polymers of glucose, differing from each other in the nature of the glycoside linkage, the amount of branching, and molecular weight (molar mass).

Starch, a polymer of α-D-glucose, consists of two parts: *amylose*, a straight-chain polymer of α-glucose [see Fig. 22.32(a)], and *amylopectin*, a highly branched polymer of α-glucose with a molecular weight that is 10 to 20 times that of amylose. Branching occurs when a third glycoside linkage attaches a branch to the main polymer chain.

Starch, the carbohydrate reservoir in plants, is the form in which glucose is stored by the plant for later use as cellular fuel. Glucose is stored in this high-molecular-weight form because it results in less stress on the plant's internal structure by osmotic

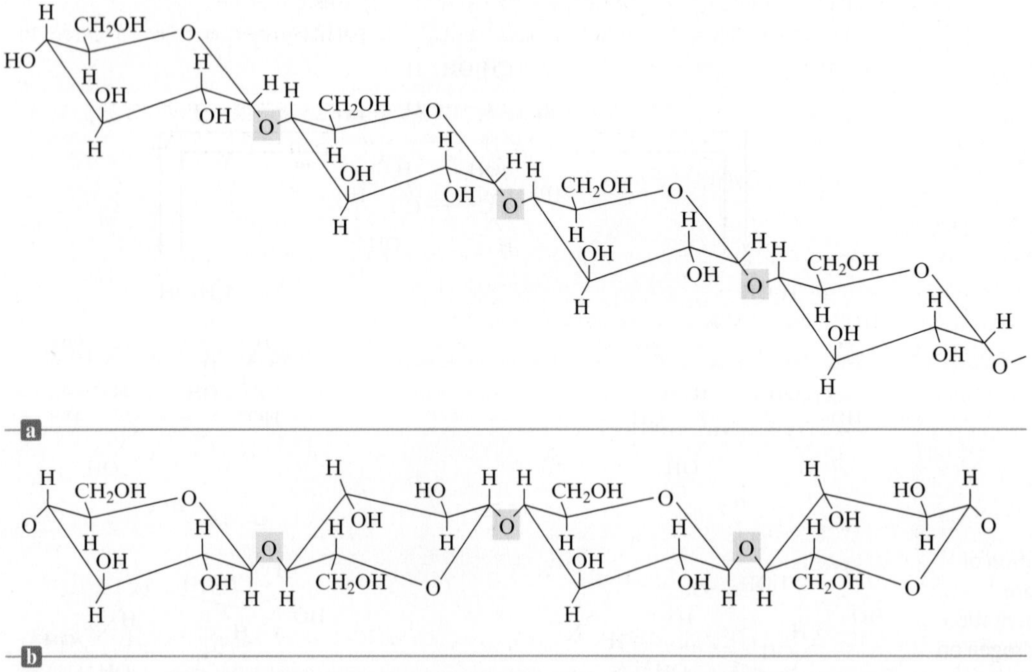

a

b

Figure 22.32 | (a) The polymer amylose is a major component of starch and is made up of α-D-glucose monomers. (b) The polymer cellulose, which consists of β-D-glucose monomers.

pressure. Recall from Section 11.6 that it is the concentration of solute molecules (or ions) that determines the osmotic pressure. Combining the individual glucose molecules into one large chain keeps the concentration of solute molecules relatively low, minimizing the osmotic pressure.

Cellulose, the major structural component of woody plants and natural fibers (such as cotton), is a polymer of β-D-glucose and has the structure shown in Fig. 22.32(b). Note that the β-glycoside linkages in cellulose give the glucose rings a different relative orientation than is found in starch. Although this difference may seem minor, it has very important consequences. The human digestive system contains α-glycosidases, enzymes that can catalyze breakage of the α-glycoside bonds in starch. These enzymes are not effective on the β-glycoside bonds of cellulose, presumably because the different structure results in a poor fit between the enzyme's active site and the carbohydrate. The enzymes necessary to cleave β-glycoside linkages, the β-glycosidases, are found in bacteria that exist in the digestive tracts of termites, cows, deer, and many other animals. Thus, unlike humans, these animals can derive nutrition from cellulose.

Glycogen, the main carbohydrate reservoir in animals, has a structure similar to that of amylopectin but with more branching. It is this branching that is thought to facilitate the rapid breakdown of glycogen into glucose when energy is required.

Nucleic Acids

Life is possible only because each cell, when it divides, can transmit the vital information about how it works to the next generation. It has been known for a long time that this process involves the chromosomes in the nucleus of the cell. Only since 1953, however, have scientists understood the molecular basis of this intriguing cellular "talent."

The substance that stores and transmits the genetic information is a polymer called **deoxyribonucleic acid (DNA)**, a huge molecule with a molecular weight as high as several billion grams per mole. Together with other similar nucleic acids called the **ribonucleic acids (RNA)**, DNA is also responsible for the synthesis of the various proteins needed by the cell to carry out its life functions. The RNA molecules, which are found in the cytoplasm outside the nucleus, are much smaller than DNA polymers, with molecular weights of only 20,000 to 40,000 grams per mole.

The monomers of the nucleic acids, called **nucleotides**, are composed of three distinct parts:

1. a *five-carbon sugar,* deoxyribose in DNA and ribose in RNA (Fig. 22.33)
2. a *nitrogen-containing organic base* of the type shown in Fig. 22.34
3. a *phosphoric acid molecule* (H_3PO_4)

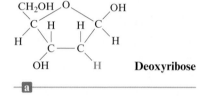

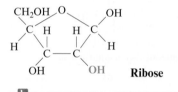

Figure 22.33 | The structure of the pentoses (a) deoxyribose and (b) ribose. Deoxyribose is the sugar molecule present in DNA; ribose is found in RNA.

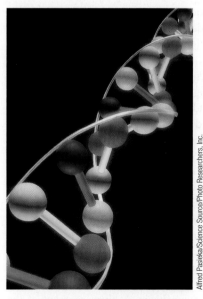

A computer image of the base pairs of DNA. The blue lines represent the sugar–phosphate backbone and the colored bars represent the hydrogen bonding between the base pairs.

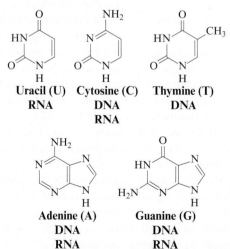

Figure 22.34 | The organic bases found in DNA and RNA.

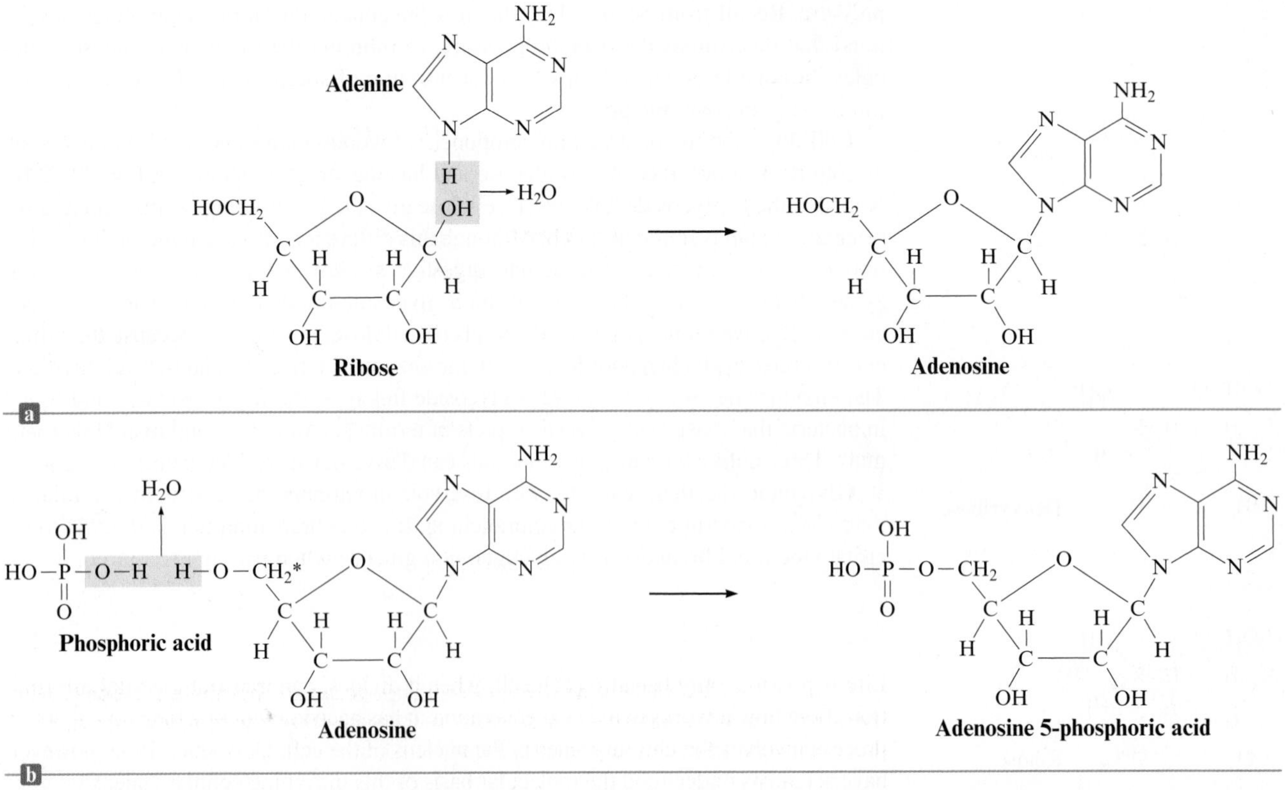

Figure 22.35 | (a) Adenosine is formed by the reaction of adenine with ribose. (b) The reaction of phosphoric acid with adenosine to form the ester adenosine 5-phosphoric acid, a nucleotide. (At biological pH, the phosphoric acid would not be fully protonated as is shown here.)

The base and the sugar combine as shown in Fig. 22.35(a) to form a unit that in turn reacts with phosphoric acid to create the nucleotide, which is an ester [Fig. 22.35(b)]. The nucleotides become connected through condensation reactions that eliminate water to give a polymer of the type represented in Fig. 22.36; such a polymer can contain a *billion* units.

The key to DNA's functioning is its *double-helical structure with complementary bases on the two strands*. The bases form hydrogen bonds to each other, as shown in Fig. 22.37. Note that the structures of cytosine and guanine make them perfect partners for hydrogen bonding, and they are *always* found as pairs on the two strands of DNA. Thymine and adenine form similar hydrogen-bonding pairs.

There is much evidence to suggest that the two strands of DNA unwind during cell division and that new complementary strands are constructed on the unraveled strands (Fig. 22.38). Because the bases on the strands always pair in the same way—cytosine with guanine and thymine with adenine—each unraveled strand serves as a template for attaching the complementary bases (along with the rest of the nucleotide). This process results in two double-helix DNA structures that are identical to the original one. Each new double strand contains one strand from the original DNA double helix and one newly synthesized strand. This replication of DNA allows for the transmission of genetic information as the cells divide.

The other major function of DNA is **protein synthesis**. A given segment of the DNA, called a **gene**, contains the code for a specific protein. These codes transmit the primary structure of the protein (the sequence of amino acids) to the construction "machinery" of the cell. There is a specific code for each amino acid in the protein, which ensures that the correct amino acid will be inserted as the protein chain grows. A code consists of a set of three bases called a **codon**.

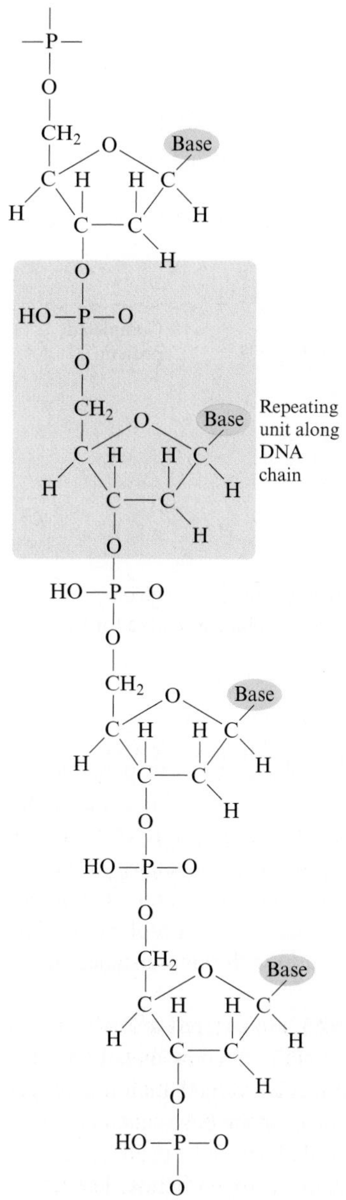

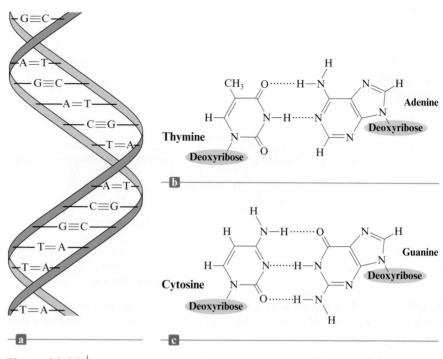

Figure 22.37 (a) The DNA double helix contains two sugar–phosphate backbones, with the bases from the two strands hydrogen-bonded to each other. The complementarity of the (b) thymine-adenine and (c) cytosine-guanine pairs.

Figure 22.36 A portion of a typical nucleic acid chain. Note that the backbone consists of sugar–phosphate esters.

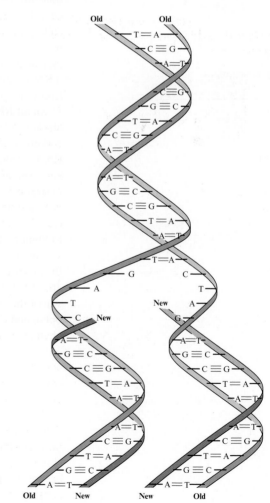

Figure 22.38 During cell division the original DNA double helix unwinds and new complementary strands are constructed on each original strand.

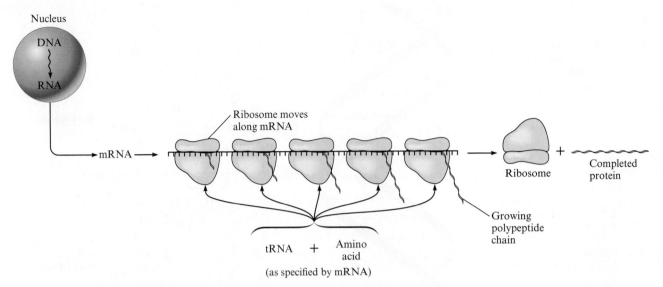

Figure 22.39 | The mRNA molecule, constructed from a specific gene on the DNA, is used as the pattern to construct a given protein with the assistance of ribosomes. The tRNA molecules attach to specific amino acids and put them in place as called for by the codons on the mRNA.

DNA stores the genetic information, while RNA molecules are responsible for transmitting this information to the ribosomes, where protein synthesis actually occurs. This complex process involves, first, the construction of a special RNA molecule called **messenger RNA (mRNA)**. The mRNA is built in the cell nucleus on the appropriate section of DNA (the gene); the double helix is "unzipped," and the complementarity of the bases is used in a process similar to that used in DNA replication. The mRNA then migrates into the cytoplasm of the cell where, with the assistance of the ribosomes, the protein is synthesized.

Small RNA fragments, called **transfer RNA (tRNA)**, are tailored to find specific amino acids and then to attach them to the growing protein chain as dictated by the codons in the mRNA. Transfer RNA has a lower molecular weight than messenger RNA. It consists of a chain of 75 to 80 nucleotides, including the bases adenine, cytosine, guanine, and uracil, among others. The chain folds back onto itself in various places as the complementary bases along the chain form hydrogen bonds. The tRNA decodes the genetic message from the mRNA, using a complementary triplet of bases called an **anticodon**. The nature of the anticodon governs which amino acid will be brought to the protein under construction.

The protein is built in several steps. First, a tRNA molecule brings an amino acid to the mRNA [the anticodon of the tRNA must complement the codon of the mRNA (Fig. 22.39)]. Once this amino acid is in place, another tRNA moves to the second codon site of the mRNA with its specific amino acid. The two amino acids link via a peptide bond, and the tRNA on the first codon breaks away. The process is repeated down the chain, always matching the tRNA anticodon with the mRNA codon.

For review

Key terms

biomolecule
organic chemistry

Section 22.1
hydrocarbons
saturated
unsaturated
alkanes
normal (straight-chain or unbranched) hydrocarbons
structural isomerism
combustion reaction
substitution reaction
dehydrogenation reaction
cyclic alkanes

Section 22.2
alkenes
cis–trans isomerism
alkynes
addition reaction
hydrogenation reaction
halogenation
polymerization

Section 22.3
aromatic hydrocarbons
phenyl group

Section 22.4
hydrocarbon derivatives
functional group
alcohols
phenol
carbonyl group
ketones
aldehydes
carboxylic acids
carboxyl group
ester
amines

Section 22.5
polymers
thermoset polymer
thermoplastic polymer
crosslinking
vulcanization
addition polymerization
free radical
condensation polymerization
copolymer
homopolymer

Hydrocarbons

> Compounds composed of mostly carbon and hydrogen atoms that typically contain chains or rings of carbon atoms
> Alkanes
>> Contain compounds with only C—C single bonds
>> Can be represented by the formula C_nH_{2n+2}
>> Are said to be saturated because each carbon present is bonded to the maximum number of atoms (4)
>> The carbon atoms are described as being sp^3 hybridized
>> Their structural isomerism involves the formation of branched chains
>> React with O_2 to form CO_2 and H_2O (called a combustion reaction)
>> Undergo substitution reactions
> Alkenes
>> Contain one or more C=C double bonds
>> Simplest alkene is C_2H_4 (ethylene), which is described as containing sp^2 hybridized carbon atoms
>> Restricted rotation about the C=C bonds in alkenes can lead to *cis–trans* isomerism
>> Undergo addition reactions
> Alkynes
>> Contain one or more C≡C triple bonds
>> Simplest example is C_2H_2 (acetylene), described as containing sp-hybridized carbon atoms
>> Undergo addition reactions
> Aromatic hydrocarbons
>> Contain rings of carbon atoms with delocalized π electrons
>> Undergo substitution reactions rather than addition reactions

Hydrocarbon derivatives

> Contain one or more functional groups
> Alcohols: contain the —OH group
> Aldehydes: contain a $\overset{\diagdown}{\underset{H}{C}}$=O group
> Ketones: contain the $\diagdown$C=O group
> Carboxylic acids: contain the —C$\overset{\diagup O}{\diagdown OH}$ group

Polymers

> Large molecules formed from many small molecules (called monomers)
>> Addition polymerization: monomers add together by a free radical mechanism
>> Condensation polymerization: monomers connect by splitting out a small molecule, such as water

Key terms

dimer
polyester
isotactic chain
syndiotactic chain
atactic chain
polystyrene
polyvinyl chloride (PVC)

Section 22.6

proteins
fibrous proteins
globular proteins
α-amino acids
side chains
dipeptide
peptide linkage
polypeptide
primary structure
secondary structure
α-helix
pleated sheet
random-coil arrangement
tertiary structure
disulfide linkage
denaturation
carbohydrates
monosaccharides (simple sugars)
pentoses
hexoses
sucrose
disaccharide
glycoside linkage
starch
cellulose
glycogen
deoxyribonucleic acid (DNA)
ribonucleic acid (RNA)
nucleotides
protein synthesis
gene
codon
messenger RNA (mRNA)
transfer RNA (tRNA)
anticodon

Proteins

> A class of natural polymers with molar masses ranging from 600 to 1,000,000

> Fibrous proteins form the structural basis of muscle, hair, and cartilage

> Globular proteins perform many biologic functions, including transport and storage of oxygen, catalysis of biologic reactions, and regulation of biological systems

> Building blocks of proteins (monomers) are α-amino acids, which connect by a condensation reaction to form a peptide linkage

> Protein structure
> > Primary: the order of amino acids in the chain
> > Secondary: the arrangement of the protein chain
> > α-helix
> > pleated sheet
> > Tertiary structure: the overall shape of the protein

Carbohydrates

> Contain carbon, hydrogen, and oxygen

> Serve as food sources for most organisms

> Monosaccharides are most commonly five-carbon and six-carbon polyhydroxy ketones and aldehydes
> > Monosaccharides combine to form more complex carbohydrates, such as sucrose, starch, and cellulose

Genetic processes

> When a cell divides, the genetic information is transmitted via deoxyribonucleic acid (DNA), which has a double helical structure
> > During cell division, the double helix unravels and a new polymer forms along each strand of the original DNA
> > The genetic code is carried by organic bases that hydrogen-bond to each other in specific pairs in the interior of the DNA double helix

Review questions   *Answers to the Review Questions can be found on the Student website (accessible from* **www.cengagebrain.com**).

1. What is a hydrocarbon? What is the difference between a saturated hydrocarbon and an unsaturated hydrocarbon? Distinguish between normal and branched hydrocarbons. What is an alkane? What is a cyclic alkane? What are the two general formulas for alkanes? What is the hybridization of carbon atoms in alkanes? What are the bond angles in alkanes? Why are cyclopropane and cyclobutane so reactive?

   The normal (unbranched) hydrocarbons are often referred to as *straight-chain hydrocarbons*. What does this name refer to? Does it mean that the carbon atoms in a straight-chain hydrocarbon really have a linear arrangement? Explain. In the shorthand notation for cyclic alkanes, the hydrogens are usually omitted. How do you determine the number of hydrogens bonded to each carbon in a ring structure?

2. What is an alkene? What is an alkyne? What are the general formulas for alkenes and alkynes, assuming one multiple bond in each? What are the bond angles in alkenes and alkynes? Describe the bonding in alkenes and alkynes using C_2H_4 and C_2H_2 as your examples. Why is there restricted rotation in alkenes and alkynes? Is the general formula for a cyclic alkene C_nH_{2n}? If not, what is the general formula, assuming one multiple bond?

3. What are aromatic hydrocarbons? Benzene exhibits resonance. Explain. What are the bond angles in benzene? Give a detailed description of the bonding in benzene. The π electrons in benzene are delocalized, while the π electrons in simple alkenes and alkynes are localized. Explain the difference.

4. Summarize the nomenclature rules for alkanes, alkenes, alkynes, and aromatic compounds. Correct the following false statements regarding nomenclature of hydrocarbons.

   a. The root name for a hydrocarbon is based on the shortest continuous chain of carbon atoms.

   b. The suffix used to name all hydrocarbons is -*ane*.

   c. Substituent groups are numbered so as to give the largest numbers possible.

   d. No number is required to indicate the positions of double or triple bonds in alkenes and alkynes.

   e. Substituent groups get the lowest number possible in alkenes and alkynes.

   f. The *ortho-* term in aromatic hydrocarbons indicates the presence of two substituent groups bonded to carbon-1 and carbon-3 in benzene.

5. What functional group distinguishes each of the following hydrocarbon derivatives?

   a. halohydrocarbons

   b. alcohols

   c. ethers

   d. aldehydes

   e. ketones

   f. carboxylic acids

   g. esters

   h. amines

   Give examples of each functional group. What prefix or suffix is used to name each functional group? What are the bond angles in each? Describe the bonding in each functional group. What is the difference between a primary, secondary, and tertiary alcohol? For the

functional groups in a–h, when is a number required to indicate the position of the functional group? Carboxylic acids are often written as RCOOH. What does —COOH indicate and what does R indicate? Aldehydes are sometimes written as RCHO. What does —CHO indicate?

6. Distinguish between isomerism and resonance. Distinguish between structural and geometric isomerism. When writing the various structural isomers, the most difficult task is identifying which are different isomers and which are identical to a previously written structure—that is, which are compounds that differ only by the rotation of a carbon single bond. How do you distinguish between structural isomers and those that are identical?

   Alkenes and cycloalkanes are structural isomers of each other. Give an example of each using C_4H_8. Another common feature of alkenes and cycloalkanes is that both have restricted rotation about one or more bonds in the compound, so both can exhibit *cis–trans* isomerism. What is required for an alkene or cycloalkane to exhibit *cis–trans* isomerism? Explain the difference between *cis* and *trans* isomers.

   Alcohols and ethers are structural isomers of each other, as are aldehydes and ketones. Give an example of each to illustrate. Which functional group in Table 22.4 can be structural isomers of carboxylic acids?

   What is optical isomerism? What do you look for to determine whether an organic compound exhibits optical isomerism? 1-Bromo-1-chloroethane is optically active whereas 1-bromo-2-chloroethane is not optically active. Explain.

7. What type of intermolecular forces do hydrocarbons exhibit? Explain why the boiling point of *n*-heptane is greater than that of *n*-butane. A general rule for a group of hydrocarbon isomers is that as the amount of branching increases, the boiling point decreases. Explain why this would be true.

   The functional groups listed in Table 22.4 all exhibit London dispersion forces, but they also usually exhibit additional dipole–dipole forces. Explain why this is the case for each functional group. Although alcohols and ethers are structural isomers of each other, alcohols always boil at significantly higher temperatures than similar-size ethers. Explain. What would you expect when comparing the boiling points of similar-size carboxylic acids to esters? $CH_3CH_2CH_3$, CH_3CH_2OH, CH_3CHO, and $HCOOH$ all have about the same molar mass, but they boil at very different temperatures. Why? Place these compounds in order by increasing boiling point.

8. Distinguish between substitution and addition reactions. Give an example of each type of reaction. Alkanes and aromatics are fairly stable compounds. To make them

react, a special catalyst must be present. What catalyst must be present when reacting Cl_2 with an alkane or with benzene? Adding Cl_2 to an alkene or alkyne does not require a special catalyst. Why are alkenes and alkynes more reactive than alkanes and aromatic compounds? All organic compounds can be combusted. What is the other reactant in a combustion reaction, and what are the products, assuming the organic compound contains only C, H, and perhaps O?

The following are some other organic reactions covered in Section 22.4. Give an example to illustrate each type of reaction.

a. Adding H_2O to an alkene (in the presence of H^+) yields an alcohol.

b. Primary alcohols are oxidized to aldehydes, which can be further oxidized to carboxylic acids.

c. Secondary alcohols are oxidized to ketones.

d. Reacting an alcohol with a carboxylic acid (in the presence of H^+) produces an ester.

9. Define and give an example of each of the following.

a. addition polymer

b. condensation polymer

c. copolymer

d. homopolymer

e. polyester

f. polyamide

Distinguish between a thermoset polymer and a thermoplastic polymer. How do the physical properties of polymers depend on chain length and extent of chain branching? Explain how crosslinking agents are used to change the physical properties of polymers. Isotactic polypropylene makes stronger fibers than atactic polypropylene. Explain. In which polymer, polyethylene or polyvinyl chloride, would you expect to find the stronger intermolecular forces (assuming the average chain lengths are equal)?

10. Give the general formula for an amino acid. Some amino acids are labeled hydrophilic and some are labeled hydrophobic. What do these terms refer to? Aqueous solutions of amino acids are buffered solutions. Explain. Most of the amino acids in Fig. 22.18 are optically active. Explain. What is a peptide bond? Show how glycine, serine, and alanine react to form a tripeptide. What is a protein, and what are the monomers in proteins? Distinguish between the primary, secondary, and tertiary structures of a protein. Give examples of the types of forces that maintain each type of structure. Describe how denaturation affects the function of a protein.

11. What are carbohydrates, and what are the monomers in carbohydrates? The monosaccharides in Table 22.8 are all optically active. Explain. What is a *disaccharide?* Which monosaccharide units make up the disaccharide sucrose? What do you call the bond that forms between the monosaccharide units? What forces are responsible for the solubility of starch in water? What is the difference between starch, cellulose, and glycogen?

12. Describe the structural differences between DNA and RNA. The monomers in nucleic acids are called nucleotides. What are the three parts of a nucleotide? The compounds adenine, guanine, cytosine, and thymine are called the nucleic acid bases. What structural features in these compounds make them bases? DNA exhibits a double-helical structure. Explain. Describe how the complementary base pairing between the two individual strands of DNA forms the overall double-helical structure. How is complementary base pairing involved in the replication of the DNA molecule during cell division? Describe how protein synthesis occurs. What is a codon, and what is a gene? The deletion of a single base from a DNA molecule can constitute a fatal mutation, whereas substitution of one base for another is often not as serious a mutation. Explain.

A blue question or exercise number indicates that the answer to that question or exercise appears at the back of this book and a solution appears in the *Solutions Guide*, as found on PowerLecture.

Questions

1. A confused student was doing an isomer problem and listed the following six names as different structural isomers of C_7H_{16}.

a. 1-sec-butylpropane

b. 4-methylhexane

c. 2-ethylpentane

d. 1-ethyl-1-methylbutane

e. 3-methylhexane

f. 4-ethylpentane

How many different structural isomers are actually present in these six names?

2. For the following formulas, what types of isomerism could be exhibited? For each formula, give an example that illustrates the specific type of isomerism. The types of isomerism are structural, geometric, and optical.

a. C_6H_{12} **b.** $C_5H_{12}O$ **c.** $C_6H_4Br_2$

3. What is wrong with the following names? Give the correct name for each compound.

a. 2-ethylpropane

b. 5-iodo-5,6-dimethylhexane

c. *cis*-4-methyl-3-pentene

d. 2-bromo-3-butanol

4. The following organic compounds cannot exist. Why?
   a. 2-chloro-2-butyne
   b. 2-methyl-2-propanone
   c. 1,l-dimethylbenzene
   d. 2-pentanal
   e. 3-hexanoic acid
   f. 5,5-dibromo-1-cyclobutanol

5. If you had a group of hydrocarbons, what structural features would you look at to rank the hydrocarbons in order of increasing boiling point?

6. Which of the functional groups in Table 22.4 can exhibit hydrogen bonding intermolecular forces? Can CH_2CF_2 exhibit hydrogen bonding? Explain.

7. A polypeptide is also called a polyamide. Nylon is also an example of a polyamide. What is a polyamide? Consider a polyhydrocarbon, a polyester, and a polyamide. Assuming average chain lengths are equal, which polymer would you expect to make the strongest fibers and which polymer would you expect to make the weakest fibers? Explain.

8. Give an example reaction that would yield the following products. Name the organic reactant and product in each reaction.
   a. alkane
   b. monohalogenated alkane
   c. dihalogenated alkane
   d. tetrahalogenated alkane
   e. monohalogenated benzene
   f. alkene

9. Give an example reaction that would yield the following products as major organic products. See Exercises 22.62 and 22.65 for some hints. For oxidation reactions, just write oxidation over the arrow and don't worry about the actual reagent.
   a. primary alcohol   e. ketone
   b. secondary alcohol   f. carboxylic acid
   c. tertiary alcohol   g. ester
   d. aldehyde

10. What is polystyrene? The following processes result in a stronger polystyrene polymer. Explain why in each case.
    a. addition of catalyst to form syndiotactic polystyrene
    b. addition of 1,3-butadiene and sulfur
    c. producing long chains of polystyrene
    d. addition of a catalyst to make linear polystyrene

11. Answer the following questions regarding the formation of polymers.
    a. What structural features must be present in a monomer in order to form a homopolymer polyester?
    b. What structural features must be present in the monomers in order to form a copolymer polyamide? (*Hint:* Nylon is an example of a polyamide. When the monomers link together to form nylon, an amide functional group results from each linkage.)
    c. What structural features must be present in a monomer that can form both an addition polymer and a condensation polymer?

12. In Section 22.6, three important classes of biologically important natural polymers are discussed. What are the three classes, what are the monomers used to form the polymers, and why are they biologically important?

Exercises

In this section similar exercises are paired.

Hydrocarbons

13. Draw the five structural isomers of hexane (C_6H_{14}).

14. Name the structural isomers in Exercise 13.

15. Draw all the structural isomers for C_8H_{18} that have the following root name (longest carbon chain). Name the structural isomers.
    a. heptane   b. butane

16. Draw all the structural isomers for C_8H_{18} that have the following root name (longest carbon chain). Name the structural isomers.
    a. hexane   b. pentane

17. Draw a structural formula for each of the following compounds.
    a. 2-methylpropane
    b. 2-methylbutane
    c. 2-methylpentane
    d. 2-methylhexane

18. Draw a structural formula for each of the following compounds.
    a. 2,2-dimethylheptane   c. 3,3-dimethylheptane
    b. 2,3-dimethylheptane   d. 2,4-dimethylheptane

19. Draw the structural formula for each of the following.
    a. 3-isobutylhexane
    b. 2,2,4-trimethylpentane, also called *isooctane*. This substance is the reference (100 level) for octane ratings.
    c. 2-*tert*-butylpentane
    d. The names given in parts a and c are incorrect. Give the correct names for these hydrocarbons.

20. Draw the structure for 4-ethyl-2,3-diisopropylpentane. This name is incorrect. Give the correct systematic name.

21. Name each of the following.

    a.
$$CH_3-\overset{\overset{\displaystyle CH_3}{|}}{\underset{\underset{\displaystyle CH_3}{|}}{C}}-CH_2-\overset{\overset{\displaystyle}{}}{\underset{\underset{\displaystyle CH_3}{|}}{CH}}-CH_2-CH_3$$

    b.
$$\underset{\underset{\displaystyle CH_3}{|}}{CH_2}-CH_2-CH_2-\underset{\underset{\displaystyle CH_3}{|}}{CH}-CH_2-CH_2-\underset{\underset{\displaystyle CH_3}{|}}{CH_2}$$

    c.
$$CH_3-\overset{\overset{\displaystyle CH_3}{|}}{\underset{\underset{\displaystyle CH_3}{|}}{C}}-CH_2-\overset{\overset{\displaystyle CH_3}{|}}{\underset{\underset{\displaystyle CH_3}{|}}{C}}-CH_3$$

d.

$$CH_3-\underset{\underset{\displaystyle CH_2-CH_3}{|}}{\overset{\overset{\displaystyle CH_2-CH_3}{|}}{C}}-CH_2-CH_2-CH_2-CH_2-CH_3$$

22. Name each of the following cyclic alkanes, and indicate the formula of the compound.

a. 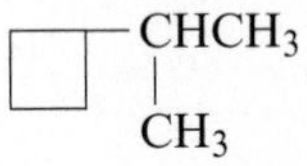 [cyclobutane ring]—CHCH$_3$ with —CH$_3$ below

b. 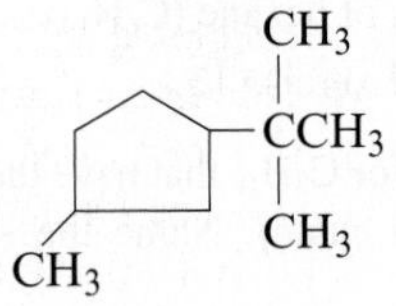 [cyclopentane ring] with CH$_3$, —CCH$_3$, CH$_3$, and CH$_3$ substituents

c. 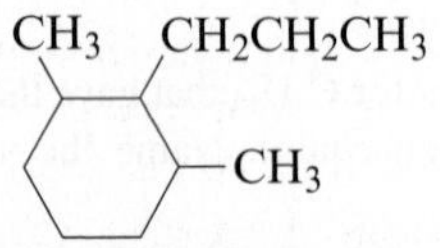 CH$_3$ CH$_2$CH$_2$CH$_3$ [cyclohexane ring] —CH$_3$

23. Give two examples of saturated hydrocarbons. How many other atoms are bonded to each carbon in a saturated hydrocarbon?

24. Draw the structures for two examples of *unsaturated* hydrocarbons. What structural feature makes a hydrocarbon unsaturated?

25. Name each of the following alkenes.

a. CH$_2$=CH—CH$_2$—CH$_3$

b.
$$CH_3-CH=CH-\underset{\underset{\displaystyle CHCH_3}{}}{CH}\overset{\overset{\displaystyle CH_2CH_3}{|}}{}$$

c.
$$CH_3CH_2\underset{\underset{\displaystyle}{}}{CH}\overset{\overset{\displaystyle CH_3}{|}}{}-CH=CH-\underset{\underset{\displaystyle CH_3}{|}}{CHCH_3}$$

26. Name each of the following alkenes or alkynes.

a.
$$CH_3-\overset{\overset{\displaystyle CH_3}{|}}{C}=\overset{\overset{\displaystyle CH_3}{|}}{C}-CH_3$$

b.
$$\overset{\overset{\displaystyle CH_3}{|}}{C}\equiv C-\overset{\overset{\displaystyle CH_3}{|}}{CH}-CH_2-CH_3$$

c.
$$CH_2=C-CH-CH_3$$
$$\;\;\;\;\;|\;\;\;|$$
$$\;\;\;\;CH_3\;CH_2-CH_3$$

27. Give the structure for each of the following.

a. 3-hexene
b. 2,4-heptadiene
c. 2-methyl-3-octene

28. Give the structure for each of the following.

a. 4-methyl-1-pentyne
b. 2,3,3-trimethyl-1-hexene
c. 3-ethyl-4-decene

29. Give the structure of each of the following aromatic hydrocarbons.

a. *o*-ethyltoluene
b. *p*-di-*tert*-butylbenzene
c. *m*-diethylbenzene
d. 1-phenyl-2-butene

30. Cumene is the starting material for the industrial production of acetone and phenol. The structure of cumene is

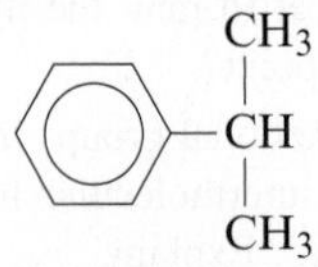

[benzene ring]—CH with CH$_3$ above and CH$_3$ below

Give the systematic name for cumene.

31. Name each of the following.

a.
$$Cl-CH_2-CH_2-\underset{\underset{\displaystyle Cl}{|}}{CH}-CH_3$$

b. CH$_3$CH$_2$CH$_2$CCl$_3$

c.
$$\underset{\underset{\displaystyle CH_3}{}}{\overset{\overset{\displaystyle CH_3}{}}{}}CCl-CH-CH-CH_3$$
$$\;\;\;\;\;\;\;\;\;\;\;\;\;|\;\;\;\;|$$
$$\;\;\;\;\;\;\;\;\;\;\;\;Cl\;\;CH_2-CH_3$$

d. CH$_2$FCH$_2$F

32. Name each of the following compounds.

a.
$$CH_3CHCH=CH_2$$
$$\;\;\;\;|$$
$$\;\;\;Cl$$

b. 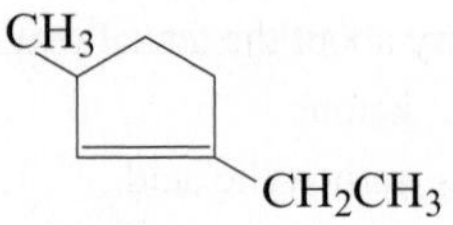 CH$_3$ [cyclopentene ring] —CH$_2$CH$_3$

c. 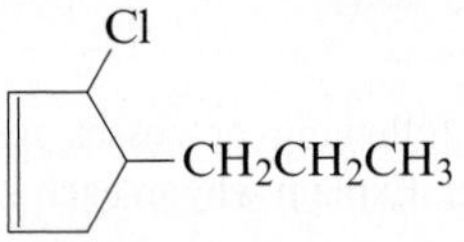 Cl [cyclopentene ring] —CH$_2$CH$_2$CH$_3$

d. 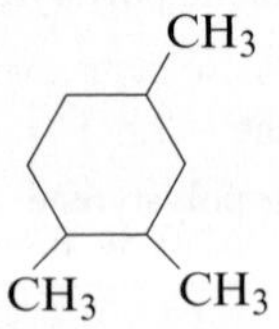 CH$_3$ [cyclohexane ring] with CH$_3$ and CH$_3$ substituents

e. 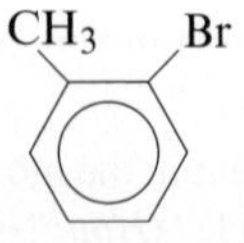 CH$_3$ Br [benzene ring]

f. 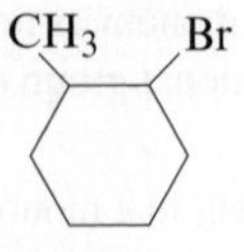 CH$_3$ Br [cyclohexane ring]

g. 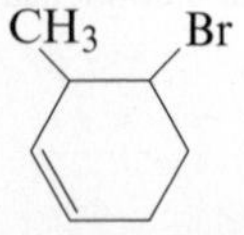 CH$_3$ Br [cyclohexene ring]

Isomerism

33. There is only one compound that is named 1,2-dichloroethane, but there are two distinct compounds that can be named 1,2-dichloroethene. Why?

34. Consider the following four structures.

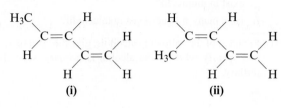

(i)   **(ii)**

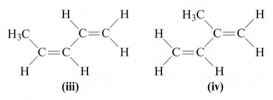

(iii)   **(iv)**

a. Which of these compounds would have the same physical properties (melting point, boiling point, density, and so on)?

b. Which of these compounds are *trans* isomers?

c. Which of these compounds do not exhibit *cis–trans* isomerism?

35. Which of the compounds in Exercises 25 and 27 exhibit *cis–trans* isomerism?

36. Which of the compounds in Exercises 26 and 28 exhibit *cis–trans* isomerism?

37. Draw all the structural isomers of C_5H_{10}. Ignore any cyclic isomers.

38. Which of the structural isomers in Exercise 37 exhibit *cis–trans* isomerism?

39. Draw all the structural and geometrical (*cis–trans*) isomers of C_3H_5Cl.

40. Draw all the structural and geometrical (*cis–trans*) isomers of bromochloropropene.

41. Draw all structural and geometrical (*cis–trans*) isomers of C_4H_7F. Ignore any cyclic isomers.

42. *Cis–trans* isomerism is also possible in molecules with rings. Draw the *cis* and *trans* isomers of 1,2-dimethylcyclohexane. In Exercise 41, you drew all of the noncyclic structural and geometric isomers of C_4H_7F. Now draw the cyclic structural and geometric isomers of C_4H_7F.

43. Draw the following.

a. *cis*-2-hexene

b. *trans*-2-butene

c. *cis*-2,3-dichloro-2-pentene

44. Name the following compounds.

a.
$$CH_3 \quad\quad Br$$
$$C=C$$
$$H \quad\quad H$$

b.
$$CH_3 \quad\quad CH_2CH_3$$
$$C=C$$
$$CH_3CH_2 \quad\quad CH_2CH_2CH_3$$

c.
$$I$$
$$|$$
$$CH_3CHCH_2 \quad\quad H$$
$$C=C$$
$$CH_3CH_2CH_2 \quad\quad I$$

45. If one hydrogen in a hydrocarbon is replaced by a halogen atom, the number of isomers that exist for the substituted compound depends on the number of types of hydrogen in the original hydrocarbon. Thus there is only one form of chloroethane (all hydrogens in ethane are equivalent), but there are two isomers of propane that arise from the substitution of a methyl hydrogen or a methylene hydrogen. How many isomers can be obtained when one hydrogen in each of the compounds named below is replaced by a chlorine atom?

a. *n*-pentane   c. 2,4-dimethylpentane

b. 2-methylbutane   d. methylcyclobutane

46. There are three isomers of dichlorobenzene, one of which has now replaced naphthalene as the main constituent of mothballs.

a. Identify the *ortho,* the *meta,* and the *para* isomers of dichlorobenzene.

b. Predict the number of isomers for trichlorobenzene.

c. It turns out that the presence of one chlorine atom on a benzene ring will cause the next substituent to add *ortho* or *para* to the first chlorine atom on the benzene ring. What does this tell you about the synthesis of *m*-dichlorobenzene?

d. Which of the isomers of trichlorobenzene will be the hardest to prepare?

Functional Groups

47. Identify each of the following compounds as a carboxylic acid, ester, ketone, aldehyde, or amine.

a. Anthraquinone, an important starting material in the manufacture of dyes:

b.

c.

d.

48. Identify the functional groups present in the following compounds.

a.

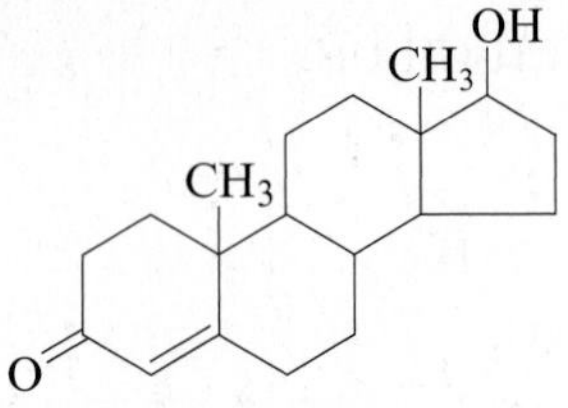

Testosterone

b. CH₃O

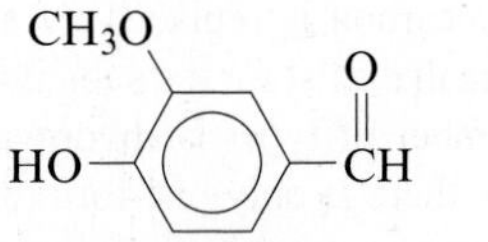

Vanillin

c.

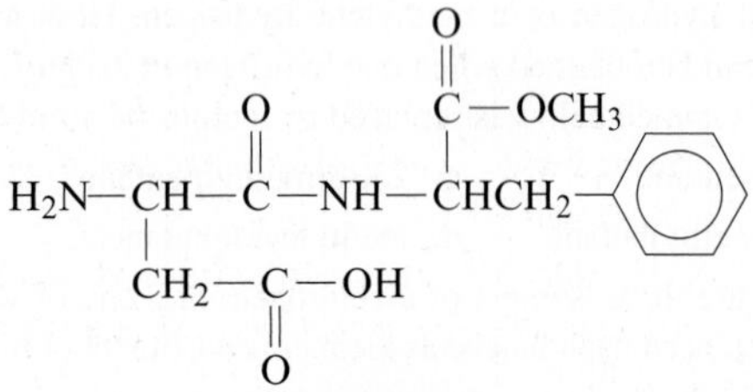

Aspartame

49. Mimosine is a natural product found in large quantities in the seeds and foliage of some legume plants and has been shown to cause inhibition of hair growth and hair loss in mice.

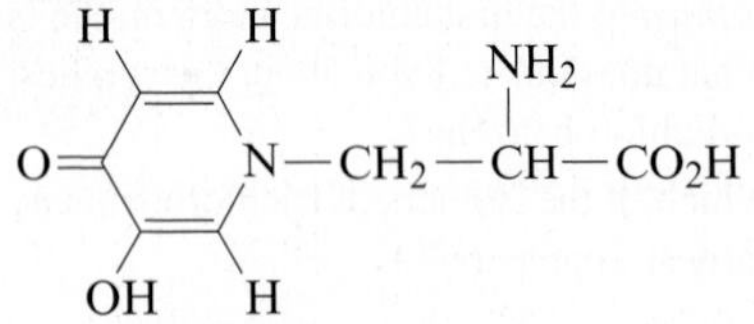

Mimosine, $C_8H_{10}N_2O_4$

a. What functional groups are present in mimosine?

b. Give the hybridization of the eight carbon atoms in mimosine.

c. How many σ and π bonds are found in mimosine?

50. Minoxidil ($C_9H_{15}N_5O$) is a compound produced by Pharmacia Company that has been approved as a treatment of some types of male pattern baldness.

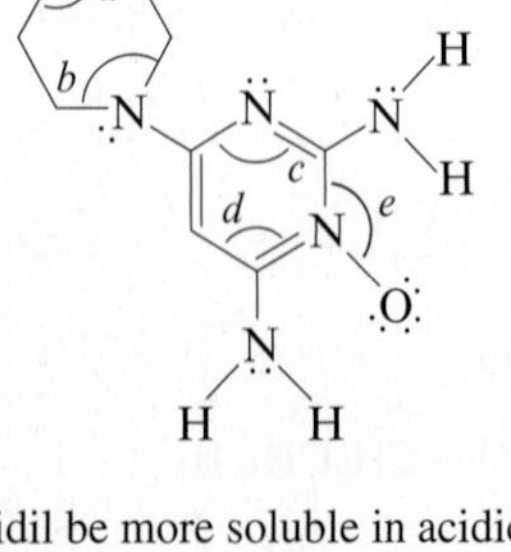

a. Would minoxidil be more soluble in acidic or basic aqueous solution? Explain.

b. Give the hybridization of the five nitrogen atoms in minoxidil.

c. Give the hybridization of each of the nine carbon atoms in minoxidil.

d. Give approximate values of the bond angles marked *a*, *b*, *c*, *d*, and *e*.

e. Including all the hydrogen atoms, how many σ bonds exist in minoxidil?

f. How many π bonds exist in minoxidil?

51. For each of the following alcohols, give the systematic name and specify whether the alcohol is primary, secondary, or tertiary.

a.

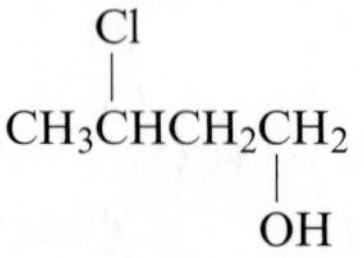

b.

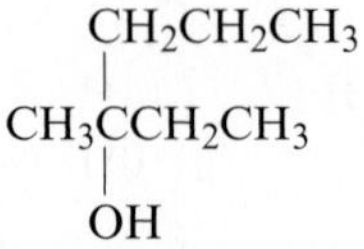

c.

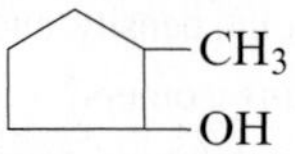

52. Draw structural formulas for each of the following alcohols. Indicate whether the alcohol is primary, secondary, or tertiary.

a. 1-butanol **c.** 2-methyl-1-butanol

b. 2-butanol **d.** 2-methyl-2-butanol

53. Name all the alcohols that have the formula $C_5H_{12}O$. How many ethers have the formula $C_5H_{12}O$?

54. Name all the aldehydes and ketones that have the formula $C_5H_{10}O$.

55. Name the following compounds.

a.

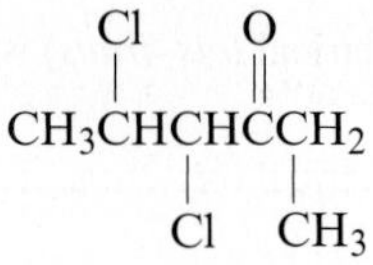

b.

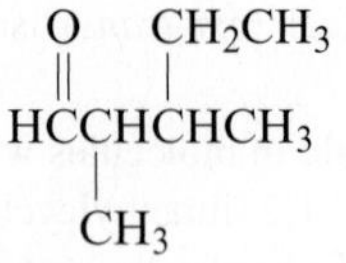

c.

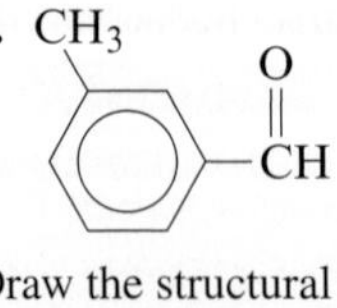

56. Draw the structural formula for each of the following.

a. formaldehyde (methanal)

b. 4-heptanone

c. 3-chlorobutanal

d. 5,5-dimethyl-2-hexanone

57. Name the following compounds.

a.

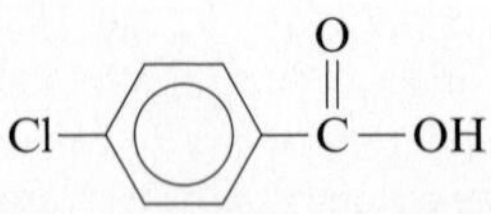

b.

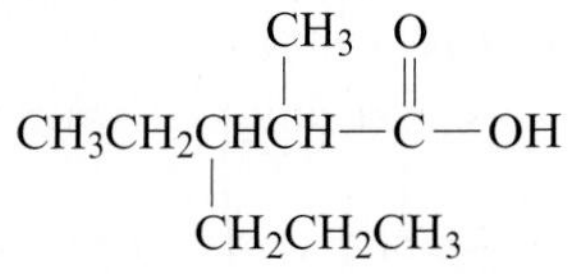

$$CH_3CH_2\overset{\displaystyle CH_3}{\underset{\displaystyle CH_2CH_2CH_3}{CHCH}}-\overset{\displaystyle O}{\overset{\displaystyle \|}{C}}-OH$$

c. HCOOH

58. Draw a structural formula for each of the following.

a. 3-methylpentanoic acid

b. ethyl methanoate

c. methyl benzoate

d. 3-chloro-2,4-dimethylhexanoic acid

59. Which of the following statements is(are) *false*? Explain why the statement(s) is(are) *false*.

a. $CH_3CH_2CH_2\overset{\displaystyle O}{\overset{\displaystyle \|}{C}}OCH_3$ is a structural isomer of pentanoic acid.

b. $HCCH_2CH_2CHCH_3$ ($\overset{\displaystyle O}{\overset{\displaystyle \|}{}}$ ... $\overset{\displaystyle CH_3}{\overset{\displaystyle |}{}}$) is a structural isomer of 2-methyl-3-pentanone.

c. $CH_3CH_2OCH_2CH_2CH_3$ is a structural isomer of 2-pentanol.

d. $CH_2\!=\!CHCHCH_3$ ($\overset{\displaystyle |}{OH}$) is a structural isomer of 2-butenal.

e. Trimethylamine is a structural isomer of $CH_3CH_2CH_2NH_2$.

60. Draw the isomer(s) specified. There may be more than one possible isomer for each part.

a. a cyclic compound that is an isomer of *trans*-2-butene

b. an ester that is an isomer of propanoic acid

c. a ketone that is an isomer of butanal

d. a secondary amine that is an isomer of butylamine

e. a tertiary amine that is an isomer of butylamine

f. an ether that is an isomer of 2-methyl-2-propanol

g. a secondary alcohol that is an isomer of 2-methyl-2-propanol

Reactions of Organic Compounds

61. Complete the following reactions.

a. $CH_3CH\!=\!CHCH_3 + H_2 \xrightarrow{Pt}$

b. $CH_2\!=\!CHCHCH\!=\!CH + 2Cl_2 \longrightarrow$
  $\quad\quad\overset{\displaystyle |}{CH_3}\quad\overset{\displaystyle |}{CH_3}$

c. $+\ Cl_2 \xrightarrow{FeCl_3}$

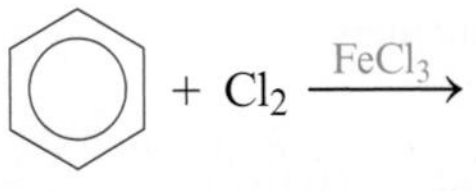

d. $CH_3C\!=\!CH_2 + O_2 \xrightarrow{Spark}$
  $\quad\quad\overset{\displaystyle |}{CH_3}$

62. Reagents such as HCl, HBr, and HOH (H_2O) can add across carbon–carbon double and triple bonds, with H forming a bond to one of the carbon atoms in the multiple bond and Cl, Br, or OH forming a bond to the other carbon atom in the multiple bond. In some cases, two products are possible. For the major organic product, the addition occurs so that the hydrogen atom in the reagent attaches to the carbon atom in the multiple bond that already has the greater number of hydrogen atoms bonded to it. With this rule in mind, draw the structure of the major product in each of the following reactions.

a. $CH_3CH_2CH\!=\!CH_2 + H_2O \xrightarrow{H^+}$

b. $CH_3CH_2CH\!=\!CH_2 + HBr \longrightarrow$

c. $CH_3CH_2C\!\equiv\!CH + 2HBr \longrightarrow$

d.

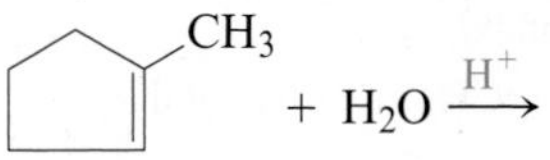

(cyclopentene with CH_3) $+ H_2O \xrightarrow{H^+}$

e.

$$\underset{CH_3}{\overset{CH_3CH_2}{}}C\!=\!C\underset{H}{\overset{CH_3}{}} + HCl \longrightarrow$$

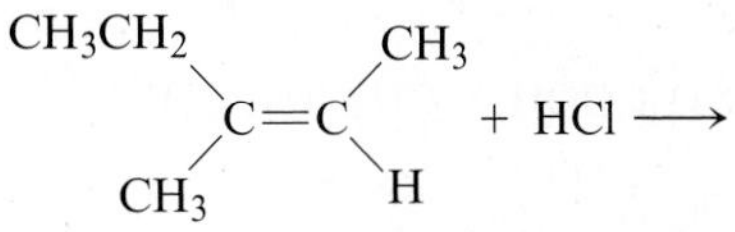

63. When toluene ($C_6H_5CH_3$) reacts with chlorine gas in the presence of iron(III) catalyst, the product is a mixture of the *ortho* and *para* isomers of $C_6H_4ClCH_3$. However, when the reaction is light-catalyzed with no Fe^{3+} catalyst present, the product is $C_6H_5CH_2Cl$. Explain.

64. Why is it preferable to produce chloroethane by the reaction of $HCl(g)$ with ethene than by the reaction of $Cl_2(g)$ with ethane? (See Exercise 62.)

65. Using appropriate reactants, alcohols can be oxidized into aldehydes, ketones, and/or carboxylic acids. Primary alcohols can be oxidized into aldehydes, which can then be oxidized into carboxylic acids. Secondary alcohols can be oxidized into ketones, while tertiary alcohols do not undergo this type of oxidation. Give the structure of the product(s) resulting from the oxidation of each of the following alcohols.

a. 3-methyl-1-butanol

b. 3-methyl-2-butanol

c. 2-methyl-2-butanol

d.

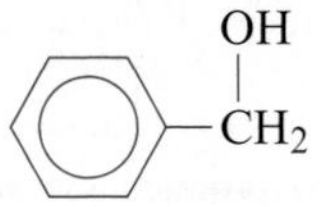

(benzene ring)$-\overset{\displaystyle OH}{\overset{\displaystyle |}{CH_2}}$

e.

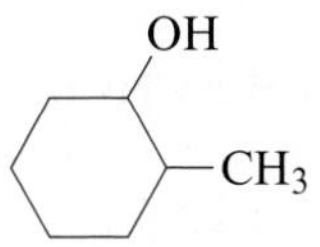

(cyclohexane ring with OH)$-CH_3$

f.

HO, OH, CH_3, CH_2, OH on cyclohexane ring

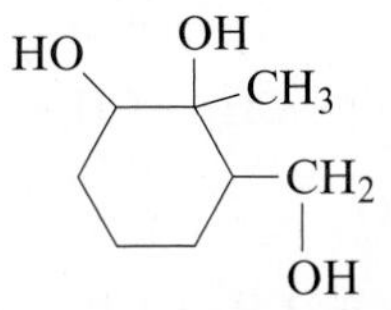

66. Oxidation of an aldehyde yields a carboxylic acid:

$$R-\overset{\displaystyle O}{\overset{\displaystyle \|}{CH}} \xrightarrow{[ox]} R-\overset{\displaystyle O}{\overset{\displaystyle \|}{C}}-OH$$

Draw the structures for the products of the following oxidation reactions.

a. propanal $\xrightarrow{[ox]}$

b. 2,3-dimethylpentanal $\xrightarrow{[ox]}$

c. 3-ethylbenzaldehyde $\xrightarrow{[ox]}$

67. How would you synthesize each of the following?

  a. 1,2-dibromopropane from propene

  b. acetone (2-propanone) from an alcohol

  c. *tert*-butyl alcohol (2-methyl-2-propanol) from an alkene (See Exercise 62.)

  d. propanoic acid from an alcohol

68. What tests could you perform to distinguish between the following pairs of compounds?

  a. $CH_3CH_2CH_2CH_3$, $CH_2{=}CHCH_2CH_3$

  b.

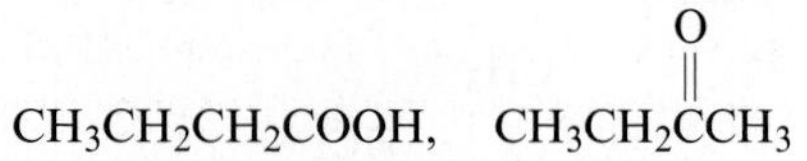

$$CH_3CH_2CH_2COOH, \quad CH_3CH_2\overset{\overset{\displaystyle O}{\|}}{C}CH_3$$

  c.

$$CH_3CH_2CH_2OH, \quad CH_3\overset{\overset{\displaystyle O}{\|}}{C}CH_3$$

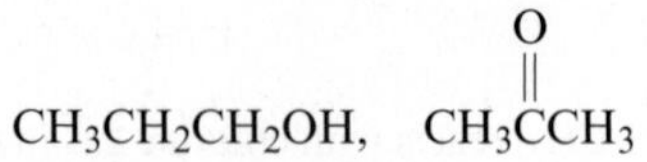

  d. $CH_3CH_2NH_2$, CH_3OCH_3

69. How would you synthesize the following esters?

  a. *n*-octylacetate

  b.

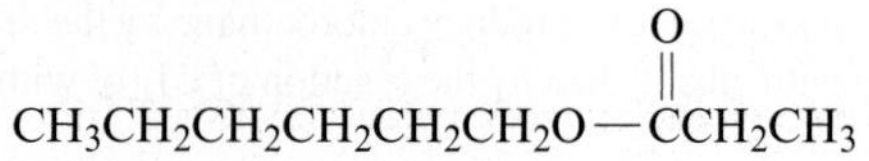

$$CH_3CH_2CH_2CH_2CH_2CH_2O{-}\overset{\overset{\displaystyle O}{\|}}{C}CH_2CH_3$$

70. Complete the following reactions.

  a. $CH_3CO_2H + CH_3OH \rightarrow$

  b. $CH_3CH_2CH_2OH + HCOOH \rightarrow$

Polymers

71. Kel-F is a polymer with the structure

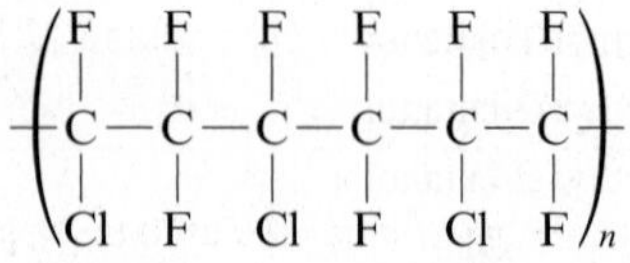

What is the monomer for Kel-F?

72. What monomer(s) must be used to produce the following polymers?

  a.

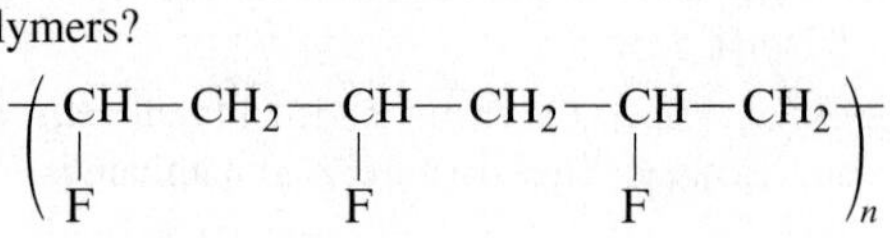

  b.

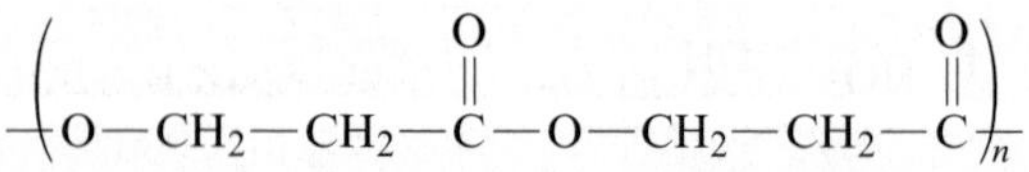

  c.

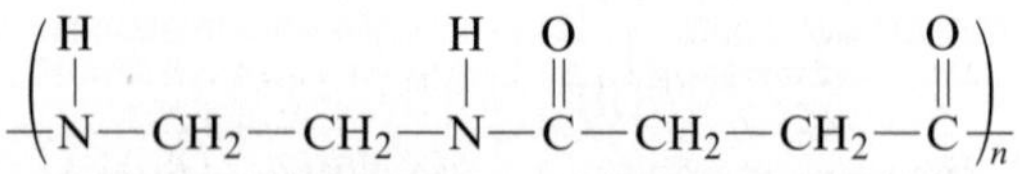

  d.

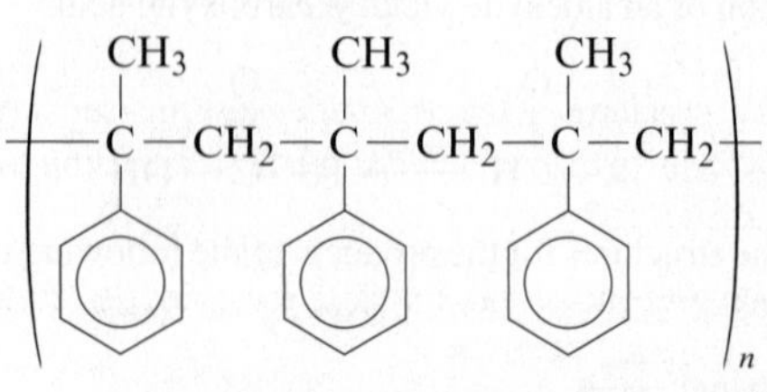

  e.

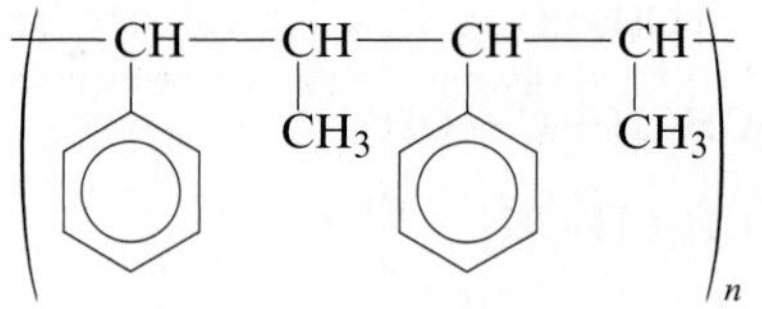

  f.

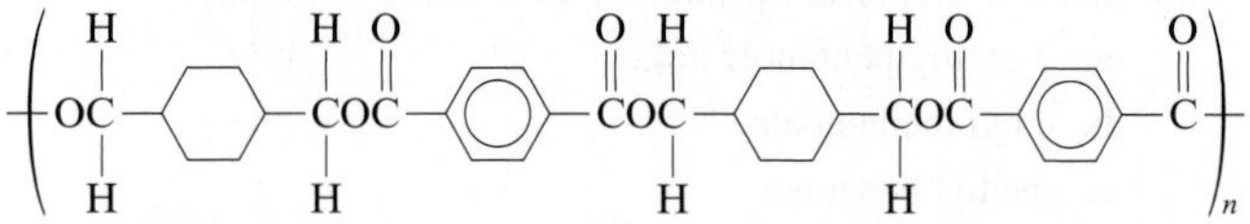

(This polymer is Kodel, used to make fibers of stain-resistant carpeting.)

Classify these polymers as condensation or addition polymers. Which are copolymers?

73. "Super glue" contains methyl cyanoacrylate,

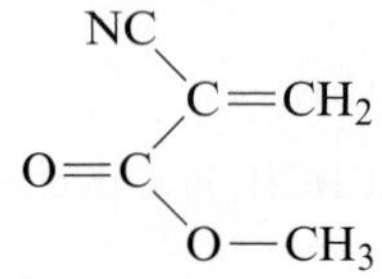

which readily polymerizes upon exposure to traces of water or alcohols on the surfaces to be bonded together. The polymer provides a strong bond between the two surfaces. Draw the structure of the polymer formed by methyl cyanoacrylate.

74. Isoprene is the repeating unit in natural rubber. The structure of isoprene is

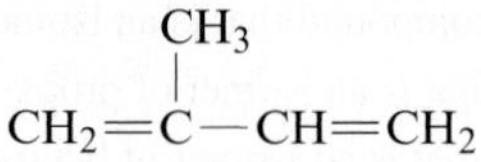

  a. Give a systematic name for isoprene.

  b. When isoprene is polymerized, two polymers of the form

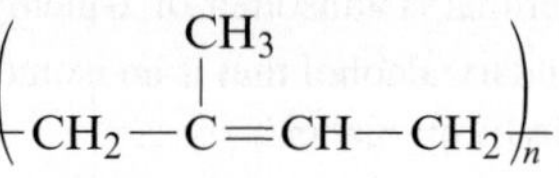

are possible. In natural rubber, the *cis* configuration is found. The polymer with the *trans* configuration about the double bond is called gutta percha and was once used in the manufacture of golf balls. Draw the structure of natural rubber and gutta percha showing three repeating units and the configuration about the carbon–carbon double bonds.

75. Kevlar, used in bulletproof vests, is made by the condensation copolymerization of the monomers

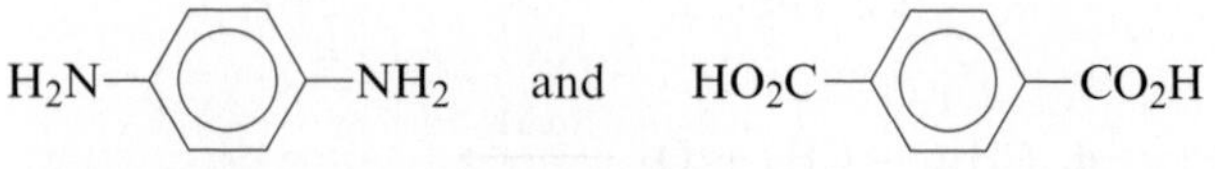

Draw the structure of a portion of the Kevlar chain.

76. The polyester formed from lactic acid,

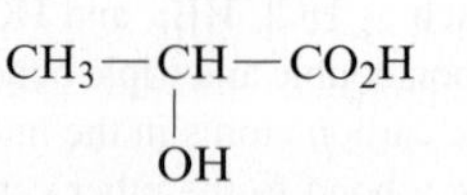

is used for tissue implants and surgical sutures that will dissolve in the body. Draw the structure of a portion of this polymer.

77. Polyimides are polymers that are tough and stable at temperatures of up to 400°C. They are used as a protective coating on the quartz fibers used in fiber optics. What monomers were used to make the following polyimide?

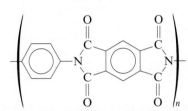

78. The Amoco Chemical Company has successfully raced a car with a plastic engine. Many of the engine parts, including piston skirts, connecting rods, and valve-train components, were made of a polymer called *Torlon:*

What monomers are used to make this polymer?

79. Polystyrene can be made more rigid by copolymerizing styrene with divinylbenzene:

$$CH{=}CH_2$$

$$CH{=}CH_2$$

How does the divinylbenzene make the copolymer more rigid?

80. Polyesters containing double bonds are often crosslinked by reacting the polymer with styrene.

a. Draw the structure of the copolymer of

$$HO{-}CH_2CH_2{-}OH \quad \text{and} \quad HO_2C{-}CH{=}CH{-}CO_2H$$

b. Draw the structure of the crosslinked polymer (after the polyester has been reacted with styrene).

81. Which of the following polymers would be stronger or more rigid? Explain your choices.

a. The copolymer of ethylene glycol and terephthalic acid or the copolymer of 1,2-diaminoethane and terephthalic acid (1,2-diaminoethane = $NH_2CH_2CH_2NH_2$)

b. The polymer of $HO{-}(CH_2)_6{-}CO_2H$ or that of

$$HO{-}\bigcirc{-}CO_2H$$

c. Polyacetylene or polyethylene (The monomer in polyacetylene is ethyne.)

82. Poly(lauryl methacrylate) is used as an additive in motor oils to counter the loss of viscosity at high temperature. The structure is

$$\left(\begin{array}{c}CH_3\\ {-}C{-}CH_2{-}\\ \\ C\\ O \quad O{-}(CH_2)_{11}{-}CH_3\end{array}\right)_n$$

The long hydrocarbon chain of poly(lauryl methacrylate) makes the polymer soluble in oil (a mixture of hydrocarbons with mostly 12 or more carbon atoms). At low temperatures the polymer is coiled into balls. At higher temperatures the balls uncoil and the polymer exists as long chains. Explain how this helps control the viscosity of oil.

Natural Polymers

83. Which of the amino acids in Fig. 22.18 contain the following functional groups in their R group?

a. alcohol   **c.** amine

b. carboxylic acid   **d.** amide

84. When pure crystalline amino acids are heated, decomposition generally occurs before the solid melts. Account for this observation. (*Hint:* Crystalline amino acids exist as $H_3\overset{+}{N}CRHCOO^-$, called *zwitterions*.)

85. Aspartame, the artificial sweetener marketed under the name NutraSweet, is a methyl ester of a dipeptide:

$$H_2N{-}CH{-}\overset{\overset{\textstyle O}{\|}}{C}{-}NH{-}CH{-}CH_2{-}\bigcirc$$
with CH_2CO_2H and CO_2CH_3

a. What two amino acids are used to prepare aspartame?

b. There is concern that methanol may be produced by the decomposition of aspartame. From what portion of the molecule can methanol be produced? Write an equation for this reaction.

86. Glutathione, a tripeptide found in virtually all cells, functions as a reducing agent. The structure of glutathione is

$$^-OOCCHCH_2CH_2\overset{\overset{\textstyle O}{\|}}{C}NHCH\overset{\overset{\textstyle O}{\|}}{C}NHCH_2COO^-$$
with $\overset{+}{N}H_3$ and CH_2SH

What amino acids make up glutathione?

87. Draw the structures of the two dipeptides that can be formed from serine and alanine.

88. Draw the structures of the tripeptides gly–ala–ser and ser–ala–gly. How many other tripeptides are possible using these three amino acids?

89. Write the sequence of all possible tetrapeptides composed of the following amino acids.

a. two phenylalanines and two glycines

b. two phenylalanines, glycine, and alanine

90. How many different pentapeptides can be formed using five different amino acids?

91. Give an example of amino acids that could give rise to the interactions pictured in Fig. 22.24 that maintain the tertiary structures of proteins.

92. What types of interactions can occur between the side chains of the following amino acids that would help maintain the tertiary structure of a protein?

    a. cysteine and cysteine
    b. glutamine and serine
    c. glutamic acid and lysine
    d. proline and leucine

93. Oxygen is carried from the lungs to tissues by the protein hemoglobin in red blood cells. Sickle cell anemia is a disease resulting from abnormal hemoglobin molecules in which a valine is substituted for a single glutamic acid in normal hemoglobin. How might this substitution affect the structure of hemoglobin?

94. Over 100 different kinds of mutant hemoglobin molecules have been detected in humans. Unlike sickle cell anemia (see Exercise 93), not all of these mutations are as serious. In one nonlethal mutation, glutamine substitutes for a single glutamic acid in normal hemoglobin. Rationalize why this substitution is nonlethal.

95. Draw cyclic structures for D-ribose and D-mannose.

96. Indicate the chiral carbon atoms found in the monosaccharides D-ribose and D-mannose.

97. In addition to using *numerical* prefixes in the general names of sugars to indicate how many carbon atoms are present, we often use the prefixes *keto-* and *aldo-* to indicate whether the sugar is a ketone or an aldehyde. For example, the monosaccharide fructose is frequently called a *ketohexose* to emphasize that it contains six carbons as well as the ketone functional group. For each of the monosaccharides shown in Table 22.8 classify the sugars as aldohexoses, aldopentoses, ketohexoses, or ketopentoses.

98. Glucose can occur in three forms: two cyclic forms and one open-chain structure. In aqueous solution, only a tiny fraction of the glucose is in the open-chain form. Yet tests for the presence of glucose depend on reaction with the aldehyde group, which is found only in the open-chain form. Explain why these tests work.

99. What are the structural differences between α- and β-glucose? These two cyclic forms of glucose are the building blocks to form two different polymers. Explain.

100. Cows can digest cellulose, but humans can't. Why not?

101. Which of the amino acids in Fig. 22.18 contain more than one chiral carbon atom? Draw the structures of these amino acids and indicate all chiral carbon atoms.

102. Why is glycine not optically active?

103. Which of the noncyclic isomers of bromochloropropene are optically active?

104. How many chiral carbon atoms does the following structure have?

105. Part of a certain DNA sequence is G–G–T–C–T–A–T–A–C. What is the complementary sequence?

106. The codons (words) in DNA (that specify which amino acid should be at a particular point in a protein) are three bases long. How many such three-letter words can be made from the four bases adenine, cytosine, guanine, and thymine?

107. Which base will hydrogen-bond with uracil within an RNA molecule? Draw the structure of this base pair.

108. Tautomers are molecules that differ in the position of a hydrogen atom. A tautomeric form of thymine has the structure

    If this tautomeric form, rather than the stable form of thymine, were present in a strand of DNA during replication, what would be the result?

109. The base sequences in mRNA that code for certain amino acids are

    | Glu: | GAA, GAG |
    | Val: | GUU, GUC, GUA, GUG |
    | Met: | AUG |
    | Trp: | UGG |
    | Phe: | UUU, UUC |
    | Asp: | GAU, GAC |

    These sequences are complementary to the sequences in DNA.

    a. Give the corresponding sequences in DNA for the amino acids listed above.
    b. Give a DNA sequence that would code for the peptide trp–glu–phe–met.
    c. How many different DNA sequences can code for the tetrapeptide in part b?
    d. What is the peptide that is produced from the DNA sequence T–A–C–C–T–G–A–A–G?
    e. What other DNA sequences would yield the same tripeptide as in part d?

110. The change of a single base in the DNA sequence for normal hemoglobin can encode for the abnormal hemoglobin giving rise to sickle cell anemia. Which base in the codon for glu in DNA is replaced to give the codon(s) for val? (See Exercises 93 and 109.)

Additional Exercises

111. Draw the following incorrectly named compounds and name them correctly.

a. 2-ethyl-3-methyl-5-isopropylhexane

b. 2-ethyl-4-*tert*-butylpentane

c. 3-methyl-4-isopropylpentane

d. 2-ethyl-3-butyne

112. In the presence of light, chlorine can substitute for one (or more) of the hydrogens in an alkane. For the following reactions, draw the possible monochlorination products.

a. 2,2-dimethylpropane + Cl_2 $\xrightarrow{hv}$

b. 1,3-dimethylcyclobutane + Cl_2 $\xrightarrow{hv}$

c. 2,3-dimethylbutane + Cl_2 $\xrightarrow{hv}$

113. Polychlorinated dibenzo-*p*-dioxins (PCDDs) are highly toxic substances that are present in trace amounts as by-products of some chemical manufacturing processes. They have been implicated in a number of environmental incidents—for example, the chemical contamination at Love Canal and the herbicide spraying in Vietnam. The structure of dibenzo-*p*-dioxin, along with the customary numbering convention, is

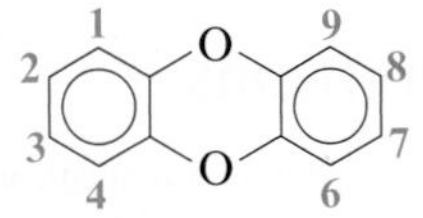

The most toxic PCDD is 2,3,7,8-tetrachloro-dibenzo-*p*-dioxin. Draw the structure of this compound. Also draw the structures of two other isomers containing four chlorine atoms.

114. Consider the following five compounds.

a. $CH_3CH_2CH_2CH_2CH_3$

b.

$CH_3CH_2CH_2CH_2$ with OH

c. $CH_3CH_2CH_2CH_2CH_2CH_3$

d.

$CH_3CH_2CH_2CH$ with O (double bond)

e.

CH_3CCH_3 with CH_3 groups

The boiling points of these five compounds are 9.5°C, 36°C, 69°C, 76°C, and 117°C. Which compound boils at 36°C? Explain.

115. The two isomers having the formula C_2H_6O boil at −23°C and 78.5°C. Draw the structure of the isomer that boils at −23°C and of the isomer that boils at 78.5°C.

116. Ignoring ring compounds, which isomer of $C_2H_4O_2$ should boil at the lowest temperature?

117. Explain why methyl alcohol is soluble in water in all proportions, while stearyl alcohol [$CH_3(CH_2)_{16}OH$] is a waxy solid that is not soluble in water.

118. Is octanoic acid more soluble in 1 *M* HCl, 1 *M* NaOH, or pure water? Explain. Drugs such as morphine ($C_{17}H_{19}NO_3$) are often

treated with strong acids. The most commonly used form of morphine is morphine hydrochloride ($C_{17}H_{20}ClNO_3$). Why is morphine treated in this way? (*Hint:* Morphine is an amine.)

119. Consider the compounds butanoic acid, pentanal, *n*-hexane, and 1-pentanol. The boiling points of these compounds (in no specific order) are 69°C, 103°C, 137°C, and 164°C. Match the boiling points to the correct compound.

120. A compound containing only carbon and hydrogen is 85.63% C by mass. Reaction of this compound with H_2O produces a secondary alcohol as the major product and a primary alcohol as the minor product. (See Exercise 62.) If the molar mass of the hydrocarbon is between 50 and 60 g/mol, name the compound.

121. Three different organic compounds have the formula C_3H_8O. Only two of these isomers react with $KMnO_4$ (a strong oxidizing agent). What are the names of the products when these isomers react with excess $KMnO_4$?

122. Consider the following polymer:

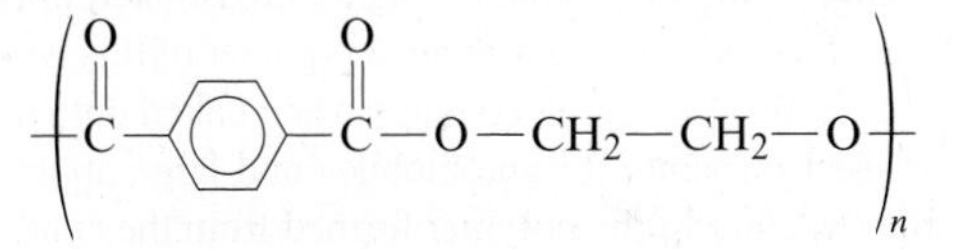

Is this polymer a homopolymer or a copolymer, and is it formed by addition polymerization or condensation polymerization? What is (are) the monomer(s) for this polymer?

123. Nylon is named according to the number of C atoms between the N atoms in the chain. Nylon-46 has 4 C atoms, then 6 C atoms, and this pattern repeats. Nylon-6 always has 6 carbon atoms in a row. Speculate as to why nylon-46 is stronger than nylon-6. (*Hint:* Consider the strengths of interchain forces.)

124. The polymer nitrile is a copolymer made from acrylonitrile and butadiene; it is used to make automotive hoses and gaskets. Draw the structure of nitrile. (*Hint:* See Table 22.7.)

125. *Polyaramid* is a term applied to polyamides containing aromatic groups. These polymers were originally made for use as tire cords but have since found many other uses.

a. Kevlar is used in bulletproof vests and many high-strength composites. The structure of Kevlar is

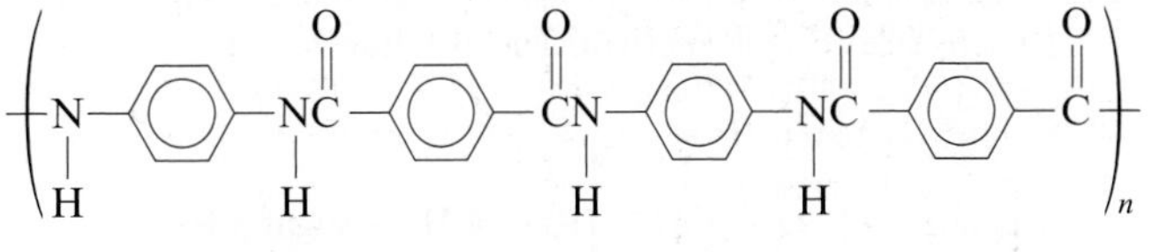

Which monomers are used to make Kevlar?

b. Nomex is a polyaramid used in fire-resistant clothing. It is a copolymer of

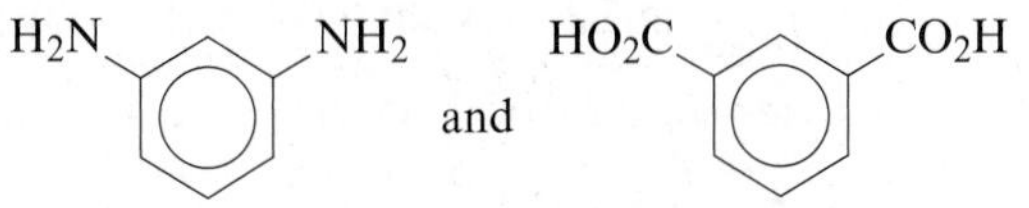

Draw the structure of the Nomex polymer. How do Kevlar and Nomex differ in their structures?

126. When acrylic polymers are burned, toxic fumes are produced. For example, in many airplane fires, more passenger deaths have been caused by breathing toxic fumes than by the fire itself. Using polyacrylonitrile as an example, what would you expect to be one of the most toxic, gaseous combustion products created in the reaction?

127. Ethylene oxide,

$$CH_2-CH_2$$
$$\diagdown O \diagup$$

is an important industrial chemical. Although most ethers are unreactive, ethylene oxide is quite reactive. It resembles C_2H_4 in its reactions in that addition reactions occur across the C—O bond in ethylene oxide.

   a. Why is ethylene oxide so reactive? (*Hint:* Consider the bond angles in ethylene oxide as compared with those predicted by the VSEPR model.)

   b. Ethylene oxide undergoes addition polymerization, forming a polymer used in many applications requiring a nonionic surfactant. Draw the structure of this polymer.

128. Another way of producing highly crosslinked polyesters is to use glycerol. Alkyd resins are a polymer of this type. The polymer forms very tough coatings when baked onto a surface and is used in paints for automobiles and large appliances. Draw the structure of the polymer formed from the condensation of

$$CH_2-CH-CH_2 \ \text{and}$$
$$\ \ | \ \ \ \ \ | \ \ \ \ \ |$$
$$OH \ \ OH \ \ OH$$
$$\text{Glycerol}$$

(phthalic acid with two CO_2H groups)
Phthalic acid

Explain how crosslinking occurs in this polymer.

129. Monosodium glutamate (MSG) is commonly used as a flavoring in foods. Draw the structure of MSG.

130. a. Use bond energies (see Table 8.4) to estimate ΔH for the reaction of two molecules of glycine to form a peptide linkage.

   b. Would you predict ΔS to favor the formation of peptide linkages between two molecules of glycine?

   c. Would you predict the formation of proteins to be a spontaneous process?

131. The reaction to form a phosphate–ester linkage between two nucleotides can be approximated as follows:

$$\overset{O}{\underset{O}{\overset{\|}{Sugar-O-P}}-OH + HO-CH_2-sugar \longrightarrow}$$

$$\overset{O}{\underset{O}{\overset{\|}{Sugar-O-P}}-O-CH_2-sugar \ + \ H_2O}$$

Would you predict the formation of a dinucleotide from two nucleotides to be a spontaneous process?

132. Considering your answers to Exercises 130 and 131, how can you justify the existence of proteins and nucleic acids in light of the second law of thermodynamics?

133. All amino acids have at least two functional groups with acidic or basic properties. In alanine, the carboxylic acid group has $K_a = 4.5 \times 10^{-3}$ and the amino group has $K_b = 7.4 \times 10^{-5}$. Three ions of alanine are possible when alanine is dissolved in water. Which of these ions would predominate in a solution with $[H^+] = 1.0 \ M$? In a solution with $[OH^-] = 1.0 \ M$?

134. The average molar mass of one base pair of nucleotides in DNA is approximately 600 g/mol. The spacing between successive base pairs is about 0.34 nm, and a complete turn in the helical structure of DNA occurs about every 3.4 nm. If a DNA molecule has a molar mass of 4.5×10^9 g/mol, approximately how many complete turns exist in the DNA α-helix structure?

135. When heat is added to proteins, the hydrogen bonding in the secondary structure is disrupted. What are the algebraic signs of ΔH and ΔS for the denaturation process?

136. In glycine, the carboxylic acid group has $K_a = 4.3 \times 10^{-3}$ and the amino group has $K_b = 6.0 \times 10^{-5}$. Use these equilibrium constant values to calculate the equilibrium constants for the following.

   a. $^+H_3NCH_2CO_2^- + H_2O \rightleftharpoons H_2NCH_2CO_2^- + H_3O^+$

   b. $H_2NCH_2CO_2^- + H_2O \rightleftharpoons H_2NCH_2CO_2H + OH^-$

   c. $^+H_3NCH_2CO_2H \rightleftharpoons 2H^+ + H_2NCH_2CO_2^-$

ChemWork Problems

These multiconcept problems (and additional ones) are found interactively online with the same type of assistance a student would get from an instructor.

137. Name each of the following alkanes.

   a. $CH_3-CH_2-CH_2-CH_2-CH_3$

   b.
$$\overset{CH_3}{\underset{|}{CH_3-CH-CH-CH_2-CH}} \ \overset{CH_3}{\underset{|}{}}$$
$$\underset{|}{\overset{|}{CH_2}} \qquad\qquad \underset{}{CH_3}$$
$$CH_3$$

   c.
$$\overset{CH_3}{\underset{|}{CH-CH_3}}$$
$$CH_3-CH_2-CH_2-CH-CH-CH_2-CH_2-CH_3$$
$$\underset{CH_3-CH_2}{\overset{|}{}}$$

138. Name each of the following alkanes.

   a.
$$\overset{Br}{\underset{|}{CH_3-C-CH-CH_2-CH_3}}$$
$$\underset{Br \ \ CH_3}{\overset{| \ \ \ |}{}}$$

   b.
$$\overset{CH_3}{\underset{|}{}} \qquad \overset{CH_2-CH_3}{\underset{|}{}}$$
$$CH_3-CH-CH-CH$$
$$\underset{I \qquad\quad CH_2-CH_3}{\overset{| \qquad\quad |}{}}$$

c. $CH_3-CH-CH_2-CH-CH_2-CH_2-CH_2-CH_2-CH_3$
     | |
    (F above)
     $\overset{\displaystyle F}{|}$

c. $CH_3-\underset{\underset{\displaystyle CH_3}{|}}{CH}-CH_2-\overset{\overset{\displaystyle F}{|}}{CH}-CH_2-CH_2-CH_2-CH_2-CH_3$

139. Name each of the following cyclic alkanes.

a. (cyclopropane with I substituent)

b. (cyclohexane with CH$_3$, CH$_3$, Cl, CH$_3$ substituents)

c. Cl (cyclopentane with CH$_2$CH$_3$)

d. (cyclopentene with CH$_3$ and CH$_2$CH$_3$)

140. Name each of the following alkenes and alkynes.

a. $CH_3CH_2-\overset{\overset{\displaystyle CH_3}{|}}{C}=CH_2$

b. $CH_2=\overset{\overset{\displaystyle CH_3}{|}}{C}-CH_2-\overset{\overset{\displaystyle CH_3}{|}}{C}=CH_2$

c. $CH_3CH_2CH-CH=CH-CH_2\overset{\overset{\displaystyle CH_3}{|}}{CH}$
     | |
   CH_2CH_3 CH_3

d. $CH_3CH_2CH_2CH_2\underset{\underset{\displaystyle Br}{|}}{CH}-C\equiv CH$

e. $CH_3-\underset{\underset{\underset{\underset{\displaystyle Cl}{|}}{\displaystyle CH_2CH_2}}{|}}{\overset{\overset{\displaystyle CH_3}{|}}{C}}-C\equiv C-\overset{\overset{\displaystyle CH_3}{|}}{CH}-CH_3$

f. $HC\equiv C-\underset{\underset{\displaystyle CH_3}{|}}{CH}-\underset{\underset{\displaystyle CH_2CH_2CH_2CH_3}{|}}{\overset{\overset{\displaystyle CH_2CH_3}{|}}{CH}}$

141. a. Name each of the following alcohols.

$HO-CH_2CH_2CH_2CH_2CH_3$

$HO-\underset{\underset{\displaystyle CH_3}{|}}{\overset{\overset{\displaystyle CH_3}{|}}{C}}-CH_2-\underset{\underset{\displaystyle CH_3}{|}}{\overset{\overset{\displaystyle OH}{|}}{CH}}$

b. Name each of the following alcohols, including the stereochemistry if *cis–trans* isomers are possible.

(cyclohexane with two OH groups)

$CH_2=CH-\underset{\underset{\displaystyle OH}{|}}{CH}-\underset{\underset{\displaystyle CH_3}{|}}{\overset{\overset{\displaystyle CH_3}{|}}{CH}}$

142. Name each of the following organic compounds.

a. $CH_3CH_2CH_2CH_2CH_2CH_2\overset{\overset{\displaystyle O}{||}}{C}H$

b. $CH_3\underset{\underset{\displaystyle CH_3}{|}}{\overset{\overset{\displaystyle CH_3CH_2}{|}}{CH}}CH_2CHCH_2\overset{\overset{\displaystyle O}{||}}{C}H$

c. $CH_3CH_2\overset{\overset{\displaystyle O}{||}}{C}CH_2CH_2CH_3$

d. $CH_3CH_2\overset{\overset{\displaystyle O}{||}}{C}\underset{\underset{\displaystyle CH_3}{|}}{CH}-\underset{\underset{\displaystyle CH_3}{|}}{\overset{\overset{\displaystyle CH_3}{|}}{CH}}$

e. $CH_3CH_2\underset{\underset{\displaystyle CH_3}{|}}{\overset{\overset{\displaystyle CH_3}{|}}{CH}}CHCH_2\overset{\overset{\displaystyle O}{||}}{C}OH$

143. Esterification reactions are carried out in the presence of a strong acid such as H_2SO_4. A carboxylic acid is warmed with an alcohol, and an ester and water are formed. You may have made a fruity-smelling ester in the lab when studying organic functional groups. Name the carboxylic acid that is necessary to complete the following esterification reaction.

$?$ + $HO-\underset{\underset{\displaystyle CH_3}{|}}{\overset{\overset{\displaystyle CH_3}{|}}{C}}-CH_3$ $\xrightarrow{\text{conc. acid}}$ $CH_3-\underset{\underset{\displaystyle Cl}{|}}{CH}-\overset{\overset{\displaystyle O}{||}}{C}-O-\underset{\underset{\displaystyle CH_3}{|}}{\overset{\overset{\displaystyle CH_3}{|}}{C}}-CH_3$

$+ H_2O$

144. Rank these organic compounds in terms of increasing water solubility (from least water soluble to most water soluble).

$$CH_3CH_2CH_2\overset{\overset{\displaystyle O}{\|}}{C}OH$$

A

$$CH_3CH_2CH=\overset{\overset{\displaystyle CH_3}{|}}{\underset{\underset{\displaystyle CH_3}{|}}{C}}$$

B

$$CH_3CH_2CH_2\overset{\overset{\displaystyle O}{\|}}{C}CH_3$$

C

Challenge Problems

145. The isoelectric point of an amino acid is the pH at which the molecule has no net charge. For glycine, that point would be the pH at which virtually all glycine molecules are in the form $^+H_3NCH_2CO_2^-$. This form of glycine is amphoteric since it can act as both an acid and a base. If we assume that the principal equilibrium at the isoelectric point has the best acid reacting with the best base present, then the reaction is

$$2\,^+H_3NCH_2CO_2^- \rightleftharpoons H_2NCH_2CO_2^- + \,^+H_3NCH_2CO_2H \quad (i)$$

Assuming this reaction is the principal equilibrium, then the following relationship must hold true:

$$[H_2NCH_2CO_2^-] = [\,^+H_3NCH_2CO_2H] \quad (ii)$$

Use this result and your answer to part c of Exercise 136 to calculate the pH at which equation (ii) is true. This pH will be the isoelectric point of glycine.

146. In 1994 chemists at Texas A & M University reported the synthesis of a non-naturally occurring amino acid (*C & E News*, April 18, 1994, pp. 26–27):

$$\underset{\underset{\displaystyle CO_2H}{}}{\overset{\overset{\displaystyle H_2N}{}}{C}}\!\!-\!\!\underset{\underset{\displaystyle CH_2SCH_3}{}}{\overset{\overset{\displaystyle CH_2\;H}{}}{C}}$$

a. To which naturally occurring amino acid is this compound most similar?

b. A tetrapeptide, phe–met–arg–phe—NH$_2$, is synthesized in the brains of rats addicted to morphine and heroin. (The

$$\text{—NH}_2 \text{ indicates that the peptide ends in } -\overset{\overset{\displaystyle O}{\|}}{C}-NH_2$$

instead of —CO$_2$H.) The TAMU scientists synthesized a similar tetrapeptide, with the synthetic amino acid above replacing one of the original amino acids. Draw a structure for the tetrapeptide containing the synthetic amino acid.

c. Indicate the chiral carbon atoms in the synthetic amino acid.

147. The structure of tartaric acid is

$$HO_2C-\overset{\overset{\displaystyle OH}{|}}{CH}-\overset{\overset{\displaystyle OH}{|}}{CH}-CO_2H$$

a. Is the form of tartaric acid pictured below optically active? Explain.

$$\underset{\underset{\displaystyle HOOC}{}}{}\overset{\overset{\displaystyle OH\;\;OH}{}}{\underset{\underset{\displaystyle H\;\;\;\;H}{}}{C-C}}\,COOH$$

(*Note:* The dashed lines show groups behind the plane of the page. The wedges show groups in front of the plane.)

b. Draw the optically active forms of tartaric acid.

148. Consider a sample of a hydrocarbon at 0.959 atm and 298 K. Upon combusting the entire sample in oxygen, you collect a mixture of gaseous carbon dioxide and water vapor at 1.51 atm and 375 K. This mixture has a density of 1.391 g/L and occupies a volume four times as large as that of the pure hydrocarbon. Determine the molecular formula of the hydrocarbon and name it.

149. Mycomycin, a naturally occurring antibiotic produced by the fungus *Nocardia acidophilus,* has the molecular formula $C_{13}H_{10}O_2$ and the systematic name 3,5,7,8-tridecatetraene-10,12-diynoic acid. Draw the structure of mycomycin.

150. Sorbic acid is used to prevent mold and fungus growth in some food products, especially cheeses. The systematic name for sorbic acid is 2,4-hexadienoic acid. Draw structures for the four geometrical isomers of sorbic acid.

151. Consider the following reactions. For parts b–d, see Exercise 62.

a. When C_5H_{12} is reacted with $Cl_2(g)$ in the presence of ultraviolet light, four different monochlorination products form. What is the structure of C_5H_{12} in this reaction?

b. When C_4H_8 is reacted with H_2O, a tertiary alcohol is produced as the major product. What is the structure of C_4H_8 in this reaction?

c. When C_7H_{12} is reacted with HCl, 1-chloro-1-methylcyclohexane is produced as the major product. What are the two possible structures for C_7H_{12} in this reaction?

d. When a hydrocarbon is reacted with H_2O and the major product of this reaction is then oxidized, acetone (2-propanone) is produced. What is the structure of the hydrocarbon in this reaction?

e. When $C_5H_{12}O$ is oxidized, a carboxylic acid is produced. What are the possible structures for $C_5H_{12}O$ in this reaction?

152. Polycarbonates are a class of thermoplastic polymers that are used in the plastic lenses of eyeglasses and in the shells of bicycle helmets. A polycarbonate is made from the reaction of bisphenol A (BPA) with phosgene ($COCl_2$):

BPA

$\xrightarrow{\text{Catalyst}}$ polycarbonate + $2n$HCl

Phenol (C_6H_5OH) is used to terminate the polymer (stop its growth).

a. Draw the structure of the polycarbonate chain formed from the above reaction.

b. Is this reaction a condensation or an addition polymerization?

153. A urethane linkage occurs when an alcohol adds across the carbon–nitrogen double bond in an isocyanate:

| Alcohol | Isocyanate | A urethane |

Polyurethanes (formed from the copolymerization of a diol with a diisocyanate) are used in foamed insulation and a variety of other construction materials. What is the structure of the polyurethane formed by the following reaction?

154. ABS plastic is a tough, hard plastic used in applications requiring shock resistance. The polymer consists of three monomer units: acrylonitrile (C_3H_3N), butadiene (C_4H_6), and styrene (C_8H_8).

a. Draw two repeating units of ABS plastic assuming that the three monomer units react in a 1:1:1 mole ratio and react in the same order as the monomers listed above.

b. A sample of ABS plastic contains 8.80% N by mass. It took 0.605 g Br_2 to react completely with a 1.20-g sample of ABS plastic. What is the percent by mass of acrylonitrile, butadiene, and styrene in this polymer sample?

c. ABS plastic does not react in a 1:1:1 mole ratio among the three monomer units. Using the results from part b, determine the relative numbers of the monomer units in this sample of ABS plastic.

155. Stretch a rubber band while holding it gently to your lips. Then slowly let it relax while still in contact with your lips.

a. What happens to the temperature of the rubber band on stretching?

b. Is the stretching an exothermic or endothermic process?

c. Explain the above result in terms of intermolecular forces.

d. What is the sign of ΔS and ΔG for stretching the rubber band?

e. Give the molecular explanation for the sign of ΔS for stretching.

156. Alcohols are very useful starting materials for the production of many different compounds. The following conversions, starting with 1-butanol, can be carried out in two or more steps. Show the steps (reactants/catalysts) you would follow to carry out the conversions, drawing the formula for the organic product in each step. For each step, a major product must be produced. (See Exercise 62.) (*Hint:* In the presence of H^+, an alcohol is converted into an alkene and water. This is the exact reverse of the reaction of adding water to an alkene to form an alcohol.)

a. 1-butanol $\longrightarrow$ butane

b. 1-butanol $\longrightarrow$ 2-butanone

157. A chemical "breathalyzer" test works because ethanol in the breath is oxidized by the dichromate ion (orange) to form acetic acid and chromium(III) ion (green). The balanced reaction is

$$3C_2H_5OH(aq) + 2Cr_2O_7^{2-}(aq) + 2H^+(aq) \longrightarrow$$
$$3HC_2H_3O_2(aq) + 4Cr^{3+}(aq) + 11H_2O(l)$$

You analyze a breathalyzer test in which 4.2 mg $K_2Cr_2O_7$ was reduced. Assuming the volume of the breath was 0.500 L at 30.°C and 750. mm Hg, what was the mole percent alcohol of the breath?

158. Estradiol is a female hormone with the following structure:

How many chiral carbon atoms are in estradiol?

Integrative Problems

These problems require the integration of multiple concepts to find the solutions.

159. An organometallic compound is one containing at least one metal–carbon bond. An example of an organometallic species is (CH_3CH_2)MBr, which contains a metal–ethyl bond.

a. If M^{2+} has the electron configuration [Ar]$3d^{10}$, what is the percent by mass of M in (CH_3CH_2)MBr?

b. A reaction involving (CH_3CH_2)MBr is the conversion of a ketone to an alcohol as illustrated here:

How does the hybridization of the starred carbon atom change, if at all, in going from reactants to products?

c. What is the systematic name of the product? (*Hint:* In this shorthand notation, all the C—H bonds have been eliminated and the lines represent C—C bonds, unless shown differently. As is typical of most organic compounds, each carbon atom has four bonds to it and the oxygen atoms have only two bonds.)

160. Helicenes are extended fused polyaromatic hydrocarbons that have a helical or screw-shaped structure.

    a. A 0.1450-g sample of solid helicene is combusted in air to give 0.5063 g CO_2. What is the empirical formula of this helicene?

    b. If a 0.0938-g sample of this helicene is dissolved in 12.5 g of solvent to give a 0.0175 M solution, what is the molecular formula of this helicene?

    c. What is the balanced reaction for the combustion of this helicene?

Marathon Problems

These problems are designed to incorporate several concepts and techniques into one situation.

161. For each of the following, fill in the blank with the correct response. All of these fill-in-the-blank problems pertain to material covered in the sections on alkanes, alkenes and alkynes, aromatic hydrocarbons, and hydrocarbon derivatives.

    a. The first "organic" compound to be synthesized in the laboratory, rather than being isolated from nature, was _____, which was prepared from _____.

    b. An organic compound whose carbon–carbon bonds are all single bonds is said to be _____.

    c. The general orientation of the four pairs of electrons around the carbon atoms in alkanes is _____.

    d. Alkanes in which the carbon atoms form a single unbranched chain are said to be _____ alkanes.

    e. Structural isomerism occurs when two molecules have the same number of each type of atom but exhibit different arrangements of the _____ between those atoms.

    f. The systematic names of all saturated hydrocarbons have the ending _____ added to a root name that indicates the number of carbon atoms in the molecule.

    g. For a branched hydrocarbon, the root name for the hydrocarbon comes from the number of carbon atoms in the _____ continuous chain in the molecule.

    h. The positions of substituents along the hydrocarbon framework of a molecule are indicated by the _____ of the carbon atom to which the substituents are attached.

    i. The major use of alkanes has been in _____ reactions, as a source of heat and light.

    j. With very reactive agents, such as the halogen elements, alkanes undergo _____ reactions, whereby a new atom replaces one or more hydrogen atoms of the alkane.

    k. Alkenes and alkynes are characterized by their ability to undergo rapid, complete _____ reactions, by which other atoms attach themselves to the carbon atoms of the double or triple bond.

    l. Unsaturated fats may be converted to saturated fats by the process of _____.

    m. Benzene is the parent member of the group of hydrocarbons called _____ hydrocarbons.

    n. An atom or group of atoms that imparts new and characteristic properties to an organic molecule is called a _____ group.

    o. A _____ alcohol is one in which there is only one hydrocarbon group attached to the carbon atom holding the hydroxyl group.

    p. The simplest alcohol, methanol, is prepared industrially by the hydrogenation of _____.

    q. Ethanol is commonly prepared by the _____ of certain sugars by yeast.

    r. Both aldehydes and ketones contain the _____ group, but they differ in where this group occurs along the hydrocarbon chain.

    s. Aldehydes and ketones can be prepared by _____ of the corresponding alcohol.

    t. Organic acids, which contain the _____ group, are typically weak acids.

    u. The typically sweet-smelling compounds called _____ result from the condensation reaction of an organic acid with an _____.

162. Choose one of the following terms to match the description given in statements (1)–(17). All of the following pertain to proteins or carbohydrates.

    a. aldohexose   **g.** disaccharides   **m.** ketohexoses

    b. saliva   **h.** disulfide   **n.** oxytocin

    c. cellulose   **i.** globular   **o.** pleated sheet

    d. CH_2O **j.** glycogen   **p.** polypeptide

    e. cysteine   **k.** glycoside linkage   **q.** primary

    f. denaturation   **l.** hydrophobic   structure

    (1) polymer consisting of many amino acids
    (2) linkage that forms between two cysteine species
    (3) peptide hormone that triggers milk secretion
    (4) proteins with roughly spherical shape
    (5) sequence of amino acids in a protein
    (6) silk protein secondary structure
    (7) water-repelling amino acid side chain
    (8) amino acid responsible for permanent wave in hair
    (9) breakdown of a protein's tertiary and/or secondary structure
    (10) animal polymer of glucose
    (11) —C—O—C— bond between rings in disaccharide sugars
    (12) empirical formula leading to the name carbohydrate
    (13) where enzymes catalyzing the breakdown of glycoside linkages are found
    (14) six-carbon ketone sugars
    (15) structural component of plants, polymer of glucose
    (16) sugars consisting of two monomer units
    (17) six-carbon aldehyde sugars

163. For each of the following, fill in the blank with the correct response(s). All of the following pertain to nucleic acids.

    a. The substance in the nucleus of the cell that stores and transmits genetic information is DNA, which stands for _____.

    b. The basic repeating monomer units of DNA and RNA are called _____.

    c. The pentose deoxyribose is found in DNA, whereas _____ is found in RNA.

d. The basic linkage in DNA or RNA between the sugar molecule and phosphoric acid is a phosphate _____ linkage.

e. The bases on opposite strands of DNA are said to be _____ to each other, which means the bases fit together specifically by hydrogen bonding to one another.

f. In a strand of normal DNA, the base _____ is always found paired with the base adenine, whereas _____ is always found paired with cytosine.

g. A given segment of the DNA molecule, which contains the molecular coding for a specific protein to be synthesized, is referred to as a _____.

h. During protein synthesis, _____ RNA molecules attach to and transport specific amino acids to the appropriate position on the pattern provided by _____ RNA molecules.

i. The codes specified by _____ are responsible for assembling the correct primary structure of proteins.

Appendixes

| Mathematical Procedures

A1.1 | Exponential Notation

The numbers characteristic of scientific measurements are often very large or very small; thus it is convenient to express them using powers of 10. For example, the number 1,300,000 can be expressed as 1.3×10^6, which means multiply 1.3 by 10 six times, or

$$1.3 \times 10^6 \times 1.3 \times \underbrace{10 \times 10 \times 10 \times 10 \times 10 \times 10}_{10^6 = 1 \text{ million}}$$

Note that each multiplication by 10 moves the decimal point one place to the right:

$$1.3 \times 10 = 13.$$
$$13 \times 10 = 130.$$
$$130 \times 10 = 1300.$$
$$\vdots$$

Thus the easiest way to interpret the notation 1.3×10^6 is that it means move the decimal point in 1.3 to the right six times:

$$1.3 \times 10^6 = 1\,3\,0\,0\,0\,0\,0 = 1,300,000$$
$$\underset{1\,2\,3\,4\,5\,6}{}$$

Using this notation, the number 1985 can be expressed as 1.985×10^3. Note that the usual convention is to write the number that appears before the power of 10 as a number between 1 and 10. To end up with the number 1.985, which is between 1 and 10, we had to move the decimal point three places to the left. To compensate for that, we must multiply by 10^3, which says that to get the intended number we start with 1.985 and move the decimal point three places to the right; that is:

$$1.985 \times 10^3 = 1\,9\,8\,5.$$
$$\underset{1\,2\,3}{}$$

Some other examples are given below.

Number	Exponential Notation
5.6	5.6×10^0 or 5.6×1
39	3.9×10^1
943	9.43×10^2
1126	1.126×10^3

So far we have considered numbers greater than 1. How do we represent a number such as 0.0034 in exponential notation? We start with a number between 1 and 10 and *divide* by the appropriate power of 10:

$$0.0034 = \frac{3.4}{10 \times 10 \times 10} = \frac{3.4}{10^3} = 3.4 \times 10^{-3}$$

Division by 10 moves the decimal point one place to the *left*. Thus the number

$$0.\,0\,0\,0\,0\,0\,0\,1\,4$$

$$7\ 6\ 5\ 4\ 3\ 2\ 1$$

can be written as 1.4×10^7.

To summarize, we can write any number in the form

$$N \times 10^{\pm n}$$

where N is between 1 and 10 and the exponent n is an integer. If the sign preceding n is positive, it means the decimal point in N should be moved n places to the right. If a negative sign precedes n, the decimal point in N should be moved n places to the left.

Multiplication and Division

When two numbers expressed in exponential notation are multiplied, the initial numbers are multiplied and the exponents of 10 are added:

$$(M \times 10^m)(N \times 10^n) = (MN) \times 10^{m+n}$$

For example (to two significant figures, as required),

$$(3.2 \times 10^4)(2.8 \times 10^3) = 9.0 \times 10^7$$

When the numbers are multiplied, if a result greater than 10 is obtained for the initial number, the decimal point is moved one place to the left and the exponent of 10 is increased by 1:

$$(5.8 \times 10^2)(4.3 \times 10^8) = 24.9 \times 10^{10}$$

$$= 2.49 \times 10^{11}$$

$$= 2.5 \ \times 10^{11} \quad \text{(two significant figures)}$$

Division of two numbers expressed in exponential notation involves normal division of the initial numbers and *subtraction* of the exponent of the divisor from that of the dividend. For example,

$$\underbrace{\frac{4.8 \times 10^8}{2.1 \times 10^3}}_{\text{Divisor}} = \frac{4.8}{2.1} \times 10^{(8-3)} = 2.3 \times 10^5$$

If the initial number resulting from the division is less than 1, the decimal point is moved one place to the right and the exponent of 10 is decreased by 1. For example,

$$\frac{6.4 \times 10^3}{8.3 \times 10^5} = \frac{6.4}{8.3} \times 10^{(3-5)} = 0.77 \times 10^{-2}$$

$$= 7.7 \times 10^{-3}$$

Addition and Subtraction

To add or subtract numbers expressed in exponential notation, *the exponents of the numbers must be the same*. For example, to add 1.31×10^5 and 4.2×10^4, we must rewrite one number so that the exponents of both are the same. The number 1.31×10^5 can be written 13.1×10^4, since moving the decimal point one place to the right can be compensated for by decreasing the exponent by 1. Now we can add the numbers:

$$13.1 \times 10^4$$
$$\underline{+\ 4.2 \times 10^4}$$
$$17.3 \times 10^4$$

In correct exponential notation the result is expressed as 1.73×10^5.

To perform addition or subtraction with numbers expressed in exponential notation, only the initial numbers are added or subtracted. The exponent of the result is the same as those of the numbers being added or subtracted. To subtract 1.8×10^2 from 8.99×10^3, we write

$$
\begin{array}{r}
8.99 \times 10^3 \\
-0.18 \times 10^3 \\
\hline
8.81 \times 10^3
\end{array}
$$

Powers and Roots

When a number expressed in exponential notation is taken to some power, the initial number is taken to the appropriate power and the exponent of 10 is multiplied by that power:

$$(N \times 10^n)^m = N^m \times 10^{m \cdot n}$$

For example,*

$$
\begin{aligned}
(7.5 \times 10^2)^3 &= 7.5^3 \times 10^{3 \cdot 2} \\
&= 422 \times 10^6 \\
&= 4.22 \times 10^8 \\
&= 4.2 \times 10^8 \quad \text{(two significant figures)}
\end{aligned}
$$

When a root is taken of a number expressed in exponential notation, the root of the initial number is taken and the exponent of 10 is divided by the number representing the root:

$$\sqrt{N \times 10^n} = (n \times 10^n)^{1/2} = \sqrt{N} \times 10^{n/2}$$

For example,

$$
\begin{aligned}
(2.9 \times 10^6)^{1/2} &= \sqrt{2.9} \times 10^{6/2} \\
&= 1.7 \times 10^3
\end{aligned}
$$

Because the exponent of the result must be an integer, we may sometimes have to change the form of the number so that the power divided by the root equals an integer. For example,

$$
\begin{aligned}
\sqrt{1.9 \times 10^3} = (1.9 \times 10^3)^{1/2} &= (0.19 \times 10^4)^{1/2} \\
&= \sqrt{0.19} \times 10^2 \\
&= 0.44 \times 10^2 \\
&= 4.4 \times 10^1
\end{aligned}
$$

In this case, we moved the decimal point one place to the left and increased the exponent from 3 to 4 to make $n/2$ an integer.

The same procedure is followed for roots other than square roots. For example,

$$
\begin{aligned}
\sqrt[3]{6.9 \times 10^5} = (6.9 \times 10^5)^{1/3} &= (0.69 \times 10^6)^{1/3} \\
&= \sqrt[3]{0.69} \times 10^2 \\
&= 0.88 \times 10^2 \\
&= 8.8 \times 10^1
\end{aligned}
$$

and

$$
\begin{aligned}
\sqrt[3]{4.6 \times 10^{10}} = (4.6 \times 10^{10})^{1/3} &= (46 \times 10^9)^{1/3} \\
&= \sqrt[3]{46} \times 10^3 \\
&= 3.6 \times 10^3
\end{aligned}
$$

*Refer to the instruction booklet for your calculator for directions concerning how to take roots and powers of numbers.

A1.2 | Logarithms

A logarithm is an exponent. Any number N can be expressed as follows:

$$N = 10^x$$

For example,

$$1000 = 10^3$$
$$100 = 10^2$$
$$10 = 10^1$$
$$1 = 10^0$$

The common, or base 10, logarithm of a number is the power to which 10 must be taken to yield the number. Thus, since $1000 = 10^3$,

$$\log 1000 = 3$$

Similarly,

$$\log 100 = 2$$
$$\log 10 = 1$$
$$\log 1 = 0$$

For a number between 10 and 100, the required exponent of 10 will be between 1 and 2. For example, $65 = 10^{1.8129}$; that is, $\log 65 = 1.8129$. For a number between 100 and 1000, the exponent of 10 will be between 2 and 3. For example, $650 = 10^{2.8129}$ and $\log 650 = 2.8129$.

A number N greater than 0 and less than 1 can be expressed as follows:

$$N = 10^{-x} = \frac{1}{10^x}$$

For example,

$$0.001 = \frac{1}{1000} = \frac{1}{10^3} = 10^{-3}$$

$$0.01 = \frac{1}{100} = \frac{1}{10^2} = 10^{-2}$$

$$0.1 = \frac{1}{10} = \frac{1}{10^1} = 10^{-1}$$

Thus

$$\log 0.001 = -3$$
$$\log 0.01 = -2$$
$$\log 0.1 = -1$$

Although common logs are often tabulated, the most convenient method for obtaining such logs is to use an electronic calculator. On most calculators the number is first entered and then the log key is punched. The log of the number then appears in the display.* Some examples are given below. You should reproduce these results on your calculator to be sure that you can find common logs correctly.

Number	Common Log
36	1.56
1849	3.2669
0.156	−0.807
1.68×10^{-5}	−4.775

*Refer to the instruction booklet for your calculator for the exact sequence to obtain logarithms.

Note that the number of digits after the decimal point in a common log is equal to the number of significant figures in the original number.

Since logs are simply exponents, they are manipulated according to the rules for exponents. For example, if $A = 10^x$ and $B = 10^y$, then their product is

$$A \cdot B = 10^x \cdot 10^y = 10^{x+y}$$

and

$$\log AB = x + y = \log A + \log B$$

For division, we have

$$\frac{A}{B} = \frac{10^x}{10^y} = 10^{x-y}$$

and

$$\log \frac{A}{B} = x - y = \log A - \log B$$

For a number raised to a power, we have

$$A^n = (10^x)^n = 10^{nx}$$

and

$$\log A^n = nx = n \log A$$

It follows that

$$\log \frac{1}{A^n} = \log A^{-n} = -n \log A$$

or, for $n = 1$,

$$\log \frac{1}{A} = -\log A$$

When a common log is given, to find the number it represents, we must carry out the process of exponentiation. For example, if the log is 2.673, then $N = 10^{2.673}$. The process of exponentiation is also called taking the antilog, or the inverse logarithm. This operation is usually carried out on calculators in one of two ways. The majority of calculators require that the log be entered first and then the keys (INV) and (LOG) pressed in succession. For example, to find $N = 10^{2.673}$ we enter 2.673 and then press (INV) and (LOG). The number 471 will be displayed; that is, $N = 471$. Some calculators have a (10^x) key. In that case, the log is entered first and then the (10^x) key is pressed. Again, the number 471 will be displayed.

Natural logarithms, another type of logarithm, are based on the number 2.7183, which is referred to as e. In this case, a number is represented as $N = e^x = 2.7183^x$. For example,

$$N = 7.15 = e^x$$

$$\ln 7.15 = x = 1.967$$

To find the natural log of a number using a calculator, the number is entered and then the (ln) key is pressed. Use the following examples to check your technique for finding natural logs with your calculator:

Number (e^x)	Natural Log (x)
784	6.664
1.61×10^3	7.384
1.00×10^{-7}	−16.118
1.00	0

If a natural logarithm is given, to find the number it represents, exponentiation to the base e (2.7183) must be carried out. With many calculators this is done using a key marked $\boxed{e^x}$ (the natural log is entered, with the correct sign, and then the $\boxed{e^x}$ key is pressed). The other common method for exponentiation to base e is to enter the natural log and then press the $\boxed{INV}$ and $\boxed{ln}$ keys in succession. The following examples will help you check your technique:

ln $N(x)$	$N(e^x)$
3.256	25.9
−5.169	5.69×10^{-3}
13.112	4.95×10^{5}

Since natural logarithms are simply exponents, they are also manipulated according to the mathematical rules for exponents given earlier for common logs.

A1.3 | Graphing Functions

In interpreting the results of a scientific experiment, it is often useful to make a graph. If possible, the function to be graphed should be in a form that gives a straight line. The equation for a straight line (a *linear equation*) can be represented by the general form

$$y = mx + b$$

where y is the *dependent variable*, x is the *independent variable*, m is the *slope*, and b is the *intercept* with the y axis.

To illustrate the characteristics of a linear equation, the function $y = 3x + 4$ is plotted in Fig. A.1. For this equation $m = 3$ and $b = 4$. Note that the y intercept occurs when $x = 0$. In this case the intercept is 4, as can be seen from the equation ($b = 4$).

The slope of a straight line is defined as the ratio of the rate of change in y to that in x:

$$m = \text{slope} = \frac{\Delta y}{\Delta x}$$

For the equation $y = 3x + 4$, y changes three times as fast as x (since x has a coefficient of 3). Thus the slope in this case is 3. This can be verified from the graph. For the triangle shown in Fig. A.1,

$$\Delta y = 34 - 10 = 24 \quad \text{and} \quad \Delta x = 10 - 2 = 8$$

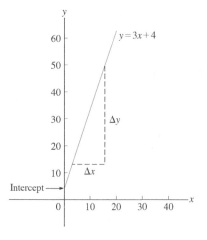

Figure A.1 | Graph of the linear equation $y = 3x + 4$.

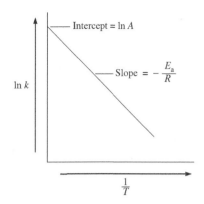

Figure A.2 | Graph of ln k versus $1/T$.

Table A.1 | Some Useful Linear Equations in Standard Form

Equation ($y = mx + b$)	What Is Plotted (y vs. x)	Slope (m)	Intercept (b)	Section in Text
$[A] = -kt + [A]_0$	$[A]$ vs. t	$-k$	$[A]_0$	12.4
$\ln[A] = -kt + \ln[A]_0$	$\ln[A]$ vs. t	$-k$	$\ln[A]_0$	12.4
$\dfrac{1}{[A]} = kt + \dfrac{1}{[A]_0}$	$\dfrac{1}{[A]}$ vs. t	k	$\dfrac{1}{[A]_0}$	12.4
$\ln P_{vap} = -\dfrac{\Delta H_{vap}}{R}\left(\dfrac{1}{T}\right) + C$	$\ln P_{vap}$ vs. $\dfrac{1}{T}$	$-\dfrac{\Delta H_{vap}}{R}$	C	10.8

Thus

$$\text{Slope} = \frac{\Delta y}{\Delta x} = \frac{24}{8} = 3$$

The preceding example illustrates a general method for obtaining the slope of a line from the graph of that line. Simply draw a triangle with one side parallel to the y axis and the other parallel to the x axis as shown in Fig. A.1. Then determine the lengths of the sides to give y and x, respectively, and compute the ratio $\Delta y/\Delta x$.

Sometimes an equation that is not in standard form can be changed to the form $y = mx + b$ by rearrangement or mathematical manipulation. An example is the equation $k = Ae^{-E_a/RT}$ described in Section 12.7, where A, E_a, and R are constants; k is the dependent variable; and $1/T$ is the independent variable. This equation can be changed to standard form by taking the natural logarithm of both sides,

$$\ln k = \ln Ae^{-E_a/RT} = \ln A + \ln e^{-E_a/RT} = \ln A - \frac{E_a}{RT}$$

noting that the log of a product is equal to the sum of the logs of the individual terms and that the natural log of $e^{-E_a/RT}$ is simply the exponent $-E_a/RT$. Thus, in standard form, the equation $k = Ae^{-E_a/RT}$ is written

$$\underbrace{\ln k}_{y} = \underbrace{-\frac{E_a}{R}}_{m}\underbrace{\left(\frac{1}{T}\right)}_{x} + \underbrace{\ln A}_{b}$$

A plot of ln k versus $1/T$ (Fig. A.2) gives a straight line with slope $-E_a/R$ and intercept ln A.

Other linear equations that are useful in the study of chemistry are listed in standard form in Table A.1.

A1.4 | Solving Quadratic Equations

A *quadratic equation,* a polynomial in which the highest power of x is 2, can be written as

$$ax^2 + bx + c = 0$$

One method for finding the two values of x that satisfy a quadratic equation is to use the *quadratic formula:*

$$x = \frac{-b \pm \sqrt{b^2 - 4ac}}{2a}$$

where a, b, and c are the coefficients of x^2, x, and the constant, respectively. For example, in determining [H$^+$] in a solution of $1.0 \times 10^{-4}\ M$ acetic acid the following expression arises:

$$1.8 \times 10^{-5} = \frac{x^2}{1.0 \times 10^{-4} - x}$$

which yields

$$x^2 + (1.8 \times 10^{-5})x - 1.8 \times 10^{-9} = 0$$

where $a = 1$, $b = 1.8 \times 10^{-5}$, and $c = -1.8 \times 10^{-9}$. Using the quadratic formula, we have

$$x = \frac{-b \pm \sqrt{b^2 - 4ac}}{2a}$$

$$= \frac{-1.8 \times 10^{-5} \pm \sqrt{3.24 \times 10^{-10} - (4)(1)(-1.8 \times 10^{-9})}}{2(1)}$$

$$= \frac{-1.8 \times 10^{-5} \pm \sqrt{3.24 \times 10^{-10} + 7.2 \times 10^{-9}}}{2}$$

$$= \frac{-1.8 \times 10^{-5} \pm \sqrt{7.5 \times 10^{-9}}}{2}$$

$$= \frac{-1.8 \times 10^{-5} \pm 8.7 \times 10^{-5}}{2}$$

Thus

$$x = \frac{6.9 \times 10^{-5}}{2} = 3.5 \times 10^{-5}$$

and

$$x = \frac{-10.5 \times 10^{-5}}{2} = -5.2 \times 10^{-5}$$

Note that there are two roots, as there always will be, for a polynomial in x^2. In this case x represents a concentration of H$^+$ (see Section 14.3). Thus the positive root is the one that solves the problem, since a concentration cannot be a negative number.

A second method for solving quadratic equations is by *successive approximations,* a systematic method of trial and error. A value of x is guessed and substituted into the equation everywhere x (or x^2) appears, except for one place. For example, for the equation

$$x^2 + (1.8 \times 10^{-5})x - 1.8 \times 10^{-9} = 0$$

we might guess $x = 2 \times 10^{-5}$. Substituting that value into the equation gives

$$x^2 + (1.8 \times 10^{-5})(2 \times 10^{-5}) - 1.8 \times 10^{-9} = 0$$

or

$$x^2 = 1.8 \times 10^{-9} - 3.6 \times 10^{-10} = 1.4 \times 10^{-9}$$

Thus

$$x = 3.7 \times 10^{-5}$$

Note that the guessed value of x (2×10^5) is not the same as the value of x that is calculated (3.7×10^{-5}) after inserting the estimated value. This means that $x = 2 \times 10^{-5}$ is not the correct solution, and we must try another guess. We take the calculated value (3.7×10^{-5}) as our next guess:

$$x^2 + (1.8 \times 10^{-5})(3.7 \times 10^{-5}) - 1.8 \times 10^{-9} = 0$$
$$x^2 = 1.8 \times 10^{-9} - 6.7 \times 10^{-10} = 1.1 \times 10^{-9}$$

Thus

$$x = 3.3 \times 10^{-5}$$

Now we compare the two values of x again:

Guessed: $x = 3.7 \times 10^{-5}$

Calculated: $x = 3.3 \times 10^{-5}$

These values are closer but not close enough. Next we try 3.3×10^{-5} as our guess:

$$x^2 + (1.8 \times 10^{-5})(3.3 \times 10^{-5}) - 1.8 \times 10^{-9} = 0$$
$$x^2 = 1.8 \times 10^{-9} - 5.9 \times 10^{-10} = 1.2 \times 10^{-9}$$

Thus

$$x = 3.5 \times 10^{-5}$$

Again we compare:

Guessed: $x = 3.3 \times 10^{-5}$

Calculated: $x = 3.5 \times 10^{-5}$

Next we guess $x = 3.5 \times 10^{-5}$ to give

$$x^2 + (1.8 \times 10^{-5})(3.5 \times 10^{-5}) - 1.8 \times 10^{-9} = 0$$
$$x^2 = 1.8 \times 10^{-9} - 6.3 \times 10^{-10} = 1.2 \times 10^{-9}$$

Thus

$$x = 3.5 \times 10^{-5}$$

Now the guessed value and the calculated value are the same; we have found the correct solution. Note that this agrees with one of the roots found with the quadratic formula in the first method.

To further illustrate the method of successive approximations, we will solve Example 14.17 using this procedure. In solving for $[H^+]$ for 0.010 M H_2SO_4, we obtain the following expression:

$$1.2 \times 10^{-2} = \frac{x(0.010 + x)}{0.010 - x}$$

which can be rearranged to give

$$x = (1.2 \times 10^{-2})\left(\frac{0.010 - x}{0.010 + x}\right)$$

We will guess a value for x, substitute it into the right side of the equation, and then calculate a value for x. In guessing a value for x, we know it must be less than 0.010, since a larger value will make the calculated value for x negative and the guessed and calculated values will never match. We start by guessing $x = 0.005$.

The results of the successive approximations are shown in the following table:

Trial	Guessed Value for x	Calculated Value for x
1	0.0050	0.0040
2	0.0040	0.0051
3	0.00450	0.00455
4	0.00452	0.00453

Note that the first guess was close to the actual value and that there was oscillation between 0.004 and 0.005 for the guessed and calculated values. For trial 3, an average of these values was used as the guess, and this led rapidly to the correct value (0.0045 to the correct number of significant figures). Also, note that it is useful to carry extra digits until the correct value is obtained. That value can then be rounded off to the correct number of significant figures.

The method of successive approximations is especially useful for solving polynomials containing x to a power of 3 or higher. The procedure is the same as for quadratic equations: Substitute a guessed value for x into the equation for every x term but one, and then solve for x. Continue this process until the guessed and calculated values agree.

A1.5 | Uncertainties in Measurements

Like all the physical sciences, chemistry is based on the results of measurements. Every measurement has an inherent uncertainty, so if we are to use the results of measurements to reach conclusions, we must be able to estimate the sizes of these uncertainties.

For example, the specification for a commercial 500-mg acetaminophen (the active painkiller in Tylenol) tablet is that each batch of tablets must contain 450 to 550 mg of acetaminophen per tablet. Suppose that chemical analysis gave the following results for a batch of acetaminophen tablets: 428 mg, 479 mg, 442 mg, and 435 mg. How can we use these results to decide if the batch of tablets meets the specification? Although the details of how to draw such conclusions from measured data are beyond the scope of this text, we will consider some aspects of how this is done. We will focus here on the types of experimental uncertainty, the expression of experimental results, and a simplified method for estimating experimental uncertainty when several types of measurement contribute to the final result.

Types of Experimental Error

There are two types of experimental uncertainty (error). A variety of names are applied to these types of errors:

$$\text{Precision} \longleftrightarrow \text{random error} \equiv \text{indeterminate error}$$
$$\text{Accuracy} \longleftrightarrow \text{systematic error} \equiv \text{determinate error}$$

The difference between the two types of error is well illustrated by the attempts to hit a target shown in Fig. 1.7 in Chapter 1.

Random error is associated with every measurement. To obtain the last significant figure for any measurement, we must always make an estimate. For example, we interpolate between the marks on a meter stick, a buret, or a balance. The precision of replicate measurements (repeated measurements of the same type) reflects the size of the random errors. Precision refers to the reproducibility of replicate measurements.

The accuracy of a measurement refers to how close it is to the true value. An inaccurate result occurs as a result of some flaw (systematic error) in the measurement: the presence of an interfering substance, incorrect calibration of an instrument, operator error, and so on. The goal of chemical analysis is to eliminate systematic error, but random errors can only be minimized. In practice, an experiment is almost always done to find an unknown value (the true value is not known—someone is trying to obtain that value by doing the experiment). In this case the precision of several replicate determinations is used to assess the accuracy of the result. The results of the replicate experiments are expressed as an average (which we assume is close to the true value) with an error limit that gives some indication of how close the average value may be to the true value. The error limit represents the uncertainty of the experimental result.

Expression of Experimental Results

If we perform several measurements, such as for the analysis for acetaminophen in painkiller tablets, the results should express two things: the average of the measurements and the size of the uncertainty.

There are two common ways of expressing an average: the mean and the median. The mean ($\bar{x}$) is the arithmetic average of the results, or

$$\text{Mean} = \bar{x} = \sum_{i=1}^{n} \frac{x_i}{n} = \frac{x_1 + x_2 + \cdots + x_n}{n}$$

where Σ means take the sum of the values. The mean is equal to the sum of all the measurements divided by the number of measurements. For the acetaminophen results given previously, the mean is

$$\bar{x} = \frac{428 + 479 + 442 + 435}{4} = 446 \text{ mg}$$

The median is the value that lies in the middle among the results. Half the measurements are above the median and half are below the median. For results of 465 mg, 485 mg, and 492 mg, the median is 485 mg. When there is an even number of results, the median is the average of the two middle results. For the acetaminophen results, the median is

$$\frac{442 + 435}{2} = 438 \text{ mg}$$

There are several advantages to using the median. If a small number of measurements is made, one value can greatly affect the mean. Consider the results for the analysis of acetaminophen: 428 mg, 479 mg, 442 mg, and 435 mg. The mean is 446 mg, which is larger than three of the four weights. The median is 438 mg, which lies near the three values that are relatively close to one another.

In addition to expressing an average value for a series of results, we must express the uncertainty. This usually means expressing either the precision of the measurements or the observed range of the measurements. The range of a series of measurements is defined by the smallest value and the largest value. For the analytical results on the acetaminophen tablets, the range is from 428 mg to 479 mg. Using this range, we can express the results by saying that the true value lies between 428 mg and 479 mg. That is, we can express the amount of acetaminophen in a typical tablet as 446 ± 33 mg, where the error limit is chosen to give the observed range (approximately).

The most common way to specify precision is by the standard deviation, s, which for a small number of measurements is given by the formula

$$s = \left[\frac{\sum_{i=1}^{n} (x_i - \bar{x})^2}{n - 1} \right]^{1/2}$$

where x_i is an individual result, $\bar{x}$ is the average (either mean or median), and n is the total number of measurements. For the acetaminophen example, we have

$$s = \left[\frac{(428 - 446)^2 + (479 - 446)^2 + (442 - 446)^2 + (435 - 446)^2}{4 - 1} \right]^{1/2} = 23$$

Thus we can say the amount of acetaminophen in the typical tablet in the batch of tablets is 446 mg with a sample standard deviation of 23 mg. Statistically this means that any additional measurement has a 68% probability (68 chances out of 100) of being between 423 mg ($446 - 23$) and 469 mg ($446 + 23$). Thus the standard deviation is a measure of the precision of a given type of determination.

The standard deviation gives us a means of describing the precision of a given type of determination using a series of replicate results. However, it is also useful to be able to estimate the precision of a procedure that involves several measurements by combining the precisions of the individual steps. That is, we want to answer the following question: How do the uncertainties propagate when we combine the results of several different types of measurements? There are many ways to deal with the propagation of uncertainty. We will discuss only one simple method here.

A Simplified Method for Estimating Experimental Uncertainty

To illustrate this method, we will consider the determination of the density of an irregularly shaped solid. In this determination we make three measurements. First, we measure the mass of the object on a balance. Next, we must obtain the volume of the solid. The easiest method for doing this is to partially fill a graduated cylinder with a liquid and record the volume. Then we add the solid and record the volume again. The difference in the measured volumes is the volume of the solid. We can then calculate the density of the solid from the equation

$$D = \frac{M}{V_2 - V_1}$$

where M is the mass of the solid, V_1 is the initial volume of liquid in the graduated cylinder, and V_2 is the volume of liquid plus solid. Suppose we get the following results:

$$M = 23.06 \text{ g}$$
$$V_1 = 10.4 \text{ mL}$$
$$V_2 = 13.5 \text{ mL}$$

The calculated density is

$$\frac{23.06 \text{ g}}{13.5 \text{ mL} - 10.4 \text{ mL}} = 7.44 \text{ g/mL}$$

Now suppose that the precision of the balance used is ± 0.02 g and that the volume measurements are precise to ± 0.05 mL. How do we estimate the uncertainty of the density? We can do this by assuming a worst case. That is, we assume the largest uncertainties in all measurements, and see what combinations of measurements will give the largest and smallest possible results (the greatest range). Since the density is the mass divided by the volume, the largest value of the density will be that obtained using the largest possible mass and the smallest possible volume:

$$D_{max} = \frac{23.08}{13.45 - 10.45} = 7.69 \text{ g/mL}$$

Largest possible mass = 23.06 + .02

Smallest possible V_2 Largest possible V_1

The smallest value of the density is

$$D_{min} = \frac{23.04}{13.35 - 10.35} = 7.20 \text{ g/mL}$$

Smallest possible mass

Largest possible V_2 Smallest possible V_1

Thus the calculated range is from 7.20 to 7.69 and the average of these values is 7.44. The error limit is the number that gives the high and low range values when added and subtracted from the average. Therefore, we can express the density as 7.44 ± 0.25 g/mL, which is the average value plus or minus the quantity that gives the range calculated by assuming the largest uncertainties.

Analysis of the propagation of uncertainties is useful in drawing qualitative conclusions from the analysis of measurements. For example, suppose that we obtained the preceding results for the density of an unknown alloy and we want to know if it is one of the following alloys:

Alloy A: $D = 7.58$ g/mL

Alloy B: $D = 7.42$ g/mL

Alloy C: $D = 8.56$ g/mL

We can safely conclude that the alloy is not C. But the values of the densities for alloys A and B are both within the inherent uncertainty of our method. To distinguish between A and B, we need to improve the precision of our determination: The obvious choice is to improve the precision of the volume measurement.

The worst-case method is very useful in estimating uncertainties when the results of several measurements are combined to calculate a result. We assume the maximum uncertainty in each measurement and calculate the minimum and maximum possible result. These extreme values describe the range and thus the error limit.

| APPENDIX 2 | The Quantitative Kinetic Molecular Model |

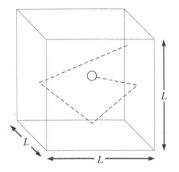

Figure A.3 | An ideal gas particle in a cube whose sides are of length L. The particle collides elastically with the walls in a random, straight-line motion.

We have seen that the kinetic molecular model successfully accounts for the properties of an ideal gas. This appendix will show in some detail how the postulates of the kinetic molecular model lead to an equation corresponding to the experimentally obtained ideal gas equation.

Recall that the particles of an ideal gas are assumed to be volumeless, to have no attraction for each other, and to produce pressure on their container by colliding with the container walls.

Suppose there are n moles of an ideal gas in a cubical container with sides each of length L. Assume each gas particle has a mass m and that it is in rapid, random, straight-line motion colliding with the walls, as shown in Fig. A.3. The collisions will be assumed to be *elastic*—no loss of kinetic energy occurs. We want to compute the force on the walls from the colliding gas particles and then, since pressure is force per unit area, to obtain an expression for the pressure of the gas.

Before we can derive the expression for the pressure of a gas, we must first discuss some characteristics of velocity. Each particle in the gas has a particular velocity u that can be divided into components u_x, u_y, and u_z, as shown in Fig. A.4. First, using u_x and u_y and the Pythagorean theorem, we can obtain u_{xy} as shown in Fig. A.4(c):

$$u_{xy}^2 = u_x^2 + u_y^2$$

Hypotenuse of   Sides of
right triangle   right triangle

Then, constructing another triangle as shown in Fig. A.4(c), we find

$$u^2 = u_{xy}^2 + u_z^2$$

or

$$u^2 = u_x^2 + u_y^2 + u_z^2$$

Now let's consider how an "average" gas particle moves. For example, how often does this particle strike the two walls of the box that are perpendicular to the x axis? It is important to realize that only the x component of the velocity affects the particle's impacts on these two walls, as shown in Fig. A.5(a). The larger the x component of the velocity, the faster the particle travels between these two walls, and the more impacts per unit of time it will make on these walls. Remember, the pressure of the gas is due to these collisions with the walls.

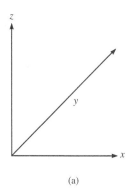

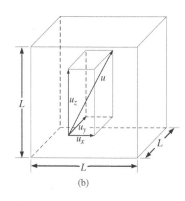

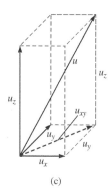

(a)   (b)   (c)

Figure A.4 | (a) The Cartesian coordinate axes.

(b) The velocity u of any gas particle can be broken down into three mutually perpendicular components: u_x, u_y, and u_z. This can be represented as a rectangular solid with sides u_x, u_y, and u_z and body diagonal u.

(c) In the xy plane,

$$u_x^2 + u_y^2 = u_{xy}^2$$

by the Pythagorean theorem. Since u_{xy} and u_x are also perpendicular,

$$u^2 = u_{xy}^2 + u_z^2 = u_x^2 + u_y^2 + u_z^2$$

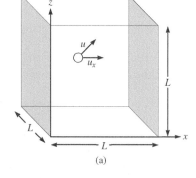

(a)

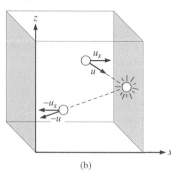

(b)

Figure A.5 | (a) Only the x component of the gas particle's velocity affects the frequency of impacts on the shaded walls, the walls that are perpendicular to the x axis.
(b) For an elastic collision, there is an exact reversal of the x component of the velocity and of the total velocity. The change in momentum (final − initial) is then

$$-mu_x - mu_x = -2mu_x$$

The collision frequency (collisions per unit of time) with the two walls that are perpendicular to the x axis is given by

$$(\text{Collision frequency})_x = \frac{\text{velocity in the } x \text{ direction}}{\text{distance between the walls}}$$

$$= \frac{u_x}{L}$$

Next, what is the force of a collision? Force is defined as mass times acceleration (change in velocity per unit of time):

$$F = ma = m\left(\frac{\Delta u}{\Delta t}\right)$$

where F represents force, a represents acceleration, Δu represents a change in velocity, and Δt represents a given length of time.

Since we assume that the particle has constant mass, we can write

$$F = \frac{m\Delta u}{\Delta t} = \frac{\Delta(mu)}{\Delta t}$$

The quantity mu is the momentum of the particle (momentum is the product of mass and velocity), and the expression $F = \Delta(mu)/\Delta t$ implies that force is the change in momentum per unit of time. When a particle hits a wall perpendicular to the x axis, as shown in Fig. A.5(b), an elastic collision results in an *exact reversal* of the x component of velocity. That is, the *sign,* or direction, of u_x reverses when the particle collides with one of the walls perpendicular to the x axis. Thus the final momentum is the *negative,* or opposite, of the initial momentum. Remember that an elastic collision means that there is no change in the *magnitude* of the velocity. The change in momentum in the x direction is then

Change in momentum $= \Delta(mu_x) =$ final momentum − initial momentum

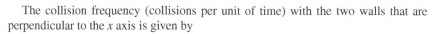

$$= -mu_x - mu_x$$

Final momentum in x direction   Initial momentum in x direction

$$= -2mu_x$$

But we are interested in the force the gas particle exerts on the walls of the box. Since we know that every action produces an equal but opposite reaction, the change in momentum with respect to the wall on impact is $-(-2mu_x)$, or $2mu_x$.

Recall that since force is the change in momentum per unit of time,

$$\text{Force}_x = \frac{\Delta(mu_x)}{\Delta t}$$

for the walls perpendicular to the x axis.

This expression can be obtained by multiplying the change in momentum per impact by the number of impacts per unit of time:

$$\text{Force}_x = (2mu_x)\left(\frac{u_x}{L}\right) = \text{change in momentum per unit of time}$$

Change in momentum per impact   Impacts per unit of time

That is,

$$\text{Force}_x = \frac{2mu_x^2}{L}$$

So far we have considered only the two walls of the box perpendicular to the x axis. We can assume that the force on the two walls perpendicular to the y axis is given by

$$\text{Force}_y = \frac{2mu_y^2}{L}$$

and that on the two walls perpendicular to the z axis by

$$\text{Force}_z = \frac{2mu_z^2}{L}$$

Since we have shown that

$$u^2 = u_x^2 + u_y^2 + u_z^2$$

the total force on the box is

$$\text{Force}_{\text{TOTAL}} = \text{force}_x + \text{force}_y + \text{force}_z$$
$$= \frac{2mu_x^2}{L} + \frac{2mu_y^2}{L} + \frac{2mu_z^2}{L}$$
$$= \frac{2m}{L}(u_x^2 + u_y^2 + u_z^2) = \frac{2m}{L}(u^2)$$

Now since we want the average force, we use the average of the square of the velocity $(\overline{u^2})$ to obtain

$$\overline{\text{Force}}_{\text{TOTAL}} = \frac{2m}{L}(\overline{u^2})$$

Next, we need to compute the pressure (force per unit of area)

$$\text{Pressure due to "average" particle} = \frac{\overline{\text{force}}_{\text{TOTAL}}}{\text{area}_{\text{TOTAL}}}$$
$$= \frac{\dfrac{2m\overline{u^2}}{L}}{6L^2} = \frac{m\overline{u^2}}{3L^3}$$

The 6 sides of the cube   Area of each side

Since the volume V of the cube is equal to L^3, we can write

$$\text{Pressure} = P = \frac{m\overline{u^2}}{3V}$$

So far we have considered the pressure on the walls due to a single, "average" particle. Of course, we want the pressure due to the entire gas sample. The number of particles in a given gas sample can be expressed as follows:

$$\text{Number of gas particles} = nN_A$$

where n is the number of moles and N_A is Avogadro's number.

The total pressure on the box due to n moles of a gas is therefore

$$P = nN_A\frac{m\overline{u^2}}{3V}$$

Next we want to express the pressure in terms of the kinetic energy of the gas molecules. Kinetic energy (the energy due to motion) is given by $\frac{1}{2}mu^2$, where m is the mass and u is the velocity. Since we are using the average of the velocity squared $(\overline{u^2})$, and since $m\overline{u^2} = 2(\frac{1}{2}m\overline{u^2})$, we have

$$P = \left(\frac{2}{3}\right)\frac{nN_A(\frac{1}{2}m\overline{u^2})}{V}$$

or

$$\frac{PV}{n} = \left(\frac{2}{3}\right)N_A(\tfrac{1}{2}m\overline{u^2})$$

Thus, based on the postulates of the kinetic molecular model, we have been able to derive an equation that has the same form as the ideal gas equation,

$$\frac{PV}{n} = RT$$

This agreement between experiment and theory supports the validity of the assumptions made in the kinetic molecular model about the behavior of gas particles, at least for the limiting case of an ideal gas.

APPENDIX 3 | Spectral Analysis

Although volumetric and gravimetric analyses are still very commonly used, spectroscopy is the technique most often used for modern chemical analysis. *Spectroscopy* is the study of electromagnetic radiation emitted or absorbed by a given chemical species. Since the quantity of radiation absorbed or emitted can be related to the quantity of the absorbing or emitting species present, this technique can be used for quantitative analysis. There are many spectroscopic techniques, as electromagnetic radiation spans a wide range of energies to include X rays; ultraviolet, infrared, and visible light; and microwaves, to name a few of its familiar forms. We will consider here only one procedure, which is based on the absorption of visible light.

If a liquid is colored, it is because some component of the liquid absorbs visible light. In a solution the greater the concentration of the light-absorbing substance, the more light absorbed, and the more intense the color of the solution.

The quantity of light absorbed by a substance can be measured by a *spectrophotometer,* shown schematically in Fig. A.6. This instrument consists of a source that emits all wavelengths of light in the visible region (wavelengths of ~400 to 700 nm); a monochromator, which selects a given wavelength of light; a sample holder for the solution being measured; and a detector, which compares the intensity of incident light

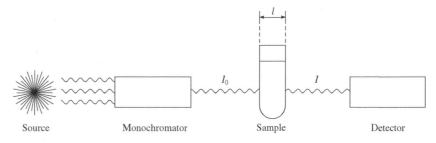

Figure A.6 | A schematic diagram of a simple spectrophotometer. The source emits all wavelengths of visible light, which are dispersed using a prism or grating and then focused, one wavelength at a time, onto the sample. The detector compares the intensity of the incident light (I_o) to the intensity of the light after it has passed through the sample (I).

I_0 to the intensity of light after it has passed through the sample I. The ratio I/I_0, called the *transmittance,* is a measure of the fraction of light that passes through the sample. The amount of light absorbed is given by the *absorbance A,* where

$$A = -\log \frac{I}{I_0}$$

The absorbance can be expressed by the *Beer–Lambert law:*

$$A = \epsilon l c$$

where ϵ is the molar absorptivity or the molar extinction coefficient (in L/mol · cm), l is the distance the light travels through the solution (in cm), and c is the concentration of the absorbing species (in mol/L). The Beer–Lambert law is the basis for using spectroscopy in quantitative analysis. If ϵ and l are known, measuring A for a solution allows us to calculate the concentration of the absorbing species in the solution.

Suppose we have a pink solution containing an unknown concentration of $Co^2(aq)$ ions. A sample of this solution is placed in a spectrophotometer, and the absorbance is measured at a wavelength where ϵ for $Co^{2+}(aq)$ is known to be 12 L/mol · cm. The absorbance A is found to be 0.60. The width of the sample tube is 1.0 cm. We want to determine the concentration of $Co^{2+}(aq)$ in the solution. This problem can be solved by a straightforward application of the Beer–Lambert law,

$$A = \epsilon l c$$

where

$$A = 0.60$$

$$\epsilon = \frac{12 \text{ L}}{\text{mol} \cdot \text{cm}}$$

$$l = \text{light path} = 1.0 \text{ cm}$$

Solving for the concentration gives

$$c = \frac{A}{\epsilon l} = \frac{0.06}{\left(12\dfrac{\text{L}}{\text{mol} \cdot \text{cm}}\right)(1.0 \text{ cm})} = 5.0 \times 10^{-2} \text{ mol/L}$$

To obtain the unknown concentration of an absorbing species from the measured absorbance, we must know the product ϵl, since

$$c = \frac{A}{\epsilon l}$$

We can obtain the product ϵl by measuring the absorbance of a solution of *known* concentration, since

Measured using a
↙ spectrophotometer

$$\epsilon l = \frac{A}{c}$$

↖ Known from making up
the solution

However, a more accurate value of the product ϵl can be obtained by plotting A versus c for a series of solutions. Note that the equation $A = \epsilon l c$ gives a straight line with slope ϵl when A is plotted against c.

For example, consider the following typical spectroscopic analysis. A sample of steel from a bicycle frame is to be analyzed to determine its manganese content. The procedure involves weighing out a sample of the steel, dissolving it in strong acid, treating the resulting solution with a very strong oxidizing agent to convert all the manganese to permanganate ion (MnO_4^-), and then using spectroscopy to determine the concentration of the intensely purple MnO_4^- ions in the solution. To do this, however, the value of ϵl for MnO_4^- must be determined at an appropriate wavelength. The absorbance values for four solutions with known MnO_4^- concentrations were measured to give the following data:

Solution	Concentration of MnO_4^- (mol/L)	Absorbance
1	7.00×10^{-5}	0.175
2	1.00×10^{-4}	0.250
3	2.00×10^{-4}	0.500
4	3.50×10^{-4}	0.875

A plot of absorbance versus concentration for the solutions of known concentration is shown in Fig. A.7. The slope of this line (change in A/change in c) is 2.48×10^3 L/mol. This quantity represents the product ϵl.

A sample of the steel weighing 0.1523 g was dissolved and the unknown amount of manganese was converted to MnO_4^- ions. Water was then added to give a solution with a final volume of 100.0 mL. A portion of this solution was placed in a spectro-

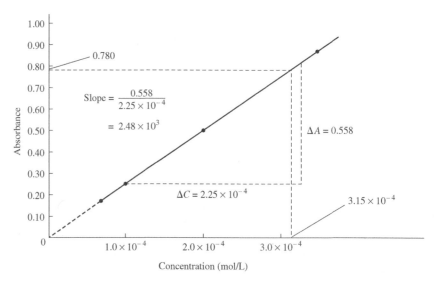

Figure A.7 | A plot of absorbance versus concentration of MnO_4^- in a series of solutions of known concentration.

photometer, and its absorbance was found to be 0.780. Using these data, we want to calculate the percent manganese in the steel. The MnO_4^- ions from the manganese in the dissolved steel sample show an absorbance of 0.780. Using the Beer–Lambert law, we calculate the concentration of MnO_4^- in this solution:

$$c = \frac{A}{\epsilon l} = \frac{0.780}{2.48 \times 10^3 \text{ L/mol}} \times 3.15 \times 10^{-4} \text{ mol/L}$$

There is a more direct way for finding c. Using a graph such as that in Fig. A.7 (often called a *Beer's law plot*), we can read the concentration that corresponds to $A = 0.780$. This interpolation is shown by dashed lines on the graph. By this method, $c = 3.15 \times 10^{-4}$ mol/L, which agrees with the value obtained above.

Recall that the original 0.1523-g steel sample was dissolved, the manganese was converted to permanganate, and the volume was adjusted to 100.0 mL. We now know that $[MnO_4]$ in that solution is 3.15×10^{-4} M. Using this concentration, we can calculate the total number of moles of MnO_4^- in that solution:

$$\text{mol } MnO_4^- = 100.0 \text{ mL} \times \frac{1 \text{ L}}{1000 \text{ mL}} \times 3.15 \times 10^{-4} \frac{\text{mol}}{\text{L}}$$

$$= 3.15 \times 10^{-5} \text{ mol}$$

Since each mole of manganese in the original steel sample yields a mole of MnO_4, that is,

$$1 \text{ mol Mn} \xrightarrow{\text{Oxidation}} 1 \text{ mol } MnO_4^-$$

the original steel sample must have contained 3.15×10^{-5} mole of manganese. The mass of manganese present in the sample is

$$3.15 \times 10^{-5} \text{ mol Mn} \times \frac{54.938 \text{ g of Mn}}{1 \text{ mol Mn}} = 1.73 \times 10^{-3} \text{ g of Mn}$$

Since the steel sample weighed 0.1523 g, the present manganese in the steel is

$$\frac{1.73 \times 10^{-3} \text{ g of Mn}}{1.523 \times 10^{-1} \text{ g of sample}} \times 100 = 1.14\%$$

This example illustrates a typical use of spectroscopy in quantitative analysis. The steps commonly involved are as follows:

1. Preparation of a calibration plot (a Beer's law plot) by measuring the absorbance values of a series of solutions with known concentrations

2. Measurement of the absorbance of the solution of unknown concentration

3. Use of the calibration plot to determine the unknown concentration

---

APPENDIX 4	Selected Thermodynamic Data

Note: All values are assumed precise to at least ± 1.

Substance and State	ΔH_f° (kJ/mol)	ΔG_f° (kJ/mol)	ΔS° (J/K · mol)	Substance and State	ΔH_f° (kJ/mol)	ΔG_f° (kJ/mol)	ΔS° (J/K · mol)
Aluminum				Barium			
Al(s)	0	0	28	Ba(s)	0	0	67
Al$_2$O$_3$(s)	−1676	−1582	51	BaCO$_3$(s)	−1219	−1139	112
Al(OH)$_3$(s)	−1277			BaO(s)	−582	−552	70
AlCl$_3$(s)	−704	−629	111	Ba(OH)$_2$(s)	−946		

(continued)

Appendix Four (continued)

Substance and State	ΔH_f° (kJ/mol)	ΔG_f° (kJ/mol)	ΔS° (J/K · mol)
Barium, *continued*			
$BaSO_4(s)$	−1465	−1353	132
Beryllium			
$Be(s)$	0	0	10
$BeO(s)$	−599	−569	14
$Be(OH)_2(s)$	−904	−815	47
Bromine			
$Br_2(l)$	0	0	152
$Br_2(g)$	31	3	245
$Br_2(aq)$	−3	4	130
$Br^-(aq)$	−121	−104	82
$HBr(g)$	−36	−53	199
Cadmium			
$Cd(s)$	0	0	52
$CdO(s)$	−258	−228	55
$Cd(OH)_2(s)$	−561	−474	96
$CdS(s)$	−162	−156	65
$CdSO_4(s)$	−935	−823	123
Calcium			
$Ca(s)$	0	0	41
$CaC_2(s)$	−63	−68	70
$CaCO_3(s)$	−1207	−1129	93
$CaO(s)$	−635	−604	40
$Ca(OH)_2(s)$	−987	−899	83
$Ca_3(PO_4)_2(s)$	−4126	−3890	241
$CaSO_4(s)$	−1433	−1320	107
$CaSiO_3(s)$	−1630	−1550	84
Carbon			
$C(s)$ (graphite)	0	0	6
$C(s)$ (diamond)	2	3	2
$CO(g)$	−110.5	−137	198
$CO_2(g)$	−393.5	−394	214
$CH_4(g)$	−75	−51	186
$CH_3OH(g)$	−201	−163	240
$CH_3OH(l)$	−239	−166	127
$H_2CO(g)$	−116	−110	219
$HCOOH(g)$	−363	−351	249
$HCN(g)$	135.1	125	202
$C_2H_2(g)$	227	209	201
$C_2H_4(g)$	52	68	219
$CH_3CHO(g)$	−166	−129	250
$C_2H_5OH(l)$	−278	−175	161
$C_2H_6(g)$	−84.7	−32.9	229.5
$C_3H_6(g)$	20.9	62.7	266.9
$C_3H_8(g)$	−104	−24	270
$C_2H_4O(g)$ (ethylene oxide)	−53	−13	242
$CH_2=CHCN(g)$	185.0	195.4	274
$CH_3COOH(l)$	−484	−389	160
$C_6H_{12}O_6(s)$	−1275	−911	212
CCl_4	−135	−65	216
Chlorine			
$Cl_2(g)$	0	0	223
$Cl_2(aq)$	−23	7	121

Substance and State	ΔH_f° (kJ/mol)	ΔG_f° (kJ/mol)	ΔS° (J/K · mol)
Chlorine, *continued*			
$Cl^-(aq)$	−167	−131	57
$HCl(g)$	−92	−95	187
Chromium			
$Cr(s)$	0	0	24
$Cr_2O_3(s)$	−1128	−1047	81
$CrO_3(s)$	−579	−502	72
Copper			
$Cu(s)$	0	0	33
$CuCO_3(s)$	−595	−518	88
$Cu_2O(s)$	−170	−148	93
$CuO(s)$	−156	−128	43
$Cu(OH)_2(s)$	−450	−372	108
$CuS(s)$	−49	−49	67
Fluorine			
$F_2(g)$	0	0	203
$F2(aq)$	−333	−279	−14
$HF(g)$	−271	−273	174
Hydrogen			
$H_2(g)$	0	0	131
$H(g)$	217	203	115
$H^+(aq)$	0	0	0
$OH^-(aq)$	−230	−157	−11
$H_2O(l)$	−286	−237	70
$H_2O(g)$	−242	−229	189
Iodine			
$I_2(s)$	0	0	116
$I_2(g)$	62	19	261
$I_2(aq)$	23	16	137
$I^-(aq)$	−55	−52	106
Iron			
$Fe(s)$	0	0	27
$Fe_3C(s)$	21	15	108
$Fe_{0.95}O(s)$ (wustite)	−264	−240	59
FeO	−272	−255	61
$Fe_3O_4(s)$ (magnetite)	−1117	−1013	146
$Fe_2O_3(s)$ (hematite)	−826	−740	90
$FeS(s)$	−95	−97	67
$FeS_2(s)$	−178	−166	53
$FeSO_4(s)$	−929	−825	121
Lead			
$Pb(s)$	0	0	65
$PbO_2(s)$	−277	−217	69
$PbS(s)$	−100	−99	91
$PbSO_4(s)$	−920	−813	149
Magnesium			
$Mg(s)$	0	0	33
$MgCO_3(s)$	−1113	−1029	66
$MgO(s)$	−602	−569	27
$Mg(OH)_2(s)$	−925	−834	64
Manganese			
$Mn(s)$	0	0	32

Substance and State	ΔH_f° (kJ/mol)	ΔG_f° (kJ/mol)	ΔS° (J/K · mol)
Manganese, *continued*			
$MnO(s)$	−385	−363	60
$Mn_3O_4(s)$	−1387	−1280	149
$Mn_2O_3(s)$	−971	−893	110
$MnO_2(s)$	−521	−466	53
$MnO_4^-(aq)$	−543	−449	190
Mercury			
$Hg(l)$	0	0	76
$Hg_2Cl_2(s)$	−265	−211	196
$HgCl_2(s)$	−230	−184	144
$HgO(s)$	−90	−59	70
$HgS(s)$	−58	−49	78
Nickel			
$Ni(s)$	0	0	30
$NiCl_2(s)$	−316	−272	107
$NiO(s)$	−241	−213	38
$Ni(OH)_2(s)$	−538	−453	79
$NiS(s)$	−93	−90	53
Nitrogen			
$N_2(g)$	0	0	192
$NH_3(g)$	−46	−17	193
$NH_3(aq)$	−80	−27	111
$NH_4^+(aq)$	−132	−79	113
$NO(g)$	90	87	211
$NO_2(g)$	34	52	240
$N_2O(g)$	82	104	220
$N_2O_4(g)$	10	98	304
$N_2O_4(l)$	−20	97	209
$N_2O_5(s)$	−42	134	178
$N_2H_4(l)$	51	149	121
$N_2H_3CH_3(l)$	54	180	166
$HNO_3(aq)$	−207	−111	146
$HNO_3(l)$	−174	−81	156
$NH_4ClO_4(s)$	−295	−89	186
$NH_4Cl(s)$	−314	−203	96
Oxygen			
$O_2(g)$	0	0	205
$O(g)$	249	232	161
$O_3(g)$	143	163	239
Phosphorus			
$P(s)$ (white)	0	0	41
$P(s)$ (red)	−18	−12	23
$P(s)$ (black)	−39	−33	23
$P_4(g)$	59	24	280
$PF_5(g)$	−1578	−1509	296
$PH_3(g)$	5	13	210
$H_3PO_4(s)$	−1279	21119	110
$H_3PO_4(l)$	−1267	—	—
$H_3PO_4(aq)$	−1288	−1143	158
$P_4O_{10}(s)$	−2984	2−2698	229
Potassium			
$K(s)$	0	0	64
$KCl(s)$	−436	−408	83

Substance and State	ΔH_f° (kJ/mol)	ΔG_f° (kJ/mol)	ΔS° (J/K · mol)
Potassium, *continued*			
$KClO_3(s)$	−391	−290	143
$KClO_4(s)$	−433	−304	151
$K_2O(s)$	−361	−322	98
$K_2O_2(s)$	−496	−430	113
$KO_2(s)$	−283	−238	117
$KOH(s)$	−425	−379	79
$KOH(aq)$	−481	2−440	9.20
Silicon			
$SiO_2(s)$ (quartz)	−911	−856	42
$SiCl_4(l)$	−687	−620	240
Silver			
$Ag(s)$	0	0	43
$Ag^+(aq)$	105	77	73
$AgBr(s)$	−100	−97	107
$AgCN(s)$	146	164	84
$AgCl(s)$	−127	−110	96
$Ag_2CrO_4(s)$	−712	−622	217
$AgI(s)$	−62	−66	115
$Ag_2O(s)$	−31	−11	122
$Ag_2S(s)$	−32	−40	146
Sodium			
$Na(s)$	0	0	51
$Na^+(aq)$	−240	−262	59
$NaBr(s)$	−360	−347	84
$Na_2CO_3(s)$	−1131	−1048	136
$NaHCO_3(s)$	−948	−852	102
$NaCl(s)$	−411	−384	72
$NaH(s)$	−56	−33	40
$NaI(s)$	−288	−282	91
$NaNO_2(s)$	−359		
$NaNO_3(s)$	−467	−366	116
$Na_2O(s)$	−416	−377	73
$Na_2O_2(s)$	−515	−451	95
$NaOH(s)$	−2427	−381	64
$NaOH(aq)$	−470	−419	50
Sulfur			
$S(s)$ (rhombic)	0	0	32
$S(s)$ (monoclinic)	0.3	0.1	33
$S^{2-}(aq)$	33	86	215
$S_8(g)$	102	50	431
$SF_6(g)$	−1209	−1105	292
$H_2S(g)$	−21	−34	206
$SO_2(g)$	−297	−300	248
$SO_3(g)$	−396	−371	257
$SO_4^{2-}(aq)$	−909	−745	20
$H_2SO_4(l)$	−814	−690	157
$H_2SO_4(aq)$	−909	−745	20
Tin			
$Sn(s)$ (white)	0	0	52
$Sn(s)$ (gray)	22	0.1	44
$SnO(s)$	−285	−257	56
$SnO_2(s)$	−581	−520	52

(continued)

Appendix Four (continued)

Substance and State	ΔH_f° (kJ/mol)	ΔG_f° (kJ/mol)	ΔS° (J/K · mol)
Tin, *continued*			
$Sn(OH)_2(s)$	−561	−492	155
Titanium			
$TiCl_4(g)$	−763	−727	355
$TiO_2(s)$	−945	−890	50
Uranium			
$U(s)$	0	0	50
$UF_6(s)$	−2137	−2008	228
$UF_6(g)$	−2113	−2029	380
$UO_2(s)$	−1084	−1029	78
$U_3O_8(s)$	−3575	−3393	282
$UO_3(s)$	−1230	−1150	99

Substance and State	ΔH_f° (kJ/mol)	ΔG_f° (kJ/mol)	ΔS° (J/K · mol)
Xenon			
$Xe(g)$	0	0	170
$XeF_2(g)$	−108	−48	254
$XeF_4(s)$	−251	−121	146
$XeF_6(g)$	−294		
$XeO_3(s)$	402		
Zinc			
$Zn(s)$	0	0	42
$ZnO(s)$	−348	−318	44
$Zn(OH)_2(s)$	−642		
$ZnS(s)$ (wurtzite)	−193		
$ZnS(s)$ (zinc blende)	−206	−201	58
$ZnSO_4(s)$	−983	−874	120

APPENDIX 5 | Equilibrium Constants and Reduction Potentials

A5.1 | Values of K_a for Some Common Monoprotic Acids

Name	Formula	Value of K_a
Hydrogen sulfate ion	HSO_4^-	1.2×10^{-2}
Chlorous acid	$HClO_2$	1.2×10^{-2}
Monochloracetic acid	$HC_2H_2ClO_2$	1.35×10^{-3}
Hydrofluoric acid	HF	7.2×10^{-4}
Nitrous acid	HNO_2	4.0×10^{-4}
Formic acid	HCO_2H	1.8×10^{-4}
Lactic acid	$HC_3H_5O_3$	1.38×10^{-4}
Benzoic acid	$HC_7H_5O_2$	6.4×10^{-5}
Acetic acid	$HC_2H_3O_2$	1.8×10^{-5}
Hydrated aluminum(III) ion	$[Al(H_2O)_6]^{3+}$	1.4×10^{-5}
Propanoic acid	$HC_3H_5O_2$	1.3×10^{-5}
Hypochlorous acid	$HOCl$	3.5×10^{-8}
Hypobromous acid	$HOBr$	2×10^{-9}
Hydrocyanic acid	HCN	6.2×10^{-10}
Boric acid	H_3BO_3	5.8×10^{-10}
Ammonium ion	NH_4^+	5.6×10^{-10}
Phenol	HOC_6H_5	1.6×10^{-10}
Hypoiodous acid	HOI	2×10^{-11}

A5.2 | Stepwise Dissociation Constants for Several Common Polyprotic Acids

Name	Formula	K_{a_1}	K_{a_2}	K_{a_3}
Phosphoric acid	H_3PO_4	7.5×10^{-3}	6.2×10^{-8}	4.8×10^{-13}
Arsenic acid	H_3AsO_4	5.5×10^{-3}	1.7×10^{-7}	5.1×10^{-12}
Carbonic acid	H_2CO_3	4.3×10^{-7}	5.6×10^{-11}	
Sulfuric acid	H_2SO_4	Large	1.2×10^{-2}	
Sulfurous acid	H_2SO_3	1.5×10^{-2}	1.0×10^{-7}	
Hydrosulfuric acid	H_2S	1.0×10^{-7}	$\sim 10^{-19}$	
Oxalic acid	$H_2C_2O_4$	6.5×10^{-2}	6.1×10^{-5}	
Ascorbic acid (vitamin C)	$H_2C_6H_6O_6$	7.9×10^{-5}	1.6×10^{-12}	
Citric acid	$H_3C_6H_5O_7$	8.4×10^{-4}	1.8×10^{-5}	4.0×10^{-6}

A5.3 | Values of K_b for Some Common Weak Bases

Name	Conjugate Formula	Acid	K_b
Ammonia	NH_3	NH_4^+	1.8×10^{-5}
Methylamine	CH_3NH_2	$CH_3NH_3^+$	4.38×10^{-4}
Ethylamine	$C_2H_5NH_2$	$C_2H_5NH_3^+$	5.6×10^{-4}
Diethylamine	$(C_2H_5)_2NH$	$(C_2H_5)_2NH_2^+$	1.3×10^{-3}
Triethylamine	$(C_2H_5)_3N$	$(C_2H_5)_3NH^+$	4.0×10^{-4}
Hydroxylamine	$HONH_2$	$HONH_3^+$	1.1×10^{-8}
Hydrazine	H_2NNH_2	$H_2NNH_3^+$	3.0×10^{-6}
Aniline	$C_6H_5NH_2$	$C_6H_5NH_3^+$	3.8×10^{-10}
Pyridine	C_5H_5N	$C_5H_5NH^+$	1.7×10^{-9}

A5.4 | K_{sp} Values at 25°C for Common Ionic Solids

Ionic Solid	K_{sp} (at 25°C)	Ionic Solid	K_{sp} (at 25°C)	Ionic Solid	K_{sp} (at 25°C)
Fluorides		Hg_2CrO_4*	2×10^{-9}	$Ni(OH)_2$	1.6×10^{-16}
BaF_2	2.4×10^{-5}	$BaCrO_4$	8.5×10^{-11}	$Zn(OH)_2$	4.5×10^{-17}
MgF_2	6.4×10^{-9}	Ag_2CrO_4	9.0×10^{-12}	$Cu(OH)_2$	1.6×10^{-19}
PbF_2	4×10^{-8}	$PbCrO_4$	2×10^{-16}	$Hg(OH)_2$	3×10^{-26}
SrF_2	7.9×10^{-10}			$Sn(OH)_2$	3×10^{-27}
CaF_2	4.0×10^{-11}	Carbonates		$Cr(OH)_3$	6.7×10^{-31}
		$NiCO_3$	1.4×10^{-7}	$Al(OH)_3$	2×10^{-32}
Chlorides		$CaCO_3$	8.7×10^{-9}	$Fe(OH)_3$	4×10^{-38}
$PbCl_2$	1.6×10^{-5}	$BaCO_3$	1.6×10^{-9}	$Co(OH)_3$	2.5×10^{-43}
AgCl	1.6×10^{-10}	$SrCO_3$	7×10^{-10}		
Hg_2Cl_2*	1.1×10^{-18}	$CuCO_3$	2.5×10^{-10}	Sulfides	
		$ZnCO_3$	2×10^{-10}	MnS	2.3×10^{-13}
Bromides		$MnCO_3$	8.8×10^{-11}	FeS	3.7×10^{-19}
$PbBr_2$	4.6×10^{-6}	$FeCO_3$	2.1×10^{-11}	NiS	3×10^{-21}
AgBr	5.0×10^{-13}	Ag_2CO_3	8.1×10^{-12}	CoS	5×10^{-22}
Hg_2Br_2*	1.3×10^{-22}	$CdCO_3$	5.2×10^{-12}	ZnS	2.5×10^{-22}
		$PbCO_3$	1.5×10^{-15}	SnS	1×10^{-26}
Iodides		$MgCO_3$	1×10^{-5}	CdS	1.0×10^{-28}
PbI_2	1.4×10^{-8}	Hg_2CO_3*	9.0×10^{-15}	PbS	7×10^{-29}
AgI	1.5×10^{-16}			CuS	8.5×10^{-45}
Hg_2I_2*	4.5×10^{-29}	Hydroxides		Ag_2S	1.6×10^{-49}
		$Ba(OH)_2$	5.0×10^{-3}	HgS	1.6×10^{-54}
Sulfates		$Sr(OH)_2$	3.2×10^{-4}		
$CaSO_4$	6.1×10^{-5}	$Ca(OH)_2$	1.3×10^{-6}	Phosphates	
Ag_2SO_4	1.2×10^{-5}	AgOH	2.0×10^{-8}	Ag_3PO_4	1.8×10^{-18}
$SrSO_4$	3.2×10^{-7}	$Mg(OH)_2$	8.9×10^{-12}	$Sr_3(PO_4)_2$	1×10^{-31}
$PbSO_4$	1.3×10^{-8}	$Mn(OH)_2$	2×10^{-13}	$Ca_3(PO_4)_2$	1.3×10^{-32}
$BaSO_4$	1.5×10^{-9}	$Cd(OH)_2$	5.9×10^{-15}	$Ba_3(PO_4)_2$	6×10^{-39}
		$Pb(OH)_2$	1.2×10^{-15}	$Pb_3(PO_4)_2$	1×10^{-54}
Chromates		$Fe(OH)_2$	1.8×10^{-15}		
$SrCrO_4$	3.6×10^{-5}	$Co(OH)_2$	2.5×10^{-16}		

*Contains Hg_2^{2+} ions. $K_{sp} = [Hg_2^{2+}][X^-]^2$ for Hg_2X_2 salts.

A5.5 | Standard Reduction Potentials at 25°C (298 K) for Many Common Half-Reactions

Half-Reaction	$\mathscr{E}°$ (V)	Half-Reaction	$\mathscr{E}°$ (V)
$F_2 + 2e^- \rightarrow 2F^-$	2.87	$O_2 + 2H_2O + 4e^- \rightarrow 4OH^-$	0.40
$Ag^{2+} + e^- \rightarrow Ag^+$	1.99	$Cu^{2+} + 2e^- \rightarrow Cu$	0.34
$Co^{3+} + e^- \rightarrow Co^{2+}$	1.82	$Hg_2Cl_2 + 2e^- \rightarrow 2Hg + 2Cl^-$	0.34
$H_2O_2 + 2H^+ + 2e^- \rightarrow 2H_2O$	1.78	$AgCl + e^- \rightarrow Ag + Cl^-$	0.22
$Ce^{4+} + e^- \rightarrow Ce^{3+}$	1.70	$SO_4^{2-} + 4H^+ + 2e^- \rightarrow H_2SO_3 + H_2O$	0.20
$PbO_2 + 4H^+ + SO_4^{2-} + 2e^- \rightarrow PbSO_4 + 2H_2O$	1.69	$Cu^{2+} + e^- \rightarrow Cu^+$	0.16
$MnO_4^- + 4H^+ + 3e^- \rightarrow MnO_2 + 2H_2O$	1.68	$2H^+ + 2e^- \rightarrow H_2$	0.00
$2e^- + 2H^+ + IO_4^- \rightarrow IO_3^- + H_2O$	1.60	$Fe^{3+} + 3e^- \rightarrow Fe$	-0.036
$MnO_4^- + 8H^+ + 5e^- \rightarrow Mn^{2+} + 4H_2O$	1.51	$Pb^{2+} + 2e^- \rightarrow Pb$	-0.13
$Au^{3+} + 3e^- \rightarrow Au$	1.50	$Sn^{2+} + 2e^- \rightarrow Sn$	-0.14
$PbO_2 + 4H^+ + 2e^- \rightarrow Pb^{2+} + 2H_2O$	1.46	$Ni^{2+} + 2e^- \rightarrow Ni$	-0.23
$Cl_2 + 2e^- \rightarrow 2Cl^-$	1.36	$PbSO_4 + 2e^- \rightarrow Pb + SO_4^{2-}$	-0.35
$Cr_2O_7^{2-} + 14H^+ + 6e^- \rightarrow 2Cr^{3+} + 7H_2O$	1.33	$Cd^{2+} + 2e^- \rightarrow Cd$	-0.40
$O_2 + 4H^+ + 4e^- \rightarrow 2H_2O$	1.23	$Fe^{2+} + 2e^- \rightarrow Fe$	-0.44
$MnO_2 + 4H^+ + 2e^- \rightarrow Mn^{2+} + 2H_2O$	1.21	$Cr^{3+} + e^- \rightarrow Cr^{2+}$	-0.50
$IO_3^- + 6H^+ + 5e^- \rightarrow \frac{1}{2}I_2 + 3H_2O$	1.20	$Cr^{3+} + 3e^- \rightarrow Cr$	-0.73
$Br_2 + 2e^- \rightarrow 2Br^-$	1.09	$Zn^{2+} + 2e^- \rightarrow Zn$	-0.76
$VO_2^+ + 2H^+ + e^- \rightarrow VO^{2+} + H_2O$	1.00	$2H_2O + 2e^- \rightarrow H_2 + 2OH^-$	-0.83
$AuCl_4^- + 3e^- \rightarrow Au + 4Cl^-$	0.99	$Mn^{2+} + 2e^- \rightarrow Mn$	-1.18
$NO_3^- + 4H^+ + 3e^- \rightarrow NO + 2H_2O$	0.96	$Al^{3+} + 3e^- \rightarrow Al$	-1.66
$ClO_2 + e^- \rightarrow ClO_2^-$	0.954	$H_2 + 2e^- \rightarrow 2H^-$	-2.23
$2Hg^{2+} + 2e^- \rightarrow Hg_2^{2+}$	0.91	$Mg^{2+} + 2e^- \rightarrow Mg$	-2.37
$Ag^+ + e^- \rightarrow Ag$	0.80	$La^{3+} + 3e^- \rightarrow La$	-2.37
$Hg_2^{2+} + 2e^- \rightarrow 2Hg$	0.80	$Na^+ + e^- \rightarrow Na$	-2.71
$Fe^{3+} + e^- \rightarrow Fe^{2+}$	0.77	$Ca^{2+} + 2e^- \rightarrow Ca$	-2.76
$O_2 + 2H^+ + 2e^- \rightarrow H_2O_2$	0.68	$Ba^{2+} + 2e^- \rightarrow Ba$	-2.90
$MnO_4^- + e^- \rightarrow MnO_4^{2-}$	0.56	$K^+ + e^- \rightarrow K$	-2.92
$I_2 + 2e^- \rightarrow 2I^-$	0.54	$Li^+ + e^- \rightarrow Li$	-3.05
$Cu^+ + e^- \rightarrow Cu$	0.52		

APPENDIX 6 | SI Units and Conversion Factors

Length

SI unit: meter (m)
1 meter = 1.0936 yards
1 centimeter = 0.39370 inch
1 inch = 2.54 centimeters (exactly)
1 kilometer = 0.62137 mile
1 mile = 5280 feet
= 1.6093 kilometers
1 angstrom = 10^{-10} meter
= 100 picometers

Mass

SI unit: kilogram (kg)
1 kilogram = 1000 grams
= 2.2046 pounds
1 pound = 453.59 grams
= 0.45359 kilogram
= 16 ounces
1 ton = 2000 pounds
= 907.185 kilograms
1 metric ton = 1000 kilograms
= 2204.6 pounds
1 atomic mass unit = 1.66056×10^{-27} kilograms

Volume

SI unit: cubic meter (m^3)
1 liter = 10^{-3} m^3
= 1 dm^3
= 1.0567 quarts
1 gallon = 4 quarts
= 8 pints
= 3.7854 liters
1 quart = 32 fluid ounces
= 0.94633 liter

Temperature

SI unit: kelvin (K)
0 K = $-273.15°C$
= $-459.67°F$
K = °C + 273.15
$°C = \dfrac{5}{9}°F - 32$
$°F = \dfrac{9}{5}°C + 32$

Energy

SI unit: joule (J)
1 joule = 1 kg · m^2/s^2
= 0.23901 calorie
= 9.4781×10^{-4} btu
(British thermal unit)
1 calorie = 4.184 joules
= 3.965×10^{-3} btu
1 btu = 1055.06 joules
= 252.2 calories

Pressure

SI unit: pascal (Pa)
1 pascal = 1 N/m^2
= 1 kg/m · s^2
1 atmosphere = 101.325 kilopascals
= 760 torr (mm Hg)
= 14.70 pounds per
square inch
1 bar = 10^5 pascals

Glossary

Accuracy the agreement of a particular value with the true value. (1.4)

Acid a substance that produces hydrogen ions in solution; a proton donor. (2.8; 4.2; 4.8)

Acid–base indicator a substance that marks the end point of an acid–base titration by changing color. (15.5)

Acid dissociation constant (K_a) the equilibrium constant for a reaction in which a proton is removed from an acid by H_2O to form the conjugate base and H_3O^+. (14.1)

Acid rain a result of air pollution by sulfur dioxide. (5.10)

Acidic oxide a covalent oxide that dissolves in water to give an acidic solution. (14.10)

Actinide series a group of 14 elements following actinium in the periodic table, in which the $5f$ orbitals are being filled. (7.11; 20.1)

Activated complex (transition state) the arrangement of atoms found at the top of the potential energy barrier as a reaction proceeds from reactants to products. (12.6)

Activation energy the threshold energy that must be overcome to produce a chemical reaction. (12.6)

Addition polymerization a type of polymerization in which the monomers simply add together to form the polymer, with no other products. (22.5)

Addition reaction a reaction in which atoms add to a carbon–carbon multiple bond. (22.2)

Adsorption the collection of one substance on the surface of another. (12.7)

Air pollution contamination of the atmosphere, mainly by the gaseous products of transportation and production of electricity. (5.10)

Alcohol an organic compound in which the hydroxyl group is a substituent on a hydrocarbon. (22.4)

Aldehyde an organic compound containing the carbonyl group bonded to at least one hydrogen atom. (22.4)

Alkali metal a Group 1A metal. (2.7; 20.2)

Alkaline earth metal a Group 2A metal. (2.7; 20.4)

Alkane a saturated hydrocarbon with the general formula C_nH_{2n+2}. (22.1)

Alkene an unsaturated hydrocarbon containing a carbon–carbon double bond. The general formula is C_nH_{2n}. (22.2)

Alkyne an unsaturated hydrocarbon containing a triple carbon–carbon bond. The general formula is C_nH_{2n-2}. (22.2)

Alloy a substance that contains a mixture of elements and has metallic properties. (10.4)

Alloy steel a form of steel containing carbon plus other metals such as chromium, cobalt, manganese, and molybdenum. (21.8)

Alpha (α) particle a helium nucleus. (19.1)

Alpha-particle production a common mode of decay for radioactive nuclides in which the mass number changes. (19.1)

Amine an organic base derived from ammonia in which one or more of the hydrogen atoms are replaced by organic groups. (14.6; 22.4)

α-Amino acid an organic acid in which an amino group and an R group are attached to the carbon atom next to the carboxyl group. (22.6)

Amorphous solid a solid with considerable disorder in its structure. (10.3)

Ampere the unit of electric current equal to one coulomb of charge per second. (18.8)

Amphoteric substance a substance that can behave either as an acid or as a base. (14.2)

Angular momentum quantum number (ℓ) the quantum number relating to the shape of an atomic orbital, which can assume any integral value from 0 to $n-1$ for each value of n. (7.6)

Anion a negative ion. (2.6)

Anode the electrode in a galvanic cell at which oxidation occurs. (18.2)

Antibonding molecular orbital an orbital higher in energy than the atomic orbitals of which it is composed. (9.2)

Aqueous solution a solution in which water is the dissolving medium or solvent. (4)

Aromatic hydrocarbon one of a special class of cyclic unsaturated hydrocarbons, the simplest of which is benzene. (22.3)

Arrhenius concept a concept postulating that acids produce hydrogen ions in aqueous solution, while bases produce hydroxide ions. (14.1)

Arrhenius equation the equation representing the rate constant as $k = Ae^{-E_a/RT}$, where A represents the product of the collision frequency and the steric factor, and $e^{-E_a/RT}$ is the fraction of collisions with sufficient energy to produce a reaction. (12.6)

Atactic chain a polymer chain in which the substituent groups such as CH_3 are randomly distributed along the chain. (22.5)

Atmosphere the mixture of gases that surrounds the earth's surface. (5.10)

Atomic number the number of protons in the nucleus of an atom. (2.5; 18)

Atomic radius half the distance between the nuclei in a molecule consisting of identical atoms. (7.12)

Atomic solid a solid that contains atoms at the lattice points. (10.3)

Atomic weight the weighted average mass of the atoms in a naturally occurring element. (2.3)

Aufbau principle the principle stating that as protons are added one by one to the nucleus to build up the elements, electrons are similarly added to hydrogen-like orbitals. (7.11)

Autoionization the transfer of a proton from one molecule to another of the same substance. (14.2)

Avogadro's law equal volumes of gases at the same temperature and pressure contain the same number of particles. (5.2)

Avogadro's number the number of atoms in exactly 12 grams of pure ^{12}C, equal to 6.022×10^{23}. (3.3)

Ball-and-stick model a molecular model that distorts the sizes of atoms but shows bond relationships clearly. (2.6)

Band model a molecular model for metals in which the electrons are assumed to travel around the metal crystal in molecular orbitals formed from the valence atomic orbitals of the metal atoms. (10.4)

Barometer a device for measuring atmospheric pressure. (5.1)

Base a substance that produces hydroxide ions in aqueous solution; a proton acceptor. (4.8)

Basic oxide an ionic oxide that dissolves in water to produce a basic solution. (14.10)

Basic oxygen process a process for producing steel by oxidizing and removing the impurities in iron using a high-pressure blast of oxygen. (21.8)

Battery a group of galvanic cells connected in series. (18.6)

Beta (β) particle an electron produced in radioactive decay. (19.1)

Beta-particle production a decay process for radioactive nuclides in which the mass number remains constant and the atomic number changes. The net effect is to change a neutron to a proton. (19.1)

Bidentate ligand a ligand that can form two bonds to a metal ion. (21.3)

Bimolecular step a reaction involving the collision of two molecules. (12.5)

Binary compound a two-element compound. (2.8)

Binding energy (nuclear) the energy required to decompose a nucleus into its component nucleons. (19.5)

Biomolecule a molecule responsible for maintaining and/or reproducing life. (22)

Blast furnace a furnace in which iron oxide is reduced to iron metal by using a very strong blast of hot air to produce carbon monoxide from coke, and then using this gas as a reducing agent for the iron. (21.8)

Bond energy the energy required to break a given chemical bond. (8.1)

Bond length the distance between the nuclei of the two atoms connected by a bond; the distance where the total energy of a diatomic molecule is minimal. (8.1)

Bond order the difference between the number of bonding electrons and the number of antibonding electrons, divided by two. It is an index of bond strength. (9.2)

Bonding molecular orbital an orbital lower in energy than the atomic orbitals of which it is composed. (9.2)

Bonding pair an electron pair found in the space between two atoms. (8.9)

Borane a covalent hydride of boron. (20.5)

Boyle's law the volume of a given sample of gas at constant temperature varies inversely with the pressure. (5.2)

Breeder reactor a nuclear reactor in which fissionable fuel is produced while the reactor runs. (19.6)

Brønsted–Lowry model a model proposing that an acid is a proton donor, and a base is a proton acceptor. (14.1)

Buffered solution a solution that resists a change in its pH when either hydroxide ions or protons are added. (15.2)

Buffering capacity the ability of a buffered solution to absorb protons or hydroxide ions without a significant change in pH;

determined by the magnitudes of [HA] and [A$^-$] in the solution. (15.3)

Calorimetry the science of measuring heat flow. (6.2)

Capillary action the spontaneous rising of a liquid in a narrow tube. (10.2)

Carbohydrate a polyhydroxyl ketone, a polyhydroxyl aldehyde, or a polymer composed of these. (22.6)

Carbon steel an alloy of iron containing up to about 1.5% carbon. (21.8)

Carboxyhemoglobin a stable complex of hemoglobin and carbon monoxide that prevents normal oxygen uptake in the blood. (21.7)

Carboxyl group the —COOH group in an organic acid. (14.2; 22.4)

Carboxylic acid an organic compound containing the carboxyl group; an acid with the general formula RCOOH. (22.4)

Catalyst a substance that speeds up a reaction without being consumed. (12.7)

Cathode the electrode in a galvanic cell at which reduction occurs. (18.2)

Cathode rays the "rays" emanating from the negative electrode (cathode) in a partially evacuated tube; a stream of electrons. (2.4)

Cathodic protection a method in which an active metal, such as magnesium, is connected to steel to protect it from corrosion. (18.6)

Cation a positive ion. (2.6)

Cell potential (electromotive force) the driving force in a galvanic cell that pulls electrons from the reducing agent in one compartment to the oxidizing agent in the other. (18.2)

Ceramic a nonmetallic material made from clay and hardened by firing at high temperature; it contains minute silicate crystals suspended in a glassy cement. (10.5)

Chain reaction (nuclear) a self-sustaining fission process caused by the production of neutrons that proceed to split other nuclei. (19.6)

Charles's law the volume of a given sample of gas at constant pressure is directly proportional to the temperature in kelvins. (5.2)

Chelating ligand (chelate) a ligand having more than one atom with a lone pair that can be used to bond to a metal ion. (21.3)

Chemical bond the force or, more accurately, the energy that holds two atoms together in a compound. (2.6)

Chemical change the change of substances into other substances through a reorganization of the atoms; a chemical reaction. (1.9)

Chemical equation a representation of a chemical reaction showing the relative numbers of reactant and product molecules. (3.8)

Chemical equilibrium a dynamic reaction system in which the concentrations of all reactants and products remain constant as a function of time. (13)

Chemical formula the representation of a molecule in which the symbols for the elements are used to indicate the types of atoms present and subscripts are used to show the relative numbers of atoms. (2.6)

Chemical kinetics the area of chemistry that concerns reaction rates. (12)

Chemical stoichiometry the calculation of the quantities of material consumed and produced in chemical reactions. (3)

Chirality the quality of having nonsuperimposable mirror images. (21.4)

Chlor–alkali process the process for producing chlorine and sodium hydroxide by electrolyzing brine in a mercury cell. (18.9)

Chromatography the general name for a series of methods for separating mixtures by using a system with a mobile phase and a stationary phase. (1.9)

Coagulation the destruction of a colloid by causing particles to aggregate and settle out. (11.8)

Codons organic bases in sets of three that form the genetic code. (22.6)

Colligative properties properties of a solution that depend only on the number, and not on the identity, of the solute particles. (11.5)

Collision model a model based on the idea that molecules must collide to react; used to account for the observed characteristics of reaction rates. (12.7)

Colloid (colloidal dispersion) a suspension of particles in a dispersing medium. (11.8)

Combustion reaction the vigorous and exothermic reaction that takes place between certain substances, particularly organic compounds, and oxygen. (22.1)

Common ion effect the shift in an equilibrium position caused by the addition or presence of an ion involved in the equilibrium reaction. (15.1)

Complete ionic equation an equation that shows all substances that are strong electrolytes as ions. (4.6)

Complex ion a charged species consisting of a metal ion surrounded by ligands. (16.3; 21.1)

Compound a substance with constant composition that can be broken down into elements by chemical processes. (1.9)

Concentration cell a galvanic cell in which both compartments contain the same components, but at different concentrations. (18.5)

Condensation the process by which vapor molecules re-form a liquid. (10.8)

Condensation polymerization a type of polymerization in which the formation of a small molecule, such as water, accompanies the extension of the polymer chain. (22.5)

Condensation reaction a reaction in which two molecules are joined, accompanied by the elimination of a water molecule. (20.3)

Condensed states of matter liquids and solids. (10.1)

Conjugate acid the species formed when a proton is added to a base. (14.1)

Conjugate acid–base pair two species related to each other by the donating and accepting of a single proton. (14.1)

Conjugate base what remains of an acid molecule after a proton is lost. (14.1)

Continuous spectrum a spectrum that exhibits all the wavelengths of visible light. (7.3)

Control rods rods in a nuclear reactor composed of substances that absorb neutrons. These rods regulate the power level of the reactor. (19.6)

Coordinate covalent bond a metal–ligand bond resulting from the interaction of a Lewis base (the ligand) and a Lewis acid (the metal ion). (21.3)

Coordination compound a compound composed of a complex ion and counter ions sufficient to give no net charge. (21.3)

Coordination isomerism isomerism in a coordination compound in which the composition of the coordination sphere of the metal ion varies. (21.4)

Coordination number the number of bonds formed between the metal ion and the ligands in a complex ion. (21.3)

Copolymer a polymer formed from the polymerization of more than one type of monomer. (22.5)

Core electron an inner electron in an atom; one not in the outermost (valence) principal quantum level. (7.11)

Corrosion the process by which metals are oxidized in the atmosphere. (18.7)

Coulomb's law $E = 2.31 \times 10^{-19}\left(\dfrac{Q_1Q_2}{r}\right)$, where E is the energy of interaction between a pair of ions, expressed in joules; r is the distance between the ion centers in nm; and Q_1 and Q_2 are the numerical ion charges. (8.1)

Counterions anions or cations that balance the charge on the complex ion in a coordination compound. (21.3)

Covalent bonding a type of bonding in which electrons are shared by atoms. (2.6; 8.1)

Critical mass the mass of fissionable material required to produce a self-sustaining chain reaction. (19.6)

Critical point the point on a phase diagram at which the temperature and pressure have their critical values; the endpoint of the liquid–vapor line. (10.9)

Critical pressure the minimum pressure required to produce liquefaction of a substance at the critical temperature. (10.9)

Critical reaction (nuclear) a reaction in which exactly one neutron from each fission event causes another fission event, thus sustaining the chain reaction. (19.6)

Critical temperature the temperature above which vapor cannot be liquefied no matter what pressure is applied. (10.9)

Crosslinking the existence of bonds between adjacent chains in a polymer, thus adding strength to the material. (22.5)

Crystal field model a model used to explain the magnetism and colors of coordination complexes through the splitting of the d orbital energies. (21.6)

Crystalline solid a solid with a regular arrangement of its components. (10.3)

Cubic closest packed (ccp) structure a solid modeled by the closest packing of spheres with an *abcabc* arrangement of layers; the unit cell is face-centered cubic. (10.4)

Cyanidation a process in which crushed gold ore is treated with an aqueous cyanide solution in the presence of air to dissolve the gold. Pure gold is recovered by reduction of the ion to the metal. (21.8)

Cyclotron a type of particle accelerator in which an ion introduced at the center is accelerated in an expanding spiral path by the use of alternating electrical fields in the presence of a magnetic field. (19.3)

Cytochromes a series of iron-containing species composed of heme and a protein. Cytochromes are the principal electron-transfer molecules in the respiratory chain. (21.7)

Dalton's law of partial pressures for a mixture of gases in a container, the total pressure exerted is the sum of the pressures that each gas would exert if it were alone. (5.5)

Degenerate orbitals a group of orbitals with the same energy. (7.7)

Dehydrogenation reaction a reaction in which two hydrogen atoms are removed from adjacent carbons of a saturated hydrocarbon, giving an unsaturated hydrocarbon. (22.1)

Denaturation the breaking down of the three-dimensional structure of a protein resulting in the loss of its function. (22.6)

Denitrification the return of nitrogen from decomposed matter to the atmosphere by bacteria that change nitrates to nitrogen gas. (20.2)

Density a property of matter representing the mass per unit volume. (1.8)

Deoxyribonucleic acid (DNA) a huge nucleotide polymer having a double-helical structure with complementary bases on the two strands. Its major functions are protein synthesis and the storage and transport of genetic information. (22.6)

Desalination the removal of dissolved salts from an aqueous solution. (11.6)

Dialysis a phenomenon in which a semipermeable membrane allows transfer of both solvent molecules and small solute molecules and ions. (11.6)

Diamagnetism a type of magnetism, associated with paired electrons, that causes a substance to be repelled from the inducing magnetic field. (9.3)

Differential rate law an expression that gives the rate of a reaction as a function of concentrations; often called the rate law. (12.2)

Diffraction the scattering of light from a regular array of points or lines, producing constructive and destructive interference. (7.2)

Diffusion the mixing of gases. (5.7)

Dilution the process of adding solvent to lower the concentration of solute in a solution. (4.3)

Dimer a molecule formed by the joining of two identical monomers. (22.5)

Dipole–dipole attraction the attractive force resulting when polar molecules line up so that the positive and negative ends are close to each other. (10.1)

Dipole moment a property of a molecule whose charge distribution can be represented by a center of positive charge and a center of negative charge. (8.3)

Direct reduction furnace a furnace in which iron oxide is reduced to iron metal using milder reaction conditions than in a blast furnace. (21.8)

Disaccharide a sugar formed from two monosaccharides joined by a glycoside linkage. (22.6)

Disproportionation reaction a reaction in which a given element is both oxidized and reduced. (20.7)

Distillation a method for separating the components of a liquid mixture that depends on differences in the ease of vaporization of the components. (1.9)

Disulfide linkage an S—S bond that stabilizes the tertiary structure of many proteins. (22.6)

Double bond a bond in which two pairs of electrons are shared by two atoms. (8.8)

Downs cell a cell used for electrolyzing molten sodium chloride. (18.9)

Dry cell battery a common battery used in calculators, watches, radios, and portable audio players. (18.6)

Dual nature of light the statement that light exhibits both wave and particulate properties. (7.2)

Effusion the passage of a gas through a tiny orifice into an evacuated chamber. (5.7)

Electrical conductivity the ability to conduct an electric current. (4.2)

Electrochemistry the study of the interchange of chemical and electrical energy. (18)

Electrolysis a process that involves forcing a current through a cell to cause a nonspontaneous chemical reaction to occur. (18.8)

Electrolyte a material that dissolves in water to give a solution that conducts an electric current. (4.2)

Electrolytic cell a cell that uses electrical energy to produce a chemical change that would otherwise not occur spontaneously. (18.8)

Electromagnetic radiation radiant energy that exhibits wavelike behavior and travels through space at the speed of light in a vacuum. (7.1)

Electron a negatively charged particle that moves around the nucleus of an atom. (2.4)

Electron affinity the energy change associated with the addition of an electron to a gaseous atom. (7.12)

Electron capture a process in which one of the inner-orbital electrons in an atom is captured by the nucleus. (19.1)

Electron spin quantum number a quantum number representing one of the two possible values for the electron spin; either $+\frac{1}{2}$ or $-\frac{1}{2}$. (7.8)

Electronegativity the tendency of an atom in a molecule to attract shared electrons to itself. (8.2)

Element a substance that cannot be decomposed into simpler substances by chemical or physical means. (1.9)

Elementary step a reaction whose rate law can be written from its molecularity. (12.5)

$E = mc^2$ Einstein's equation proposing that energy has mass; E is energy, m is mass, and c is the speed of light. (7.2)

Empirical formula the simplest whole-number ratio of atoms in a compound. (3.6)

Enantiomers isomers that are nonsuperimposable mirror images of each other. (21.4)

End point the point in a titration at which the indicator changes color. (4.8)

Endothermic refers to a reaction where energy (as heat) flows into the system. (6.1)

Energy the capacity to do work or to cause heat flow. (6.1)

Enthalpy a property of a system equal to $E + PV$, where E is the internal energy of the system, P is the pressure of the system, and V is the volume of the system. At constant pressure the change in enthalpy equals the energy flow as heat. (6.2)

Enthalpy (heat) of fusion the enthalpy change that occurs to melt a solid at its melting point. (10.8)

Entropy a thermodynamic function that measures randomness or disorder. (17.1)

Enzyme a large molecule, usually a protein, that catalyzes biological reactions. (12.7)

Equilibrium constant the value obtained when equilibrium concentrations of the chemical species are substituted in the equilibrium expression. (13.2)

Equilibrium expression the expression (from the law of mass action) obtained by multiplying the product concentrations and dividing by the multiplied reactant concentrations, with each concentration raised to a power represented by the coefficient in the balanced equation. (13.2)

Equilibrium point (thermodynamic definition) the position where the free energy of a reaction system has its lowest possible value. (17.8)

Equilibrium position a particular set of equilibrium concentrations. (13.2)

Equivalence point (stoichiometric point) the point in a titration when enough titrant has been added to react exactly with the substance in solution being titrated. (4.9; 15.4)

Ester an organic compound produced by the reaction between a carboxylic acid and an alcohol. (22.4)

Exothermic refers to a reaction where energy (as heat) flows out of the system. (6.1)

Exponential notation expresses a number as $N \times 10^M$, a convenient method for representing a very large or very small number and for easily indicating the number of significant figures. (1.5)

Faraday a constant representing the charge on one mole of electrons; 96,485 coulombs. (18.4)

Filtration a method for separating the components of a mixture containing a solid and a liquid. (1.9)

First law of thermodynamics the energy of the universe is constant; same as the law of conservation of energy. (6.1)

Fission the process of using a neutron to split a heavy nucleus into two nuclei with smaller mass numbers. (19.6)

Flotation process a method of separating the mineral particles in an ore from the gangue that depends on the greater wettability of the mineral pieces. (21.8)

Formal charge the charge assigned to an atom in a molecule or polyatomic ion derived from a specific set of rules. (8.12)

Formation constant (stability constant) the equilibrium constant for each step of the formation of a complex ion by the addition of an individual ligand to a metal ion or complex ion in aqueous solution. (16.3)

Formula equation an equation representing a reaction in solution showing the reactants and products in undissociated form, whether they are strong or weak electrolytes. (4.6)

Fossil fuel coal, petroleum, or natural gas; consists of carbon-based molecules derived from decomposition of once-living organisms. (6.5)

Frasch process the recovery of sulfur from underground deposits by melting it with hot water and forcing it to the surface by air pressure. (20.6)

Free energy a thermodynamic function equal to the enthalpy (H) minus the product of the entropy (S) and the Kelvin temperature (T); $G = H - TS$. Under certain conditions the change in free energy for a process is equal to the maximum useful work. (17.4)

Free radical a species with an unpaired electron. (22.5)

Frequency the number of waves (cycles) per second that pass a given point in space. (7.1)

Fuel cell a galvanic cell for which the reactants are continuously supplied. (18.6)

Functional group an atom or group of atoms in hydrocarbon derivatives that contains elements in addition to carbon and hydrogen. (22.4)

Fusion the process of combining two light nuclei to form a heavier, more stable nucleus. (19.6)

Galvanic cell a device in which chemical energy from a spontaneous redox reaction is changed to electrical energy that can be used to do work. (18.2)

Galvanizing a process in which steel is coated with zinc to prevent corrosion. (18.7)

Gamma (γ) ray a high-energy photon. (19.1)

Gangue the impurities (such as clay or sand) in an ore. (21.8)

Geiger–Müller counter (Geiger counter) an instrument that measures the rate of radioactive decay based on the ions and electrons produced as a radioactive particle passes through a gas-filled chamber. (19.4)

Gene a given segment of the DNA molecule that contains the code for a specific protein. (22.6)

Geometrical (cis–trans) isomerism isomerism in which atoms or groups of atoms can assume different positions around a rigid ring or bond. (21.4; 22.2)

Glass an amorphous solid obtained when silica is mixed with other compounds, heated above its melting point, and then cooled rapidly. (10.5)

Glass electrode an electrode for measuring pH from the potential difference that develops when it is dipped into an aqueous solution containing H^+ ions. (18.5)

Glycoside linkage a C—O—C bond formed between the rings of two cyclic monosaccharides by the elimination of water. (22.6)

Graham's law of effusion the rate of effusion of a gas is inversely proportional to the square root of the mass of its particles. (5.7)

Greenhouse effect a warming effect exerted by the earth's atmosphere (particularly CO_2 and H_2O) due to thermal energy retained by absorption of infrared radiation. (6.5)

Ground state the lowest possible energy state of an atom or molecule. (7.4)

Group (of the periodic table) a vertical column of elements having the same valence electron configuration and showing similar properties. (2.7)

Haber process the manufacture of ammonia from nitrogen and hydrogen, carried out at high pressure and high temperature with the aid of a catalyst. (3.10; 20.2)

Half-life (of a radioactive sample) the time required for the number of nuclides in a radioactive sample to reach half of the original value. (19.2)

Half-life (of a reactant) the time required for a reactant to reach half of its original concentration. (12.4)

Half-reactions the two parts of an oxidation–reduction reaction, one representing oxidation, the other reduction. (4.10; 17.1)

Halogen a Group 7A element. (2.7; 20.7)

Halogenation the addition of halogen atoms to unsaturated hydrocarbons. (22.2)

Hard water water from natural sources that contains relatively large concentrations of calcium and magnesium ions. (20.4)

Heat energy transferred between two objects due to a temperature difference between them. (6.1)

Heat capacity the amount of energy required to raise the temperature of an object by one degree Celsius. (6.2)

Heat of fusion the enthalpy change that occurs to melt a solid at its melting point. (10.8)

Heat of hydration the enthalpy change associated with placing gaseous molecules or ions in water; the sum of the energy needed to expand the solvent and the energy released from the solvent–solute interactions. (11.2)

Heat of solution the enthalpy change associated with dissolving a solute in a solvent; the sum of the energies needed to expand both solvent and solute in a solution and the energy released from the solvent–solute interactions. (11.2)

Heat of vaporization the energy required to vaporize one mole of a liquid at a pressure of one atmosphere. (10.8)

Heating curve a plot of temperature versus time for a substance where energy is added at a constant rate. (10.8)

Heisenberg uncertainty principle a principle stating that there is a fundamental limitation to how precisely both the position and momentum of a particle can be known at a given time. (7.5)

Heme an iron complex. (21.7)

Hemoglobin a biomolecule composed of four myoglobin-like units (proteins plus heme) that can bind and transport four oxygen molecules in the blood. (21.7)

Henderson–Hasselbalch equation an equation giving the relationship between the pH of an acid–base system and the concentrations of base and acid:

$$pH = pK_a + \log\left(\frac{[\text{base}]}{[\text{acid}]}\right). \quad (15.2)$$

Henry's law the amount of a gas dissolved in a solution is directly proportional to the pressure of the gas above the solution. (11.3)

Hess's law in going from a particular set of reactants to a particular set of products, the enthalpy change is the same whether the reaction takes place in one step or in a series of steps; in summary, enthalpy is a state function. (6.3)

Heterogeneous equilibrium an equilibrium involving reactants and/or products in more than one phase. (13.4)

Hexagonal closest packed (hcp) structure a structure composed of closest packed spheres with an *ababab* arrangement of layers; the unit cell is hexagonal. (10.4)

Homogeneous equilibrium an equilibrium system where all reactants and products are in the same phase. (13.4)

Homopolymer a polymer formed from the polymerization of only one type of monomer. (22.5)

Hund's rule the lowest energy configuration for an atom is the one having the maximum number of unpaired electrons allowed by the Pauli exclusion principle in a particular set of degenerate orbitals, with all unpaired electrons having parallel spins. (7.11)

Hybrid orbitals a set of atomic orbitals adopted by an atom in a molecule different from those of the atom in the free state. (9.1)

Hybridization a mixing of the native orbitals on a given atom to form special atomic orbitals for bonding. (9.1)

Hydration the interaction between solute particles and water molecules. (4.1)

Hydride a binary compound containing hydrogen. The hydride ion, H^-, exists in ionic hydrides. The three classes of hydrides are covalent, interstitial, and ionic. (20.3)

Hydrocarbon a compound composed of carbon and hydrogen. (22.1)

Hydrocarbon derivative an organic molecule that contains one or more elements in addition to carbon and hydrogen. (22.4)

Hydrogen bonding unusually strong dipole–dipole attractions that occur among molecules in which hydrogen is bonded to a highly electronegative atom. (10.1)

Hydrogenation reaction a reaction in which hydrogen is added, with a catalyst present, to a carbon–carbon multiple bond. (22.2)

Hydrohalic acid an aqueous solution of a hydrogen halide. (20.7)

Hydrometallurgy a process for extracting metals from ores by use of aqueous chemical solutions. Two steps are involved: selective leaching and selective precipitation. (21.8)

Hydronium ion the H_3O^+ ion; a hydrated proton. (14.1)

Hypothesis one or more assumptions put forth to explain the observed behavior of nature. (1.2)

Ideal gas law an equation of state for a gas, where the state of the gas is its condition at a given time; expressed by $PV = nRT$, where P = pressure, V = volume, n = moles of the gas, R = the universal gas constant, and T = absolute temperature. This equation expresses behavior approached by real gases at high T and low P. (5.3)

Ideal solution a solution whose vapor pressure is directly proportional to the mole fraction of solvent present. (11.4)

Indicator a chemical that changes color and is used to mark the end point of a titration. (4.8; 15.5)

Integrated rate law an expression that shows the concentration of a reactant as a function of time. (12.2)

Interhalogen compound a compound formed by the reaction of one halogen with another. (20.7)

Intermediate a species that is neither a reactant nor a product but that is formed and consumed in the reaction sequence. (12.5)

Intermolecular forces relatively weak interactions that occur between molecules. (10.1)

Internal energy a property of a system that can be changed by a flow of work, heat, or both; $\Delta E = q + w$, where ΔE is the change in the internal energy of the system, q is heat, and w is work. (6.1)

Ion an atom or a group of atoms that has a net positive or negative charge. (2.6)

Ion exchange (water softening) the process in which an ion-exchange resin removes unwanted ions (for example, Ca^{2+} and Mg^{2+}) and replaces them with Na^+ ions, which do not interfere with soap and detergent action. (20.4)

Ion pairing a phenomenon occurring in solution when oppositely charged ions aggregate and behave as a single particle. (11.7)

Ion-product (dissociation) constant (K_w) the equilibrium constant for the auto-ionization of water; $K_w = [H^+][OH^-]$. At 25°C, $K_w = 1.0 \times 10^{-14}$. (14.2)

Ion-selective electrode an electrode sensitive to the concentration of a particular ion in solution. (18.5)

Ionic bonding the electrostatic attraction between oppositely charged ions. (2.6; 8.1)

Ionic compound (binary) a compound that results when a metal reacts with a nonmetal to form a cation and an anion. (8.1)

Ionic solid a solid containing cations and anions that dissolves in water to give a solution containing the separated ions, which are mobile and thus free to conduct an electric current. (2.6; 10.3)

Irreversible process any real process. When a system undergoes the changes State 1 → State 2 → State 1 by any real pathway, the universe is different than before the cyclic process took place in the system. (17.9)

Isoelectronic ions ions containing the same number of electrons. (8.4)

Isomers species with the same formula but different properties. (21.4)

Isotactic chain a polymer chain in which the substituent groups such as CH_3 are all arranged on the same side of the chain. (22.5)

Isotonic solutions solutions having identical osmotic pressures. (11.6)

Isotopes atoms of the same element (the same number of protons) with different numbers of neutrons. They have identical atomic numbers but different mass numbers. (2.5; 18)

Ketone an organic compound containing the carbonyl group

bonded to two carbon atoms. (22.4)

Kinetic energy $(\frac{1}{2}mv^2)$ energy due to the motion of an object; dependent on the mass of the object and the square of its velocity. (6.1)

Kinetic molecular theory (KMT) a model that assumes that an ideal gas is composed of tiny particles (molecules) in constant motion. (5.6)

Lanthanide contraction the decrease in the atomic radii of the lanthanide series elements, going from left to right in the periodic table. (21.1)

Lanthanide series a group of 14 elements following lanthanum in the periodic table, in which the $4f$ orbitals are being filled. (7.11; 20.1; 21.1)

Lattice a three-dimensional system of points designating the positions of the centers of the components of a solid (atoms, ions, or molecules). (10.3)

Lattice energy the energy change occurring when separated gaseous ions are packed together to form an ionic solid. (8.5)

Law of conservation of energy energy can be converted from one form to another but can be neither created nor destroyed. (6.1)

Law of conservation of mass mass is neither created nor destroyed. (1.2; 2.2)

Law of definite proportion a given compound always contains exactly the same proportion of elements by mass. (2.2)

Law of mass action a general description of the equilibrium condition; it defines the equilibrium constant expression. (13.2)

Law of multiple proportions a law stating that when two elements form a series of compounds, the ratios of the masses of the second element that combine with one gram of the first element can always be reduced to small whole numbers. (2.2)

Leaching the extraction of metals from ores using aqueous chemical solutions. (21.8)

Lead storage battery a battery (used in cars) in which the anode is lead, the cathode is lead coated with lead dioxide, and the electrolyte is a sulfuric acid solution. (18.6)

Le Châtelier's principle if a change is imposed on a system at equilibrium, the position of the equilibrium will shift in a direction that tends to reduce the effect of that change. (13.7)

Lewis acid an electron-pair acceptor. (14.11)

Lewis base an electron-pair donor. (14.11)

Lewis structure a diagram of a molecule showing how the valence electrons are arranged among the atoms in the molecule. (8.10)

Ligand a neutral molecule or ion having a lone pair of electrons that can be used to form a bond to a metal ion; a Lewis base. (21.3)

Lime–soda process a water-softening method in which lime and soda ash are added to water to remove calcium and magnesium ions by precipitation. (14.6)

Limiting reactant (limiting reagent) the reactant that is completely consumed when a reaction is run to completion. (3.11)

Line spectrum a spectrum showing only certain discrete wavelengths. (7.3)

Linear accelerator a type of particle accelerator in which a changing electrical field is used to accelerate a positive ion along a linear path. (19.3)

Linkage isomerism isomerism involving a complex ion where the ligands are all the same but the point of attachment of at least one of the ligands differs. (21.4)

Liquefaction the transformation of a gas into a liquid. (20.1)

Localized electron (LE) model a model that assumes that a molecule is composed of atoms that are bound together by sharing pairs of electrons using the atomic orbitals of the bound atoms. (8.9)

London dispersion forces the forces, existing among noble gas atoms and nonpolar molecules, that involve an accidental dipole that induces a momentary dipole in a neighbor. (10.1)

Lone pair an electron pair that is localized on a given atom; an electron pair not involved in bonding. (8.9)

Magnetic quantum number (m_ℓ) the quantum number relating to the orientation of an orbital in space relative to the other orbitals with the same ℓ quantum number. It can have integral values between ℓ and $-\ell$, including zero. (7.6)

Main-group (representative) elements elements in the groups labeled 1A, 2A, 3A, 4A, 5A, 6A, 7A, and 8A in the periodic table. The group number gives the sum of the valence s and p electrons. (7.11; 18.1)

Major species the components present in relatively large amounts in a solution. (14.3)

Manometer a device for measuring the pressure of a gas in a container. (5.1)

Mass the quantity of matter in an object. (1.3)

Mass defect the change in mass occurring when a nucleus is formed from its component nucleons. (19.5)

Mass number the total number of protons and neutrons in the atomic nucleus of an atom. (2.5; 18)

Mass percent the percent by mass of a component of a mixture (11.1) or of a given element in a compound. (3.5)

Mass spectrometer an instrument used to determine the relative masses of atoms by the deflection of their ions on a magnetic field. (3.2)

Matter the material of the universe. (1.9)

Messenger RNA (mRNA) a special RNA molecule built in the cell nucleus that migrates into the cytoplasm and participates in protein synthesis. (22.6)

Metal an element that gives up electrons relatively easily and is lustrous, malleable, and a good conductor of heat and electricity. (2.7)

Metalloids (semimetals) elements along the division line in the periodic table between metals and nonmetals. These elements exhibit both metallic and nonmetallic properties. (7.13; 20.1)

Metallurgy the process of separating a metal from its ore and preparing it for use. (20.1; 21.8)

Millimeters of mercury (mm Hg) a unit of pressure, also called a torr, 760 mm Hg = 760 torr = 101,325 Pa = 1 standard atmosphere. (5.1)

Mineral a relatively pure compound as found in nature. (21.8)

Model (theory) a set of assumptions put forth to explain the observed behavior of matter. The models of chemistry usually involve assumptions about the behavior of individual atoms or molecules. (1.2)

Moderator a substance used in a nuclear reactor to slow down the neutrons. (19.6)

Molal boiling-point elevation constant a constant characteristic of a particular solvent that gives the change in boiling point as a function of solution molality; used in molecular weight determinations. (11.5)

Molal freezing-point depression constant a constant characteristic of a particular solvent that gives the change in freezing point as a function of the solution molality; used in molecular weight determinations (11.5)

Molality the number of moles of solute per kilogram of solvent in a solution. (11.1)

Molar heat capacity the energy required to raise the temperature of one mole of a substance by one degree Celsius. (6.2)

Molar mass the mass in grams of one mole of molecules or formula units of a substance; also called *molecular weight.* (3.4)

Molar volume the volume of one mole of an ideal gas; equal to 22.42 liters at STP. (5.4)

Molarity moles of solute per volume of solution in liters. (4.3; 11.1)

Mole (mol) the number equal to the number of carbon atoms in exactly 12 grams of pure ^{12}C: Avogadro's number. One mole represents 6.022×10^{23} units. (3.3)

Mole fraction the ratio of the number of moles of a given component in a mixture to the total number of moles in the mixture. (5.5; 11.1)

Mole ratio (stoichiometry) the ratio of moles of one substance to moles of another substance in a balanced chemical equation. (3.9)

Molecular formula the exact formula of a molecule, giving the types of atoms and the number of each type. (3.6)

Molecular orbital (MO) model a model that regards a molecule as a collection of nuclei and electrons, where the electrons are assumed to occupy orbitals much as they do in atoms, but having the orbitals extend over the entire molecule. In this model the electrons are assumed to be delocalized rather than always located between a given pair of atoms. (9.2; 10.4)

Molecular orientations (kinetics) orientations of molecules during collisions, some of which can lead to reaction while others cannot. (12.6)

Molecular solid a solid composed of neutral molecules at the lattice points. (10.3)

Molecular structure the three-dimensional arrangement of atoms in a molecule. (8.13)

Molecularity the number of species that must collide to produce the reaction represented by an elementary step in a reaction mechanism. (12.5)

Molecule a bonded collection of two or more atoms of the same or different elements. (2.6)

Monodentate (unidentate) ligand a ligand that can form one bond to a metal ion. (21.3)

Monoprotic acid an acid with one acidic proton. (14.2)

Monosaccharide (simple sugar) a polyhydroxy ketone or aldehyde containing from three to nine carbon atoms. (22.6)

Myoglobin an oxygen-storing biomolecule consisting of a heme complex and a proton. (21.7)

Natural law a statement that expresses generally observed behavior. (1.2)

Nernst equation an equation relating the potential of an electrochemical cell to the concentrations of the cell components:

$$\mathscr{E} = \mathscr{E}^\circ - \frac{0.0591}{n} \log(Q) \text{ at } 25°C \text{ (18.5)}$$

Net ionic equation an equation for a reaction in solution, where strong electrolytes are written as ions, showing only those components that are directly involved in the chemical change. (4.6)

Network solid an atomic solid containing strong directional covalent bonds. (10.5)

Neutralization reaction an acid–base reaction. (4.8)

Neutron a particle in the atomic nucleus with mass virtually equal to the proton's but with no charge. (2.5; 19)

Nitrogen cycle the conversion of N_2 to nitrogen-containing compounds, followed by the return of nitrogen gas to the atmosphere by natural decay processes. (20.2)

Nitrogen fixation the process of transforming N_2 to nitrogen-containing compounds useful to plants. (20.2)

Nitrogen-fixing bacteria bacteria in the root nodules of plants that can convert atmospheric nitrogen to ammonia and other nitrogen-containing compounds useful to plants. (20.2)

Noble gas a Group 8A element. (2.7; 20.8)

Node an area of an orbital having zero electron probability. (7.7)

Nonelectrolyte a substance that, when dissolved in water, gives a nonconducting solution. (4.2)

Nonmetal an element not exhibiting metallic characteristics. Chemically, a typical nonmetal accepts electrons from a metal. (2.7)

Normal boiling point the temperature at which the vapor pressure of a liquid is exactly one atmosphere. (10.8)

Normal melting point the temperature at which the solid and liquid states have the same vapor pressure under conditions where the total pressure on the system is one atmosphere. (10.8)

Normality the number of equivalents of a substance dissolved in a liter of solution. (11.1)

Nuclear atom an atom having a dense center of positive charge (the nucleus) with electrons moving around the outside. (2.4)

Nuclear transformation the change of one element into another. (19.3)

Nucleon a particle in an atomic nucleus, either a neutron or a proton. (19)

Nucleotide a monomer of the nucleic acids composed of a five-carbon sugar, a nitrogen-containing base, and phosphoric acid. (22.6)

Nucleus the small, dense center of positive charge in an atom. (2.4)

Nuclide the general term applied to each unique atom; represented by $^A_Z X$, where X is the symbol for a particular element. (19)

Octet rule the observation that atoms of nonmetals tend to form the most stable molecules when they are surrounded by eight electrons (to fill their valence orbitals). (8.10)

Open hearth process a process for producing steel by oxidizing and removing the impurities in molten iron using external heat and a blast of air or oxygen. (21.8)

Optical isomerism isomerism in which the isomers have opposite effects on plane-polarized light. (21.4)

Orbital a specific wave function for an electron in an atom. The square of this function gives the probability distribution for the electron. (7.5) *d*-**Orbital splitting** a splitting of the *d* orbitals of the metal ion in a complex such that the orbitals pointing at the ligands have higher energies than those pointing between the ligands. (21.6)

Order (of reactant) the positive or negative exponent, determined by experiment, of the reactant concentration in a rate law. (12.2)

Organic acid an acid with a carbon-atom backbone; often contains the carboxyl group. (14.2)

Organic chemistry the study of carbon-containing compounds (typically chains of carbon atoms) and their properties. (22)

Osmosis the flow of solvent into a solution through a semipermeable membrane. (11.6)

Osmotic pressure (π) the pressure that must be applied to a solution to stop osmosis; $\pi = MRT$. (11.6)

Ostwald process a commercial process for producing nitric acid by the oxidation of ammonia. (20.2)

Oxidation an increase in oxidation state (a loss of electrons). (4.9; 17.1)

Oxidation–reduction (redox) reaction a reaction in which one or more electrons are transferred. (4.9; 17.1)

Oxidation states a concept that provides a way to keep track of electrons in oxidation–reduction reactions according to certain rules. (4.9; 21.3)

Oxidizing agent (electron acceptor) a reactant that accepts electrons from another reactant. (4.9; 17.1)

Oxyacid an acid in which the acidic proton is attached to an oxygen atom. (14.2)

Ozone O_3, the form of elemental oxygen in addition to the much more common O_2. (20.5)

Paramagnetism a type of induced magnetism, associated with unpaired electrons, that causes a substance to be attracted into the inducing magnetic field. (9.3)

Partial pressures the independent pressures exerted by different gases in a mixture. (5.5)

Particle accelerator a device used to accelerate nuclear particles to very high speeds. (19.3)

Pascal the SI unit of pressure; equal to newtons per meter squared. (5.1)

Pauli exclusion principle in a given atom no two electrons can have the same set of four quantum numbers. (7.8)

Peptide linkage the bond resulting from the condensation reaction between amino acids; represented by:

$$\underset{\overset{|}{\text{C}}}{\overset{\text{O}}{\parallel}}\quad\underset{\overset{|}{\text{N}}}{\overset{\text{H}}{|}}$$

—C—N— (22.6)

Percent dissociation the ratio of the amount of a substance that is dissociated at equilibrium to the initial concentration of the substance in a solution, multiplied by 100. (14.5)

Percent yield the actual yield of a product as a percentage of the theoretical yield. (3.11)

Periodic table a chart showing all the elements arranged in columns with similar chemical properties. (2.7)

pH curve (titration curve) a plot showing the pH of a solution being analyzed as a function of the amount of titrant added. (15.4)

pH scale a log scale based on 10 and equal to $-\log[H^+]$; a convenient way to represent solution acidity. (14.3)

Phase diagram a convenient way of representing the phases of a substance in a closed system as a function of temperature and pressure. (10.9)

Phenyl group the benzene molecule minus one hydrogen atom. (22.3)

Photochemical smog air pollution produced by the action of light on oxygen, nitrogen oxides, and unburned fuel from auto exhaust to form ozone and other pollutants. (5.10)

Photon a quantum of electromagnetic radiation. (7.2)

Physical change a change in the form of a substance, but not in its chemical composition; chemical bonds are not broken in a physical change. (1.9)

Pi (π) bond a covalent bond in which parallel *p* orbitals share an electron pair occupying the space above and below the line joining the atoms. (9.1)

Planck's constant the constant relating the change in energy for a system to the frequency of the electromagnetic radiation absorbed or emitted; equal to 6.626×10^{-34} J $\cdot$ s. (7.2)

Polar covalent bond a covalent bond in which the electrons are not shared equally because one atom attracts them more strongly than the other. (8.1)

Polar molecule a molecule that has a permanent dipole moment. (4.1)

Polyatomic ion an ion containing a number of atoms. (2.6)

Polyelectronic atom an atom with more than one electron. (7.9)

Polymer a large, usually chainlike molecule built from many small molecules (monomers). (22.5)

Polymerization a process in which many small molecules (monomers) are joined together to form a large molecule. (22.2)

Polypeptide a polymer formed from amino acids joined together by peptide linkages. (22.6)

Polyprotic acid an acid with more than one acidic proton. It dissociates in a stepwise manner, one proton at a time. (14.7)

Porous disk a disk in a tube connecting two different solutions in a galvanic cell that allows ion flow without extensive mixing of the solutions. (18.2)

Porphyrin a planar ligand with a central ring structure and various substituent groups at the edges of the ring. (21.7)

Positional probability a type of probability that depends on the number of arrangements in space that yield a particular state. (17.1)

Positron production a mode of nuclear decay in which a particle is formed having the same mass as an electron but opposite charge. The net effect is to change a proton to a neutron. (19.1)

Potential energy energy due to position or composition. (6.1)

Precipitation reaction a reaction in which an insoluble substance forms and separates from the solution. (4.5)

Precision the degree of agreement among several measurements of the same quantity; the reproducibility of a measurement. (1.4)

Primary structure (of a protein) the order (sequence) of amino acids in the protein chain. (22.6)

Principal quantum number (n) the quantum number relating to the size and energy of an orbital; it can have any positive integer value. (7.6)

Probability distribution the square of the wave function indicating the probability of finding an electron at a particular point in space. (7.5)

Product a substance resulting from a chemical reaction. It is shown to the right of the arrow in a chemical equation. (3.8)

Protein a natural high-molecular-weight polymer formed by condensation reactions between amino acids. (22.6)

Proton a positively charged particle in an atomic nucleus. (2.5; 19)

Pure substance a substance with constant composition. (1.9)

Pyrometallurgy recovery of a metal from its ore by treatment at high temperatures. (21.8)

Quantization the concept that energy can occur only in discrete units called *quanta*. (7.2)

Rad a unit of radiation dosage corresponding to 10^{-2} J of energy deposited per kilogram of tissue (from *radiation absorbed dose*). (19.7)

Radioactive decay (radioactivity) the spontaneous decomposition of a nucleus to form a different nucleus. (2.4; 19.1)

Radiocarbon dating (carbon-14 dating) a method for dating ancient wood or cloth based on the rate of radioactive decay of the nuclide $^{14}_{6}C$. (19.4)

Radiotracer a radioactive nuclide, introduced into an organism for diagnostic purposes, whose pathway can be traced by monitoring its radioactivity. (19.4)

Random error an error that has an equal probability of being high or low. (1.4)

Raoult's law the vapor pressure of a solution is directly proportional to the mole fraction of solvent present. (11.4)

Rate constant the proportionality constant in the relationship between reaction rate and reactant concentrations. (12.2)

Rate of decay the change in the number of radioactive nuclides in a sample per unit time. (19.2)

Rate-determining step the slowest step in a reaction mechanism, the one determining the overall rate. (12.5)

Rate law (differential rate law) an expression that shows how the rate of reaction depends on the concentration of reactants. (12.2)

Reactant a starting substance in a chemical reaction. It appears to the left of the arrow in a chemical equation. (3.8)

Reaction mechanism the series of elementary steps involved in a chemical reaction. (12.5)

Reaction quotient, Q a quotient obtained by applying the law of mass action to initial concentrations rather than to equilibrium concentrations. (13.5)

Reaction rate the change in concentration of a reactant or product per unit time. (12.1)

Reactor core the part of a nuclear reactor where the fission reaction takes place. (19.6)

Reducing agent (electron donor) a reactant that donates electrons to another substance to reduce the oxidation state of one of its atoms. (4.9; 17.1)

Reduction a decrease in oxidation state (a gain of electrons). (4.9; 17.1)

Rem a unit of radiation dosage that accounts for both the energy of the dose and its effectiveness in causing biological damage (from *roentgen equivalent for man*). (19.7)

Resonance a condition occurring when more than one valid Lewis structure can be written for a particular molecule. The actual electronic structure is not represented by any one of the Lewis structures but by the average of all of them. (8.12)

Reverse osmosis the process occurring when the external pressure on a solution causes a net flow of solvent through a semipermeable membrane from the solution to the solvent. (11.6)

Reversible process a cyclic process carried out by a hypothetical pathway, which leaves the universe exactly the same as it was before the process. No real process is reversible. (17.9)

Ribonucleic acid (RNA) a nucleotide polymer that transmits the genetic information stored in DNA to the ribosomes for protein synthesis. (22.6)

Roasting a process of converting sulfide minerals to oxides by heating in air at temperatures below their melting points. (21.8)

Root mean square velocity the square root of the average of the squares of the individual velocities of gas particles. (5.6)

Salt an ionic compound. (14.8)

Salt bridge a U-tube containing an electrolyte that connects the two compartments of a galvanic cell, allowing ion flow without extensive mixing of the different solutions. (18.1)

Scientific method the process of studying natural phenomena, involving observations, forming laws and theories, and testing of theories by experimentation. (1.2)

Scintillation counter an instrument that measures radioactive decay by sensing the flashes of light produced in a substance by the radiation. (19.4)

Second law of thermodynamics in any spontaneous process, there is always an increase in the entropy of the universe. (17.2)

Secondary structure (of a protein) the three-dimensional structure of the protein chain (for example, α-helix, random coil, or pleated sheet). (22.6)

Selective precipitation a method of separating metal ions from an aqueous mixture by using a reagent whose anion forms a precipitate with only one or a few of the ions in the mixture. (4.7; 16.2)

Semiconductor a substance conducting only a slight electric current at room temperature, but showing increased conductivity at higher temperatures. (10.5)

Semipermeable membrane a membrane that allows solvent but not solute molecules to pass through. (11.6)

SI system International System of units based on the metric system and units derived from the metric system. (1.3)

Side chain (of amino acid) the hydrocarbon group on an amino acid represented by H, CH_3, or a more complex substituent. (22.6)

Sigma (σ) bond a covalent bond in which the electron pair is shared in an area centered on a line running between the atoms. (9.1)

Significant figures the certain digits and the first uncertain digit of a measurement. (1.4)

Silica the fundamental silicon–oxygen compound, which has the empirical formula SiO_2, and forms the basis of quartz and certain types of sand. (10.5)

Silicates salts that contain metal cations and polyatomic silicon–oxygen anions that are usually polymeric. (10.5)

Single bond a bond in which one pair of electrons is shared by two atoms. (8.8)

Smelting a metallurgical process that involves reducing metal ions to the free metal. (21.8)

Solubility the amount of a substance that dissolves in a given volume of solvent at a given temperature. (4.2)

Solubility product constant the constant for the equilibrium expression representing the dissolving of an ionic solid in water. (16.1)

Solute a substance dissolved in a liquid to form a solution. (4.2; 11.1)

Solution a homogeneous mixture. (1.9)

Solvent the dissolving medium in a solution. (4.2)

Somatic damage radioactive damage to an organism resulting in its sickness or death. (19.7)

Space-filling model a model of a molecule showing the relative sizes of the atoms and their relative orientations. (2.6)

Specific heat capacity the energy required to raise the temperature of one gram of a substance by one degree Celsius. (6.2)

Spectator ions ions present in solution that do not participate directly in a reaction. (4.6)

Spectrochemical series a listing of ligands in order based on their ability to produce d-orbital splitting. (21.6)

Spontaneous fission the spontaneous splitting of a heavy nuclide into two lighter nuclides. (19.1)

Spontaneous process a process that occurs without outside intervention. (17.1)

Standard atmosphere a unit of pressure equal to 760 mm Hg. (5.1)

Standard enthalpy of formation the enthalpy change that accompanies the formation of one mole of a compound at 25°C from its elements, with all substances in their standard states at that temperature. (6.4)

Standard free energy change the change in free energy that will occur for one unit of reaction if the reactants in their standard states are converted to products in their standard states. (17.6)

Standard free energy of formation the change in free energy that accompanies the formation of one mole of a substance from its constituent elements with all reactants and products in their standard states. (17.6)

Standard hydrogen electrode a platinum conductor in contact with 1 M H$^+$ ions and bathed by hydrogen gas at one atmosphere. (18.3)

Standard reduction potential the potential of a half-reaction under standard state conditions, as measured against the potential of the standard hydrogen electrode. (18.3)

Standard solution a solution whose concentration is accurately known. (4.3)

Standard state a reference state for a specific substance defined according to a set of conventional definitions. (6.4)

Standard temperature and pressure (STP) the condition 0°C and 1 atmosphere of pressure. (5.4)

Standing wave a stationary wave as on a string of a musical instrument; in the wave mechanical model, the electron in the hydrogen atom is considered to be a standing wave. (7.5)

State function (property) a property that is independent of the pathway. (6.1)

States (of matter) the three different forms in which matter can exist; solid, liquid, and gas. (1.9)

Stereoisomerism isomerism in which all the bonds in the isomers are the same but the spatial arrangements of the atoms are different. (21.4)

Steric factor the factor (always less than 1) that reflects the fraction of collisions with orientations that can produce a chemical reaction. (12.6)

Stoichiometric quantities quantities of reactants mixed in exactly the correct amounts so that all are used up at the same time. (3.10)

Strong acid an acid that completely dissociates to produce an H$^+$ ion and the conjugate base. (4.2; 14.2)

Strong base a metal hydroxide salt that completely dissociates into its ions in water. (4.2; 14.6)

Strong electrolyte a material that, when dissolved in water, gives a solution that conducts an electric current very efficiently. (4.2)

Structural formula the representation of a molecule in which the relative positions of the atoms are shown and the bonds are indicated by lines. (2.6)

Structural isomerism isomerism in which the isomers contain the same atoms but one or more bonds differ. (21.4; 22.1)

Subcritical reaction (nuclear) a reaction in which less than one neutron causes another fission event and the process dies out. (19.6)

Sublimation the process by which a substance goes directly from the solid to the gaseous state without passing through the liquid state. (10.8)

Subshell a set of orbitals with a given azimuthal quantum number. (7.6)

Substitution reaction (hydrocarbons) a reaction in which an atom, usually a halogen, replaces a hydrogen atom in a hydrocarbon. (22.1)

Supercooling the process of cooling a liquid below its freezing point without its changing to a solid. (10.8)

Supercritical reaction (nuclear) a reaction in which more than one neutron from each fission event causes another fission event. The process rapidly escalates to a violent explosion. (19.6)

Superheating the process of heating a liquid above its boiling point without its boiling. (10.8)

Superoxide a compound containing the O$_2^-$ anion. (19.2)

Surface tension the resistance of a liquid to an increase in its surface area. (10.2)

Surroundings everything in the universe surrounding a thermodynamic system. (6.1)

Syndiotactic chain a polymer chain in which the substituent groups such as CH$_3$ are arranged on alternate sides of the chain. (22.5)

Syngas synthetic gas, a mixture of carbon monoxide and hydrogen, obtained by coal gasification. (6.6)

System (thermodynamic) that part of the universe on which attention is to be focused. (6.1)

Systematic error an error that always occurs in the same direction. (1.4)

Tempering a process in steel production that fine-tunes the proportions of carbon crystals and cementite by heating to intermediate temperatures followed by rapid cooling. (21.8)

Termolecular step a reaction involving the simultaneous collision of three molecules. (12.5)

Tertiary structure (of a protein) the overall shape of a protein, long and narrow or globular, maintained by different types of intramolecular interactions. (22.6)

Theoretical yield the maximum amount of a given product that can be formed when the limiting reactant is completely consumed. (3.11)

Theory a set of assumptions put forth to explain some aspect of the observed behavior of matter. (1.2)

Thermal pollution the oxygen-depleting effect on lakes and rivers of using water for industrial cooling and returning it to its natural source at a higher temperature. (11.3)

Thermodynamic stability (nuclear) the potential energy of a particular nucleus as compared to the sum of the potential energies of its component protons and neutrons. (19.1)

Thermodynamics the study of energy and its interconversions. (6.1)

Thermoplastic polymer a substance that when molded to a certain shape under appropriate conditions can later be remelted. (22.5)

Thermoset polymer a substance that when molded to a certain shape under pressure and high temperatures cannot be softened again or dissolved. (22.5)

Third law of thermodynamics the entropy of a perfect crystal at 0 K is zero. (17.5)

Titration a technique in which one solution is used to analyze another. (4.8)

Torr another name for millimeter of mercury (mm Hg). (5.1)

Transfer RNA (tRNA) a small RNA fragment that finds specific amino acids and attaches them to the protein chain as dictated by the codons in mRNA. (22.6)

Transition metals several series of elements in which inner orbitals (*d* or *f* orbitals) are being filled. (7.11; 20.1)

Transuranium elements the elements beyond uranium that are made artificially by particle bombardment. (19.3)

Triple bond a bond in which three pairs of electrons are shared by two atoms. (8.8)

Triple point the point on a phase diagram at which all three states of a substance are present. (10.9)

Tyndall effect the scattering of light by particles in a suspension. (11.8)

Uncertainty (in measurement) the characteristic that any measurement involves estimates and cannot be exactly reproduced. (1.4)

Unimolecular step a reaction step involving only one molecule. (12.5)

Unit cell the smallest repeating unit of a lattice. (10.3)

Unit factor method an equivalence statement between units used for converting from one unit to another. (1.6)

Universal gas constant the combined proportionality constant in the ideal gas law; 0.08206 L · atm/K · mol or 8.3145 J/K · mol. (5.3)

Valence electrons the electrons in the outermost principal quantum level of an atom. (7.11)

Valence shell electron-pair repulsion (VSEPR) model a model whose main postulate is that the structure around a given atom in a molecule is determined principally by minimizing electron-pair repulsions. (8.13)

Van der Waals equation a mathematical expression for describing the behavior of real gases. (5.8)

van't Hoff factor the ratio of moles of particles in solution to moles of solute dissolved. (11.7)

Vapor pressure the pressure of the vapor over a liquid at equilibrium. (10.8)

Vaporization (evaporization) the change in state that occurs when a liquid evaporates to form a gas. (10.8)

Viscosity the resistance of a liquid to flow. (10.2)

Volt the unit of electrical potential defined as one joule of work per coulomb of charge transferred. (18.2)

Voltmeter an instrument that measures cell potential by drawing electric current through a known resistance. (18.2)

Volumetric analysis a process involving titration of one solution with another. (4.8)

Vulcanization a process in which sulfur is added to rubber and the mixture is heated, causing crosslinking of the polymer chains and thus adding strength to the rubber. (22.5)

Wave function a function of the coordinates of an electron's position in three-dimensional space that describes the properties of the electron. (7.5)

Wave mechanical model a model for the hydrogen atom in which the electron is assumed to behave as a standing wave. (7.5)

Wavelength the distance between two consecutive peaks or troughs in a wave. (7.1)

Weak acid an acid that dissociates only slightly in aqueous solution. (4.2; 14.2)

Weak base a base that reacts with water to produce hydroxide ions to only a slight extent in aqueous solution. (4.2; 14.6)

Weak electrolyte a material that, when dissolved in water, gives a solution that conducts only a small electric current. (4.2)

Weight the force exerted on an object by gravity. (1.3)

Work force acting over a distance. (6.1)

X-ray diffraction a technique for establishing the structure of crystalline solids by directing X rays of a single wavelength at a crystal and obtaining a diffraction pattern from which interatomic spaces can be determined. (10.3)

Zone of nuclear stability the area encompassing the stable nuclides on a plot of their positions as a function of the number of protons and the number of neutrons in the nucleus. (19.1)

Zone refining a metallurgical process for obtaining a highly pure metal that depends on continuously melting the impure material and recrystallizing the pure metal. (21.8)

Answers to Selected Exercises

The answers listed here are from the *Complete Solutions Guide,* in which rounding is carried out at each intermediate step in a calculation in order to show the correct number of significant figures for that step. Therefore, an answer given here may differ in the last digit from the result obtained by carrying extra digits throughout the entire calculation and rounding at the end (the procedure you should follow).

Chapter 11

11. 9.74 M **13.** 160 mL   **15.** 4.5 M **17.** As the temperature increases, the gas molecules will have a greater average kinetic energy. A greater fraction of the gas molecules in solution will have kinetic energy greater than the attractive forces between the gas molecules and the solvent molecules. More gas molecules will escape to the vapor phase, and the solubility of the gas will decrease.   **19.** The levels of the liquids in each beaker will become constant when the concentration of solute is the same in both beakers. Because the solute is less volatile, the beaker on the right will have a larger volume when the concentrations become equal. Water will initially condense in this beaker in a larger amount than solute is evaporating, while the net change occurring initially in the other beaker is for water to evaporate in a larger amount than solute is condensing. Eventually the rate that solute and H_2O leave and return to each beaker will become equal when the concentrations become equal.   **21.** No. For an ideal solution, $\Delta H_{soln} = 0$.   **23.** Normality is the number of equivalents per liter of solution. For an acid or a base, an equivalent is the mass of acid or base that can furnish 1 mole of protons (if an acid) or accept 1 mole of protons (if a base). A proton is an H^+ ion. Molarity is defined as the moles of solute per liter of solution. When the number of equivalents equals the number of moles of solute, then normality = molarity. This is true for acids that only have one acidic proton in them and for bases that accept only one proton per formula unit. Examples of acids where equivalents = moles solute are HCl, HNO_3, HF, and $HC_2H_3O_2$. Examples of bases where equivalents = moles solute are NaOH, KOH, and NH_3. When equivalents $\neq$ moles solute, then normality $\neq$ molarity. This is true for acids that donate more than one proton (H_2SO_4, H_3PO_4, H_2CO_3, etc.) and for bases that react with more than one proton per formula unit [$Ca(OH)_2$, $Ba(OH)_2$, $Sr(OH)_2$, etc.].   **25.** Only statement b is true. A substance freezes when the vapor pressures of the liquid and solid phases are the same. When a solute is added to water, the vapor pressure of the solution at 0°C is less than the vapor pressure of the solid; the net result is for any ice present to convert to liquid in order to try to equalize the vapor pressures (which never can occur at 0°C). A lower temperature is needed to equalize the vapor pressures of water and ice, hence the freezing point is depressed. For statement a, the vapor pressure of a solution is directly related to the mole fraction of solvent (not solute) by Raoult's law. For statement c, colligative properties depend on the number of solute particles present and not on the identity of the solute. For statement d, the boiling point of water is increased because the sugar solute decreases the vapor pressure of the water; a higher temperature is required for the vapor pressure of the solution to equal the external pressure so boiling can occur.   **27.** Isotonic solutions are those that have identical osmotic pressures. Crenation and hemolysis refer to phenomena that occur when red blood cells are bathed in solutions having a mismatch in osmotic pressure between the inside and the outside of the cell. When red blood cells are in a solution having a higher osmotic pressure than that of the cells, the cells shrivel as there is a net transfer of water out of the cells. This is called crenation. Hemolysis occurs when the red blood cells are bathed in a solution having lower osmotic pressure than that inside the cell. Here, the cells rupture as there is a net transfer of water to inside the red blood cells.   **29.** 1.06 g/mL; 0.0180 mole fraction H_3PO_4, 0.9820 mole fraction H_2O; 0.981 mol/L; 1.02 mol/kg   **31.** HCl: 12 M, 17 m, 0.23; HNO_3: 16 M, 37 m, 0.39; H_2SO_4: 18 M, 200 m, 0.76; $HC_2H_3O_2$: 17 M, 2000 m, 0.96; NH_3: 15 M, 23 m, 0.29   **33.** 35%; 0.39; 7.3 m; 3.1 M **35.** 10.1% by mass; 2.45 mol/kg   **37.** 23.9%; 1.6 m, 0.028, 4.11 N **39.** $NaI(s) \rightarrow Na^+(aq) + I^-(aq)$ $\Delta H_{soln} = -8$ kJ/mol   **41.** The attraction of water molecules for Al^{3+} and OH^- cannot overcome the larger lattice energy of $Al(OH)_3$.   **43. a.** CCl_4; **b.** H_2O; **c.** H_2O; **d.** CCl_4; **e.** H_2O; **f.** H_2O; **g.** CCl_4 **45. a.** NH_3; **b.** CH_3CN; **c.** CH_3COOH **47.** As the length of the hydrocarbon chain increases, the solubility decreases because the nonpolar hydrocarbon chain interacts poorly with the polar water molecules.   **49.** 1.04×10^{-3} mol/L · atm; 1.14×10^{-3} mol/L   **51.** 50.0 torr   **53.** 0.918   **55.** 23.2 torr at 25°C; 70.0 torr at 45°C   **57. a.** 290 torr; **b.** 0.69   **59.** $X_{methanol} = X_{propanol} = 0.500$ **61.** solution c   **63. a.** These two molecules are named acetone (CH_3COCH_3) and water. As discussed in Section 11.4 on nonideal solutions, acetone–water solutions exhibit negative deviations from Raoult's law. Acetone and water have the ability to hydrogen bond with each other, which gives the solution stronger intermolecular forces compared to the pure states of both solute and solvent. In the pure state, acetone cannot hydrogen bond with itself, so the middle diagram illustrating negative deviations from Raoult's law is the correct choice for acetone–water solutions. **b.** These two molecules are named ethanol (CH_3CH_2OH) and water. Ethanol–water solutions show positive deviations from Raoult's law. Both substances can hydrogen bond in the pure state, and they can continue this in solution. However, the solute–solvent interactions are weaker for ethanol–water solutions due to the significant nonpolar part of ethanol (CH_3—CH_2 is the nonpolar part of ethanol). This nonpolar part of ethanol weakens the intermolecular forces in solution, so the first diagram illustrating positive deviations from Raoult's law is the correct choice for ethanol–water solutions. **c.** These two molecules are named heptane (C_7H_{16}) and hexane (C_6H_{14}). Heptane and hexane are very similar nonpolar substances; both are composed entirely of nonpolar C—C bonds and relatively nonpolar C—H bonds, and both have a similar size and shape. Solutions of heptane and hexane should be ideal, so the third diagram illustrating no deviation from Raoult's law is the correct choice for heptane–hexane solutions. **d.** These two molecules are named heptane (C_7H_{16}) and water. The interactions between the nonpolar heptane molecules and the polar water molecules will certainly be weaker in solution compared to the pure solvent and pure solute interactions. This results in positive deviations from Raoult's law (the first diagram).   **65.** 101.5°C   **67.** 14.8 g $C_3H_8O_3$ **69.** $T_f = -29.9$°C, $T_b = 108.2$°C   **71.** 6.6×10^{-2} mol/kg; 590 g/mol (610 g/mol if no rounding of numbers)   **73. a.** $\Delta T = 2.0 \times 10^{-5}$°C, $\pi = 0.20$ torr; **b.** Osmotic pressure is better for determining the molar mass of large molecules. A temperature change of 10^{-5}°C is very difficult to measure. A change in height of a column of mercury by 0.2 mm is not as hard to measure precisely.   **75.** 2.51×10^5 g/mol   **77.** Dissolve 210 g sucrose in some water and dilute to 1.0 L in a volumetric flask. To get 0.62 ± 0.01 mol/L, we need 212 ± 3 g sucrose.   **79. a.** 0.010 m Na_3PO_4 and 0.020 m KCl; **b.** 0.020 m HF; **c.** 0.020 m $CaBr_2$ **81. a.** $T_f = -13$°C; $T_b = 103.5$°C; **b.** $T_f = -4.7$°C; $T_b = 101.3$°C   **83.** 1.67   **85. a.** $T_f = -0.28$°C; $T_b = 100.077$°C; **b.** $T_f = -0.37$°C; $T_b = 100.10$°C   **87.** 2.63 (0.0225 m), 2.60 (0.0910 m), 2.57 (0.278 m); $i_{average} = 2.60$ **89. a.** yes; **b.** no   **91.** Benzoic acid is capable of hydrogen bonding, but a significant part of benzoic acid is the nonpolar benzene ring. In benzene, a hydrogen bonded dimer forms.

The dimer is relatively nonpolar and thus more soluble in benzene than in water. Benzoic acid would be more soluble in a basic solution because of the reaction $C_6H_5CO_2H + OH^- \rightarrow C_6H_5CO_2^- + H_2O$. By removing the proton from benzoic acid, an anion forms, and like all anions, the species becomes more soluble in water.   **93. a.** 26.6 kJ/mol; **b.** -657 kJ/mol   **95. a.** Water boils when the vapor pressure equals the pressure above the water. In an open pan, $P_{atm} \approx 1.0$ atm. In a pressure cooker, $P_{inside} > 1.0$ atm and water boils at a higher temperature. The higher the cooking temperature, the faster the cooking time. **b.** Salt dissolves in water, forming a solution with a melting point lower than that of pure water ($\Delta T_f = K_f m$). This happens in water on the surface of ice. If it is not too cold, the ice melts. This process won't occur if the ambient temperature is lower than the depressed freezing point of the salt solution. **c.** When water freezes from a solution, it freezes as pure water, leaving behind a more concentrated salt solution. **d.** On the CO_2 phase diagram, the triple point is above 1 atm and $CO_2(g)$ is the stable phase at 1 atm and room temperature. $CO_2(l)$ can't exist at normal atmospheric pressures, which explains why dry ice sublimes rather than boils. In a fire extinguisher, $P > 1$ atm and $CO_2(l)$ can exist. When CO_2 is released from the fire extinguisher, $CO_2(g)$ forms as predicted from the phase diagram. **e.** Adding a solute to a solvent increases the boiling point and decreases the freezing point of the solvent. Thus, the solvent is a liquid over a wider range of temperatures when a solute is dissolved.   **97.** 0.600   **99.** $P_{ideal} = 188.6$ torr; $X_{acetone} = 0.512$, $X_{methanol} = 0.488$; the actual vapor pressure of the solution is smaller than the ideal vapor pressure, so this solution exhibits a negative deviation from Raoult's law. This occurs when solute–solvent attractions are stronger than for the pure substances.   **101.** 776 g/mol   **103.** $C_2H_4O_3$; 151 g/mol (exp.); 152.10 g/mol (calc.); $C_4H_8O_6$ **105.** 1.97% NaCl   **107. a.** 100.77°C; **b.** 23.1 mm Hg; **c.** Assume an ideal solution; assume no ions form ($i = 1$); assume the solute is nonvolatile.   **109.** 639 g/mol; 33.7 torr

119. 30.% A: $\chi_A = \dfrac{0.30y}{0.70x + 0.30y}$, $\chi_B = 1 - \chi_A$;

50.% A: $\chi_A = \dfrac{y}{x + y}$, $\chi_B = 1 - \dfrac{y}{x + y}$;

80.% A: $\chi_A = \dfrac{0.80y}{0.20x + 0.80y}$, $\chi_B = 1 - \chi_A$;

30.% A: $\chi_A^V = \dfrac{0.30x}{0.30x + 0.70y}$, $\chi_B^V = 1 - \dfrac{0.30x}{0.30x + 0.70y}$;

50.% A: $\chi_A^V = \dfrac{x}{x + y}$, $\chi_B^V = 1 - \chi_A^V$;

80.% A: $\chi_A^V = \dfrac{0.80x}{0.80x + 0.20y}$, $\chi_B^V = 1 - \chi_A^V$

121. 72.5% sucrose and 27.5% NaCl by mass; 0.313   **123.** 0.050   **125.** 44% naphthalene, 56% anthracene   **127.** -0.20°C, 100.056°C   **129. a.** 46 L; **b.** No; a reverse osmosis system that applies 8.0 atm can only purify water with solute concentrations less than 0.32 mol/L. Salt water has a solute concentration of $2(0.60\ M) = 1.2\ M$ ions. The solute concentration of salt water is much too high for this reverse osmosis unit to work.   **131.** $i = 3.00$; $CdCl_2$

Chapter 12

11. In a unimolecular reaction, a single reactant molecule decomposes to products. In a bimolecular reaction, two molecules collide to give products. The probability of the simultaneous collision of three molecules with enough energy and orientation is very small, making termolecular steps very unlikely.   **13.** All of these choices would affect the rate of the reaction, but only b and c affect the rate by affecting the value of the rate constant k. The value of the rate constant is dependent on temperature. It also depends on the activation energy. A catalyst will change the value of k because the activation energy changes. Increasing the concentration (partial pressure) of either O_2 or NO does not affect the value of k, but it does increase the rate of the reaction because both concentrations appear in the rate law.   **15.** The average rate decreases with time because the reverse reaction occurs more frequently as the concentration of products increases. Initially, with no products present, the rate of the forward reaction is at its fastest; but as time goes on, the rate gets slower and slower since products are converting back into reactants. The instantaneous rate will also decrease with time. The only rate that is constant is the initial rate. This is the instantaneous rate taken at $t \approx 0$. At this time, the amount of products is insignificant and the rate of the reaction only depends on the rate of the forward reaction.   **17.** When the rate doubles as the concentration quadruples, the order is 1/2. For a reactant that has an order of −1, the rate will decrease by a factor of 1/2 when the concentrations are doubled.   **19.** Two reasons are: **1)** the collision must involve enough energy to produce the reaction; i.e., the collision energy must equal or exceed the activation energy. **2)** the relative orientation of the reactants must allow formation of any new bonds necessary to produce products.   **21.** Enzymes are very efficient catalysts. As is true for all catalysts, enzymes speed up a reaction by providing an alternative pathway for reactants to convert to products. This alternative pathway has a smaller activation energy and, hence, a faster rate. Also true is that catalysts are not used up in the overall chemical reaction. Once an enzyme comes in contact with the correct reagent, the chemical reaction quickly occurs, and the enzyme is then free to catalyze another reaction. Because of the efficiency of the reaction step, only a relatively small amount of enzyme is needed to catalyze a specific reaction, no matter how complex the reaction.   **23.** P_4: 6.0×10^{-4} mol/L · s; H_2: 3.6×10^{-3} mol/L · s   **25. a.** average rate of decomposition of $H_2O_2 = 2.31 \times 10^{-5}$ mol/L · s, rate of production of $O_2 = 1.16 \times 10^{-5}$ mol/L · s; **b.** average rate of decomposition of $H_2O_2 = 1.16 \times 10^{-5}$ mol/L · s, rate of production of O_2 $= 5.80 \times 10^{-6}$ mol/L · s   **27. a.** mol/L · s; **b.** mol/L · s; **c.** s^{-1}; **d.** L/mol · s; **e.** L^2/mol^2 · s   **29. a.** rate $= k[NO]^2[Cl_2]$; **b.** 1.8×10^2 L^2/mol^2 · min   **31. a.** rate $= k[NOCl]^2$; **b.** 6.6×10^{-29} cm^3/molecules · s; **c.** 4.0×10^{-8} L/mol · s   **33. a.** rate $= k[I^-][OCl^-]$; **b.** 3.7 L/mol · s; **c.** 0.083 mol/L · s   **35. a.** first order in Hb and first order in CO; **b.** rate $= k[Hb][CO]$; **c.** 0.280 L/μmol · s; **d.** 2.26 μmol/L · s   **37.** rate $= k[H_2O_2]$; $\ln[H_2O_2] = -kt + \ln[H_2O_2]_0$;

$k = 8.3 \times 10^{-4}$ s^{-1}; 0.037 M **39.** rate $= k[NO_2]^2$; $\dfrac{1}{[NO_2]} = kt + \dfrac{1}{[NO_2]_0}$;

$k = 2.08 \times 10^{-4}$ L/mol · s; 0.131 M **41. a.** rate $= k$; $[C_2H_5OH] = -kt +$ $[C_2H_5OH]_0$; because slope $= -k$, $k = 4.00 \times 10^{-5}$ mol/L · s; **b.** 156 s;

c. 313 s   **43.** rate $= k[C_4H_6]^2$; $\dfrac{1}{[C_4H_6]} = kt + \dfrac{1}{[C_4H_6]_0}$; $k = 1.4 \times 10^{-2}$

L/mol · s   **45.** second order; 0.1 M **47. a.** $\frac{1}{[A]} = -kt + [A]_0$; **b.** 1.0×10^{-2} s; **c.** 2.5×10^{-4} M **49.** 9.2×10^{-3} s^{-1}; 75 s   **51.** 150. s   **53.** 12.5 s   **55. a.** 1.1×10^{-2} M; **b.** 0.025 M **57.** 1.6 L/mol · s   **59. a.** rate $= k[CH_3NC]$; **b.** rate $= k[O_3][NO]$; **c.** rate $= k[O_3]$; **d.** rate $= k[O_3][O]$ **61.** Rate $= k[C_4H_9Br]$; $C_4H_9Br + 2H_2O \rightarrow C_4H_9OH + Br^- + H_3O^+$; the intermediates are $C_4H_9^+$ and $C_4H_9OH_2^+$.

63.

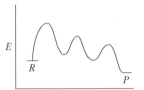

65. 341 kJ/mol   **67.** The graph of $\ln(k)$ versus $1/T$ is linear with slope $= -E_a/R = -1.2 \times 10^4$ K; $E_a = 1.0 \times 10^2$ kJ/mol   **69.** 9.5×10^{-5} L/mol · s   **71.** 51°C   **73.** $H_3O^+(aq) + OH^-(aq) \rightarrow 2H_2O(l)$ should have the faster rate. H_3O^+ and OH^- will be electrostatically attracted to each other; Ce^{4+} and Hg_2^{2+} will repel each other (so E_a is much larger).   **75. a.** NO; **b.** NO$_2$; **c.** 2.3   **77.** CH_2D—CH_2D should be the product. If the mechanism is possible, then the reaction must be $C_2H_4 + D_2 \rightarrow CH_2DCH_2D$. If we got this product, then we could conclude that this is a possible mechanism. If we got some other product, e.g., CH_3CHD_2, then we would conclude that the mechanism is wrong. Even though this mechanism correctly predicts the products of the reaction, we cannot say conclusively that this is the correct mechanism; we might be able to conceive of other mechanisms that would give the same product as our proposed one.   **79.** The rate depends on the number of reactant molecules adsorbed on the surface of the catalyst. This quantity is proportional to the concentration of reactant. However, when all the catalyst surface sites are occupied, the rate becomes independent of the concentration of reactant.   **81.** 215°C   **83. a.** 20 min; **b.** 30 min; **c.** 15 min   **85.** 427 s

87. 1.0×10^2 kJ/mol   **89.** 6.58×10^{-6} mol/L · s   **91. a.** 25 kJ/mol; **b.** 12 s;

c. T	Interval	$54 - 2$(Intervals)
21.0°C	16.3 s	21°C
27.8°C	13.0 s	28°C
30.0°C	12 s	30.°C

This rule of thumb gives excellent agreement to two significant figures. **93. a.** 115 L^3/mol^3 · s; **b.** 87.0 s; **c.** [A] $= 1.27 \times 10^{-5}$ M, [B] $= 1.00$ M **103.** rate $= \dfrac{k[I^-][OCl^-]}{[OH^-]}$; $k = 6.0 \times 10^1$ s^{-1} **105. a.** first order with respect to both reactants; **b.** rate $= k[NO][O_3]$; **c.** $k' = 1.8$ s^{-1}; $k'' = 3.6$ s^{-1}; **d.** $k = 1.8 \times 10^{-14}$ cm^3/molecules · s   **107. a.** For a three-step reaction with the first step limiting, the energy-level diagram could be

Note that the heights of the second and third humps must be lower than the first-step activation energy. However, the height of the third hump could be higher than the second hump. One cannot determine this absolutely from the information in the problem.

b.

$F_2 \rightarrow 2F$	slow
$F + H_2 \rightarrow HF + H$	fast
$H + F \rightarrow HF$	fast
$F_2 + H_2 \rightarrow 2HF$	

c. F_2 was the limiting reactant.   **109. a.** [B] $\gg$ [A] so that [B] can be considered constant over the experiments. (This gives us a pseudo-order rate law equation.) **b.** Rate $= k[A]^2[B]$, $k = 0.050$ L^2/mol^2 · s   **111.** Rate $= k[A][B]^2$, $k = 1.4 \times 10^{-2}$ L^2/mol^2 · s   **113.** 2.20×10^{-5} s^{-1}; 5.99×10^{21} molecules **115.** 1.3×10^{-5} s^{-1}; 112 torr

Chapter 13

11. No, equilibrium is a dynamic process. Both the forward and reverse reactions are occurring at equilibrium, just at equal rates. Thus the forward and reverse reactions will distribute ^{14}C atoms between CO and CO_2.   **13.** A large value for K ($K \gg 1$) indicates there are relatively large concentrations of product gases and/or solutes as compared with the concentrations of reactant gases and/or solutes at equilibrium. A reaction with a large K value is a good source of products.   **15.** 4 molecules H_2O, 2 molecules CO, 4 molecules H_2, and 4 molecules CO_2 are present at equilibrium.   **17.** K and K_p are equilibrium constants as determined by the law of mass action. For K, the units used for concentrations are mol/L, and for K_p, partial pressures in units of atm are used (generally). Q is called the reaction quotient. Q has the exact same form as K or K_p, but instead of equilibrium concentrations, initial concentrations are used to calculate the Q value. Q is of use when it is compared to the K value. When $Q = K$ (or when $Q_p = K_p$), the reaction is at equilibrium. When $Q \neq K$, the reaction is not at equilibrium and one can determine what has to be the net change for the system to get to equilibrium.   **19.** We always try to make good assumptions that simplify the math. In some problems, we can set up the problem so that the net change, x, that must occur to reach equilibrium is a small number. This comes in handy when you have expressions like $0.12 - x$ or $0.727 + 2x$. When x is small, we assume that it makes little difference when subtracted from or added to some relatively big number. When this is true, $0.12 - x \approx 0.12$ and $0.727 + 2x \approx 0.727$. If the assumption holds by the 5% rule, then the assumption is assumed valid. The 5% rule refers to x (or $2x$ or $3x$, etc.) that was assumed small compared to some number. If x (or $2x$ or $3x$, etc.) is less than 5% of the number the assumption was made against, then the assumption will be assumed valid. If the 5% rule fails to work, one can generally use a math procedure called the method of successive approximations to solve the quadratic or cubic equation.

21. a. $K = \dfrac{[NO]^2}{[N_2][O_2]}$; **b.** $K = \dfrac{[NO_2]^2}{[N_2O_4]}$; **c.** $K = \dfrac{[SiCl_4][H_2]^2}{[SiH_4][Cl_2]^2}$;

d. $K = \dfrac{[PCl_3]^2[Br_2]^3}{[PBr_3]^2[Cl_2]^3}$ **23. a.** 0.11; **b.** 77; **c.** 8.8; **d.** 1.7×10^{-4} **25.** 4.0×10^6

27. 1.7×10^{-5} **29.** 6.3×10^{-13} **31.** 4.6×10^3

33. a. $K = \dfrac{[H_2O]}{[NH_3]^2[CO_2]}$, $K_p = \dfrac{P_{H_2O}}{P_{NH_3}^2 \times P_{CO_2}}$; **b.** $K = [N_2][Br_2]^3$,

$K_p = P_{N_2} \times P_{Br_2}^3$; **c.** $K = [O_2]^3$, $K_p = P_{O_2}^3$; **d.** $K = \dfrac{[H_2O]}{[H_2]}$, $K_p = \dfrac{P_{H_2O}}{P_{H_2}}$

35. only reaction d **37.** 8.0×10^9 **39. a.** not at equilibrium; $Q > K$, shift left; **b.** at equilibrium; **c.** not at equilibrium; $Q > K$, shift left **41. a.** decrease; **b.** no change; **c.** no change; **d.** increase **43.** 0.16 mol **45.** 3.4 **47.** 0.056 **49.** $[N_2]_0 = 10.0\ M$, $[H_2]_0 = 11.0\ M$ **51.** $[SO_3] = [NO] = 1.06\,M$; $[SO_2] = [NO_2] = 0.54\,M$ **53.** 7.8×10^{-2} atm **55.** $P_{SO_2} = 0.38$ atm; $P_{O_2} = 0.44$ atm; $P_{SO_3} = 0.12$ atm **57. a.** $[NO] = 0.032\ M$, $[Cl_2] = 0.016\ M$, $[NOCl] = 1.0\ M$; **b.** $[NO] = [NOCl] = 1.0\ M$, $[Cl_2] = 1.6 \times 10^{-5}\,M$; **c.** $[NO] = 8.0 \times 10^{-3}\,M$, $[Cl_2] = 1.0\,M$, $[NOCl] = 2.0\,M$ **59.** $[CO_2] = 0.39\ M$, $[CO] = 8.6 \times 10^{-3}\ M$, $[O_2] = 4.3 \times 10^{-3}\ M$ **61.** 0.27 atm **63. a.** no effect; **b.** shifts left; **c.** shifts right **65. a.** right; **b.** right; **c.** no effect; **d.** left; **e.** no effect **67. a.** left; **b.** right; **c.** left; **d.** no effect; **e.** no effect; **f.** right **69.** increase **71.** 2.6×10^{81} **73.** 6.74×10^{-6} **75. a.** 0.379 atm; **b.** 0.786 **77.** $[Fe^{3+}] = 2 \times 10^{-4}\ M$, $[SCN^-] = 0.08\ M$, $[FeSCN^{2+}] = 0.020\ M$ **79.** 1.43×10^{-2} atm **81.** pink **83.** Added OH^- reacts with H^+ to produce H_2O. As H^+ is removed, the reaction shifts right to produce more H^+ and CrO_4^{2-}. Because more CrO_4^{2-} is produced, the solution turns yellow. **85.** $9.0 \times 10^{-3}\ M$ **87.** 0.50 **89.** $3.0 \times 10^{-6}\ M$ **99.** $P_{CO} = 0.58$ atm, $P_{CO_2} = 1.65$ atm **101.** $[NOCl] = 2.0\,M$, $[NO] = 0.050\,M$, $[Cl_2] = 0.025\ M$ **103.** 2.1×10^{-3} atm **105.** $P_{NO_2} = 0.704$ atm, $P_{N_2O_4} = 0.12$ atm **107. a.** Assuming mol/L units for concentrations, $K = 8.16 \times 10^{47}$; **b.** 33.9 L **109.** 0.63 **111.** 0.240 atm **113. a.** 2.33×10^{-4}; **b.** The argon gas will increase the volume of the container. This is because the container is a constant-pressure system, and if the number of moles increases at constant T and P, the volume must increase. An increase in volume will dilute the concentrations of all gaseous reactants and gaseous products. Because there are more moles of product gases than reactant gases (3 moles vs. 2 moles), the dilution will decrease the numerator of K more than the denominator will decrease. This causes $Q < K$ and the reaction shifts right to get back to equilibrium. Because temperature was unchanged, the value of K will not change. K is a constant as long as temperature is constant. **115.** 192 g; 1.3 atm **117.** 33

Chapter 14

19. as an acid: $HCO_3^-(aq) + H_2O(l) \rightleftharpoons CO_3^{2-}(aq) + H_3O^+(aq)$; as a base: $HCO_3^-(aq) + H_2O(l) \rightleftharpoons H_2CO_3(aq) + OH^-(aq)$; as an acid: $H_2PO_4^-(aq) + H_2O(l) \rightleftharpoons HPO_4^{2-}(aq) + H_3O^+(aq)$; as a base: $H_2PO_4^-(aq) + H_2O(l) \rightleftharpoons H_3PO_4(aq) + OH^-(aq)$. **21.** b, c, and d **23.** 10.78 (4 significant figures); 6.78 (3 significant figures); 0.78 (2 significant figures); a pH value is a logarithm. The numbers to the left of the decimal place identify the power of 10 to which $[H^+]$ is expressed in scientific notation—for example, 10^{-11}, 10^{-7}, 10^{-1}. The number of decimal places in a pH value identifies the number of significant figures in $[H^+]$. In all three pH values, the $[H^+]$ should be expressed only to two significant figures since these pH values have only two decimal places.

25.

NH_3	+	NH_3	$\rightleftharpoons$	NH_2^-	+	NH_4^+
Acid		Base		Conjugate Base		Conjugate Acid

One of the NH_3 molecules acts as a base and accepts a proton to form NH_4^+. The other NH_3 molecule acts as an acid and donates a proton to form NH_2^-. NH_4^+ is the conjugate acid of the NH_3 base. In the reverse reaction, NH_4^+ donates a proton. NH_2^- is the conjugate base of the NH_3 acid. In the reverse reaction, NH_2^- accepts a proton. Conjugate acid–base pairs differ only by an H^+ in the formula. **27. a.** These would be 0.10 M solutions of strong acids like HCl, HBr, HI, HNO$_3$, H$_2$SO$_4$, or HClO$_4$. **b.** These are salts of the conjugate acids of the bases in Table 14.3. These conjugate acids are all weak acids. Three examples would be 0.10 M solutions of NH$_4$Cl, CH$_3$NH$_3$NO$_3$, and C$_2$H$_5$NH$_3$Br. Note that the anions used to form these salts are conjugate bases of strong acids; this is because they have no acidic or basic properties in

water (with the exception of HSO_4^-, which has weak acid properties). **c.** These would be 0.10 M solutions of strong bases like LiOH, NaOH, KOH, RbOH, CsOH, Ca(OH)$_2$, Sr(OH)$_2$, and Ba(OH)$_2$. **d.** These are salts of the conjugate bases of the neutrally charged weak acids in Table 14.2. The conjugate bases of weak acids are weak bases themselves. Three examples would be 0.10 M solutions of NaClO$_2$, KC$_2$H$_3$O$_2$, and CaF$_2$. The cations used to form these salts are Li$^+$, Na$^+$, K$^+$, Rb$^+$, Cs$^+$, Ca^{2+}, Sr^{2+}, and Ba^{2+} since these cations have no acidic or basic properties in water. Notice that these are the cations of the strong bases that you should memorize. **e.** There are two ways to make a neutral salt. The easiest way is to combine a conjugate base of a strong acid (except for HSO_4^-) with one of the cations from the strong bases. These ions have no acidic/basic properties in water so salts of these ions are neutral. Three examples would be 0.10 M solutions of NaCl, KNO$_3$, and SrI$_2$. Another type of strong electrolyte that can produce neutral solutions are salts that contain an ion with weak acid properties combined with an ion of opposite charge having weak base properties. If the K_a for the weak acid ion is equal to the K_b for the weak base ion, then the salt will produce a neutral solution. The most common example of this type of salt is ammonium acetate, NH$_4$C$_2$H$_3$O$_2$. For this salt, K_a for NH$_4^+$ = K_b for C$_2$H$_3$O$_2^-$ = 5.6×10^{-10}. This salt, at any concentration, produces a neutral solution. **29. a.** $H_2O(l) + H_2O(l) \rightleftharpoons H_3O^+(aq) + OH^-(aq)$ or

$H_2O(l) \rightleftharpoons H^+(aq) + OH^-(aq)$ $K = K_w = [H^+][OH^-]$

b. $HF(aq) + H_2O(l) \rightleftharpoons F^-(aq) + H_3O^+(aq)$ or

$HF(aq) \rightleftharpoons H^+(aq) + F^-(aq)$ $K = K_a = \dfrac{[H^+][F^-]}{[HF]}$

c. $C_5H_5N(aq) + H_2O(l) \rightleftharpoons C_5H_5NH^+(aq) + OH^-(aq)$

$$K = K_b = \dfrac{[C_5H_5NH^+][OH^-]}{[C_5H_5N]}$$

31. a. This expression holds true for solutions of strong acids having a concentration greater than $1.0 \times 10^{-6}\ M$. For example, 0.10 M HCl, 7.8 M HNO$_3$, and $3.6 \times 10^{-4}\ M$ HClO$_4$ are solutions where this expression holds true. **b.** This expression holds true for solutions of weak acids where the two normal assumptions hold. The two assumptions are that the contribution of H^+ from water is negligible and that the acid is less than 5% dissociated in water (from the assumption that x is small compared to some number). This expression will generally hold true for solutions of weak acids having a K_a value less than 1×10^{-4}, as long as there is a significant amount of weak acid present. Three example solutions are 1.5 M HC$_2$H$_3$O$_2$, 0.10 M HOCl, and 0.72 M HCN. **c.** This expression holds true for strong bases that donate 2 OH^- ions per formula unit. As long as the concentration of the base is above $5 \times 10^{-7}\ M$, this expression will hold true. Three examples are $5.0 \times 10^{-3}\ M$ Ca(OH)$_2$, $2.1 \times 10^{-4}\ M$ Sr(OH)$_2$, and $9.1 \times 10^{-5}\ M$ Ba(OH)$_2$. **d.** This expression holds true for solutions of weak bases where the two normal assumptions hold. The assumptions are that the OH^- contribution from water is negligible and that the base is less than 5% ionized in water. For the 5% rule to hold, you generally need bases with $K_b < 1 \times 10^{-4}$ and concentrations of weak base greater than 0.10 M. Three examples are 0.10 M NH$_3$, 0.54 M C$_6$H$_5$NH$_2$, and 1.1 M C$_5$H$_5$N. **33.** One reason HF is a weak acid is that the H—F bond is unusually strong and thus, is difficult to break. This contributes to the reluctance of the HF molecules to dissociate in water. **35. a.** $HClO_4(aq) + H_2O(l) \rightarrow H_3O^+(aq) + ClO_4^-(aq)$ or $HClO_4(aq) \rightarrow H^+(aq) + ClO_4^-(aq)$; water is commonly omitted from K_a reactions. **b.** $CH_3CH_2CO_2H(aq) \rightleftharpoons H^+(aq) + CH_3CH_2CO_2^-(aq)$; **c.** $NH_4^+(aq) \rightleftharpoons H^+(aq) + NH_3(aq)$ **37. a.** H_2O, base; H_2CO_3, acid; H_3O^+, conjugate acid; HCO_3^-, conjugate base; **b.** $C_5H_5NH^+$, acid; H_2O, base; C_5H_5N, conjugate base; H_3O^+, conjugate acid; **c.** HCO_3^-, base; $C_5H_5NH^+$, acid; H_2CO_3, conjugate acid; C_5H_5N, conjugate base **39. a.** HClO$_4$, strong acid; **b.** HOCl, weak acid; **c.** H$_2$SO$_4$, strong acid; **d.** H$_2$SO$_3$, weak acid **41.** $HClO_4 > HClO_2 > NH_4^+ > H_2O$ **43. a.** HCl; **b.** HNO$_2$; **c.** HCN since it has a larger K_a value. **45. a.** $1.0 \times 10^{-7}\ M$, neutral; **b.** 12 M, basic; **c.** $8.3 \times 10^{-16}\ M$, acidic; **d.** $1.9 \times 10^{-10}\ M$, acidic **47. a.** endothermic; **b.** $[H^+] = [OH^-] = 2.34 \times 10^{-7}\ M$ **49.** [45] **a.** pH = pOH = 7.00; **b.** pH = 15.08, pOH = -1.08; **c.** pH = -1.08, pOH = 15.08; **d.** pH = 4.27, pOH = 9.73 [46] **a.** pH = 14.18, pOH = -0.18; **b.** pH = -0.44, pOH = 14.44; **c.** pH = pOH = 7.00; **d.** pH = 10.86, pOH = 3.14 **51. a.** pH = 6.88, pOH = 7.12, $[H^+] = 1.3 \times 10^{-7}\ M$, $[OH^-] = 7.6 \times 10^{-8}\ M$, acidic; **b.** pH = 0.92, pOH = 13.08, $[H^+] = 0.12\ M$,

$[OH^-] = 8.4 \times 10^{-14}\,M$, acidic; **c.** pH = 10.89, pOH = 3.11, $[H^+] = 1.3 \times 10^{-11}\,M$, $[OH^-] = 7.8 \times 10^{-4}\,M$, basic; **d.** pH = pOH = 7.00, $[H^+] = [OH^-] = 1.0 \times 10^{-7}\,M$, neutral **53.** pOH = 11.9, $[H^+] = 8 \times 10^{-3}\,M$, $[OH^-] = 1 \times 10^{-12}\,M$, acidic **55. a.** H^+, ClO_4^-, H_2O; 0.602; **b.** H^+, NO_3^-, H_2O; 0.602 **57. a.** 1.00; **b.** −0.70; **c.** 7.00 **59.** $3.2 \times 10^{-3}\,M$ **61.** Add 4.2 mL of 12 *M* HCl to water with mixing; add enough water to bring the solution volume to 1600 mL. **63. a.** HNO_2 and H_2O, 2.00; **b.** $HC_2H_3O_2$ and H_2O, 2.68 **65.** $[H^+] = [F^-] = 3.5 \times 10^{-3}\,M$, $[OH^-] = 2.9 \times 10^{-12}\,M$, $[HF] = 0.017\,M$, 2.46 **67.** $[HC_3H_5O_2] = 0.099\,M$, $[C_3H_5O_2^-] = [H^+] = 1.1 \times 10^{-3}\,M$, $[OH^-] = 9.1 \times 10^{-12}$, pH = 2.96, 1.1% dissociated. **69.** 1.96 **71.** 1.57; $5.9 \times 10^{-9}\,M$ **73. a.** 0.60%; **b.** 1.9%; **c.** 5.8%; **d.** Dilution shifts equilibrium to the side with the greater number of particles (% dissociation increases). **e.** $[H^+]$ also depends on initial concentration of weak acid. **75.** 1.4×10^{-4} **77.** 0.16 **79.** 0.024 *M* **81.** 6×10^{-3}

83. a. $NH_3(aq) + H_2O(l) \rightleftharpoons NH_4^+(aq) + OH^-(aq)$ $K_b = \dfrac{[NH_4^+][OH^-]}{[NH_3]}$;

b. $C_5H_5N(aq) + H_2O(l) \rightleftharpoons C_5H_5NH^+(aq) + OH^-(aq)$

$$K_b = \dfrac{[C_5H_5NH^+][OH^-]}{[C_5H_5N]}$$

85. $NH_3 > C_5H_5N > H_2O > NO_3^-$ **87. a.** $C_6H_5NH_2$; **b.** $C_6H_5NH_2$; **c.** OH^-; **d.** CH_3NH_2 **89. a.** 13.00; **b.** 7.00; **c.** 14.30 **91. a.** K^+, OH^-, and H_2O, 0.015 *M*, 12.18; **b.** Ba^{2+}, OH^-, and H_2O, 0.030 *M*, 12.48 **93.** 0.16 g **95.** NH_3 and H_2O, $1.6 \times 10^{-3}\,M$, 11.20 **97. a.** $[OH^-] = 8.9 \times 10^{-3}\,M$, $[H^+] = 1.1 \times 10^{-12}\,M$, 11.96; **b.** $[OH^-] = 4.7 \times 10^{-5}\,M$, $[H^+] = 2.1 \times 10^{-10}\,M$, 9.68 **99.** 12.00 **101. a.** 1.3%; **b.** 4.2%; **c.** 6.4% **103.** 1.0×10^{-9}

105. $H_2SO_3(aq) \rightleftharpoons HSO_3^-(aq) + H^+(aq)$ K_{a_1} reaction

$HSO_3^-(aq) \rightleftharpoons SO_3^{2-}(aq) + H^+(aq)$ K_{a_2} reaction

107. 3.00 **109.** 4.00; $1.0 \times 10^{-19}\,M$ **111.** −0.30 **113.** $HCl > NH_4Cl > KNO_3 > KCN > KOH$ **115.** OCl^- **117. a.** $[OH^-] = [H^+] = 1.0 \times 10^{-7}\,M$, pH = 7.00; **b.** $[OH^-] = 2.4 \times 10^{-5}\,M$, $[H^+] = 4.2 \times 10^{-10}\,M$, pH = 9.38 **119. a.** 5.82; **b.** 10.95 **121.** $[HN_3] = [OH^-] = 2.3 \times 10^{-6}\,M$, $[Na^+] = 0.010\,M$, $[N_3^-] = 0.010\,M$, $[H^+] = 4.3 \times 10^{-9}\,M$ **123.** NaF **125.** 3.66 **127.** 3.08 **129. a.** neutral; **b.** basic; $NO_2^- + H_2O \rightleftharpoons HNO_2 + OH^-$; **c.** acidic; $C_5H_5NH^+ \rightleftharpoons C_5H_5N + H^+$; **d.** acidic because NH_4^+ is a stronger acid than NO_2^- is a base; $NH_4^+ \rightleftharpoons NH_3 + H^+$; $NO_2^- + H_2O \rightleftharpoons HNO_2 + OH^-$; **e.** basic; $OCl^- + H_2O \rightleftharpoons HOCl + OH^-$; **f.** basic because OCl^- is a stronger base than NH_4^+ is an acid; $OCl^- + H_2O \rightleftharpoons HOCl + OH^-$, $NH_4^+ \rightleftharpoons NH_3 + H^+$ **131. a.** $HIO_3 < HBrO_3$; as the electronegativity of the central atom increases, acid strength increases. **b.** $HNO_2 < HNO_3$; as the number of oxygen atoms attached to the central atom increases, acid strength increases. **c.** $HOI < HOCl$; same reasoning as in part a. **d.** $H_3PO_3 < H_3PO_4$; same reasoning as in part b. **133. a.** $H_2O < H_2S < H_2Se$; acid strength increases as bond energy decreases. **b.** $CH_3CO_2H < FCH_2CO_2H < F_2CHCO_2H < F_3CCO_2H$; as the electronegativity of the neighboring atoms increases, acid strength increases. **c.** $NH_4^+ < HONH_3^+$; same reasoning as in part b. **d.** $NH_4^+ < PH_4^+$; same reasoning as in part a. **135. a.** basic; $CaO(s) + H_2O(l) \rightarrow Ca(OH)_2(aq)$; **b.** acidic; $SO_2(g) + H_2O(l) \rightarrow H_2SO_3(aq)$; **c.** acidic; $Cl_2O(g) + H_2O(l) \rightarrow 2HOCl(aq)$ **137. a.** $B(OH)_3$, acid; H_2O, base; **b.** Ag^+, acid; NH_3, base; **c.** BF_3, acid; F^-, base **139.** $Al(OH)_3(s) + 3H^+(aq) \rightarrow Al^{3+}(aq) + 3H_2O(l)$; $Al(OH)_3(s) + OH^-(aq) \rightarrow Al(OH)_4^-(aq)$ **141.** Fe^{3+}; because it is smaller with a greater positive charge, Fe^{3+} will be more strongly attracted to a lone pair of electrons from a Lewis base. **143.** 990 mL H_2O **145.** NH_4Cl **147. a.** $Hb(O_2)_4$ in lungs, HbH_4^{4+} in cells; **b.** Decreasing $[CO_2]$ will decrease $[H^+]$, favoring $Hb(O_2)_4$ formation. Breathing into a bag raises $[CO_2]$. **c.** $NaHCO_3$ lowers the acidity from accumulated CO_2. **149.** $4.2 \times 10^{-2}\,M$ **151. a.** 2.62; **b.** 2.4%; **c.** 8.48 **153. a.** HA is a weak acid. Most of the acid is present as HA molecules; only one set of H^+ and A^- ions is present. In a strong acid, all of the acid would be dissociated into H^+ and A^- ions. **b.** 2.2×10^{-3} **155.** 9.2×10^{-7} **157. a.** 1.66; **b.** Fe^{2+} ions will produce a less acidic solution (higher pH) due to the lower charge on Fe^{2+} as compared with Fe^{3+}. As the charge on a metal ion increases, acid strength of the hydrated ion increases. **159.** acidic; $HSO_4^- \rightleftharpoons SO_4^{2-} + H^+$; 1.54 **161. a.** H_2SO_3; **b.** $HClO_3$; **c.** H_3PO_3; NaOH and KOH are ionic compounds composed of either Na^+ or K^+ cations and OH^- anions. When soluble ionic compounds dissolve in water, they form the ions from which they are formed. The acids in this problem are all covalent compounds. When these acids dissolve in water, the covalent bond between oxygen and hydrogen breaks to form H^+ ions. **171.** 7.20. **173.** 4540 mL **175.** 4.17 **177.** 0.022 *M* **179.** 917 mL of water evaporated **181.** PO_4^{3-}, $K_b = 0.021$; HPO_4^{2-}, $K_b = 1.6 \times 10^{-7}$; $H_2PO_4^-$, $K_b = 1.3 \times 10^{-12}$; from the K_b values, PO_4^{3-} is the strongest base. **183. a.** basic; **b.** acidic; **c.** basic; **d.** acidic; **e.** acidic **185.** 1.0×10^{-3} **187.** 5.4×10^{-4} **189.** 2.5×10^{-3}

Chapter 15

9. When an acid dissociates, ions are produced. A common ion is when one of the product ions in a particular equilibrium is added from an outside source. For a weak acid dissociating to its conjugate base and H^+, the common ion would be the conjugate base; this would be added by dissolving a soluble salt of the conjugate base into the acid solution. The presence of the conjugate base from an outside source shifts the equilibrium to the left so less acid dissociates. **11.** The more weak acid and conjugate base present, the more H^+ and/or OH^- that can be absorbed by the buffer without significant pH change. When the concentrations of weak acid and conjugate base are equal (so that pH = pK_a), the buffer system is equally efficient at absorbing either H^+ or OH^-. If the buffer is overloaded with weak acid or with conjugate base, then the buffer is not equally efficient at absorbing either H^+ or OH^-. **13. a.** Let's call the acid HA, which is a weak acid. When HA is present in the beakers, it exists in the undissociated form, making it a weak acid. A strong acid would exist as separate H^+ and A^- ions. **b.** beaker c → beaker a → beaker e → beaker b → beaker d **c.** pH = pK_a when a buffer solution is present that has equal concentrations of the weak acid and conjugate base. This is beaker e. **d.** The equivalence point is when just enough OH^- has been added to exactly react with all of the acid present initially. This is beaker b. **e.** Past the equivalence point, the pH is dictated by the concentration of excess OH^- added from the strong base. We can ignore the amount of hydroxide added by the weak conjugate base that is also present. This is beaker d. **15.** The three key points to emphasize in your sketch are the initial pH, the pH at the halfway point to equivalence, and the pH at the equivalence point. For the two weak bases titrated, pH = pK_a at the halfway point to equivalence (50.0 mL HCl added) because [weak base] = [conjugate acid] at this point. For the initial pH, the strong base has the highest pH (most basic), while the weakest base has the lowest pH (least basic). At the equivalence point, the strong base titration has pH = 7.0. The weak bases titrated have acidic pHs at the equivalence point because the conjugate acids of the weak bases titrated are the major species present. The weakest base has the strongest conjugate acid so its pH will be lowest (most acidic) at the equivalence point.

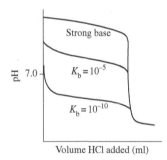

Volume HCl added (ml)

17. Only the third beaker represents a buffer solution. A weak acid and its conjugate base both must be present in large quantities in order to have a buffer solution. This is only the case in the third beaker. The first beaker represents a beaker full of strong acid that is 100% dissociated. The second beaker represents a weak acid solution. In a weak acid solution, only a small fraction of the acid is dissociated. In this representation, 1/10 of the weak acid has dissociated. The only B^- present in this beaker is from the dissociation of the weak acid. A buffer solution has B^- added from another source. **19.** When strong acid or strong base is added to a sodium bicarbonate/sodium carbonate buffer mixture, the strong acid/base is neutralized. The reaction goes to completion resulting in the strong acid/base being replaced with a weak acid/base. This results in a new buffer solution. The reactions are $H^+(aq) + CO_3^{2-}(aq) \rightarrow HCO_3^-(aq)$; $OH^-(aq) + HCO_3^-(aq) \rightarrow CO_3^{2-}(aq) + H_2O(l)$ **21. a.** 2.96; **b.** 8.94; **c.** 7.00; **d.** 4.89   **23.** 1.1% vs. 1.3×10^{-2}% dissociated; the presence of $C_3H_5O_2^-$ in solution 21d greatly inhibits the dissociation of $HC_3H_5O_2$. This is called the *common ion effect.*   **25. a.** 1.70; **b.** 5.49; **c.** 1.70; **d.** 4.71   **27. a.** 4.29; **b.** 12.30; **c.** 12.30; **d.** 5.07   **29.** solution d; solution d is a buffer solution that resists pH changes.   **31.** 3.40   **33.** 3.48; 3.22   **35. a.** 5.14; **b.** 4.34; **c.** 5.14; **d.** 4.34   **37.** 4.37   **39. a.** 7.97; **b.** 8.73; both solutions have an initial pH = 8.77. The two solutions differ in their buffer capacity. Solution b with the larger concentrations has the greater capacity to resist pH change.   **41.** 15 g   **43. a.** 0.19; **b.** 0.59; **c.** 1.0; **d.** 1.9   **45.** 1.3×10^{-2} M **47.** HOCl; there are many possibilities. One possibility is a solution with [HOCl] = 1.0 M and [NaOCl] = 0.35 M.   **49.** 8.18; add 0.20 mole NaOH   **51.** solution d   **53. a.** 1.0 mole; **b.** 0.30 mole; **c.** 1.3 moles   **55. a.** ~22 mL base added; **b.** buffer region is from ~1 mL to ~21 mL base added. The maximum buffering region would be from ~5 mL to ~17 mL base added with the halfway point to equivalence (~11 mL) as the best buffer point. **c.** ~11 mL base added; **d.** 0 mL base added; **e.** ~22 mL base added (the stoichiometric point); **f.** any point after the stoichiometric point (volume base added > ~22 mL)   **57. a.** 0.699; **b.** 0.854; **c.** 1.301; **d.** 7.00; **e.** 12.15   **59. a.** 2.72; **b.** 4.26; **c.** 4.74; **d.** 5.22; **e.** 8.79; **f.** 12.15

61.

Volume (mL)	pH
0.0	2.43
4.0	3.14
8.0	3.53
12.5	3.86
20.0	4.46
24.0	5.24
24.5	5.6
24.9	6.3
25.0	8.28
25.1	10.3
26.0	11.30
28.0	11.75
30.0	11.96

See *Solutions Guide* for pH plot.

63.

Volume (mL)	pH
0.0	11.11
4.0	9.97
8.0	9.58
12.5	9.25
20.0	8.65
24.0	7.87
24.5	7.6
24.9	6.9
25.0	5.28
25.1	3.7
26.0	2.71
28.0	2.24
30.0	2.04

See *Solutions Guide* for pH plot.

65. a. 4.19, 8.45; **b.** 10.74; 5.96; **c.** 0.89, 7.00   **67.** 2.1×10^{-6} **69. a.** yellow; **b.** 8.0; **c.** blue   **71.** phenolphthalein   **73.** Phenol red is one possible indicator for the titration in Exercise 57. Phenolphthalein is one possible indicator for the titration in Exercise 59.   **75.** Phenolphthalein is one possible indicator for Exercise 61. Bromcresol green is one possible indicator for Exercise 63.   **77.** The pH is between 5 and 8.   **79. a.** yellow; **b.** green; **c.** yellow; **d.** blue   **81.** $pOH = pK_b + \log\dfrac{[\text{acid}]}{[\text{base}]}$ **83. a.** The optimal buffer pH is about 8.1; **b.** 0.083 at pH = 7.00; 8.3 at pH = 9.00; **c.** 8.08; 7.95   **85. a.** potassium fluoride + HCl; **b.** benzoic acid + NaOH; **c.** acetic acid + sodium acetate; **d.** HOCl + NaOH; **e.** ammonium chloride + NaOH   **87. a.** $1.1 \approx 1$; **b.** A best buffer has approximately equal concentrations of weak acid and conjugate base, so pH $\approx pK_a$ for a best buffer. The pK_a

value for an $H_3PO_4/H_2PO_4^-$ buffer is $-\log(7.5 \times 10^{-3}) = 2.12$. A pH of 7.15 is too high for an $H_3PO_4/H_2PO_4^-$ buffer to be effective. At this high of a pH, there would be so little H_3PO_4 present that we could hardly consider it a buffer; this solution would not be effective in resisting pH changes, especially when a strong base is added.   **89. a.** 1.8×10^9; **b.** 5.6×10^4; **c.** 1.0×10^{14} **91.** 4.4 L   **93.** 65 mL   **95.** 180. g/mol; 3.3×10^{-4}; assumed acetylsalicylic acid is a weak monoprotic acid.   **97.** 0.210 M **99.** 1.74×10^{-8} **109.** 49 mL   **111.** 3.9 L   **113. a.** 200.0 mL; **b. i.** H_2A, H_2O; **ii.** H_2A, HA^-, H_2O, Na^+; **iii.** HA^-, H_2O, Na^+; **iv.** HA^-, A^{2-}, H_2O, Na^+; **v.** A^{2-}, H_2O, Na^+; **vi.** A^{2-}, H_2O, Na^+, OH^-; **c.** $K_{a_1} = 1 \times 10^{-4}$; $K_{a_2} = 1 \times 10^{-8}$ **115. a.** The major species present at the various points after H^+ reacts completely are A: CO_3^{2-}, H_2O, Na^+; B: $CO_3^{2-}, HCO_3^-, H_2O, Cl^-, Na^+$; C: $HCO_3^-, H_2O, Cl^-, Na^+$; D: $HCO_3^-, CO_2 (H_2CO_3), H_2O, Cl^-, Na^+$; E: $CO_2 (H_2CO_3), H_2O, Cl^-, Na^+$; F: $H^+, CO_2 (H_2CO_3), H_2O, Cl^-, Na^+$; **b.** B: pH = 10.25; D: pH = 6.37   **117.** pH $\approx$ 5.0; $K_a \approx 1 \times 10^{-10}$ **119.** 3.00   **121.** 2.78

Chapter 16

9. K_{sp} values can only be compared to determine relative solubilities when the salts produce the same number of ions. Here, Ag_2S and CuS do not produce the same number of ions when they dissolve, so each has a different mathematical relationship between the K_{sp} value and the molar solubility. To determine which salt has the larger molar solubility, you must do the actual calculations and compare the two molar solubility values.   **11. i.** This is the result when you have a salt that breaks up into two ions. Examples of these salts would be AgCl, $SrSO_4$, $BaCrO_4$, and $ZnCO_3$. **ii.** This is the result when you have a salt that breaks up into three ions, either two cations and one anion or one cation and two anions. Some examples are SrF_2, Hg_2I_2, and Ag_2SO_4. **iii.** This is the result when you have a salt that breaks up into four ions, either three cations and one anion (Ag_3PO_4) or one cation and three anions (ignoring the hydroxides, there are no examples of this type of salt in Table 16.1). **iv.** This is the result when you have a salt that breaks up into five ions, either three cations and two anions [$Sr_3(PO_4)_2$] or two cations and three anions (no examples of this type of salt are in Table 16.1).   **13.** For the K_{sp} reaction of a salt dissolving into its respective ions, the common ion would be if one of the ions in the salt was added from an outside source. When a common ion is present, the K_{sp} equilibrium shifts to the left resulting in less of the salt dissolving into its ions.   **15.** Some people would automatically think that an increase in temperature would increase the solubility of a salt. This is not always the case because some salts show a decrease in solubility as temperature increases. The two major methods used to increase solubility of a salt both involve removing one of the ions in the salt by reaction. If the salt has an anion with basic properties, adding H^+ will increase the solubility of the salt because the added H^+ will react with the basic ion, thus removing it from solution. More salt dissolves in order to make up for the lost ion. Some examples of salts with basic ions are AgF, $CaCO_3$, and $Al(OH)_3$. The other way to remove an ion is to form a complex ion. For example, the Ag^+ ion in silver salts forms the complex ion $Ag(NH_3)_2^+$ as ammonia is added. Silver salts increase their solubility as NH_3 is added because the Ag^+ ion is removed through complex ion formation.   **17.** In 2.0 M NH_3, the soluble complex ion $Ag(NH_3)_2^+$ forms, which increases the solubility of AgCl(s). The reaction is $AgCl(s) + 2NH_3 \rightleftharpoons Ag(NH_3)_2 + Cl^-$. In 2.0 M NH_4NO_3, NH_3 is only formed by the dissociation of the weak acid NH_4^+. There is not enough NH_3 produced by this reaction to dissolve AgCl(s) by the formation of the complex ion.   **19. a.** $AgC_2H_3O_2(s) \rightleftharpoons Ag^+ (aq) + C_2H_3O_2^-(aq)$; $K_{sp} = [Ag^+][C_2H_3O_2^-]$; **b.** $Al(OH)_3(s) \rightleftharpoons Al^{3+} (aq) + 3OH^-(aq)$; $K_{sp} = [Al^{3+}][OH^-]^3$; **c.** $Ca_3(PO_4)_2(s) \rightleftharpoons 3Ca^{2+}(aq) + 2PO_4^{3-}(aq)$; $K_{sp} = [Ca^{2+}]^3[PO_4^{3-}]^2$ **21. a.** 2.3×10^{-9}; **b.** 8.20×10^{-19} **23.** 1.4×10^{-8} **25.** 3.92×10^{-5} **27. a.** 1.6×10^{-5} mol/L; **b.** 9.3×10^{-5} mol/L; **c.** 6.5×10^{-7} mol/L   **29.** 0.89 g   **31.** 1.3×10^{-4} mol/L   **33.** 2×10^{-11} mol/L   **35. a.** CaF_2; **b.** $FePO_4$ **37. a.** 4×10^{-17} mol/L; **b.** 4×10^{-11} mol/L; **c.** 4×10^{-29} mol/L   **39. a.** 1.4×10^{-2} mol/L; **b.** 1.2×10^{-3} mol/L; **c.** 3.9×10^{-3} mol/L   **41.** 2.3×10^{-11} mol/L   **43.** 3.5×10^{-10} **45.** If the anion in the salt can act as a base in water, then the solubility of the salt will increase as the solution becomes more acidic. Added H^+ will react with the base, forming the conjugate acid. As the basic anion is removed, more of the salt will dissolve to replenish the basic anion. The salts with basic anions are Ag_3PO_4, $CaCO_3$, $CdCO_3$, and $Sr_3(PO_4)_2$. Hg_2Cl_2 and PbI_2 do not have any pH dependence because Cl^- and I^- are terrible bases (the conjugate bases of strong acids).

$$Ag_3PO_4(s) + H^+(aq) \longrightarrow 3Ag^+(aq) + HPO_4^{2-}(aq) \xrightarrow{\text{excess } H^+}$$
$$3Ag^+(aq) + H_3PO_4(aq)$$

$$CaCO_3(s) + H^+(aq) \longrightarrow Ca^{2+}(aq) + HCO_3^-(aq) \xrightarrow{\text{excess } H^+}$$
$$Ca^{2+}(aq) + H_2CO_3(aq) [H_2O(l) + CO_2(g)]$$

$$CdCO_3(s) + H^+(aq) \longrightarrow Cd^{2+}(aq) + HCO_3^-(aq) \xrightarrow{\text{excess } H^+}$$
$$Cd^{2+}(aq) + H_2CO_3(aq) [H_2O(l) + CO_2(g)]$$

$$Sr_3(PO_4)_2(s) + 2H^+(aq) \longrightarrow 3Sr^{2+}(aq) + 2HPO_4^{2-}(aq) \xrightarrow{\text{excess } H^+}$$
$$3Sr^{2+}(aq) + 2H_3PO_4(aq)$$

47. 1.5×10^{-19} g   **49.** No precipitate forms.   **51.** $PbF_2(s)$ will not form.   **53.** $[K^+] = 0.160\ M$, $[C_2O_4^{2-}] = 3.3 \times 10^{-7}\ M$, $[Ba^{2+}] = 0.0700\ M$, $[Br^-] = 0.300\ M$ **55.** 7.5×10^{-6} mol/L   **57.** $[AgNO_3] > 5.6 \times 10^{-5}\ M$ **59.** $PbS(s)$ will form first, followed by $Pb_3(PO_4)_2(s)$, and $PbF_2(s)$ will form last.

61. a.

$$Ni^{2+} + CN^- \rightleftharpoons NiCN^+$$
$$NiCN^+ + CN^- \rightleftharpoons Ni(CN)_2$$
$$Ni(CN)_2 + CN^- \rightleftharpoons Ni(CN)_3^-$$
$$\underline{Ni(CN)_3^- + CN^- \rightleftharpoons Ni(CN)_4^{2-}}$$
$$Ni^{2+} + 4CN^- \rightleftharpoons Ni(CN)_4^{2-}$$

b.

$$V^{3+} + C_2O_4^{2-} \rightleftharpoons VC_2O_4^+$$
$$VC_2O_4^+ + C_2O_4^{2-} \rightleftharpoons V(C_2O_4)_2^-$$
$$\underline{V(C_2O_4)_2^- + C_2O_4^{2-} \rightleftharpoons V(C_2O_4)_3^{3-}}$$
$$V^{3+} + 3C_2O_4^{2-} \rightleftharpoons V(C_2O_4)_3^{3-}$$

63. 1.0×10^{42} **65.** $Hg^{2+}(aq) + 2I^-(aq) \rightleftharpoons HgI_2(s)$ (orange precipitate); $HgI_2(s) + 2I^-(aq) \rightleftharpoons HgI_4^{2-}(aq)$ (soluble complex ion)   **67.** $3.3 \times 10^{-32}\ M$ **69. a.** $1.0 \times 10^{-3}\ M$; **b.** $2.0 \times 10^{-7}\ M$; **c.** $8.0 \times 10^{-15}\ M$ **71. a.** 1.2×10^{-8} mol/L; **b.** 1.5×10^{-4} mol/L; **c.** The presence of NH_3 increases the solubility of AgI. Added NH_3 removes Ag^+ from solution by forming the complex ion $Ag(NH_3)_2^+$. As Ag^+ is removed, more AgI will dissolve to replenish the Ag^+ concentration.   **73.** 4.7×10^{-2} mol/L   **75.** Test tube 1: added Cl^- reacts with Ag^+ to form the silver chloride precipitate. The net ionic equation is $Ag^+(aq) + Cl^-(aq) \rightarrow AgCl(s)$. Test tube 2: added NH_3 reacts with Ag^+ ions to form the soluble complex ion $Ag(NH_3)_2^+$. As this complex ion forms, Ag^+ is removed from solution, which causes $AgCl(s)$ to dissolve. When enough NH_3 is added, then all of the silver chloride precipitate will dissolve. The equation is $AgCl(s) + 2NH_3(aq) \rightarrow Ag(NH_3)_2^+(aq) + Cl^-(aq)$. Test tube 3: added H^+ reacts with the weak base NH_3 to form NH_4^+. As NH_3 is removed, Ag^+ ions are released to solution, which can then react with Cl^- to re-form $AgCl(s)$. The equations are $Ag(NH_3)_2^+(aq) + 2H^+(aq) \rightarrow Ag^+(aq) + 2NH_4^+(aq)$ and $Ag^+(aq) + Cl^-(aq) \rightarrow AgCl(s)$.   **77.** 1.7 g AgCl; 5×10^{-9} mol/L   **79.** 2.7×10^{-5} mol/L; the solubility of hydroxyapatite will increase as a solution gets more acidic, since both phosphate and hydroxide can react with H^+. 6×10^{-8} mol/L; the hydroxyapatite in the tooth enamel is converted to the less soluble fluorapatite by fluoride-treated water. The less soluble fluorapatite will then be more difficult to dissolve, making teeth less susceptible to decay.   **81.** Addition of more than 9.0×10^{-6} g $Ca(NO_3)_2$ should start precipitation of $CaF_2(s)$.   **83.** 7.0×10^{-8} **85.** 6.2×10^5 **87. a.** 6.7×10^{-6} mol/L; **b.** 1.2×10^{-13} mol/L; **c.** $Pb(OH)_2(s)$ will not form since $Q < K_{sp}$.   **89. a.** 1.6×10^{-6}; **b.** 0.056 mol/L   **91.** $Ba(OH)_2$, pH = 13.34; $Sr(OH)_2$, pH = 12.93; $Ca(OH)_2$, pH = 12.15   **99. a.** 0.33 mol/L; **b.** 0.33 M; **c.** $4.8 \times 10^{-3}\ M$ **101. a.** 7.1×10^{-7} mol/L; **b.** 8.7×10^{-3} mol/L; **c.** The presence of NH_3 increases the solubility of AgBr. Added NH_3 removes Ag^+ from solution by forming the complex ion, $Ag(NH_3)_2^+$. As Ag^+ is removed, more $AgBr(s)$ will dissolve to replenish the Ag^+ concentration. **d.** 0.41 g AgBr; **e.** Added HNO_3 will have no effect on the $AgBr(s)$ solubility in pure water. Neither H^+ nor NO_3^- react with Ag^+ or Br^- ions. Br^- is the conjugate base of the strong acid HBr, so it is a terrible base. However, added

HNO_3 will reduce the solubility of $AgBr(s)$ in the ammonia solution. NH_3 is a weak base ($K_b = 1.8 \times 10^{-5}$). Added H^+ will react with NH_3 to form NH_4^+. As NH_3 is removed, a smaller amount of the $Ag(NH_3)_2^+$ complex ion will form, resulting in a smaller amount of $AgBr(s)$ that will dissolve.   **103.** 5.7×10^{-2} mol/L   **105.** 3 M **107.** $4.8 \times 10^{-11}\ M$ **109. a.** 5.8×10^{-4} mol/L; **b.** greater; F^- is a weak base ($K_b = 1.4 \times 10^{-11}$), so some of the F^- is removed by reaction with water. As F^- is removed, more SrF_2 will dissolve; **c.** 3.5×10^{-3} mol/L   **111.** pH = 0.70; 0.20 M Ba^{2+}; 23 g $BaSO_4(s)$

Chapter 17

11. Living organisms need an external source of energy to carry out these processes. For all processes combined, ΔS_{univ} must be greater than zero (the 2nd law).   **13.** As any process occurs, ΔS_{univ} will increase; ΔS_{univ} cannot decrease. Time also goes in one direction, just as ΔS_{univ} goes in one direction.   **15.** This reaction is kinetically slow but thermodynamically favorable ($\Delta G < 0$). Thermodynamics only tells us if a reaction can occur. To answer the question will the reaction occur, one also needs to consider the kinetics (speed of reaction). The ultraviolet light provides the activation energy for this slow reaction to occur.   **17.** $\Delta S_{surr} = -\Delta H/T$; heat flow ($\Delta H$) into or out of the system dictates ΔS_{surr}. If heat flows into the surroundings, the random motions of the surroundings increase, and the entropy of the surroundings increases. The opposite is true when heat flows from the surroundings into the system (an endothermic reaction). Although the driving force described here really results from the change in entropy of the surroundings, it is often described in terms of energy. Nature tends to seek the lowest possible energy.   **19.** Note that these substances are not in the solid state but are in the aqueous state; water molecules are also present. There is an apparent increase in ordering when these ions are placed in water. The hydrating water molecules must be in a highly ordered state when surrounding these anions.   **21.** One can determine ΔS° and ΔH° for the reaction using the standard entropies and standard enthalpies of formation in Appendix 4, then use the equation $\Delta G^\circ = \Delta H^\circ - T\Delta S^\circ$. One can also use the standard free energies of formation in Appendix 4. And finally, one can use Hess's law to calculate ΔG°. Here, reactions having known ΔG° values are manipulated to determine ΔG° for a different reaction. For temperatures other than 25°C, ΔG° is estimated using the $\Delta G^\circ = \Delta H^\circ - T\Delta S^\circ$ equation. The assumptions made are that the ΔH° and ΔS° values determined from Appendix 4 data are temperature independent. We use the same ΔH° and ΔS° values as determined when $T = 25$°C, then plug in the new temperature in kelvins into the equation to estimate ΔG° at the new temperature.   **23.** The light source for the first reaction is necessary for kinetic reasons. The first reaction is just too slow to occur unless a light source is available. The kinetics of a reaction are independent of the thermodynamics of a reaction. Even though the first reaction is more favorable thermodynamically (assuming standard conditions), it is unfavorable for kinetic reasons. The second reaction has a negative ΔG° value and is a fast reaction, so the second reaction occurs very quickly. When considering if a reaction will occur, thermodynamics and kinetics must both be considered.   **25.** a, b, c
27. Possible arrangements for one molecule:

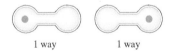

1 way   1 way

Both are equally probable.
Possible arrangements for two molecules:

1 way   2 ways   1 way
               Most probable

Possible arrangement for three molecules:

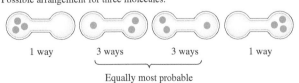

1 way   3 ways   3 ways   1 way

Equally most probable

29. We draw all of the possible arrangements of the two particles in the three levels.

2 kJ	—	—	x	—	x	xx
1 kJ	—	x	—	xx	x	—
0 kJ	xx	x	x	—	—	—
Total E =	0 kJ	1 kJ	2 kJ	2 kJ	3 kJ	4 kJ

The most likely total energy is 2 kJ.   **31. a.** H_2 at 100°C and 0.5 atm; **b.** N_2 at STP; **c.** $H_2O(l)$ **33. a.** negative; **b.** positive   **35.** Spontaneous ($\Delta G < 0$) for b, c, d   **37.** 89.3 J/K · mol   **39. a.** yes ($\Delta G < 0$); **b.** 196 K   **41. a.** negative; **b.** negative; **c.** negative; **d.** positive   **43. a.** $C_{graphite}(s)$; **b.** $C_2H_5OH(g)$; **c.** $CO_2(g)$ **45. a.** negative, -186 J/K; **b.** positive, 187 J/K; **c.** hard to predict since $\Delta n = 0$; 138 J/K   **47.** 262 J/K · mol   **49. a.** ΔH and ΔS are both positive; **b.** $S_{rhombic}$ **51. a.** ΔH and ΔS are both negative; **b.** low temperatures   **53. a.** $\Delta H° = -803$ kJ, $\Delta S° = -4$ J/K, $\Delta G° = -802$ kJ; **b.** $\Delta H° = 2802$ kJ, $\Delta S° = -262$ J/K, $\Delta G° = 2880.$ kJ; **c.** $\Delta H° = -416$ kJ, $\Delta S° = -209$ J/K, $\Delta G° = -354$ kJ; **d.** $\Delta H° = -176$ kJ, $\Delta S° = -284$ J/K, $\Delta G° = -91$ kJ   **55.** -5.40 kJ; 328.6 K; $\Delta G°$ is negative below 328.6 K.   **57.** -817 kJ   **59.** -731 kJ/mol   **61. a.** 53 kJ; **b.** No, the reaction is not spontaneous at standard concentrations and 298 K. **c.** $T > 630$ K   **63.** $CH_4(g) + CO_2(g) \rightarrow CH_3CO_2H(l)$, $\Delta H° = -16$ kJ, $\Delta S° = -240.$ J/K, $\Delta G° = 56$ kJ; $CH_3OH(g) + CO(g) \rightarrow CH_3CO_2H(l)$, $\Delta H° = -173$ kJ, $\Delta S° = -278$ J/K, $\Delta G° = -90.$ kJ; the second reaction is preferred. It should be run at temperatures below 622 K.   **65.** -188 kJ   **67. a.** shifts right; **b.** no shift since the reaction is at equilibrium; **c.** shifts left   **69.** -198 kJ; 5.07×10^{34} **71.** 8.72; 0.0789   **73.** 140 kJ   **75. a.** 2.22×10^5; **b.** 94.3   **77.** -71 kJ/mol   **79.** $\Delta H° = 1.1 \times 10^5$ J/mol; $\Delta S° = 330$ J/K · mol; the major difference in the plot is the slope of the line. An endothermic process has a negative slope for the $\ln(K)$ versus $1/T$ plot, whereas an exothermic process has a positive slope.   **81.** 447 J/K · mol   **83.** decreases; ΔS will be negative since 2 moles of gaseous reactants form 1 mole of gaseous product. For ΔG to be negative, ΔH must be negative (exothermic). For exothermic reactions, K decreases as T increases, so the ratio of the partial pressure of PCl_5 to the partial pressure of PCl_3 will decrease.   **85.** 43.7 K   **87.** 7.0×10^{-4} **89.** 60   **91.** ΔS is more favorable (less negative) for reaction 2 than for reaction 1, resulting in $K_2 > K_1$. In reaction 1, seven particles in solution form one particle. In reaction 2, four particles form one, which results in a smaller decrease in positional probability than for reaction 1.   **93.** 725 K   **103. a.** Vessel 1: At 0°C, this system is at equilibrium, so $\Delta S_{univ} = 0$ and $\Delta S = \Delta S_{surr}$. Because the vessel is perfectly insulated, $q = 0$ so $\Delta S_{surr} = 0 = \Delta S_{sys}$. **b.** Vessel 2: The presence of salt in water lowers the freezing point of water to a temperature below 0°C. In vessel 2 the conversion of ice into water will be spontaneous at 0°C, so $\Delta S_{univ} > 0$. Because the vessel is perfectly insulated, $\Delta S_{surr} = 0$. Therefore, ΔS_{sys} must be positive ($\Delta S > 0$) in order for ΔS_{univ} to be positive.   **105.** $\Delta H° = 286$ kJ; $\Delta G° = 326$ kJ; $K = 7.22 \times 10^{-58}$; $P_{O_3} = 3.3 \times 10^{-41}$ atm; this partial pressure represents one molecule of ozone per 9.5×10^{17} L of air. Equilibrium is probably not maintained under the conditions because the concentration of ozone is not large enough to maintain equilibrium.   **107. a.**

$$k_f = A \exp\left(\frac{-E_a}{RT}\right) \text{ and } k_r = A \exp\left(\frac{-(E_a - \Delta G°)}{RT}\right),$$

$$\frac{k_f}{k_r} = \exp\left(\frac{-E_a}{RT} + \frac{(E_a - \Delta G°)}{RT}\right) = \exp\left(\frac{-\Delta G°}{RT}\right)$$

From $\Delta G° = -RT \ln K$, K also equals the same expression: $K = \exp\left(\frac{-\Delta G°}{RT}\right)$, so $K = \frac{k_f}{k_r}$. **b.** A catalyst increases the value of the rate constant (increases rate) by lowering the activation energy. For the equilibrium constant K to remain constant, both k_f and k_r must increase by the same factor. Therefore, a catalyst must increase the rate of both the forward and the reverse reactions.   **109. a.** 0.333; **b.** $P_A = 1.50$ atm; $P_B = 0.50$ atm; **c.** $\Delta G = \Delta G° + RT \ln(P_B/P_A) = 2722 \text{ J} - 2722 \text{ J} = 0$ **111.** greater than 7.5 torr   **113.** 6 M **115.** 61 kJ/mol   **117.** -4.1 kJ/mol

Chapter 18

15. Oxidation: increase in oxidation number, loss of electrons; reduction: decrease in oxidation number, gain of electrons   **17.** Reactions a, b, and c are oxidation–reduction reactions.

	Oxidizing Agent	Reducing Agent	Substance Oxidized	Substance Reduced
a.	H_2O	CH_4	$CH_4(C)$	$H_2O(H)$
b.	$AgNO_3$	Cu	Cu	$AgNO_3(Ag)$
c.	HCl	Zn	Zn	$HCl(H)$

19. Electrochemistry is the study of the interchange of chemical and electrical energy. A redox (oxidation–reduction) reaction is a reaction in which one or more electrons are transferred. In a galvanic cell, a spontaneous redox reaction occurs which produces an electric current. In an electrolytic cell, electricity is used to force a nonspontaneous redox reaction to occur.   **21.** Magnesium is an alkaline earth metal; Mg will oxidize to Mg^{2+}. The oxidation state of hydrogen in HCl is $+1$. To be reduced, the oxidation state of H must decrease. The obvious choice for the hydrogen product is $H_2(g)$ where hydrogen has a zero oxidation state. The balanced reaction is: $Mg(s) + 2HCl(aq) \rightarrow MgCl_2(aq) + H_2(g)$. Mg goes from the 0 to the $+2$ oxidation state by losing two electrons. Each H atom goes from the $+1$ to the 0 oxidation state by gaining one electron. Since there are two H atoms in the balanced equation, then a total of two electrons are gained by the H atoms. Hence, two electrons are transferred in the balanced reaction. When the electrons are transferred directly from Mg to H^+, no work is obtained. In order to harness this reaction to do useful work, we must control the flow of electrons through a wire. This is accomplished by making a galvanic cell, which separates the reduction reaction from the oxidation reaction in order to control the flow of electrons through a wire to produce a voltage.   **23.** An extensive property is one that depends on the amount of substance. The free energy change for a reaction depends on whether 1 mole of product is produced or 2 moles of product is produced or 1 million moles of product is produced. This is not the case for cell potentials, which do not depend on the amount of substance. The equation that relates ΔG to E is $\Delta G = -nFE$. It is the n term that converts the intensive property E into the extensive property ΔG. n is the number of moles of electrons transferred in the balanced reaction that ΔG is associated with.   **25.** A potential hazard when jump-starting a car is that the electrolysis of $H_2O(l)$ can occur. When $H_2O(l)$ is electrolyzed, the products are the explosive gas mixture of $H_2(g)$ and $O_2(g)$. A spark produced during jump-starting a car could ignite any $H_2(g)$ and $O_2(g)$ produced. Grounding the jumper cable far from the battery minimizes the risk of a spark nearby the battery where $H_2(g)$ and $O_2(g)$ could be collecting.   **27.** You need to know the identity of the metal so you know which molar mass to use. You need to know the oxidation state of metal ion in the salt so the moles of electrons transferred can be determined. And finally, you need to know the amount of current and the time the current was passed through the electrolytic cell. If you know these four quantities, then the mass of metal plated out can be calculated.   **29. a.** $3I^-(aq) + 2H^+(aq) + ClO^-(aq) \rightarrow I_3^-(aq) + Cl^-(aq) + H_2O(l)$; **b.** $7H_2O(l) + 4H^+(aq) + 3As_2O_3(s) + 4NO_3^-(aq) \rightarrow 4NO(g) + 6H_3AsO_4(aq)$; **c.** $16H^+(aq) + 2MnO_4^-(aq) + 10Br^-(aq) \rightarrow 5Br_2(l) + 2Mn^{2+}(aq) + 8H_2O(l)$; **d.** $8H^+(aq) + 3CH_3OH(aq) + Cr_2O_7^{2-}(aq) \rightarrow 2Cr^{3+}(aq) + 3CH_2O(aq) + 7H_2O(l)$ **31. a.** $2H_2O(l) + Al(s) + MnO_4^-(aq) \rightarrow Al(OH)_4^-(aq) + MnO_2(s)$; **b.** $2OH^-(aq) + Cl_2(g) \rightarrow Cl^-(aq) + OCl^-(aq) + H_2O(l)$; **c.** $OH^-(aq) + H_2O(l) + NO_2^-(aq) + 2Al(s) \rightarrow NH_3(g) + 2AlO_2^-(aq)$ **33.** $4NaCl(aq) + 2H_2SO_4(aq) + MnO_2(s) \rightarrow 2Na_2SO_4(aq) + MnCl_2(aq) + Cl_2(g) + 2H_2O(l)$ **35.** The reducing agent causes reduction to occur; it does this by containing the species that is oxidized. Oxidation occurs at the anode, so the reducing agent will be in the anode compartment. The oxidizing agent causes oxidation to occur; it does this by containing the species that is reduced. Reduction occurs at the cathode, so the oxidizing agent will be in the cathode compartment. Electron flow is always from the anode compartment to the cathode compartment.   **37.** See Fig. 18.3 of the text for a typical galvanic cell. The anode compartment contains the oxidation half-reaction compounds/ions, and the cathode compartment contains the reduction half-reaction compounds/ions. The electrons flow from the anode to the cathode. For each of the following answers, all solutes are 1.0 M and all gases are at 1.0 atm. **a.** $7H_2O(l) + 2Cr^{3+}(aq) + 3Cl_2(g) \rightarrow Cr_2O_7^{2-}(aq) + 6Cl^-(aq) + 14H^+(aq)$; cathode: Pt electrode; Cl_2 bubbled into solution, Cl^- in solution; anode: Pt electrode; Cr^{3+}, H^+, and $Cr_2O_7^{2-}$ in solution; **b.** $Cu^{2+}(aq) + Mg(s) \rightarrow Cu(s) + Mg^{2+}(aq)$; cathode: Cu electrode; Cu^{2+} in solution; anode: Mg electrode; Mg^{2+} in solution **39. a.** 0.03 V; **b.** 2.71 V **41.** See Exercise 37 for a description of a galvanic cell. For each of the following answers, all solutes are 1.0 M and all gases are at 1.0 atm. In the salt bridge, cations flow to the cathode and anions flow to the anode. **a.** $Cl_2(g) + 2Br^-(aq) \rightarrow Br_2(aq) + 2Cl^-(aq)$, $\mathscr{E}° = 0.27$ V; cathode: Pt electrode; $Cl_2(g)$ bubbled in, Cl^- in solution; anode: Pt electrode; Br_2 and Br^- in solution; **b.** $3H_2O(l) + 5IO_4^-(aq) + 2Mn^{2+}(aq) \rightarrow 5IO_3^-(aq) + 2MnO_4^-(aq) + 6H^+(aq)$, $\mathscr{E}° = 0.09$ V; cathode: Pt electrode; IO_4^-, IO_3^-, and H_2SO_4 (as a source of H^+) in solution; anode: Pt electrode; Mn^{2+}, MnO_4^-, and H_2SO_4 in solution **43.** 37a. $Pt|Cr^{3+}$ (1.0 M), H^+ (1.0 M), $Cr_2O_7^{2-}$ (1.0 M)$||Cl_2$ (1.0 atm)$|Cl^-$ (1.0 M)$|$Pt; 37b. $Mg|Mg^{2+}$ (1.0 M)$||Cu^{2+}$ (1.0 M)$|$Cu; 41a. $Pt|Br^-$ (1.0 M), Br_2 (1.0 M)$||Cl_2$ (1.0 atm)$|Cl^-$ (1.0 M)$|$Pt; 41b. $Pt|Mn^{2+}$ (1.0 M), MnO_4^- (1.0 M), H^+ (1.0 M)$||IO_4^-$ (1.0 M), H^+ (1.0 M), IO_3^- (1.0 M) $|$Pt **45. a.** $Au^{3+}(aq) + 3Cu^+(aq) \rightarrow 3Cu^{2+}(aq) + Au(s)$, $\mathscr{E}° = 1.34$ V; **b.** $2VO_2^+(aq) + 4H^+(aq) + Cd(s) \rightarrow Cd^{2+}(aq) + 2VO^{2+}(aq) + 2H_2O(l)$, $\mathscr{E}° = 1.40$ V **47. a.** $16H^+ + 2MnO_4^- + 10I^- \rightarrow 5I_2 + 2Mn^{2+} + 8H_2O$, $\mathscr{E}°_{cell} = 0.97$ V, spontaneous; **b.** $16H^+ + 2MnO_4^- + 10F^- \rightarrow 5F_2 + 2Mn^{2+} + 8H_2O$, $\mathscr{E}°_{cell} = -1.36$ V, not spontaneous **49.** $\mathscr{E}° = 0.41$ V, $\Delta G° = -79$ kJ **51.** 45a. -388 kJ; 45b. $-270.$ kJ **53.** -0.829 V; the two values agree to two significant figures. **55.** 1.24 V **57.** $K^+ < H_2O < Cd^{2+} < I_2 < AuCl_4^- < IO_3^-$ **59. a.** no; **b.** yes; **c.** yes; **61. a.** Of the species available, Ag^+ would be the best oxidizing agent because it has the largest $\mathscr{E}°$ value. **b.** Of the species available, Zn would be the best reducing agent because it has the largest $-\mathscr{E}°$ value. **c.** $SO_4^{2-}(aq)$ can oxidize $Pb(s)$ and $Zn(s)$ at standard conditions. When $SO_4^{2-}(aq)$ is coupled with these reagents, $\mathscr{E}_{cell}$ is positive. **d.** $Al(s)$ can reduce $Ag^+(aq)$ and $Zn^{2+}(aq)$ at standard conditions because $\mathscr{E}°_{cell} > 0$. **63. a.** $Cr_2O_7^{2-}$, O_2, MnO_2, IO_3^-; **b.** $PbSO_4$, Cd^{2+}, Fe^{2+}, Cr^{3+}, Zn^{2+}, H_2O **65.** $ClO^-(aq) + 2NH_3(aq) \rightarrow Cl^-(aq) + N_2H_4(aq) + H_2O(l)$, $\mathscr{E}°_{cell} = 1.00$ V; because $\mathscr{E}°_{cell}$ is positive for this reaction, at standard conditions ClO^- can spontaneously oxidize NH_3 to the somewhat toxic N_2H_4. **67. a.** larger; **b.** smaller **69.** Electron flow is always from the anode to the cathode. For the cells with a nonzero cell potential, we will identify the cathode, which means the other compartment is the anode. **a.** 0; **b.** 0.018 V; compartment with $[Ag^+] = 2.0$ M is cathode; **c.** 0.059 V; compartment with $[Ag^+] = 1.0$ M is cathode; **d.** 0.26 V; compartment with $[Ag^+] = 1.0$ M is cathode; **e.** 0 **71.** 2.12 V **73.** 1.09 V **75.** [37]. **a.** $\Delta G° = -20$ kJ; 1×10^3; **b.** $\Delta G° = -523$ kJ; 5.12×10^{91}; [41]. **a.** $\Delta G° = -52$ kJ; 1.4×10^9; **b.** $\Delta G° = -90$ kJ; 2×10^{15} **77. a.** $Fe^{2+}(aq) + Zn(s) \rightarrow Zn^{2+}(aq) + Fe(s)$ $\mathscr{E}°_{cell} = 0.32$ V; **b.** -62 kJ; 6.8×10^{10}; **c.** 0.20 V **79. a.** 0.23 V; **b.** 1.2×10^{-5} M **81.** 0.16 V, copper is oxidized. **83.** 1.7×10^{-30} **85. a.** no reaction; **b.** $Cl_2(g) + 2I^-(aq) \rightarrow I_2(s) + 2Cl^-(aq)$, $\mathscr{E}°_{cell} = 0.82$ V; $\Delta G° = -160$ kJ; $K = 5.6 \times 10^{27}$; **c.** no reaction; **d.** $4Fe^{2+}(aq) + 4H^+(aq) + O_2(g) \rightarrow 4Fe^{3+}(aq) + 2H_2O(l)$, $\mathscr{E}°_{cell} = 0.46$ V; $\Delta G° = -180$ kJ; $K = 1.3 \times 10^{31}$; **87.** 0.151 V; -29.1 kJ **89.** 5.1×10^{-20} **91.** -0.14 V **93. a.** 30. hours; **b.** 33 s; **c.** 1.3 hours **95. a.** 16 g; **b.** 25 g; **c.** 71 g; **d.** 4.9 g **97.** Bi **99.** 9.12 L F_2 (anode), 29.2 g K (cathode) **101.** 1×10^5 A **103.** 1.14×10^{-2} M **105.** Au followed by Ag followed by Ni followed by Cd **107.** Cathode: $2H_2O + 2e^- \rightarrow H_2(g) + 2OH^-$; anode: $2H_2O \rightarrow O_2(g) + 4H^+ + 4e^-$ **109. a.** cathode: $Ni^{2+} + 2e^- \rightarrow Ni$; anode: $2Br^- \rightarrow Br_2 + 2e^-$; **b.** cathode: $Al^{3+} + 3e^- \rightarrow Al$; anode: $2F^- \rightarrow F_2 + 2e^-$; **c.** cathode: $Mn^{2+} + 2e^- \rightarrow Mn$; anode: $2I^- \rightarrow I_2 + 2e^-$ **111. a.** cathode: $Ni^{2+} + 2e^- \rightarrow Ni$; anode: $2Br^- \rightarrow Br_2 + 2e^-$; **b.** cathode: $2H_2O + 2e^- \rightarrow H_2 + 2OH^-$; anode: $2H_2O \rightarrow O_2 + 4H^+ + 4e^-$; **c.** cathode: $2H_2O + 2e^- \rightarrow H_2 + 2OH^-$; anode: $2I^- \rightarrow I_2 + 2e^-$ **113.** 0.250 mol **115. a.** 0.10 V, SCE is anode; **b.** 0.53 V, SCE is anode; **c.** 0.02 V, SCE is cathode; **d.** 1.90 V, SCE is cathode; **e.** 0.47 V, SCE is cathode **117. a.** decrease; **b.** increase; **c.** decrease; **d.** decrease; **e.** same **119. a.** $\Delta G° = -582$ kJ; $K = 3.45 \times 10^{102}$; $\mathscr{E}° = 1.01$ V; **b.** -0.65 V **121.** Aluminum has the ability to form a durable oxide coating over its surface. Once the HCl dissolves this oxide coating, Al is exposed to H^+ and is easily oxidized to Al^{3+}. Thus, the Al foil disappears after the oxide coating is dissolved. **123.** 1.14 V **125.** $w_{max} = -13{,}200$ kJ; the work done can be no larger than the free energy change. If the process were reversible all of the free energy released would go into work, but this does not occur in any real process. Fuel cells are more efficient in converting chemical energy to electrical energy; they are also less massive. Major disadvantage: They are expensive. **127.** 0.98 V **129.** 7.44×10^4 A **131.** To produce 1.0 kg Al by the Hall-Heroult process requires 5.4×10^4 kJ of energy. To melt 1.0 kg Al requires 4.0×10^2 kJ of energy. It is feasible to recycle Al by melting the metal because, in theory, it takes less than 1% of the energy required to produce the same amount of Al by the Hall-Heroult process. **133.** +3 **141.** $\mathscr{E}° = \frac{T\Delta S°}{nF} - \frac{\Delta H°}{nF}$; if we graph $\mathscr{E}°$ versus T, we should get a straight line ($y = mx + b$). The slope of the line is equal to $\Delta S°/nF$ and the y-intercept is equal to $-\Delta H°/nF$. $\mathscr{E}°$ will have little temperature dependence for cell reactions with $\Delta S°$ close to zero. **143.** 9.8×10^{-6} **145.** 2.39×10^{-7} **147. a.** ± 0.02 pH units; $\pm 6 \times 10^{-6}$ M H^+; **b.** ± 0.001 V **149. a.** 0.16 V; **b.** 8.6 mol **151.** $[Ag^+] = 4.6 \times 10^{-18}$ M; $[Ni^{2+}] = 1.5$ M **153.** 0.64 V **155. a.** 5.77×10^{10}; **b.** -12.2 kJ/mol **157.** Osmium(IV) nitrate; $[Ar]4s^13d^{10}$

Chapter 19

1. The characteristic frequencies of energies emitted in a nuclear reaction suggest that discrete energy levels exist in the nucleus. The extra stability of certain numbers of nucleons and the predominance of nuclei with even numbers of nucleons suggest that the nuclear structure might be described by using quantum numbers. **3.** β-particle production has the net effect of turning a neutron into a proton. Radioactive nuclei having too many neutrons typically undergo β-particle decay. Positron production has the net effect of turning a proton into a neutron. Nuclei having too many protons typically undergo positron decay. **5.** The transuranium elements are the elements having more protons than uranium. They are synthesized by bombarding heavier nuclei with neutrons and positive ions in a particle accelerator. **7.** $\Delta E = \Delta mc^2$; The key difference is the mass change when going from reactants to products. In chemical reactions, the mass change is indiscernible. In nuclear processes, the mass change is discernible. It is the conversion of this discernible mass change into energy that results in the huge energies associated with nuclear processes. **9.** The temperatures of fusion reactions are so high that all physical containers would be destroyed. At these high temperatures, most of the electrons are stripped from the atoms. A plasma of gaseous ions is formed that can be controlled by magnetic fields. **11. a.** ${}^3_1H \rightarrow {}^3_2He + {}^{\ 0}_{-1}e$; **b.** ${}^8_3Li \rightarrow {}^8_4Be + {}^{\ 0}_{-1}e$, ${}^8_4Be \rightarrow 2\ {}^4_2He$; overall reaction: ${}^8_3Li \rightarrow 2\ {}^4_2He + {}^{\ 0}_{-1}e$; **c.** ${}^7_4Be + {}^{\ 0}_{-1}e \rightarrow {}^7_3Li$; **d.** ${}^8_5B \rightarrow {}^8_4Be + {}^{\ 0}_{+1}e$ **13. a.** ${}^{234}_{90}Th$; this is α-particle production. **b.** ${}^{\ 0}_{-1}e$; this is β-particle production. **15. a.** ${}^{68}_{31}Ga + {}^{\ 0}_{-1}e \rightarrow {}^{68}_{30}Zn$; **b.** ${}^{62}_{29}Cu \rightarrow {}^{\ 0}_{+1}e + {}^{62}_{28}Ni$; **c.** ${}^{212}_{87}Fr \rightarrow {}^4_2He + {}^{208}_{85}At$; **d.** ${}^{129}_{51}Sb \rightarrow {}^{\ 0}_{-1}e + {}^{129}_{52}Te$ **17.** 7 α particles; 4 β particles **19. a.** ${}^{241}_{95}Am \rightarrow {}^4_2He + {}^{237}_{93}Np$; **b.** ${}^{209}_{83}Bi$; **c.** The intermediate radionuclides are ${}^{237}_{93}Np$, ${}^{233}_{91}Pa$, ${}^{233}_{92}U$, ${}^{229}_{90}Th$, ${}^{225}_{88}Ra$, ${}^{225}_{89}Ac$, ${}^{221}_{87}Fr$, ${}^{217}_{85}At$, ${}^{213}_{83}Bi$, ${}^{213}_{84}Po$, and ${}^{209}_{82}Pb$. **21.** ${}^{53}_{26}Fe$ has too many protons. It will undergo positron production, electron capture, and/or alpha-particle production. ${}^{59}_{26}Fe$ has too many neutrons and will undergo beta-particle production. (See Table 19.2 of the text.) **23. a.** ${}^{249}_{98}Cf + {}^{18}_{8}O \rightarrow {}^{263}_{106}Sg + 4\ {}^1_0n$; **b.** ${}^{259}_{104}Rf$ **25.** 690 hours **27.** ${}^{81}Kr$ is most stable since it has the longest half-life. ${}^{73}Kr$ is "hottest" since it decays very rapidly due to its very short half-life. ${}^{73}Kr$, 81s; ${}^{74}Kr$, 34.5 min; ${}^{76}Kr$, 44.4 h; ${}^{81}Kr$, 6.3×10^5 yr **29.** 15 μg ${}^{47}CaCO_3$ should be ordered at a minimum. **31.** 19.6% **33.** The fraction that remains is 0.0041, or 0.41%. **35.** 26 g **37.** 2.3 counts per minute per gram of C. No; for a 10.-mg C sample, it would take roughly 40 min to see a single disintegration. This is too long to wait, and the background radiation would probably be much greater than the ${}^{14}C$ activity. Thus ${}^{14}C$ dating is not practical for very small samples. **39.** 3.8×10^9 yr **41.** 4.3×10^6 kg/s **43.** ${}^{232}Pu$, -1.715×10^{14} J/mol; ${}^{231}Pa$, -1.714×10^{14} J/mol **45.** ${}^{12}C$: 1.230×10^{-12} J/nucleon; ${}^{235}U$: 1.2154×10^{-12} J/nucleon; since ${}^{56}Fe$ is the most stable known nucleus, the binding energy per nucleon for ${}^{56}Fe$ would be larger than that of ${}^{12}C$ or ${}^{235}U$. (See Fig. 19.9 of the text.) **47.** 6.01513 u **49.** -2.0×10^{10} J/g of hydrogen nuclei **51.** The Geiger–Müller tube has a certain response time. After the gas in the tube ionizes to produce a "count," some time must elapse for the gas to return to an electrically neutral state. The response of the tube levels off because, at high activities, radioactive particles are entering the tube faster than the tube can respond to them. **53.** Water is produced in this reaction by removing an OH group from one substance and an H from the other substance. There are two ways to do this:

i.

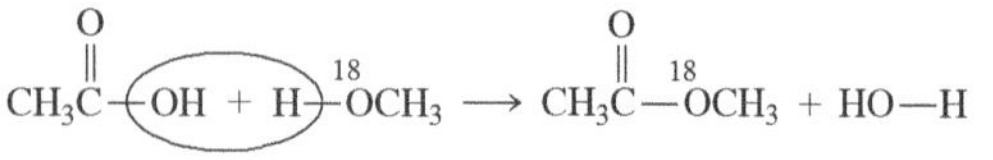

ii.

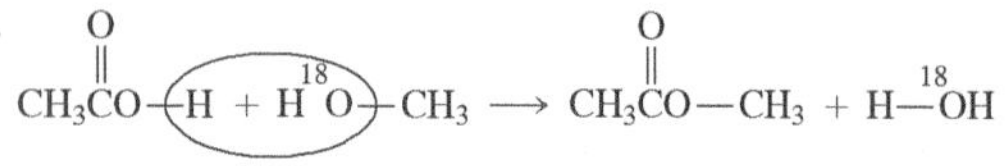

Because the water produced is not radioactive, methyl acetate forms by the first reaction, where all of the oxygen-18 ends up in methyl acetate. **55.** 2 neutrons; 4 β particles **57.** Strontium. Xe is chemically unreactive and not readily incorporated into the body. Sr can be easily oxidized to Sr^{2+}. Strontium is in the same family as calcium and could be absorbed and concentrated in the body in a fashion similar to Ca^{2+}. The chemical properties determine where radioactive material may be concentrated in the body or how easily it may be excreted. **59. a.** unstable; beta production; **b.** stable; **c.** unstable; positron production or electron capture; **d.** unstable, positron production, electron capture, or alpha production. **61.** 3800 decays/s **63.** The third-life will be the time required for the number of nuclides to reach one-third of the original value ($N_0/3$). The third-life of this nuclide is 49.8 years. **65.** 1975 **67.** 900 g ^{235}U **69.** 7×10^5 m/s; 8×10^{-16} J/nuclei; **71.** All evolved $O_2(g)$ comes from water. **79.** 77% ^{238}U and 23% ^{235}U **81.** Assuming that (1) the radionuclide is long lived enough that no significant decay occurs during the time of the experiment, and (2) the total activity is uniformly distributed only in the rat's blood, $V = 10$. mL. **83. a.** $^{12}_6C$; **b.** ^{13}N, ^{13}C, ^{14}N, ^{15}O, and ^{15}N; **c.** -5.950×10^{11} J/mol ^1H **85.** 4.3×10^{-29} **87.** $^{249}_{97}Bk + ^{22}_{10}Ne \rightarrow ^{267}_{107}Bh + 4\ ^1_0n$; 62.7 s; $[Rn]7s^25f^{14}6d^5$

Chapter 20

1. The gravity of the earth cannot keep the light H_2 molecules in the atmosphere. **3.** The acidity decreases. Solutions of Be^{2+} are acidic, while solutions of the other M^{2+} ions are neutral. **5.** For Groups 1A–3A, the small size of H (as compared to Li), Be (as compared to Mg), and B (as compared to Al) seems to be the reason why these elements have nonmetallic properties, while others in Groups 1A–3A are strictly metallic. The small size of H, Be, and B also causes these species to polarize the electron cloud in nonmetals, thus forcing a sharing of electrons when bonding occurs. For Groups 4A–6A, a major difference between the first and second members of a group is the ability to form π bonds. The smaller elements form stable π bonds, while the larger elements do not exhibit good overlap between parallel p orbitals and, in turn, do not form strong π bonds. For Group 7A, the small size of F as compared to Cl is used to explain the low electron affinity of F and the weakness of the F—F bond. **7.** In order to maximize hydrogen bonding interactions in the solid phase, ice is forced into an open structure. This open structure is why $H_2O(s)$ is less dense than $H_2O(l)$. **9.** Group 1A and 2A metals are all easily oxidized. They must be produced in the absence of materials (H_2O, O_2) that are capable of oxidizing them. **11. a.** $\Delta H° = 207$ kJ, $\Delta S° = 216$ J/K; **b.** $T > 958$ K **13.** $4Li(s) + O_2(g) \rightarrow 2Li_2O(s)$; $2Li(s) + S(s) \rightarrow Li_2S(s)$; $2Li(s) + Cl_2(g) \rightarrow 2LiCl(s)$; $12Li(s) + P_4(s) \rightarrow 4Li_3P(s)$; $2Li(s) + H_2(g) \rightarrow 2LiH(s)$; $2Li(s) + 2H_2O(l) \rightarrow 2LiOH(aq) + H_2(g)$; $2Li(s) + 2HCl(aq) \rightarrow 2LiCl(aq) + H_2(g)$ **15.** When lithium reacts with excess oxygen, Li_2O forms, which is composed of Li^+ and O_2^- ions. This is called an oxide salt. When sodium reacts with oxygen, Na_2O_2 forms, which is composed of Na^+ and O_2^{2-} ions. This is called a peroxide salt. When potassium (or rubidium or cesium) reacts with oxygen, KO_2 forms, which is composed of K^+ and O_2^- ions. For your information, this is called a superoxide salt. So the three types of alkali metal oxides that can form differ in the oxygen anion part of the formula (O^{2-} vs. O_2^{2-} vs. O_2^-). Each of these anions have unique bonding arrangements and oxidation states. **17.** The small size of the Li^+ cation results in a much greater attraction to water. The attraction to water is not so great for the other alkali metal ions. Thus lithium salts tend to absorb water. **19.** $CaCO_3(s) + H_2SO_4(aq) \rightarrow CaSO_4(aq) + H_2O(l) + CO_2(g)$ **21.** 3.84×10^6 g Ba **23.** Beryllium has a small size and a large electronegativity as compared with the other alkaline earth metals. The electronegativity of Be is so high that it does not readily give up electrons to nonmetals,

as is the case for the other alkaline earth metals. Instead, Be has significant covalent character in its bonds; it prefers to share valence electrons rather than give them up to form ionic bonds. **25.** element 113: $[Rn]7s^25f^{14}6d^{10}7p^1$; element 113 would fall below Tl in the periodic table. We would expect element 113, like Tl, to form $+1$ and $+3$ oxidation states in its compounds. **27.** $B_2H_6(g) + 3O_2(g) \rightarrow 2B(OH)_3(s)$ **29.** $2Ga(s) + 3F_2(g) \rightarrow 2GaF_3(s)$; $4Ga(s) + 3O_2(g) \rightarrow 2Ga_2O_3(s)$; $2Ga(s) + 3S(s) \rightarrow Ga_2S_3(s)$; $2Ga(s) + 6HCl(aq) \rightarrow 2GaCl_3(aq) + 3H_2(g)$ **31.** An amphoteric substance is one that can behave as either an acid or a base. Al_2O_3 dissolves in both acidic and basic solutions. The reactions are $Al_2O_3(s) + 6H^+(aq) \rightarrow 2Al^{3+}(aq) + 3H_2O(l)$ and $Al_2O_3(s) + 2OH^-(aq) + 3H_2O(l) \rightarrow 2Al(OH)_4^-(aq)$. **33.** Compounds containing Si—Si single and multiple bonds are rare, unlike compounds of carbon. The bond strengths of the Si—Si and C—C single bonds are similar. The difference in bonding properties must be for other reasons. One reason is that silicon does not form strong pi bonds, unlike carbon. Another reason is that silicon forms particularly strong sigma bonds to oxygen, resulting in compounds with Si—O bonds instead of Si—Si bonds. **35.** $O=C=O$ The darker green orbitals about carbon are sp hybrid orbitals. The lighter green orbitals about each oxygen are sp^2 hybrid orbitals, and the gold orbitals about all of the atoms are unhybridized p atomic orbitals. In each double bond in CO_2, one σ bond and one π bond exist. The two carbon–oxygen σ bonds are formed from overlap of sp hybrid orbitals from carbon with an sp^2 hybrid orbital from each oxygen. The two carbon–oxygen π bonds are formed from side-to-side overlap of the unhybridized p atomic orbitals from carbon with an unhybridized p atomic orbital from each oxygen, as illustrated in the figure. **37. a.** $SiO_2(s) + 2C(s) \rightarrow Si(s) + 2CO(g)$; **b.** $SiCl_4(l) + 2Mg(s) \rightarrow Si(s) + 2MgCl_2(s)$; **c.** $Na_2SiF_6(s) + 4Na(s) \rightarrow Si(s) + 6NaF(s)$ **39.** 2:1 **41.** Nitrogen's small size does not provide room for all four oxygen atoms, making NO_4^{3-} unstable. Phosphorus is larger so PO_4^{3-} is more stable. To form NO_3^-, a pi bond must form. Phosphorus doesn't form strong pi bonds as readily as nitrogen. **43.** $2Bi_2S_3(s) + 9O_2(g) \rightarrow 2Bi_2O_3(s) + 6SO_2(g)$; $2Bi_2O_3(s) + 3C(s) \rightarrow 4Bi(s) + 3CO_2(g)$; $2Sb_2S_3(s) + 9O_2(g) \rightarrow 2Sb_2O_3(s) + 6SO_2(g)$; $2Sb_2O_3(s) + 3C(s) \rightarrow 4Sb(s) + 3CO_2(g)$

45.

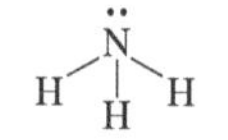

Trigonal bipyramid; dsp^3

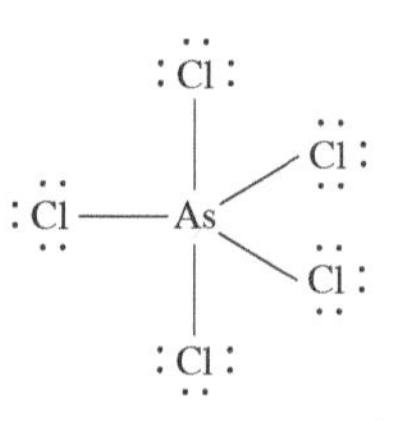

Trigonal bipyramid; dsp^3

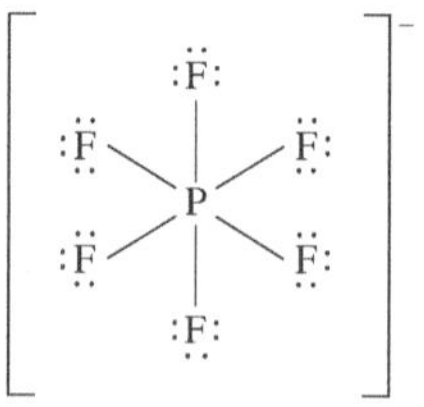

Octahedral; d^2sp^3

Nitrogen does not have low-energy d orbitals it can use to expand its octet. Both NF_5 and NCl_6^- would require nitrogen to have more than 8 valence electrons around it; this never happens. **47.** $\frac{1}{2}N_2(g) + \frac{1}{2}O_2(g) \rightarrow NO(g)$; $\Delta G° = \Delta G°_{f(NO)}$; NO (and some other oxides of nitrogen) have weaker bond's as compared with the triple bond of N_2 and the double bond of O_2. Because of this, NO (and some other oxides of nitrogen) has a higher (positive) standard free energy of formation as compared to the relatively stable N_2 and O_2 molecules. **49.** The pollution provides nitrogen and phosphorus nutrients so the algae can grow. The algae consume dissolved oxygen, causing fish to die. **51.** The acidic hydrogens in the oxyacids of phosphorus all are bonded to oxygen. The hydrogens bonded directly to phosphorus are not acidic. H_3PO_4 has three oxygen-bonded hydrogens, and it is a triprotic acid. H_3PO_3 has only two of the hydrogens bonded to oxygen, and it is a diprotic acid. The third oxyacid of phosphorus, H_3PO_2, has only one of the hydrogens bonded to an oxygen; it is a monoprotic acid. **53.** 821 nm **55.** $H_2SeO_4(aq) + 3SO_2(g) \rightarrow Se(s) + 3SO_3(g) + H_2O(l)$ **57.** In the upper atmosphere, O_3 acts as a filter for ultraviolet (UV) radiation:

$$O_3 \xrightarrow{h\nu} O_2 + O$$

O_3 is also a powerful oxidizing agent. It irritates the lungs and eyes, and at high concentration, it is toxic. The smell of a "spring thunderstorm" is O_3 formed during lightning discharges. Toxic materials don't necessarily smell bad. For example, HCN smells like almonds. **59.** The M.O. electron configuration of O_2 has two unpaired electrons in the degenerate pi antibonding (π_{2p}^*) orbitals. A substance with unpaired electrons is paramagnetic. **61.** From the following Lewis structure, each oxygen atom has a tetrahedral arrangement of electron pairs. Therefore, bond angles are $\approx 109.5°$, and each O is sp^3 hybridized.

$$:\!\ddot{F}\!-\!\ddot{O}\!-\!\ddot{O}\!-\!\ddot{F}\!:$$

Formal charge: 0   0   0   0
Oxidation state: -1 $+1$ $+1$ -1

Oxidation states are more useful. We are forced to assign $+1$ as the oxidation state for oxygen. Oxygen is very electronegative, and $+1$ is not a stable oxidation state for this element.

63.

V-shaped; $<109.5°$

See-saw; $\approx 120°$, $\approx 90°$ Octahedral; $90°$

OF_4 would not form. This compound would require oxygen to have more than 8 valence electrons around it. This never occurs for oxygen; oxygen does not have low-energy d orbitals it can use to expand its octet. **65.** The oxyacid strength increases as the number of oxygens in the formula increases. Therefore, the order of the oxyacids from weakest to strongest acid is $HOCl < HClO_2 < HClO_3 < HClO_4$. **67. a.** IO_4^-; **b.** IO_3^-; **c.** IF_2^-; **d.** IF_4^-; **e.** IF_6^- **69.** Helium is unreactive and doesn't combine with any other elements. It is a very light gas and would easily escape the earth's gravitational pull as the planet was formed. **71.** One would expect RnF_2, RnF_4, and maybe RnF_6 to form in fashion similar to XeF_2, XeF_4, and XeF_6. The chemistry of radon is difficult to study because radon isotopes are all radioactive. The hazards of dealing with radioactive materials are immense. **73.** $N_2H_4(l) + O_2(g) \rightarrow N_2(g) + 2H_2O(g)$; $\Delta H = -590.$ kJ **75.** If the compound contained Ga(II), it would be paramagnetic and if the compound contained Ga(I) and Ga(III), it would be diamagnetic. Paramagnetic compounds have an apparent greater mass in a magnetic field. **77.** 9.3×10^{-5} mol/L **79. a.** $AgCl(s) \xrightarrow{h\nu} Ag(s) + Cl$; the reactive chlorine atom is trapped in the crystal. When light is removed, Cl reacts with silver atoms to re-form AgCl; i.e., the reverse reaction occurs. In pure AgCl, the Cl atoms escape, making the reverse reaction impossible. **b.** Over time, chlorine is lost and the dark silver metal is permanent. **81.** 6.5×10^{27} atoms **83.** Strontium and calcium are both alkaline earth metals, so both have similar chemical properties. Since milk is a good source of calcium, strontium could replace some calcium in milk without much difficulty. **85.** $+6$ oxidation state: SO_4^{2-}, SO_3, SF_6; $+4$ oxidation state: SO_3^{2-}, SO_2, SF_4; $+2$ oxidation state: SCl_2; 0 oxidation state: S_8 and all other elemental forms of sulfur; -2 oxidation state: H_2S, Na_2S **97.** Ca; 12.698 **99.** I **101.** For the reaction

$$\underset{:\ddot{O}:}{\overset{:\ddot{O}}{N}}\!-\!\ddot{N}\!=\!\ddot{O}: \longrightarrow NO_2 + NO$$

the activation energy must in some way involve the breaking of a nitrogen–nitrogen single bond. For the reaction

$$\underset{:\ddot{O}:}{\overset{:\ddot{O}}{N}}\!-\!\ddot{N}\!=\!\ddot{O}: \longrightarrow O_2 + N_2O$$

at some point nitrogen–oxygen bonds must be broken. N—N single bonds (160 kJ/mol) are weaker than N—O single bonds (201 kJ/mol). In addition, resonance structures indicate that there is more double-bond character in the N—O bonds than in the N—N bond. Thus NO_2 and NO are preferred by kinetics because of the lower activation energy. **103.** 5.89 **105.** 20. g **107. a.** 7.1 g; **b.** 979 nm; this electromagnetic radiation is not visible to humans; it is in the infrared region of the electromagnetic radiation spectrum; **c.** n-type **109. a.** $+6$; **b.** 4.42

Chapter 21

5. The lanthanide elements are located just before the $5d$ transition metals. The lanthanide contraction is the steady decrease in the atomic radii of the lanthanide elements when going from left to right across the periodic table. As a result of the lanthanide contraction, the sizes of the $4d$ and $5d$ elements are very similar. This leads to a greater similarity in the chemistry of the $4d$ and $5d$ elements in a given vertical group. **7.** No; both the *trans* and the *cis* forms of $Co(NH_3)_4Cl_2^+$ have mirror images that are superimposable. For the *cis* form, the mirror image only needs a 90° rotation to produce the original structure. Hence, neither the *trans* nor the *cis* form is optically active. **9.** $Fe_2O_3(s) + 6H_2C_2O_4(aq) \rightarrow 2Fe(C_2O_4)_3^{3-}(aq) + 3H_2O(l) + 6H^+(aq)$; the oxalate anion forms a soluble complex ion with iron in rust (Fe_2O_3), which allows rust stains to be removed.

11. a.
$\uparrow \quad \uparrow \quad \uparrow$

$\uparrow\downarrow \quad \uparrow\downarrow$
small Δ

$CoCl_4^{2-}$ is an example of a weak-field case having three unpaired electrons.

b.
$- \quad -$

$\uparrow\downarrow \quad \uparrow\downarrow \quad \uparrow\downarrow$
large Δ

CN^- is a strong-field ligand so $Co(CN)_6^{3-}$ will be a low-spin case having zero unpaired electrons.

13. From Table 21.16, the red octahedral $Co(H_2O)_6^{2+}$ complex ion absorbs blue-green light ($\lambda \approx 490$ nm), whereas the blue tetrahedral $CoCl_4^{2-}$ complex ion absorbs orange light ($\lambda \approx 600$ nm). Because tetrahedral complexes have a d-orbital splitting much less than octahedral complexes, one would expect the tetrahedral complex to have a smaller energy difference between split d orbitals. This translates into longer-wavelength light absorbed ($E = hc/\lambda$) for tetrahedral complex ions compared to octahedral complex ions. Information from Table 21.16 confirms this. **15.** SCN^-, NO_2^-, and OCN^- can form linkage isomers; all are able to bond to the metal ion in two different ways. **17.** Sc^{3+} has no electrons in d orbitals. Ti^{3+} and V^{3+} have d electrons present. The color of transition metal complexes results from electron transfer between split d orbitals. If no d electrons are present, no electron transfer can occur, and the compounds are not colored. **19.** At high altitudes, the oxygen content of air is lower, so less oxyhemoglobin is formed which diminishes the transport of oxygen in the blood. A serious illness called high-altitude sickness can result from the decrease of O_2 in the blood. High-altitude acclimatization is the phenomenon that occurs in the human body in response to the lower amounts of oxyhemoglobin in the blood. This response is to produce more hemoglobin, and, hence, increase the oxyhemoglobin in the blood. High-altitude acclimatization takes several weeks to take hold for people moving from lower altitudes to higher altitudes. **21. a.** Ni: $[Ar]4s^23d^8$; **b.** Cd: $[Kr]5s^24d^{10}$; **c.** Zr: $[Kr]5s^24d^2$; **d.** Os: $[Xe]6s^24f^{14}5d^6$ **23. a.** Ti: $[Ar]4s^23d^2$; Ti^{2+}: $[Ar]3d^2$; Ti^{4+}: $[Ne]3s^23p^6$ or $[Ar]$; **b.** Re: $[Xe]6s^24f^{14}5d^5$; Re^{2+}: $[Xe]4f^{14}5d^5$; Re^{3+}: $[Xe]4f^{14}5d^4$; **c.** Ir: $[Xe]6s^24f^{14}5d^7$; Ir^{2+}: $[Xe]4f^{14}5d^7$; Ir^{3+}: $[Xe]4f^{14}5d^6$ **25. a.** Fe^{3+}: $[Ar]3d^5$; **b.** Ag^+: $[Kr]4d^{10}$; **c.** Ni^{2+}: $[Ar]3d^8$; **d.** Cr^{3+}: $[Ar]3d^3$ **27. a.** molybdenum(IV) sulfide, molybdenum(VI) oxide; **b.** MoS_2, +4; MoO_3, +6; $(NH_4)_2Mo_2O_7$, +6; $(NH_4)_6Mo_7O_{24} \cdot 4H_2O$, +6   **29.** NH_3 is a weak base that produces OH^- ions in solution. The white precipitate is $Cu(OH)_2(s)$. $Cu^{2+}(aq) + 2OH^-(aq) \rightarrow Cu(OH)_2(s)$; with excess NH_3 present, Cu^{2+} forms a soluble complex ion, $Cu(NH_3)_4^{2+}$. $Cu(OH)_2(s) + 4NH_3(aq) \rightarrow Cu(NH_3)_4^{2+}(aq) + 2OH^-(aq)$ **31.** $[Co(NH_3)_6]I_3$: 3 moles of AgI; $[Pt(NH_3)_4I_2]I_2$: 2 moles of AgI; $Na_2[PtI_6]$: 0 moles of AgI; $[Cr(NH_3)_4I_2]I$: 1 mole of AgI   **33. a.** hexacyanomanganate(II) ion; **b.** cis-tetraamminedichlorocobalt(III) ion; **c.** pentaamminechlorocobalt(II) ion   **35. a.** hexaamminecobalt(II) chloride; **b.** hexaaquacobalt(III) iodide; **c.** potassium tetrachloroplatinate(II); **d.** potassium hexachloroplatinate(II); **e.** pentaamminechlorocobalt(III) chloride; **f.** triamminetrinitrocobalt(III)   **37. a.** $K_2[CoCl_4]$; **b.** $[Pt(H_2O)(CO)_3]Br_2$; **c.** $Na_3[Fe(CN)_2(C_2O_4)_2]$; **d.** $[Cr(NH_3)_3Cl(H_2NCH_2CH_2NH_2)]I_2$

39. a.

cis   trans

b.

cis   trans

c.

d.

$en = N\frown N = NH_2CH_2CH_2NH_2$

41. monodentate   bidentate   bridging

43. a. 2; **b.** 3; **c.** 4; **d.** 4

45.

47. $Cr(acac)_3$ and cis-$Cr(acac)_2(H_2O)_2$ are optically active.   **49.** With five electrons each in a different orbital, diagram (a) is for the weak-field $Fe(H_2O)_6^{3+}$ complex ion. With three electrons, diagram (b) is for the $Cr(NH_3)_5Cl^{2+}$ complex ion. With six electrons all paired up, diagram (c) is for the strong-field $Co(NH_3)_4Br_2^+$ complex ion.

51. a. Fe^{2+}

High spin   Low spin

b. Fe^{3+} **c.** Ni^{2+}

High spin

53. weak field   **55. a.** 0; **b.** 2; **c.** 2   **57.** $Co(CN)_6^{3-} < Co(en)_3^{3+} < Co(H_2O)_6^{3+} < CoI_6^{3-}$ **59.** The violet complex ion absorbs yellow-green light ($\lambda \approx 570$ nm), the yellow complex ion absorbs blue light ($\lambda \approx 450$ nm), and the green complex ion absorbs red light ($\lambda \approx 650$ nm). The violet complex ion is $Cr(H_2O)_6^{3+}$, the yellow complex ion is $Cr(NH_3)_6^{3+}$, and the green complex ion is $Cr(H_2O)_4Cl_2^+$.   **61.** $CoBr_4^{2-}$ is a tetrahedral complex ion, while $CoBr_6^{4-}$ is an octahedral complex ion. Since tetrahedral d-orbital splitting is less than one-half the octahedral d-orbital splitting, the octahedral complex ion ($CoBr_6^{4-}$) will absorb higher-energy light, which will have a shorter wavelength than 3.4×10^{-6} m ($E = hc/\lambda$).   **63.** 5   **65. a.** -11 kJ; **b.** $\Delta H° = 172.5$ kJ; $\Delta S° = 176$ J/K; $T > 980.$ K   **67.** Fe_2O_3: iron has a +3 oxidation state; Fe_3O_4: iron has a +8/3 oxidation state. The three iron ions in Fe_3O_4 must have a total charge of +8. The only combination that works is to have two Fe^{3+} ions and one Fe^{2+} ion per formula unit. This makes sense from the other formula for magnetite, $FeO \cdot Fe_2O_3$. FeO has an Fe^{2+} ion and Fe_2O_3 has two Fe^{3+} ions.   **69.** $8CN^-(aq) + 4Ag(s) + O_2(g) + 2H_2O(l) \rightarrow 4Ag(CN)_2^-(aq) + 4OH^-(aq)$ **71.** The molecular formula is $EuC_{15}H_{21}O_6$. Because each $acac^-$ is $C_5H_7O_2^-$, an abbreviated molecular formula is $Eu(acac)_3$.   **73.** 0.66 V; -130 kJ; 2.2×10^{22} **75.** There are four geometrical isomers (labeled i–iv). Isomers iii and iv are optically active, and the nonsuperimposable mirror images are shown.

i.   **ii.**

iii.

optically active   mirror   mirror image of iii (nonsuperimposable)

iv.

optically active   mirror   mirror image of iv (nonsuperimposable)

77. Octahedral Cr^{2+} complexes should be used. Cr^{2+}: $[Ar]3d^4$; high-spin (weak-field) Cr^{2+} complexes have four unpaired electrons and low-spin (strong-field) Cr^{2+} complexes have two unpaired electrons. Ni^{2+}: $[Ar]3d^8$; octahedral Ni^{2+} complexes will always have two unpaired electrons, whether high or low spin. Therefore, Ni^{2+} complexes cannot be used to distinguish weak- from strong-field ligands by examining magnetic properties. Alternatively, the ligand field strengths can be measured using visible spectra. Either Cr^{2+} or Ni^{2+} complexes can be used for this method.   **79. a.** $[Co(C_5H_5N)_6]Cl_3$; **b.** $[Cr(NH_3)_5I]I_2$; **c.** $[Ni(NH_2CH_2CH_2NH_2)_3]Br_2$; **d.** $K_2[Ni(CN)_4]$; **e.** $[Pt(NH_3)_4Cl_2][PtCl_4]$ **81.** 60   **89.** $Ni^{2+} = d^8$; if ligands

A and B produced very similar crystal fields, the *cis*-[NiA$_2$B$_4$]$^{2+}$ complex ion would give the following octahedral crystal field diagram for a d^8 ion:

$$\underline{\uparrow} \quad \underline{\uparrow}$$
$$\underline{\uparrow\downarrow} \quad \underline{\uparrow\downarrow} \quad \underline{\uparrow\downarrow}$$

This is paramagnetic.

Because it is given that the complex ion is diamagnetic, the A and B ligands must produce different crystal fields, giving a unique *d*-orbital splitting diagram that would result in a diamagnetic species.   **91. a.** -0.26 V; **b.** From standard reduction potentials, Co^{3+} ($\mathscr{E}° = 1.82$ V) is a much stronger oxidizing agent than Co(en)$_3^{3+}$ ($\mathscr{E}° = -0.26$ V); **c.** In aqueous solution, Co^{3+} forms the hydrated transition metal complex, Co(H$_2$O)$_6^{3+}$. In both complexes, Co(H$_2$O)$_6^{3+}$ and Co(en)$_3^{3+}$, cobalt exists as Co^{3+}, which has 6 *d* electrons. If we assume a strong-field case for each complex, then the *d*-orbital splitting diagram for each has the six electrons paired in the lower-energy t_{2g} orbitals. When each complex ion gains an electron, the electron enters the higher-energy e_g orbitals. Since en is a stronger-field ligand than H$_2$O, then the *d*-orbital splitting is larger for Co(en)$_3^{3+}$, and it takes more energy to add an electron to Co(en)$_3^{3+}$ than to Co(H$_2$O)$_6^{3+}$. Therefore, it is more favorable for Co(H$_2$O)$_6^{3+}$ to gain an electron than for Co(en)$_3^{3+}$ to gain an electron.   **93.** No, in all three cases six bonds are formed between Ni^{2+} and nitrogen. So ΔH values should be similar. $\Delta S°$ for formation of the complex ion is most negative for 6 NH$_3$ molecules reacting with a metal ion (7 independent species become 1). For penten reacting with a metal ion, 2 independent species become 1, so $\Delta S°$ is the least negative. Thus the chelate effect occurs because the more bonds a chelating agent can form to the metal, the less unfavorable $\Delta S°$ becomes for the formation of the complex ion, and the larger the formation constant.

95.
$$\underline{\quad} \qquad d_{z^2}$$
$$\underline{\quad} \quad \underline{\quad} \qquad d_{x^2-y^2}, d_{xy}$$
$$\underline{\quad} \quad \underline{\quad} \qquad d_{xz}, d_{yz}$$

97. The coordinate system for the complex ion is shown below. From the coordinate system, the CN$^-$ ligands are in a square planar arrangement. Since CN$^-$ produces a much stronger crystal field, the diagram will most resemble that of a square planar complex:

$$
\begin{array}{c}
\text{NC} \diagdown \quad | \quad \diagup \text{CN} \cdots y \\
\quad \quad \text{Ni} \\
\text{NC} \diagup \quad | \quad \diagdown \text{CN} \cdots x \\
\quad \text{NH}_3 \\
\quad \quad \vdots \\
\quad \quad z
\end{array}
$$

$$
\begin{array}{ll}
d_{x^2-y^2} & \underline{\quad} \\
d_{z^2} & \text{\textHarpoonup\textHarpoondown} \\
d_{xy} & \text{\textHarpoonup\textHarpoondown} \\
d_{xz} & \text{\textHarpoonup\textHarpoondown} \; \text{\textHarpoonup\textHarpoondown} \;\; d_{yz}
\end{array}
$$

With the NH$_3$ ligands on the *z* axis, we will assume that the d_{z^2} orbital is destabilized more than the d_{xy} orbital. This may or may not be the case.
99. [Cr(NH$_3$)$_5$I]I$_2$; octahedral

Chapter 22

1. a. 1-sec-butylpropane

$$
\begin{array}{c}
\text{CH}_2\text{CH}_2\text{CH}_3 \\
| \\
\text{CH}_3\text{CHCH}_2\text{CH}_3
\end{array}
$$

3-methylhexane is correct.
c. 2-ethylpentane

$$
\begin{array}{c}
\text{CH}_3\text{CHCH}_2\text{CH}_2\text{CH}_3 \\
| \\
\text{CH}_2\text{CH}_3
\end{array}
$$

3-methylhexane is correct.

b. 4-methylhexane

$$
\begin{array}{c}
\text{CH}_3 \\
| \\
\text{CH}_3\text{CH}_2\text{CH}_2\text{CHCH}_2\text{CH}_3
\end{array}
$$

3-methylhexane is correct.
d. 1-ethyl-1-methylbutane

$$
\begin{array}{c}
\text{CH}_2\text{CH}_3 \\
| \\
\text{CHCH}_2\text{CH}_2\text{CH}_3 \\
| \\
\text{CH}_3
\end{array}
$$

3-methylhexane is correct.

e. 3-methylhexane

$$
\begin{array}{c}
\text{CH}_3\text{CH}_2\text{CHCH}_2\text{CH}_2\text{CH}_3 \\
| \\
\text{CH}_3
\end{array}
$$

f. 4-ethylpentane

$$
\begin{array}{c}
\text{CH}_3\text{CH}_2\text{CH}_2\text{CHCH}_3 \\
| \\
\text{CH}_2\text{CH}_3
\end{array}
$$

3-methylhexane is correct.

All six of these are the same compound. They only differ from each other by rotations about one or more carbon–carbon single bonds. Only one isomer of C$_7$H$_{16}$ is present in all of these names: 3-methylhexane.

3. a.
$$
\begin{array}{c}
\text{CH}_3\text{CHCH}_3 \\
| \\
\text{CH}_2\text{CH}_3
\end{array}
$$

The longest chain is 4 carbons long. The correct name is 2-methylbutane.

b.
$$
\begin{array}{c}
\quad\quad\quad\quad \text{I} \quad \text{CH}_3 \\
\quad\quad\quad\quad | \quad\;\; | \\
\text{CH}_3\text{CH}_2\text{CH}_2\text{CH}_2\overset{\displaystyle|}{\text{C}}-\text{CH}_2 \\
\quad\quad\quad\quad\quad\quad\;\; | \\
\quad\quad\quad\quad\quad\quad\; \text{CH}_3
\end{array}
$$

The longest chain is 7 carbons long and we would start the numbering system at the other end for lowest possible numbers. The correct name is 3-iodo-3-methylheptane.

c.
$$
\begin{array}{c}
\quad\quad\quad\quad\quad \text{CH}_3 \\
\quad\quad\quad\quad\quad\; | \\
\text{CH}_3\text{CH}_2\text{CH}=\text{C}-\text{CH}_3
\end{array}
$$

This compound cannot exhibit *cis–trans* isomerism since one of the double-bonded carbons has the same two groups (CH$_3$) attached. The numbering system should also start at the other end to give the double bond the lowest possible number. 2-methyl-2-pentene is correct.

d.
$$
\begin{array}{c}
\text{Br OH} \\
| \quad | \\
\text{CH}_3\text{CHCHCH}_3
\end{array}
$$

The OH functional group gets the lowest number. 3-bromo-2-butanol is correct.

5. Hydrocarbons are nonpolar substances exhibiting only London dispersion forces. Size and shape are the two most important structural features relating to the strength of London dispersion forces. For size, the bigger the molecule (the larger the molar mass), the stronger the London dispersion forces and the higher the boiling point. For shape, the more branching present in a compound, the weaker the London dispersion forces and the lower the boiling point.   **7.** The amide functional group is:

$$
\begin{array}{c}
\;\;\; \text{O} \;\;\; \text{H} \\
\;\;\; \| \;\;\;\; | \\
-\text{C}-\text{N}-
\end{array}
$$

When the amine end of one amino acid reacts with the carboxylic acid end of another amino acid, the two amino acids link together by forming an amide functional group. A polypeptide has many amino acids linked together, with each linkage made by the formation of an amide functional group. Because all linkages result in the presence of the amide functional group, the resulting polymer is called a polyamide. For nylon, the monomers also link together by forming the amide functional group (the amine end of one monomer reacts with a carboxylic acid end of another monomer to give the amide functional group linkage). Hence nylon is also a polyamide. The correct order of strength is polyhydrocarbon < polyester < polyamide. The difference in strength is related to the types of intermolecular forces present. All of these polymers have London dispersion forces. However, polyhydrocarbons only have London dispersion forces. The polar ester group in polyesters and the polar amide group in polyamides give rise to additional dipole forces. The polyamide has the ability to form relatively strong hydrogen bonding interactions, hence why it would form the strongest fibers.

9. a.
$$
\text{CH}_2=\text{CH}_2 + \text{H}_2\text{O} \xrightarrow{\text{H}^+} \begin{array}{c} \text{OH} \;\; \text{H} \\ | \quad\;\; | \\ \text{CH}_2-\text{CH}_2 \end{array} \;\; 1° \text{ alcohol}
$$

b.
$$
\text{CH}_3\text{CH}=\text{CH}_2 + \text{H}_2\text{O} \xrightarrow{\text{H}^+} \begin{array}{c} \text{OH} \;\; \text{H} \\ | \quad\;\; | \\ \text{CH}_3\text{CH}-\text{CH}_2 \end{array} \;\; 2° \text{ alcohol}
$$
$$
\text{major product}
$$

c.

$$CH_3C{=}CH_2 + H_2O \xrightarrow{H^+} CH_3\overset{\overset{\displaystyle OH}{|}}{C}{-}\overset{\overset{\displaystyle H}{|}}{CH_2} \quad 3° \text{ alcohol}$$
with CH_3 below first carbon

major product

d.

$$CH_3CH_2OH \xrightarrow{\text{oxidation}} CH_3\overset{\overset{\displaystyle O}{\|}}{C}H \quad \text{aldehyde}$$

e.

$$CH_3\overset{\overset{\displaystyle OH}{|}}{C}HCH_3 \xrightarrow{\text{oxidation}} CH_3\overset{\overset{\displaystyle O}{\|}}{C}CH_3 \quad \text{ketone}$$

f.

$$CH_3CH_2CH_2OH \xrightarrow{\text{oxidation}} CH_3CH_2\overset{\overset{\displaystyle O}{\|}}{C}{-}OH \quad \text{carboxylic acid}$$

or

$$CH_3CH_2\overset{\overset{\displaystyle O}{\|}}{C}H \xrightarrow{\text{oxidation}} CH_3CH_2\overset{\overset{\displaystyle O}{\|}}{C}{-}OH$$

g.

$$CH_3OH + HO\overset{\overset{\displaystyle O}{\|}}{C}CH_3 \longrightarrow CH_3{-}O{-}\overset{\overset{\displaystyle O}{\|}}{C}CH_3 + H_2O \quad \text{ester}$$

11. a. A polyester forms when an alcohol functional group reacts with a carboxylic acid functional group. The monomer for a homopolymer polyester must have an alcohol functional group and a carboxylic acid functional group present in the structure. **b.** A polyamide forms when an amine functional group reacts with a carboxylic acid functional group. For a copolymer polyamide, one monomer would have at least two amine functional groups present and the other monomer would have at least two carboxylic acid functional groups present. For polymerization to occur, each monomer must have two reactive functional groups present. **c.** To form an addition polymer, a carbon–carbon double bond must be present. To form a polyester, the monomer would need the alcohol and carboxylic acid functional groups present. To form a polyamide, the monomer would need the amine and carboxylic acid functional groups present. The two possibilities are for the monomer to have a carbon–carbon double bond, an alcohol functional group, and a carboxylic acid functional group all present, or to have a carbon–carbon double bond, an amine functional group, and a carboxylic acid functional group present.

13. $CH_3{-}CH_2{-}CH_2{-}CH_2{-}CH_2{-}CH_3$

$$CH_3{-}\overset{\overset{\displaystyle CH_3}{|}}{C}H{-}CH_2{-}CH_2{-}CH_3$$

$$CH_3{-}CH_2{-}\overset{\overset{\displaystyle CH_3}{|}}{C}H{-}CH_2{-}CH_3$$

$$CH_3{-}\overset{\overset{\displaystyle CH_3}{|}}{\underset{\underset{\displaystyle CH_3}{|}}{C}}{-}CH_2{-}CH_3$$

$$CH_3{-}\overset{\overset{\displaystyle CH_3}{|}}{C}H{-}\overset{\overset{\displaystyle CH_3}{|}}{C}H{-}CH_3$$

15. a.

$$CH_3\overset{\overset{\displaystyle CH_3}{|}}{C}HCH_2CH_2CH_2CH_2CH_3$$
2-methylheptane

$$CH_3CH_2\overset{\overset{\displaystyle CH_3}{|}}{C}HCH_2CH_2CH_3$$
3-methylheptane

$$CH_3CH_2CH_2\overset{\overset{\displaystyle CH_3}{|}}{C}HCH_2CH_2CH_3$$
4-methylheptane

b.

$$CH_3{-}\overset{\overset{\displaystyle CH_3}{|}}{\underset{\underset{\displaystyle CH_3}{|}}{C}}{-}\overset{\overset{\displaystyle CH_3}{|}}{\underset{\underset{\displaystyle CH_3}{|}}{C}}{-}CH_3$$
2,2,3,3-tetramethylbutane

17. a.

$$CH_3\overset{\overset{\displaystyle CH_3}{|}}{C}HCH_3$$

b.

$$CH_3\overset{\overset{\displaystyle CH_3}{|}}{C}HCH_2CH_3$$

c.

$$CH_3\overset{\overset{\displaystyle CH_3}{|}}{C}HCH_2CH_2CH_3$$

d.

$$CH_3\overset{\overset{\displaystyle CH_3}{|}}{C}HCH_2CH_2CH_2CH_3$$

19. a. $CH_3CH_2{-}\overset{\overset{\displaystyle {|}}{C}}H{-}CH_2CH_2CH_3$
with $CH_2{-}CH{-}CH_3$ and CH_3 branches

b.

$$CH_3{-}\overset{\overset{\displaystyle CH_3}{|}}{\underset{\underset{\displaystyle CH_3}{|}}{C}}{-}CH_2{-}\overset{\overset{\displaystyle {|}}{C}H}{-}CH_3 \quad (CH_3)$$

c. $CH_3{-}CH{-}CH_2CH_2CH_3$

$$CH_3{-}\overset{\overset{\displaystyle {|}}{\underset{\underset{\displaystyle CH_3}{|}}{C}}}{-}CH_3$$

d. 4-ethyl-2-methylheptane; 2,2,3-trimethylhexane
21. a. 2,2,4-trimethylhexane; **b.** 5-methylnonane; **c.** 2,2,4,4-tetramethylpentane; **d.** 3-ethyl-3-methyloctane
23. $CH_3{-}CH_2{-}CH_2{-}CH_3$;

$$\begin{array}{c} H \quad H \\ | \quad | \\ H{-}C{-}C{-}H \\ | \quad | \\ H{-}C{-}C{-}H \\ | \quad | \\ H \quad H \end{array}$$

Each carbon is bonded to four other atoms.
25. a. 1-butene; **b.** 4-methyl-2-hexene; **c.** 2,5-dimethyl-3-heptene
27. a. $CH_3CH_2CH{=}CHCH_2CH_3$; **b.** $CH_3CH{=}CHCH{=}CHCH_2CH_3$;
c. $CH_3CHCH{=}CHCH_2CH_2CH_2CH_3$
with CH_3 below.

29. a.

benzene ring with CH_3 and CH_2CH_3

b.

$$H_3C{-}\overset{\overset{\displaystyle CH_3}{|}}{\underset{\underset{\displaystyle CH_3}{|}}{C}}{-}\langle benzene \rangle{-}\overset{\overset{\displaystyle CH_3}{|}}{\underset{\underset{\displaystyle CH_3}{|}}{C}}{-}CH_3$$

c.

benzene ring with CH_2CH_3 and CH_2CH_3

d.

benzene ring with $CH_2CH{=}CHCH_3$

31. a. 1,3-dichlorobutane; **b.** 1,1,1-trichlorobutane; **c.** 2,3-dichloro-2,4-dimethylhexane; **d.** 1,2-difluoroethane **33.** $CH_2Cl{-}CH_2Cl$,1,2-dichloroethane: There is free rotation about the $C{-}C$ single bond that doesn't lead to different compounds. $CHCl{=}CHCl$, 1,2-dichloroethene: There is no rotation about the $C{=}C$ double bond. This creates the *cis* and *trans* isomers, which are different compounds. **35.** [25], compounds b and c; [27], all compounds
37. $CH_2{=}CHCH_2CH_2CH_3$ $CH_3CH{=}CHCH_2CH_3$

$$CH_2{=}\overset{\overset{\displaystyle {|}}{\underset{\underset{\displaystyle CH_3}{|}}{C}}}CH_2CH_3 \quad CH_3\overset{\overset{\displaystyle {|}}{\underset{\underset{\displaystyle CH_3}{|}}{C}}}{=}CHCH_3$$

$$CH_3\overset{\overset{\displaystyle {|}}{\underset{\underset{\displaystyle CH_3}{|}}{C}}}HCH{=}CH_2$$

39.

$$CH_3-CHCH=CH_2$$

(Structures: three structures — a $C=C$ with Cl and CH$_3$, CH$_2$CH=CH$_2$, Cl substituents; a second $C=C$ isomer; and a cyclopropane with Cl)

41. (Ten alkene structures)

$$CH_2=CHCHCH_3 \ (F)$$

$$CH_2=CCH_3 \ (CH_3, F)$$

$$CH_2=CCH_2CH_3 \ (F)$$

$$CH=CCH_3 \ (F, CH_3)$$

$$CH_2=CHCH_2CH_2 \ (F)$$

43. a.

$$CH_3 \quad CH_2CH_2CH_3$$
$$C=C$$
$$H \qquad H$$

b.

$$CH_3 \qquad H$$
$$C=C$$
$$H \qquad CH_3$$

c.

$$CH_3 \qquad CH_2CH_3$$
$$C=C$$
$$Cl \qquad Cl$$

45. a. 3 monochloro isomers of *n*-pentane; **b.** 4 monochloro isomers of 2-methylbutane; **c.** 3 monochloro isomers of 2,4-dimethylpentane; **d.** 4 monochloro isomers of methylcyclobutane   **47. a.** ketone; **b.** aldehyde; **c.** carboxylic acid; **d.** amine

49. a.

(structure labeled with: amine, ketone, alcohol, carboxylic acid)

b. 5 carbons in ring and the carbon in —CO$_2$H: sp^2; the other two carbons: sp^3; **c.** 24 sigma bonds, 4 pi bonds   **51. a.** 3-chloro-1-butanol, primary; **b.** 3-methyl-3-hexanol, tertiary; **c.** 2-methylcyclopentanol, secondary   **53.** 1-pentanol; 2-pentanol; 3-pentanol; 2-methyl-1-butanol; 2-methyl-2-butanol; 3-methyl-2-butanol; 3-methyl-1-butanol; 2,2-dimethyl-1-propanol; 6 ethers   **55. a.** 4,5-dichloro-3-hexanone; **b.** 2,3-dimethylpentanal; **c.** 3-methylbenzaldehyde or *m*-methylbenzaldehyde   **57. a.** 4-chlorobenzoic acid or *p*-chlorobenzoic acid; **b.** 3-ethyl-2-methylhexanoic acid; **c.** methanoic acid (common name = formic acid)   **59.** Only statement d is false.

$$O$$
$$\|$$
2-butenal: $HCCH=CHCH_3$.

The formula of 2-butenal is C_4H_6O, while the alcohol has a formula of C_4H_8O.

61. a. $CH_3CH_2CH_2CH_3$

b.

$$Cl \ Cl \qquad Cl \ Cl$$
$$| \ | \qquad | \ |$$
$$CH_2CHCHCHCH$$
$$| \ |$$
$$CH_3 \ CH_3$$

c. (benzene ring with Cl) $+ \ HCl$

d. $C_4H_8(g) + 6O_2(g) \longrightarrow 4CO_2(g) + 4H_2O(g)$

63. For the iron-catalyzed reaction, one of the *ortho* or *para* hydrogens in benzene is replaced by chlorine. When an iron catalyst is not present, the benzene hydrogens are unreactive. To substitute for an alkane hydrogen, light must be present. For toluene, the light-catalyzed reaction substitutes a chlorine for a hydrogen in the methyl group attached to the benzene ring.

65. a.

$$O \qquad\qquad O$$
$$\| \qquad\qquad \|$$
$$H-C-CH_2CHCH_3 + HO-C-CH_2CHCH_3$$
$$| \qquad\qquad |$$
$$CH_3 \qquad\qquad CH_3$$

b.

$$O$$
$$\|$$
$$CH_3-C-CHCH_3$$
$$|$$
$$CH_3$$

c. No reaction

d. (benzaldehyde + benzoic acid structures)

e. (cyclohexanone with CH$_3$)

f. (two cyclohexanone structures with OH and CH$_3$ and C—H / C—OH)

67. a. $CH_3CH=CH_2 + Br_2 \rightarrow CH_3CHBrCH_2Br$;

b.

$$OH \qquad\qquad\qquad O$$
$$| \qquad\qquad\qquad \|$$
$$CH_3-CH-CH_3 \xrightarrow{oxidation} CH_3-C-CH_3;$$

c.

$$CH_3 \qquad\qquad\qquad\qquad CH_3$$
$$| \qquad\qquad\qquad\qquad\qquad |$$
$$CH_2=C-CH_3 + H_2O \xrightarrow{H^+} CH_2-C-CH_3;$$
$$\qquad\qquad\qquad\qquad\qquad\qquad | \quad |$$
$$\qquad\qquad\qquad\qquad\qquad\qquad H \quad OH$$

d.

$$\qquad\qquad\qquad\qquad\qquad\qquad O$$
$$\qquad\qquad\qquad\qquad\qquad\qquad \|$$
$$CH_3CH_2CH_2OH \xrightarrow{KMnO_4} CH_3CH_2C-OH$$

69. a.

$$O$$
$$\|$$
$$CH_3C-OH + HOCH_2(CH_2)_6CH_3 \longrightarrow$$
$$\qquad\qquad\qquad O$$
$$\qquad\qquad\qquad \|$$
$$\qquad\qquad CH_3C-O-CH_2(CH_2)_6CH_3 + H_2O;$$

b.

$$O$$
$$\|$$
$$CH_3CH_2C-OH + HOCH_2(CH_2)_4CH_3 \longrightarrow$$
$$\qquad\qquad\qquad O$$
$$\qquad\qquad\qquad \|$$
$$\qquad\qquad CH_3CH_2C-O-CH_2(CH_2)_4CH_3 + H_2O$$

71. $CFCl=CF_2$

73.

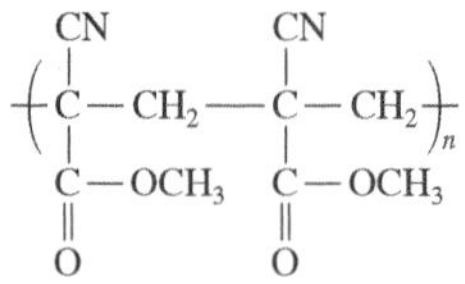

75.

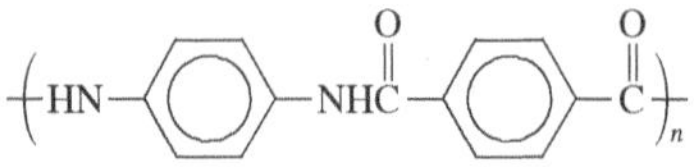

77.

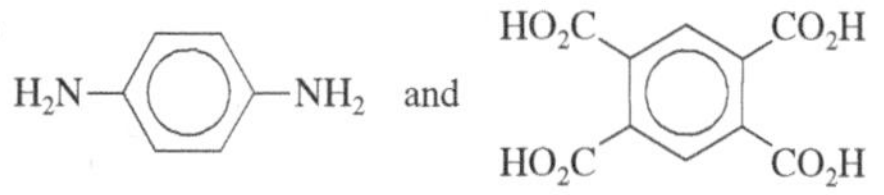

and

79. Divinylbenzene has two reactive double bonds that are used during formation of the polymer. The key is for the double bonds to insert themselves into two different polymer chains. When this occurs, the two chains are bonded together (are crosslinked). The chains cannot move past each other because of the crosslinks making the polymer more rigid.   **81. a.** The polymer from 1,2-diaminoethane and terephthalic acid is stronger because of the possibility of hydrogen bonding between chains. **b.** The polymer of

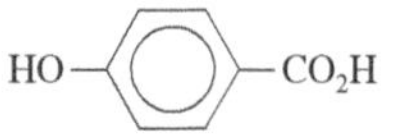

is more rigid because the chains are stiffer due to the rigid benzene rings in the chains. **c.** Polyacetylene is $nHC{\equiv}CH \rightarrow +(CH{=}CH)_n$. Polyacetylene is more rigid because the double bonds in the chains make the chains stiffer. **83. a.** serine; tyrosine; threonine; **b.** aspartic acid; glutamic acid; **c.** histidine; lysine; arginine; tryptophan; **d.** glutamine; asparagine   **85. a.** aspartic acid and phenylalanine; **b.** Aspartame contains the methyl ester of phenylalanine. This ester can hydrolyze to form methanol, $RCO_2CH_3 + H_2O \rightleftharpoons RCO_2H + CH_3OH$.
87.

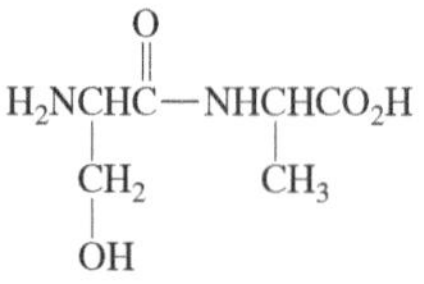

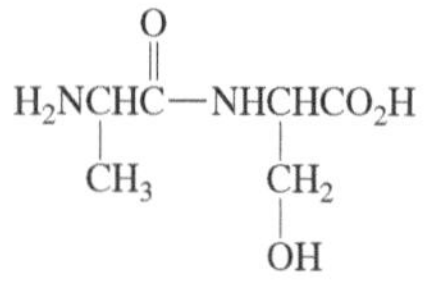

89. a. Six tetrapeptides are possible. From NH_2 to CO_2H end: phe–phe–gly–gly, gly–gly–phe–phe, gly–phe–phe–gly, phe–gly–gly–phe, phe–gly–phe–gly, gly–phe–gly–phe; **b.** Twelve tetrapeptides are possible. From NH_2 to CO_2H end: phe–phe–gly–ala, phe–phe–ala–gly, phe–gly–phe–ala, phe–gly–ala–phe, phe–ala–phe–gly, phe–ala–gly–phe, gly–phe–phe–ala, gly–phe–ala–phe, gly–ala–phe–phe, ala–phe–phe–gly, ala–phe–gly–phe, ala–gly–phe–phe   **91.** Ionic: his, lys, or arg with asp or glu; hydrogen bonding: ser, glu, tyr, his, arg, asn, thr, asp, gln, or lys with any amino acid; covalent: cys with cys; London dispersion: all amino acids with nonpolar R groups (gly, ala, pro, phe, ile, trp, met, leu, val); dipole–dipole: tyr, thr, and ser with each other   **93.** Glutamic acid has a polar R group and valine has a nonpolar R group. The change in polarity of the R groups could affect the tertiary structure of hemoglobin and affect the ability of hemoglobin to bond to oxygen.

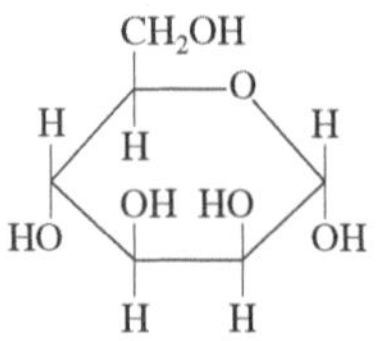

95.

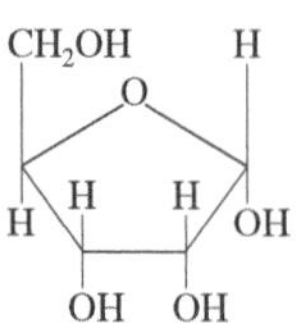

97. aldohexose: glucose, mannose, galactose; aldopentose: ribose, arabinose; ketohexose: fructose; ketopentose: ribulose   **99.** They differ in the orientation of a hydroxy group on a particular carbon. Starch is composed from α-D-glucose, and cellulose is composed from β-D-glucose.

101. The chiral carbons are marked with asterisks.

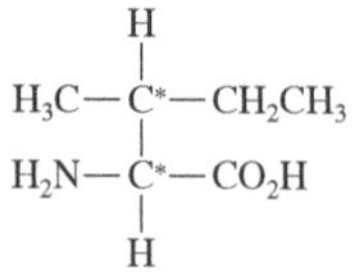

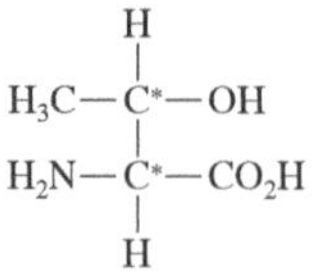

103.

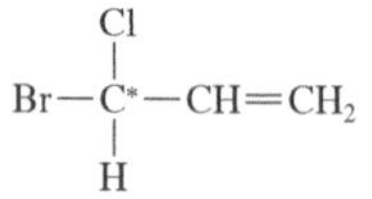

is optically active. The chiral carbon is marked with an asterisk.
105. C–C–A–G–A–T–A–T–G   **107.** Uracil will H-bond to adenine.

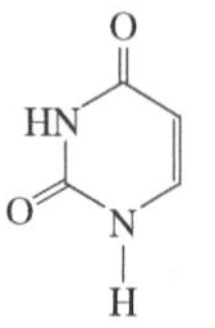

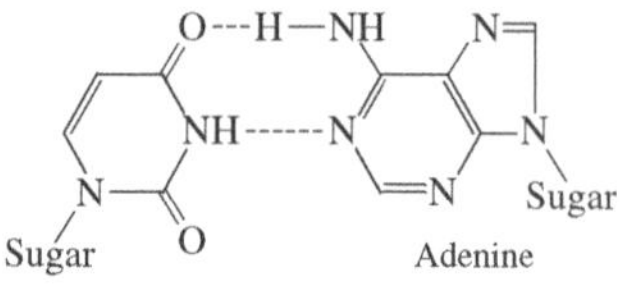

109. a. glu: CTT, CTC; val: CAA, CAG, CAT, CAC; met: TAC; trp: ACC; phe: AAA, AAG; asp: CTA, CTG; **b.** ACC–CTT–AAA–TAC or ACC–CTC–AAA–TAC or ACC–CTT–AAG–TAC or ACC–CTC–AAG– TAC; **c.** four (see answer to part b); **d.** met–asp–phe; **e.** TAC–CTA–AAG; TAC–CTA–AAA; TAC–CTG–AAA   **111. a.** 2,3,5,6-tetramethyloctane; **b.** 2,2,3,5-tetramethylheptane; **c.** 2,3,4-trimethylhexane; **d.** 3-methyl-1-pentyne
113.

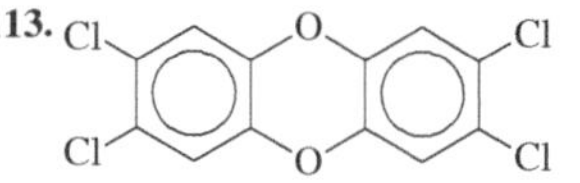

There are many possibilities for isomers. Any structure with four chlorines replacing four hydrogens in any of the numbered positions would be an isomer; i.e., 1,2,3,4-tetrachloro-dibenzo-p-dioxin is a possible isomer. **115.** $-23°C$: $CH_3{-}O{-}CH_3$; $78.5°C$: $CH_3{-}CH_2{-}OH$ **117.** Alcohols consist of two parts, the polar OH group and the nonpolar hydrocarbon chain attached to the OH group. As the length of the nonpolar hydrocarbon chain increases, the solubility of the alcohol decreases. In methyl alcohol (methanol), the polar OH group can override the effect of the nonpolar CH_3 group, and methyl alcohol is soluble in water. In stearyl alcohol, the molecule consists mostly of the long nonpolar hydrocarbon chain, so it is insoluble in water.   **119.** n-hexane, 69°C; pentanal, 103°C; 1-pentanol, 137°C; butanoic acid, 164°C.   **121.** 2-propanone; propanoic acid   **123.** In nylon, hydrogen-bonding interactions occur due to the presence of N—H bonds in the polymer. For a given polymer chain length, there are more N—H groups in nylon-46 as compared to nylon-6. Hence, nylon-46 forms a stronger polymer compared to nylon-6 due to the increased hydrogen-bonding interactions.
125. a.

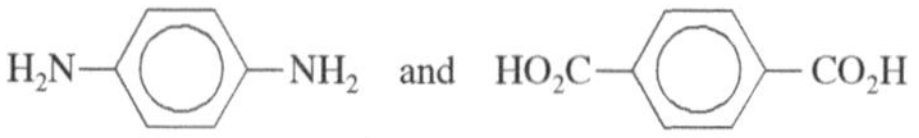

and

b. Repeating unit:

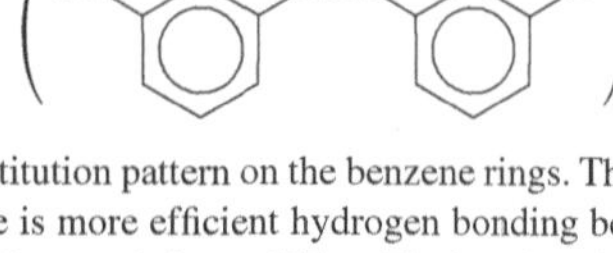

The two polymers differ in the substitution pattern on the benzene rings. The Kevlar chain is straighter, and there is more efficient hydrogen bonding between Kevlar chains than between Nomex chains.   **127. a.** The bond angles in the ring are about 60°. VSEPR predicts bond angles close to 109°. The bonding electrons are much closer together than they prefer, resulting in strong electron–electron repulsions. Thus ethylene oxide is unstable (reactive). **b.** The ring opens up during polymerization and the monomers link together through the formation of O—C bonds.

$$+O-CH_2CH_2-O-CH_2CH_2-O-CH_2CH_2+_n$$

129. 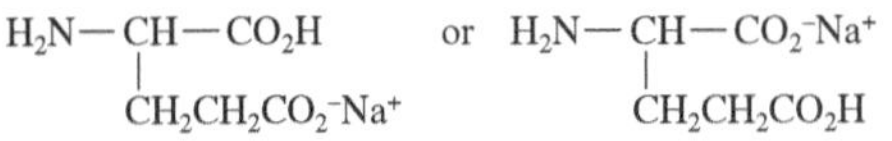 H_2N—CH—CO_2H   or   H_2N—CH—CO_2⁻Na⁺
                 | |
         CH_2CH_2CO_2⁻Na⁺   CH_2CH_2CO_2H

The first structure is MSG, which is impossible for you to predict.   **131.** $\Delta G = \Delta H - T\Delta S$; for the reaction, we break a P–O and O–H bond and form a P–O and O–H bond, so $\Delta H \approx 0$ based on bond dissociation energies. ΔS for this process is negative (unfavorable) because positional probability decreases. Thus, $\Delta G > 0$ due to the unfavorable ΔS term, and the reaction is not expected to be spontaneous.

133.

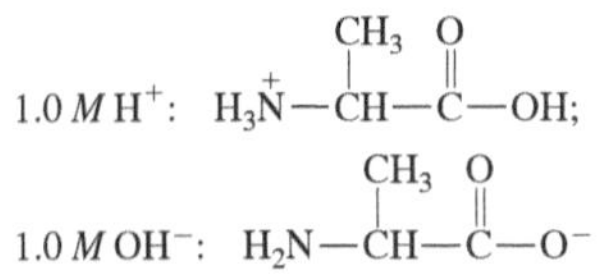

1.0 M H⁺:   H_3N⁺—CH—C—OH;
                          | ||
                        CH_3 O

1.0 M OH⁻:   H_2N—CH—C—O⁻
                         | ||
                       CH_3 O

135. Both ΔH and ΔS are positive values.   **145.** 6.07   **147. a.** No; the mirror image is superimposable.

b.

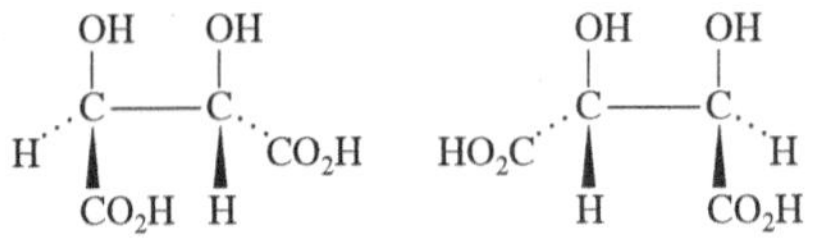

149.

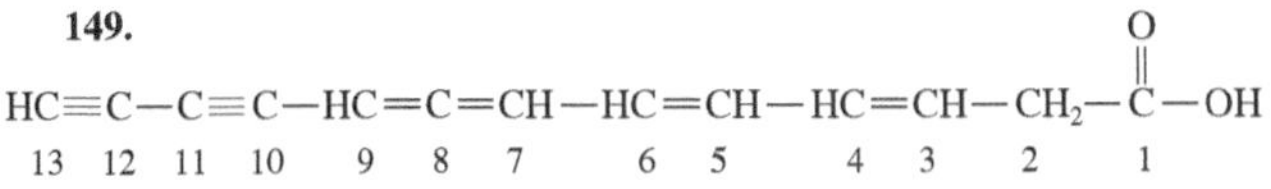

HC≡C—C≡C—HC=C=CH—HC=CH—HC=CH—CH_2—C—OH
13 12 11 10   9   8   7   6   5   4   3   2   1 ‖
                                                                    O

151. a. 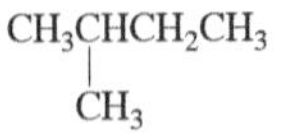 CH_3CHCH_2CH_3
                      |
                     CH_3

b. 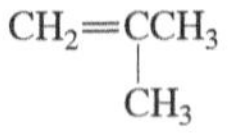 CH_2=CCH_3
                  |
                 CH_3

c.

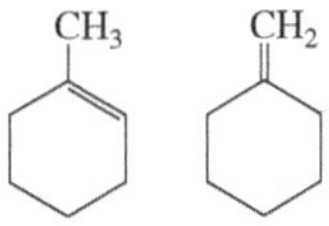

d. CH_2=CHCH_3

e. 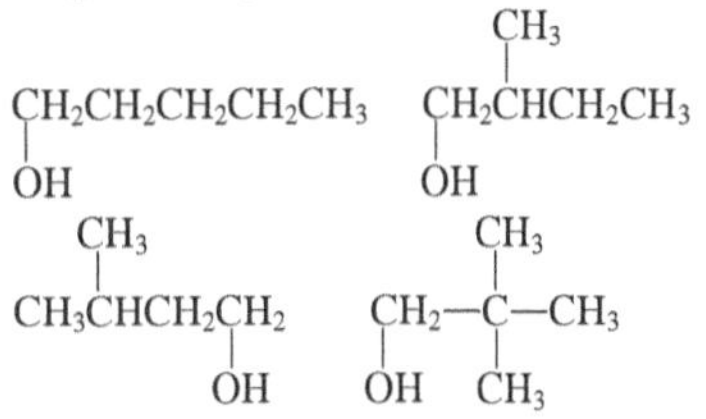 CH_2CH_2CH_2CH_2CH_3   CH_2CHCH_2CH_3
         | |
        OH   OH

CH_3CHCH_2CH_2   CH_2—C—CH_3
        | | | |
       CH_3   OH   OH   CH_3

with CH_3 substituents shown above.

153.

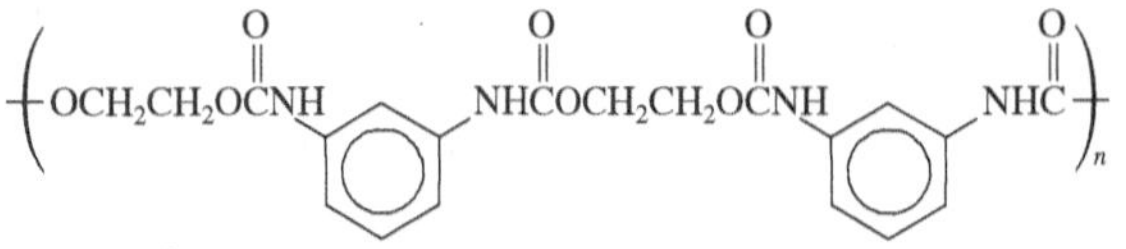

155. a. The temperature of the rubber band increases when it is stretched; **b.** exothermic (heat is released); **c.** As the chains are stretched, they line up more closely resulting in stronger London dispersion forces between the chains. Heat is released as the strength of the intermolecular forces increases. **d.** ΔG is positive and ΔS is negative; **e.** The structure of the stretched polymer chains is more ordered (has a smaller positional probability) than in unstretched rubber. Therefore, entropy decreases as the rubber band is stretched.   **157.** 0.11%   **159. a.** 37.50%; **b.** The hybridization changes from sp^2 to sp^3; **c.** 3,4-dimethyl-3-hexanol

Index

Table 2.4 | Common Type II Cations

Ion	Systematic Name
Fe^{3+}	Iron(III)
Fe^{2+}	Iron(II)
Cu^{2+}	Copper(II)
Cu^+	Copper(I)
Co^{3+}	Cobalt(III)
Co^{2+}	Cobalt(II)
Sn^{4+}	Tin(IV)
Sn^{2+}	Tin(II)
Pb^{4+}	Lead(IV)
Pb^{2+}	Lead(II)
Hg^{2+}	Mercury(II)
Hg_2^{2+}*	Mercury(I)
Ag^+	Silver[†]
Zn^{2+}	Zinc[†]
Cd^{2+}	Cadmium[†]

*Note that mercury(I) ions always occur bound together to form Hg_2^{2+} ions.
[†]Although these are transition metals, they form only one type of ion, and a Roman numeral is not used.

Table 2.5 | Common Polyatomic Ions

Ion	Name	Ion	Name
Hg_2^{2+}	Mercury(I)	NCS^- or SCN^-	Thiocyanate
NH_4^+	Ammonium	CO_3^{2-}	Carbonate
NO_2^-	Nitrite	HCO_3^-	Hydrogen carbonate
NO_3^-	Nitrate		(bicarbonate is a widely
SO_3^{2-}	Sulfite		used common name)
SO_4^{2-}	Sulfate	ClO^- or OCl^-	Hypochlorite
HSO_4^-	Hydrogen sulfate	ClO_2^-	Chlorite
	(bisulfate is a widely	ClO_3^-	Chlorate
	used common name)	ClO_4^-	Perchlorate
OH^-	Hydroxide	$C_2H_3O_2^-$	Acetate
CN^-	Cyanide	MnO_4^-	Permanganate
PO_4^{3-}	Phosphate	$Cr_2O_7^{2-}$	Dichromate
HPO_4^{2-}	Hydrogen phosphate	CrO_4^{2-}	Chromate
$H_2PO_4^-$	Dihydrogen phosphate	O_2^{2-}	Peroxide
		$C_2O_4^{2-}$	Oxalate
		$S_2O_3^{2-}$	Thiosulfate

Table 4.1 | Simple Rules for the Solubility of Salts in Water

1. Most nitrate (NO_3^-) salts are soluble.
2. Most salts containing the alkali metal ions (Li^+, Na^+, K^+, Cs^+, Rb^+) and the ammonium ion (NH_4^+) are soluble.
3. Most chloride, bromide, and iodide salts are soluble. Notable exceptions are salts containing the ions Ag^+, Pb^{2+}, and Hg_2^{2+}.
4. Most sulfate salts are soluble. Notable exceptions are $BaSO_4$, $PbSO_4$, Hg_2SO_4, and $CaSO_4$.
5. Most hydroxides are only slightly soluble. The important soluble hydroxides are NaOH and KOH. The compounds $Ba(OH)_2$, $Sr(OH)_2$, and $Ca(OH)_2$ are marginally soluble.
6. Most sulfide (S^{2-}), carbonate (CO_3^{2-}), chromate (CrO_4^{2-}), and phosphate (PO_4^{3-}) salts are only slightly soluble, except for those containing the cations in Rule 2.

Table 8.4 | Average Bond Energies (kJ/mol)

	Single Bonds							Multiple Bonds	
H—H	432	N—H	391	I—I	149			C=C	614
H—F	565	N—N	160	I—Cl	208			C≡C	839
H—Cl	427	N—F	272	I—Br	175			O=O	495
H—Br	363	N—Cl	200					C=O*	745
H—I	295	N—Br	243	S—H	347			C≡O	1072
		N—O	201	S—F	327			N=O	607
C—H	413	O—H	467	S—Cl	253			N=N	418
C—C	347	O—O	146	S—Br	218			N≡N	941
C—N	305	O—F	190	S—S	266			C≡N	891
C—O	358	O—Cl	203					C=N	615
C—F	485	O—I	234						
C—Cl	339			Si—Si	340				
C—Br	276	F—F	154	Si—H	393				
C—I	240	F—Cl	253	Si—C	360				
C—S	259	F—Br	237	Si—O	452				
		Cl—Cl	239						
		Cl—Br	218						
		Br—Br	193						

*C=O(CO$_2$) = 799

Table 8.5 | Bond Lengths and Bond Energies for Selected Bonds

Bond	Bond Type	Bond Length (pm)	Bond Energy (kJ/mol)
C—C	Single	154	347
C=C	Double	134	614
C≡C	Triple	120	839
C—O	Single	143	358
C=O	Double	123	745
C—N	Single	143	305
C=N	Double	138	615
C≡N	Triple	116	891